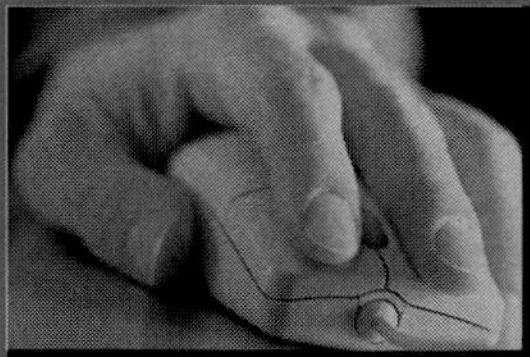

The Microbiology Place

Log in.

Organize.

Motivate & assess.

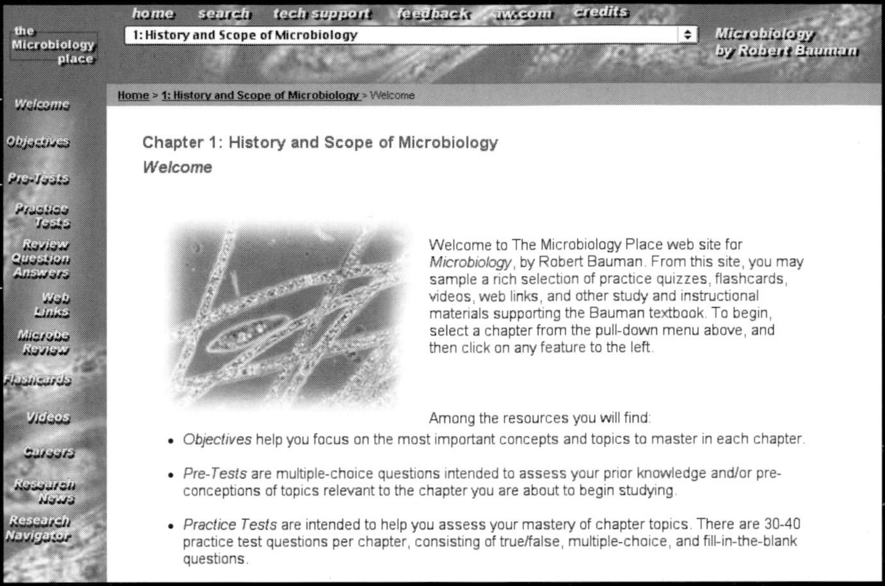
home search tech support feedback aw.com credits
the Microbiology place
1: History and Scope of Microbiology
Microbiology by Robert Bauman
Home > 1: History and Scope of Microbiology > Welcome
Welcome
Objectives
Pre-Tests
Practice Tests
Review Question Answers
Web Links
Microbe Review
Flashcards
Videos
Careers
Research News
Research Navigator
Chapter 1: History and Scope of Microbiology
Welcome
Welcome to The Microbiology Place web site for Microbiology, by Robert Bauman. From this site, you may sample a rich selection of practice quizzes, flashcards, videos, web links, and other study and instructional materials supporting the Bauman textbook. To begin, select a chapter from the pull-down menu above, and then click on any feature to the left.
Among the resources you will find:
Objectives help you focus on the most important concepts and topics to master in each chapter.
Pre-Tests are multiple-choice questions intended to assess your prior knowledge and/or pre-conceptions of topics relevant to the chapter you are about to begin studying.
Practice Tests are intended to help you assess your mastery of chapter topics. There are 30-40 practice test questions per chapter, consisting of true/false, multiple-choice, and fill-in-the-blank questions.

To help teach microbiology more effectively, we've arranged for you and your students to enjoy access to several media resources. Included in these resources are tools for instructors and student learning and assessment. Go to www.microbiologyplace.com.

What your system needs to use these media resources:

WINDOWS™
266 MHz Intel Pentium processor or greater
Windows 98 or higher
64 MB Minimum RAM
800 X 600 screen resolution
Browser: Internet Explorer 5.0 or higher; Netscape 4.7, 6.2, or higher
Plug-Ins: Shockwave Player 8, Flash Player 5, QuickTime 4
Internet Connection: 56k modem minimum for website
NOTE: Use of Netscape 6.0 is not recommended due to a known compatibility issue between Netscape 6.0 and the Flash and Shockwave plug-ins.

MACINTOSH™
266 MHz Minimum CPU
OS 8.6 or higher
64 MB Minimum RAM
800 x 600 screen resolution,thousands of colors
Browser: OS 8-9: Netscape 4.7, 6.2, or higher;(Web only: Internet Explorer 5 or higher); OS X: Internet Explorer 5.2 or higher
Plug-Ins: Shockwave Player 8, Flash Player 5, QuickTime 4
Internet Connection: 56k modem minimum for website
NOTE: Use of Netscape 6.0 is not recommended due to a known compatibility issue between Netscape 6.0 and the Flash and Shockwave plug-ins. Also, Mac OS X users should run this CD-ROM on Internet Explorer only due to known compatibility issues between OS X and Netscape.

Got technical questions?

For technical support, please visit www.aw.com/techsupport and complete the appropriate online form. Technical support is available Monday-Friday, 9 a.m. to 6 p.m. Eastern Time (US and Canada).

Here's your personal ticket to success:

How to log on to www.microbiologyplace.com:

1. Go to www.microbiologyplace.com.
2. Click ***Microbiology* by Robert W. Bauman**.
3. Click "Register."
4. Scratch off the silver foil coating below to reveal your pre-assigned access code.
5. Enter your pre-assigned access code exactly as it appears below.
6. Complete the online registration form to create your own personal user Login Name and Password.
7. Once your personal Login Name and Password are confirmed by email, go back to www.microbiologyplace.com, type in your new Login Name and Password, and click "Log In."

Your Access Code is:

If there is no silver foil covering the access code above, the code may no longer be valid. In that case, please contact your Benjamin Cummings sales representative.

Important: Please read the License Agreement, located on the launch screen before using the The Microbiology Place website. By using the website, you indicate that you have read, understood and accepted the terms of this agreement.

Microbiology

Brief Edition

ROBERT W. BAUMAN, Ph.D.
Amarillo College

Contributions by:

Elizabeth Machunis-Masuoka, Ph.D.
University of Virginia

Ian Tizard, Ph.D.
Texas A&M University

PEARSON
Benjamin Cummings

San Francisco Boston New York
Cape Town Hong Kong London Madrid Mexico City
Montreal Munich Paris Singapore Sydney Tokyo Toronto

Publisher *Daryl Fox*
Executive Editors *Leslie Berriman, Serina Beauparlant*
Project Editor *Barbara Yien*
Managing Editor *Wendy Earl*
Production Editor *Sharon Montooth*
Text Development Editor *Laura Bonazzoli*
Art Development Editor *Laura Southworth*
Principal Artist *Kenneth X. Probst*
Artists *Darwen Hennings, Cassio Lynm, Precision Graphics, J.B. Woolsey & Associates*
Copyeditor *Alan Titche*
Art and Photo Coordinator *Kelly Murphy*
Photo Researchers *Diane Austin, Kathleen Olson*
Text Designer *tani hasegawa*
Cover Designer *Lillian Carr*
Contributing Editors *Claire Brassert, Kay Ueno, Amy Teeling*
Market Development Manager *Cheryl Cechvala*
Executive Marketing Manager *Lauren Harp*
Publishing Assistants *Ryan Shaw, Ziki Dekel*
Manufacturing Buyer *Stacey Weinberger*
Compositor *GTS Graphics*

Cover Image: T4 bacteriophages bursting from an *E. coli* bacterium. Kenneth X. Probst
Photograph by Lee D. Simon, Photo Researchers, Inc.

Text, photography, and illustration credits appear following the Appendices.

Library of Congress Cataloging-in-Publication Data

Bauman, Robert W.
Microbiology / Robert W. Bauman ; contributions by Elizabeth Machunis-Masuoka, Ian Tizard. — Brief ed.
p. cm.
Includes index.
ISBN 0-8053-7676-3
1. Microbiology. I. Machunis-Masuoka, Elizabeth. II. Tizard, Ian R.
III.
Title.
QR41.2.B382 2004
579—dc22 2004052728

PEARSON
Benjamin Cummings

ISBN 0-8053-7676-3

1 2 3 4 5 6 7 8 9 10—VHC—07 06 05 04

www.aw.com/bc

I dedicate this labor of love to my students,
who urged me to write this book, and to my family,
who have graciously shared their lives with
"the book."

Robert W. Bauman

ABOUT THE AUTHOR

Robert W. Bauman is a professor of biology and chairman of the Department of Biological Sciences at Amarillo College in Amarillo, Texas. He teaches microbiology and human anatomy and physiology. In 2004, the students of Amarillo College selected Dr. Bauman as the recipient of the John S. Mead Faculty Excellence Award. He received an M.A. degree in botany from the University of Texas at Austin, and a Ph.D. in biology from Stanford University. His research interests have included the morphology and ecology of freshwater algae, the cell biology of marine algae (particularly the deposition of cell walls and intercellular communication), and environmentally triggered chromogenesis in butterflies. He is a member of the American Society of Microbiology (ASM), Texas Community College Teacher's Association (TCCTA), American Association for the Advancement of Science (AAAS), and The Lepidopterist's Society. When he is not writing books, he enjoys spending time with his family: gardening, hiking, camping, playing games (with lots of laughter), and reading classics out loud by a crackling fire.

ABOUT THE ILLUSTRATION TEAM

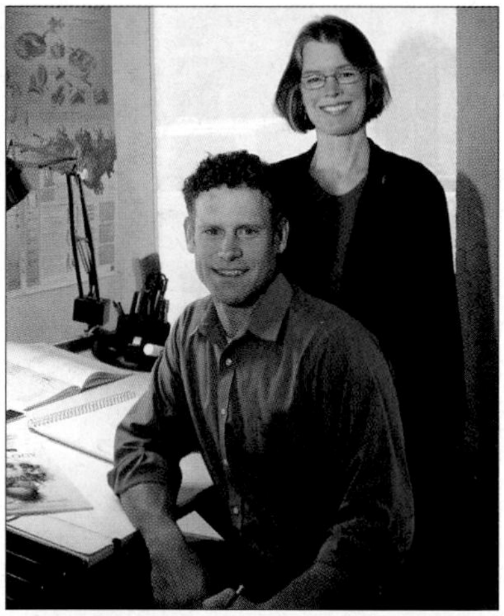

Kenneth Xavier Probst, the principal artist, received his M.A. in Medical and Biological Illustration from the Johns Hopkins University School of Medicine and has been a staff scientific illustrator for Benjamin Cummings since 2000. In addition to his work as an educator, fine artist, graphic and industrial designer, Kenneth has received recognition for his work as a medical illustrator, including Certificates of Merit from the Association of Medical Illustrators and a Vesalius Trust Research Grant.

Laura Southworth is a freelance art development editor and scientific illustrator in Forestville, California. She began her career while obtaining a B.S. degree in biology from California Polytechnic State University, San Luis Obispo. For over twenty years she has been designing and creating scientific art for publications in many disciplines, including biology, human anatomy and physiology, entomology, microbiology, and molecular biology.

Richard A. Robison photographed many of the unique micrographs found in this text. He is an Associate Professor of microbiology at Brigham Young University in Provo, Utah. He earned a Ph.D. in microbiology from BYU in 1988, with graduate minors in chemistry and biochemistry. His current research interests center around frank bacterial pathogens of humans and animals. Dr. Robison is director of a biosafety level-3 facility and collaborates with scientists at several national and private laboratories. Photographing the microscopic world has been a favorite hobby of his for over 20 years.

PREFACE

I have taught microbiology to undergraduates for over 15 years, and have witnessed firsthand how students struggle with the same topics and concepts year after year. Increasingly, I also find that students come into class with various levels of preparation—while some have strong science backgrounds, others lack a foundation in chemistry and/or biology—making it a challenge to decide how to gear the course. In setting out to create this textbook, my goal was to write an introductory text that helped address these concerns and that was also engaging and enjoyable to read, weaving in the many exciting and interesting microbiology-related topics that appear daily in the news and in research journals. I wished to highlight the positive effects of microorganisms in our lives, along with the medically important microorganisms that cause disease. Lastly, I aimed to write a text that explains complex topics (especially metabolism, genetics, and immunology) in a way beginning students can understand, while still presenting a thorough, comprehensive, and accurate overview of microbiology.

To achieve these goals, a great deal of thought and research, and the feedback of literally hundreds of students and instructors (listed at the end of this preface), went into the development of this textbook. From the beginning I was determined *not* to write a book that would be just like every other microbiology textbook, but instead one that improved upon conventional explanations and illustrations in substantive and effective ways. To this end, I have collaborated with educators, students, editors, and top scientific illustrators in the field to develop this text. The following sections provide an outline of the book's main themes and features.

A Note about the Brief Edition

This Brief Edition was produced in response to instructor requests for a shorter, more affordable paperback version of the original hardcover edition of *Microbiology*. It is identical to the hardcover edition with the exception that it does not include the eight chapters covering pathogens and environmental microbiology that are found in the hardcover edition, making it ideal for use by instructors who prefer to cover these topics with their own instructional materials. The Brief Edition is fully supported by a comprehensive suite of instructor and student supplements, described on page xi.

A Groundbreaking Art Program That Engages as it Teaches

This book's stunning art program features a number of innovations and pedagogical tools designed to help students understand the material. **Half-micrograph, half-illustration figures (Figures 1 and 2)** depict microbial structures in a three-dimensional, visually striking way that draws students' interest while also teaching them what these structures really look like. **Orientation figures (Figure 3)** remind students of "the big picture". **Step-by-step diagrams (Figure 3)** break complex processes into smaller, more manageable pieces; the step numbers are also color-coded to correspond to step numbers in the text. **Figure legend questions (Figure 4)** engage students by asking them to apply what they've learned in the text to answer questions about topics featured in the art. **Consistent use of icons and colors (Figure 4)** from figure to figure makes it easier for students to recognize structures and processes throughout the book. **Ruler bars** and TEM / SEM / LM designations provide readers added context with which to study the art.

Half-micrograph, half-illustration figures help students visualize microbial structures as they really are.

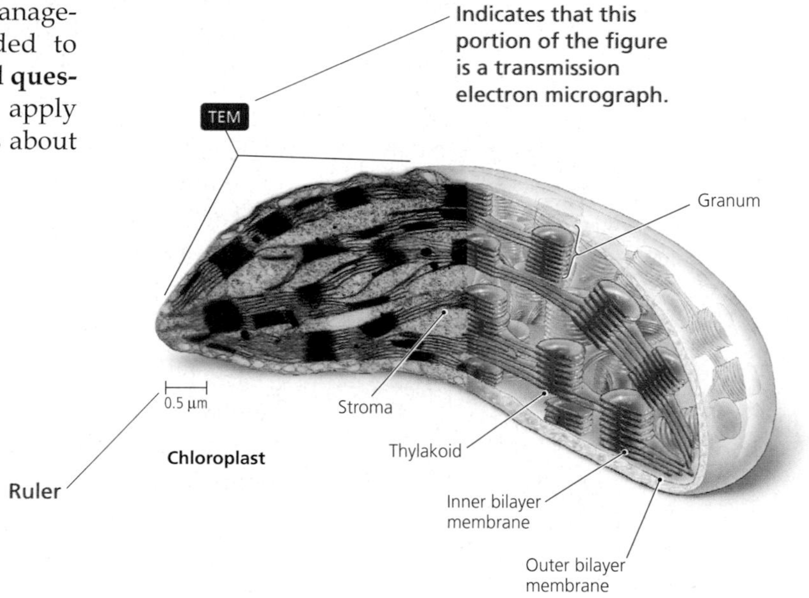

Figure 1

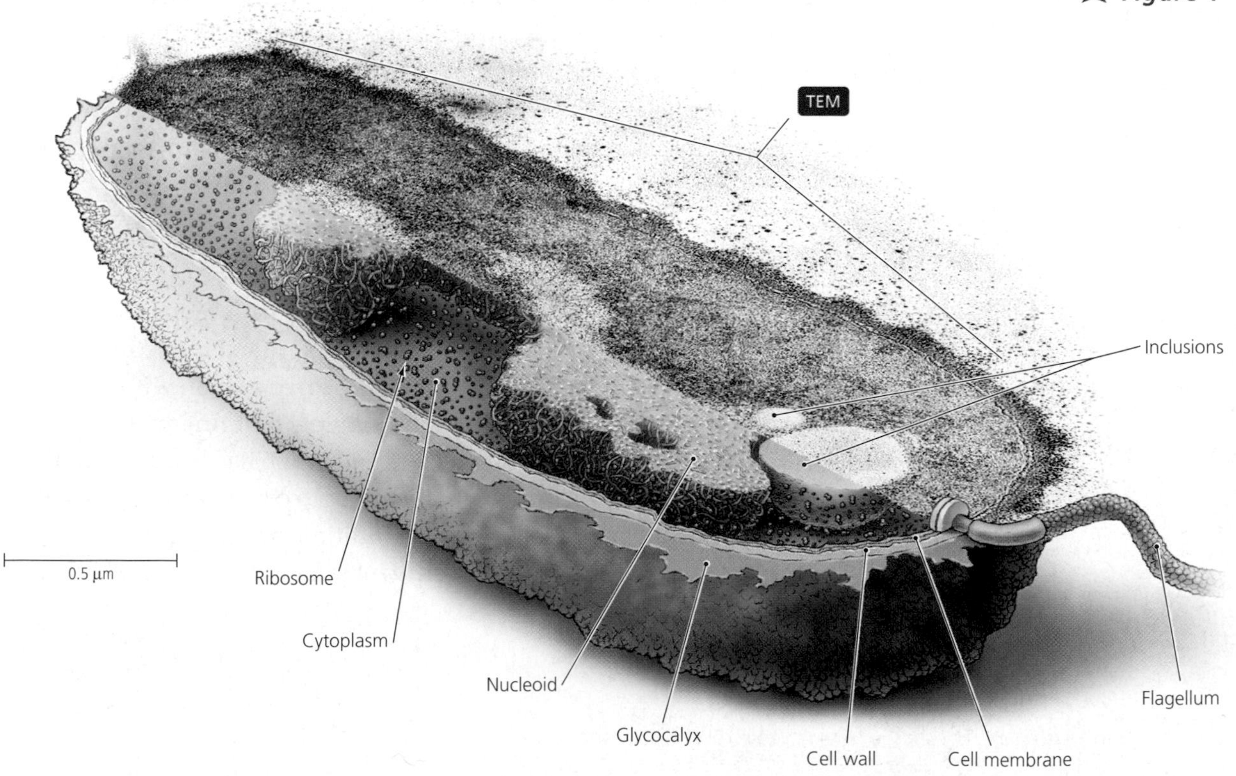

Figure 2

➤ *Figure 3*

Orientation figures help students keep "the big picture" in mind.

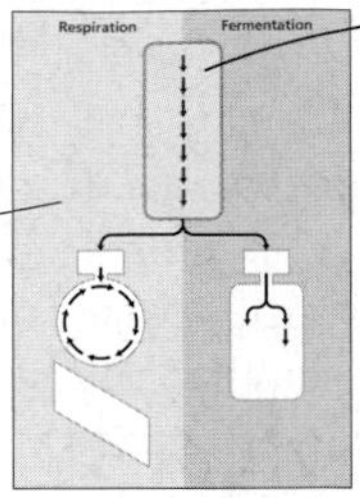

Step-by-step diagrams help students understand complex topics. The step numbers are color-coded to match step numbers in the text.

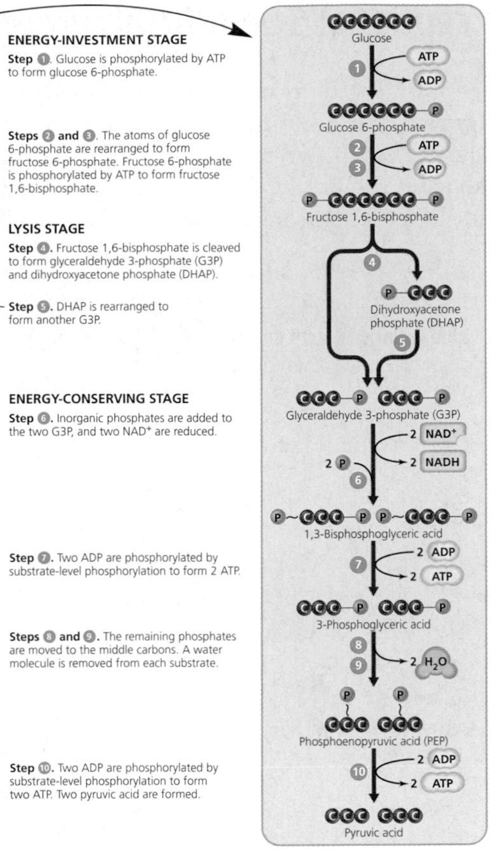

ENERGY-INVESTMENT STAGE
Step 1. Glucose is phosphorylated by ATP to form glucose 6-phosphate.
Steps 2 and 3. The atoms of glucose 6-phosphate are rearranged to form fructose 6-phosphate. Fructose 6-phosphate is phosphorylated by ATP to form fructose 1,6-bisphosphate.
LYSIS STAGE
Step 4. Fructose 1,6-bisphosphate is cleaved to form glyceraldehyde 3-phosphate (G3P) and dihydroxyacetone phosphate (DHAP).
Step 5. DHAP is rearranged to form another G3P.
ENERGY-CONSERVING STAGE
Step 6. Inorganic phosphates are added to the two G3P, and two NAD+ are reduced.
Step 7. Two ADP are phosphorylated by substrate-level phosphorylation to form 2 ATP.
Steps 8 and 9. The remaining phosphates are moved to the middle carbons. A water molecule is removed from each substrate.
Step 10. Two ADP are phosphorylated by substrate-level phosphorylation to form two ATP. Two pyruvic acid are formed.
Glucose
ATP
ADP
Glucose 6-phosphate
Fructose 1,6-bisphosphate
Dihydroxyacetone phosphate (DHAP)
Glyceraldehyde 3-phosphate (G3P)
2 NAD+
2 NADH
1,3-Bisphosphoglyceric acid
3-Phosphoglyceric acid
2 H2O
Phosphoenopyruvic acid (PEP)
Pyruvic acid

▼ *Figure 4*

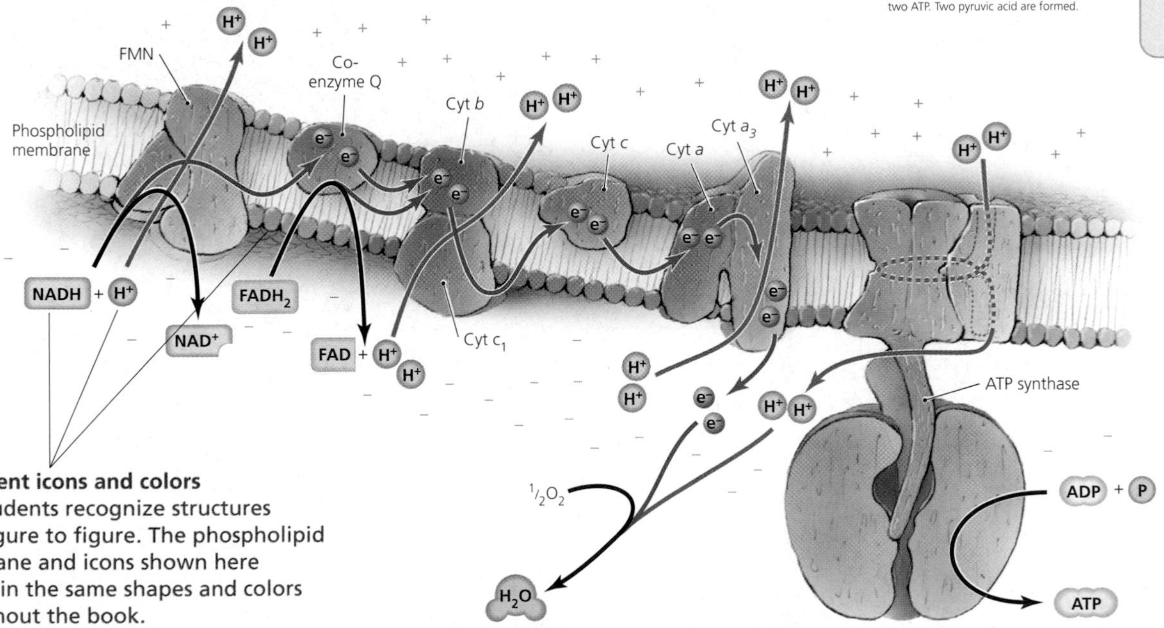

Consistent icons and colors help students recognize structures from figure to figure. The phospholipid membrane and icons shown here appear in the same shapes and colors throughout the book.

▲ ***Figure 5.21***

One possible arrangement of an electron transport chain. Electron transport chains are located in the cytoplasmic membranes of prokaryotes, and in the inner membranes of mitochondria in eukaryotes. The exact types and sequences of carrier molecules in electron transport chains vary among organisms. As electrons move down the chain (red arrows), their energy is used to pump protons (H^+) across the membrane. The protons then flow through ATP synthase (ATPase), which synthesizes ATP. Approximately one molecule of ATP is generated for every two protons that cross the membrane. *Why is it essential that all the electron carriers of an electron transport chain be membrane bound?*

Figure 5.21 *The carrier molecules of electron transport chains are membrane bound for at least two reasons: (1) Carrier molecules need to be in fairly close proximity so they can pass electrons between each other. (2) The goal of the chain is the pumping of* H^+ *across a membrane to create a proton gradient; without a membrane, there could be no gradient.*

Figure legend questions ask students to apply concepts learned in the text to answer questions about the art. Answers to the questions appear upside-down.

Exceptional Scientific Accuracy

Exceptional attention has been paid to **scientific accuracy (Figure 5)** throughout the art program. Many proteins, for instance, are drawn to look like their actual shapes based on data from the Protein Data Bank from the Research Collaboratory for Structural Bioinformatics. Microbial structures are drawn from actual micrographs. And, where appropriate, photos are used in place of (or in addition to) illustrations in order to more accurately depict a topic or process **(Figure 6).**

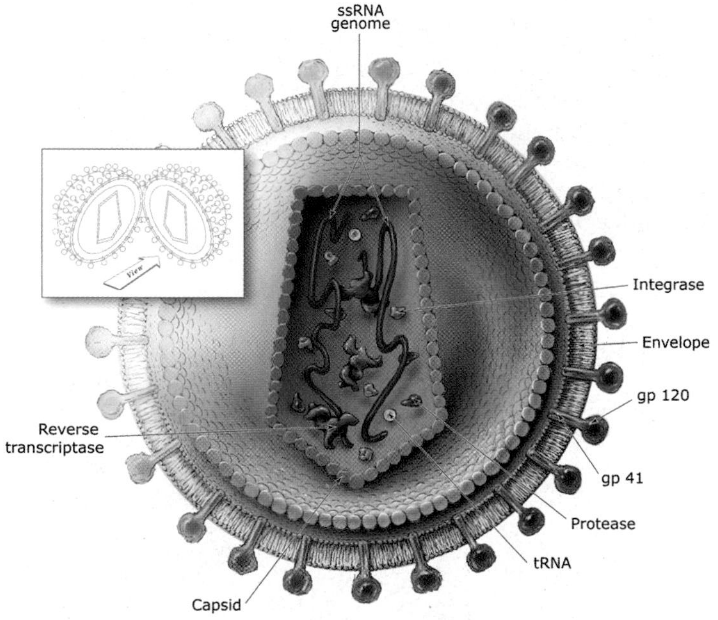

▼ ***Figure 5***
Scientific accuracy
The reverse transcriptase depicted in this illustration of HIV was based on data from the Protein Data Bank.

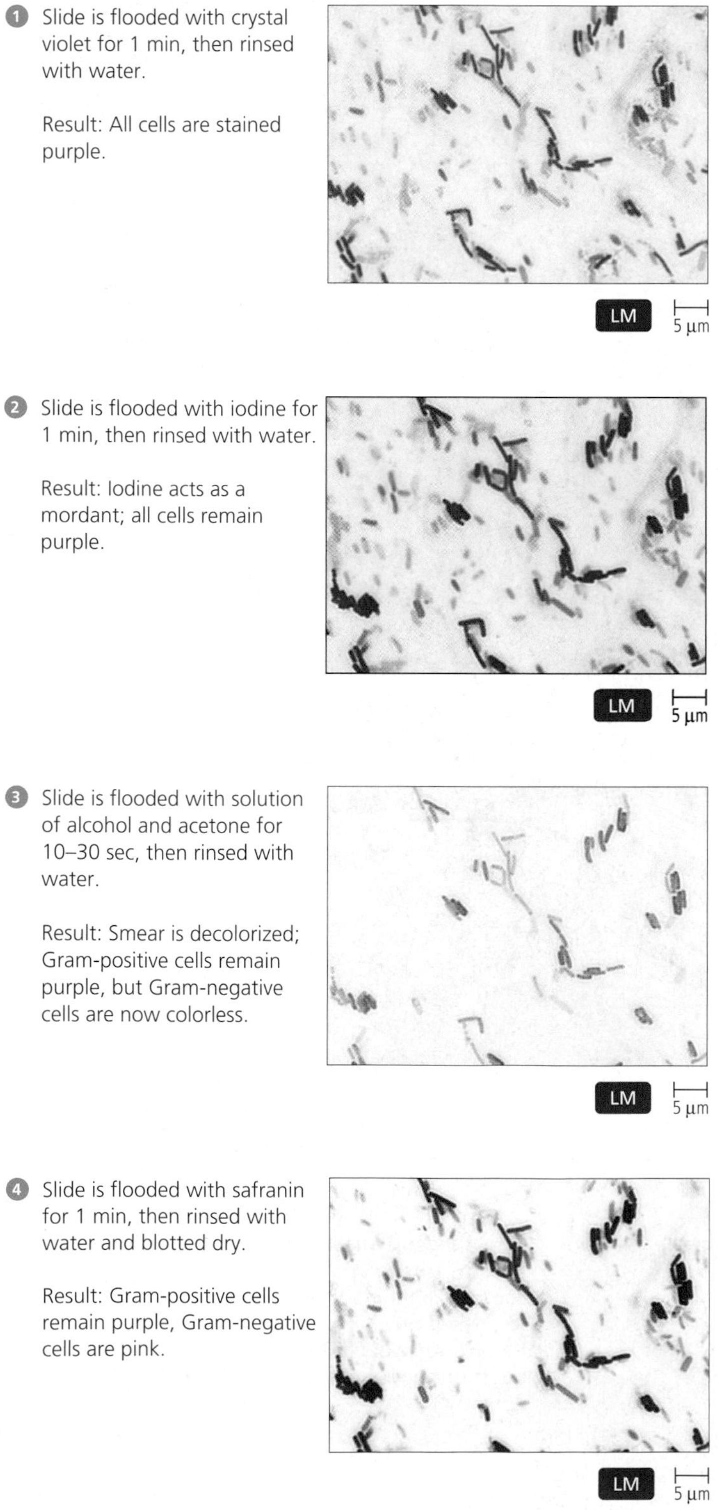

1 Slide is flooded with crystal violet for 1 min, then rinsed with water.
Result: All cells are stained purple.
LM 5 µm
2 Slide is flooded with iodine for 1 min, then rinsed with water.
Result: Iodine acts as a mordant; all cells remain purple.
LM 5 µm
3 Slide is flooded with solution of alcohol and acetone for 10–30 sec, then rinsed with water.
Result: Smear is decolorized; Gram-positive cells remain purple, but Gram-negative cells are now colorless.
LM 5 µm
4 Slide is flooded with safranin for 1 min, then rinsed with water and blotted dry.
Result: Gram-positive cells remain purple, Gram-negative cells are pink.
LM 5 µm

◀ ***Figure 6***
High-quality micrographs are found throughout the text, and are also available to instructors in electronic format. Here, four same-field micrographs depict the Gram staining procedure.

Coverage of Timely and High-Interest Topics from Both Research and Everyday Life

The text includes several features designed to draw students' interest and bring the wonders of microbiology to life. **Chapter-opening vignettes (Figure 7)** pique the reader's curiosity about the chapter at hand, covering such topics as the use of microbes in genetically modified foods (Ch. 8) and the problem of so-called "superbugs" that are resistant to our most powerful antibiotics (Ch. 10). **Highlight boxes (Figure 8)** focus on fascinating topics in microbiology, such as microbial control in the preparation of sushi (Ch. 9) and the medicinal and cosmetic uses of Botox (Ch. 11). These boxes demonstrate the relevance of applied microbiology to everyday life, encouraging interest in the field. Meanwhile, **New Frontiers boxes (Figure 9)** introduce students to cutting-edge research and future challenges related to microbiology. Topics include bioterrorism (Ch. 1), biofilms (Chs. 3 and 6), and the genetic modification of mosquitoes to curb the spread of malaria (Ch. 8). A complete list of box topics is included on page xxvi.

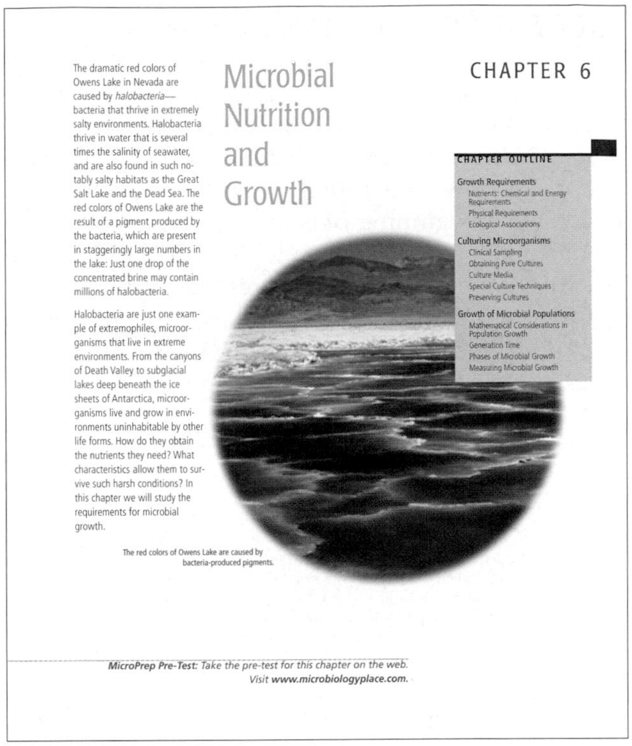

CHAPTER 6

Microbial Nutrition and Growth

The dramatic red colors of Owens Lake in Nevada are caused by *halobacteria*—bacteria that thrive in extremely salty environments. Halobacteria thrive in water that is several times the salinity of seawater, and are also found in such notably salty habitats as the Great Salt Lake and the Dead Sea. The red colors of Owens Lake are the result of a pigment produced by the bacteria, which are present in staggeringly large numbers in the lake: Just one drop of the concentrated brine may contain millions of halobacteria.

Halobacteria are just one example of extremophiles, microorganisms that live in extreme environments. From the canyons of Death Valley to subglacial lakes deep beneath the ice sheets of Antarctica, microorganisms live and grow in environments uninhabitable by other life forms. How do they obtain the nutrients they need? What characteristics allow them to survive such harsh conditions? In this chapter we will study the requirements for microbial growth.

CHAPTER OUTLINE

Growth Requirements
- Nutrients: Chemical and Energy Requirements
- Physical Requirements
- Ecological Associations

Culturing Microorganisms
- Clinical Sampling
- Obtaining Pure Cultures
- Culture Media
- Special Culture Techniques
- Preserving Cultures

Growth of Microbial Populations
- Mathematical Considerations in Population Growth
- Generation Time
- Phases of Microbial Growth
- Measuring Microbial Growth

The red colors of Owens Lake are caused by bacteria-produced pigments.

MicroPrep Pre-Test: *Take the pre-test for this chapter on the web. Visit* ***www.microbiologyplace.com.***

▲ ***Figure 7***
Chapter openers include striking photos and interesting vignettes to draw readers in.

Highlight 9.2 Sushi: A Case Study in Microbial Control

Sushi—it either makes you squirm or salivate! Although technically the term refers to rice, it's generally understood to refer to bite-sized slices of raw fish served with rice. Long a staple of Japanese cuisine, sushi has become popular throughout the United States. But isn't eating raw fish dangerous? What methods of microbial control are applied in the preparation of sushi?

It's true that raw fish can contain harmful microorganisms. Such parasites as the kind of roundworms called anasakids are commonly found in fish and can cause gastrointestinal symptoms. Accordingly, before fish can be served raw, the Food and Drug Administration requires that it be frozen at −20°C (−4°F) for 7 days, or at −37°C (−35°F) for 15 hours. Unfortunately, freezing does *not* kill all bacterial or viral pathogens. Consumers should be aware that they always assume some risk of food poisoning caused by such bacteria as *S. aureus, E. coli,* and species of *Salmonella* and *Vibrio* whenever they eat any kind of raw seafood, including raw oysters, seviche, or carpaccio.

Even though diners tend to fixate on the safety of raw fish, cooked rice left sitting at room temperature is also vulnerable to the growth of pathogens. To counter this potential problem, sushi rice is prepared with vinegar, which acidifies the rice. At pH values below

4.6, rice becomes too acidic to support the growth of pathogens. Sushi bars can also reduce the risk posed by pathogens by keeping restaurant temperatures cool. Additionally, it is believed that wasabi, the fiery horseradish-like green paste commonly eaten with sushi, contains antimicrobial properties, although its mechanism of action is not well understood.

With these antimicrobial precautions in place, the vast majority of diners consume sushi safely meal after meal (although pregnant women and people with compromised immune systems should avoid all raw seafood). When properly prepared, sushi is beautiful, low in calories, and a source of heart-healthy omega 3-fatty acids—and very delicious.

◀ ***Figure 8***
Highlight boxes focus on fascinating topics in microbiology.

New Frontiers 8.1 Building a Better Mosquito

What if scientists could genetically engineer mosquitoes to be incapable of transmitting deadly diseases, such as malaria? Researchers are attempting to do just that. One possible strategy involves using transposons to carry genes into mosquitoes that would boost their production of defensive proteins that destroy pathogens. Mosquitoes naturally produce such proteins, but not in high enough doses to kill off all pathogens. Scientists theorize that if they can genetically engineer and breed mosquitoes that destroy pathogens, these mosquitoes could then be released into the environment to breed with wild mosquitoes, creating subsequent populations that can no longer spread disease.

To date, researchers have successfully used transposons as vectors to insert defense-boosting genes into *Aedes* mosquitoes, which transmit yellow fever. While this is a promising achievement, *Aedes* is only one of over 100 disease-transmitting mosquito species. Recombinant technologies that work on *Aedes* may not necessarily work on, for instance, *Anopheles gambia* mosquitoes, which are the primary carriers of the most deadly form of malaria. In addition, some worry that unleashing transgenic mosquitoes into the wild may have adverse consequences. Given that malaria alone kills over 2 million people each year, however, the quest for a safer mosquito is one that scientists are compelled to continue.

Aedes mosquito

◀ ***Figure 9***
New Frontiers boxes highlight cutting-edge research as well as future challenges in microbiology.

Student-Focused Pedagogy

Every page of this textbook was written with students' specific needs in mind. To help students be better prepared in class, **Online "MicroPrep" Pre-tests** are available for each chapter at www.microbiologyplace.com. The pre-tests are cited at the beginning of each chapter, and are designed to be taken before a given chapter is read in order to (1) review relevant topics introduced in previous chapters (or in fundamental chemistry/biology courses), and (2) test student's preconceptions and/or knowledge concerning the chapter's topics. **Learning Objectives (Figure 10)** are integrated throughout each chapter to give students a preview of topics covered. **Integrated pronunciation and etymology guides (Figure 10)** help students pronounce and remember vocabulary. **Critical Thinking questions (Figure 10),** which are both integrated throughout the text and listed at the end of each chapter, encourage students to apply what they have learned to a novel situation or to a case study. **End-of-chapter summaries** and **review questions (Figure 11)** help students review the material.

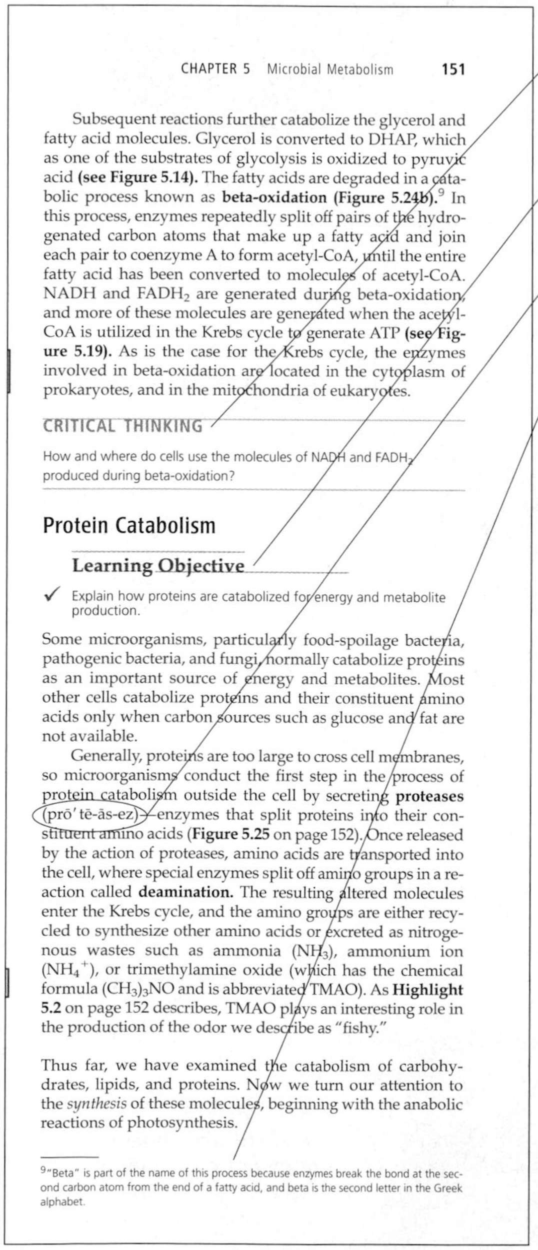

CHAPTER 5 Microbial Metabolism 151

Subsequent reactions further catabolize the glycerol and fatty acid molecules. Glycerol is converted to DHAP, which as one of the substrates of glycolysis is oxidized to pyruvic acid **(see Figure 5.14).** The fatty acids are degraded in a catabolic process known as **beta-oxidation (Figure 5.24b).**[9] In this process, enzymes repeatedly split off pairs of the hydrogenated carbon atoms that make up a fatty acid and join each pair to coenzyme A to form acetyl-CoA, until the entire fatty acid has been converted to molecules of acetyl-CoA. NADH and $FADH_2$ are generated during beta-oxidation, and more of these molecules are generated when the acetyl-CoA is utilized in the Krebs cycle to generate ATP **(see Figure 5.19).** As is the case for the Krebs cycle, the enzymes involved in beta-oxidation are located in the cytoplasm of prokaryotes, and in the mitochondria of eukaryotes.

CRITICAL THINKING

How and where do cells use the molecules of NADH and $FADH_2$ produced during beta-oxidation?

Protein Catabolism

Learning Objective

✓ Explain how proteins are catabolized for energy and metabolite production.

Some microorganisms, particularly food-spoilage bacteria, pathogenic bacteria, and fungi, normally catabolize proteins as an important source of energy and metabolites. Most other cells catabolize proteins and their constituent amino acids only when carbon sources such as glucose and fat are not available.

Generally, proteins are too large to cross cell membranes, so microorganisms conduct the first step in the process of protein catabolism outside the cell by secreting **proteases** (prō'tē-ās-ez)—enzymes that split proteins into their constituent amino acids (**Figure 5.25** on page 152). Once released by the action of proteases, amino acids are transported into the cell, where special enzymes split off amino groups in a reaction called **deamination.** The resulting altered molecules enter the Krebs cycle, and the amino groups are either recycled to synthesize other amino acids or excreted as nitrogenous wastes such as ammonia (NH_3), ammonium ion (NH_4^+), or trimethylamine oxide (which has the chemical formula $(CH_3)_3NO$ and is abbreviated TMAO). As **Highlight 5.2** on page 152 describes, TMAO plays an interesting role in the production of the odor we describe as "fishy."

Thus far, we have examined the catabolism of carbohydrates, lipids, and proteins. Now we turn our attention to the *synthesis* of these molecules, beginning with the anabolic reactions of photosynthesis.

[9]"Beta" is part of the name of this process because enzymes break the bond at the second carbon atom from the end of a fatty acid, and beta is the second letter in the Greek alphabet.

Figure 10

Critical thinking questions are integrated throughout the text, as well as at the end of each chapter.

Learning Objectives throughout each chapter keep the student focused.

Integrated pronunciation guides help students pronounce key terms.

Etymology guides help students retain word definitions and enhance comprehension.

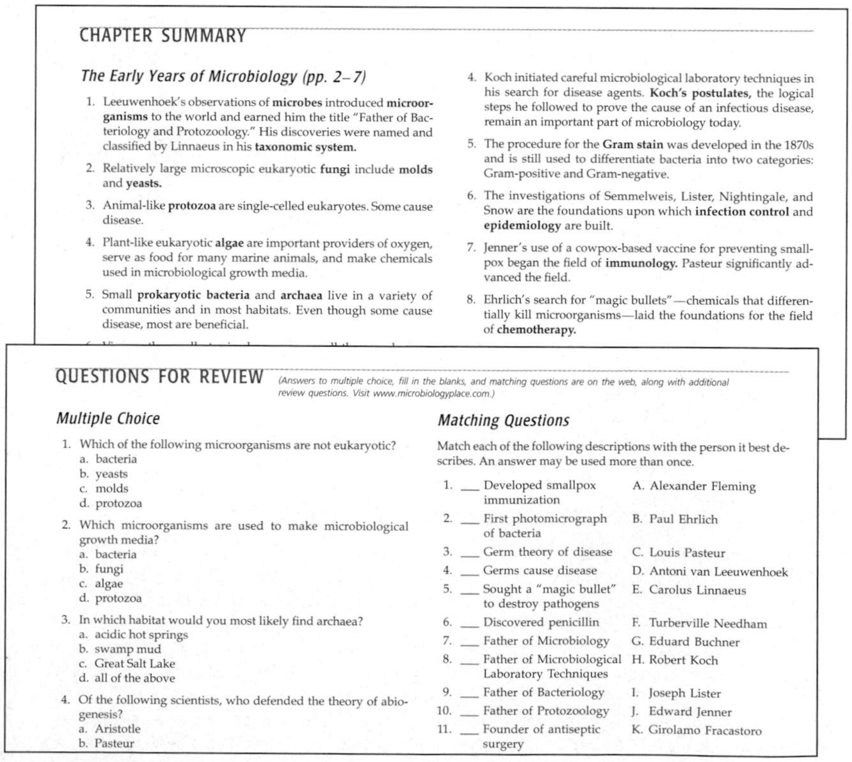

CHAPTER SUMMARY

The Early Years of Microbiology (pp. 2–7)

1. Leeuwenhoek's observations of **microbes** introduced **microorganisms** to the world and earned him the title "Father of Bacteriology and Protozoology." His discoveries were named and classified by Linnaeus in his **taxonomic system.**
2. Relatively large microscopic eukaryotic **fungi** include **molds** and **yeasts.**
3. Animal-like **protozoa** are single-celled eukaryotes. Some cause disease.
4. Plant-like eukaryotic **algae** are important providers of oxygen, serve as food for many marine animals, and make chemicals used in microbiological growth media.
5. Small **prokaryotic bacteria** and **archaea** live in a variety of communities and in most habitats. Even though some cause disease, most are beneficial.
4. Koch initiated careful microbiological laboratory techniques in his search for disease agents. **Koch's postulates,** the logical steps he followed to prove the cause of an infectious disease, remain an important part of microbiology today.
5. The procedure for the **Gram stain** was developed in the 1870s and is still used to differentiate bacteria into two categories: Gram-positive and Gram-negative.
6. The investigations of Semmelweis, Lister, Nightingale, and Snow are the foundations upon which **infection control** and **epidemiology** are built.
7. Jenner's use of a cowpox-based vaccine for preventing smallpox began the field of **immunology.** Pasteur significantly advanced the field.
8. Ehrlich's search for "magic bullets"—chemicals that differentially kill microorganisms—laid the foundations for the field of **chemotherapy.**

QUESTIONS FOR REVIEW (Answers to multiple choice, fill in the blanks, and matching questions are on the web, along with additional review questions. Visit www.microbiologyplace.com.)

Multiple Choice

1. Which of the following microorganisms are not eukaryotic?
   a. bacteria
   b. yeasts
   c. molds
   d. protozoa
2. Which microorganisms are used to make microbiological growth media?
   a. bacteria
   b. fungi
   c. algae
   d. protozoa
3. In which habitat would you most likely find archaea?
   a. acidic hot springs
   b. swamp mud
   c. Great Salt Lake
   d. all of the above
4. Of the following scientists, who defended the theory of abiogenesis?
   a. Aristotle
   b. Pasteur

Matching Questions

Match each of the following descriptions with the person it best describes. An answer may be used more than once.

1. ___ Developed smallpox immunization	A. Alexander Fleming
2. ___ First photomicrograph of bacteria	B. Paul Ehrlich
3. ___ Germ theory of disease	C. Louis Pasteur
4. ___ Germs cause disease	D. Antoni van Leeuwenhoek
5. ___ Sought a "magic bullet" to destroy pathogens	E. Carolus Linnaeus
6. ___ Discovered penicillin	F. Turberville Needham
7. ___ Father of Microbiology	G. Eduard Buchner
8. ___ Father of Microbiological Laboratory Techniques	H. Robert Koch
9. ___ Father of Bacteriology	I. Joseph Lister
10. ___ Father of Protozoology	J. Edward Jenner
11. ___ Founder of antiseptic surgery	K. Girolamo Fracastoro

Figure 11

End-of-chapter summaries and **review questions** help students master the material.

A Powerful Media and Supplements Package

This text is accompanied by **The Microbiology Place CD-ROM/Website (Figure 12).** The Microbiology Place (www.microbiologyplace.com) hosts the MicroPrep Pre-tests, along with over 40 practice test questions for each chapter of the text, review questions that utilize micrographs and other art from the text, weblinks, electronic flashcards, answers to questions in the text, information about careers in microbiology, and more. It also contains links to **Research Navigator,** a powerful online research tool with access to three exclusive databases of credible and reliable source material, including the *Journal of Applied Microbiology,* the *Annual Review of Microbiology, Science,* and the *New York Times* science pages. Available in web and/or CD format.

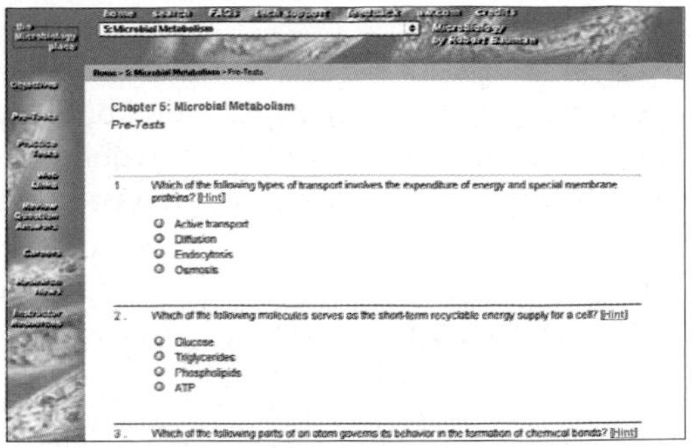

▲ *Figure 12*
The Microbiology Place website/CD-ROM includes MicroPrep Pre-Tests to aid student preparation, along with practice quizzes, web links, flashcards, and more.

Other supplements available with this text include the following:

For the Instructor

Instructor's Resource CD-ROM 0-8053-7657-7
This cross-platform CD-ROM provides PowerPoint® lecture outlines, digital videos of microorganisms, and electronic files of the line art, photographs, and micrographs from the book, as well as hundreds of additional images relating to pathogens, diseases, and environmental microbiology (approximately 800 images, **Figure 13**). All images are identified by figure number and are available in both PowerPoint® and jpeg formats, enabling instructors to quickly and easily prepare powerful, visually engaging classroom presentations. In addition, several key illustrations are provided in a customizable, step-by-step format—allowing instructors maximum flexibility in deciding how to present complex topics within PowerPoint®.

Instructor's Manual and Test Bank 0-8053-7658-5
The Instructor's Manual contains expanded chapter summaries, a visual guide to the art, tips for teaching the course, a guide to The Microbiology Place Website, and more. The Test Bank contains approximately 1800 multiple-choice, matching, true/false, fill-in-the-blank, and essay questions, with their answers.

Transparency Acetates 0-8053-7661-5
Features approximately 400 illustrations, including all illustrations from the text and hundreds of additional illustrations relating to pathogens, diseases, and environmental microbiology.

Computerized Test Bank CD-ROM 0-8053-7659-3
This easy-to-use, cross-platform testing program enables instructors to view and edit electronic questions from the Test Bank, create multiple tests, and print them in a variety of formats.

Biological Agents of Disease/HIV Slide Set for Microbiology 0-8053-7671-2
An additional 100 clinical photos of microbes and pathologies, all from sources other than the text.

▲ *Figure 13*
The Instructor's Resource CD-ROM includes lecture outlines, digital videos of microorganisms, and approximately 800 illustrations and photos from the Bauman text.

CourseCompass™
CourseCompass™ is a nationally hosted online course management system offering preloaded book-specific content, including testing and assessment, weblinks, illustrations, and photos. To view a demonstration, visit www.coursecompass.com.

For the Student

Study Guide 0-8053-7662-3
by Elizabeth Machunis-Masuoka
Students can master key concepts and earn a better grade with the help of the valuable tools found in this study guide. The study guide includes expanded chapter summaries, a guide to The Microbiology Place website, critical thinking questions, and a variety of self-test questions with answers.

Microbiology: A Photographic Atlas for the Laboratory
0-8053-2732-0
by Steve K. Alexander and Dennis Strete
A full-color atlas of approximately 400 high-quality color photographs that demonstrate the results of laboratory procedures and the morphology of important microorganisms.

Laboratory Experiments in Microbiology, Seventh Edition
0-8053-7673-9
by Ted R. Johnson and Christine L. Case
This lab manual includes 57 experiments that demonstrate basic microbiological techniques and applications.

Microbiology Coloring Book 0-06-041925-3
by I. Edward Alcamo and Lawrence M. Elson
This microbiology coloring atlas asks the reader to color a series of figures that convey microbiological principles and processes. An efficient review of all areas pertinent to a microbiology course, it simplifies the learning process and provides visually appealing figures that can be used for future study. The act of coloring helps students remember the structure and processes of microbiology, and how they relate to one another.

Microbiology: A Laboratory Manual, Sixth Edition
0-8053-7648-8
by James Cappuccino and Natalie Sherman
This versatile laboratory manual can be used with any undergraduate microbiology text and course. Known for its brief laboratory activities, minimal equipment requirements, and competitive price, the manual includes a variety of experiments selected to assist in the teaching of basic principles and techniques. Each of the 77 experiments includes an overview, a purpose, an in-depth discussion of the principle involved, easy-to-follow procedures, and lab reports with review and critical thinking questions. Comprehensive introductory material and laboratory safety instructions are provided.

The Microbe Files: Cases in Microbiology for the Undergraduate
by Marjorie K. Cowan
The first of its kind, *The Microbe Files* provides allied health and non-major microbiology students a fascinating series of short cases that help them apply what they have learned in the course. It's available in two versions:
(with answers) 0-8053-4927-8
(without answers) 0-8053-4928-6

Reviewers, Class Testers, and Focus Group Participants

I wish to thank the hundreds of instructors and students who reviewed and/or class-tested early drafts of this textbook. I also wish to thank the faculty and students who participated in focus groups during this book's development. Your comments have informed this book from beginning to end, and I am deeply grateful.

Reviewers and/or Class Testers

Ghava Ahmed
Seton Hall University

Craig Almeida
Stonehill College

Robin Almeida
University of North Carolina—Greensboro

John Amaral
University of San Francisco

Dawn Anderson
Berea College

Gail Baker
LaGuardia Community College

Susan Baxley
Troy State University—Montgomery

Keith Belcher
Austin Peay State University

Clinton Benjamin
Lower Columbia College

Margaret Beucher
Brigham Young University

Benjie Blair
Jacksonville State University

Pat Blaney
Brevard Community College—Cocoa

Susan Bornstein-Forst
Marian College

Kathryn Brooks
Michigan State University

Linda Bruslind
Oregon State University

Dave Bryan
Cincinnati State Technical Community College

Gregory W. Buck
Texas A&M University—Corpus Christi

George Bullerjahn
Bowling Green State University

Jackie Butler
Grayson County Community College

Christie Canady
Middle Georgia College

Jeff Cavanaugh
Troy State University—Montgomery

Elaine Charters
Xavier University

Kathy Clark
College of Southern Idaho

L. Clendaniel
Delaware Technical College—Georgetown Campus

Candice Coffin
Briar Cliff University

Don Collier
Calhoun Community College

Janet Cooper
Rockhurst University

Rhena Cooper
North Idaho College

Elaine Cox
Bossier Parish Community College

Lorraine Cramer
University of North Carolina—Chapel Hill

Victoria Crnekovic
Parkland College

Tucker Crum
Wellesley College

Don Dailey
Austin Peay State University

Roxanne Davenport
Tulsa Community College

Kelley Davis
University of Missouri, Columbia

Louis Debetaz
Angelina College

Paul DeLange
Kettering College of the Medical Arts

Paul Demchick
Barton College

Charles Denny
University of South Carolina—Sumter

Don Deters
Bowling Green State University

Beverly Dixon
California State University—Hayward

Diane Dixon
Southeast Oklahoma State University

Sandra Dubowsky
Allen County Community College

Khrys Duddleston
University of Alaska—Anchorage

Frances Duncan
Pensacola Jr. College

David Elmendorf
University of Central Oklahoma

Debby Filler
Anoka Ramsey Community College

John Friede
Villanova University

Denise Friedman
Hudson Valley Community College

David Fulford
Edinboro University of Pennsylvania

Karen Fulford
Morris Brown College

Fred Funk
Northern Arizona University

Rhonda Gamble
Mineral Area College

Beth Gaydos
San Jose City College

Aiah Gbakima
Morgan State University

David Gilmore
Arkansas State University

Michael Griffin
Angelo State University

Dale Hall
Southwestern Community College

Hershell Hanks
Collin County Community College

Ernie Hannig
University of Texas—Dallas

Ann Hanson
University of Maine

Elaine Hardwick
Minnesota State University—Mankato

Julie Harlass
Montgomery College

Donna Harman
Lubbock Christian University

Randall Harris
William Carey College

Sharon Harris
Columbia Basin Community College

Donna Hazelwood
Dakota State University

Alice Helm
University of Illinois—Urbana-Champaign

Daniel Herman
Grand Valley State University

Elizabeth Hoffman
Ashland Community College

Carolyn Holcroft
Foothill College

James Holda
University of Akron

Georgia Ineichen
Hinds Community College

Janice Ito
Leeward Community College

Bob Iwan
Anoka Ramsey Community College

Ruper Iyer
Wharton County Jr. College

Edythe Jones
Claflin University

Hinrich Kaiser
LaSierra University

Kevin Kaiser
Cape Fear Community College

Phyllis Katz
Endicott College

Karen Keller
Allegany College of Maryland

Karen Kendall-Fite
Columbia State Community College

Karen Kesterson
University of Nevada—Las Vegas
Harry Kestler
Lorain County Community College
Peggy Knittel
Eastern Wyoming College
Dubear Kroening
University of Wisconsin—Fox Valley
Karl Kvistberg
Nicolet College
Carol Lauzon
California State University—Hayward
Michael Lee
Joliet Jr. College
Jettie Lights
Columbus State University
Gayle LoPiccolo
Montgomery College
Candi Lostroh
Simmons College
Bonnie Lustigman
Montclair State University
Dawn Madl
Northeast Wisconsin Technical College
Reza Marvdashti
San Jacinto College, North
Luis Materon
University of Texas—Pan American
Kim Maznicki
Seminole Community College
Donald McGarey
Kennesaw State University
Elizabeth McPherson
University of Tennessee—Knoxville
Robert McReynolds
San Jacinto College, Central
Gina Marie Morris
Frank Phillips College
Quian Moss
Des Moines Community College
Rita Moyes
Texas A&M University
John Myers
Owens Community College
Linda Nazitski
Tunxis Community College
Mary Jane Niles
University of San Francisco
Catherine O'Brien
San Jacinto College, South
Karen Palin
Bates College
Robin Patterson
Butler Community College
Bobbie Pettriess
Wichita State University
Marcia Pierce
Eastern Kentucky University
Laraine Powers
East Tennessee State University
Davis Pritchett
University of Louisiana—Monroe
John Pritchett
Florence Darlington Technical College
Charlie Purvis
Jefferson Community College, Southwest
Kirsten Raines
San Jacinto College, South
Kimberly Raun
Wharton County Jr. College
Sabina Rech
San Jose State University
Thomas Reilly
Ball State University
Clifford Renk
Florida Gulf Coast University
Jean Revie
South Mountain Community College
Fayette Reynolds
Berkshire Community College
Tim Rhoads
South Georgia College
Kathleen Richardson
Portland Community College
Shelly Robertson
University of Wyoming
Jackie Rosby
Milwaukee Area Technical College
Ed Rowland
Ohio University
Michael Ruhl
Vermont Regional Jr. College
Barbara Rundell
College of DuPage
Jerry Sanders
University of Michigan—Flint
Pramila Sen
Houston Community College, Central Campus
Victoria Sharpe
Blinn College—Bryan Campus
Patty Shields
University of Maryland
Brian R. Shmaefsky
Kingwood College
Linda Simpson
University of North Carolina—Charlotte
Luci Nell Simpson
Weatherford College
Cheryl Smith
Pennsylvania College of Technology
Sally Smith
Okefenokee Technical College
Kris Snow
Fox Valley Technical College
Walter Spangenberg
Spokane Falls Community College
Robert Speed
Wallace Community College
Kathy Steinart
Bellevue Community College
Kathleen Steinert
Bellevue Community College
Gail Stewart
Camden County College
John Story
Northwest Arkansas Community College
Julie Sutherland
College of DuPage
Jeffrey Taylor
North Central State College
Monica Tischler
Benedictine University
Stephen Torosian
University of New Hampshire
James Urban
Kansas State University
Margaret Waldman
Allegany College of Maryland
Burton Webb
Indiana University School of Medicine
Larry Weiskirch
Onondaga Community College
S. Williams
Delaware Technical College—Georgetown Campus
Christina Wilson
Southern Illinois University, Edwardsville
T. Yee Wong
University of Memphis
Patricia Yokell
New Hampshire Technical Institute
Kathy Zarilla
Durham Technical Community College

Faculty Focus Group Participants

Bill Boyko
Sinclair Community College
Elizabeth Hoffman
Ashland Community College
Richard Karp
University of Cincinnati
Marcia Pierce
Eastern Kentucky University
Charlie Purvis
Jefferson Community College, Southwest
Galen Renwick
Indiana University, Southeast
Bill Staddon
Eastern Kentucky University

Student Focus Group Participants

Instructor: Pat Johnson
Palm Beach Community College
Sharon Cordero
Palm Beach Community College
Samantha Dill
Palm Beach Community College
Melanie Radd
Palm Beach Community College
Joshua Suttle
Palm Beach Community College
Instructor: Kathryn Brooks
Michigan State University
Traci Klomparens
Michigan State University
Katy Loyd
Michigan State University
Nicole Westover
Michigan State University
Instructor: Joy Voorhees
Capital University
Christina Tompkins
Capital University

Acknowledgments

Writing a textbook was a larger undertaking than I imagined when I began. Though my name is on the cover, this book has truly been a team effort. I am deeply grateful to Daryl Fox, Serina Beauparlant, Leslie Berriman, and Barbara Yien of Benjamin Cummings, and to the team they gathered to make this book happen. Daryl's support for this book never wavered, and his willingness to take risks in support of innovation is much appreciated. Serina and Barbara helped develop the "vision" for this book, constantly coming up with ideas for making it more effective and compelling. As project editor, Barbara also had the Herculean task of coordinating everything and keeping us all on track. Thank you, Barbara—"Mack" and I will miss the daily phone calls. Laura Bonazzoli and Alan Titche edited the manuscript thoroughly and meticulously—credit for my successes is in large part due to them, while any errors or problems are mine alone. Laura Southworth developed many of this book's complex "process" figures and ensured that all the art in the text was styled consistently. The incomparable Ken Probst is responsible for creating this book's amazingly beautiful biological illustrations, including the cover image. Kay Ueno and Amy Teeling provided editorial guidance in the early stages of this book. Elizabeth Machunis-Masuoka, Randall Harris, Temma Al-Mukhtar, Michelle Bauman, Laura Bonazzoli, and Ian Tizard all made valuable writing contributions to the text and its supplements. Wendy Earl, Sharon Montooth, and Kelly Murphy pulled out all the stops to meet the book's aggressive production schedule. Diane Austin and Kathleen Olson did a superb job researching the photos for this text and Rich Robison supplied many of the text's wonderful and unique micrographs. tani hasegawa created the book's elegant design. Thanks also to Tracey Hines, Nan Kemp, Ziki Dekel, Ryan Shaw, and Blythe Robbins for their administrative and research assistance. Finally, Cheryl Cechvala, Lauren Harp, and the Benjamin Cummings book reps have done a terrific job of keeping in touch with the professors and students who provided so many wonderful suggestions for this textbook. I am especially grateful to Larry Weiskirch of Onondaga Community College and Mary Jane Niles of the University of San Francisco for their expertise and advice on the immunology chapters.

On the home front, I am grateful for Nan Kemp, Dr. Nichol Dolby, and Art Schneider at Amarillo College; Chris Gilbert, former student, racquetball partner, and dear friend; Tracey Hays, business consultant and partner in what really matters; Dr. Vance Esler, Larry Latham, Mike Isley, and Dr. Dustin Anderson, lunch buddies—all of them were always supportive and helpful. My "secretarial staff," Michelle, Jennie, Elizabeth, and Jeremy Bauman, were always there to type, file, surf the web, run to the FedEx box, or rub my feet and neck. Many men may be blessed, but I am blessed above all; you, my family, are the best. Finally, thanks to JC, who got me through it all.

Robert W. Bauman
Microbiology
Benjamin Cummings
1301 Sansome Street
San Francisco, CA 94111

BRIEF EDITION CONTENTS-AT-A-GLANCE

CONTENTS

CHAPTER 16
Specific Defense: The Immune Response 462

CHAPTER 17
Immunization and Immune Testing 487

CHAPTER 18
Hypersensitivities, Autoimmune Diseases, and Immune Deficiencies 509

BOXES

New Frontiers

Highlights

Case Studies

CHAPTER 1

A Brief History of Microbiology

Life as we know it would not exist without microorganisms. Plants depend on microorganisms to help them obtain the nitrogen they need for survival. Animals such as cows and sheep need microorganisms in order to digest the cellulose in their plant-based diets. Our ecosystems rely on microorganisms to enrich soil, degrade wastes, and support life. We use microorganisms to make wine and cheese and to develop vaccines and antibiotics. The human body is home to billions of microorganisms, many of which help keep us healthy. Microorganisms are not only an essential part of our lives, they are quite literally a part of us.

Of course, some microorganisms do cause disease, from the common cold to AIDS. The threats of bioterrorism and new or re-emerging infectious diseases are real. This textbook explores all the roles—both harmful and beneficial—that microorganisms play in our lives, as well as their sophisticated structures and processes. We begin with a look at the history of microbiology, starting with the invention of crude microscopes that revealed, for the first time, the existence of this miraculous, miniature world.

CHAPTER OUTLINE

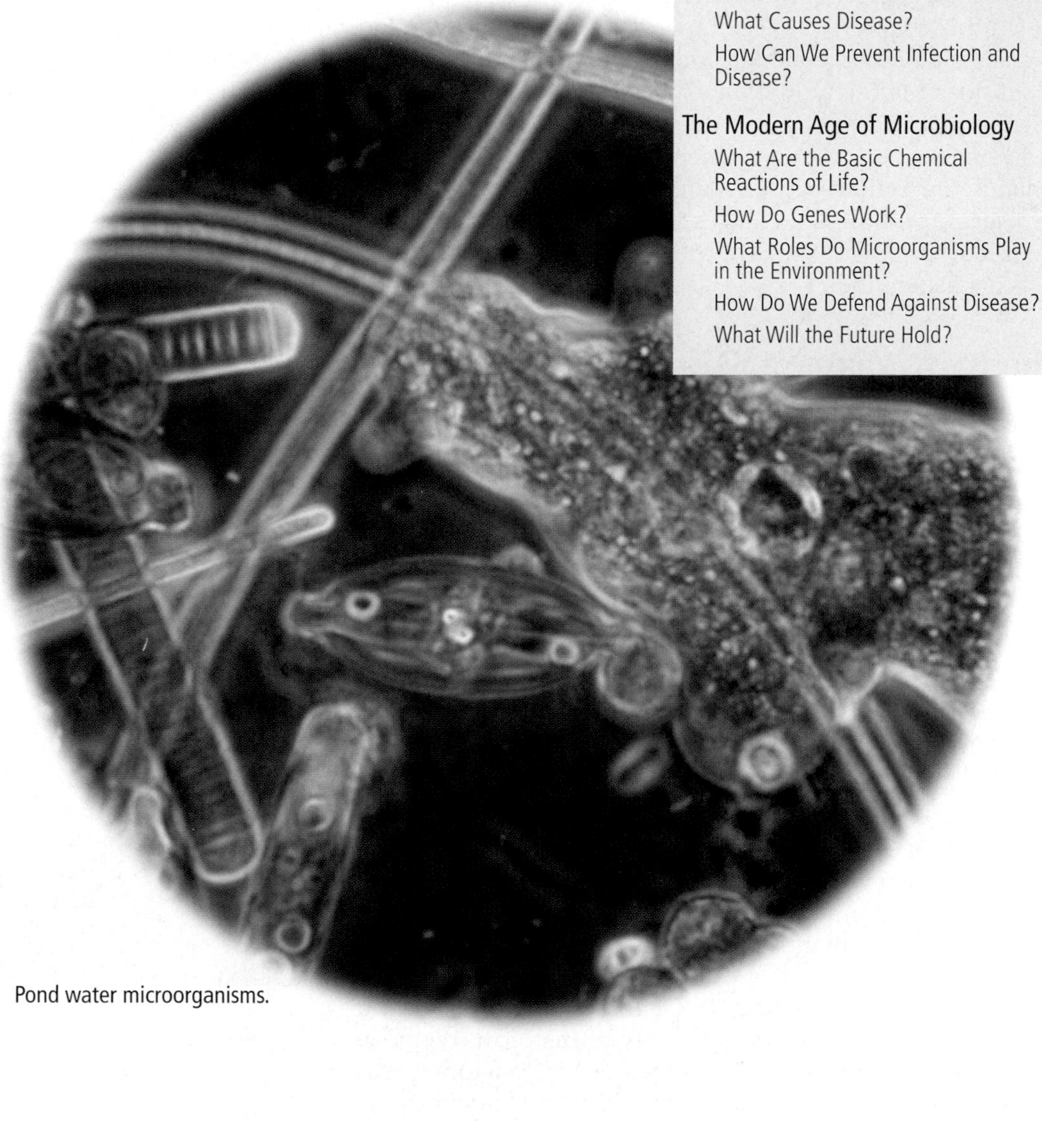

Pond water microorganisms.

MicroPrep Pre-Test: *Take the pre-test for this chapter on the web. Visit* ***www.microbiologyplace.com****.*

Science is the study of nature that proceeds by posing questions about observations. Why are there seasons? What is the function of the nodules at the base of this plant? Why does this bread taste sour? What does plaque from between teeth look like when magnified? Why are so many crows dying this winter? What causes disease?

Many early written records show that people have always asked questions like these. For example, the Greek physician Hippocrates (ca. 460–ca. 377 B.C.) wondered whether there is a link between environment and disease, and the Greek historian Thucydides (ca. 460–ca. 404 B.C.) questioned why he and other survivors of the plague could have intimate contact with victims and not fall ill again. For many centuries the answers to these and other fundamental questions about the nature of life remained largely unanswered. But then, about 350 years ago, the invention of the microscope began to provide some clues.

In this chapter we'll see how one man's determination to answer a fundamental question about the nature of life—What does life really look like?—led to the birth of a new science called *microbiology.* We'll then see how the search for answers to other questions, such as those concerning spontaneous generation, why fermentation occurs, and the cause of disease, prompted advances in this new science. Finally, we'll look briefly at some of the key questions microbiologists are asking today.

Figure 1.1
Antoni van Leeuwenhoek, the Father of Protozoology and Bacteriology, who reported the existence of protozoa in 1674 and of bacteria in 1676. *Why did Leeuwenhoek discover protozoa before bacteria?*

Figure 1.1 *Protozoa are generally larger than bacteria.*

The Early Years of Microbiology

The early years of microbiology brought the first observations of microbial life and the initial efforts to organize them into logical classifications.

What Does Life Really Look Like?

Learning Objectives

- ✓ Describe the world-changing scientific contributions of Leeuwenhoek.
- ✓ Define microbes in the words of Leeuwenhoek and as we know them today.

Only a few people have changed the world of science forever. We've all heard of Galileo, Newton, and Einstein, but the list also includes Antoni van Leeuwenhoek (lā′vĕn-huk; 1632–1723), a Dutch tailor, merchant, and lens grinder, and the man who first discovered the microbial world **(Figure 1.1).**

Leeuwenhoek was born in Delft, the Netherlands, and lived most of his 90 years in the city of his birth. What set Leeuwenhoek apart from other men of his generation was an insatiable curiosity coupled with an almost stubborn desire to do everything for himself. His journey to fame began simply enough, when as a tailor he needed to examine the quality of cloth. Rather than merely buying one of the magnifying lenses already available, he learned to grind glass and made his own **(Figure 1.2).** Soon he began asking the

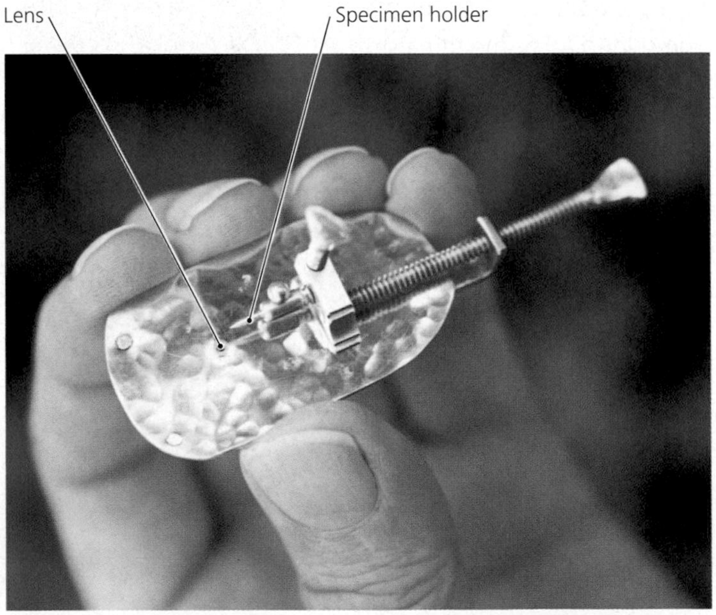

Figure 1.2
Leeuwenhoek's microscope. This simple device is little more than a magnifying glass with screws for manipulating the specimen; yet with it, Leeuwenhoek changed the way we see our world. The lens, which is convex on both sides, is about the size of a pinhead. The object to be viewed was mounted either directly on the specimen holder or inside a small glass tube, which was then mounted on the specimen holder.

question "What does it really look like?" of everything in his world: the stinger of a bee, the brain of a fly, the leg of a louse, a drop of blood, flakes of his own skin. To find answers, he spent hours examining, reexamining, and recording every detail of each object he observed.

Making and looking through his simple microscopes, most really no more than magnifying glasses, became the overwhelming passion of his life. His enthusiasm and dedication are evident from the fact that he sometimes personally extracted the metal for his microscope from ore. Further, he often made a new microscope for each specimen, which remained mounted so that he could view it again and again. Then one day, he turned a lens onto a drop of water. We don't know what he expected to see, but certainly he saw more than he had anticipated. As he reported to the Royal Society of London[1] in 1674, he was surprised and delighted by

> *some green streaks, spirally wound serpent-wise, and orderly arranged. . . . Among these there were, besides, very many little animalcules, some were round, while others a bit bigger consisted of an oval. On these last, I saw two little legs near the head, and two little fins at the hind most end of the body. . . . And the motion of most of these animalcules in the water was so swift, and so various, upwards, downwards, and round about, that 'twas wonderful to see.*

Leeuwenhoek had discovered a previously unknown microbial world, which today we know to be populated with tiny animals, fungi, algae, and single-celled protozoa **(Figure 1.3).** In a later report to the Royal Society, he noted that

> *the number of these animals in the scurf [plaque] of a man's teeth, are so many that I believe they exceed the number of men in a kingdom. . . . I found too many living animals therein, that I guess there might have been in a quantity of matter no bigger than the 1/100 part of a [grain of] sand.*

From the figure accompanying this report and the precise description of the size of these organisms from between his teeth, we know that Leeuwenhoek was reporting the existence of bacteria. By the end of the 19th century, Leeuwenhoek's "beasties," as he sometimes dubbed them, were called **microbes,** and today we also know them as **microorganisms.** Both terms include all organisms that are too small to be seen without a microscope.

Because of the quality of his microscopes, his profound observational skills, his detailed reports over a 50-year period, and his report of the discovery of many types of microorganisms, Antoni van Leeuwenhoek was elected to the Royal Society in 1680. He and Isaac Newton were probably the most famous scientists of their time.

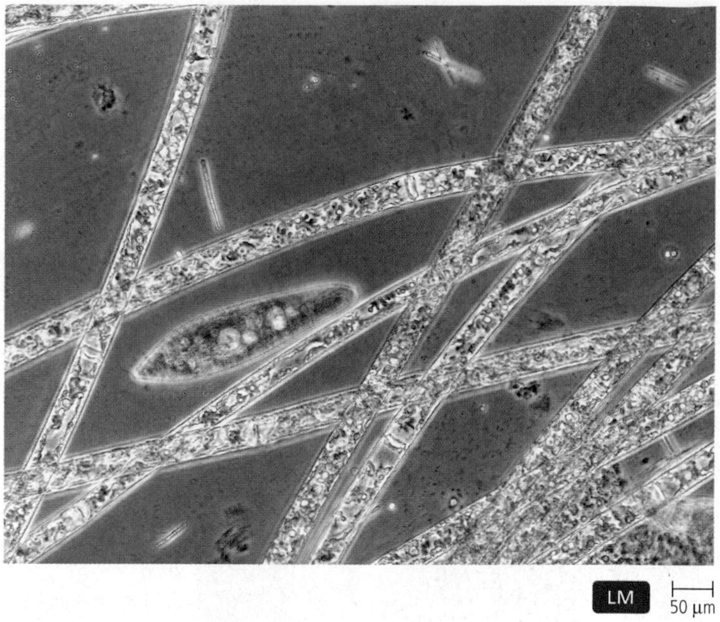

Figure 1.3
The microbial world. Leeuwenhoek reported seeing a scene very much like this, one full of numerous, fantastic, cavorting creatures.

How Can Microbes Be Classified?

Learning Objectives

- ✓ List five groups of microorganisms.
- ✓ Explain why protozoa, algae, and nonmicrobial parasitic worms are studied in microbiology.
- ✓ Differentiate between prokaryotic and eukaryotic organisms.

Shortly after Leeuwenhoek made his discoveries, the Swedish botanist Carolus Linnaeus (1707–1778) developed a **taxonomic system**—that is, a system for naming plants and animals and grouping similar organisms together. For instance, Linnaeus and other scientists of the period grouped all organisms into either the animal kingdom or the plant kingdom. Today, biologists still use this basic system, but they have modified Linnaeus's scheme by adding categories that more realistically reflect the relationships among organisms. For example, scientists no longer classify yeasts, molds, and mushrooms as plants, but instead as fungi. We examine taxonomic schemes in more detail in Chapter 4.

The microorganisms that Leeuwenhoek described can be grouped into five basic categories: fungi, protozoa, algae, prokaryotes, and small animals. The only microbes not described by Leeuwenhoek are *viruses,*[2] which are too small to

[1]The Royal Society of London for the Promotion of Natural Knowledge, granted a royal charter in 1662, is one of the older and more prestigious scientific groups in Europe.
[2]Technically, viruses are not "organisms" because they neither replicate themselves nor carry on the chemical reactions of living things. See Chapter 3 for a fuller discussion of this issue.

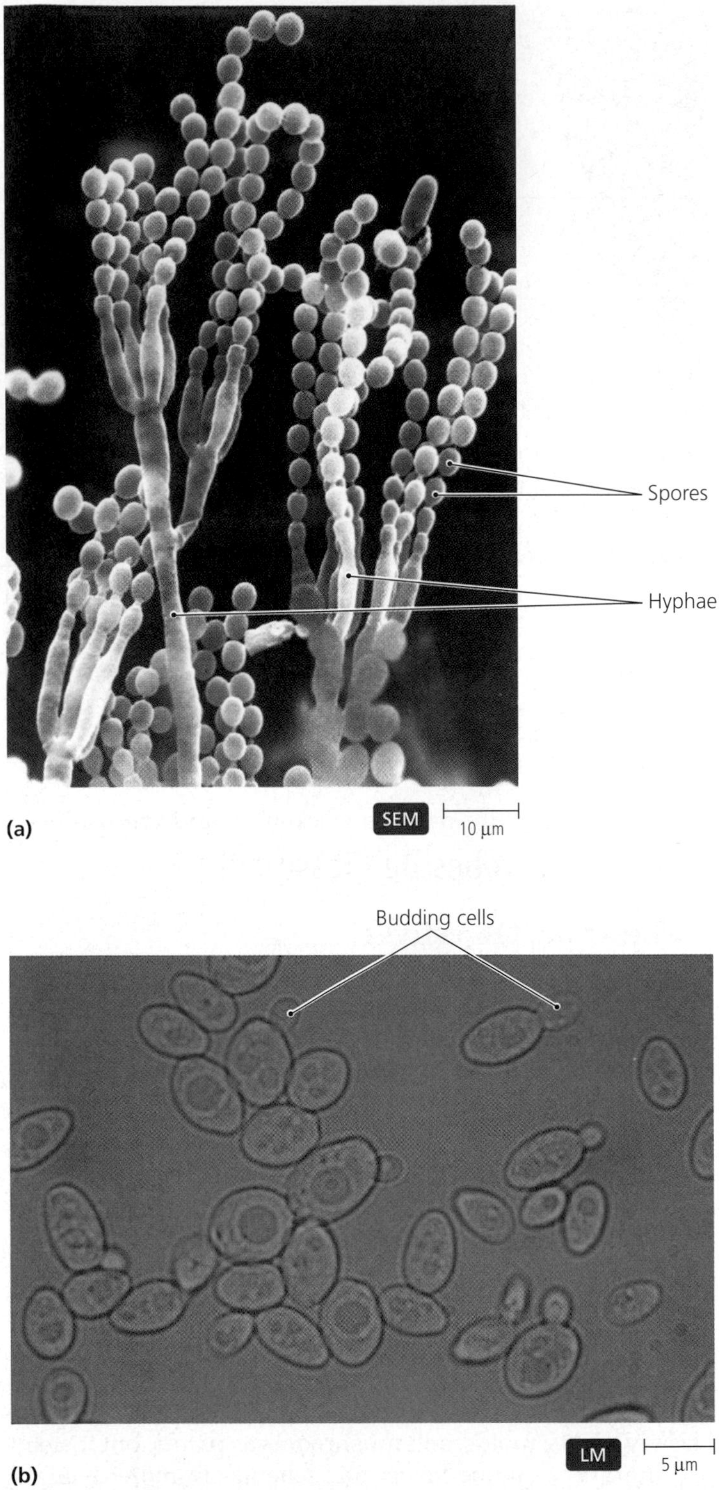

▲ *Figure 1.4*

Fungi. **(a)** The mold *Penicillium chrysogenum,* which produces penicillin, has long filamentous hyphae that intertwine to form its body. It reproduces by spores. **(b)** The yeast *Saccharomyces cerevisiae.* Yeasts are round to oval and typically reproduce by budding.

be seen without an electron microscope. We briefly consider organisms in the first four categories in the following sections.

Fungi

Fungi (fŭn′jī)[3] are organisms whose cells are **eukaryotic;**[4] that is, each of their cells contains a nucleus composed of genetic material surrounded by a distinct membrane. Fungi are different from plants because they obtain their food from other organisms (rather than making it by photosynthesis). They differ from animals by having cell walls.

Microscopic fungi include some molds and yeasts. **Molds** are typically multicellular organisms that grow as long filaments, called *hyphae,* that intertwine to make up the body of the mold. Molds reproduce by sexual and asexual spores, which are cells that produce a new individual without fusing with another cell **(Figure 1.4a).** The cottony growths on cheese, bread, and jams are examples of molds. *Penicillium chrysogenum* (pen-i-sil′ē-ŭm krī-so′jĕn-ŭm) is a mold that produces penicillin.

Yeasts are unicellular and typically oval to round. They reproduce asexually by *budding,* a process in which a daughter cell grows off the mother cell. Some yeasts also produce sexual spores. An example of a useful yeast is *Saccharomyces cerevisiae* (sak-ă-rō-mī′sēz se-ri-vis′ē-ē; **Figure 1.4b**) which causes bread to rise and produces alcohol from sugar. *Candida albicans* (kan′did-ă al′bi-kanz) is a yeast that causes most cases of yeast infections in women.

Fungi and their significance in the environment, in food production, and as agents of human disease are discussed in Chapter 12.

Protozoa

Protozoa are single-celled eukaryotes that are similar to animals in their nutritional needs and cellular structure. In fact, *protozoa* is Greek for "first animals." Most protozoa are capable of locomotion, and one way scientists categorize protozoa is according to their locomotive structures: *pseudopodia,*[5] *cilia,*[6] or *flagella.*[7] Pseudopodia are extensions of a cell that flow in the direction of travel **(Figure 1.5a).** Cilia are numerous short, hairlike protrusions of a cell that beat rhythmically to propel the protozoan through its environment **(Figure 1.5b).** Flagella are also extensions of a cell, but are fewer, longer, and more whiplike than cilia **(Figure 1.5c).** Some protozoa, such as the malaria-causing *Plasmodium* (plaz-mō′dē-ŭm), are nonmotile in their mature forms.

Protozoa typically live freely in water, but some live inside animal hosts, where they can cause disease. Most protozoa reproduce asexually, though some are sexual as well. The protozoa are discussed further in Chapter 12.

---

[3]Plural of the Latin *fungus,* meaning mushroom.
[4]From Greek *eu,* meaning true, and *karyon,* meaning kernel (which in this case refers to the nucleus of a cell).
[5]From Greek *pseudes,* meaning false, and *podos,* meaning foot.
[6]Plural of the Latin *cilium,* meaning eyelid.
[7]Plural of the Latin *flagellum,* meaning whip.

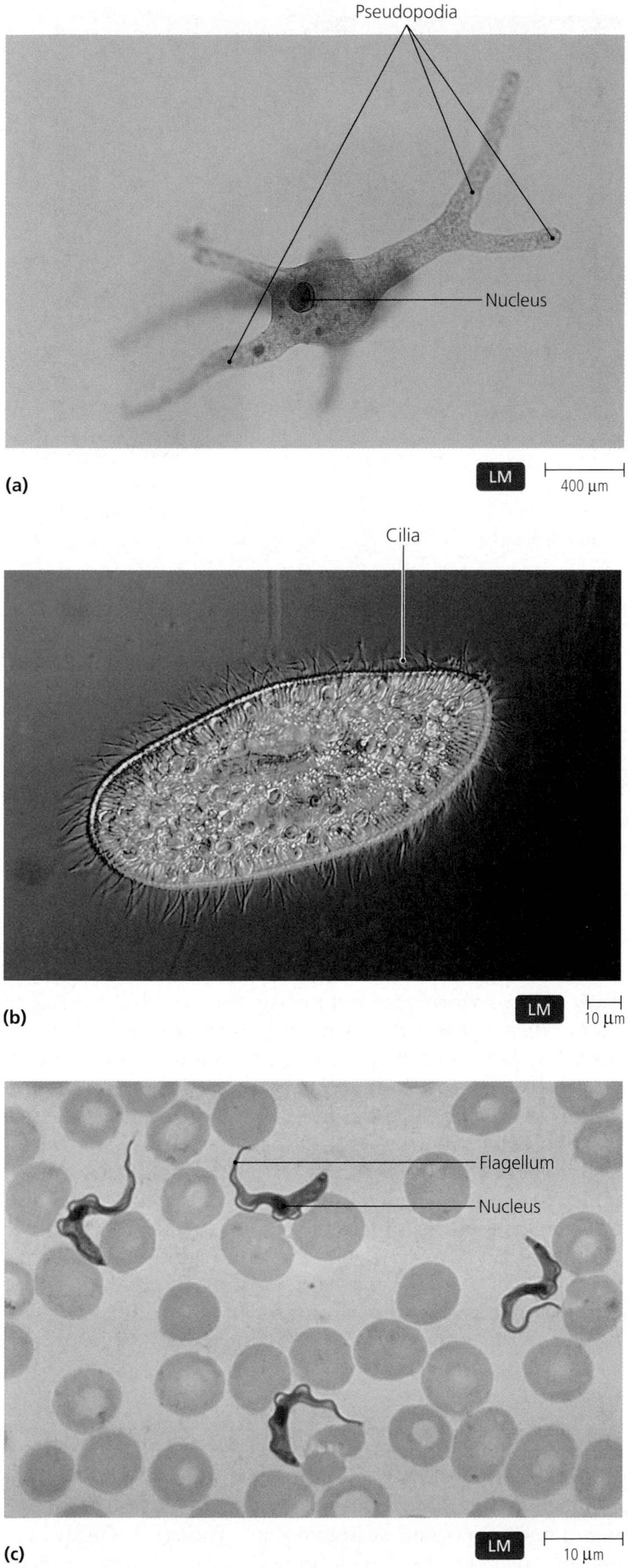

Figure 1.5
Locomotive structures of protozoa. **(a)** Pseudopodia, which are cellular extensions used for locomotion and feeding, as seen in *Amoeba proteus.* **(b)** Cilia, which are short, motile, hairlike extrusions over the entire surface of the cell, as seen in *Paramecium.* **(c)** Flagella, which are whiplike extensions that are less numerous and longer than cilia, as seen in *Trypanosoma brucei,* the protozoan that causes African sleeping sickness.
How do cilia and flagella differ?

Figure 1.5 *Cilia are short, numerous, and cover the cell, whereas flagella are long and relatively few in number.*

Algae

Algae[8] are unicellular or multicellular *photosynthetic* organisms (**Figure 1.6** on page 6); that is, like plants they make their own food from carbon dioxide and water using energy from sunlight. They differ from plants in the relative simplicity of their reproductive structures. Algae are categorized on the basis of their pigmentation, their storage products, and the composition of their cell walls.

Large algae, commonly called seaweeds and kelps, are common in the world's oceans. Chemicals from their gelatinous cell walls are used as thickeners and emulsifiers in many food and cosmetic products, as well as in microbiological laboratory media.

Unicellular algae are common in freshwater ponds, streams, and lakes, and in the oceans as well. They are the major food of small aquatic and marine animals and provide most of the world's oxygen as a by-product of photosynthesis. The glasslike cell walls of diatoms provide grit for many polishing compounds. Other aspects of the biology of algae are discussed in Chapter 12.

Prokaryotes

Prokaryotes[9] are unicellular microbes that lack nuclei. There are two kinds of prokaryotes—**bacteria** and **archaea.** Bacterial cell walls are composed of a polysaccharide called *peptidoglycan,* though some bacteria lack cell walls. The cell walls of archaea lack peptidoglycan and instead are composed of other polymers. Members of both groups reproduce asexually. Chapters 4 and 11 examine other differences between bacteria and archaea.

Most bacteria and archaea are much smaller than eukaryotic cells (**Figure 1.7** on page 6). They live singly or in pairs, chains, or clusters in almost every habitat containing sufficient moisture. Archaea are found in extreme environments, such as the highly saline Mono Lake in California, acidic hot springs in Yellowstone National Park, and oxygen-depleted mud at the bottom of swamps.

[8]Plural of the Latin *alga,* meaning seaweed.
[9]From Greek *pro,* meaning before, and *karyon,* meaning kernel.

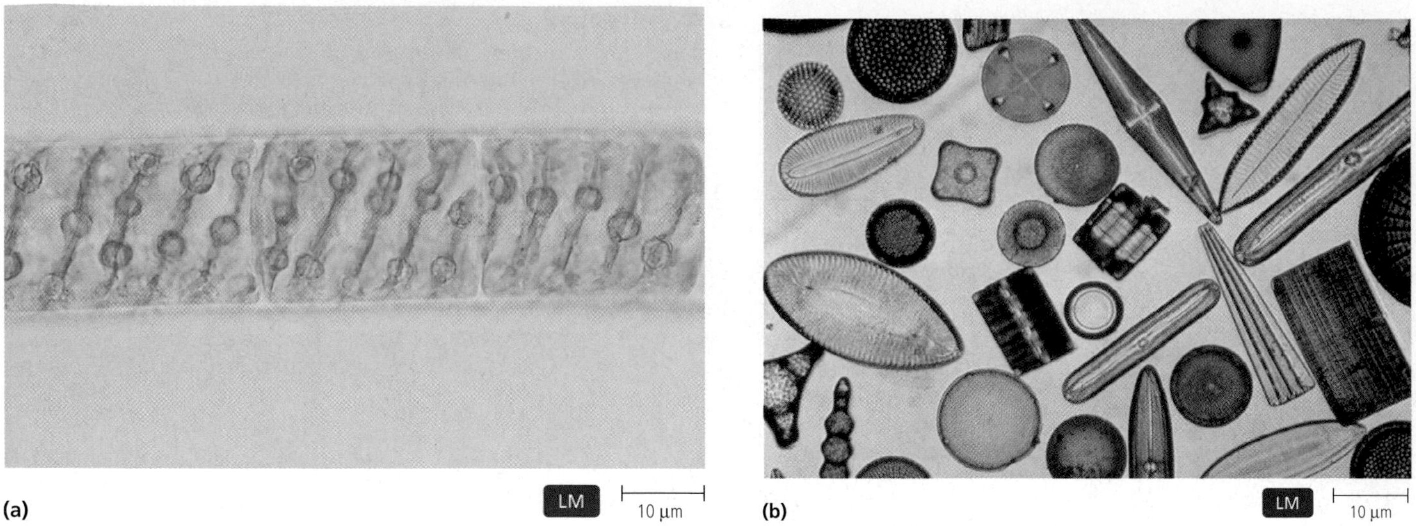

▲ ***Figure 1.6***

Algae. **(a)** *Spirogyra* sp. These microscopic algae grow as chains of cells containing helical photosynthetic organs. **(b)** Diatoms. These beautiful algae have glasslike cell walls.

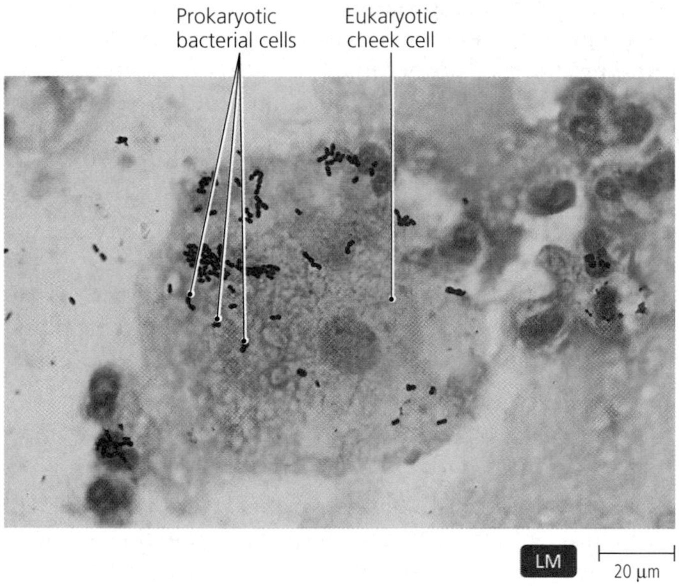

▲ ***Figure 1.7***

Cells of the bacterium *Streptococcus*, which Gram stain purple, surrounding a (eukaryotic) human cheek cell, which Gram stains pink. Notice the size difference.

Though bacteria often have a poor reputation in our world, the great majority of bacteria do not cause disease in animals, humans, or crops. Indeed, bacteria are beneficial to us in many ways. For example, dead plants and animals are degraded by bacteria (and fungi) to release phosphorus, sulfur, nitrogen, and carbon back into the air, soil, and water to be used by new generations of organisms. Without microbial recyclers, the world would be buried under the petrified corpses of uncountable dead organisms.

CRITICAL THINKING

A few bacteria produce disease because they derive nutrition from human cells and produce toxic wastes. Algae do not cause disease. Why not?

Other Organisms of Importance to Microbiologists

Microbiologists also study parasitic worms, which range in size from microscopic forms **(Figure 1.8)** to adult tapeworms over 7 meters (approximately 23 feet) in length. Even though most of these worms are not microscopic as adults, many of them cause diseases that were studied by early microbiologists. Further, laboratory technicians diagnose infections of parasitic worms by finding microscopic eggs and immature stages in blood, fecal, urine, and lymph specimens.

The only type of microbes that remained hidden from Leeuwenhoek and other early microbiologists are viruses, which are much smaller than the smallest prokaryote and are not visible by light microscopy **(Figure 1.9).** Viruses could not be seen until the electron microscope was invented in 1932. All viruses are acellular (not composed of cells) obligatory parasites composed of small amounts of genetic material (either DNA or RNA) surrounded by a protein coat. The general characteristics of viruses are discussed in Chapter 13.

Leeuwenhoek first reported the existence of microorganisms in 1674, but microbiology did not develop significantly as a field of study for almost two centuries. There were a number of reasons for this delay. First, Leeuwenhoek was a suspicious and secretive man. Though he built over 400 microscopes, he never trained an apprentice, and he never sold or gave away a microscope. In fact, he never let *anyone*—not his family or such distinguished visitors as the Czar of Russia and the Queen of England—so much as peek through his very best instruments. When Leeuwenhoek died, the secret of creating superior microscopes was lost. It took almost 100 years for scientists to make microscopes of equivalent quality.

Another reason that microbiology was slow to develop as a science is that scientists in the 1700s considered microbes to be curiosities of nature and insignificant to human affairs. But in the late 1800s, scientists began to adopt a new philosophy, one that demanded experimental proof rather than mere acceptance of traditional knowledge. This fresh philosophical foundation, accompanied by improved microscopes, new laboratory techniques, and a drive to answer a series of pivotal questions, propelled microbiology to the forefront as a scientific discipline.

The Golden Age of Microbiology

Learning Objective

✓ List and answer four questions that propelled research in what is called the "Golden Age of Microbiology."

For about 50 years, during what is now called "The Golden Age of Microbiology," scientists and the blossoming field of microbiology were driven by the search for answers to the following four questions:

1. Is spontaneous generation of microbial life possible?
2. What causes fermentation?
3. What causes disease?
4. How can we prevent infection and disease?

Competition among scientists who were striving to be the first to answer these questions drove exploration and discovery in microbiology during the late 1800s and early 1900s. These scientists' discoveries and the fields of study they initiated continue to shape the course of microbiological research today.

In the next sections we consider these questions, and how the great scientists accumulated the experimental evidence that answered them.

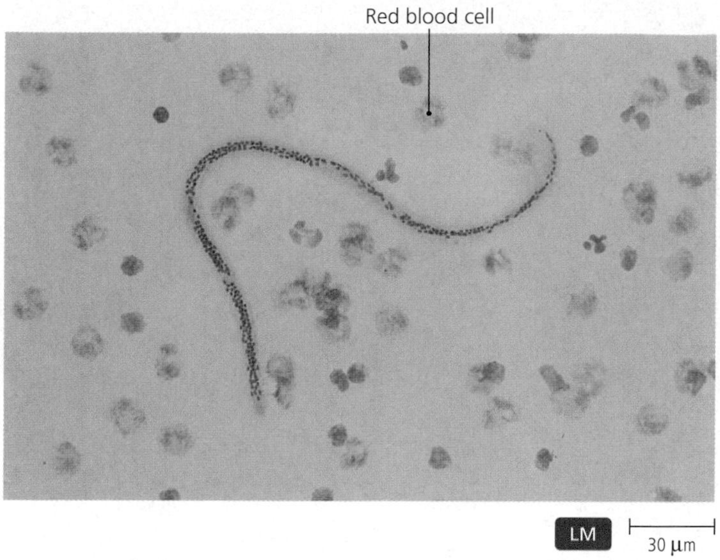

▲ ***Figure 1.8***
An immature stage of a parasitic worm in blood.

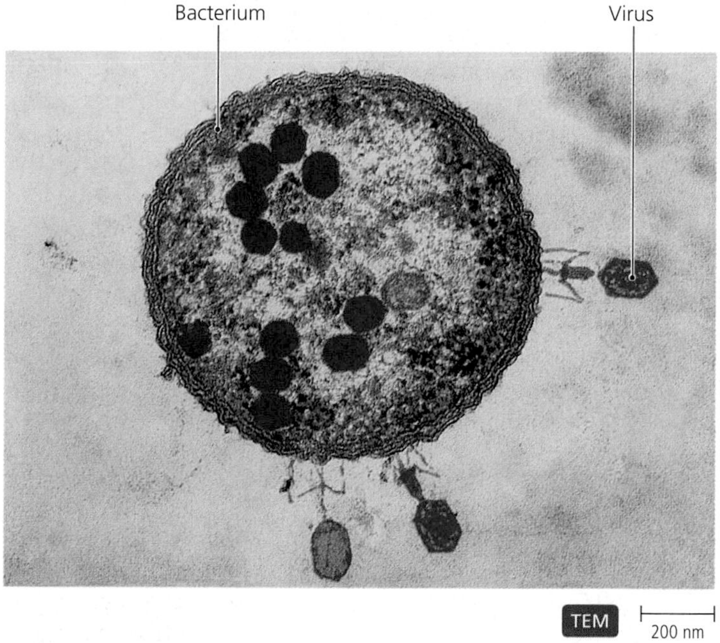

▲ ***Figure 1.9***
Viruses, which are acellular obligatory parasites, are too small to be seen with a light microscope. This photo is a colorized electron microscope image of viruses that infect bacteria. Notice how small the viruses are compared to the bacterium.

Is Spontaneous Generation of Microbial Life Possible?

Learning Objectives

✓ Identify the scientists who argued in favor of spontaneous generation.
✓ Compare and contrast the investigations of Redi, Needham, Spallanzani, and Pasteur concerning spontaneous generation.
✓ List four steps in the scientific method of investigation.

A dry lakebed has lain under the relentless North African desert sun for eight long months. The cracks in the baked, parched mud are wider than a man's hand. There is no sign of life anywhere in the scorched terrain. With the abruptness characteristic of desert storms, rain falls in a torrent, and a raging flood of roiling water and mud crashes down the dry streambed and fills the lake. Within hours, what had been a lifeless, dry mudflat becomes a pool of water teeming with billions of shrimp; by the next day it is home to hundreds of toads. Where did these animals come from?

▲ ***Figure 1.10***

Redi's experiments. When the flask remained unsealed, maggots covered the meat within a few days. When the flask was sealed, flies were kept away and no maggots appeared on the meat. When the flask opening was covered with gauze, flies were kept away and no maggots appeared on the meat, although a few maggots appeared on top of the gauze.

Many scientists and philosophers of ages past thought that living things can arise via three processes: through asexual reproduction, through sexual reproduction, or from nonliving matter. The appearance of shrimp and toads in the mud of what so recently was a dry lakebed was seen as an example of the third process, which came to be known as *abiogenesis*[10] or **spontaneous generation.** The theory of spontaneous generation as promulgated by Aristotle was widely accepted for over 1900 years because it seemed to explain a variety of commonly observed phenomena, such as the appearance of maggots on spoiling meat. However, the validity of the theory came under challenge in the 17th century.

Redi's Experiments

In the late 1600s, the Italian physician Francesco Redi (1626–1697) demonstrated by a series of experiments that when decaying meat was kept isolated from flies, maggots never developed, while meat exposed to flies was soon infested **(Figure 1.10).** As a result of experiments such as these, scientists began to doubt Aristotle's theory and adopt the view that animals only come from other animals.

Needham's Experiments

The debate over spontaneous generation was rekindled when Leeuwenhoek discovered microbes and showed that they appeared after a few days in freshly collected rainwater. Though scientists agreed that larger animals could not arise spontaneously, they disagreed about Leeuwenhoek's "wee animalcules"; surely they did not have parents, did they? They must arise spontaneously.

The proponents of spontaneous generation pointed to the careful demonstrations of British investigator John T. Needham (1713–1781). He boiled beef gravy and infusions[11] of plant material in vials, which he then tightly sealed with corks. Some days later, Needham observed that the vials were cloudy, and examination revealed an abundance of "microscopical [*sic*] animals of most dimensions." As he explained it, there must be a "life force" that causes inanimate matter to spontaneously come to life, since he had heated the vials sufficiently to kill everything. Needham's experiments so impressed the Royal Society that they elected him a member.

Spallanzani's Experiments

Then, in 1799, the Italian scientist Lazzaro Spallanzani (1729–1799) reported results that contradicted Needham's findings. Spallanzani boiled some infusions for almost an hour and sealed the vials by melting their slender necks closed. His infusions remained clear, unless he broke the seal and exposed the infusion to air, after which they became cloudy with microorganisms. He concluded three things:

- Needham had either failed to heat his vials sufficiently to kill all microbes, or he had not sealed them tightly enough.
- Microorganisms exist in the air and can contaminate experiments.
- Spontaneous generation of microorganisms does not occur; all living things arise from other living things.

---

[10]From Greek *a*, meaning not, *bios*, meaning life, and *genein*, meaning to produce.
[11]Infusions are broths made by steeping plant or animal material in water.

Although Spallanzani's experiments would appear to have settled the controversy once and for all, it proved difficult to dethrone a theory that had held sway for 2000 years, especially when so notable a scientist as Aristotle had propounded it. One of the criticisms of Spallanzani's work was that his sealed vials did not allow enough air for organisms to thrive; another objection was that his prolonged heating destroyed the "life force." The debate continued until the French chemist Louis Pasteur **(Figure 1.11)** conducted experiments that finally laid the theory of spontaneous generation to rest.

Pasteur's Experiments

Louis Pasteur (1822–1895) was an indefatigable worker who pushed himself as hard as he pushed others. As he wrote his sisters, "To *will* is a great thing dear sisters, for Action and Work usually follow Will, and almost always Work is accompanied by Success. These three things, Work, Will, Success, fill human existence. Will opens the door to success both brilliant and happy; Work passes these doors, and at the end of the journey Success comes to crown one's efforts." When his wife complained about his long hours in the laboratory, he replied, "I will lead you to fame."

Pasteur's determination and hard work are apparent in his investigations of spontaneous generation. Like Spallanzani, he boiled infusions long enough to kill everything. But instead of sealing the flasks, he bent their necks into an S-shape, which allowed air to enter while preventing the introduction of dust and microbes into the broth (**Figure 1.12a** on page 10).

Crowded for space and lacking funds, he improvised an incubator in the space under a staircase. Day after day he crawled on hands and knees into this incommodious space and examined his flasks for the cloudiness that would indicate the presence of living organisms. In 1861, he reported that his "swan-necked flasks" remained free of microbes even 18 months later. Since they contained all the nutrients (including air) known to be required by living things, he concluded, "Never will spontaneous generation recover from the mortal blow of this simple experiment."

Pasteur followed this experiment with demonstrations that microbes in the air were the "parents" of Needham's microorganisms. He broke the necks off some flasks, exposing the liquid in them directly to the air, and he carefully tilted others so that the liquid touched the dust that had accumulated in their necks **(Figure 1.12b).** The next day, all of these flasks were cloudy with microbes. He concluded that the microbes in the liquid were the progeny of microbes that had been on the dust particles.

The Scientific Method

The debate over spontaneous generation led to the development of a generalized **scientific method** by which questions are answered through observations of the outcomes of carefully controlled experiments, instead of by conjecture or according to the opinions of any authority figure. The scientific method, which provides a *framework* for conducting an investigation rather than a rigid set of specific "rules," consists of four basic steps (**Figure 1.13** on page 11):

▲ ***Figure 1.11***
Louis Pasteur, who disproved spontaneous generation and came to be known as the Father of Microbiology.

1. A group of observations leads a scientist to ask a question about some phenomenon.
2. The scientist generates a hypothesis—that is, a potential answer to the question.
3. The scientist designs and conducts an experiment to test the hypothesis.
4. Based on the observed results of the experiment, the scientist either accepts, rejects, or modifies the hypothesis.

As shown in Figure 1.13, the scientist then returns to earlier steps in the method, either modifying hypotheses and then testing them, or repeatedly testing accepted hypotheses, until the evidence for a hypothesis is convincing. Accepted hypotheses that explain many observations and are repeatedly verified by numerous scientists over many years may become *theories* or *laws*.

Note that in order for experiments (and their results) to be accepted as valid by the scientific community, they must include appropriate *control groups*—groups that are treated exactly the same as the other groups in the experiment, except for the one variable that the experiment is designed to test. In Pasteur's experiments on spontaneous generation, for example, his "control flasks" contained a sterile infusion composed of all the nutrients living things need, as well as air made available through the flasks' "swan necks." His "experimental flasks" for testing his hypothesis—that microbes would reach (and subsequently grow in) the infusion through contact with dust particles—were exposed to exactly the same conditions, *plus* contact with the dust in the

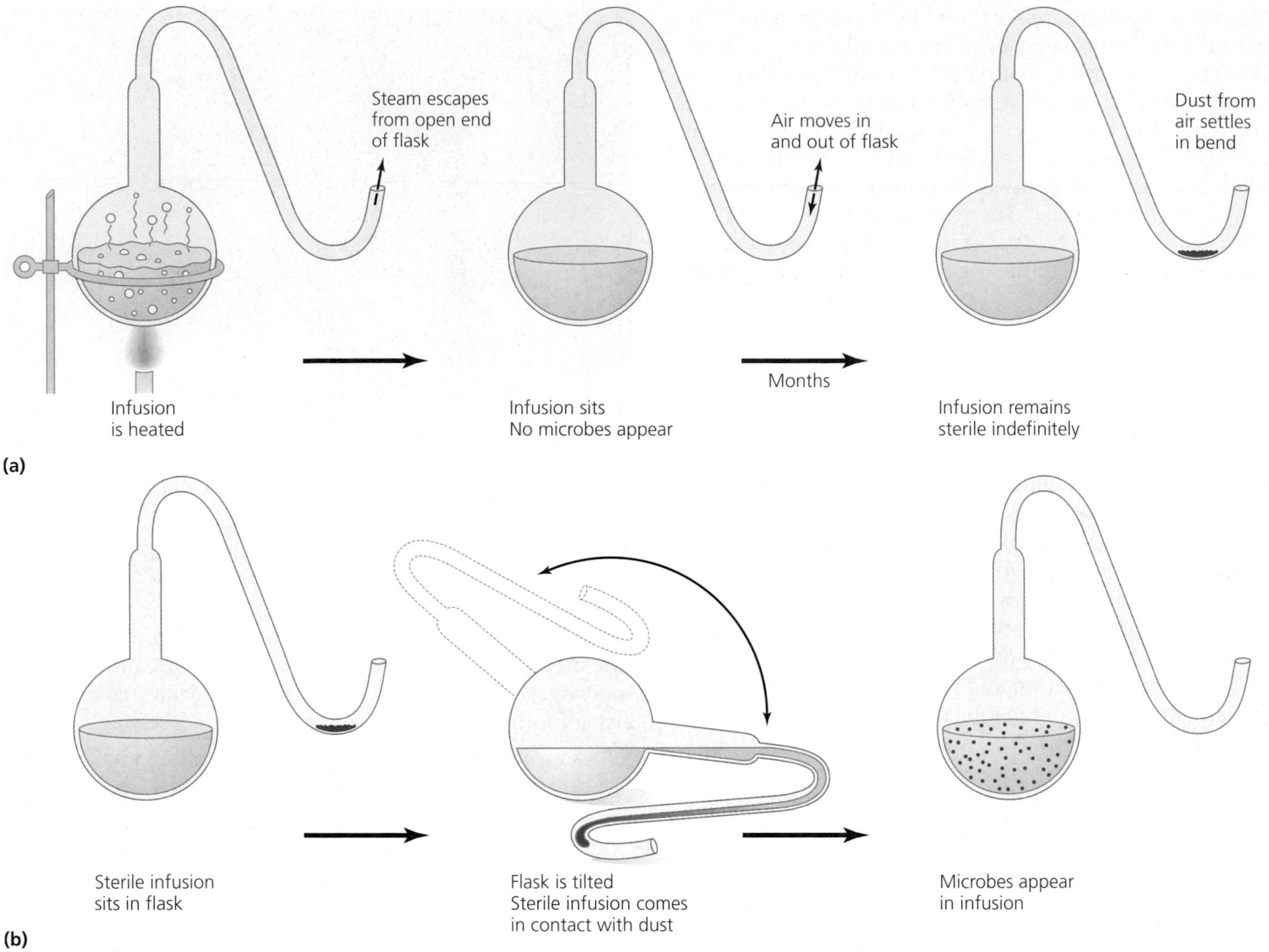
Steam escapes
from open end
of flask
Air moves in
and out of flask
Dust from
air settles
in bend
Months
Infusion
is heated
Infusion sits
No microbes appear
Infusion remains
sterile indefinitely
(a)
Sterile infusion
sits in flask
Flask is tilted
Sterile infusion comes
in contact with dust
Microbes appear
in infusion
(b)

▲ ***Figure 1.12***

Pasteur's experiments with "swan-necked flasks." **(a)** As long as the flask remained upright, no microbial growth appeared in the infusion. **(b)** When the flask was tilted so that some dust from the bend in the neck was carried back into the flask, the infusion became cloudy with microbes within a day.

bend in the neck. Because exposure to the dust was the *only* difference between the control and experimental groups, Pasteur was able to conclude that the microbes growing in the infusion arrived on the dust particles.

What Causes Fermentation?

Learning Objectives

- ✓ Discuss the significance of Pasteur's fermentation experiments to our world today.
- ✓ Explain why Pasteur is known as the Father of Microbiology.
- ✓ Identify the scientist whose experiments led to the field of biochemistry and the study of metabolism.

The controversy over spontaneous generation was largely a philosophical exercise among men who conducted research to gain basic scientific knowledge. They had no practical goals in mind—except, perhaps, personal aggrandizement in the form of financial support, honor, and prestige. However, the second question that moved microbial studies forward in the 1800s had tremendous practical applications.

Our story resumes in 19th-century France where spoiled, acidic wine was threatening the livelihood of many grape growers. The initial question was, "Why is the wine spoiled?" but this led to a more fundamental question, "What causes the fermentation of grape juice into wine?" These questions were so important to vintners that they funded research concerning fermentation, hoping scientists could develop methods to promote the production of alcohol and prevent spoilage by acid during fermentation.

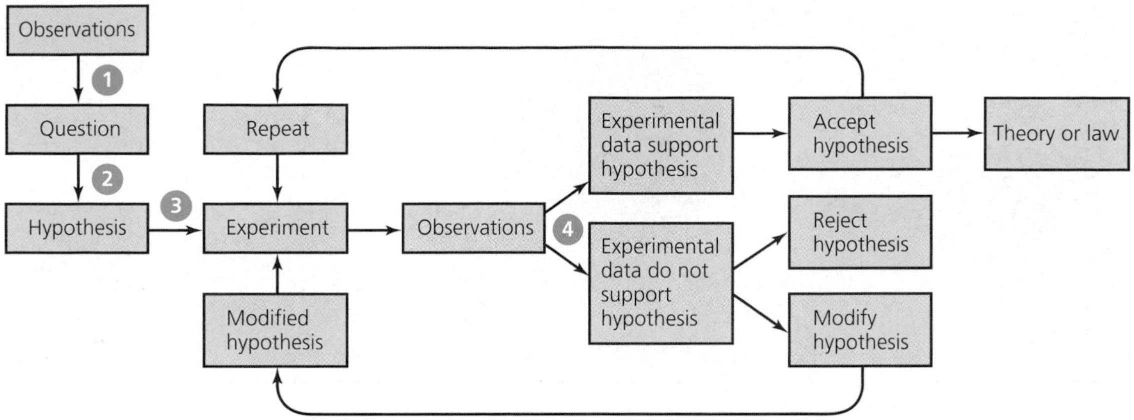

▲ ***Figure 1.13***

The scientific method, which forms the framework for scientific research. ① Observations of natural phenomena lead a scientist to ask a question. ② The scientist generates a hypothesis to explain the observations. ③ The scientist designs and conducts an experiment to test the hypothesis. ④ Observations of the experiment's results either support the hypothesis, which is then provisionally accepted, or fail to support the hypothesis. Unsupported hypotheses may be either rejected or modified and retested. A hypothesis that continues to be accepted after repeated testing may be called a theory or a law.

Pasteur's Experiments

Scientists of the 1800s used the word *fermentation* to mean not only the formation of alcohol from sugar, but also other chemical reactions such as the formation of lactic acid, the putrefaction of meat, and the decomposition of waste. Many scientists asserted that air caused fermentation reactions, others insisted that living organisms were responsible.

The debate over the cause of fermentation reactions was linked to the debate over spontaneous generation. Some scientists proposed that the yeasts observed amidst alcohol fermentation products were nonliving globules of chemicals and gases. Others thought that yeasts were alive and were spontaneously generated during fermentation. Still others asserted that yeasts not only were living organisms, but also caused fermentation.

Pasteur conducted a series of careful observations and experiments that answered the question "What causes fermentation?" First, he observed yeast cells growing and budding in grape juice and conducted experiments showing that they arise only from other yeast cells. Then, by sealing some sterile flasks containing grape juice and yeast, and by leaving others open to the air, he demonstrated that yeast could grow with or without oxygen; that is, he discovered that yeasts are *facultative anaerobes*[12]—organisms that can live with or without oxygen. Finally, by introducing bacteria and yeast cells into different flasks of sterile grape juice, he proved that bacteria ferment grape juice to produce acids, and that yeast cells ferment grape juice to produce alcohol (**Figure 1.14** on page 12).

Pasteur's discovery that *anaerobic* bacteria fermented grape juice into acids suggested a method for preventing the spoilage of wine. His name became a household word when he developed *pasteurization,* a process of heating the grape juice just enough to kill most contaminating bacteria without changing the juice's basic qualities, so that it could then be inoculated with yeast to ensure that alcohol fermentation occurred. Pasteur thus began the field of **industrial microbiology** (or **biotechnology**) in which microbes are intentionally used to manufacture products (**Table 1.1** on page 13). Today pasteurization is used routinely on milk to eliminate pathogens that cause diseases such as bovine tuberculosis and brucellosis; it is also used to eliminate pathogens in juices and other beverages.

These are just a few of the many experiments Pasteur conducted with microbes. While a few of Pasteur's successes can be attributed to the superior microscopes available in the late 1800s, his genius is clearly evident in his carefully designed and straightforward experiments. Because of his many, varied, and significant accomplishments in working with microbes, Pasteur may be considered the Father of Microbiology.

Buchner's Experiments

Studies on fermentation began with the idea that fermentation reactions were strictly chemical and did not involve living organisms. This idea was supplanted by Pasteur's work showing that fermentation proceeded only when living cells were present, and that different types of microorganisms growing under varied conditions produced different end products.

In 1897, the German scientist Eduard Buchner (1860–1917) resurrected the chemical explanation by showing

[12]From Greek *an,* meaning not, *aer,* meaning air (i.e., oxygen), and *bios,* meaning life.

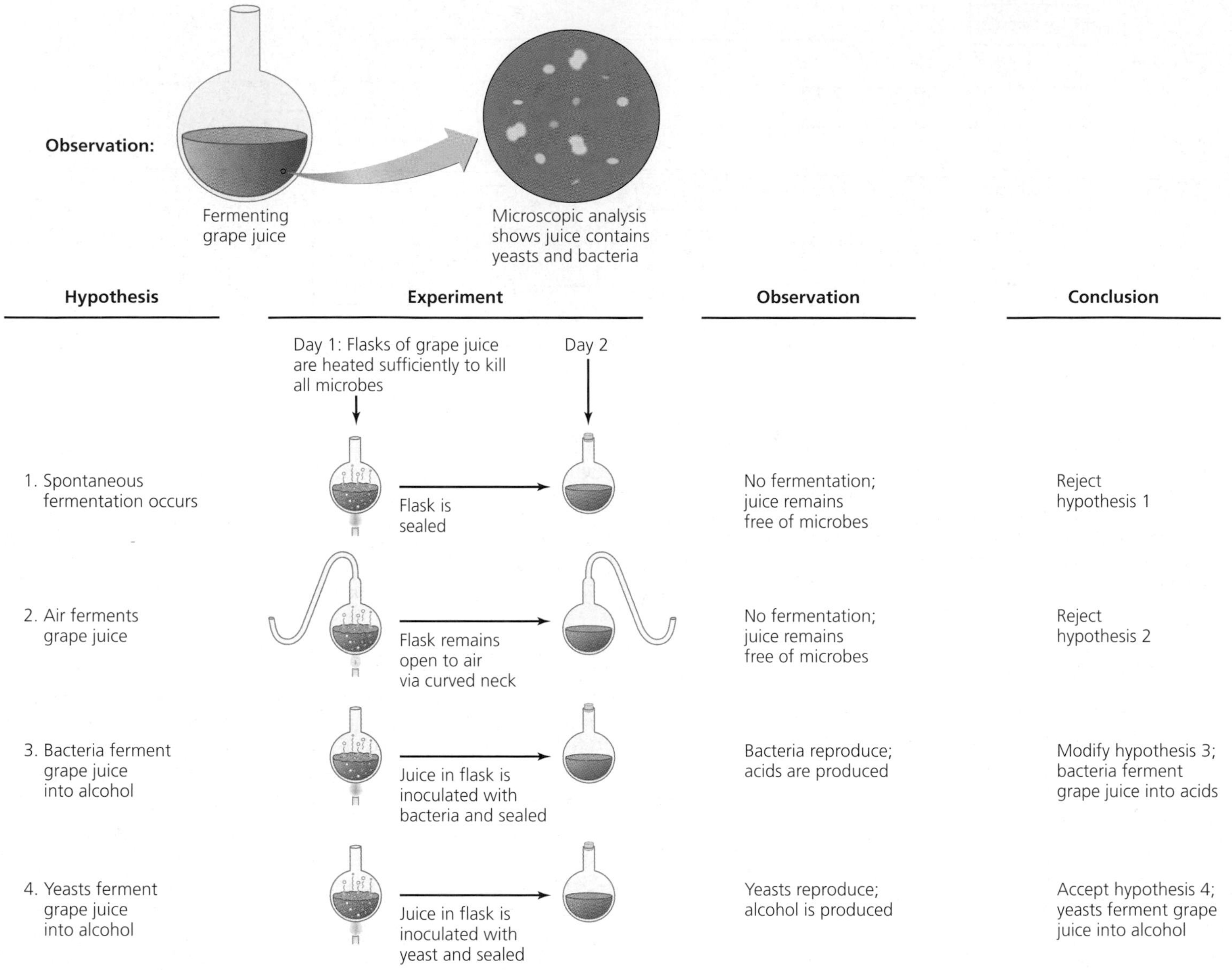

Observation:
Fermenting grape juice
Microscopic analysis shows juice contains yeasts and bacteria
Hypothesis
Experiment
Observation
Conclusion
Day 1: Flasks of grape juice are heated sufficiently to kill all microbes
Day 2
1. Spontaneous fermentation occurs
Flask is sealed
No fermentation; juice remains free of microbes
Reject hypothesis 1
2. Air ferments grape juice
Flask remains open to air via curved neck
No fermentation; juice remains free of microbes
Reject hypothesis 2
3. Bacteria ferment grape juice into alcohol
Juice in flask is inoculated with bacteria and sealed
Bacteria reproduce; acids are produced
Modify hypothesis 3; bacteria ferment grape juice into acids
4. Yeasts ferment grape juice into alcohol
Juice in flask is inoculated with yeast and sealed
Yeasts reproduce; alcohol is produced
Accept hypothesis 4; yeasts ferment grape juice into alcohol

▲ ***Figure 1.14***

How Pasteur applied the scientific method in investigating the nature of fermentation. After observing that fermenting grape juice contained both yeasts and bacteria, Pasteur hypothesized that these organisms cause fermentation. Upon eliminating the possibility that fermentation could occur spontaneously or be caused by air (hypotheses 1 and 2), he concluded that fermentation requires the presence of living cells. The results of additional experiments (those testing hypotheses 3 and 4) indicated that bacteria ferment grape juice to produce acids, and that yeasts ferment grape juice to produce alcohol. *Which of Pasteur's flasks was the control?*

Figure 1.14 *The sealed flask that remained free of microorganisms served as the control.*

that fermentation does not require living cells **(Figure 1.15).** Buchner's experiments demonstrated the presence of *enzymes,* which are cell-produced proteins that promote chemical reactions. Buchner's work began the field of **biochemistry** and the study of **metabolism,** a term that refers to the sum of all chemical reactions within an organism.

CRITICAL THINKING

How might the debate over spontaneous generation have been different if Buchner had conducted his experiments in 1857 instead of 1897?

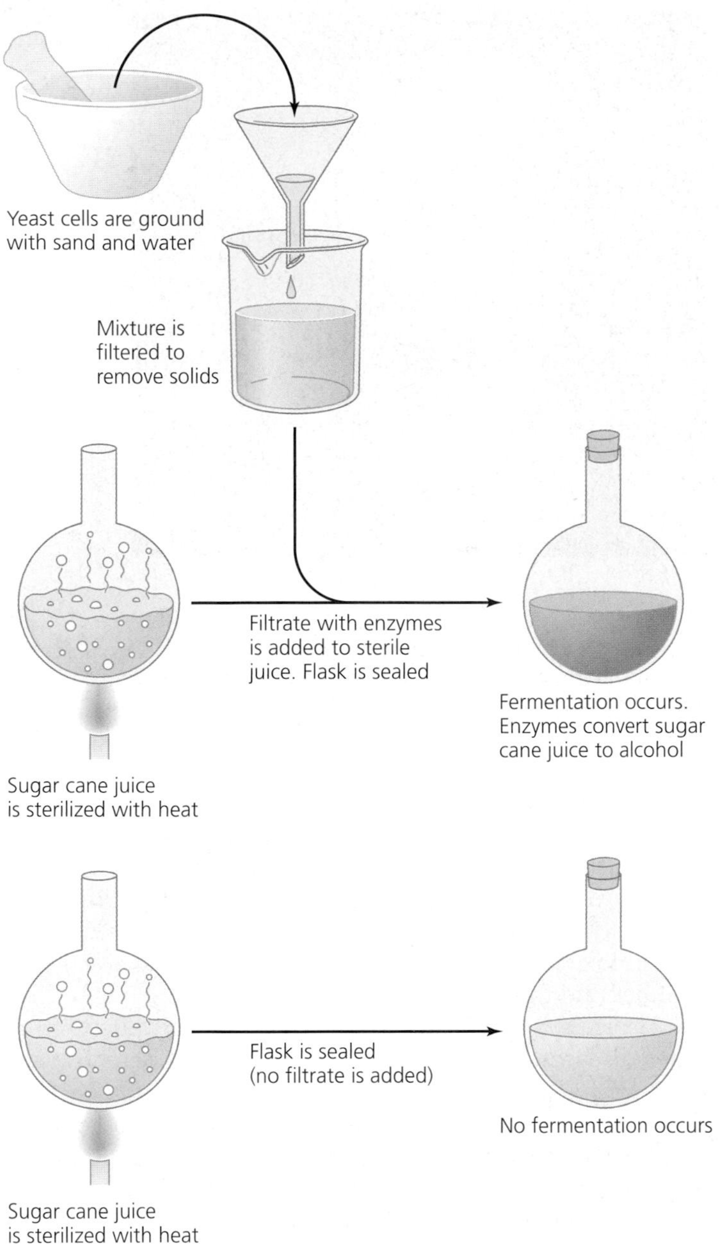
Yeast cells are ground with sand and water
Mixture is filtered to remove solids
Filtrate with enzymes is added to sterile juice. Flask is sealed
Fermentation occurs. Enzymes convert sugar cane juice to alcohol
Sugar cane juice is sterilized with heat
Flask is sealed (no filtrate is added)
No fermentation occurs
Sugar cane juice is sterilized with heat

▲ ***Figure 1.15***

Buchner's demonstration of acellular fermentation. After he ground yeast cells in water with quartz sand, the filtered liquid was capable of fermenting sterilized sugar cane juice into alcohol. This experiment started the field of biochemistry. *What substances from the crushed yeast cells carried out fermentation?*

Table 1.1 Some Industrial Uses of Microbes

Product or Process	Contribution of Microorganism
Foods and Beverages	
Cheese	Flavoring and ripening produced by bacteria and fungi; flavors dependent on the source of milk and the type of microorganism
Alcoholic beverages	Alcohol produced by bacteria or yeast by fermentation of sugars in fruit juice or grain
Soy sauce	Produced by fungal fermentation of soybeans
Vinegar	Produced by bacterial fermentation of sugar
Yogurt	Produced by bacteria growing in skim milk
Sour cream	Produced by bacteria growing in cream
Artificial sweetener	Amino acids synthesized by bacteria from sugar
Bread	Rising of dough produced by action of yeast; sourdough results from bacterial-produced acids
Other Products	
Antibiotics	Produced by bacteria and fungi
Human growth hormone, human insulin	Produced by genetically engineered bacteria
Laundry enzymes	Isolated from bacteria
Vitamins	Isolated from bacteria
Diatomaceous earth (used in polishes and buffing compounds)	Composed of cell walls of microscopic algae
Pest control chemicals	Insect pests killed or inhibited by bacterial pathogens
Drain opener	Protein-digesting and fat-digesting enzymes produced by bacteria

What Causes Disease?

Learning Objectives

✓ List at least seven contributions made by Koch to the field of microbiology.

✓ List the four steps that must be taken to prove the cause of an infectious disease.

✓ Describe the contribution of Gram to the field of microbiology.

Figure 1.15 *Enzymes released from the yeast cells carried out fermentation.*

You are a physician in London, and it is August, 1854. It is past midnight, and you have been visiting patients since before dawn. As you enter the room of your next patient you observe with frustration and despair that this case is like hundreds of others you and your colleagues have attended in the neighborhood over the past month.

A five-year-old boy with a vacant stare lies in bed listlessly. As you watch, he is suddenly gripped by severe abdominal cramps, and his gastrointestinal tract empties in an explosion of watery diarrhea. The voided fluid is clear, colorless, odorless, and streaked with thin flecks of white mucus, reminiscent of water poured off a pot of cooking rice. His anxious mother changes his bedclothes as his father gives him a sip of water, but it is of little use. With a heavy heart you confirm the parents' fear—their child has cholera, and there is nothing you can do. He will likely die before morning. As you despondently turn to go, the question that has haunted you for two months is foremost in your mind: What causes such a disease?

The third question that propelled the advance of microbiology concerned disease, defined generally as any abnormal condition in the body. Prior to the 1800s, disease was attributed to various factors, including evil spirits, sin, imbalances in body fluids, and foul vapors. Although the Italian philosopher Girolamo Fracastoro (1478–1553) conjectured as early as 1546 that "germs[13] of contagion" cause disease, the idea that germs might be invisible living organisms awaited the invention of Leeuwenhoek's microscope 130 years later.

Pasteur's discovery that bacteria are responsible for spoiling wine led naturally to his hypothesis in 1857 that microorganisms are also responsible for diseases. This idea came to be known as the **germ theory of disease.** Since a particular disease is typically accompanied by the same symptoms in all affected individuals, early investigators suspected that diseases such as cholera, tuberculosis, and anthrax are each caused by a specific germ, called a **pathogen.**[14] Today we know that some diseases are genetic and that allergic reactions and environmental toxins cause others, so the germ theory applies only to *infectious*[15] *diseases.*

Just as Pasteur was the chief investigator in disproving spontaneous generation and determining the cause of fermentation, so investigations in **etiology**[16] (the study of causation of disease) were dominated by Robert Koch (1843–1910) **(Figure 1.16).**

▲ ***Figure 1.16***
Robert Koch, who was instrumental in modifying the scientific method to prove that a given pathogen caused a specific disease.

Koch's Experiments

Koch was a country doctor in Germany when he began a race with Pasteur to discover the cause of anthrax, which is a potentially fatal disease, primarily of animals, in which toxins produce ulceration of the skin. Anthrax caused untold financial losses to farmers and ranchers in the 1800s, and the disease can be spread to humans. (See **New Frontiers 1.1** on page 22.)

Koch carefully examined the blood of infected animals, and in every case he identified a rod-shaped bacterium[17] that formed chains. He observed the formation of resting stages (endospores) within the bacterial cells and showed that the spores always produced anthrax when they were injected into mice. This was the first time that a bacterium was proven to cause a disease. As a result of his successful work on anthrax, Koch was able to move to Berlin and was given facilities and funding to continue his research.

Heartened by his success, Koch turned his attention to other diseases. He had been fortunate when he chose anthrax for his initial investigations, because anthrax bacteria are quite large and easily identified with the microscopes of that time. However, most bacteria are very small, and different types exhibit few or no visible differences. Koch puzzled how to distinguish among these bacteria.

He solved the problem by taking specimens (for instance, blood, pus, or sputum) from disease victims and then smearing the specimens onto a solid surface such as a slice of potato or a gelatin medium. He then waited for bacteria and fungi present in the specimen to multiply and form distinct colonies **(Figure 1.17).** Koch hypothesized that each colony consisted of the progeny of a single cell. He then inoculated samples from each colony into laboratory animals to see which caused disease. Koch's method of isolation is a standard technique in microbiological and medical labs to this day, though a gel called *agar,* derived from red seaweed, is used instead of gelatin or potato.

[13]From Latin *germen,* meaning sprout.
[14]From Greek *pathos,* meaning disease, and *genein,* meaning to produce.
[15]From Latin *inficere,* meaning to taint (i.e., with a pathogen).
[16]From Greek *aitia,* meaning cause, and *logos,* meaning word or study.
[17]Now known as *Bacillus anthracis*—Latin for "the rod of anthrax."

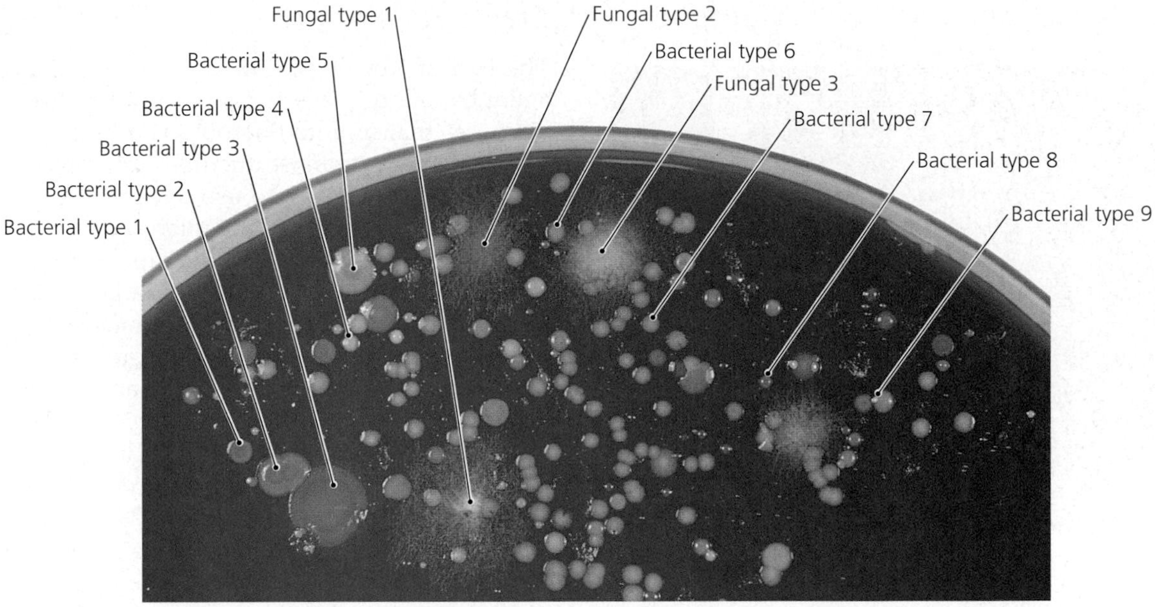

▲ ***Figure 1.17***

Bacterial and fungal colonies on a solid surface (agar). Differences in colony size, shape, and color indicate the presence of different species of bacteria and fungi. Such differences allowed Koch to isolate specific types of microbes that could be tested for their ability to cause disease.

Koch and his colleagues are also responsible for many other advances in laboratory microbiology, including the following:

- Simple staining techniques for bacterial cells and flagella
- The first photomicrograph of bacteria
- The first photograph of bacteria in diseased tissue
- Techniques for estimating the number of bacteria in a solution based on the number of colonies that form after inoculation onto a solid surface
- The use of steam to sterilize growth media
- The use of Petri[18] dishes to hold solid growth media
- Aseptic laboratory techniques such as transferring bacteria between media using a platinum wire that had been heat-sterilized in a flame
- Elucidation of bacteria as distinct species

For these achievements, Koch is considered the Father of the Microbiological Laboratory.

CRITICAL THINKING

French microbiologists, led by Pasteur, tried to isolate a single bacterium by diluting liquid media until only a single type of bacterium could be microscopically observed in a sample of the diluted medium. What advantages does Koch's method have over the French method?

Koch's Postulates

After discovering the anthrax bacterium, Koch continued to search for disease agents. In two pivotal scientific publications in 1882 and 1884, he announced that the cause of tuberculosis was a rod-shaped bacterium, *Mycobacterium tuberculosis* (mī-kō-bak-tēr'ē-ŭm tū-ber-kyū-lō'sis). In 1905 he received the Nobel Prize in Physiology or Medicine for this work.

In his publications on tuberculosis, Koch elucidated a series of steps that must be taken to prove the cause of any infectious disease. These steps, now known as **Koch's postulates,** are one of his more important contributions to microbiology. The postulates, which we discuss in more detail in Chapter 14, are the following:

1. The suspected causative agent must be found in every case of the disease and be absent from healthy hosts.
2. The agent must be isolated and grown outside the host.
3. When the agent is introduced to a healthy, susceptible host, the host must get the disease.
4. The same agent must be reisolated from the diseased experimental host.

We use the term *suspected causative agent* because "agent" can refer to a fungus, protozoan, bacterium, virus, or other pathogen. It is "suspected" until the postulates have

[18]Named for Richard Petri, Koch's assistant, who invented them in 1887.

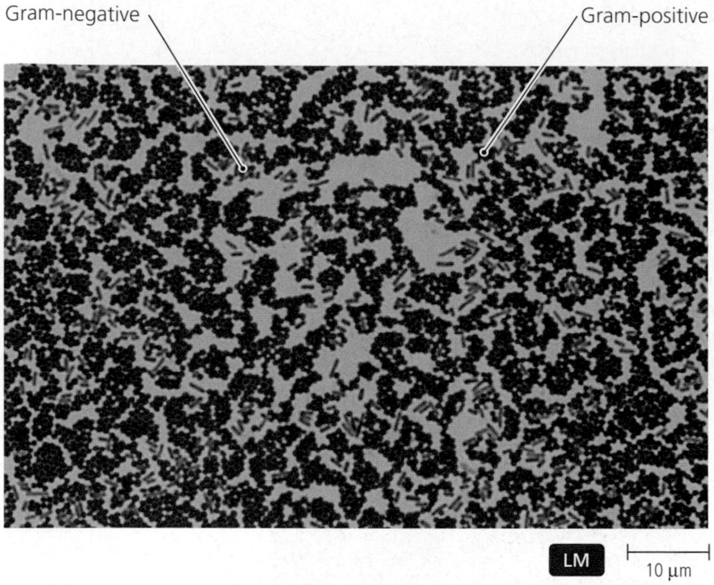

▲ *Figure 1.18*

Results of Gram staining. Gram-positive cells (in this case *Staphylococcus aureus*) are purple; Gram-negative cells (in this case *Escherichia coli*) are pink.

been fulfilled. There are practical and ethical limits in the application of Koch's postulates, but in almost every case they must be satisfied before the cause of an infectious disease is proven.

CRITICAL THINKING

Why aren't Koch's postulates always useful in proving the cause of a given disease? Consider a variety of diseases, such as cholera, pneumonia, Alzheimer's, AIDS, Down syndrome, and lung cancer.

During microbiology's "golden years," other scientists used Koch's postulates, as well as laboratory techniques introduced by Koch and Pasteur, to discover the causes of most protozoan and bacterial diseases, as well as some viral diseases. For example, Charles Laveran (1845–1922) showed that a protozoan is the cause of malaria, and Edwin Klebs (1834–1913) described the bacterium that causes diphtheria. Dmitri Ivanowski (1864–1920) and Martinus Beijerinck (1851–1931) discovered that a certain disease in tobacco plants is caused by a pathogen that passes through filters with such extremely small pores that bacteria cannot pass through. Beijerinck, recognizing that the pathogen was not bacterial, called it a *filterable virus*. Now such pathogens are simply called *viruses*. As previously noted, viruses couldn't be seen until electron microscopes were invented in 1932. The American physician Walter Reed (1851–1902) proved in 1900 that viruses can cause such diseases as yellow fever in humans. Chapter 13 deals with the science of *virology*.

A partial list of scientists and the pathogens they discovered is provided in **Table 1.2**.

Gram's Stain

The first of Koch's postulates demands that the suspected agent be found in every case of a given disease, which presupposes that minute microbes can be seen and identified. However, because most microbes are colorless and difficult to see, scientists began to use dyes to stain them and make them more visible under the microscope.

Though Koch reported a simple staining technique in 1877, the Danish scientist Christian Gram (1853–1938) developed a more important staining technique in 1884. His procedure, which involves the application of a series of dyes, leaves some microbes purple and others pink. We now label the first group of cells as *Gram-positive* and the second as *Gram-negative,* and we use the Gram procedure to separate bacteria into these two large groups **(Figure 1.18).**

The **Gram stain** is still the most widely used staining technique. It is one of the first steps carried out in any laboratory where bacteria are being identified, and is one of the procedures you will learn in micro lab. The full procedure is discussed in Chapter 4 (page 109).

How Can We Prevent Infection and Disease?

Learning Objectives

- ✓ Identify four health care practitioners who did pioneering research in the areas of public health microbiology and epidemiology.
- ✓ Name two scientists whose work with vaccines began the field of immunology.
- ✓ Describe the quest for a "magic bullet."

The last great question that drove microbiological research during the "Golden Age" was how to prevent infectious diseases. Though some methods of preventing or limiting disease were discovered even before it was understood that microorganisms caused contagious diseases, great advances occurred only after Pasteur and Koch showed that life comes from life and that microorganisms can cause diseases.

In the mid-1800s, modern principles of hygiene, such as those involving sewage and water treatment, personal cleanliness, and pest control, were not widely practiced. Typically, medical personnel and health care facilities lacked adequate cleanliness. *Nosocomial*[19] *infections*—infections acquired in a health care setting—were rampant. For example, surgical patients frequently succumbed to gangrene acquired while under their doctor's care, and many women who gave birth in hospitals died from puerperal[20] fever. Four health care practitioners who were especially instrumental in changing the way health care is delivered were Semmelweis, Lister, Nightingale, and Snow.

[19]From Greek *nosos,* meaning disease, and *komein,* meaning to care for (relating to a hospital).

[20]From Latin *puerperus,* meaning childbirth.

Table 1.2 Some Notable Scientists of the "Golden Age of Microbiology" and the Agents of Human Disease They Discovered

Scientist	Year	Disease	Agent
Robert Koch	1876	Anthrax	*Bacillus anthracis* (bacterium)
Albert Neisser	1879	Gonorrhea	*Neisseria gonorrheae* (bacterium)
Charles Laveran	1880	Malaria	*Plasmodium* species (protozoa)
Carl Eberth	1880	Typhoid fever	*Salmonella typhi* (bacterium)
Robert Koch	1882	Tuberculosis	*Mycobacterium tuberculosis* (bacterium)
Edwin Klebs	1883	Diphtheria	*Corynebacterium diphtheriae* (bacterium)
Theodore Escherich	1884	Traveler's diarrhea Bladder infection	*Escherichia coli* (bacterium)
Albert Fraenkel	1884	Pneumonia	*Streptococcus pneumoniae* (bacterium)
Robert Koch	1884	Cholera	*Vibrio cholerae* (bacterium)
David Bruce	1887	Undulant fever (brucellosis)	*Brucella melitensis* (bacterium)
Anton Weichselbaum	1887	Meningococcal meningitis	*Neisseria meningitidis* (bacterium)
A. A. Gartner	1888	Salmonellosis (form of food poisoning)	*Salmonella* species (bacterium)
Shibasaburo Kitasato	1889	Tetanus	*Clostridium tetani* (bacterium)
Dmitri Ivanowski and Martinus Beijerinck	1892 1898	Tobacco mosaic disease	*Tobamovirus* tobacco mosaic virus
William Welsh and George Nuttall	1892	Gas gangrene	*Clostridium perfringens* (bacterium)
Alexandre Yersin and Shibasaburo Kitasato	1894	Bubonic plague	*Yersinia pestis* (bacterium)
Kiyoshi Shiga	1898	Shigellosis (a type of severe diarrhea)	*Shigella dysenteriae* (bacterium)
Walter Reed	1900	Yellow fever	*Flavivirus* yellow fever virus
Robert Forde and Joseph Dutton	1902	African sleeping sickness	*Trypanosoma brucei gambiense* (protozoan)

Semmelweis and Handwashing

Ignaz Semmelweis (1818–1865) was a physician on the obstetric ward of a teaching hospital in Vienna. In about 1848 he observed that women giving birth in the wing where medical students were trained died from puerperal fever at a rate 20 times higher than the mortality rates of either women attended by midwives in an adjoining wing or women who gave birth at home.

Though Pasteur had not yet elaborated his germ theory of disease, Semmelweis hypothesized that medical students carried "cadaver particles" from their autopsy studies into the delivery rooms, and that these "particles" resulted in puerperal fever. Semmelweis gained support for his hypothesis when a doctor who sliced his finger during an autopsy died after showing symptoms similar to those of puerperal fever. Today we know that the primary cause of puerperal fever are bacteria in the genus *Streptococcus* (strep-tō-kok'ŭs; see Figure 1.7), which are usually harmless on the skin or in the mouth but cause severe complications when they enter the blood.

Semmelweis began requiring medical students to wash their hands with chlorinated lime water, a substance long used to eliminate the smell of cadavers. Mortality in the subsequent year dropped from 18.3% to 1.3%. Despite his success, Semmelweis was ridiculed by the director of the hospital and eventually forced to leave. He returned to his native Hungary, where his insistence on handwashing met with general approval when it continued to produce higher patient survival rates.

Though his impressive record made it easier for later doctors to institute changes, Semmelweis was unsuccessful in gaining support for his method from most European doctors. He became severely depressed and was committed to a mental hospital, where he died from an infection of *Streptococcus,* the very organism he had fought for so long.

Lister's Antiseptic Technique

Shortly after Semmelweis was rejected in Vienna, the English physician Joseph Lister (1827–1912) modified and advanced the idea of *antisepsis*[21] in health care settings. As a surgeon, Lister was aware of the dreadful consequences that

[21]From Greek *anti,* meaning against, and *sepein,* meaning putrefaction.

resulted from the infection of wounds. Therefore, he began spraying wounds, surgical incisions, and dressings with carbolic acid (phenol), a chemical that had previously proven effective in reducing odor and decay in sewage. Like Semmelweis, he initially met with some resistance, but when he showed that it reduced deaths among his patients by two-thirds, his method was accepted into common practice. In this manner, Lister vindicated Semmelweis, became the founder of antiseptic surgery, and opened new fields of research into antisepsis and disinfection.

Nightingale and Nursing

Florence Nightingale (1820–1910) **(Figure 1.19)** was a dedicated English nurse who succeeded in introducing cleanliness and other antiseptic techniques into nursing practice. She was instrumental in setting standards of hygiene that saved innumerable lives during the Crimean War of 1854–56. One of her first requisitions in the military hospital was for 200 scrubbing brushes, which she and her assistants used diligently in the squalid wards. She next arranged for each patient's filthy clothes and dressings to be replaced or cleaned at a different location, thus removing many sources of infection. She thoroughly documented statistical comparisons to show that poor food and unsanitary conditions in the hospitals were responsible for the deaths of many soldiers.

After the war, Nightingale returned to England, where she actively exerted political pressure to reform hospitals and implement public health policies. Perhaps her greatest achievements were in nursing education. For example, she founded the Nightingale School for Nurses—the first of its kind in the world.

▲ ***Figure 1.19***
Florence Nightingale, the founder of modern nursing, was influential in introducing antiseptic technique into nursing practice.

Snow and Epidemiology

Another English physician, John Snow (1813–1858), also played a key role in setting standards for good public hygiene to prevent the spread of infectious diseases. Snow had been studying the propagation of cholera and suspected that the disease was spread by a contaminating agent in water. In 1854, he mapped the occurrence of cholera cases during an epidemic in London and showed that they centered around a public water supply on Broad Street. When, after his recommendation, the water pump was dismantled, the cholera epidemic abated.

Though Snow did not know the cause of cholera, his careful documentation of the epidemic highlighted the critical need for adequate sewage treatment and a pure water supply. His study was the foundation for two branches of microbiology—**infection control** and **epidemiology,**[22] which is the study of the occurrence, distribution, and spread of disease in humans.

Jenner's Vaccine

In about 1789, the English physician Edward Jenner (1749–1823) tested the hypothesis that a mild disease called cowpox provided protection against potentially fatal smallpox. After he intentionally inoculated his son with pus collected from a milkmaid's cowpox lesion, the boy developed cowpox, which, of course, he survived. When Jenner then attempted to infect his son with smallpox, he found that the boy had become immune[23] to it. (Note that experiments that intentionally expose human subjects to deadly pathogens are considered unethical today.) In 1798 Jenner reported similar results from additional experiments, demonstrating the validity of the procedure he named *vaccination* after *Vaccinia,*[24] the virus that causes cowpox. Jenner invented vaccination (the term *immunization* is often used synonymously today), established a safe treatment for preventing smallpox, and began the field of **immunology**—the study of the body's specific defenses against pathogens. Chapters 16–18 discuss immunology.

Pasteur later capitalized on Jenner's work by producing weakened strains of various pathogens for use in preventing the serious diseases they cause. In honor of Jenner's work with cowpox, Pasteur used the term *vaccine* to refer to all weakened, protective strains of pathogens. He subsequently developed successful vaccines against fowl cholera, anthrax, and rabies.

Ehrlich's "Magic Bullets"

Gram's discovery that bacteria could be differentiated into two types by staining suggested to the German microbiologist

[22]From Greek *epi,* meaning upon, *demos* meaning people, and *logos* meaning word or study.
[23]From Latin *immunis,* meaning free.
[24]From Latin *vacca,* meaning cow.

Paul Ehrlich (1854–1915) that chemicals could be used to kill microorganisms differentially. To investigate this idea, Ehrlich undertook an exhaustive survey of chemicals to find a "magic bullet" that would destroy pathogens while remaining nontoxic to humans. By 1908, he had discovered chemicals active against trypanosomes (the protozoan parasites that cause sleeping sicknesses) and against *Treponema pallidum* (trep-ō-nē′mă pal′li-dŭm), the causative agent of syphilis. His discoveries began the branch of medical microbiology known as **chemotherapy.**

In summary, the Golden Age of Microbiology was a time when researchers proved that living things come from other living things, that microorganisms can cause fermentation and disease, and that certain procedures and chemicals can limit, prevent, and cure infectious diseases. These discoveries were made by scientists who applied the scientific method to biological investigation, and they led to an explosion of knowledge in a number of scientific disciplines **(Figure 1.20).**

The Modern Age of Microbiology

Learning Objective

✓ List four major questions that drive microbiological investigations today.

The vast increase in the number of microbiological investigations and in scientific knowledge during the 1800s opened new fields of science, including disciplines called environmental science, immunology, epidemiology, chemotherapy, and genetic engineering (**Table 1.3,** page 20). Microorganisms played a significant role in the development of these disciplines because microorganisms are relatively easy to grow, take up little space, and are available by the trillions. Much of what has been learned about microbes also applies to other organisms, including humans. In the rest of this text we examine advances made in these branches of microbiology, though it would require thousands of books this size to deal with all that is known.

Once the developing science of microbiology had successfully answered questions about spontaneous generation, fermentation, and disease, additional questions arose in each branch of the new science. Since the early 20th century, microbiologists have worked to answer these new questions. In this section we briefly consider some of the 20th century's overarching questions in both basic and applied research. The chapter concludes with a look at some of the questions that might propel microbiological research for the next 50 years.

What Are the Basic Chemical Reactions of Life?

Biochemistry is the study of metabolism—that is, the chemical reactions that occur in living organisms. Biochemistry

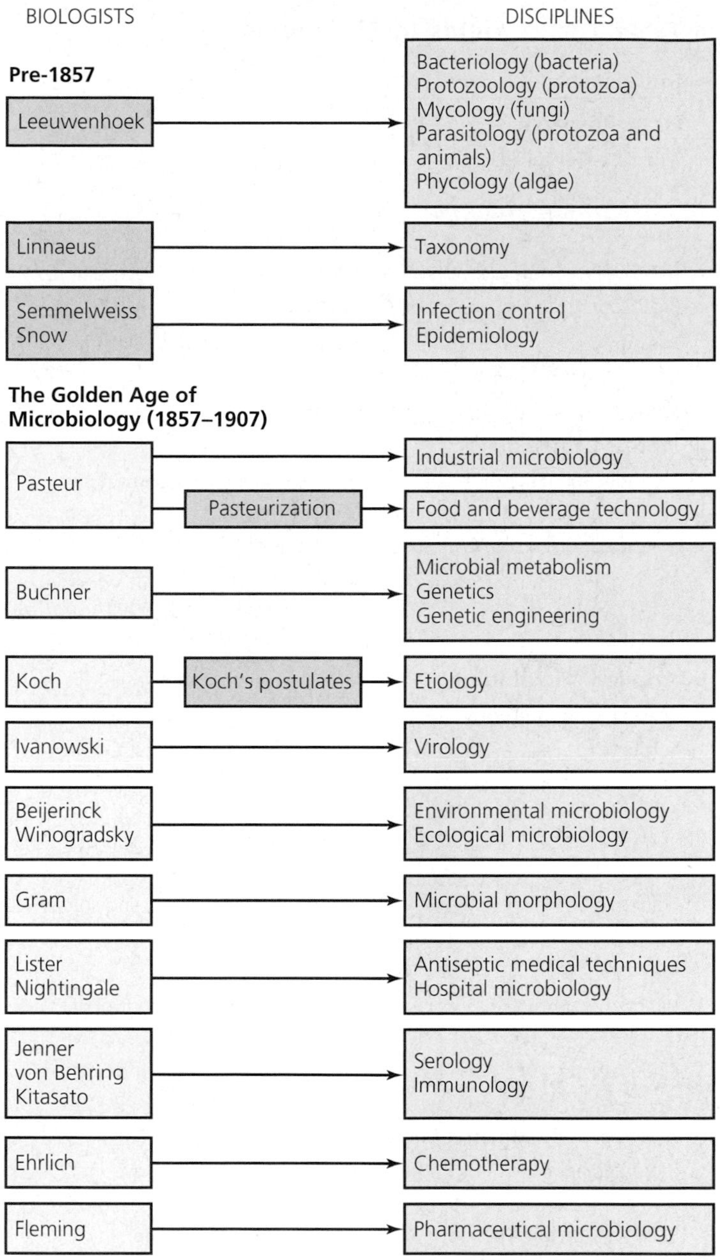

▲ ***Figure 1.20***

Some of the many scientific disciplines and applications that arose from the pioneering work of scientists just before and during the "Golden Age of Microbiology."

began with Pasteur's work on fermentation by yeast and bacteria, and with Buchner's discovery of enzymes in yeast extract, but by the early 1900s many scientists thought that the metabolic reactions of microbes had little to do with the metabolism of plants and animals.

In contrast, microbiologists Albert Kluyver (1888–1956) and his student C. B. van Niel (1897–1985) proposed that basic biochemical reactions are shared by all living things, that these reactions are relatively few in number, and that their primary feature is the transfer of electrons and hydrogen ions. In adopting this view, scientists could use

Table 1.3 Fields of Microbiology

Disciplines	Subject(s) of Study
Basic Research	
Microbe-Centered	
Bacteriology	Bacteria and archaea
Phycology	Algae
Mycology	Fungi
Protozoology	Protozoa
Parasitology	Parasitic protozoa and parasitic animals
Virology	Viruses
Process-Centered	
Microbial metabolism	Biochemistry: chemical reactions within cells
Microbial genetics	Functions of DNA and RNA
Environmental microbiology	Relationships between microbes, and among microbes, other organisms, and their environment
Applied Microbiology	
Medical Microbiology	
Serology	Antibodies in blood serum, particularly as an indicator of infection
Immunology	Body's defenses against specific diseases
Epidemiology	Frequency, distribution, and spread of disease
Etiology	Causes of disease
Infection control	Hospital hygiene and control of nosocomial infections
Chemotherapy	Development and use of drugs to treat infectious diseases
Applied Environmental Microbiology	
Bioremediation	Use of microbes to remove pollutants
Public health microbiology	Sewage treatment, water purification, and control of insects that spread disease
Agricultural microbiology	Use of microbes to control insect pests
Industrial Microbiology (Biotechnology)	
Food and beverage technology	Reduction or elimination of harmful microbes in food and drink
Pharmaceutical microbiology	Manufacture of vaccines and antibiotics
Recombinant DNA technology	Alteration of genes in microbes to synthesize useful products

microbes as model systems to answer questions about metabolism in all organisms. Research during the 20th century validated this approach to understanding basic metabolic processes, but scientists have also documented an amazing metabolic diversity. Chapter 5 discusses basic metabolic processes, and Chapter 6 considers metabolic diversity.

Basic biochemical research has many practical applications, including:

- The design of herbicides and pesticides that are specific in their action and have no long-term adverse effects on the environment.
- The diagnosis of illnesses and the monitoring of a patient's responses to treatment. For example, physicians routinely monitor liver disease by measuring blood levels of certain enzymes and products of liver metabolism.
- The treatment of metabolic diseases. One example is treating phenylketonuria, a disease resulting from the inability to properly metabolize the amino acid phenylalanine, by eliminating foods containing phenylalanine from the diet.
- The design of drugs to treat leukemia, gout, bacterial infections, malaria, herpes, AIDS, asthma, and heart attacks.

CRITICAL THINKING

Albert Kluyver said, "From elephant to . . . bacterium—it is all the same!" What did he mean?

How Do Genes Work?

Genetics, the scientific study of inheritance, started in the mid-1800s as an offshoot of botany, but scientists studying microbes made most of the great advances in this discipline.

Microbial Genetics

While working with the bacterium *Streptococcus pneumoniae* (strep-tō-kok'ŭs nū-mō'nē-ī), Oswald Avery (1877–1955), Colin MacLeod (1909–1972), and Maclyn McCarty (1911–) determined that genes are contained in molecules of DNA. In 1958, George Beadle (1903–1989) and Edward Tatum (1909–1975), working with the bread mold *Neurospora crassa* (nū-ros'pōr-ă kras'ă), established that a gene's activity is related to the function of the specific protein coded by that gene. Other researchers, also working with microbes, determined the exact way in which genetic information is translated into a protein, the rates and mechanisms of genetic mutation, and the methods by which cells control genetic expression. We examine all of these aspects of microbial genetics in Chapter 7.

Over the past 40 years, advances in microbial genetics developed into several new disciplines that are among the faster growing areas of scientific research today, including *molecular biology, recombinant DNA technology,* and *gene therapy.*

Molecular Biology

Molecular biology combines aspects of biochemistry, cell biology, and genetics to explain cell function at the molecular level. Molecular biologists are particularly concerned with *genome*[25] *sequencing*. Using techniques perfected on microorganisms, molecular biologists have sequenced the genomes of many organisms, including humans and some of their pathogens. It is hoped that a fuller understanding of the genomes of organisms will result in practical ways to limit disease, repair genetic defects, and enhance agricultural yield.

The American Nobel laureate Linus Pauling (1901–1994) proposed in 1965 that gene sequences could provide a means of understanding possible evolutionary relationships and processes, establishing taxonomic categories that more closely reflect these relationships, and identifying the existence of microbes that have never been cultured in a laboratory. Two examples illustrate such uses of gene sequencing data:

- In the 1970s, Carl Woese (1928–) discovered that significant differences in nucleic acid sequences among organisms clearly reveal that cells belong to one of *three* major groups—bacteria, archaea, or eukaryotes—and not merely two groups (prokaryotes and eukaryotes) as previously thought.
- Scientists showed in 1990 that cat-scratch fever is caused by a bacterium that cannot be cultured. The bacterium was discovered by recognition of the sequence of a portion of its ribonucleic acid that differs from all other known ribonucleic acid sequences.

Recombinant DNA Technology

Molecular biology is applied in **recombinant DNA technology,**[26] commonly called *genetic engineering*, which was first developed using microbial models. Geneticists manipulate genes in microbes, plants, and animals for practical applications. For instance, once scientists have inserted the gene for human blood-clotting factor into *Escherichia coli* (esh-ĕ-rik′ē-ă kō′lī), the bacterium produces the factor in a pure form. This technology is a boon to hemophiliacs, who previously depended on clotting factor isolated from donated blood which was possibly contaminated by life-threatening viral pathogens.

Gene Therapy

An exciting new area of study is the use of recombinant DNA technology for **gene therapy,** a process that involves inserting a missing gene or repairing a defective one in human cells. This procedure uses harmless viruses to insert a desired gene into host cells, where it is incorporated into a chromosome and begins to function normally. Chapter 8 examines recombinant DNA technology and gene therapy in more detail.

What Roles Do Microorganisms Play in the Environment?

Learning Objectives

✓ Identify the field of microbiology that studies the role of microorganisms in the environment.

✓ Name the fastest growing scientific disciplines in microbiology today.

Ever since Koch and Pasteur, most research in microbiology has focused on pure cultures of individual species; however, microorganisms are not alone in the "real world." Instead, they live in natural microbial communities in the soil, water, the human body, and other habitats, and these communities play critical roles in such processes as the production of vitamins and *bioremediation*—the use of living bacteria, fungi, and algae to detoxify polluted environments.

Microbial communities also play an essential role in the decay of dead organisms and the recycling of chemicals such as carbon, nitrogen, and sulfur. Martinus Beijerinck discovered bacteria capable of converting nitrogen gas (N_2) from the air into nitrate (NO_3), the form of nitrogen used by plants, and the Russian microbiologist Sergei Winogradsky (1856–1953) elucidated the role of microorganisms in the recycling of sulfur. Together these two microbiologists developed laboratory techniques for isolating and growing environmentally important microbes.

Another role of microbes in the environment is the causation of disease. Although most microorganisms are not pathogenic, some microbes do pose threats to human health. In the following section, and in Chapters 15 and 16, we'll examine how the human body defends against microbial pathogens.

CRITICAL THINKING

The ability of farmers around the world to produce crops such as corn, wheat, and rice is often limited by the lack of nitrogen-based fertilizer. How might scientists use Beijerinck's discovery to increase world supplies of grain?

How Do We Defend Against Disease?

Why do some people get sick during the flu season while their close friends and family remain well? The germ theory of disease showed not only that microorganisms can cause diseases, but also that the body can defend itself—otherwise, everyone would be sick most of the time.

The work of Jenner and Pasteur on vaccines showed that the body can protect itself from repeated diseases by the same organism. The German bacteriologist Emil von

[25]A genome is the total genetic information of an organism.

[26]Recombinant DNA is DNA composed of genes from more than one organism.

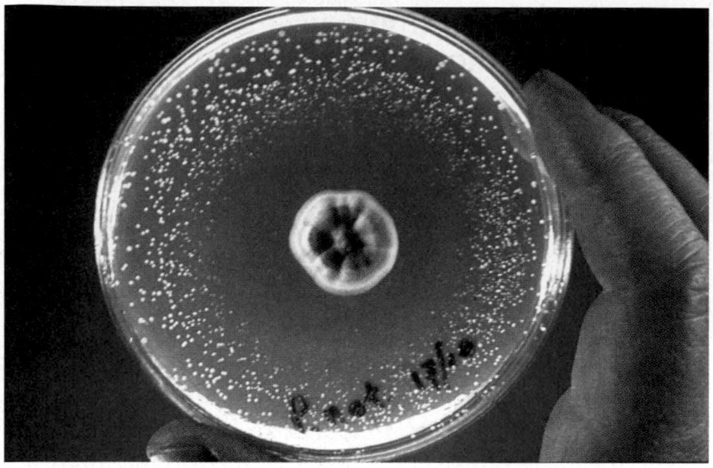

▲ *Figure 1.21*

The bacteriocidal effects of penicillin on a bacterial "lawn" in a Petri dish. The clear ring surrounding the fungus colony, which is producing the antibiotic, is where the penicillin killed the bacteria.

Behring (1854–1917) and the Japanese microbiologist Shibasaburo Kitasato (1856–1931), working in Koch's laboratory, reported the existence in the blood of chemicals and cells that fight infection. Their studies developed into the fields of *serology,* the study of blood serum[27]—specifically, the chemicals in the liquid portion of blood that fight disease—and *immunology,* the study of the body's defense against specific pathogens. Chapters 16–18 cover these aspects of microbiology, which are of utmost importance to physicians, nurses, and other health care practitioners.

Ehrlich introduced the idea of a "magic bullet" that would kill pathogens, but it wasn't until Alexander Fleming (1881–1955) discovered penicillin **(Figure 1.21)** in 1929 and Gerhard Domagk (1895–1964) discovered sulfa drugs in 1935 that medical personnel finally had drugs effective against a wide range of bacteria. We study chemotherapy, and some physical and chemical agents used to control microorganisms in the environment, in Chapters 9 and 10.

[27]Latin, meaning whey. Serum is the liquid that remains after blood coagulates.

New Frontiers 1.1 Bioterrorism

At 2 A.M. on October 2, 2001, a 63-year-old photo editor named Bob Stevens was admitted to JFK Medical Center in Florida, exhibiting fever, confusion, and low blood pressure. Within a few hours, he had lost consciousness. The diagnosis, confirmed on October 4, was both startling and unexpected: inhalational anthrax, the first documented case on U.S. soil since 1976. On October 5, Stevens died. At the time, no one yet knew that Stevens had contracted anthrax after handling a piece of mail containing anthrax spores—and that his death was the result of a bioterrorist attack that would ultimately claim four more lives, infect 13 other people, and sound alarms throughout a nation still reeling from the events of September 11.

Bioterrorism—the use of viruses, bacteria, fungi, or their toxins for threatening or harmful purposes—is not new. British soldiers sent blankets containing smallpox to enemy tribes during the French and Indian War, and the use of biological agents in both world wars (such as the dropping of plague-infected fleas in cities to spread disease) has been documented. However, bioterrorism is arguably a greater threat today than ever before. The technology to cultivate biological agents, and to effectively disseminate them for purposes of mass destruction, is more advanced than ever. And the roster of potential biological agents now includes newly discovered viruses such as Ebola and bacteria that have been genetically modified to resist antibiotics (including drug-resistant strains of *Bacillus anthracis,* the bacterium that causes anthrax).

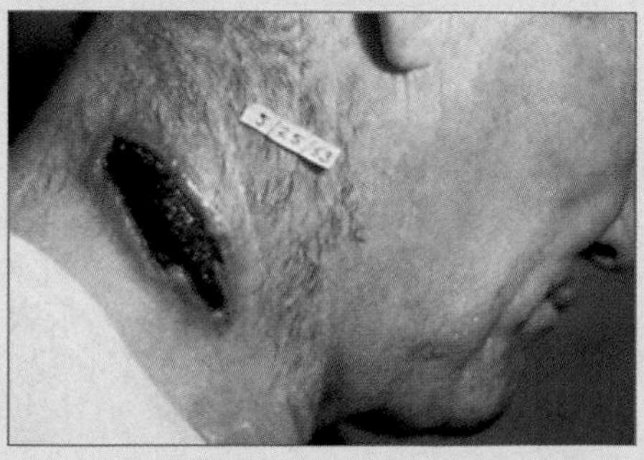

Skin lesion caused by *Bacillus anthracis*

Combating bioterrorism requires a joint effort by microbiologists, epidemiologists, immunologists, molecular biologists, and health care workers in governmental agencies and the private sector. The challenges include (first and foremost) preventing such an attack; detecting that an outbreak has occurred; distinguishing between a natural outbreak and a bioterrorist act; and identifying the agent(s) responsible. Plans must then be made for containing the agent's spread, as well as for treating victims expediently and effectively.

Microbiology plays an important role in addressing all these challenges. For example, scientists are in the process of developing *biosensors,* which are handheld devices powered by genetically altered immune cells from mice and jellyfish that can quickly detect harmful microbial agents in the air. Health care workers must apply their knowledge of microbiology to diagnosing and treating patients. And of course, the development of new vaccines, antibiotics, and germicides to defend against biological attacks will necessarily be informed by microbiological research.

Fortunately, transforming microorganisms into effective weapons for mass destruction remains difficult. Pathogens and biotoxins often lose their virulence or toxicity during storage. Even after deployment, various technical and environmental factors—such as dilution in air and water or deactivation by sunlight—further curb their effectiveness. Nonetheless, the threat of bioterrorism remains a real one. Mounting a defense is critical and has the added benefit of increasing our ability to respond to what is likely an even greater threat: the emergence of new infectious diseases, or the reemergence of naturally occurring ones.

What Will the Future Hold?

The science of microbiology is built on asking and answering questions. What began with the curiosity of a dedicated lens grinder in the Netherlands has come far in the past 350 years and has expanded into disciplines as diverse as immunology, recombinant DNA technology, and bioremediation. However, the adage remains true: *The more questions we answer, the more questions we have.*

What will microbiologists discover next? Among the questions for the next 50 years are the following:

- What is it about the physiology of life forms known only by their nucleic acid sequences that prevents those life forms from being grown in the laboratory?
- Can bacteria and archaea be used in ultraminiature technologies such as living computer circuit boards?
- How can an understanding of microbial communities help us understand communities of larger organisms, including humans?
- What genetic sequences make some microbes pathogenic, and what can we do at a genetic level to defend against these pathogens?
- Does life exist beyond planet Earth, and if so, what are its features?
- How can we reduce the threat of infectious diseases, especially those that can be used by bioterrorists? (See New Frontiers 1.1, page 22.)

CHAPTER SUMMARY

The Early Years of Microbiology (pp. 2–7)

1. Leeuwenhoek's observations of **microbes** introduced **microorganisms** to the world and earned him the title "Father of Bacteriology and Protozoology." His discoveries were named and classified by Linnaeus in his **taxonomic system.**
2. Relatively large microscopic **eukaryotic fungi** include **molds** and **yeasts.**
3. Animal-like **protozoa** are single-celled eukaryotes. Some cause disease.
4. Plant-like eukaryotic **algae** are important providers of oxygen, serve as food for many marine animals, and make chemicals used in microbiological growth media.
5. Small **prokaryotic bacteria** and **archaea** live in a variety of communities and in most habitats. Even though some cause disease, most are beneficial.
6. Viruses, the smallest microbes, are so small they can be seen only by using an electron microscope.
7. Parasitic worms, the largest organisms studied by microbiologists, are often visible without a microscope, although their immature stages are microscopic.

The Golden Age of Microbiology (pp. 7–19)

1. The study of the Golden Age of Microbiology includes a look at the men who proposed or refuted the theory of **spontaneous generation**: Aristotle, Redi, Needham, Spallanzani, and Pasteur (the Father of Microbiology). The **scientific method** that emerged then remains the accepted sequence of study today.
2. The study of fermentation by Pasteur and Buchner led to the fields of **industrial microbiology (biotechnology)** and **biochemistry,** and to the study of **metabolism.**
3. Koch, Pasteur, and others proved that **pathogens** cause infectious diseases, an idea that is known as the **germ theory of disease. Etiology** is the study of the causation of diseases.
4. Koch initiated careful microbiological laboratory techniques in his search for disease agents. **Koch's postulates,** the logical steps he followed to prove the cause of an infectious disease, remain an important part of microbiology today.
5. The procedure for the **Gram stain** was developed in the 1870s and is still used to differentiate bacteria into two categories: Gram-positive and Gram-negative.
6. The investigations of Semmelweis, Lister, Nightingale, and Snow are the foundations upon which **infection control** and **epidemiology** are built.
7. Jenner's use of a cowpox-based vaccine for preventing smallpox began the field of **immunology.** Pasteur significantly advanced the field.
8. Ehrlich's search for "magic bullets"—chemicals that differentially kill microorganisms—laid the foundations for the field of **chemotherapy.**

The Modern Age of Microbiology (pp. 19–23)

1. Microbiology in the modern age has focused on answering questions regarding **biochemistry,** which is the study of metabolism; microbial genetics, which is the study of inheritance in microorganisms; and **molecular biology,** which involves investigations of cell function at the molecular level.
2. Scientists have applied knowledge from basic research to answer questions in **recombinant DNA technology** and **gene therapy.**
3. The study of microorganisms in their natural environment is **environmental microbiology.**
4. The discovery of chemicals in the blood that are active against specific pathogens advanced immunology and began the field of serology.
5. Advancements in chemotherapy were made in the 1900s with the discovery of numerous substances, such as penicillin and sulfa drugs, that inhibit pathogens.

QUESTIONS FOR REVIEW

(Answers to multiple choice, fill in the blanks, and matching questions are on the web, along with additional review questions. Visit www.microbiologyplace.com.)

Multiple Choice

1. Which of the following microorganisms are not eukaryotic?
   a. bacteria
   b. yeasts
   c. molds
   d. protozoa
2. Which microorganisms are used to make microbiological growth media?
   a. bacteria
   b. fungi
   c. algae
   d. protozoa
3. In which habitat would you most likely find archaea?
   a. acidic hot springs
   b. swamp mud
   c. Great Salt Lake
   d. all of the above
4. Of the following scientists, who defended the theory of abiogenesis?
   a. Aristotle
   b. Pasteur
   c. Needham
   d. Spallanzani
5. Which of the following scientists hypothesized that a bacterial colony arises from a single bacterial cell?
   a. Antoni van Leeuwenhoek
   b. Louis Pasteur
   c. Robert Koch
   d. Richard Petri
6. Which scientist first hypothesized that medical personnel can infect patients with pathogens?
   a. Edward Jenner
   b. Joseph Lister
   c. John Snow
   d. Ignaz Semmelweis

Fill in the Blanks

Fill in the blanks with the name(s) of the scientist(s) whose investigations led to the following fields of study in microbiology.

1. Environmental microbiology ________ and ________
2. Biochemistry ________ and ________
3. Chemotherapy ________
4. Immunology ________
5. Public health microbiology ________
6. Etiology ________
7. Epidemiology ________
8. Biotechnology ________
9. Food microbiology ________

Matching Questions

Match each of the following descriptions with the person it best describes. An answer may be used more than once.

1. ___ Developed smallpox immunization	A. Alexander Fleming
2. ___ First photomicrograph of bacteria	B. Paul Ehrlich
3. ___ Germ theory of disease	C. Louis Pasteur
4. ___ Germs cause disease	D. Antoni van Leeuwenhoek
5. ___ Sought a "magic bullet" to destroy pathogens	E. Carolus Linnaeus
6. ___ Discovered penicillin	F. John T. Needham
7. ___ Father of Microbiology	G. Eduard Buchner
8. ___ Father of Microbiological Laboratory Techniques	H. Robert Koch
9. ___ Father of Bacteriology	I. Joseph Lister
10. ___ Father of Protozoology	J. Edward Jenner
11. ___ Founder of antiseptic surgery	K. Girolamo Fracastoro

Short Answer

1. Why was the theory of spontaneous generation a hindrance to the development of the field of microbiology?
2. Discuss the significant difference between the flasks used by Pasteur and Spallanzani. How did Pasteur's investigation settle the dispute about spontaneous generation?
3. List four types of microorganisms from smallest to largest.
4. Defend this statement: "The investigations of Antoni van Leeuwenhoek changed the world forever."
5. Why would a *macro*scopic tapeworm be studied in *micro*biology?
6. Describe what has been called "The Golden Age of Microbiology" with reference to four major questions that propelled scientists during that period.
7. List four major questions that drive microbiological investigations today.
8. Refer to the four steps in the scientific method in describing Pasteur's fermentation experiments.
9. List Koch's postulates, and explain why they are significant.

CRITICAL THINKING

1. If Robert Koch had become interested in a viral disease such as influenza instead of anthrax (caused by a bacterium), how might his list of lifetime accomplishments be different? Why?
2. In 1911, the Polish scientist Kasimir Funk proposed that a limited diet of polished white rice (rice without the husks) caused beriberi, a disease of the central nervous system. Even though history has proven him correct—beriberi is caused by a thiamine deficiency, which in his day resulted from unsophisticated milling techniques that removed the thiamine-rich husks—Funk was criticized by his contemporaries, who told him to find the microbe that caused beriberi. Explain how the prevailing scientific philosophy of the day shaped Funk's detractors' point of view.
3. *Haemophilus influenzae* does not cause flu, but it received its name because it was once thought to be the cause. Explain how a proper application of Koch's postulates would have prevented this error in nomenclature.
4. Just before winter break in early December, your roommate stocks the refrigerator with a gallon of milk, but both of you leave before opening it. When you return in January, the milk has soured. Your roommate is annoyed because the milk was pasteurized and thus should not have spoiled. Explain why your roommate's position is unreasonable.
5. Design an experiment to prove that microbes don't spontaneously generate in milk.
6. The British General Board of Health concluded in 1855 that the Broad Street cholera epidemic (see page 18) resulted from fermentation of "nocturnal clouds of vapor" from the polluted Thames River. How could an epidemiologist prove or disprove this claim?

CHAPTER 2

The Chemistry of Microbiology

CHAPTER OUTLINE

Is there microbial life on other planets? From telescopic observations and space probes we have identified the chemicals found on our nearest planetary neighbors. Venus, for example, has a thick atmosphere of CO_2, sulfur, and sulfuric acid (H_2SO_4), and its soil contains a significant amount of iron. Microorganisms could potentially live on Venus, since carbon (essential for all life forms) is available in Venus's atmosphere, and we know some microbes use iron and sulfur for energy. To date, probes have found no evidence of life on Venus or any other planet, but the tantalizing possibility remains.

What environmental conditions are needed to support microbial life on Earth? How can we develop treatments against harmful bacteria and viruses? How do we differentiate one microorganism from another? The answers to these questions, and to many other important questions in microbiology, have their basis in chemistry.

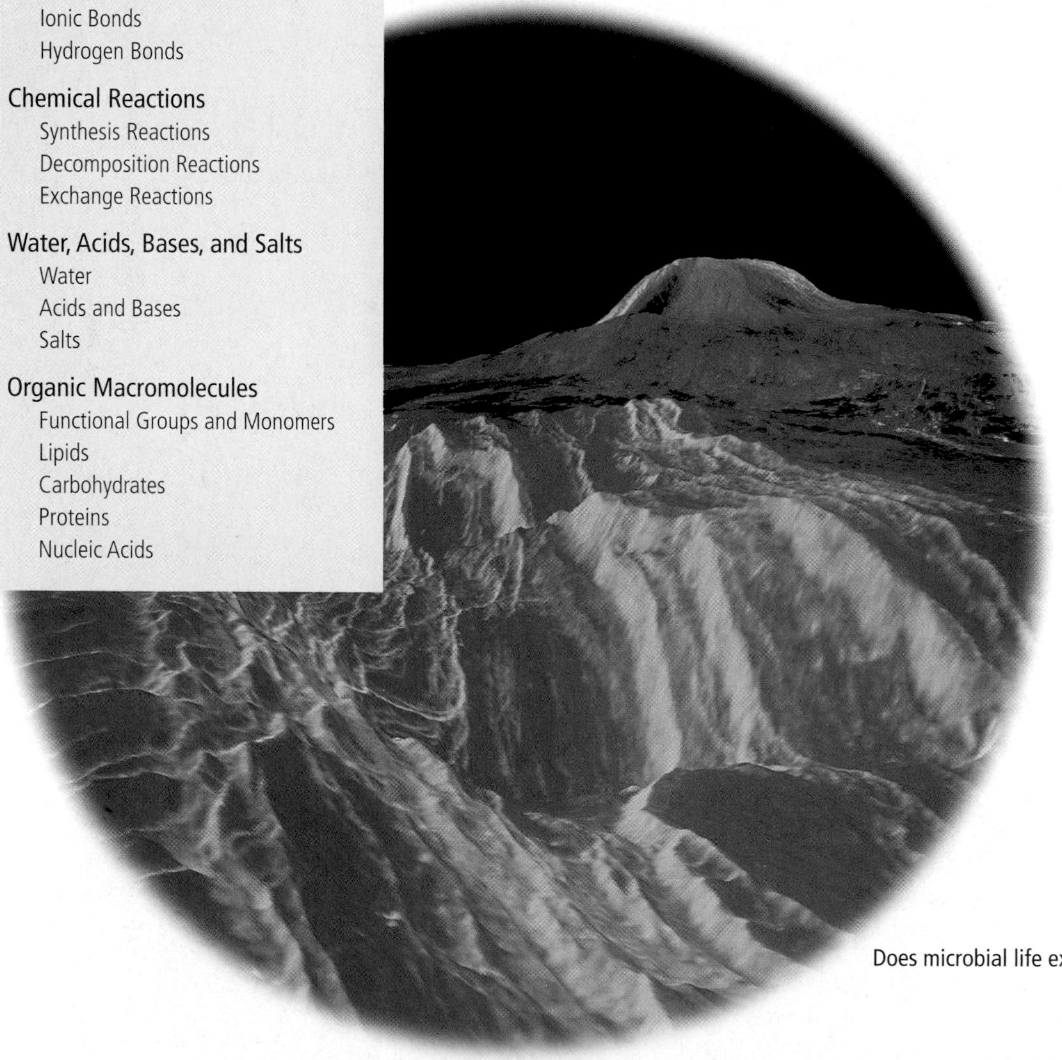

Does microbial life exist on Venus?

MicroPrep Pre-Test: *Take the pre-test for this chapter on the web. Visit* ***www.microbiologyplace.com***

Learning some basic concepts of chemistry will enable you to understand more fully the variety of interactions between microorganisms and their environment—which includes you. If you plan a career in health care, you will find microbial chemistry involved in the diagnosis of disease, the response of the immune system, the growth and identification of pathogenic microorganisms in the laboratory, and the function and selection of antibiotics. Even preserving your own health is aided by an understanding of the fundamentals of chemistry.

In this chapter we study atoms, which are the basic units of chemistry, and we consider how atoms react with one another to form chemical bonds and molecules. Then we examine the three major categories of chemical reactions. The chapter concludes with a look at the molecules of greatest importance to life: water, acids, bases, lipids, carbohydrates, proteins, nucleic acids, and ATP.

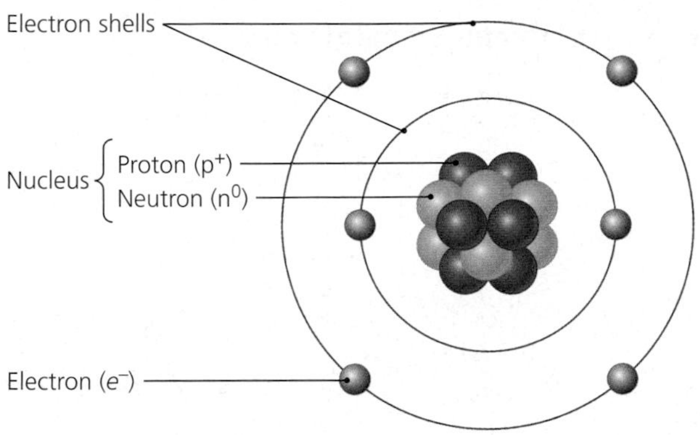

▲ ***Figure 2.1***

An example of a Bohr model of atomic structure. This drawing is not to scale; for the electrons to be shown in scale with the greatly magnified nucleus, they would have to occupy orbits located many miles from the nucleus. Put another way, the volume of an entire atom is 100 trillion times the volume of its nucleus.

Atoms

Learning Objective

✓ Define *matter, atom,* and *element,* and explain how these terms relate to one another.

Matter is defined as anything that takes up space and has mass.[1] The smallest chemical units of matter are **atoms.** Atoms are extremely small, and only the very largest of them can be seen using the most powerful microscopes. Therefore, scientists have developed various models to conceptualize and illustrate the structure of atoms.

Atomic Structure

Learning Objective

✓ Draw and label an atom, showing the parts of the nucleus and orbiting electrons.

In 1913, the Danish physicist Niels Bohr proposed a simple model in which negatively charged subatomic particles called **electrons** orbit a centrally located nucleus like planets in a miniature solar system **(Figure 2.1).** A nucleus is composed of uncharged **neutrons** and positively charged **protons.** (The only exception to this description is the nucleus of a normal hydrogen atom, which is composed of only a single proton and no neutrons.) The number of electrons in an atom typically equals the number of protons, so overall atoms are electrically neutral.

An **element** is matter that is composed of a single type of atom. For example, gold is an element because it consists of only gold atoms. In contrast, the ink in your pen is not an element because it is composed of many different kinds of atoms.

Elements differ from one another in their **atomic number,** which is the number of protons in their nuclei. For example, the atomic numbers of hydrogen, carbon, and oxygen are 1, 6, and 8, respectively, because all hydrogen nuclei contain a single proton, all carbon nuclei have six protons, and all oxygen nuclei have eight protons.

The **atomic mass** of an atom (sometimes called its *atomic weight*) is the sum of the masses of its protons, neutrons, and electrons. Protons and neutrons each have a mass of approximately 1 *atomic mass unit,*[2] which is also called a *dalton.*[3] An electron is much less massive, with a mass of about 0.00054 dalton. Electrons are often ignored in discussions of atomic mass because their contribution to the overall mass is negligible. Therefore, the sum of the number of protons and neutrons approximates the atomic mass of an atom.

There are 93 naturally occurring elements known;[4] however, organisms typically utilize only about 20 elements, each of which has its own symbol that is derived from its English or Latin name (**Table 2.1** on page 28).

Isotopes

Learning Objective

✓ List at least four ways that radioactive isotopes are useful.

Every atom of an element has the same number of protons, but atoms of a given element can differ in the number of neutrons in their nuclei. Atoms that differ in this way are called **isotopes.** For example, there are three naturally

[1] *Mass* and *weight* are sometimes confused. Mass is the quantity of material in something, whereas weight is the effect of gravity on mass. Even though an astronaut is weightless in space, his mass is the same in space as on Earth.

[2] An atomic mass unit (dalton) is 1/597,728,630,000,000,000,000,000, or 1.673×10^{-24} grams.

[3] Named for John Dalton, the British chemist who helped develop atomic theory around 1800.

[4] For many years, scientists thought that there were only 92 naturally occurring elements, but natural plutonium was discovered in Africa in 1997.

Table 2.1 Common Elements of Life

Element	Symbol	Atomic Number	Atomic Mass*	Biological Significance
Hydrogen	H	1	1	Component of organic molecules and water; H^+ released by acids
Boron	B	5	11	Essential for plant growth
Carbon	C	6	12	Backbone of organic molecules
Nitrogen	N	7	14	Component of amino acids, proteins, and nucleic acids
Oxygen	O	8	16	Component of many organic molecules and water; OH^- released by bases; necessary for aerobic metabolism
Sodium (Natrium)	Na	11	23	Principal cation outside cells
Magnesium	Mg	12	24	Component of many energy-transferring enzymes
Silicon	Si	14	28	Component of cell wall of diatoms
Phosphorus	P	15	31	Component of nucleic acids and ATP
Sulfur	S	16	32	Component of proteins
Chlorine	Cl	17	35	Principal anion outside cells
Potassium (Kalium)	K	19	39	Principal cation inside cells; essential for nerve impulses
Calcium	Ca	20	40	Utilized in many intercellular signaling processes; essential for muscular contraction
Manganese	Mn	25	54	Component of some enzymes
Iron (Ferrum)	Fe	26	56	Component of energy-transferring proteins; transports oxygen in the blood of many animals
Cobalt	Co	27	59	Component of vitamin B_{12}
Copper (Cuprum)	Cu	29	64	Component of some enzymes
Zinc	Zn	30	65	Component of some enzymes
Molybdenum	Mo	42	96	Component of some enzymes
Iodine	I	53	127	Component of thyroid hormones

*Rounded to nearest whole number.

occurring isotopes of carbon, each having six protons and six electrons **(Figure 2.2).** Over 95% of carbon atoms also have six neutrons. Because these atoms have six protons and six neutrons, the atomic mass of this isotope is about 12 daltons, and it is known as carbon-12, symbolized as ^{12}C. Atoms of carbon-13 (^{13}C) have seven neutrons per nucleus, and ^{14}C atoms each have eight neutrons.

Unlike the first two isotopes, the nucleus of ^{14}C is unstable due to the ratio of its protons and neutrons. Unstable atomic nuclei release energy and subatomic particles such as neutrons, protons, and electrons in a process called *radioactive decay*. Atoms that undergo radioactive decay are *radioactive isotopes*. Radioactive decay and radioactive isotopes play important roles in microbiological research, medical diagnosis, the treatment of disease, and the complete destruction of contaminating microbes (sterilization) of medical equipment and chemicals.

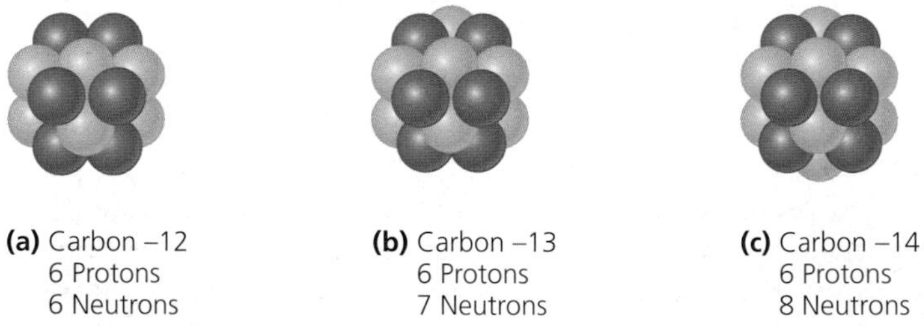

▲ *Figure 2.2*

Nuclei of the three naturally occurring isotopes of carbon. Each isotope also has six electrons, which are not shown. *What are the atomic number and atomic mass of each of these isotopes?*

Electron Configurations

While the nuclei of atoms determine their identities, nuclei of different atoms almost never come close enough together

Figure 2.2 The atomic number of all three is 6; their atomic masses are 12, 13, and 14, respectively.

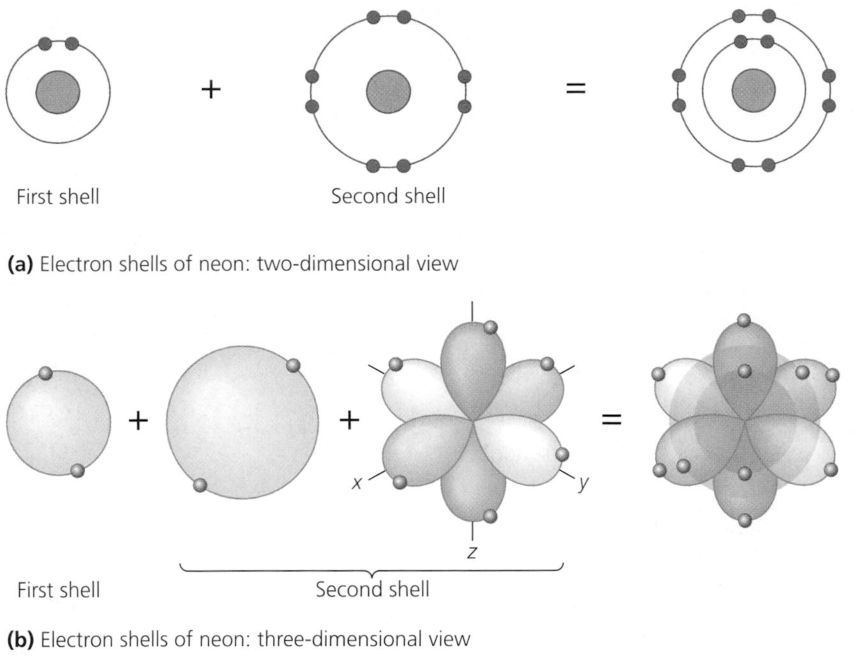

Figure 2.3

Electron configurations. **(a)** Two-dimensional model (Bohr diagram) of the electron shells of neon. **(b)** Three-dimensional model of the electron shells of neon. In this model, the first shell is a small sphere, whereas the second shell consists of a larger sphere plus three pairs of ellipses that extend from the nucleus at right angles. Larger shells (not shown) are even more complex.

to interact.[5] Typically, only the electrons of atoms interact, which is why it is electrons that determine an atom's *chemical behavior*. Thus, because all of the isotopes of carbon (for example) have the same number of electrons, all these isotopes behave the same way in chemical reactions, even though their nuclei are different.

Scientists know that electrons do not really orbit the nucleus in a two-dimensional circle, as indicated by a Bohr diagram **(Figure 2.3a)**, but instead in three-dimensional *electron shells* or *clouds* that assume unique shapes dependent on the energy of the electrons **(Figure 2.3b)**. More accurately put, an electron shell depicts the *probable* locations of electrons at a given time; nevertheless, it is simpler and more convenient to draw electron shells as circles.

Each electron shell can hold only a certain, maximum number of electrons. For example, the first shell (the one nearest the nucleus) can accommodate a maximum of two electrons, and the second shell can hold no more than eight electrons. Atoms of hydrogen and helium have one and two electrons, respectively; thus, these two elements have only a single electron shell. A lithium atom, which has three electrons, has two shells.

Atoms with more than 10 electrons require more shells. The third shell holds up to eight electrons when it is the outermost shell, though its capacity increases to 18 when the fourth shell contains two electrons. Heavier atoms have even more shells, but these atoms do not play significant roles in the processes of life.

It is the electrons in the outermost shell of atoms, called *valence electrons*, that actually interact with other atoms. **Figure 2.4** on page 30 depicts the electron configurations of atoms of some elements important to microbial life. Notice that except for helium, atoms of all elements in a given column have the same number of valence electrons. Helium is placed in the far right-hand column with the other inert gases, because its outer shell is full, though it has two rather than eight valence electrons.

Chemical Bonds

Learning Objectives

- ✓ Describe the configuration of electrons in a stable atom.
- ✓ Contrast molecules and compounds.

Outer electron shells are stable when they contain eight electrons (except for the first electron shell, which is stable with only two electrons, since that is its maximum number). When atoms' outer shells are not filled with eight electrons, they either have room for more electrons or have "extra" electrons, depending on whether it is easier for them to gain electrons or lose electrons. For example, an oxygen atom, with six electrons in its outer shell, has two "unfilled spaces" (see Figure 2.4), because it requires less energy for the oxygen atom to gain two electrons than to lose six electrons. A calcium atom, by contrast, has two "extra" electrons in its outer (fourth) shell, because it requires less energy to lose these two electrons than to gain six new ones. When a calcium atom loses two electrons, its third shell, which is then its outer shell, is full and stable with eight electrons.

As previously noted, an atom's outermost electrons are called valence electrons, and thus the outermost shell of an atom is the *valence shell*. An atom's **valence,**[6] defined as its combining capacity, is considered to be positive if its valence shell has extra electrons to give up, and to be negative if its valence shell has spaces to fill. Thus a calcium atom, with two electrons in its valence shell, has a valence of $+2$, whereas an oxygen atom, with two spaces to fill in its valence shell, has a valence of -2.

Atoms combine with one another by either sharing or transferring valence electrons in such a way as to fill their valence shells. Such interactions between atoms are called **chemical bonds.** Two or more atoms held together by chemical bonds form a **molecule.** If a molecule contains atoms of more than one element, it is a **compound.** Two hydrogen atoms bonded together form a hydrogen molecule, which is not a compound because only one element is involved.

[5]Except during nuclear reactions, such as occur in nuclear power plants.
[6]From Latin *valentia*, meaning strength.

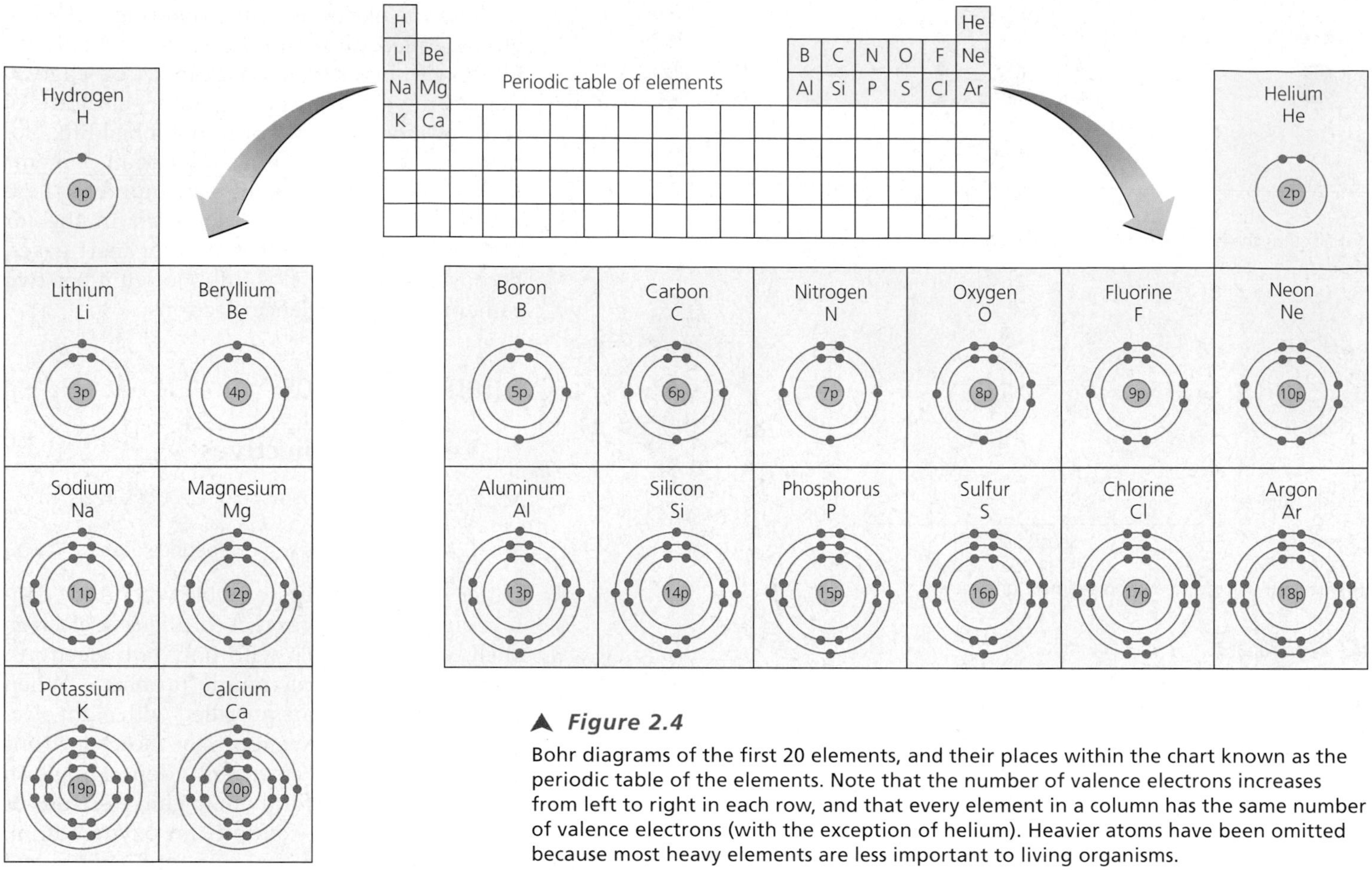

▲ ***Figure 2.4***

Bohr diagrams of the first 20 elements, and their places within the chart known as the periodic table of the elements. Note that the number of valence electrons increases from left to right in each row, and that every element in a column has the same number of valence electrons (with the exception of helium). Heavier atoms have been omitted because most heavy elements are less important to living organisms.

However, two hydrogen atoms bonded to an oxygen atom form a molecule of water (H_2O), which is a compound.

In this section, we discuss the three principal types of chemical bonds: *nonpolar covalent bonds, polar covalent bonds,* and *ionic bonds.* We also consider *hydrogen bonds,* which are weak forces that act with polar covalent bonds to give certain large chemicals their characteristic three-dimensional shapes.

CRITICAL THINKING

Neon (atomic mass 10) and argon (atomic mass 18) are *inert* elements, which means that they very rarely form chemical bonds. Give the electron configuration of their atoms and explain why these elements are inert.

Nonpolar Covalent Bonds

Learning Objective

✓ Contrast nonpolar covalent, polar covalent, and ionic bonds.

A **covalent**[7] **bond** is the sharing of a pair of electrons by two atoms. Consider, for example, what happens when two hydrogen atoms approach one another. Each hydrogen atom consists of a single proton orbited by a single electron. Since the valence shell of each atom requires two electrons to be filled, each atom shares its single electron with the other, forming a hydrogen molecule in which both atoms have full shells **(Figure 2.5a).** Similarly, two oxygen atoms can share electrons, but they must share *two* pairs of electrons for their valence shells to be full **(Figure 2.5b).** Since two pairs of electrons are involved, oxygen atoms form two covalent bonds, or a *double covalent bond,* with one another.

The attraction of an atom for electrons is called its **electronegativity.** The more electronegative an atom, the greater the pull its nucleus exerts on electrons. Note in **Figure 2.6** on page 32, which displays the electronegativities of atoms of several elements, that electronegativities tend to increase from left to right in the chart. The reason is that elements toward the right of the chart have more protons and thus exert a greater pull on electrons. Electronegativities of elements decrease from top to bottom in the chart because of the increasing distance between the nucleus and the valence shell as elements get larger.

Atoms with equal or nearly equal electronegativities, such as two hydrogen atoms or a hydrogen and a carbon, share electrons equally or nearly equally. In chemistry and

[7]From Latin *co,* meaning with or together, and *valentia,* meaning strength.

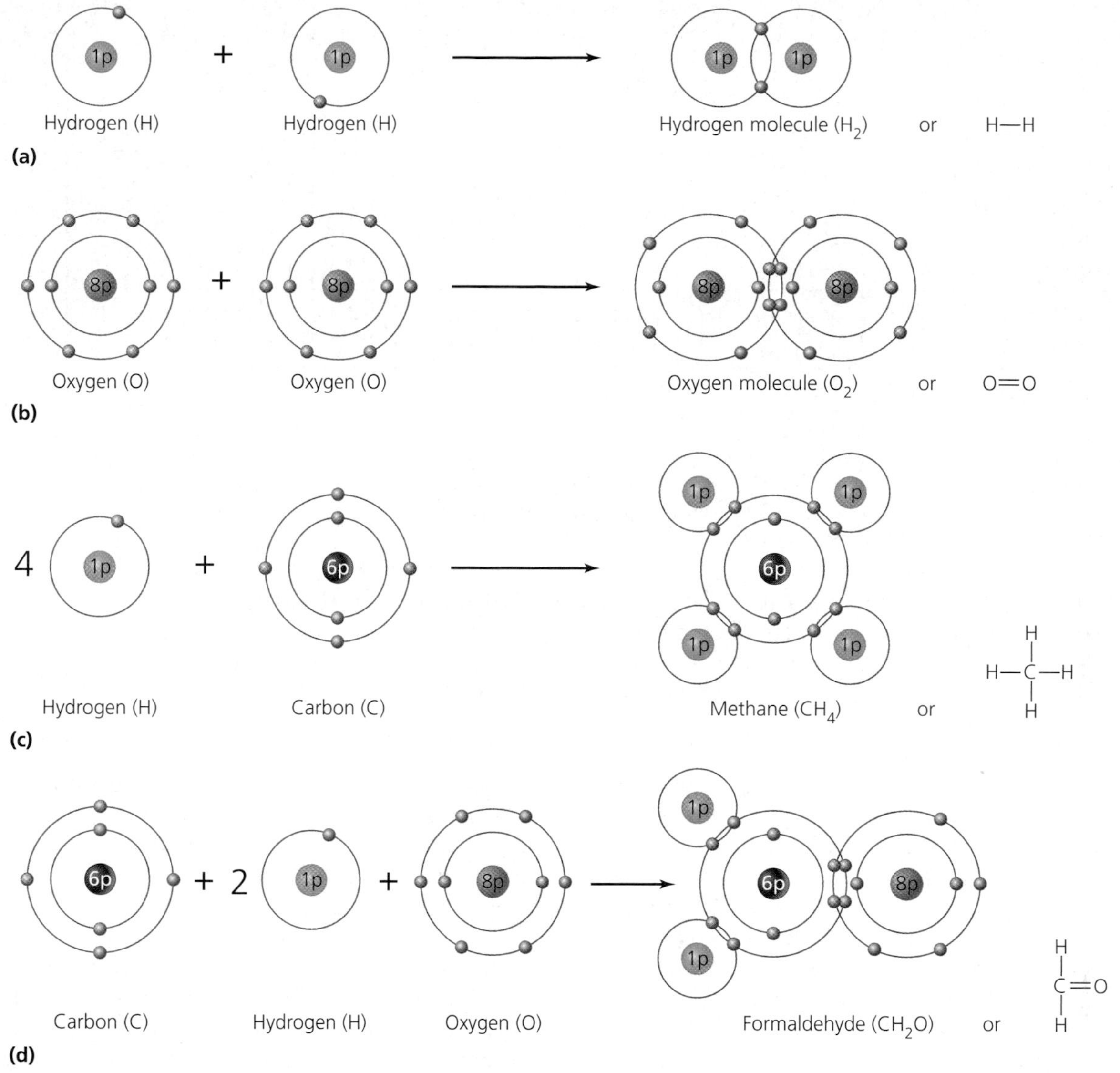

▲ ***Figure 2.5***

Four molecules formed by covalent bonds. **(a)** Hydrogen. Each hydrogen atom needs another electron to have a full valence shell. The two atoms share their electrons, forming a covalent bond. **(b)** Oxygen. Oxygen atoms have six electrons in their valence shells; thus, they need two electrons each. When they share with each other, two covalent bonds are formed. Note that the valence electrons of oxygen atoms are in the second shell. **(c)** A methane molecule, which has four single covalent bonds. **(d)** Formaldehyde. The carbon atom forms a double bond with the oxygen atom and single bonds with two hydrogen atoms. *Which of these molecules are also compounds? Why?*

Figure 2.5 *Methane and formaldehyde molecules are also compounds because they are composed of more than one element.*

physics, "poles" are opposed forces, such as north and south magnetic poles or positive and negative terminals of a battery. In the case of atoms with similar electronegativities, the shared electrons tend to spend an equal amount of time around each nucleus of the pair, and no poles exist; therefore, the bond between them is a **nonpolar covalent bond.** All the covalent bonds illustrated in Figure 2.5 are nonpolar.

A hydrogen molecule can be symbolized a number of ways:

H—H H:H H_2

In the first symbol, the dash represents the chemical bond between the atoms. In the second symbol, the dots represent the electron pair of the covalent bond. These two symbols

I	II											III	IV	V	VI	VII	Inert gases
H 2.1																	He 0.0
Li 1.0	Be 1.5											B 2.0	C 2.5	N 3.0	O 3.5	F 4.0	Ne 0.0
Na 0.9	Mg 1.2											Al 1.5	Si 1.8	P 2.1	S 2.5	Cl 3.0	Ar 0.0
K 0.8	Ca 1.0	Sc 1.3	Ti 1.5	V 1.6	Cr 1.6	Mn 1.5	Fe 1.8	Co 1.8	Ni 1.8	Cu 1.9	Zn 1.6	Ga 1.6	Ge 1.8	As 2.0	Se 2.4	Br 2.8	Kr 0.0

▲ ***Figure 2.6***

Electronegativity values of selected elements. The values are expressed according to the Pauling scale, named for the Nobel Prize-winning chemist Linus Pauling, who based the scale on bond energies. Pauling chose to compare the electronegativity of each element to that of fluorine, to which he assigned a value of 4.0.

are known as *structural formulas.* In the third symbol, known as a *molecular formula,* the subscript "2" indicates the number of hydrogen atoms that are bonded, not the number of shared electrons. Each of these symbols indicates the same thing—two hydrogen atoms are sharing a pair of electrons.

Many atoms need more than one electron to fill their valence shell. For instance, a carbon atom has four valence electrons and needs to gain four more if it is to have eight in its valence shell **(see Figure 2.5c).** As before, a line in the structural formula represents a covalent bond formed from the sharing of two electrons. Note that two covalent bonds are formed between an oxygen atom and a carbon atom in formaldehyde **(see Figure 2.5d).** This fact is represented by a double line, which indicates that the carbon atom shares four electrons with the oxygen atom.

Carbon atoms are critical to life. Since a carbon atom has four electrons in its valence shell, it has equal tendency to either lose four electrons or gain four electrons. Either event produces a full outer shell. The result is that carbon atoms tend to share electrons and form four covalent bonds with one another, and with many other types of atoms. Each carbon atom in effect acts as a four-way intersection where different components of a molecule can attach. One result of this feature is that carbon atoms can form very large chains that constitute the "backbone" of many biologically important molecules. Carbon chains can be branched or unbranched, and some even close back on themselves to form rings. Compounds that contain carbon and hydrogen atoms are called **organic compounds.** Among the many biologically important organic compounds are proteins and carbohydrates, which are discussed later in the chapter.

CRITICAL THINKING

An article in the local newspaper about gangrene states that the tissue-destroying toxin, lecithinase, is "an organic compound." But many people consider "organic" chemicals to mean something is good. Explain the apparent contradiction.

Polar Covalent Bonds

Learning Objective

✓ Explain the relationship between electronegativity and the polarity of a covalent bond.

If two covalently bound atoms have significantly different electronegativities, their electrons will not be shared equally. Instead, the electron pair will spend more time orbiting the nucleus of the atom with greater electronegativity. This type of bond, in which there is unequal sharing of electrons, is a **polar covalent bond.** An example of a molecule with polar covalent bonds is water **(Figure 2.7a).**

Because oxygen is more electronegative than hydrogen, the electrons spend more time near the oxygen nucleus than near the hydrogen nuclei, and thus the oxygen atom acquires a partial negative charge (symbolized as δ^-). The hydrogen nuclei each have a corresponding partial positive charge (δ^+). The covalent bonds between the oxygen atom and two hydrogen atoms are called polar because they have opposite electrical charges.

Polar covalent bonds can form between many different elements, but the most important polar covalent bonds for life are those that involve hydrogen because they allow hydrogen bonding, which we discuss shortly.

Both nonpolar and polar covalent bonds form angles between atoms such that the distances between electron orbits are maximized. The bond angle for water is shown in **Figure 2.7b.** However, it is more convenient to simply draw molecules as if all the atoms were in one plane, as in **Figure 2.7c.**

CRITICAL THINKING

The deadly poison hydrogen cyanide has the chemical formula H—C≡N. Describe the bonds between carbon and hydrogen, and between carbon and nitrogen, in terms of the number of electrons involved.

Triple covalent bonds are stronger and more difficult to break than single covalent bonds. Explain the reason why by referring to the stability of a valence shell that contains eight electrons.

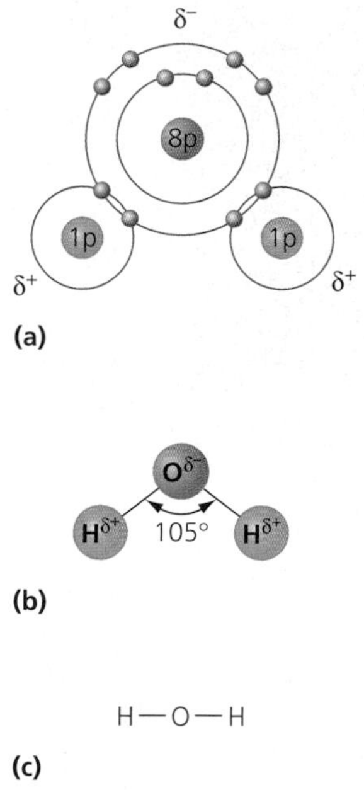

▲ ***Figure 2.7***

Polar covalent bonding in a water molecule. **(a)** A Bohr model of a water molecule, which has two polar covalent bonds. When the electronegativities of two atoms are significantly different, the shared electrons of covalent bonds spend more time around the more electronegative atom, giving it a partial negative charge (δ^-). **(b)** The bond angle in a water molecule. Atoms maximize the distances between electron orbitals in polar and nonpolar covalent bonds. **(c)** A common, convenient way of representing water as a polar molecule without showing bond angles.

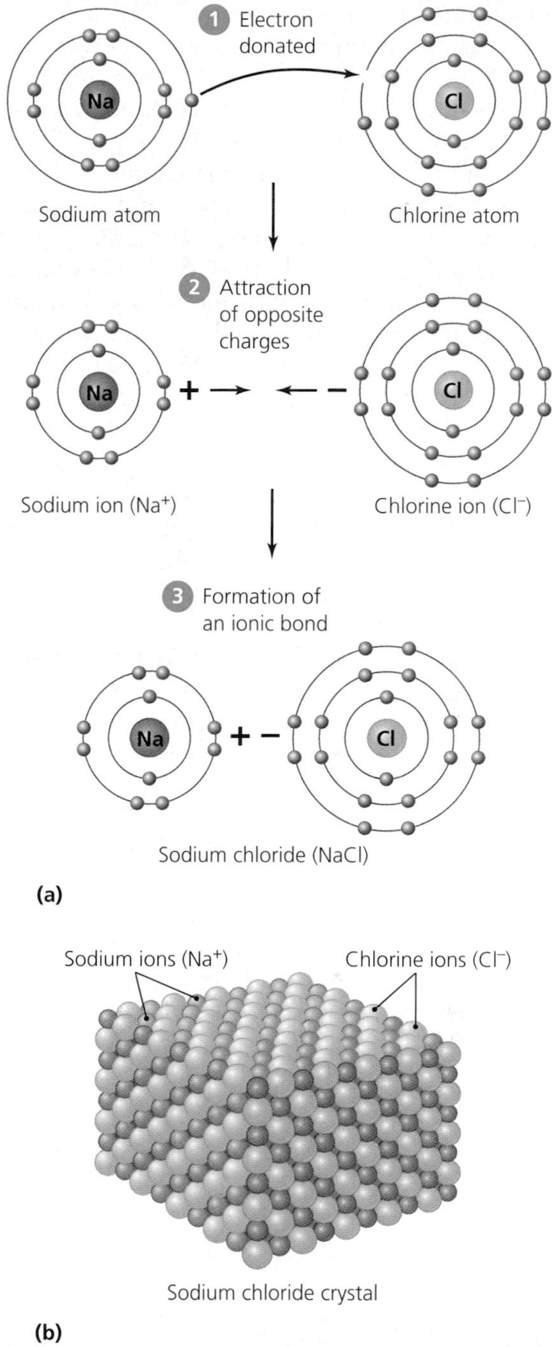

▲ ***Figure 2.8***

The interaction of sodium and chlorine to form an ionic bond. **(a)** Because the chlorine atom has a much higher electronegativity than the sodium atom, it "strips" the valence electron from sodium ①, forming a positively charged sodium ion and a negatively charged chlorine ion ②. The ion's opposite charges attract one another, forming an ionic bond ③. **(b)** The arrangement of sodium and chlorine ions in a sodium chloride (NaCl) crystal.

Ionic Bonds

Learning Objective

✓ Define ionization using the terms *cation* and *anion*.

Consider what happens when two atoms with vastly different electronegativities—for example, sodium, with one electron in its valence shell and an electronegativity of 0.9, and chlorine, with seven electrons in its valence shell and an electronegativity of 3.0—come together. Chlorine has such a higher electronegativity that it very strongly attracts sodium's valence electron, and the result is that the sodium loses that electron to chlorine (step ① in **Figure 2.8a**).

Now that the chlorine atom has one more electron than it has protons, it has a full negative charge, and the sodium atom, which has lost an electron, now has a full positive charge (step 2). An atom or group of atoms that has either a full negative charge or a full positive charge is called an *ion*. Positively charged ions are called **cations,** whereas negatively charged ions are called **anions.**

Because of their opposite charges, cations and anions attract each other and form what is termed an **ionic bond** (step 3). They form crystalline ionic compounds known as **salts,** such as sodium chloride (NaCl), also known as table salt **(Figure 2.8b),** and potassium chloride (KCl, sodium-free table salt). Ionic bonds differ from covalent bonds in that ions do not share electrons. Instead, the bond is formed from the attraction of opposite electrical charges.

The polar bonds of water molecules interfere with the ionic bonds of salts, causing *dissociation* (also called *ionization*) **(Figure 2.9).** This occurs as the partial negative charge on the oxygen atom of water attracts cations, and the partial positive charge on hydrogen atoms attracts anions. Obviously, the presence of polar bonds interferes with the attraction between the cation and anion.

When cations and anions disassociate from one another and become surrounded by water molecules (are hydrated), they are called **electrolytes** because they can conduct electricity through the solution. Electrolytes are critical for life because they stabilize a variety of compounds, act as electron carriers, and allow electrical gradients to exist within cells. We examine these functions of electrolytes in later chapters.

In nature, chemical bonds range from nonpolar bonds to polar bonds to ionic bonds. The important thing to remember is that electrons are shared between atoms in covalent bonds, and transferred from one atom to another in ionic bonds.

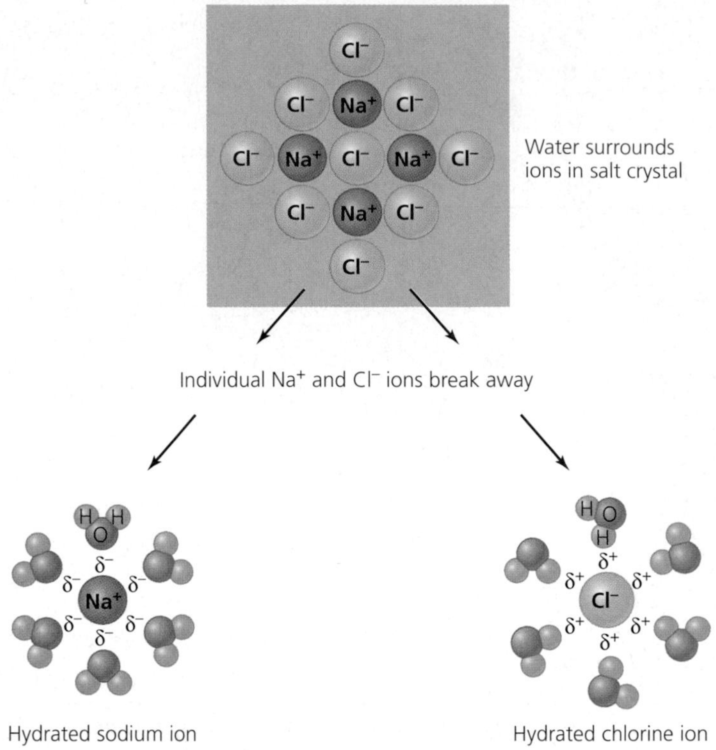

Figure 2.9

Dissociation of NaCl in water. When water surrounds the ions in a NaCl crystal, the partial charges on water molecules are attracted to charged ions, and the water molecules hydrate the ions by surrounding them. The partial negative charges on oxygen atoms are attracted to cations (in this case, the sodium ions), and the partial positive charges on hydrogen atoms are attracted to anions (the chlorine ions). Because the ions no longer attract one another, the salt crystal dissolves. Hydrated ions are called electrolytes.

CRITICAL THINKING

According to the chart in Figure 2.6, what type of bond (nonpolar covalent, polar covalent, or ionic) would you expect between chlorine and potassium? Between carbon and nitrogen? Between phosphorus and oxygen? Explain your reasoning in each case.

Hydrogen Bonds

Learning Objective

✓ Describe hydrogen bonds, and discuss their importance in living organisms.

As we have seen, hydrogen atoms bind to oxygen atoms by means of polar covalent bonds, which results in partial positive charges on the hydrogen atoms. Hydrogen atoms form polar covalent bonds with atoms of other elements as well.

The electrical attraction between a partially charged hydrogen atom and a full or partial negative charge on either a different region of the same molecule or another molecule is called a **hydrogen bond (Figure 2.10).** Hydrogen bonds can be likened to weak ionic bonds in that they arise from the

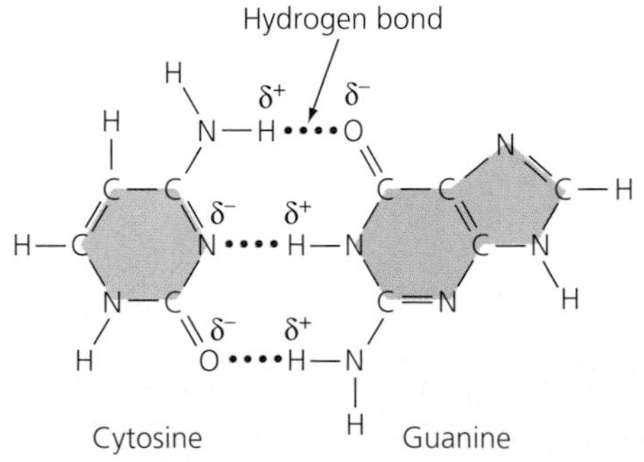

Figure 2.10

Hydrogen bonds can hold together portions of the same molecule, or hold two different molecules together. In this case, three hydrogen bonds are holding molecules of cytosine and guanine together.

Table 2.2 Characteristics of Chemical Bonds

Type of Bond	Description	Relative Strength
Nonpolar covalent bond	Pair of electrons is nearly equally shared between two atoms	Strong
Polar covalent bond	Electrons spend more time around the more electronegative of two atoms	Strong
Ionic bond	Electrons are stripped from a cation by an anion	Weaker than covalent in aqueous environments
Hydrogen bond	Partial positive charges on hydrogen atoms are attracted to full and partial negative charges on other molecules or other regions of the same molecule	Weaker than ionic

attraction of positive and negative charges. Notice also that although they are a consequence of polar covalent bonds between hydrogen atoms and other, more electronegative atoms, hydrogen bonds themselves are not covalent bonds—they do not involve the sharing of electrons.

As we have seen, covalent bonds are essential for life because they strongly link atoms together to form molecules. Hydrogen bonds, though weaker than covalent bonds, are also essential. The cumulative effect of numerous hydrogen bonds is to stabilize the three-dimensional shapes of large molecules. For example, the familiar double-helix shape of DNA is due to the stabilizing effects of thousands of hydrogen bonds holding the molecule together. Exact shape is critical for the functioning of enzymes, antibodies, intercellular chemical messengers, and the recognition of target cells by pathogens. Further, because hydrogen bonds are weak, they can be overcome when necessary. For example, the two complementary halves of a DNA molecule are held together primarily by hydrogen bonds, and can be separated for DNA replication and other processes (see Figure 7.5 on page 206).

Table 2.2 summarizes the characteristics of chemical bonds.

Chemical Reactions

Learning Objective

✓ Describe three general types of chemical reactions found in living things.

You are already familiar with many consequences of chemical reactions: You add yeast to bread dough and it rises; enzymes in your laundry detergent remove grass stains; and gasoline burned in your car releases energy to speed you on your way. What exactly is happening in these reactions? What is the precise definition of a chemical reaction?

We have discussed how bonds are formed via the sharing of electrons or the attraction of positive and negative charges. Scientists define **chemical reactions** as the making or breaking of such chemical bonds. All chemical reactions begin with **reactants**—the atoms, ions, or molecules that exist at the beginning of a reaction. Similarly, all chemical reactions result in **products**—the atoms, ions, or molecules left after the reaction is complete.

Reactants and products may have very different physical and chemical characteristics. For example, hydrogen and oxygen are gases, and both have very different properties from water. However, the numbers and types of atoms never change in a chemical reaction; atoms are neither destroyed nor created, only rearranged.

Now let's turn our attention to three general categories of chemical reactions that occur in microorganisms: *synthesis, decomposition,* and *exchange reactions.*

Synthesis Reactions

Learning Objectives

✓ Give an example of a synthesis reaction that involves the formation of a water molecule.

✓ Contrast endothermic and exothermic chemical reactions.

Synthesis reactions involve the formation of larger, more complex molecules. Synthesis reactions can be expressed symbolically as:

$$\text{Reactant} + \text{Reactant} \rightarrow \text{Product(s)}$$

The arrow indicates the direction of the reaction and the formation of new chemical bonds. For example, algae make their own glucose (sugar) using the following reaction:

$$6\,H_2O + 6\,CO_2 \rightarrow C_6H_{12}O_6 + 6\,O_2$$

The reaction is read, "Six molecules of water plus six molecules of carbon dioxide yield one molecule of glucose and six molecules of oxygen." Notice that the total number and kind of atoms are the same on both sides of the reaction.

An important type of synthesis reaction is **dehydration synthesis,** in which two smaller molecules are joined together by a covalent bond, and a water molecule is also formed (**Figure 2.11a** on page 36). The word *dehydration* in the name of this type of reaction refers to the fact that one of the products is a water molecule formed when a hydrogen ion (H^+) from one reactant combines with a hydroxyl ion (OH^-) from another reactant.

Synthesis reactions require energy to break bonds in the reactants and to form new bonds to make products. Reactions

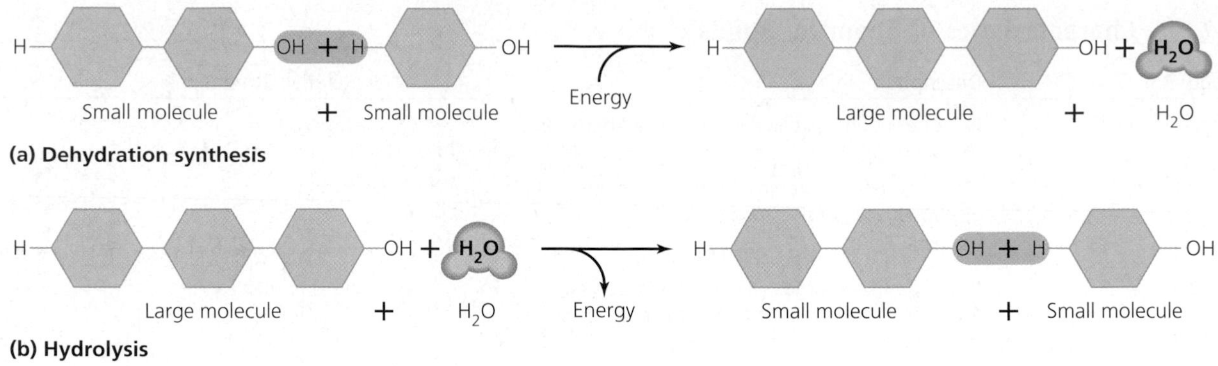

▲ ***Figure 2.11***

Two types of chemical reactions in living things. **(a)** Dehydration synthesis. In this energy-requiring reaction, a hydroxyl ion (OH^-) removed from one reactant and a hydrogen ion (H^+) removed from another reactant combine to form hydrogen hydroxide (HOH), which is water. **(b)** Hydrolysis, an energy-yielding reaction that is the reverse of a dehydration synthesis reaction. *What are the scientific words meaning "energy-requiring" and "energy-yielding"?*

Figure 2.11 Endothermic means "energy requiring," and exothermic means "energy releasing."

that require energy are said to be **endothermic**[8] **reactions** because they trap energy within new molecular bonds. As we will see in Chapter 6, an energy supply for fueling synthesis reactions is one common requirement of all living things.

Taken together, all of the synthesis reactions in an organism are called **anabolism.**

Decomposition Reactions

Learning Objective

✓ Give an example of a decomposition reaction that involves breaking the bonds of a water molecule.

Decomposition reactions are the reverse of synthesis reactions in that they break bonds within larger molecules to form smaller atoms, ions, and molecules. These reactions release energy and are therefore **exothermic.**[9] In general, decomposition reactions can be represented by the following formula:

$$\text{Reactant} \rightarrow \text{Product} + \text{Product}$$

An example of a biologically important decomposition reaction is the aerobic decomposition of glucose to form carbon dioxide and water:

$$C_6H_{12}O_6 + 6\,O_2 \rightarrow 6\,H_2O + 6\,CO_2$$

Note that this reaction is exactly the reverse of the synthesis reaction in algae that we examined previously. Synthesis and decomposition reactions are often reversible in living things.

A common type of decomposition reaction is **hydrolysis,**[10] the reverse of dehydration synthesis **(Figure 2.11b).** In hydrolytic reactions, a covalent bond in a large molecule is broken, and the ionic components of water (H^+ and OH^-) are added to the products.

Collectively, all of the decomposition reactions in an organism are called **catabolism.**

Exchange Reactions

Learning Objective

✓ Compare exchange reactions to synthesis and decomposition reactions.

Exchange reactions (also called *transfer reactions*) have features similar to both synthesis and decomposition reactions. For instance, they involve breaking and forming covalent bonds, and they involve both endothermic and exothermic steps. As the name suggests, atoms are moved from one molecule to another. In general, these reactions can be represented as either

$$A + BC \rightarrow AB + C$$

or

$$AB + CD \rightarrow AD + BC$$

An important exchange reaction within organisms is the phosphorylation of glucose:

$$\underset{\text{Glucose}}{C_6H_{12}O_6} + \underset{\text{Adenosine triphosphate}}{\text{A-Ⓟ-Ⓟ-Ⓟ}} \rightarrow \underset{\text{Glucose phosphate}}{C_6H_{12}O_6\text{-Ⓟ}} + \underset{\text{Adenosine diphosphate}}{\text{A-Ⓟ-Ⓟ}}$$

The sum of all of the chemical reactions in an organism, both catabolic and anabolic, is called **metabolism.** We examine metabolism in more detail in Chapter 5.

[8]From Greek *endon,* meaning within, and *thermos,* meaning heat (energy).
[9]From Greek *exo,* meaning outside, and *thermos,* meaning heat (energy).
[10]From Greek *hydor,* meaning water, and *lysis,* meaning loosing.

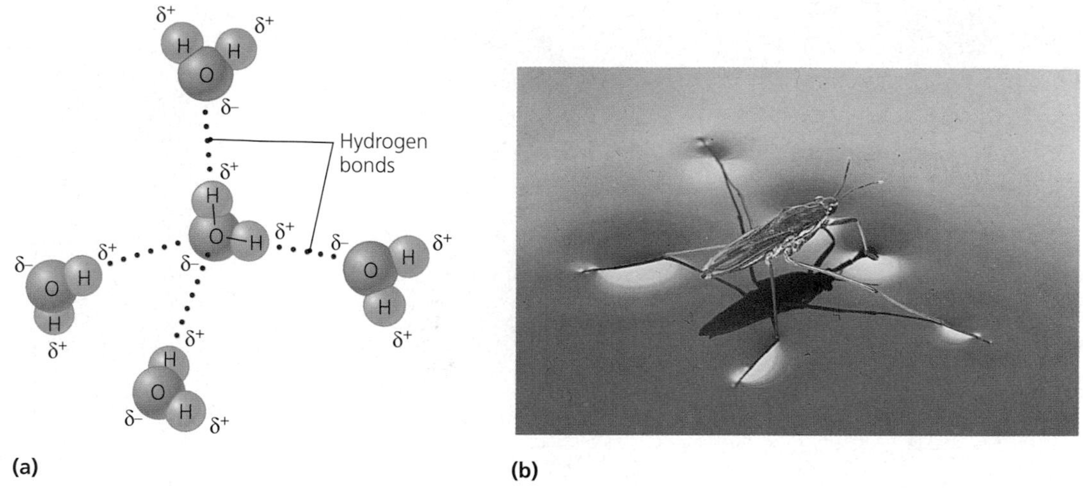

Figure 2.12 The cohesiveness of liquid water. **(a)** Water molecules are cohesive because hydrogen bonds cause them to stick to one another. **(b)** One result of cohesiveness in water is surface tension, which can be strong enough to support the weight of insects known as water striders.

Water, Acids, Bases, and Salts

As previously noted, living things depend on organic compounds, those that contain carbon and hydrogen atoms. Living things also require a variety of **inorganic chemicals,** which typically lack carbon. Such inorganic substances include water, oxygen molecules, metal ions, and many acids, bases, and salts. In this section we examine the characteristics of some of these inorganic substances.

Water

Learning Objective

✓ Describe five qualities of water that make it vital to life.

Water is the most abundant substance in organisms, constituting 50–99% of their mass. Most of the special characteristics that make water vital result from the fact that a water molecule has two polar covalent bonds, which allows hydrogen bonding between water molecules and their neighbors. Among the special properties of water are the following:

1. Water molecules are cohesive; that is, they tend to stick to one another through hydrogen bonding **(Figure 2.12).** This property generates many special characteristics of water, including *surface tension,* which allows water to form a thin layer on the surface of cells. This aqueous layer is necessary for the transport of dissolved materials into and out of a cell.
2. Water is an excellent *solvent;* that is, it dissolves salts and other electrically charged molecules because it is attracted to both positive and negative charges (see Figure 2.9).
3. Water remains a liquid across a wider range of temperatures than other molecules of its size. This is critical because living things require water in liquid form.
4. Water can absorb significant amounts of heat energy without itself changing temperature. Further, when heated water molecules eventually evaporate, they take much of this absorbed energy with them. These properties moderate temperature fluctuations that would otherwise damage organisms.
5. Water molecules participate in many chemical reactions within cells, both as reactants in hydrolysis and as products of dehydration synthesis.

CRITICAL THINKING

How can hydrogen bonding between water molecules help explain water's ability to absorb large amounts of energy before evaporating?

Acids and Bases

Learning Objective

✓ Contrast acids, bases, and salts, and explain the role of buffers.

As we have seen, the polar bonds of water molecules dissociate salts into their component cations and anions. A similar process occurs with substances known as acids and bases.

An **acid** is a substance that dissociates into one or more hydrogen ions (H^+) and one or more anions (**Figure 2.13a** on page 38). Acids can be inorganic molecules such as hydrochloric acid (HCl) and sulfuric acid (H_2SO_4), or organic molecules such as amino acids and nucleic acids. Familiar organic acids are found in lemon juice, black coffee, and tea. Of course, the anions of organic acids contain carbon, while those of inorganic acids do not.

A **base** is a molecule that binds with H^+ when dissolved in water. Some bases dissociate into cations and *hydroxyl ions* (OH^-) **(Figure 2.13b),** which then combine with hydrogen ions to form water molecules:

$$H^+ + OH^- \rightarrow H_2O$$

Other bases, such as household ammonia (NH_3), directly accept hydrogen ions and become compound ions such as NH_4^+ (ammonium). Another common household base is baking soda (sodium bicarbonate, $NaHCO_3$).

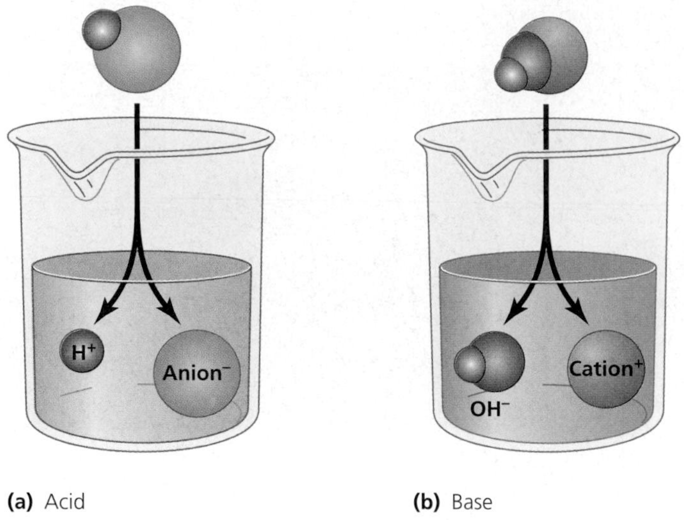

Figure 2.13
Acids and bases. **(a)** Acids dissociate in water into hydrogen ions and anions. **(b)** Many bases dissociate into hydroxyl ions and cations.

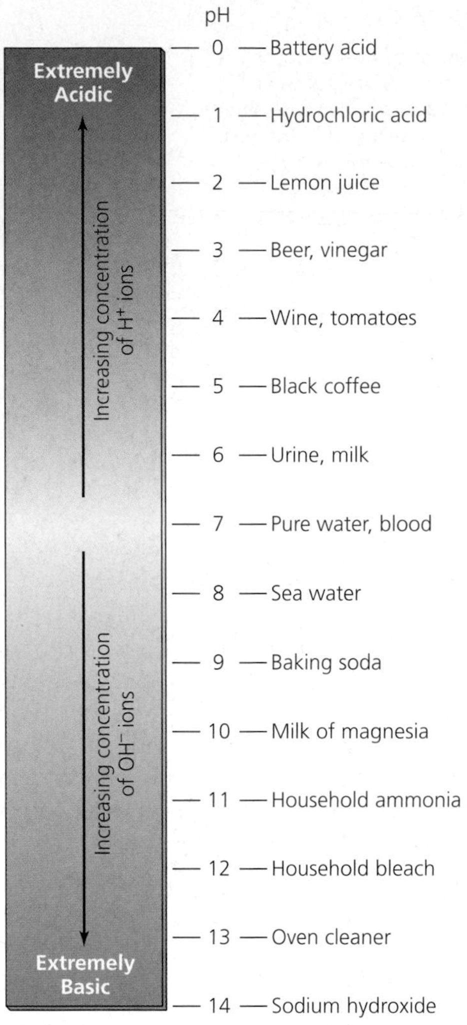

Figure 2.14
The pH scale. Values below 7 are acidic; values above 7 are basic.

Metabolism requires a relatively constant balance of acids and bases because hydrogen ions and hydroxyl ions are involved in many chemical reactions. Further, many complex molecules such as proteins lose their functional shapes when acidity changes. If the concentration of either hydrogen ions or hydroxyl ions deviates too far from normal, metabolism ceases.

The concentration of hydrogen ions in a solution is expressed using a logarithmic **pH scale (Figure 2.14).** The term *pH* comes from *potential hydrogen,* which is the negative of the logarithm of the concentration of hydrogen ions. In this logarithmic scale, it is important to notice that acidity increases as pH values decrease, and that each decrease by a whole number in pH indicates a 10-fold increase in acidity (hydrogen ion concentration). For example, a glass of grapefruit juice, which has a pH of 3.0, contains 10 times as many hydrogen ions as the same volume of tomato juice, which has a pH of 4.0. Similarly, tomato juice is 1000 times more acidic than pure water, which has a pH of 7.0 (neutral). Water is neutral because it dissociates into one hydrogen cation and one hydroxyl anion:

$$H_2O \rightarrow H^+ + OH^-$$

Alkaline (basic) substances have pH values greater than 7.0. They reduce the number of free hydrogen ions by combining with them. For bases that produce hydroxyl ions, the concentration of hydroxyl ions is inversely related to the concentration of hydrogen ions.

Organisms can tolerate only a certain, relatively narrow pH range. Fluctuations outside an organism's preferred range inhibit its metabolism and may even be fatal. Most organisms contain natural **buffers**—substances, such as proteins, that prevent drastic changes in internal pH. In a laboratory culture, the metabolic activity of microorganisms can change the pH of microbial growth solutions as nutrients are taken up and wastes are released; therefore, pH buffers may be added to them (see Table 6.5 on page 184). One common buffer used in microbiological media is KH_2PO_4, which exists as either a weak acid or a weak base, depending on the pH of its environment. Under acidic conditions, KH_2PO_4 is a base that combines with H^+, neutralizing the acidic environment; in alkaline conditions, however, KH_2PO_4 acts as an acid, releasing hydrogen ions.

Microorganisms differ in their ability to tolerate various ranges of pH. Most grow best when the pH is between 6.5 and 8.5. Photosynthetic bacteria known as *cyanobacteria* grow well in more basic solutions. Fungi generally tolerate acidic environments better than most prokaryotes, though acid-loving prokaryotes, called *acidophiles,* require acidic conditions. Two pathogenic bacteria that tolerate acidic conditions in the human body, allowing them to grow where other bacteria cannot, are *Propionibacterium acnes* (prō-pē-on-i-bak-tēr′ē-ŭm ak′nēz), which contributes to acne in the skin, which normally has a pH of about 4.0; and *Helicobacter pylori* (hel′ĭ-kō-bak′ter pī′lō-rē),[11] a curved bacterium that has been shown to cause ulcers in the stomach, which can have a pH as low as 1.5 when acid is being actively secreted. **Highlight 2.1** focuses on how the use of antacids may increase the survival rates of certain disease-causing bacteria in the stomach.

Microorganisms can change the pH of their environment by utilizing acids and bases and by producing acidic or basic wastes. For example, fermentative microorganisms form organic acids from the decomposition of sugar, and the bacterium *Thiobacillus* (thī-ō-bă-sil′ŭs) can reduce the pH of its environment to 0.0. Acid produced by this bacterium in mine water dissolves enough uranium and copper from low-grade ore to make some mines profitable.

Scientists measure pH with a pH meter or with test papers impregnated with chemicals (such as litmus or phenol red) that change color in response to pH. In a microbiological laboratory, changes in color of such pH indicators incorporated into microbial growth media are commonly used to distinguish among bacterial genera.

Salts

As we have seen, a salt is a compound that dissociates in water into cations and anions other than H^+ and OH^-. Acids and hydroxyl-yielding bases neutralize each other during exchange reactions that produce water and salt. For instance, milk of magnesia (magnesium hydroxide), is an antacid used to neutralize excess stomach acid. The chemical reaction is

$$\underset{\text{Magnesium hydroxide}}{Mg(OH)_2} + \underset{\text{Hydrochloric acid}}{2\,HCl} \rightarrow \underset{\text{Magnesium chloride (a salt)}}{MgCl_2} + \underset{\text{Water}}{2\,H_2O}$$

Cations and anions of salts are electrolytes. A cell uses electrolytes to create electrical differences between its inside and outside, to transfer electrons from one location to another, and as important components of many enzymes. Certain organisms also use salts such as calcium carbonate ($CaCO_3$) to provide structure and support for their cells.

CRITICAL THINKING

Why are two molecules of hydrochloric acid neutralized by only one molecule of magnesium hydroxide?

Organic Macromolecules

Inorganic molecules play important roles in an organism's metabolism; however, water excluded, they compose only about 1.5% of its mass. Inorganic molecules are typically too small and too simple to constitute an organism's basic structures, or to perform the complicated chemical reactions required of life. These functions are fulfilled by organic molecules, which are generally larger and much more complex.

Highlight 2.1 Raw Oysters and Antacids: A Deadly Mix?

Raw oyster lovers, beware. Researchers have found evidence that taking antacids may make you more susceptible to becoming ill from *Vibrio vulnificus,* a bacterium commonly ingested by eating tainted raw oysters. The highly acidic environment of the stomach kills many bacteria before they are able to cause disease, but taking antacids reduces the stomach's acidity, creating conditions in which more bacteria survive. Researchers found that the presence of antacids in a simulated gastric environment significantly increased the survival rate of *V. vulnificus,* which is the leading cause of food-poisoning fatalities in the United States.

Reference: Koo, J. *et al.* 2001. Antacid increases survival of *Vibrio vulnificus* and *Vibrio vulnificus* phage in a gastrointestinal model. *Applied and Environmental Microbiology* 67:2895–2902.

Functional Groups and Monomers

Learning Objective

✓ Define a "functional group" as it relates to organic chemistry.

As we have seen, organic molecules contain carbon and hydrogen atoms, and each carbon atom can form four covalent bonds with other atoms (see Figure 2.5c and d). Carbon atoms that are linked together in branched chains, unbranched chains, and rings provide the basic frameworks of organic molecules.

Atoms of other elements are bound to these carbon frameworks to form an unlimited number of compounds. Besides carbon and hydrogen, the most common elements in organic compounds are oxygen, nitrogen, phosphorus, and sulfur. Other elements, such as iron, copper, molybdenum, manganese, zinc, and iodine, are important in some proteins.

Atoms often appear in certain common arrangements called **functional groups.** For example, —NH_2, the amino functional group, is found in all amino acids, and —OH, the hydroxyl functional group,[12] is common to all alcohols. When a class of organic molecules is discussed, the letter **R** (for *residue*) designates atoms in the compound that vary from one molecule to another. The symbol R—OH, therefore,

[11] The name *pylori* refers to the pylorus, a region of the stomach.

[12] Note that the hydroxyl functional group is not the same thing as the hydroxyl *ion* because the former is covalently bonded to a carbon atom.

Table 2.3 Functional Groups of Organic Molecules, and Some Classes of Compounds in Which They Are Found

Structure	Name	Class of Compounds
—OH	Hydroxyl	Alcohol Monosaccharide Amino acid
$R{-}CH_2{-}O{-}CH_2{-}R'$	Ether	Disaccharide Polysaccharide
$R{-}C({=}O){-}R'$	Internal carbonyl (a carbon atom on each side)	Ketone Carbohydrate
$R{-}C({=}O){-}H$	Terminal carbonyl (a carbon atom on only one side)	Aldehyde
$R{-}C({=}O){-}O{-}H$	Carboxyl	Amino acid Nucleic acid
$R{-}CH{-}NH_2$	Amino	Amino acid Protein
$R{-}C({=}O){-}O{-}R'$	Ester	Fat Wax
$R{-}CH_2{-}SH$	Sulfhydryl	Amino acid Protein
$R{-}CH_2{-}O{-}P(OH)_2{=}O$	Organic phosphate	Phospholipid Nucleotide ATP

represents the general formula for an alcohol. **Table 2.3** describes some common functional groups of organic molecules.

There is a great variety of organic compounds, but certain basic types are used by all organisms. These molecules, known as *macromolecules* because they are very large, are lipids, carbohydrates, proteins, and nucleic acids. Proteins, carbohydrates, and nucleic acids are composed of simpler subunits known as **monomers,**[13] which are basic building blocks. The monomers of these macromolecules are joined together to form chains of monomers called **polymers.**[14] Some macromolecular polymers are composed of hundreds of thousands of monomers.

Lipids

Learning Objectives

- ✓ Describe the structure of a triglyceride molecule, and compare it to that of a phospholipid.
- ✓ Distinguish among saturated, unsaturated, and polyunsaturated fatty acids.

Lipids are a diverse group of organic macromolecules not composed of monomers. They have one common trait—they are **hydrophobic;**[15] that is, they are insoluble in water. Lipids have little or no affinity for water because they are composed almost entirely of carbon and hydrogen atoms linked by nonpolar covalent bonds. Because these bonds are nonpolar, they have no attraction to the polar bonds of water molecules. To look at it another way, the polar water molecules are attracted to each other and exclude the nonpolar lipid molecules. There are four major groups of lipids in cells: fats, phospholipids, waxes, and steroids.

Fats

Organisms make **fats** in dehydration synthesis reactions that form *esters* between three chainlike fatty acids and an alcohol named *glycerol* **(Figure 2.15a).** Fats are also called *triglycerides* because they contain three fatty acid molecules linked to a molecule of glycerol.

The three fatty acids in a fat molecule may be identical or different from one another, but each usually has 12 to 20 carbon atoms. An important difference among fatty acids is the presence and location of double bonds between the carbon atoms **(Figure 2.15b).** When the carbon atoms are linked solely by single bonds, every carbon atom, with the exception of the terminal ones, is covalently linked to two hydrogen atoms. Such a fatty acid is **saturated** with hydrogen. In contrast, **unsaturated fatty acids** contain at least one double bond between adjacent carbon atoms, and therefore contain

13From Greek *mono,* meaning one, and *meris,* meaning part.
14From Greek *poly,* meaning many, and *meris,* meaning part.
15From Greek *hydor,* meaning water, and *phobos,* meaning fear.

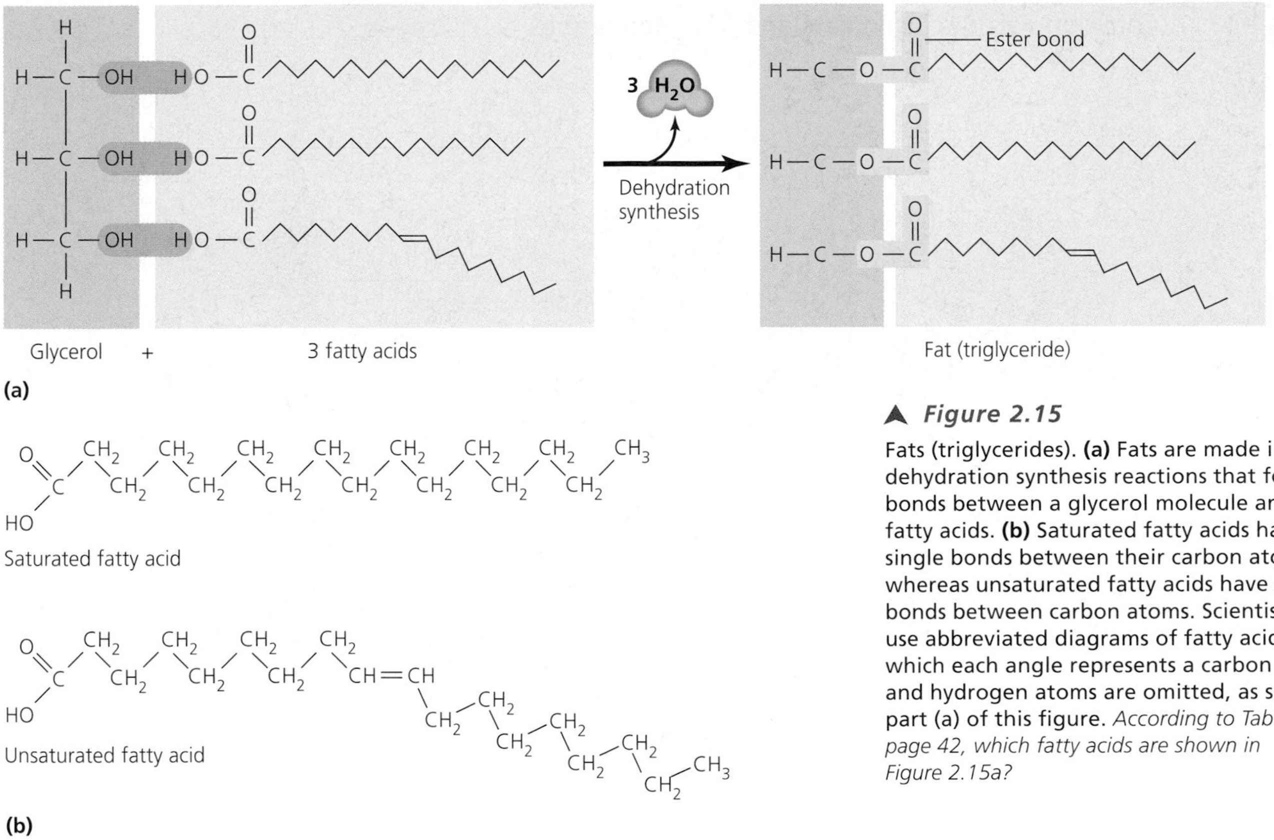

Figure 2.15
Fats (triglycerides). **(a)** Fats are made in dehydration synthesis reactions that form ester bonds between a glycerol molecule and three fatty acids. **(b)** Saturated fatty acids have only single bonds between their carbon atoms, whereas unsaturated fatty acids have double bonds between carbon atoms. Scientists often use abbreviated diagrams of fatty acids in which each angle represents a carbon atom, and hydrogen atoms are omitted, as seen in part (a) of this figure. *According to Table 2.4 on page 42, which fatty acids are shown in Figure 2.15a?*

***Figure 2.15** Stearic acid, palmitic acid, and oleic acid.*

at least one carbon atom bound to only a single hydrogen atom. If several double bonds exist in even one fatty acid of a molecule of fat, then it is a **polyunsaturated** fat.

Saturated fats (composed of saturated fatty acids), like those found in animals, are usually solid at room temperature because their fatty acids can pack close together. Unsaturated fatty acids, by contrast, are bent at every double bond, and so cannot pack tightly; they remain liquid at room temperature. Most fats in plants are unsaturated or polyunsaturated. **Table 2.4** on page 42 compares the structures and melting points of four common fatty acids.

Fats contain an abundance of energy stored in their carbon–carbon covalent bonds. Indeed, the primary role of fats in organisms is to store energy. As we see in Chapter 5, fats can be catabolized to provide energy for movement, synthesis, and transport.

Phospholipids

Phospholipids are similar to fats, but they contain two fatty acid chains instead of three. In phospholipids, the third carbon atom of glycerol is linked to a phosphate (PO_4) functional group instead of to a fatty acid (**Figure 2.16a** on page 43). Like fats, different phospholipids contain different fatty acids. Small organic groups linked to the phosphate group provide additional variety.

Part of a phospholipid molecule is attracted to water; that is, whereas the fatty acid "tail" portion of the molecule is nonpolar and thus hydrophobic, the phospholipid "head" is polar and is thus **hydrophilic**[16] **(Figure 2.16b)**. As a result, phospholipids placed in a watery environment will always self-assemble into forms that keep the fatty acid tails away from water. Two such forms are possible **(Figure 2.16c):** micelles, which have a single layer of phospholipid and resemble balls, and phospholipid bilayers, which resemble a two-ply bag.

The fatty acids tails, which are hydrophobic, congregate in the water-free interior of micelles and bilayers. The polar phosphate heads orient toward the water because they are hydrophilic. Phospholipid bilayers make up the outer membranes of all cells, as well as the internal membranes of plant, fungal, and animal cells.

Waxes

Waxes contain one long-chain fatty acid linked covalently to a long-chain alcohol by an ester bond. Waxes do not have a hydrophilic head; thus, they are completely water insoluble. Certain microorganisms, such as *Mycobacterium tuberculosis* (mī′kō-bak-tēr′ē-ŭm tū-ber-kyū-lō′sis), are surrounded by a waxy wall, making them resistant to drying. Some marine

[16]From Greek *philos,* meaning love.

Table 2.4 Common Fatty Acids in Fats and Cell Membranes

Number of Carbon Atoms: Double Bonds	Type of Fatty Acid	Structure and Formula	Common Name	Melting Point
16 : 0	Saturated	$CH_3(CH_2)_{14}COOH$ 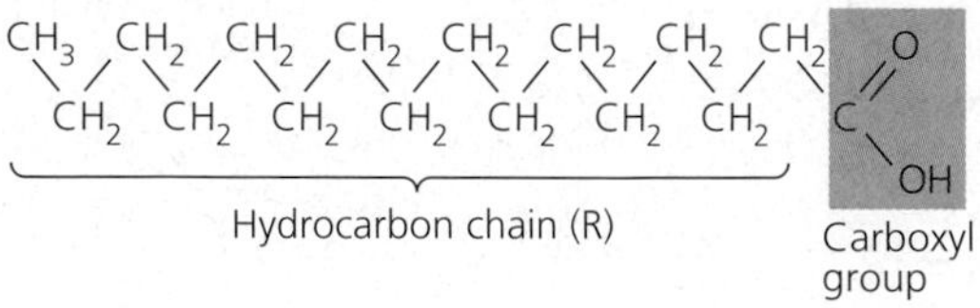	Palmitic acid	63°C
18 : 0	Saturated	$CH_3(CH_2)_{16}COOH$ 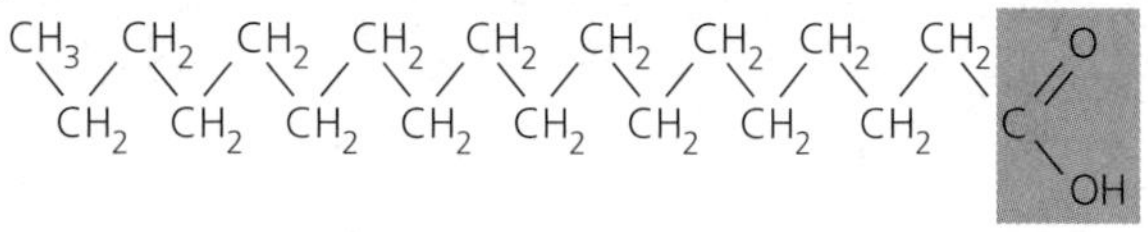	Stearic acid	70°C
18 : 1	Monounsaturated	$CH_3(CH_2)_7CH{=}CH(CH_2)_7COOH$ 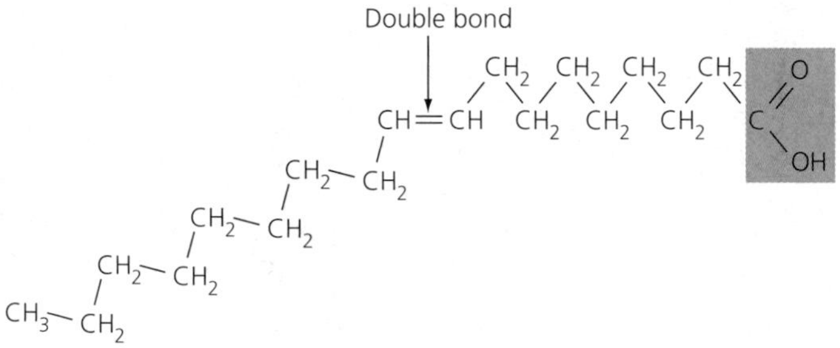	Oleic acid	16°C
18 : 2	Polyunsaturated	$CH_3(CH_2)_4(CH{=}CHCH_2)_2(CH_2)_6COOH$ 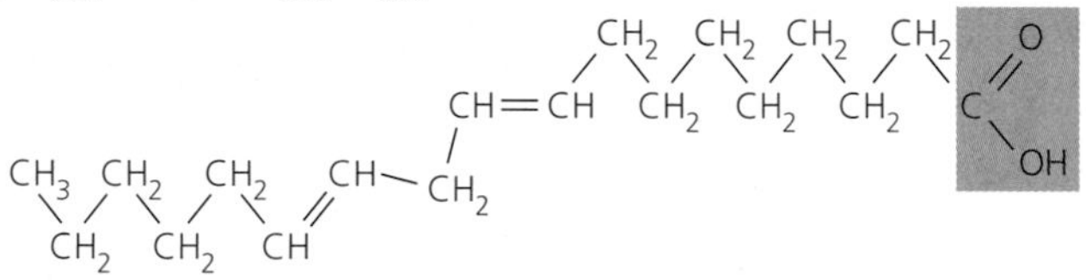	Linoleic acid	−5°C

microbes use waxes instead of fats as energy storage molecules.

Steroids

A final group of lipids are **steroids.** Steroids consist of four rings (containing five or six carbon atoms) that are fused to one another and attached to various side chains and functional groups (**Figure 2.17a** on page 44). Steroids play many roles in human metabolism. Some act as hormones; another steroid, *cholesterol,* is familiar to you as an undesirable component of food. However, cholesterol is also an essential part of the phospholipid bilayer membrane surrounding all animal cells. Cells of fungi, plants, and one group of bacteria (mycoplasmas) have similar sterol molecules in their membranes. Sterols, which are steroids with an —OH functional group, interfere with the tight packing of the fatty acid chains of phospholipids **(Figure 2.17b).** This keeps the membranes fluid and flexible at low temperatures. Without steroids such as cholesterol, the membranes of cells would become stiff and inflexible in the cold.

CRITICAL THINKING

We have seen that it is important that biological membranes remain flexible. Most bacteria lack sterols in their membranes and instead incorporate unsaturated phospholipids in the membranes to resist tight packing. Reexamine Table 2.4. Which fatty acid might best protect the membranes of an ice-dwelling bacterium?

Carbohydrates

Learning Objective

✓ Discuss the roles of carbohydrates in living systems.

Carbohydrates are organic molecules composed solely of atoms of carbon, hydrogen, and oxygen. Most carbohydrate

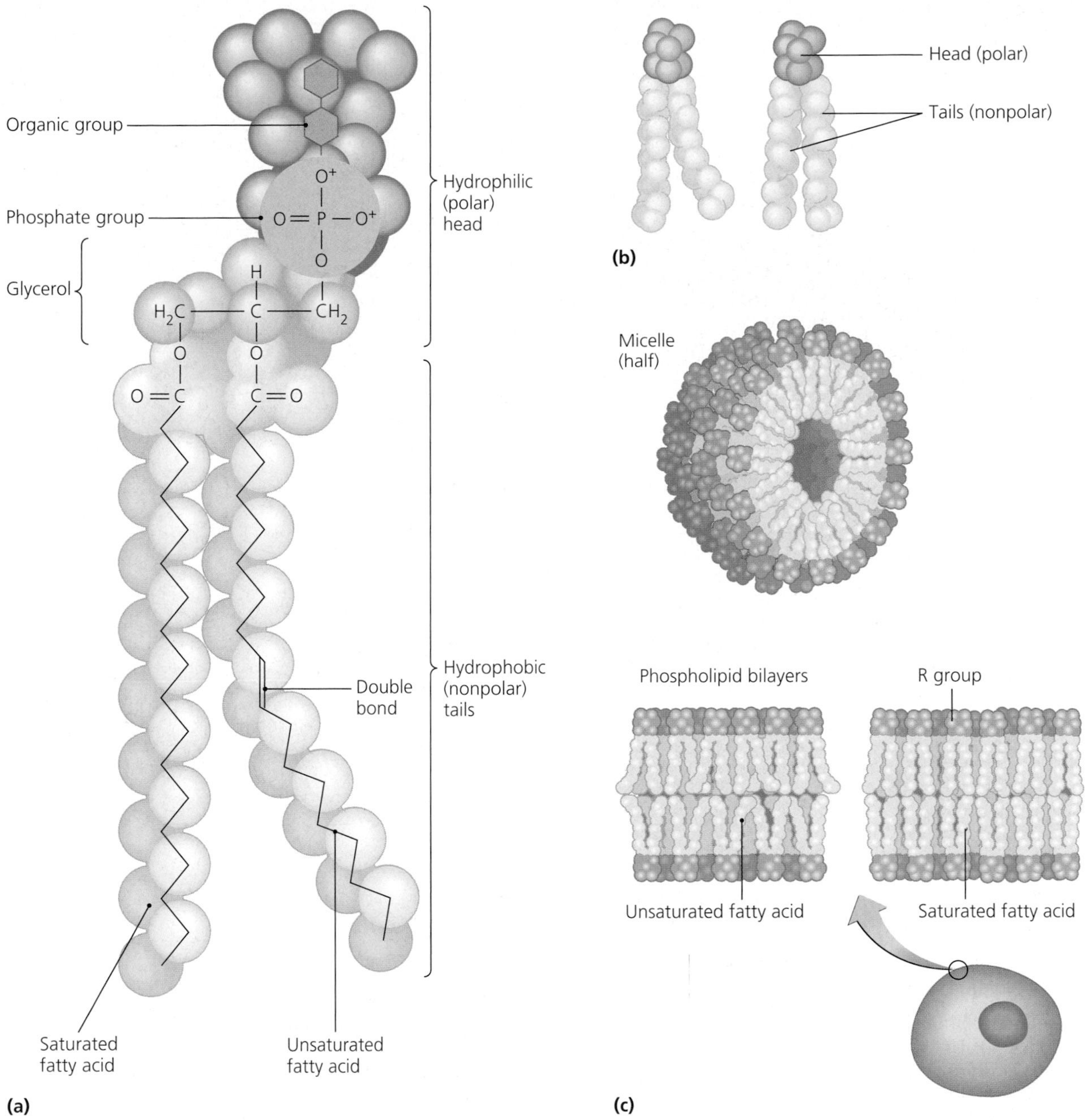

Figure 2.16

Phospholipids. **(a)** A phospholipid is composed of a hydrophilic (polar) "head," which is composed of glycerol and a phosphate group, and two hydrophobic (nonpolar) fatty acid "tails." **(b)** The symbols commonly used to represent phospholipids. **(c)** In water, phospholipids self-assemble into spherical micelles or bilayers. Note that phospholipids containing unsaturated fatty acids do not pack together as tightly as those containing saturated fatty acids.

compounds contain an equal number of oxygen and carbon atoms, and twice as many hydrogen atoms as carbon atoms, so the general formula for a carbohydrate is $(CH_2O)_n$, where n indicates the number of CH_2O units.

Carbohydrates play many important roles in organisms. Large carbohydrates such as starch and glycogen are used for the long-term storage of chemical energy, while a smaller carbohydrate molecule—glucose—serves as a ready energy source in most cells. Carbohydrates also form part of the backbones of DNA and RNA, and others are converted routinely into amino acids. Additionally, polymers of carbohydrate form the cell walls of fungi, plants, algae, and prokaryotes, and are involved in intercellular interactions between animal cells. For example, specific carbohydrates found on the surfaces of white blood cells determine which cells interact in immune responses against pathogens.

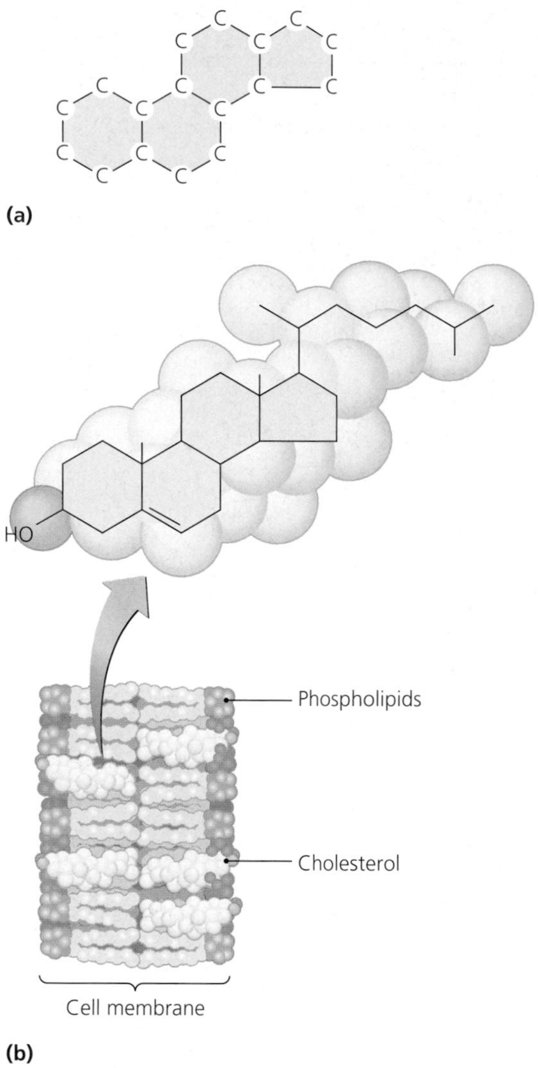

▲ ***Figure 2.17***

Steroids. **(a)** Steroids are lipids characterized by four "fused" carbon rings. **(b)** The steroid cholesterol functions in animal and protozoan cell membranes to prevent packing of phospholipids, which keeps the membranes fluid at low temperatures.

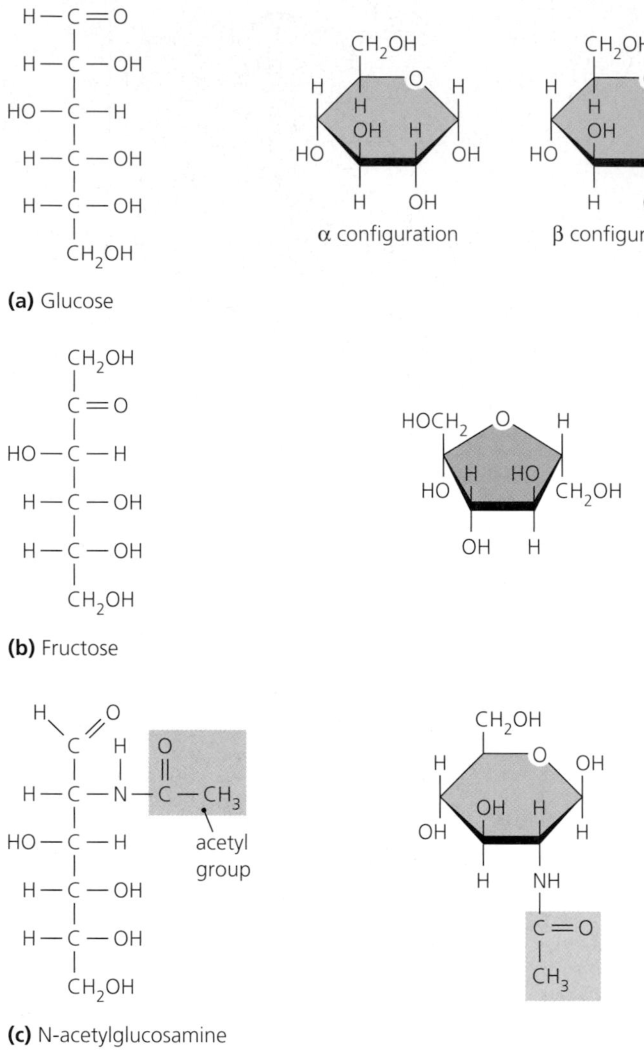

▲ ***Figure 2.18***

Monosaccharides (simple sugars). Although simple sugars may exist as either linear molecules (at left) or rings (at right), energy dynamics generally favor ring forms. **(a)** Glucose, a hexose, is the primary energy source for cellular metabolism and an important monomer in many larger carbohydrates. Alpha and beta ring configurations exist. **(b)** Fructose, or fruit sugar, a hexose. **(c)** *N*-acetylglucosamine (NAG), a monomer in many bacterial cell walls.

Monosaccharides

The simplest carbohydrates are **monosaccharides**[17] or simple sugars **(Figure 2.18).** The general names for the classes of monosaccharides are formed from a prefix indicating the number of carbon atoms, and the suffix *-ose*. For example, *pentoses* are sugars with five carbon atoms, and *hexoses* are sugars with six carbon atoms. Pentoses and hexoses are particularly important in the metabolism of cells. For example, deoxyribose, which is the sugar component of DNA, is a pentose. Glucose is a hexose and the primary energy molecule of cells, and fructose is a hexose found in fruit.

Monosaccharides may exist as linear molecules, but due to energy dynamics they usually take cyclic (ring) forms. In some cases, more than one cyclic structure may exist. For example, glucose can assume an alpha (α) configuration or a beta (β) configuration (see Figure 2.18a). As we will see, these configurations play important roles in the formation of different polymers.

Disaccharides

When two monosaccharide molecules are linked together via dehydration synthesis, the result is a **disaccharide.** For

[17]From Greek *sakcharon,* meaning sugar.

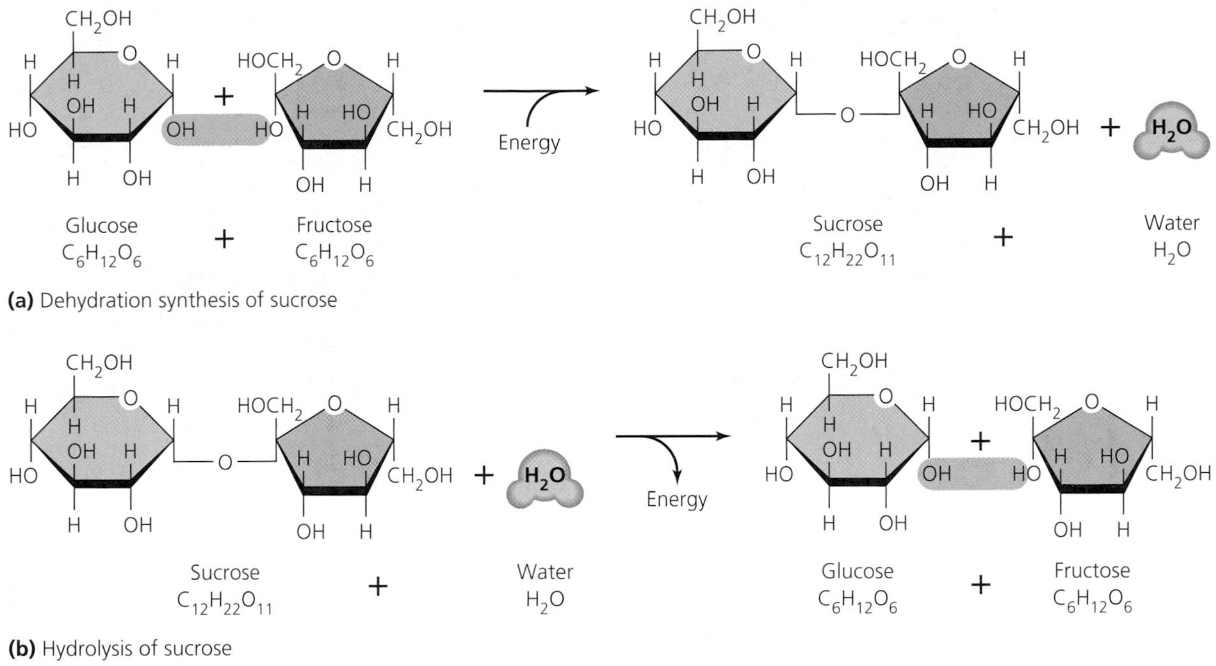

▲ *Figure 2.19*
Disaccharides. **(a)** Formation of the disaccharide sucrose via dehydration synthesis. **(b)** Breakdown of sucrose via hydrolysis.

example, the linkage of two hexoses, glucose and fructose, forms sucrose (table sugar) and a molecule of water **(Figure 2.19a).** Other disaccharides include maltose (malt sugar) and lactose (milk sugar). Disaccharides can be broken down via hydrolysis into their constituent monosaccharides **(Figure 2.19b).**

Polysaccharides

Polysaccharides are polymers composed of hundreds or thousands of monosaccharides that have been covalently linked in dehydration synthesis reactions. Even polysaccharides that contain only glucose monomers can be quite diverse because they can differ according to their monosaccharide monomer configurations (either alpha or beta) and their shapes (either branched or unbranched). Cellulose, the main constituent of the cell walls of plants and green algae, is a long unbranched molecule that contains only β-monomers of glucose linked between carbons 1 and 4 of alternating monomers; such bonds are termed β-1,4 bonds (**Figure 2.20a** on page 46). Amylose, a starch (a storage compound in plants), has only α-1,4 bonds and is unbranched **(Figure 2.20b);** amylopectin, another plant starch, contains mostly α-1,4 monomers, but also has a few α-1,6 bonds and is branched **(Figure 2.20c).** Glycogen, a storage molecule formed in the liver and muscle cells of animals, is a highly branched molecule with both α-1,4 and α-1,6 bonds **(Figure 2.20d). Highlight 2.2** on page 47 explores the roles of these bond types, and other factors as well, in the digestibility of cellulose, the most abundant polysaccharide on Earth.

The cell walls of bacteria are composed of *peptidoglycan,* which is made of polysaccharides and amino acids (see Figure 3.12 on page 68). Polysaccharides may also be linked to lipids to form lipopolysaccharides, which can form cell markers such as those involved in the ABO blood typing system in humans.

Proteins

Learning Objectives

✓ Describe five general functions of proteins in organisms.
✓ Sketch and label four levels of protein structure.

The most complex organic compounds are **proteins,** which are composed mostly of carbon, hydrogen, oxygen, nitrogen, and sulfur. Proteins perform many functions in cells, including:

- Structure. Proteins are structural components in cell walls, in membranes, and within cells themselves. Proteins are also the primary structural material of hair, nails, the outer cells of skin, muscle, and flagella and cilia (which act to move microorganisms through their environment).
- Enzymatic catalysis. *Catalysts* are chemicals that enhance the speed or likelihood of a chemical reaction. Catalysts in cells, which are typically proteins, are called *enzymes.*
- Regulation. Some proteins regulate cell function by stimulating or hindering either the action of other proteins or the expression of genes. Hormones are examples of regulatory proteins.

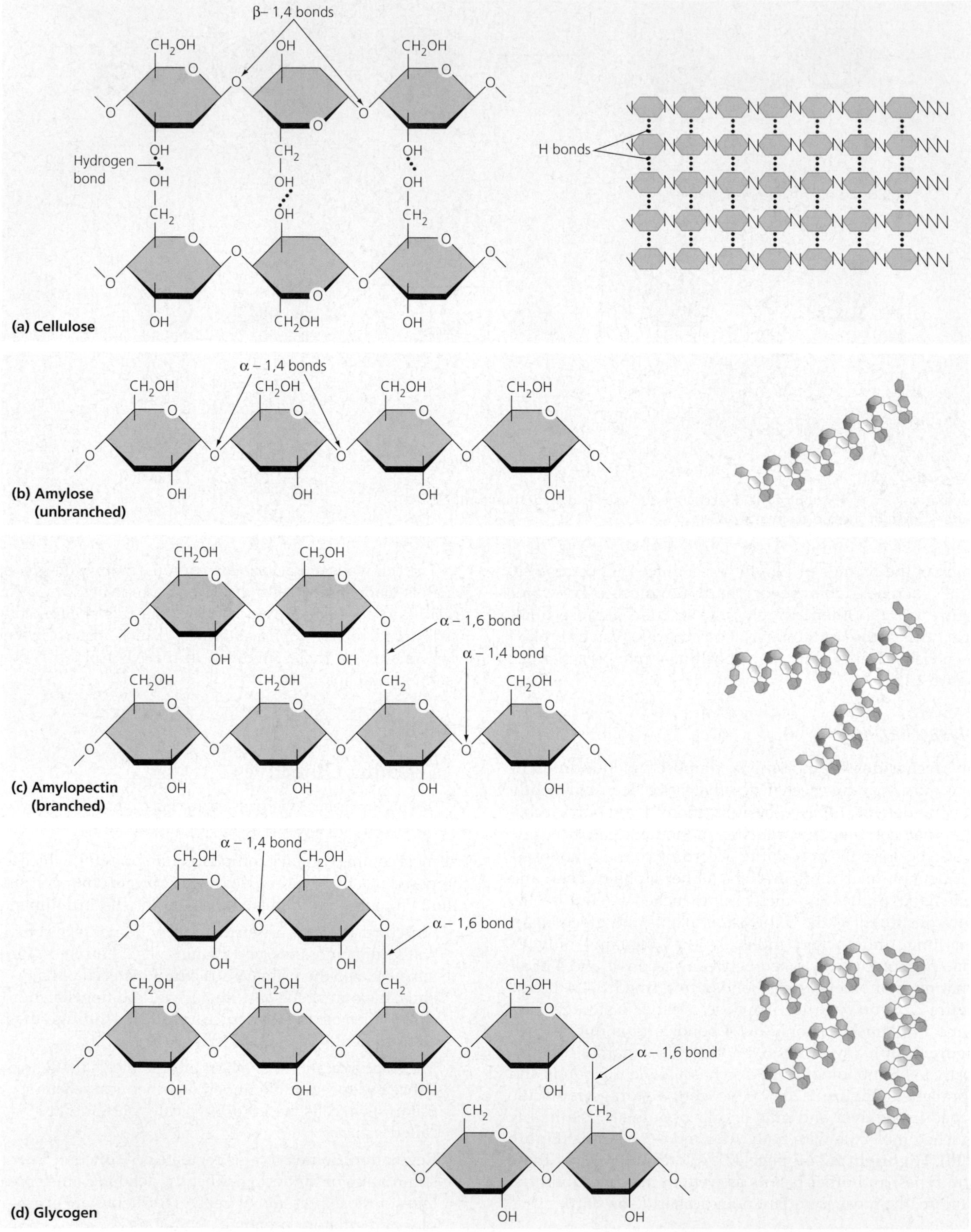

β- 1,4 bonds
CH_2OH
OH
O
Hydrogen bond
CH_2
H bonds
(a) Cellulose
α – 1,4 bonds
(b) Amylose (unbranched)
α – 1,6 bond
α – 1,4 bond
(c) Amylopectin (branched)
(d) Glycogen

◀ ***Figure 2.20***

Polysaccharides. All four polysaccharides shown here are composed solely of glucose but differ in the configuration of the glucose monomers and the amount of branching. **(a)** Cellulose, the major structural material in plants, is unbranched and contains only β-1,4 bonds. **(b)** Amylose is an unbranched plant starch with only α-1,4 bonds. **(c)** Amylopectin is a branched plant starch with mostly α-1,4 bonds but also a few α-1,6 bonds. **(d)** Glycogen, a highly branched storage molecule in animals, is composed of glucose monomers linked by α-1,4 or α-1,6 bonds.

- Transportation. As we will see in Chapter 3, certain proteins act as channels and "pumps" that move substances into or out of cells.
- Defense and offense. *Antibodies* and *complement* are examples of proteins that defend your body against microorganisms, and some bacteria produce proteins called *bacteriocins* that kill other bacteria.

A protein's function is dependent on its shape, which is determined by the molecular structures of its constituent parts.

Highlight 2.2 Why Can't We Digest Wood?

Why is it that termites can digest wood, but humans cannot? The answer obviously involves biochemistry, but it involves bacteria as well.

Wood consists mainly of cellulose, a constituent of plant cell walls. A comparison of the structures of cellulose, plant starches, and animal glycogen in Figure 2.20 reveals that the glucose monomers of cellulose are joined with covalent bonds called β-1,4 bonds, but the monomers in starches and glycogen are linked with α-1,4 bonds and α-1,6 bonds. Animals possess enzymes that recognize and hydrolyze alpha bonds, but they lack *cellulase*, the enzyme that recognizes the shape of β-1,4 bonds. Therefore, animals (including humans) cannot digest cellulose.

How, then, do termites survive on wood, and herbivores such as horses and cattle thrive on grass, when their diets are composed primarily of cellulose? Their guts provide an ideal environment for billions of cellulose-digesting bacteria, which secrete cellulase. This enzyme hydrolyzes cellulose into glucose, which the animals can then absorb.

Amino Acids

Proteins are polymers composed of monomers called **amino acids.** Amino acids contain a basic amino group (—NH_2), an acidic carboxyl group (—COOH), and a hydrogen atom, all attached to the same carbon atom, which is known as the *α-carbon* **(Figure 2.21).** A fourth bond attaches the α-carbon to a side group (—R) that varies among different amino acids. The side group may be a single hydrogen atom, various chains, or various complex ring structures. There are hundreds of amino acids, but most organisms use only 21 amino acids to build proteins. The different side groups affect the way amino acids interact with one another within a given protein, as well as how the protein interacts with other molecules. A change in an amino acid's side group may seriously interfere with a protein's normal function.

Because amino acids contain both an acidic carboxyl group and a basic amino group, they have both positive and negative charges and are easily soluble in water. The French physicist Jean Baptiste Biot (1774–1862) discovered that aqueous solutions of organic molecules such as amino acids and simple sugars bend light rays passing through

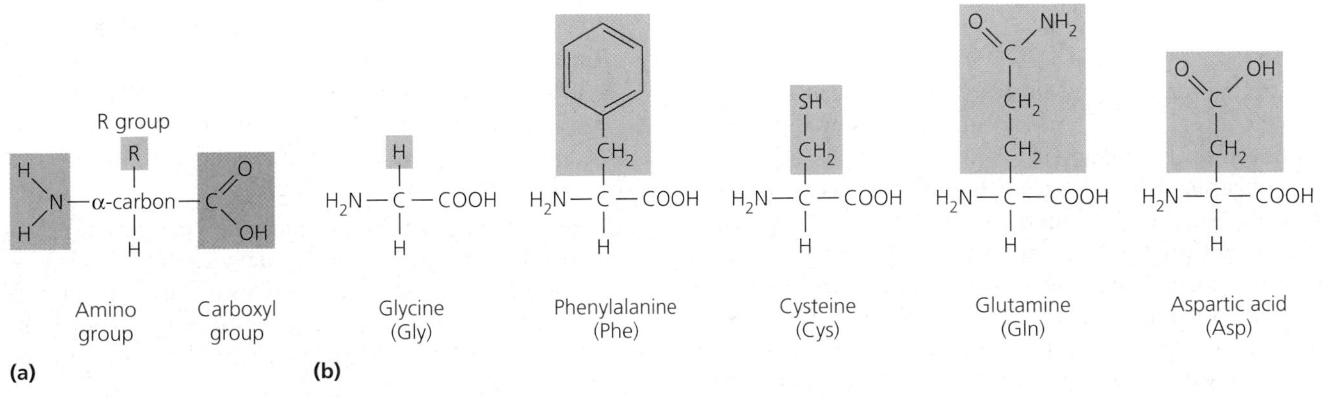

▲ ***Figure 2.21***

Amino acids. **(a)** The basic structure of amino acids. The central α-carbon is attached to an amino group, a hydrogen atom, a carboxyl group, and a side group (—R group) that varies among amino acids. **(b)** Some selected amino acids, with their side groups highlighted. Note that each amino acid has a distinctive three-letter abbreviation.

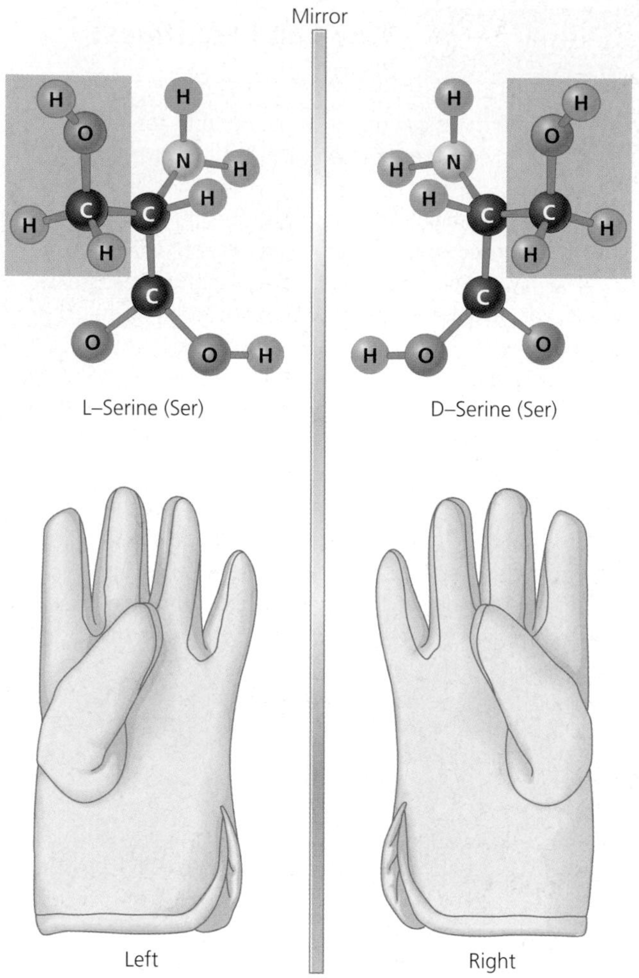

Figure 2.22

Stereoisomers, molecules that are mirror images of one another. When dissolved in water, D isomers bend light clockwise, and L forms bend light counterclockwise. Just as a right-handed glove does not fit a left hand, so a D stereoisomer cannot be substituted for an L stereoisomer in metabolic reactions.

the solution. Some molecules, known as *D forms,*[18] bend light rays clockwise; other molecules bend light rays counterclockwise and are known as *L forms.*[19]

Many organic molecules exist as both D and L forms that are *stereoisomers* of one another; that is; they have the same atoms and functional groups but are mirror images of each other **(Figure 2.22).** Amino acids in proteins are almost always L forms—except for glycine, which does not have a stereoisomer. Interestingly, organisms almost always use D sugars in metabolism and polysaccharides. The rare stereoisomers—D amino acids and L sugars—do exist in some bacterial cell walls and in some antibiotics.

CRITICAL THINKING

Why isn't there a stereoisomer of glycine?

Peptide Bonds

Cells link amino acids together in chains that somewhat resemble beads on a necklace. By a dehydration synthesis reaction, a covalent bond is formed between the carbon of the carboxyl group of one amino acid, and the nitrogen of the amino group of the next amino acid in the chain **(Figure 2.23).** As we study in detail in Chapter 7, cells follow the organism's genetic instructions to link amino acids together in precise sequences.

Scientists refer to covalent bonds between amino acids by the special name **peptide**[20] **bonds.** A molecule composed of two amino acids linked together by a single peptide bond is called a dipeptide (see Figure 2.23); longer chains of amino acids are called *polypeptides.*

Protein Structure

Proteins are unbranched polypeptides composed of hundreds to thousands of amino acids linked together in specific patterns as determined by genes. The structure of a protein molecule is directly related to its function; therefore, an understanding of protein structure is critical to understanding certain specific chemical reactions, the action of antibiotics, and specific defense against pathogens. Every protein has at least three levels of structure, and some proteins have four levels.

1. Primary structure. The primary structure of a protein is its sequence of amino acids **(Figure 2.24a).** Cells use many different types of amino acids in proteins, though not every protein contains all types. The primary structures of proteins vary widely in length and amino acid sequence.

   A change in a single amino acid can drastically affect a protein's overall structure and function, though this is not always the case. For instance, the replacement of valine by alanine in position 136 of the primary structure of a particular sheep brain protein, called cellular prion (prē'on) protein, may result in a disease called *scrapie.* The altered protein is believed to have spread into cows, causing *mad cow disease,* and from cows into humans, causing *new variant Creutzfeldt-Jakob* (kroyts-felt-yah-kŭp) *disease.*[21] However, numerous substitutions can be made in other, noncritical regions of cellular prion protein, with no ill effects.
2. Secondary structure. Ionic and hydrogen bonds, and hydrophobic and hydrophilic characteristics, cause many polypeptide chains to fold into either coils, which are called *α-helices,* or accordionlike structures called *β-pleated sheets* **(Figure 2.24b).** Proteins are typically

[18]From Latin *dexter,* meaning on the right.
[19]From Latin *laevus,* meaning on the left.
[20]From *peptone,* the name given to short chains of amino acids resulting from the partial digestion of protein.
[21]Named for the two German neurobiologists who first described the disease.

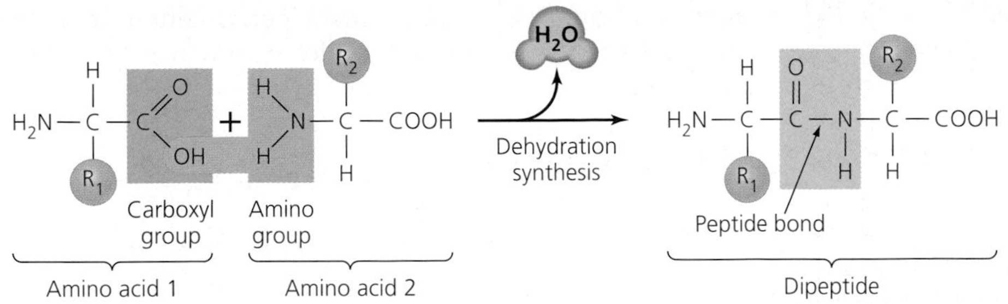

Figure 2.23
The linkage of amino acids by peptide bonds via a dehydration reaction. In this reaction, removal of a hydroxyl group from amino acid 1 and a hydrogen atom from amino acid 2 results in the production of a dipeptide—which is two amino acids linked by a single peptide bond—and a molecule of water.

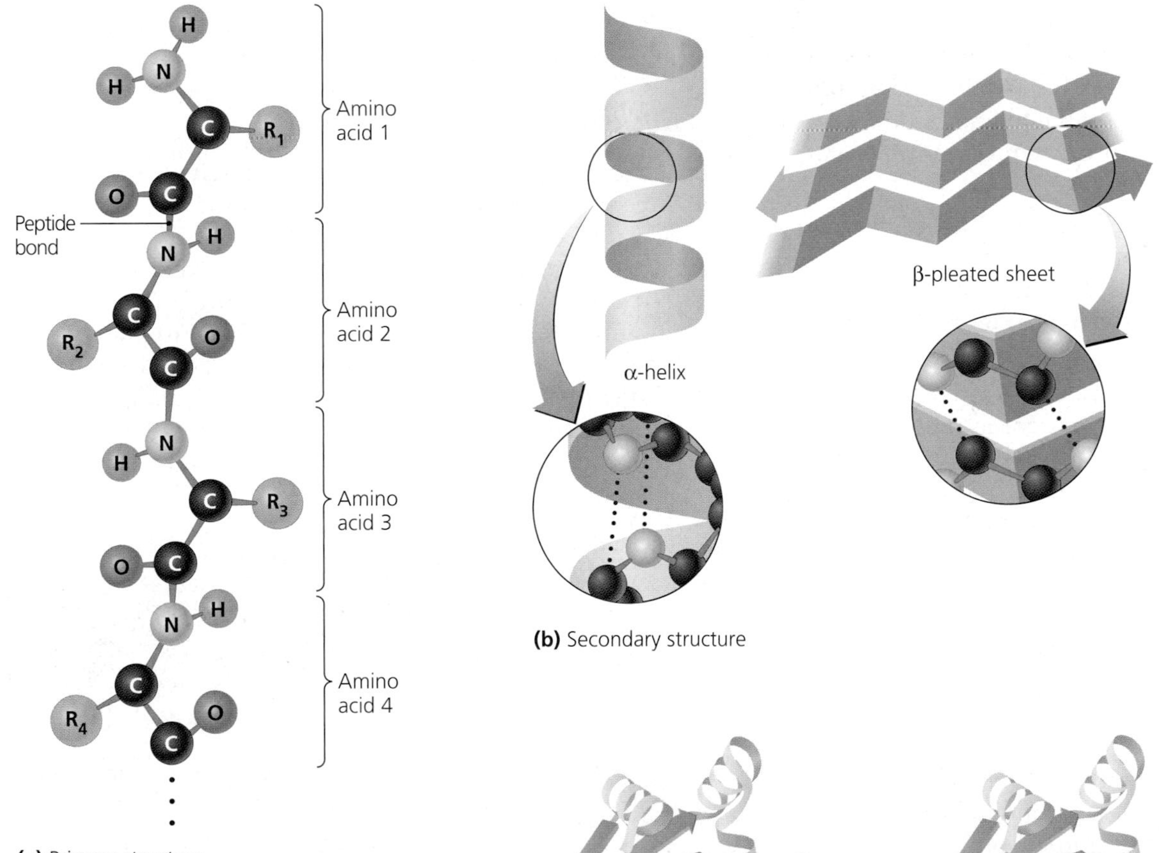

(a) Primary structure

(b) Secondary structure

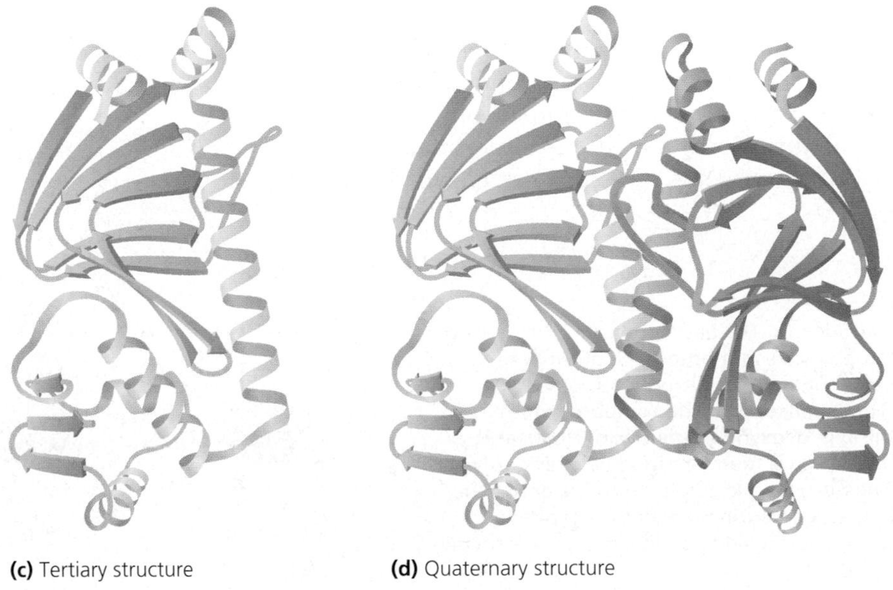

(c) Tertiary structure

(d) Quaternary structure

Figure 2.24
Levels of protein structure. **(a)** A protein's primary structure is the sequence of amino acids in a polypeptide. **(b)** Secondary structure arises as a result of interactions, such as hydrogen bonding, between regions of the polypeptide. Secondary structure takes two basic shapes: α-helices and β-pleated sheets. **(c)** The more complex tertiary structure is a three-dimensional shape defined by further hydrogen bonding as well as covalent bonding between sulfur atoms of adjoining cysteine. **(d)** Those proteins that are composed of more than one polypeptide chain have a quaternary structure.

composed of both α-helices and β-pleated sheets linked by short sequences of amino acids that do not show such secondary structure. Because of its primary structure, the protein that causes new variant Creutzfeldt-Jakob disease has β-pleated sheets in locations where the normal protein has α-helices.

3. Tertiary structure. Polypeptides further fold into complex three-dimensional shapes that are not repetitive like α-helices and β-pleated sheets **(Figure 2.24c),** but are uniquely designed to accomplish the function of the protein. Scientists are only beginning to understand the interactions that determine tertiary structure, but it is clear that covalent bonds between —R groups of amino acids, hydrogen bonds, ionic bonds, and other molecular interactions are important. For instance, nonpolar (hydrophobic) side chains fold into the interior of molecules, away from the presence of water.

   Some proteins form strong covalent bonds between sulfur atoms of cysteine molecules that are brought into proximity by the folding of the polypeptide. These *disulfide bridges* are critical in maintaining tertiary structure of many proteins.

4. Quaternary structure. Some proteins are composed of two or more polypeptide chains linked together by disulfide bridges or other bonds. The overall shape of such a protein may be globular **(Figure 2.24d)** or fibrous (threadlike).

Organisms may further modify proteins by combining them with other organic or inorganic molecules. For instance, *glycoproteins* are proteins covalently bound with carbohydrates, *lipoproteins* are proteins bonded with lipids, *metalloproteins* contain metallic ions, and *nucleoproteins* are proteins bonded with nucleic acids.

Since protein function is determined by protein shape, anything that severely interrupts shape also disrupts function. As we have seen, shape and function can be altered by amino acid substitution. Additionally, physical and chemical factors such as heat, changes in pH, and salt concentration can interfere with hydrogen and ionic bonding, which in turn disrupts the three-dimensional structure of proteins. This process is called **denaturation.** Denaturation can be permanent or temporary (if the denatured protein is able to return to its original shape again).

Nucleic Acids

Learning Objectives

- ✓ Describe the basic structure of a nucleotide.
- ✓ Compare and contrast DNA and RNA.
- ✓ Contrast the structures of ATP, ADP, and AMP.

The nucleic acids **deoxyribonucleic acid (DNA)** and **ribonucleic acid (RNA)** are vital as the genetic material of cells and viruses. Moreover, RNA, acting as an enzyme, binds amino acids together to form polypeptides. DNA and RNA are both unbranched macromolecular polymers that differ primarily in the structures of their monomers, which we discuss next.

Nucleotides

Each monomer of nucleic acids, a **nucleotide,** consists of three parts **(Figure 2.25a):** (1) phosphoric acid ionized *phosphate* (PO_4^{2-}); (2) a pentose sugar, either deoxyribose or ribose **(Figure 2.25b);** and (3) one of five cyclic (ring-shaped) nitrogenous bases: **adenine (A), guanine (G), cytosine (C), thymine (T),** or **uracil (U) (Figure 2.25c).** Adenine and guanine are double-ringed molecules of a class called *purines,* whereas cytosine, thymine, and uracil have single rings and are *pyrimidines.* DNA contains A, G, C, and T bases, whereas RNA contains A, G, C, and U bases. As their names suggest,

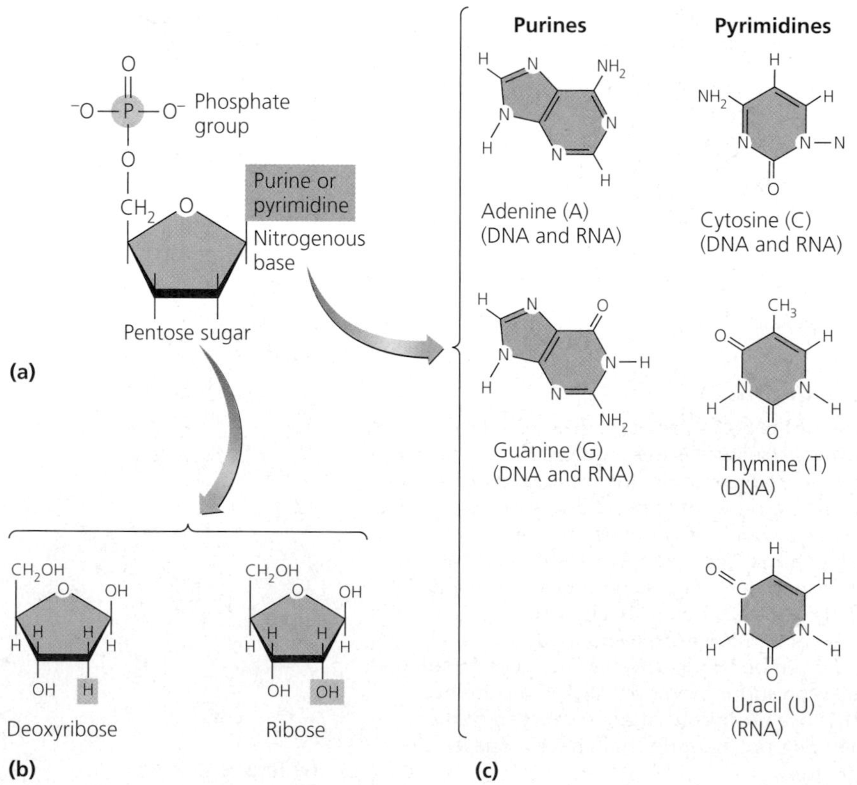

Figure 2.25 ➤

Nucleotides. **(a)** The basic structure of nucleotides, each of which is composed of a phosphate, a pentose sugar, and a nitrogenous base. **(b)** The pentose sugars deoxyribose, which is found in deoxyribonucleic acid (DNA), and ribose, which is found in ribonucleic acid (RNA). **(c)** The nitrogenous bases, which are either the double-ringed purines adenosine or guanine, or the single-ringed pyrimidines cytosine, thymine, or uracil.

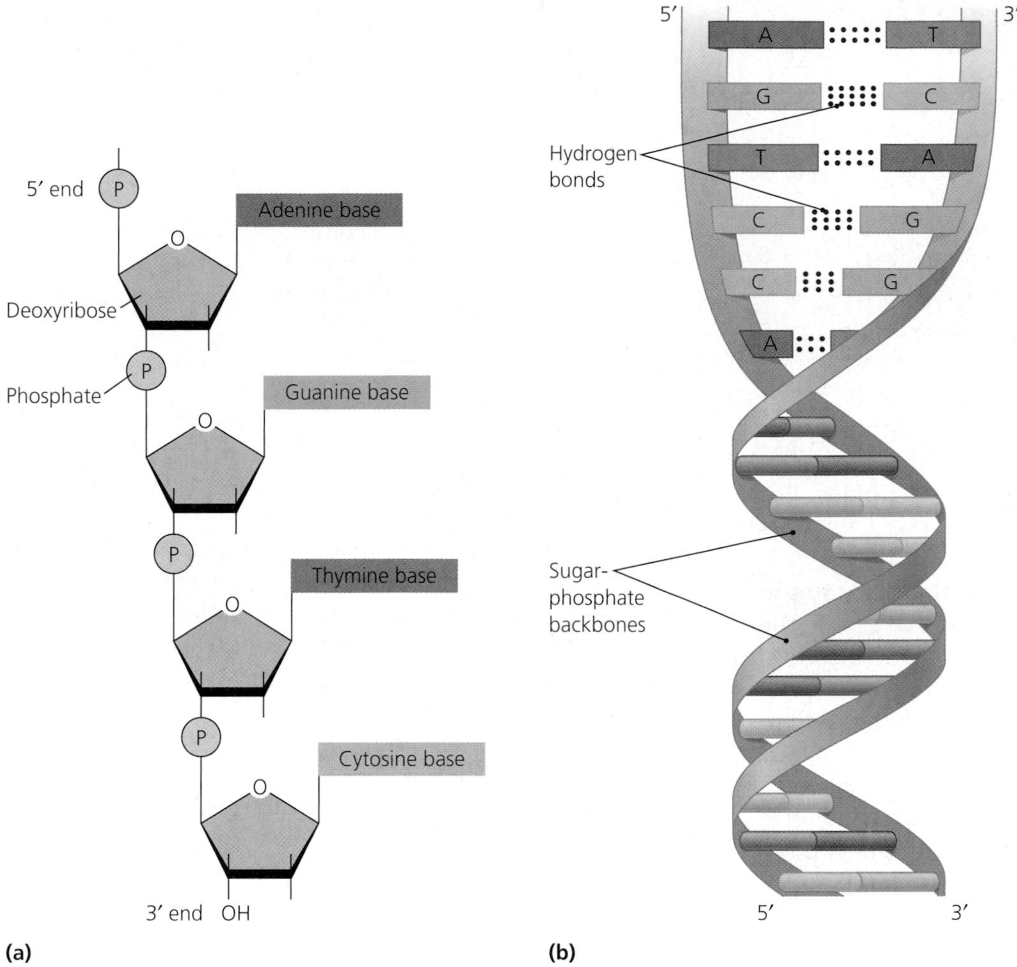

Figure 2.26

Basic nucleic acid structure. **(a)** Nucleotides are polymerized to form chains in which the nitrogenous bases extend from a sugar-phosphate backbone like the teeth of a comb. **(b)** Specific pairs of nitrogenous bases form hydrogen bonds between adjacent nucleotide chains to form the familiar DNA double helix. *How can you determine that the molecule in (a) is DNA and not RNA?*

Figure 2.26 It is DNA because its nucleotides have deoxyribose sugar and because some of them have thymine bases (not uracil, as in RNA).

DNA nucleotides contain deoxyribose, and RNA nucleotides contain ribose.

Each nucleotide is also named for the base it contains. Thus a nucleotide made with uracil is a uracil RNA nucleotide (or uracil *ribonucleotide*), and a nucleotide made with adenine and ribose is an adenine ribonucleotide. Likewise, a nucleotide composed of adenine and deoxyribose is an adenine DNA nucleotide (or adenine *deoxyribonucleotide*).

CRITICAL THINKING

A textbook states that only five nucleotide bases are found in cells, but a laboratory worker correctly reports that she has isolated eight different nucleotides. Explain why both are correct.

Nucleic Acid Structure

Nucleic acids, like polysaccharides and proteins, are polymers. They are composed of nucleotides linked by covalent bonds between the phosphate of one nucleotide and the sugar of the next. Polymerization results in a linear spine composed of alternating sugars and phosphates, with bases extending from it rather like the teeth of a comb **(Figure 2.26a)**. The ends of a chain of nucleotides are different. At one end, called the 5′ end[22] (five prime end), carbon 5′ of the sugar is attached to a phosphate group. At the other end (3′ end), carbon 3′ of the sugar is not attached to a phosphate group.

The atoms of the bases in nucleotides are arranged in such a manner that hydrogen bonds readily form between specific bases of two adjacent nucleic acid chains. Three hydrogen bonds form between an adjacent pair composed of cytosine (C) and guanine (G), whereas two hydrogen bonds form between an adjacent pair composed of adenine (A) and thymine (T) in DNA **(Figure 2.26b)**, or between an adjacent pair composed of adenine (A) and uracil (U) in RNA. Hydrogen bonds do not readily form between other combinations of nucleotide bases; for example, adenine does not readily pair with cytosine, guanine, or another adenine nucleotide.

In cells and most viruses that use DNA as a genome, DNA molecules are double stranded. The two strands of

[22]Carbon atoms in organic molecules are commonly identified by numbers. In a nucleotide, carbon atoms 1, 2, 3, etc. belong to the base, and carbon atoms 1′, 2′, 3′, etc. belong to the sugar.

Table 2.5 Comparison of Nucleic Acids

Characteristic	DNA	RNA
Sugar	Deoxyribose	Ribose
Purine nucleotides	A and G	A and G
Pyrimidine nucleotides	T and C	U and C
Number of strands	Double stranded in cells and in most DNA viruses; single stranded in parvoviruses	Single stranded in cells and in most RNA viruses; double stranded in reoviruses
Function	Genetic material of all cells and DNA viruses	Protein synthesis in all cells; genetic material of RNA viruses

DNA are complementary to one another; that is, the specificity of nucleotide base pairing ensures that opposite strands are composed of complementary nucleotides. For instance, if one strand has the sequence AATGCT, then its complement has TTACGA.

The two strands are also *antiparallel*; that is, they run in opposite directions. One strand runs from the 3′ end to the 5′ end, while its complement runs in the opposite direction from its 5′ end to its 3′ end. Though hydrogen bonds are relatively weak bonds, thousands of them exist at normal temperatures, forming a stable, double-stranded DNA molecule, which looks much like a ladder: The two deoxyribose-phosphate chains are the side rails, and base pairs form the rungs. Hydrogen bonding also twists the phosphate-deoxyribose backbones into a helix (see Figure 2.26b). Thus, typical DNA is a double helix.

Nucleic Acid Function

DNA is the genetic material of all organisms and of many viruses; it carries instructions for the synthesis of RNA molecules and proteins. By controlling the synthesis of enzymes and regulatory proteins, DNA controls the synthesis of all other molecules in an organism. Genetic instructions are carried in the sequence of nucleotides that make up the nucleic acid. Even though only four kinds of bases are found in DNA (A, T, G, and C), they can be sequenced in distinctive patterns that create genetic diversity and code for an infinite number of proteins, just as an alphabet of only four letters could spell a very large number of words. Cells replicate their DNA molecules and pass copies to their descendents, ensuring that each has the instructions necessary for life.

Ribonucleic acids play several roles in the synthesis of proteins, including catalyzing the synthesis of proteins. RNA molecules also function as structural components of ribosomes, and in place of DNA as the genome of RNA viruses.

Table 2.5 compares and contrasts RNA and DNA. We examine the synthesis and function of DNA and RNA in detail in Chapter 7.

ATP

Phosphate in nucleotides and other molecules is a highly reactive functional group and can form covalent bonds with other phosphate groups to make diphosphate and triphosphate molecules. Such molecules made from ribose nucleotides are important in many metabolic reactions. The names of these molecules indicate the nucleotide base and the number of phosphate groups they contain. Thus, cells make adenosine monophosphate (AMP) from the nitrogenous base adenine, ribose sugar, and one phosphate group; adenosine diphosphate (ADP), which has two phosphate groups; and **adenosine triphosphate** (ă-den′ō-sēn trī-fos′fāt) or **ATP**, which has three phosphate groups **(Figure 2.27).**

ATP is the principal, short-term, recyclable energy supply for cells. When the phosphate bonds of ATP are broken,

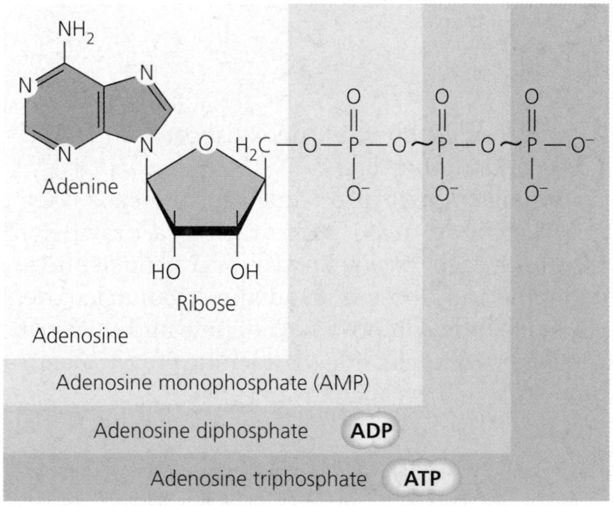

▲ ***Figure 2.27***

Adenosine triphosphate (ATP), the primary short-term, recyclable energy supply for cells. Energy is stored in "high-energy" bonds between the phosphate groups. *What is the relationship between AMP and adenine ribonucleotide?*

Figure 2.27 AMP and adenine ribonucleotide are two names for the same thing.

a significant amount of energy is released; in fact, more energy is released from phosphate bonds than is released from most other covalent bonds. For this reason, the phosphate–phosphate bonds of ATP are known as *high-energy bonds,* and to show these specialized bonds, ATP is often symbolized as

$$A-Ⓟ\sim Ⓟ\sim Ⓟ$$

Energy is released when ATP is converted to ADP, and when phosphate is removed from ADP to form AMP, though the latter reaction is not as common in cells. Energy released from the phosphate bonds of ATP is used for important life-sustaining activities, such as synthesis reactions, locomotion, and transportation of substances into and out of cells.

Cells also use ATP as a structural molecule in the formation of *coenzymes.* As we will see in Chapter 5, coenzymes such as *flavin adenine dinucleotide, nicotinamide nucleotide,* and *coenzyme* A function in many metabolic reactions.

A cell's supply of ATP is limited; therefore, an important part of cellular metabolism is replenishing ATP stores. We discuss the important ATP-generating reactions in Chapter 5.

CHAPTER SUMMARY

Atoms (pp. 27–29)

1. **Atoms** contain negatively charged **electrons** spinning around a nucleus composed of uncharged **neutrons** and positively charged **protons.**
2. An **element** is matter composed of a single type of atom.
3. The number of protons in the nucleus of an atom is its **atomic number.** The sum of the masses of its protons, neutrons, and electrons is an atom's **atomic mass,** which is estimated by adding the number of neutrons and protons (because electrons have little mass).
4. **Isotopes** are atoms of an element that differ only in the numbers of neutrons they contain.

Chemical Bonds (pp. 29–35)

1. The region of space occupied by electrons is an electron shell. The number of electrons in the outermost or **valence shell** of an atom determines the atom's reactivity. Most valence shells hold a maximum of eight electrons. Sharing or transferring valence electrons to fill a valence shell results in **chemical bonds.**
2. A chemical bond results when two atoms share a pair of electrons. The **electronegativity** of each of the atoms, which is the strength of their attraction for electrons, determines whether the bond between them will be a **nonpolar covalent bond** (equal sharing of electrons), a **polar covalent bond** (unequal sharing of electrons), or an **ionic bond** (giving up of electrons from one atom to another).
3. If a **molecule** contains atoms of more than one element, it is a **compound. Organic compounds** are those that contain carbon and hydrogen atoms.
4. An **anion** is an atom with an extra electron and thus a negative charge. A **cation** has lost an electron and thus has a positive charge. Ionic bonding between the two types of ions makes **salt.** When salts dissolve in water, their ions are called **electrolytes.**
5. **Hydrogen bonds** are relatively weak but important chemical bonds. They hold molecules in specific shapes and confer unique properties to water molecules.

Chemical Reactions (pp. 35–36)

1. **Chemical reactions** result from the making or breaking of chemical bonds in a process in which **reactants** are changed into **products.**
2. **Synthesis reactions** form larger, more complex molecules. In **dehydration synthesis,** a molecule of water is removed from the reactants as the larger molecule is formed. **Endothermic reactions** require energy. **Anabolism** is the sum of all synthesis reactions in an organism.
3. **Decomposition reactions** break larger molecules into smaller molecules and are **exothermic** because they release energy. **Hydrolysis** is a decomposition reaction that uses water as one of the reactants. The sum of all decomposition reactions in an organism is called **catabolism.**
4. **Exchange reactions** involve exchanging atoms between reactants.
5. **Metabolism** is the sum of all anabolic and catabolic chemical reactions in an organism.

Water, Acids, Bases, and Salts (pp. 37–39)

1. **Inorganic** molecules typically lack carbon.
2. Water is a vital inorganic compound because of its properties as a solvent, its liquidity, its great capacity to absorb heat, and its participation in chemical reactions.
3. **Acids** release hydrogen ions. **Bases** release hydroxyl anions. The relative strength of each is assessed on a logarithmic **pH scale,** which measures the hydrogen ion concentration in a substance.
4. **Buffers** are substances that prevent drastic changes in pH.

Organic Macromolecules (pp. 39–53)

1. Certain groups of atoms in common arrangements, called **functional groups,** are found in organic macromolecules. **Monomers** are simple subunits that can be covalently linked to form chainlike **polymers.**
2. **Lipids,** which include fats, phospholipids, waxes, and steroids, are **hydrophobic** (insoluble in water) macromolecules.

3. **Fat** molecules are formed from a glycerol and three chainlike fatty acids. **Saturated fatty acids** contain more hydrogen in their structural formulas than **unsaturated fatty acids,** which contain double bonds between some carbon atoms. If several double bonds exist in the fatty acids of a molecule of fat, it is a **polyunsaturated** fat.
4. **Phospholipids** contain two fatty acid chains and a phosphate functional group. The phospholipid head is **hydrophilic,** whereas the fatty acid portion of the molecule is hydrophobic.
5. **Waxes** contain a long-chain fatty acid covalently linked to a long-chain alcohol. Waxes, which are water insoluble, are components of cell walls and are sometimes used as energy storage molecules.
6. **Steroid** lipids such as cholesterol help maintain the structural integrity of membranes as temperature fluctuates.
7. **Carbohydrates** such as **monosaccharides, disaccharides,** and **polysaccharides** serve as energy sources, structural molecules, and recognition sites during intercellular interactions.
8. **Proteins** are structural components of cells, enzymatic catalysts, regulators of various activities, molecules involved in the transportation of substances, and molecules involved in an organism's defense. They are composed of **amino acids** linked by **peptide bonds,** and they possess primary, secondary, tertiary, and (sometimes) quaternary structures that affect their function. **Denaturation** of a protein disrupts its structure and subsequently its function.
9. **Deoxyribonucleic acid (DNA)** and **ribonucleic acid (RNA)** are unbranched macromolecular polymers of **nucleotides,** each composed either of deoxyribose or ribose sugar, phosphoric acid, and a nitrogenous base. Five different bases exist: **adenine, guanine, cytosine, thymine,** and **uracil.** DNA contains A, G, C, and T nucleotides. RNA uses U nucleotides instead of T nucleotides.
10. The structure of nucleic acids allows for genetic diversity, correct copying of genes for their passage on to the next generation, and the accurate synthesis of proteins.
11. **Adenosine triphosphate (ATP),** which is related to adenine nucleotide, is the most important short-term energy storage molecule in cells. It is also incorporated into the structure of many coenzymes.

QUESTIONS FOR REVIEW

(Answers to multiple choice and fill in the blanks are on the web, along with additional review questions. Visit www.microbiologyplace.com.)

Multiple Choice

1. Which of the following structures have no electrical charge?
   a. protons
   b. neutrons
   c. electrons
   d. ions
2. The atomic mass of an atom most closely approximates the sum of the masses of all its
   a. protons.
   b. electrons.
   c. isotopes.
   d. protons and neutrons.
3. One isotope of iodine differs from another in
   a. the number of protons.
   b. the number of electrons.
   c. the number of neutrons.
   d. atomic number.
4. Which of the following is *not* an organic compound?
   a. monosaccharide
   b. protein
   c. water
   d. steroid
5. Which of the following terms most correctly describes the bonds in a molecule of water?
   a. nonpolar covalent bond
   b. polar covalent bond
   c. ionic bond
   d. hydrogen bond
6. In water, cations and anions of salts disassociate from one another and become surrounded by water molecules. In this state, the ions are also called
   a. electrically negative.
   b. ionically bonded.
   c. electrolytes.
   d. hydrogen bonds.
7. Which of the following can be most accurately described as a decomposition reaction?
   a. $C_6H_{12}O_6 + 6\,O_2 \rightarrow 6\,H_2O + 6\,CO_2$
   b. glucose + ATP $\rightarrow$ glucose phosphate + ADP
   c. $6\,H_2O + 6\,CO_2 \rightarrow C_6H_{12}O_6 + 6\,O_2$
   d. $A + BC \rightarrow AB + C$
8. Which of the following statements about a carbonated cola beverage with a pH of 2.9 is true?
   a. It has a relatively high concentration of hydrogen ions.
   b. It has a relatively low concentration of hydrogen ions.
   c. It has equal amounts of hydroxyl and hydrogen ions.
   d. Cola is a buffered solution.
9. Proteins are polymers of
   a. amino acids.
   b. fatty acids.
   c. nucleic acids.
   d. monosaccharides.
10. Which of the following are hydrophobic organic molecules?
    a. proteins
    b. carbohydrates
    c. lipids
    d. nucleic acids

Fill in the Blanks

1. The outermost electron shell of an atom is known as the ________ shell.
2. The type of chemical bond between atoms with nearly equal electronegativities is called a ________ ________ bond.
3. The principal short-term energy storage molecule in cells is ________.
4. The most common long-term energy storage molecule is ________.
5. Groups of atoms such as NH_2 or OH that appear in certain common arrangements are called ________.
6. The reverse of dehydration synthesis is ________.
7. Reactions that release energy are called ________ reactions.
8. All chemical reactions begin with reactants and result in new molecules called ________.

Short Answer

1. List three main types of chemical bonds, and give an example of each.
2. Name five properties of water that are vital to life.
3. Describe the difference(s) among saturated fatty acids, unsaturated fatty acids, and polyunsaturated fatty acids.
4. What is the difference between atomic oxygen and molecular oxygen?
5. Explain how the polarity of water molecules makes water an excellent solvent.

CRITICAL THINKING

1. Anthrax is caused by a bacterium, *Bacillus anthracis*, a pathogen that avoids defenses against disease by synthesizing an outer glycoprotein covering made from D-glutamic acid. This covering is not digestible by white blood cells that normally engulf bacteria. Why is the covering indigestible?
2. Dehydrogenation is a chemical reaction in which a saturated fat is converted to an unsaturated fat. Explain why the name for this reaction is an appropriate one.
3. Two freshmen disagree about an aspect of chemistry. The nursing major insists that H^+ is the symbol for a hydrogen ion. The physics major insists that H^+ is the symbol for a proton. How can you help them resolve their disagreement?
4. When an egg white is heated, it changes from liquid to solid. When gelatin is cooled, it changes from liquid to solid. Both gelatin and egg white are proteins. From what you have learned about proteins, why can the gelatin be changed back to liquid but the cooked egg cannot?
5. When amino acids are synthesized in a test tube, D and L forms occur in equal amounts. However, cells use only L forms in their proteins. Occasionally, meteorites are found to contain amino acids. Based on these facts, how could NASA scientists determine whether the amino acids recovered from space are evidence of Earth-like extraterrestrial life or of nonmetabolic processes?

CHAPTER 3

Cell Structure and Function

CHAPTER OUTLINE

Can a microbe be a magnet? The answer is yes, if it is a magnetobacterium.

Magnetobacteria are microorganisms with an unusual feature: cellular structures called *magnetosomes.* Magnetosomes are stored deposits (also called inclusions) of the mineral magnetite. These deposits allow magnetobacteria to respond to the lines of the Earth's magnetic field, much like a compass. In the Southern Hemisphere, magnetobacteria exist as south-seeking varieties; in the Northern Hemisphere, they exist as north-seeking varieties.

How do these bacteria benefit from magnetosomes? Magnetobacteria prefer environments with little or no oxygen, such as those that exist far below the surfaces of land and sea. The magnetosomes are drawn toward the underground magnetic poles, helping magnetobacteria move toward the oxygen-scarce regions that are their favored homes.

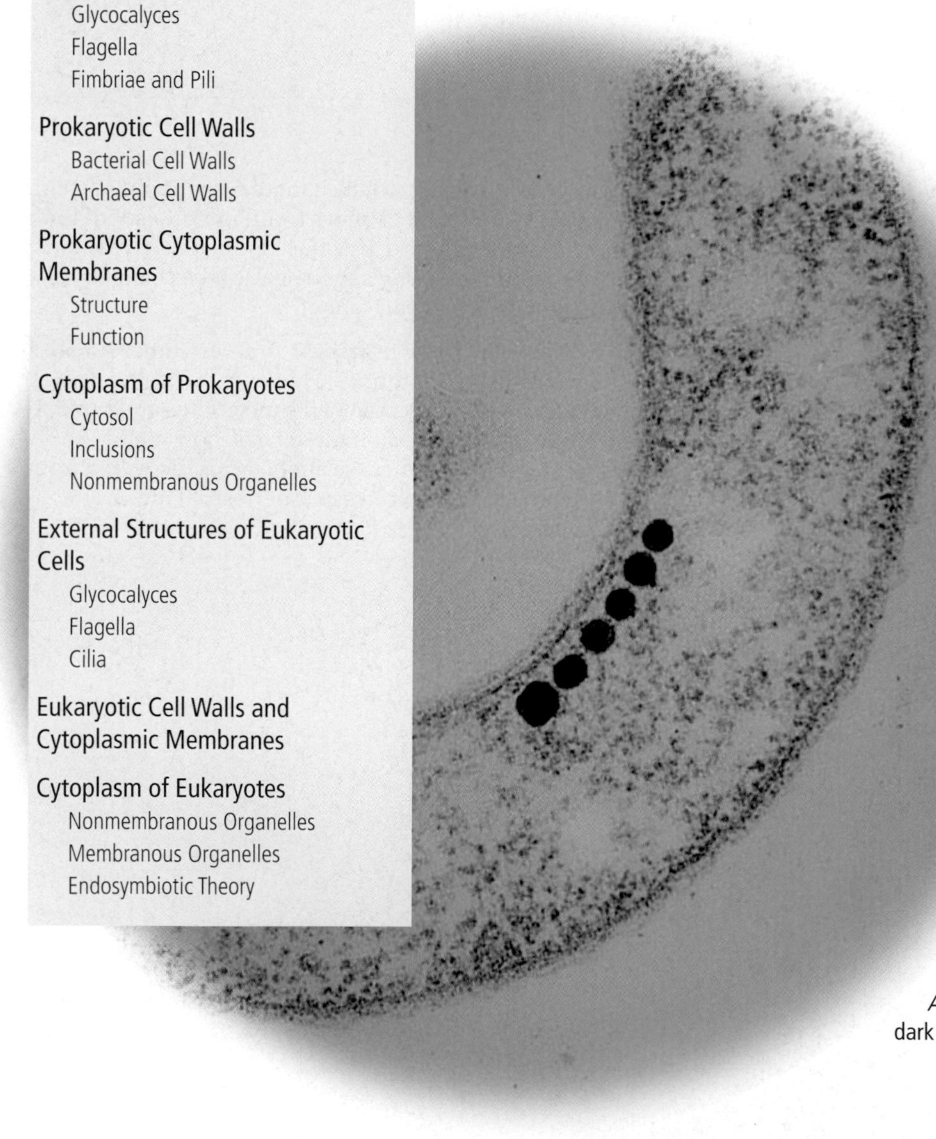

Aquaspirillum magnetotacticum, a magnetobacterium. The dark circles make up the magnetosome.

MicroPrep Pre-Test: *Take the pre-test for this chapter on the web. Visit* ***www.microbiologyplace.com.***

All living things—including our bodies and the bacterial, protozoan, and fungal pathogens that attack us—are composed of living cells. If we want to understand disease and its treatment, therefore, we must first understand the life of cells. How pathogens attack our cells, how our bodies defend themselves, how current medical treatments assist our bodies in recovering—all of these activities have their basis in the biology of our, and our pathogens', cells.

In this chapter, we will examine cells and the structures within cells. We will discuss similarities and differences between the two major kinds of cells, eukaryotic and prokaryotic. The differences are particularly important because they allow researchers to develop treatments that inhibit or kill pathogens without adversely affecting a patient's own cells. We will also learn about cellular structures that allow pathogens to evade the body's defenses and cause disease.

Processes of Life

Learning Objective

✓ Describe the four major processes of living cells.

As we discussed in Chapter 1, microbiology is the study of particularly small living things. What we have not yet discussed is the question of how we define life. Scientists once thought that living things were composed of special organic chemicals, such as glucose and amino acids, that carried a "life force" found only in living organisms. These organic chemicals were thought to be formed only by living things and to be very different from the inorganic chemicals of nonliving things.

The idea that organic chemicals could come only from living organisms had to be abandoned in 1828, when Friedrich Wöhler (1800–1882) synthesized urea, an organic molecule, using only inorganic substrates in his laboratory. Today we know that all living things contain both organic and inorganic chemicals, and that many organic chemicals can be made from inorganic chemicals by laboratory processes. If organic chemicals can be made even in the absence of life, what is the difference between a living thing and a nonliving thing? What is life?

At first this may seem a simple question. After all, you can usually tell when something is alive. However, defining "life" itself is difficult, so biologists generally avoid setting a definition, preferring instead to describe characteristics common to all living things. Biologists can agree that all living things share at least four processes of life: growth, reproduction, responsiveness, and metabolism.

- **Growth.** Living things can grow; that is, they can increase in size.
- **Reproduction.** Organisms normally have the ability to reproduce themselves. Reproduction means that they increase in number, producing more organisms like themselves. Reproduction may be accomplished asexually (alone) or sexually with a mate. Note that reproduction is an increase in number, whereas growth is an increase in size. Growth and reproduction often occur simultaneously. We consider several methods of reproduction when we examine microorganisms in detail in Chapters 11–13.
- **Responsiveness.** All living things respond to their environment. They have the ability to change internal and/or external properties in reaction to changing conditions around or within them. Many organisms also have the ability to move toward or away from environmental stimuli—a response called *taxis.*
- **Metabolism.** Metabolism can be defined as the ability of organisms to take in nutrients from outside themselves and use the nutrients in a series of controlled chemical reactions to provide the energy and structures needed to grow, reproduce, and be responsive. Metabolism is a unique process of living things; nonliving things cannot metabolize. Cells store metabolic energy in the chemical bonds of *adenosine triphosphate* (ă-den′ō-sēn trī-fos′fāt) or *ATP.* The major processes of microbial metabolism, including the generation of ATP, are discussed in Chapters 5–7.

Table 3.1 shows how these characteristics, along with cell structure, relate to various kinds of microbes.

Table 3.1 Characteristics of Life and Their Distribution in Microbes

Characteristic	Prokaryotes	Eukaryotes	Viruses
Growth: increase in size	Occurs in all	Occurs in all	Does not occur
Reproduction: increase in number	Occurs in all	Occurs in all	Occurs only inside a host cell
Responsiveness: ability to react to environmental stimuli	Occurs in all	Occurs in all	Occurs in some viruses as a reaction to host cells
Metabolism: controlled chemical reactions of organisms	Occurs in all	Occurs in all	Uses host cell's metabolism
Cellular structure: membrane-bound structure capable of all of the above functions	Present in all	Present in all	Lacks cytoplasmic membrane or cellular structure

Highlight 3.1 Are Viruses Alive?

An ongoing discussion in microbiology concerns the nature of viruses. Are they alive?

Viruses are noncellular; that is, they lack cell membranes, cell walls, and most other cellular components. However, they do contain genetic material in the form of either DNA or RNA. (No virus has both DNA and RNA.) While many viruses use DNA molecules for their genes, others such as HIV and poliovirus use RNA for genes instead of DNA. No cells use RNA molecules for their genes.

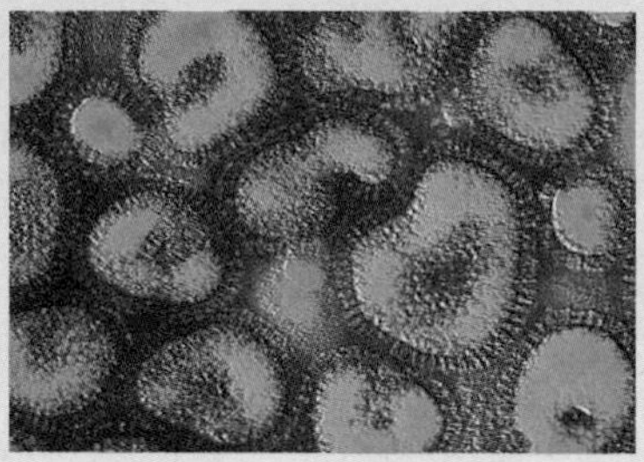

Influenzaviruses

Viruses have some other characteristics of living cells. For instance, some demonstrate responsiveness to their environment when they inject their genetic material into susceptible host cells. However, viruses lack most characteristics of life: They are unable to grow, reproduce, or metabolize outside of a host cell, although once they enter a cell they take control of the cell's metabolism and cause it to make more viruses. This takeover typically leads to the death of the cell and results in disease in an organism.

CRITICAL THINKING

In your opinion, are viruses alive? What factors led you to your conclusion?

Organisms may not exhibit these processes at all times. For instance, in some organisms, reproduction may be postponed or curtailed by age or disease, or in humans at least, by choice. Likewise, the rate of metabolism may be reduced, as occurs in a seed, a hibernating animal, or a bacterial endospore,[1] and growth often stops when an animal reaches a certain size. However, microorganisms typically grow, reproduce, respond, and metabolize as long as conditions are suitable. The proper conditions for the metabolism and growth of various types of microorganisms are discussed in Chapter 6.

A group of pathogens essential to the debate regarding "What constitutes life?" are the viruses. **Highlight 3.1** discusses the question of whether or not viruses should be considered living things.

Eukaryotic and Prokaryotic Cells: An Overview

Learning Objective

✓ Compare and contrast prokaryotic and eukaryotic cells.

In the 1800s, two German biologists, Theodor Schwann (1804–1881) and Matthias Schleiden (1810–1882) developed the theory that all living things are composed of cells. *Cells* are living entities, surrounded by a membrane, that are capable of growing, reproducing, responding, and metabolizing. The smallest living things are single-celled microorganisms.

There are many different kinds of cells **(Figure 3.1).** Some cells are free-living, independent organisms; others live together in colonies or form the bodies of multicellular organisms. Cells also exist in various sizes, from the smallest bacteria to bird eggs, which are the largest of cells. All cells are described as either *prokaryotes* (prō-kar′ē-ōts) or *eukaryotes* (yū-kar′ē-ōts).

The distinctive structural feature of **prokaryotes** is not what they have, but what they lack. The light microscope reveals that prokaryotes do not have a membrane surrounding their genetic material (DNA); that is, they do not have a nucleus (**Figure 3.2a** on page 60). The word *prokaryote* comes from Greek words meaning "before nucleus." Moreover, electron microscopy has revealed that prokaryotes also lack various types of internal membrane-bound structures present in eukaryotic cells. Not only are prokaryotes simpler than eukaryotes; they are also typically smaller—approximately 1.0 μm in diameter, as compared to 10–100 μm for eukaryotic cells (**Figure 3.3** on page 61). (See **Highlight 3.2** for some interesting exceptions concerning the size of prokaryotes.)

There are two types of prokaryotic cells: *bacteria* and *archaea.* These cell types fundamentally differ in the nucleotide base sequences of their ribosomal RNA, in the type of lipids in their cytoplasmic membranes, and in the chemistry of their cell walls. Chapter 11 discusses these differences in more detail.

Eukaryotes have a membrane surrounding their DNA, forming a nucleus (see **Figure 3.2b**). Indeed, the term *eukaryote* comes from Greek words meaning "true nucleus." Besides the nuclear membrane, eukaryotes have numerous other internal membranes that compartmentalize cellular functions. These membrane-bound compartments are **organelles**—specialized structures that act like tiny organs to carry on the various functions of the cell. Organelles and their functions are discussed later in this chapter. The cells of algae, protozoa, fungi, animals, and plants are eukaryotic. Eukaryotes are usually larger and more complex than prokaryotes.

[1]Endospores are resting stages, produced by some bacteria, that are tolerant of environmental extremes.

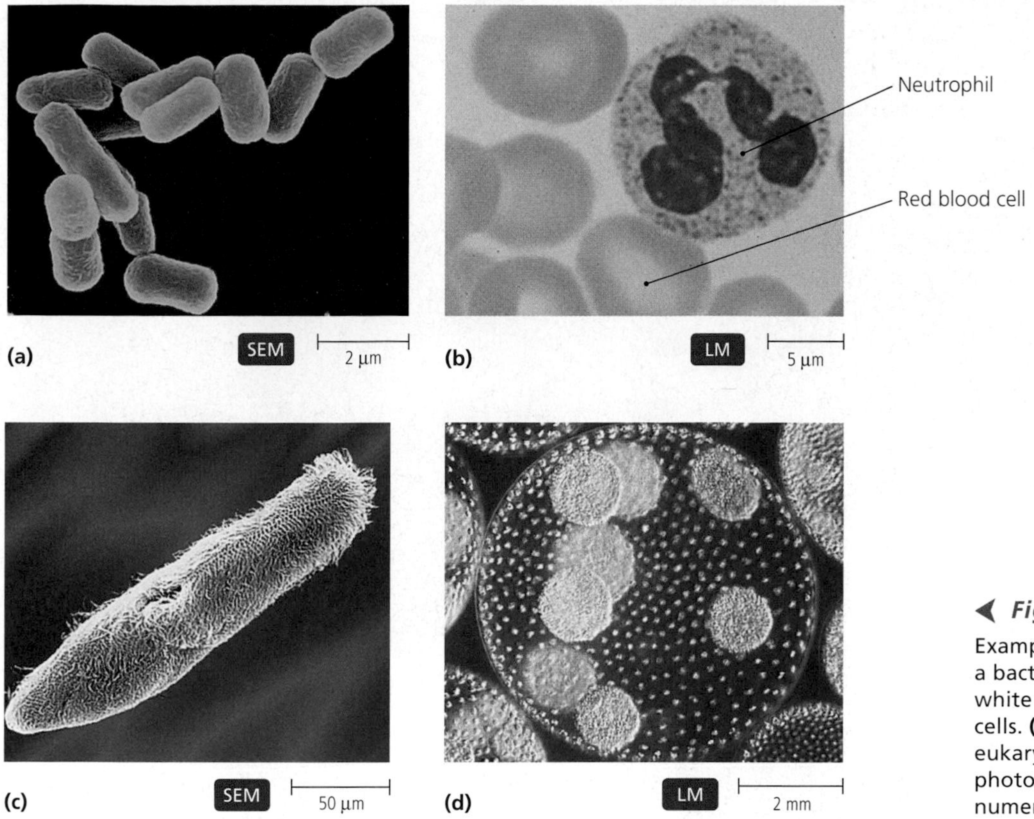

Figure 3.1

Examples of types of cells. **(a)** *Escherichia coli*, a bacterial cell. **(b)** A neutrophil, a defensive white blood cell, amidst human red blood cells. **(c)** *Paramecium*, a single-celled eukaryote. **(d)** *Volvox*, a colonial, photosynthetic eukaryote composed of numerous small cells and daughter colonies.

Highlight 3.2 Unusual Giants

Most prokaryotes are small compared to eukaryotes. The dimensions of a typical bacterial cell—for example, a cell of *Escherichia coli*—are 1.0 μm × 2.0 μm. Typical eukaryotic cells are often thousands of micrometers in diameter.

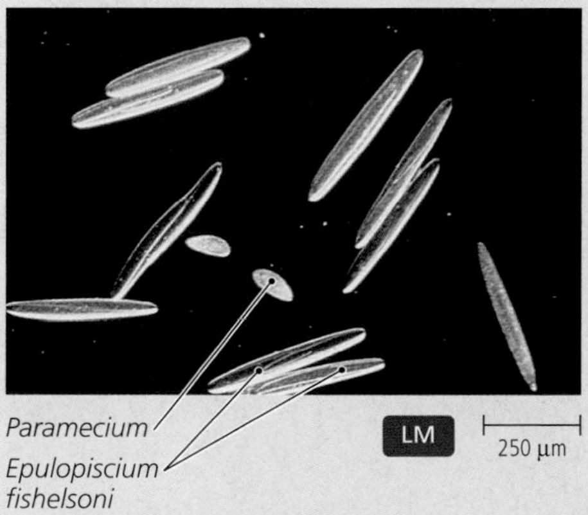

For many years biologists thought that the small size of prokaryotes was a necessary result of the nature of their cells. Without internal compartments, any prokaryotes that were as large as eukaryotes could not isolate and control the metabolic reactions of life; nor could they efficiently distribute nutrients and eliminate wastes.

In 1985 scientists located a new unicellular organism from the intestines of a surgeonfish. Its cells were 0.6 mm (600 μm) long, a size visible to the unaided eye. Microscopic examination revealed the presence of a fine covering of what appeared to be cilia, which are a feature of eukaryotic cells only. The new organism was named *Epulopiscium fishelsoni,*[a] and it was assumed that such a large ciliated cell was eukaryotic. A surprise came when these cells were examined with an electron microscope. They contained no nuclei or other eukaryotic organelles, and the "cilia" more closely resembled short bacterial flagella. *Epulopiscium,* as it turned out, is indeed a giant prokaryote. (Compare the size of *Epulopiscium* with that of *Paramecium,* a eukaryote, in the box photo.)

It is not, however, the largest prokaryote. In 1997, Heide Shultz, a graduate student of marine microbiology in Bremen, Germany, discovered an even larger bacterium, named *Thiomargarita namibiensis,*[b] which lives in chains of 2–50 cells in ocean sediments off the coast of Namibia, Africa. The spherical cells of *Thiomargarita* can grow to 750 μm in diameter, or about the size of the period at the end of this sentence. Such large cells have volumes over 66 times those of the largest *Epulopiscium.* If a single cell of *Thiomargarita* were the size of an African elephant, then *Epulopiscium* would be the size of a cow, and *Escherichia* would be the size of an ant.

[a]*Epulopiscium* in Latin means "guest at a banquet of fish."

[b]This name means "sulfur pearl of Namibia," because granules of sulfur stored in the cytoplasm cause the cells to glisten white, like a string of pearls.

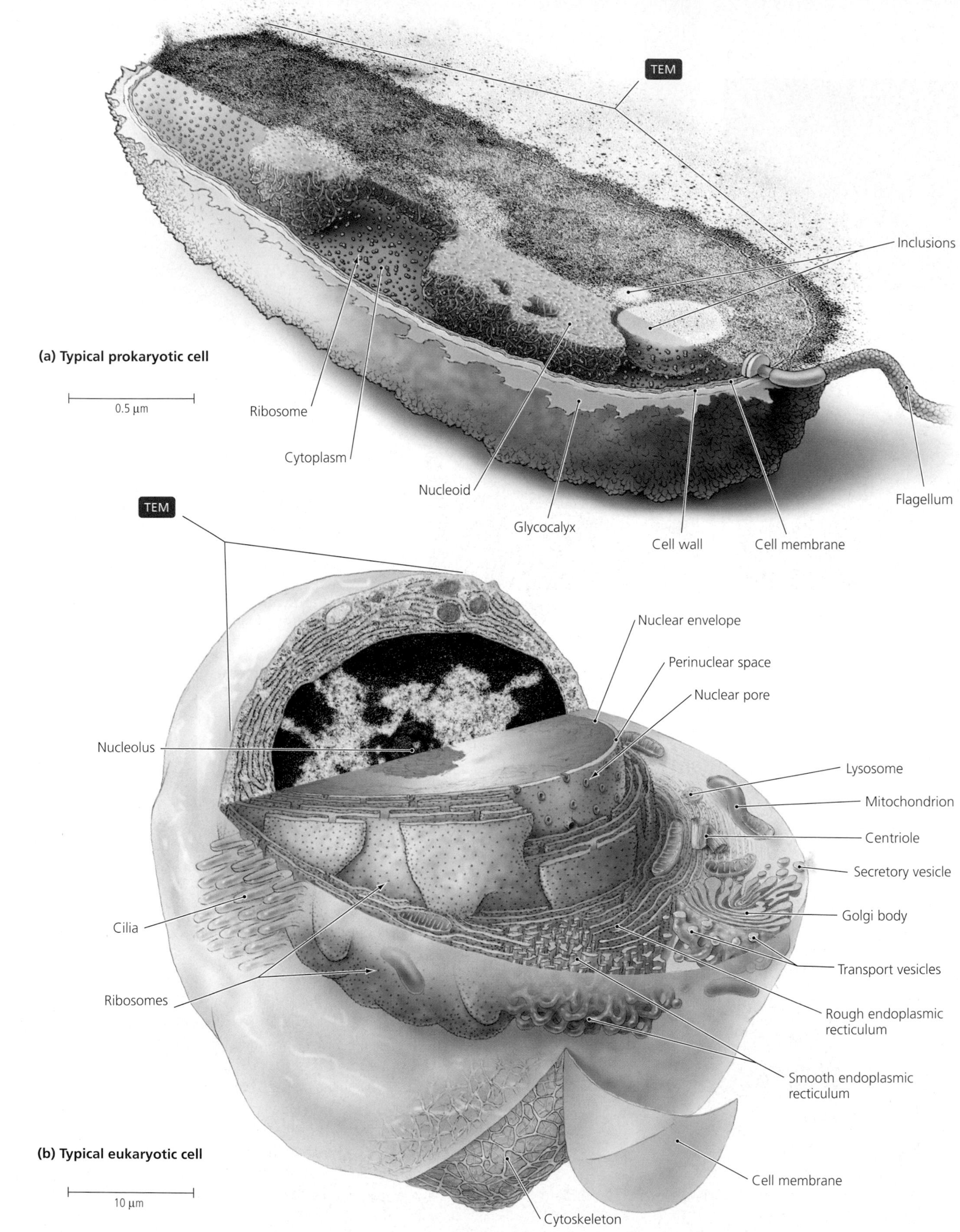

▲ ***Figure 3.2***

Comparison of cells. **(a)** A prokaryote. **(b)** A eukaryote. Note differences in magnification.
What major difference between prokaryotes and eukaryotes was visible to early microscopists?

Figure 3.2 *Eukaryotes contain nuclei, which are visible with light microscopy, whereas prokaryotes lack nuclei.*

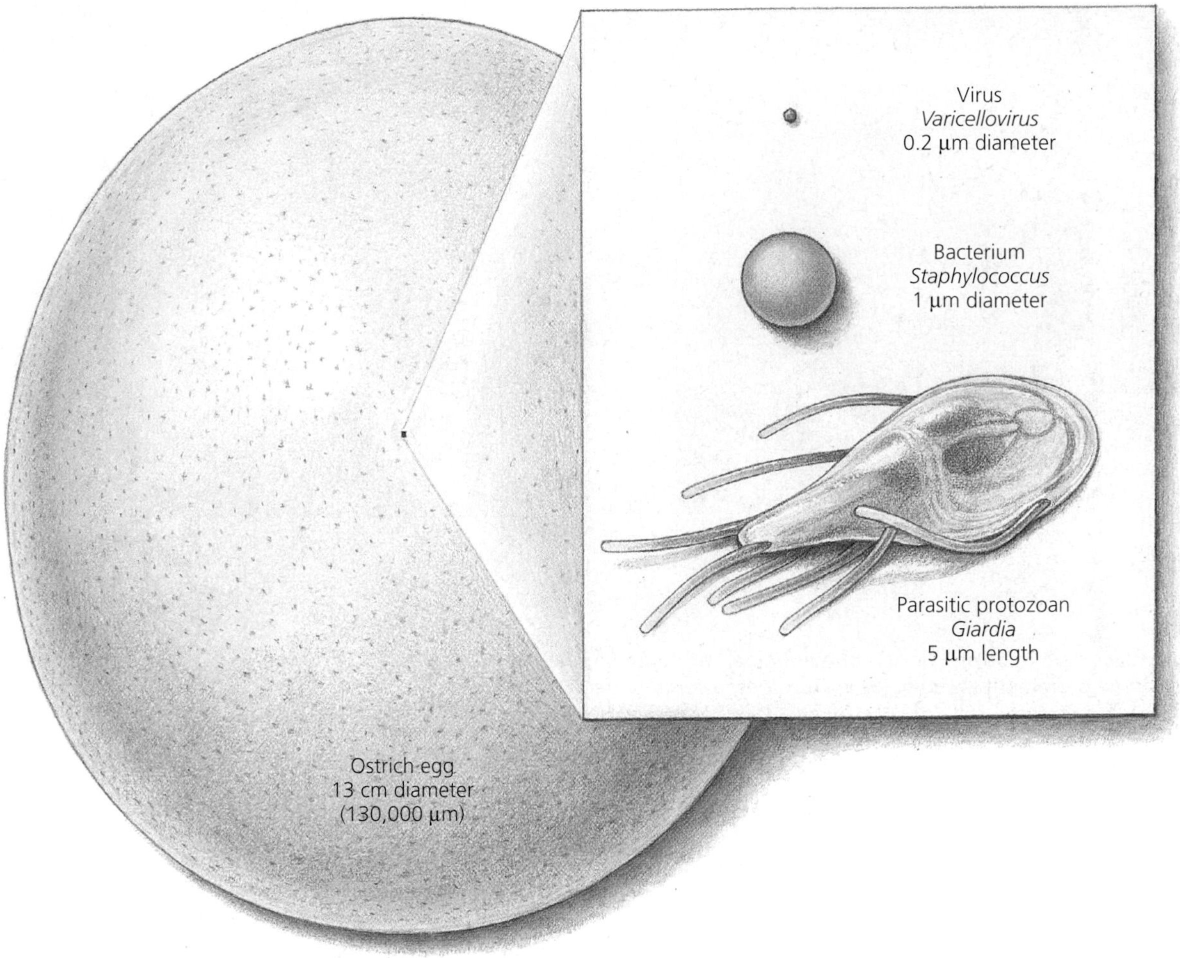

*At this scale, the inset box on the egg would actually be too small to be visible. (Width of box would be about 0.006 mm.)

▲ ***Figure 3.3***
Approximate size of various types of cells. Note that *Staphylococcus*, a prokaryote, is smaller than *Giardia*, a eukaryote. A chicken pox virus is shown only for comparison; viruses are not cellular.

Although there are many kinds of cells, they all share the characteristic processes of life as previously described, as well as certain physical features. In this chapter, we will distinguish between prokaryotic and eukaryotic "versions" of physical features common to cells, including (1) external structures, (2) the cell wall, (3) the cytoplasmic membrane, and (4) the cytoplasm. We will also discuss features unique to each type. Further details of prokaryotes and eukaryotes, their classification, and their ability to cause disease are discussed in Chapters 11 and 12.

External Structures of Prokaryotic Cells

Many cells have special external features that enable them to respond to other cells and their environment. In prokaryotes, these features include glycocalyces, flagella, fimbriae, and pili.

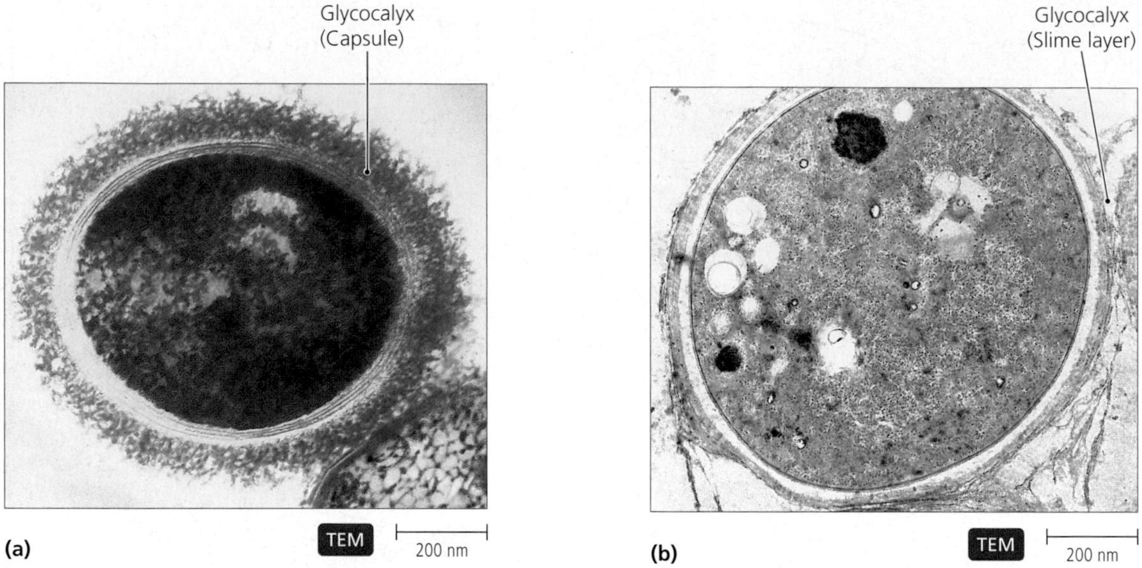

▲ ***Figure 3.4***

Glycocalyces. **(a)** Micrograph of *Streptococcus pneumoniae*, the common cause of pneumonia, showing a prominent capsule. **(b)** *Bacteroides*, a common fecal bacterium, has a slime layer surrounding the cell. *What advantage does a glycocalyx provide a cell?*

Figure 3.4 A glycocalyx provides protection from drying and from being devoured; it may also help attach cells to one another and to surfaces in the environment.

Glycocalyces

Learning Objectives

✓ Describe the composition, function, and relevance to human health of glycocalyces.

✓ Distinguish between capsules and slime layers.

Some cells have a gelatinous, sticky substance that surrounds the outside of the cell. This substance is known as a **glycocalyx** (plural: *glycocalyces*), which literally means "sugar cup." The glycocalyx may be composed of polysaccharides, polypeptides, or both. These chemicals are produced inside the cell and extruded onto the cell's surface.

When the glycocalyx of a prokaryote is composed of organized repeating units of organic chemicals firmly attached to the cell surface, the glycocalyx is called a **capsule (Figure 3.4a).** A loose, water-soluble glycocalyx is called a **slime layer (Figure 3.4b).** Capsules and slime layers protect cells from desiccation (drying).

The presence of a glycocalyx is a feature of numerous pathogenic prokaryotes. Their glycocalyces play an important role in the ability of these cells both to survive and to cause disease. Slime layers are often *viscous* (sticky), providing one means by which prokaryotes attach to surfaces. For example, they enable oral bacteria to colonize the teeth, where they produce acid and cause decay. Since the chemicals in many capsules are similar to those normally found in the body, they may prevent bacteria from being recognized and devoured by defensive cells of the host. For example, the capsules of *Streptococcus pneumoniae* (strep-tō-kok-ŭs nū-mō'nē-ē) and *Klebsiella pneumoniae* (kleb-sē-el'ă nū-mō'nē-ē) enable these prokaryotes to avoid destruction by defensive cells in the respiratory tract and cause pneumonia. Unencapsulated strains of these bacteria do not cause disease because the body's defensive cells destroy them.

Flagella

Learning Objectives

✓ Discuss the structure and function of bacterial flagella.

✓ List and describe four prokaryotic flagellar arrangements.

A cell's motility may enable it to flee from a harmful environment or move toward a favorable environment such as one where food or light is available. The structures responsible for this microbial movement are flagella. **Flagella** (singular: *flagellum*) are long, whiplike structures that extend beyond the surface of the cell and glycocalyx and propel the cell through its environment.

Structure

Bacterial flagella are composed of three parts: a long, thin *filament*, a *hook*, and a *basal body* **(Figure 3.5).** The filament is the whiplike shaft, about 20 nm in diameter, that extends out into the cell's environment. It is composed of many identical globular molecules of a protein called *flagellin*. Molecules of flagellin are arranged in chains and form a helix around a

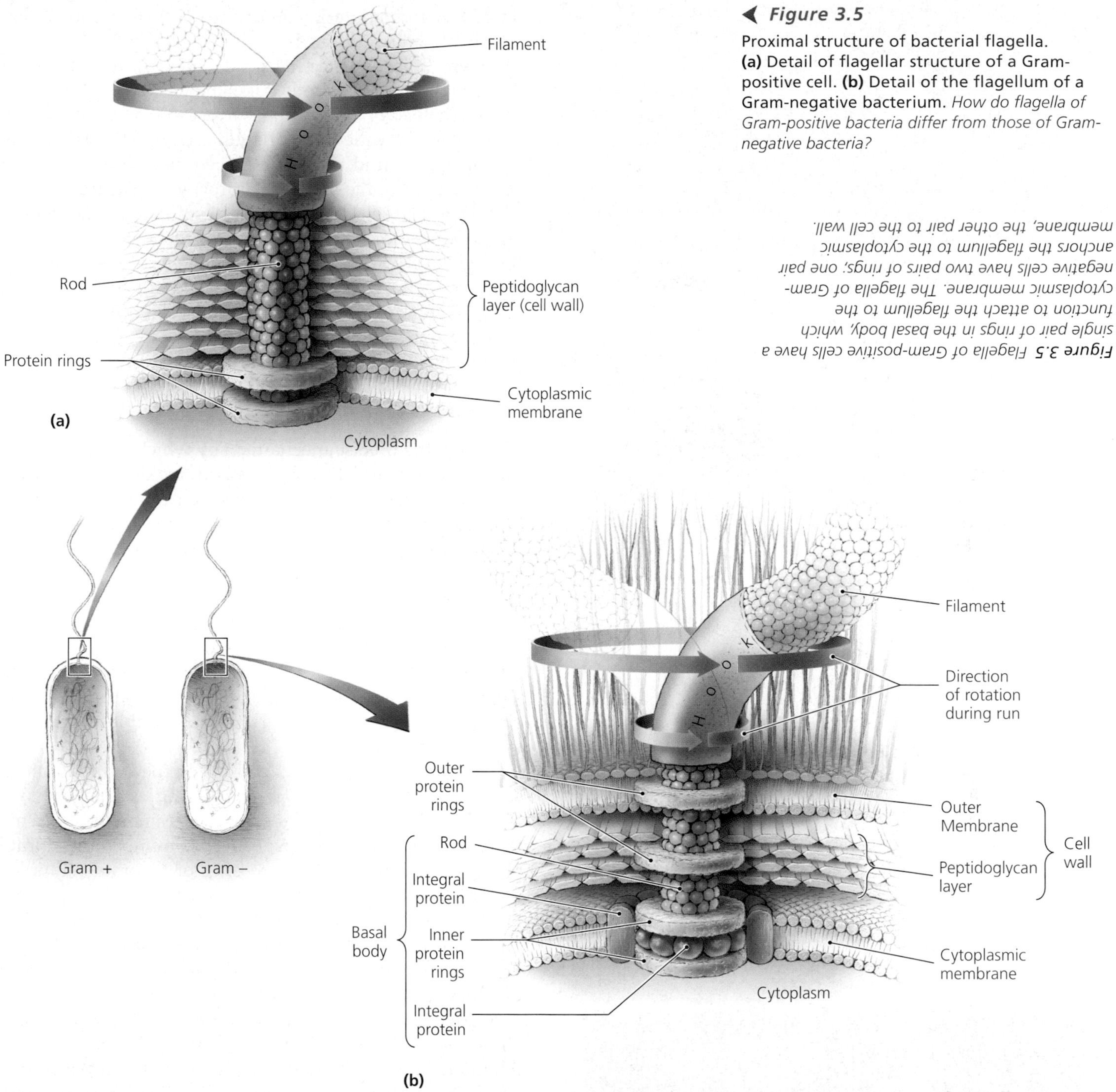

Figure 3.5

Proximal structure of bacterial flagella. **(a)** Detail of flagellar structure of a Gram-positive cell. **(b)** Detail of the flagellum of a Gram-negative bacterium. *How do flagella of Gram-positive bacteria differ from those of Gram-negative bacteria?*

Figure 3.5 *Flagella of Gram-positive cells have a single pair of rings in the basal body, which function to attach the flagellum to the cytoplasmic membrane. The flagella of Gram-negative cells have two pairs of rings; one pair anchors the flagellum to the cytoplasmic membrane, the other pair to the cell wall.*

hollow core. No membrane covers the filament of bacterial flagella. At its base the filament inserts into a curved structure, the hook, which is composed of a different protein. The basal body, which is composed of still different proteins, anchors the filament and hook to the cell wall by means of a rod and a series of either two or four rings. Together the hook, rod, and rings allow the filament to rotate 360°. Differences in the proteins associated with bacterial flagella vary enough to allow classification of species into groups (strains) called *serovars.*

Arrangement

Bacteria may have one of several flagellar arrangements (**Figure 3.6** on page 64). Cells with a single flagellum are **monotrichous.**[2] **Lophotrichous**[3] cells have a tuft of flagella at only one end of the cell, whereas **amphitrichous**[4] cells have flagella at both ends. Flagella that cover the surface of

[2]From Greek *monos,* meaning one, and *trichos,* meaning a hair.
[3]From Greek *lophos,* meaning tuft.
[4]From Greek *amphi,* meaning both.

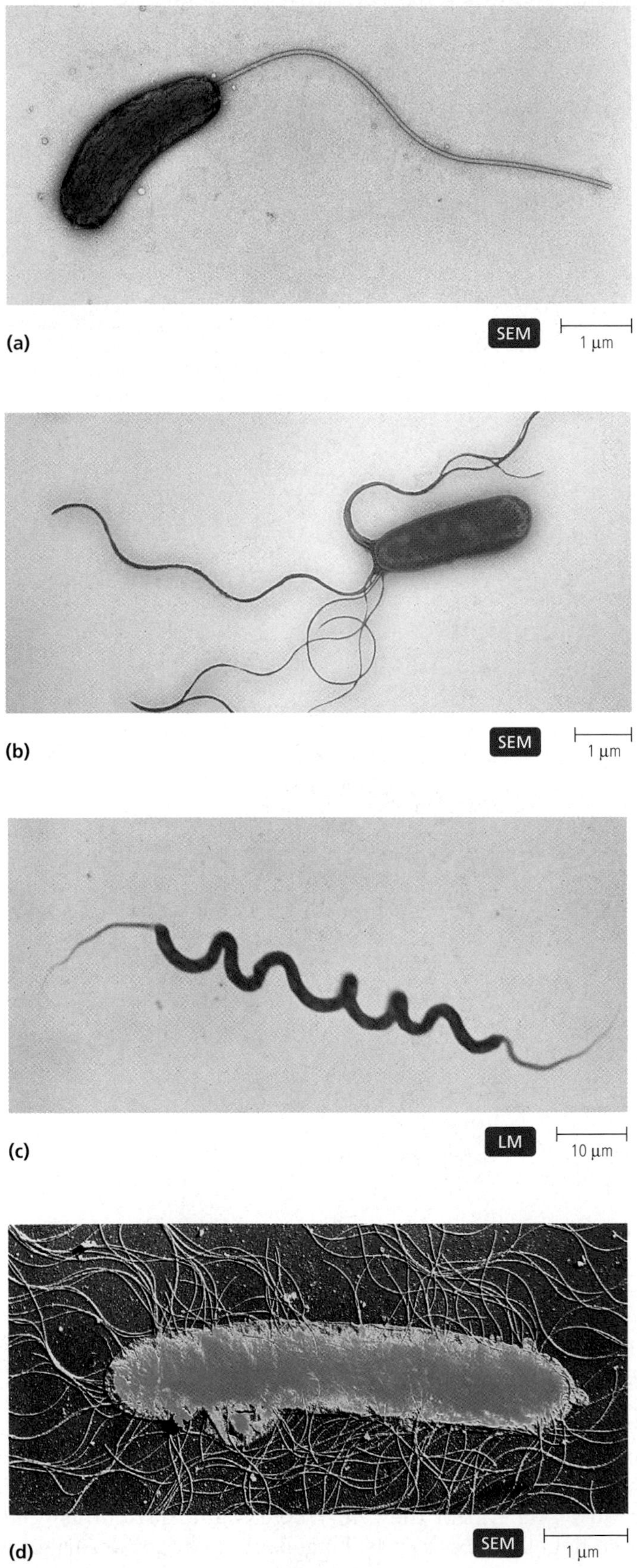

▲ *Figure 3.6*

Micrographs of basic arrangements of bacterial flagella. **(a)** Monotrichous. **(b)** Lophotrichous. **(c)** Amphitrichous. **(d)** Peritrichous.

the cell are termed **peritrichous,**[5] in contrast to **polar** flagella, which are only at the ends.

Some spiral-shaped bacteria, called *spirochetes* (spī′rō-kēts),[6] have amphitrichous flagella that spiral tightly around the cell instead of protruding into the surrounding medium. These flagella, called **endoflagella,** form an **axial filament** that wraps around the cell between its cell membrane and an outer membrane found in spirochetes **(Figure 3.7).** Rotation of endoflagella evidently causes the axial filament to rotate around the cell, causing the spirochete to "corkscrew" through its medium. *Treponema pallidum* (trep-ō-nē′mă pal′li-dŭm), the agent of syphilis, and *Borrelia burgdorferi* (bō-rē′lē-a bŭrg-dor′fer-ē), the cause of Lyme disease, are notable spirochetes. Some scientists think the corkscrew motility of these pathogens allows them to invade viscous human tissues.

Function

Although the precise mechanism by which bacterial flagella move is not completely understood, we do know that they rotate 360° like the shaft of an electric motor rather than whipping from side to side. This rotation propels the bacterium through the environment. Rotation of a flagellum may be either clockwise or counterclockwise and can be reversed by the cell. If more than one flagellum is present, the flagella align and rotate together as a unit.

Bacteria move with a series of "runs" punctuated by "tumbles" **(Figure 3.8).** Runs are movements of a cell in a single direction for some time; tumbles are abrupt, random, changes in direction. Both runs and tumbles occur in response to stimuli.

Receptors for light or chemicals on the surface of the cell send signals to the flagella, which then adjust their speed and direction of rotation. A bacterium can position itself in a more favorable environment by varying the number and duration of runs and tumbles. The presence of favorable stimuli increases the number of runs and decreases the number of tumbles; as a result, the cell tends to move toward an attractant. Unfavorable stimuli increase the number of tumbles, which increases the likelihood that it will move in another direction, away from a repellant.

Movement in response to a stimulus is termed **taxis.** The stimulus may be either light **(phototaxis)** or a chemical **(chemotaxis).** Movement toward a favorable stimulus is *positive taxis,* whereas movement away from an unfavorable stimulus is *negative taxis.* For example, movement toward a nutrient would be *positive chemotaxis.*

Fimbriae and Pili

Learning Objective

✓ Compare and contrast the structures and functions of fimbriae, pili, and flagella.

[5]From Greek *peri,* meaning around.
[6]From Greek *speira,* meaning coil, and *chaeta,* meaning hair.

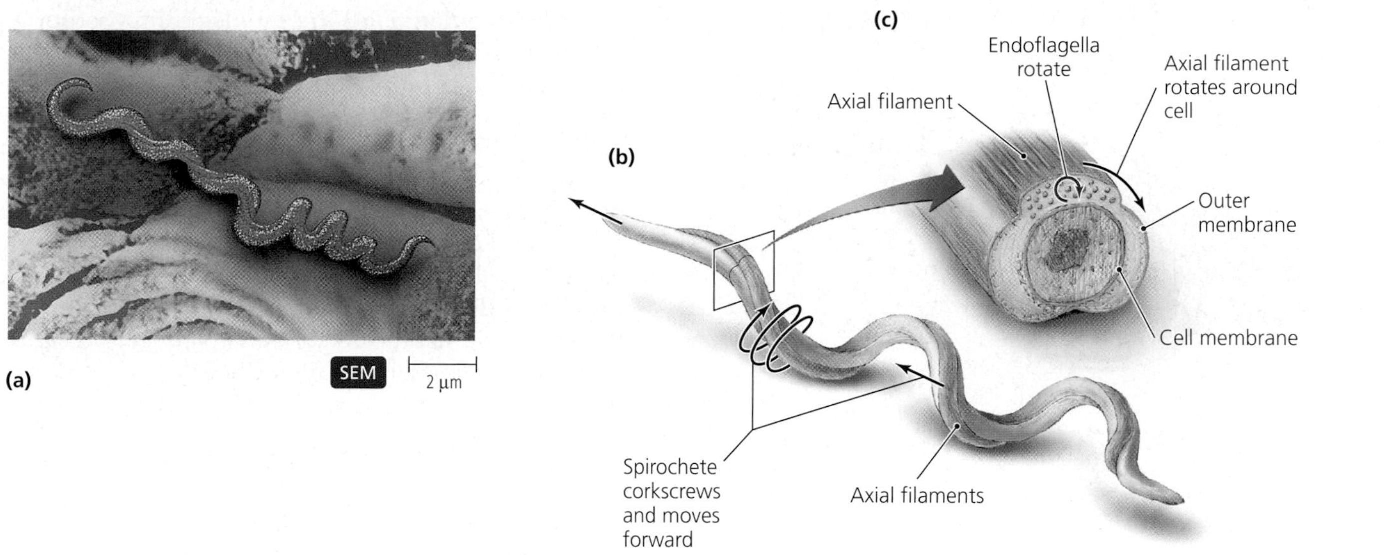

▲ *Figure 3.7*

Axial filament. **(a)** Scanning electron micrograph of *Treponema pallidum*, a spirochete. **(b)** Diagram of axial filament wrapped around a spirochete. **(c)** Cross-section of the spirochete, which reveals that the axial filament is composed of endoflagella.

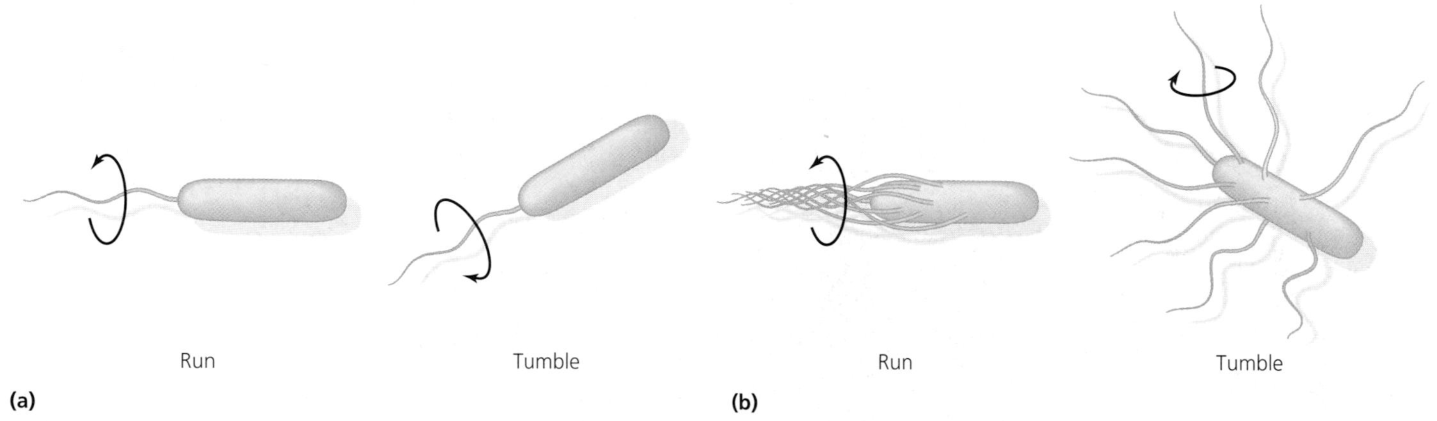

▲ *Figure 3.8*

Patterns of bacterial motility. **(a)** Motion of a monotrichous bacterium. **(b)** Motion of a peritrichous bacterium. In peritrichous bacteria, runs occur when all of the flagella rotate counterclockwise and become bundled. Tumbles occur when the flagella rotate clockwise, become unbundled, and the cell spins randomly. *What triggers a bacterial flagellum to rotate counterclockwise, producing a run?*

Figure 3.8 Favorable environmental conditions induce runs.

Some prokaryotes have nonmotile extensions called fimbriae and pili. Some Gram-negative bacteria (defined shortly) use **fimbriae**—sticky, proteinaceous, bristlelike projections—to adhere to one another and to substances in the environment. There may be hundreds of fimbriae per cell. They are usually shorter than flagella (**Figure 3.9** on page 66). An example of a prokaryote with fimbriae is *Neisseria gonorrhoeae* (nī-sē′rē-ă go-nōr-rē′ē), the bacterium that causes gonorrhea. Pathogens must be able to adhere to their hosts if they are to survive and cause disease. This bacterium is able to colonize the mucous membrane of the reproductive tract by attaching with fimbriae. *Neisseria* cells that lack fimbriae are nonpathogenic.

Fimbriae also serve an important function in **biofilms,** slimy masses of bacteria adhering to a substrate by means of fimbriae and glycocalyces. It has been estimated that about 99% of bacteria in nature exist in biofilms. Biofilms are a topic of intense interest to researchers because of the roles they play in human diseases and in industry. See the **New Frontiers** 3.1 on page 67.

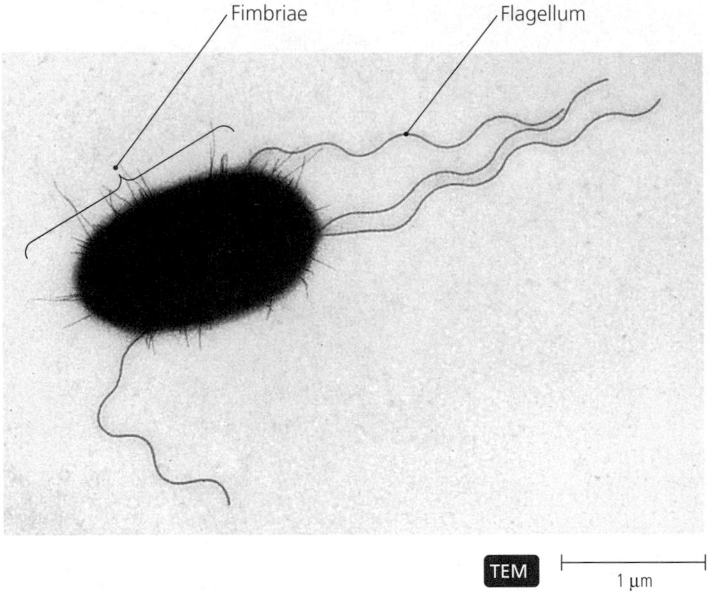

▲ ***Figure 3.9***
Fimbriae. *Escherichia coli* has fimbriae and flagella.

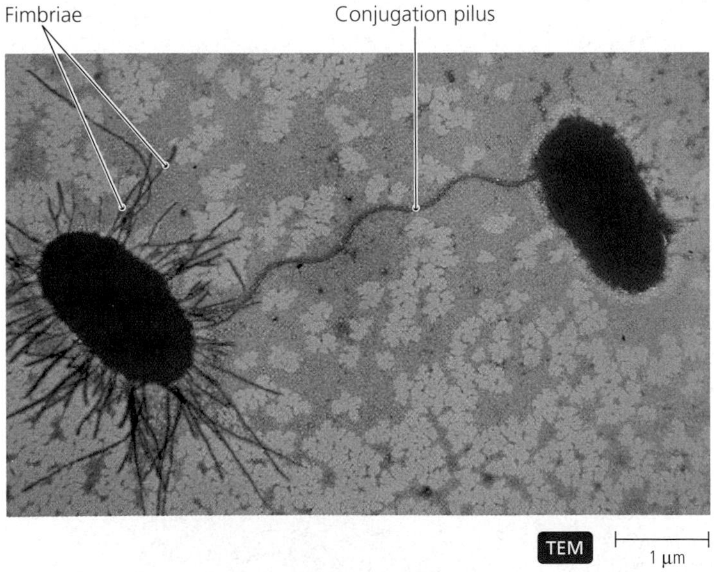

▲ ***Figure 3.10***
Pilus. Two *Escherichia coli* cells are connected by a conjugation (sex) pilus. *How are pili different from prokaryotic flagella?*

Figure 3.10 Prokaryotic flagella are flexible structures that rotate to propel the cell; pili are hollow tubes that mediate transfer of DNA from one cell to another.

Pili (singular: *pilus*) are tubules composed of a protein called *pilin*. Pili are longer than fimbriae, but usually shorter than flagella. Typically only one to ten pili are present per cell in bacteria that have them. Bacteria use pili to move across a substrate or toward another bacterium via a process that appears to be similar to the use of a grappling hook. The bacterium extrudes a pilus, which attaches to the substrate or to another bacterium, then the bacterium pulls itself along the pilus toward the attachment point. So-called **conjugation pili** mediate the transfer of DNA from one cell to the other via a process termed *conjugation* **(Figure 3.10),** which is discussed in detail in Chapter 7.

Prokaryotic Cell Walls

The cells of most prokaryotes are surrounded by a **cell wall** that provides structure and shape to the cell and protects it from osmotic forces. In addition, a cell wall assists some cells in attaching to other cells or in eluding antimicrobial drugs. Note that animal cells do not have walls, a difference that plays a key role in treatment of many bacterial diseases with certain types of antibiotics. For example, penicillin attacks the cell wall of bacteria but is harmless to human cells, which lack walls.

Recall that there are two basic types of prokaryotes: bacteria and archaea. One difference between these prokaryotes involves the chemistry of their cell walls. Scientists have studied bacterial walls more extensively than those of archaea, so we will concentrate our discussion on bacterial cell walls.

Bacterial Cell Walls

Learning Objectives

- ✓ Compare and contrast the cell walls of Gram-positive and Gram-negative prokaryotes in terms of structure and Gram staining.
- ✓ Describe the clinical implications of the structure of the Gram-negative cell wall.
- ✓ Compare and contrast the cell walls of acid-fast bacteria with typical Gram-positive cell walls.

Most bacteria have a cell wall, composed of **peptidoglycan,** a complex polysaccharide. Peptidoglycan is composed of two regularly alternating sugars, called N-*acetylglucosamine (NAG)* and N-*acetylmuramic acid (NAM),* which are structurally similar to glucose **(Figure 3.11).** NAG and NAM are covalently linked in long chains in which NAG alternates with NAM (**Figure 3.12** on page 68). These chains are the "glycan" portions of peptidoglycan.

Chains of NAG and NAM are attached to other chains by crossbridges of four amino acids (tetrapeptides). These crossbridges are the "peptido" portion of peptidoglycan. Depending on the bacterium, these tetrapeptide bridges are either covalently bonded to one another or are held together by *short connecting* chains of amino acids as shown in Figure 3.12. Peptidoglycan covers the entire surface of the cell.

New Frontiers 3.1 Understanding Biofilms

They form plaque on teeth; they are the slime on rocks in rivers and streams; and they can cause disease, clog drains, or aid in hazardous-waste cleanup. They are **biofilms**—organized, layered systems of bacteria attached to a surface. Understanding biofilms holds the key to many important clinical and industrial applications.

Biofilm on a tooth surface SEM 5 μm

Recent research has demonstrated that bacteria in biofilms behave in significantly different ways from individual, free-floating bacteria. For example, as a free-floating cell, the soil bacterium *Pseudomonas putida* propels itself through water with its flagella; however, once it becomes part of a biofilm, it turns off the genes for flagellar proteins and starts synthesizing pili instead. In addition, the genes for antibiotic resistance in *P. putida* are more active within cells in a biofilm than within a free-floating cell.

Bacteria in a biofilm also communicate via chemical messages that help them organize and form three-dimensional structures. The architecture of a biofilm provides protection that free-floating bacteria lack. For example, the lower concentrations of oxygen found in the interior of biofilms thwart the effectiveness of some antibiotics. Moreover, the presence of multiple bacterial species in most biofilms increases the likelihood that some bacteria within the biofilm community will remain resistant to a given antibiotic.

The more we understand biofilms, the more readily we can reduce their harmful effects or put them to good use. Biofilms account for two-thirds of bacterial infections in humans, including gum disease and the serious lung infections suffered by cystic fibrosis patients. Biofilms are also the culprits in many industrial problems, including corroded pipes and clogged water filters, which cause millions of dollars of damage each year. Fortunately, not all biofilms are detrimental; some have shown potential as aids in preventing and controlling certain kinds of pollution.

1. Slimebuster. *Nature* 408:284–286.
2. The Center for Biofilm Engineering at Montana State University, Bozeman: www.erc.montana.edu

The two basic types of bacterial cell walls are **Gram-positive** cell walls and **Gram-negative** cell walls, and bacterial cells are described as being either "Gram-positive" or "Gram-negative" depending on the kind of wall they have. Gram-positive and Gram-negative cells are distinguished by the use of the Gram staining procedure, which is described in Chapter 4. The Gram staining procedure was invented long before the structure and chemical nature of prokaryotic cell walls were known.

Gram-Positive Cell Walls

Gram-positive cell walls have a relatively thick layer of peptidoglycan that also contains unique polysaccharides called *teichoic* (tī-kō′ĭk)[7] *acids.* Some teichoic acids are covalently linked to lipids, forming *lipoteichoic acids* that anchor the peptidoglycan to the cell membrane (**Figure 3.13a** on page 69). The thick cell wall of a Gram-positive organism retains the crystal violet dye used in the Gram staining procedure, so the stained cells appear purple under magnification.

Some additional chemicals are associated with the walls of some Gram-positive cells. For example, species of *Mycobacterium* (mī′kō-bak-tēr′ē-ŭm), which include the causative agents of tuberculosis and leprosy, have walls with up to 60% mycolic acid, a waxy lipid. Mycolic acid helps these cells survive desiccation and makes them difficult to stain with regular water-based dyes. Special staining procedures such as the acid-fast stain have been developed for these Gram-positive cells with large amounts of waxy lipids, so they are called *acid-fast bacteria* (see Chapter 4).

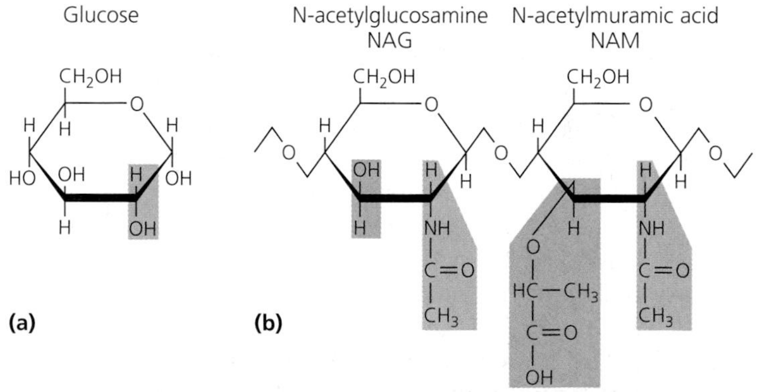

▲ *Figure 3.11*

Comparison of the structures of glucose, NAG, and NAM. **(a)** Glucose. **(b)** *N*-acetylglucosamine (NAG) and *N*-acetylmuramic acid (NAM) molecules linked as in peptidoglycan. Blue shading indicates the differences between glucose and the other two sugars. Orange boxes highlight the difference between NAG and NAM.

[7]From Greek *teichos,* meaning wall.

Figure 3.12 ➤

Structure of peptidoglycan. Peptidoglycan is composed of chains of NAG and NAM linked by tetrapeptide crossbridges and connecting chains of amino acids to form a relatively rigid structure. The amino acids of the crossbridges differ among bacterial species.

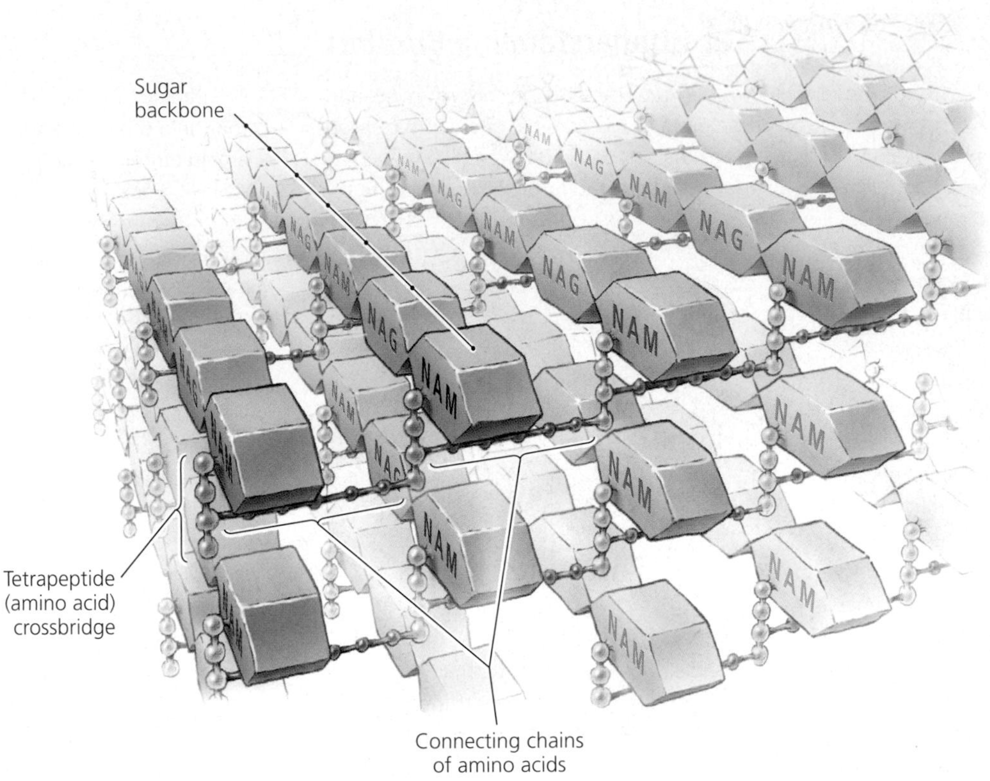

Gram-Negative Cell Walls

Gram-negative cell walls have only a thin layer of peptidoglycan **(Figure 3.13b),** but outside this layer is a bilayer membrane composed of phospholipids, channel proteins (called *porins*), and **lipopolysaccharide (LPS).** LPS is a union of lipid with sugar. The lipid portion of LPS is known as **lipid A.** The erroneous idea that lipid A is *inside* Gram-negative cells led to the use of the term *endotoxin*[8] for this chemical. Lipid A is released from dead cells when the cell wall disintegrates, and it may trigger fever, vasodilation, inflammation, shock, and blood clotting in humans. Because killing large numbers of Gram-negative cells with antimicrobial drugs releases large amounts of lipid A, which might threaten the patient more than the live bacteria, any internal infection by Gram-negative bacteria is cause for concern.

The Gram-negative cell wall can also be an impediment to the treatment of disease. The outer membrane may prevent the movement of penicillin to the underlying peptidoglycan, thus rendering the drug ineffectual against many Gram-negative pathogens.

Between the cell membrane and the outer membrane of Gram-negative organisms is a **periplasmic space** (see Figure 3.13b). The periplasmic space contains the peptidoglycan and *periplasm*, the name given to the gel between the membranes of Gram-negative cells. Periplasm contains water, nutrients, and substances secreted by the cell, such as digestive enzymes and proteins involved in specific transport. The enzymes function to catabolize large nutrient molecules into smaller molecules that can be absorbed or transported into the cell.

Because the cell walls of Gram-positive and Gram-negative organisms differ, the Gram stain is an important diagnostic tool. After the Gram staining procedure Gram-negative cells appear pink, and Gram-positive cells appear purple.

[8]From Greek *endo*, meaning inside, and *toxikon*, meaning poison.

Bacteria without Cell Walls

A few bacteria, such as *Mycoplasma pneumoniae* (mī'kō-plaz-mă nū-mō'nē-ē), lack cell walls entirely. In the past, these bacteria were often mistaken for viruses because of their small size and lack of walls. However, they do have other features of prokaryotic cells, including prokaryotic ribosomes (discussed later in the chapter).

Archaeal Cell Walls

Learning Objective

✓ Contrast types of archaeal cell walls with one another and with bacterial cell walls.

The archaea have walls containing a variety of specialized polysaccharides and proteins, but no peptidoglycan. The absence of peptidoglycan is one reason scientists classify archaea in a separate group from bacteria.

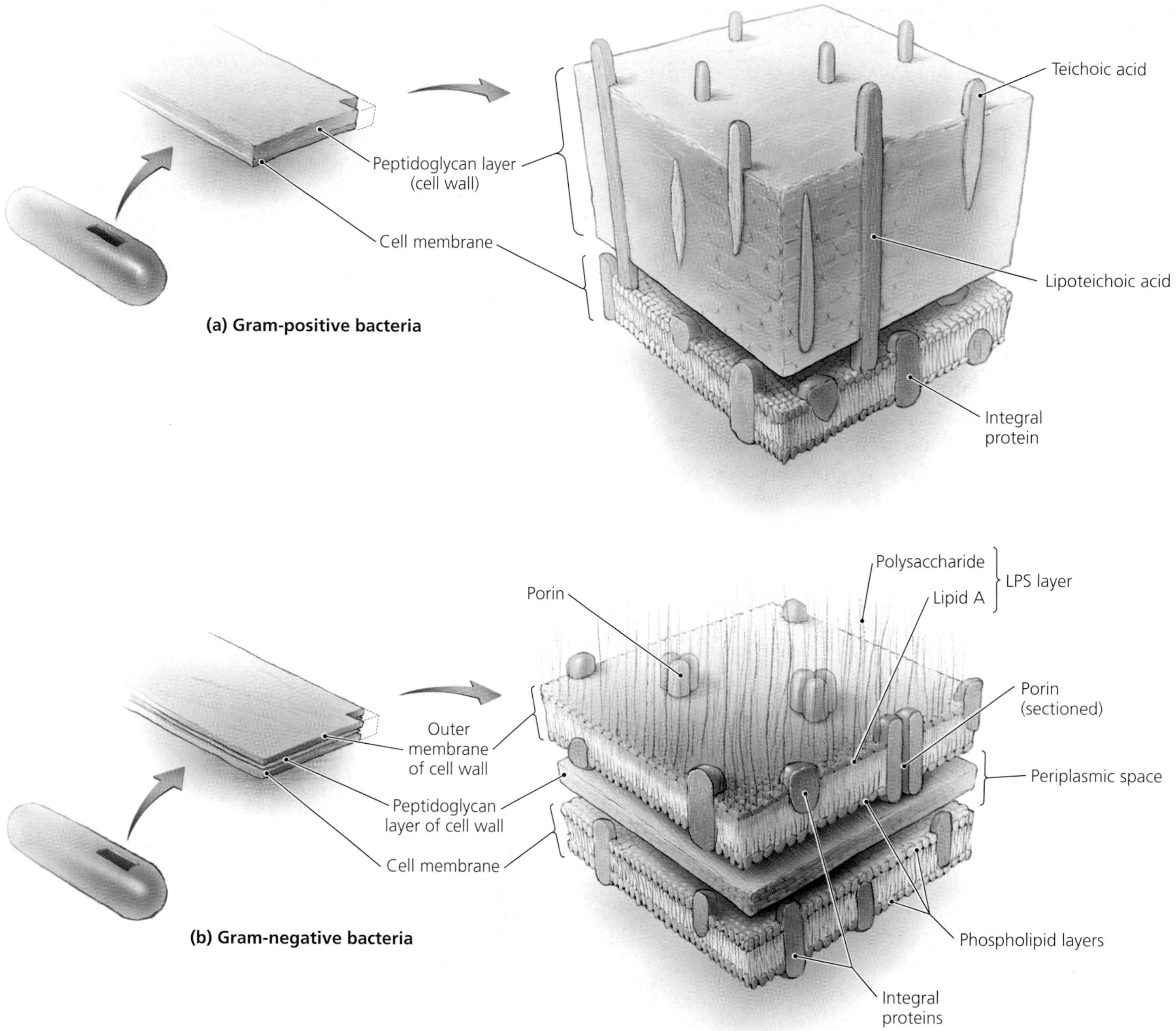

▲ ***Figure 3.13***

Comparison of cell walls of Gram-positive and Gram-negative bacteria. **(a)** The Gram-positive cell wall has a thick layer of peptidoglycan and teichoic acids that anchor the wall to the cell membrane. **(b)** The Gram-negative cell wall has a thin layer of peptidoglycan and an outer membrane composed of lipopolysaccharide (LPS), phospholipids, and proteins. *What effects can lipid A have on human physiology?*

Nevertheless, archaeal cells share some characteristics with bacteria. Like Gram-positive bacteria, Gram-positive archaea have a thick wall and Gram stain purple. Gram-negative archaeal cells have a layer of protein covering the wall, rather than a lipid membrane as seen in Gram-negative bacteria, but they do stain pink.

CRITICAL THINKING

After a man infected with the bacterium *Escherichia coli* was treated with the correct antibiotic for this pathogen, the bacterium was no longer found in the man's blood, but his symptoms of fever and inflammation worsened. What caused the man's response to the treatment? Why was his condition worsened by the treatment?

Figure 3.13 *Lipid A can cause shock, blood clotting, and fever in humans.*

Prokaryotic Cytoplasmic Membranes

Beneath the glycocalyx and the cell wall is the **cytoplasmic membrane.** The cytoplasmic membrane may also be referred to as the *cell membrane* or the *plasma membrane.*

Structure

Learning Objectives

- ✓ Diagram a phospholipid bilayer, and explain its significance in reference to a cytoplasmic membrane.
- ✓ Explain the fluid mosaic model of membrane structure.

Cytoplasmic membranes are composed of lipids and associated proteins. Bacterial membranes contain phospholipids **(Figure 3.14);** by contrast, archaeal membranes are composed of lipids that lack phosphate groups and have branched hydrocarbons linked to glycerol by ether linkages instead of ester linkages (see Table 2.3). Some archaea—particularly those that thrive in very hot water—have a single layer of lipid composed of two glycerol groups linked with branched hydrocarbon chains. Some bacterial membranes also contain sterol-like molecules, called *hopanoids,* that help stabilize the membrane. Because scientists are most familiar with bacterial cytoplasmic membranes, the remainder of our discussion will concentrate on these structures.

The structure of a bacterial cytoplasmic membrane is referred to as a **phospholipid bilayer** (see Figure 3.14). As discussed in Chapter 2, a phospholipid molecule is bipolar; that is, the two ends of the molecule are different. The phosphate heads of each phospholipid molecule are *hydrophilic;*[9] that is, they are attracted to water at the two surfaces of the membrane. The hydrocarbon tails of each phospholipid molecule are *hydrophobic*[10] and huddle together with other tails in the interior of the membrane, away from water. Phospholipids placed in a watery environment naturally form a bilayer because of their bipolar nature.

About half of a bacterial cytoplasmic membrane is composed of proteins, many of which are *integral proteins* inserted between the phospholipids. Some integral proteins

[9]From Greek *hydro,* meaning water, and *philos,* meaning love.
[10]From Greek *hydro,* meaning water, and *phobos,* meaning fear.

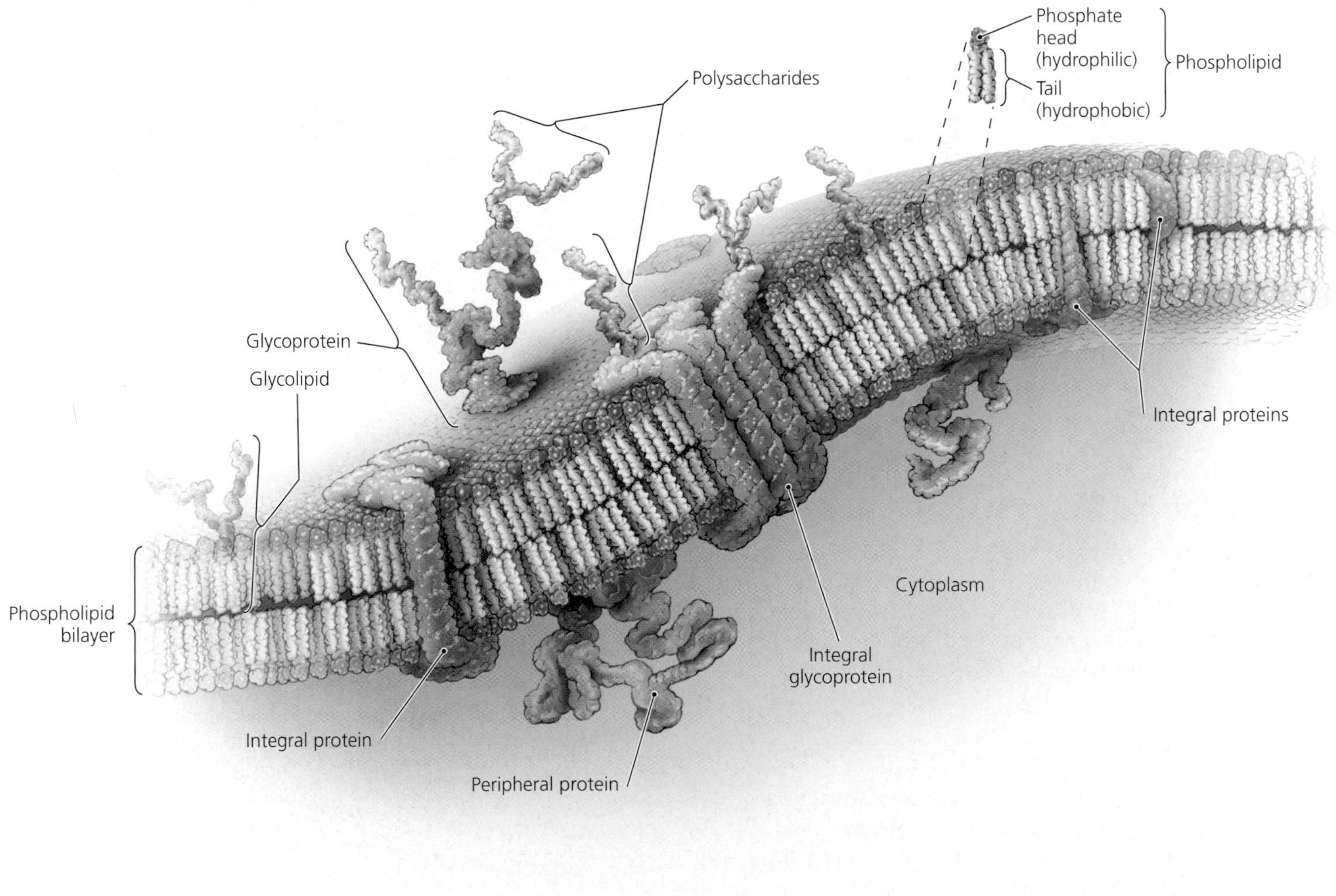

▲ ***Figure 3.14***
The structure of a prokaryotic cytoplasmic membrane: a phospholipid bilayer.

penetrate the entire bilayer; others are found in only half the bilayer. *Peripheral proteins* are loosely attached to the membrane on one side or the other. Some of the membrane proteins have polysaccharide groups attached to them. Proteins chemically bound to polysaccharides are called *glycoproteins*. Proteins of cell membranes may act as recognition proteins, enzymes, receptors, carriers, or channels.

The **fluid mosaic model** describes our current understanding of membrane structure. The term *mosaic* indicates that the membrane proteins are arranged in a way that resembles the tiles in a mosaic, and *fluid* indicates that the proteins and lipids are free to flow laterally within the membrane.

Function

Learning Objectives

- ✓ Describe the functions of the cytoplasmic membrane as they relate to permeability.
- ✓ Compare and contrast the passive and active processes by which materials cross the cytoplasmic membrane.
- ✓ Define osmosis, and distinguish among isotonic, hypertonic, and hypotonic solutions.

A cytoplasmic membrane does more than separate the contents of the cell from the outside environment. In all prokaryotic cells, the cytoplasmic membrane controls the passage of substances into and out of the cell. Nutrients are brought into the cell, and wastes are removed. The membrane also functions in energy production, and to harvest light energy in photosynthetic prokaryotes. Energy production and photosynthesis are discussed in Chapter 5.

In its function of controlling the contents of the cell, the cytoplasmic membrane is **selectively permeable;** that is, it allows some substances to cross it while preventing the crossing of others. How does the membrane exert control over the contents of the cell, and the substances that move across it?

A phospholipid bilayer is naturally impermeable to most substances. Large molecules cannot cross through it; ions and molecules with an electrical charge are repelled by it; and hydrophilic substances cannot easily cross its hydrophobic interior. However, cytoplasmic membranes, unlike plain phospholipid bilayers in a scientist's test tube, contain proteins, and these proteins allow substances to cross the membrane by functioning as pores, channels, or carriers.

Movement across the cytoplasmic membrane occurs either by passive or active processes. Passive processes do not require the expenditure of a cell's ATP energy store, whereas active processes require the expenditure of ATP, either directly or indirectly. Active and passive processes will be discussed shortly, but first you must understand another feature of selectively permeable cytoplasmic membranes: their ability to maintain a *concentration gradient.*

Membranes enable a cell to concentrate chemicals on one side of the membrane or the other. The difference in concentration of a chemical on the two sides of a membrane is its **concentration gradient** (also known as a *chemical gradient*).

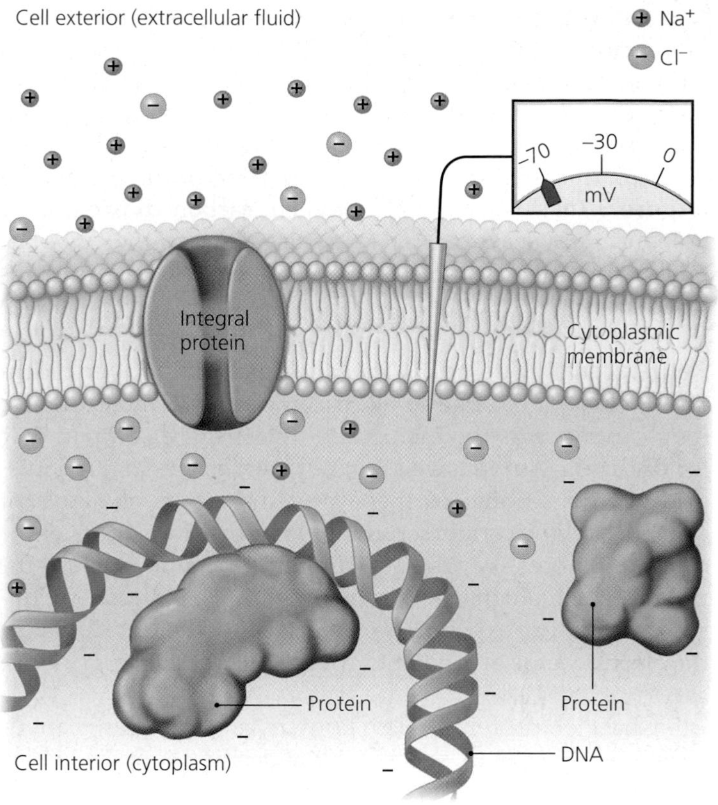

▲ ***Figure 3.15***

Electrical potential of a cell membrane. The electrical potential exists across a membrane because there are more negative charges inside the cell than outside it.

Because many of the substances that have concentration gradients across cell membranes are electrically charged chemicals, a corresponding **electrical gradient,** or voltage, exists across the membrane **(Figure 3.15).** For example, a greater concentration of negatively charged proteins exists inside the membrane, while positively charged sodium ions are more concentrated outside the membrane. One result of the segregation of electrical charges by a membrane is that the interior of a cell is usually electrically negative compared to the exterior. This tends to repel negatively charged chemicals and attract positively charged substances into the cell.

Together the chemical and electrical gradients are called the **electrochemical gradient.** A cytoplasmic membrane can use the energy inherent in its electrochemical gradient to transport substances into or out of a cell.

Passive Processes

In passive processes, the electrochemical gradient provides a source of energy; the cell does not expend its ATP energy reserve. Passive processes include diffusion, facilitated diffusion, and osmosis.

Diffusion **Diffusion** is the net movement of a chemical down its concentration gradient—that is, from an area of higher concentration to an area of lower concentration. It requires no energy output by the cell, a common feature of all passive processes. In fact, diffusion occurs even in the absence of cells or their membranes. In the case of diffusion into or out of cells, only chemicals that are small or lipid soluble can diffuse through the lipid portion of the membrane **(Figure 3.16a).** For example, oxygen, carbon dioxide, alcohol, and fatty acids can freely diffuse through the cytoplasmic membrane, but molecules such as glucose and proteins cannot.

Facilitated Diffusion The phospholipid bilayer blocks the movement of large or electrically charged molecules, so they cannot cross the membrane unless there is a pathway for diffusion. As we have seen, cell membranes contain integral proteins. Some of these proteins act as channels or carriers to allow certain molecules to diffuse into or out of the cell. This process is called **facilitated diffusion** because the proteins facilitate the process by providing a pathway for diffusion. The cell expends no energy in facilitated diffusion; the electrochemical gradient provides all of the energy necessary.

Channel proteins, also known as *permeases,* can be either nonspecific or specific. Nonspecific channels, which are common in prokaryotes, allow the passage of a wide range of chemicals that have the right size or electrical charge **(Figure 3.16b).** Specific channel proteins, which are more common among eukaryotic cells, carry only specific substrates **(Figure 3.16c).** A specific carrier has a binding site that is selective for one substance.

CRITICAL THINKING

A scientist, who is studying passive movement of chemicals across the cell membrane of *Salmonella typhi* (sal'mō-nel'ă tī 'fē), measures the rate at which two chemicals diffuse into a cell as a function of external concentration. The results are shown in the figure. Chemical A diffuses into the cell more rapidly than does B at lower external concentrations, but the rate levels off as the external concentration increases. The rate of diffusion of chemical B continues to increase as the external concentration increases.

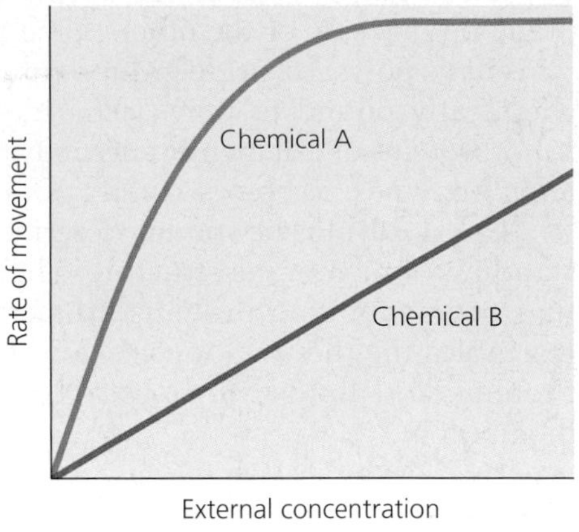

1. How can you explain the differences in the diffusion rates of chemicals A and B?
2. Why does the diffusion rate of chemical A taper off?
3. How could the cell increase the diffusion rate of chemical A?
4. How could the cell increase the diffusion rate of chemical B?

Osmosis When discussing simple and facilitated diffusion, we considered a solution in terms of the solutes (dissolved materials) it contains, because it is those solutes that move in and out of the cell. In contrast, with osmosis it is useful to consider the concentration of the solvent, which in organisms is always water. **Osmosis** is the special name given to the diffusion of water across a selectively permeable membrane—that is, across a membrane that is permeable to water molecules, but not to all solutes that are present, such as proteins, amino acids, salts, or glucose **(Figure 3.16d).** Because they cannot freely penetrate the membrane, such solutes cannot diffuse no matter how unequal their concentrations on either side of the membrane may be. Instead, what diffuses is the water, which crosses from the side of the membrane that contains a higher concentration of water (lower concentration of solute) to the side that contains a lower concentration of water

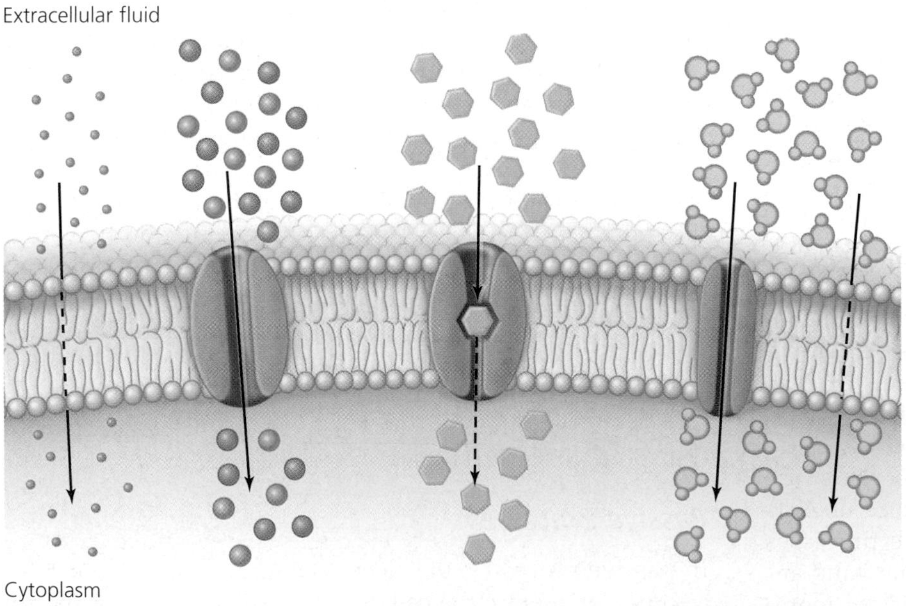

(a) Diffusion through the phospholipid bilayer

(b) Facilitated diffusion through a nonspecific channel protein

(c) Facilitated diffusion through a channel protein specific for one chemical; binding of substrate causes shape change in channel protein

(d) Osmosis, the diffusion of water through a specific channel protein or through a phospholipid bilayer

◀ ***Figure 3.16***

Passive processes of movement across a cell membrane. Passive processes always involve movement down an electrochemical gradient. **(a)** Diffusion. **(b)** Facilitated diffusion through a nonspecific channel protein. **(c)** Facilitated diffusion through a specific channel protein. **(d)** Osmosis through a nonspecific channel protein or through a phospholipid bilayer.

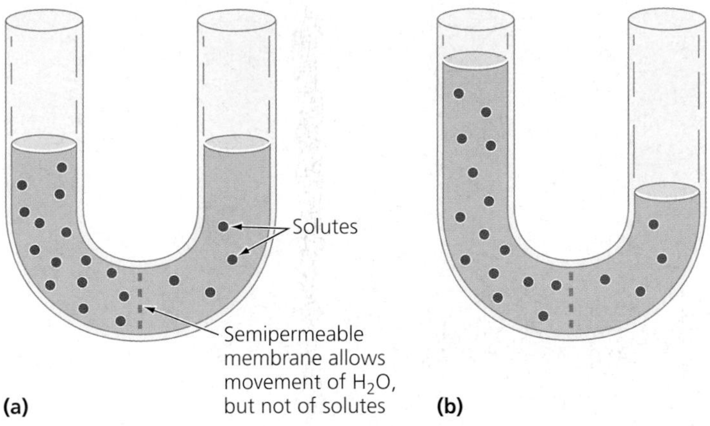

▲ ***Figure 3.17***

Osmosis, the diffusion of water across a selectively permeable membrane. **(a)** A membrane separates two solutions of different concentrations in a U-shaped tube. The membrane is permeable to water, but not to the solute. **(b)** After time has passed, water has moved down its concentration gradient until water pressure prevented the osmosis of any additional water. *Which side of the tube more closely represents a living cell?*

Figure 3.17 The left-hand side represents the cell, because cells are typically hypertonic to their environment.

(higher concentration of solute). Osmosis continues until equilibrium is reached, or until the pressure of water is equal to the force of osmosis **(Figure 3.17).**

We commonly classify solutions according to their concentrations of solutes and water. When solutions on either side of a selectively permeable membrane have the same concentration of solutes and water, the two solutions are said to be **isotonic.**[11] In an isotonic situation, neither side of a selectively permeable membrane will experience a net loss or gain of water (**Figure 3.18a** on page 74).

When the concentrations of solutions are unequal, they are named according to their relative concentration of *solutes.* The solution with the higher concentration of solutes is said to be **hypertonic**[12] to the other. The solution with a lower concentration of solutes is **hypotonic**[13] in comparison. Note that the terms *hypertonic* and *hypotonic* refer to the concentration of solute, even though osmosis refers to the movement of the *solvent,* which, in cells, is water. The terms *isotonic, hypertonic,* and *hypotonic* are relative. For example, a glass of tap water is isotonic to another glass of the same water, but it is hypertonic compared to distilled water, and hypotonic when compared to seawater. In biology, the three terms are traditionally used relative to the interior of cells.

Obviously, a higher concentration of solutes necessarily means a lower concentration of water; that is, a hypertonic solution has a lower concentration of water than does a hypotonic solution. Like other chemicals, water moves down its concentration gradient from a hypotonic solution into a hypertonic solution. A cell placed in a hypertonic solution will therefore lose water and shrivel **(Figure 3.18b).**

On the other hand, water will diffuse into a cell placed in a hypotonic solution because the cell has a higher solutes-to-water concentration. As water moves into the cell, water pressure against its cytoplasmic membrane increases, and the cell expands **(Figure 3.18c).** One function of a cell wall is to resist further osmosis and prevent cells from bursting.

It is useful to compare solutions to the concentration of solutes in a patient's blood cells. Isotonic saline solutions administered to a patient have the same percent dissolved solute (in this case, salt) as do the patient's blood cells. Thus, the patient's intracellular and extracellular environments remain in equilibrium when an isotonic saline solution is administered. However, if the patient is infused with a hypertonic solution, water will move out of the patient's cells, and the cells will shrivel, a condition called *plasmolysis*. Conversely, if a patient is infused with a hypotonic solution, water will move into the patient's cells, which will swell and possibly burst.

CRITICAL THINKING

Solutions hypertonic to bacteria and fungi are used for food preservation. For instance, jams and jellies are hypertonic with sugar, and pickles are hypertonic with salt. How do hypertonic solutions kill bacteria and fungi that would otherwise spoil these foods?

Active Processes

As stated previously, active processes require the cell to expend energy to move materials across the cytoplasmic membrane against their electrochemical gradient. This is analogous to moving water uphill. Energy stored in molecules of ATP is used for active transport. As we will see, ATP may be utilized directly during transport, or indirectly at some other site and at some other time. Active processes in prokaryotes include *active transport* by means of carrier proteins and a special process termed *group translocation.*

Active Transport Like facilitated diffusion, **active transport** utilizes transmembrane carrier proteins; however, the functioning of active transport carrier proteins requires the cell to expend ATP to transport molecules across the membrane. Some carrier proteins are referred to as *gated channels* or *ports* because they are controlled. When the cell is in need of a substance, the carrier protein becomes functional (the gate "opens"). At other times the gate is "closed."

If only one substance is transported at a time, the carrier protein is called a *uniport* (**Figure 3.19a** on page 75). In contrast, *antiports* simultaneously transport two chemicals, but in opposite directions; that is, one substance is transported into the cell at the same time that a second substance is transferred

[11]From Greek *isos,* meaning equal, and *tonos,* meaning tone.
[12]From Greek *hyper,* meaning more or over.
[13]From Greek *hypo,* meaning less or under.

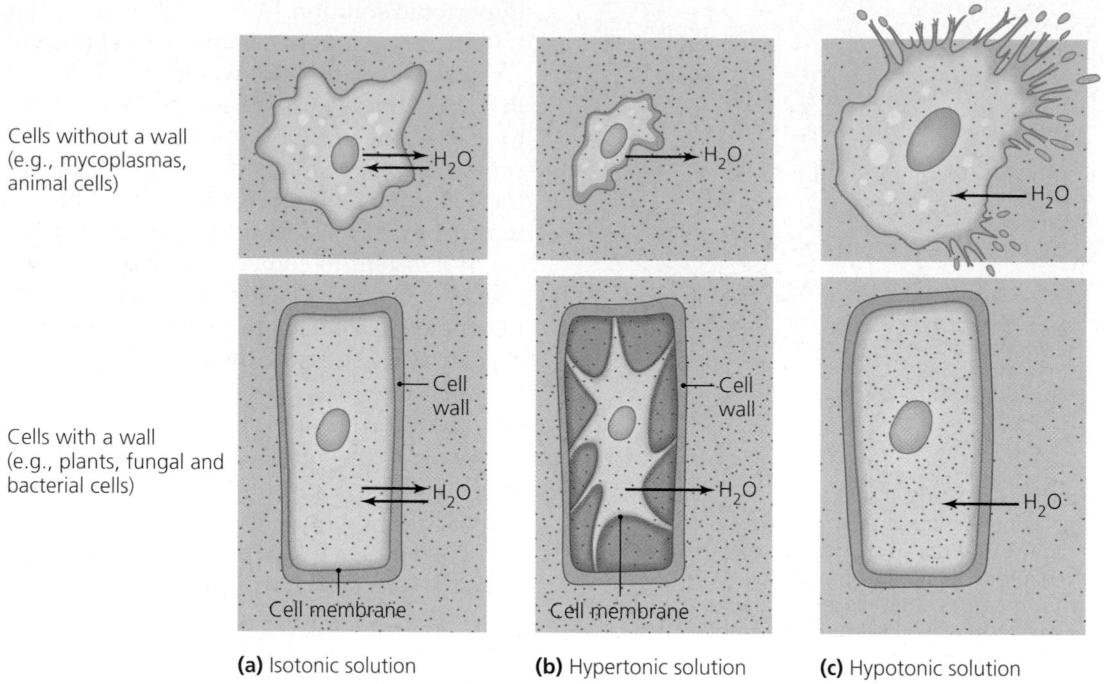

▲ ***Figure 3.18***

Effects of isotonic, hypertonic, and hypotonic solutions on cells. **(a)** Cells in isotonic solutions experience no net movement of water. **(b)** Cells in hypertonic solutions shrink due to the net movement of water out of the cell. **(c)** Cells in hypotonic solutions undergo a net gain of water. Animal cells burst because they lack a cell wall; in cells with a cell wall, the pressure of water pushing against the interior of the wall eventually stops the movement of water into the cell.

out of the cell **(Figure 3.19b).** In other types of active transport, two substances move together in the same direction across the membrane by means of a single carrier protein. Such carrier proteins are known as *symports* **(Figure 3.19c).**

In all cases, active transport moves substances against their electrochemical gradient. Typically, the carrier protein breaks down ATP into ADP and inorganic phosphate during transport. ATP releases energy that is used to move the chemical against its electrochemical gradient across the membrane.

With symports and antiports, one chemical's electrochemical gradient may provide the energy needed to transport the second chemical. For example, H^+ moving into a cell down its electrochemical gradient by facilitated diffusion provides energy to carry glucose into the cell, against the glucose gradient. The two processes are linked by a symport. However, ATP is still utilized for transport because the H^+ gradient was previously established by the active pumping of H^+ to the outside of the cell by an ATP-dependent H^+ uniport. The use of ATP is thus separated in time and space from the active transport of glucose, but ATP was still expended.

Group Translocation **Group translocation** is an active process that occurs only in some prokaryotes. In group translocation, the substance being actively transported across the membrane is chemically changed during transport **(Figure 3.20).** The membrane is impermeable to the altered substance, trapping it inside the cell. Group translocation is very efficient at bringing substances into a cell. It can operate efficiently even if the external concentration of the chemical being transported is as low as 1 part per million (ppm).

One well-studied example of group translocation is the accumulation of glucose inside a bacterial cell. As glucose is transported across the bacterial cell membrane, it is phosphorylated; that is, a phosphate group is added to the glucose. The glucose is changed into glucose-6-phosphate, a sugar that cannot cross back out, but can be utilized in the ATP-producing metabolism of the cell. Other carbohydrates, fatty acids, purines, and pyrimidines are also brought into bacterial cells by group translocation. A summary of prokaryotic transport processes is shown in **Table 3.2** on page 76.

Cytoplasm of Prokaryotes

Cytoplasm is the general term used to describe the semiliquid, gelatinous material inside a cell. Cytoplasm is semitransparent, fluid, elastic, and aqueous. It is composed of cytosol, inclusions, ribosomes, and a cytoskeleton.

Figure 3.19 ▶
Mechanisms of active transport.
(a) Via a uniport. **(b)** Via an antiport. **(c)** Via a uniport coupled with a symport. *What is the usual source of energy for active transport?*

Figure 3.19 ATP is the usual source of energy for active transport processes.

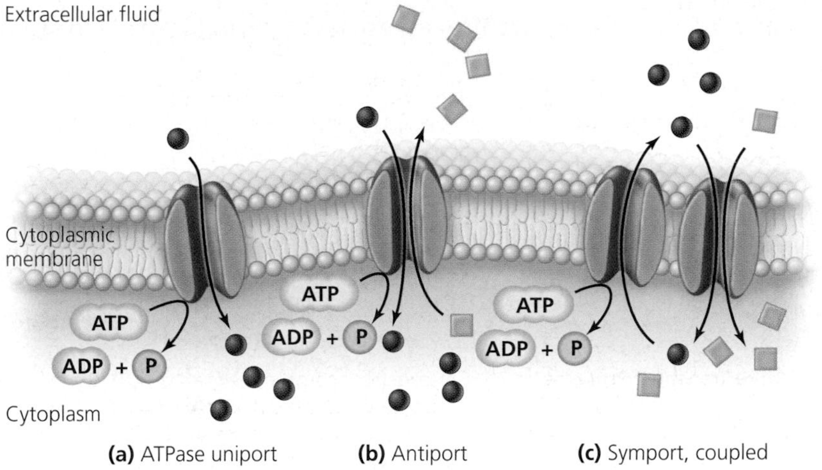

▼ ***Figure 3.20***
Group translocation. This process involves a chemical change in a substance as it is being transported. This figure depicts glucose being transported into a bacterial cell via group translocation.

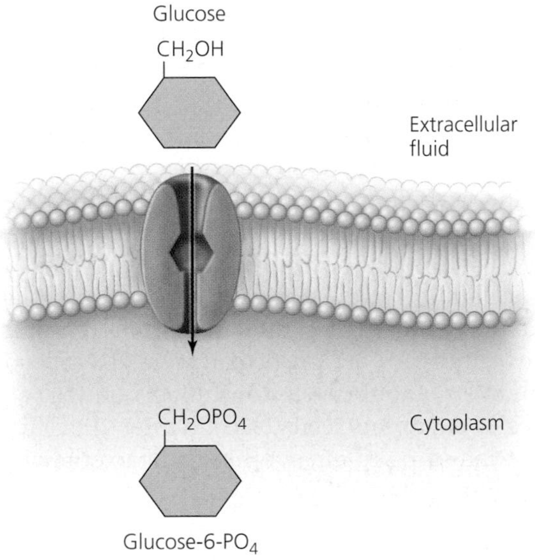

Cytosol

Learning Objective

✓ Describe cytoplasm and its basic contents.

The liquid portion of the cytoplasm is called **cytosol.** It is mostly water, but it also contains dissolved and suspended substances, including ions, carbohydrates, proteins (mostly enzymes), lipids, and wastes. The cytosol of prokaryotes also contains the cell's DNA in a region called the **nucleoid.** Most prokaryotes have a single, circular chromosome that is not surrounded by a nuclear membrane. *Vibrio cholerae* (vib′rē-ō kol′er-ē), the bacterium that causes cholera, is an unusual prokaryote in that it has two chromosomes.

The cytosol is the site of some chemical reactions. For example, enzymes within the cytosol function to produce amino acids.

Inclusions

Deposits, called **inclusions,** are often found within the cytosol of prokaryotes. Inclusions may include reserve deposits of lipids, starch, or compounds containing nitrogen, phosphate, or sulfur (**Figure 3.21** on page 76). Such chemicals may be taken in and stored in the cytosol when nutrients are in abundance and then utilized when nutrients are scarce. The presence of inclusions is diagnostic for several pathogenic bacteria.

Many aquatic cyanobacteria (blue-green photosynthetic prokaryotes) contain inclusions called *gas vesicles* that store gases in protein sacs. The gases buoy the cells to the surface and into the light needed for photosynthesis. Other interesting inclusions are the small crystals of magnetite stored by *magnetobacteria,* featured at the beginning of this chapter.

Nonmembranous Organelles

Learning Objective

✓ Describe the structure and function of ribosomes and the cytoskeleton.

As previously noted, prokaryotes lack the membrane-bound organelles found in eukaryotic cells. However, two types of *nonmembranous organelles* are found in direct contact with the cytosol in prokaryotic cytoplasm. Some scientists do not consider them to be true organelles because they lack a membrane, but in this text they are considered a type of organelle. Nonmembranous organelles in prokaryotes include ribosomes and the cytoskeleton.

Ribosomes

Ribosomes are the sites of protein synthesis in cells. Prokaryotic cells have thousands of ribosomes in their cytoplasm, which gives cytoplasm a grainy appearance. The approximate size of ribosomes—and indeed other cellular

Table 3.2 Transport Processes Across Prokaryotic Cytoplasmic Membranes

	Description	Substances Transported
Passive Transport Processes	Processes require no use of energy by the cell; the electrochemical gradient provides energy.	
Diffusion	Molecules move down their electrochemical gradient through the phospholipid bilayer of the membrane.	Oxygen, carbon dioxide, lipid-soluble chemicals
Facilitated diffusion	Molecules move down their electrochemical gradient through channels or carrier proteins.	Glucose, fructose, urea, some vitamins
Osmosis	Water molecules move down their concentration gradient across a selectively permeable membrane.	Water
Active Transport Processes	Cell expends energy in the form of ATP to move a substance against its electrochemical gradient. Some active processes are linked to a passive process and use the energy of the passive process to carry a substance across the membrane.	
Active transport	ATP-dependent carrier proteins bring substances into cell.	Na^+, K^+, Ca^{2+}, H^+, Cl^-
Group translocation	The substance is chemically altered during transport; found only in some prokaryotes.	Glucose, mannose, fructose

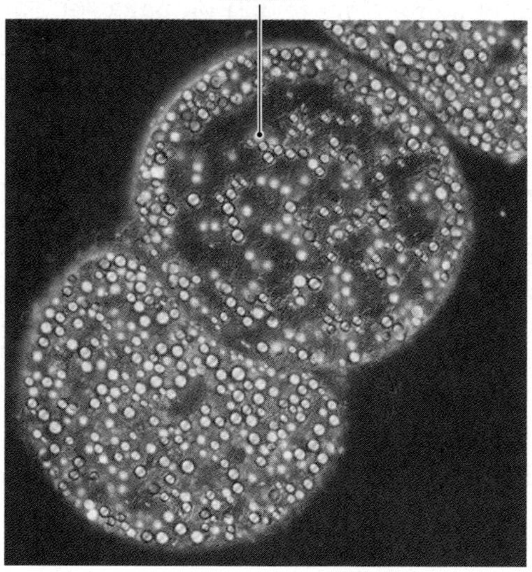

▲ ***Figure 3.21***
Cellular inclusions. Shown here are sulfur globules in the purple sulfur bacterium *Chromatium buderi.*

structures—is expressed in Svedbergs (S),[14] and is determined by their sedimentation rate—the rate at which they move to the bottom of a test tube during centrifugation. As you might expect, large, compact, heavy particles sediment faster than small, loosely packed, or light ones, and are assigned a higher number. Prokaryotic ribosomes are 70S; in contrast, the larger ribosomes of eukaryotes are 80S.

All ribosomes are composed of two subunits, each of which is composed of protein and a type of RNA called **ribosomal RNA (rRNA).** The subunits of prokaryotic 70S ribosomes are a smaller 30S subunit and a larger 50S subunit; the 30S subunit contains protein and a single rRNA molecule, whereas the 50S subunit has two rRNA molecules. Because sedimentation rates are dependent not only on mass and size but also on shape, the sedimentation rates of subunits do not add up to the sedimentation rate of a whole ribosome.

Many antibiotics act on prokaryotic 70S ribosomes or their subunits without deleterious effects on the larger 80S ribosomes of eukaryotic cells (see Chapter 10). This is why antibiotics can stop protein synthesis in bacteria without affecting protein synthesis in a patient.

Cytoskeleton

Most cells contain an internal network of fibers called a **cytoskeleton** that plays a role in forming a cell's basic shape. Prokaryotes were long thought to lack cytoskeletons, but recent research has revealed that rod-shaped prokaryotes have a simple one **(Figure 3.22).** Spherical prokaryotes appear to lack cytoskeletons.

To this point, we have discussed basic features of bacterial and archaeal prokaryotic cells. Chapter 11 discusses the classification of prokaryotic organisms in more detail. Next we turn our attention to eukaryotic cells.

External Structures of Eukaryotic Cells

Eukaryotic cells have many external structural similarities to prokaryotic cells, and some uniquely eukaryotic features as well. In this section we discuss the structure and function of eukaryotic glycocalyces, flagella, and cilia.

[14] Svedberg units are named for Theodor Svedberg, Nobel Prize winner and inventor of the ultracentrifuge.

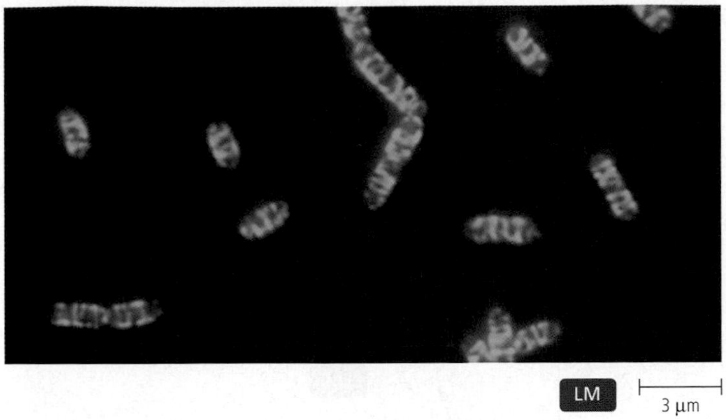

▲ *Figure 3.22*
The simple helical cytoskeleton of the rod-shaped prokaryote *Bacillus subtilis*. This cytoskeleton, which is composed of only a single protein, has been stained with a fluorescent dye.

Glycocalyces

Learning Objective

✓ Describe the composition, function, and importance of eukaryotic glycocalyces.

Animal and protozoan cells lack cell walls, but they have sticky carbohydrate glycocalyces that are anchored to their cytoplasmic membranes by covalently bonding to membrane proteins and lipids. The functions of eukaryotic glycocalyces, which are never as structurally organized as prokaryotic capsules, include helping to anchor animal cells to each other, strengthening the cell surface, providing some protection against dehydration, and functioning in cell-to-cell recognition and communication. Glycocalyces are absent in eukaryotic cells that have cell walls.

Flagella

Learning Objective

✓ Compare and contrast the structure and function of prokaryotic and eukaryotic flagella.

Structure and Arrangement

Flagella of eukaryotes (**Figure 3.23a** on page 78) differ structurally from flagella of prokaryotes. The shaft of a eukaryotic flagellum is composed of molecules of a globular protein called *tubulin* arranged in chains to form hollow *microtubules* **(Figure 3.23c).** Nine pairs of microtubules surround two microtubules in the center. This "9 + 2" arrangement of microtubules is common to all flagellated eukaryotic cells, whether they are found in protozoa, algae, animals, or plants. The filaments of eukaryotic flagella are anchored to the cell by a basal body, but no hook connects the two parts, as in prokaryotes. The basal body has *triplets* of microtubules instead of pairs, and there are no microtubules in the center, so it has a "9 + 0" arrangement of microtubules.

Another difference between the flagella of eukaryotes and prokaryotes is that eukaryotic flagella are surrounded by an extension of the cell membrane and are filled with cytosol. Thus, the flagella of eukaryotes are inside the cell, not extensions outside the cell. Eukaryotic flagella may be single or multiple and are generally found at one pole of the cell.

Function

The flagella of eukaryotes also move differently from those of prokaryotes. Rather than rotating like prokaryotic flagella, those of eukaryotes undulate rhythmically (**Figure 3.24a** on page 79). Some eukaryotic flagella push the cell through the medium (as occurs in animal sperm), whereas others pull the cell through the medium (as occurs in many protozoa). Positive and negative phototaxis and chemotaxis are seen in eukaryotic cells, but such cells do not move in runs and tumbles.

Cilia

Learning Objectives

✓ Describe the structure and function of cilia.
✓ Compare and contrast eukaryotic cilia and flagella.

Some eukaryotic cells move by means of hairlike structures called **cilia,** which cover the surface of the cell and are shorter and more numerous than flagella (see **Figure 3.23b**). No prokaryotic cells have cilia. Like flagella, cilia are composed primarily of tubulin microtubules, which are arranged in a "9 + 2" arrangement of pairs in their shafts and a "9 + 0" arrangement of triplets in their basal bodies (see **Figure 3.23c**).

A single cell may have hundreds or even thousands of cilia. Cilia beat rhythmically, much like a swimmer doing a butterfly stroke **(Figure 3.24b).** Coordinated beating of cilia propels single-celled eukaryotes through their environment. Cilia are also used within multicellular eukaryotes to move substances in the local environment past the surface of the cell. For example, such movement of cilia helps cleanse the human respiratory tract of dust and microorganisms.

Eukaryotic Cell Walls and Cytoplasmic Membranes

Learning Objectives

✓ Compare and contrast prokaryotic and eukaryotic cell walls and cytoplasmic membranes.
✓ Contrast exocytosis and endocytosis.
✓ Describe the role of pseudopodia in eukaryotic cells.

The eukaryotic cells of fungi, algae, and plants have cell walls. Recall that glycocalyces are absent from eukaryotes with cell walls; instead, the cell wall takes on one of the functions of a glycocalyx by providing protection from the environment. The wall also provides shape and support against osmotic pressure. Eukaryotic cell walls are composed of various polysaccharides, but not the peptidoglycan seen in the walls of most bacteria.

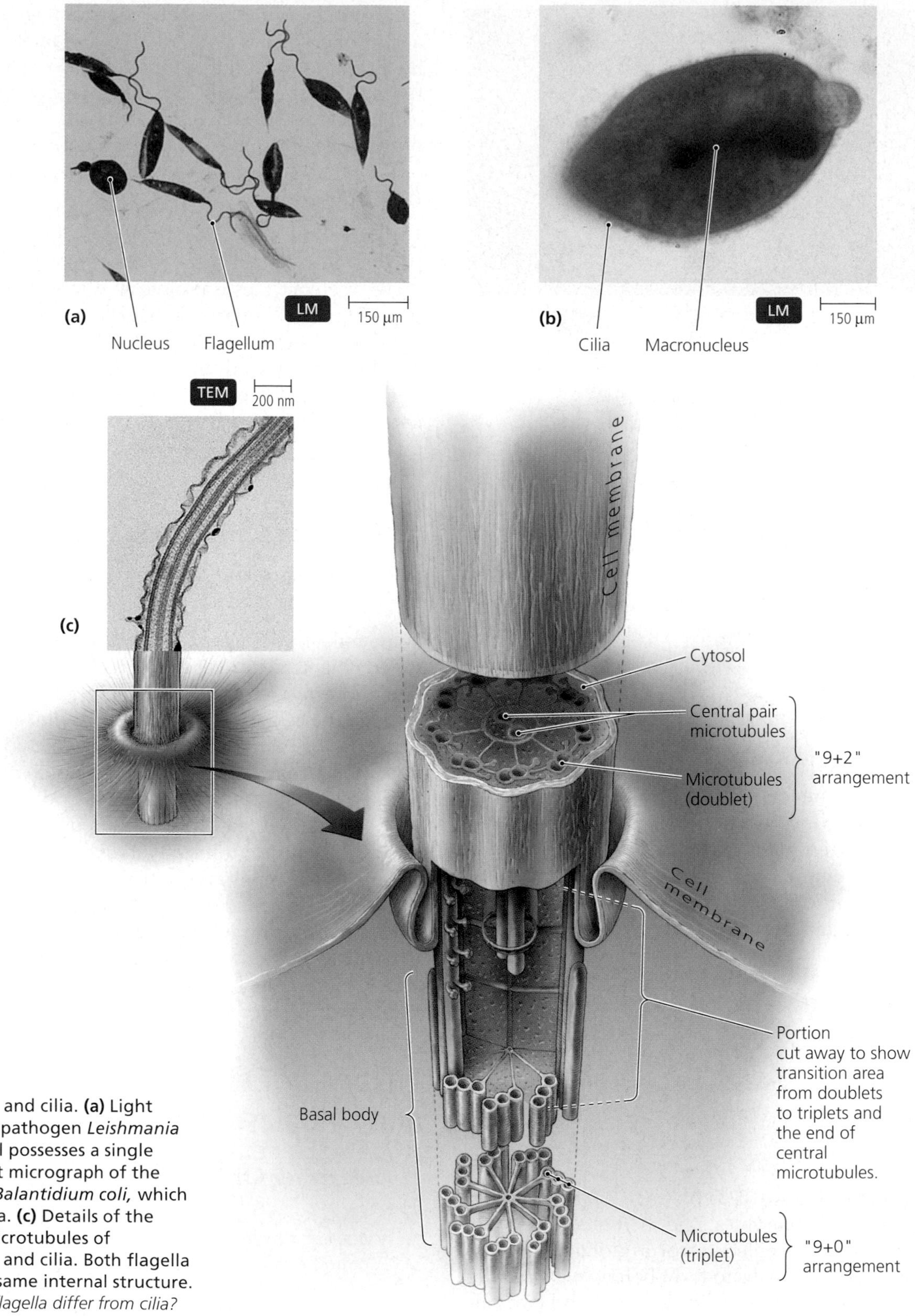

Figure 3.23 ➤

Eukaryotic flagella and cilia. **(a)** Light micrograph of the pathogen *Leishmania donovani;* each cell possesses a single flagellum. **(b)** Light micrograph of the human pathogen *Balantidium coli,* which is covered with cilia. **(c)** Details of the arrangement of microtubules of eukaryotic flagella and cilia. Both flagella and cilia have the same internal structure. *How do eukaryotic flagella differ from cilia?*

Figure 3.23 *Flagella are longer and less numerous than cilia.*

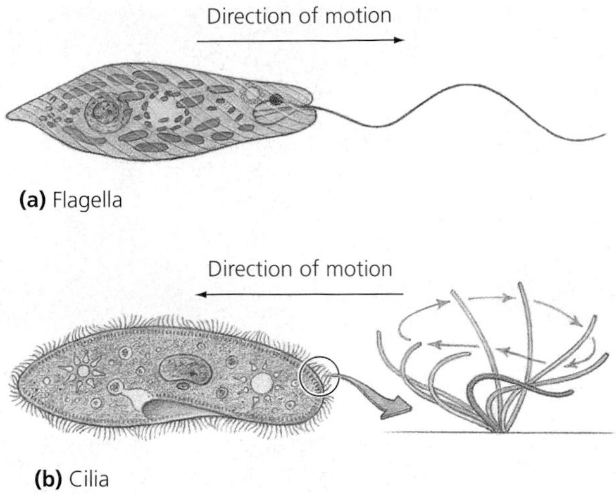

▲ ***Figure 3.24***

Movement of eukaryotic flagella and cilia. **(a)** Eukaryotic flagella undulate in waves that begin at one end and traverse the length of the flagellum. **(b)** Cilia move with a power stroke followed by a return stroke. In the power stroke a cilium is stiff; it relaxes during the return stroke. *How is the movement of eukaryotic flagella different from that of prokaryotic flagella?*

Figure 3.24 Eukaryotic flagella undulate in a wave that moves down the flagellum; the flagella of prokaryotes rotate about the basal body.

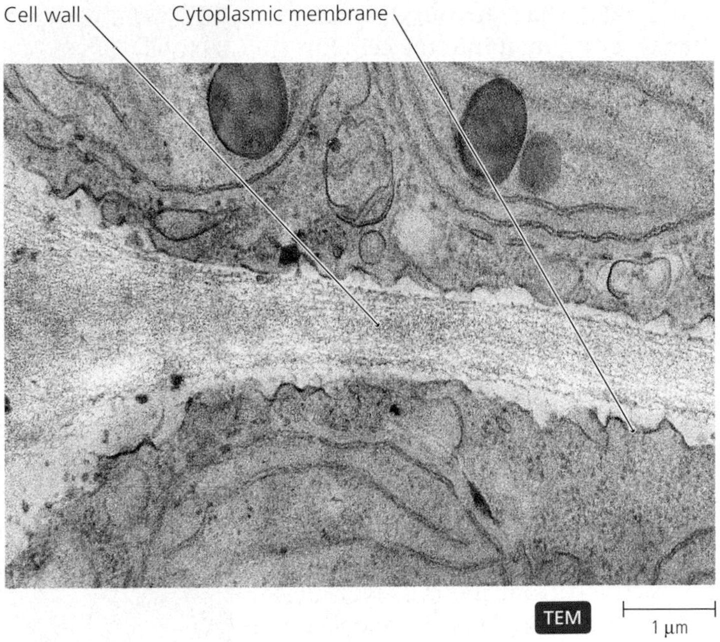

▲ ***Figure 3.25***

A eukaryotic cell wall, as seen in the red alga *Gelidium*. This cell wall is composed of layers of the polysaccharide called agar. *What is the function of a cell wall?*

Figure 3.25 The cell wall provides support, protection, and resistance to osmotic forces.

The walls of plant cells are composed of *cellulose*, a polysaccharide that is familiar to you as paper and dietary fiber. Fungi also have walls of polysaccharides, including cellulose, *chitin*, and/or *glucomannan*. The walls of algae **(Figure 3.25)** are composed of a variety of polysaccharides and other chemicals, depending on the type of alga. These chemicals include cellulose, *agar*, *carrageenan*, *silicates*, *algin*, calcium carbonate, or a combination of these substances. Fungi and algae are discussed in more detail in Chapter 12.

All eukaryotic cells have cytoplasmic membranes **(Figure 3.26).** (Botanists call the cytoplasmic membrane of an algal or plant cell a *plasmalemma*.) A eukaryotic cytoplasmic membrane, like those of bacteria, is a fluid mosaic of phospholipids and proteins, which act as recognition molecules, enzymes, receptors, carriers, or channels. Additionally, within multicellular animals some membrane proteins serve to anchor cells to each other.

Unlike most bacterial membranes, eukaryotic membranes contain steroid lipids *(sterols)*, such as cholesterol in animal cells, that help maintain membrane fluidity. Paradoxically, at high temperatures sterols stabilize a phospholipid bilayer by making it less fluid, but at low temperatures sterols have the opposite effect—they prevent phospholipid packing, making the membrane more fluid.

Like its prokaryotic counterpart, a eukaryotic cytoplasmic membrane controls the movement of materials into and out of the cell. Eukaryotic cytoplasmic membranes use both passive processes (diffusion, facilitated diffusion, and osmosis; see Figure 3.16) and active transport (see Figure 3.19). Eukaryotic membranes do not perform group translocation, which occurs only in some prokaryotes, but many perform another type of active transport—**endocytosis** (**Figure 3.27** on page 80), which involves physical manipulation of the cytoplasmic membrane around the cytoskeleton. Endocytosis occurs when the membrane distends to form **pseudopodia**

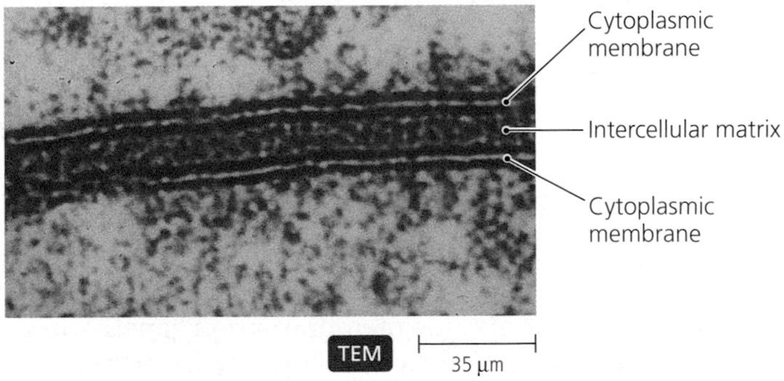

▲ ***Figure 3.26***

Eukaryotic cytoplasmic membrane. Note that this micrograph depicts the cytoplasmic membranes of two adjoining cells.

(false feet) that surround a substance, bringing it into the cell. Endocytosis is termed **phagocytosis** if a solid is brought into the cell, and **pinocytosis** if only liquid is brought into the cell. Nutrients brought into a cell by endocytosis are then enclosed in a *food vesicle*. Vesicles and digestion of the nutrients they contain are discussed in more detail shortly. The process of phagocytosis is more fully discussed in Chapter 15 as it relates to the defense of the body against disease.

Exocytosis, another eukaryotic process, is the reverse of endocytosis in that it enables substances to be exported from the cell. Not all eukaryotic cells can perform endocytosis or exocytosis.

Some eukaryotes also use pseudopodia as a means of locomotion. The cell extends a pseudopod and then the cytoplasm streams into it, a process called *amoeboid action* **(Figure 3.28).**

Table 3.3 lists some of the features of endocytosis and exocytosis.

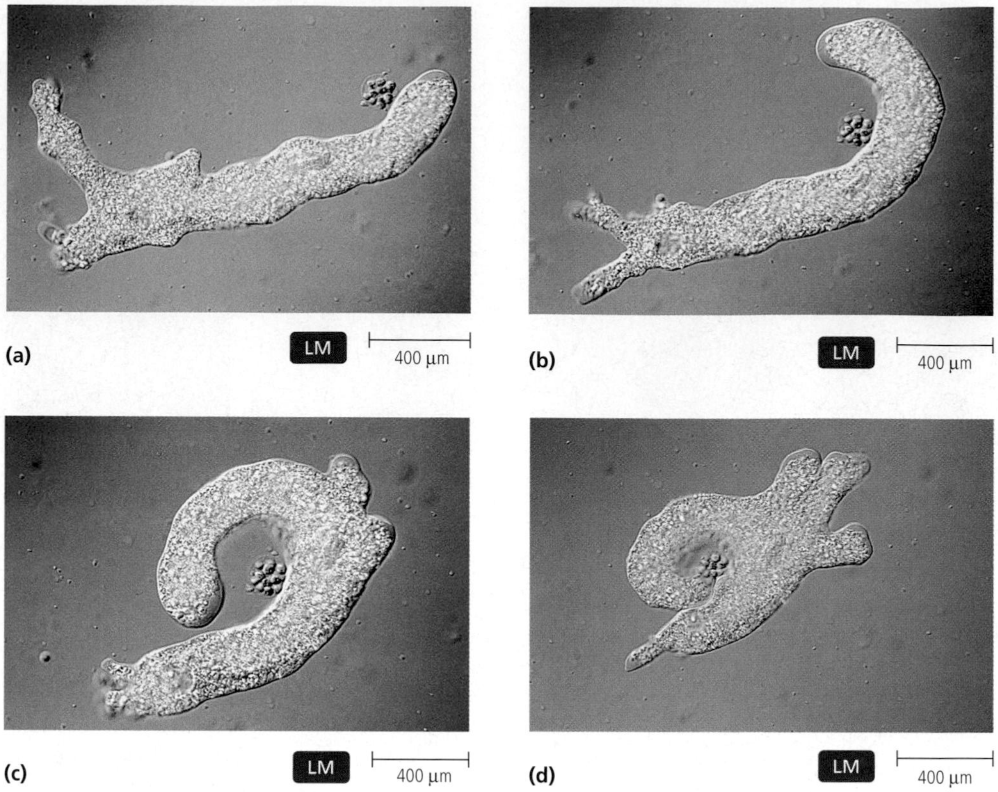

▲ ***Figure 3.27***

Endocytosis. Pseudopodia extend to surround solid and/or liquid nutrients, which become incorporated into a food vesicle inside the cytoplasm. *What is the difference between phagocytosis and pinocytosis?*

Figure 3.27 Phagocytosis is endocytosis of a solid; pinocytosis is endocytosis of a liquid.

Cytoplasm of Eukaryotes

Learning Objectives

✓ Compare and contrast the cytoplasm of prokaryotes and eukaryotes.

✓ Identify nonmembranous and membranous organelles.

The cytoplasm of eukaryotic cells is more complex than that of either bacteria or archaea. The most distinctive difference is the presence of numerous membranous organelles in eukaryotes. However, before we discuss these membranous organelles, we will briefly consider the nonmembranous organelles in eukaryotes.

Nonmembranous Organelles

Learning Objectives

✓ Describe the structure and function of ribosomes, cytoskeletons, and centrioles.

✓ Compare and contrast the ribosomes of prokaryotes and eukaryotes.

✓ List and describe the three filaments of a eukaryotic cytoskeleton.

Here we discuss three nonmembranous organelles found in eukaryotes: ribosomes and cytoskeleton (both of which are also present in prokaryotes), and centrioles (which are present only in certain kinds of eukaryotic cells).

Ribosomes

The cytosol of eukaryotes, like that of prokaryotes, is a semitransparent fluid composed primarily of water containing dissolved and suspended proteins, ions, carbohydrates, lipids, and wastes. Within the cytosol of eukaryotic cells are ribosomes that are larger than prokaryotic ribosomes; instead of 70S ribosomes, eukaryotic ribosomes are 80S and are composed of 60S and 40S subunits. (Chapter 10 discusses many antibiotics that inhibit protein synthesis by 70S prokaryotic ribosomes without adversely affecting protein synthesis by the 80S ribosomes in eukaryotic cells.) In addition to the ribosomes found within the cytosol, many eukaryotic ribosomes are attached to the membranes of the *endoplasmic reticulum* (discussed shortly).

Table 3.3 Active Transport Processes Found Only in Eukaryotes: Endocytosis and Exocytosis

	Description	Substances Transported
Endocytosis: phagocytosis and pinocytosis	Substances are surrounded by pseudopodia and brought into the cell. Phagocytosis involves solid substances; pinocytosis involves liquids.	Bacteria, viruses, aged and dead cells; liquid nutrients in extracellular solutions
Exocytosis	Vesicles containing substances are fused with cytoplasmic membrane, dumping their contents to the outside.	Wastes, secretions

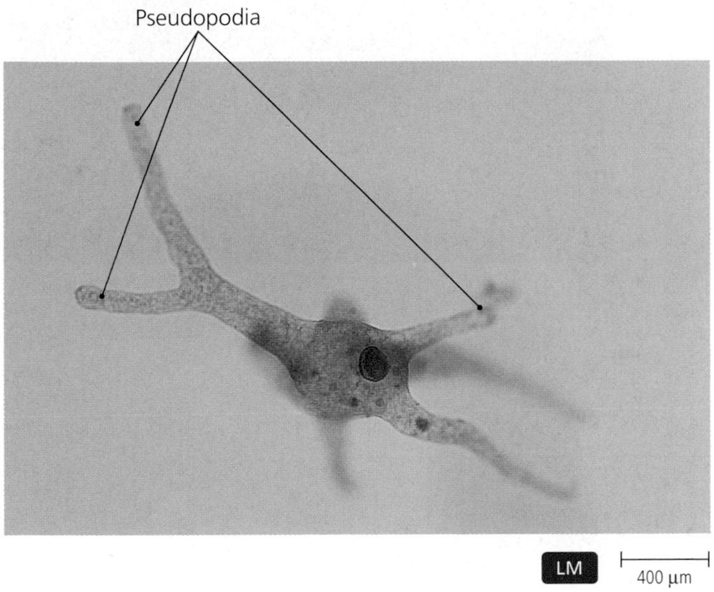

▲ ***Figure 3.28***
Amoeboid action. Pseudopodia move a cell to a new location. This micrograph is of a cell caught in motion. Pseudopodia protrude from the surface, cytoplasm flows into them, and the cell progresses across the surface.

Cytoskeleton

Eukaryotic cells contain an extensive cytoskeleton composed of an internal network of fibers and tubules. The eukaryotic cytoskeleton acts to anchor organelles, functions in cytoplasmic streaming and in movement of organelles within the cytosol, enables contraction of the cell, moves the cell membrane during endocytosis and amoeboid action, and provides the basic shape of many cells.

The eukaryotic cytoskeleton is made up of *tubulin microtubules* (also found in flagella, cilia, and centrioles), thinner *microfilaments* composed of *actin*, and *intermediate filaments* composed of various proteins (**Figure 3.29** on page 82).

Centrioles and Centrosome

Animal cells and some fungal cells contain two **centrioles,** which lie at right angles to each other near the nucleus, in a region of the cytoplasm called the **centrosome** (**Figure 3.30** on page 82). Plants, algae, fungi, and prokaryotes lack both centrioles and centrosomes. Centrioles are composed of nine *triplets* of microtubules arranged in a way that resembles the "9 + 0" arrangement seen at the base of eukaryotic flagella and cilia.

Centrioles appear to play a role in *mitosis* (nuclear division), *cytokinesis* (cell division), and in the formation of flagella and cilia. However, because many eukaryotic cells that lack centrioles, such as brown algal sperm and numerous one-celled algae, are still able to form flagella and undergo mitosis and cytokinesis, the function of centrioles is the subject of ongoing research.

Membranous Organelles

Learning Objectives

- ✓ Discuss the function of each of the following membranous organelles: *nucleus, ER, Golgi body, lysosome, peroxisome, vesicle, vacuole, mitochondrion,* and *chloroplast.*
- ✓ Label the structures associated with each of the membranous organelles.

Eukaryotic cells contain a variety of organelles that are surrounded by phospholipid bilayer membranes similar to the cytoplasmic membrane. These membranous organelles include the nucleus, endoplasmic reticulum, Golgi body, lysosomes, peroxisomes, vacuoles, vesicles, mitochondria, and chloroplasts. Prokaryotic cells lack these structures.

Nucleus

The **nucleus** is usually spherical to ovoid and is often the largest organelle in a cell[15] (**Figure 3.31** on page 83). Some cells have a single nucleus; others are multinucleate, while still others lose their nuclei. The nucleus is often referred to as "the control center of the cell" because it contains most of the cell's genetic instructions in the form of DNA. Cells that lose their nuclei, such as mammalian red blood cells, can survive for only a few months.

Just as the semiliquid portion of the cell is called cytoplasm, the semiliquid matrix of the nucleus is called **nucleoplasm.** Within the nucleoplasm may be one or more **nucleoli** (singular: *nucleolus*), which are specialized

[15]Historically, the nucleus was not considered an organelle because it is large and not considered part of the cytoplasm.

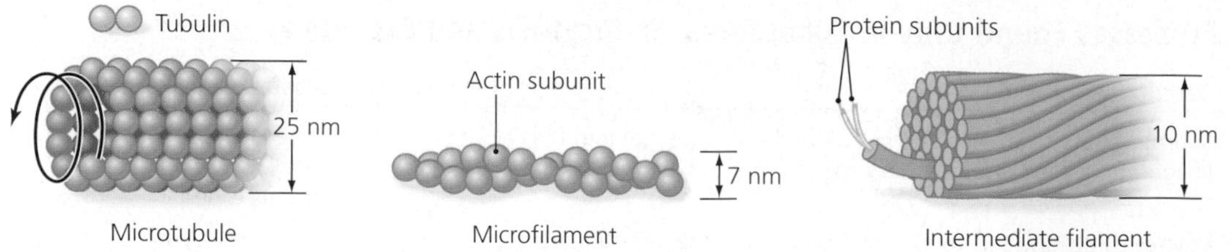

(a)

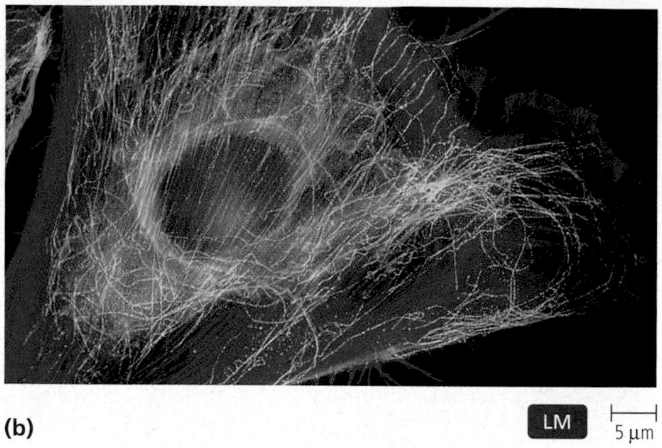

(b)

▲ ***Figure 3.29***

Eukaryotic cytoskeleton. The cytoskeleton of eukaryotic cells serves to anchor organelles, provides a "track" for the movement of organelles throughout the cell, and provides shape to animal cells. Their cytoskeletons are composed of microtubules, microfilaments, and intermediate filaments. **(a)** Artist's rendition of cytoskeleton filaments. **(b)** Various elements of the cytoskeleton shown here have been stained with different fluorescent dyes.

(a)

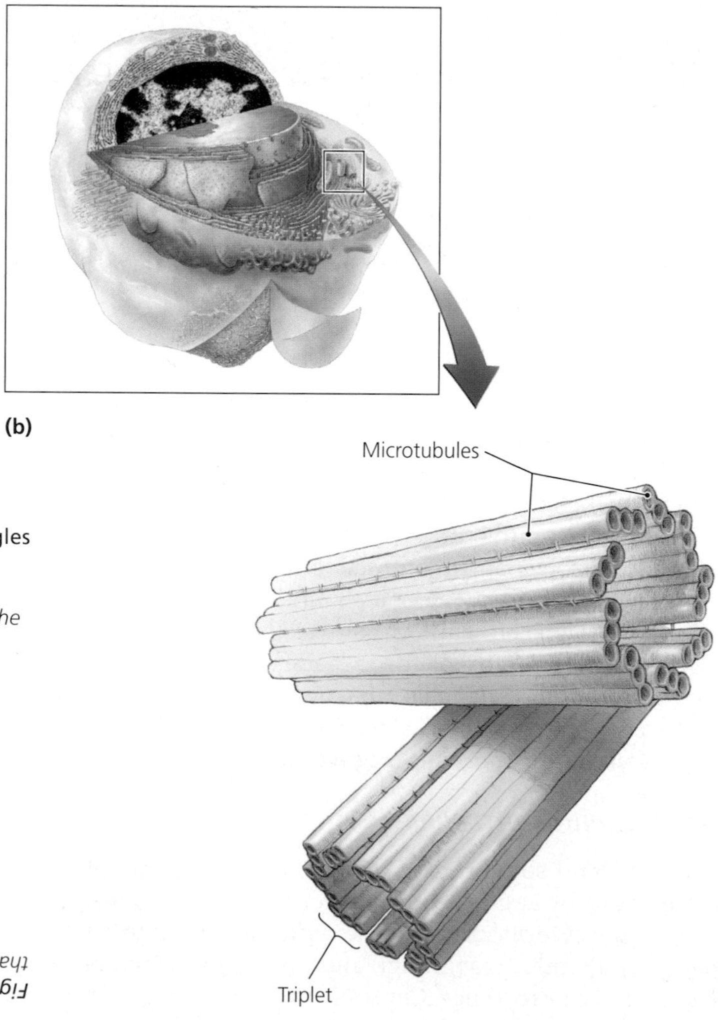

(b)

▲ ***Figure 3.30***

Centrosome. A centrosome is composed of two centrioles at right angles to one another; each centriole has nine triplets of microtubules. **(a)** Transmission electron micrographs of centrosomes and centrioles. **(b)** Artist's rendition of a centrosome. *How do centrioles compare with the basal body and shafts of eukaryotic flagella and cilia (see Figure 3.23)?*

***Figure 3.30** Centrioles have the same "9 + 0" arrangement of microtubules that is found in the basal bodies of eukaryotic cilia and flagella.*

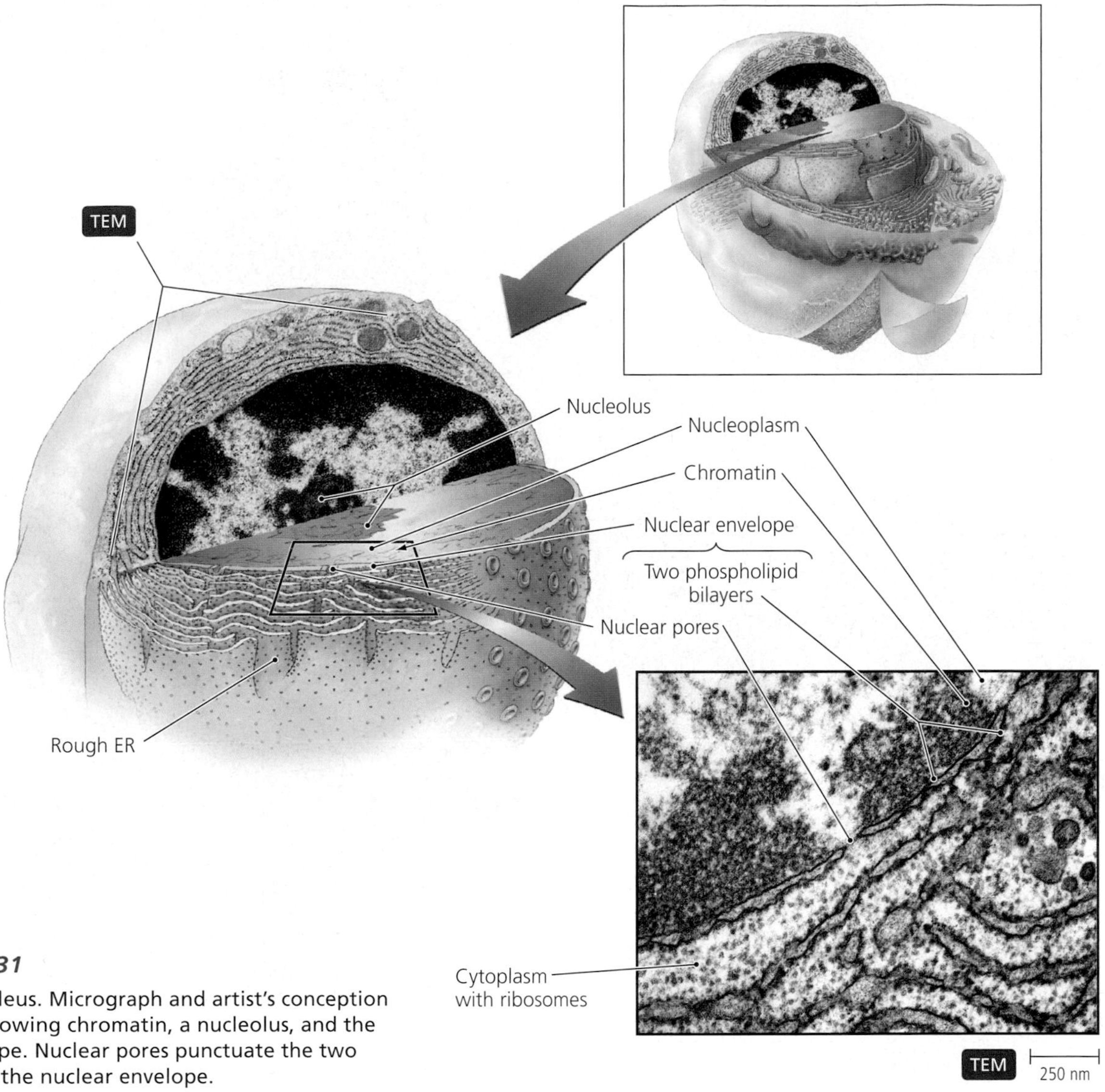

▲ ***Figure 3.31***

Eukaryotic nucleus. Micrograph and artist's conception of a nucleus showing chromatin, a nucleolus, and the nuclear envelope. Nuclear pores punctuate the two membranes of the nuclear envelope.

regions where RNA is synthesized. The nucleoplasm also contains **chromatin,** which is a threadlike mass of DNA associated with special proteins called *histones* that play a role in packaging nuclear DNA. During mitosis, the process of eukaryotic nuclear division, chromatin becomes visible as *chromosomes*. Mitosis is discussed in more detail in Chapter 12.

Surrounding the nucleus is a double membrane called the **nuclear envelope,** which is composed of two phospholipid bilayers, for a total of four phospholipid layers (see Figure 3.31). The nuclear envelope contains **nuclear pores** that function to control the import and export of substances through the envelope.

Endoplasmic Reticulum

Continuous with the outer membrane of the nuclear envelope is a netlike arrangement of hollow tubules called **endoplasmic reticulum (ER)** (**Figure 3.32** on page 84). The ER traverses the cytoplasm of eukaryotic cells. Endoplasmic reticulum functions as a transport system and is found in two forms: **smooth endoplasmic reticulum (SER)** and **rough endoplasmic reticulum (RER).** SER plays a role in lipid synthesis as well as transport. Rough endoplasmic reticulum is rough because ribosomes adhere to its outer surface. Proteins produced by ribosomes on the RER are inserted into the lumen (central canal) of the RER and transported throughout the cell.

Golgi Body

A **Golgi**[16] **body** is like the shipping department of a cell: it receives, processes, and packages large molecules for export from the cell (**Figure 3.33** on page 84). The Golgi body packages secretions in sacs called **secretory vesicles,** which then fuse with the cytoplasmic membrane before dumping their

[16]Golgi was an Italian histologist who first described the organelle in 1898. This organelle is also known as a Golgi complex or Golgi apparatus, and in plants and algae as a dictyosome.

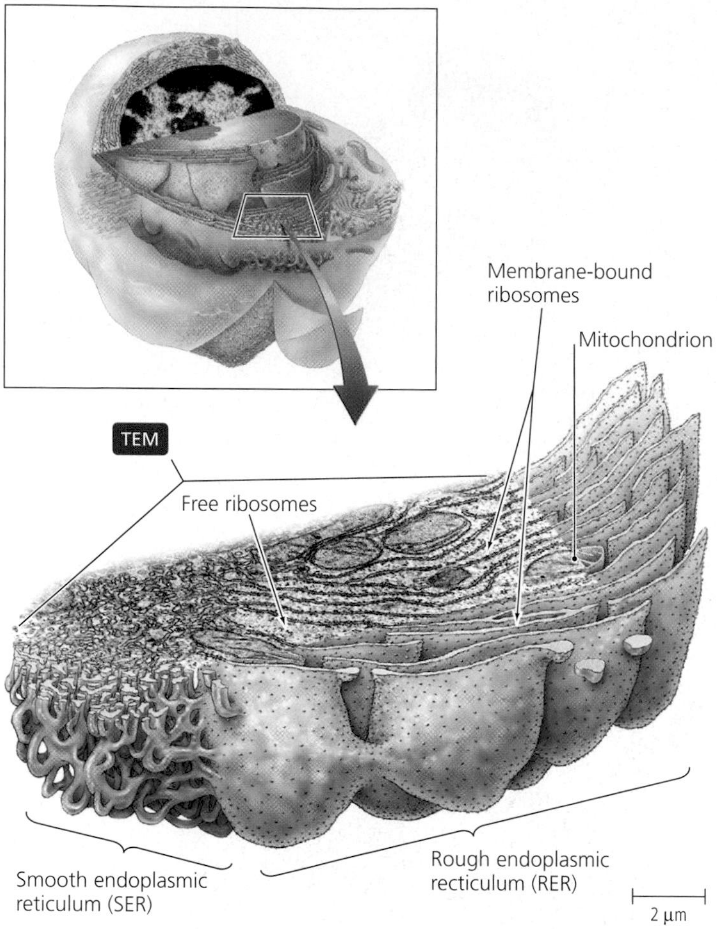

▲ ***Figure 3.32***

Endoplasmic reticulum. ER functions in transport throughout the cell. Ribosomes are on the surface of rough ER; smooth ER lacks ribosomes.

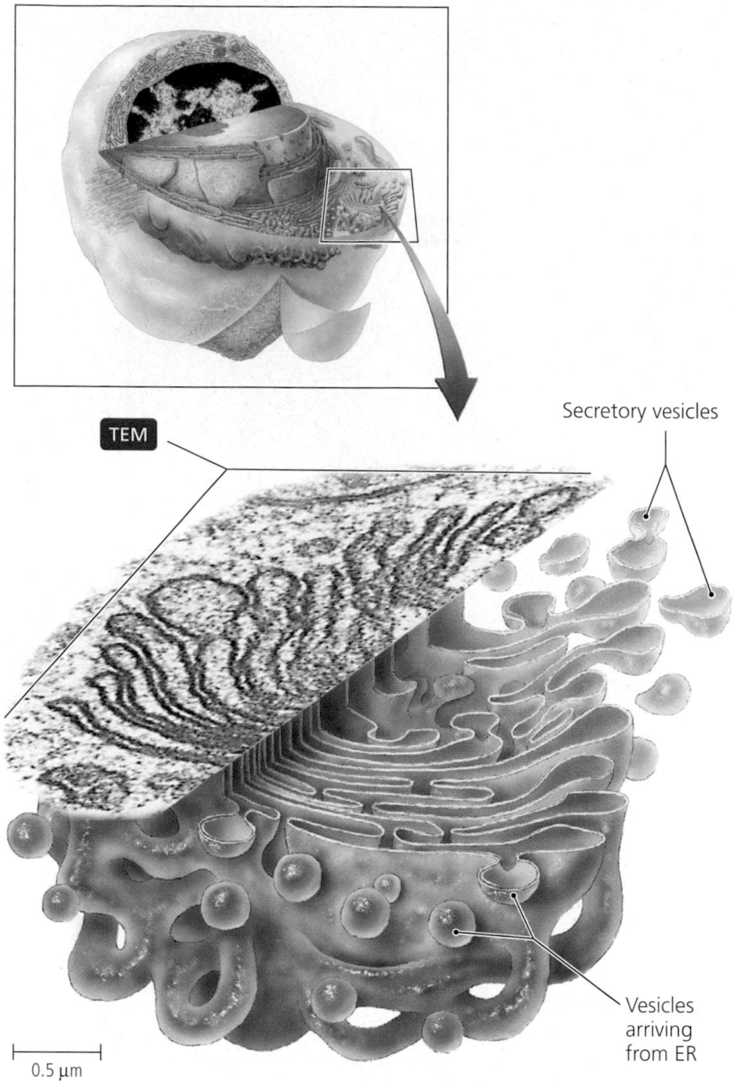

▲ ***Figure 3.33***

Golgi body. A Golgi body is composed of flattened sacs. Proteins synthesized by ribosomes on RER are transported via vesicles to a Golgi body. The Golgi then modifies the proteins and sends them via secretory vesicles to the cytoplasmic membrane, where they can be secreted from the cell by exocytosis.

contents outside the cell via exocytosis. Golgi bodies are composed of a series of flattened hollow sacs that are circumscribed by a phospholipid bilayer.

Lysosomes, Peroxisomes, Vacuoles, and Vesicles

Lysosomes, peroxisomes, vacuoles, and vesicles are membranous sacs that function to store and transfer chemicals within eukaryotic cells. Both **vesicle** and **vacuole** are general terms for such sacs. Large vacuoles are found in plant and algal cells that store starch, lipids, and other substances in the center of the cell. Often a central vacuole is so large that the rest of the cytoplasm is pressed against the cell wall in a thin layer **(Figure 3.34).**

Lysosomes, which are found in animal cells, contain catabolic enzymes that damage the cell if they are released from their packaging into the cytosol. The enzymes are used during the self-destruction of old, damaged, and diseased cells, and to digest nutrients that have been phagocytized. For example, the digestive enzymes in lysosomes are utilized by white blood cells to destroy phagocytized pathogens **(Figure 3.35).**

Peroxisomes are vesicles that contain *oxidase* and *catalase*, which are enzymes that degrade poisonous metabolic wastes (such as free radicals and hydrogen peroxide) resulting from some oxygen-dependent reactions. Peroxisomes are found in all eukaryotic cells but are especially prominent in the kidney and liver cells of mammals.

Mitochondria

Mitochondria are spherical to elongated structures found in most eukaryotic cells (**Figure 3.36** on page 86). Like nuclei, they have two membranes, each composed of a phos-

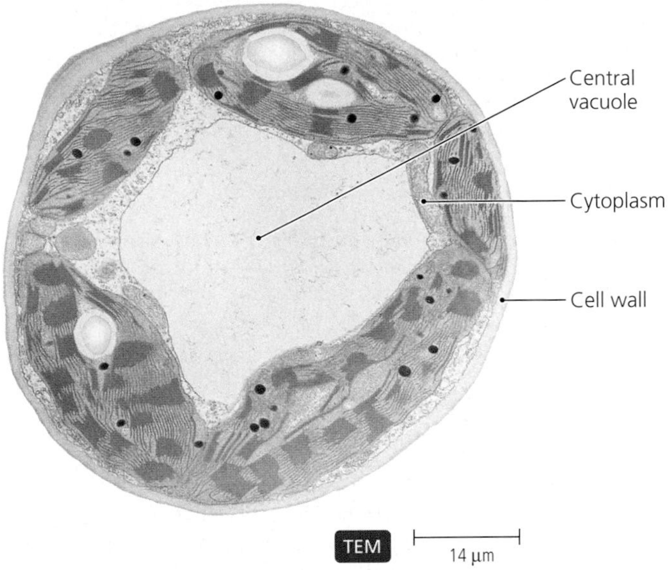

▲ ***Figure 3.34***

Vacuole. The large central vacuole of a plant cell, which constitutes a storehouse for the cell, presses the cytoplasm against the cell wall.

pholipid bilayer. The inner membrane is folded into numerous **cristae** that increase the inner membrane's surface area. Mitochondria are often called the "powerhouses of the cell" because their cristae produce most of the ATP in the cell. The chemical reactions that produce ATP are discussed in Chapter 5.

The interior matrix of a mitochondrion contains small ("prokaryotic") 70S ribosomes and a circular molecule of DNA. This DNA contains genes for some RNA molecules and for a few mitochondrial polypeptides that are manufactured by mitochondrial ribosomes; however, most mitochondrial proteins are coded by nuclear DNA and synthesized by cytoplasmic ribosomes.

Chloroplasts

Chloroplasts are light-harvesting structures found in photosynthetic eukaryotes (**Figure 3.37** on page 86). Like mitochondria and the nucleus, chloroplasts have two phospholipid bilayer membranes and DNA. Further, like mitochondria, chloroplasts can synthesize a few polypeptides with their own 70S ribosomes. The pigments of chloroplasts gather light energy to produce ATP and form sugar from carbon dioxide. Numerous membranous sacs called *thylakoids* form an extensive surface area for the biochemical and photochemical reactions of chloroplasts. The fluid between the thylakoids and the inner membrane is called the *stroma*.

Photosynthetic prokaryotes lack chloroplasts and instead have infoldings of their cytoplasmic membranes called *photosynthetic lamellae*. The details of photosynthesis are discussed in Chapter 5.

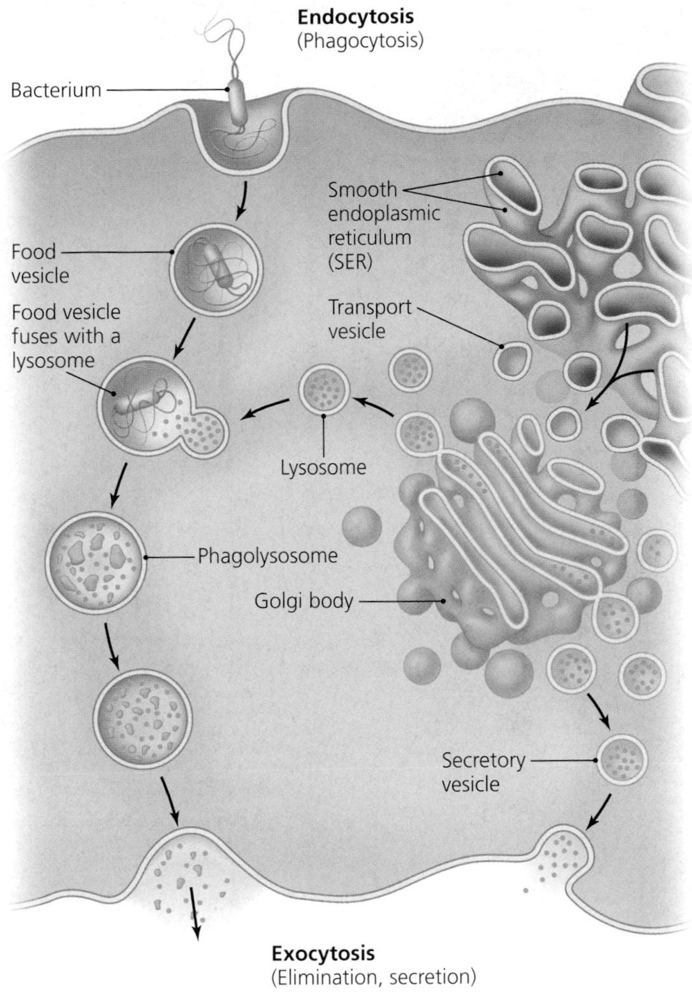

▲ ***Figure 3.35***

The roles of vesicles in the destruction of a phagocytized pathogen within a white blood cell. Before endocytosis, vesicles from the SER deliver digestive enzymes to the Golgi body, which then packages them into lysosomes. During endocytosis, phagocytized particles (in this case, a bacterium) are enclosed within a food vesicle, which then fuses with a lysosome to form a phagolysosome. Once digestion within the phagolysosome is complete, the resulting wastes can be expelled from the cell via exocytosis.

The functions of the nonmembranous and membranous organelles, and their distribution among prokaryotic and eukaryotic cells, are summarized in **Table 3.4** on page 87.

Endosymbiotic Theory

Learning Objectives

- ✓ Describe the endosymbiotic theory of the origin of mitochondria, chloroplasts, and eukaryotic cells.
- ✓ List evidence for and against the endosymbiotic theory.

Mitochondria and chloroplasts are semiautonomous; that is, they divide independently of the cell but remain dependent

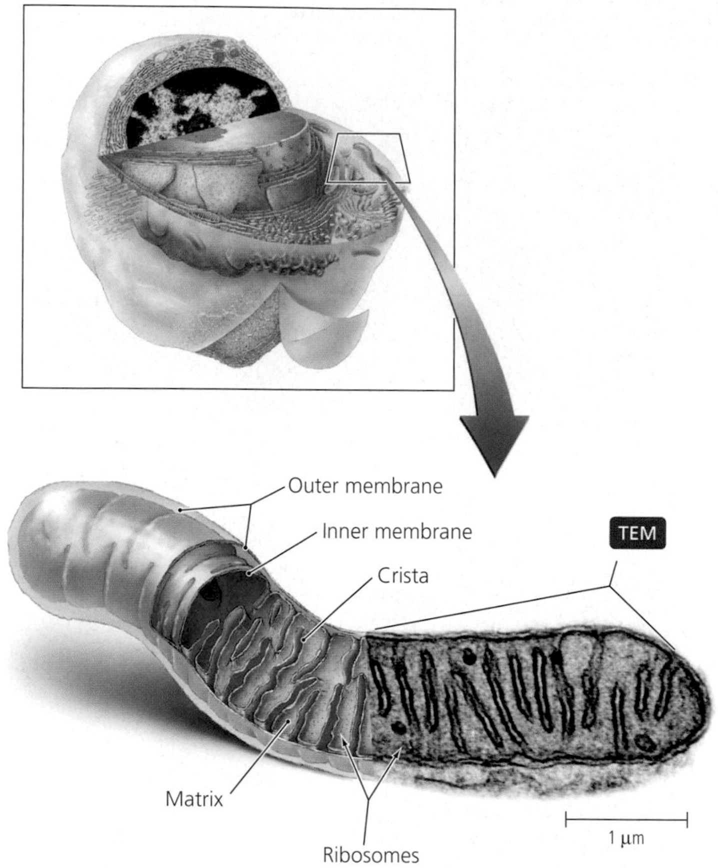

Figure 3.36
Mitochondria. Note the double membrane. The inner membrane is folded into cristae that increase its surface area. *What is the importance of the increased surface area of the inner membrane that results from having cristae?*

Figure 3.36 The chemicals involved in aerobic ATP production are located on the inner membranes of mitochondria. Increased surface area provides more space for more chemicals.

Figure 3.37
Chloroplasts. Chloroplasts have an ornate internal structure designed to harvest light energy for photosynthesis.

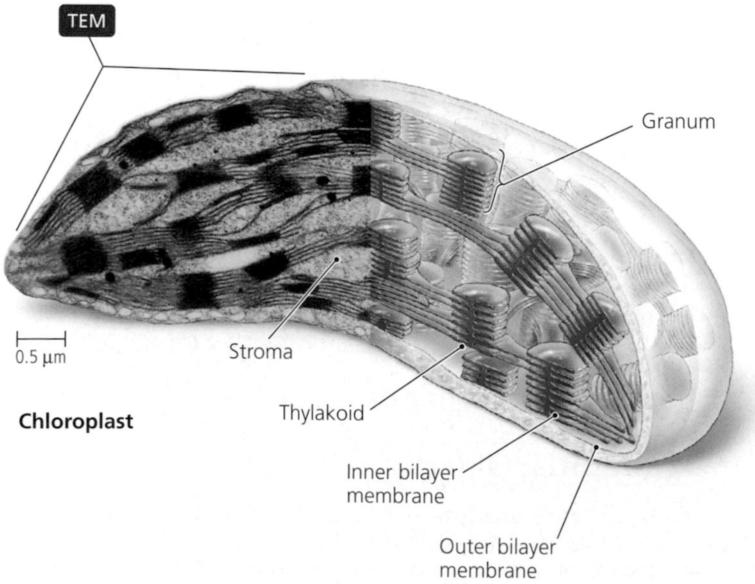

on the cell for most of their proteins. As we have seen, both mitochondria and chloroplasts contain a small amount of DNA and 70S ribosomes, and each can produce a few polypeptides with its own ribosomes. The presence of circular DNA, 70S ribosomes, and two bilipid membranes in these semiautonomous organelles led scientist Lynn Margulis (1938–) to propose the **endosymbiotic[17] theory** for the formation of eukaryotic cells, which suggests that eukaryotes formed from the phagocytosis of small aerobic[18] prokaryotes by larger anaerobic prokaryotes. The smaller prokaryotes were not destroyed by the larger cells, but instead became internal parasites that remained surrounded by a vesicular membrane of the host.

According to the theory, the parasites eventually lost the ability to exist independently, but they retained a portion of their DNA, some ribosomes, and their cytoplasmic membranes. During the same time, the larger cell became dependent on the parasites for aerobic ATP production. According to the theory, the aerobic prokaryotes eventually evolved into mitochondria. Margulis described a similar scenario for the origin of chloroplasts from phagocytized photosynthetic prokaryotes. The theory provides an explanation for the presence of 70S ribosomes and circular DNA within mitochondria and chloroplasts, and it accounts for the presence of their two membranes.

The endosymbiotic theory is not universally accepted, however, partly because it does not explain all of the facts. For example, the theory provides no explanation for the two membranes of the nuclear envelope; nor does it explain why only a few polypeptides of mitochondria and chloroplasts are made in the organelles while the bulk of their proteins come from nuclear DNA and cytoplasmic ribosomes.

Table 3.5 summarizes features of prokaryotic and eukaryotic cells.

CRITICAL THINKING

Eukaryotic cells are almost always larger than prokaryotic cells. What structures might allow for their larger size?

[17] From Greek *endo*, meaning inside, and *symbiosis*, meaning to live with.
[18] Aerobic means requiring oxygen; anaerobic is the opposite.

Table 3.4 Nonmembranous and Membranous Organelles of Cells

	General Function	Prokaryotes	Eukaryotes
Nonmembranous Organelles			
Ribosomes	Protein synthesis	Present in all	Present in all
Cytoskeleton	Shape in prokaryotes; support, cytoplasmic streaming, and endocytosis in eukaryotes	Present in some	Present in all
Centrioles	Appear to play a role in mitosis, cytokinesis, and flagella and cilia formation in animal cells	Absent in all	Present in animals
Membranous Organelles	Sequester chemical reactions within the cell		
Nucleus	"Control center" of the cell	Absent in all	Present in all
Endoplasmic reticulum	Transport within the cell, lipid synthesis	Absent in all	Present in all
Golgi bodies	Exocytosis, secretion	Absent in all	Present in some
Lysosomes	Breakdown of nutrients, self-destruction of damaged or aged cells	Absent in all	Present in some
Peroxisomes	Neutralization of toxins	Absent in all	Present in some
Vacuoles	Storage	Absent in all	Present in some
Vesicles	Storage, digestion, transport	Absent in all	Present in all
Mitochondria	Aerobic ATP production	Absent in all	Present in most
Chloroplasts	Photosynthesis	Absent in all	Present in plants and algae

Table 3.5 Comparison of Prokaryotic and Eukaryotic Cells

Characteristic	Prokaryotes	Eukaryotes
Size	Commonly 0.2–2.0 μm in diameter	Commonly 10–100 μm in diameter
Nucleus	Absent	Present
Membrane-bound organelles	Absent	Present; include ER, Golgi bodies, lysosomes, mitochondria, and chloroplasts
Glycocalyx	Present as organized capsule or as unorganized slime layer	Present surrounding some animal cells
Motility	Some have simple flagella composed of a filament, hook, and basal body that rotates	Some have complex flagella or cilia composed of a "9 + 2" arrangement of microtubules that undulate; others move by means of amoeboid action
Flagella	Present	Present
Cilia	Absent	Present
Fimbriae and pili	Present	Absent
Cell wall	Present in most; in bacteria, typically composed of peptidoglycan	Present in most; composed of cellulose, algin, agar, carrageenan, silicate, glucomannan, or chitin
Cytoplasmic membrane	Lacks carbohydrates; usually lacks sterols	Contains glycolipids, glycoproteins, and sterols
Cytosol	Present	Present
Inclusions	Present	Present
Ribosomes	Smaller (70S)	Larger (80S) in cytoplasm, but 70S in mitochondria and chloroplasts
Chromosomes	Commonly single, circular, and lacking histones	Typically more than one, linear, and containing histones

CHAPTER SUMMARY

Processes of Life (pp. 57–58)

1. All living things have some common features, including **growth,** an increase in size; **reproduction,** an increase in number; **responsiveness,** reactions to environmental stimuli; **metabolism,** controlled chemical reactions in an organism; and cellular structure.
2. Viruses do not grow, self reproduce, or metabolize.

Eukaryotic and Prokaryotic Cells: An Overview (pp. 58–61)

1. All cells can be classified as either **prokaryotic** or **eukaryotic.** Prokaryotic cells lack a nucleus and membrane-bound organelles. Bacteria and archaea are prokaryotic. Eukaryotes have internal, membrane-bound **organelles,** including nuclei. Animals, plants, algae, fungi, and protozoa are eukaryotic.

External Structures of Prokaryotic Cells (pp. 61–66)

1. Cells share common structural features. These include external structures, cell walls, cytoplasmic membranes, and cytoplasm. The external structures of prokaryotic cells include glycocalyces, flagella, fimbriae, and pili.
2. **Glycocalyces** are sticky external sheaths of cells. They may be loosely attached **slime layers** or firmly attached **capsules.** Glycocalyces prevent cells from drying out. Capsules protect cells from phagocytosis by other cells, and slime layers enable cells to stick to each other and to surfaces in their environment.
3. A **flagellum** is a long, whiplike protrusion of some cells composed of a basal body, hook, and filament. Flagella allow cells to move toward favorable conditions such as nutrients or light, or move away from unfavorable stimuli such as poisons.
4. The flagella of prokaryotes may be single **(monotrichous),** grouped at one end of the cell **(lophotrichous),** be at both ends of the cell **(amphitrichous),** or cover the cell **(peritrichous). Endoflagella,** which are special flagella of spirochetes, form **axial filaments,** located in the periplasmic space.
5. **Taxis** is movement that may be either a positive response or a negative response to light **(phototaxis)** or chemicals **(chemotaxis).**
6. **Fimbriae** are nonmotile extensions of some bacterial cells that function along with glycocalyces to adhere cells to one another and to environmental surfaces. A mass of such bacteria on a surface is termed a **biofilm.**
7. **Pili,** which are also called **conjugation pili,** are hollow, nonmotile tubes of protein that connect some prokaryotic cells. They mediate the movement of DNA from one cell to another. Not all bacteria have fimbriae or pili.

Prokaryotic Cell Walls (pp. 66–69)

1. Most prokaryotic cells have **cell walls** that provide shape and support against osmotic pressure. Cell walls are composed of polysaccharide chains.
2. The cell walls of bacteria are composed of a large interconnected molecule of **peptidoglycan.** Peptidoglycan is composed of alternating sugar molecules called *N*-acetylglucosamine (NAG) and *N*-acetylmuramic acid (NAM). **Gram-positive** cells have thick layers of peptidoglycan. **Gram-negative** cells have thin layers of peptidoglycan and an external wall membrane with a **periplasmic space** between. This wall membrane contains **lipopolysaccharide (LPS),** which contains **lipid A.** During an infection with Gram-negative bacteria, lipid A can accumulate in the blood, causing shock, fever, and blood clotting. Acid-fast bacteria have waxy lipids in their cell walls.
3. Cell walls of archaea lack peptidoglycan.

Prokaryotic Cytoplasmic Membranes (pp. 70–74)

1. A **cytoplasmic membrane** is typically composed of lipid molecules arranged in a double layer configuration called a **phospholipid bilayer.** Proteins associated with the membrane vary in location and function and are able to flow laterally within the membrane. The **fluid mosaic model** is descriptive of the current understanding of the membrane.
2. Archaea do not have phospholipid membranes, and some have a single layer of lipid instead of a bilayer.
3. The **selectively permeable** cytoplasmic membrane prevents the passage of some substances while allowing other substances to pass through protein pores or channels, sometimes requiring carrier molecules. The relative concentrations inside and outside the cell of chemicals **(concentration gradients)** and of electrical charges **(electrical gradients)** create an **electrochemical gradient** across the membrane. The electrochemical gradient has a predictable effect on the passage of substances through the membrane.
4. Passive processes that move chemicals across the cytoplasmic membrane require no energy expenditure by the cell. Molecular size and concentration gradients determine the rate of simple **diffusion. Facilitated diffusion** depends on the electrochemical gradient and carriers within the membrane that allow certain substances to pass through the membrane. **Osmosis** specifically refers to the diffusion of water molecules across a selectively permeable membrane.
5. The concentrations of solutions can be compared. **Hypertonic** solutions have a higher concentration of solutes than **hypotonic** solutions, which have a lower concentration of solutes. Two **isotonic solutions** have the same concentrations of solutes.
6. Active transport processes require cell energy from ATP. **Active transport** moves a substance against its electrochemical gradient via carrier proteins. These carriers may move two substances in the same direction at once *(symports)* or move substances in opposite directions *(antiports).* **Group translocation** occurs in prokaryotes; during it the substance being transported is chemically altered in transit.

Cytoplasm of Prokaryotes (pp. 74–76)

1. **Cytoplasm** is composed of the liquid **cytosol** inside a cell plus nonmembranous organelles and inclusions. **Inclusions** in the cytosol are deposits of various substances.
2. Both prokaryotic and eukaryotic cells contain nonmembranous organelles.
3. The **nucleoid** is the nuclear region in prokaryotic cytosol. It has no nuclear envelope and usually contains a single circular molecule of DNA.
4. Inclusions include reserve deposits of lipids, starch, or compounds containing nitrogen, phosphate, or sulfur. Inclusions called gas vesicles store gases.
5. **Ribosomes,** composed of protein and **ribosomal RNA (rRNA),** are nonmembranous organelles, found in both prokaryotes and eukaryotes, that function to make proteins. The 70S ribosomes of prokaryotes are smaller than the 80S ribosomes of eukaryotes.
6. The **cytoskeleton** is a network of fibrils that appears to provide the basic shape of rod-shaped prokaryotes.

External Structures of Eukaryotic Cells (pp. 76–77)

1. Eukaryotic animal and some protozoan cells lack cell walls but have glycocalyces that prevent desiccation, provide support, and enable cells to stick together.
2. Some eukaryotic cells have long, whiplike flagella that differ from the flagella of prokaryotes. They have no hook and the basal bodies and shafts are composed arrangements of microtubules. Further, eukaryotic flagella are surrounded by the cytoplasmic membrane.
3. Some eukaryotic cells are covered with **cilia,** which have the same structure as eukaryotic flagella but are much shorter and more numerous.

Eukaryotic Cell Walls and Cytoplasmic Membranes (pp. 77–80)

1. Fungal, plant, algal, and some protozoan cells have cell walls composed of polysaccharides. Cell walls provide support, shape, and protection from osmotic forces.
2. Fungal cell walls are composed of chitin or other polysaccharides. Plant cell walls are composed of cellulose. Algal cell walls contain agar, carrageenan, algin, cellulose, or other chemicals.
3. Eukaryotic cytoplasmic membranes contain steroids such as cholesterol, which act to strengthen and solidify the membranes when temperatures rise and provide fluidity when temperatures fall.
4. Some eukaryotic cells transport substances into the cytoplasm via **endocytosis,** which is an active process requiring the expenditure of energy by the cell. In endocytosis, **pseudopodia**—movable extensions of the cytoplasm and membrane of the cell—surround a substance and move it into the cell. When solids are brought into the cell, endocytosis is called **phagocytosis;** the incorporation of liquids by endocytosis is called **pinocytosis.**
5. **Exocytosis** is the active export of substances out of a cell.

Cytoplasm of Eukaryotes (pp. 80–87)

1. Eukaryotic cytoplasm is characterized by membranous organelles, particularly a nucleus. It also contains nonmembranous organelles and cytosol.
2. The 80S ribosomes of eukaryotic cells are composed of 60S and 40S subunits. They are found free in the cytosol and attached to endoplasmic reticulum. The ribosomes within mitochondria and chloroplasts are 70S.
3. The eukaryotic cytoskeleton is composed of microtubules, intermediate filaments, and microfilaments. It provides an infrastructure and aids in movement of cytoplasm and organelles.
4. **Centrioles,** which are nonmembranous organelles in animal cells only, are found in a region of the cytoplasm called the **centrosome** and are composed of triplets of microtubules in a "9 + 0" arrangement. Centrioles appear to function in the formation of flagella and cilia and in cell division.
5. The **nucleus,** a membranous structure in eukaryotic cells, contains **nucleoplasm** in which are found one or more **nucleoli** and **chromatin.** Chromatin consists of the multiple strands of DNA and associated histone proteins that become obvious as chromosomes during mitosis. **Nuclear pores** penetrate the four phospholipid layers of the **nuclear envelope** (membrane).
6. The **endoplasmic reticulum** functions as a transport system. It can be **rough ER (RER),** which has ribosomes on its surface, or **smooth ER (SER),** which lacks ribosomes.
7. The **Golgi body** is a series of flattened hollow sacs surrounded by phospholipid bilayers. It packages large molecules destined for export from the cell in **secretory vesicles,** which release these molecules from the cell via exocytosis.
8. **Vesicles** and **vacuoles** are general terms for membranous sacs that store or carry substances. More specifically, **lysosomes** of animal cells contain digestive enzymes, and **peroxisomes** contain enzymes that neutralize poisonous free radicals and hydrogen peroxide.
9. Four phospholipid layers surround **mitochondria,** site of production of most ATP in a eukaryotic cell. The inner bilayer is folded into **cristae,** which greatly increase the surface area available for chemicals that generate ATP.
10. Photosynthetic eukaryotes possess **chloroplasts,** which are organelles containing membranous thylakoids that provide increased surface area for photosynthetic reactions.
11. The **endosymbiotic theory** has been suggested to explain why mitochondria and chloroplasts have 70S ribosomes, circular DNA, and two membranes. The theory states that the ancestors of these organelles were prokaryotic cells that were internalized by other prokaryotes and then lost the ability to exist outside their host—thus forming early eukaryotes.

QUESTIONS FOR REVIEW

(Answers to multiple choice and matching questions are on the web, along with additional review questions. Visit www.microbiologyplace.com.)

Multiple Choice Questions

1. A cell may allow a large or charged chemical to move across the cytoplasmic membrane, down the chemical's electrochemical gradient, in a process called
   a. active transport.
   b. facilitated diffusion.
   c. endocytosis.
   d. pinocytosis.
2. Which of the following statements concerning growth and reproduction is *not* true?
   a. Growth and reproduction may occur simultaneously in living organisms.
   b. A living organism must reproduce to be considered alive.
   c. Living things may stop growing and reproducing, yet still be alive.
   d. Normally, living organisms have the ability to grow and reproduce themselves.
3. A "9 + 2" arrangement of microtubules is seen in
   a. prokaryotic flagella.
   b. nucleoids.
   c. eukaryotic flagella.
   d. Golgi bodies.
4. Which of the following is most associated with diffusion?
   a. symports
   b. antiports
   c. carrier proteins
   d. endocytosis
5. Which of the following is *not* associated with prokaryotic organisms?
   a. nucleoid
   b. glycocalyx
   c. cilia
   d. circular DNA
6. Which of the following is true of Svedbergs?
   a. They are not exact but are useful for comparisons.
   b. They are abbreviated "sv."
   c. They are prokaryotic in nature but exhibit some eukaryotic characteristics.
   d. They are an expression of sedimentation rate during high-speed centrifugation.
7. Which of the following statements is true?
   a. The cell walls of bacteria are composed of peptidoglycan.
   b. Peptidoglycan is a fatty acid.
   c. Gram-positive cell walls have a relatively thin layer of peptidoglycan anchored to the cytoplasmic membrane by teichoic acids.
   d. Peptidoglycan is found mainly in the cell walls of fungi, algae, and plants.
8. Which of the following is *not* a function of a glycocalyx?
   a. It forms pseudopodia for faster mobility of an organism.
   b. It can protect a bacterial cell from drying out.
   c. It hides the bacterial cell from other cells.
   d. It allows a bacterium to stick to a host.
9. Bacterial flagella
   a. are anchored to the cell by a basal body.
   b. are composed of many identical globular proteins arranged around a central core.
   c. are surrounded by an extension of the cytoplasmic membrane.
   d. are composed of tubulin in hollow microtubules in a "9 + 2" arrangement.
10. Which cellular structure is important in classifying a bacterial species as Gram-positive or Gram-negative?
    a. flagella
    b. cell wall
    c. cilia
    d. glycocalyx
11. A cell is moving uric acid across the cell membrane against its electrochemical gradient. Which of the following statements is true?
    a. The exterior of the cell is probably electrically negative compared to the interior of the cell.
    b. The acid probably moves by a passive means such as facilitated diffusion.
    c. The acid moves by an active process such as active transport.
    d. The movement of the acid requires phagocytosis.
12. Gram-positive cells
    a. have a thick cell wall, which retains crystal violet dye.
    b. contain teichoic acids in their cell walls.
    c. appear purple after Gram staining.
    d. all of the above

Matching Questions

Match the structures on the left with the descriptions on the right. A letter may be used more than once or not at all, and more than one letter may be correct for each blank.

1. ___ Glycocalyx	A. Bristlelike projections found in quantities of 100 or more
2. ___ Flagella	B. Long whip
3. ___ Axial filaments	C. Responsible for conjugation
4. ___ Cilia	D. "Sugar cup" composed of polysaccharides and/or polypeptides
5. ___ Fimbriae	E. Short, numerous, nonmotile projections
6. ___ Pili	F. Responsible for motility of spirochetes
	G. Not used for motility
	H. Made of tubulin in eukaryotes
	I. Made of flagellin in prokaryotes

Match the term on the left with its description on the right. Only one description is intended for each term.

7. ___ Ribosome	A. Site of protein synthesis
8. ___ Cytoskeleton	B. Contains enzymes to neutralize hydrogen peroxide
9. ___ Centriole	C. Functions as the transport system within a eukaryotic cell

10. ___ Nucleus D. Allows contraction of the cell
11. ___ Mitochondrion E. Site of most DNA in eukaryotes
12. ___ Chloroplast F. Contains microtubules in "9 + 0" arrangement
13. ___ ER G. Light-harvesting organelle
14. ___ Golgi body H. Packages large molecules for export from the cell
15. ___ Peroxisome I. Its internal membranes are sites for ATP production

Short Answer

1. A local newspaper writer has contacted you, an educated microbiology student from a respected college. He wants to obtain scientific information for an article he is writing about "life" and poses the following query: "What is the difference between a living thing and a nonliving thing?" Knowing that he will edit your material to fit the article, give an intelligent, scientific response.
2. Define cytosol.
3. The term *fluid mosaic* has been used in describing the cytoplasmic membrane. How does each word of that phrase accurately describe our current understanding of a cell membrane?
4. Describe (or draw) an example of diffusion down a concentration gradient.
5. Sketch, name, and describe four flagellar arrangements in prokaryotes.
6. What is the difference between growth and reproduction?
7. Compare bacterial cells and algal cells, giving at least four similarities and four differences.
8. Contrast a cell of *Streptococcus pyogenes* (a bacterium) with the unicellular protozoan *Entamoeba histolytica*, listing at least eight differences.
9. Differentiate among pili, fimbriae, prokaryotic flagella, eukaryotic flagella, and cilia using sketches and descriptive labels.
10. Can nonliving things metabolize? Explain your answer.
11. Contrast prokaryotic and eukaryotic cells by filling in the following table.

Characteristic	Prokaryotes	Eukaryotes
Size		
Presence of nucleus		
Presence of membrane-bound organelles		
Structure of flagella		
Chemicals in cell walls		
Type of ribosomes		
Structure of chromosomes		

12. What is the function of glycocalyces and fimbriae in forming a biofilm?
13. What factors may prevent a molecule from moving across a cell membrane?
14. Compare and contrast three types of passive transport across a cell membrane.
15. Contrast the following active processes for transporting materials into or out of a cell: active transport, group translocation, endocytosis, exocytosis.
16. Contrast symports and antiports.
17. Describe the endosymbiotic theory. What evidence supports the theory? Which features of eukaryotic cells are not explained by the theory?

CRITICAL THINKING

1. A scientist develops a chemical that prevents Golgi bodies from functioning. Contrast the specific effects the chemical would have on human cells versus bacterial cells.
2. Methylene blue binds to DNA. What structures in a yeast cell would be stained by this dye?
3. A new chemotherapeutic drug kills bacteria, but not humans. Discuss the possible ways the drug may act selectively on bacterial cells.
4. Some bacterial toxins cause cells lining the digestive tract to secrete ions, making the contents of the tract hypertonic. What effect does this have on a patient's water balance?
5. A researcher carefully inserts an electrode into a plant cell. He determines that the electrical charge across the cytoplasmic membrane is −70 millivolts. Then he slips the electrode deeper into the cell across another membrane and measures an electrical charge of −90 millivolts compared to the outside. What relatively large organelle is surrounded by the second membrane? Explain your answer.
6. Does Figure 3.2a on page 60 best represent a Gram-positive or a Gram-negative cell? Defend your answer.
7. An electron micrograph of a newly discovered cell shows long projections with a basal body in the cell wall. What kind of projections are these? How might this cell behave in its environment because of the presence of this structure?
8. An entry in a recent scientific journal reports positive phototaxis in a newly described species. What condition could you create in the lab to encourage the growth of these organisms?
9. A medical microbiological lab report indicates that sample X contained a biofilm, and that one species in the biofilm was identified as *Neisseria gonorrhoeae.* Is this strain of *Neisseria* likely to be pathogenic? Why or why not?

CHAPTER 4

Microscopy, Staining, and Classification

CHAPTER OUTLINE

Units of Measurement

Microscopy
- General Principles of Microscopy
- Light Microscopy
- Electron Microscopy
- Probe Microscopy

Staining
- Preparing Specimens for Staining
- Principles of Staining
- Simple Stains
- Differential Stains
- Special Stains
- Staining for Electron Microscopy

Classification and Identification of Microorganisms
- Linnaeus, Whittaker, and Taxonomic Categories
- Domains
- Taxonomic and Identifying Characteristics
- Taxonomic Keys

When we want to weigh ourselves, we just step on a bathroom scale. But what can we use to measure the mass of something as delicate as the cell wall of a microorganism? The answer is an interference microscope, which uses a split beam of light to form vertical light and dark bands across a specimen (see the photo). Where light rays have traveled through a specimen and slowed, the pattern of light and dark bands is shifted, and the amount of shift is directly proportional to the change in speed. That change in speed can then be correlated with the specimen's density; the denser the specimen, the slower light travels through it. In this way, we can "weigh" a cell wall—or a nucleus or a chloroplast, or even a single chromosome.

This chapter explores the illuminating world of microscopes, and the amazing lessons about microbial life that we continue to learn from them.

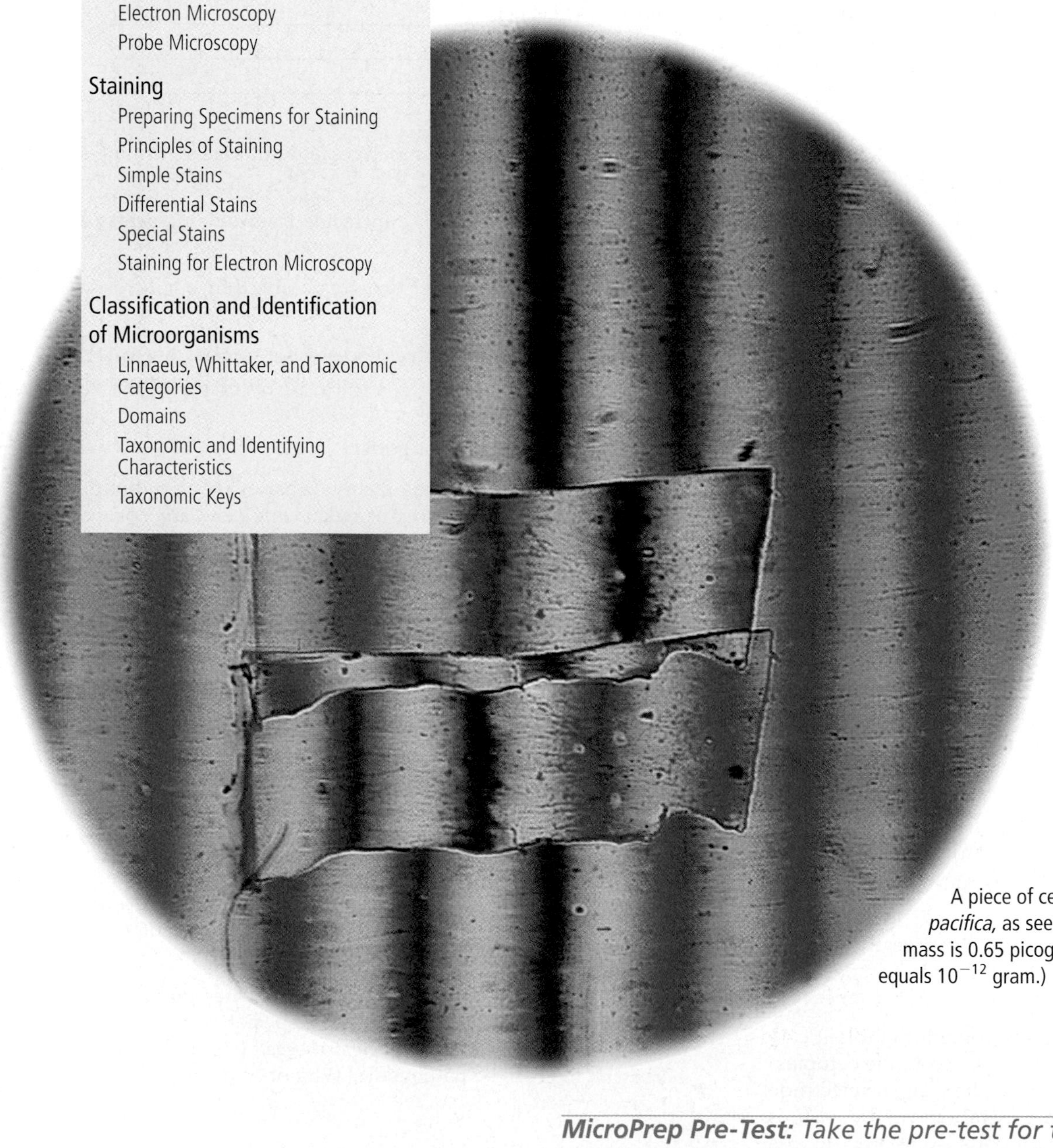

A piece of cell wall from the marine red alga *Griffithsia pacifica,* as seen through an interference microscope. Its mass is 0.65 picogram per square micrometer. (A picogram equals 10^{-12} gram.)

MicroPrep Pre-Test: *Take the pre-test for this chapter on the web. Visit* ***www.microbiologyplace.com***

Either the well was very deep, or she fell very slowly, for she had plenty of time as she went down to look about her, and to wonder what was going to happen next. . . ."Curiouser and curiouser!" cried Alice. . . .
—*Alice's Adventures in Wonderland* by Lewis Carroll

Like Alice falling into Wonderland or traveling through a looking glass, scientists have entered the marvelous microbial world through advances in microscopy. With the invention of new laboratory techniques and the construction of new instruments, biologists are still discovering "curiouser and curiouser" wonders about the microbial world. In this chapter we will discuss some of the techniques microbiologists use to enter that world. We begin with a discussion of metric units as they relate to measuring the size of microbes. We then examine the instruments and staining techniques used in microbiology. Finally, we consider the classification schemes used to categorize the inhabitants of the microbial wonderland.

Units of Measurement

Learning Objectives

✓ Identify the two primary metric units used to measure the diameters of microbes.

✓ List the metric units of length in order, from meter to nanometer.

Microorganisms are small. This may seem an obvious statement, but it is one that cannot be taken for granted. Exactly how small are they? How can we measure the width and length of microbes?

Typically, a unit of measurement is smaller than the object being measured. For example, we measure a person's height in feet or inches, not in miles. Likewise, the diameter of a dime is measured in fractions of an inch, not in feet. So, measuring the size of a microbe requires units that are smaller than even the smallest interval on a ruler marked with English units (typically 1/16 inch). Even smaller units, such as 1/64 inch or 1/128 inch, become quite cumbersome and very difficult to use when we are dealing with microorganisms.

So that they can work with units that are simpler and in standard use the world over, scientists use metric units of measurement. Unlike the English system, the metric system is a decimal system, so each unit is one-tenth the size of the next largest unit. Using the metric system, even extremely small units are much easier to use than the fractions involved in the English system.

The unit of length in the metric system is the *meter (m),* which is slightly longer than a yard. One-tenth of a meter is a *decimeter (dm),* and one-hundredth of a meter is a *centimeter (cm),* which is equivalent to about a third of an inch. One tenth of a centimeter is a *millimeter (mm),* which is the thickness of a dime. A millimeter is still too large to measure the size of most microorganisms, but in the metric system we continue to divide by multiples of 10 until we have a unit appropriate for use. Thus, one-thousandth of a millimeter is a *micrometer (μm),* which is small enough to be useful in measuring the size of cells. One-thousandth of a micrometer is a *nanometer (nm),* a unit used to measure the smallest cellular organelles and viruses. A nanometer is one-billionth of a meter.

Table 4.1 presents these metric units and some English equivalents. Refer to Figure 3.3 on page 61 for a visual size comparison of a typical eukaryotic cell, prokaryotic cell, and virus particle.

Microscopy

Learning Objective

✓ Define microscopy.

Microscopy[1] refers to the use of light or electrons to magnify objects. The science of microbiology began when Leeuwenhoek used primitive microscopes to observe and report the

[1]From Greek *micro,* meaning small, and *skopein,* meaning to view.

Table 4.1 Metric Units of Length

Metric Unit (abbreviation)	Meaning of Prefix	Metric Equivalent	U.S. Equivalent	Representative Microbiological Application of the Unit
Meter (m)	—[a]	1 m	39.37 in (about a yard)	Length of pork tapeworm, *Taenia solium* (e.g., 1.8 m–8.0 m)
Decimeter (dm)	1/10	0.1 m = 10^{-1} m	3.94 in	—[b]
Centimeter (cm)	1/100	0.01 m = 10^{-2} m	0.39 in; 1 in = 2.54 cm	Diameter of a mushroom cap (e.g., 12 cm)
Millimeter (mm)	1/1000	0.001 m = 10^{-3} m	—	Diameter of a bacterial colony (e.g., 2.3 mm); length of a tick (e.g., 5.7 mm)
Micrometer (μm)	1/1,000,000	0.000001 m = 10^{-6} m	—	Diameter of white blood cells (e.g., 5 μm–25 μm)
Nanometer (nm)	1/1,000,000,000	0.000000001 m = 10^{-9} m	—	Diameter of a poliovirus (e.g., 25 nm)

[a]The meter is the standard metric unit of length.
[b]Decimeters are rarely used; centimeters are much more commonly used units.

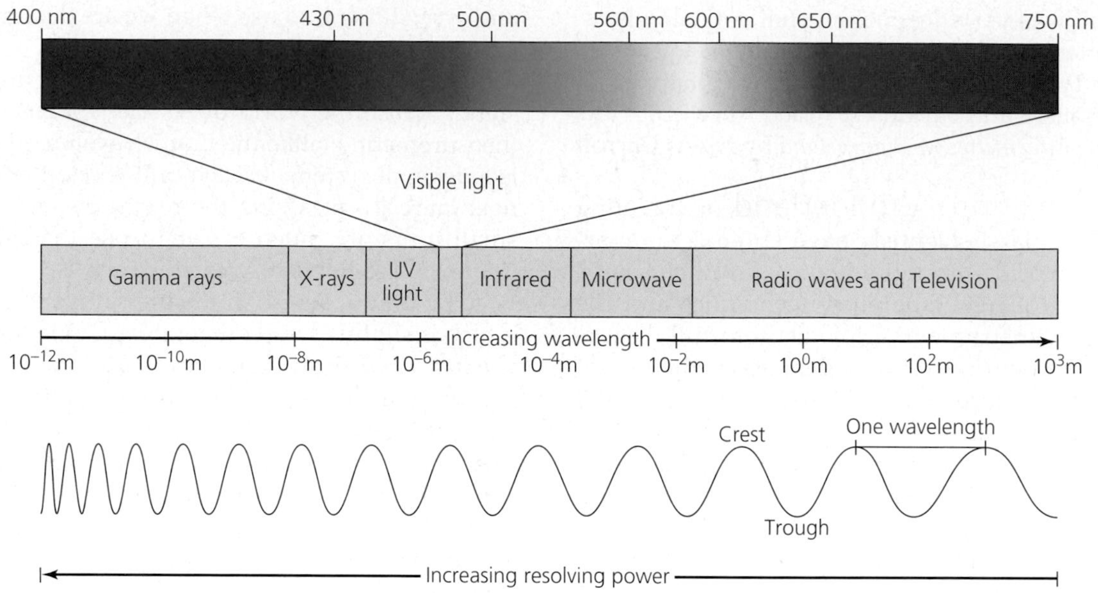

▲ ***Figure 4.1***

The electromagnetic spectrum. Visible light comprises a narrow band of wavelengths of radiation. Visible and ultraviolet light are used in microscopy.

existence of microorganisms. Since that time, scientists and engineers have developed a variety of light and electron microscopes.

General Principles of Microscopy

Learning Objectives

- ✓ Explain the relevance of electromagnetic radiation to microscopy.
- ✓ Define *empty magnification.*
- ✓ List and explain two factors that determine resolving power.
- ✓ Discuss the relationship between contrast and staining in microscopy.

General principles involved in both light and electron microscopy include the wavelength of radiation, the magnification of an image, the resolving power of the instrument, and contrast in the specimen.

Wavelength of Radiation

Visible light is one part of a spectrum of electromagnetic radiation that includes X-rays, microwaves, and radio waves **(Figure 4.1).** Note that beams of radiation may be referred to as either rays or waves. These various forms of radiation differ in **wavelength**—the distance between two corresponding parts of a wave. The human eye discriminates among different wavelengths of visible light and sends patterns of nerve impulses to the brain, which interprets the impulses as different colors. For example, we see wavelengths of 400 nm as violet and wavelengths of 650 nm as red. White light, composed of many colors (wavelengths), has an average wavelength of 550 nm.

In Chapter 2 we learned that electrons are negatively charged particles that orbit the nuclei of atoms. Besides being particulate, moving electrons act as waves with wavelengths dependent upon the voltage of an electron beam. For example, the wavelength of electrons at 10,000 volts (V) is 0.01 nm, whereas the wavelength of electrons at 1,000,000 V is 0.001 nm. As we will see, using radiation of smaller wavelengths results in enhanced microscopy.

Magnification

Magnification is the apparent increase in size of an object. It is indicated by a number and "×", which is read "times." For example, the *Paramecium* in Figure 4.15c is magnified 2000× or 2000 times. Magnification results when a beam of radiation *refracts* (bends) as it passes through a lens. Curved glass lenses refract light, and magnetic fields (magnetic lenses) refract electron beams. Let's consider the magnifying power of a glass lens that is convex on both sides.

A lens refracts light because it is *optically dense* compared to the surrounding medium (such as air); that is, light travels more slowly through the lens than through air. This can be compared to a car moving from a paved road onto a dirt road at an angle. As the right front tire leaves the pavement, it has less traction and slows down. Since the other wheels continue at their original speed, the car will veer toward the dirt, and the line of travel will bend to the right. Likewise, the leading edge of a light beam slows as it enters glass, and the beam bends **(Figure 4.2a).** Light also bends as it leaves the glass and reenters the air.

Because of its curvature, a lens refracts light rays that pass through its periphery more than light rays that pass

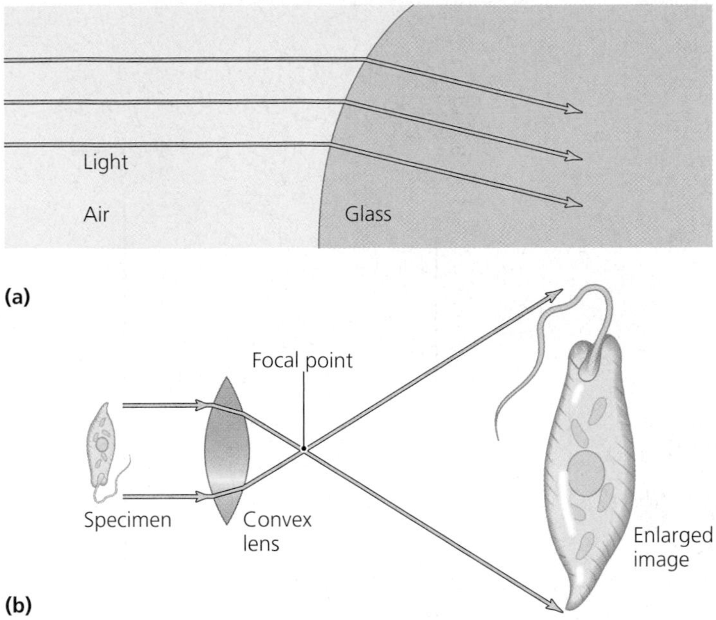

Figure 4.2

Light refraction and image magnification by a convex glass lens. **(a)** Light passing through a lens refracts (bends) because light rays slow down as they enter the glass, and light at the leading edge of a beam that strikes the glass at an angle slows first. **(b)** A convex lens focuses light on a focal point. The image is enlarged as light rays pass the focal point and spread apart.

through its center. The lens focuses light rays on a *focal point*. Importantly, for the purpose of microscopy, light rays spread apart as they travel past the focal point and produce an enlarged image **(Figure 4.2b)**. The degree to which the image is enlarged depends on the thickness of the lens, its curvature, and the speed of light through its substance.

Microscopists could combine lenses to obtain an image magnified millions of times, but the image would be faint and blurry. Such magnification is said to be *empty magnification*. The properties that determine the clarity of an image, which in turn determines the useful magnification of a microscope, are *resolution* and *contrast*.

Resolution

Resolution, which is also called *resolving power*, is the ability to distinguish between objects that are close together. An optometrist's eye chart is a test of resolution at a distance of 20 feet (6.1 m). Leeuwenhoek's microscopes had a resolving power of about 1 μm; that is, he could distinguish between objects if they were more than about 1 μm apart, and objects closer together than 1 μm appeared as a single object. The closer together that two objects are still distinguishable as separate objects, the better the resolution. Modern microscopes have fivefold better resolution than Leeuwenhoek's; they can distinguish between objects as close together as 0.2 μm. **Figure 4.3** on page 96 illustrates the size of various objects that can be resolved by the unaided human eye and by various types of microscopes.

Why do modern microscopes have better resolution than Leeuwenhoek's microscopes? A principle of microscopy is that resolution distance is dependent on (1) the wavelength of the light or electron beam and (2) the **numerical aperture** of the lens, which refers to the ability of a lens to gather light.

Resolution distance is calculated using the following formula:

$$\text{Resolution distance} = \frac{0.61 \times \text{wavelength}}{\text{numerical aperture}}$$

The resolution of today's microscopes is greater than that of Leeuwenhoek's microscopes because modern microscopes use shorter wavelength radiation, such as blue light or electron beams, and because they have lenses with larger numerical apertures.

Contrast

Contrast refers to differences in intensity between two objects, or between an object and its background. Contrast is important in determining resolution. For example, although you can easily distinguish between two golf balls lying side by side on a putting green 15 m away, at that distance it is much more difficult to distinguish between them if they are lying on a white towel.

Most microorganisms are colorless and have very little contrast whether one uses light or electrons. One way to increase the contrast between microorganisms and their background is to stain them. Stains and staining techniques are covered later in the chapter. As we will see, the use of light that is in *phase*—that is, in which all of the waves' crests and troughs are aligned—can also enhance contrast.

Light Microscopy

Learning Objectives

- ✓ Contrast simple and compound microscopes.
- ✓ Compare and contrast bright-field microscopy, dark-field microscopy, and phase microscopy.
- ✓ Compare and contrast fluorescent and confocal microscopes.

Several classes of microscopes use various types of light to examine microscopic specimens. The most common microscopes are *bright-field microscopes,* in which the background (or *field*) is illuminated. In *dark-field microscopes,* the specimen is made to appear light against a dark background. *Phase microscopes* use the alignment or misalignment of light waves to achieve the desired contrast between a living specimen and its background. *Fluorescent microscopes* use invisible ultraviolet light to cause specimens to radiate visible light, a phenomenon called *fluorescence*. Microscopes that use lasers to illuminate fluorescent chemicals in a thin plane of a specimen are called *confocal microscopes*. Next we examine each of these kinds of light microscope in turn.

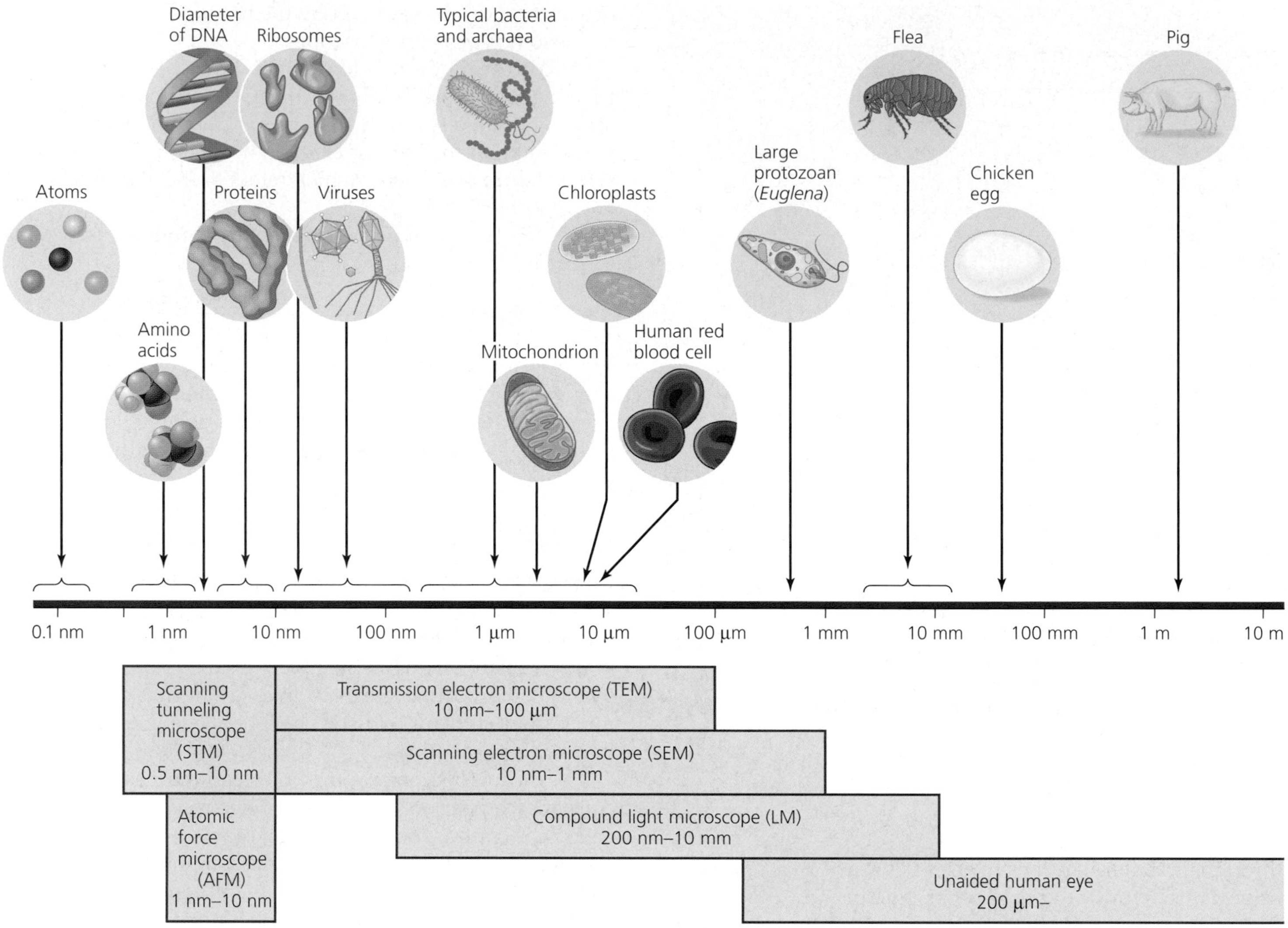

▲ ***Figure 4.3***

The limits of resolution (and some representative objects within those ranges) of the human eye and of various types of microscopes.

Bright-Field Microscopes

There are two basic types of bright-field microscopes: *simple microscopes* and *compound microscopes.*

Simple Microscopes As we learned in Chapter 1, Leeuwenhoek first reported his observations of microorganisms using a simple microscope in 1673. A **simple microscope,** which contains a single magnifying lens, is more similar to a magnifying glass than to a modern microscope (see Figure 1.2). Though Leeuwenhoek did not invent the microscope, he was the finest lens grinder of his day and produced microscopes of exceptional quality. They were capable of approximately 300× magnification and achieved excellent clarity, far surpassing other microscopes of his time.

Compound Microscopes Simple microscopes have been replaced in modern laboratories by compound microscopes. A **compound microscope** uses a series of lenses for magnification **(Figure 4.4a).** Many scientists, including Galileo Galilei, made compound microscopes as early as 1590, but it was not until about 1830 that scientists developed compound microscopes that exceeded the clarity and magnification of Leeuwenhoek's simple microscope.

In a basic compound microscope, magnification is achieved as light rays pass through a specimen and into an **objective lens,** which is the lens immediately above the object being magnified **(Figure 4.4b).** An objective lens is really a series of lenses that not only create a magnified image, but are also engineered to reduce aberrations in the shape and color of the image. Most light microscopes used in biology have three or four objective lenses mounted on a **revolving**

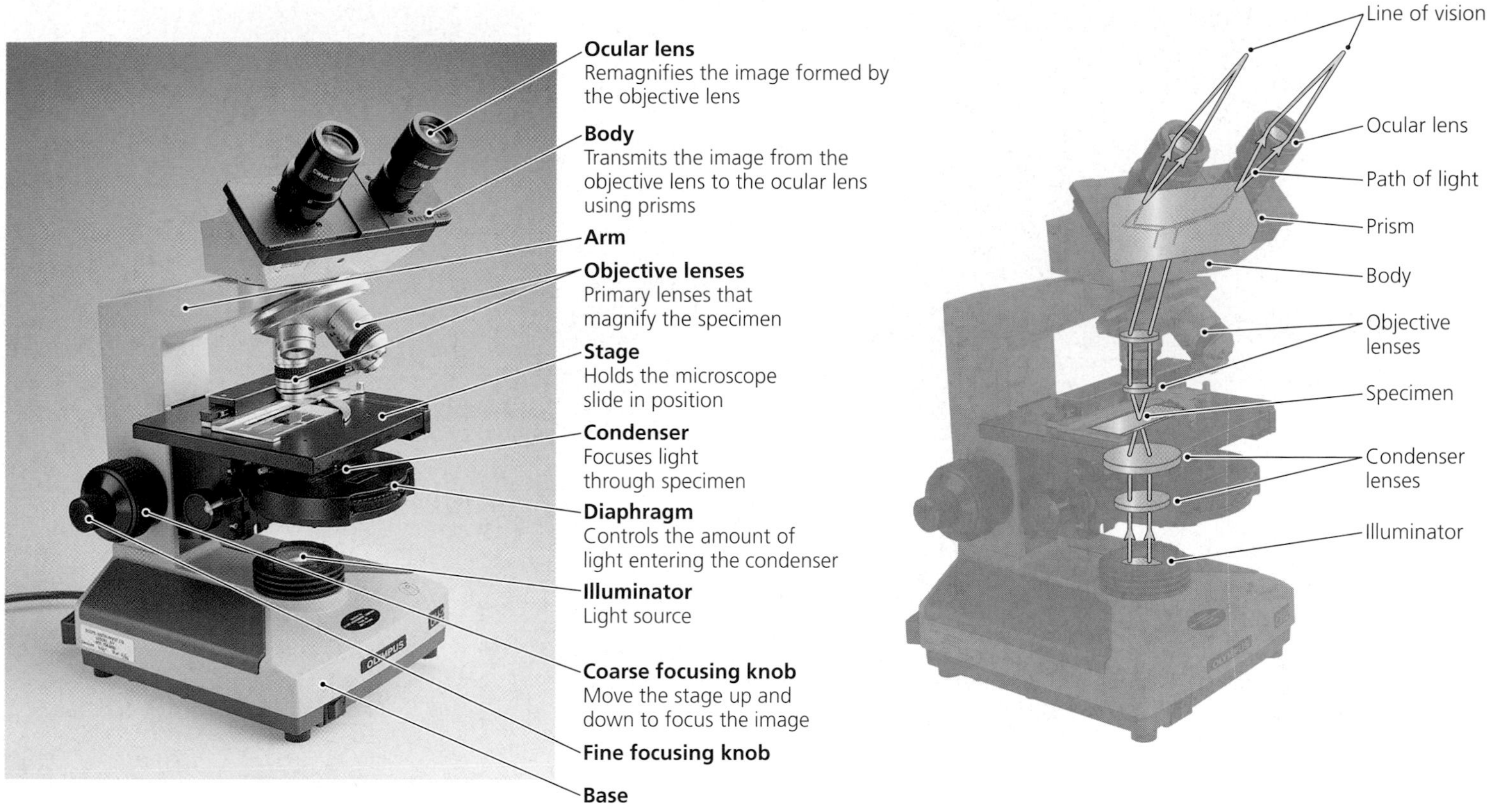

(a) (b)

▲ ***Figure 4.4***

A bright-field, compound light microscope. **(a)** The parts of a compound microscope, which uses a series of lenses to produce an image up to 2000× magnification. **(b)** The path of light in a compound microscope; light travels from bottom to top. *Why can't light microscopes produce clear images that are magnified 10,000×?*

Figure 4.4 While it is possible for a light microscope to produce an image that is magnified 10,000×, magnification above 2000× is empty magnification because pairs of objects in the specimen are too close together to resolve with even the shortest-wavelength (blue) light.

nosepiece. The objective lenses on a typical microscope are: *scanning objective lens* (4×), *low power objective lens* (10×), *high power lens* or *high dry objective lens* (40×), and *oil immersion objective lens* (100×).

Not only does an oil immersion lens increase magnification, it also increases resolution. As we have seen, light refracts as it travels from air into glass and also from glass into air; therefore, some of the light passing out of a glass slide is bent so much that it bypasses the lens (**Figure 4.5a** on page 98). Placing *immersion oil* between the slide and an *oil immersion objective lens* enables the lens to capture this light because light travels through immersion oil at the same speed as through glass. Since light is traveling at a uniform speed through the slide, the immersion oil, and the glass lens, it does not refract **(Figure 4.5b).** Immersion oil increases the numerical aperture, which increases resolution, because more light rays are gathered into the lens to produce the image. Obviously, the space between the slide and the lens can be filled with oil only if the distance between the lens and the specimen, called the *working distance,* is small.

An objective lens bends the light rays, which then pass up through the hollow body of the microscope and through one or two **ocular lenses,** which are the lenses closest to the eyes. Microscopes with a single ocular lens are *monocular,* and those with two are *binocular*. Ocular lenses magnify the image created by the objective lens, typically another 10 times (10×).

The **total magnification** of a compound microscope is determined by multiplying the magnification of the objective lens by the magnification of the ocular lens. Thus, total magnification using a 10× ocular lens and a 10× low power objective lens is 100×. Using the same ocular and a 100× oil immersion objective produces 1000× magnification. Some light microscopes, using higher magnification oil immersion objective lenses and ocular lenses, can achieve 2000× magnification, but this is the limit of useful magnification for

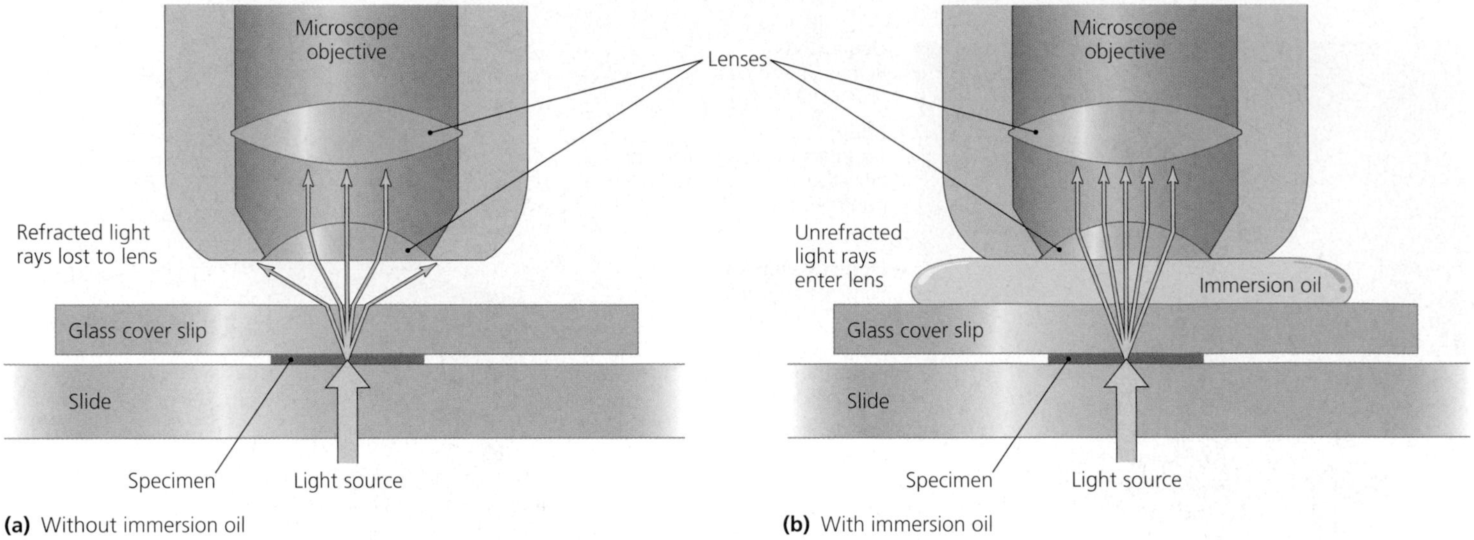

▲ ***Figure 4.5***

The effect of immersion oil on resolution. **(a)** Without immersion oil, light is refracted as it moves from the cover glass into the air. Part of the scattered light misses the objective lens. **(b)** With immersion oil. Because light travels through the oil at the same speed as it does through glass, no light is refracted as it leaves the specimen, and more light enters the lens, which increases resolution.

light microscopes because their resolution is restricted by the wavelength of visible light.

Modern compound microscopes also have a **condenser lens** (or lenses), which directs light through the specimen, as well as one or more mirrors or prisms that deflect the path of the light rays from an objective lens to the ocular lens (see Figure 4.4b). Some microscopes have mirrors or prisms that direct light to a camera through a special tube. Photographs of a microscopic image are called **micrographs (Figure 4.6).**

Dark-Field Microscopes

Pale objects are best observed with **dark-field microscopes.** These microscopes utilize a *dark-field stop* in the condenser that prevents light from directly entering the objective lens **(Figure 4.7).** Instead, light rays are reflected inside the condenser so that they pass into the slide at such an oblique angle that they miss the objective lens. Only those light rays that are scattered by the specimen enter the objective lens and are seen, so the specimen appears light against a dark background (see Figure 4.6b). This increases contrast and enables observation of more details than are visible in bright-field microscopy. Dark-field microscopes are especially useful for examining small or colorless cells.

Phase Microscopes

Scientists use **phase microscopes** to examine living microorganisms or specimens that would be damaged or altered by attaching them to slides or staining them. Basically, phase microscopes treat one set of light rays differently from another set of light rays.

Light rays are said to be *in phase* when their crests and troughs are aligned, and *out of phase* when their crests and troughs are not aligned (**Figure 4.8a** on page 100). Light rays that are in phase reinforce one another, producing a brighter image, and light rays that are out of phase interfere with one another and produce a darker image. Light rays passing through a specimen naturally slow down and are shifted about 1/4 wavelength out of phase. A special filter called a *phase plate,* which is mounted in the phase objective lens, retards these rays another 1/4 wavelength, so that they are 1/2 wavelength out of phase with their neighbors. When the phase microscope lens brings the two sets of rays together, troughs of one wave interfere with the crests of the other—because they are out of phase **(Figure 4.8b)**—and contrast is created. There are two types of phase microscopes: phase-contrast and differential interference contrast microscopes.

Phase-Contrast Microscopes The simplest phase microscopes, **phase-contrast microscopes,** produce sharply defined images in which fine structures can be seen in living cells (see Figure 4.6c). These microscopes are particularly useful for observing cilia and flagella.

Differential Interference Contrast Microscopes **Differential interference contrast microscopes** (also called Nomarski[2] microscopes) create phase interference patterns. They also use prisms that split light beams into their component wavelengths (colors). This significantly increases

[2]After the French physicist Georges Nomarski, who invented the differential interference contrast microscope.

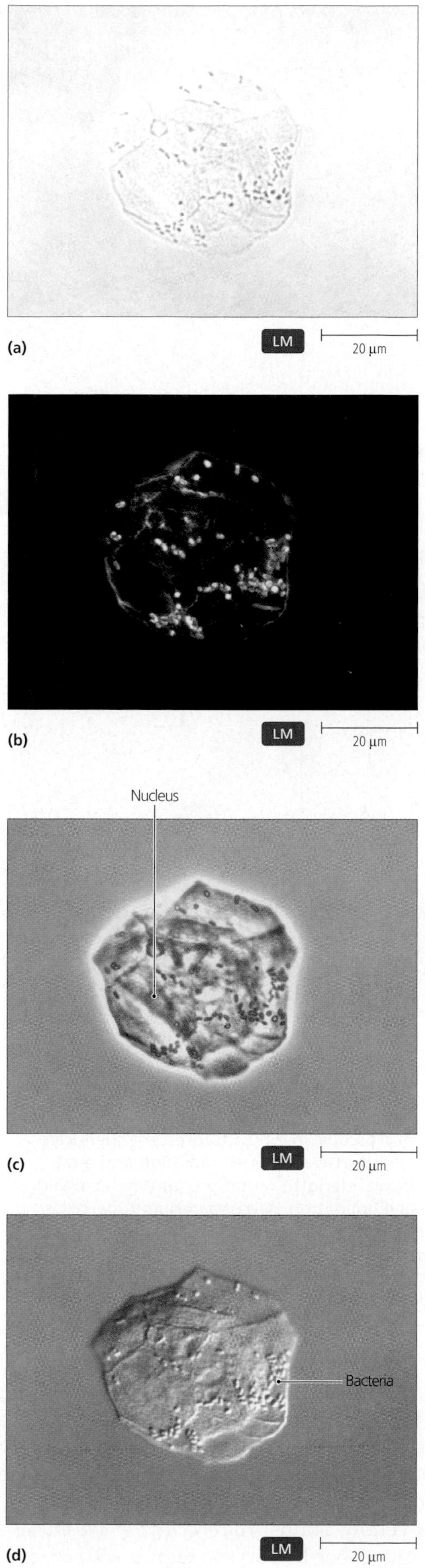

Figure 4.6

Four kinds of micrographs, in this case of bacteria and a human cheek cell. **(a)** Bright-field microscopy reveals some internal structures. **(b)** Dark-field microscopy increases contrast between some internal structures and between the edges of the cell and the surrounding medium. **(c)** Phase-contrast microscopy provides greater resolution of internal structures. **(d)** Differential interference contrast (Nomarski) microscopy produces a three-dimensional effect.

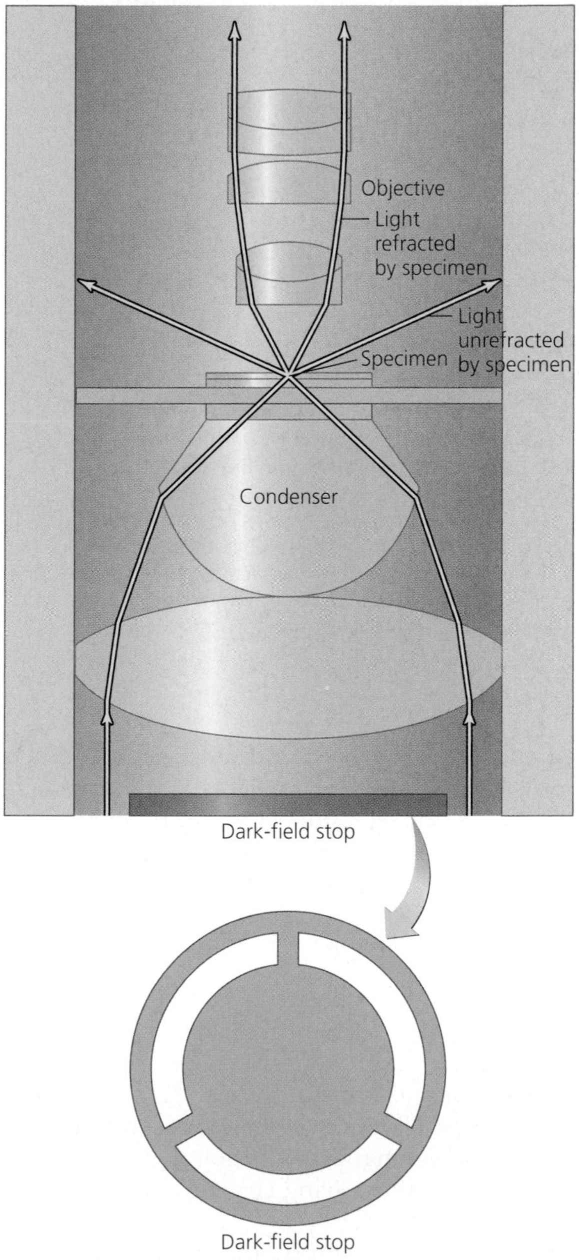

Figure 4.7

The light path in a dark-field microscope. A dark-field stop prevents light from entering the specimen directly; only light rays that are scattered by the specimen reach the objective lens and can be seen. The resulting high-contrast image—a brightly lit specimen against a dark background—enhances resolution.

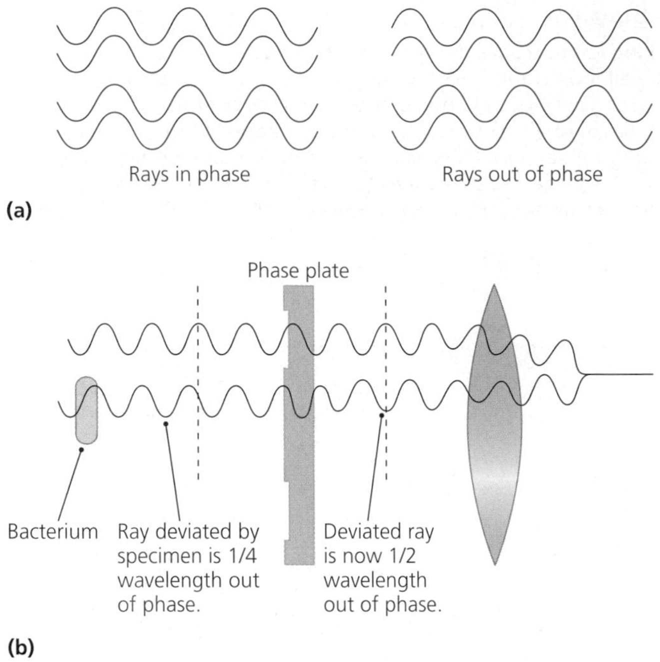

▲ ***Figure 4.8***

Principles of phase microscopy. **(a)** Light rays that are in phase are aligned; their crests and troughs reinforce one another to produce a brighter image. Rays that are out of phase interfere with one another and produce a darker image. **(b)** Together, the regions of a specimen and a phase plate built into a phase objective lens slow some of the light rays such that they are 1/2 wavelength out of phase. These rays interfere with light rays that bypass the specimen, which produces contrast between the field and various regions of the specimen (see Figure 4.6c and d).

contrast and gives the image a dramatic three-dimensional or shadowed appearance, almost as if light were striking the specimen from one side (see Figure 4.6d). This technique also produces unnatural colors, which enhance contrast.

Fluorescent Microscopes

Molecules that absorb energy from invisible radiation (such as ultraviolet light) and then radiate the energy back as a longer, visible wavelength are said to be *fluorescent*. **Fluorescent microscopes** use an ultraviolet (UV) light source to fluoresce objects. UV light increases resolution because it has a shorter wavelength than visible light, and contrast is improved because fluorescing structures are visible against a black background.

Some cells—for example, the pathogen *Pseudomonas aeruginosa* (soo-dō-mō'nas ā-rū-ji-nō'să)—and some cellular molecules (such as chlorophyll in photosynthetic organisms) are naturally fluorescent. Other cells and cellular structures can be stained with fluorescent dyes. When these dyes are bombarded with ultraviolet light, they emit visible light and show up as bright orange, green, yellow, or other colors against a black background (see Figure 3.29b on page 82).

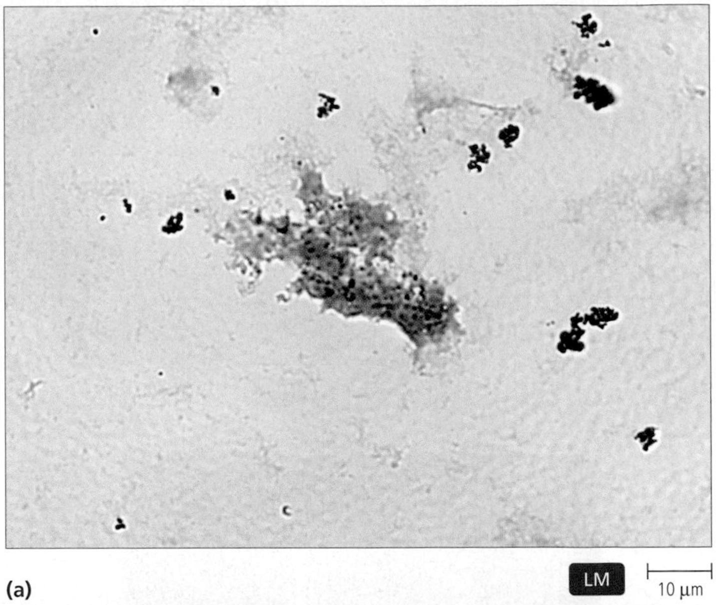

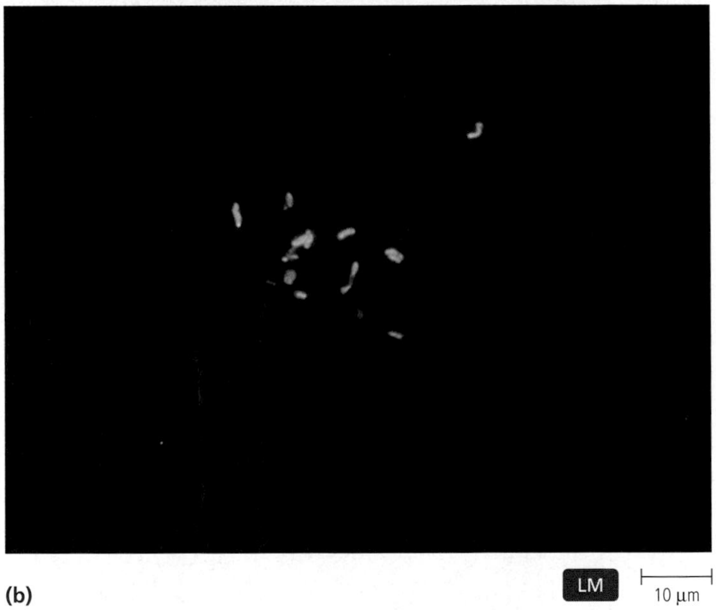

▲ ***Figure 4.9***

Fluorescent microscopy, in which fluorescent chemicals absorb invisible short-wavelength radiation and emit visible (longer wavelength) radiation. **(a)** When viewed under normal illumination, *Mycobacterium tuberculosis* cells stained with the fluorescent dye auramine O are invisible amid the mucus and debris in a sputum smear. **(b)** When the same smear is viewed under UV light, the bacteria fluoresce and are clearly visible.

Some fluorescent dyes are specific for certain cells. For example, the dye fluorescein isothiocyanate attaches to cells of *Bacillus anthracis* (ba-sil'ŭs an-thrā'sis), the causative agent of anthrax, and appears apple green when viewed in a fluorescent microscope. Another dye, auramine O, which fluoresces yellow, stains *Mycobacterium tuberculosis* (mī'kō-bak-tēr'ē-ŭm tū-ber-kyū-lō'sis; **Figure 4.9**).

Fluorescent microscopy is also used in a process called *immunofluorescence.* First, fluorescent dyes are covalently linked to Y-shaped immune system proteins called *antibodies* **(Figure 4.10a).** Given the opportunity, these dye-tagged antibodies will bind specifically to complementary-shaped *antigens,* which are molecules (or portions of molecules) that are present, for example, on the surface of microbial cells. When viewed under UV light, a microbial specimen that has bound dye-tagged antibodies becomes visible **(Figure 4.10b).** In addition to identifying pathogens, including those that cause syphilis, rabies, and Lyme disease, scientists use immunofluorescence to locate and make visible a variety of proteins of interest.

Confocal Microscopes

Even though **confocal**[3] **microscopes** also use fluorescent dyes or fluorescent antibodies, these microscopes use ultraviolet lasers to illuminate the fluorescent chemicals in only a single plane that is no thicker than 1.0 μm; the rest of the specimen remains dark and out of focus. Visible light emitted by the dyes passes through a pinhole aperture, which helps eliminate blurring that can occur with other types of microscopes and increases resolution by up to 40%. Each image from a confocal microscope is thus an "optical slice" through the specimen, as if it had been thinly cut. Once individual images are digitized, a computer is used to construct a three-dimensional representation, which can be rotated and viewed from any direction (**Figure 4.11** on page 102). Confocal microscopes have been particularly useful for examining the relationships among various organisms within complex microbial communities called *biofilms.* Regular light microscopy cannot produce clear images within a living biofilm, and removing surface layers from a biofilm would change the dynamics of the community.

Electron Microscopy

Learning Objective

✓ Contrast transmission electron microscopes with scanning electron microscopes in terms of how they work, the images they produce, and the advantages of each.

Even with the most expensive phase microscope using the best oil immersion lens with the highest numerical aperture, resolution is still limited by the wavelength of visible light. Because the shortest visible radiation (violet) has a wavelength of about 400 nm, structures closer together than about 200 nm cannot be distinguished using a light microscope. By contrast, electrons traveling as waves have wavelengths between 0.01 nm and 0.001 nm, which is one ten-thousandth to one hundred-thousandth the wavelength of visible light. The resolving power of electron microscopes is therefore much greater than that of light microscopes, and with greater resolving power comes the possibility of

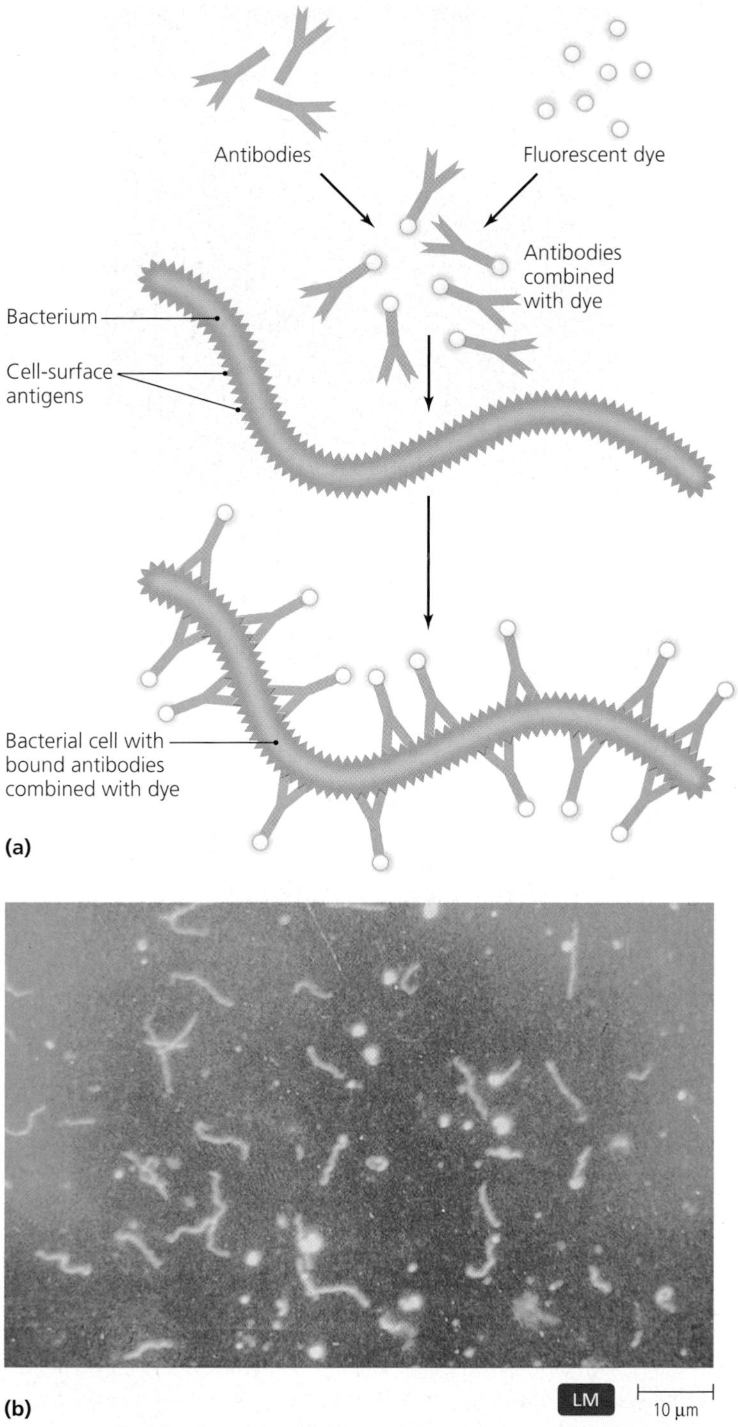

▲ ***Figure 4.10***

Immunofluorescence. **(a)** After a fluorescent dye is covalently linked to an antibody, the dye-antibody combination binds to the antibody's target, making the target visible under fluorescent microscopy. **(b)** Immunofluorescent staining of *Treponema pallidum,* the causative agent of syphilis. The bacteria are brightly colored against a dark background.

[3]From *coinciding focal* points of (laser) light.

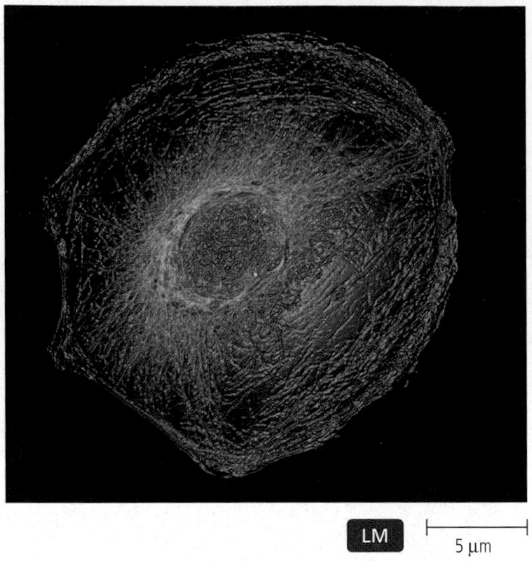

▲ ***Figure 4.11***

Confocal microscopy. Images are computer generated and created from many individual "optical slices" through a fluorescently-stained specimen. Here a cell's nucleus (purple) and cytoskeleton (red and green) are clearly seen.

greater magnification. **Highlight 4.1** describes the invention of the electron microscope in the 1930s.

Generally, electron microscopes magnify objects 10,000× to 100,000×, though magnification millions of times with good resolution is possible. Electron microscopes provide detailed views of the smallest bacteria, viruses, internal cellular structures, and even molecules and large atoms. Cellular structures that can be seen only by using electron microscopy are referred to as a cell's *ultrastructure*. Ultrastructural details cannot be made visible by light microscopy because they are too small to be resolved.

There are two general types of electron microscopes: *transmission electron microscopes* and *scanning electron microscopes*.

Transmission Electron Microscopes

A **transmission electron microscope (TEM)** generates a beam of electrons that ultimately produces an image on a fluorescent screen **(Figure 4.12a).** The path of electrons is similar to the path of light in a light microscope. From their source, the electrons pass through the specimen, through magnetic fields (instead of glass lenses) that manipulate and focus the beam, and then onto a fluorescent screen that absorbs electrons, thereby changing some of their energy into visible light **(Figure 4.12b).** Dense areas of the specimen block electrons from striking the screen, resulting in a dark area on the screen. In regions where the specimen is less dense, the screen fluoresces more brightly. As with light microscopy, contrast and resolution can be enhanced through the use of electron-dense stains, which are discussed later. The brightness of each region of the screen corresponds to the number of electrons striking it. Therefore, the image on the screen is composed of light and dark areas much like a photographic negative.

The screen can be folded out of the way to enable the electrons to strike a photographic film, located in the base of the microscope. Prints made from the film are called *transmission electron micrographs* or *TEM images*. Such images can be colorized to emphasize certain features **(Figure 4.13).**

Matter, including air, absorbs electrons, so the column of a transmission electron microscope must be a vacuum, and the specimen must be very thin. Before thicker specimens such as whole cells can be examined, they must be dehydrated, embedded in plastic, and cut to a thickness of about 100 nm with a diamond or glass knife mounted in a slicing machine called an *ultramicrotome*. Such thin sections are placed on a small copper grid and inserted into the microscope through an air lock.

Because the vacuum and slicing of the specimens are required, electron microscopes cannot be used to study living organisms. Unfortunately dehydration and sectioning can introduce shrinkage, distortion, and other *artifacts,* which

Highlight 4.1 "One of the Most Important Inventions of the Century"

By the beginning of the 20th century, microscopists had reached an impasse. They could not increase the useful magnification of microscopes because resolution is dependent on the wavelength of light. In 1931, Ernst Ruska, an engineering student at the Technical University in Berlin, posited that electrons, which had been shown to sometimes act like waves with very short wavelengths, might be focused on a specimen, providing the means for greater resolution and greater useful magnification. But one major obstacle to this idea remained: the lack of a lens that could focus electron waves. So Ruska set out to produce one.

In 1931, Ruska built the first magnetic lens that could focus a beam of electrons, much as a glass lens focuses light waves. In 1933, using a series of magnetic lenses, he created the first electron microscope. This prototype was capable of magnification of only a few hundred times, and it would be another decade before a working electron microscope would become a reality.

Still, Ruska's surpassing idea—the principle that electrons could be used for microscopy—opened the door for modern microbiologists to study the ultrastructure of cells. Calling the electron microscope "one of the most important inventions of the century," the Nobel Society awarded Ernst Ruska the Nobel Prize for physics more than 50 years later, in 1986.

CRITICAL THINKING

Electron microscopy reveals the ultrastructure of a cell. What cellular structures are considered ultrastructural?

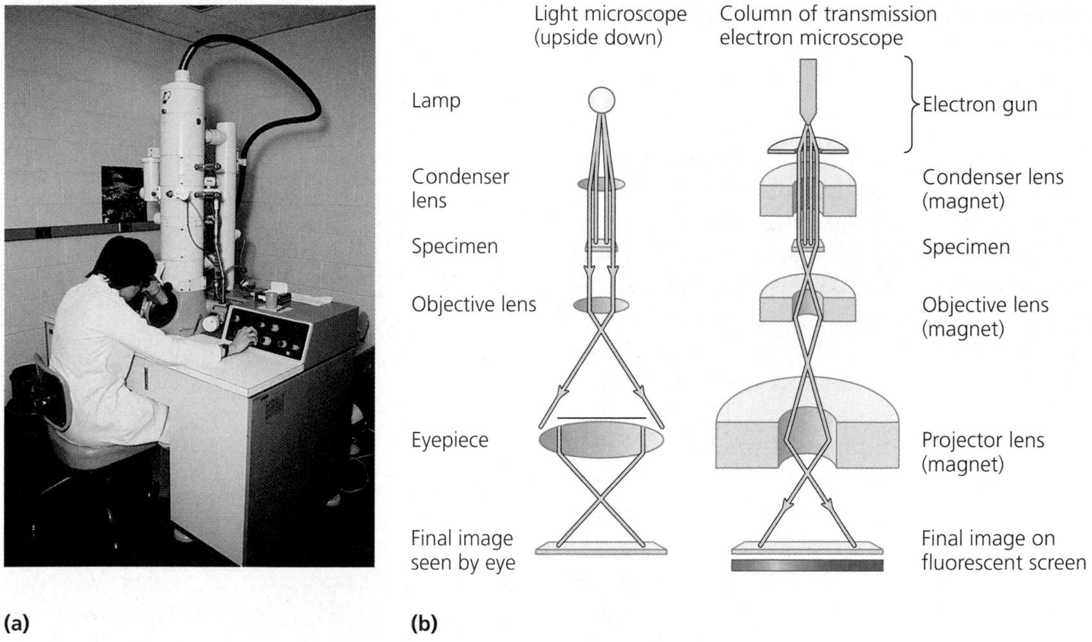

Figure 4.12

A transmission electron microscope (TEM). **(a)** TEMs are much larger than light microscopes. **(b)** The path of electrons through a TEM (at right), as compared to the path of light through a light microscope (at left, drawn upside down to facilitate the comparison). *Why must air be evacuated from the column of an electron microscope?*

Figure 4.12 *Air would absorb electrons, so there would be no radiation to produce an image.*

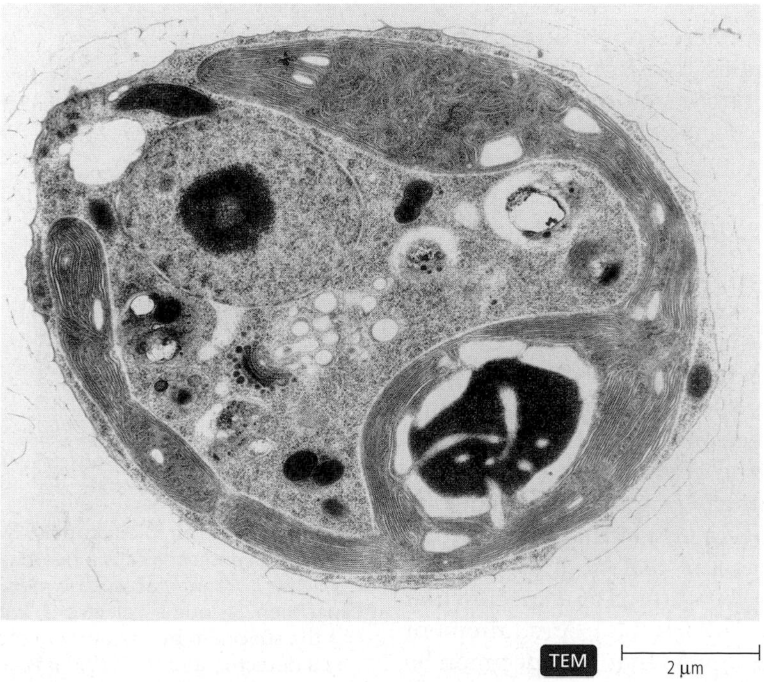

Figure 4.13

Transmission electron image. A transmission electron micrograph reveals much internal detail not visible by light microscopy.

are structures that appear in a TEM image but are not present in natural specimens.

Scanning Electron Microscopes

A **scanning electron microscope (SEM)** also uses magnetic fields within a vacuum tube to manipulate a beam of electrons, called primary electrons **(Figure 4.14a).** However, rather than passing electrons through a specimen, the SEM rapidly focuses them back and forth across the specimen's surface, which has previously been coated with a metal such as platinum or gold **(Figure 4.14b).** The primary electrons knock electrons off the surface of the specimen, and these scattered secondary electrons pass through a detector and a photomultiplier, producing an amplified signal that is displayed on a monitor. Typically, scanning microscopes are used to magnify up to 10,000× with a resolution of about 20 nm.

One advantage of scanning microscopy over transmission microscopy is that whole specimens can be observed, because sectioning is not required. Scanning electron micrographs can be beautifully realistic and three-dimensional **(Figure 4.15).** Two disadvantages of a scanning electron microscope are that it magnifies only the external surface of a specimen, and, like TEM, it requires a vacuum and thus can examine only dead organisms.

Probe Microscopy

Learning Objective

✓ Describe two variations of probe microscopes.

A relatively recent advance in microscopy utilizes minuscule, pointed, electronic probes to magnify more than 100,000,000 times. There are two variations of probe microscopes: *scanning tunneling microscopes* and *atomic force microscopes*.

Scanning Tunneling Microscopes

A **scanning tunneling microscope (STM)** passes a metallic probe, sharpened to end in a single atom, back and forth across and slightly above the surface of a specimen. Rather than scattering a beam of electrons into a detector, as in scanning electron microscopy, a scanning tunneling microscope measures the flow of electrons to and from the probe and the specimen's surface. The amount of electron flow, called a *tunneling current,* is directly proportional to the distance from the probe to the specimen's surface. A scanning tunneling microscope can measure distances as small as 0.01 nm and reveal details on the surface of a specimen at the atomic level (**Figure 4.16a** on page 106). A requirement for scanning tunneling microscopy is that the specimen be electrically conductive.

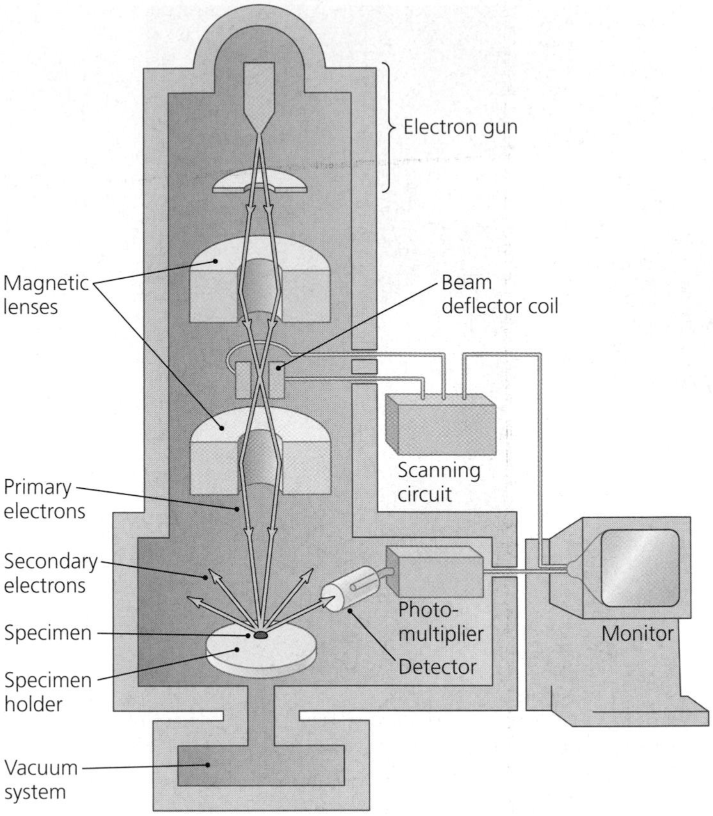

(a)

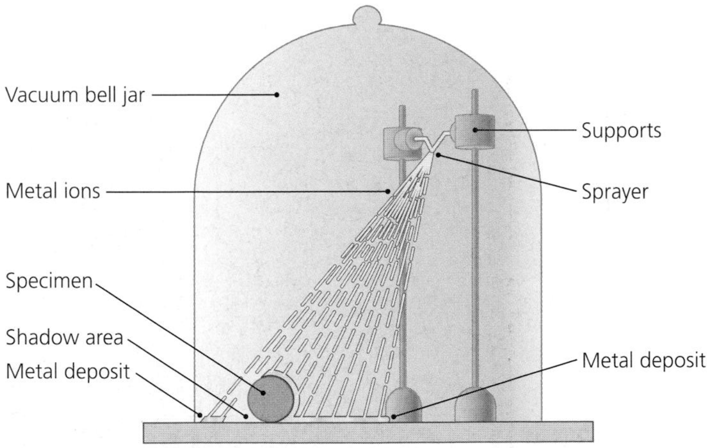

(b)

▲ *Figure 4.14*

Scanning electron microscope (SEM). **(a)** The SEM uses magnetic lenses to focus a beam of primary electrons, which are scanned across the metal-coated surface of a specimen. Secondary electrons, knocked off the surface of the specimen by the primary electrons, are collected by a detector, and their signal is amplified and displayed on a monitor. **(b)** The coating of a specimen with a thin layer of metal ions in preparation for its examination in the SEM.

(a) Diatom SEM 10 μm

(b) *Aspergillus* SEM 50 μm

(c) *Paramecium* SEM 25 μm

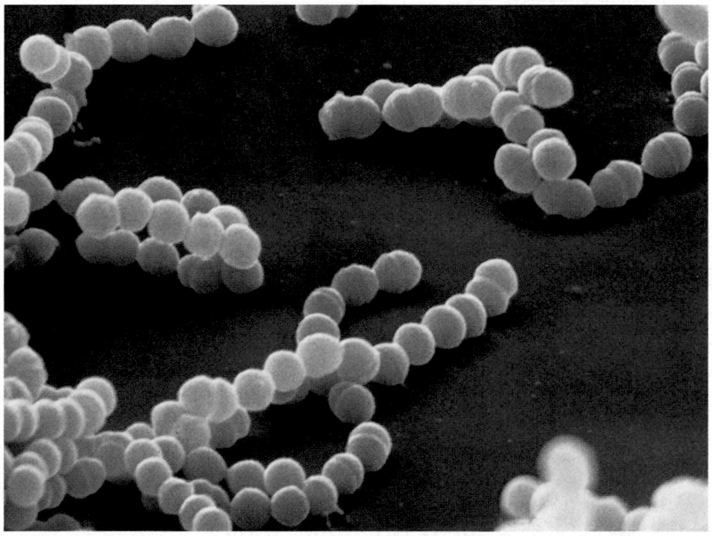

(d) *Streptococcus* SEM 1 μm

▲ ***Figure 4.15***
SEM images.

Atomic Force Microscopes

An **atomic force microscope (AFM)** also uses a pointed probe, but it traverses the tip of the probe lightly on the surface of the specimen, rather than at a distance. This might be likened to the way a person reads Braille. Deflection of a laser beam aimed at the probe's tip measures vertical movements, which when translated by a computer reveals the atomic topography.

Unlike tunneling microscopes, atomic force microscopes can magnify specimens that do not conduct electrons. They can also magnify living specimens, because neither an electron beam nor a vacuum is required (see **Figure 4.16b**). These microscopes have been used to magnify the surfaces of bacteria, viruses, proteins, and amino acids. Recent studies using atomic force microscopes have provided insights into the molecular folding of lipopolysaccharides on the surfaces of Gram-negative pathogenic bacteria.

Table 4.2 on page 107 summarizes the features of the various types of microscopes.

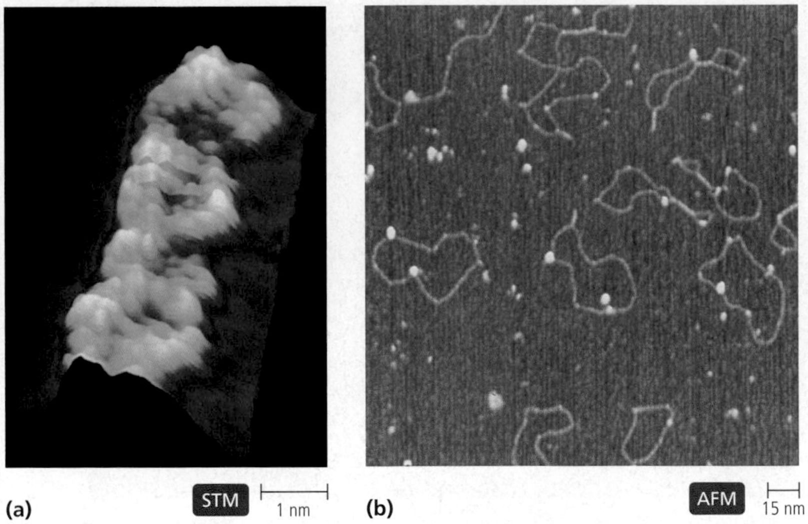

Figure 4.16

Probe microscopy. **(a)** Scanning tunneling microscopes reveal surface detail; in this case, three turns of a DNA double helix. **(b)** Plasmid DNA being digested by an enzyme, viewed through atomic force microscopy.

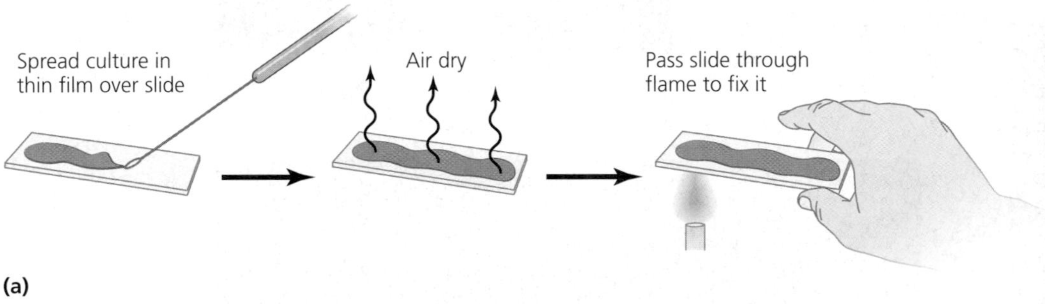

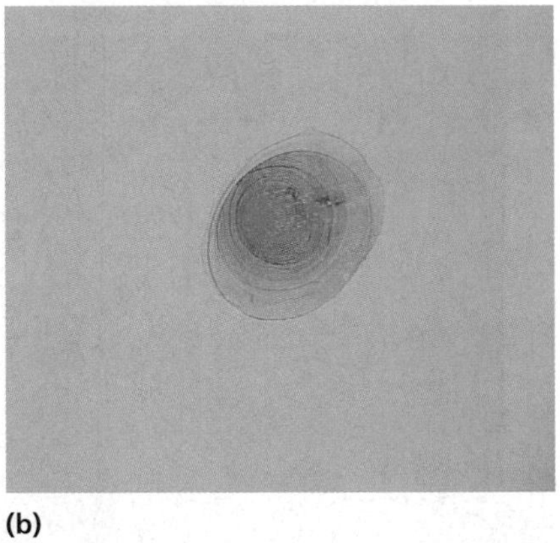

Figure 4.17

Preparing a specimen for staining. **(a)** Microorganisms are spread in liquid across the surface of a slide using a circular motion. After drying in the air, the smear is passed through the flame of a Bunsen burner to fix the cells to the glass. Alternatively, chemical fixation can be used. **(b)** A properly prepared smear lacks clumps and is thinner toward the edges than in the middle. *Why must a smear be fixed to the slide?*

Staining

Earlier we discussed the difficulty of resolving two distant white golf balls viewed against a white background. If the balls were painted black, they could be distinguished more readily from the background, and from one another. This illustrates why staining increases contrast and resolution.

Most microorganisms are colorless and difficult to view with bright-field microscopes. Microscopists use stains to make microorganisms and their parts more visible because stains increase contrast between structures, and between a specimen and its background. Electron microscopy also requires the treatment of specimens with stains or coatings to enhance contrast.

In this section we examine how scientists prepare specimens for staining and how stains work, and we consider five kinds of stains used for light microscopy. We conclude with a look at staining for electron microscopy.

Preparing Specimens for Staining

Learning Objective

✓ Explain the purposes of a smear, heat fixation, and chemical fixation in the preparation of a specimen for microscopic viewing.

Many investigations of microorganisms, especially those seeking to identify pathogens, begin with light microscopic observation of stained specimens. **Staining** simply means coloring specimens with stains, which are also called dyes.

Before microbiologists stain microorganisms, they must place them on and then firmly attach them to a microscope slide **(Figure 4.17).** Typically, this involves making a *smear* and *fixing* it to the slide. If the organisms are growing in a liquid, a small drop is spread across the surface of the slide. If the organisms are growing on a solid surface, such as an agar plate, then they are mixed into a small drop of water on

Figure 4.17 Fixation causes the specimen to adhere to the glass so that it does not easily wash off during staining.

Table 4.2 Comparison of Types of Microscopes

Type of Microscope	Typical Image	Description of Image	Special Features	Typical Uses
Light Microscopes		Useful magnification 1× to 2000×; resolution to 200 nm	Use visible light; shorter, blue wavelengths provide better resolution	
Bright-field		Colored or clear specimen against bright background	Simple to use; relatively inexpensive; stained specimens often required	Observation of killed stained specimens and naturally colored live ones; also used to count microorganisms
Dark-field		Bright specimen against dark background	Use a special filter in the condenser that prevents light from directly passing through a specimen; only light scattered by the specimen is visible	Observation of living, colorless, unstained organisms
Phase-contrast	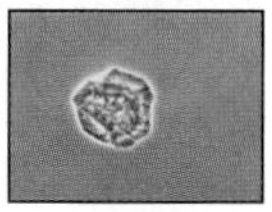	Specimen has light and dark areas	Use a special condenser that splits a polarized light beam into two beams, one of which passes through the specimen, and one of which bypasses the specimen; the beams are then rejoined before entering the oculars; contrast in the image results from the interactions of the two beams	Observation of internal structures of living microbes
Differential interference contrast (Nomarski)	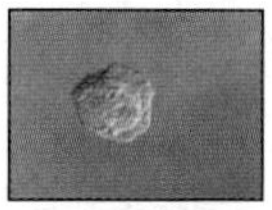	Image appears three-dimensional	Use two separate beams instead of a split beam; false color and a three-dimensional effect result from interactions of light beams and lenses; no staining required	Observation of internal structures of living microbes
Fluorescent		Brightly colored fluorescent structures against dark background	An ultraviolet light source causes fluorescent natural chemicals or dyes to emit visible light	Localization of specific chemicals or structures; used as an accurate and quick diagnostic tool for detection of pathogens
Confocal	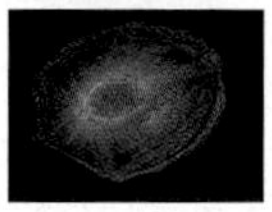	Single plane of structures or cells that have been specifically stained with fluorescent dyes	Use a laser to fluoresce only one plane of the specimen at a time	Detailed observation of structures of cells within communities
Electron Microscopes		Typical magnification 1000× to 100,000×; resolution to 0.001 nm	Use electrons traveling as waves with short wavelengths; require specimens to be in a vacuum, so cannot be used to examine living microbes	
Transmission	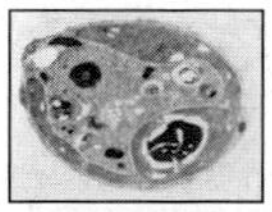	Monotone, two-dimensional, highly magnified images; may be color-enhanced	Produce two-dimensional image of ultrastructure of cells	Observation of internal ultrastructural detail of cells and observation of viruses and small bacteria
Scanning		Monotone, three-dimensional, surface images; may be color-enhanced	Produce three-dimensional view of the surface of microbes and cellular structures	Observation of the surface details of structures
Probe Microscopes		Magnification greater than 100,000,000× with resolving power greater than that of electron microscopes	Use microscopic probes that move over the surface of a specimen	
Scanning tunneling	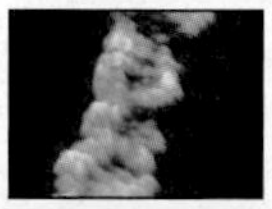	Individual molecules and atoms visible	Measures the flow of electrical current between the tip of a probe and the specimen to produce an image of the surface at atomic level	Observation of the surface of objects; provide extremely fine detail, high magnification, and great resolution
Atomic force	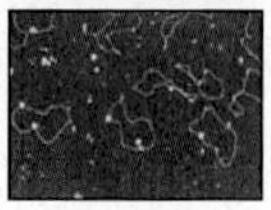	Individual molecules and atoms visible	Measure the deflection of a laser beam aimed at the tip of a probe that travels across the surface of the specimen	Observation of living specimens at the molecular and atomic levels

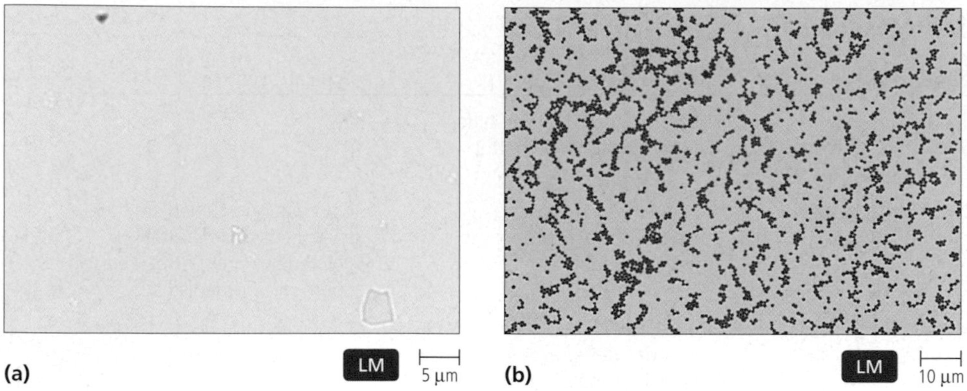

Figure 4.18

Simple stains, which are simply and quickly performed, increase contrast, and allow determination of size, shape, and arrangement of cells. **(a)** Unstained *Staphylococcus aureus.* **(b)** Methylene blue simple stain of *Staphylococcus aureus* (cocci) and *Escherichia coli* (rods). Note that all cells, no matter their type, stain the same color with a simple stain because only one dye is used.

the slide. Either way, the thin film of organisms on the slide is called a **smear.**

The smear is air dried and then attached or fixed to the surface of the slide using either heat or chemicals. Robert Koch developed **heat fixation** more than a hundred years ago. In heat fixation, the slide is gently heated by passing the slide, smear up, through the flame of a Bunsen burner. Alternately, **chemical fixation** involves applying methyl alcohol (methanol) or formalin (a solution of formaldehyde in water) to the smear for one minute. Desiccation (drying) and fixation kill the microorganisms, attach them firmly to the slide, and generally preserve their shape and size. It is important to smear and fix specimens properly so that they are not lost during staining.

Specimens prepared for electron microscopy are also dried, because water vapor from a wet specimen would stop the electron beam. As we have seen, transmission electron microscopy requires that the desiccated sample also be sliced very thin, generally before staining.

Principles of Staining

Learning Objective

✓ Describe the uses of acidic and basic dyes, mentioning ionic bonding and pH.

Dyes used as microbiological stains for light microscopy are usually salts. As we saw in Chapter 2, a salt is composed of a positively charged *cation* and a negatively charged *anion.* At least one of the two ions in the molecular makeup of dyes is colored; this colored portion of a dye is known as the *chromophore.* Chromophores typically bind to chemicals via covalent, ionic, or hydrogen bonds. For example, methylene blue chloride is composed of a cationic chromophore, methylene blue, and a chloride anion. Since methylene blue is positively charged, it ionically bonds to negatively charged molecules in cells, including DNA and many proteins. In contrast, anionic dyes, for example, eosin, bind to positively charged molecules such as some amino acids.

Anionic chromophores are also called **acidic dyes** because they stain alkaline structures and work best in acidic (low pH) environments. Positively charged, cationic chromophores are called **basic dyes** because they combine with and stain acidic structures; further, they work best under basic (higher pH) conditions. In microbiology, basic dyes are used more commonly than acidic dyes because most cells are negatively charged. Acidic dyes are used in negative staining, which is discussed shortly.

Some stains do not form bonds with cellular chemicals, but rather function because of their solubility characteristics. For example, Sudan Black selectively stains membranes because it is lipid soluble and accumulates in phospholipid bilayers.

Simple Stains

Learning Objective

✓ Describe the simple, Gram, acid-fast, and endospore staining procedures.

Simple stains are composed of a single basic dye such as crystal violet, safranin, or methylene blue. They are "simple" because they involve no more than soaking the smear in the dye for 30–60 seconds and then rinsing off the slide with water. (A properly fixed specimen will remain attached to the slide despite this treatment.) After carefully blotting the slide dry, the microbiologist observes the smear under the microscope. Simple stains are used to determine size, shape, and arrangement of cells **(Figure 4.18).**

Differential Stains

Most stains used in microbiology are **differential stains,** which use more than one dye so that different cells, chemicals,

Figure 4.19 ➤

The Gram staining procedure. A specimen is smeared and fixed to a slide. The classical procedure consists of four steps. Gram-positive cells (in this case, *Bacillus cereus*) remain purple throughout the procedure; Gram-negative cells (here, *Escherichia coli*) end up pink.

1 Slide is flooded with crystal violet for 1 min, then rinsed with water.

Result: All cells are stained purple.

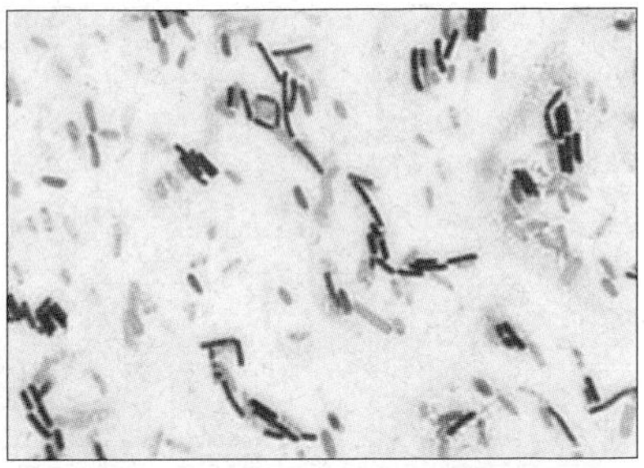

2 Slide is flooded with iodine for 1 min, then rinsed with water.

Result: Iodine acts as a mordant; all cells remain purple.

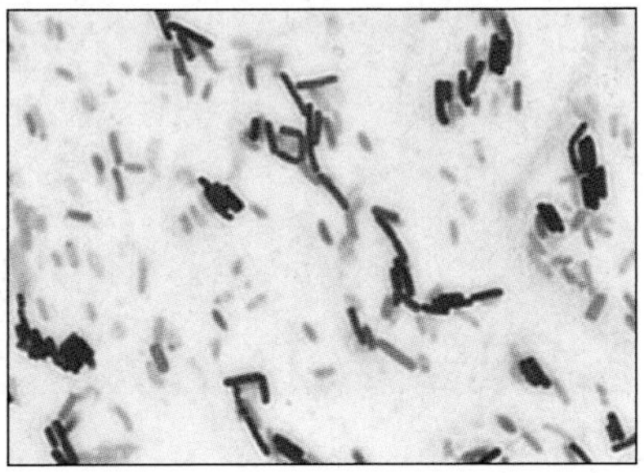

3 Slide is flooded with solution of alcohol and acetone for 10–30 sec, then rinsed with water.

Result: Smear is decolorized; Gram-positive cells remain purple, but Gram-negative cells are now colorless.

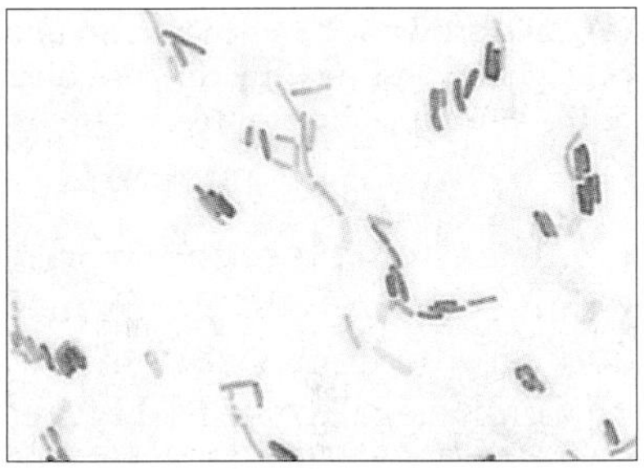

4 Slide is flooded with safranin for 1 min, then rinsed with water and blotted dry.

Result: Gram-positive cells remain purple, Gram-negative cells are pink.

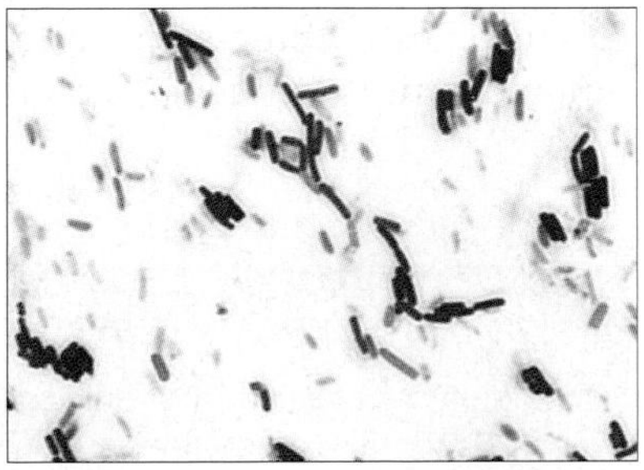

or structures can be distinguished when microscopically examined. Common differential stains are the *Gram stain*, the *acid-fast stain*, and the *endospore stain*.

Gram Stain

In 1884, the Danish scientist Hans Christian Gram developed the most frequently used differential stain, which now bears his name. The **Gram stain** differentiates between two large groups of microorganisms: purple-staining Gram-positive cells, and pink-staining Gram-negative cells. Recall from Chapter 3 that these cells differ significantly in the chemical and physical structures of their cell walls (see Figure 3.13 on page 69). Typically, a Gram stain is the first step a medical laboratory technologist performs to identify bacterial pathogens.

Let's examine the Gram staining procedure as it was originally developed, and as it is typically performed today, over a hundred years later. For the purposes of our discussion, we'll assume that a smear has been made on a slide and heat fixed, and that the smear contains both Gram-positive and Gram-negative colorless bacteria. The classical Gram staining procedure has the following four steps **(Figure 4.19):**

1. Flood the smear with the basic dye crystal violet for 1 min, and then rinse with water. Crystal violet, which is called the **primary stain,** colors all cells.
2. Flood the smear with an iodine solution for 1 min, and then rinse with water. Iodine is a **mordant,** a substance that binds to a dye and makes it less soluble. After this step, all cells remain purple.
3. Rinse the smear with a solution of ethanol and acetone for 10–30 sec, and then rinse with water. This solution, which acts as a **decolorizing agent,** breaks down the thin cell wall of Gram-negative cells, allowing the stain and mordant to be washed away; these cells are now colorless. Gram-positive cells, with their thicker cell walls, remain purple.
4. Flood the smear with safranin for 1 min, and then rinse with water. This red **counterstain** provides a contrasting color to the primary stain. Although all types of cells may absorb safranin, the resulting pink color is masked by the darker purple dye already in Gram-positive cells. After this step, Gram-negative cells now appear pink, whereas Gram-positive cells remain purple.

After the final step, the slide is blotted dry in preparation for microscopy.

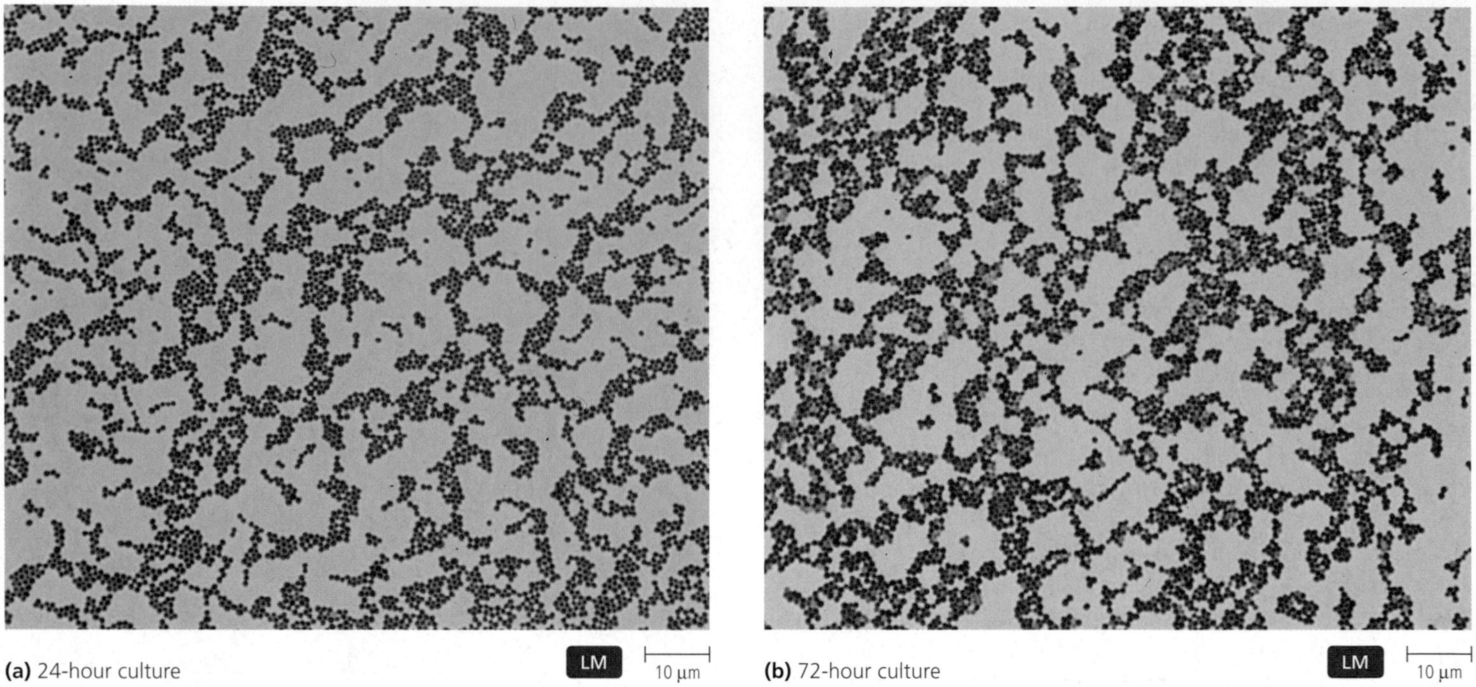

(a) 24-hour culture **(b)** 72-hour culture

▲ ***Figure 4.20***

Effects of time on Gram staining of *Staphylococcus aureus,* a Gram-positive organism. **(a)** Cells grown for 24 hours stain uniformly purple. **(b)** In this 72-hour culture, older, presumably less healthy cells are bleached by the decolorizer, lose their purple color, and stain pink. This creates a false Gram-negative reaction in many cells.

The Gram procedure works best with young cells. Older Gram-positive cells bleach more easily than younger cells and can therefore stain pink, which makes them appear to be Gram-negative cells **(Figure 4.20).** Therefore, smears for Gram-staining should come from freshly grown bacteria.

Microscopists have developed minor variations on Gram's original procedure. For example, 95% ethanol may be used to decolorize instead of Gram's ethanol-acetone mixture. In a three-step variation, safranin dissolved in ethanol simultaneously decolorizes and counterstains.

New Frontiers 4.1 focuses on a miniature sensor that is being developed to differentiate between Gram-positive and Gram-negative bacteria.

Acid-Fast Stain

The **acid-fast stain** is another important differential stain because it stains cells of the genera *Mycobacterium* and *Nocardia* (nō-kar′dē-ă), which cause many human diseases, including tuberculosis, leprosy, and other lung and skin infections. Cells of these bacteria have large amounts of waxy lipid in

New Frontiers 4.1 "Smart" Bandages of the Future

Imagine a bandage that not only covers an infected cut on your finger, but also identifies whether the infectious agent is Gram-positive or Gram-negative. The use of such "smart" bandages may not be far off. Researchers at the University of Rochester have developed a miniature sensor that can differentiate between

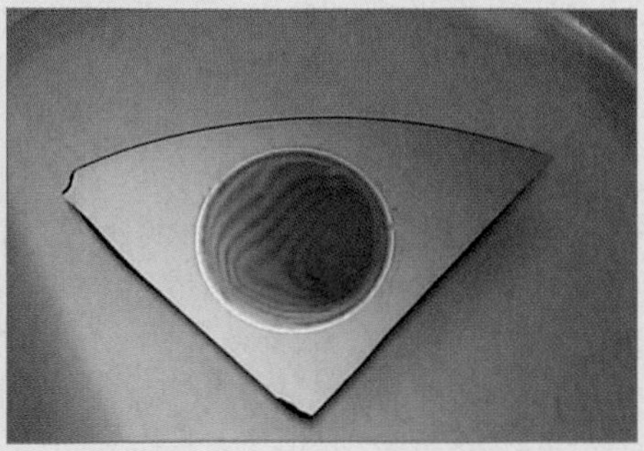

Gram-positive and Gram-negative bacteria even more reliably than the traditional Gram staining procedure. The sensor (see photo) is a porous silicon chip that can detect lipid A, which is present only on the outer membrane of Gram-negative bacteria (see Figure 3.13 on p. 69). When lipid A molecules bind to the organic receptor molecules that coat the sensor, they cause a color change on the sensor that can be used to differentiate between the two main groups of bacteria. While the sensor is not yet practical for household use, a version of it could someday be incorporated in bandages to rapidly identify infectious agents, or be used in meat packaging plants to detect contaminants.

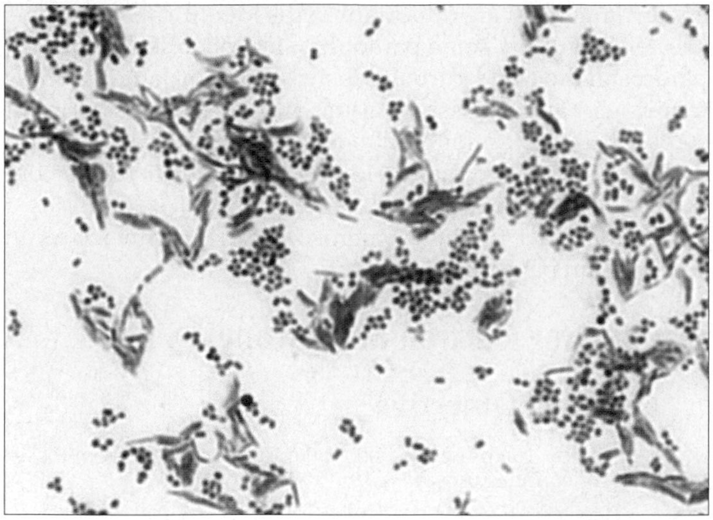

▲ ***Figure 4.21***

Ziehl-Neelsen acid-fast stain. Acid-fast cells such as these rod-shaped *Mycobacterium bovis* cells stain pink or red. Non-acid-fast cells—in this case, *Streptococcus pyogenes*—stain blue. *Why isn't the Gram stain utilized to stain* Mycobacterium*?*

Figure 4.21 *Cell walls of* Mycobacterium *are composed of waxy materials that repel the water-based dyes of the Gram stain.*

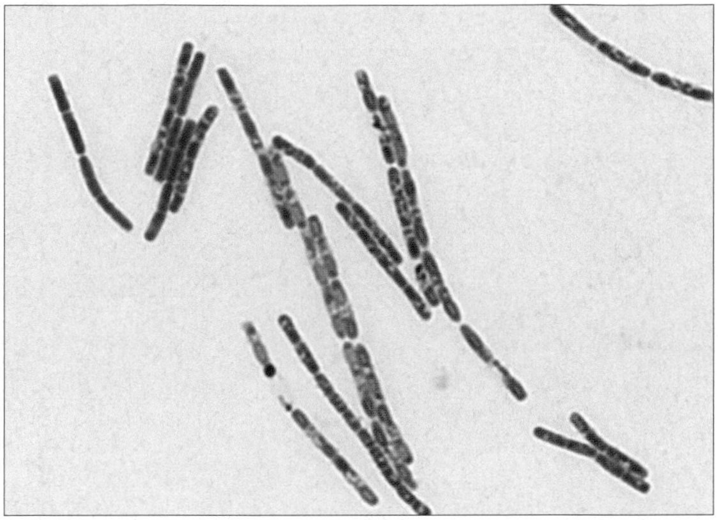

▲ ***Figure 4.22***

Schaeffer-Fulton endospore stain of *Bacillus anthracis*. The nearly impermeable spore wall retains the green dye during decolorization. Vegetative cells, which lack spores, pick up the counterstain and appear red. *Why don't the spores stain red as well?*

Figure 4.22 *Heat from steam is used to drive the green primary stain into the endospores. Counterstaining is performed at room temperature, and the thick, impermeable walls of the endospores resist the counterstain.*

their cell walls, so they do not readily stain with simple water-based dyes such as methylene blue and crystal violet, or with the Gram stain.

Modern microbiological laboratories commonly use a variation of the acid-fast stain developed by Franz Ziehl (1857–1926) and Friedrich Neelsen (1854–1894) in 1883. Their procedure is as follows:

1. Cover the smear with a small piece of tissue paper to retain the dye during the procedure.
2. Flood the slide with the red primary stain, carbolfuchsin, for several minutes while warming it over steaming water. In this procedure, heat is used to drive the stain through the waxy wall and into the cell, where it remains trapped.
3. Remove the tissue paper, cool the slide, and then decolorize the smear by rinsing it with a solution of hydrochloric acid (pH < 1.0). The bleaching action of hydrochloric acid removes color from both non-acid-fast cells and the background. Acid-fast cells retain their red color because the acid cannot penetrate the waxy wall. The name of the procedure is derived from this step; that is, the cells are colorfast in acid.
4. Counterstain with methylene blue, which stains only bleached, non-acid-fast cells.

The Ziehl-Neelsen acid-fast staining procedure results in pink acid-fast cells, which can be differentiated from blue non-acid-fast cells, including human cells and tissue **(Figure 4.21).** The presence of *acid-fast rods (AFRs)* in sputum is indicative of mycobacterial infection.

CRITICAL THINKING

In what ways are the Gram stain and the acid-fast staining procedures similar? In the acid-fast procedure, what takes the place of Gram's iodine mordant?

Endospore Stain

Some bacteria—notably those of the genera *Bacillus* and *Clostridium* (klos-trid′ē-ŭm), which contain species that cause such diseases as anthrax, gangrene, and tetanus—produce **endospores.** These dormant, highly resistant cells form inside the cytoplasm of the bacteria and can survive environmental extremes such as desiccation, heat, and harmful chemicals. Endospores cannot be stained by normal staining procedures because their walls are practically impermeable to all chemicals. The **Schaeffer-Fulton endospore stain** uses heat to drive the primary stain, *malachite green,* into the endospore. After cooling, the slide is decolorized with water and counterstained with safranin. This staining procedure results in green-stained endospores and red-colored vegetative cells **(Figure 4.22).**

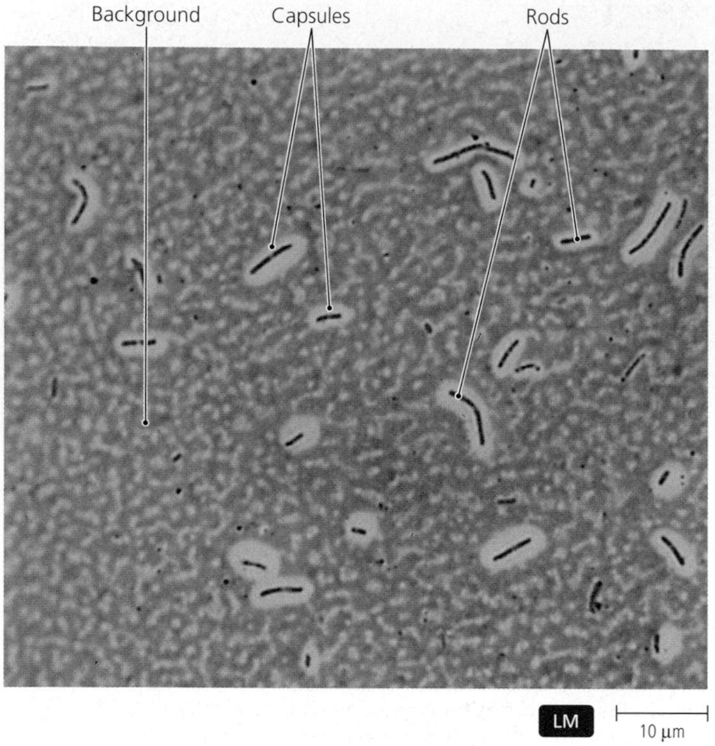

▲ *Figure 4.23*

Negative (capsule) stain of *Klebsiella pneumoniae*. Notice that the acidic dye stains the background and does not penetrate the capsule. The cells have been counterstained with a different (basic) dye.

Special Stains

Special stains are simple stains designed to reveal special microbial structures. There are three types of special stains: *negative stains, flagellar stains,* and *fluorescent stains* (which we already discussed in the section on fluorescent microscopy).

Negative (Capsule) Stain

Most dyes used to stain bacterial cells, such as crystal violet, methylene blue, malachite green, and safranin, are basic dyes. These dyes stain cells by attaching to negatively charged molecules within them.

Acidic dyes, by contrast, are repulsed by the negative charges on the surface of cells and therefore do not stain them. Such stains are called **negative stains** because they stain the background and leave cells colorless. Eosin and nigrosin are examples of acidic dyes used for negative staining. A counterstain may be added to color the cells.

Negative stains are used primarily to reveal the presence of negatively charged bacterial capsules. Therefore, they are also called **capsule stains.** Encapsulated cells appear to have a halo surrounding them **(Figure 4.23).**

Flagellar Stain

Bacterial flagella are extremely thin and thus normally invisible with light microscopy, but their presence, number, and arrangement are important in the identification of some species, including some pathogens. Flagellar stains, such as pararosanaline and carbolfuchsin, and mordants, such as tannic acid and potassium alum, are applied in a series of steps. These molecules bind to the flagella, increase their diameter, and change their color, all of which increases contrast and makes them visible **(Figure 4.24).**

Stains used for light microscopy are summarized in **Table 4.3** on page 114.

Staining for Electron Microscopy

Learning Objective

✓ Explain how stains used for electron microscopy differ from those used for light microscopy.

Laboratory technicians increase contrast and resolution for transmission electron microscopy by using stains, just as they do for light microscopy. However, stains used for transmission electron microscopy are not colored dyes, but instead chemicals containing atoms of heavy metals such as lead, osmium, tungsten, and uranium, which absorb electrons. Electron-dense stains may bind to molecules within specimens, or they may stain the background. The latter type of negative staining is used to provide contrast for extremely small specimens such as viruses and molecules.

Stains for electron microscopy can be general, in that they stain most objects to some degree, or they may be highly specific. For example, osmium tetraoxide (OsO_4) has an affinity for lipids and is thus used to enhance the contrast of membranes. Electron-dense stains can also be linked to antibodies to provide an even greater degree of staining specificity because antibodies bind only to their specific target molecules.

Classification and Identification of Microorganisms

Learning Objective

✓ Discuss the purposes of classification and identification of organisms.

Biologists classify organisms for several reasons: to bring a sense of order and organization to the variety and diversity of living things, to enhance communication, to make predictions about the structure and function of similar organisms, and to uncover and understand potential evolutionary connections. They sort organisms on the basis of mutual similarities into nonoverlapping groups called **taxa.**[4] As we discussed in Chapter 1, **taxonomy**[5] is the science of classifying and naming organisms. Taxonomy consists of *classification,* which is assigning organisms to taxa based upon

[4]From Greek *taxis,* meaning order.
[5]From *taxis* and Greek *nomos,* meaning rule.

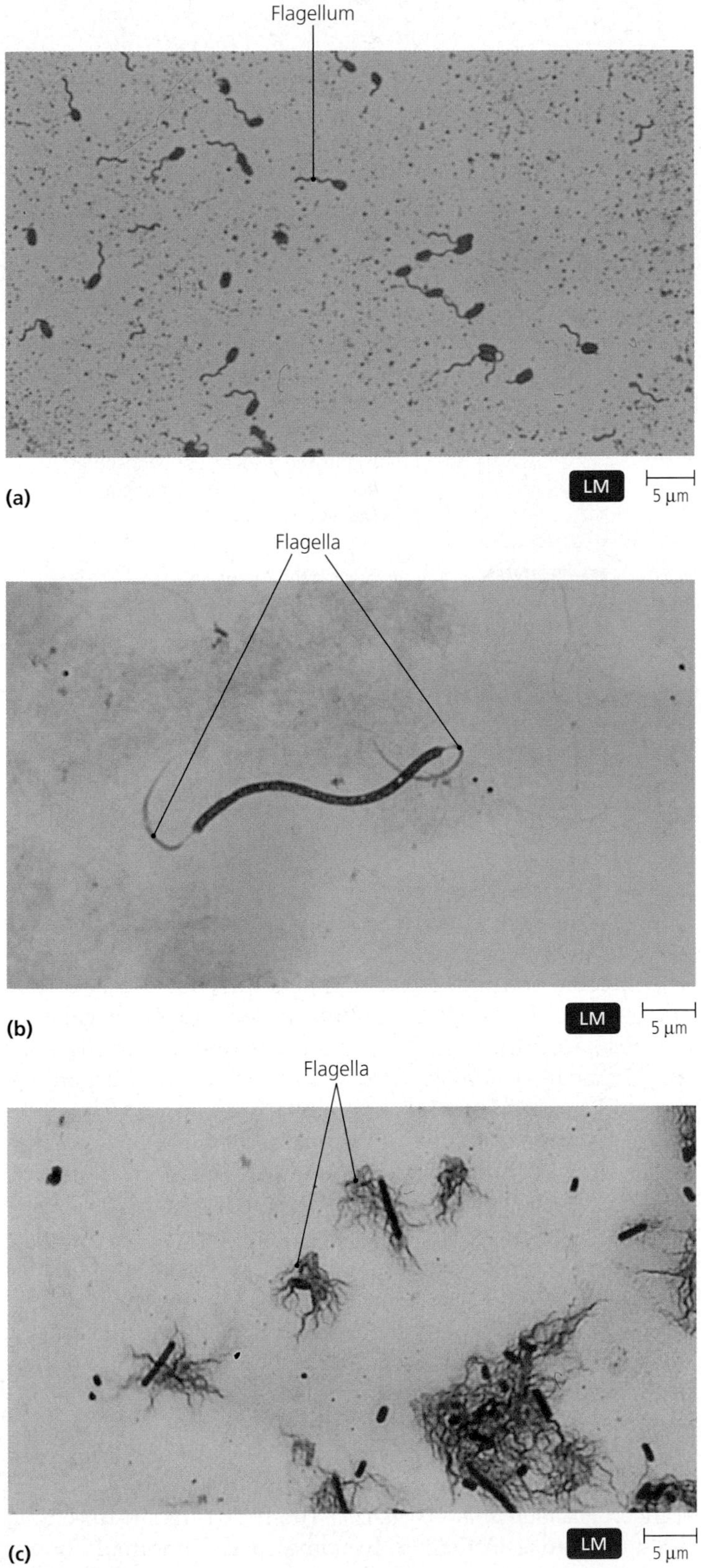

▲ ***Figure 4.24***

Flagellar stain. Various bacteria have different numbers and arrangements of flagella, features that are important in identifying some species. **(a)** *Pseudomonas aeruginosa* has a single flagellum. **(b)** *Spirullum volutans* has flagella at each end. **(c)** *Proteus vulgaris* is covered with many flagella.

similarities; *nomenclature,* which is concerned with the rules of naming organisms; and *identification,* which is the practical science of determining that an isolated individual or population belongs to a particular taxon. In this book we concentrate on classification and identification.

Since all members of any given taxon share certain common features, taxonomy enables scientists to both organize large amounts of information about organisms and make predictions based on knowledge of similar organisms. For example, if one member of a taxon is important in recycling nitrogen in the environment, it is likely that others in the group will play a similar ecological role. Likewise, a clinician might suggest a treatment against one pathogen based on what has been effective against another pathogen in the same taxon.

Identification of organisms is an essential part of taxonomy because it enables scientists to communicate effectively and be confident that they are discussing the same organism. Further, identification is often essential for treatment of groups of diseases such as meningitis and pneumonia, which can be caused by pathogens as different as fungi, bacteria, and viruses.

In this section we examine the historical basis of taxonomy, consider modern advances in this field, and briefly consider various taxonomic methods. This chapter also presents a general overview of the taxonomy of prokaryotes (which are considered in greater detail in Chapter 11); of animals, protozoa, fungi, and algae (Chapter 12); and of viruses, viroids, and prions (Chapter 13).

Linnaeus, Whittaker, and Taxonomic Categories

Learning Objectives

- ✓ Discuss the difficulties in defining species of microorganisms.
- ✓ List the hierarchy of taxa from general to specific.
- ✓ Define binomial nomenclature.
- ✓ Describe a few modifications of the Linnaean system of taxonomy.

Our current system of taxonomy began in 1758 with the publication of *Systema Naturae* by the Swedish botanist Carolus Linnaeus. Until his time, the names of organisms were often strings of descriptive terms that varied from country to country and from one scientist to another. Linnaeus provided a system that standardized the naming and classification of organisms based on characteristics they have in common. He grouped similar organisms that can successfully interbreed into categories called **species.**

The definition of species as "a group of organisms that interbreed to produce viable offspring" works relatively well for more complex, sexually reproducing organisms, but it is not satisfactory for asexual organisms, including most microorganisms. As a result, for asexual organisms many scientists define a species as a collection of *strains*—populations of cells that arose from a single cell—that share many stable properties and differ from other strains. Not surprisingly, this definition sometimes results in disagreements and inconsistencies in the classification of microbial life.

Table 4.3 Stains Used for Light Microscopy

Type of Stain	Examples	Results	Typical Images	Representative Uses
Simple stain (uses a single dye)	Crystal violet Methylene blue	Uniform purple stain Uniform blue stain		Reveals size, morphology, and arrangement of cells
Differential stains (use two or more dyes to differentiate between cells or structures)	Gram stain	Gram-positive cells are purple; gram-negative cells are pink	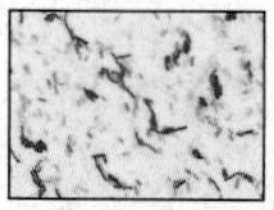	Differentiates between Gram-positive and Gram-negative bacteria, which is typically the first step in their identification
	Ziehl-Neelsen acid-fast stain	Pink to red acid-fast cells and blue non-acid-fast cells	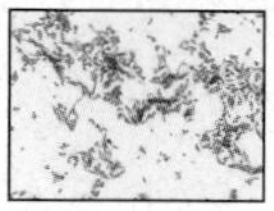	Distinguishes the genera *Mycobacterium* and *Nocardia* from other bacteria
	Schaeffer-Fulton endospore stain	Green endospores and pink to red vegetative cells		Highlights the presence of endospores produced by species in the genera *Bacillus* and *Clostridium*
Special stains	Negative stain for capsules	Background is dark, cells unstained or stained with simple stain	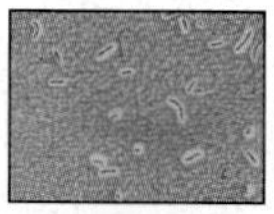	Reveals bacterial capsules
	Flagellar stain	Bacterial flagella become visible		Allows determination of number and location of bacterial flagella

In Linnaeus's system, which forms the basis of modern taxonomy, similar species are grouped into **genera,**[6] and similar genera into still larger taxonomic categories. That is, genera sharing common features are grouped together to form **families;** similar families are grouped into **orders;** orders are grouped into **classes;** classes into **phyla,**[7] and phyla into **kingdoms (Figure 4.25).**

All these categories, including species and genera, are taxa, which are hierarchical; that is, each successive taxon has a broader description than the preceding one, and each taxon includes all the taxa beneath it. The rules of nomenclature require that all taxa have Latin or Latinized names, in part because the language of science during Linnaeus's time was Latin, and in part because using Latin ensures that no country or ethnic group has priority in the language of taxonomy. The name *Chondrus crispus* (kon'drŭs krisp'ŭs) describes the exact same algal species all over the world, despite the fact that in England its common name is Irish moss, in Ireland it is carragheen, in North America it is curly moss, and it isn't really a moss at all!

When new microscopic, genetic, or biochemical techniques identify new or more detailed characteristics of organisms, a taxon may be split into two or more taxa. Alternatively, several taxa may be lumped together into a single taxon. For example, the genus "*Diplococcus*" has been united (synonymized) with the genus *Streptococcus* (strep-tō-kok'ŭs), and the name *Diplococcus pneumoniae* has been changed to reflect this synonymy—to *Streptococcus pneumoniae.*

Linnaeus assigned each species a descriptive name consisting of its genus name and a **specific epithet.** The genus name is always a noun, and it is written first and capitalized. The specific epithet always contains only lowercase letters and is usually an adjective. Both names, together called a *binomial,* are either printed in italics or underlined. Because the Linnaean system assigns two names to every organism, it is said to use **binomial[8] nomenclature.**

Consider the following examples of binomials. *Enterococcus faecalis* (en'ter-ō-kok'ŭs fē-kă'lis)[9] is a fecal bacterium. Whereas *Enterococcus faecium* (fē-sē'ŭm) is in the same genus, it is a different species because of certain differing characteristics, so it is given a different specific epithet. Humans are classified as *Homo sapiens* (hō'mō sā'pē-enz); the genus name means "man" and the specific epithet means "wise." The genus *Homo* contains no other living species, but scientists have assigned some fossil remains to *Homo neanderthalensis* (nē-an'der-thol-en'sis).

Note that even though binomials are often descriptive of an organism, sometimes they can be misleading. For instance, *Haemophilus influenzae* (hē-mof'i-lŭs in-flū-en'zē) does not cause influenza. In some cases, binomials honor

[6]Plural of Latin *genus,* meaning race or birth.
[7]Plural of *phylum,* from Greek *phyllon,* meaning tribe. The term *phylum* is used for animals and bacteria; the corresponding taxon in mycology and botany is called a *division.*
[8]From Greek *bi,* meaning two, and *nomos,* meaning rule.
[9]From Greek *enteron,* meaning intestine, and *kokkos,* meaning berry.

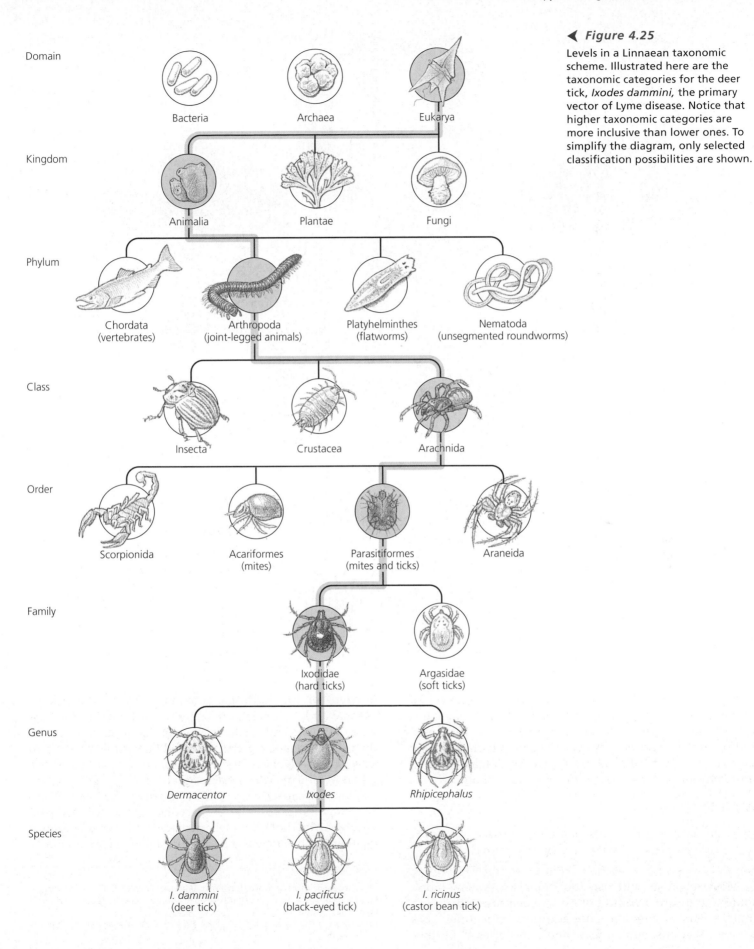

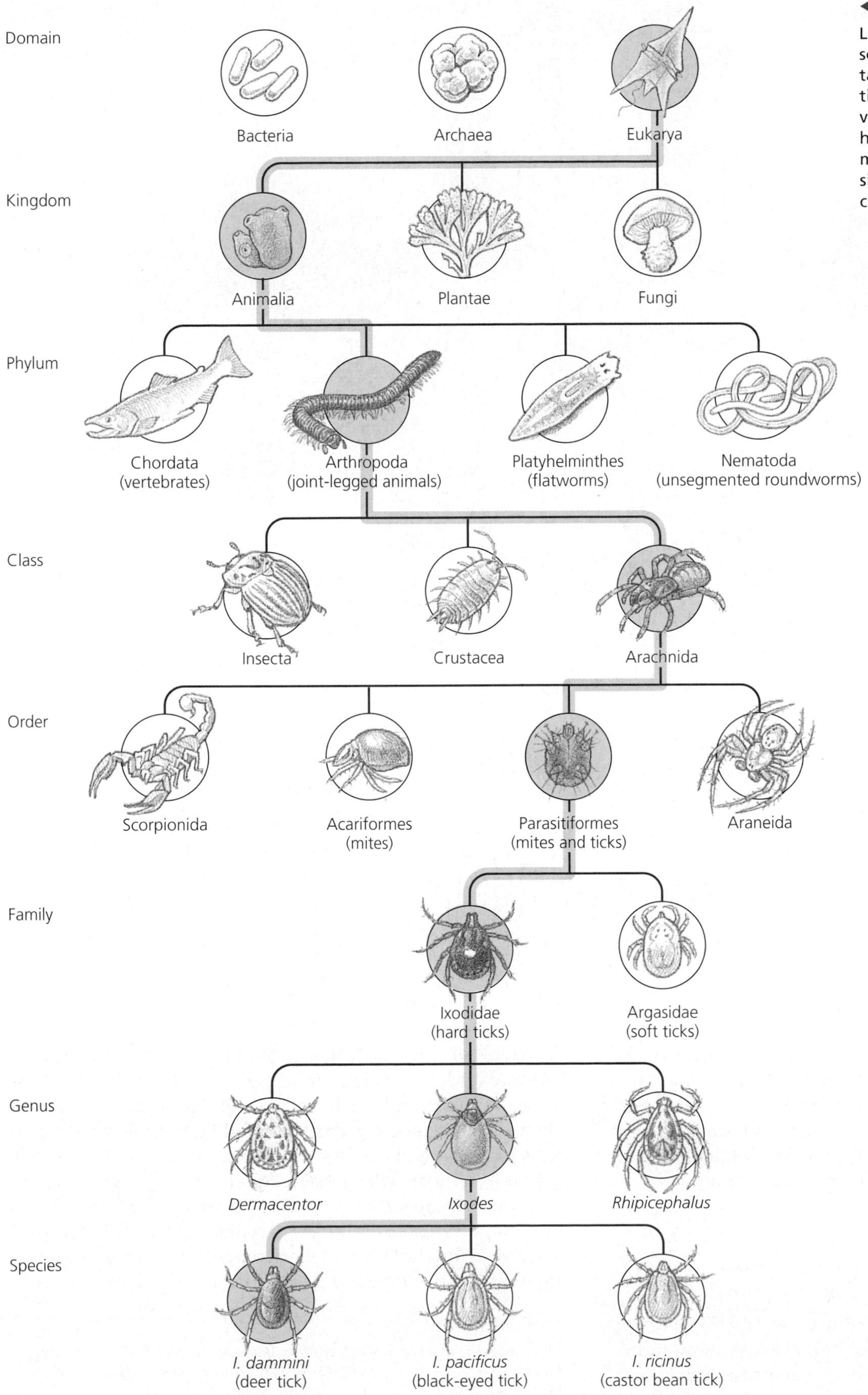

Figure 4.25

Levels in a Linnaean taxonomic scheme. Illustrated here are the taxonomic categories for the deer tick, *Ixodes dammini,* the primary vector of Lyme disease. Notice that higher taxonomic categories are more inclusive than lower ones. To simplify the diagram, only selected classification possibilities are shown.

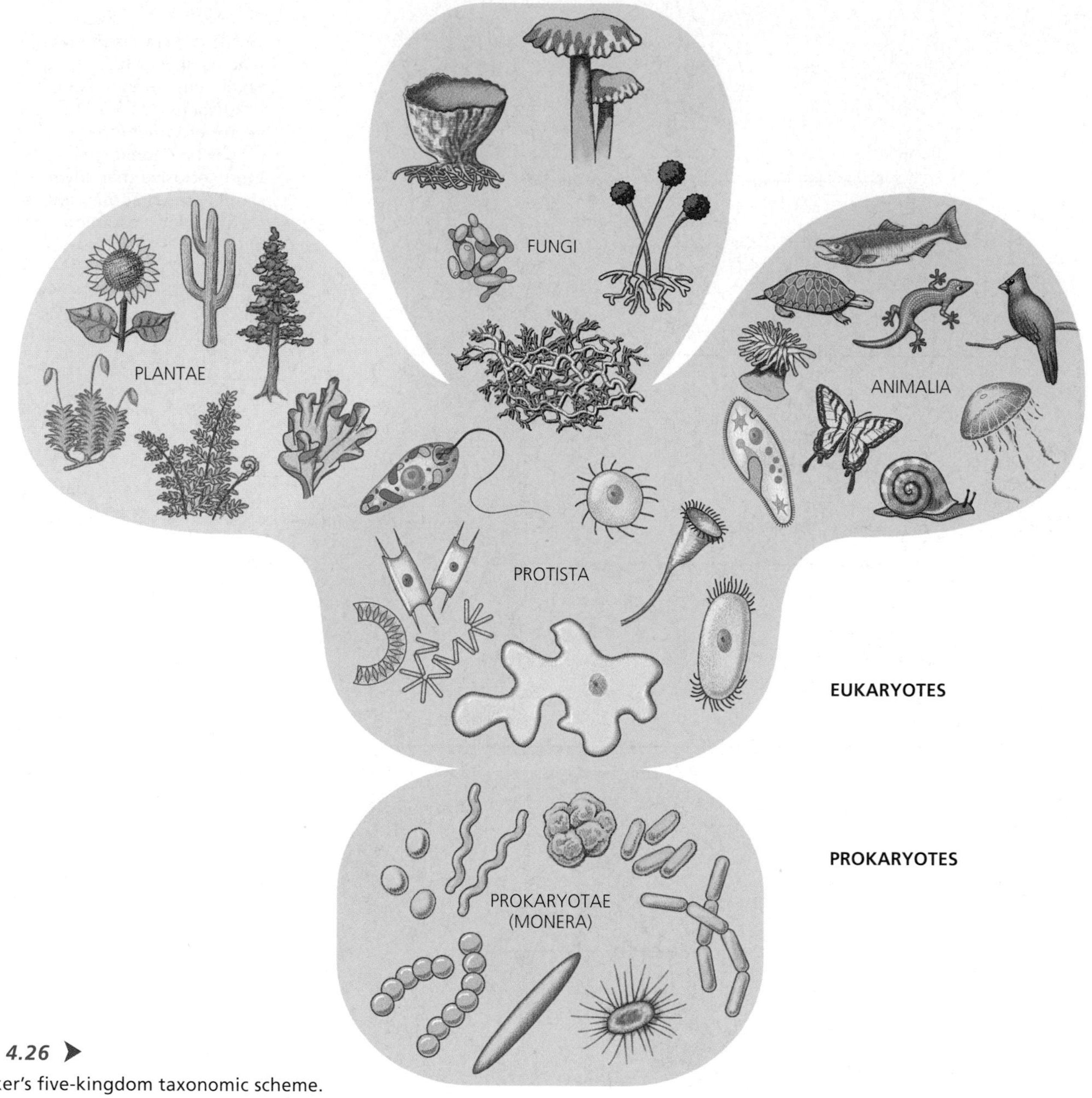

Figure 4.26 ➤
Whittaker's five-kingdom taxonomic scheme.

people. Examples include *Pasteurella haemolytica* (pas-ter-el′ ă hē-mō-lit′i-kă), a bacterium named after the microbiologist Louis Pasteur; *Escherichia coli* (esh-ĕ-rik′ē-ă kō-lī), a bacterium named after the physician Theodor Escherich; and *Isabella abottae* (iz-a-bel′ah ab′ot-tē), a marine alga named after the phycologist and taxonomist Isabella Abbott.

CRITICAL THINKING

Examine the binomials of the species discussed in this section. What do the genus names and specific epithets indicate about the organisms?

Most scientists still use the Linnaean system today, though significant modifications have been adopted. For example, scientists sometimes use additional categories, such as tribes, sections, subfamilies, and subspecies.[10] Further, Linnaeus divided all organisms into only two kingdoms (Plantae and Animalia), and he did not know of the existence of viruses and smaller pathogens. As scientists learned more about organisms, they adopted taxonomic schemes to reflect these advances in knowledge. For example, in 1959, Robert Whittaker (1920–1980) proposed a widely accepted taxonomic approach based on five kingdoms: Animalia, Plantae, Fungi, Protista,[11] and Prokaryotae (also called Monera) **(Figure 4.26).** Though widely accepted, this scheme groups together in a single kingdom (Protista) such obviously

[10]Subspecies, which are also called *varieties* or *strains,* differ only slightly from each other. Not all taxonomists agree on how much difference between two microbes constitutes a strain versus a species.

[11]The Kingdom Protista consists of all algae, slime molds, and single-celled animal-like organisms.

Highlight 4.2 Woese's Taxonomic Revolution

Carl Woese labored for years to understand the taxonomic relationships among prokaryotic cells. Morphology (shape) and biochemical tests did not provide enough information to classify prokaryotes fully, so for over a decade Woese painstakingly sequenced the nucleotides of prokaryotic ribosomal RNA (rRNA) in an effort to unravel the relationships among these organisms.

Then, in 1976, he sequenced rRNA from an odd group of prokaryotes that produce methane gas as a metabolic waste. Woese was surprised when their rRNA did not contain nucleotide sequences characteristic of bacteria. At first he thought he had made a mistake in handling the samples, but repeated testing showed that methanogens, as they are called, were not like other prokaryotic or eukaryotic organisms. They were something new to science, a third branch of life. Woese had discovered the Archaea, ushering microbiologists into a new and curious wonderland of microbes.

Ribosomal nucleotide sequences suggest that there may be at least 50 kingdoms of Bacteria and three kingdoms of Archaea. Further, scientists examining substances such as human saliva, water, soil, and rock regularly discover novel nucleotide sequences. When these sequences are compared to known sequences stored in a computer database, they cannot be associated with any previously identified organism. This suggests that many curious new forms of microbial life have never been grown in a laboratory and still await discovery.

CRITICAL THINKING

In 1999, microbiologists at Stanford University announced the discovery of 31 new species of bacteria that thrive between the teeth and gums of humans. The bacteria could be not be grown in the researcher's laboratories, nor were any of them ever observed via any kind of microscopy.

If they couldn't culture them or see them, how could the researchers know they had discovered 31 new species? If they couldn't examine the cells for the presence of a nucleus, how did they determine that the organisms were prokaryotes and not eukaryotes?

disparate organisms as massive brown seaweeds (kelps) and unicellular protozoa. Further, because it does not address the taxonomy of viruses nor all the differences between microorganisms, scientists have proposed other taxonomic schemes that have from four to more than 50 kingdoms.

Sometimes students are upset that taxonomists do not agree about all the taxonomic categories or the species they contain, but it must be realized that classification of organisms reflects the state of our current knowledge and theories. Taxonomists change their schemes to accommodate new information, and not every expert agrees with every proposed modification.

Another significant development in taxonomy is a shift in its basic goal—from Linnaeus's goal of classifying and naming organisms as a means of cataloging them, to the more modern goal of understanding the relationships among groups of organisms. Linnaeus based his taxonomic scheme primarily on organisms' structural similarities, and whereas such terms as *genus* and *family* may suggest the existence of some common lineage, he and his contemporaries thought of species as divinely created and generally immutable entities. However, when Charles Darwin (1809–1882) propounded his theory of the evolution of species by natural selection (a century after Linnaeus published his pivotal work on the taxonomy of plants), taxonomists came to consider that common ancestry largely explains the similarities among organisms in the various taxa. Today, most taxonomists agree that a major goal of modern taxonomy is to reflect a *phylogenetic*[12] *hierarchy;* that is, that the ways in which organisms are grouped reflect their derivation from common ancestors. The spatial arrangement of kingdoms seen in Figure 4.26 (and the placement of smaller taxa within them, as well) reflects this phylogenetic approach to taxonomy.

Taxonomists' efforts to classify organisms according to their ancestry has resulted in reduced emphasis on comparisons of physical and chemical traits, and in greater emphasis on comparisons of their genetic material. Such work has led to a proposal to add a new, most inclusive taxon: the *domain.*

Domains

Learning Objective

✓ List and describe the three domains proposed by Carl Woese.

Spurred primarily by the work of Carl Woese (1928–) and his collaborators, taxonomists are now comparing the nucleotide sequences of the smaller ribosomal RNA (rRNA) subunits of both prokaryotes and eukaryotes. Because these rRNA molecules are present in all cells and are crucial to protein synthesis, changes in their nucleotide sequences are presumably very rare. By sequencing the nucleotides in rRNA, Woese and his coworkers discovered three basic type of ribosomes **(Highlight 4.2),** which led them to propose a new classification scheme in which a new taxon, called a **domain,** contains (or in some schemes replaces) the Linnaean taxon of kingdom. The three domains identified by Woese—**Eukarya, Bacteria,** and **Archaea**—are based on three basic types of cells as determined by ribosomal nucleotide sequences.

Domain Eukarya includes all eukaryotic cells, all of which contain eukaryotic rRNA sequences. The domains Bacteria and Archaea include all prokaryotic cells. They contain bacterial and archaeal rRNA sequences, respectively,

[12]From Greek *phyllon,* meaning tribe, and Latin *genus,* meaning birth (i.e., origin of a group).

which differ significantly from one another and from those in eukaryotic cells. In addition to differences in rRNA sequences, cells of the three domains differ in many other characteristics, including the lipids in their cell membranes, transfer RNA (tRNA) molecules, and sensitivity to antibiotics. The taxonomy of organisms within the three domains is discussed in Chapters 11 and 12, which cover prokaryotes and eukaryotes, respectively.

Ribosomal nucleotide sequences have also given microbiologists a new way to define prokaryotic species. Some scientists propose that prokaryotes whose rRNA sequence differs from that of other prokaryotes by more than 3% be classified as a distinct species. While this definition has the advantage of being precise, not all taxonomists agree with it.

Taxonomic and Identifying Characteristics

Learning Objective

✓ Describe five procedures used by taxonomists to identify and classify microorganisms.

The criteria and laboratory techniques used for classifying and identifying microorganisms are quite numerous and include macroscopic and microscopic examination of physical characteristics, differential staining characteristics, growth characteristics, microorganisms' interactions with antibodies, microorganisms' susceptibilities to viruses, nucleic acid analysis, biochemical tests, and organisms' environmental requirements, including the temperature and pH ranges of their various types of habitats. Clearly, then, microbial taxonomy is too broad a subject to cover in one chapter, and thus the details of the criteria for the classification of major groups are provided in subsequent chapters.

It is important to note that even though scientists may use a given technique to either classify or identify microorganisms, the criteria used for identifying a particular organism are not always the same as those that were used to classify it. For example, even though laboratory technologists distinguish the genus *Escherichia* from other bacterial genera by its inability to utilize citric acid (citrate) as a sole carbon source, this identifying characteristic was not vital in the classification of *Escherichia.*

Bergey's Manual of Determinative Bacteriology, first published in 1923 and now in its 9th edition (1994), contains information used for the identification of prokaryotes. *Bergey's Manual of Systematic Bacteriology* (2nd edition, 2000) is a similar reference work that is used for classification. Its scheme is based on ribosomal RNA sequences, which presumably reflect phylogenetic (evolutionary) relationships. Each of these volumes is known as "Bergey's Manual."

Linnaeus did not know of the existence of viruses and thus did not include them in his original taxonomic hierarchy; nor are viruses assigned to any of the five kingdoms or three domains because viruses are acellular and generally lack rRNA. Virologists do classify viruses into families and genera, but higher taxa are poorly defined for viruses. Viral taxonomy is discussed further in Chapter 13.

With this background, let's turn now to some brief discussions of five types of information microbiologists commonly use to distinguish among microorganisms: physical characteristics, biochemical tests, serological tests, phage typing, and analysis of nucleic acids.

Physical Characteristics

Many physical characteristics are used to identify microorganisms. Scientists can usually identify protozoa, fungi, algae, and parasitic worms based solely upon their *morphology* (shape). As we see in Chapter 6, lab technologists can also use the physical appearance of a bacterial colony[13] to help identify microorganisms. As we have discussed, stains are used to view the size and shape of individual bacterial cells and to show the presence or absence of identifying features such as endospores and flagella.

Linnaeus categorized prokaryotic cells into two genera based upon two prevalent shapes. He classified spherical prokaryotes in the genus "*Coccus*,"[14] and he placed rod-shaped cells in the genus *Bacillus.*[15] However, subsequent studies have revealed that there are vast differences between many of the thousands of spherical and rod-shaped prokaryotes, and thus physical characteristics alone are not sufficient for the classification of prokaryotes. Instead, taxonomists rely primarily on genetic differences as revealed by metabolic dissimilarities, and more and more frequently on rRNA sequences.

CRITICAL THINKING

Why is the genus name *"Coccus"* placed within quotation marks, but the genus name *Bacillus* is not?

Biochemical Tests

Microbiologists distinguish many prokaryotes that are similar in microscopic appearances and staining characteristics on the basis of differences in their ability to utilize or produce certain chemicals. Biochemical tests include procedures that determine an organism's ability to ferment various carbohydrates; utilize various substrates such as specific amino acids, starch, citrate, and gelatin; or produce waste products such as hydrogen sulfide (H_2S) gas **(Figure 4.27).** Differences in fatty acid composition of bacteria are also used to distinguish between bacteria. Obviously, biochemical tests can be used to identify only those microbes that can be grown under laboratory conditions.

Laboratory technicians utilize biochemical tests to identify pathogens, allowing physicians to prescribe appropriate treatments. Many tests require that the microorganisms be *cultured* (grown) for 12 to 24 hours, though this time can be greatly reduced by the use of rapid identification tools. Such

[13]A group of bacteria that has arisen from a single cell grown on a solid laboratory medium.

[14]From Greek *kokkos,* meaning berry. This genus name has been supplanted by many genera, including *Staphylococcus, Micrococcus,* and *Streptococcus.*

[15]From Latin *bacillum,* meaning small rod.

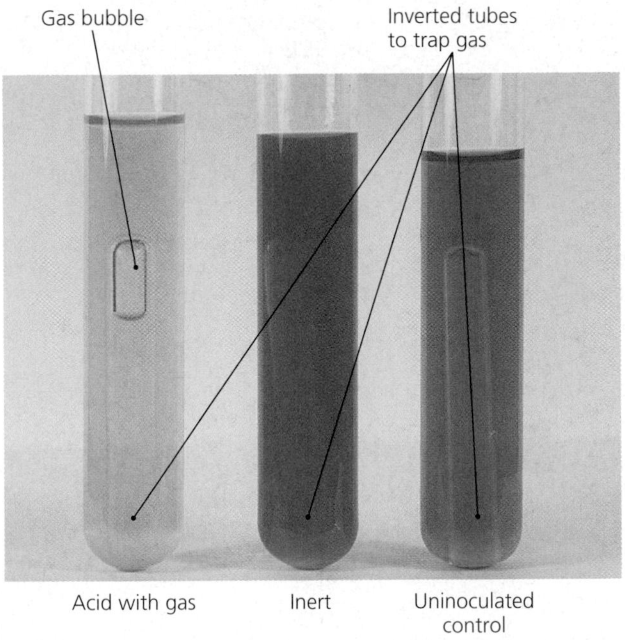

(a)

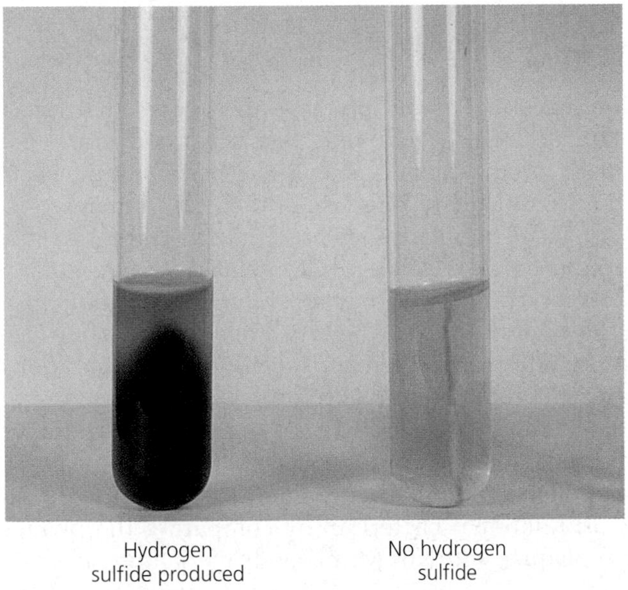

(b)

▲ ***Figure 4.27***

Two biochemical tests for identifying bacteria. **(a)** A carbohydrate utilization test. At left is a tube in which the bacteria have metabolized a particular carbohydrate to produce acid (which changes the color of a pH indicator, phenol red, to red) and gas, as indicated by the gas bubble. (Other species of bacteria produce acid but no gas.) At center is a tube inoculated with bacteria that are "inert" with respect to this test (they do not metabolize the carbohydrate). At right is a control tube into which no bacteria were inoculated. **(b)** A hydrogen sulfide (H_2S) test. Bacteria that produce H_2S are identified by the black precipitate formed by the reaction of the H_2S with iron present in the medium.

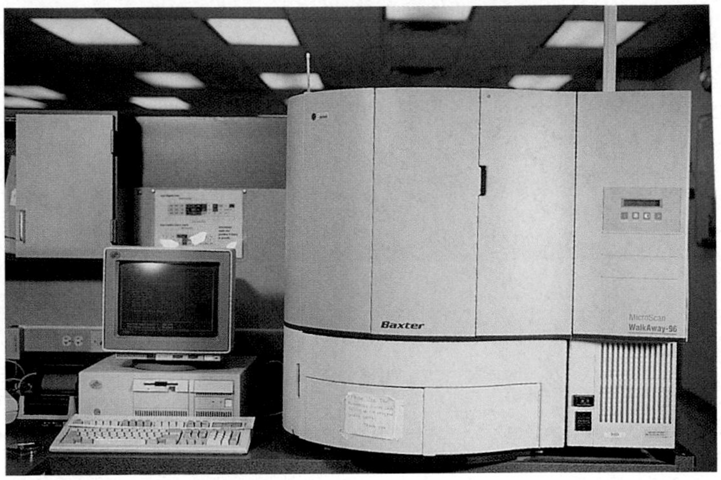

(a) MicroScan instrument

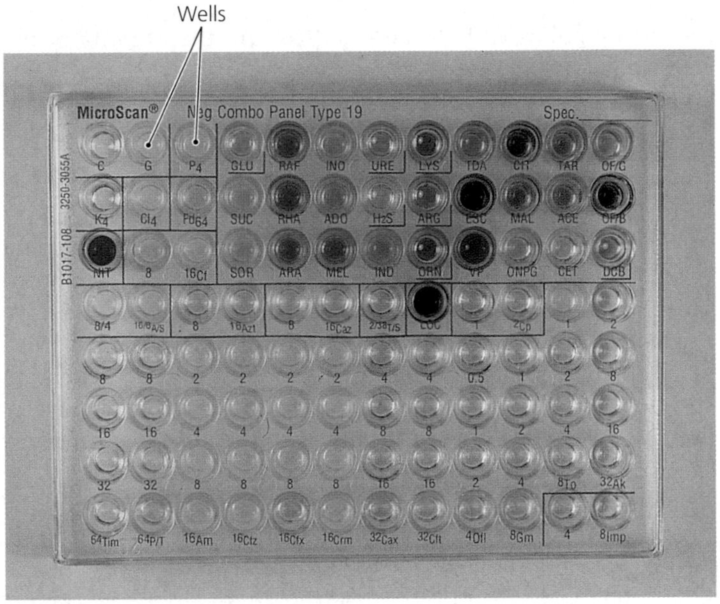

(b) MicroScan® panel

▲ ***Figure 4.28***

One tool for the rapid identification of bacteria, the automated MicroScan® system. **(a)** The MicroScan® instrument. **(b)** A MicroScan® panel, a plate containing numerous wells, each the site of a particular biochemical test. The instrument ascertains the identity of the organism by reading the pattern of colors in the wells after the biochemical tests have been performed.

tools exist for many groups of medically important pathogens, such as Gram-negative bacteria in the family Enterobacteriaceae, Gram-positive bacteria, yeasts, and filamentous fungi. Automated systems for identifying pathogens, such as the one shown in **Figure 4.28a,** read the results of a whole battery of biochemical tests performed in a plastic plate containing numerous small wells **(Figure 4.28b).** A color change in a well indicates the presence of a particular metabolic reaction, and the machine reads the

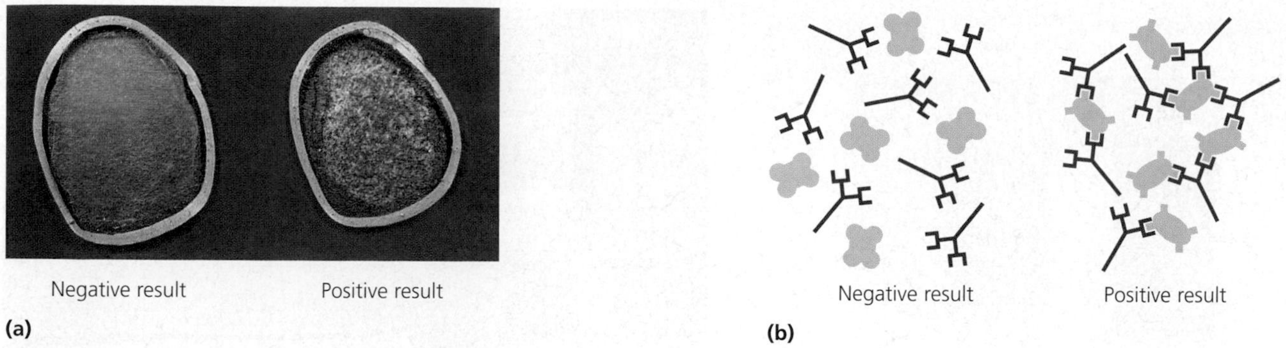

▲ ***Figure 4.29***
An agglutination test, one type of serological test. **(a)** In a positive agglutination test, visible clumps are formed by the binding of antibodies to their target antigens present on cells. **(b)** The processes involved in agglutination tests. In a negative result, antibody binding cannot occur because its specific target is not present; in a positive result, specific binding does occur. Note that agglutination occurs because each antibody molecule can bind to two antigen molecules.

pattern of colors in the plate to ascertain the identity of the pathogen.

Serological Tests

In the narrowest sense, *serology* is the study of serum, the liquid portion of blood after the clotting factors have been removed, and an important site of antibodies. In its most practical application, serology is the study of antigen-antibody reactions in laboratory settings. As we have seen, antibodies are immune system proteins that bind very specifically to target antigens. Antibodies and antigens are discussed in detail in Chapter 16; in this section we briefly consider the use of serological testing to identify microorganisms.

Many microorganisms are *antigenic;* that is, within a host organism they trigger an immune response that results in the production of antibodies. Suppose, for example, that a scientist injects a sample of *Borrelia burgdorferi* (bō-rē′lē-ă bŭrg-dōr′fer-ē), the bacterium that causes Lyme disease, into a rabbit. The bacterium has many surface proteins and carbohydrates that are antigenic because they are foreign to the rabbit. The rabbit responds to these foreign antigens by producing antibodies against them. These antibodies can be isolated from the rabbit's serum and concentrated into a solution known as an **antiserum.** Antisera bind to the antigens that triggered their production.

In a procedure called an **agglutination test,** antiserum is mixed with a sample that potentially contains its target cells. If the antigenic cells are present, antibodies in the antiserum will clump (*agglutinate*) the antigen **(Figure 4.29).** Other antigens, and therefore other organisms, remain unaffected because antibodies are highly specific for their targets.

Antisera can be used to distinguish among species, and even among strains of the same species. For example, a particularly pathogenic strain of *Escherichia coli* was classified and is identified by the presence of both antigen number 157 on its cell wall (designated O157) and antigen number 7 on its flagella (designated H7). This pathogen, known as *E. coli* O157:H7, has caused several deaths in the United States over the past few years. Other serological tests, such as *enzyme-linked immunosorbent assay (ELISA)* and *western blotting,* are discussed in Chapter 17.

Phage Typing

Bacteriophages (or simply **phages**) are viruses that infect and usually destroy bacterial cells. Just as antibodies are specific for their target antigens, phages are specific for the hosts they can infect. **Phage typing,** like serological testing, works because of such specificity. One bacterial strain may be susceptible to a particular phage while a related strain is not.

In phage typing, a technician spreads a solution containing the bacterium to be identified across a solid surface of growth medium and then adds small drops of solutions containing different types of bacteriophage. Wherever a specific phage is able to infect and kill bacteria, the resulting lack of bacterial growth produces within the bacterial lawn a clear area called a **plaque (Figure 4.30).** A microbiologist can identify an unknown bacterium by comparing the phages that form plaques with known phage-bacteria interactions.

Analysis of Nucleic Acids

As we have discussed, the sequence of nucleotides in nucleic acid molecules provides a powerful tool for the classification and identification of microbes. In many cases, nucleic acid analysis has confirmed classical taxonomic hierarchies. In other cases, as in Woese's discovery of domains, curious new organisms and relationships that were not obvious from classical methodologies have come to light. Techniques of nucleotide sequencing and comparison, such as *DNA base composition, polymerase chain reaction (PCR),* and *nucleic acid hybridization,* are best understood after we have discussed microbial genetics; therefore, these techniques are discussed in Chapter 8.

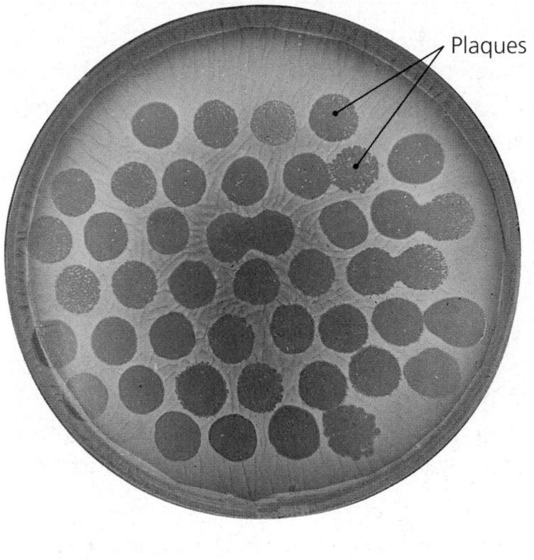

▲ *Figure 4.30*

Phage typing. Drops containing type A bacteriophages were added to this plate after its entire surface was inoculated with an unknown strain of *Salmonella.* After 12 hours of bacterial growth, clear zones, called plaques, developed where the phages killed bacteria. Given the great specificity of phage A for infecting and killing its host, the strain of bacterium can be identified as *Salmonella typhi* type A.

Determination of the proportion of a cell's DNA that is guanine and cytosine, a quantity referred to as the cell's *GC ratio,* has also become a part of prokaryotic taxonomy. Scientists express the ratio as follows:

$$\frac{G + C}{A + T + G + C} \times 100\%$$

GC ratios vary from 20% to 80% among prokaryotes, and often (but not always) organisms that share other characteristics have similar GC ratios. Organisms that were once thought to be closely related but have widely different GC ratios are invariably not as closely related as thought.

Taxonomic Keys

As we have seen, taxonomists, medical clinicians, and researchers can use a wide variety of information, including morphology, chemical characteristics, and results from biochemical, serological, and phage typing tests, in their efforts to identify microorganisms, including pathogens. But how can all these characteristics and results be organized so that they can be used efficiently to identify an unknown organism? All this information is often arranged in **dichotomous keys,** which contain a series of paired statements worded so that only one of two "either/or" choices applies to any particular organism **(Figure 4.31).** Based on which of the two

▼ *Figure 4.31*

Use of a dichotomous taxonomic key. The example presented here involves identification of the genera of potentially pathogenic, intestinal bacteria. **(a)** A sample key. To use it, choose the one statement in a pair that applies to the organism to be identified, and then either refer to another key (as indicated by words) or go to the appropriate place within this key (as indicated by a number). **(b)** A flowchart that shows the various paths that might be followed in using the key presented in part a. Highlighted is the path taken when the bacterium in question is *Escherichia*.

1a. Gram-positive cells................................ Gram-positive bacteria
1b. Gram-negative cells.............................. 2

2a. Rod-shaped cells.................................. 3
2b. Non-rod-shaped cells............................ Cocci and pleomorphic bacteria

3a. Can tolerate oxygen............................... 4
3b. Cannot tolerate oxygen.......................... Obligate anaerobes

4a. Ferments lactose.................................. 5
4b. Cannot ferment lactose......................... Non-lactose fermenters

5a. Can use citric acid as a sole carbon source....................................... 6
5b. Cannot use citric acid alone.................. 8

6a. Produces hydrogen sulfide gas............... *Salmonella*
6b. Does not produce hydrogen sulfide gas.. 7

7a. Produces acetoin.................................. *Enterobacter*
7b. Does not produce acetoin....................... *Citrobacter*

8a. Produces gas from glucose..................... *Escherichia*
8b. Does not produce gas from glucose....... *Shigella*

(a)

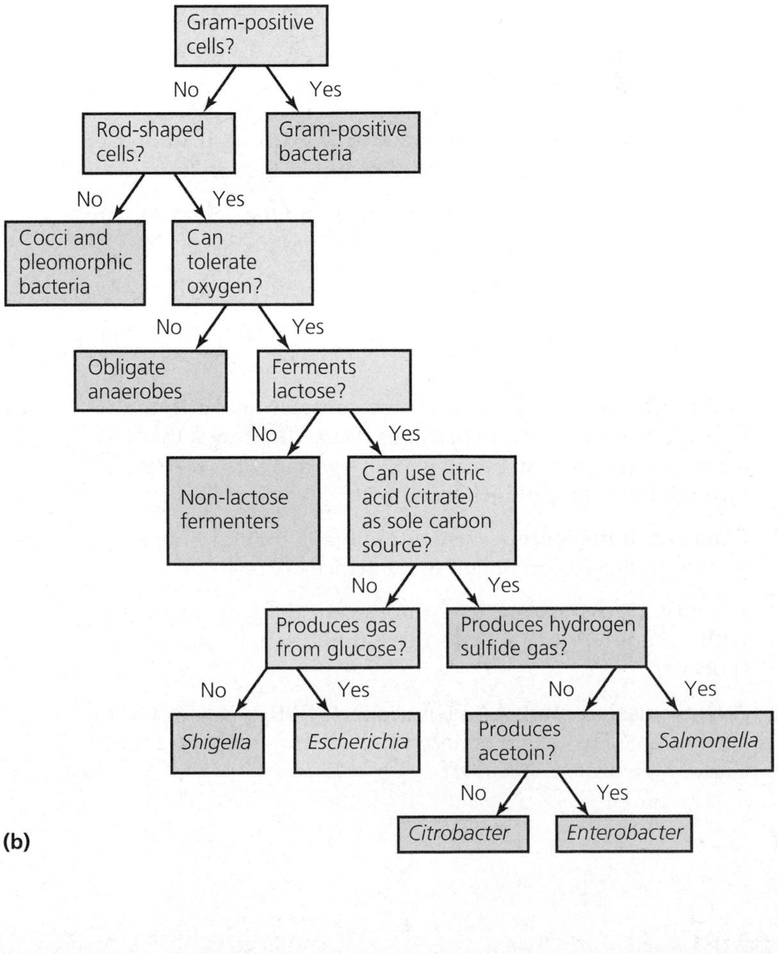

(b)

statements applies, the key either directs the user to another pair of statements, or it provides the name of the organism in question. Note that more than one key can be created to enable the identification of a given set of organisms, but all such keys involve mutually exclusive, "either/or" choices that send the user along a path that leads to the identity of the unknown organism.

CRITICAL THINKING

A clinician obtains a specimen of urine from a patient suspected to have bladder infection. From the specimen she cultures a Gram-negative, rod-shaped bacterium that ferments lactose in the presence of oxygen, utilizes citrate, and produces acetoin but not hydrogen sulfide. Using the key presented in Figure 4.31, identify the genus of the infective bacterium.

CHAPTER SUMMARY

Units of Measurement (p. 93)

1. The metric system is a decimal system in which each unit is one-tenth the size of the next largest unit.
2. The basic unit of length in the metric system is the **meter.**

Microscopy (pp. 93–105)

1. **Microscopy** refers to the passage of light or electrons of various **wavelengths** through lenses to **magnify** objects and provide **resolution** and contrast so that those objects can be viewed and studied.
2. **Immersion oil** is used to fill the space between the specimen and a lens to reduce light refraction and thus increase the **numerical aperture** and resolution.
3. Staining techniques and polarized light may be used to enhance **contrast** between the object and its background.
4. **Simple microscopes** contain a single magnifying lens, whereas **compound microscopes** use a series of lenses for magnification.
5. The lens closest to the object being magnified is the **objective lens,** several of which are mounted on a **revolving nosepiece.** The lenses closest to the eyes are **ocular lenses. Condenser lenses** lie beneath the stage and direct light though the slide.
6. The magnifications of the objective lens and the ocular lens are multiplied together to give **total magnification.**
7. A photograph of a microscopic image is a **micrograph.**
8. **Dark-field microscopes** provide a dark background for small or colorless specimens.
9. **Phase microscopes,** such as **phase-contrast** and **differential interference contrast microscopes,** cause light rays that pass through a specimen to be out of phase with light rays that pass through the field, producing contrast.
10. **Fluorescent microscopes** use ultraviolet light and fluorescent dyes to fluoresce specimens and enhance contrast.
11. A **confocal microscope** uses fluorescent dyes in conjunction with computers to provide three-dimensional images of a specimen.
12. A **transmission electron microscope (TEM)** provides an image produced by the transmission of electrons through a thinly sliced, dehydrated specimen.
13. A **scanning electron microscope (SEM)** provides a three-dimensional image by scattering electrons from the surface of a specimen.
14. Minuscule electronic probes are used in **scanning tunneling microscopes (STM)** and in **atomic force microscopes (AFM)** to reveal details at the atomic level.

Staining (pp. 106–112)

1. **Staining** organisms with dyes for light microscopy involves making a **smear** or thin film of the specimens on a slide and then either passing the slide through a flame **(heat fixation)** or applying a chemical **(chemical fixation)** to attach the specimens to the slide. **Acidic dyes** or **basic dyes** are used to stain different portions of an organism to aid viewing and identification.
2. **Simple stains** involve the simple process of soaking the smear with one dye and then rinsing with water. **Differential stains** such as the **Gram stain, acid-fast stain,** and **endospore stain** use more than one dye to differentiate different cells, chemicals, or structures.
3. The Gram stain procedure includes use of a **primary stain,** a **mordant,** a **decolorizing agent,** and a **counterstain** that results in either purple (Gram-positive) or pink (Gram-negative) organisms, depending on the chemical structures of their cell walls.
4. The acid-fast stain is used to differentiate cells with waxy cell walls. **Endospores** are stained by the **Schaeffer-Fulton endospore stain** procedure.
5. Stains that stain the background and leave the cells colorless are called **negative** (or **capsule**) **stains.**

Classification and Identification of Microorganisms (pp. 112–122)

1. **Taxa** are nonoverlapping groups of organisms that are studied and named in **taxonomy.** Carolus Linnaeus invented a system of taxonomy, grouping similar interbreeding organisms into **species,** species into **genera,** genera into **families,** families into **orders,** orders into **classes,** classes into **phyla,** and phyla into **kingdoms.**
2. Linnaeus gave each species a descriptive name consisting of a genus name and **specific epithet.** This practice of naming organisms with two names is called **binomial nomenclature.**
3. Carl Woese proposed the existence of three taxonomic **domains** based on three cell types revealed by rRNA sequencing: **Eukarya, Bacteria,** and **Archaea.**

4. Taxonomists rely primarily on genetic differences revealed by metabolic dissimilarities to classify organisms. Species or strains within species may be distinguished by using **antisera, agglutination tests,** nucleic acid analysis, or **phage typing** with **bacteriophages,** in which unknown bacteria are identified by observing **plaques** (regions of a bacterial lawn where the phage has killed bacterial cells).
5. Microbiologists use **dichotomous keys,** which involve stepwise choices between paired characteristics, to assist them in identifying microbes.

QUESTIONS FOR REVIEW

(Answers to multiple choice and fill in the blanks are on the web, along with additional review questions. Visit www.microbiologyplace.com.)

Multiple Choice

1. Which of the following is smallest?
   a. decimeter
   b. millimeter
   c. nanometer
   d. micrometer
2. A nanometer is ________ than a micrometer.
   a. 10 times larger
   b. 10 times smaller
   c. 1000 times larger
   d. 1000 times smaller
3. Resolution is best described as
   a. the ability to view something that is small.
   b. the ability to magnify a specimen.
   c. the ability to distinguish between two adjacent objects.
   d. the difference between two waves of electromagnetic radiation.
4. Curved glass lenses ________ light.
   a. refract
   b. bend
   c. magnify
   d. both a and b
5. Which of the following factors is important in making an image appear larger?
   a. the thickness of the lens
   b. the curvature of the lens
   c. the speed of the light passing through the lens
   d. all of the above
6. Which of the following is different between light microscopy and transmission electron microscopy?
   a. magnification
   b. resolution
   c. wavelengths
   d. all of the above
7. Which of the following types of microscopes produces a three-dimensional image with a shadowed appearance?
   a. simple microscope
   b. Nomarski microscope
   c. fluorescent microscope
   d. transmission electron microscope
8. Which of the following microscopes combines the greatest magnification with the best resolution?
   a. confocal microscope
   b. probe microscope
   c. TEM
   d. SEM
9. Negative stains such as eosin are also called
   a. capsule stains.
   b. endospore stains.
   c. simple stains.
   d. acid-fast stains.
10. In the binomial system of nomenclature, which term is always written in lowercase letters?
   a. kingdom
   b. domain
   c. genus
   d. specific epithet

Fill in the Blanks

1. If an objective magnifies 40×, and each binocular lens magnifies 15×, the total magnification of the object being viewed is ________.
2. The type of fixation developed by Koch for bacteria is ________.
3. Immersion oil ________ (increases/decreases) the numerical aperture, which ________ (increases/decreases) resolution because ________ (more/fewer) light rays are involved.
4. ________ refers to differences in intensity between two objects.
5. Cationic chromophores such as methylene blue ionically bond to ________ (positively/negatively) charged chemicals such as DNA and proteins.

Short Answer

1. Explain how the principle "electrons travel as waves" applies to microscopy.
2. Critique the following definition of magnification given by a student on a microbiology test: "Magnification makes things bigger."
3. Why can electron microscopes magnify only dead organisms?
4. Put the following substances in the order they are used in a Gram stain: counterstain, decolorizing agent, mordant, primary stain.
5. Why is Latin used in taxonomic nomenclature?
6. Give three characteristics of a "specific epithet."
7. How does the study of the nucleotide sequences of ribosomal RNA fit into a discussion of taxonomy?
8. An atomic force microscope can magnify a living cell, whereas electron microscopes and scanning tunneling microscopes cannot. What requirement of electron and scanning tunneling microscopes precludes the imaging of living specimens?

CRITICAL THINKING

1. Miki came home from microbiology lab with very green fingers and a bad grade. When asked about this, she replied that she was doing a Gram stain but it never worked the way the book said it should. Boone overheard the conversation and said that she must have used the wrong chemicals. What dye was she probably using, and what structure does that chemical normally stain?
2. Why is the definition of species as "successfully interbreeding organisms" not satisfactory for most microorganisms?
3. With the exception of the discovery of new organisms, is it logical to assume that taxonomy as we currently know it will stay the same? Why or why not?
4. A novice microbiology student incorrectly explains that immersion oil increases the magnification of his microscope. What is the function of immersion oil?

CHAPTER 5

Microbial Metabolism

The next time you bite into a delicious piece of chocolate, consider this: Microbial metabolism played a key role in how that chocolate came to taste so good.

Chocolate comes from cacao seeds, found inside the pods of *Theobroma cacao* trees. After the pods are split open, the seeds and the surrounding pulp are scooped out and placed in heaps on top of plantain or banana leaves. The heaps are then covered and left to ferment for 2–7 days. Fermentation occurs as microorganisms—including yeast and several kinds of bacteria—grow on the fleshy, sugary pulp. Bathed in fermenting pulp, the cacao seeds (which start off tasting bitter) begin to develop the flavors and colors that we associate with chocolate. After fermentation, the seeds are dried, roasted, and then processed further by chocolate manufacturers before becoming one of our favorite desserts.

In this chapter we will learn about many metabolic processes of microorganisms, including fermentation.

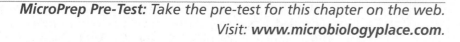

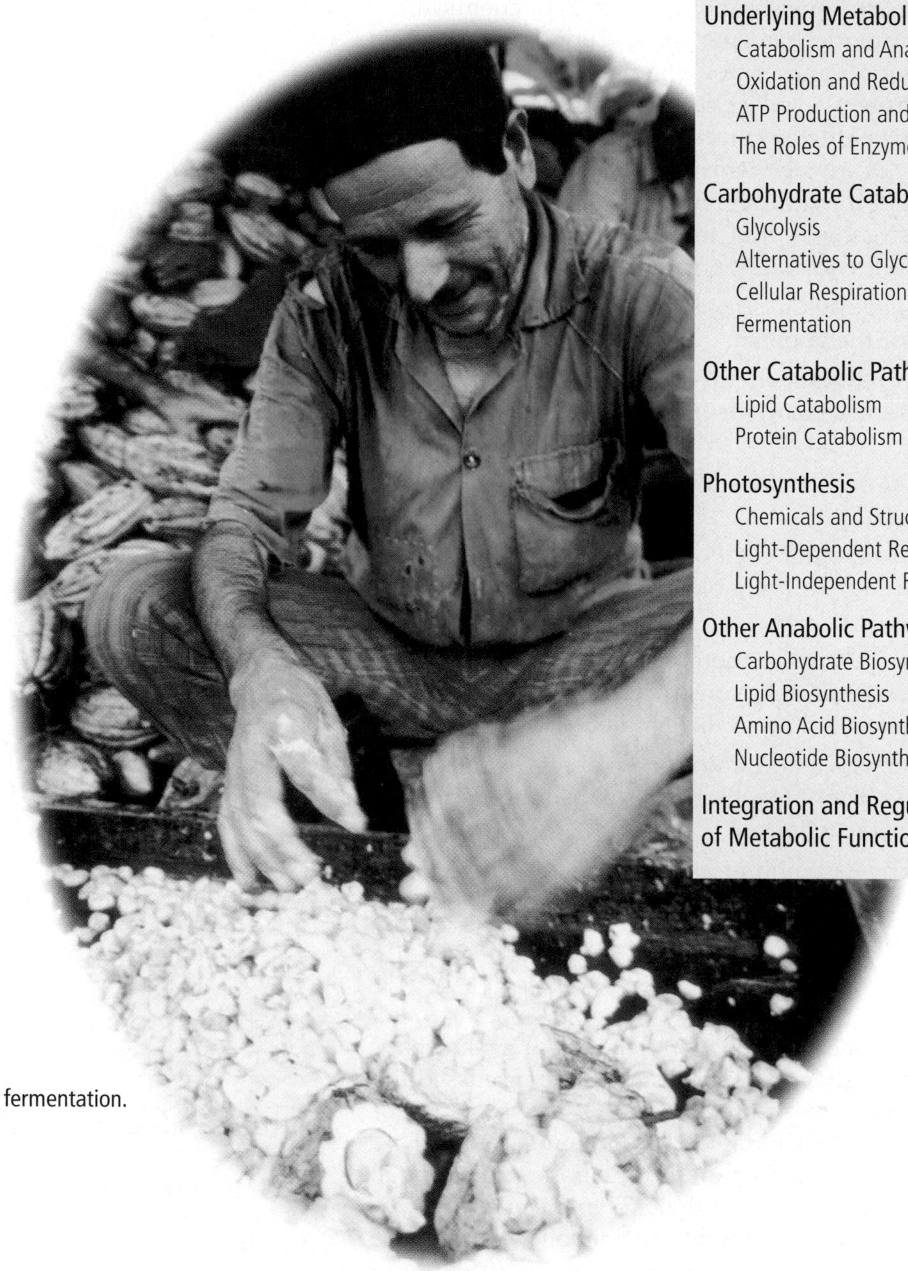

A worker prepares cacao seeds for fermentation.

CHAPTER OUTLINE

MicroPrep Pre-Test: *Take the pre-test for this chapter on the web.*
Visit: ***www.microbiologyplace.com.***

How do pathogens acquire energy and nutrients at the expense of a patient's health? How does grape juice turn into wine, and how does yeast cause bread to rise? How do disinfectants, antiseptics, and antimicrobial drugs work? When laboratory personnel perform biochemical tests to identify unknown microorganisms and help diagnose disease, what exactly are they doing?

The answers to all of these questions require an understanding of microbial **metabolism,**[1] the collection of controlled biochemical reactions that takes place within the cells of an organism. While it is true that metabolism in its entirety is complex, consisting of thousands of chemical reactions and control mechanisms, the reactions are nevertheless elegantly logical and can be understood in a simplified form. In this chapter we will concern ourselves only with the central metabolic pathways and energy metabolism.

Your study of metabolism will be manageable if you keep in mind that the ultimate function of metabolism is to reproduce the organism, and that metabolic processes are guided by the following eight elementary statements:

- Every cell acquires *nutrients,* which are the chemicals necessary for metabolism.
- Metabolism requires energy from light or from the *catabolism* (kă-tab'ō-lizm), or breakdown, of acquired nutrients.
- Energy is stored in the chemical bonds of *adenosine triphosphate (ATP).*
- Using *enzymes,* cells catabolize nutrient molecules to form elementary building blocks called *precursor metabolites.*
- Using these precursor metabolites, energy from ATP, and other enzymes, cells construct larger building blocks in *anabolic* (an-ă-bol'ik), or biosynthetic, reactions.
- Cells use enzymes and additional energy from ATP to link building blocks together to form macromolecules in *polymerization* reactions.
- Cells grow by assembling macromolecules into cellular structures such as ribosomes, membranes, and cell walls.
- Cells typically reproduce once they have doubled in size.

We will discuss each aspect of metabolism in the chapters that most directly apply. For instance, we discussed the first step of metabolism—the active and passive transport of nutrients into cells—in Chapter 3. In this chapter we will examine the importance of enzymes in catabolic and anabolic reactions, study the three ways that ATP molecules are synthesized, and show that catabolic and anabolic reactions are linked. We will also examine the catabolism of nutrient molecules; the anabolic reactions involved in the synthesis of carbohydrates, lipids, amino acids, and nucleotides; and a few ways that cells control their metabolic activities. Genetic control of metabolism and the polymerization of DNA, RNA, and proteins are discussed in Chapter 7, and the specifics of cell division are covered in Chapters 11 and 12.

[1]From Greek *metabole,* meaning change.

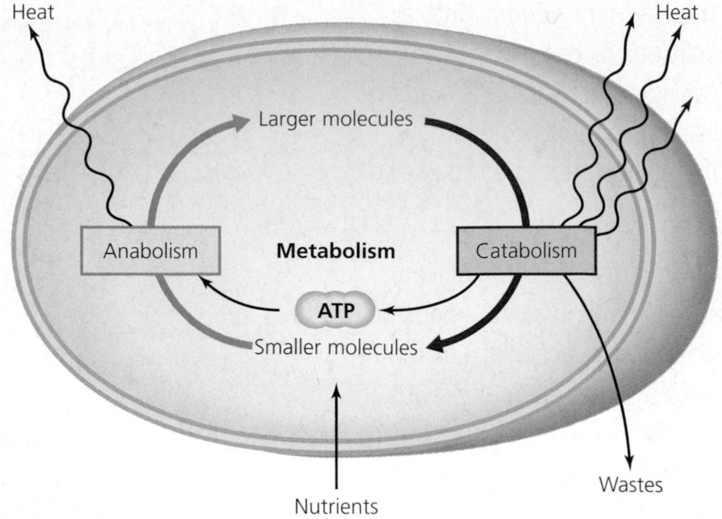

▲ ***Figure 5.1***

Metabolism is composed of catabolic and anabolic reactions. Some energy released in catabolism is stored in ATP molecules, but most is lost as heat. Anabolic reactions require energy, typically provided by ATP. There is some heat loss in anabolism as well. The products of catabolism provide many of the building blocks for anabolic reactions; other anabolic building blocks must be acquired directly as nutrients.

Basic Chemical Reactions Underlying Metabolism

In the following sections we will examine the basic concepts of catabolism, anabolism, and a special class of reactions called *oxidation-reduction reactions.* The latter involve the transfer of electrons between molecules. Then we will turn our attention briefly to the synthesis of ATP and energy storage before we discuss the organic catalysts called *enzymes,* which make metabolism possible.

Catabolism and Anabolism

Learning Objective

✓ Distinguish among metabolism, anabolism, and catabolism.

Metabolism, which is all of the chemical reactions in an organism, can be divided into two major classes of reactions: **catabolism** and **anabolism (Figure 5.1).** A series of reactions is called a *pathway.* Cells have *catabolic pathways,* which break larger molecules into smaller products, and *anabolic pathways,* which synthesize large molecules from the smaller products of catabolism. Even though catabolic and anabolic pathways are intimately linked in cells, it is often useful to study the two types of pathways as if they were separate.

When catabolic pathways break down large molecules, they release energy; that is, catabolic pathways are *exergonic* (ek-ser-gon'ik). Cells store some of this released energy in the bonds of ATP, though much of it is lost as heat. Another result of the breakdown of large molecules by catabolic pathways is the production of numerous smaller molecules, some of which are **precursor metabolites** of anabolism. Some organisms, such as *Escherichia coli* (esh-ĕ-rik'ē-ă kō'lī), can synthesize everything in their cells from these precursor metabolites; other organisms must acquire some anabolic building blocks as nutrients. Note that catabolic *pathways,* but not necessarily *individual* catabolic *reactions,* produce ATP and metabolites; a given catabolic pathway may produce ATP, or metabolites, or both. An example of a catabolic pathway is the breakdown of lipids into glycerol and fatty acids.

Anabolic pathways are functionally the opposite of catabolic pathways in that they synthesize macromolecules and cellular structures. Because building anything requires energy, anabolic pathways are *endergonic* (en-der-gon'ik); that is, they require more energy than they release. The energy required for anabolic pathways usually comes from ATP molecules produced during catabolism. An example of an anabolic pathway is the synthesis of lipids for cell membranes from glycerol and fatty acids.

To summarize, a cell's metabolism involves both catabolic pathways that break down macromolecules to supply molecular building blocks and energy in the form of ATP, and anabolic pathways that use the building blocks and ATP to synthesize macromolecules needed for growth and reproduction.

Oxidation and Reduction Reactions

Learning Objective

✓ Contrast reduction and oxidation reactions.

Many metabolic reactions involve the transfer of electrons from a molecule that donates an electron (called an *electron donor*) to a molecule that accepts an electron (called an *electron acceptor*). Such electron transfers are called **oxidation-reduction reactions** or **redox reactions.** The reactions in which electrons are accepted are *reduction reactions,* whereas the reactions in which electrons are donated are *oxidation reactions* **(Figure 5.2).** Electron acceptors are said to be *reduced* because their gain in electrons reduces their overall electrical charge (that is, they are more negatively charged). Molecules that donate electrons are said to be *oxidized* because frequently their electrons are donated to oxygen atoms.

Reduction and oxidation reactions are always coupled (as represented in Figure 5.2) because every electron that is gained by one molecule must be donated by some other molecule. A molecule may be reduced by gaining either a simple electron or an electron that is part of a hydrogen atom—which, as we saw in Chapter 2, is composed of one proton and one electron. (**New Frontiers 5.1** on page 128 describes an interesting example of how some prokaryotes are able to reduce gold dissolved in solution.) In contrast, a molecule may be oxidized in one of three ways: by losing a simple electron, by losing a hydrogen atom, or by gaining an oxygen atom. Because biological oxidations often involve the loss of hydrogen atoms, such reactions are also called *dehydrogenation* (dē-hī'drō-jen-ā'shŭn) *reactions.*

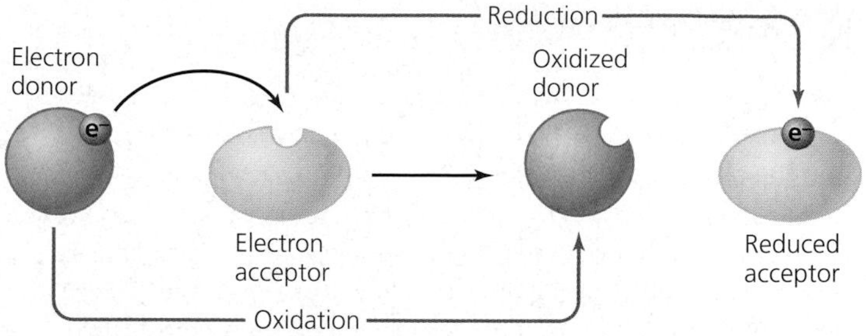

▲ ***Figure 5.2***

Oxidation-reduction or redox reactions. When electrons are transferred from donor molecules to acceptor molecules, donors become oxidized and acceptors become reduced. *Why are acceptor molecules said to be reduced when they are gaining electrons?*

Figure 5.2 "Reduction" refers to the overall electrical charge on a molecule. Because electrons have a negative charge, the gain of an electron reduces the molecule's overall charge.

Electrons rarely exist freely in cytoplasm; instead, they orbit atomic nuclei. Therefore, cells use electron carrier molecules to carry electrons (often in hydrogen atoms) from one location in a cell to another. Three important electron carrier molecules, which are derived from vitamins, are **nicotinamide adenine dinucleotide (NAD^+), nicotinamide adenine dinucleotide phosphate ($NADP^+$),** and **flavin adenine dinucleotide (FAD).** Cells use each of these molecules in specific metabolic pathways to carry pairs of electrons. One of the electrons carried by either NAD^+ or $NADP^+$ is part of a hydrogen atom, forming NADH and NADPH. FAD carries two electrons as hydrogen atoms ($FADH_2$). Many metabolic pathways, including those that synthesize ATP, require such electron carrier molecules.

ATP Production and Energy Storage

Learning Objective

✓ Compare and contrast the three types of ATP phosphorylation.

Nutrients contain energy, but that energy is spread throughout their chemical bonds and generally is not concentrated enough for use in anabolic reactions. During catabolism, organisms release energy from nutrients that can then be concentrated and stored in high-energy phosphate bonds of molecules such as ATP. This happens by a general process called *phosphorylation* (fos'fōr-i-lā'shŭn), in which inorganic phosphate (PO_4^{2-}) is added to a substrate. For example,

New Frontiers 5.1 Gold-Mining Microbes

Gold, as found in nature, exists in two forms: gold-ore deposits, usually found near the Earth's crust, and gold dissolved in solution, as found in thermal springs and in seawater. Dissolved gold (which is gold in its oxidized form) is largely useless to humans; it cannot be converted easily or inexpensively into the valuable objects that we produce from solid gold (which is gold in its reduced form). Even though gold in either form is toxic when ingested by most living things, scientists have recently discovered that certain bacteria and archaea can metabolize dissolved gold. When placed in a solution containing gold, these microorganisms reduce the dissolved gold and shed flecks of solid gold as metabolic waste.

Solid gold is gold in its reduced form.

Scientists have long known that some prokaryotes can transfer electrons from an electron donor (commonly hydrogen) to metals such as iron and uranium, in the process reducing these metals. This knowledge led a team of researchers from the University of Massachusetts, Amherst, headed by microbiologist Derek Lovley, to hypothesize that such microorganisms might also be able to transfer electrons to gold in its dissolved form, thereby reducing it and precipitating solid gold. Focusing on microorganisms known for their ability to reduce iron, Lovley's team found that although not all iron-reducing microbes could also reduce gold, some could, including *Pyrobaculum islandicum* and *Pyrococcus furiosus* (archaea), and *Thermotoga maritima* and *Shewanella algae* (bacteria). Lovley's research thus suggests that microorganisms may play a role in the formation of some gold-ore deposits.

Entrepreneurial minds may wonder if this research also has practical—that is, potentially profitable—applications. While it is true that a great deal of dissolved gold is found in thermal springs and oceans, the gold is very dilute—only minute amounts are present in very large volumes of water. Moreover, were someone to perfect a way of using microorganisms to convert dissolved gold to great quantities of solid gold, they would be wise to keep it to themselves: So much solid gold could become available that its market value would plunge dramatically.

Reference: Kashefi, K., J. M. Tor, K. P. Nevin, and D. R. Lovley. 2001. Reductive Precipitation of Gold by Assimilatory Fe(III)-Reducing Bacteria and Archaea. *Applied and Environmental Microbiology* 67(7):3275–9.

cells phosphorylate adenosine diphosphate (ADP), which has two phosphate groups, to form adenosine triphosphate (ATP), which has three phosphate groups **(Figure 5.3).**

As we will examine in the following sections, cells phosphorylate ADP to form ATP in three specific ways:

- *Substrate-level phosphorylation* (see page 139), which involves the transfer of phosphate to ADP from another phosphorylated organic compound
- *Oxidative phosphorylation* (see page 147), in which energy from redox reactions of *respiration* (described shortly) is used to attach inorganic phosphate to ADP
- *Photophosphorylation* (see page 154), in which light energy is used to phosphorylate ADP with inorganic phosphate

We will investigate each of these in more detail as we proceed through the chapter.

In summary, after ADP is phosphorylated to produce ATP, anabolic pathways use some energy of ATP by breaking a phosphate bond (which re-forms ADP). Thus the cyclical interconversion of ADP and ATP functions somewhat like rechargeable batteries: ATP molecules store energy from light (in photosynthetic organisms) and from catabolic reactions and then release stored energy to drive cellular processes (including anabolic reactions, active transport, and movement). Re-formed ADP molecules can be "recharged" to ATP again and again (Figure 5.3).

The Roles of Enzymes in Metabolism

Learning Objectives

✓ Draw a table listing the six basic types of enzymes, their activities, and an example of each.

✓ Describe the components of a holoenzyme, and contrast protein and RNA enzymes.

✓ Define *activation energy, enzyme, apoenzyme, cofactor, coenzyme, active site,* and *substrate,* and describe their roles in enzyme activity.

✓ Describe how temperature, pH, substrate concentration, and competitive and noncompetitive inhibition affect enzyme activity.

As we saw in Chapter 2, reactions occur when chemical bonds are broken or formed between atoms. In catabolic reactions, a bond must be destabilized before it will break, whereas in anabolic reactions reactants must collide with sufficient energy before bonds will form between them. In anabolism, increasing either the concentrations of reactants or ambient temperatures will increase the number of collisions and produce more chemical reactions; however, in living organisms, neither reactant concentration nor temperature is usually high enough to ensure that bonds will form. Therefore, the chemical reactions of life depend upon *catalysts,* which are chemicals that increase the likelihood of a reaction but are not permanently changed in the process. Organic catalysts are known as **enzymes.**

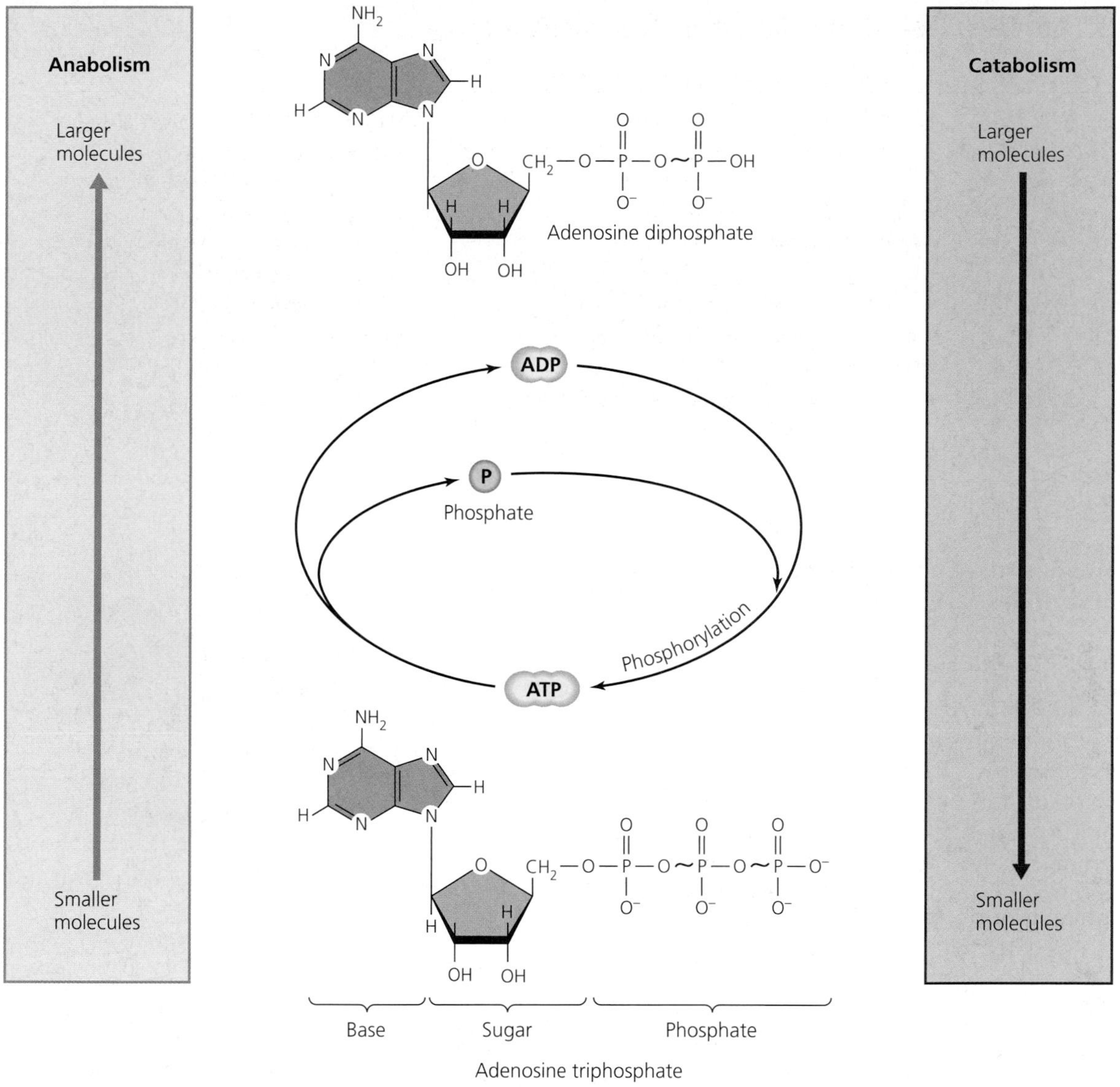

▲ ***Figure 5.3***

Phosphorylation of ADP to form ATP. Cells add phosphate via a high-energy bond to ADP, making ATP. During anabolism, ATP gives up its energy and phosphate to become ADP, which is then recycled to ATP.

Naming and Classifying Enzymes

The names of enzymes usually end with the suffix "-ase," and the name of each enzyme often incorporates the name of that enzyme's **substrate,** which is the molecule the enzyme acts upon. Based on their mode of action, enzymes can be grouped into six basic categories:

1. *Hydrolases* catabolize molecules by adding water in a decomposition process known as *hydrolysis.* Hydrolases are used primarily in the depolymerization of macromolecules.
2. *Isomerases*[2] rearrange the atoms within a molecule but do not add or remove anything (so they are neither catabolic nor anabolic).
3. *Ligases* or *polymerases* join two molecules together (and are thus anabolic). They often use energy supplied by ATP.
4. *Lyases* split large molecules (and are thus catabolic) without using water in the process.
5. *Oxidoreductases* remove electrons from (oxidize) or add electrons to (reduce) various substrates. They are used in both catabolic and anabolic pathways.
6. *Transferases* transfer functional groups, such as an amino group (NH_2), a phosphate group, or a two-carbon (acetyl) group, between molecules. Transferases can be anabolic.

[2]An isomer is a compound with the same molecular formula as another molecule, but with a different arrangement of atoms.

Table 5.1 Enzyme Classification Based on Reaction Types

Class	Type of Reaction Catalyzed	Example
Hydrolase	Hydrolysis (catabolic)	Lipase—breaks down lipid molecules
Isomerase	Rearrangement of atoms within a molecule (neither catabolic nor anabolic)	Phosphoglucoisomerase—converts glucose 6-phosphate into fructose 6-phosphate during glycolysis
Ligase or polymerase	Joining two or more chemicals together (anabolic)	Acetyl-CoA synthetase—combines acetate and coenzyme A to form acetyl-CoA for the Krebs cycle
Lyase	Splitting a chemical into smaller parts without using water (catabolic)	Fructose 1,6-bisphosphate aldolase—splits fructose 1,6-bisphosphate into G3P and DHAP during glycolysis
Oxidoreductase	Transfer of electrons or hydrogen atoms from one molecule to another	Lactic acid dehydrogenase—oxidizes lactic acid to form pyruvic acid during fermentation
Transferase	Moving a functional group from one molecule to another (may be anabolic)	Hexokinase—transfers phosphate from ATP to glucose in the first step of glycolysis

Table 5.1 summarizes these types of enzymes and gives examples of each.

The Makeup of Enzymes

Many enzymes are composed entirely of protein, but others are composed of protein portions, called **apoenzymes** (ap′ō-en-zīms), that are inactive if they are not bound to one or more nonprotein substances called cofactors. **Cofactors** are either inorganic ions (such as iron, magnesium, zinc, or copper ions) or certain organic molecules called coenzymes. All **coenzymes** (ko-en′zīms) are either vitamins or contain vitamins, which are organic molecules that are required for metabolism but cannot be synthesized by certain organisms (especially mammals). Some apoenzymes bind with inorganic cofactors, some bind with coenzymes, and some bind with both. The binding of an apoenzyme and its cofactor(s) forms an active enzyme, called a **holoenzyme** (hol-ō-en′zīm; **Figure 5.4**).

Table 5.2 lists several examples of inorganic cofactors and organic cofactors (coenzymes). Note that three important coenzymes are the electron carriers NAD^+, $NADP^+$, and FAD, which, as we have seen, carry electrons in hydrogen atoms from place to place within cells. We will examine more closely the roles of these coenzymes in the generation of ATP later in the chapter.

Not all enzymes are proteinaceous; some are RNA molecules called **ribozymes.** In eukaryotes, ribozymes process other RNA molecules by removing sections of RNA and splicing the remaining pieces together. Recently, researchers have discovered that the functional core of a ribosome is a ribozyme; therefore, given that ribosomes make all proteins, ribosomal enzymes make protein enzymes.

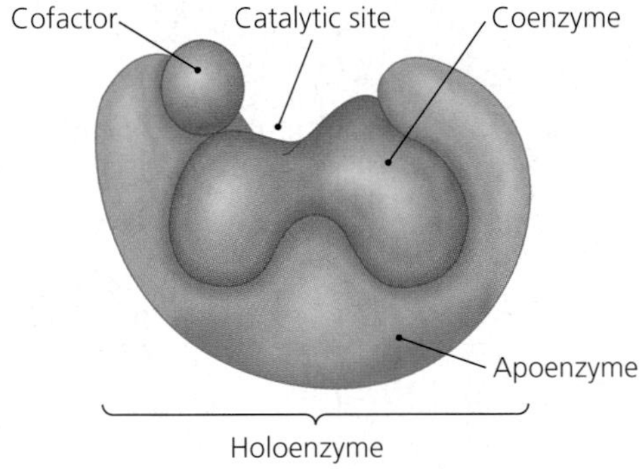

▲ ***Figure 5.4***

Makeup of a protein enzyme. The combination of a proteinaceous apoenzyme with one or more cofactors forms a holoenzyme, which is the active form of the enzyme. The label "cofactor" represents either an inorganic ion or a coenzyme, which is an organic cofactor derived from a vitamin. The apoenzyme is inactive unless it is bound to its cofactors. *Name four metal ions that can act as cofactors.*

Figure 5.4 *Iron, magnesium, zinc and copper ions can act as cofactors.*

Enzyme Activity

Within cells, enzymes catalyze reactions by lowering the **activation energy,** which is the amount of energy needed to trigger a chemical reaction **(Figure 5.5).** Whereas heat can provide energy to trigger reactions, the temperatures needed to reach activation energy for most metabolic reactions are often too high to allow cells to survive, so enzymes are needed if metabolism is to occur. This is true regardless of whether the enzyme is a protein or RNA, or whether the chemical reaction is anabolic or catabolic.

The activity of enzymes depends on the closeness of fit between the functional sites of an enzyme and its substrate. The shape of an enzyme's functional site, called its **active site,** is complementary to the shape of the substrate. Generally, the shapes and locations of only a few amino acids or nucleotides determines the shape of an enzyme's active site. A change in a single component—for instance, through mutation—can render an enzyme less effective or even completely nonfunctional.

Table 5.2 Representative Cofactors of Enzymes

Cofactors	Example of Use in Enzymatic Activity	Substance Transferred in Enzymatic Activity	Vitamin Source (of Coenzyme)
Inorganic (metal ion)			
Magnesium (Mg^{2+})	Forms bond with ADP during phosphorylation	Phosphate	None
Organic (coenzymes)			
Nicotinamide adenine dinucleotide (NAD^+)	Carrier of reducing power	Two electrons and a hydrogen ion	Niacin
Nicotinamide adenine dinucleotide phosphate ($NADP^+$)	Carrier of reducing power	Two electrons and a hydrogen ion	Niacin
Flavin adenine dinucleotide (FAD)	Carrier of reducing power	Two hydrogen atoms	Riboflavin
Tetrahydrofolate	Used in synthesis of nucleotides and some amino acids	One-carbon molecule	Folic acid
Coenzyme A	Formation of acetyl-CoA in Krebs cycle and beta-oxidation	Two-carbon molecule	Pantothenic acid
Pyridoxal phosphate	Transaminations in the synthesis of amino acids	Amine group	Pyridoxine
Thiamine pyrophosphate	Decarboxylation of pyruvic acid	Aldehyde group (CHO)	Thiamine

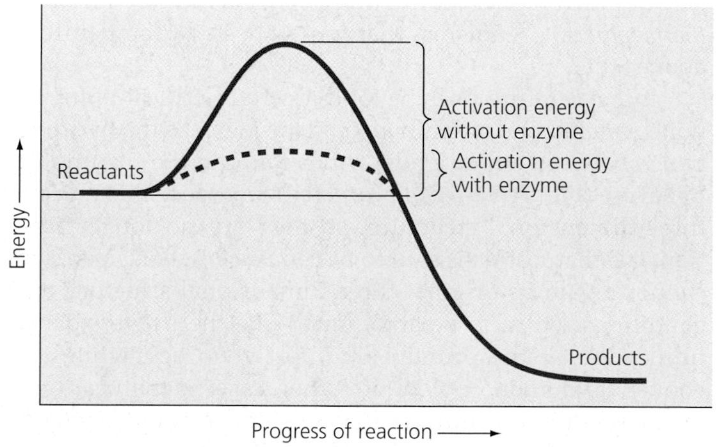

Figure 5.5

The effect of enzymes on chemical reactions. Enzymes catalyze reactions by lowering the activation energy—that is, the energy needed to trigger the reaction.

This *enzyme-substrate specificity,* which is critical to enzyme activity, has been likened to the fit between a lock and key. This analogy is not completely apt because enzymes change shape slightly when they bind to their substrate, almost as if a lock could grasp its key once it has been inserted. This latter description of enzyme-substrate specificity is called the *induced-fit model* (**Figure 5.6** on page 132).

In some cases, several different enzymes possess active sites that are complementary to various portions of a single substrate molecule. For example, an important precursor metabolite called phosphoenolpyruvic acid (PEP) is the substrate for at least five enzymes; depending on the enzyme involved, various products are produced from PEP. For instance, in one catabolic pathway PEP is converted to pyruvic acid, whereas in a particular anabolic pathway PEP is converted to the amino acid phenylalanine.

Although the exact ways that enzymes lower activation energy are not known, it appears that several mechanisms are involved. Some enzymes appear to bring reactants into sufficiently close proximity to enable a bond to form, whereas other enzymes change the shape of a reactant, inducing a bond to be broken. In any case, enzymes increase the likelihood that bonds will form or break.

The activity of enzymes is believed to follow the process illustrated in **Figure 5.7** on page 133, which depicts the catabolic lysis of a molecule called fructose 1,6-bisphosphate:

1. An enzyme associates with a specific substrate molecule having a shape that is complementary to that enzyme's active site.
2. The enzyme and its substrate bind to form a temporary intermediate compound called an *enzyme-substrate complex.* The binding of the substrate induces the enzyme to fit the shape of the substrate even more closely.
3. Bonds within the substrate are broken, forming two

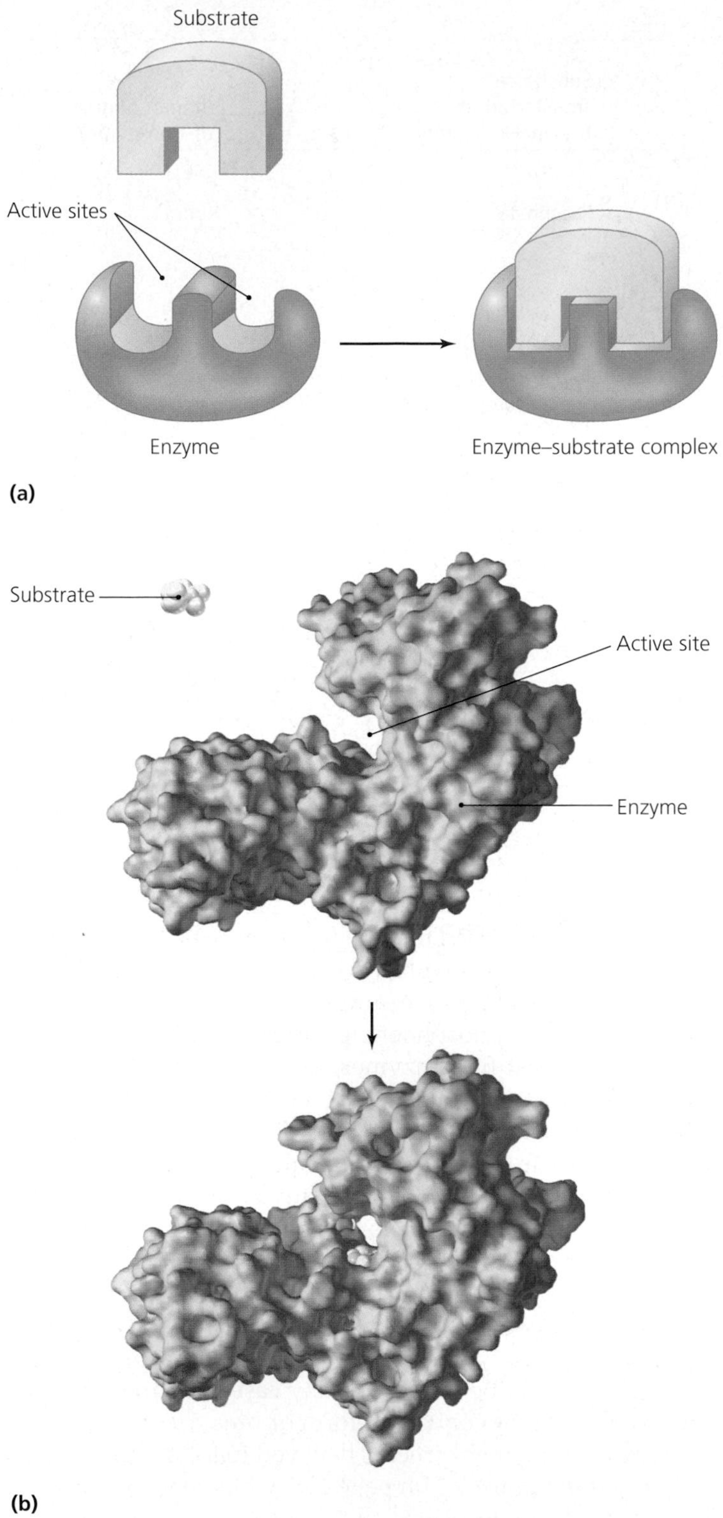

▲ *Figure 5.6*

(a) The induced-fit model of enzyme-substrate interaction. The enzyme's active site is generally complementary to the shape of its substrate, but a perfect fit between them does not occur until the substrate and enzyme bind to form an enzyme-substrate complex. **(b)** Space-filling models of an enzyme binding to a substrate.

(and in some other reactions, more than two) products. (Note that in an anabolic reaction, instead of the breakage of a bond, two reactants are linked together to form a single product.)

4. The enzyme disassociates from the newly formed molecules, which diffuse away from the site of the reaction, and the enzyme resumes its original configuration and is ready to associate with another substrate molecule.

Many factors influence the rate of enzymatic reactions, including temperature, pH, enzyme and substrate concentrations, and the presence of inhibitors.

Temperature As mentioned, higher temperatures tend to increase the rate of most chemical reactions because molecules are moving faster and collide more frequently, which encourages bonds to form or break. However, this is not entirely true of enzymatic reactions, because the active sites of enzymes change shape as temperature changes. If the temperature rises too high or falls too low, an enzyme is often no longer able to achieve a fit with its substrate.

Each enzyme has an optimal temperature for its activity (**Figure 5.8a** on page 134). The optimum temperature for the enzymes in the human body is about 37°C, which is normal body temperature. Part of the reason certain pathogens can cause disease in humans is that the optimal temperature for the enzymes in those microorganisms is also 37°C. The enzymes of some other microorganisms, however, function best at much higher temperatures; this is the case for *hyperthermophiles,* organisms that grow best at temperatures above 80°C.

If temperature rises beyond a certain critical point, the noncovalent bonds within an enzyme (such as the hydrogen bonds between amino acids) will break, and the enzyme will **denature** (**Figure 5.9**). Denatured enzymes lose their specific three-dimensional structure, so they are no longer functional. Denaturation is said to be *permanent* when an enzyme cannot regain its original three-dimensional structure once conditions return to normal, much like the irreversible solidification of the protein albumin when egg whites are cooked and then cooled. In other cases denaturation is *reversible*—the denatured enzyme's noncovalent bonds reform upon the return of normal conditions.

CRITICAL THINKING

Explain why hyperthermophiles do not cause disease in humans.

pH Extremes of pH also denature enzymes when ions released from acids and bases interfere with hydrogen bonding and distort and disrupt an enzyme's secondary and tertiary structures. Therefore, each enzyme has an optimal pH (**Figure 5.8b**).

Changing the pH provides a way to control the growth of unwanted microorganisms by denaturing their proteins. For example, vinegar (acetic acid, pH 3.0) acts as a preserva-

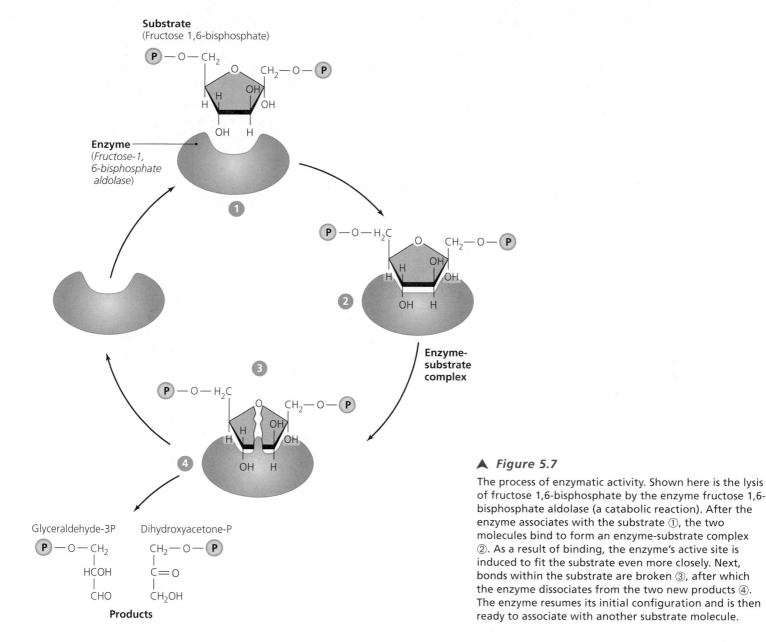

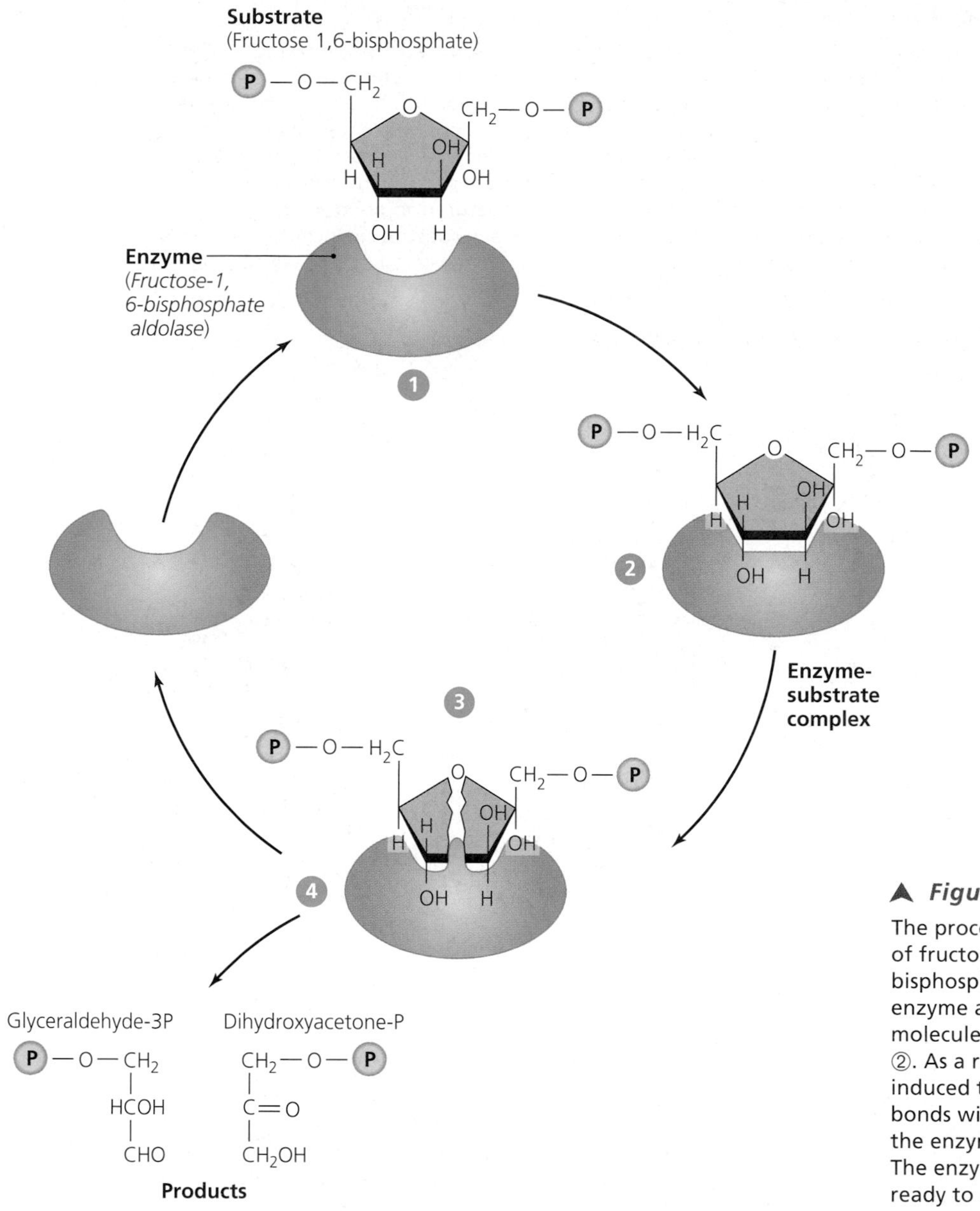

▲ ***Figure 5.7***

The process of enzymatic activity. Shown here is the lysis of fructose 1,6-bisphosphate by the enzyme fructose 1,6-bisphosphate aldolase (a catabolic reaction). After the enzyme associates with the substrate ①, the two molecules bind to form an enzyme-substrate complex ②. As a result of binding, the enzyme's active site is induced to fit the substrate even more closely. Next, bonds within the substrate are broken ③, after which the enzyme dissociates from the two new products ④. The enzyme resumes its initial configuration and is then ready to associate with another substrate molecule.

tive in dill pickles, and ammonia (pH 11.5) can be used as a disinfectant.

CRITICAL THINKING

In addition to extremes in temperature and pH, other chemical and physical agents, including ionizing radiation, alcohol, enzymes, and heavy-metal ions, denature proteins. For example, the first antimicrobial drug, salvarsan, contained the heavy metal arsenic and was used to inhibit the enzymes of the bacterium *Treponema pallidum,* the causative agent of syphilis.

Given that both human and bacterial enzymes are denatured by heavy metals, how was salvarsan used to treat syphilis without poisoning the patient? Why is syphilis no longer treated with arsenic-containing compounds?

Enzyme and Substrate Concentration Another factor that determines the rate of enzymatic activity within cells is the concentration of substrate present (**Figure 5.8c**). As substrate concentration increases, enzymatic activity increases as more and more enzyme active sites bind more and more substrate molecules. Eventually, when all enzyme active sites have bound substrate, the enzymes have reached their saturation point, and the addition of more substrate will not increase the rate of enzymatic activity.

Obviously, the rate of enzymatic activity is also affected by the concentration of enzyme within cells. In fact, one way that organisms regulate their metabolism is by controlling the quantity and timing of enzyme synthesis. In other words, many enzymes are produced in the amounts and at the times they are needed to maintain metabolic activity. Chapter 7 discusses the role of genetic mechanisms in the regulation of enzyme synthesis. Additionally, eukaryotic

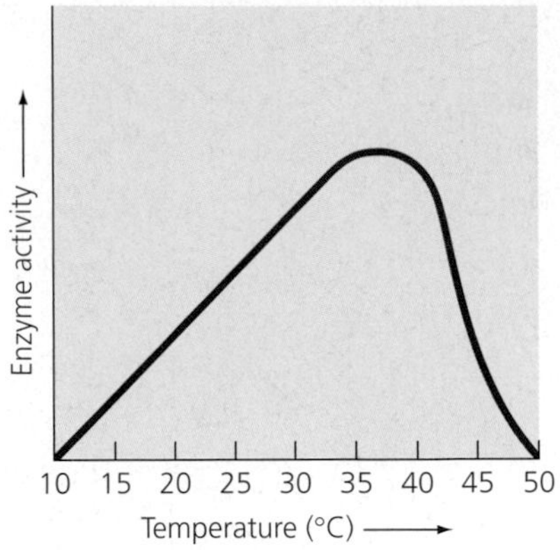

(a) Temperature

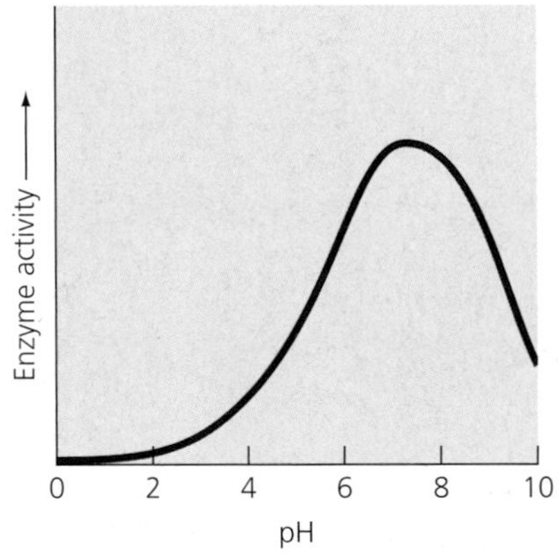

(b) pH

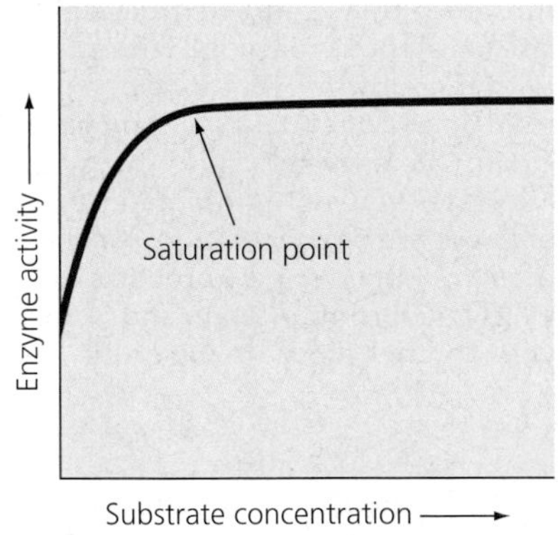

(c) Substrate concentration

Figure 5.8

The effects of temperature, pH, and substrate concentration on enzyme activity. **(a)** Rising temperature enhances enzymatic activity to a point, but above some optimal temperature the enzyme denatures and loses function. **(b)** Enzymes typically have some optimal pH, at which point enzymatic activity reaches a maximum. **(c)** At lower substrate concentrations, enzyme activity increases as the substrate concentration increases and as more and more active sites are utilized. At the substrate concentration at which all active sites are utilized, termed the saturation point, enzymatic activity reaches a maximum, and any additional increase in substrate concentration has no effect on enzyme activity. *What is the optimal pH of the enzyme shown in part (b)?*

Figure 5.8 The enzyme's optimal pH is approximately 7.

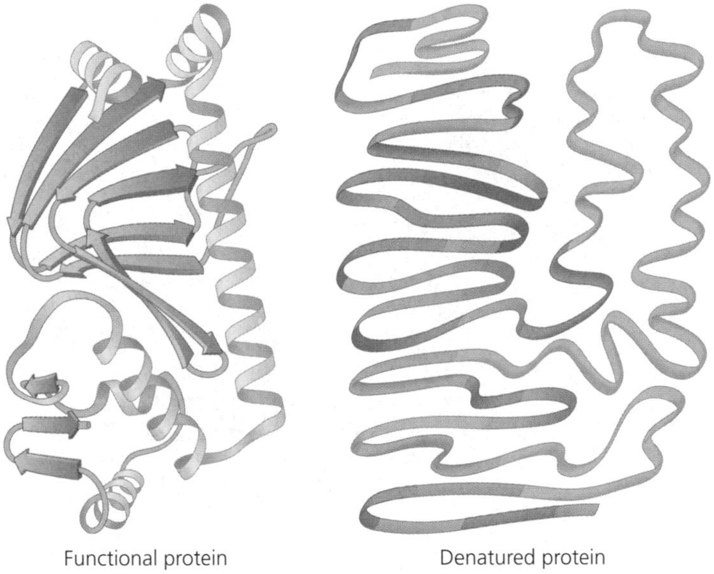

Figure 5.9

Denaturation of protein enzymes. Breakage of noncovalent bonds (such as hydrogen bonds) causes the protein to lose its secondary and tertiary structure and become denatured; as a result, the enzyme is no longer functional.

cells control some enzymatic activities by compartmentalizing enzymes inside membranes so that certain metabolic reactions proceed physically separated from the rest of the cell. For example, white blood cells catabolize phagocytized pathogens using enzymes packaged within lysosomes.

Inhibitors Enzymatic activity can be influenced by a variety of inhibitory substances that block an enzyme's active site. Enzymatic inhibitors, which may be either competitive or noncompetitive, do not denature enzymes.

Competitive inhibitors are shaped such that they fit into an enzyme's active site and thus prevent the normal substrate from binding **(Figure 5.10a).** However, such inhibitors do not undergo a chemical reaction to form products. Competitive inhibitors can bind permanently or reversibly to an active site. Permanent binding results in permanent loss of enzymatic activity; reversible competition can be overcome by an increase in the concentration of substrate molecules, which increases the likelihood that active sites will be filled with substrate instead of inhibitor **(Figure 5.10b).**

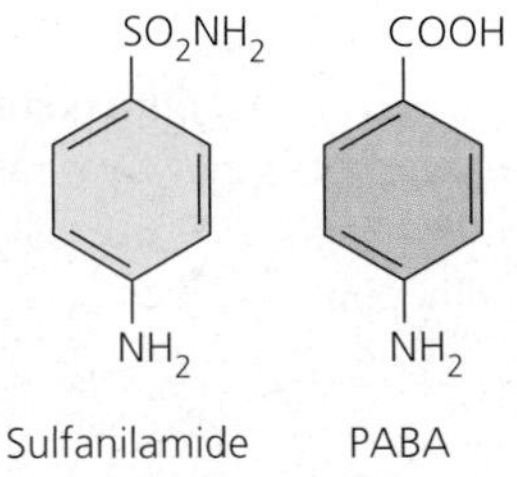

A good example of competitive inhibition is the action of sulfanilamide, which has a shape similar to that of paraaminobenzoic acid (PABA).

Sulfanilamide has great affinity for the active site of an enzyme required in the conversion of PABA

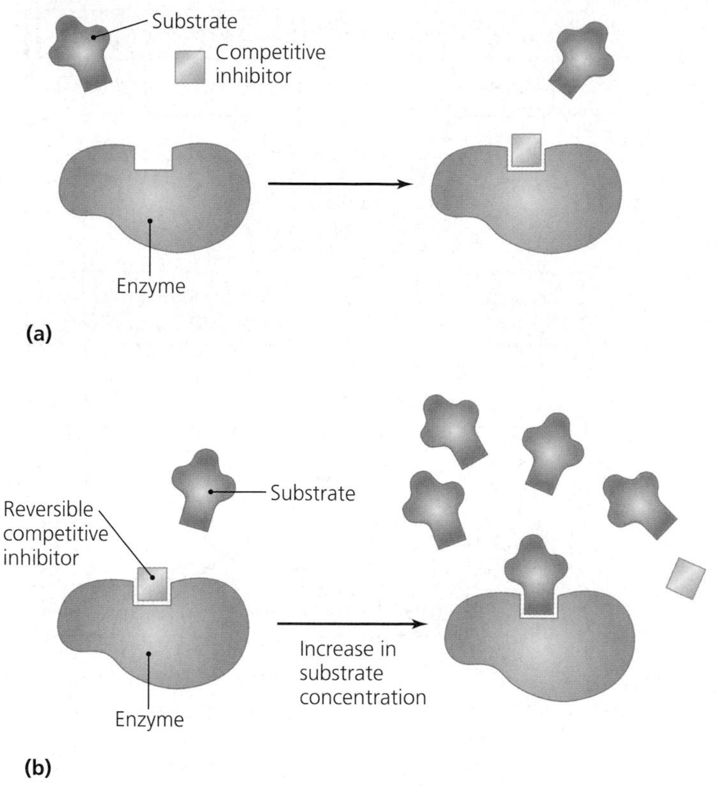

Figure 5.10

Competitive inhibition of enzyme activity. **(a)** Inhibitory molecules, which are similar in shape to substrate molecules, compete for and block active sites. **(b)** Reversible inhibition can be overcome by an increase in substrate concentration.

into folic acid, which is essential for DNA synthesis. Once sulfanilamide is bound to the enzyme, it stays bound. As a result, it prevents synthesis of folic acid. Sulfanilamide effectively inhibits bacteria that make folic acid from PABA without harming people because humans lack the necessary enzymes; they must acquire folic acid in their diets.

Noncompetitive inhibitors do not bind to the active site but instead prevent enzymatic activity by binding to an *allosteric* (al-ō-stār′ik) *site* located elsewhere on the enzyme. Binding at an allosteric site alters the shape of the active site so that substrate cannot be bound. Allosteric control of enzyme activity can take two forms: inhibitory and excitatory. *Allosteric (noncompetitive) inhibition* halts enzymatic activity in the manner just described **(Figure 5.11a).** In *excitatory allosteric control,* the binding of certain activator molecules (such as a heavy-metal ion cofactor) to an allosteric site causes a change in shape of the active site, which activates an otherwise inactive enzyme **(Figure 5.11b).** Some enzymes have several allosteric sites, both inhibitory and excitatory, which allows their function to be closely regulated.

Cells often control the action of enzymes through **feedback inhibition** (also called *negative feedback* or *end-product inhibition*). Allosteric feedback inhibition functions in much the way a thermostat controls a heater. As the room gets warmer, a sensor inside the thermostat changes shape and sends an electrical signal that turns off the flame or electrical coil in the heater. Similarly, in metabolic feedback inhibition, the end-product of a series of reactions is an allosteric inhibitor of an enzyme in an earlier part of the pathway (**Figure 5.12a** on page 136). Because the product of each reaction in the pathway is the substrate for the next reaction, inhibition of the first enzyme in the series inhibits the entire pathway, thereby saving the cell energy. For example, in *Escherichia coli*, the presence of the amino acid isoleucine allosterically inhibits the first enzyme in the metabolic pathway that produces isoleucine. In this manner, the bacterium prevents the accumulation of isoleucine (and intermediate products) when the amino acid is available from the environment. When environmental isoleucine is depleted, the first metabolic enzyme is no longer inhibited, and isoleucine production resumes.

Feedback inhibition can occur in even more complex ways. For instance, even though the first step in the synthesis of the amino acids tyrosine, phenylalanine, and tryptophan is the same—the linkage of phosphoenolpyruvic acid (PEP) and erythrose 4-phosphate to form 3-deoxy-arabinoheptulosonic acid 7-phosphate (DAHAP)—three different

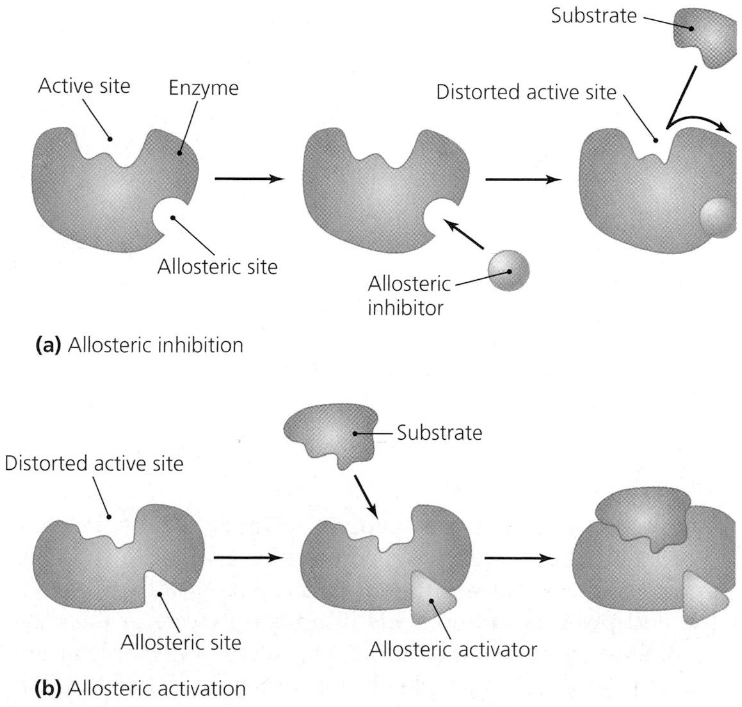

Figure 5.11

Allosteric control of enzyme activity. **(a)** Allosteric (noncompetitive) inhibition results from a change in the shape of the active site when an inhibitor binds to an allosteric site. **(b)** Allosteric activation results when the binding of an activator molecule to an allosteric site causes a change in the active site that makes it capable of binding substrate.

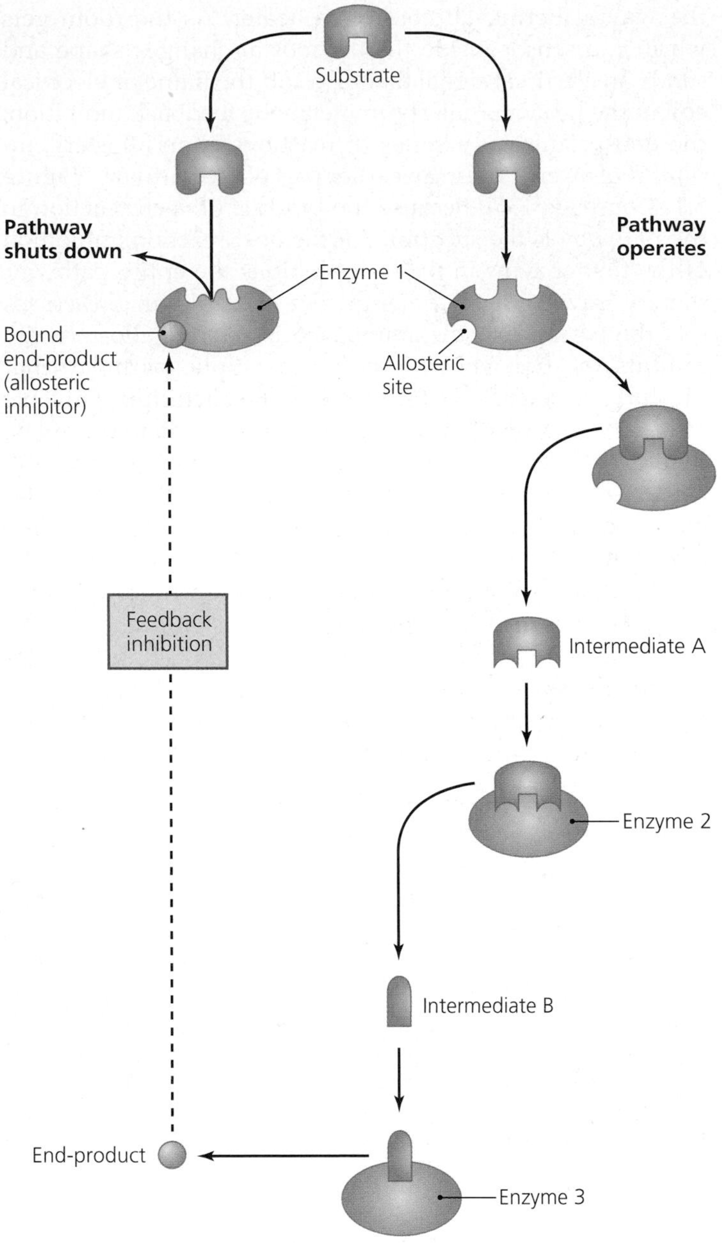

(a)

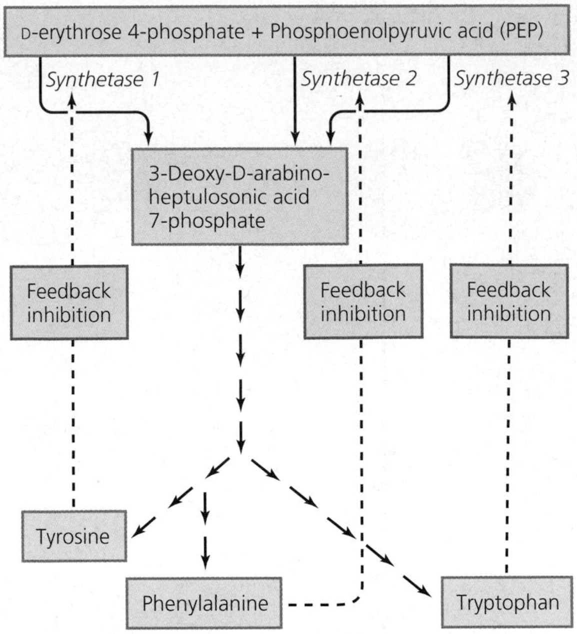

(b)

Figure 5.12

Feedback inhibition. **(a)** The end-product of a metabolic pathway allosterically inhibits the initial step, shutting down the pathway. **(b)** In this example, each of the three end-products (tyrosine, phenylalanine, and tryptophan) inhibits a different synthetase enzyme. An excess of all three end-products is needed to completely hinder the synthesis of 3-deoxy-arabino-heptulosonic acid 7-phosphate.

synthetase enzymes are involved **(Figure 5.12b).** In other words, enzymatic reactions convert DAHAP into the three amino acids via three different pathways. As shown, each of the end-product amino acids inhibits only one of the three synthetase enzymes. Therefore, an excess of all three amino acids is needed to completely hinder synthesis of DAHAP and thereby stop the production of these amino acids.

To this point we have viewed the concept of metabolism as a collection of chemical reactions (pathways) that can be categorized as either catabolic (breaking down) or anabolic (building up). Because enzymes are required to lower the activation energy of these reactions, we examined these catalysts in some detail.

Energy is also critical to metabolism, so we examined redox reactions as a means of transferring energy within cells. We saw, for example, that redox reactions and carrier molecules are used to transfer energy from catabolic pathways to ATP, a molecule that stores energy in cells.

We will now consider how cells acquire and utilize metabolites, which are used to synthesize the macromolecules necessary for growth and, eventually, reproduction—the ultimate goal of metabolism. We will also consider in more detail the phosphorylation of ADP to make ATP.

Carbohydrate Catabolism

Learning Objective

✓ In general terms, describe the three stages of aerobic glucose catabolism (glycolysis, the Krebs cycle, and the electron transport chain), including their substrates, products, and net energy production.

Organisms oxidize carbohydrates as their primary energy source for anabolic reactions. They use glucose most commonly, though other sugars, amino acids, and fats are also utilized, often by first converting them into glucose. Glucose is catabolized via one of two processes: either via *cellular*

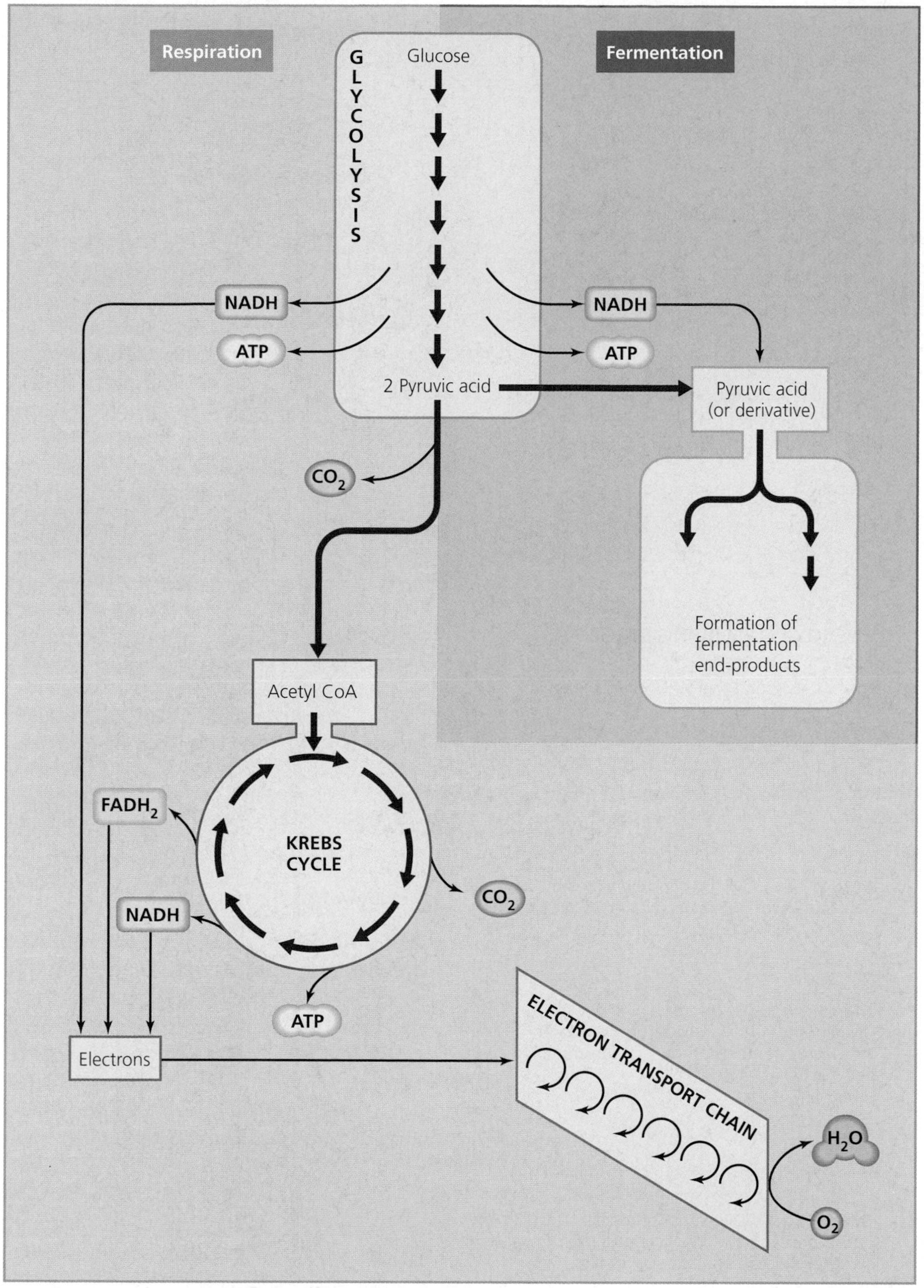

◀ ***Figure 5.13***

Summary of glucose catabolism. Glucose catabolism begins with glycolysis, which forms pyruvic acid and two molecules of both ATP and NADH. Two pathways branch from pyruvic acid: respiration and fermentation. In aerobic respiration, the Krebs cycle and the electron transport chain completely oxidize pyruvic acid to CO_2 and H_2O, in the process synthesizing many molecules of ATP. Fermentation results in the incomplete oxidation of pyruvic acid to form organic fermentation products.

respiration[3]—a process that involves the complete breakdown of glucose to carbon dioxide and water—or via *fermentation*, which results in organic waste products.

As shown in **Figure 5.13**, both cellular respiration and fermentation begin with *glycolysis* (glī-kol′i-sis), a process that catabolizes a single molecule of glucose to two molecules of pyruvic acid (also called pyruvate) and results in a small amount of ATP production. Respiration then continues via the *Krebs cycle* and the *electron transport chain*, which results in a significant amount of ATP production. Fermentation involves the conversion of pyruvic acid into other organic compounds. Because it lacks the Krebs cycle and electron transport chain, fermentation results in the production of much less ATP than does respiration.

[3]Cellular respiration is often referred to simply as *respiration*, which should not be confused with breathing, also called respiration.

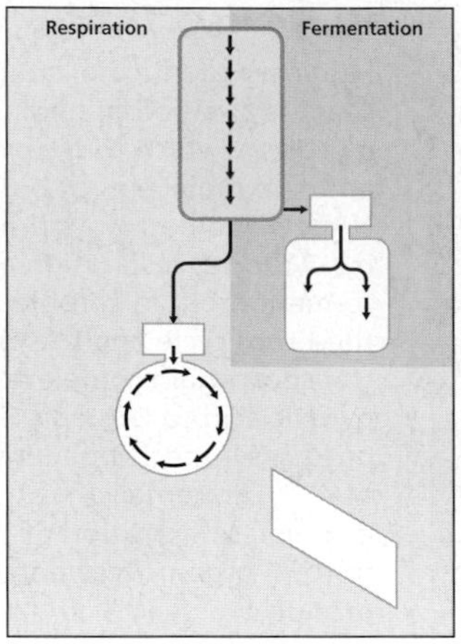

ENERGY-INVESTMENT STAGE

Step 1. Glucose is phosphorylated by ATP to form glucose 6-phosphate.

Steps 2 and 3. The atoms of glucose 6-phosphate are rearranged to form fructose 6-phosphate. Fructose 6-phosphate is phosphorylated by ATP to form fructose 1,6-bisphosphate.

LYSIS STAGE

Step 4. Fructose 1,6-bisphosphate is cleaved to form glyceraldehyde 3-phosphate (G3P) and dihydroxyacetone phosphate (DHAP).

Step 5. DHAP is rearranged to form another G3P.

ENERGY-CONSERVING STAGE

Step 6. Inorganic phosphates are added to the two G3P, and two NAD^+ are reduced.

Step 7. Two ADP are phosphorylated by substrate-level phosphorylation to form two ATP.

Steps 8 and 9. The remaining phosphates are moved to the middle carbons. A water molecule is removed from each substrate.

Step 10. Two ADP are phosphorylated by substrate-level phosphorylation to form two ATP. Two pyruvic acid are formed.

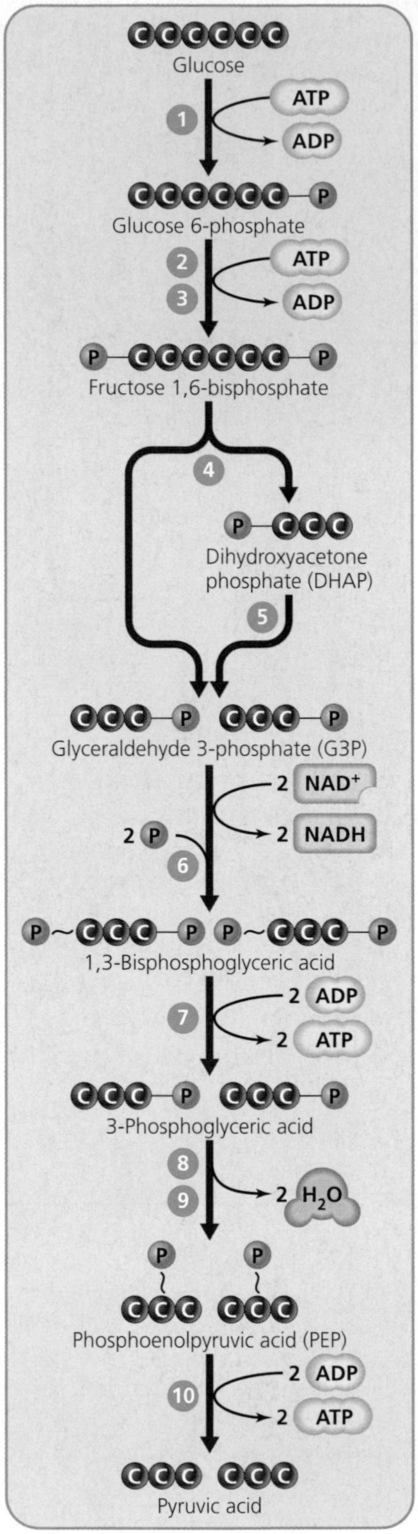

▲ ***Figure 5.14***

Glycolysis (also known as the Embden-Meyerhof pathway), in which glucose is cleaved and ultimately transformed into two molecules of pyruvic acid. Four ATPs are formed and two ATPs are used, so a net gain of two ATPs results. Two molecules of NAD^+ are reduced to NADH.

The following is a simplified discussion of glucose catabolism. Details of the substrates and enzymes involved are provided in Appendix A. To help understand the basic reactions in each of the pathways of glucose catabolism, pay special attention to three things: the number of carbon atoms in each of the intermediate products, the relative numbers of ATP molecules produced in each pathway, and the changes in the coenzymes NAD^+ and FAD as they are reduced and then oxidized back to their original forms.

Glycolysis

Glycolysis,[4] also called the *Embden-Meyerhof pathway* after the scientists who discovered it, is the first step in the catabolism of glucose via both respiration and fermentation. Glycolysis occurs in most cells. In general, as its name implies, glycolysis involves the splitting of a six-carbon glucose molecule into two three-carbon sugar molecules. When these three-carbon molecules are oxidized to pyruvic acid, some of the energy released is stored in molecules of ATP.

Glycolysis, which occurs in the cytoplasm, can be divided into three stages involving a total of 10 steps **(Figure 5.14),** each of which is catalyzed by its own enzyme. The three stages of glycolysis are:

1. Energy-investment stage (steps 1–3). As with money, one must invest before a profit can be made. In this case, the energy in two molecules of ATP is invested to phosphorylate a six-carbon glucose molecule and rearrange its atoms to form fructose 1,6-bisphosphate.
2. Lysis stage (steps 4 and 5). Fructose 1,6-bisphosphate is cleaved into glyceraldehyde 3-phosphate (G3P)[5] and dihydroxyacetone phosphate (DHAP). Each of these compounds contains three carbon atoms and is freely convertible into the other.
3. Energy-conserving stage (steps 6–10). G3P is oxidized to pyruvic acid, yielding two ATP molecules. DHAP is converted to G3P and also oxidized to pyruvic acid, yielding another two ATP molecules, for a total of four ATP molecules.

Our study of glycolysis provides our first opportunity to study *substrate-level phosphorylation* (see steps 1, 7, and 10 in Figure 5.14). Let's examine this important process more closely by considering the 10th and final step of glycolysis.

Each of the two phosphoenolpyruvic acid (PEP) molecules produced in step 9 of glycolysis is a three-carbon compound containing a high-energy phosphate bond. In the presence of a specific holoenzyme (which requires a Mg^{2+} cofactor), the high-energy phosphate in PEP (one substrate) is transferred to an ADP molecule (a second substrate) to form ATP **(Figure 5.15);** the direct transfer of the phosphate between the two substrates is the reason the process is called **substrate-level phosphorylation.** A variety of substrate-level phosphorylations occur in metabolism. As you might expect, each type has its own enzyme that recognizes both its substrate molecule and ADP.

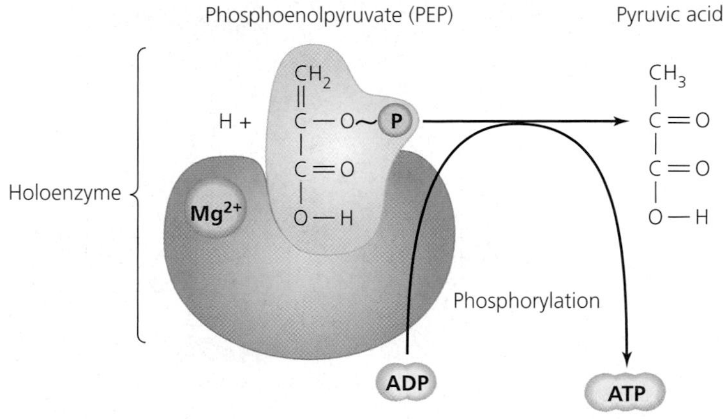

▲ *Figure 5.15*

Substrate-level phosphorylation, in which high-energy phosphate bonds are transferred from one substrate to another. *What role does Mg^{2+} play in this reaction?*

Figure 5.15 Mg^{2+} is a cofactor of the enzyme.

In glycolysis, two ATP molecules are invested by substrate-level phosphorylation to prime glucose for lysis, and four molecules of ATP are produced, by substrate-level phosphorylation. Therefore, a net gain of two ATP molecules occurs for each molecule of glucose that is oxidized to pyruvic acid. Glycolysis also yields two molecules of NADH.

Alternatives to Glycolysis

Learning Objective

✓ Compare the pentose phosphate pathway and the Entner-Doudoroff pathway with glycolysis in terms of energy production and products.

The initial part of the catabolism of glucose can also proceed via two alternate pathways: the pentose phosphate pathway and the Entner-Doudoroff pathway. Though they yield fewer molecules of ATP than glycolysis, these alternate pathways reduce coenzymes and yield different substrate metabolites that are needed in anabolic pathways. Next we briefly examine each of these alternate pathways.

Pentose Phosphate Pathway

The **pentose phosphate pathway** is named for the phosphorylated pentose (five-carbon) sugars—ribulose, xylulose, and ribose—that are formed from glucose 6-phosphate by enzymes in the pathway (**Figure 5.16** on page 140). The pentose phosphate pathway is primarily used for the production of precursor metabolites used in anabolic reactions, including the synthesis of nucleotides for nucleic acids, of certain amino acids, and of glucose by *photosynthesis*

[4]From Greek *glykys,* meaning sweet, and *lysein,* meaning to loosen.

[5]G3P is also known as phosphoglyceraldehyde or PGAL.

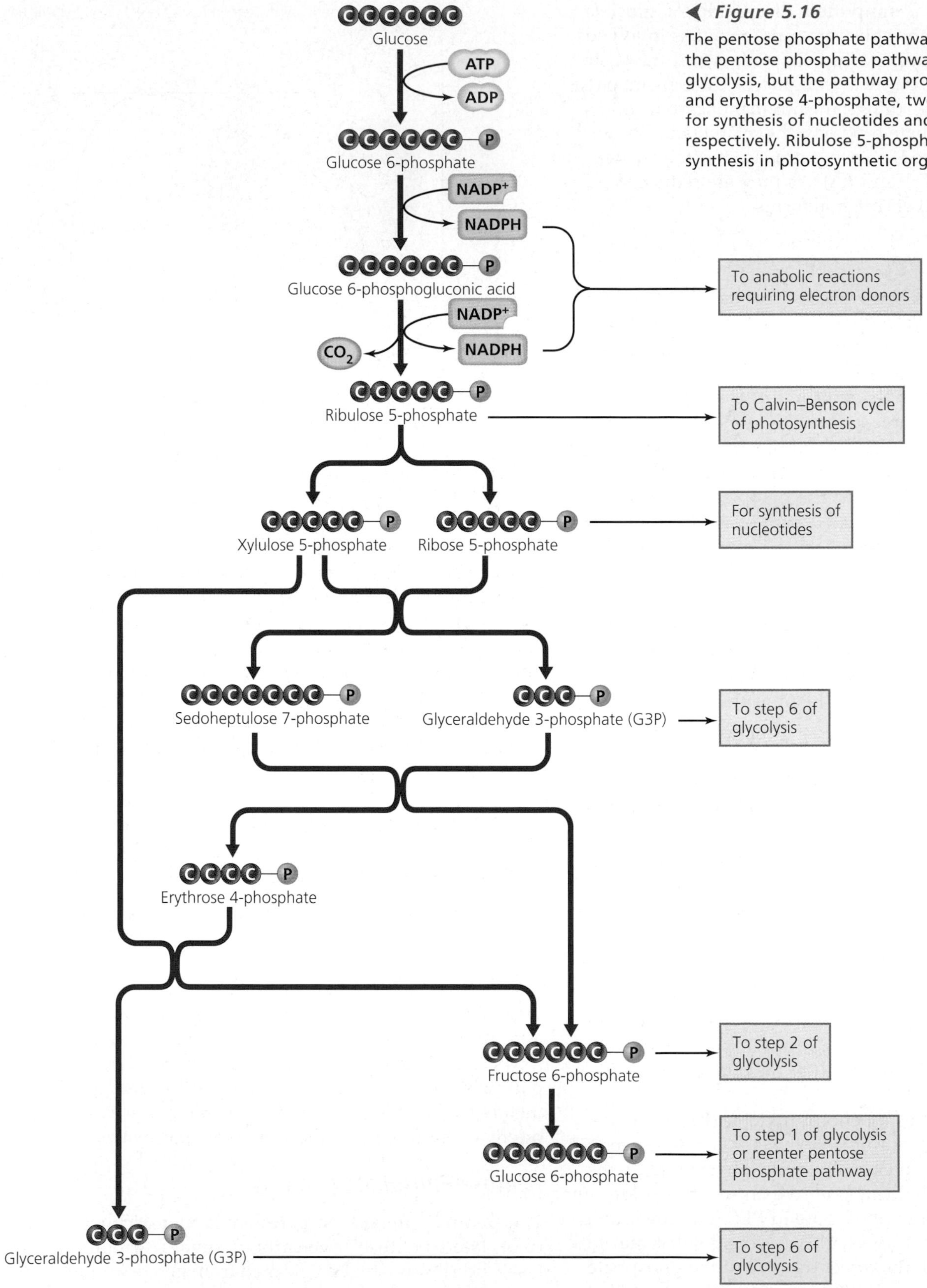

Figure 5.16

The pentose phosphate pathway. Energy captured during the pentose phosphate pathway is less than that of glycolysis, but the pathway produces ribose 5-phosphate and erythrose 4-phosphate, two metabolites necessary for synthesis of nucleotides and certain amino acids, respectively. Ribulose 5-phosphate, necessary for glucose synthesis in photosynthetic organisms, is also produced.

(described in a later section). The pathway also reduces two molecules of $NADP^+$ to NADPH and nets a single molecule of ATP from each molecule of glucose. NADPH is a necessary coenzyme for anabolic enzymes that synthesize DNA nucleotides, steroids, and fatty acids.

CRITICAL THINKING

Examine the biosynthetic pathway for the production of the amino acids tryptophan, tyrosine, and phenylalanine in Figure 5.12b. Where do the initial reactants (PEP and erythrose 4-phosphate) originate?

Entner-Doudoroff Pathway

Most bacteria use glycolysis and the pentose phosphate pathway, but a few substitute the **Entner-Doudoroff pathway (Figure 5.17)** for glycolysis. This pathway, named for its discoverers, is a series of reactions that catabolize glucose to pyruvic acid using different enzymes from those used in either glycolysis or the pentose phosphate pathway.

Among organisms, only a very few bacteria use the Entner-Doudoroff pathway. These include the Gram-negative *Pseudomonas aeruginosa* (soo-dō-mō′nas ā-rŭ-ji-nō′să), and the Gram-positive bacterium *Enterococcus faecalis* (en-te-rō-kok′kus fē-kā′lis). Like the pentose phosphate pathway, the Entner-Doudoroff pathway nets only a single molecule of ATP for each molecule of glucose, but it does yield precursor metabolites and NADPH. The latter is unavailable from glycolysis.

CRITICAL THINKING

Even though *Pseudomonas aeruginosa* and *Enterococcus faecalis* usually grow harmlessly in the body, they can cause disease. Because these bacteria use the Entner-Doudoroff pathway instead of glycolysis to catabolize glucose, investigators sometimes use clinical tests that provide evidence of the Entner-Doudoroff pathway to identify the presence of these potential pathogens.

Suppose you were able to identify the presence of any specific organic compound. Name a substrate molecule you would find in *Pseudomonas* and *Enterococcus* cells, but not in human cells.

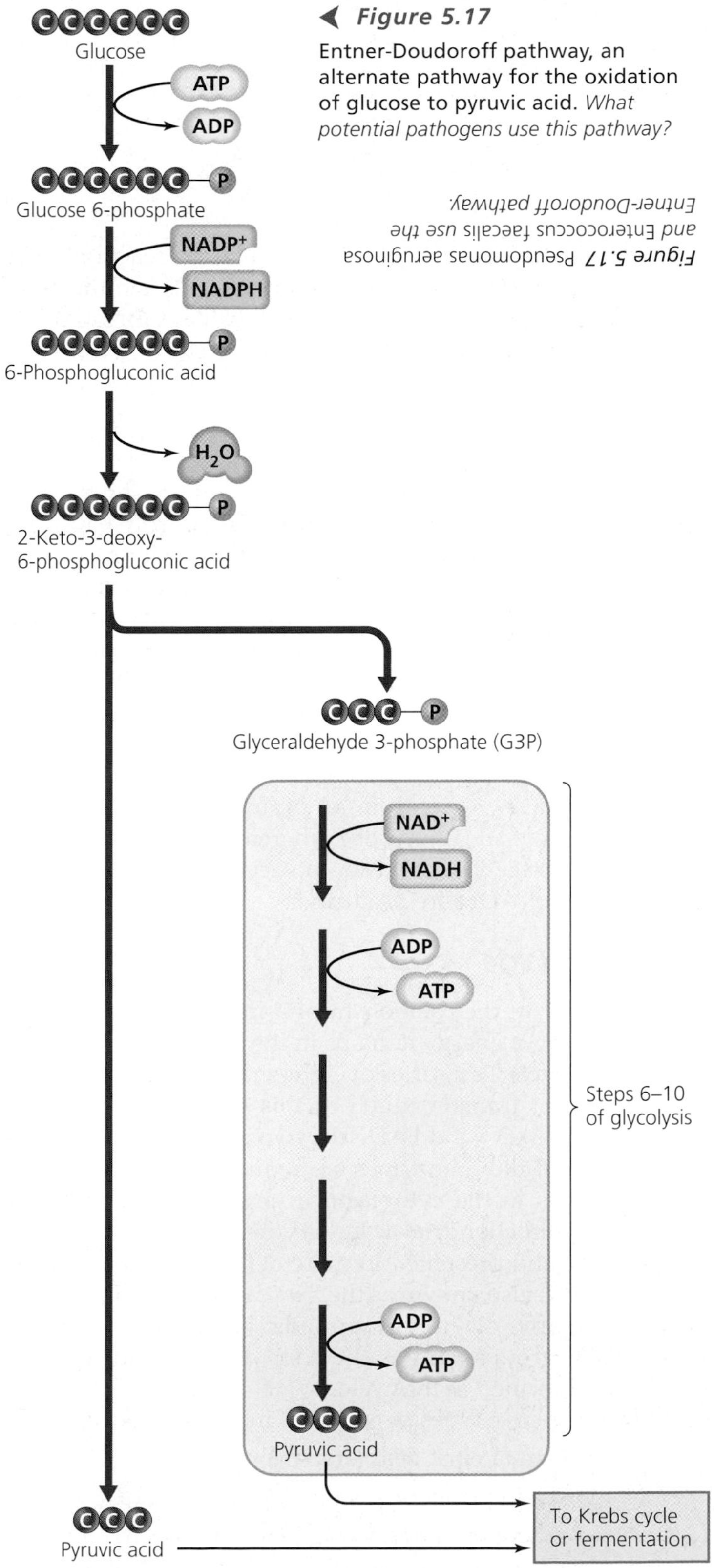

◀ *Figure 5.17* Entner-Doudoroff pathway, an alternate pathway for the oxidation of glucose to pyruvic acid. *What potential pathogens use this pathway?*

Figure 5.17 Pseudomonas aeruginosa *and* Enterococcus faecalis *use the Entner-Doudoroff pathway.*

Cellular Respiration

Learning Objectives

- ✓ Discuss the roles of acetyl-CoA, the Krebs cycle, and electron transport in carbohydrate catabolism.
- ✓ Contrast electron transport in aerobic and anaerobic respiration.
- ✓ Identify four classes of carriers in electron transport chains.
- ✓ Describe the role of chemiosmosis in oxidative phosphorylation of ATP.

After glucose has been oxidized via glycolysis or one of the alternate pathways, a cell uses the resultant pyruvic acid molecules to complete either cellular respiration or fermentation (which we will discuss in a later section). Our topic here—**cellular respiration**—is a metabolic process that involves the complete oxidation of substrate molecules and

Figure 5.18 ➤
Formation of acetyl-CoA. The responsible enzyme acts in a stepwise manner to ① remove CO_2 from pyruvic acid, ② attach the remaining two-carbon acetate to coenzyme A, and ③ reduce a molecule of NAD^+ to NADH.

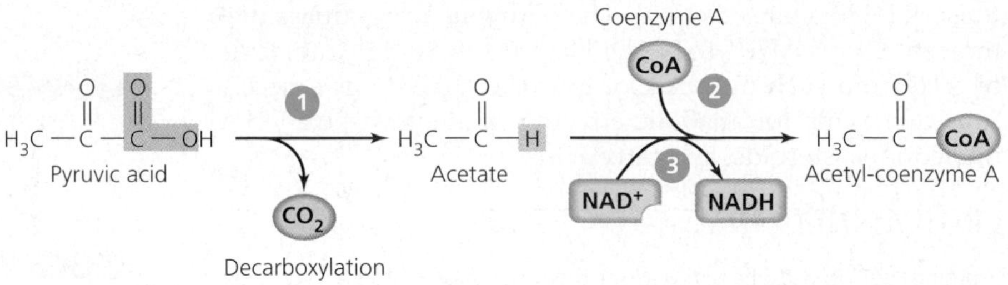

then production of ATP by a series of redox reactions. The three stages of cellular respiration are: (1) synthesis of acetyl-CoA, (2) the Krebs cycle, and (3) a final series of redox reactions, which collectively constitute an electron transport chain.

Synthesis of Acetyl-CoA

Before pyruvic acid (generated by glycolysis, the pentose phosphate pathway, and the Entner-Doudoroff pathway) can enter the Krebs cycle for respiration, it must first be converted to *acetyl-coenzyme A* or **acetyl-CoA** (as′e-til kō-ā′; see Figure 5.13). Enzymes remove one carbon from pyruvic acid as CO_2 and join the remaining two-carbon acetate to *coenzyme A* with a high-energy bond **(Figure 5.18).** The removal of CO_2, called *decarboxylation,* requires a coenzyme derived from the vitamin thiamine. One molecule of NADH is also produced during this reaction.

Recall that two molecules of pyruvic acid were derived from each molecule of glucose. Therefore, at this stage, two molecules of acetyl-CoA, two molecules of CO_2, and two molecules of NADH are produced.

The Krebs Cycle

At this point in the catabolism of a molecule of glucose, a great amount of energy remains in the bonds of acetyl-CoA. The **Krebs cycle**[6] is a series of eight enzymatically catalyzed reactions that transfer much of this stored energy to the coenzymes NAD^+ and FAD. The two carbons in acetate are oxidized, and the coenzymes are reduced. The Krebs cycle, which occurs in the cytoplasm in prokaryotes and in the matrix of mitochondria in eukaryotes, is diagrammed in **Figure 5.19** and presented in more detail in Appendix A on page A-6. It is also known as the *tricarboxylic acid (TCA) cycle,* because many of its compounds have three carboxyl (—COOH) groups, and as the *citric acid cycle,* for the first compound formed in the cycle.

There are five types of reactions in the Krebs cycle:

- Anabolism of citric acid (step 1)
- Isomerization reactions (steps 2, 7, and 8)
- Redox reactions (steps 3, 4, 6, and 8)
- Decarboxylations (steps 3 and 4)
- Substrate-level phosphorylation (step 5)

[6]Named for biochemist Sir Hans Krebs (1900–1981), who elucidated its reactions in the 1940s.

In the first step of the Krebs cycle, the splitting of the high-energy bond between acetate and coenzyme A releases enough energy to enable the binding of the freed two-carbon acetate to a four-carbon compound called oxaloacetic acid, forming the six-carbon compound citric acid.

As you study Figure 5.19, notice that after isomerization (step 2), the decarboxylations of the Krebs cycle release two molecules of CO_2 for each acetyl-CoA that enters (steps 3 and 4). Thus, for every two carbon atoms that enter the cycle, two are lost to the environment. At this juncture in the respiration of a molecule of glucose, six carbon atoms have been lost to the environment: two as CO_2 molecules produced in decarboxylation of two molecules of pyruvic acid to form two acetyl-CoA molecules, and four in CO_2 molecules produced in decarboxylations in the *two* turns through the Krebs cycle. (One molecule of acetyl-CoA enters the cycle at a time.)

A small amount of ATP is also produced in the Krebs cycle. For every two molecules of acetyl-CoA that pass through the Krebs cycle, two molecules of ATP are generated by substrate-level phosphorylation (step 5). A molecule of guanosine triphosphate (GTP), which is similar to ATP, serves as an intermediary in this process.

Redox reactions reduce FAD to $FADH_2$ (step 6) and NAD^+ to NADH (steps 3, 4, and 8), so that for every two molecules of acetyl-CoA that move through the cycle, six molecules of NADH and two of $FADH_2$ are formed. In the Krebs cycle, little energy is captured directly in high-energy phosphate bonds, but much energy is transferred via electrons to NADH and $FADH_2$. These coenzymes are the most important molecules of respiration because they carry a large amount of energy that is subsequently used to phosphorylate ADP to ATP.

CRITICAL THINKING

We have examined the total ATP, NADH, and $FADH_2$ production in the Krebs cycle for each molecule of glucose coming through Embden-Meyerhof glycolysis. How many of each of these molecules would be produced if the Entner-Doudoroff pathway were used instead of glycolysis?

Electron Transport

Some scientists estimate that each day an average human synthesizes his or her own weight in ATP molecules and uses them for metabolism, responsiveness, growth, and cell reproduction. ATP turnover in prokaryotes is relatively as

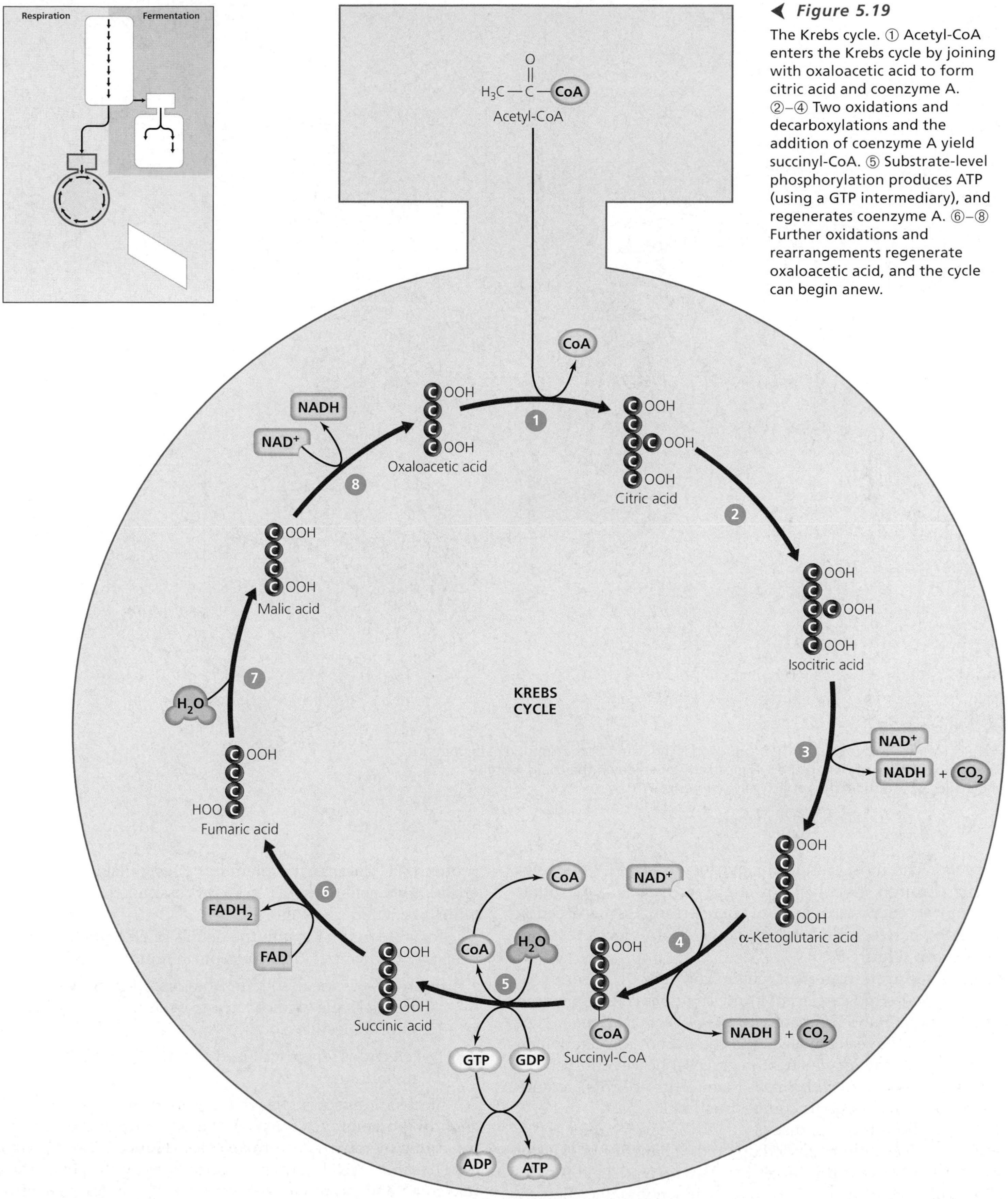

Figure 5.19

The Krebs cycle. ① Acetyl-CoA enters the Krebs cycle by joining with oxaloacetic acid to form citric acid and coenzyme A. ②–④ Two oxidations and decarboxylations and the addition of coenzyme A yield succinyl-CoA. ⑤ Substrate-level phosphorylation produces ATP (using a GTP intermediary), and regenerates coenzyme A. ⑥–⑧ Further oxidations and rearrangements regenerate oxaloacetic acid, and the cycle can begin anew.

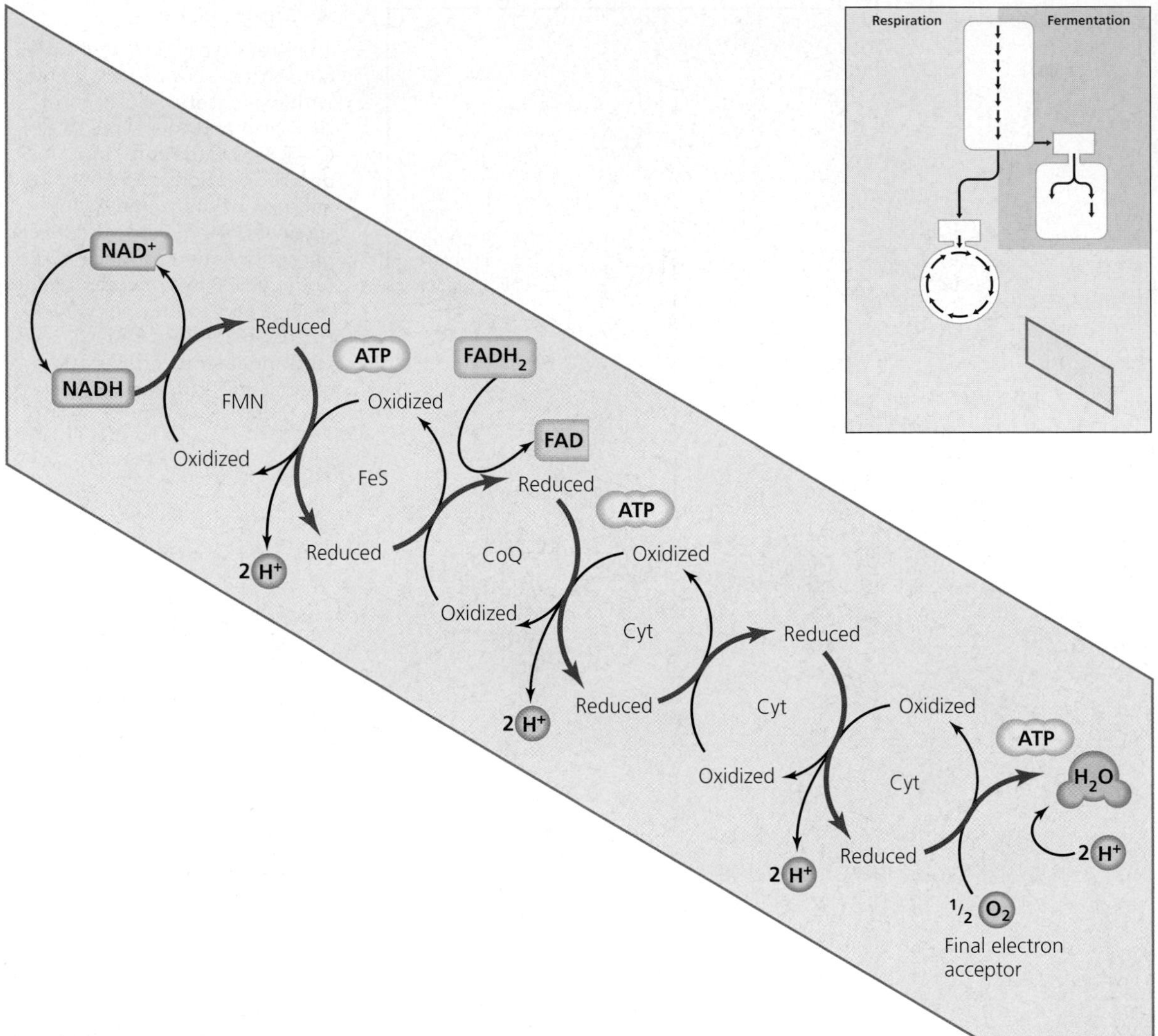

▲ ***Figure 5.20***

An electron transport chain. ATP production is indicated at the approximate point in the chain that energy is captured as electrons move down the chain, but the molecules of the chain do not actually synthesize ATP.

copious. The most significant production of ATP does not occur through glycolysis or the Krebs cycle, but rather through the stepwise release of energy from a series of redox reactions between molecules known as an **electron transport chain (Figure 5.20).**

An electron transport chain consists of a series of membrane-bound carrier molecules that pass electrons from one to another and ultimately to a *final electron acceptor.* Typically, as we have seen, electrons come from the catabolism of an organic molecule such as glucose; however, microorganisms called *chemolithotrophs* (kem′ō-lith′-ō-trōfs) acquire electrons from inorganic sources such as H_2, NO^{2-}, or Fe^{2+}. (Chemolithotrophs are discussed further in Chapter 6.) In any case, electrons are passed down the chain like buckets in a fire brigade to the final acceptor. As with a bucket brigade, the final step of electron transport is irreversible. Energy from the electrons is used to actively transport (pump) protons (H^+) across the membrane, establishing a *proton gradient* that generates ATP via a process called *chemiosmosis,* which we will discuss shortly.

To avoid getting lost in the details of electron transport, keep the following critical concepts in mind:

- Electrons pass sequentially from one membrane-bound carrier molecule to another, and eventually to a final acceptor molecule.
- The electrons' energy is used to pump protons across the membrane.

Electron transport chains are located in the inner mitochondrial membranes (cristae) of eukaryotes and in the cytoplasmic membrane of prokaryotes **(Figure 5.21).** Though NADH and $FADH_2$ donate electrons as hydrogen atoms (electrons and protons), many carrier molecules pass only the electrons down the chain. There are four categories of carrier molecules in electron transport chains.

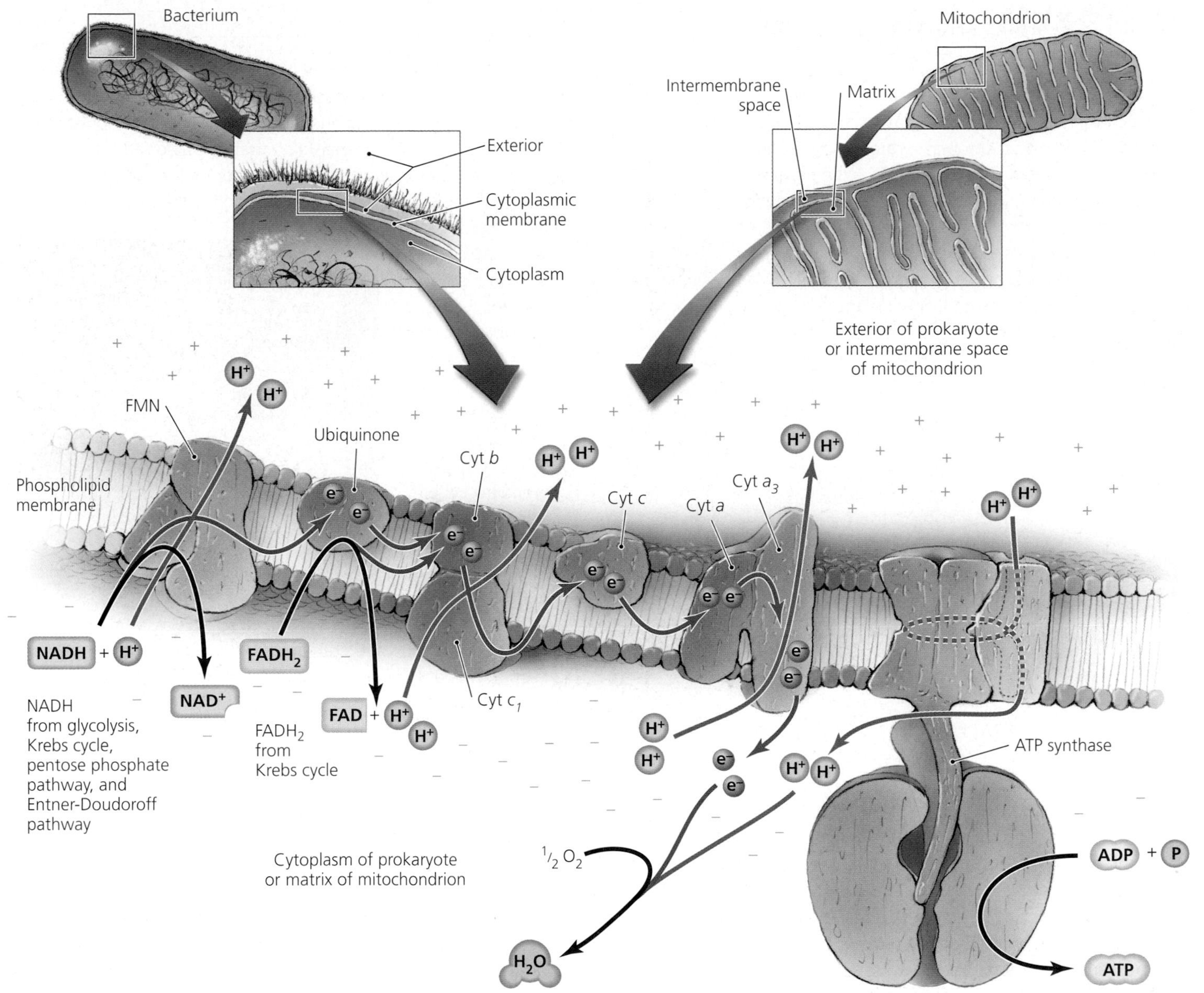

▲ ***Figure 5.21***

One possible arrangement of an electron transport chain. Electron transport chains are located in the cytoplasmic membranes of prokaryotes, and in the inner membranes of mitochondria in eukaryotes. The exact types and sequences of carrier molecules in electron transport chains vary among organisms. As electrons move down the chain (red arrows), their energy is used to pump protons (H^+) across the membrane. The protons then flow through ATP synthase (ATPase), which synthesizes ATP. Approximately one molecule of ATP is generated for every two protons that cross the membrane. *Why is it essential that all the electron carriers of an electron transport chain be membrane bound?*

Figure 5.21 The carrier molecules of electron transport chains are membrane bound for at least two reasons: (1) Carrier molecules need to be in fairly close proximity so they can pass electrons between each other. (2) The goal of the chain is the pumping of H^+ across a membrane to create a proton gradient; without a membrane, there could be no gradient.

Highlight 5.1 Glowing Bacteria

Bacteria in the genus *Photobacterium* possess an interesting electron transport chain that generates light instead of ATP. These organisms can switch the flow of electrons from a standard electron transport chain (composed of cytochromes, proton pumps, and O_2 as the final electron acceptor) to an alternate chain. Whereas the standard chain establishes a proton gradient that is used to synthesize ATP, the alternate chain uncouples electron transport and ATP; instead of transferring electrons to proton pumps, the alternate chain shunts the electrons to the coenzyme flavin mononucleotide (FMN). Then, in the presence of an enzyme called luciferase and a long-chain hydrocarbon, the alternate chain emits light as it transfers electrons to O_2 (see the box diagram). The exact mechanism of bioluminescence is not known, but both FMN and the hydrocarbon are oxidized as oxygen is reduced.

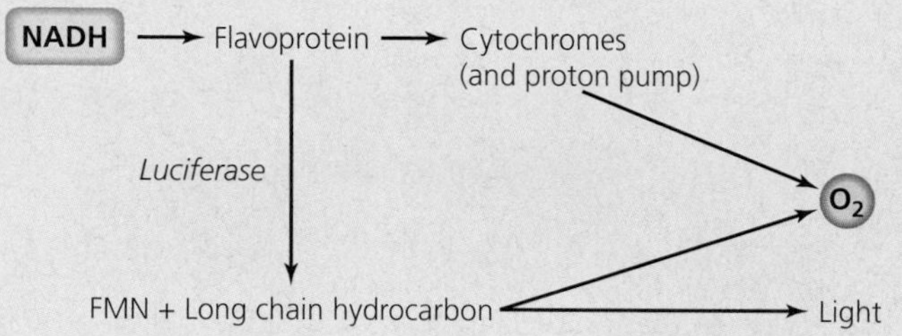

Interestingly, free-living *Photobacterium* are usually not bioluminescent. It is primarily when these bacteria colonize the tissues of marine animals such as squid and fish that they use their light-generating pathway. The animals gain from the association because the light produced serves as an attractant for mates and a warning against predators. One species, the "flashlight fish" (*Photoblepharon palpebratus*), has a special organ near its mouth that is specially adapted for the growth of luminescent bacteria. Enough light is generated from millions of bacteria that the fish can navigate over coral reefs at night and attract prey to their light. The light organ even has a membrane that descends like an eyelid to control the amount of light emitted.

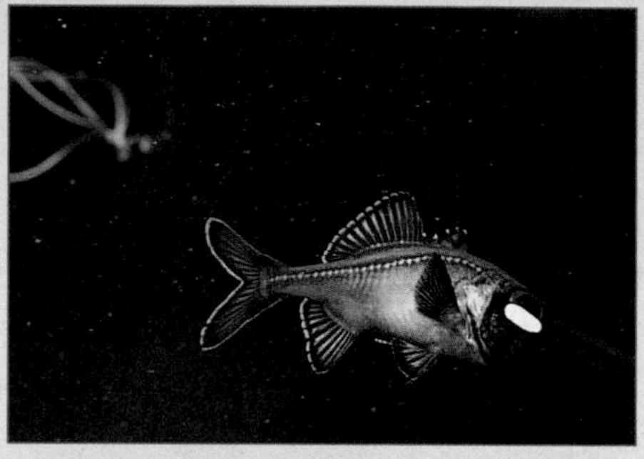

Photoblepharon palpebratus

It is not clear what the bacteria gain from this association. Presumably, the protection and nutrients the bacteria gain from the fish make up for the enormous metabolic cost the bacteria incur in the form of lost ATP synthesis.

They are:

- *Flavoproteins* are integral membrane proteins, many of which contain flavin, a coenzyme derived from riboflavin (vitamin B_2). One form of flavin is flavin mononucleotide (FMN), which is the initial carrier molecule of electron transport chains of mitochondria. The familiar FAD is a coenzyme for other flavoproteins. Like all carrier molecules in the electron transport chain, flavoproteins alternate between the reduced and oxidized states.
- *Ubiquinones* (yū-bik′wi-nōns) are lipid-soluble, nonprotein carriers that are so named because they are ubiquitous in all cells. Ubiquinones are derived from vitamin K. In mitochondria, the ubiquinone is called *coenzyme Q*.
- *Metal-containing proteins* are a mixed group of integral proteins with a wide-ranging number of iron, sulfur, and copper atoms that can alternate between the reduced and oxidized states. Iron-sulfur proteins occur in various places in electron transport chains of various organisms. Copper proteins are found only in electron transport chains involved in photosynthesis (discussed shortly).
- *Cytochromes* (sī′tō-krōms) are integral proteins associated with *heme,* which is the same iron-containing, nonprotein, pigmented molecule found in the hemoglobin of blood. Iron can alternate between a reduced (Fe^{2+}) state and an oxidized (Fe^{3+}) state. Cytochromes are identified by letters and numbers based on the order in which they were identified, so their sequence in electron transport chains does not always seem logical.

The carrier molecules in electron transport chains are diverse—bacteria typically have different carrier molecules arranged in different sequences than do archaea or the mitochondria of eukaryotes. Some prokaryotes, including *E. coli,* can even vary their carrier molecules under different environmental conditions. Even among bacteria, the makeup of carrier molecules can be variable. For example, the pathogens *Neisseria* (nī-se′rē-ă) and *Pseudomonas* contain two cytochromes, *a* and a_3—together called *cytochrome oxidase*—that oxidize cytochrome *c;* such bacteria are said to be *oxidase positive.* In contrast, other bacterial pathogens, such as *Escherichia, Salmonella,* and *Proteus* (prō′tē-ŭs), lack cytochrome oxidase and are thus considered to be *oxidase negative.*

Electrons carried by NADH enter the transport chain at a flavoprotein, and those carried by $FADH_2$ are introduced via a ubiquinone. This explains why more molecules of ATP are generated from NADH than from $FADH_2$. Researchers do not agree on which carrier molecules are the actual proton pumps, nor on the number of protons that are pumped. Figure 5.21 shows one possibility.

In some organisms, the final electron acceptors are oxygen atoms, which, with the addition of hydrogen ions, generate H_2O; these organisms conduct **aerobic**[7] **respiration** and are called *aerobes*. Other organisms, called *anaerobes,*[8] use other inorganic molecules (or rarely an organic molecule) instead of oxygen as the final electron acceptor and perform **anaerobic respiration.** The anaerobic bacterium *Desulfovibrio* (dē′sul-fō-vib-rē-ō), for example, reduces sulfate (SO_4^{2-}) to hydrogen sulfide gas (H_2S), whereas other anaerobes in the genera *Bacillus* (ba-sil′ŭs) and *Pseudomonas* utilize nitrate (NO_3^-) to produce nitrite ions (NO_2^-), nitrous oxide (N_2O), or nitrogen gas (N_2). Some prokaryotes—particularly archaea called methanogens—reduce carbonate (CO_3^{2-}) to methane gas (CH_4). **Highlight 5.1** describes an unusual bacterial electron transport system that produces light instead of ATP.

Laboratory technologists test for products of anaerobic respiration, such as nitrite, to aid in identification of some species of bacteria. As discussed more fully in Chapter 25, anaerobic respiration is also critical for the recycling of nitrogen and sulfur in nature.

CRITICAL THINKING

Cyanide is a potent poison because it irreversibly blocks cytochrome a_3. What effect would its action have on the rest of the electron transport chain? What would be the redox state (reduced or oxidized) of coenzyme Q in the presence of cyanide?

In summary, glycolysis, the pentose phosphate pathway, the Entner-Doudoroff pathway, and the Krebs cycle strip electrons, which carry energy, from glucose molecules and transfer them to molecules of NADH and $FADH_2$. In turn, NADH and $FADH_2$ pass the electrons to an electron transport chain. As the electrons move down the electron transport chain, proton pumps use the electrons' energy to actively transport protons (hydrogen ions) across the membrane.

Recall, however, that the significance of electron transport is not merely that it pumps protons across a membrane, but that it ultimately results in the synthesis of ATP. We turn now to the process by which cells synthesize ATP using the concentration gradient produced by the large concentration of protons on one side of a membrane.

Chemiosmosis

Chemiosmosis is a general term for the use of ion gradients to generate ATP; that is, ATP is synthesized utilizing energy released by the flow of ions down their electrochemical gradient across a membrane. The term should not be confused with osmosis of water. To understand chemiosmosis, we need to review several concepts from Chapter 3 concerning diffusion and phospholipid membranes.

Recall that chemicals diffuse from areas of high concentration to areas of low concentration, and toward an electrical charge opposite their own. We call the composite of differences in concentration and charge an *electrochemical gradient.* Chemicals diffuse down their electrochemical gradients. Recall as well that membranes of cells and organelles are impermeable to most chemicals unless a specific protein channel allows their passage across the membrane. A membrane maintains an electrochemical gradient by keeping one or more chemicals in a higher concentration on one side. The blockage of diffusion creates potential energy, like water behind a dam.

Chemiosmosis uses the potential energy of an electrochemical gradient to phosphorylate ADP into ATP. Even though chemiosmosis is a general principle with relevance to both *oxidative phosphorylation* and *photophosphorylation,* here we consider it as it relates to oxidative phosphorylation.

As we have seen, cells use the energy released in the redox reactions of electron transport chains to actively transport protons (H^+) across a membrane. Theoretically, an electron transport chain pumps three pairs of protons for each pair of electrons contributed by NADH, and pumps two pairs of protons for each electron pair delivered by $FADH_2$. This difference results from the fact that $FADH_2$ delivers electrons farther down the chain than does NADH; therefore, energy carried by $FADH_2$ is used to transport one-third fewer protons (see Figure 5.20). Because lipid bilayers are impermeable to protons, the transport of protons to one side of the membrane creates an electrochemical gradient known as a **proton gradient,** which has potential energy known as a *proton motive force.*

Hydrogen ions, propelled by the proton motive force, flow down their electrochemical gradient through protein channels, called **ATP synthases (ATPases),** that phosphorylate molecules of ADP to ATP (see Figure 5.21). Such phosphorylation is called **oxidative phosphorylation** because the proton gradient is created by the oxidation of components of an electron transport chain.

In the past, scientists attempted to calculate the exact number of ATP molecules synthesized per pair of electrons that travel down an electron transport chain. However, it is now apparent that phosphorylation and oxidation are not directly coupled. In other words, chemiosmosis does not require exact constant relationships among the number of molecules of NADH and $FADH_2$ reduced, the number of electrons that move down an electron transport chain, and the number of molecules of ATP that are synthesized. Additionally, cells use proton gradients for other cellular processes, including active transport and bacterial flagellar motion, so not every transported electron results in ATP production.

Nevertheless, about 34 molecules of ADP per molecule of glucose are oxidatively phosphorylated to ATP via chemiosmosis: three from each of the 10 molecules of NADH generated from glycolysis, the synthesis of acetyl-CoA, and the Krebs cycle, and two from each of the two

[7]From Greek *aer,* meaning air (that is, oxygen), and *bios,* meaning life.
[8]The Greek prefix *an* means not.

molecules of $FADH_2$ generated in the Krebs cycle. Given that glycolysis produces a net two molecules of ATP by substrate-level phosphorylation, and that the Krebs cycle produces two more, the complete aerobic oxidation of one molecule of glucose by a prokaryote can theoretically yield a net total of 38 molecules of ATP **(Table 5.3).** The theoretical net maximum for eukaryotic cells is generally given as 36 molecules of ATP because the energy from two ATP molecules is required to transport NADH generated in glycolysis into the mitochondria.

CRITICAL THINKING

Suppose you could insert a tiny pH probe into the space between mitochondrial membranes. Would the pH be above or below 7.0? Why?

Fermentation

Learning Objectives

- ✓ Describe fermentation, and contrast it with respiration.
- ✓ Identify three useful end-products of fermentation, and explain how fermentation reactions are used in the identification of bacteria.
- ✓ Discuss the use of biochemical tests for metabolic enzymes and products in the identification of bacteria.

Sometimes cells cannot completely oxidize glucose by cellular respiration. For instance, they may lack sufficient final electron acceptors, as is the case, for example, for an aerobic bacterium in the anaerobic environment of the colon. Electrons cannot flow down an electron transport chain unless oxidized carrier molecules are available to receive them. Our bucket brigade analogy can help clarify this point.

Suppose the last person in the brigade did not throw the water, but instead held onto two full buckets. What would happen? The entire brigade would soon consist of firefighters holding full buckets of water. The analogous situation occurs in an electron transport chain: All the carrier molecules are forced to remain in their reduced states when there is not a final electron acceptor. Without the movement of electrons down the chain, protons cannot be transported, the proton motive force is lost, and oxidative phosphorylation of ADP to ATP ceases. Without sufficient ATP, a cell is unable to anabolize, grow, or divide.

ATP could be synthesized in glycolysis and the Krebs cycle by substrate-level phosphorylation. After all, together these pathways produce four molecules of ATP per molecule of glucose. However, careful consideration reveals that glycolysis and the Krebs cycle require a continual supply of oxidized NAD^+ molecules (see Figures 5.14 and 5.19). In respiration, electron transport produces the required NAD^+, but without a final electron acceptor, this source of NAD^+ ceases to be available. A cell in such a predicament must use an alternate source of NAD^+ provided by alternative metabolic pathways, called *fermentation pathways.*

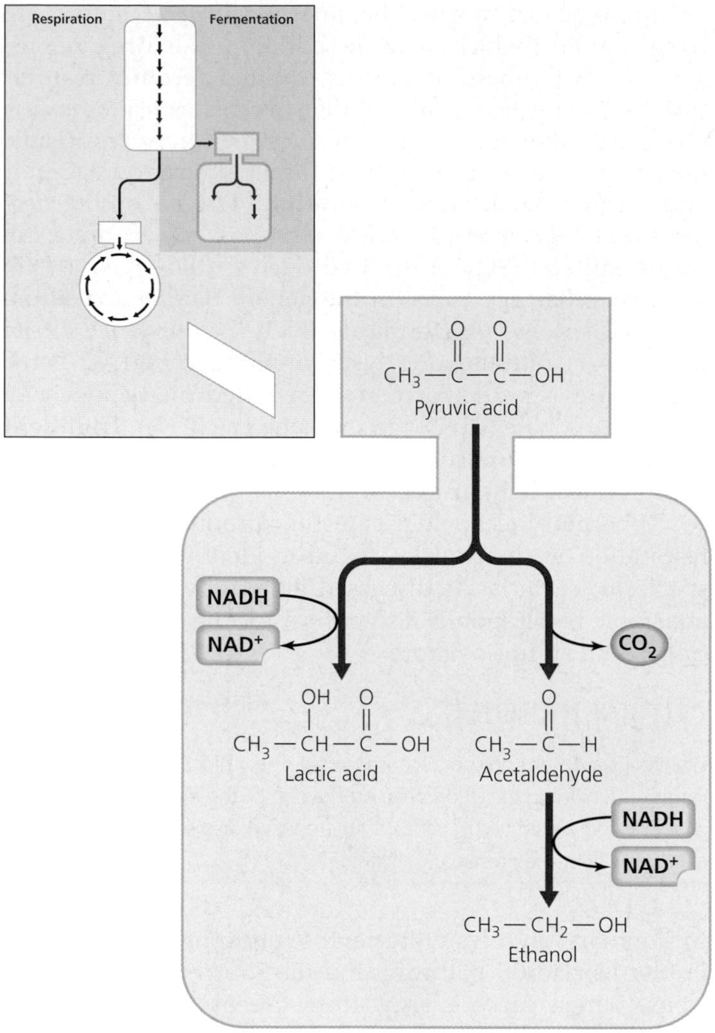

▲ *Figure 5.22*

Fermentation. In the simplest fermentation reaction, NADH reduces pyruvic acid (from glycolysis) to form lactic acid. Another simple fermentation pathway involves a decarboxylation reaction and reduction to form ethanol.

In the vernacular, fermentation refers to the production of alcohol from sugar, but in microbiology, fermentation has an expanded meaning. **Fermentation** is the partial oxidation of sugar (or other metabolites) to release energy using an organic molecule as an electron acceptor rather than an electron transport chain. In other words, fermentation pathways are metabolic reactions that oxidize NADH to NAD^+ while reducing organic molecules, which are the final electron acceptors. In contrast, as previously noted, aerobic respiration reduces oxygen, and anaerobic respiration reduces other inorganic chemicals such as sulfate and nitrate or (rarely) an organic molecule. **Figure 5.22** illustrates two common fermentation pathways that reduce pyruvic acid to lactic acid and ethanol, oxidizing NADH in the process.

Table 5.3 Summary of Prokaryotic Aerobic Respiration of One Molecule of Glucose

Pathway	ATP Produced	ATP Used	NADH Produced	$FADH_2$ Produced	Comments
Glycolysis	4	2	2	0	Two molecules of water produced
Synthesis of acetyl-CoA and Krebs cycle	2	0	8	2	Two molecules of water used, six molecules of CO_2 produced
Aerobic electron transport chain	34	0	0	0	Ten molecules of NADH, two molecules of $FADH_2$, and six molecules of O_2 used, six molecules of water produced
Total	40	2			
Net total	**38**				

The essential function of fermentation is the regeneration of NAD^+ for glycolysis, so that ADP molecules can be phosphorylated to ATP. Even though fermentation pathways are not as energetically efficient as respiration because much of the potential energy stored in glucose remains in the bonds of fermentation products, the major benefit of fermentation is that it allows ATP production to continue in the absence of cellular respiration. Microorganisms with fermentation pathways can colonize anaerobic environments, which are unavailable to strictly aerobic organisms. **Table 5.4** on page 150 compares fermentation to aerobic and anaerobic respiration with respect to four crucial aspects of these processes.

Microorganisms produce a variety of fermentation products depending on the enzymes and substrates available to each. Though fermentation products are wastes to the cells that make them, many are useful to humans, including ethanol (drinking alcohol), acetic acid (vinegar), and lactic acid (used in the production of cheese, sauerkraut, and pickles) (**Figure 5.23** on page 150).

Other fermentation products are harmful to human health and industry. For example, fermentation products of the bacterium *Clostridium perfringens* (klos-tri'dē-um per-frin'jens) are involved in the necrosis (death) of muscle tissue associated with gangrene. Also, as you may recall from Chapter 1, Pasteur discovered that bacterial contaminants in grape juice fermented the sugar into unwanted products such as acetic acid and lactic acid, which spoiled the wine.

Laboratory personnel routinely use the detection of fermentation products in the identification of microbes. For example, *Proteus* ferments glucose but not lactose, whereas *Escherichia* and *Enterobacter* ferment both. Further, glucose fermentation by *Escherichia* produces mixed acids (acetic, lactic, succinic, and formic), while *Enterobacter* produces 2,3-butanediol. Common fermentation tests contain a carbohydrate and a pH indicator, which is a molecule that changes color as the pH changes. An organism that utilizes the carbohydrate causes a change in pH, causing the pH indicator to change color.

In this section on carbohydrate catabolism, we have spent some time examining glycolysis, alternatives to glycolysis, the Krebs cycle, and electron transport because these pathways are central to metabolism. They generate all of the precursor metabolites and most of the ATP needed for anabolism. We have seen that some ATP is generated in respiration by substrate-level phosphorylation (in both glycolysis and the Krebs cycle), and that most ATP is generated by oxidative phosphorylation via chemiosmosis utilizing the reducing power of NADH and $FADH_2$. We also saw that some microorganisms use fermentation to provide an alternate source of NAD^+ when conditions are not suitable for respiration.

Thus far we have concentrated on the catabolism of glucose as a representative carbohydrate, but microorganisms can also use other molecules as energy sources. In the next section we will examine catabolic pathways that utilize lipids and proteins.

Other Catabolic Pathways

Lipid and protein molecules contain abundant energy in their chemical bonds and can also be converted into precursor metabolites. These molecules are first catabolized to produce their constituent monomers, which serve as substrates in glycolysis and the Krebs cycle.

Lipid Catabolism

Learning Objective

✓ Explain how lipids are catabolized for energy and metabolite production.

The most common lipids involved in ATP and metabolite production are fats, which, as we saw in Chapter 2, consist of glycerol and fatty acids. In the first step of fat catabolism, enzymes called *lipases* hydrolyze the bonds attaching the glycerol to the fatty acid chains (**Figure 5.24a** on page 151).

Table 5.4 Comparison of Aerobic Respiration, Anaerobic Respiration, and Fermentation

	Aerobic Respiration	Anaerobic Respiration	Fermentation
Oxygen required	Yes	No	No
Type of phosphorylation	Substrate-level and oxidative	Substrate-level and oxidative	Substrate-level
Final electron (hydrogen) acceptor	Oxygen	NO_3^-, SO_4^{2-}, or CO_3^{2-}	Organic molecules
Potential molecules of ATP produced	36–38	2–36	2

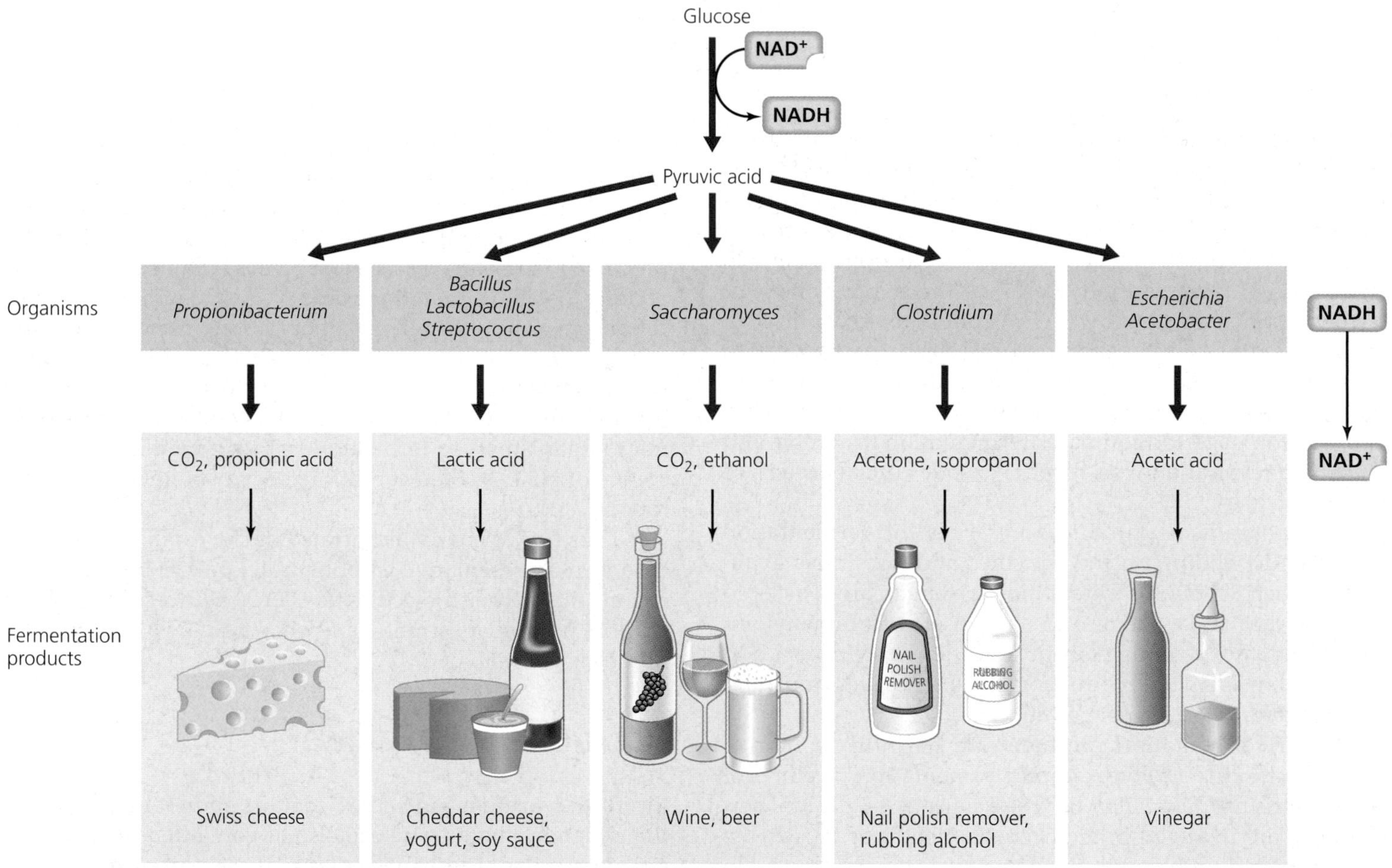

▲ ***Figure 5.23***

Representative fermentation products and the organisms that produce them. All of the organisms are bacteria except *Saccharomyces,* which is a yeast. *Why does Swiss cheese have holes, but cheddar does not?*

Figure 5.23 *Swiss cheese is a product of Propionibacterium, which ferments pyruvic acid to produce CO_2; bubbles of this gas cause the holes in the cheese. Cheddar is a product of Lactobacillus, which does not produce gas.*

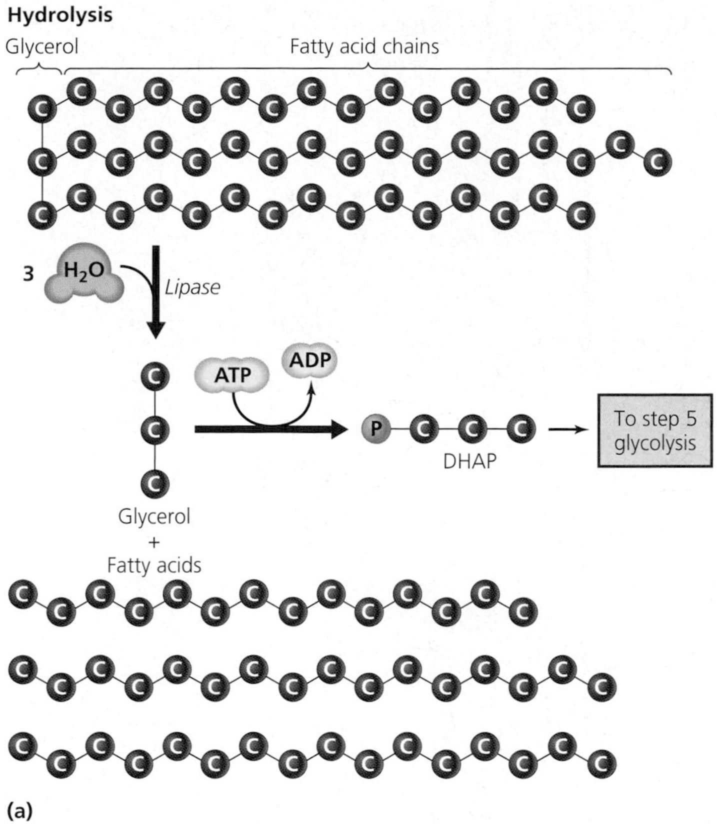

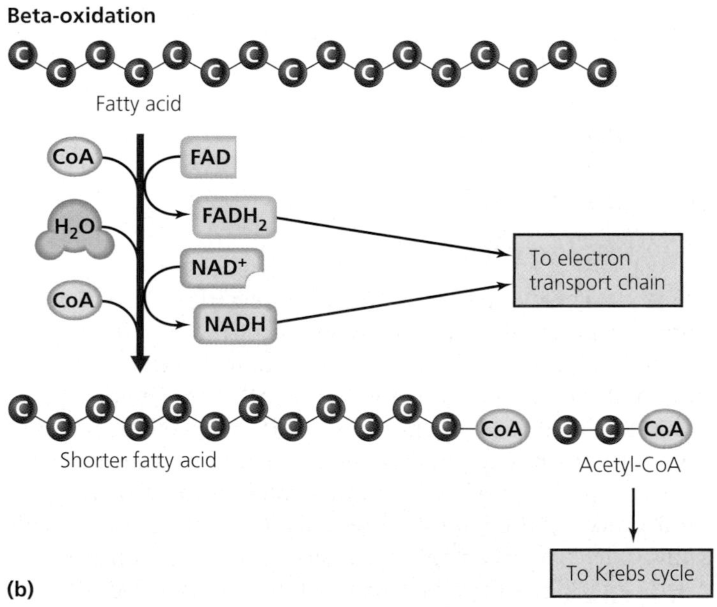

▲ ***Figure 5.24***

Catabolism of a fat molecule. **(a)** Lipase breaks fats into glycerol and three fatty acids by hydrolysis. Glycerol is converted to DHAP, which can be catabolized via glycolysis and the Krebs cycle. **(b)** Fatty acids are catabolized via beta-oxidation reactions that produce molecules of acetyl-CoA and reduced coenzymes (NADH and $FADH_2$).

Subsequent reactions further catabolize the glycerol and fatty acid molecules. Glycerol is converted to DHAP, which as one of the substrates of glycolysis is oxidized to pyruvic acid (see Figure 5.14). The fatty acids are degraded in a catabolic process known as **beta-oxidation (Figure 5.24b).**[9] In this process, enzymes repeatedly split off pairs of the hydrogenated carbon atoms that make up a fatty acid and join each pair to coenzyme A to form acetyl-CoA, until the entire fatty acid has been converted to molecules of acetyl-CoA. NADH and $FADH_2$ are generated during beta-oxidation, and more of these molecules are generated when the acetyl-CoA is utilized in the Krebs cycle to generate ATP (see Figure 5.19). As is the case for the Krebs cycle, the enzymes involved in beta-oxidation are located in the cytoplasm of prokaryotes, and in the mitochondria of eukaryotes.

CRITICAL THINKING

How and where do cells use the molecules of NADH and $FADH_2$ produced during beta-oxidation?

Protein Catabolism

Learning Objective

✓ Explain how proteins are catabolized for energy and metabolite production.

Some microorganisms, particularly food-spoilage bacteria, pathogenic bacteria, and fungi, normally catabolize proteins as an important source of energy and metabolites. Most other cells catabolize proteins and their constituent amino acids only when carbon sources such as glucose and fat are not available.

Generally, proteins are too large to cross cell membranes, so prokaryotes often conduct the first step in the process of protein catabolism outside the cell by secreting **proteases** (prō′ tē-ās-ez)—enzymes that split proteins into their constituent amino acids (**Figure 5.25** on page 152). Once released by the action of proteases, amino acids are transported into the cell, where special enzymes split off amino groups in a reaction called **deamination.** The resulting altered molecules enter the Krebs cycle, and the amino groups are either recycled to synthesize other amino acids or excreted as nitrogenous wastes such as ammonia (NH_3), ammonium ion (NH_4^+), or trimethylamine oxide (which has the chemical formula $(CH_3)_3NO$ and is abbreviated TMAO). As **Highlight 5.2** on page 152 describes, TMAO plays an interesting role in the production of the odor we describe as "fishy."

Thus far, we have examined the catabolism of carbohydrates, lipids, and proteins. Now we turn our attention to the *synthesis* of these molecules, beginning with the anabolic reactions of photosynthesis.

[9]"Beta" is part of the name of this process because enzymes break the bond at the second carbon atom from the end of a fatty acid, and beta is the second letter in the Greek alphabet.

Figure 5.25 ➤

Protein catabolism in microbes. Hydrolysis of proteins by secreted proteases releases amino acids, which are deaminated after uptake to produce molecules used as substrates in the Krebs cycle. R indicates the side group, which varies among amino acids.

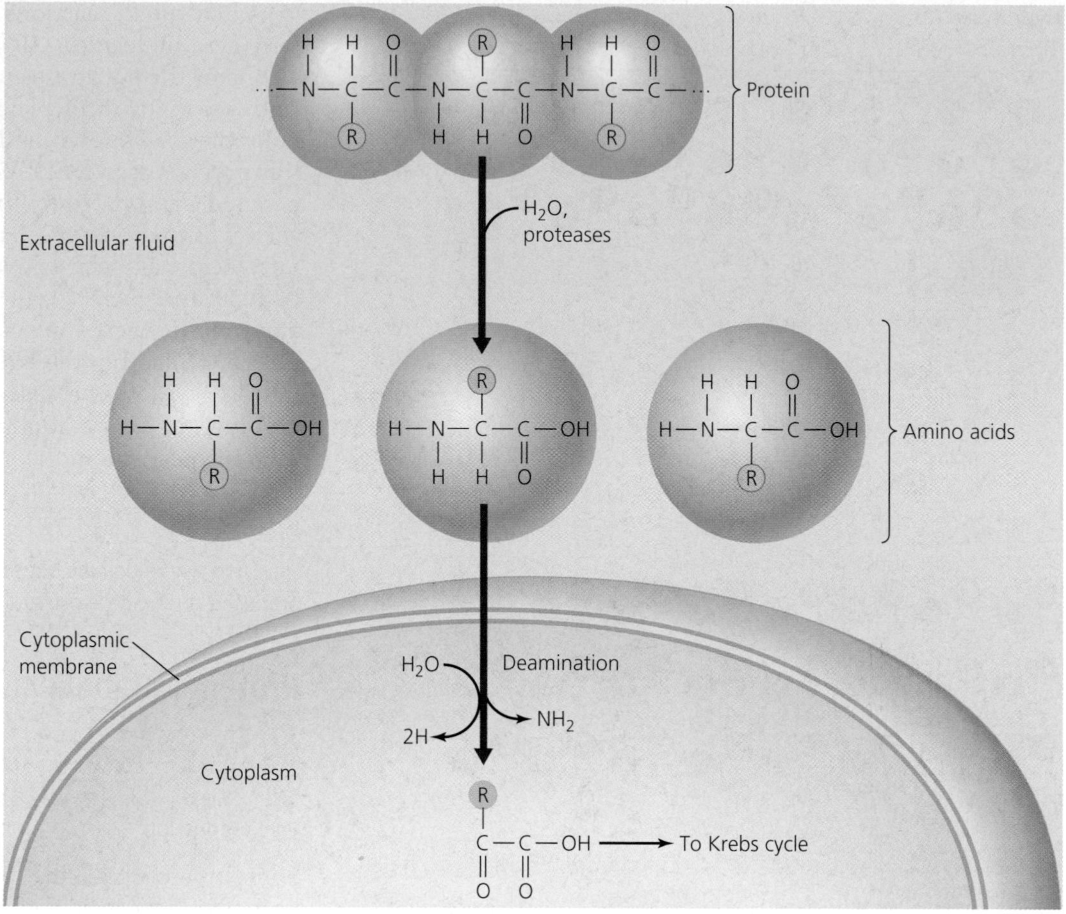

Highlight 5.2 What's That Fishy Smell?

Ever wonder why fish at the supermarket sometimes have that "fishy" smell? The reason relates to bacterial metabolism.

Trimethylamine oxide (TMAO) is a nitrogenous waste product of fish metabolism used to dispose of excess nitrogen (as amine groups) produced in the catabolism of amino acids. TMAO is an odorless substance that does not affect the smell, appearance, or taste of fresh fish. However, some bacteria use TMAO as a final electron acceptor in anaerobic respiration, reducing TMAO to trimethylamine (TMA), a compound with a very definite—and, for most people, unpleasant—"fishy" odor. A human nose is able to detect even a few molecules of TMA. Because bacterial degradation of fish and the production of TMA begin as soon as a fish is dead, a fish that smells fresh probably is fresh, or was frozen while still fresh.

Photosynthesis

Learning Objective

✓ Define photosynthesis.

Many organisms use only organic molecules as a source of energy and metabolites, but where do they acquire organic molecules? Ultimately every food chain begins with anabolic pathways in organisms that synthesize their own organic molecules from inorganic carbon dioxide. Most of these organisms capture light energy from the sun and use it to drive the synthesis of carbohydrates from CO_2 and H_2O by a process called **photosynthesis.** Cyanobacteria, purple sulfur bacteria, green sulfur bacteria, green nonsulfur bacteria, purple nonsulfur bacteria, a few protozoa, algae, and green plants are photosynthetic.

CRITICAL THINKING

Photosynthetic organisms are rarely pathogenic. Why?

▼ ***Figure 5.26***

Photosynthetic structures in a prokaryote. **(a)** Chlorophyll molecules are grouped together in thylakoid membranes to form photosystems. Chlorophyll contains a light-absorbing active site that is connected to a long hydrocarbon tail. The chlorophyll shown here is bacteriochlorophyll *a*; other chlorophyll types differ in the side chains protruding from the central ring. **(b)** Prokaryotic thylakoids are infoldings of the cytoplasmic membrane.

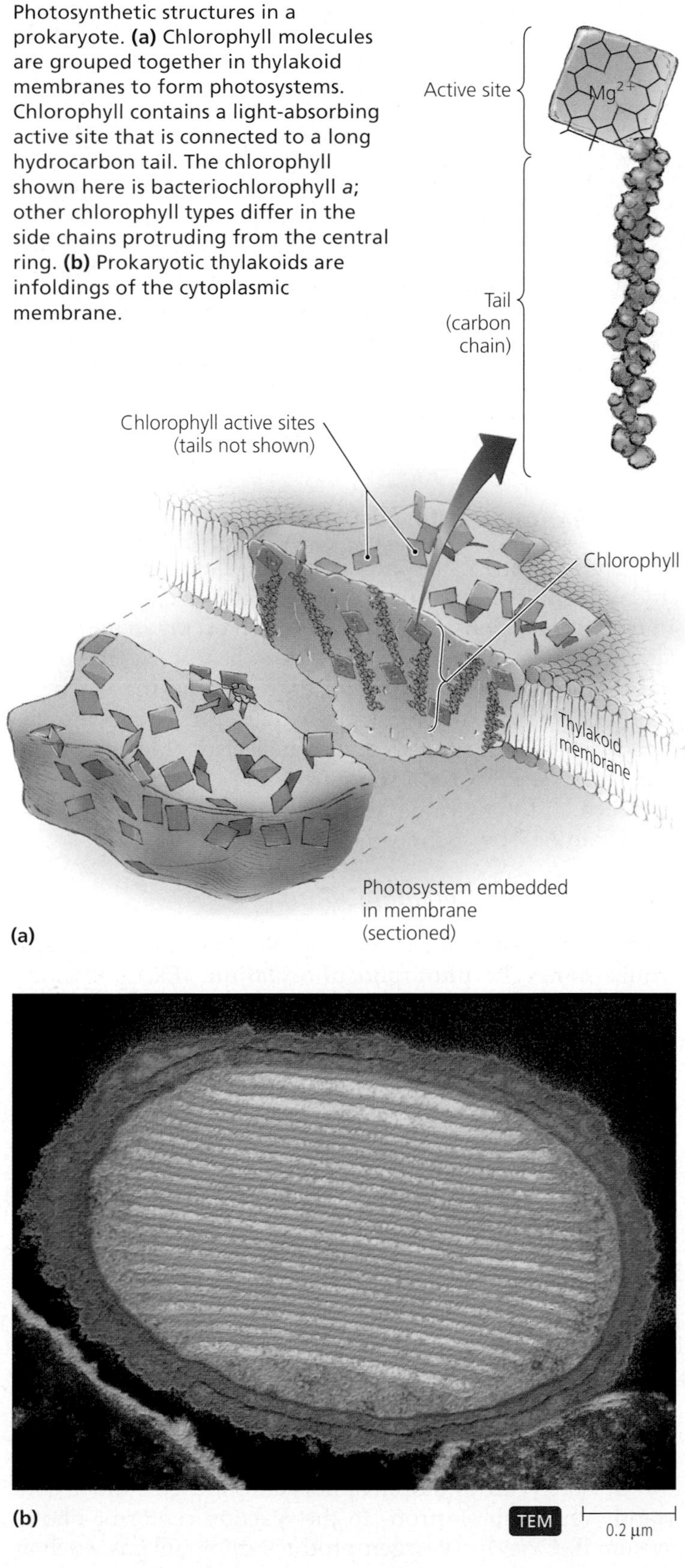

Chemicals and Structures

Learning Objective

✓ Compare and contrast the basic chemicals and structures involved in photosynthesis in prokaryotes and eukaryotes.

Photosynthetic organisms capture light energy with pigment molecules, the most important of which are **chlorophylls.** Chlorophyll molecules are composed of a hydrocarbon tail attached to a light-absorbing *active site* centered around a magnesium ion (Mg^{2+}) **(Figure 5.26a).** Active sites are structurally similar to the cytochrome molecules found in electron transport chains, except chlorophylls use Mg^{2+} rather then Fe^{2+}. Chlorophylls, typically designated with letters—for example, chlorophyll *a,* chlorophyll *b,* and bacteriochlorophyll *a*—vary slightly in the lengths and structures of their hydrocarbon tails and in the atoms that extend from their active sites. Green plants, algae, photosynthetic protozoa, and cyanobacteria principally use chlorophyll *a,* whereas green and purple bacteria use bacteriochlorophylls.

The slight structural differences among chlorophylls cause them to absorb light of different wavelengths. For example, chlorophyll *a* from algae best absorbs light with wavelengths of about 425 nm and 660 nm (violet and red), whereas bacteriochlorophyll *a* from purple bacteria best absorbs light with wavelengths of about 350 nm and 880 nm (ultraviolet and infrared). Because they best use light with differing wavelengths, algae and purple bacteria successfully occupy different ecological niches.

Cells arrange numerous molecules of chlorophyll and other pigments within a protein matrix to form light-harvesting matrices called **photosystems** that are embedded in cellular membranes called **thylakoids.** Thylakoids of photosynthetic prokaryotes are invaginations of the cytoplasmic membranes **(Figure 5.26b).** The thylakoids of eukaryotes appear to be formed from infoldings of the inner membranes of chloroplasts, though thylakoid membranes and the inner membranes are not connected in mature chloroplasts (see Figure 3.37). Thylakoids of chloroplasts are arranged in stacks called *grana.* An outer chloroplast membrane surrounds the grana, and the space between the outer membrane and the thylakoid membrane is known as the *stroma.*

There are two types of photosystems, named photosystem I (PS I) and photosystem II (PS II), in the order of their discovery. Photosystems absorb light energy and use redox reactions to store the energy in molecules of ATP and NADPH. Because they depend upon light energy, these reactions of photosynthesis are classified as **light-dependent reactions.** Photosynthesis also involves **light-independent reactions** that actually synthesize glucose from carbon dioxide and water. Historically, these reactions were called *light* and *dark reactions,* but this older terminology implies that light-independent reactions occur only in the dark, which is not the case. We first consider the light-dependent reactions of the two photosystems.

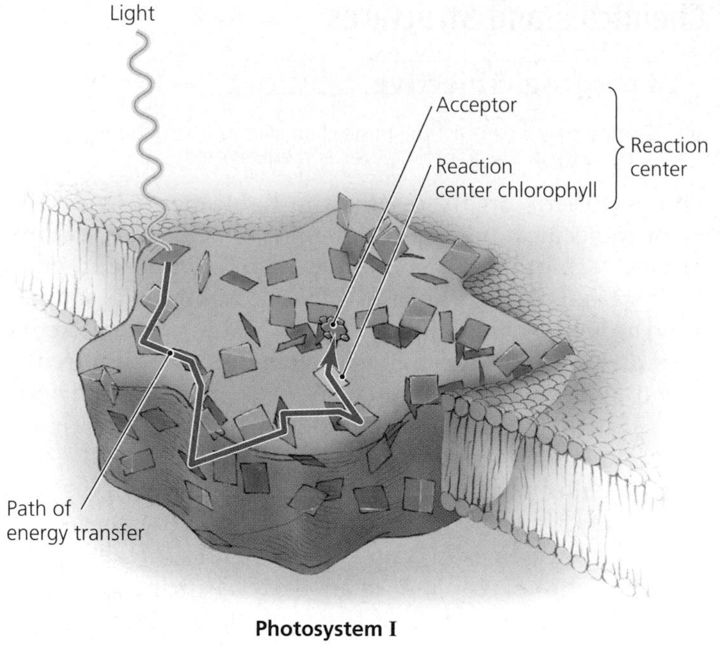

Figure 5.27

Reaction center of a photosystem. Light energy absorbed by pigments anywhere in the photosystem is transferred to the reaction center chloroplast, where it is used to excite electrons for delivery to the reaction center's electron acceptor.

Figure 5.28

The light-dependent reactions of photosynthesis: cyclic and noncyclic photophosphorylation. The red arrows indicate the flow of electrons excited by light energy. **(a)** In cyclic photophosphorylation, electrons excited by light striking photosystem I travel from its reaction center, down an electron transport chain, and then return to the reaction center. The proton gradient produced drives the phosphorylation of ADP by ATP synthase (chemiosmosis). **(b)** In noncyclic photophosphorylation, light striking photosystem II excites electrons that are passed down an electron transport chain to photosystem I, simultaneously establishing a proton gradient that is used to phosphorylate ADP to ATP. Light energy collected by photosystem I further excites the electrons, which are used to reduce $NADP^+$ to NADPH via an electron transport chain. In oxygenic organisms, new electrons are provided by the lysis of H_2O (bottom left).

Light-Dependent Reactions

Learning Objectives

- ✓ Describe the components and function of the two photosystems, PS I and PS II.
- ✓ Contrast cyclic and noncyclic photophosphorylation.

The pigments of photosystem I absorb light energy and transfer it to a neighboring molecule within the photosystem until the energy eventually arrives at a special chlorophyll molecule called the **reaction center chlorophyll (Figure 5.27).** Light energy from hundreds of such transfers excites electrons in the reaction center chlorophyll, which passes its excited electrons to the reaction center's electron acceptor, which is the initial carrier of an electron transport chain. As electrons move down the chain, their energy is used to pump protons across the membrane, creating a proton motive force. In prokaryotes, protons are pumped out of the cell; in eukaryotes they are pumped from the stroma into the interior of the thylakoids.

The proton motive force is used in chemiosmosis, in which, you should recall, protons flow down their electrochemical gradient through ATPases, which generate ATP. In photosynthesis, this process is called *photophosphorylation,* which can be either cyclic or noncyclic.

Cyclic Photophosphorylation

Electrons moving from one carrier molecule to another in a thylakoid must eventually pass to a final electron acceptor. In **cyclic photophosphorylation,** which occurs in all photosynthetic organisms, the final electron acceptor is the original reaction center chlorophyll that donated the electrons **(Figure 5.28a).** In other words, when light energy excites electrons in PS I, they pass down an electron transport chain and return to PS I. The energy from the electrons is used to establish a proton gradient that drives the phosphorylation of ADP to ATP by chemiosmosis.

Noncyclic Photophosphorylation

Some photosynthetic bacteria and all plants and algae also utilize **noncyclic photophosphorylation.** (Exceptions are green and purple sulfur bacteria.) Noncyclic photophosphorylation, which requires both PS I and PS II, not only generates molecules of ATP, but also reduces molecules of coenzyme $NADP^+$ to NADPH **(Figure 5.28b).**

When light energy excites electrons of PS II, they are passed to PS I through an electron transport chain. (Note that photosystem II occurs first in the pathway, and photosystem I second, a result of the fact that the photosystems were named for the order in which they were discovered, not the order in which they operate.) PS I further energizes the electrons with additional light energy and transfers them through an electron transport chain to $NADP^+$, which is thereby reduced to NADPH. Hydrogen ions in NADPH come from the stroma or cytosol. NADPH subsequently participates in the synthesis of glucose in the light-independent reactions, which we will examine in the next section.

In noncyclic photophosphorylation, a cell must constantly replenish electrons to the reaction center of photosystem II. *Oxygenic* (oxygen-producing) organisms, such as algae, green plants, and cyanobacteria, derive electrons from the dissociation of H_2O (see Figure 5.28b). In these organisms, two molecules of water give up their electrons, producing molecular oxygen (O_2) as a waste product during

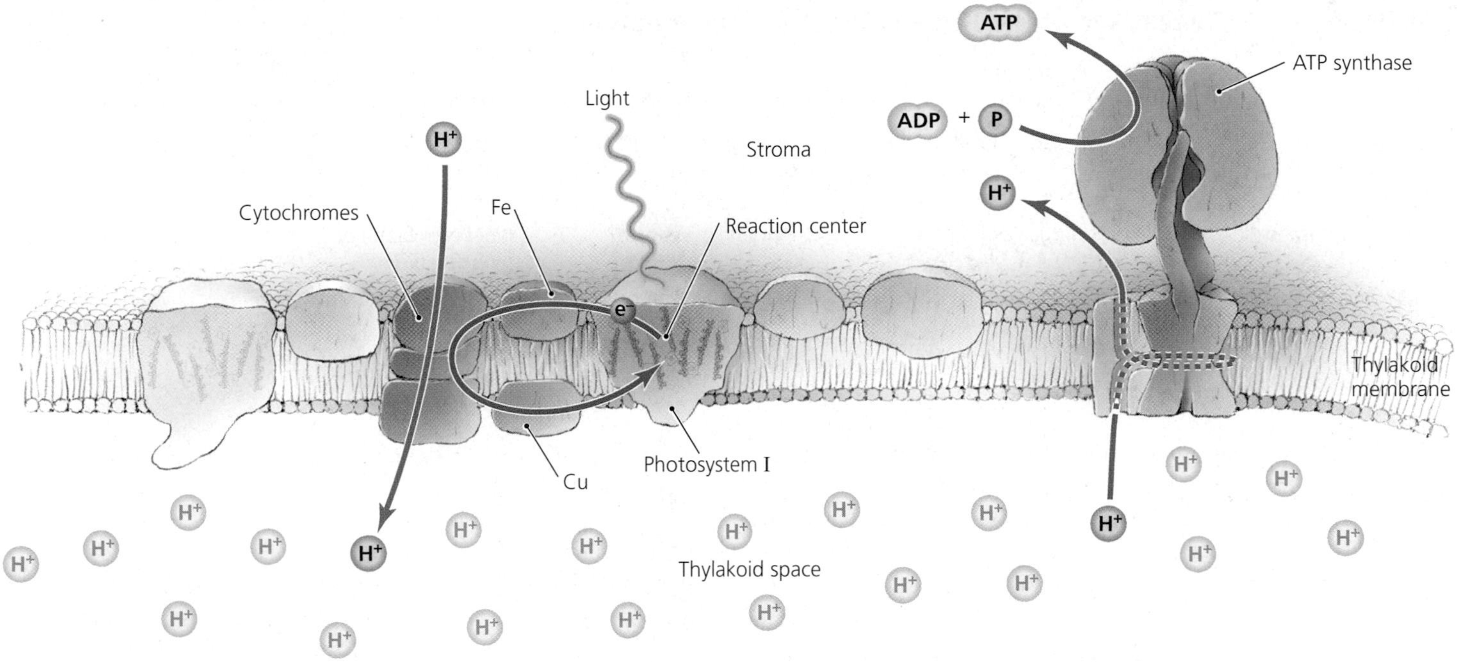

(a) Cyclic photophosphorylation

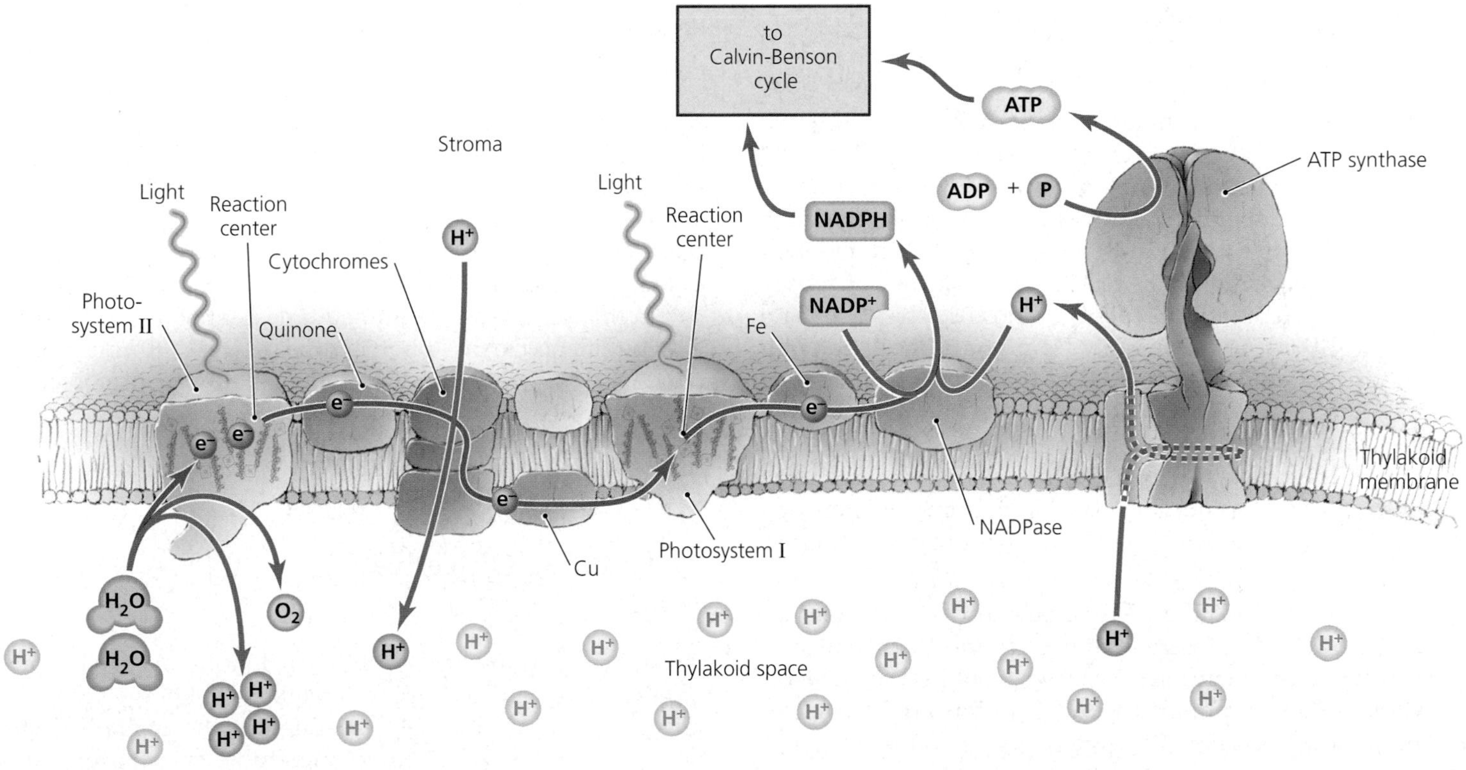

(b) Noncyclic photophosphorylation

photosynthesis. *Anoxygenic* photosynthetic bacteria get electrons from inorganic compounds such as H_2S, which results in a nonoxygen waste such as sulfur.

Table 5.5 on page 156 compares photophosphorylation to substrate-level and oxidative phosphorylation.

CRITICAL THINKING

We have seen that of the two ways that ATP is generated via chemiosmosis—photophosphorylation and oxidative phosphorylation—the former can be cyclical, but the latter is never cyclical. Why can't oxidative phosphorylation be cyclical; that is, why aren't electrons passed back to the molecules that donated them?

Table 5.5 A Comparison of the Three Types of Phosphorylation

	Source of Phosphate	Source of Energy	Location in Eukaryotic Cell	Location in Prokaryotic Cell
Substrate-level phosphorylation	Organic molecule	High-energy bond between donor and phosphate	Cytosol and mitochondrial matrix	Cytosol
Oxidative phosphorylation	Inorganic phosphate (PO_4^{2-})	Proton motive force	Inner membrane of mitochondrion	Cell membrane
Photophosphorylation	Inorganic phosphate (PO_4^{2-})	Proton motive force	Thylakoid of chloroplast	Thylakoid of cell membrane

To this point, we have examined the use of photosynthetic pigments and thylakoid structure to harvest light energy to produce both ATP and reducing power in the form of NADPH. Next we examine the light-independent reactions of photosynthesis.

Light-Independent Reactions

Learning Objectives

- ✓ Contrast the light-dependent and light-independent reactions of photosynthesis.
- ✓ Describe the reactants and products of the Calvin-Benson cycle.

Light-independent reactions of photosynthesis do not require light directly; instead, they use ATP and NADPH generated by the light-dependent reactions. The key reaction of the light-independent pathway of photosynthesis is **carbon fixation** by the **Calvin-Benson cycle,**[10] which involves the attachment of molecules of CO_2 to molecules of a five-carbon organic compound called ribulose 1,5-bisphosphate (RuBP). RuBP is derived initially from phosphorylation of a precursor metabolite produced by the pentose phosphate pathway (see Figure 5.16).

The Calvin-Benson cycle is reminiscent of the Krebs cycle in that the substrates of the cycle are regenerated. It is helpful to notice the number of carbon atoms during each part of the three steps of the Calvin-Benson cycle **(Figure 5.29):**

1. Fixation of CO_2. Enzymes attach three molecules of carbon dioxide (3 carbon atoms) to three molecules of RuBP (15 carbon atoms), which are split to form six molecules of 3-phosphoglyceric acid (18 carbon atoms).
2. Reduction. NADPH reduces the six molecules of 3-phosphoglyceric acid to form six molecules of glyceraldehyde 3-phosphate (G3P) (18 C). These reactions require six molecules each of ATP and NADPH generated by the light-dependent reactions.
3. Regeneration of RuBP. The cell regenerates three molecules of RuBP (15 C) from five molecules of G3P (15 C); it also uses the remaining molecule of glyceraldehyde 3-phosphate to synthesize glucose by reversing the reactions of glycolysis.

In summary, ATP and NADPH from the light-dependent reactions drive the synthesis of glucose from CO_2 in the light-independent reactions of the Calvin-Benson cycle. For every three molecules of CO_2 that enter the Calvin-Benson cycle, a molecule of glyceraldehyde 3-phosphate (G3P) leaves. Glycolysis is subsequently reversed to anabolically synthesize glucose 6-phosphate from two molecules of G3P. A more detailed account of the Calvin-Benson cycle is given in Appendix A on page A-7.

The processes of oxygenic photosynthesis and aerobic respiration complement one another to complete both a carbon cycle and an oxygen cycle. During the synthesis of glucose in oxygenic photosynthesis, water and carbon dioxide are used, and oxygen is released as a waste product; in aerobic respiration, oxygen serves as the final electron acceptor in the oxidation of glucose to carbon dioxide and water.

In this section we examined an essential anabolic pathway for life on Earth—photosynthesis, which produces glucose. Next we will consider anabolic pathways involved in the synthesis of other organic molecules.

Other Anabolic Pathways

Learning Objective

- ✓ Define amphibolic reaction.

Anabolic reactions are synthesis reactions. As such, they require energy and a source of metabolites. Energy for anabolism is provided by ATP generated in the catabolic reactions of aerobic respiration, anaerobic respiration, and fermentation, and by the initial redox reactions of photosynthesis. Glycolysis, the Krebs cycle, and the pentose phosphate pathway provide 12 basic precursor metabolites from which all macromolecules and cellular structures can be made. Some microorganisms, such as *E. coli*, can synthesize all 12 precursors (**Table 5.6** on page 160); other organisms, such as humans, must acquire some precursors in their diets.

Many anabolic pathways are the reversal of the catabolic pathways we have discussed; therefore, much of the

[10]Named for the men who elucidated its pathways.

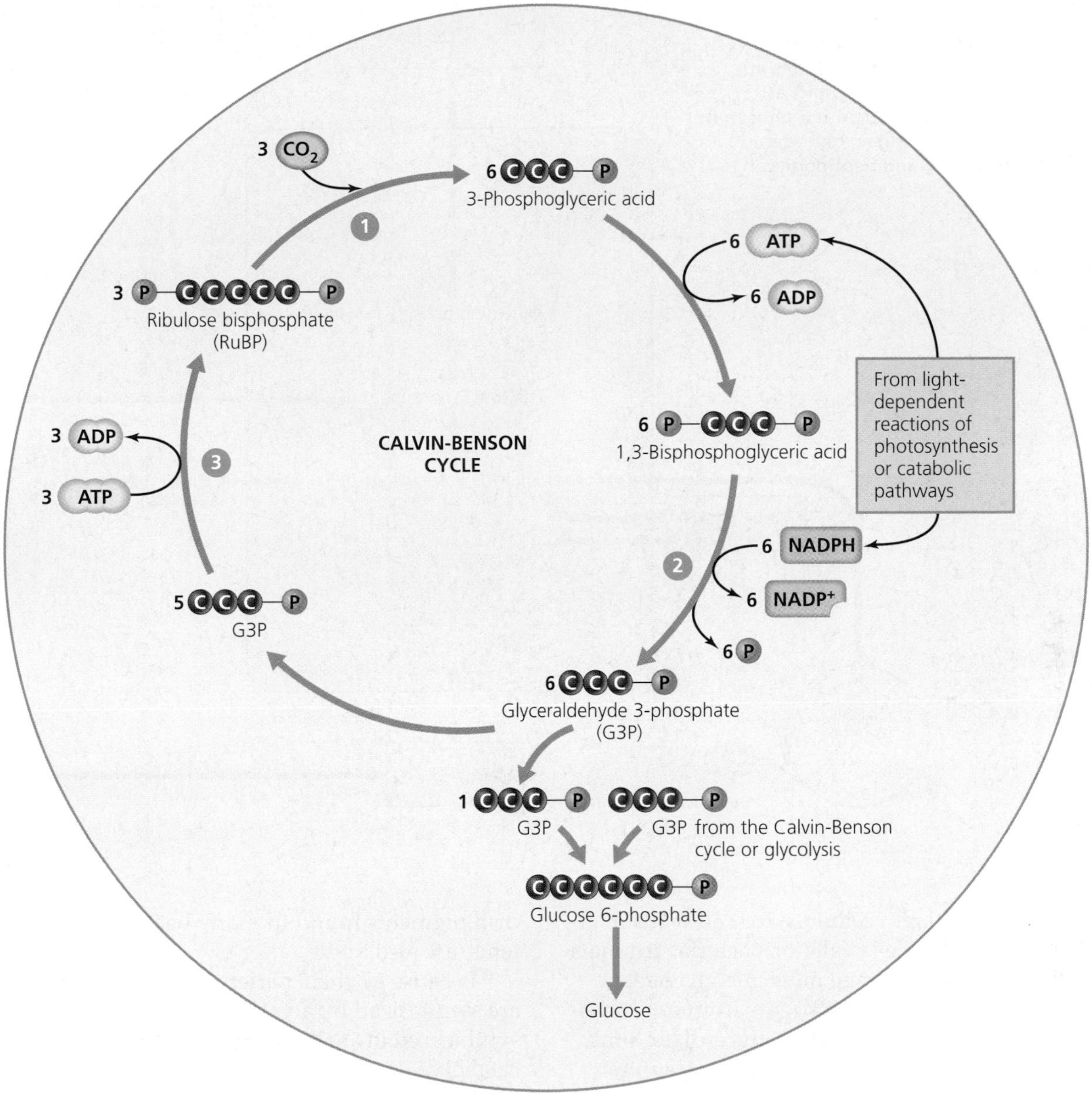

▲ ***Figure 5.29***

Simplified diagram of the Calvin-Benson cycle ①. Three molecules of RuBP combine with three molecules of CO_2 ②. The resulting molecules are reduced to form six molecules of G3P ③. Five molecules of G3P are converted to three molecules of RuBP, which completes the cycle. Two turns of the cycle also yield two molecules of G3P, which are polymerized to synthesize glucose 6-phosphate.

material that follows has *in a sense* been discussed in the sections on catabolism. Reactions that can proceed in either direction—toward catabolism or toward anabolism—are said to be **amphibolic.** The following sections discuss the synthesis of carbohydrates, lipids, amino acids, and nucleotides. The polymerizations of amino acids into proteins, and of nucleotides into RNA and DNA, are closely linked to genetics, so they are covered in Chapter 7.

Carbohydrate Biosynthesis

Learning Objective

✓ Describe the biosynthesis of carbohydrates.

As we have seen, anabolism begins in photosynthetic organisms with carbon fixation by the enzymes of the Calvin-Benson cycle to form molecules of G3P. Enzymes use G3P as the starting point for synthesizing sugars, complex

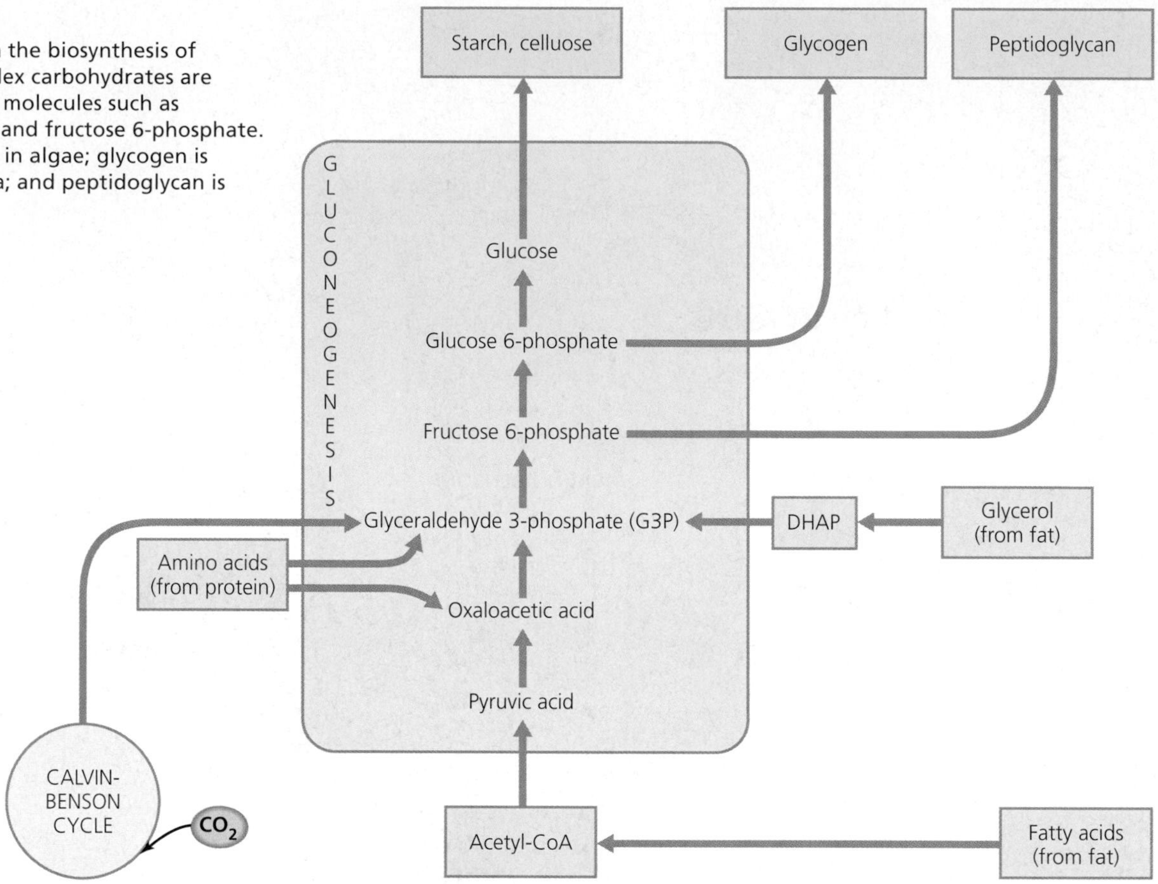

Figure 5.30 The role of gluconeogenesis in the biosynthesis of complex carbohydrates. Complex carbohydrates are synthesized from simple sugar molecules such as glucose, glucose 6-phosphate, and fructose 6-phosphate. Starch and cellulose are found in algae; glycogen is found in animals and protozoa; and peptidoglycan is found in bacteria.

polysaccharides such as starch, cellulose for cell walls in algae, and peptidoglycan for cell walls of bacteria. Animals and protozoa synthesize the storage molecule glycogen.

Some cells are able to synthesize sugars from noncarbohydrate precursors such as amino acids, glycerol, and fatty acids by pathways collectively called *gluconeogenesis*[11] (glū′kō-nē-ō-jen′ĕ-sis; **Figure 5.30**). Most of the reactions of gluconeogenesis are amphibolic, using enzymes of glycolysis in reverse, but four of the reactions require unique enzymes. Gluconeogenesis is highly endergonic and can proceed only if there is an adequate supply of energy.

Lipid Biosynthesis

Learning Objective

✓ Describe the biosynthesis of lipids.

As we saw in Chapter 2, lipids are a diverse group of organic molecules that function as energy-storage compounds and as components of membranes. *Carotenoids,* which are reddish pigments found in many bacterial and plant photosystems, are also lipids.

Because of their variety, it is not surprising that lipids are synthesized by a variety of routes. For example, fats are synthesized in anabolic reactions that are the reverse of their catabolism—cells polymerize glycerol and three fatty acids **(Figure 5.31).** Glycerol is derived from G3P generated by the Calvin-Benson cycle and glycolysis; the fatty acids are produced by the linkage of two-carbon acetyl-CoA molecules to one another by a sequence of endergonic reactions that effectively reverse the catabolic reactions of beta-oxidation. Other lipids, such as steroids, are synthesized in complex pathways involving polymerizations and isomerizations of sugar and amino acid metabolites.

Mycobacterium tuberculosis (mī′kō-bak-tēr′ē-ŭm tū-ber-kyū-lō-sis), the pathogen that causes tuberculosis, makes copious amounts of a waxy lipid called *mycolic acid,* which is incorporated into its cell wall. The arduous, energy-intensive process of making long lipid chains explains why this organism grows slowly, and why tuberculosis requires a long course of treatment with antimicrobial drugs.

[11] From Greek *glykys,* meaning sweet, *neo,* meaning new, and *genesis,* meaning generate.

Figure 5.31 ➤
Biosynthesis of fat, a lipid. A fat molecule is synthesized from glycerol and three molecules of fatty acid, the precursors of which are produced in glycolysis.

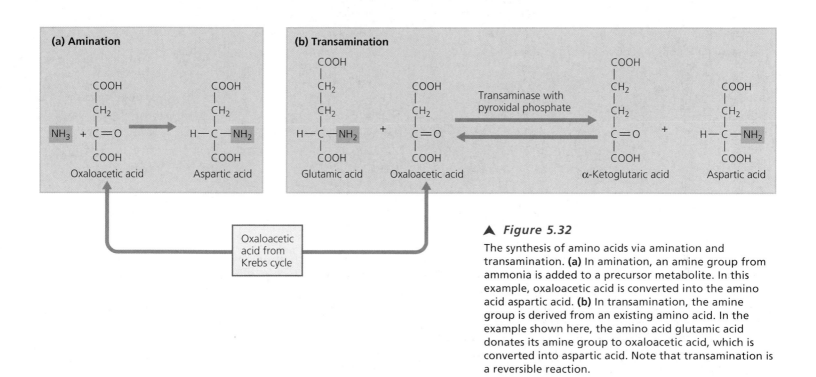

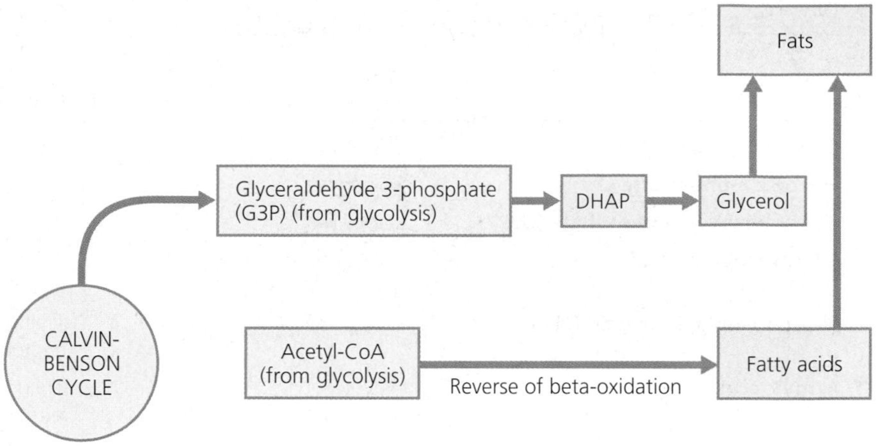

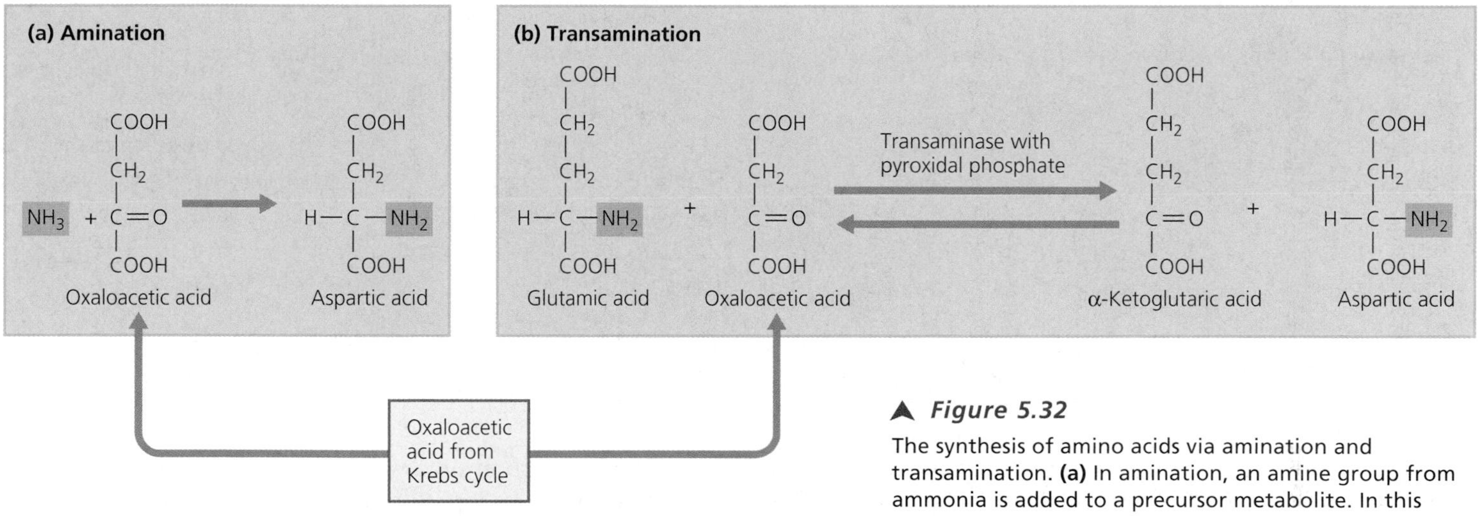

▲ **Figure 5.32**
The synthesis of amino acids via amination and transamination. **(a)** In amination, an amine group from ammonia is added to a precursor metabolite. In this example, oxaloacetic acid is converted into the amino acid aspartic acid. **(b)** In transamination, the amine group is derived from an existing amino acid. In the example shown here, the amino acid glutamic acid donates its amine group to oxaloacetic acid, which is converted into aspartic acid. Note that transamination is a reversible reaction.

Amino Acid Biosynthesis

Learning Objective

✓ Describe the biosynthesis of amino acids.

Amino acids are synthesized from precursor metabolites derived from glycolysis, the Krebs cycle, and the pentose phosphate pathway, and from other amino acids. Some organisms (such as *E. coli,* and most plants and algae) synthesize all their amino acids from precursor metabolites. Other organisms, including humans, must acquire so-called *essential amino acids* in their diets because they cannot synthesize them. One extreme example is *Lactobacillus* (lak-tō-bă-sil'ŭs), a bacterium that ferments milk and produces some cheeses. This microorganism cannot synthesize any of its amino acids; it acquires all of them by catabolizing proteins in its environment.

Precursor metabolites are converted to amino acids by the addition of an amine group. This process is called **amination** when the amine group comes from ammonia (NH_3); an example is the formation of aspartic acid from NH_3 and the Krebs cycle intermediate oxaloacetic acid **(Figure 5.32a).** Amination reactions are the reverse of the catabolic deamination reactions we discussed previously. More commonly, however, a cell moves the amine group from one amino acid and adds it to a metabolite, producing a different amino

Table 5.6 The Twelve Precursor Metabolites

	Pathway that Generates the Metabolite	Examples of Macromolecule Synthesized from Metabolite[a]	Examples of Functional Use
Glucose 6-phosphate	Glycolysis	Lipopolysaccharide	Outer membrane of cell wall
Fructose 6-phosphate	Glycolysis	Peptidoglycan	Cell wall
Glyceraldehyde 3-phosphate (G3P)	Glycolysis	Glycerol portion of lipids	Fats–energy storage
Phosphoglyceric acid	Glycolysis	Amino acids: cysteine, glycine, and serine	Enzymes
Phosphoenolpyruvic acid (PEP)	Glycolysis	Amino acids: phenylalanine, tryptophan, and tyrosine	Enzymes
Pyruvic acid	Glycolysis	Amino acids: alanine, leucine, and valine	Enzymes
Ribose 5-phosphate	Pentose phosphate pathway	DNA, RNA, amino acid and histidine	Genome, enzymes
Erythrose 4-phosphate	Pentose phosphate pathway	Amino acids: phenylalanine, tryptophan, and tyrosine	Enzymes
Acetyl-CoA	Krebs cycle	Fatty acid portion of lipids	Cell membrane
α-Ketoglutaric acid	Krebs cycle	Amino acids: arginine, glutamic acid, glutamine, and proline	Enzymes
Succinyl-CoA	Krebs cycle	Heme	Cytochrome electron carrier
Oxaloacetate	Krebs cycle	Amino acids: aspartic acid, asparagine, isoleucine, lysine, methionine, and threonine	Enzymes

[a]Examples given apply to the bacterium *E. coli.*

acid. This process is called **transamination** because the amine group is transferred from one amino acid to another **(Figure 5.32b)**. All transamination enzymes use a coenzyme, *pyridoxal phosphate,* which is derived from vitamin B_6.

Ribozymes of ribosomes polymerize amino acids into proteins. This energy-demanding process is examined in Chapter 7 because it is intimately linked with genetics.

CRITICAL THINKING

What class of enzyme is involved in amination reactions? What class of enzyme catalyzes transaminations?

Nucleotide Biosynthesis

Learning Objective

✓ Describe the biosynthesis of nucleotides.

Recall from Chapter 2 that the building blocks of nucleic acids are nucleotides, each of which consists of a five-carbon sugar, a phosphate group, and a purine or pyrimidine base (see Figure 2.25a). Nucleotides are produced from precursor metabolites of glycolysis and the Krebs cycle **(Figure 5.33):**

- The five-carbon sugars—ribose in RNA and deoxyribose in DNA—are derived from ribose 5-phosphate from the pentose phosphate pathway.
- The phosphate group is derived ultimately from ATP.
- Purines and pyrimidines are synthesized in a series of ATP-requiring reactions from the amino acids glutamine and aspartic acid derived from Krebs cycle intermediates, ribose 5-phosphate and folic acid (which is a vitamin for humans but can be synthesized by many bacteria and protozoa).

The anabolic reactions by which polymerases polymerize nucleotides to form DNA and RNA are discussed in Chapter 7.

Integration and Regulation of Metabolic Functions

Learning Objectives

✓ Describe interrelationships between catabolism and anabolism in terms of ATP and substrates.

✓ Discuss regulation of metabolic activity.

As we have seen, catabolic and anabolic reactions interact with one another in several ways. First, ATP molecules

Figure 5.33 ➤
The biosynthesis of nucleotides.

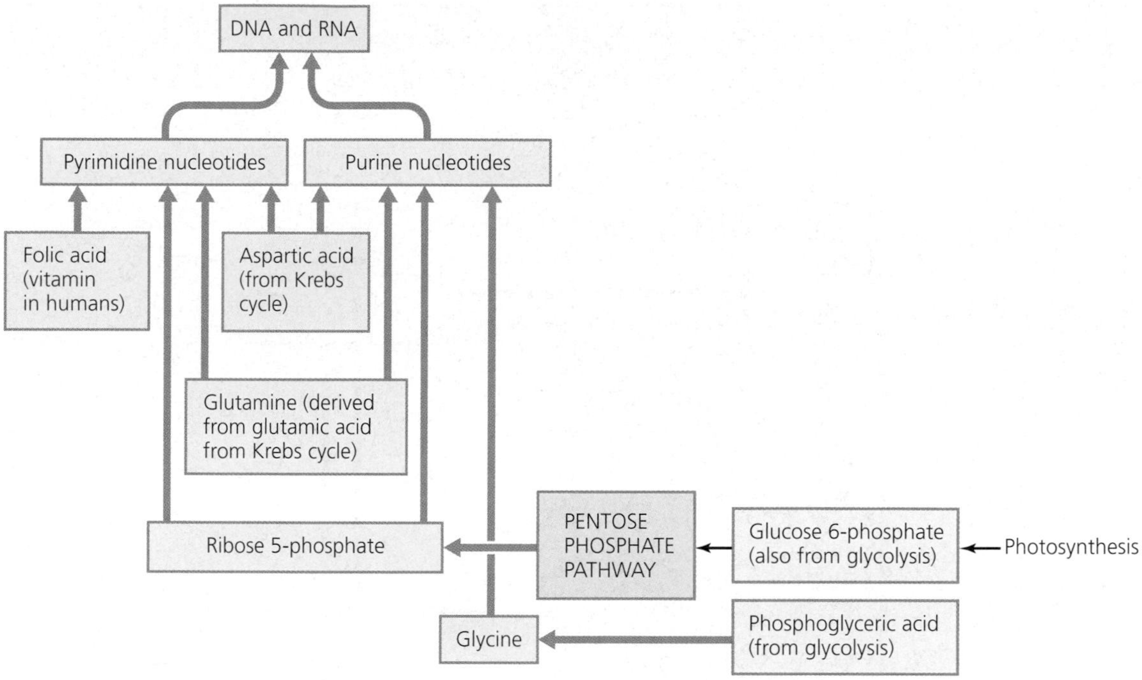

produced by catabolism are used to drive anabolic reactions. Second, precursor metabolites produced by catabolic pathways are used as substrates for anabolic reactions. Additionally, most metabolic pathways are amphibolic; they function as part of either catabolism or anabolism as needed.

Cells regulate metabolism in a variety of ways to maximize efficiency in growth and reproductive rate. Among the mechanisms involved are the following:

- Cells synthesize or degrade channel and transport proteins to increase or decrease the concentration of chemicals in the cytosol and organelles.
- Cells often synthesize the enzymes needed to catabolize a particular substrate only when that substrate is available. For instance, the enzymes of beta-oxidation are not produced when there are no fatty acids to catabolize.
- If two energy sources are available, cells catabolize the more energy efficient of the two. For example, a bacterium growing in the presence of both glucose and lactose will produce enzymes only for the transport and catabolism of glucose. Once the supply of glucose is depleted, lactose-utilizing proteins are produced.
- Cells synthesize the metabolites they need, but they typically cease synthesis if a metabolite is available as a nutrient. For instance, bacteria grown with an excess of aspartic acid will cease the amination of oxaloacetic acid (see Figure 5.32).
- Eukaryotic cells keep metabolic processes from interfering with each other by isolating particular enzymes within membrane-bounded organelles. For example, proteases sequestered within lysosomes digest phagocytized proteins without destroying vital proteins in the cytosol.
- Cells use inhibitory and excitatory allosteric sites on enzymes to control the activity of enzymes (see Figure 5.11).
- Feedback inhibition slows or stops anabolic pathways when the product is in abundance (see Figure 5.12).
- Cells regulate catabolic and anabolic pathways that use the same substrate molecules by requiring different coenzymes for each. For instance, NADH is used almost exclusively with catabolic enzymes, whereas NADPH is typically used for anabolism.

Note that these regulatory mechanisms are generally of two types: *control of gene expression,* in which cells control the amount and timing of protein (enzyme) production, and *control of metabolic expression,* in which cells control the activity of proteins (enzymes) once they have been produced.

Numerous interrelationships among the metabolic pathways discussed in this chapter are diagrammed schematically in **Figure 5.34** on page 162.

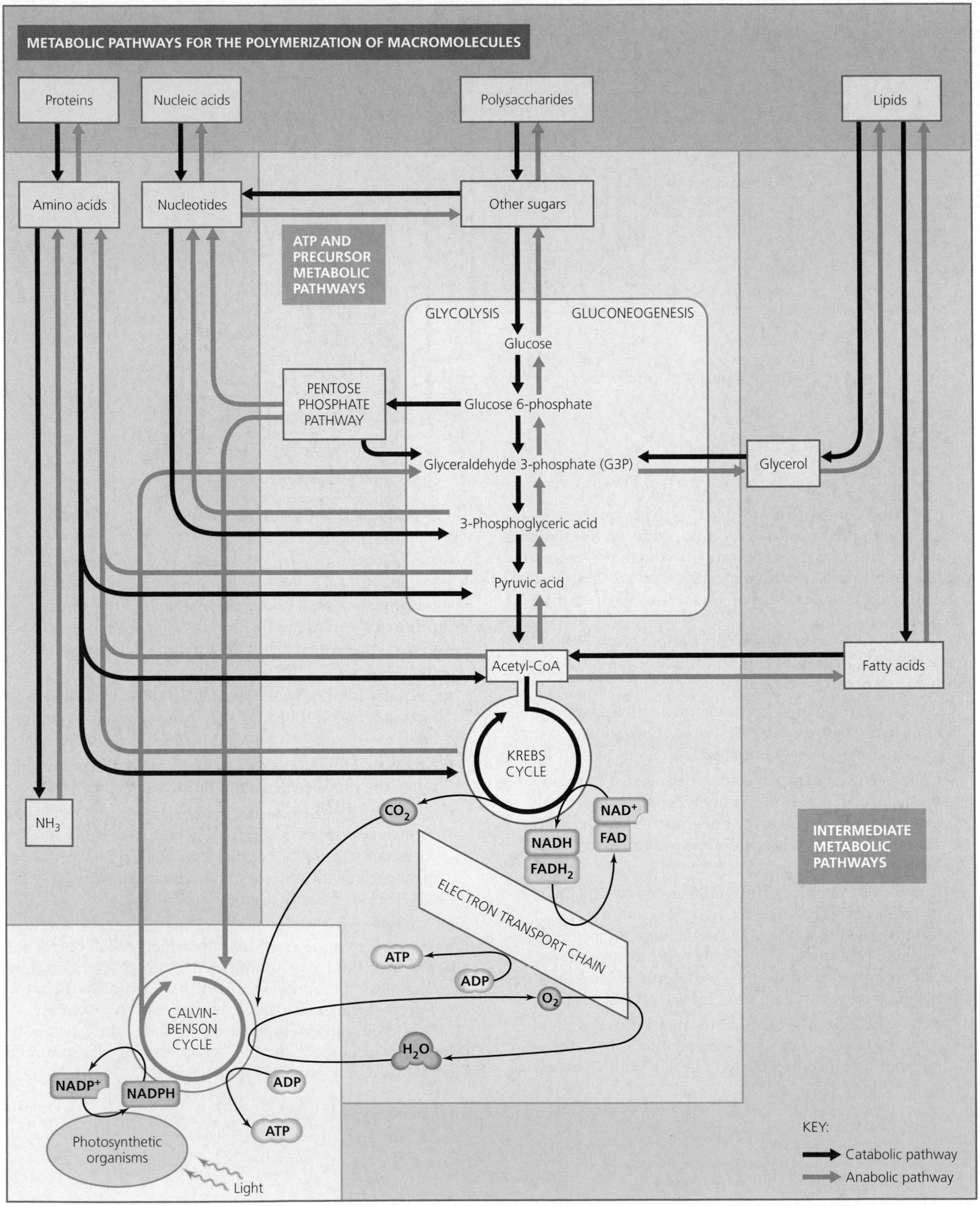
METABOLIC PATHWAYS FOR THE POLYMERIZATION OF MACROMOLECULES
Proteins
Nucleic acids
Polysaccharides
Lipids
Amino acids
Nucleotides
Other sugars
ATP AND PRECURSOR METABOLIC PATHWAYS
GLYCOLYSIS
GLUCONEOGENESIS
Glucose
PENTOSE PHOSPHATE PATHWAY
Glucose 6-phosphate
Glyceraldehyde 3-phosphate (G3P)
Glycerol
3-Phosphoglyceric acid
Pyruvic acid
Acetyl-CoA
Fatty acids
KREBS CYCLE
NH$_3$
CO$_2$
NAD$^+$
FAD
NADH
FADH$_2$
INTERMEDIATE METABOLIC PATHWAYS
ELECTRON TRANSPORT CHAIN
ATP
ADP
O$_2$
H$_2$O
CALVIN-BENSON CYCLE
NADP$^+$
NADPH
ADP
ATP
Photosynthetic organisms
Light
KEY:
Catabolic pathway
Anabolic pathway

◀ *Figure 5.34*

Integration of cellular metabolism. Cells possess three major categories of metabolic pathways: pathways for the polymerization of macromolecules (proteins, nucleic acids, polysaccharides, and lipids), intermediate pathways, and ATP and precursor pathways (glycolysis, Krebs cycle, the pentose phosphate pathway, and the Entner-Doudoroff pathway [not shown]). Cells of photosynthetic organisms also have the Calvin-Benson cycle.

CHAPTER SUMMARY

Basic Chemical Reactions Underlying Metabolism (pp. 126–136)

1. **Metabolism** is the sum of complex biochemical reactions within the cells of an organism, including **catabolism,** which breaks down molecules and releases energy, and **anabolism,** which synthesizes molecules and uses energy.
2. **Precursor metabolites,** often produced in catabolic reactions, are used to synthesize all other organic compounds.
3. **Reduction** reactions are those in which the electrical charge on a chemical is made more negative by the addition of electrons. The molecule that donates an electron is **oxidized.** If the electron is part of a hydrogen atom, an oxidation reaction is also called dehydrogenation. Oxidation and reduction reactions always occur in pairs called **oxidation-reduction (redox) reactions**.
4. Three important electron carrier molecules are **nicotinamide adenine dinucleotide (NAD^+), nicotinamide adenine dinucleotide phosphate ($NADP^+$)**, and **flavin adenine dinucleotide (FAD).**
5. Phosphorylation is the addition of phosphate to a molecule. Three types of phosphorylation form ATP: **Substrate-level phosphorylation** involves the transfer of phosphate to ADP from another phosphorylated organic compound. In **oxidative phosphorylation,** energy from redox reactions of respiration is used to attach inorganic phosphate (PO_4^{2-}) to ADP. **Photophosphorylation** is the phosphorylation of ADP with inorganic phosphate using energy from light.
6. Catalysts increase the rates of chemical reactions and are not permanently changed in the process. **Enzymes,** which are organic catalysts, are often named for their **substrates,** the molecules upon which they act. Enzymes can be classified according to their mode of action into hydrolases, isomerases, ligases (polymerases), lyases, oxidoreductases, or transferases.
7. **Apoenzymes** are the protein portions of protein enzymes that may require one or more **cofactors** such as inorganic ions or organic cofactors called **coenzymes.** The combination of both apoenzyme and its cofactors is a **holoenzyme.** RNA molecules functioning as enzymes are called **ribozymes.**
8. **Activation energy** is the amount of energy required to initiate a chemical reaction.
9. Substrates fit into the specifically shaped **active sites** of the enzymes that catalyze their reactions.
10. Enzymes may be **denatured** by physical and chemical factors such as heat and pH. Denaturation may be reversible or permanent.
11. Enzyme activity proceeds at a rate proportional to the concentration of substrate molecules until all the active sites are filled to saturation.
12. **Competitive inhibitors** block active sites and thereby block enzyme activity. In allosteric inhibition, **noncompetitive inhibitors** attach to an allosteric site on an enzyme, altering the active site so that it is no longer functional.
13. **Feedback inhibition** (negative feedback) occurs when the final product of a series of reactions is an allosteric inhibitor of some previous step in the series. Thus, accumulation of the end-product "feeds back" into the series a signal that stops the process.

Carbohydrate Catabolism (pp. 136–149)

1. **Glycolysis** involves the splitting of a glucose molecule in a three-stage, 10-step process that ultimately results in two molecules of pyruvic acid and a net gain of two ATP and two NADH molecules.
2. The **pentose phosphate** and **Entner-Doudoroff pathways** are alternative means for the catabolism of glucose that yield fewer ATP molecules than does glycolysis. However, they produce precursor metabolites not produced in glycolysis.
3. **Cellular respiration** is a metabolic process that involves the complete oxidation of substrate molecules and the production of ATP following a series of redox reactions.
4. Two carbons from pyruvic acid join coenzyme A to form acetyl-coenzyme A (**acetyl-CoA**), which then enters the **Krebs cycle,** a series of eight enzymatic steps that transfer energy from acetyl-CoA to coenzymes NAD^+ and FAD.
5. An **electron transport chain** is a series of redox reactions that pass electrons from one membrane-bound carrier to another, and then to a final electron acceptor. The energy from these electrons is used to pump protons across the membrane.
6. The four classes of carrier molecules in electron transport systems are flavoproteins, ubiquinones, metal-containing proteins, and cytochromes.
7. Aerobes use oxygen atoms as final electron acceptors of their electron transport chains in a process known as **aerobic respiration,** whereas anaerobes use other inorganic molecules (such as NO_3^-, SO_4^{2-}, and CO_3^{2-}, or rarely an organic molecule) as the final electron acceptor in **anaerobic respiration.**

8. In **chemiosmosis,** ions flow down their electrochemical gradient across a membrane through **ATP synthase (ATPase)** to synthesize ATP.
9. A **proton gradient** is an electrochemical gradient of hydrogen ions across a membrane. It has potential energy known as a proton motive force.
10. Oxidative phosphorylation and photophosphorylation use chemiosmosis.
11. **Fermentation** is the partial oxidation of sugar to release energy using an organic molecule rather than an electron transport chain as the final electron acceptor. End-products of fermentation, which are often useful to humans and aid in identification of microbes in the laboratory, include acids, alcohols, and gases.

Other Catabolic Pathways (pp. 149–152)

1. Lipids and proteins can be catabolized into smaller molecules, which can be used as substrates for glycolysis and the Krebs cycle.
2. **Beta-oxidation** is a catabolic process in which enzymes split pairs of hydrogenated carbon atoms from a fatty acid and join them to coenzyme A to form acetyl-CoA.
3. **Proteases** secreted by microorganisms digest proteins outside the microbes' cell walls. The resulting amino acids are moved into the cell and used in anabolism, or **deaminated** and catabolized for energy.

Photosynthesis (pp. 152–156)

1. **Photosynthesis** is a process in which light energy is captured by pigment molecules called **chlorophylls** (bacteriochlorophylls in some bacteria) and transferred to ATP and metabolites. **Photosystems** are networks of light-absorbing chlorophyll molecules and other pigments held within a protein matrix on membranes called **thylakoids.**
2. The redox reactions of photosynthesis are classified as **light-dependent reactions** and **light-independent reactions.**
3. A **reaction center chlorophyll** is a special chlorophyll molecule in a photosystem in which electrons are excited by light energy and passed to an acceptor molecule of an electron transport chain.
4. In **cyclic photophosphorylation,** the electrons return to the original reaction center after passing down the electron transport chain.
5. In **noncyclic photophosphorylation,** photosystem II works with photosystem I, and the electrons are used to reduce $NADP^+$ to NADPH. In oxygenic photosynthesis, cyanobacteria, algae, and green plants replenish electrons to the reaction center by dissociation of H_2O molecules, resulting in the release of O_2 molecules. Anoxygenic bacteria derive electrons from inorganic compounds such as H_2S, producing waste such as sulfur.
6. In the light-independent pathway of photosynthesis, **carbon fixation** occurs in the **Calvin-Benson cycle,** in which CO_2 is reduced to produce glucose.

Other Anabolic Pathways (pp. 156–160)

1. **Amphibolic reactions** are metabolic reactions that are reversible.
2. Some cells are able to synthesize glucose from amino acids, glycerol, and fatty acids via a process called gluconeogenesis.
3. **Amination** reactions involve adding an amine group from ammonia to a metabolite to make an amino acid. **Transamination** occurs when an amine group is transferred from one amino acid to another.
4. Nucleotides are synthesized from precursor metabolites produced by glycolysis, the Krebs cycle, and the pentose phosphate pathway.

Integration and Regulation of Metabolic Functions (pp. 160–163)

1. Cells regulate metabolism by control of gene expression or metabolic expression. They control the latter in a variety of ways, including synthesizing or degrading channel proteins and enzymes, sequestering reactions in membrane-bounded organelles (seen only in eukaryotes), and feedback inhibition.

QUESTIONS FOR REVIEW

(Answers to multiple choice, matching questions, and fill in the blanks are on the web, along with additional review questions. Visit www.microbiologyplace.com.)

Multiple Choice

For each of the phrases in Questions 1–7, indicate the type of metabolism referred to, using the following choices:

a. anabolism only
b. both anabolism and catabolism (amphibolic)
c. catabolism only

1. Breaks a large molecule into smaller ones
2. Includes dehydration synthesis reactions
3. Is exergonic
4. Is endergonic
5. Involves the production of cell membrane constituents
6. Includes hydrolytic reactions
7. Includes metabolism
8. Redox reactions
   a. transfer energy.
   b. transfer electrons.
   c. involve oxidation and reduction.
   d. are involved in all of the above.
9. A reduced molecule
   a. has gained electrons.
   b. has become more positive in charge.

c. has lost electrons.
d. is an electron donor.

10. Activation energy
   a. is the amount of energy required during an activity such as flagellar motion.
   b. requires the addition of nutrients in the presence of water.
   c. is lowered by the action of organic catalysts.
   d. results from the movement of molecules.

11. Coenzymes
   a. are types of apoenzymes.
   b. are proteins.
   c. are inorganic cofactors.
   d. are organic cofactors.

12. Which of the following statements best describes ribozymes?
   a. Ribozymes are proteins that aid in the production of ribosomes.
   b. Ribozymes are nucleic acids that produce ribose sugars.
   c. Ribozymes store enzymes in ribosomes.
   d. Ribozymes process RNA molecules in eukaryotes.

13. Which of the following does *not* affect the function of enzymes?
   a. ubiquinone
   b. substrate concentration
   c. temperature
   d. competitive inhibitors

14. Most oxidation reactions in bacteria involve the
   a. removal of hydrogen ions and electrons.
   b. removal of oxygen.
   c. addition of hydrogen ions and electrons.
   d. addition of hydrogen ions.

15. Under ideal conditions, the fermentation of one glucose molecule by a bacterium allows a net gain of how many ATP molecules?
   a. 2
   b. 4
   c. 38
   d. 0

16. Under ideal conditions, the complete aerobic oxidation of one molecule of glucose by a bacterium allows a net gain of how many ATP molecules?
   a. 2
   b. 4
   c. 38
   d. 0

17. Which of the following statements is *not* true of the Entner-Doudoroff pathway?
   a. It is a series of reactions that synthesizes glucose.
   b. Its products are sometimes used to determine the presence of *Pseudomonas*.
   c. It is a pathway of chemical reactions that catabolizes glucose.
   d. It is an alternative pathway to glycolysis.

18. Reactions involved in the light-independent reactions of photosynthesis constitute the
   a. Krebs cycle.
   b. Entner-Doudoroff pathway.
   c. Calvin-Benson cycle.
   d. pentose phosphate pathway.

Matching Questions

1. ___ Occurs when energy from a compound containing phosphate reacts with ADP to form ATP
2. ___ Involves formation of ATP via reduction of coenzymes in the electron transport chain
3. ___ Begins with glycolysis
4. ___ Occurs when all active sites on substrate molecules are filled

A. Saturation
B. Oxidative phosphorylation
C. Substrate-level phosphorylation
D. Photophosphorylation
E. Carbohydrate catabolism

Fill in the Blanks

1. The final electron acceptor in cyclic photophosphorylation is ________________.
2. Two ATP molecules are used to initiate glycolysis. Enzymes generate molecules of ATP for each molecule of glucose that undergoes glycolysis. Thus a net gain of ________________ molecules of ATP is produced in glycolysis.
3. The initial catabolism of glucose occurs by glycolysis and/or the ________________ and ________________ pathways.
4. ________________ is a cyclic series of eight reactions involved in the catabolism of acetyl-CoA that yields eight molecules of NADH and two molecules of $FADH_2$.
5. The final electron *acceptor* in aerobic respiration is ________________.
6. Three common electron acceptors in anaerobic respiration are ________________, ________________, and ________________.
7. Chemolithotrophs *acquire* electrons from (organic/inorganic) ________________ compounds.
8. Complete the following chart:

Category of Enzymes	Description
	Catabolyzes substrate by adding water
Isomerase	
Ligase/polymerase	
	Moves functional groups such as an acetyl group
	Adds or removes electrons
Lyase	

Short Answer

1. Why are enzymes necessary for anabolic reactions to occur in living organisms?
2. How do organisms control the rate of metabolic activities in their cells?
3. How does a noncompetitive inhibitor at an allosteric site affect a whole pathway of enzymatic reactions?
4. Explain the mechanism of negative feedback with respect to enzyme action.

5. Facultative anaerobes can live under either aerobic or anaerobic conditions. What metabolic pathways allow these organisms to continue to harvest energy from sugar molecules in the absence of oxygen?
6. How does oxidation of a molecule occur without oxygen?
7. List at least four groups of microorganisms that are photosynthetic.
8. Why do we breathe oxygen and give off carbon dioxide?
9. Why do cyanobacteria and algae take in carbon dioxide and give off oxygen?
10. What happens to the carbon atoms in sugar catabolized by *E. coli*?
11. How do yeast cells make alcohol and cause bread to rise?
12. Where specifically does the most significant production of ATP occur in prokaryotic and eukaryotic cells?
13. Why are vitamins essential metabolic factors for microbial metabolism?
14. A laboratory scientist notices that a certain bacterium does not utilize lactose when glucose is available in its environment. Describe a cellular regulatory mechanism that would explain this observation.

CRITICAL THINKING

1. Why does an organism that uses glycolysis and the Krebs cycle also need the pentose phosphate pathway?
2. Describe how bacterial fermentation causes milk to sour.
3. *Giardia lamblia* and *Entamoeba histolytica* are protozoa that live in the colons of mammals and can cause life-threatening diarrhea. Interestingly, these microbes lack mitochondria. What kind of pathway must they have for carbohydrate catabolism?
4. Two cultures of a facultative anaerobe are grown in the same type of medium, but one is exposed to air and the other is maintained under anaerobic conditions. Which of the two cultures will contain more cells at the end of a week? Why?
5. What is the maximum number of molecules of ATP that can be generated by a bacterium after the complete aerobic oxidation of a fat molecule containing three 12-carbon chains? (Assume that all the available energy released during catabolism goes to ATP production.)
6. In terms of its effects on metabolism, why is a fever over 40°C often life threatening?
7. Using the categories defined in the text on page 129, classify each of the enzymes of glycolysis, the Krebs cycle, and transamination (see Figure 5.32 on page 159).
8. How are photophosphorylation and oxidative phosphorylation similar? How are they different?
9. Members of the pathogenic bacterial genus *Haemophilus* require NAD^+ and heme from their environment. For what purpose does *Haemophilus* use these growth factors?
10. Compare and contrast aerobic respiration, anaerobic respiration, and fermentation.
11. Scientists estimate that up to one-third of Earth's biomass is composed of methanogenic prokaryotes in ocean sediments (*Science* 295:2067–2070). Describe the metabolism of these organisms.
12. A young student was troubled by the idea that a bacterium is able to control its diverse and complex metabolic activities, even though it lacks a brain. How would you explain their metabolic control?
13. If a bacterium uses beta-oxidation to catabolize a molecule of the fatty acid arachidic acid, which contains 20 carbon atoms, how many acetyl-CoA molecules will be generated?

CHAPTER 6

Microbial Nutrition and Growth

The dramatic red colors of Owens Lake in Nevada are caused by *halobacteria*—bacteria that thrive in extremely salty environments. Halobacteria thrive in water that is several times the salinity of seawater and are also found in such notably salty habitats as the Great Salt Lake and the Dead Sea. The red colors of Owens Lake are the result of a pigment produced by the bacteria, which are present in staggeringly large numbers in the lake: Just one drop of the concentrated brine may contain millions of halobacteria.

Halobacteria are just one example of extremophiles, microorganisms that live in extreme environments. From the canyons of Death Valley to subglacial lakes deep beneath the ice sheets of Antarctica, microorganisms live and grow in environments uninhabitable by other life forms. How do they obtain the nutrients they need? What characteristics allow them to survive such harsh conditions? In this chapter we will study the requirements for microbial growth.

CHAPTER OUTLINE

The red colors of Owens Lake are caused by bacteria-produced pigments.

MicroPrep Pre-Test: *Take the pre-test for this chapter on the web. Visit* ***www.microbiologyplace.com.***

We saw in Chapter 5 that metabolism—the collection of controlled chemical reactions within cells—is a major characteristic of all living things. Recall as well that the ultimate outcome of metabolic activity is reproduction, an increase in the number of individual cells or organisms. When speaking of the reproductive activities of microbes in general, and of bacteria in particular, microbiologists typically use the term *growth,* referring to an increase in a *population* of microbes rather than to an increase in size. The result of such microbial growth is a discrete **colony,** an aggregation of cells arising from a single parent cell. Put another way, the *reproduction* of individual microorganisms results in the *growth* of a colony. Further, the common expression "The microorganisms *grow* in salt-containing media" is widely understood to mean that they metabolize and reproduce.

In this chapter we consider the characteristics of microbial growth from two different but related perspectives: We examine the requirements of microbes living in their natural settings, including their chemical, physical, and energy requirements, and we explore how microbiologists try to duplicate natural conditions to grow microorganisms in the laboratory so that they can be transported, identified, and studied. We conclude by examining bacterial population dynamics and some techniques for measuring bacterial growth.

Growth Requirements

As we discussed in Chapter 5, organisms use a variety of chemicals—called **nutrients**—for their energy needs and to build organic molecules and cellular structures. The most common of these nutrients are compounds containing necessary elements such as carbon, oxygen, nitrogen, and hydrogen (Table 6.1 on page 170). Like all organisms, microbes obtain nutrients from a variety of sources in their environment, and they must bring nutrients into their cells by the passive and active transport processes we discussed in Chapter 3. When they acquire their nutrients by living in or on another organism, they may cause disease as they interfere with their hosts' metabolism and nutrition.

Nutrients: Chemical and Energy Requirements

Learning Objectives

- ✓ Describe the roles of carbon, hydrogen, oxygen, nitrogen, trace elements, and vitamins in microbial growth and reproduction.
- ✓ Compare the four basic categories of organisms based on their carbon and energy sources.
- ✓ Distinguish among anaerobes, aerobes, aerotolerant anaerobes, facultative anaerobes, microaerophiles, and capnophiles.
- ✓ Explain how oxygen can be fatal to organisms by discussing singlet oxygen, superoxide radical, peroxide anion, and hydroxyl radical, and describe how organisms protect themselves from toxic forms of oxygen.

We begin our examination of microbial growth requirements by considering three things all cells need for metabolism: a carbon source, a source of energy, and a source of electrons or hydrogen atoms.

Sources of Carbon, Energy, and Electrons

Organisms can be categorized into two broad groups based on their source of carbon. Organisms that utilize an inorganic carbon source (that is, carbon dioxide) as their sole source of carbon are called *autotrophs* (aw'tō-trōfs), so-named because they "feed themselves."[1] More precisely, autotrophs make organic compounds from CO_2 and thus need

[1]From Greek *auto,* meaning self, and *trophe,* meaning nutrition.

Carbon Source	Energy Source: Light *(photo-)*	Energy Source: Chemical compounds *(chemo-)*
Carbon dioxide *(auto-)*	***Photoautotrophs*** • Plants, algae, and cyanobacteria use H_2O to reduce CO_2, producing O_2 as a byproduct • Photosynthetic green sulfur and purple sulfur bacteria do not use H_2O nor produce O_2	***Chemoautotrophs*** • Hydrogen, sulfur, and nitrifying bacteria
Organic compounds *(hetero-)*	***Photoheterotrophs*** • Green nonsulfur and purple nonsulfur bacteria	***Chemoheterotrophs*** • Aerobic respiration: most animals, fungi, and protozoa, and many bacteria • Anaerobic respiration: some animals, protozoa, and bacteria • Fermentation: some bacteria and yeasts

◀ ***Figure 6.1***
The four basic groups of organisms based on their carbon and energy sources. Additionally, every organism utilizes electrons from either organic molecules (organotrophs) or inorganic molecules (lithotrophs) in redox reactions.

New Frontiers 6.1 Is There Life on Other Planets?

We are accustomed to thinking of a biological world based upon photosynthesis—photosynthetic organisms provide a suitable carbon source for animals, fungi, protozoa, and most bacteria, which are then food for larger organisms, including humans. However, based on evidence from samples within the earth's crust, many living things are lithotrophic chemoautotrophic prokaryotes that acquire all of their nutritional and energy needs from inorganic chemicals such as hydrogen, sulfur, and ferrous iron (Fe^{2+}). Such microbes also fill the role of primary producers and support entire communities of animals on the bottoms of the oceans, where no light penetrates. Some scientists believe that organisms of this type were the first living things on Earth, and that they are the only things that could live on other planets in the solar system.

Additionally, scientists have recently discovered novel microorganisms living in hydrothermal waters far beneath the mountain ranges of Idaho. These chemoautolithotrophic microorganisms, 90% of which are archaea, live in an environment completely devoid of sunlight. Instead of obtaining energy from light or organic matter, they use hydrogen as an energy source. Because the habitat of these microorganisms is very similar to what we know of the environment on Mars, some scientists speculate that microbial life may be possible on the Red Planet.

Mars

not feed on organic compounds from other organisms to acquire carbon. In contrast, organisms called *heterotrophs*[2] (het′er-ō-trōfs) catabolize reduced organic molecules (such as proteins, carbohydrates, amino acids, and fatty acids) they acquire from other organisms.

Organisms can also be categorized according to whether they use chemicals or light as a source of energy for such cellular processes as anabolism, intracellular transport, and motility. Organisms that acquire energy from redox reactions involving inorganic and organic chemicals are called *chemotrophs* (kēm′ō-trōfs). (Recall from Chapter 5 that these reactions are either aerobic respiration, anaerobic respiration, or fermentation, depending on the final electron acceptor.) Organisms that use light as their energy source are called *phototrophs*[3] (fō′tō-trōfs).

Based on their carbon and energy sources, then, most organisms can be categorized into one of four basic groups: **photoautotrophs, chemoautotrophs, photoheterotrophs,** and **chemoheterotrophs (Figure 6.1).** Plants, some protozoa, and algae are photoautotrophs and animals, fungi, and other protozoa are chemoheterotrophs. Prokaryotes are most diverse, with members in all four groups.

Additionally, the cells of all organisms require electrons or hydrogen atoms. Hydrogen is the most common chemical element in cells, and it is so common in organic molecules and water that it is never a *limiting nutrient;* that is, metabolism is never interrupted by a lack of hydrogen. Hydrogen is essential for hydrogen bonding and in electron transfer. Organisms that acquire electrons (typically as part of hydrogen atoms) from the same organic molecules that provide them carbon and energy are called **organotrophs** (ōr′găn-ō-trōfs); alternatively, organisms that acquire electrons or hydrogen atoms from inorganic sources (such as H_2, NO^{2-}, H_2S, and Fe^{2+}) are called **lithotrophs**[4] (lith′ō-trōfs). **New Frontiers 6.1** focuses on lithotrophic chemoautotrophs (also called chemolithoautotrophs), which are prokaryotic organisms that acquire carbon, energy, and electrons from inorganic chemicals.

CRITICAL THINKING

The filamentous bacterium *Beggiatoa* gets its carbon from carbon dioxide and electrons and energy from hydrogen sulfide. What is its nutritional classification?

Not all organisms are easy to classify. For instance, *Euglena granulata* typically uses light energy and gets its carbon from carbon dioxide. However, when it is cultured on a suitable medium in the dark, this microbe utilizes energy and carbon solely from organic compounds. What is the nutritional classification of *Euglena?*

Oxygen Requirements

As we learned in Chapter 5, oxygen is essential for **obligate aerobes** because it serves as the final electron acceptor of electron transport chains, which produce most of the ATP in these organisms. By contrast, oxygen is a deadly poison for **obligate anaerobes.** How can oxygen be a fatal toxin for one group of organisms and yet be essential for another?

The key to understanding this apparent incongruity is understanding that neither gaseous molecular oxygen (O_2) of the atmosphere nor the oxygen that is covalently bound in compounds such as carbohydrates and water is poisonous.

---

[2]From Greek *hetero,* meaning other, and *trophe,* meaning nutrition.
[3]From Greek *photos,* meaning light, and *trophe,* meaning nutrition.
[4]From Greek *lithos,* meaning rock, and *trophe,* meaning nutrition.

Table 6.1 Elements Required by Microorganisms

Element	Proportion of Prokaryotic Cells' Dry Weight	FUNCTION
Carbon	50%	Basic structural component of organic compounds
Oxygen	20%	Component of many organic and inorganic compounds; O_2 is the final electron acceptor in aerobic respiration
Nitrogen	14%	Constituent of amino acids and nucleotides
Hydrogen	8%	Constituent of organic compounds; electrons of hydrogen atoms are used in redox reactions
Phosphorus	3%	As phosphate (PO_4^{2-}), constituent of nucleic acids, phospholipids, some coenzymes, and ATP
Sodium	1%	Cation used to maintain electrical balance of cells
Sulfur	1%	Constituent of the amino acids cysteine and methionine; disulfide bonds are important for tertiary structure of proteins
Potassium	1%	Inorganic cation involved in maintaining electrical balance inside cells; cofactor for some enzymes, including some involved in protein synthesis
Chlorine	0.5%	Principal inorganic anion involved in maintaining electrical balance of cells
Magnesium	0.5%	Component of chlorophyll; cofactor for some enzymes
Calcium	0.5%	Responsible in part for the resistance of bacterial endospores; cofactor for some enzymes; constituent of some algal cell walls and protozoan shells
Iron	0.25%	Constituent of cytochromes in electron transport chains; cofactor for some enzymes
Trace elements (cobalt, copper, manganese, molybdenum, nickel, selenium, silicon, tungsten, vanadium, and zinc)	0.25%	Cofactors of some enzymes and silicon is a constituent of some algal cell walls

Rather, the forms of oxygen that are toxic are those that are highly reactive. They are toxic for the same reason that oxygen is the final electron acceptor for aerobes: They are excellent oxidizing agents, so they steal electrons from other compounds, which in turn steal electrons from still other compounds. The resulting chain of vigorous oxidations causes irreparable damage to cells by oxidizing important compounds, including proteins and lipids.

There are four toxic forms of oxygen:

1. **Singlet oxygen ($^1O_2$)** Singlet oxygen is molecular oxygen with electrons that have been boosted to a higher energy state, typically during aerobic metabolism. Singlet oxygen is a very reactive oxidizing agent. Phagocytic cells, such as certain human white blood cells, use it to oxidize pathogens. Because singlet oxygen is also photochemically produced by the reaction of oxygen and light, phototrophic microorganisms often contain pigments called **carotenoids** (ka-rot'e-noyds) that prevent toxicity by removing the excess energy of singlet oxygen.
2. **Superoxide radical (O_2^-)** A few superoxide radicals form during the incomplete reduction of O_2 during electron transport in aerobes and during metabolism by anaerobes in the presence of oxygen. Superoxide radicals are so reactive and toxic that aerobic organisms

must produce enzymes called *superoxide dismutases* (dis'myu-tās-es) to detoxify them. These enzymes, which have active sites that contain metal ions—Zn^{2+}, Mn^{2+}, Fe^{2+}, Ni^{2+}, or Cu^{2+}, depending on the organism—combine two superoxide radicals and two protons to form hydrogen peroxide (H_2O_2) and molecular oxygen (O_2):

$$2O_2^- + 2H^+ \rightarrow H_2O_2 + O_2$$

The reason that anaerobes are susceptible to oxygen is that they lack superoxide dismutase; they die as a result of the oxidizing reactions of superoxide radicals formed in the presence of oxygen.

3. **Peroxide anion (O_2^{2-})** Hydrogen peroxide formed during reactions catalyzed by superoxide dismutase (and during other metabolic reactions) contains peroxide anion, another highly reactive oxidant. It is peroxide anion that makes hydrogen peroxide an effective antimicrobial agent. Aerobes contain either catalase or peroxidase, enzymes that detoxify peroxide anion.

   Catalase converts hydrogen peroxide to water and molecular oxygen:

$$2\,H_2O_2 \rightarrow 2\,H_2O + O_2$$

   A simple test for catalase involves adding a sample from a bacterial colony to a drop of hydrogen peroxide. The production of bubbles of oxygen indicates the presence of catalase **(Figure 6.2).**

   Peroxidase breaks down hydrogen peroxide without forming oxygen, using a reducing agent such as the coenzyme NADH:

$$H_2O_2 + NADH + H^+ \rightarrow 2\,H_2O + NAD^+$$

   Obligate anaerobes either lack both catalase and peroxidase or have only a small amount of them, so they are susceptible to the toxic action of hydrogen peroxide.

4. **Hydroxyl radical ($OH^\cdot$)** Hydroxyl radicals result from ionizing radiation and from the incomplete reduction of hydrogen peroxide:

$$H_2O_2 + e^- + H^+ \rightarrow H_2O + OH^\cdot$$

   Hydroxyl radicals are the most reactive of the four toxic forms of oxygen, but because hydrogen peroxide does not accumulate in aerobic cells (due to the action of catalase and peroxidase), the threat of hydroxyl radical is virtually eliminated in aerobic cells.

Besides the enzymes superoxide dismutase, catalase, and peroxidase, aerobes use other antioxidants, such as vitamins C and E, to protect themselves against toxic oxygen products. These antioxidants provide electrons that reduce toxic forms of oxygen.

Not all organisms are either strict **aerobes** or **anaerobes;** many organisms can live in various oxygen concentrations between these two extremes. For example, some aerobic organisms can maintain life via fermentation or anaerobic respiration, though their metabolic efficiency is often reduced in the absence of oxygen. Such organisms are called

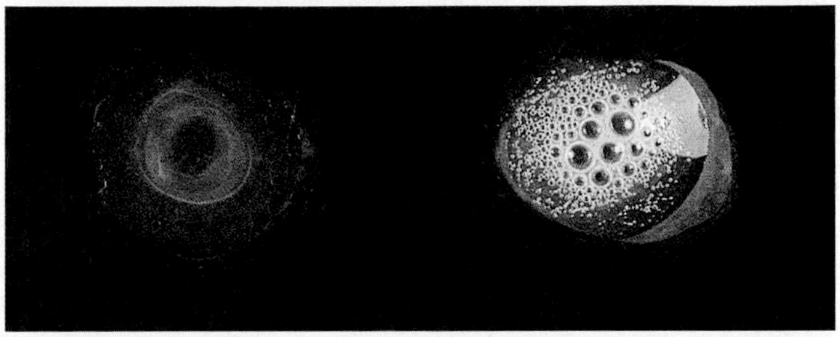

▲ *Figure 6.2*

Catalase test. The enzyme catalase converts hydrogen peroxide into water and oxygen, the latter of which can be seen as visible bubbles. *Enterococcus faecalis* **(a)** is catalase negative, whereas *Staphylococcus epidermidis* **(b)** is catalase positive.

facultative anaerobes. *Escherichia coli* (esh-ĕ-rik'ē-ă kō'lī) is an example of a facultatively anaerobic bacterium.

Aerotolerant anaerobes do not use aerobic metabolism, but they tolerate oxygen by having some of the enzymes that detoxify oxygen's poisonous forms. The lactobacilli that transform cucumbers into pickles, and milk into cheese, are aerotolerant. These organisms can be grown without the special conditions required by strict anaerobes, which will be discussed shortly.

Microaerophiles, such as the ulcer-causing pathogen *Helicobacter pylori*[5] (hel'ĭ-kō-bak'ter pī'lo-rē), are aerobes that require oxygen levels of 2% to 10%. This concentration of oxygen is found in the stomach. Microaerophiles are damaged by the 21% concentration of oxygen in the atmosphere, presumably because they have limited ability to detoxify hydrogen peroxide and superoxide radicals.

Microbial groups contain members with each of the five types of oxygen requirement. Algae, most fungi and protozoa, and many prokaryotes are obligate aerobes. A few yeasts and numerous prokaryotes are facultative anaerobes. Many prokaryotes and a few protozoa are aerotolerant, microaerophilic, or obligate anaerobes. The oxygen requirement of an organism can be identified by growing it in a medium that contains an oxygen gradient from top to bottom **(Figure 6.3).**

Nitrogen Requirements

Another essential element is nitrogen, which is contained in many organic compounds, including the amine group of amino acids and as part of nucleotide bases. Nitrogen makes up about 14% of the dry weight of microbial cells.

Nitrogen is often a growth-limiting nutrient for many organisms; that is, their anabolism ceases because they do not have sufficient nitrogen to build proteins and nucleotides.

---

[5]From semi-helical shape of the cell, and *pylorus,* which is the distal portion of the stomach.

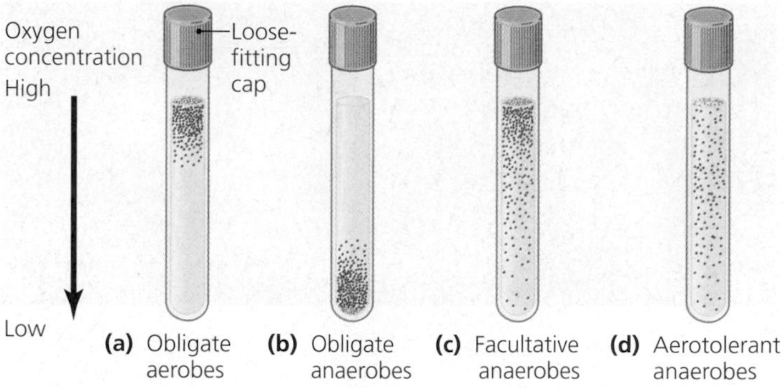

▲ ***Figure 6.3***

Using a liquid growth medium to identify the oxygen requirements of organisms. The surface of thioglycollate medium, which is exposed to atmospheric oxygen, is aerobic. Oxygen concentration decreases with depth; the bottom of the tube is anaerobic. **(a)** Obligate aerobes cannot survive below the depth to which oxygen penetrates the medium. **(b)** Obligate anaerobes cannot tolerate any oxygen. **(c)** Facultative anaerobes can grow with or without oxygen, but their ability to use aerobic respiration pathways enhances their growth near the surface. **(d)** Aerotolerant aerobes can grow equally well with or without oxygen; their growth is relatively evenly distributed throughout the medium. *Where in such a test tube would the growth zone be for a microaerophilic aerobe?*

Figure 6.3 *Microaerophiles would be found slightly below the surface, but neither directly at the surface nor in the depths of the tube.*

Organisms acquire nitrogen from organic and inorganic nutrients. For example, most photosynthetic organisms can reduce nitrate (NO_3^-) to ammonium (NH_4^+), which can then be used for biosynthesis. In addition, all cells recycle nitrogen from their amino acids and nucleotides.

Though nitrogen constitutes about 79% of the atmosphere, relatively few organisms can utilize nitrogen gas. A few bacteria, notably many cyanobacteria and *Rhizobium* (rī-zō′bē-ŭm), reduce nitrogen gas (N_2) to ammonia (NH_3) via a process called **nitrogen fixation.** Nitrogen fixation is essential for life on Earth because nitrogen-fixers provide nitrogen in a useable form to other organisms. In Chapter 11 we discuss nitrogen-fixing prokaryotes, as well as nitrifying prokaryotes—those that oxidize nitrogenous compounds to acquire electrons for electron transport.

Other Chemical Requirements

Together, carbon, hydrogen, oxygen, and nitrogen make up more than 95% of the dry weight of cells; phosphorus, sulfur, calcium, manganese, magnesium, copper, iron, and a few other elements constitute the rest. Phosphorus is a component of phospholipid membranes, DNA, RNA, ATP, and some proteins. Sulfur is a component of sulfur-containing amino acids, which bind to one another via disulfide bonds that are critical to the tertiary structure of proteins, and in vitamins such as thiamine (B_1) and biotin.

Other elements are called **trace elements** because they are required in very small ("trace") amounts (see Table 6.1 on page 170). For example, a few atoms of selenium dissolved out of the walls of glass test tubes provide the total requirement for the growth of green algae in a laboratory. Other trace elements are usually found in sufficient quantities dissolved in water. For this reason, tap water can sometimes be used instead of distilled or deionized water to grow microorganisms in the laboratory.

Some microorganisms—for example, algae and photosynthetic bacteria—are lithotrophic photoautotrophs; that is, they can synthesize all of their metabolic and structural needs from inorganic nutrients. These organisms have every enzyme and cofactor they need to produce all their cellular components. Most organisms, however, require small amounts of certain organic chemicals that they cannot synthesize, in addition to those that provide carbon and energy. These organic chemicals are called **growth factors (Table 6.2).** For example, vitamins are growth factors for some microorganisms. Recall that vitamins constitute all or part of many coenzymes and cannot be synthesized by some organisms. (Note, however, that vitamins are not growth factors for microorganisms that can manufacture them, such as *E. coli*.) Growth factors for various microbes include certain amino acids, purines, pyrimidines, cholesterol, NADH, and heme.

CRITICAL THINKING

Given that *Haemophilus ducreyi* is a chemoheterotrophic pathogen that requires heme as a growth factor, deduce how this bacterium phosphorylates most of its ADP to form ATP. Defend your answer.

Physical Requirements

Learning Objective

✓ Explain how extremes of temperature, pH, and osmotic and hydrostatic pressure limit microbial growth.

In addition to chemical nutrients, organisms have physical requirements for growth, including specific conditions of temperature, pH, osmolarity, and pressure.

Temperature

Temperature plays an important role in microbial life through its effects on the three-dimensional configurations of biological molecules. Recall that to function properly, proteins require a specific three-dimensional shape that is determined in part by temperature-sensitive hydrogen bonds, which are more likely to form at lower temperatures, and more likely to break at higher temperatures. When the hydrogen bonds break, the proteins denature and lose function. Additionally, lipids, such as those that are components of the membranes of cells and organelles, are also temperature

Table 6.2 Some Growth Factors of Microorganisms and Their Functions

Growth Factor	Function
Amino acids	Components of proteins
Cholesterol	Used by mycoplasmas (bacteria) for cell membranes
Heme	Functional portion of cytochromes in electron transport system
NADH	Electron carrier
Niacin (nicotinic acid)	Precursor of NAD^+ and $NADP^+$
Pantothenic acid	Component of coenzyme A
Para-aminobenzoic acid	Precursor of folic acid, which is involved in metabolism of one-carbon compounds and nucleic acid synthesis
Purines, pyrimidines	Components of nucleic acids
Pyroxidine (vitamin B_6)	Utilized in transamination syntheses of amino acids
Riboflavin (vitamin B_2)	Precursor of FAD
Thiamine (vitamin B_1)	Utilized in some decarboxylation reactions

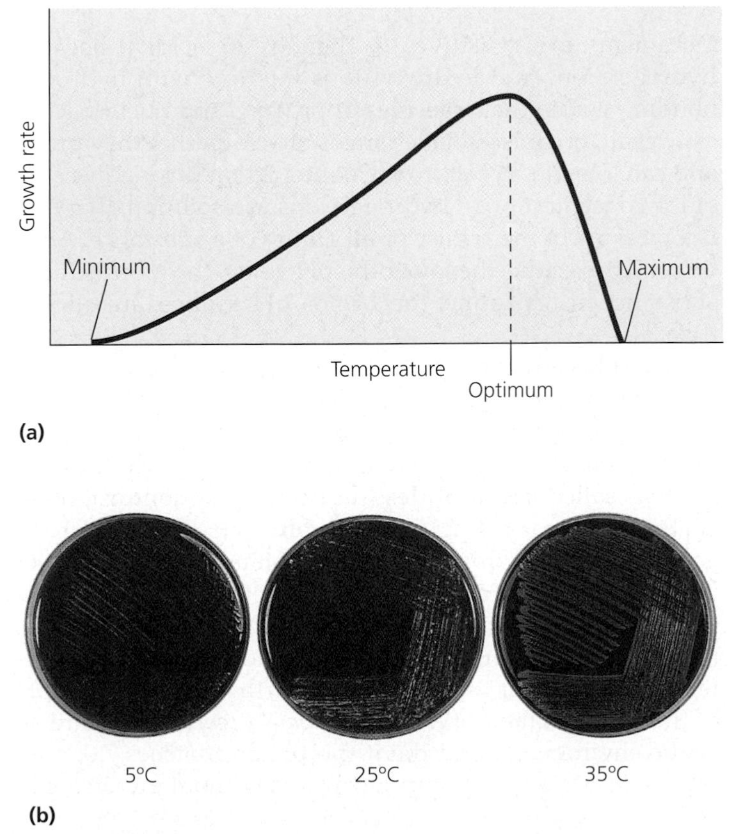

Figure 6.4

The effects of temperature on microbial growth. (a) Minimum, optimum, and maximum temperatures, as determined from a graph showing growth rate plotted against temperature. (b) Growth of *Staphylococcus aureus* on blood agar after 24 hours of incubation at three different temperatures. *If microorganisms can survive at temperatures lower than their minimum growth temperature, then why is it so-named?*

Figure 6.4 *The minimum growth temperature is defined as the lowest temperature that supports metabolism. Many organisms can survive at low temperatures without actively metabolizing.*

sensitive. If the temperature is too low, membranes become rigid and fragile; if the temperature is too high, the lipids become too fluid, and the membrane cannot contain the cell or organelle.

Thus, because temperature plays an important role in the three-dimensional structure of many types of biological molecules, different temperatures have different effects on the survival and growth of microbes **(Figure 6.4).** The lowest temperature at which an organism is able to conduct metabolism is called the *minimum growth temperature.* Note, however, that many microbes, particularly bacteria, survive temperatures far below this temperature, despite the fact that cell membranes are less fluid and transport processes are too slow to support metabolic activity. The highest temperature at which an organism continues to metabolize is called the *maximum growth temperature;* when the temperature exceeds this value, the organism's proteins are permanently

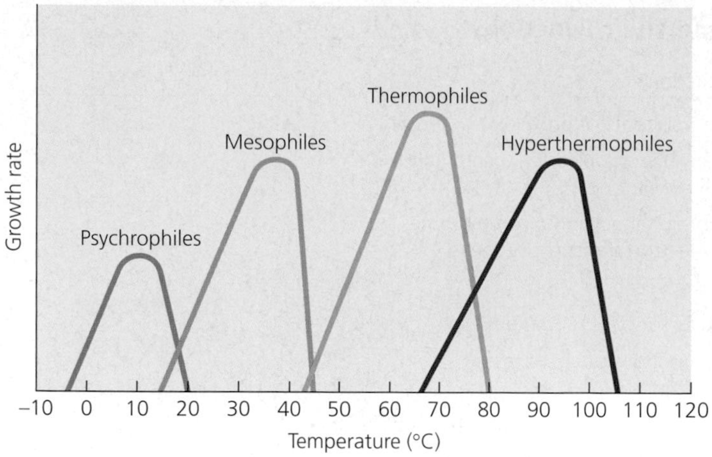

▲ ***Figure 6.5***
Four categories of microbes based on temperature ranges for growth. *Categorize the bacterium* Vibrio marinus, *which has an optimum growth temperature near 10°C.*

Figure 6.5 Vibrio marinus *is a psychrophile.*

denatured, and it dies. The temperature at which an organism's metabolic activities produce the highest growth rate is the **optimum growth temperature.** Each organism thus survives over a *temperature range,* within which its growth and metabolism are supported.

CRITICAL THINKING

Examine the graph in Figure 6.4. Note that the growth rate increases slowly until the optimum is reached, and then it declines steeply over higher temperatures. In other words, organisms tolerate a wider range of temperatures below their optimal temperature than they do above the optimum. Explain this observation.

Based on their preferred temperature ranges—the temperatures within which their metabolic activity and growth are best supported—microbes can be categorized into four overlapping groups **(Figure 6.5). Psychrophiles**[6] (sī'krō-fīls) grow best at temperatures below about 15°C and can even continue to grow at temperatures below 0°C. They die at temperatures much above 20°C. In nature, psychrophilic algae, fungi, and bacteria live in snowfields, ice, and cold water **(Figure 6.6).** They do not cause disease in humans because they cannot survive at body temperature; some do cause food spoilage in refrigerators. Psychrophiles present unique challenges to laboratory investigations, because they must be kept at cold temperatures. For example, microscope stages must be refrigerated, and the air temperatures needed to maintain living psychrophiles are uncomfortably cold for lab personnel.

Mesophiles[7] (mez'ō-fīls) are organisms that grow best in temperatures ranging from 20°C to about 40°C (see Figure 6.5), though they can survive at higher and lower temperatures. Because normal body temperature is approximately 37°C, human pathogens are mesophiles. *Thermoduric*[8] *organisms* are mesophiles that can survive brief periods at higher temperatures. Inadequate heating during pasteurization and canning can result in food spoilage by thermoduric mesophiles.

Thermophiles[9] (ther'mō-fīls) grow at temperatures above 45°C in habitats such as compost piles and hot springs. Some members of the Archaea, called **hyperthermophiles,** grow in water above 80°C; others live at temperatures above 100°C[10]. The current record-holder is *Pyrodictium,* a bacterium of submarine hot springs, which grows naturally in water at 113°C and can survive up to an hour in an autoclave at 121°C. Thermophiles and hyperthermophiles do not cause disease because they "freeze" at body temperature. **Highlight 6.1** discusses the biology of these interesting organisms.

CRITICAL THINKING

Over 100 years ago, doctors infected syphilis victims with malaria parasites to induce a high fever. Surprisingly, such treatment often cured the syphilis infection. Explain how this could occur.

pH

Organisms are sensitive to changes in acidity because hydrogen ions and hydroxyl ions interfere with hydrogen bonding within the molecules of proteins and nucleic acids; as a result, organisms have ranges of acidity that they prefer and can tolerate. Recall from Chapter 2 that pH is a measure of the concentration of hydrogen ions in a solution; that is, it is a measure of the acidity or alkalinity of a substance. A pH below 7.0 is acidic; the lower the pH value, the more acidic a substance is. Alkaline (or basic) pH values are higher than 7.0.

Most bacteria and protozoa, including most pathogens, grow best in a narrow range around a neutral pH—that is, between pH 6.5 and pH 7.5, which is also the pH range of most tissues and organs in the human body; such microbes are thus called **neutrophiles** (nū'trō-fīls). By contrast, other bacteria and many fungi are **acidophiles** (ă-sid'ō-phīls), organisms that grow best in acidic habitats. One example of acidophilic microbes are the chemoautotrophic bacteria that live in mines and in water that runs off from mine tailings (waste rock), habitats that have pHs as low as 0.0. These bacteria oxidize sulfur to sulfuric acid, further lowering the pH of their environment. Whereas *obligate acidophiles* require an acidic environment and die if the pH approaches 7.0, *acid-tolerant microbes* merely survive in acid without preferring it.

[6]From Greek *psuchros,* meaning cold, and *philos,* meaning love.
[7]From Greek *mesos,* meaning middle, and *philos,* meaning love.
[8]From Greek *therme,* meaning hot. The organisms are so-named because of their ability to endure or tolerate heat.
[9]From Greek *therme,* meaning hot, and *philos,* meaning love.
[10]Water can remain a liquid above 100°C if it has a high salt content or is under pressure, such as occurs in geysers or deep ocean troughs.

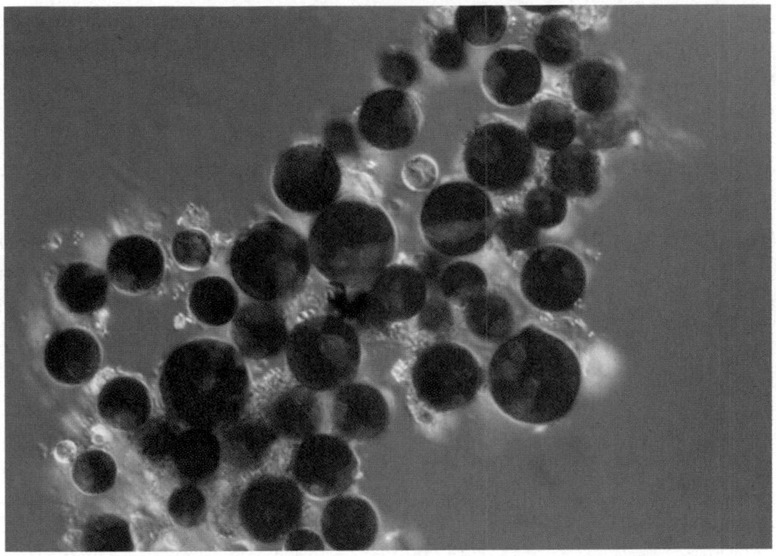

(a) (b)

Figure 6.6

An example of a psychrophile. **(a)** The alga *Chlamydomonas nivalis* colors this summertime snowbank in the Sierra Nevada mountains pink. **(b)** Microscopic view of the red-pigmented spores of *C. nivalis.*

Highlight 6.1 Some Like it Hot: Extreme Thermophiles

Extreme thermophiles, also called hyperthermophiles, live in extremely high-temperature environments, even those in excess of 100°C. Their heat-stable lipids, nucleic acids, and proteins allow them to survive—indeed thrive—at temperatures that would destroy the organic molecules of mesophiles. Hyperthermophilic organisms present unique challenges to our understanding of organic chemistry: Why don't their lipid membranes melt? How do their DNA molecules remain intact? What prevents their proteins from denaturing at such high temperatures?

The cells and chemicals of hyperthermophiles have special features that enable them to remain intact at high temperatures:

- Their cytoplasmic membranes do not contain fatty acids, which would melt at such high environmental temperatures. Instead, the membranes are monolayers (instead of bilayers) composed of hydrocarbon chains containing 40 carbon atoms bonded to glycerol phosphate. This accounts, at least in part, for their stability at high temperatures.
- The nucleic acids of extreme thermophiles appear to be stabilized by the presence of enzymes that fold DNA into unique heat-stable supercoils, by high concentrations of potassium ion, and by heat-stable proteins that bind to and stabilize DNA.
- The enzymes of thermophiles are also heat stable; they contain more hydrophobic amino acids than their counterparts in mesophiles, and form additional bonds between neighboring amino acids.

As we will see in Chapter 8, enzymes from hyperthermophiles play important roles in genetic engineering and industry. For example, DNA polymerases isolated from *Thermus aquaticus*, a thermophile found in hot springs at Yellowstone National Park, and from *Thermococcus litoralis*, a deep ocean bacterium, are used in the *polymerase chain reaction (PCR)* technique, a method of creating unlimited amounts of identical DNA strands. Additionally, enzymes from thermophiles and extreme thermophiles catalyze reactions in industrial processes that optimally operate between 50°C and 100°C.

Thermophilic bacteria, which can be distinguished by their orange-colored carotenoids, surround the Grand Prismatic Spring in Yellowstone National Park.

Many organisms produce acidic waste products that accumulate in their environment until eventually they inhibit further growth. For example, many cheeses are acidic due to lactic acid produced by fermenting bacteria and fungi. The low pH of these cheeses then acts as a preservative by preventing any further microbial growth. Other acidic foods, such as sauerkraut and dill pickles, are also kept from spoiling because most organisms cannot tolerate their low pH.

The normal acidity of certain regions of the body inhibits microbial growth and retards many kinds of infection. At one site, the vaginas of adult women, acidity results from the fermentation of carbohydrates by normal resident bacteria. If the growth of these normal residents is disrupted—for instance, by antibiotic therapy—the resulting higher pH may allow yeasts to grow and cause a yeast infection. Another site, the stomach, is inhospitable to most microbes because of the normal production of stomach acid. However, the acid-tolerant bacterium *Helicobacter pylori* neutralizes stomach acid by secreting bicarbonate and urease, an enzyme that converts urea to ammonia, which is alkaline. The growth of *Helicobacter* in certain portions of the digestive tract is the cause of most ulcers.

Alkaline conditions also inhibit the growth of most microbes, but **alkalinophiles** live in alkaline soils and water up to pH 11.5. For example, *Vibrio cholerae* (vib'rē-ō kol'er-ī), the causative agent of cholera, grows best outside of the body in water at pH 9.0.

Physical Effects of Water

Microorganisms require water; they must be in a moist environment if they are to be metabolically active. Water is not only needed to dissolve enzymes and nutrients but is also an important reactant in many metabolic reactions. Even though most cells die in the absence of water, some microorganisms—for example, the bacterium *Mycobacterium tuberculosis* (mī'kō-bak-tēr'ē-ŭm tū-ber-kyū-lō'sis)—have cell walls that retain water, allowing them to survive for months under dry conditions. Additionally, the spores and cysts of some other single-celled microbes can cease most metabolic activity for years; these cells are in essence in a state of suspended animation because they neither grow nor reproduce in their dry condition.

We now consider the physical effects of water on microbes by examining two topics: osmotic pressure and hydrostatic pressure.

Osmotic Pressure As we saw in Chapter 3, *osmosis* is the diffusion of water across a semipermeable membrane and is driven by unequal solute concentrations on the two sides of such a membrane. The *osmotic pressure* of a solution is the pressure exerted on a semipermeable membrane by a solution containing solutes (dissolved material) that cannot freely cross the membrane. Osmotic pressure is related to the concentration of dissolved molecules and ions in a solution. Solutions with greater concentrations of such solutes are *hypertonic* relative to those with a lower solute concentration, which are *hypotonic*.

Osmotic pressure can have dire effects on cells. For example, a cell placed in fresh water (a hypotonic solution relative to the cell's cytoplasm) gains water from its environment and swells to the limit of its cell wall. Cells that lack a cell wall—animal cells and some bacterial, fungal, and protozoan cells—will swell until they burst in hypotonic solutions. By contrast, a cell placed in seawater, which is a solution containing about 3.5% solutes and thus hypertonic to most cells, loses water into the surrounding saltwater. Such a cell can die from **plasmolysis** (plaz-mol'i-sis), or shriveling of its cytoplasm. Osmotic pressure accounts for the preserving action of salt in jerky and salted fish, and of sugar in jellies, preserves, and honey. In those foods, the salt and sugar are solutes that draw water out of any microbial cells that are present, preventing the growth and reproduction of the microbes.

Osmotic pressure restricts organisms to certain environments. Some microbes, called **obligate halophiles,**[11] are adapted to growth under high osmotic pressure such as exists in the Great Salt Lake and smaller salt ponds. They may grow in up to 30% salt and will burst if placed in freshwater. Other microbes are *facultative halophiles;* that is, though they do not require high salt concentrations, they can tolerate them. One potential bacterial pathogen, *Staphylococcus aureus* (staf'i-lō-kok'ŭs o'rē-ŭs), can tolerate up to 20% salt, which allows it to colonize the surface of the skin—an environment that is too salty for most microbes. *S. aureus* causes a number of different skin and mucous membrane diseases ranging from pimples, styes, and boils to life-threatening scalded skin and toxic shock syndromes.

Hydrostatic Pressure Water exerts pressure in proportion to its depth. For every additional 10 m of depth, water pressure increases 1 atmosphere (atm). Therefore, the pressure at 100 m below the surface is 10 atm—ten times greater than at the surface. Obviously, the pressure in deep ocean basins and trenches, which are thousands of meters below the surface, is tremendous. Organisms that live under such extreme pressure are called **barophiles**[12] (bar'ō-fīls). Their membranes and enzymes do not merely tolerate pressure, but are dependent on pressure to maintain their three-dimensional, functional shapes. Thus barophiles brought to the surface quickly die because their proteins denature. Obviously, barophiles cannot cause diseases in humans, plants, or animals that do not live at great depths.

---

[11] From Greek *halos,* meaning salt, and *philos,* meaning love.

[12] From Greek *baros,* meaning weight, and *philos,* meaning love.

New Frontiers 6.2 Biofilms, Quorum Sensing, and Disease Prevention

Given that many microorganisms become more harmful when they are part of a biofilm, scientists are seeking ways to prevent biofilms from forming in the first place. Researchers have learned that drugs that block cell receptors, thereby disrupting communication among bacterial cells, will inhibit biofilm formation and thus prevent disease. Such receptor-blocking drugs have successfully blocked biofilm formation and prevented disease in mice and are being considered for use in humans.

Another possible approach to preventing biofilm formation involves artificially amplifying quorum sensing while bacteria are still relatively few in number. Some pathogens hide within capsules and in blood clots, and only after they have multiplied significantly and formed a biofilm do they emerge and become "visible" to the immune system. With amplified quorum sensing, bacteria might produce certain proteins earlier in an infection cycle than normal, thus "revealing" themselves sooner to the immune system, which may then eliminate them before they can form biofilms and cause disease.

Ecological Associations

Learning Objective

✓ Describe how quorum sensing can lead to formation of a biofilm.

Organisms don't live in a purely chemical/physical (abiotic) environment—they live in association with other individuals of their own and different species. The relationships between organisms can be viewed as falling along a continuum stretching from causing harm to providing benefits.

Relationships in which one organism harms or even kills another organism are considered *antagonistic relationships.* As discussed in Chapter 1, viruses are especially clear examples of antagonistic microbes; they require a cell in which to replicate themselves and almost always kill their cellular hosts. Beneficial relationships take at least two forms: synergistic relationships and symbiotic relationships. In *synergistic relationships,* the individual members of an association cooperate such that each receives benefits that exceed those that would result if each lived by itself, even though each member could live separately. In *symbiotic relationships,* organisms live in such close nutritional or physical contact that they become interdependent, such that the members rarely (if ever) live outside the relationship. Symbiotic relationships are discussed in greater detail in Chapter 14, particularly as they relate to the production of disease.

Biofilms are examples of complex relationships among numerous individuals, which are often different species, that attach as a group to surfaces and display metabolic and structural traits different from those expressed by any of the microorganisms alone. Some scientists estimate that 65% of bacterial diseases are caused by biofilms.

Biofilms often form as a result of a process called **quorum sensing,** in which bacteria respond to the density of nearby bacteria. The bacteria secrete molecules into their environment that act as signals; the cells also possess receptors for these signal molecules. When bacterial densities are low, the concentration of signaling molecules remains low, and few receptors bind signal molecules. However, when the density of bacteria increases, the concentration of signal molecules also increases, such that more and more receptors become occupied. Once the binding of signal molecules exceeds a certain threshold amount, the expresion of previously suppressed genes is triggered, and the result is that the bacteria have new characteristics, such as the ability to form biofilms.

Dental plaque on teeth is an example of a biofilm. The ability of some strains of *Salmonella enterica* (sal'mŏ-nel'ă en-ter'i-kă), *Staphylococcus aureus,* and *Pseudomonas aeruginosa* (soo-dō-mō'nas ā-rū-ji-nō'să) to cause disease in humans is also linked to their ability to form biofilms. **New Frontiers 6.2** describes research into preventing disease by manipulating the formation of biofilms.

Culturing Microorganisms

As we saw in Chapter 1, the second of Koch's postulates for demonstrating that a certain agent causes a specific disease requires that microorganisms be isolated and cultivated. Medical laboratory personnel must also grow pathogens as a step in the diagnosis of many diseases. To cultivate or *culture* microorganisms, a sample called an **inoculum** (plural: *inocula*) is introduced into a collection of nutrients called a **medium.** Microorganisms that grow from an inoculum are also called a *culture;* thus, the word **culture** can refer to the act of cultivating microorganisms or to the microorganisms that are cultivated.

Cultures can be grown in liquid media called **broths,** or on the surface of solid media. Cultures that are visible on the surface of solid media are called colonies. Bacterial and fungal colonies often have distinctive characteristics—including color, size, shape, elevation, texture, and appearance of the colony's margin (edge)—that taken together aid in identifying the microbial species that formed the colony **(Figure 6.7).**

Microbiologists obtain inocula from a variety of sources. *Environmental specimens* are taken from such sources as ponds, streams, soil, and air; environmental sampling is discussed in

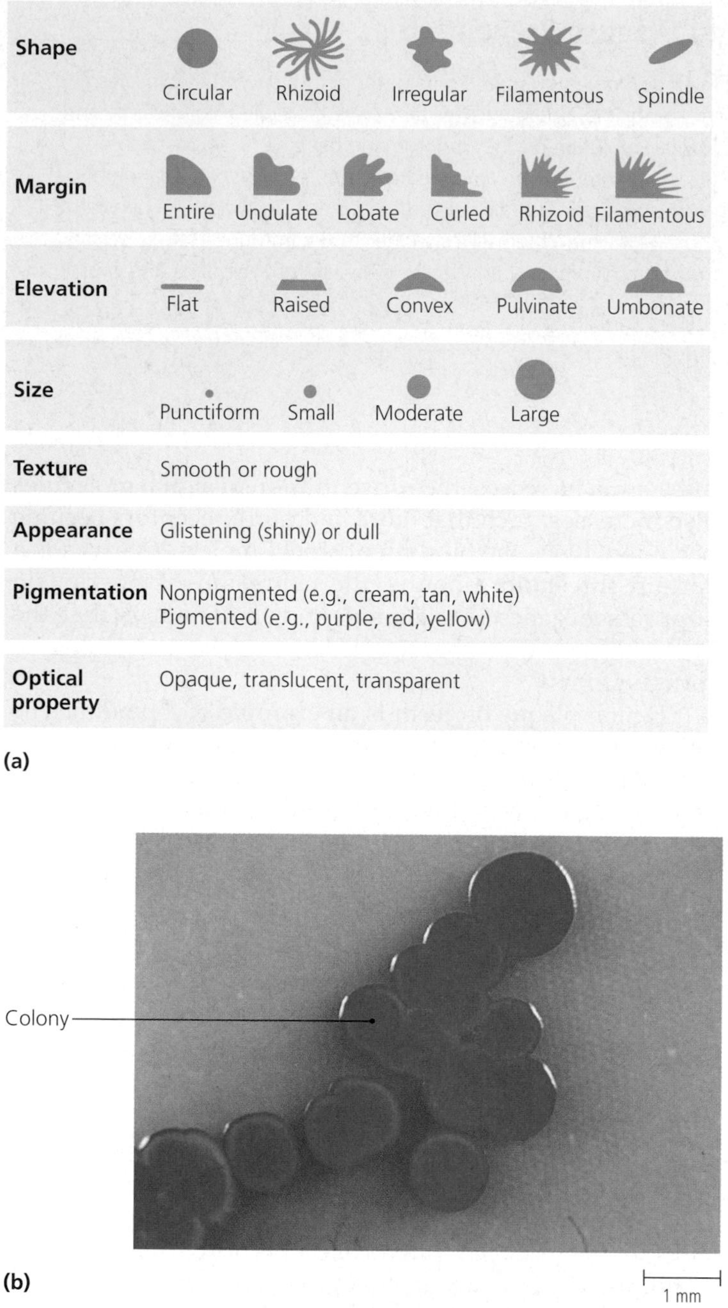

▲ ***Figure 6.7***

Characteristics of bacterial colonies. **(a)** Shape, margin, elevation (side view), size, texture, appearance, pigmentation (color), and optical properties are described by a variety of terms. **(b)** *Serratia marcescens* growing on an agar surface. These colonies are circular, entire, convex, small, smooth, shiny, red, and opaque.

some detail in Chapter 26. *Clinical specimens* are taken from patients and handled in ways that facilitate the examination of or testing for the presence of microorganisms. Finally, another source of inocula is the cultures originally grown from environmental or clinical specimens and maintained in storage in the laboratory. Next we briefly examine clinical sampling.

Clinical Sampling

Learning Objective

✓ Describe methods for collecting clinical specimens from the skin and from the respiratory, reproductive, and urinary tracts.

Diagnosis and treatment of disease often depend on the isolation and correct identification of pathogens. Physicians and other health care professionals must properly obtain samples from their patients, and must then transport them quickly and correctly to a microbiology laboratory for culture and identification. They must take care to prevent the contamination of samples with microorganisms from the environment or other regions of the patient's body, and they must prevent infecting themselves with pathogens while sampling. In this regard, the Centers for Disease Control and Prevention (CDC) has established a set of guidelines, called *universal precautions,* to protect health care professionals from contamination by pathogens.

In clinical microbiology, a **clinical specimen** is a sample of human material, such as feces, saliva, cerebrospinal fluid, or blood, that is examined or tested for the presence of microorganisms. As summarized in **Table 6.3,** health care professionals collect clinical specimens using a variety of techniques and equipment. Specimens must be properly labeled and promptly transported to a microbiological laboratory to both avoid death of the pathogens and minimize the growth of normal organisms. Clinical specimens are often transported in special *transport media* that are chemically formulated to maintain the relative abundance of different microbial species or to maintain an anaerobic environment.

Obtaining Pure Cultures

Learning Objective

✓ Describe the two most common methods by which microorganisms can be isolated for culture.

Clinical specimens are collected in order to identify a suspected pathogen, but they also contain *normal microbiota,* which are microorganisms associated with a certain area of the body (see Table 14.2) without causing diseases. As a result, the suspected pathogen in a specimen must be isolated from the normal microbiota in culture. Scientists use several techniques to isolate organisms in **pure cultures,** that is, cultures composed of cells arising from a single progenitor. The word *axenic*[13] (ā-zen′ik) is also used to refer to a pure culture. The progenitor from which a particular pure culture is derived may be either a single cell or a group of related cells; therefore, the progenitor is termed a **colony-forming unit (CFU).**

[13]From Greek *a,* meaning no, and *xenos,* meaning stranger.

Table 6.3 Clinical Specimens and the Methods Used to Collect Them

Type or Location of Specimen	Collection Method
Skin, accessible mucous membrane (including eye, outer ear, nose, throat, vagina, cervix, urethra) or open wounds	Sterile swab brushed across the surface; care should be taken not to contact neighboring tissues
Blood	Needle aspiration from vein, anticoagulants are included in the specimen transfer tube
Cerebrospinal fluid	Needle aspiration from subarachnoid space of spinal column
Stomach	Intubation, which involves inserting a tube into the stomach, often via a nostril
Urine	In aseptic collection, a catheter is inserted into the bladder through the urethra; in the "clean catch" method, initial urination washes the urethra, and the specimen is midstream urine
Lungs	Collection of sputum either dislodged by coughing or acquired via a catheter
Diseased tissue	Surgical removal (biopsy)

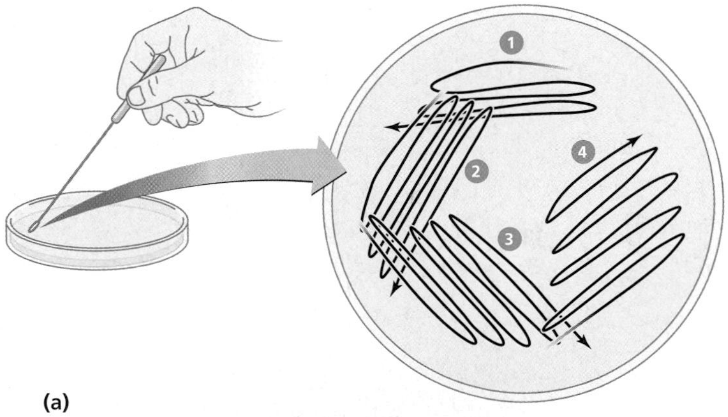

(a)

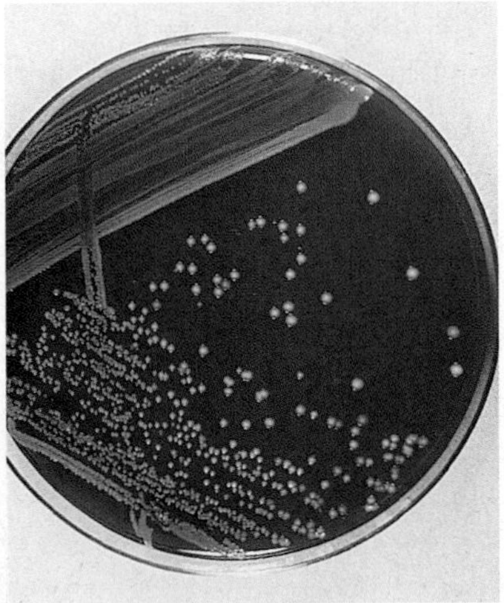
(b)

Figure 6.8 The streak plate method of isolation. **(a)** An inoculum is spread across the surface of an agar plate in a sequential pattern of streaks (as indicated by the numbers and arrows). The loop is sterilized between streaks. In streaks 2, 3, and 4, bacteria are picked up from the previous streak, diluting the number of cells each time. **(b)** A streak plate showing colonies of *Neisseria gonorrhoea*.

Of course, in all microbiological procedures care must be taken to reduce the chance of contamination as occurs when instruments or air currents carry foreign microbes into culture vessels. All media, vessels, and instruments must be **sterile**—that is, free of any microbial contaminants. Sterilization and *aseptic techniques,* which are designed to limit contamination, are discussed in Chapter 9. Now we examine two common isolation techniques: streak plates and pour plates.

Streak Plates

The most commonly used isolation technique in microbiological laboratories is the **streak plate** method. In this technique, a sterile inoculating loop (or sometimes a needle) is used to spread an inoculum across the surface of a solid medium in *Petri dishes,* which are clear, flat culture dishes with loose-fitting lids. The loop is used to lightly streak a set pattern that gradually dilutes the sample to a point that CFUs are isolated from one another **(Figure 6.8a).** After an appropriate period of time called **incubation,** colonies develop from each isolate **(Figure 6.8b).** The various types of organisms present are distinguished from one another by differences in colonial characteristics (see Figure 6.9b). Samples from each variety can then be inoculated in new media to establish axenic cultures.

CRITICAL THINKING

Using the terms in Figure 6.7a, describe the shape, margin, pigmentation, and optical properties of bacterial colonies 3 and 5 seen in Figure 6.9b.

Pour Plates

In the **pour plate** technique, CFUs are separated from one another using a series of dilutions. There are various ways to perform pour plate isolations. In one method, an initial 1-milliliter sample is mixed into 9.0 ml of medium in a test tube. After mixing, a new sample from this medium is then used to inoculate a second tube of liquid medium. The process is repeated to establish a series of dilutions **(Figure 6.9a).** Samples from the more diluted media are mixed in Petri dishes with sterile, warm medium containing *agar*—a gelling agent derived from the cell walls of red algae. After the agar cools and solidifies, and the Petri plates are incubated, isolated colonies—colonies that are separate and distinct from all others—form in the dishes from CFUs that have been separated via the dilution series **(Figure 6.9b).** One difference between this method and the streak plate technique is that colonies form both at and below the surface of the medium. As before, pure cultures can be established from distinct colonies.

Isolation techniques work well only if a relatively large number of CFUs of the organism of interest are present in the initial sample, and if the medium used supports the growth of that microbe. As discussed later, special media and enriching techniques can be used to increase the likelihood of success.

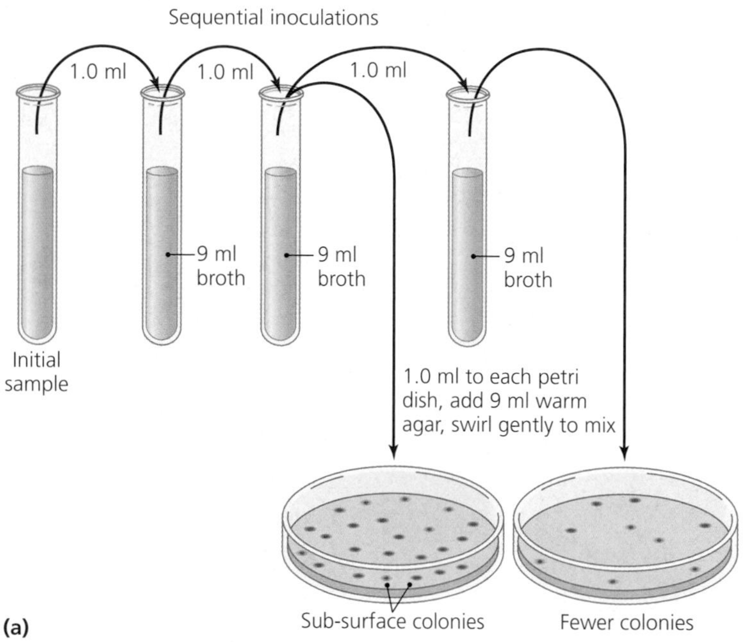

(a)

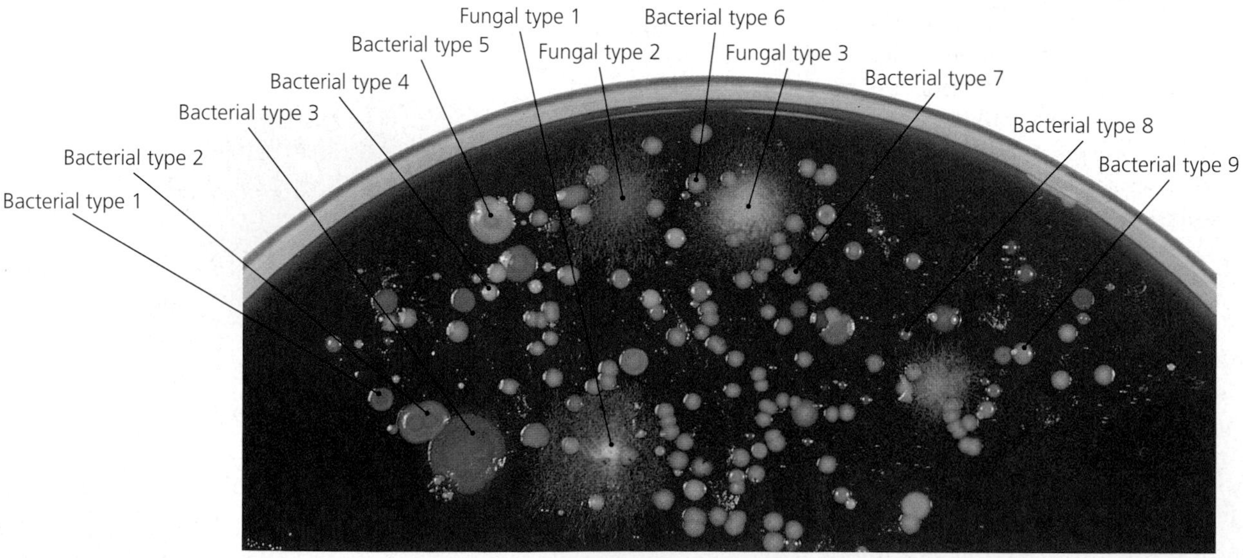

(b)

Figure 6.9

The pour plate method of isolation. **(a)** After an initial sample is diluted through a series of transfers, the final dilutions are mixed with warm agar in Petri plates. Individual CFUs form colonies in and on the agar. **(b)** A portion of a plate showing the results of isolation.

Other Isolation Techniques

Streak plates and pour plates are used primarily to establish pure cultures of bacteria, but they can also be used for some fungi, particularly yeasts. Protozoa and motile unicellular algae are not usually cultured on solid media because they do not remain in one location to form colonies. Instead, they are isolated through a series of dilutions, but remain in broth culture media. In cases of fairly large microorganisms such as the protists *Euglena* and *Amoeba,* hollow tubes with small diameters called *micropipettes* can be used to pick up a single cell, which is then used to establish a culture.

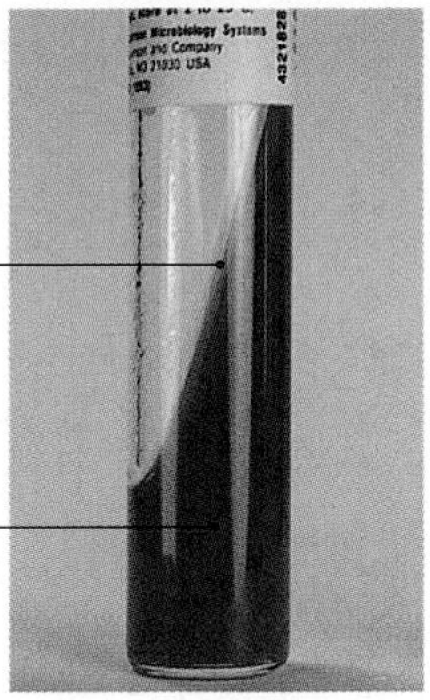

Figure 6.10
Slant tube containing solid media (in this case, acetamide agar).

Culture Media

Learning Objective

✓ Describe six types of general culture media available for bacterial culture.

Culturing microorganisms can be an exacting science. Although some microbes, such as *E. coli,* are not particular about their nutritional needs and can be grown in a variety of media, others, such as *Neisseria gonorrheae* (nī-sē′rē-ă go-nor-re′-ē′) and *Haemophilus influenzae* (hē-mof′i-lŭs in-flū-en′zē), require specific nutrients, including specific growth factors. The majority of prokaryotes have never been successfully grown in any culture medium, in part because scientists have concentrated their efforts on culturing pathogens and commercially important species. However, some pathogens, such as the syphilis bacterium *Treponema pallidum* (trep-ō-nē′mă pal′li-dŭm), have never been cultured in any laboratory medium, despite over a century of effort.

A variety of media are available for microbiological cultures, and more are developed each year to support the needs of food, water, industrial, and clinical microbiologists. Most media are available from commercial sources and come in powdered forms that require only the addition of water to make broths. A common medium, for example, is *nutrient broth,* which contains powdered beef extract and peptones (short chains of amino acids produced by enzymatic digestion of protein) dissolved in water. For some purposes broths are adequate, but if solid media are needed, dissolving about 1.5% agar into warm broth, pouring the liquid mixture into an appropriate vessel, and allowing it to cool provides a solid surface to support colonial growth. Media made solid by the addition of agar to a broth have the word *agar* in their names; thus *nutrient agar* is nutrient broth to which 1.5% agar has been added.

Agar, a complex polysaccharide, is a useful compound in microbiology for several reasons:

- Most microbes cannot digest agar; therefore, agar media remain solid even when bacteria and fungi are growing on them.
- Powdered agar dissolves in water at 100°C, a temperature at which most nutrients remain undamaged.
- Agar solidifies at temperatures below 40°C, so temperature-sensitive, sterile nutrients such as vitamins and blood can be added without detriment, and liquid agar can be poured over most bacterial cells without harming them. The latter technique plays a role in the pour plate isolation technique.
- Solid agar does not melt below 100°C; thus, it can be used to culture many thermophiles.

As we have seen, still-warm liquid agar media can be poured into Petri dishes, which once the agar solidifies are then called **Petri plates.** When warm agar media are poured into test tubes that are then placed at an angle and left to cool until the agar solidifies, the result is **slant tubes,** or **slants (Figure 6.10).** The slanted surface provides a larger surface area for aerobic microbial growth while the butt of the tube remains almost anaerobic. If the tubes are kept vertical until the agar solidifies, they are called *deeps.*

Next we examine six types of general culture media: defined media, complex media, selective media, differential media, anaerobic media, and transport media. It is important to note that these types of media are not mutually exclusive categories; that is, in some cases a given medium can be classified into more than one category.

Defined Media

If microorganisms are to grow and multiply in culture, the medium must provide essential nutrients (including an appropriate energy source for chemotrophs), water, an appropriate oxygen level, and the required physical conditions (such as the correct pH and suitable osmotic pressure and temperature). A **defined medium** (also called a **synthetic medium**) is one in which the exact chemical composition is known (**Table 6.4** on page 182). Relatively simple defined media containing inorganic salts and a source of CO_2 (such as sodium bicarbonate) are available for chemolithotrophs and phototrophs, particularly cyanobacteria and algae. Chemoheterotrophs also require organic molecules such as glucose, amino acids, and vitamins, which supply carbon and energy or are vital growth factors.

Organisms that require a relatively large number of growth factors are termed *fastidious.* Such organisms may be used as living assays for the presence of growth factors. For example, a scientist needing to know if a sample contains a certain vitamin could inoculate the sample with a fastidious

Table 6.4 Ingredients of a Representative Defined (Synthetic) Medium for Culturing *E. coli*

Glucose	1.0 g
Na_2HPO_4	16.4 g
KH_2PO_4	1.5 g
$(NH_4)_2PO_4$	2.0 g
$MgSO_4 \cdot 7H_2O$	0.2 g
$CaCl_2$	0.01 g
$FeSO_4 \cdot 7H_2O$	0.005 g
Distilled or deionized water	Enough to bring volume to 1 L

organism that requires the vitamin. If the microbe grows, the vitamin is present in the sample. The amount of growth provides an estimate of the amount of vitamin present, scant growth indicating a small amount.

Complex Media

For most clinical cultures, defined media are unnecessarily troublesome to prepare. Most chemoheterotrophs, including pathogens, are routinely grown on **complex media** that contain nutrients released by the partial digestion of yeast, beef, soy, or proteins such as casein from milk. The exact chemical composition of a complex medium is unknown because partial digestion releases many different chemicals in a variety of concentrations.

Complex media have advantages over defined media. Because a complex medium contains a variety of nutrients, including growth factors, it can support a wider variety of different microorganisms. Complex media are also used to culture organisms whose exact nutritional needs are unknown. Nutrient broth, Trypticase™ soy agar, and MacConkey agar are some common complex media. Blood is often added to complex media to provide additional growth factors, such as NADH and heme. Such a fortified medium is said to be *enriched* and can support the growth of many fastidious microorganisms.

Selective Media

Selective media typically contain substances that either favor the growth of particular microorganisms or inhibit the growth of unwanted ones. Eosin, methylene blue, and crystal violet dyes as well as bile salts are included in media to inhibit the growth of Gram-positive bacteria without affecting Gram-negatives. A high concentration of NaCl (table salt) in a medium selects for halophiles and for salt-tolerant bacteria such as the pathogen *S. aureus.* Sabouraud dextrose agar has a slightly low pH, which by inhibiting the growth of bacteria is selective for fungi **(Figure 6.11).**

A medium can also become a selective medium when a single crucial nutrient is left out of it. For example, leaving glucose out of Trypticase™ soy agar makes the resulting medium selective for organisms that can meet all their carbon requirements by catabolizing amino acids.

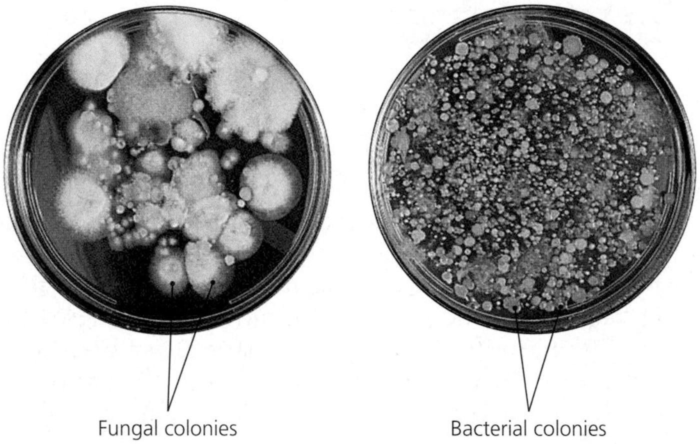

▲ ***Figure 6.11***
An example of the use of a selective medium. After inoculation with a diluted soil sample, the presence of a slightly acidic pH in Sabouraud dextrose agar (left) makes it selective for fungi by inhibiting the growth of bacteria. At right for comparison is a nutrient agar plate inoculated with an identical sample.

CRITICAL THINKING

Examine the ingredients of MacConkey agar as listed in Table 6.5 on page 184. Does this medium select for Gram-positive or Gram-negative bacteria? Explain your reasoning.
The sole carbon source in citrate medium is citric acid (citrate). Why might a laboratory microbiologist use this medium?

Differential Media

Differential media are formulated such that either the presence of visible changes in the medium or differences in the appearances of colonies helps microbiologists differentiate among different kinds of bacteria growing on the medium. Such media take advantage of the fact that different bacteria utilize the ingredients of any given medium in different ways. One example of the use of a differential medium involves the differences in organisms' utilization of the red blood cells in blood agar **(Figure 6.12).** *Streptococcus pneumoniae* (strep-tō-kok'ŭs nu-mō'nē-ī) partially

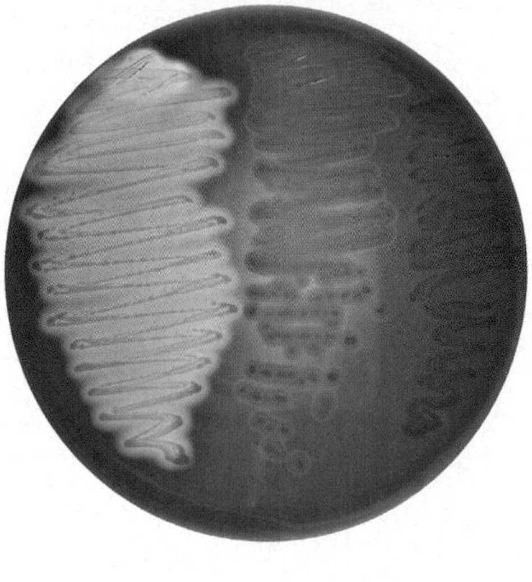

▲ *Figure 6.12*

The use of blood agar as a differential medium. *Streptococcus pyogenes* (left) completely uses red blood cells, producing a clear zone termed beta-hemolysis. *Streptococcus pneumoniae* (middle) partially uses red blood cells, producing a discoloration termed alpha-hemolysis. *Enterococcus faecalis* (right) does not use red blood cells; the lack of any change in the medium around colonies is termed gamma-hemolysis even though no red cells are hemolyzed.

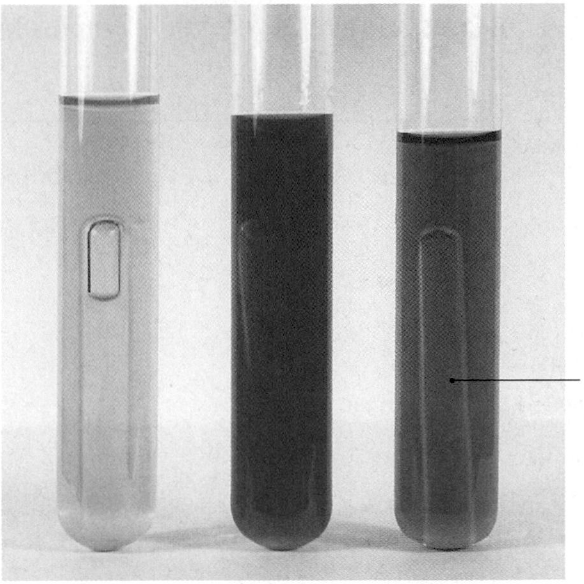

▲ *Figure 6.13*

The use of carbohydrate utilization tubes as differential media. Each tube contains a single kind of simple carbohydrate (a sugar) as a carbon source, and the dye phenol red as a pH indicator. *Escherichia coli* fermented the sugar, and the acid produced lowered the pH enough to cause the phenol red to turn yellow. This bacterium also produced gas, which is visible as a small bubble in the Durham tube. *Alcaligenes faecalis* did not ferment this carbohydrate; because no acid was produced, the color of the medium did not change.

digests (lyses) red blood cells, producing around its colonies a greenish-brown discoloration denoted *alpha-hemolysis.* By contrast, *Streptococcus pyogenes* (pī-oj′en-ēz) completely digests red blood cells, producing around its colonies clear zones termed *beta-hemolysis. Enterococcus faecalis* (en′ter-ō-kok′ŭs fē-kă′lis) does not digest red blood cells, so the agar appears unchanged, a reaction called *gamma-hemolysis* even though no lysis occurs. In some differential media, such as carbohydrate utilization broth tubes, a pH-sensitive dye changes color when bacteria metabolizing sugars produce acid waste products **(Figure 6.13).** Some common differential complex media are described in **Table 6.5** on page 184.

Many media are both selective and differential; that is, they enhance the growth of certain species that can then be distinguished from other species by variations in their effect on the medium or by the color of colonies they produce. For example, bile salts and crystal violet in MacConkey agar both inhibit the growth of Gram-positive bacteria and differentiate between lactose-fermenting and non-lactose-fermenting Gram-negative bacteria **(Figure 6.14).**

CRITICAL THINKING

Using as many of the following terms that apply—selective, differential, broth, solid, defined, complex—categorize each of the media listed in Tables 6.4 and 6.5.

Anaerobic Media

Obligate anaerobes require special culture conditions in that their cells must be protected from free oxygen. Anaerobes can be introduced with a straight inoculating wire into the anoxic (oxygen-free) depths of solid media to form a *stab culture,* but special media called **reducing media** provide better anaerobic culturing conditions. These media contain compounds, such as sodium thioglycollate, that chemically combine with free oxygen and remove it from the medium. Heat is used to drive the absorbed oxygen from the thioglycollate immediately before such a medium is inoculated.

The use of Petri plates presents special problems for the culture of anaerobes because each dish has a loose-fitting lid that allows the entry of air. For the culture of anaerobes, inoculated Petri plates are placed in sealable containers containing reducing chemicals **(Figure 6.15a).** Of course, the air-tight lids of anaerobic culture vessels must be sealed so that oxygen cannot enter. Only anaerobes that can tolerate exposure to oxygen can be cultured by this method because inoculation and transfer occur outside of the anaerobic environment. Laboratories that routinely study strict anaerobes have larger anaerobic glove boxes, which are transparent

Table 6.5 Representative Differential Complex Media

Medium and Ingredients		Use	Interpretation of Results
MacConkey Medium		For the culture and differentiation of enteric bacteria based on their ability to ferment lactose	Lactose fermenters produce red to pink colonies; nonfermenters form colorless or transparent colonies
Peptone	20.0 g		
Agar	12.0 g		
Lactose	10.0 g		
Bile salts	5.0 g		
NaCl	5.0 g		
Neutral red	0.075 g		
Crystal violet	0.001 g		
Water to bring volume to 1 L			
Eosin Methylene Blue (EMB) Agar		For the isolation, culture, and differentiation of Gram-negative enteric bacteria	Lactose-fermenting bacteria may appear with a green metallic sheen; nonfermenters appear as colorless or light purple colonies
Agar	13.5 g		
Pancreatic digest of casein	10.0 g		
Lactose	5.0 g		
Sucrose	5.0 g		
K_2HPO_4	2.0 g		
Eosin Y	0.4 g		
Methylene blue	0.065 g		
Water to bring volume to 1 L			
Triple Sugar Iron Agar		For the differentiation of enteric Gram-negative bacteria based on their fermentation of glucose, sucrose, and lactose and the production of H_2S gas	Red slant/red butt: no fermentation; yellow slant/red butt: glucose fermentation only; yellow slant/yellow butt: glucose and lactose and/or sucrose fermentation; if the butt turns black, H_2S is produced
Peptone	20.0 g		
Agar	12.0 g		
Lactose	10.0 g		
Sucrose	10.0 g		
NaCl	5.0 g		
Beef extract	3.0 g		
Yeast extract	3.0 g		
Glucose	1.0 g		
Ferric citrate	0.3 g		
$Na_2S_2O_3$	0.3 g		
Phenol red	0.025 g		
Water to bring volume to 1 L			
Blood Agar		For culture of fastidious microorganisms and differentiation of hemolytic microorganisms	Partial digestion of blood: alpha-hemolysis; complete digestion of blood: beta-hemolysis; no digestion of blood: gamma-hemolysis
Agar	15.0 g		
Pancreatic digest of casein	15.0 g		
Papaic digest of soybean meal	5.0 g		
NaCl	5.0 g		
Sterile blood	50.0 ml		
Water to bring volume to 950.0 ml (Blood is added to medium after autoclaving and cooling.)			

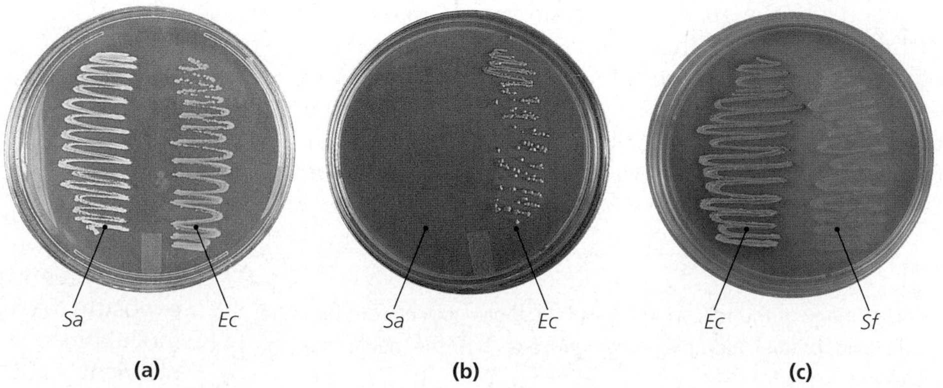

Figure 6.14 ➤ Use of MacConkey agar as a selective and differential medium. **(a)** Whereas both the Gram-positive *Staphylococcus aureus* (*Sa*) and the Gram-negative *Escherichia coli* (*Ec*) grow on nutrient agar, MacConkey agar **(b)** selects for Gram-negative bacteria and inhibits Gram-positive bacteria. **(c)** MacConkey agar also differentiates between Gram-negative bacteria on their ability to ferment lactose. The colonies of the lactose-fermenting *E. coli* are easily distinguished from those of the non-lactose-fermenting *Shigella flexneri* (*Sf*).

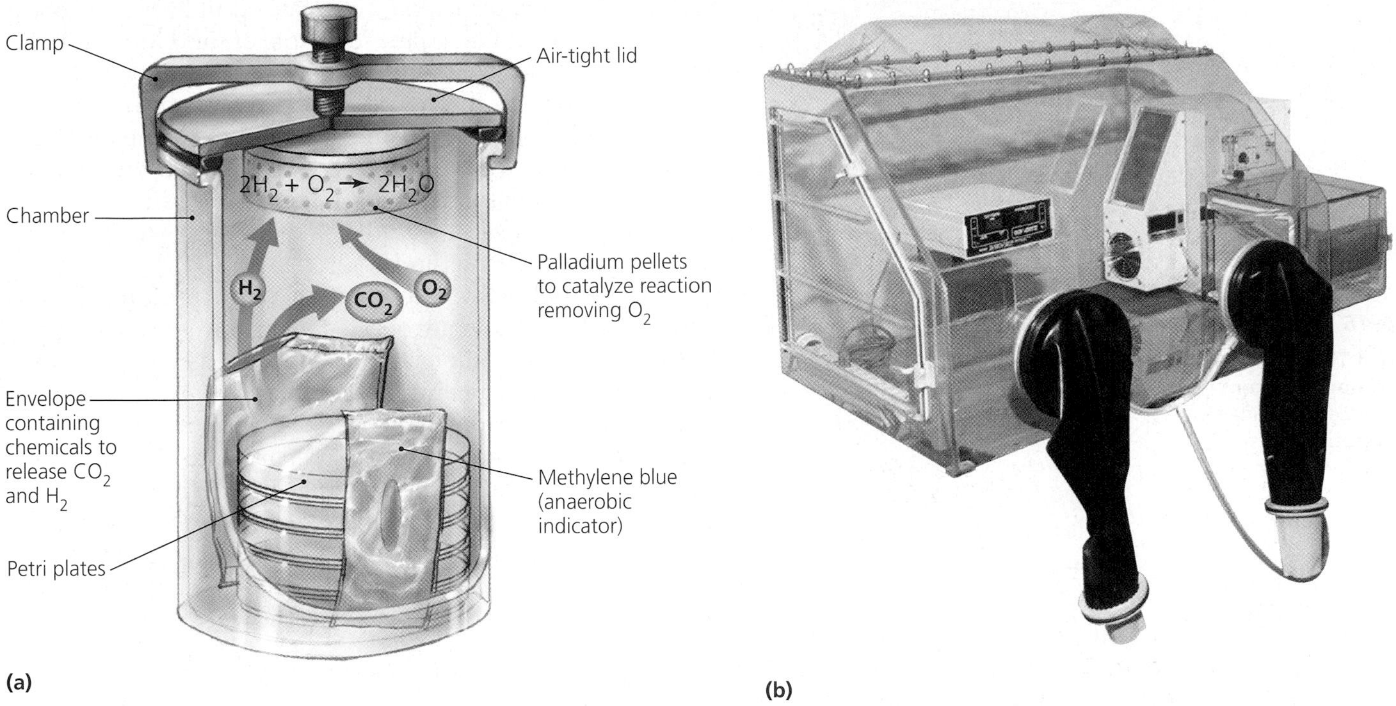

▲ ***Figure 6.15***

Anaerobic culture equipment. **(a)** The Gas-Pak anaerobic system utilizes chemicals to create an anaerobic environment inside a sealable, airtight jar. Methylene blue, which turns colorless in the absence of oxygen, indicates when the environment within the jar is anaerobic. **(b)** An anaerobic glove box, which provides an environment for the transfer and growth of strict anaerobes.

airtight chambers with special airtight rubber gloves, chemicals that remove oxygen, and air locks **(Figure 6.15b).** These chambers allow scientists to manipulate equipment and anaerobic cultures in an oxygen-free environment.

Transport Media

Hospital personnel use special **transport media** to carry clinical specimens of feces, urine, saliva, sputum, blood, and other bodily fluids in such a way as to ensure that people are not infected and that the specimens are not contaminated. Speed in transporting clinical specimens to the laboratory is extremely important because pathogens often do not survive outside the body as long as normal microbiota. Stool and other specimens are transported in buffered media designed to maintain the ratios among different microorganisms. Anaerobic clinical specimens can be transported for less than an hour in syringes from which the needles have been removed, but longer transport times require that the specimens be injected into anaerobic transport media.

Special Culture Techniques

Learning Objective

✓ Discuss the use of special culture methods including animal and cell culture, low-oxygen culture, and enrichment culture.

Not all organisms can be grown under the culture conditions we have discussed. Scientists have developed other techniques to culture many of these organisms.

Animal and Cell Culture

Microbiologists have developed animal and cell culture techniques for growing microbes for which artificial media are inadequate. The causative agents of leprosy and syphilis, for example, must be grown in animals because all attempts to grow them using standard culture techniques have been unsuccessful. *Mycobacterium leprae* (mī′kō-bak-tēr′ē-ŭm lep′-rī) is cultured in armadillos, whose internal conditions (including a relatively low body temperature) provide the conditions this microbe prefers. Rabbits meet the culture needs for *Treponema pallidum,* the bacterium that causes syphilis. Because viruses and small bacteria called rickettsias and chlamydias are obligate intracellular parasites—that is, they grow and reproduce only within living cells—bird eggs and cultures of living cells are used to culture these organisms.

Low-Oxygen Culture

As we have discussed, many types of organisms prefer oxygen conditions that are intermediate between strictly aerobic and anaerobic environments. *Carbon dioxide incubators,* machines that electronically monitor and control CO_2 levels,

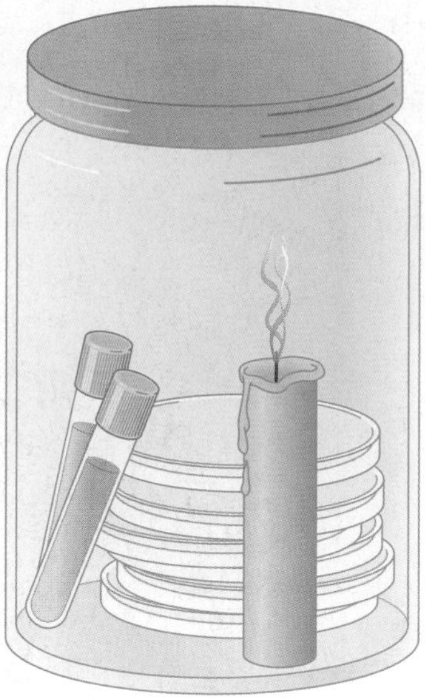

Figure 6.16 ➤

A candle jar. The burning candle consumes oxygen and releases carbon dioxide until it extinguishes itself; the resulting environment is suitable for growing microaerophiles and capnophiles.

provide atmospheres that mimic the environments of the intestinal tract, the respiratory tract, and other body tissues and thus are useful for culturing these kinds of organisms. Smaller and much less expensive alternatives to CO_2 incubators are *candle jars* **(Figure 6.16).** In these simple but effective devices, culture plates are sealed in a jar along with a lit candle; the flame consumes much of the O_2, replacing it with CO_2. The candle eventually extinguishes itself, creating an environment that is ideal for aerotolerant anaerobes, microaerophiles, and **capnophiles** which are organisms such as *Neisseria gonorrhoeae* that grow best with a relatively high concentration of carbon dioxide in addition to low oxygen levels. Remaining oxygen in the jar prevents the growth of strict anaerobes. The use of packets of chemicals that remove oxygen from the jar has largely replaced candles in modern microbiology labs.

Enrichment Culture

Bacteria that are present in small numbers may be overlooked on a streak plate or overwhelmed by faster-growing, more abundant strains. This is especially true of organisms in soil and fecal samples that contain a wide variety of microbial species. To isolate potentially important microbes that might otherwise be overlooked, microbiologists enhance the growth of less abundant organisms by a variety of techniques.

In the late 1800s, the Dutch microbiologist Martinus Beijerinck (1851–1931) introduced the most common of these methods, called simply **enrichment culture.** Enrichment cultures use a selective medium and are designed to increase very small numbers of a chosen microbe to observable levels. For example, suppose a microbiologist specializing in environmental clean up wanted to isolate an organism capable of digesting crude oil to have on hand should it be required to clean an oil-soaked beach. Even though a sample of beach sand might contain a few such organisms, it would also likely contain many millions of unwanted common bacteria. To isolate oil-utilizing microbes, the scientist would inoculate a sample of sand into a tube of selective medium containing oil as the sole carbon source, and then incubate it. Then a small amount of the culture would be transferred into a new tube of the same medium, to be incubated again. After a series of such enrichment transfers, any remaining bacteria will be oil-utilizing organisms. Different species could be isolated by either streak-plate or pour-plate methods.

Cold enrichment is another technique used to enrich a culture with cold-tolerant species, such as *Vibrio cholerae,* the bacteria that cause cholera. Stool specimens or water samples suspected of containing the bacterium are incubated in a refrigerator instead of at 37°C. Cold enrichment works because *Vibrio* cells are much less sensitive to cold than are more common fecal bacteria such as *E. coli;* therefore, *Vibrio* continues to grow in the cold while the other species are inhibited. The result of cold enrichment is a culture with a greater percentage of *Vibrio* cells than the original sample; the *Vibrio* cells can then be isolated by other methods.

CRITICAL THINKING

Beijerinck used the concept of enrichment culture to isolate aerobic and anaerobic nitrogen-fixing bacteria, sulfate-reducing bacteria, and sulfur-oxidizing bacteria. What kind of selective media could he have used for isolating each of these four types of microbes?

Preserving Cultures

Learning Objective

✓ Contrast refrigeration, deep freezing, and lyophilization as methods for preserving cultures of microbes.

Because cells metabolize, storing living cells necessitates slowing their metabolism to prevent the excessive accumulation of waste products and the exhaustion of all nutrients in a medium. **Refrigeration** is often the best technique for storing bacterial cultures for short periods of time.

Deep-freezing and lyophilization are used for long-term storage of bacterial cultures. **Deep-freezing** involves freezing the cells at temperatures from −50°C to −95°C. Deep frozen cultures can be restored years later by thawing them and placing a sample in an appropriate medium.

Lyophilization (lī-of′i-li-zā′shŭn) (freeze drying) involves removing water from a frozen culture using an intense vacuum. Under these conditions, ice directly becomes a gas (sublimates) and is removed from cells without permanently damaging cellular structures and chemicals. Lyophilized cultures can last for decades and are revived by adding lyophilized cells to liquid culture media.

Growth of Microbial Populations

Most unicellular microorganisms reproduce by *binary fission,* a process in which a cell grows to twice its normal size and

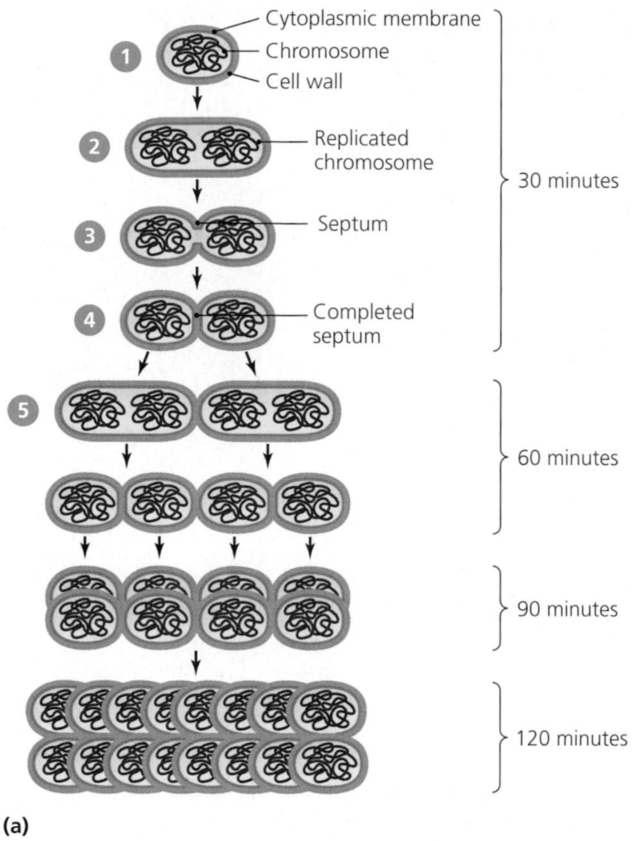

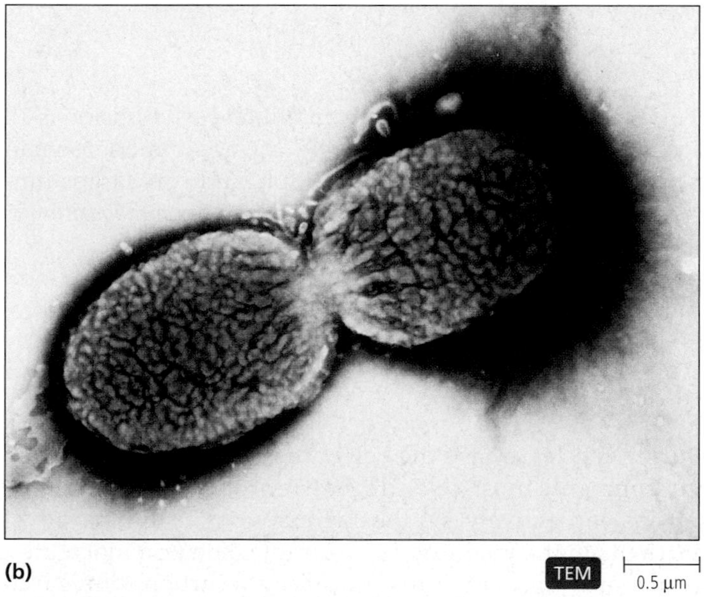

(b)

▲ ***Figure 6.17***

Binary fission. **(a)** The events in binary fission. ① The cell replicates its chromosome; the duplicated DNA molecules are attached to the cytoplasmic membrane. ② The cell elongates, and growth between attachment sites pushes the chromosomes apart. ③ The cell forms a new cell wall (a septum) across the midline. ④ The septum is completed; daughter cells may remain attached (as shown here) or may completely separate. ⑤ The process repeats. All the cells may divide in parallel planes and remain attached to form a chain, or they may divide in different planes to form a cluster (as shown here). **(b)** Transmission electron micrograph of a bacterial cell undergoing binary fission.

divides in half to produce two equally sized daughter cells. Binary fission generally involves four steps as illustrated in **Figure 6.17** for a prokaryotic cell.

1. The cell replicates its chromosome. The duplicated chromosomes are attached to the cytoplasmic membrane. (In eukaryotic cells chromosomes are attached to microtubules.)
2. The cell elongates and growth between attachment sites pushes the chromosomes apart. (Eukaryotic cells segregate their chromosomes by *mitosis,* a process described in Chapter 12.)
3. The cell forms a new cell membrane and wall (septum) across the midline.
4. When the septum is completed, the daughter cells may remain attached as shown in the figure, or they may completely separate. When the cells remain attached, further binary fissions in parallel planes produces a chain. When further divisions are in different planes, the cells become a cluster (as shown in the figure).
5. The process repeats.

Other reproductive strategies of prokaryotes and eukaryotes are discussed in Chapters 11 and 12. Here we consider only the growth of populations by binary fission, using bacterial cultures as examples. We begin with a brief discussion of the mathematics of population growth.

Mathematical Considerations in Population Growth

Learning Objective

✓ Describe logarithmic growth.

With binary fission, any given cell divides to form two cells; then each of these new cells divides in two, to make four, and then four become eight, and so on. This type of growth, called **logarithmic growth** or **exponential growth,** produces very different results from simple addition, or *arithmetic growth.* We can compare these two types of growth by considering what would happen over time to two identical hypothetical populations, as shown in **Figure 6.18.** In this case we assume that a population of hypothetical species A increases by adding one new cell every 20 minutes, whereas the cells in hypothetical species B divide by binary fission every 20 minutes. After 20 minutes, each population, which started with but a single cell, would have two cells; after 40 minutes, species A would have three cells, while species B would have four cells. At this point there is little difference in the growth of the two populations, but after 2 hours, the arithmetically growing species A would have only seven cells, whereas the logarithmically growing species B would have increased to 64 cells. Clearly, logarithmic growth can increase a population's size dramatically—after only 7 hours, species B will have reproduced over 2 million cells!

The number of cells arising from a single cell reproducing by binary fission is calculated as 2^n, where n is the number of generations, in other words, multiply two times itself n number of times. To calculate the total number of cells in a population, we multiply the original number of cells by $2n$. If, for example, species B had begun with three cells instead of one, then after 2 hours it would have 192 cells ($3 \times 2^6 = 3 \times 64 = 192$).

A visible culture of bacteria may consist of trillions of cells, so microbiologists use scientific notation to deal with the huge numbers involved. One advantage of scientific notation is that large numbers are expressed as powers of 10, making them easier to read and write. For example, consider our culture of species B. After 10 hours (30 generations) it would have 1,073,741,824 (2^{30}) cells. This large number can be rounded off and expressed more succinctly in scientific notation as 1.07×10^9. After 30 more generations, scientific notation would be the only practical way to express the huge number of cells in the culture, which would be 1.15×10^{18} (2^{60}). Such a number, if written out (the digits 115 followed by 16 zeroes), would be impractically large.

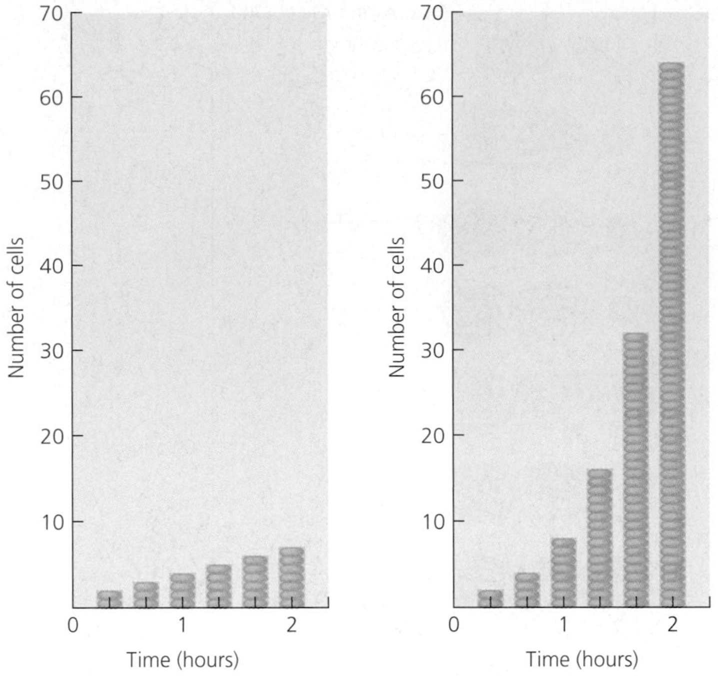

(a) Species A **(b)** Species B

▲ ***Figure 6.18***

A comparison of arithmetic and logarithmic growth. Given two initial populations consisting of a single cell each, after 2 hours the arithmetically growing species A will have seven cells, whereas the logarithmically growing species B will have 64 cells.

Generation Time

Learning Objective

✓ Explain what is meant by the generation time of bacteria.

The time required for a bacterial cell to grow and divide is its **generation time.** Viewed another way, generation time is also the time required for a population of cells to double in number. Generation times vary among populations and are dependent on chemical and physical conditions. Under optimal conditions, some bacteria (such as *E. coli* and *S. aureus*) have a generation time of 20 minutes or less. For this reason, food contaminated with only a few of these organisms can cause food poisoning if not properly refrigerated and cooked. Most bacteria have a generation time of 1–3 hours, though some slow-growing species such as *Mycobacterium leprae* require more than 10 days before they double. The math required to calculate generation time for a population is presented in Appendix B.

Phases of Microbial Growth

Learning Objectives

✓ Draw and label a bacterial growth curve.

✓ Describe what occurs at each phase of a population's growth.

A graph that plots the numbers of organisms in a growing population over time is known as a **growth curve.** When drawn using an arithmetic scale on the y-axis, a plot of exponential growth presents two problems **(Figure 6.19a):** It is difficult or impossible to distinguish numbers in early generations from the baseline, and as the population grows it becomes impossible to accommodate the graph on a single page.

The solution to these problems is to replace the arithmetic scale on the y-axis with a logarithmic (log) scale **(Figure 6.19b).** Such a log scale, in which each division is 10 times larger than the preceding one, can accommodate small numbers at the lower end of the graph, and very large numbers at the upper end. This kind of graph is *semilogarithmic,* because only one axis uses a log scale.

When bacteria are inoculated into a liquid medium, there are four distinct phases to a population's growth curve: the lag, log, stationary, and death phases **(Figure 6.20).**

Lag Phase

During the **lag phase** the cells are adjusting to their new environment; most cells do not reproduce immediately, but instead actively synthesize enzymes to utilize novel nutrients in the medium. For example, bacteria inoculated from a medium containing glucose as a carbon source into a medium containing lactose must synthesize two types of proteins: membrane proteins to transport lactose into the cell, and the enzyme lactase to catabolize the lactose. The lag phase can last less than an hour or for days depending on the species and the chemical and physical conditions of the medium.

Log Phase

Eventually, the bacteria synthesize the necessary chemicals for conducting metabolism in their new environment, and

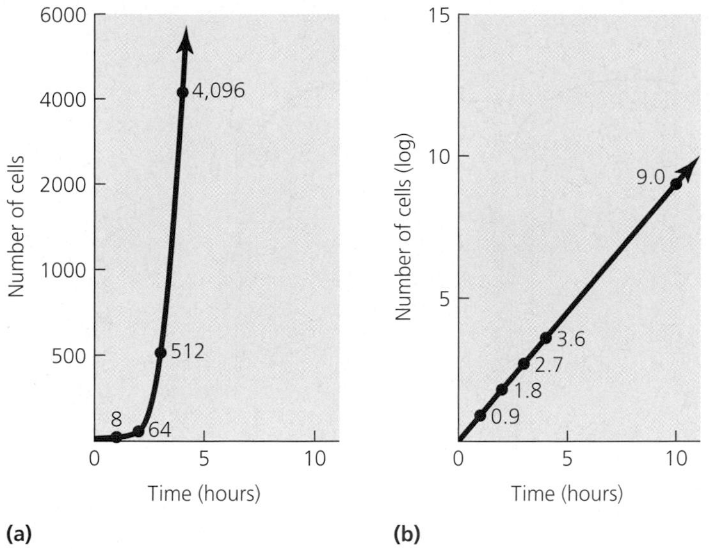

▲ ***Figure 6.19***

Two growth curves of logarithmic growth. The generation time for this *E. coli* population is 20 minutes. **(a)** An arithmetic graph. Using an arithmetic scale for the *y*-axis makes it difficult to ascertain actual numbers of cells near the beginning, and impossible to plot points after only a short time. **(b)** A semilogarithmic graph. Using a logarithmic scale for the *y*-axis solves both of these problems. Note that a plot of logarithmic population growth using a logarithmic scale produces a straight line.

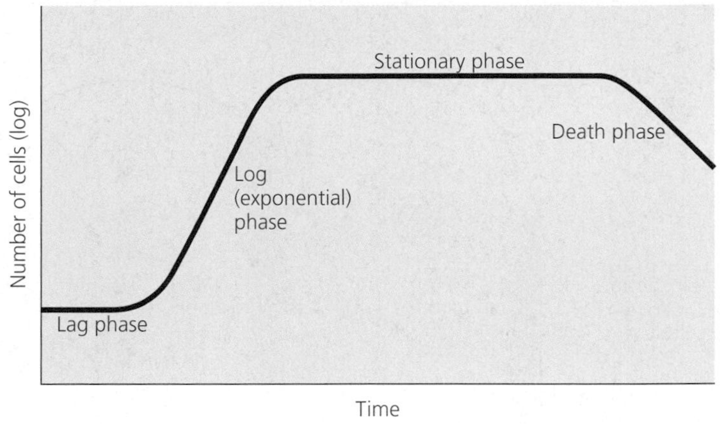

▲ ***Figure 6.20***

A typical microbial growth curve, including the four phases of population growth. *Why do cells trail behind their optimum reproductive potential during the lag phase?*

Figure 6.20 *During lag phase, cells are synthesizing the metabolic machinery and chemicals required for optimal reproduction.*

they then enter a phase of rapid chromosome replication, growth, and reproduction. This is the **log phase,** so-named because the population increases logarithmically, and the reproductive rate reaches a constant as DNA and protein syntheses are maximized. Because the metabolic rate of individual cells is at a maximum during log phase, this phase is sometimes preferred for industrial and laboratory purposes. Populations in log phase are more susceptible to antimicrobial drugs that interfere with metabolism, such as erythromycin, and to drugs that interfere with the formation of cell structures, such as the inhibition of cell wall synthesis by penicillin. Further, populations in log phase are preferred for Gram staining because most cells' walls are intact—an important characteristic for correct staining.

Stationary Phase

If bacterial growth continued at the exponential rate of the log phase, bacteria would soon overwhelm the earth. This does not occur because as nutrients are depleted and wastes accumulate, the rate of reproduction decreases. Eventually, the number of dying cells equals the number of cells being produced, and the size of the population becomes stationary—hence the name **stationary phase.** During this phase the metabolic rate of surviving cells declines.

The onset of the stationary phase can be postponed indefinitely (and thus exponential growth can be maintained indefinitely) by a special apparatus called a *chemostat,* which continually removes wastes (along with old medium and some cells) and adds fresh medium. Chemostats are used in industrial fermentation processes.

Death Phase

If nutrients are not added and wastes are not removed, a population reaches a point at which cells die at a faster rate than they are produced. Such a culture has entered the **death phase** (or *decline phase*). Bear in mind that during the death phase, some cells remain alive and continue metabolizing and reproducing, but the number of dying cells exceeds the number of new cells produced, so that eventually the population decreases to a fraction of its previous abundance. In some cases, all the cells die, while in others a few survivors may remain indefinitely. The latter case is especially true for cultures of bacteria that can develop resting structures called *endospores* (see Chapter 11).

Measuring Microbial Growth

Learning Objective

✓ Contrast direct and indirect methods of measuring bacterial growth.

We have discussed the concepts of population growth and have seen that large numbers result from logarithmic growth, but we have not discussed practical methods of determining the size of a microbial population. Because of each cell's small size and incredible rate of reproduction, it is not possible to actually count every one in a population. (For one thing, they grow so rapidly that their number changes during the count.) Therefore, laboratory personnel must estimate the number of

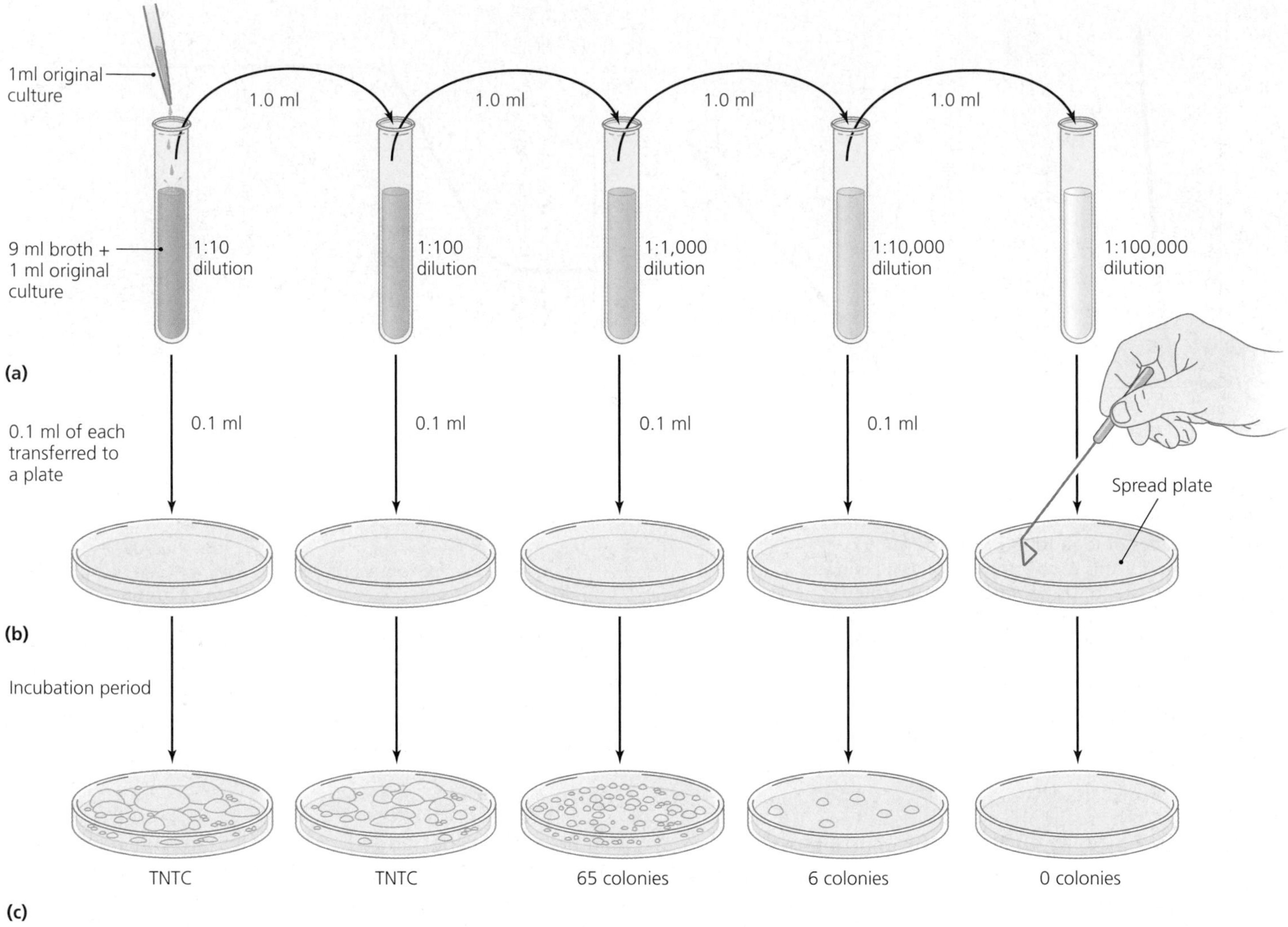

▲ ***Figure 6.21***

A viable plate count for estimating microbial population size. **(a)** Serial dilutions. A series of 10-fold dilutions is made. **(b)** Plating. A 0.1-ml sample from each dilution is poured onto a plate and spread with a sterile rod. Alternatively, 0.1 ml of each dilution can be mixed with melted agar medium and poured into plates.
(c) Counting. Plates are examined after incubation. Some plates may contain so much growth that colonies are too numerous to count (TNTC). The number of colonies is multiplied by the reciprocal of the dilution to estimate the concentration of bacteria in original culture—in this case, 65 colonies × 10,000 = 650,000 bacteria/ml.

cells in a population by counting the number in a small, representative sample and then multiplying to estimate the number in the whole specimen. For example, if there are 25 cells in a microliter (μl) sample of urine, then there are approximately 25 million cells in a liter of urine.

Estimating the number of microorganisms in a sample is useful for determining such things as the severity of urinary tract infections, the effectiveness of pasteurization and other methods of food preservation, the degree of fecal contamination of water supplies, and the effectiveness of particular disinfectants and antibiotics.

Microbiologists use either direct or indirect methods to estimate the number of cells. We begin with direct methods of measuring bacterial growth.

Direct Methods

Among the many direct techniques are viable plate counts, membrane filtration, microscopic counts, the use of electronic counters, and the most probable number method.

Viable Plate Counts What if the number of cells in even a very small sample is still too great to count? If, for example, a l-milliliter sample of milk containing 20,000 bacterial cells were plated on a Petri plate, there would be too many colonies to count. In such cases, microbiologists make a series of dilutions and count the number of colonies resulting on a spread or pour plate from each dilution. Scientists count the colonies on plates with 30–300 colonies and multiply the number by the reciprocal of the dilution to estimate the number of bacteria per ml of the original culture. This method is called a **viable plate count (Figure 6.21).**

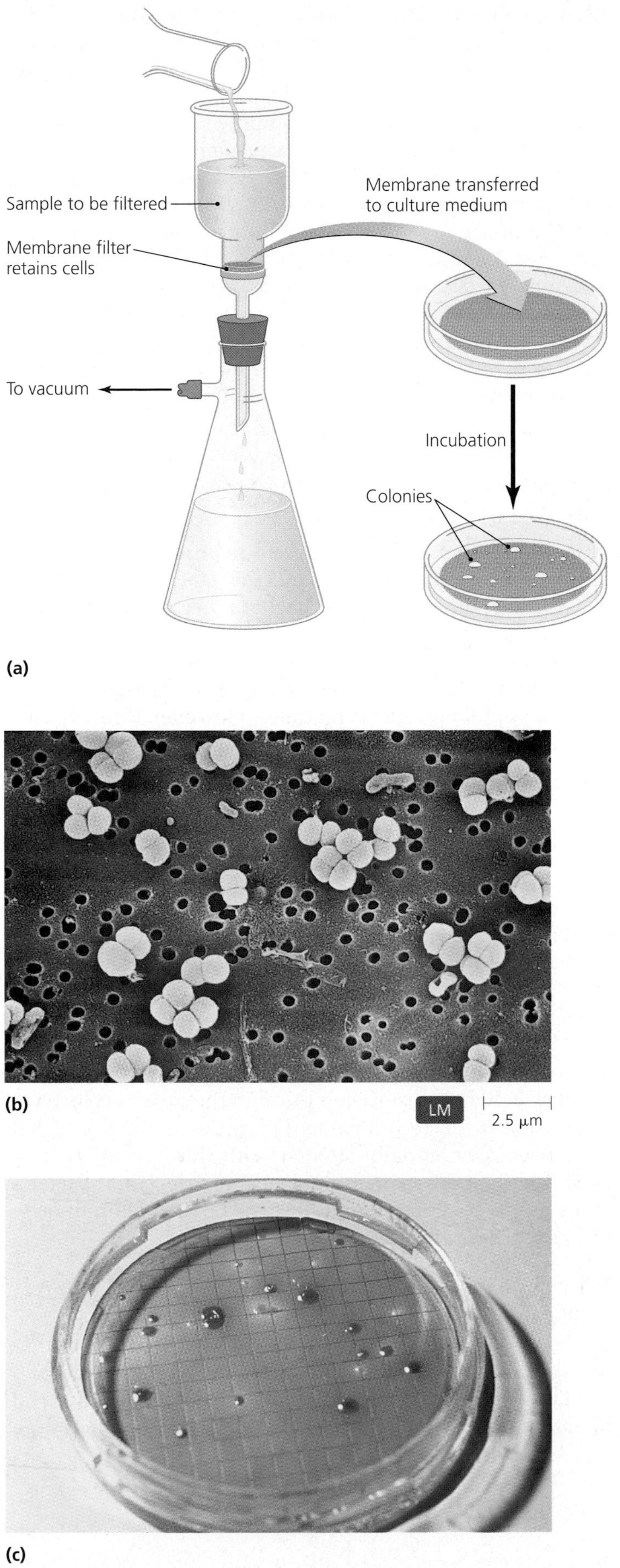

Figure 6.22

The use of membrane filtration to estimate microbial population size. **(a)** After all the bacteria in a given volume of sample are trapped on a membrane filter, the filter is transferred onto an appropriate medium and incubated. The microbial population is estimated by multiplying the number of colonies counted by the volume of sample filtered. **(b)** Bacteria trapped on the surface of a membrane filter. **(c)** Colonies growing on a solid medium after being transferred from the membrane filter. Scientists use a super-imposed grid to help them count the colonies. *If the colonies in (c) resulted from filtering 2.5 liters of stream water, what is the minimum number of bacteria per liter in the stream?*

Figure 6.22 $\frac{25 \text{ colonies}}{2.5 \text{ L}} = 10 \text{ colonies/L}$

When a plate has fewer than 30 colonies, it is not used to estimate the number of bacteria in the original sample because the chance of underestimating the population increases when the number of colonies is small. Recall that the number of colonies on a plate indicates the number of *colony-forming units* that were inoculated onto the plate. This number differs from the actual number of cells when the colony forming units are composed of more than one cell. In such cases, a viable plate count underestimates the number of cells present in the sample.

The accuracy of a viable plate count is also dependent on the homogeneity of the dilutions, the ability of the bacteria to grow on the medium used, the number of cell deaths, and the growth phase of the sample population. Thoroughly mixing each dilution, inoculating multiple plates per dilution, and using log-phase cultures minimize errors.

Membrane Filtration Viable plate counts allow scientists to estimate the number of microorganisms when the population is very large, but if the population density is very small—as is the case, for example, for fecal bacteria in a stream or lake—microbes are more accurately counted by **membrane filtration (Figure 6.22).** In this method, a large sample (perhaps as large as several liters) is poured (or drawn under a vacuum) through a membrane filter with pores small enough to trap the cells. The membrane is then transferred onto a solid medium, and the colonies present after incubation are counted. In this case, the number of colonies is equal to the number of CFUs in the original large sample.

CRITICAL THINKING

Viable plate counts are used to estimate population size when the density of microorganisms is high, whereas membrane filtration is used when the density is low. Why is a viable plate count appropriate when the density is high, but not when the density is low?

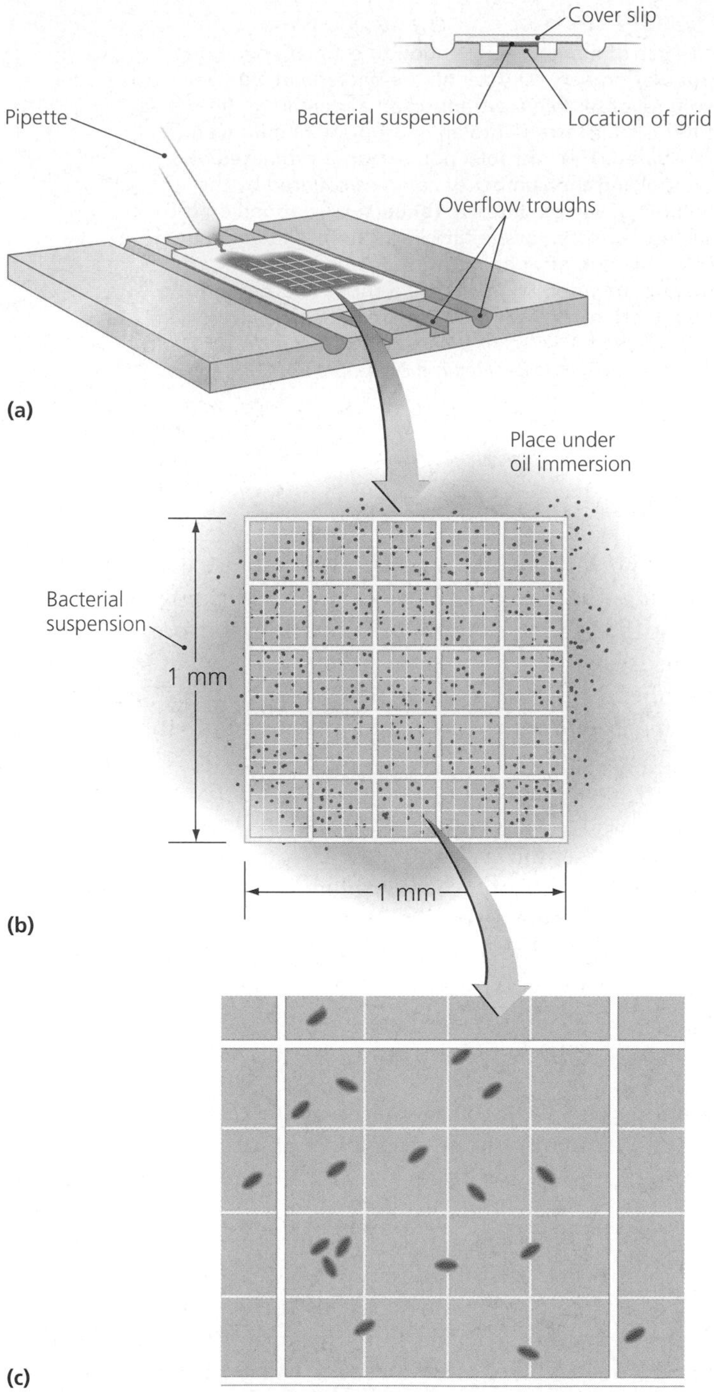

▲ *Figure 6.23*

The use of a cell counter for estimating microbial numbers. **(a)** The counter is a glass slide with an etched grid that is exactly 0.02 mm lower than the bottom of the cover slip. A bacterial suspension placed next to the cover slip through a pipette moves under the cover slip and over the grid by capillary action. **(b)** View of a 1 mm^2 portion of the grid through the microscope. Each square millimeter of the grid has 25 large squares, each of which is divided into 16 small squares. **(c)** Enlarged view of one large square containing 15 cells. The number of bacteria in several large squares is counted and averaged. The calculations involved in estimating the number of bacteria per milliliter (cm^3) of suspension are described in the text.

Microscopic Counts Microbiologists can also count microorganisms directly through a microscope rather than inoculating them onto the surface of a solid medium. In this method, particularly suitable for stained prokaryotes and relatively large eukaryotes, a sample is placed on a *cell counter* (also called a *Petroff-Hauser counting chamber*), which is a glass slide composed of an etched grid positioned beneath a glass cover slip **(Figure 6.23).** Because the cover slip is 0.02 mm above the grid, the volume of bacterial suspension over a 1 mm^2 portion of the grid is 1 mm $\times$ 1 mm $\times$ 0.02 mm = 0.02 mm^3. Each 1 mm^2 grid contains 25 large squares; a microbiologist counts the number of bacteria in several of the large squares and then calculates the mean number of bacteria per square. The number of bacteria per milliliter (cm^3) can be calculated as follows:

$$\text{mean no. of bacteria per square} \times 25 \text{ squares} = \text{no. of bacteria per } 0.02 \text{ mm}^3$$

$$\text{no. of bacteria per } 0.02 \text{ mm}^3 \times 50 = \text{no. of bacteria per mm}^3$$

$$\text{no. of bacteria per mm}^3 \times 1000 = \text{no. of bacteria per cm}^3 \text{ (ml)}$$

This means that one only needs to multiply the mean number of bacteria per square by 1,250,000 (25 $\times$ 50 $\times$ 1000) to calculate the number of bacteria per ml of bacterial suspension.

Direct microscopic counts are advantageous when there are more than 10,000,000 cells per ml or when a speedy estimate of population size is required. However, direct counts can be problematic because it is often difficult to differentiate between living and dead cells, and difficult to count motile microorganisms.

Electronic Counters A *Coulter counter* is a device that directly counts cells as they interrupt an electrical current flowing across a narrow tube held in front of an electronic detector. This device is useful for counting the larger cells of yeast, algae, and protozoa; it is less useful for bacterial counts because of debris in the media and the presence of filaments and clumps of cells.

Flow cytometry is a variation of counting with a Coulter counter. A cytometer uses a light-sensitive detector to record changes in light transmission through the tube as cells pass. Scientists use this technique to distinguish among cells that have been differentially stained with fluorescent dyes or tagged with fluorescent antibodies. They can count bacteria in a solution and even count host cells that contain fluorescently stained intracellular parasites.

Most Probable Number The **most probable number (MPN) method** is a statistical estimating technique based on the fact that the more bacteria in a sample, the more dilutions are required to reduce their number to zero.

Let's consider an example of the use of the MPN method to estimate the number of fecal bacteria contaminating a

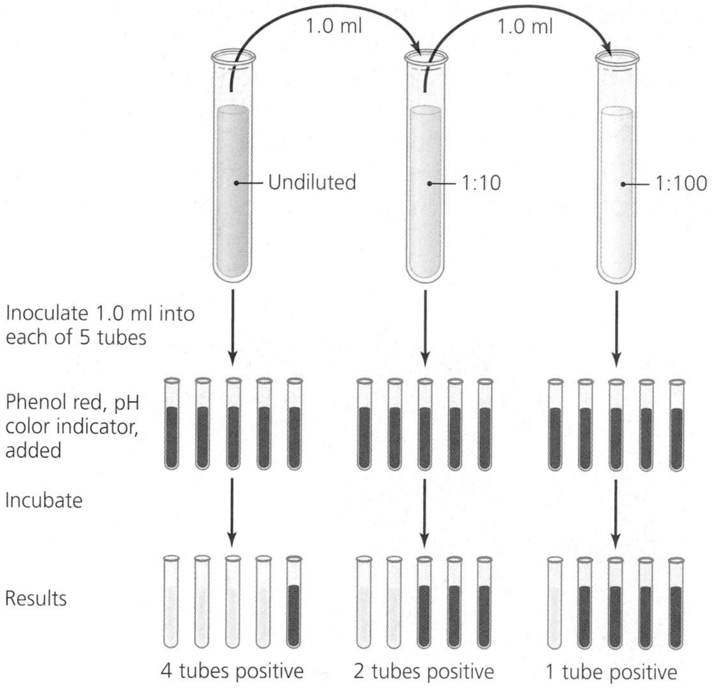

▲ ***Figure 6.24***

The most probable number (MPN) method for estimating microbial numbers. Typically, sets of five test tubes are used for each of three dilutions. After incubation, the number of tubes showing growth in each set is used to enter an MPN table (see Table 6.6), which provides an estimate of the number of cells per 100 ml of liquid. *If the results were 5, 3, 1, what would be the most probable number of microorganisms in the original broth?*

Figure 6.24 The MPN is 110 ml.

stream. A researcher inoculates a set of test tubes of a broth medium with a sample of stream water. Even though the more tubes that are used, the more accurate is the MPN method, accuracy must be balanced against the time and cost involved in inoculating and incubating numerous tubes. Typically, a set of five tubes is inoculated.

The researcher also inoculates a set of five tubes with a 1:10 dilution, and another set of five tubes with a 1:100 dilution of stream water. Thus, there are 15 test tubes—the first set of five tubes inoculated with undiluted sample, the second set with a 1:10 dilution, and the third set with a 1:100 dilution **(Figure 6.24).**

After incubation for 48 hours, the researcher counts the number of test tubes in each set that show growth as determined by some method, such as cloudiness of the broth, gas production, or pH changes. This generates three numbers, in this case—4, 2, 1—that are compared to the numbers in an MPN table (**Table 6.6** on page 195). How statisticians develop MPN tables is beyond the scope of our discussion, but they are constructed in such a way that they accurately estimate the number of cells in a culture 95% of the time. Thus, when growth occurs in four of the undiluted broth tubes, two of the 1:10 tubes, and only one of the 1:100 tubes (4,2,1), the MPN table estimates that there were 26 bacteria/100 ml of stream water.

The most probable number method is useful for counting microorganisms that will not grow on solid media, when bacterial counts are required routinely, and when samples of wastewater, drinking water, and food samples contain too few organisms to use a viable plate count. The MPN method is also used to count algal cells because algae seldom form distinct colonies on solid media.

Indirect Methods

It is not always necessary to count microorganisms to estimate population size or density. Industrial and research microbiologists use indirect methods that measure such variables as metabolic activity, dry weight, and turbidity instead of counting microorganisms, colonies, or MPN tubes.

Metabolic Activity Under standard temperature conditions, the rate at which a population of cells utilizes nutrients and produces wastes is dependent on their number. Once they establish the metabolic rate of a microorganism, scientists can indirectly estimate the number of cells in a culture by measuring changes in such things as nutrient utilization, waste production, or pH.

Dry Weight The abundance of some microorganisms, particularly filamentous microorganisms, is difficult to measure by direct methods. Instead, these organisms are filtered from their culture medium, dried, and weighed. The *dry weight method* is suitable for broth cultures, but growth cannot be followed over time because the organisms are killed during the process.

Turbidity As bacteria reproduce in a broth culture, the broth often becomes *turbid* (cloudy) **(Figure 6.25a).** Generally, the greater the bacterial population, the more turbid a broth will be. An indirect method for estimating the growth of a microbial population involves measuring changes in turbidity using a device called a *spectrophotometer* **(Figure 6.25b).**

A spectrophotometer measures the amount of light transmitted through a culture under standardized conditions **(Figure 6.25c).** The greater the concentration of bacteria within a broth, the more light will be absorbed and scattered, and the less light will pass through and strike a light-sensitive detector. Generally, transmission is inversely proportional to the population size; that is, the larger the population grows, the less light will reach the detector.

Scales on the gauge of a spectrophotometer report *percentage of transmission* and *absorbance.* These are two ways of looking at the same things, for example, 25% transmission is the same thing as 75% absorbance. Direct counts must be calibrated with transmission and absorbance readings to provide estimates of population size. Once these values are determined, spectrophotometry provides estimates of population size more quickly than any direct method.

The benefits of measuring turbidity to estimate population growth include ease of use and speed. However, the technique is useful only if the concentration of cells exceeds 1 million per milliliter; densities below this value generally do not produce turbidity. Further, the technique is accurate only if the cells are suspended uniformly in the medium. If they form either a *pellicle* (a film of cells at the surface) or a *sediment* (an accumulation of cells at the bottom), their number will be underestimated. Further, spectrophotometry does not distinguish between living and dead cells.

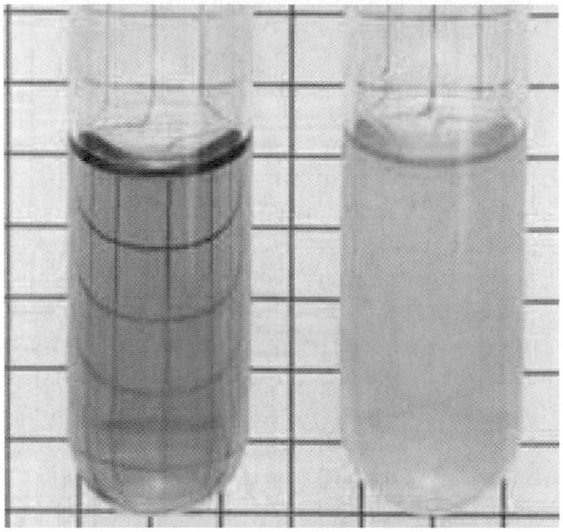

(a)

(b)

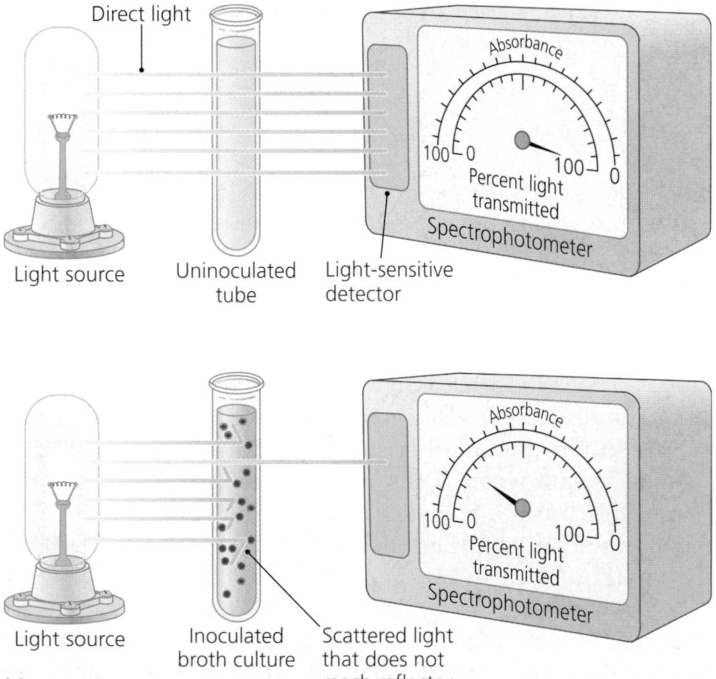

(c)

Figure 6.25 ➤
Turbidity and the use of spectrophotometry in indirectly measuring population size. **(a)** Turbidity (right), an increased optical density or cloudiness of a solution. **(b)** A spectrophotometer. **(c)** The principle of spectrophotometry. After passing a light beam through an uninoculated sample of the culture medium, the scale is set at 100% transmission. In an inoculated sample, the microbial cells absorb and scatter light, reducing the amount reaching the detector. The percentage of light transmitted is inversely proportional to population density.

Table 6.6 Most Probable Number Table

Number Out of 5 Tubes Giving Positive Results in Three Dilutions			Most Probable Number of Bacteria per 100ml	Number Out of 5 Tubes Giving Positive Results in Three Dilutions			Most Probable Number of Bacteria per 100ml
0	0	0	1.8	4	3	0	27
0	0	1	2	4	3	1	33
0	1	0	2	4	4	0	34
0	2	0	4	5	0	0	23
1	0	0	2	5	0	1	30
1	0	1	4	5	0	2	40
1	1	0	4	5	1	0	30
1	1	1	6	5	1	1	50
1	2	0	6	5	1	2	60
2	0	0	4	5	2	0	50
2	0	1	7	5	2	1	70
2	1	0	7	5	2	2	90
2	1	1	9	5	3	0	80
2	2	0	9	5	3	1	110
2	3	0	12	5	3	2	140
3	0	0	8	5	3	3	170
3	0	1	11	5	4	0	130
3	1	0	11	5	4	1	170
3	1	1	14	5	4	2	220
3	2	0	14	5	4	3	280
3	2	1	17	5	4	4	350
4	0	0	13	5	5	0	240
4	0	1	17	5	5	1	300
4	1	0	17	5	5	2	500
4	1	1	21	5	5	3	900
4	1	2	26	5	5	4	1,600
4	2	0	22	5	5	5	≥ 1,600
4	2	1	26				

CHAPTER SUMMARY

Growth Requirements (pp. 168–177)

1. A **colony,** which is a visible population of microorganisms living in one place, grows in size as the number of cells increases.
2. Chemical **nutrients** such as carbon, hydrogen, oxygen, and nitrogen are required for the growth of microbial populations.
3. **Photoautotrophs** use carbon dioxide as a carbon source and light energy from the environment to make their own food; **chemoautotrophs** use carbon dioxide as a carbon source but catabolize organic molecules for energy. **Photoheterotrophs** are photosynthetic organisms that acquire energy from light and acquire nutrients via catabolism of organic compounds; **chemoheterotrophs** use organic compounds for both energy and carbon. **Organotrophs** acquire electrons from organic sources, whereas **lithotrophs** acquire electrons from inorganic sources.
4. **Obligate aerobes** require oxygen as the final electron acceptor of the electron transport chain, whereas **obligate anaerobes** cannot tolerate oxygen and use an electron acceptor other than oxygen.
5. The four toxic forms of oxygen are **singlet oxygen ($^1O_2$),** which is neutralized by pigments called **carotenoids; superoxide radicals (O_2^-),** which are detoxified by superoxide dismutase; **peroxide anion (O_2^{2-}),** which is detoxified by catalase or peroxidase; and **hydroxyl radicals ($OH^\cdot$),** the most reactive of the toxic forms of oxygen.

6. Microbes are described in terms of their oxygen requirements and limitations as strict **aerobes,** which require oxygen; strict **anaerobes,** which cannot tolerate oxygen; as **facultative anaerobes,** which can live with or without oxygen; as **aerotolerant anaerobes,** which prefer anaerobic conditions, but can tolerate exposure to low levels of oxygen; or as **microaerophiles,** which require low levels of oxygen. A **capnophile** grows best with high CO_2 levels in addition to low oxygen levels.
7. Nitrogen, acquired from organic or inorganic sources, is an essential element for microorganisms. Some bacteria can reduce nitrogen gas into a more usable form via a process called **nitrogen fixation.**
8. In addition to the main elements found in microbes, very small amounts of **trace elements** are required. Vitamins are among the **growth factors,** which are organic chemicals required in small amounts for metabolism.
9. Though microbes survive within the limits imposed by a minimum growth temperature and a maximum growth temperature, an organism's metabolic activities produce the highest growth rate at the **optimum growth temperature.**
10. Microbes are described in terms of their temperature requirements as (from coldest to warmest) **psychrophiles, mesophiles, thermophiles,** or **hyperthermophiles.**
11. **Neutrophiles** grow best at neutral pH, **acidophiles** grow best in acidic surroundings, and **alkalinophiles** live in alkaline habitats.
12. Osmotic pressure can cause cells to die from either swelling and bursting or shriveling **(plasmolysis).** The cell walls of some microorganisms protect them from osmotic shock. **Obligate halophiles** require high osmotic pressure, whereas facultative halophiles do not require but can tolerate such conditions.
13. **Barophiles,** organisms that normally live under the extreme hydrostatic pressure at great depth below the surface of a body of water, often cannot live at the pressure found at the surface.
14. **Quorum sensing** is the process by which bacteria respond to changes in microbial density by utilizing signal and receptor molecules. **Biofilms,** which are communities of cells attached to surfaces, are one result of quorum sensing.

Culturing Microorganisms (pp. 177–186)

1. Microbiologists culture microorganisms by transferring an **inoculum** from a clinical or environmental **specimen** into a **medium** such as **broth** or solid media. The microorganisms grow into a **culture.** On solid surfaces, cultures are seen as colonies.
2. **Pure cultures** (axenic cultures) contain cells of only one species and are derived from a **colony-forming unit (CFU)** composed of a single cell or group of related cells. To obtain pure cultures, **sterile** equipment and use of aseptic techniques are critical.
3. The **streak plate** method allows CFUs to be isolated, incubated, and observed. The **pour plate** technique isolates CFUs via a series of dilutions.
4. Petri dishes that are filled with solid media are called **Petri plates. Slant tubes (slants)** are test tubes containing agar media that solidified while the tube was resting at an angle.
5. A **defined medium** (also known as **synthetic medium**) provides exact known amounts of nutrients for the growth of a particular microbe. **Complex media** contain a variety of growth factors. **Selective media** either inhibit the growth of unwanted microorganisms or favor the growth of particular microbes. Microbiologists use **differential media** to distinguish among groups of bacteria. **Reducing media** provide conditions conducive to culturing anaerobes. **Transport media** are designed to move specimens safely from one location to another while maintaining the relative abundance of organisms and preventing contamination of the specimen or environment.
6. Special culture techniques include the use of animal and cell cultures, low-oxygen cultures, **enrichment cultures,** and **cold enrichment cultures.**
7. Cultures can be preserved in the short term by **refrigeration,** and in the long term by **deep-freezing** and **lyophilization.**

Growth of Microbial Populations (pp. 186–195)

1. Bacteria grow by **logarithmic** or **exponential growth.**
2. A population of bacteria doubles during its **generation time**—the time also required for a bacterial cell to grow and divide.
3. A graph that plots the number of bacteria growing in a population over time is called a **growth curve.** When bacteria are grown in a broth and the growth curve is plotted on a semilogarithmic scale, the population's growth curve has four phases. In the **lag phase,** the organisms are adjusting to their environment. In the **log phase,** the population is most actively growing. In the **stationary phase,** new organisms are being produced at the same rate at which they are dying. In the **death phase,** the organisms are dying more quickly than they can be replaced by new organisms.
4. Direct methods for estimating population size include **viable plate counts, membrane filtration,** microscopic counts, electronic counters, flow cytometry, and the **most probable number (MPN) method.**
5. Indirect methods include measurements of metabolic activity, dry weight, and turbidity.

QUESTIONS FOR REVIEW

(Answers to Multiple Choice and Fill in the Blanks are on the Web, along with additional review questions. Visit www.microbiologyplace.com.)

Multiple Choice

1. Which of the following can grow in a Petri plate?
   a. an anaerobe
   b. a colony on an agar surface
   c. viruses on an agar surface
   d. barophiles
2. Which of the following terms best describes an organism that cannot exist in the presence of oxygen?
   a. obligate aerobe
   b. facultative aerobe
   c. obligate anaerobe
   d. facultative anaerobe
3. Superoxide dismutase
   a. causes hydrogen peroxide to become toxic.
   b. detoxifies superoxide radicals.
   c. neutralizes singlet oxygen.
   d. is missing in aerobes.
4. The most reactive of the four toxic forms of oxygen is
   a. the hydroxyl radical.
   b. the peroxide anion.
   c. the superoxide radical.
   d. singlet oxygen.
5. Microaerophiles that grow best with a high concentration of carbon dioxide in addition to a low level of oxygen are called
   a. aerotolerant.
   b. capnophiles.
   c. facultative anerobes.
   d. fastidious.
6. Which is *not* a growth factor for various microbes?
   a. cholesterol
   b. water
   c. vitamins
   d. heme
7. Organisms that may be found thriving in icy waters are described as
   a. barophiles.
   b. thermophiles.
   c. mesophiles.
   d. psychrophiles.
8. Barophiles
   a. cannot cause diseases in humans.
   b. live at normal barometric pressure.
   c. die if put under high pressure.
   d. thrive in warm air.
9. This statement, "In the laboratory, a sterile inoculating loop is moved across the agar surface in a culture dish, thinning a sample and isolating individuals," describes which of the following:
   a. broth culture
   b. pour plate
   c. streak plate
   d. dilution plate
10. In a defined medium,
    a. the exact chemical composition of the medium is known.
    b. agar is available for microbial nutrition.
    c. blood may be included.
    d. organic chemicals are excluded.
11. Which of the following is most useful in representing population growth on a graph?
    a. logarithmic reproduction of the growth curve
    b. a semilogarithmic graph using a log scale on the y-axis
    c. an arithmetic graph of the lag phase followed by a logarithmic section for the log, stationary, and death phases
    d. none of the above would best represent a population growth curve
12. Which of the following methods is best for counting fecal bacteria from a stream to determine the safety of the water for drinking?
    a. dry weight
    b. turbidity
    c. viable plate counts
    d. membrane filtration
13. A Coulter counter is
    a. a statistical estimation using five dilution tubes and a table of numbers to estimate the number of bacteria per milliliter.
    b. an indirect method of counting microorganisms.
    c. a device that directly counts microbes as they pass through a tube in front of an electronic detector.
    d. a device that directly counts microbes that are differentially stained with fluorescent dyes.
14. Lyophilization can be described as
    a. freeze-drying.
    b. deep-freezing.
    c. refrigeration.
    d. pickling.
15. Quorum sensing is
    a. the ability to respond to changes in population density.
    b. a characteristic of most bacteria.
    c. dependent on direct contact among cells.
    d. associated with colonies on an agar plate.

Fill in the Blanks

1. All cells require a source of __________, __________, and __________.
2. A toxic form of oxygen, __________ oxygen, is molecular oxygen with electrons that have been boosted to a higher energy state.
3. All cells recycle the essential element __________ from amino acids and nucleotides.
4. __________ are small organic molecules that are required in minute amounts for metabolism.
5. The lowest temperature at which a microbe continues to metabolize is called its __________.

6. Cells that shrink in hypertonic solutions such as saltwater are responding to ________________ pressure.
7. Obligate ________________ may exist in salt ponds because of their ability to withstand high osmotic pressure.

Short Answer

1. High temperature affects the shape of particular molecules. How does this affect the life of a microbe?
2. Support or refute the following statement: Microbes cannot tolerate the low pH of the human stomach.
3. Explain quorum sensing and describe how it is related to biofilm formation.
4. Why must media, vessels, and instruments be sterilized before they are used for microbiological procedures?
5. Why is agar used in microbiology?
6. What is the difference between complex media and defined media?
7. Draw and label the four distinct phases of a bacterial growth curve. Describe what is happening within the culture as it passes through the phases.
8. If there are 47 cells in a μl of sewage, how many cells are there in a liter?
9. List three indirect methods of counting microbes.
10. List five direct methods of counting microbes.
11. Explain the differences among photoautotrophs, chemoautotrophs, photoheterotrophs, chemoheterotrophs, organotrophs, and lithotrophs.

CRITICAL THINKING

1. A microbiologist describes an organism as a chemoheterotrophic, aerotolerant, mesophilic, facultatively halophilic bacillus. Describe the organism's metabolic and structural features in plain English.
2. Pasteurization is a technique that uses temperatures of about 72°C to neutralize potential pathogens in foods. What effect does this temperature have on the enzymes and cellular metabolism of pathogens? Why does the heat of pasteurization kill some microorganisms yet fail to affect thermophiles?
3. If two cultures of a facultative anaerobe were grown under identical conditions except that one was exposed to oxygen and the other was completely deprived of oxygen, what differences would you expect to see between the dry weights of the cultures? Why?

CHAPTER 7

Microbial Genetics

Do genes hold the secrets to life? If we can identify all the genes in an organism, and determine the functions of each, can we explain all of biological behavior? What genes do organisms have in common, and what genes make them unique? What genes cause certain microorganisms to be harmful or even deadly, and how can we develop drugs or techniques to target a pathogen's genes?

These are the kinds of questions that drive the dynamic world of genetic research. We now have complete genome maps, or genetic blueprints, of hundreds of viruses, bacteria, and other organisms. We even have a working draft of the human genome, which our knowledge of microbial genetics is helping us to analyze. For example, by studying the genes of the *Escherichia coli* bacterium and then identifying which genes we share, we can determine the roles these same genes play in humans. Genetically speaking, you may have much more in common with microorganisms than you think!

CHAPTER OUTLINE

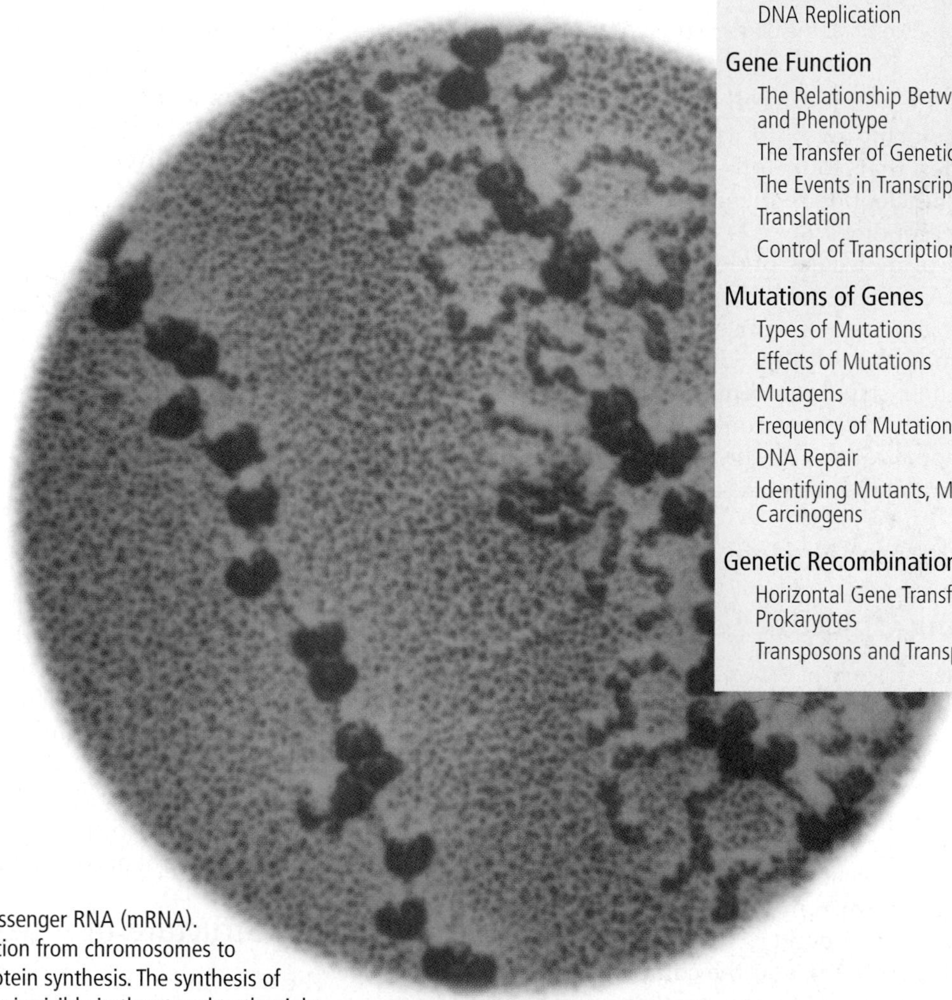

Ribosomes on two strands of messenger RNA (mRNA). mRNA carry genetic information from chromosomes to ribosomes, enabling protein synthesis. The synthesis of polypeptides is visible in the strand to the right.

MicroPrep Pre-Test: *Take the pre-test for this chapter on the web. Visit **www.microbiologyplace.com.***

Genetics is the study of inheritance and inheritable traits as expressed in an organism's genetic material. Geneticists study many aspects of inheritance, including the physical structure and function of genetic material, mutations, and the transfer of genetic material among organisms. In this chapter, we will examine these topics as they apply to microorganisms, the study of which have formed much of the basis of our understanding of human, animal, and plant genetics.

The Structure and Replication of Genomes

Learning Objective

✓ Compare and contrast the genomes of prokaryotes and eukaryotes.

The **genome** (je'nōm) of a cell or virus is its entire genetic complement, including both its **genes**—specific sequences of nucleotides that code for polypeptides or RNA molecules—and nucleotide sequences that connect genes to one another. The genomes of cells and DNA viruses are composed solely of molecules of deoxyribonucleic acid (DNA), whereas RNA viruses use ribonucleic acid instead. We will examine the genomes of viruses in more detail in Chapter 13. The remainder of this chapter focuses on *bacterial* genomes—their structure, replication, function, mutation, and repair, and how they compare and contrast with eukaryotic genomes and with the genomes of archaea. We begin by examining the structure of nucleic acids.

The Structure of Nucleic Acids

Learning Objective

✓ Describe the structure of DNA, and discuss how it facilitates the ability of DNA to act as genetic material.

As we studied in Chapter 2, nucleic acids are polymers of nucleotides, each of which contains a pentose sugar (deoxyribose in DNA, ribose in RNA), a phosphate, and one of five nitrogenous bases (guanine, cytosine, adenine, thymine, or uracil). These bases hydrogen bond in specific ways called **base pairs (bp):** In DNA, the complementary bases guanine and cytosine bond to one another with three hydrogen bonds **(Figure 7.1a)**, and the complementary bases adenine and thymine bond to one another with two hydrogen bonds **(Figure 7.1b)**. In RNA, uracil (not thymine) bonds with adenine **(Figure 7.1c)**.

Deoxyribonucleotides are linked through their sugars and phosphates to form the two backbones of a helical, double-stranded DNA (dsDNA) molecule **(Figure 7.1d)**. The carbon atoms of deoxyribose are numbered 1′ (pronounced "one prime") through 5′. One end of a DNA strand is called the 5′ end because it terminates in a phosphate group attached to a 5′ carbon; the opposite (3′) end terminates with a hydroxyl group bound to a 3′ carbon of deoxyribose. The two strands are oriented in opposite directions to each other; one strand runs in a 3′ to 5′ direction, while the other runs in a 5′ to 3′ direction. Scientists say the two strands are *antiparallel*. The base pairs extend into the middle of the molecule in a way reminiscent of the steps of a spiral staircase.

The lengths of DNA molecules are not usually given in metric units; instead the length of a DNA molecule is expressed in base pairs. For example, the genome of *Mycoplasma genitalium* (mī'kō-plaz-mă jen-ē-tal'ē-ŭm) is 580,070 bp long, making it the smallest known cellular genome.

The structure of DNA helps explain its ability to act as genetic material. First, the linear sequence of nucleotides carries the instructions for the synthesis of polypeptides and RNA molecules—in much the way a sequence of letters carries information used to form words and sentences. Secondly, the complementary structure of the two strands allows a cell to make exact copies to pass to its progeny. We will examine the genetic code and DNA replication shortly.

CRITICAL THINKING

The chromosome of *Mycobacterium tuberculosis* is 4,411,529 bp long. A scientist who isolates and counts the number of nucleotides in its DNA molecule discovers that there are 2,893,963 molecules of guanine. How many molecules of the other three nucleotides are in the original DNA?

The amount of DNA in a genome can be extraordinary, as some examples will illustrate. The bacterium *Escherichia coli* (esh-ĕ-rik'ē-ă kō'lī) is approximately 2 μm long and 1 μm in diameter, but its genome consists primarily of a 4.6×10^6 bp DNA molecule that is about 1600 μm long—800 times longer than the cell. The human genome has about 3 billion base pairs in 46 nuclear DNA molecules plus a mitochondrial DNA molecule, and it would be 1.6 meters (1,600,000 μm) long if all 47 molecules were laid end to end. Most of the human genome is packed into a nucleus that is typically only 5 μm in diameter. This is like packing 10 miles of thread into a golf ball! To understand how cells package such prodigious amounts of DNA into such small spaces, we must first understand that prokaryotes and eukaryotes package DNA in different ways. We begin by examining the structure of prokaryotic genomes.

The Structure of Prokaryotic Genomes

The DNA in prokaryotic genomes is found in two structures: chromosomes and plasmids.

Prokaryotic Chromosomes

Prokaryotic cells package the main portion of their DNA, along with associated molecules of protein and RNA, as one or two **chromosomes.**[1] Scientists have studied the genes of

[1]From Greek *chroma*, meaning color (because they typically stain darkly), and *soma*, meaning body.

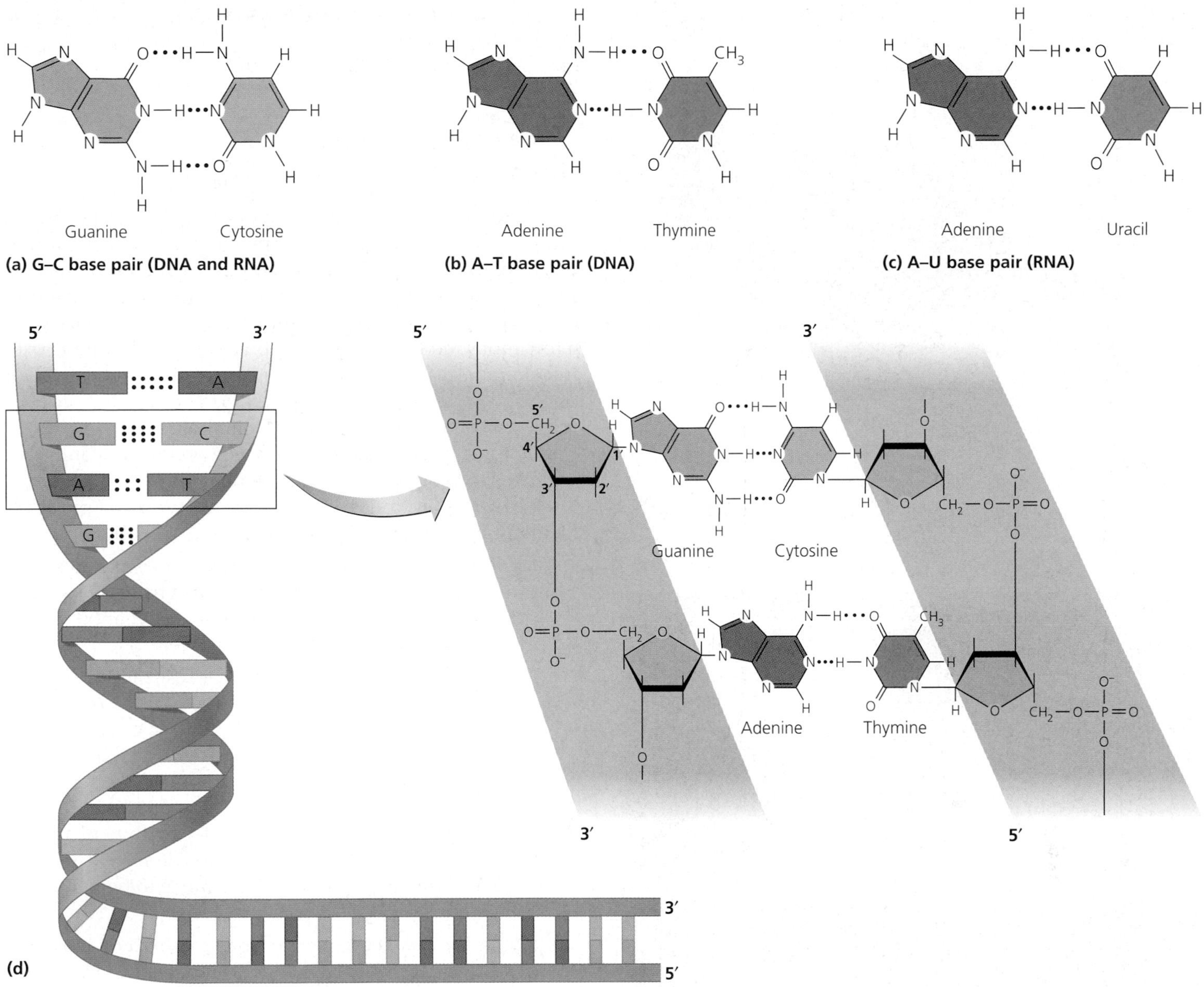

▲ *Figure 7.1*

The structure of nucleic acids, which are polymers of nucleotides consisting of a pentose sugar, a phosphate, and a nitrogenous base. **(a)** Base pairing between the complementary bases guanine (G) and cytosine (C) formed by three hydrogen bonds, found in both DNA and RNA. **(b)** Base pairing between the complementary bases adenine (A) and thymine (T) formed by two hydrogen bonds, found in DNA only. **(c)** Base pairing between adenine and uracil (U), found in RNA only. Notice the structural similarities between thymine and uracil. **(d)** Double-stranded DNA, which consists of antiparallel strands of nucleotides held to one another by the hydrogen bonding between complementary bases. *What structures do DNA nucleotides and RNA nucleotides have in common?*

Figure 7.1 Both DNA and RNA nucleotides are composed of a pentose sugar, a phosphate, and a nitrogenous base.

bacteria more than those of archaea, so the remainder of our discussion deals with bacterial genomes.

A typical bacterial chromosome (**Figure 7.2a** on page 202) consists of a circular molecule of DNA localized in a region of the cell called the **nucleoid.** No membrane surrounds the nucleoid, though the chromosome is packed in such a way that a distinct boundary is visible between the nucleoid and the rest of the cytosol. Chromosomal DNA is folded into loops that are 50,000–100,000 bp long **(Figure 7.2b)** held in place by molecules of protein and RNA. The entire chromosome is further folded and coiled like a skein of yarn into a compact mass.

For many years scientists thought that all prokaryotes had only a single circular chromosome, but we now know that some bacterial species contain two chromosomes, and that at least one member of such a pair may be linear. *Agrobacterium tumefaciens* (ag′rō-bak-ti′rē-um tū′ me-fāsh-enz), a

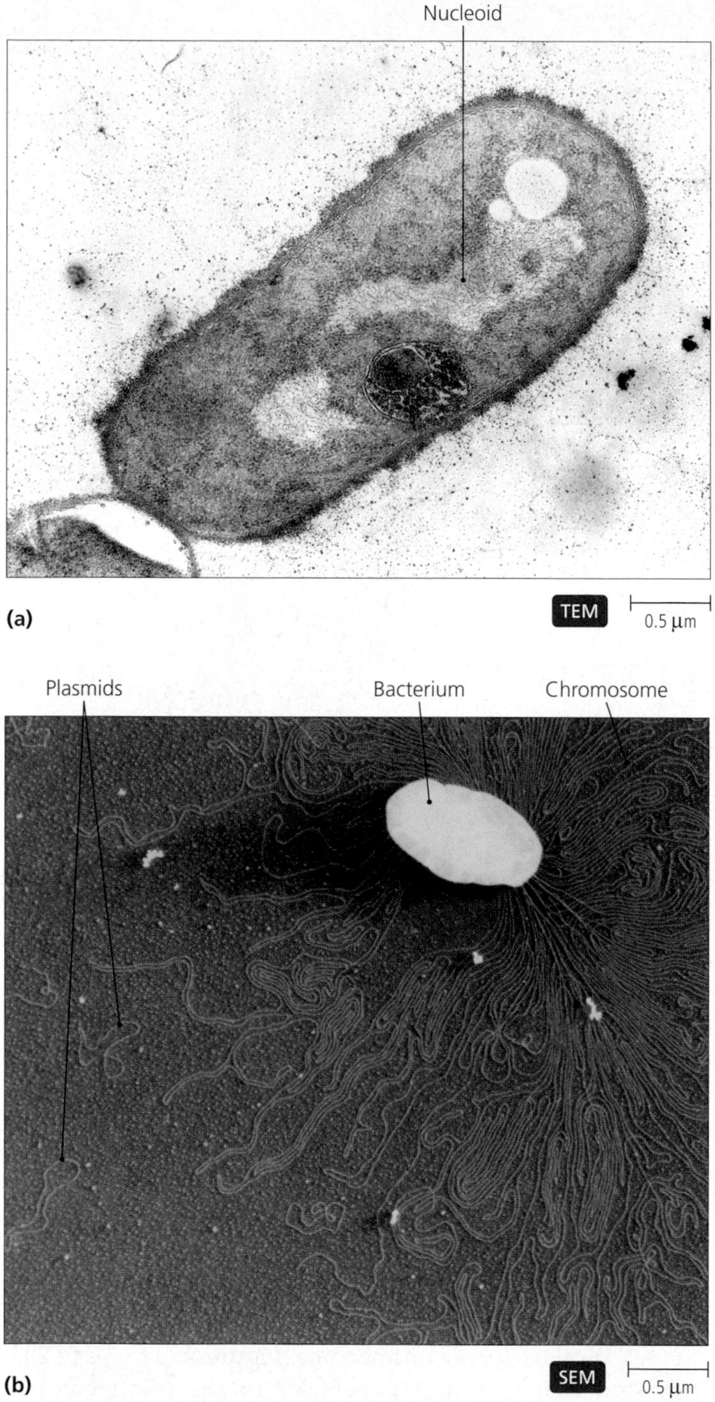

▲ ***Figure 7.2***

Bacterial genome. **(a)** Bacterial chromosomes are packaged in a region of the cytosol called the nucleoid, which is not surrounded by a membrane. **(b)** The packing of a circular bacterial chromosome into loops, as seen after the cell was gently broken open to release the chromosome. Extrachromosomal DNA in the form of plasmids is also visible.

bacterium used to transfer genes into plants, is an example of a prokaryote with two chromosomes, one circular and one linear. Another example of a prokaryote with two chromosomes is *Vibrio cholerae* (vib′rē-ō kol′er-ă), the bacterium that causes cholera.

Plasmids

Learning Objective

✓ Describe the structure and function of plasmids.

In addition to chromosomes, many bacterial cells contain one or more **plasmids,** which are small, circular molecules of DNA that replicate independently of the chromosome. Plasmids are circular and usually 1% to 5% the size of a bacterial chromosome (see Figure 7.2b), ranging in size from a few thousand bp to a few million bp. Each plasmid carries information required for its own replication, and often for one or more cellular traits. Typically, genes carried on plasmids are not essential for normal metabolism, for growth, or for cellular reproduction, but *plasmid* genes can confer advantages to the cells that carry them.

Researchers have identified many types of plasmids (sometimes also called *factors*), including the following:

- *Fertility (F) factors* carry instructions for *conjugation,* a process involved in transferring genes from one bacterial cell to another. We will consider conjugation in more detail near the end of this chapter.
- *Resistance (R) factors* carry genes for resistance to one or more antimicrobial drugs, heavy metals, or toxins. By processes we will discuss shortly, certain cells can transfer resistance factors to other cells, which then acquire resistance to the same antimicrobial chemicals. One example of the effects of an R factor involves gonorrhea, which is no longer treated with penicillin because cells of *Neisseria gonorrhoeae* (nī-sē′rē-ă gon-ō-rē′ē) acquired an R factor for resistance to penicillin from strains of *Streptococcus.*
- *Bacteriocin factors* carry genes for proteinaceous toxins called *bacteriocins,* which kill bacterial cells of the same or similar species that lack the factor. In this way a bacterium containing the plasmid can kill its competitors.
- *Virulence plasmids* carry instructions for structures, enzymes, or toxins that enable a bacterium to become pathogenic. For example, *E. coli,* a normal resident of the human gastrointestinal tract, causes diarrhea only when it carries plasmids that code for certain toxins.
- So-called *cryptic plasmids* have been observed in photographs, but scientists have as yet been unable to identify their functions.

Now that we have examined the structure of prokaryotic genomes, we turn to the structure of eukaryotic genomes.

The Structure of Eukaryotic Genomes

Eukaryotic genomes consist of both nuclear and extranuclear DNA.

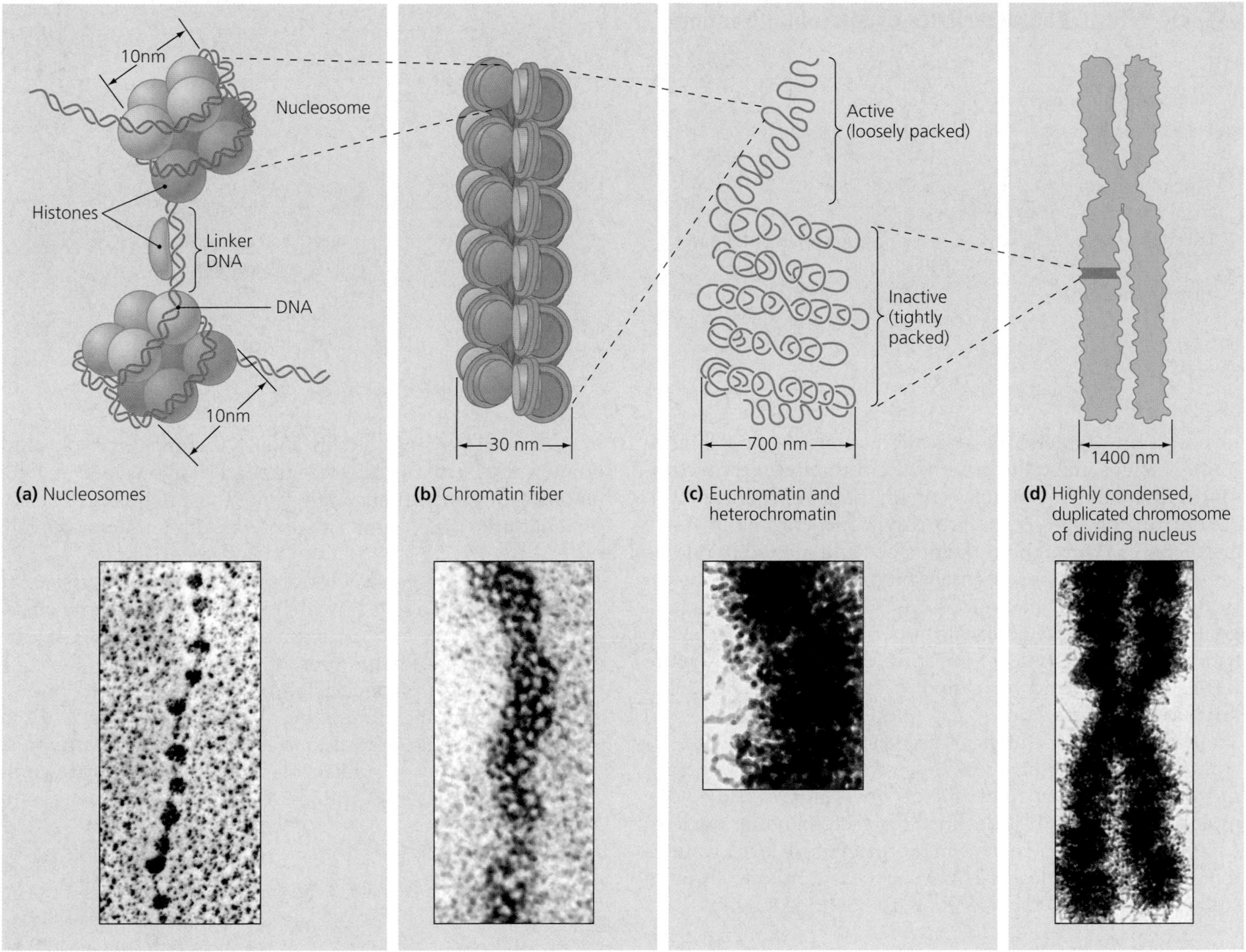

▲ ***Figure 7.3***

Eukaryotic nuclear chromosomal packaging. **(a)** Histones stabilize and package DNA to form nucleosomes connected by linker DNA. **(b)** Nucleosomes clump to form chromatin fibers. **(c)** Chromatin fibers fold and are organized into active euchromatin and inactive heterochromatin. **(d)** During nuclear division (mitosis), duplicated chromatin fully condenses into a mitotic chromosome that is visible by light microscopy. If the nucleosomes were actually the size shown in **(a)**, the chromosome in **(d)** would be 20 m (about 65 feet) long.

Nuclear Chromosomes

Learning Objective

✓ Compare and contrast prokaryotic and eukaryotic chromosomes.

Typically, eukaryotic cells have more than one nuclear chromosome in their genomes, though one species of Australian ant has a single chromosome per nucleus, and some eukaryotic cells such as mammalian red blood cells lose their chromosomes as they mature. Eukaryotic chromosomes differ from their typical prokaryotic counterparts in that they are linear (rather than circular) and are sequestered within a nucleus. As we saw in Chapter 3, the nucleus is an organelle surrounded by two membranes, which together are called the *nuclear envelope*. Given that a typical eukaryotic cell must package substantially more DNA than its prokaryotic counterpart, it is not surprising that nuclear chromosomes are more elaborate than those of prokaryotes.

The structures involved in the packaging of eukaryotic chromosomes are depicted in **Figure 7.3.** Eukaryotic chromosomes are composed of DNA and globular proteins called **histones.** DNA, which has an overall negative electrical charge, wraps around the positively charged histones to

Table 7.1 Characteristics of Microbial Genomes

	Bacteria	Archaea	Eukarya
Number of chromosomes	One or rarely two	One	With one exception, two or more
Plasmids present?	In some cells; frequently more than one per cell	In some cells	In some fungi and protozoa
Type of nucleic acid	Circular or linear dsDNA	Circular dsDNA	Linear dsDNA in nucleus; circular dsDNA in mitochondria, chloroplasts, and plasmids
Location of DNA	In nucleoid and in plasmids in cytosol	In nucleoid and in plasmids in cytosol	In nuclei, and in mitochondria, chloroplasts, and plasmids in cytosol
Histones present?	No, though chromosome is associated with a small amount of nonhistone protein	Yes	Yes

form 10-nm-diameter beads called **nucleosomes (Figure 7.3a).** Nucleosomes clump with other proteins to form **chromatin fibers** that are about 30 nm in diameter **(Figure 7.3b).** Except during *mitosis* (nuclear division), chromatin fibers are dispersed throughout the nucleus and are too thin to be resolved without the extremely high magnification of electron microscopes. In regions of the chromosome where genes are active, the chromatin fibers are loosely packed to form *euchromatin* (yū-krō′ mă-tin); inactive DNA is more tightly packed and is called *heterochromatin* (het′er-ō-krō′ mă-tin; **Figure 7.3c**).

Prior to *mitosis* (nuclear division), a cell replicates its chromosomes and then condenses them into pairs of chromosomes visible by light microscopy **(Figure 7.3d).** (One molecule of each pair is destined for each daughter nucleus. Chapter 12 discusses mitosis in more detail.) The net result is that each DNA molecule is packaged as a mitotic chromosome that is 50,000× shorter than its extended length.

Extranuclear DNA of Eukaryotes

Not all of the DNA of a eukaryotic genome is contained in its nuclear chromosomes; most eukaryotic cells also have mitochondria, and plant, algal, and some protozoan cells have chloroplasts that also contain DNA. DNA molecules of mitochondria and chloroplasts are circular and resemble the circular chromosomes of prokaryotes. Genes located on these "prokaryotic" chromosomes code for about 5% of the RNA and polypeptides required for the organelle's replication and function; nuclear DNA codes for the remaining 95% of polypeptides and RNA molecules required for mitochondrial and chloroplast function. Recall from Chapter 2 that some proteins have a quarternary structure formed from the association of individual polypeptides. Interestingly, polypeptides coded by mitochondrial or chloroplast chromosomes do not alone constitute functional proteins. Rather, they become functional only when associated with polypeptides coded by nuclear chromosomes.

In addition to the extranuclear DNA in their mitochondria, some fungi and protozoa carry plasmids. For instance, most strains of the yeast *Saccharomyces cerevisiae* (sak-ah-rō-mī′sēz se-ri-vis′ē-ē) contain about 70 copies of a plasmid known as a *2-μm circle.* Each 2-μm circle is about 6300 bp long and has four protein-encoding genes that are involved solely in replicating the plasmid and confer no other traits to the cell.

In summary, the genome of a prokaryotic cell consists of both chromosomal DNA, which is usually in a single circular chromosome, and all extrachromosomal DNA in the form of plasmids that is present. By contrast, a eukaryotic genome consists of nuclear chromosomal DNA in one or more linear chromosomes, plus all the extranuclear DNA in mitochondria, chloroplasts, and any plasmids that are present. The genomes of prokaryotes and eukaryotes are compared and contrasted in **Table 7.1.**

CRITICAL THINKING

In Chapter 3 we learned that the endosymbiotic theory proposes that mitochondria and chloroplasts evolved from prokaryotes living within other prokaryotes. What aspects of the eukaryotic genome support this theory? What aspects do not support the theory?

DNA Replication

Learning Objectives

- ✓ Describe the replication of DNA as a semiconservative process.
- ✓ Compare and contrast the synthesis of leading and lagging strands in DNA replication

DNA replication, an anabolic polymerization process, allows a cell to pass copies of its genome to its descendants. As discussed in Chapter 2, all polymerization processes require monomers (building blocks) and energy. *Triphosphate deoxyribonucleotides*—nucleotides with three phosphate groups linked together by two high-energy bonds—serve both functions in DNA replication. In other words, the building blocks of DNA carry within themselves the energy required for DNA synthesis **(Figure 7.4).** The structure of guanosine triphosphate nucleotide (dGTP), shown in Figure 7.4a, differs from that of cytidine triphosphate (dCTP), thymidine triphosphate (dTTP), and adenosine triphosphate (dATP)

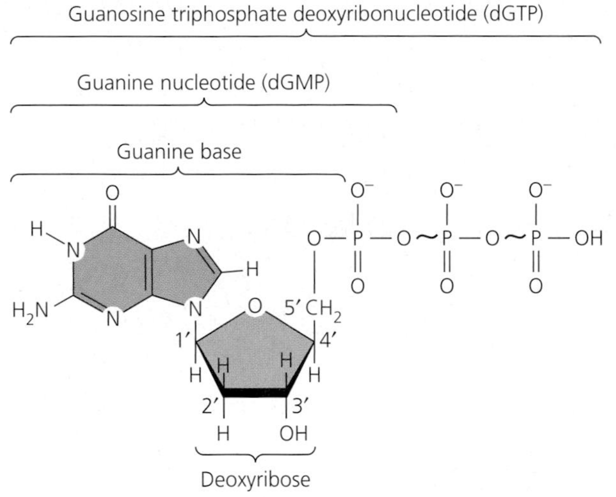

(a)

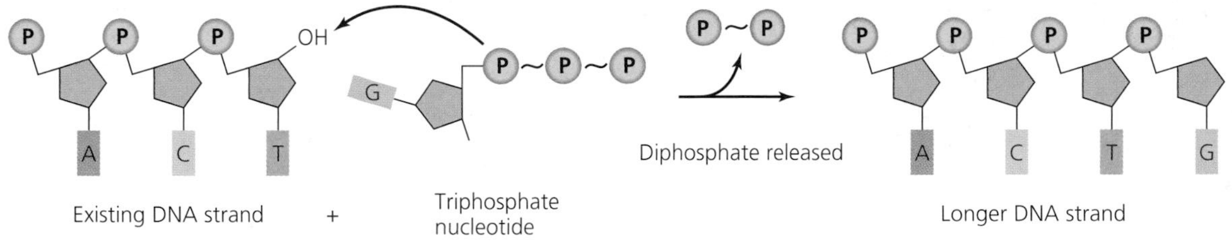

(b)

Figure 7.4
The dual role of triphosphate deoxyribonucleotides as building blocks and energy sources in DNA synthesis. **(a)** Guanosine triphosphate deoxyribonucleotide (dGTP), like all the triphosphate monomers of DNA, is a nucleotide to which two additional phosphate groups are attached. **(b)** The energy required for DNA polymerization (the addition of nucleotide building blocks to a DNA strand) is carried in each triphosphate nucleotide in the high-energy bonds between phosphate groups. *What is the difference between dGTP and guanosine triphosphate ribonucleotide (rGTP)?*

Figure 7.4 *This molecule (dGTP) contains deoxyribose; rGTP contains ribose.*

only in the kind of base present; dATP has a structure similar to that of the energy-storage molecule ATP, except that ATP is an RNA nucleotide (see Figure 2.26).

The key to DNA replication is the complementary structure of the two strands: Adenine and guanine in one strand bond with thymine and cytosine, respectively, in the other. DNA replication is a simple concept—a cell separates the two original strands and uses each as a template for the synthesis of a new complementary strand. Biologists say that DNA replication is *semiconservative* because each daughter DNA molecule is composed of one original strand and one new strand. The following sections examine DNA replication in more detail.

Initial Processes in DNA Replication

DNA replication (**Figure 7.5** on page 206) begins at a specific sequence of nucleotides called an *origin* (not shown). First, a cell removes chromosomal proteins, exposing the DNA helix. Next, enzymes called *DNA helicases* locally "unzip" the DNA molecule by breaking the hydrogen bonds between complementary nucleotide bases, which exposes the nucleotides in a *replication fork* (Figure 7.5a). Other protein molecules stabilize the single strands so that they do not rejoin while replication proceeds.

After the helicases untwist and separate the strands, a molecule of an enzyme called *DNA polymerase* (po-lim′er-ās) binds to each strand. Scientists have identified five kinds of prokaryotic DNA polymerase, and six kinds from eukaryotes. These eleven enzymes vary in their specific functions, but all of them share one important feature—they synthesize DNA by adding nucleotides only to a hydroxyl group at the 3′ end of a nucleic acid. Because of the latter feature, DNA polymerases replicate DNA in only one direction—5′ to 3′—like a jeweler stringing pearls to make a necklace, adding them one at a time, always moving from one end of the string to the other.

Because the two template strands are antiparallel, cells synthesize new strands in two different ways. One strand, called the *leading strand,* is synthesized continuously as a single long chain of nucleotides. The other strand, called the *lagging strand,* is synthesized in short segments that are later joined. We will consider synthesis of the leading strand before examining replication of the lagging strand, even though the two processes occur simultaneously.

Synthesis of the Leading Strand

The cell synthesizes the leading strand toward the replication fork in the following series of five steps, the first three of which are shown in Figure 7.5b:

1. An enzyme called *primase* synthesizes a short RNA molecule that is complementary to the template DNA strand. This *RNA primer* provides a 3′ hydroxyl group required by DNA polymerase.
2. Triphosphate deoxyribonucleotides form hydrogen bonds with their complements in the parental strand.

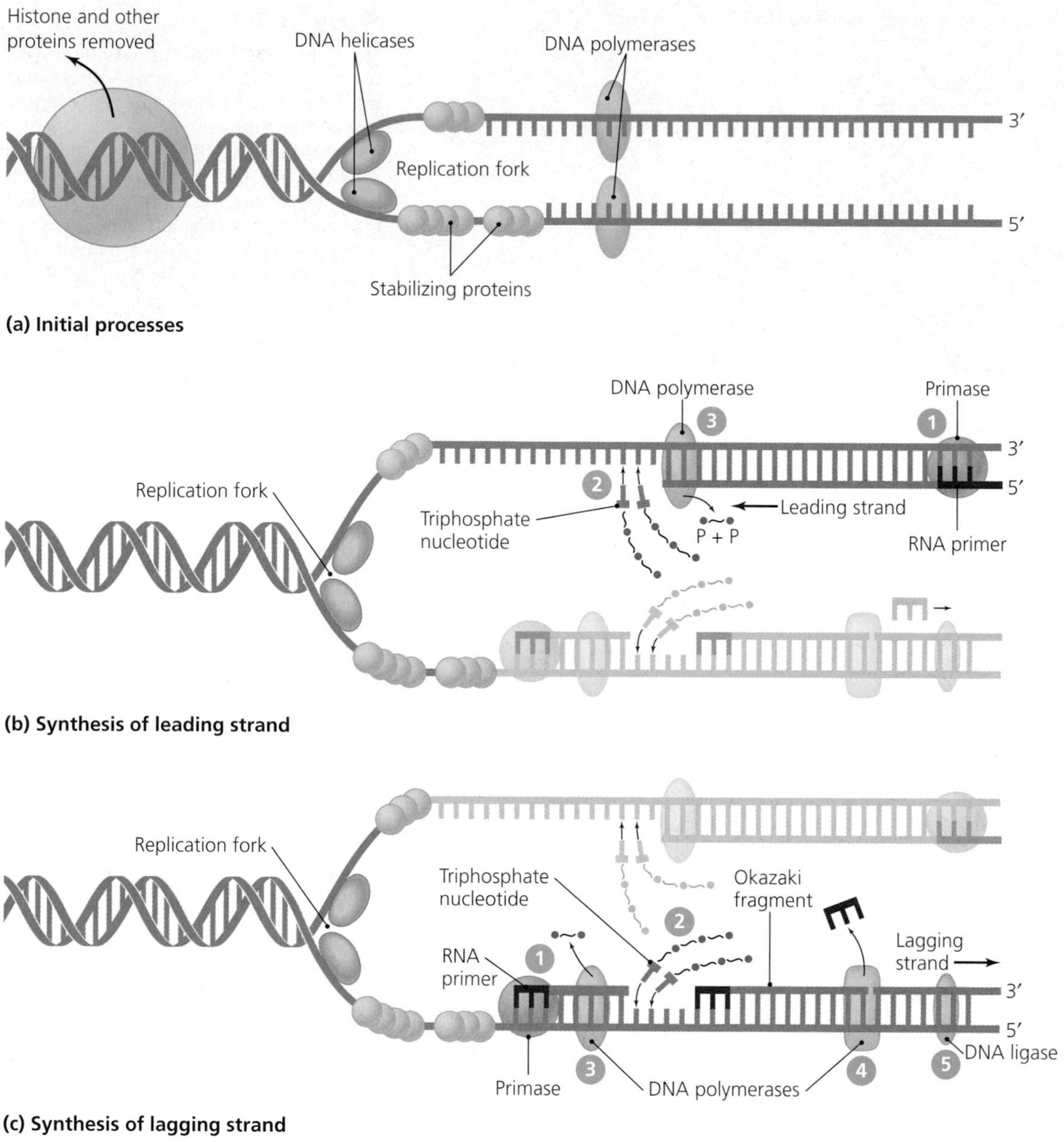

▲ *Figure 7.5*

DNA replication. **(a)** Initial processes. The cell removes histones and other proteins from the DNA molecule. Helicases unzip the double helix—breaking hydrogen bonds between complementary base pairs—to form a replication fork. **(b)** Continuous synthesis of the leading strand. DNA synthesis always moves in the 5′ to 3′ direction, so the leading strand is synthesized toward the replication fork. The numbers refer to the steps in the process, which are described in the text. (Steps 4 and 5, the proofreading function of DNA polymerase and the replacement of the RNA primer with DNA, are not shown.) **(c)** Discontinuous synthesis of the lagging strand, which proceeds moving away from the replication fork. Actual Okazaki fragments are about 1000 nucleotides long. *Why is DNA replication termed "semiconservative"?*

Figure 7.5 "Semiconservative" refers to the fact that each of the daughter molecules retains one parental strand and has one new strand; in other words, each is half new and half old.

3. Using the energy in the high-energy bonds of the triphosphate deoxyribonucleotides, DNA polymerase covalently joins them by dehydration synthesis to form the leading strand. A typical DNA polymerase can add about 500–1000 nucleotides per second to a new strand.
4. DNA polymerase also performs a proofreading function (not shown). About one out of every 100,000 nucleotides is mismatched with its template; for instance, a guanine might become incorrectly paired with a thymine. DNA polymerase recognizes most such errors and removes the incorrect nucleotides before proceeding with synthesis. Because of this proofreading function, only about one error remains for every billion (10^9) base pairs replicated.

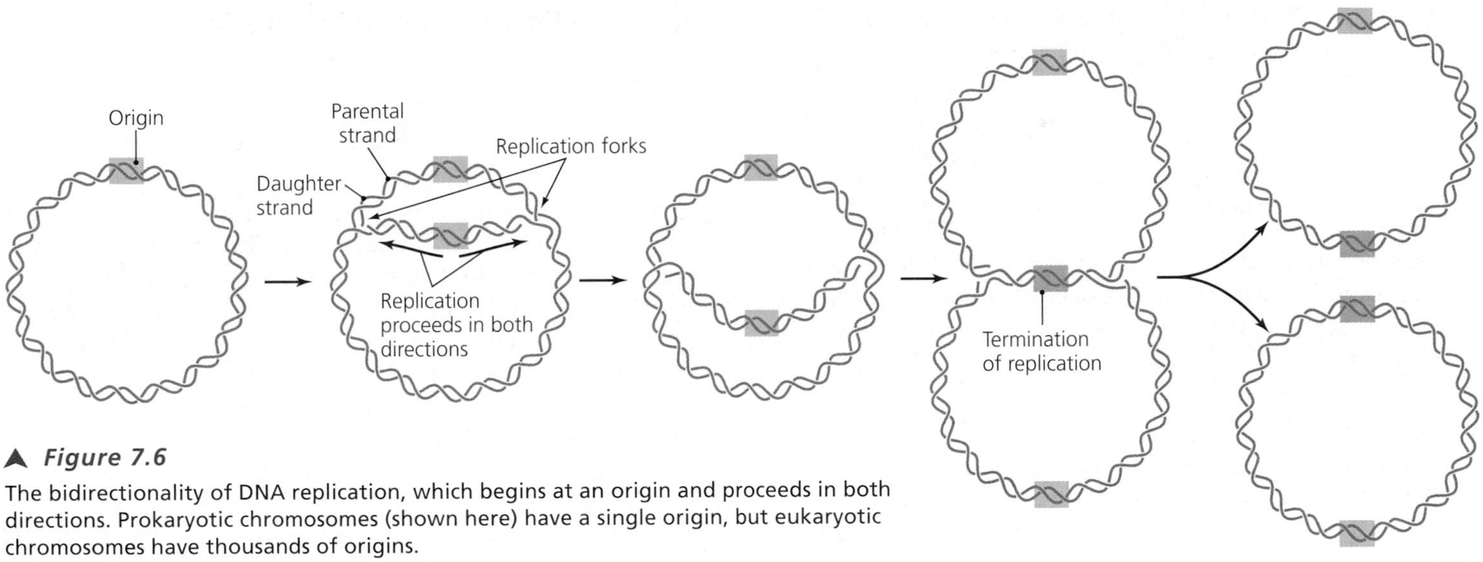

▲ ***Figure 7.6***
The bidirectionality of DNA replication, which begins at an origin and proceeds in both directions. Prokaryotic chromosomes (shown here) have a single origin, but eukaryotic chromosomes have thousands of origins.

5 Another enzyme replaces the RNA primer with DNA (not shown).

Synthesis of the Lagging Strand

Because DNA polymerase adds nucleotides only in a 5′ to 3′ direction, it moves away from the replication fork as it synthesizes the lagging strand. As a result, the lagging strand is synthesized discontinuously and always lags behind the process occurring in the leading strand. The steps in the synthesis of the lagging strand are as follows (Figure 7.5c):

1 As with the leading strand, primase synthesizes RNA primers.

2 Nucleotides pair up with their complements in the template—adenine with thymine, and cytosine with guanine.

3 DNA polymerase joins neighboring nucleotides and proofreads. In contrast to synthesis of the leading strand, however, the lagging strand is synthesized in discontinuous segments called *Okazaki fragments,* named for the Japanese scientist Reiji Okazaki (1930–1975), who first identified them. Each Okazaki fragment requires a new RNA primer and consists of about 1000 nucleotides.

4 Another type of DNA polymerase replaces the RNA primers of Okazaki fragments with DNA and further proofreads the daughter strand.

5 *DNA ligase* seals the gaps between adjacent Okazaki fragments to form a continuous DNA strand.

In summary, synthesis of the leading strand proceeds continuously toward the replication fork from a single RNA primer at the origin, following the helicases and replication fork down the DNA. The lagging strand is synthesized away from the replication fork, discontinuously as a series of Okazaki fragments, each of which begins with its own RNA primer. All the primers are eventually replaced with DNA nucleotides, and ligase joins the Okazaki fragments.

DNA replication is *semiconservative;* each daughter molecule is composed of one parental strand and one daughter strand. This replication process, which produces double-stranded daughter molecules with a nucleotide sequence identical to that in the original double helix, ensures that the integrity of an organism's genome is maintained each time it is copied.

Other Characteristics of DNA Replication

DNA replication is usually *bidirectional;* that is, DNA synthesis proceeds in both directions from the origin. In prokaryotes, the process of replication typically proceeds from a single origin, so it involves two sets of enzymes, two replication forks, two leading strands, and two lagging strands **(Figure 7.6).** By contrast, the large size of eukaryotic chromosomes necessitates thousands of origins per molecule, each generating two replication forks; otherwise, the replication of eukaryotic chromosomes would take days instead of hours.

DNA replication is further complicated by **methylation** of the daughter strands, in which a cell adds a methyl group ($—CH_3$) to one or two bases that are part of specific nucleotide sequences. Plant and animal cells methylate cytosine bases exclusively, whereas prokaryotes typically methylate adenine bases and only rarely a cytosine base.

Methylation can play a role in a variety of cellular processes, including the following:

- *Control of genetic expression.* In some cases, genes that are methylated are "turned off" and are not transcribed, whereas in other cases methylated genes are "turned on" and are transcribed.
- *Initiation of DNA replication.* In many prokaryotes, methylated nucleotide sequences play a role in initiating DNA replication.
- *Protection against viral infection.* Methylation at specific sites in the nucleotide sequence enables cells to

distinguish their DNA from viral DNA, which lacks methylation. The cells can then selectively degrade the viral DNA.

- *Repair of DNA.* The role of methylation in some DNA repair mechanisms is discussed on page 225.

CRITICAL THINKING

We have seen that the hydrogen bonds between complementary nucleotides are crucial to the structure of dsDNA because they hold the two strands together. Why couldn't the two strands be effectively linked by covalent bonds?

We have examined the physical structure of cellular genes—the specific sequences of DNA nucleotides—and how cells replicate their genes. Now we will consider how genes function, and how cells control genetic expression.

Gene Function

The first topic we must consider if we are to understand gene function is the relationship between an organism's genotype and its phenotype.

The Relationship Between Genotype and Phenotype

Learning Objective

✓ Explain how the genotype of an organism determines its phenotype.

The **genotype**[2] (jen′ō-tīp) of an organism is the actual set of genes in its genome. A genotype differs from a genome in that a genome also includes nucleotides that are not part of genes, such as the nucleotide sequences that link genes together. At the molecular level, the genotype consists of all the series of DNA nucleotides that carry instructions for an organism's life. **Phenotype**[3] (fē′nō-tīp) refers to the physical features and functional traits of an organism, including characteristics such as structures, morphology, and metabolism. For example, the shape of a cell, the presence and location of flagella, the enzymes and cytochromes of electron transport chains, and membrane receptors that trigger chemotaxis are all phenotypic traits.

Genotype determines phenotype by specifying what kinds of RNA and which structural, enzymatic, and regulatory protein molecules are produced. Though genes do not code *directly* for such molecules as phospholipids or for behaviors such as chemotaxis, ultimately phenotypic traits result from the actions of RNA and protein molecules that are themselves coded by DNA.

Not all genes are active at all times; that is, the information of a genotype is not always expressed as a phenotype. For example, *E. coli* activates genes for lactose catabolism only when it detects lactose in its environment.

The Transfer of Genetic Information

Learning Objective

✓ State the central dogma of genetics, and explain the roles of DNA and RNA in polypeptide synthesis.

Cells must continually synthesize proteins required for growth, reproduction, metabolism, and regulation. This synthesis requires that they accurately transfer the genetic information contained in DNA nucleotide sequences to the amino acid sequences of polypeptides. However, cells do not transfer the information coded in DNA directly, but first make an RNA copy of the gene. In this copying process, called **transcription,**[4] the information is copied as RNA nucleotide sequences; RNA molecules in ribosomes then synthesize polypeptides in a process called **translation.**[5] These processes make up the **central dogma** of genetics: DNA is transcribed to RNA, which is translated to form polypeptides. **Highlight 7.1** presents an analogy of the central dogma. There are a few exceptions to the central dogma. For example, some RNA viruses transcribe DNA from an RNA template—a process that is the reverse of cellular transcription.

In the following sections, we will examine the processes of transcription and translation.

The Events in Transcription

Learning Objective

✓ Describe three steps in RNA transcription, mentioning the following: DNA, RNA polymerase, promoter, 5′ to 3′ direction, and terminator.

Cells transcribe three types of RNA from DNA:

- **messenger RNA (mRNA)** molecules, which carry genetic information from chromosomes to ribosomes;
- **ribosomal RNA (rRNA)** molecules, which combine with ribosomal polypeptides to form ribosomes—the organelles that synthesize polypeptides; and
- **transfer RNA (tRNA)** molecules, which deliver amino acids to the ribosomes.

We will more closely examine the functions of each type of RNA shortly.

Whereas transcription occurs in the nucleoid region of the cytoplasm in prokaryotes, in eukaryotes it occurs in the nucleus, mitochondria, and chloroplasts. In both prokaryotes and eukaryotes, the steps of RNA transcription are similar: *initiation of transcription, elongation of the RNA transcript,* and *termination of transcription.* The events of transcription in bacteria are depicted in **Figure 7.7** on page 210.

[2]From Greek *genos,* meaning race, and *typos,* meaning type.
[3]From Greek *phainein,* meaning to show.
[4]From Latin *trans,* meaning across, and *scribere,* meaning to write—that is, to transfer in writing.
[5]From Latin *translatus,* meaning transferred.

Highlight 7.1 Understanding the Central Dogma: It's (No Longer) Greek to Me

Suppose you were trying to understand the following message (which is a portion of the oath of Hippocrates, written in the Greek alphabet):

ΔιαιτημασιτεχρησομαιεπωΦελεινκαμνοντωνκατα
δυναμινκαικρισινεμηνεπιδηληειδεκαιαδικιηειρζειν.

If the Greek alphabet is foreign to you, you might have the Greek characters *transcribed* into the familiar English alphabet as a first step in understanding the message:

Diaiteimasi te chreisomai ep ophelein kamnonton kata dunamin kai krisin emein epi deileisei de kai adikiei eirzein

Then you could begin the process of having the Greek words, now expressed in English letters, *translated* into English words:

I will prescribe treatment to the best of my ability and judgment to help the sick and never for a harmful or illicit purpose

To a ribosome in a cell, DNA would be like a foreign language written in a foreign alphabet. Thus, a cell must use processes analogous to those just described: It must first *transcribe* the "foreign alphabet" of DNA nucleotides (genes) into the more "familiar alphabet" of RNA nucleotides; then it must *translate* the message formed by these "letters" into the "words" (amino acids) that make up the "message" (a polypeptide). In this way a genotype can be expressed as a phenotype.

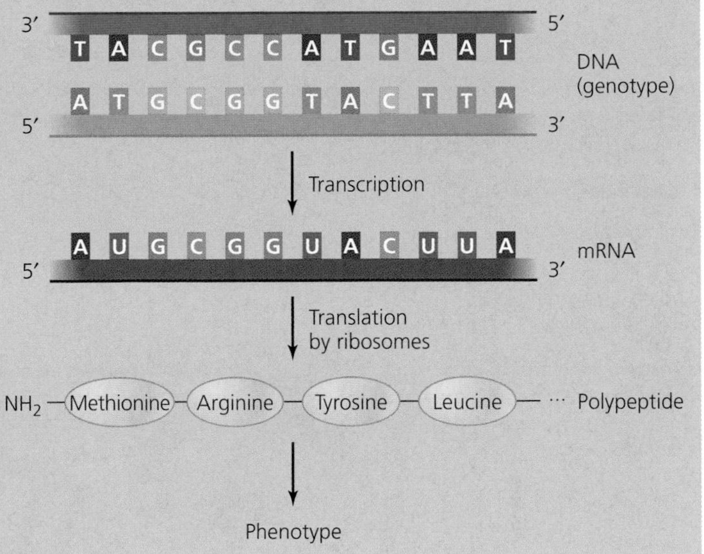

Initiation of Transcription

RNA polymerase—the enzyme that synthesizes RNA—moves by an unknown mechanism along a DNA helix until it reaches a specific nucleotide sequence called a **promoter,** which is located near the beginning of a gene and initiates transcription (Figure 7.7a). In bacteria, a portion of RNA polymerase called the *sigma factor* is necessary for recognition of the promoter. In contrast, separate protein *transcription factors* assist in binding eukaryotic RNA polymerase to promoters. Once it adheres to a promoter, RNA polymerase unwinds and unzips the DNA in the promoter region; and then continues moving along the DNA, unzipping the double helix as it moves.

RNA polymerases do not adhere equally strongly to all promoter sequences. The greater the attraction between RNA polymerase and a promoter, the more likely that transcription will proceed. One way that cells control the relative amounts of transcribed RNA is by using different promoter sequences for different genes. Ultimately, variation in promoters affects the amounts and kinds of polypeptides produced.

Elongation of the RNA Transcript

RNA transcription does not actually begin in the promoter region, but at a spot 10 nucleotides away. There, triphosphate ribonucleotides (rATP, rUTP, rGTP, and rCTP) align opposite their complements in the open DNA. RNA polymerase links together two adjacent ribonucleotide molecules using energy from the phosphate bonds of the first ribonucleotide. The enzyme then moves down the DNA strand, elongating RNA by repeating the process (Figure 7.7b). During transcription of the first 10 ribonucleotides, RNA polymerase releases its sigma factor, which causes RNA polymerase to adhere tightly to the DNA. Once this occurs, transcription proceeds to completion. The released sigma factor joins with another molecule of RNA polymerase to locate another promoter and initiate another transcription.

Many molecules of RNA polymerase may concurrently transcribe the same gene (**Figure 7.8** on page 211). In this way, a cell simultaneously produces numerous identical copies of RNA from a single gene—much in the way many identical prints can be made from a single photographic negative.

Like DNA polymerase, RNA polymerase links nucleotides in the 5′ to 3′ direction only; however, RNA polymerase differs from DNA polymerase in the following ways:

- RNA polymerase unwinds and opens DNA by itself; helicase is not required.
- RNA polymerase is slower than DNA polymerase, proceeding at a rate of about 50 nucleotides per second.
- RNA polymerase incorporates ribonucleotides instead of deoxyribonucleotides.
- Uracil nucleotides are incorporated instead of thymine nucleotides.
- RNA polymerase lacks a proofreading function, leaving a base pair error about every 10,000 nucleotides.

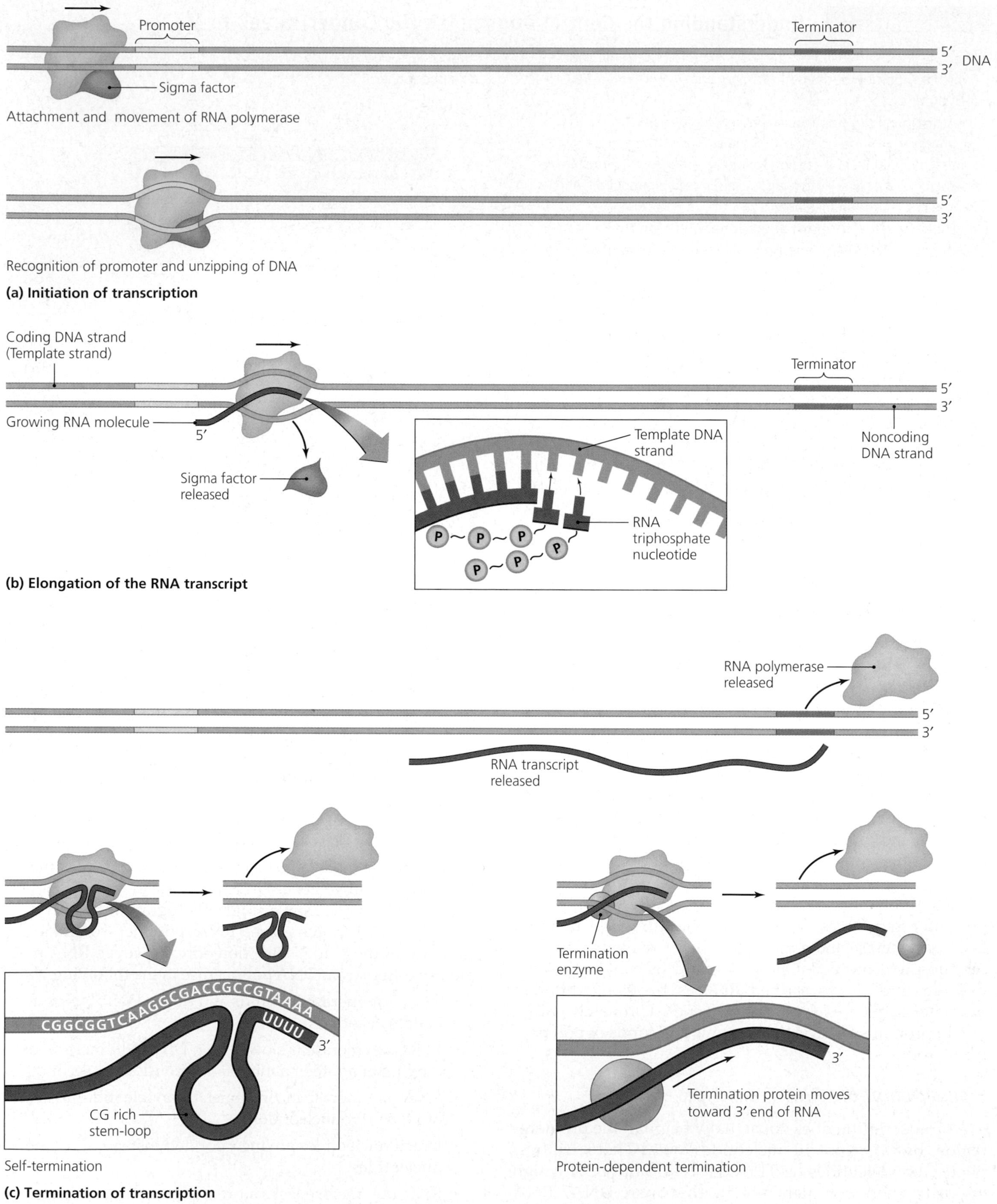
Promoter
Terminator
5′
3′
DNA
Sigma factor
Attachment and movement of RNA polymerase
5′
3′
Recognition of promoter and unzipping of DNA
(a) Initiation of transcription
Coding DNA strand
(Template strand)
Terminator
5′
3′
Growing RNA molecule
5′
Noncoding
DNA strand
Template DNA
strand
Sigma factor
released
RNA
triphosphate
nucleotide
P
(b) Elongation of the RNA transcript
RNA polymerase
released
5′
3′
RNA transcript
released
Termination
enzyme
CGGCGGTCAAGGCGACCGCCGTAAAA
UUUU
3′
CG rich
stem-loop
3′
Termination protein moves
toward 3′ end of RNA
Self-termination
Protein-dependent termination
(c) Termination of transcription

Figure 7.7

The events in the transcription of RNA in bacteria. The helical dsDNA molecule is depicted as straight, parallel strands for clarity. **(a)** Initiation of transcription. RNA polymerase attaches nonspecifically to DNA and travels down its length until it recognizes a promoter sequence. Sigma factor enhances promoter recognition in bacteria. Upon recognition of the promoter, RNA polymerase unzips the DNA molecule beginning at the promoter. **(b)** Elongation of the RNA transcript. After triphosphate ribonucleotides align with their DNA complements, RNA polymerase links them together, synthesizing RNA. The triphosphate ribonucleotides also provide the energy required for RNA synthesis. **(c)** Termination of transcription, which is effected by the release of RNA polymerase. In self-termination, the transcription of DNA terminator sequences cause the RNA to fold, loosening the grip of RNA polymerase on the DNA; in enzyme-dependent termination, a termination enzyme pushes between RNA polymerase and the DNA, releasing the polymerase. *What is the difference between a promoter sequence and an origin?*

***Figure 7.7** A promoter is a DNA sequence that initiates transcription; an origin is a point where DNA replication begins.*

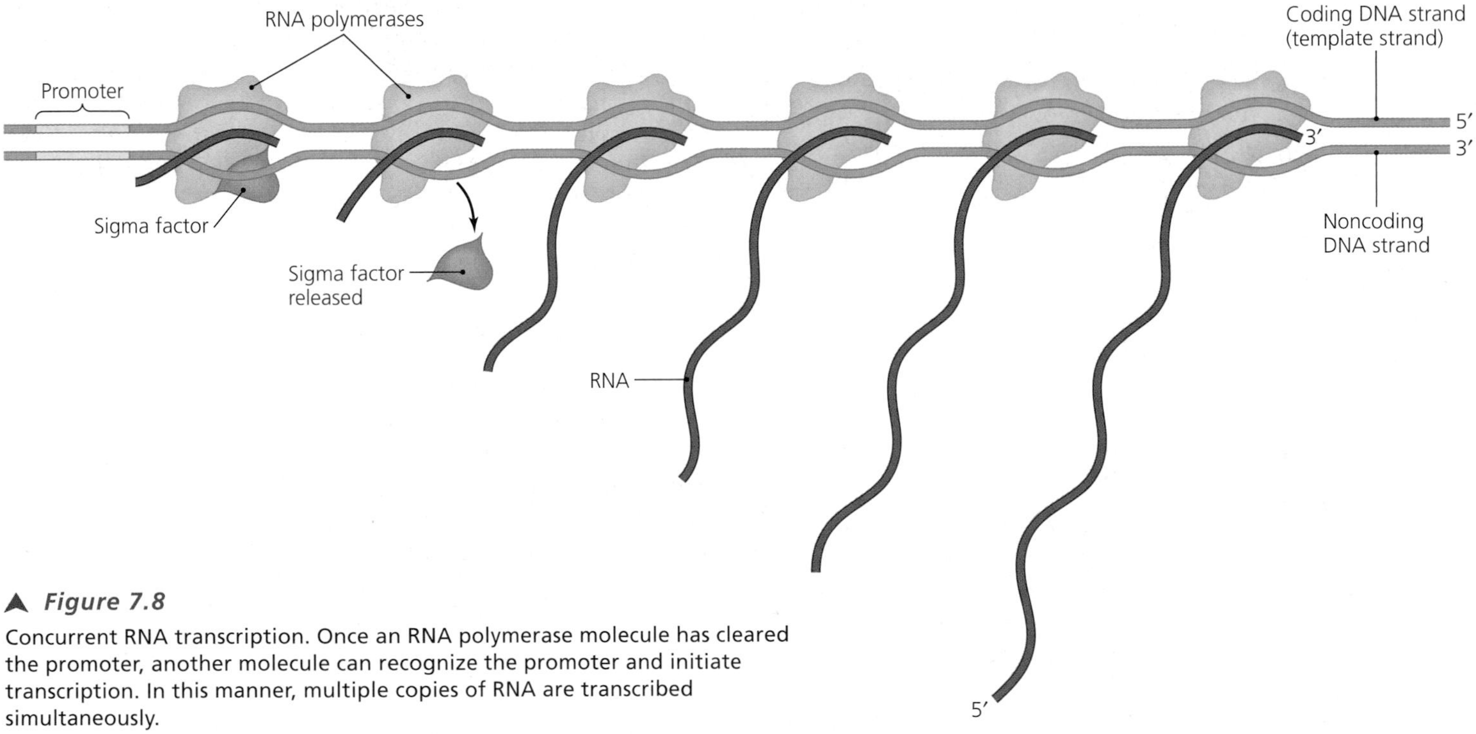

Figure 7.8

Concurrent RNA transcription. Once an RNA polymerase molecule has cleared the promoter, another molecule can recognize the promoter and initiate transcription. In this manner, multiple copies of RNA are transcribed simultaneously.

CRITICAL THINKING

On average, RNA polymerase makes one error for every 10,000 nucleotides it incorporates in RNA. By contrast, only one base pair error remains for every billion base pairs during DNA replication. Explain why the accuracy of RNA transcription is not as critical as the accuracy of DNA replication.

Termination of Transcription

Transcription terminates when RNA polymerase and the transcribed RNA are released from DNA (see Figure 7.7c). As stated previously, once RNA polymerase loses its sigma factor, it becomes tightly associated with the DNA molecule and cannot be removed easily; therefore, the termination of transcription is complicated. Scientists have elucidated two types of termination processes in bacteria: those that are self-terminating and those that depend on the action of an additional termination enzyme.

Self-Termination Self-termination occurs when RNA polymerase transcribes a **terminator** sequence of DNA composed of two symmetrical series: one that is very rich in guanine and cytosine bases, followed by a region rich in adenine bases (see Figure 7.7c, lower left). RNA polymerase slows down during transcription of the GC rich portion of the terminator because the three hydrogen bonds between each guanine and cytosine base pair makes unwinding the DNA helix more difficult. This pause in transcription provides enough time for the RNA molecule to form hydrogen bonds between its own symmetrical sequences, forming a stem and loop structure that puts tension on the union of RNA polymerase and the DNA. When RNA polymerase transcribes the adenine rich portion of the terminator, the relatively few hydrogen bonds between the adenine bases of DNA and the uracil bases of RNA cannot withstand the tension, and the RNA transcript breaks away from the DNA, releasing RNA polymerase.

Enzyme-Dependent Termination The second type of termination depends on a termination protein that binds to a specific RNA sequence near the end of an RNA transcript. The protein moves toward the 3′ end, pushing between RNA polymerase and the DNA strand and forcing them apart; this releases RNA polymerase and the RNA transcript (see Figure 7.7c, lower right).

Now that we have discussed how cells use DNA as the genetic material, maintain the integrity of their genomes through semiconservative replication, and transcribe RNA from DNA genes, we turn to the process of translation and the role of each type of RNA.

Translation

Learning Objectives

- ✓ Describe the genetic code in general, and identify the relationship between codons and amino acids.
- ✓ Describe the translation of polypeptides, identifying the roles of the three types of RNA.

Translation is the process whereby ribosomes use the genetic information of nucleotide sequences to synthesize polypeptides composed of specific amino acid sequences. As we studied in Chapter 2, some proteins are simple polypeptides, whereas other proteins are composed of several polypeptides bound together in a quaternary structure.

Ribosomes can be thought of as "polypeptide factories," so consider the following analogy between translation and a hypothetical automobile factory. Boxcars deliver preformed parts to the factory at the correct times and in the correct order to manufacture one of a large variety of automobile models, depending on instructions from corporate headquarters delivered by special courier. Similarly, molecules of tRNA (the boxcars) deliver preformed amino acids (the parts) to a ribosome (the factory), which can manufacture an infinite variety of polypeptides (the car models) by assembling amino acids (the parts) in the correct order according to the instructions from DNA (corporate headquarters) delivered via mRNA (the special courier).

How do ribosomes interpret the nucleotide sequence of mRNA to determine the correct order in which to assemble amino acids? To answer this question, we will consider the genetic code, examine in more detail the RNA molecules that participate in translation, and describe the specific steps of translation.

The Genetic Code

When geneticists in the early 20th century began to consider that DNA might be the genetic molecule, they were confronted with a problem: How can four kinds of DNA nucleotides (adenine, thymine, guanine, and cytosine) specify the 21 different amino acids commonly found in proteins? If each nucleotide coded for one amino acid, only four amino acid could be specified. Even if pairs of nucleotides served as the code—for instance, if AA, AT, and TA each specified a different amino acid—only 16 amino acids (that is, 4^2) could be accommodated. Eventually scientists showed that genes are composed of sequences of three nucleotides that specify amino acids. For example, the DNA nucleotide sequence TTT specifies the amino acid lysine, and TTA codes for asparagine. There are 64 possible arrangements of the four nucleotides in triplets (4^3)—more than enough to specify 21 amino acids.

These examples are DNA triplets, but ribosomes do not directly access genetic information on a DNA molecule. Instead, molecules of mRNA carry the code to the ribosomes; therefore, scientists define the genetic code **(Figure 7.9)** as triplets of mRNA nucleotides called **codons** (kō′ donz) that code for specific amino acids. AAA is a codon for lysine, and AAU is a codon for asparagine.

In most cases, 61 codons specify amino acids and three codons—UAA, UAG, and UGA—instruct ribosomes to stop translating; though, under some conditions, UGA codes for the 21st amino acid, selenocysteine. Codon AUG also has a dual function, acting as both a start signal and coding for an amino acid. For eukaryotic and archaeal proteins, AUG codes for methionine, while in prokaryotes, mitochondria, and chloroplasts it codes for N-formylmethionine (fMet), which is similar in structure to methionine:

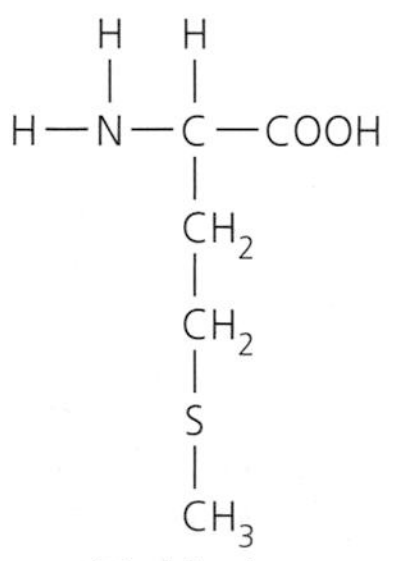

Methionine

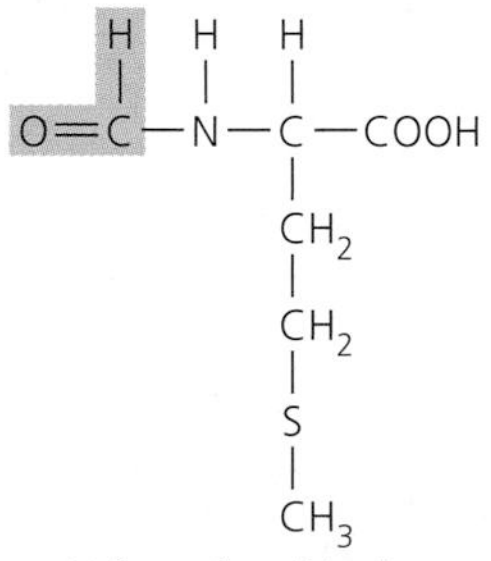

N-formylmethionine

As you examine the genetic code, notice that it is redundant; that is, more than one codon is associated with all the amino acids except methionine and tryptophan. With most redundant codons, the first two nucleotides determine the amino acid, and the third nucleotide is inconsequential. For example, the codons GUU, GUC, GUA, and GUG all specify the amino acid valine.

Interestingly, the genetic code is nearly universal; that is, with few exceptions, ribosomes in archaeal, bacterial, plant, fungal, protozoan, and animal cells use the same genetic code. Some exceptions are listed in **Table 7.2.**

First nucleotide base (5′ position)	Second nucleotide base: U	Second nucleotide base: C	Second nucleotide base: A	Second nucleotide base: G	Third nucleotide base (3′ position)
U	UUU Phenylalanine (Phe)	UCU Serine (Ser)	UAU Tyrosine (Tyr)	UGU Cysteine (Cys)	U
U	UUC	UCC	UAC	UGC	C
U	UUA Leucine (Leu)	UCA	UAA STOP	UGA STOP Selenocysteine (SeCys)	A
U	UUG	UCG	UAG STOP	UGG Tryptophan (Trp)	G
C	CUU Leucine (Leu)	CCU Proline (Pro)	CAU Histidine (His)	CGU Arginine (Arg)	U
C	CUC	CCC	CAC	CGC	C
C	CUA	CCA	CAA Glutamine (Gln)	CGA	A
C	CUG	CCG	CAG	CGG	G
A	AUU Isoleucine (Ile)	ACU Threonine (Thr)	AAU Asparagine (Asn)	AGU Serine (Ser)	U
A	AUC	ACC	AAC	AGC	C
A	AUA	ACA	AAA Lysine (Lys)	AGA Arginine (Agn)	A
A	AUG START Methionine (Met) fmethionine (fMet) in prokaryotes	ACG	AAG	AGG	G
G	GUU Valine (Val)	GCU Alanine (Ala)	GAU Aspartic acid (Asp)	GGU Glycine (Gly)	U
G	GUC	GCC	GAC	GGC	C
G	GUA	GCA	GAA Glutamic acid (Glu)	GGA	A
G	GUG	GCG	GAG	GGG	G

▲ ***Figure 7.9***

The genetic code, the complete set of mRNA codons and the amino acids for which they code. AUG is both the start codon and specifies methionine (Met) in eukaryotes and N-formylmethionine (fMet) in prokaryotes, mitochondria, and chloroplasts. Three codons (UAA, UAG, and UGA) are stop codons that do not typically specify amino acids.

Table 7.2 Some Exceptions to the Genetic Code

Codon	Usual Use	Alternate Use
AUA	Codes for isoleucine	Codes for methionine in mitochondria
UAG	STOP	Codes for glutamine in some protozoa and algae and for pyrrolysine (22nd amino acid) in some prokaryotes
CGG	Codes for arginine	Codes for tryptophan in plant mitochondria
UGA	STOP, selenocysteine	Codes for tryptophan in mitochondria and mycoplasmas (type of bacteria)

CRITICAL THINKING

A scientist isolates a molecule of mRNA with the following base sequence: AUGUACGACAUAUGCAUA. What is the sequence of amino acids in the polypeptide synthesized by a prokaryotic ribosome from this message? What would be different if the message were translated in a mitochondrion instead?

Participants in Translation

As we have discussed, transcription produces messenger RNA, transfer RNA, and ribosomal RNA—each of which is involved in translation. We now discuss each kind of RNA in turn.

Messenger RNA Messenger RNA carries genetic information (in the form of RNA nucleotide sequences) from a chromosome to ribosomes. In prokaryotes a basic mRNA molecule contains sequences of nucleotides that are recognized by ribosomes: an AUG start codon, sequential codons for other amino acids in the polypeptide, and at least one of the three stop codons. A single molecule of prokaryotic mRNA often contains start codons and instructions for more than one polypeptide arranged in series (**Figure 7.10** on page 214). Because both transcription and the subsequent events of translation occur in the cytosol of prokaryotes, prokaryotic ribosomes can begin translation before transcription is finished.

Eukaryotic mRNA differs from prokaryotic mRNA in several ways:

- Newly synthesized eukaryotic mRNA is called *pre-messenger RNA* because it contains noncoding

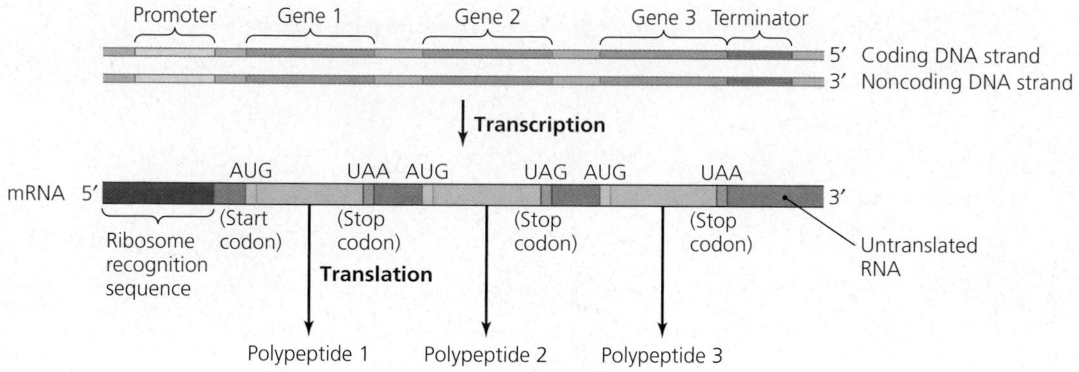

Figure 7.10 ➤

Prokaryotic mRNA. Prokaryotes typically code for several related polypeptides via a single mRNA molecule. The mRNA molecule shown here has transcripts of three genes coding three polypeptides. The transcript of each gene begins with a start codon and ends with a stop codon.

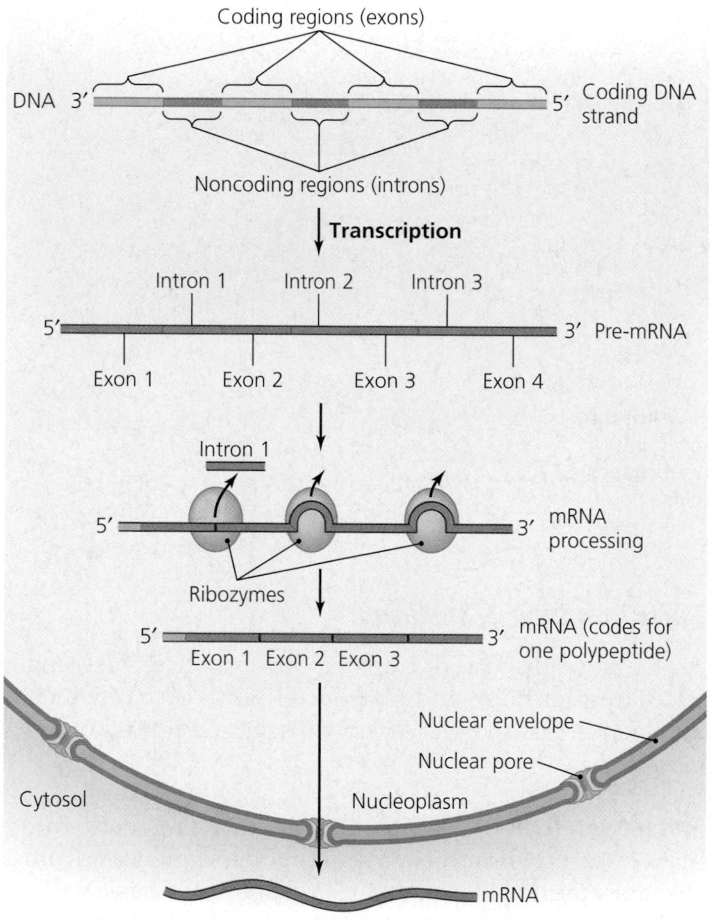

▲ ***Figure 7.11***

Eukaryotic mRNA. Within the nucleus, transcription produces pre-mRNA, which contains coding exons and noncoding introns. Ribozymes process pre-mRNA by removing introns and splicing together exons to form a molecule that codes for a single polypeptide. Eukaryotic mRNA then moves from the nucleus to the cytoplasm.

sequences called **introns,** which are removed to make functional mRNA containing only coding regions called **exons (Figure 7.11).** The "in" in *intron* refers to *intervening* sequences (that is, they lie between coding regions), whereas the "ex" in exon refers to the fact that these coding regions are *expressed.* Small RNA molecules in the nucleus act as ribozymes (ribosomal enzymes) to process pre-mRNA into mRNA; that is, they remove the introns and connect the exons to produce a functional mRNA molecule.

- A molecule of eukaryotic mRNA contains instructions for only one polypeptide.
- Eukaryotic mRNA is not translated until it is fully transcribed and processed and has left the nucleus because eukaryotic ribosomes are located in the cytoplasm. In other words, transcription and translation of a molecule of eukaryotic mRNA do not occur simultaneously.

Transfer RNA A transfer RNA (tRNA) molecule is a sequence of about 75 ribonucleotides that curves back on itself to form three main hairpin loops held in place by hydrogen bonding between complementary nucleotides. Although transfer RNA molecules can be modeled simplistically by a cloverleaf structure **(Figure 7.12a),** their three-dimensional shape is more complex. For simplicity, tRNA will be represented in subsequent figures by an icon shaped like the illustration in **Figure 7.12b.**

A molecule of tRNA transfers the correct amino acid to a ribosome during polypeptide synthesis. To this end, a tRNA has an *acceptor stem* for a specific amino acid at its 3′ end, and an **anticodon** (an-tē-kō′ don) triplet in its bottom loop (see Figure 7.12b). Specific enzymes in the cytoplasm *charge* each tRNA molecule; that is, they attach the appropriate amino acid to the acceptor stem. The existence of only one specific enzyme for each amino acid ensures that every tRNA molecule carries only one specific amino acid.

Anticodons are complementary to mRNA codons, and each acceptor stem is designed to carry one particular amino acid, which varies with the tRNA. In other words, each transfer RNA carries a specific amino acid and recognizes mRNA codons only for that amino acid. A tRNA molecule is designated by a superscript abbreviation of its amino acid. For example, $tRNA^{Phe}$ carries phenylalanine, and $tRNA^{Ser}$ transfers serine.

The fact that 62 codons specify the 21 amino acids used by cells does not mean that there must be 62 anticodons on

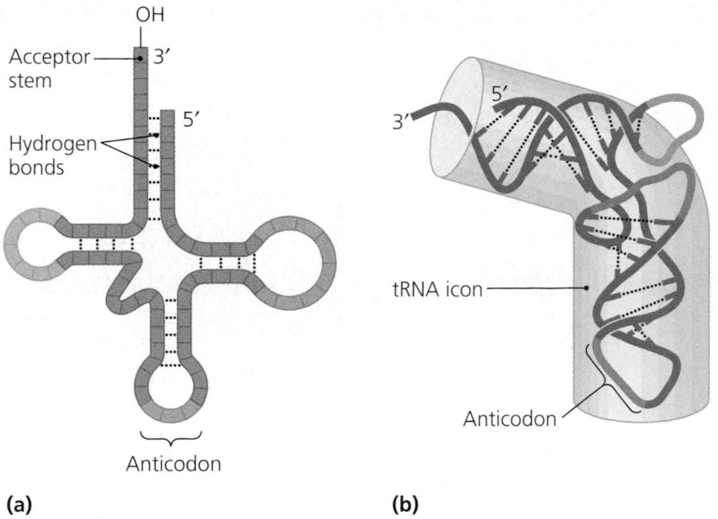

▲ *Figure 7.12*

Transfer RNA. **(a)** A two-dimensional "cloverleaf" representation of tRNA showing three hairpin loops held in place by intramolecular hydrogen bonding. **(b)** A three-dimensional drawing of the same tRNA. A specific amino acid attaches to the 3′ acceptor stem; the anticodon is a nucleotide triplet that is complementary to the mRNA codon for that amino acid.

62 different types of tRNA, because many tRNA molecules recognize more than one codon. *E. coli,* for example, has only about 40 different tRNAs. The variability in codon recognition by tRNA is due to "wobble" of the anticodon's third nucleotide. Wobble, which is a change of angle from the normal axis of the molecule, allows the third nucleotide to hydrogen bond to a nucleotide other than its usual complement. For example, a guanine nucleotide in the third position normally bonds to cytosine, but it can wobble and also pair with uracil; whether a codon has cytosine or uracil in the third position makes no difference, because the same tRNA recognizes either nucleotide in the third position. For example, the codons UUC and UUU both specify the amino acid phenylalanine because the anticodon AAG recognizes both of them. Similarly, UCC and UCU code for serine, and UAC and UAU code for tyrosine. This redundancy in the genetic code helps protect cells against the effects of errors in replication and transcription.

CRITICAL THINKING

We have seen that wobble makes the genetic code redundant in the third position for C and U. After reexamining the genetic code in Figure 7.9, state what other nucleotides in the second or third position appear to accommodate anticodon wobbling.

Ribosomes and Ribosomal RNA Prokaryotic ribosomes, which are also called 70S ribosomes based on their sedimentation rate in an ultracentrifuge, are extremely complex associations of ribosomal RNAs and polypeptides. Each ribosome is composed of two subunits: 50S and 30S **(Figure 7.13a).** The 50S subunit is in turn composed of two rRNA molecules (23S and 5S) and 33 different polypeptides, whereas the 30S subunit consists of one molecule of 16S rRNA and 21 ribosomal polypeptides. The ribosomes of mitochondria and chloroplasts are also 70S ribosomes composed of the similar subunits and polypeptides.

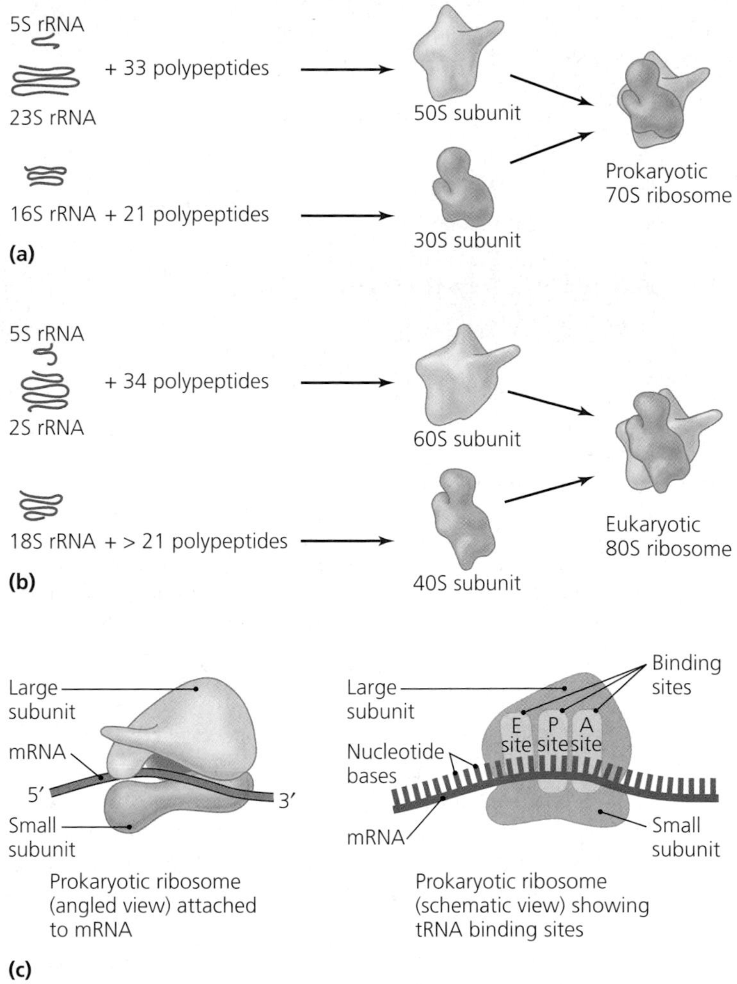

▲ *Figure 7.13*

Ribosomal structure. **(a)** The 70S prokaryotic ribosome, which is composed of polypeptides and three rRNA molecules arranged in 50S and 30S subunits. **(b)** The 80S eukaryotic ribosome, which is composed of molecules of rRNA and polypeptides, arranged in 60S and 40S subunits. **(c)** Transfer RNA binding sites in a ribosome, which is presented schematically for simplicity. The P site holds the transfer RNA attached to the growing polypeptide, whereas the A site accepts the tRNA carrying the next amino acid in the polypeptide. Transfer RNAs exit from the E site.

In contrast, both the cytosol and the rough endoplasmic reticulum (RER) of eukaryotic cells have 80S ribosomes composed of 60S and 40S subunits **(Figure 7.13b).** These subunits contain larger molecules of rRNA and more polypeptides than the corresponding prokaryotic subunits, though researchers do not agree on their exact number. The term *eukaryotic ribosome* is understood to mean only the 80S

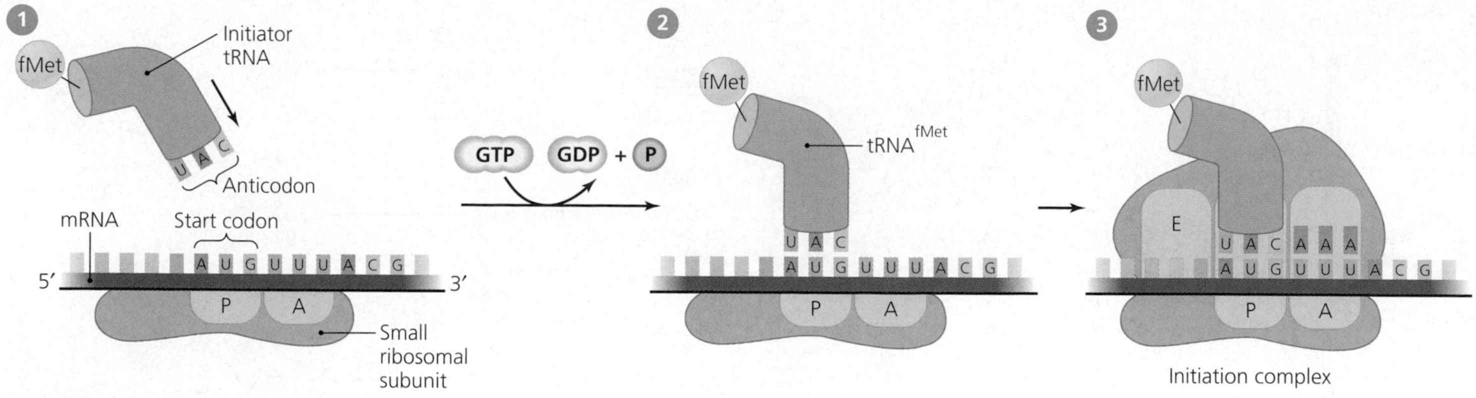

▲ ***Figure 7.14***

The initiation of translation in prokaryotes. ① The smaller ribosomal subunit attaches to mRNA at a ribosomal recognition sequence containing a start codon (AUG). ② The anticodon of $tRNA^{fMet}$ aligns with the start codon on the mRNA; energy from GTP (not shown) is used to bind the tRNA in place. ③ The larger ribosomal subunit attaches to form an initiation complex—a complete ribosome attached to mRNA.

ribosomes of the cytosol and RER. Since the ribosomes of mitochondria and chloroplasts are 70S, they are called *prokaryotic ribosomes* even though they are in eukaryotic cells.

The structural differences between prokaryotic and eukaryotic ribosomes play a crucial role in the efficacy and safety of antimicrobial drugs. Because erythromycin, for example, binds only to the 23S rRNA found exclusively in prokaryotic ribosomes, it has no effect on eukaryotic ribosomes, and thus no deleterious effect on the patient. Chapter 10 discusses antimicrobial drugs in more detail.

The smaller subunit of a ribosome is shaped to accommodate three codons at one time—that is, nine nucleotide bases of a molecule of mRNA. Each ribosome also has three tRNA binding sites that are named for their function **(Figure 7.13c):**

- The **A site** accommodates a tRNA delivering an *amino acid.*
- The **P site** holds a tRNA and the growing *polypeptide.*
- Discharged tRNAs *exit* from the **E site.**

We will now examine translation, the process whereby ribosomes actually synthesize polypeptides using amino acids delivered by tRNAs and following the genetic instructions of mRNA.

Stages of Translation

Researchers divide translation into three stages: *initiation, elongation,* and *termination.* All three stages require additional protein factors that assist the ribosomes. Initiation and elongation also require energy provided by molecules of the ribonucleotide GTP, which are free in the cytosol (that is, they are not part of an RNA molecule).

Initiation During initiation, mRNA, the two ribosomal subunits, several protein factors, and $tRNA^{fMet}$ (in prokaryotes) or $tRNA^{Met}$ (in eukaryotes) form an *initiation complex.* The events of initiation in a bacterium are as follows **(Figure 7.14):**

1. The smaller ribosomal subunit attaches to mRNA at a ribosomal recognition sequence, with the start codon at its P site.
2. $tRNA^{fMet}$ (whose anticodon is complementary to the start codon) attaches to the ribosome's P site; GTP supplies the energy required for binding.
3. The larger ribosomal subunit attaches to form a complete initiation complex.

Elongation Synthesis of a polypeptide is a cyclical process that involves the sequential addition of amino acids to a polypeptide chain growing at the P site. **Figure 7.15** illustrates several cycles of the process. The steps of each cycle occur as follows:

1. The transfer RNA whose anticodon matches the next codon—in this case, phenylalanine (Phe)—delivers its amino acid to the A site. Another protein called *elongation factor* (not shown) escorts the tRNA along with a molecule of GTP. Energy from GTP (not shown) is used to stabilize each tRNA as it is added to the A site.
2. A ribozyme in the larger ribosomal subunit forms a peptide bond by dehydration synthesis between the terminal amino acid of the polypeptide chain (in this case, N-formylmethionine) and the newly introduced amino acid. The polypeptide is now attached to the tRNA occupying the A site.
3. Using energy supplied by more GTP, the ribosome moves one codon down the mRNA. This transfers each tRNA to the adjacent binding site; that is, the first tRNA moves from the P site to the E site, and the second tRNA

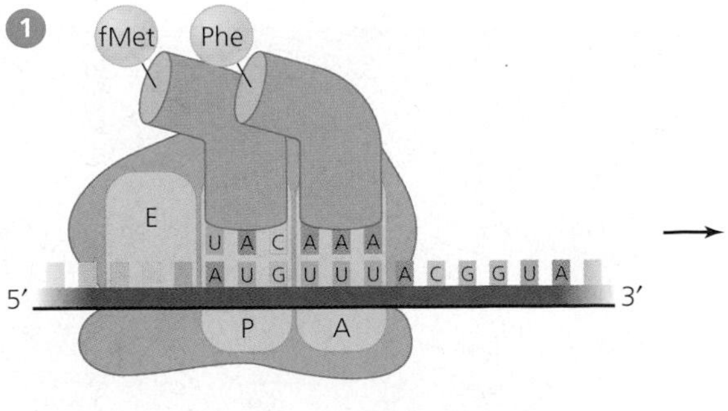

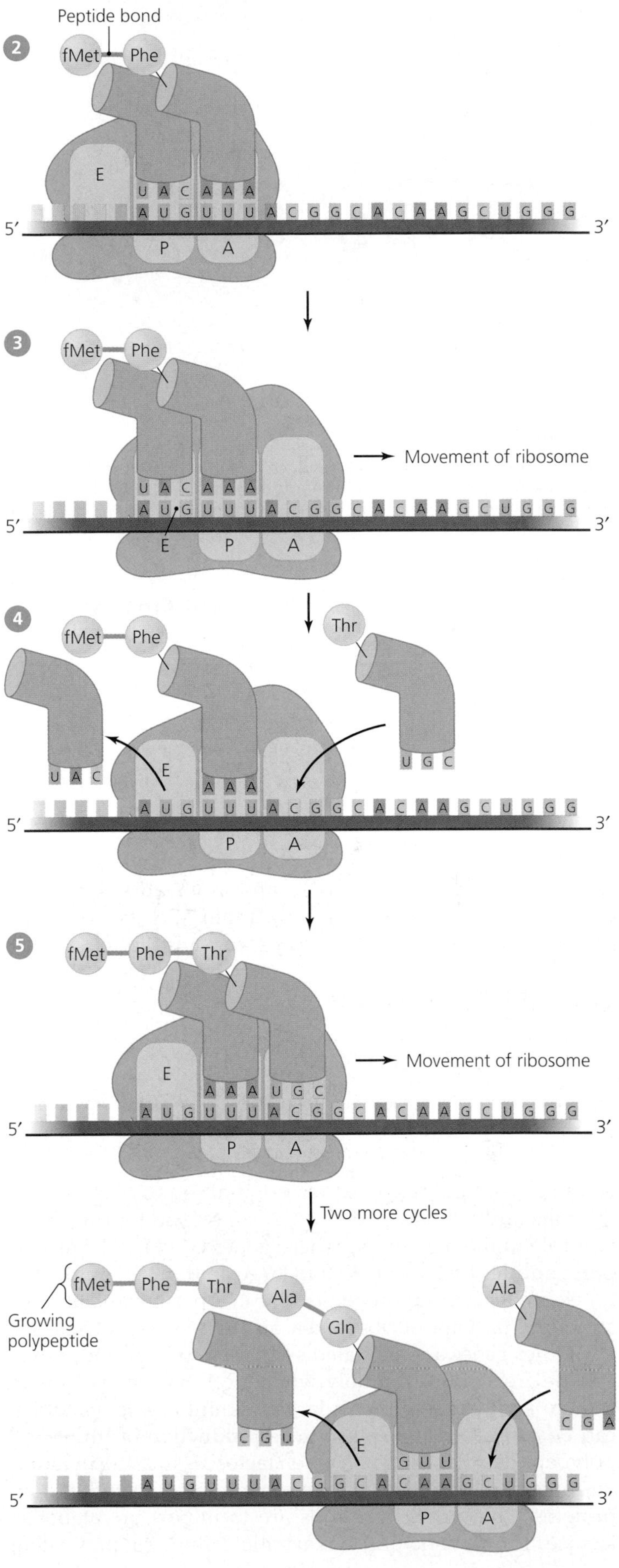

▲ ***Figure 7.15***

The elongation stage of translation. Transfer RNAs sequentially deliver amino acids as directed by the codons of the mRNA. Ribosomal RNA in the large ribosomal subunit catalyzes a peptide bond between the amino acid at the A site and the growing polypeptide at the P site. The steps in the process are described in the text. *Which component macromolecule of a ribosome—protein or rRNA—is the enzyme that actually links amino acids to form a polypeptide?*

Figure 7.15 *A ribozyme (RNA) is the active enzyme in protein synthesis.*

(with the attached polypeptide) moves to the vacated P site.

4 The ribosome releases the "empty" tRNA from the E site. In the cytosol, the appropriate enzyme recharges it with another molecule of its specific amino acid.

5 The cycle repeats, each time adding another amino acid (in this case, threonine, then alanine, and then glutamine).

As elongation proceeds, ribosomal movement exposes the start codon, allowing another ribosome to attach behind the first one. In this way, one ribosome after another attaches at the start codon and begins to translate identical polypeptide molecules from the same message. Such a group of ribosomes, called a *polyribosome*, resembles beads on a string (**Figure 7.16** on page 218).

Termination Termination does not involve tRNA; instead, proteins called *release factors* halt elongation. Though the exact mechanism is unknown, it appears that release factors somehow recognize stop codons and modify the larger ribosomal subunit in such a way as to activate another of its ribozymes, which severs the polypeptide from the final tRNA (resident at the P site). The ribosome then dissociates into its subunits.

CRITICAL THINKING

If a scientist synthesizes a DNA molecule with the nucleotide base sequence TACGGGGGAGGGGGAGGGGGA and then uses it for transcription and translation, what would be the amino acid sequence of the product? (Refer to Figure 7.9 for the genetic code.)

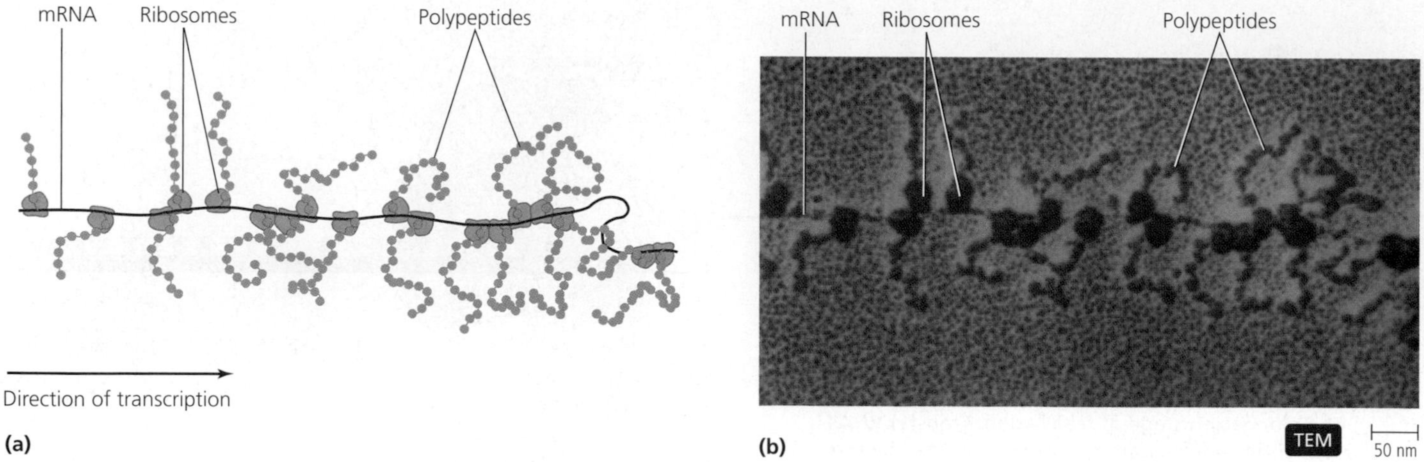

Figure 7.16

A polyribosome in a prokaryotic cell. As a ribosome moves down the mRNA, the start codon (AUG) becomes available to another ribosome. In this manner, numerous identical polypeptides are translated simultaneously from a single mRNA molecule.

Table 7.3 Comparison of Genetic Processes

Process	Purpose	Beginning Point	Ending Point
Replication	To duplicate the cell's genome	Origin	Origin or the end of the DNA molecule
Transcription	To synthesize RNA	Promoter	Terminator
Translation	To synthesize polypeptides	AUG start codon	UAA, UAG, or UGA stop codons

The processes we have examined thus far—how a cell replicates DNA, transcribes RNA, and translates RNA into polypeptides—are summarized in **Table 7.3.** Next we examine the way cells control the process of transcription.

Control of Transcription

Learning Objectives

- ✓ Explain the operon model of transcriptional control in prokaryotes.
- ✓ Contrast the regulation of an inducible operon with that of a repressible operon, and give an example of each.

About 75% of genes are expressed at all times; that is, they are constantly transcribed and translated and play a persistent role in the phenotype. These genes code for RNAs and polypeptides that are needed in large amounts by the cell—for example, integral proteins of the cytoplasmic membrane, structural proteins of ribosomes, and enzymes of glycolysis.

Other genes are regulated so that the polypeptides they encode are synthesized only when a cell has need of them. Protein synthesis requires a large amount of energy, which can be conserved if a cell forgoes production of unneeded polypeptides. For example, when lactose is missing in its environment, a bacterial cell saves energy by not synthesizing proteins whose only functions are to import or catabolize lactose. Prokaryotic and eukaryotic cells regulate protein synthesis in many ways. Cells most typically stop synthesis by stopping transcription. In a few cases, cells stop translation directly.

Proteins coded by regulated genes include a few structural proteins and many enzymes. An example of the former is found in *Trypanosoma brucei* (tri-pan'ō-sō' mă brus'ē) (the protozoan that causes African sleeping sickness), which has many genes for surface glycoproteins. Each protozoan activates only one of the genes at a time so that a population of cells contains members with a wide variety of glycoproteins. This is advantageous to *Trypanosoma* because when its host's immune system recognizes and attacks cells with one kind of glycoprotein, cells with other types may remain unrecognized and continue to reproduce.

Since much of our knowledge of regulatory mechanisms has come from the study of microorganisms such as *E. coli,* we will examine two types of enzyme regulation in this bacterium: *induction* and *repression.* But before we examine induction and repression, we must first consider a special arrangement of prokaryotic genes that plays a role in gene expression.

The Nature of Prokaryotic Operons

In 1961, Francois Jacob (1920–) and Jacques Monod (1910–1976) proposed the *operon model* to explain gene

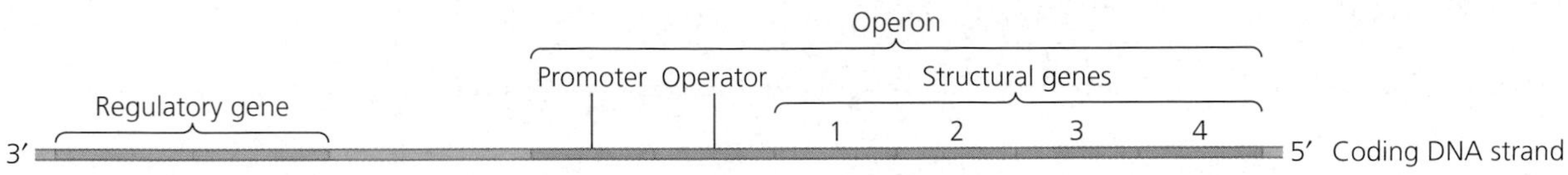

▲ ***Figure 7.17***

An operon, which consists of genes, their promoter, and a contiguous operator. The genes code for enzymes and structures such as channel and carrier proteins. A nearby regulatory gene codes for a protein that controls the operon.

regulation in bacteria. An **operon** consists of a promoter, an adjacent regulatory element called an **operator** where a *repressor protein* binds to stop transcription, and a series of genes, which code for enzymes and structures such as channel proteins **(Figure 7.17).** Operons are either repressed (turned off) or induced (turned on) by proteins coded by a regulatory gene (located elsewhere). **Inducible operons** are not usually transcribed and must be activated by *inducers.* **Repressible operons** operate in reverse fashion—they are transcribed continually until deactivated by *repressors.* To clarify these concepts, let's examine some inducible and repressible operons found in *E. coli.*

The Lactose Operon, an Inducible Operon

The *lactose (lac) operon* of *E. coli* is an inducible operon and the first operon whose structure and action were elucidated. It includes a promoter, an operator, and three genes that encode for proteins involved in the catabolism of lactose **(Figure 7.18a).** The operon is controlled by a regulatory gene that is constantly transcribed and translated to produce a

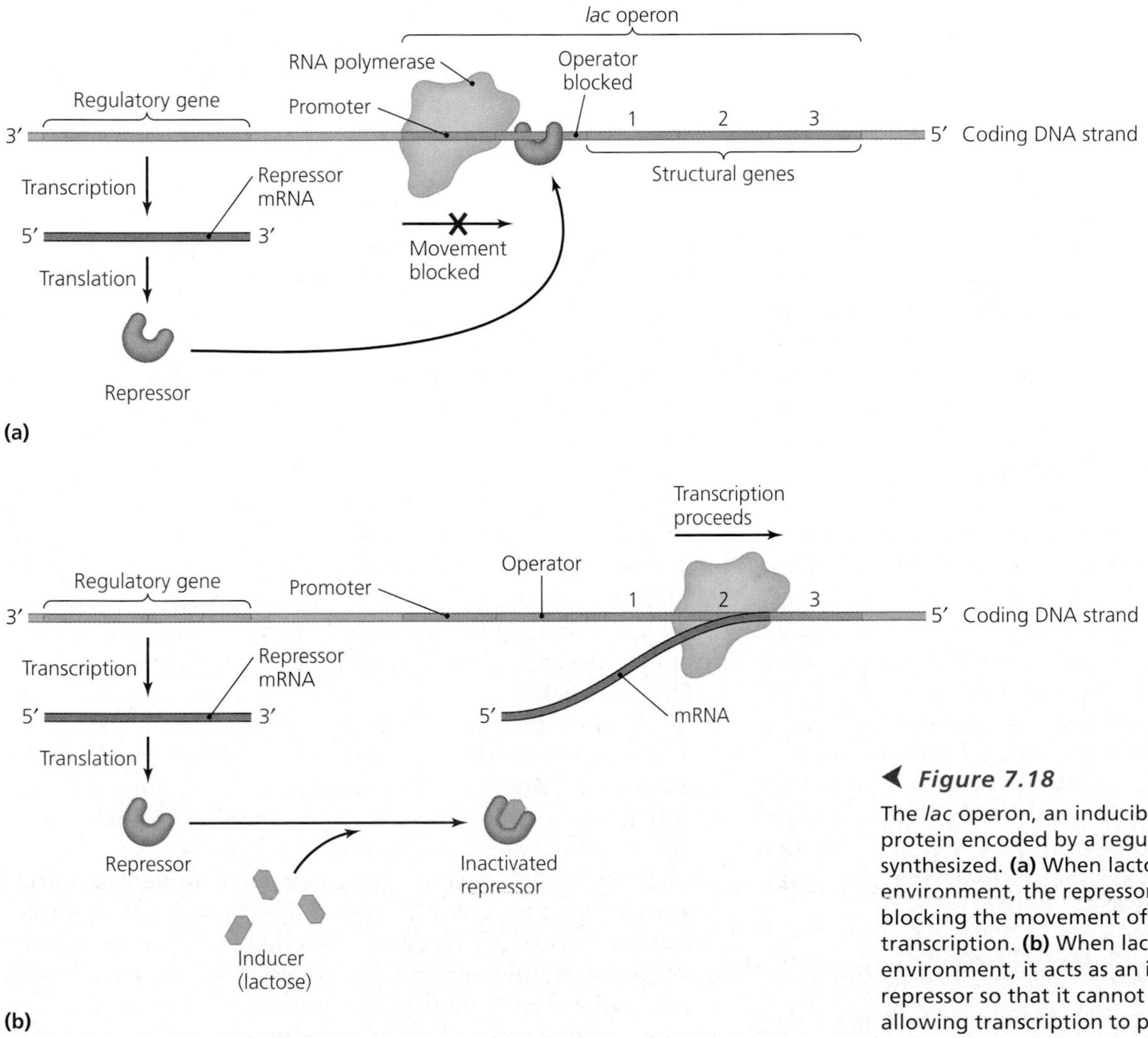

◀ ***Figure 7.18***

The *lac* operon, an inducible operon. The repressor, a protein encoded by a regulatory gene, is constantly synthesized. **(a)** When lactose is absent from the cell's environment, the repressor binds to the operator, blocking the movement of RNA polymerase and halting transcription. **(b)** When lactose is present in the cell's environment, it acts as an inducer by inactivating the repressor so that it cannot bind to the operator, allowing transcription to proceed.

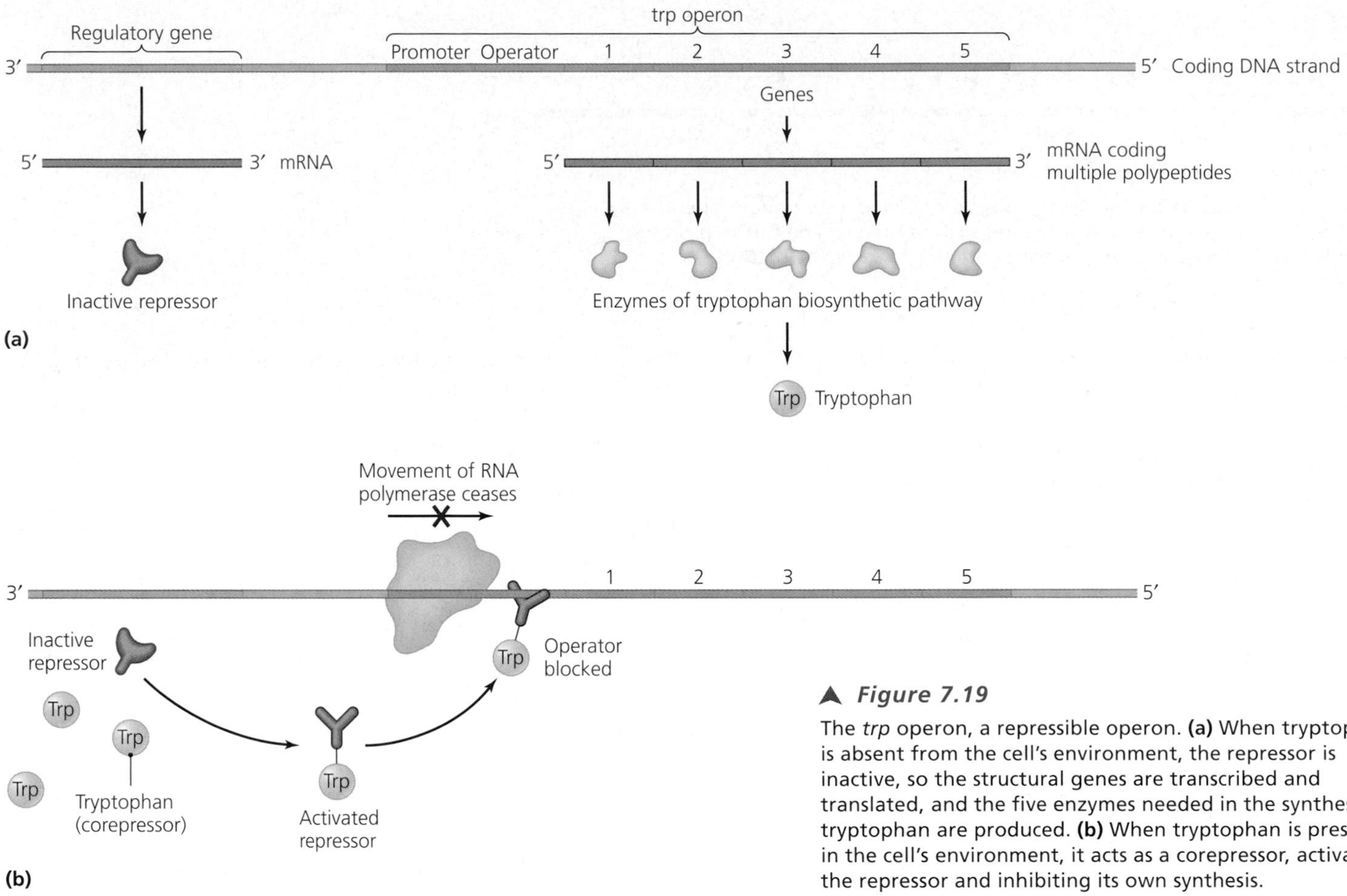

▲ ***Figure 7.19***

The *trp* operon, a repressible operon. **(a)** When tryptophan is absent from the cell's environment, the repressor is inactive, so the structural genes are transcribed and translated, and the five enzymes needed in the synthesis of tryptophan are produced. **(b)** When tryptophan is present in the cell's environment, it acts as a corepressor, activating the repressor and inhibiting its own synthesis.

repressor protein that attaches to DNA at the *lac* operator. This repressor prevents RNA polymerase from moving beyond the promoter, stopping synthesis of mRNA. Thus, the *lac* operon is usually inactive.

Whenever lactose becomes available in an *E. coli* cell's environment, it acts as an inducer and changes the quaternary structure of the repressor so that it is inactivated and can no longer attach to DNA **(Figure 7.18b).** This absence of binding allows transcription of the three structural genes to proceed—the operon has been induced and has become active. Ribosomes translate the newly synthesized mRNA to produce enzymes that catabolize lactose. Once the lactose supply has been depleted, there is no more inducer, and the repressor once again becomes active, suppressing transcription and translation of the *lac* operon. In this manner, *E. coli* cells conserve energy by synthesizing enzymes for the catabolism of lactose only when lactose is available to them.

Such inducible operons are often involved in controlling catabolic pathways whose polypeptides are not needed unless a particular nutrient is available. A different situation occurs with anabolic pathways such as those that synthesize amino acids such as tryptophan.

The Tryptophan Operon, a Repressible Operon

E. coli normally synthesizes the amino acids it needs for polypeptide synthesis; however, it can save energy by using amino acids available in its environment. In such cases, *E. coli* represses the genes for a given amino acid's synthetic pathway.

The *tryptophan (trp) operon,* which consists of a promoter, an operator, and five genes that code for the enzymes involved in the synthesis of tryptophan, is an example of such a repressible operon. Just as with the *lac* operon, a regulatory gene codes for a repressor molecule that is constantly synthesized. In contrast to inducible operons, however, the repressor of repressible operons is normally *inactive*. Thus, in the case of the repressible *trp* operon, whenever tryptophan is not present in the environment, the *trp* operon is active: The appropriate mRNA is transcribed, the enzymes for tryptophan synthesis are translated, and tryptophan is produced **(Figure 7.19a).**

When tryptophan is available, it activates the repressor by binding to it. The activated repressor then binds to the operator, halting the movement of RNA polymerase and halting transcription **(Figure 7.19b).** In other words, tryptophan acts as a *corepressor* of its own synthesis.

Biochemical analyses have revealed numerous variations of repressor-operator regulatory control. For example, scientists have discovered operons that use dual regulatory proteins, multiple operons controlled by a single repressor, and operons with multiple operators. Furthermore, some operons are not merely on-off systems, but can be fine-tuned

New Frontiers 7.1 Controlling Disease by Controlling Gene Expression

Recent advances in our understanding of transcriptional control may provide new techniques for treating and even preventing diseases caused by numerous bacteria.

For example, many strains of the bacterium *Staphylococcus* that cause food poisoning and a variety of diseases (including impetigo, ear infections, and toxic shock syndrome) are resistant to penicillin. One protein that confers such resistance is an enzyme called beta-lactamase, which deactivates molecules of penicillin by cleaving them. Normally, the beta-lactamase gene is repressed, but in the presence of penicillin the gene is induced and beta-lactamase is synthesized.

New research has provided scientists a greater understanding of how penicillin induces the beta-lactamase gene in these bacteria. Penicillin binds to a receptor protein on the bacterial cell membrane, which causes the receptor to become enzymatic. The newly formed enzyme breaks down the repressor of the beta-lactamase gene. In the absence of repressor, the gene is transcribed, beta-lactamase is produced, and penicillin is deactivated. Now that this metabolic pathway for penicillin resistance is understood, it may be possible to design drugs that block the receptor's action and leave *Staphylococcus* susceptible to penicillin.

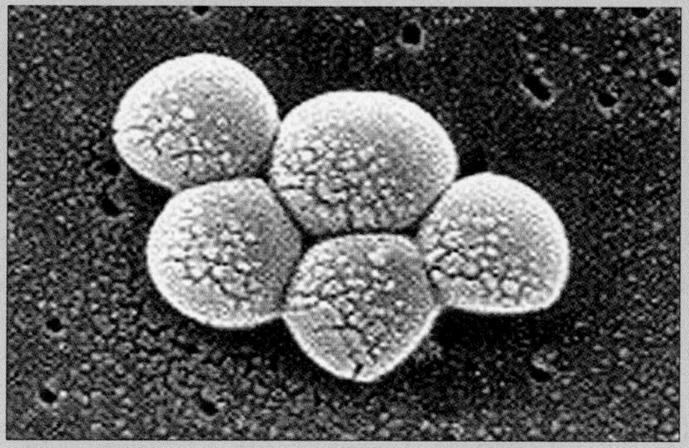

Staphylococcus aureus.

Table 7.4 The Roles of Operons in the Regulation of Transcription

Type of Regulation	Type of Metabolic Pathway Regulated	Regulating Condition
Inducible operons	Catabolic pathways	Presence of substrate of pathway
Repressible operons	Anabolic pathways	Presence of product

so that transcription rates vary with the concentration of corepressors.

Table 7.4 summarizes how the basic characteristics of inducible and repressible operons relate to the regulation of transcription. **New Frontiers 7.1** examines potential practical aspects of transcriptional control in health care.

Mutations of Genes

Learning Objective

✓ Define mutation.

The phenotype of a cell is dependent upon both the integrity and accurate control of its genes; however, the nucleotide sequences of genes are not always accurately maintained. A **mutation** is a change in the nucleotide base sequence of a genome, particularly its genes. Mutations of genes are almost always deleterious, though a few make no difference to the organism. Very rarely a mutation could lead to a protein having a novel property that improves the ability of an organism and its descendants to survive and reproduce. Mutations in asexual organisms are generally passed on to the organism's progeny, but mutations in multicellular sexual organisms are passed to offspring only if the mutation occurs in gametes or gamete-producing cells.

Types of Mutations

Learning Objective

✓ Define and describe three types of point mutations.

Mutations range from large changes in an organism's genome, such as the loss or gain of an entire chromosome, to the most common type of mutation—**point mutations**—in which just one or a few nucleotide base pairs are affected. Point mutations include base pair **substitutions, insertions,** and **deletions.** These types of mutations can be demonstrated by the following analogy. Suppose that the DNA code was represented by the letters THECATATEELK. Grouping the letters into triplets (like codons) yields THE CAT ATE ELK. The substitution of a single letter could either change the meaning of the sentence, as in THE RAT ATE ELK, or result in a meaningless phrase, such as THE CAT RTE ELK. Insertion or deletion of a letter produces more serious changes, such as TRH ECA TAT EEL K or TEC ATA TEE LK. Insertions and deletions are also called **frameshift mutations** because nucleotide triplets subsequent to the mutation are displaced, creating new sequences of codons that result in vastly altered polypeptide sequences. Frameshift mutations affect proteins much more seriously than mere substitutions because a frame shift affects all codons subsequent to the mutation.

Mutations can also involve inversion (THE ACT ATE KLE), duplication (THE CAT CAT ATE ELK ELK), or

Figure 7.20 ▶

The effects of the various types of point mutations. Base-pair substitutions can result in silent mutations **(a)**, missense mutations **(b)**, or nonsense mutations **(c)**. Frameshift insertions **(d)** and frameshift deletions **(e)** usually result in severe missense or nonsense mutations because all codons downstream from the mutation are altered.

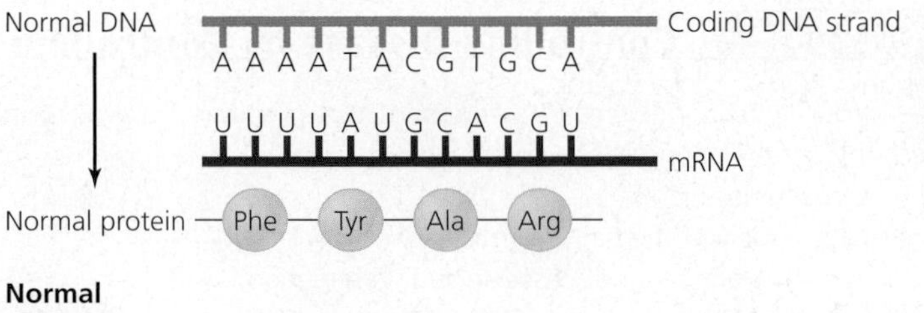

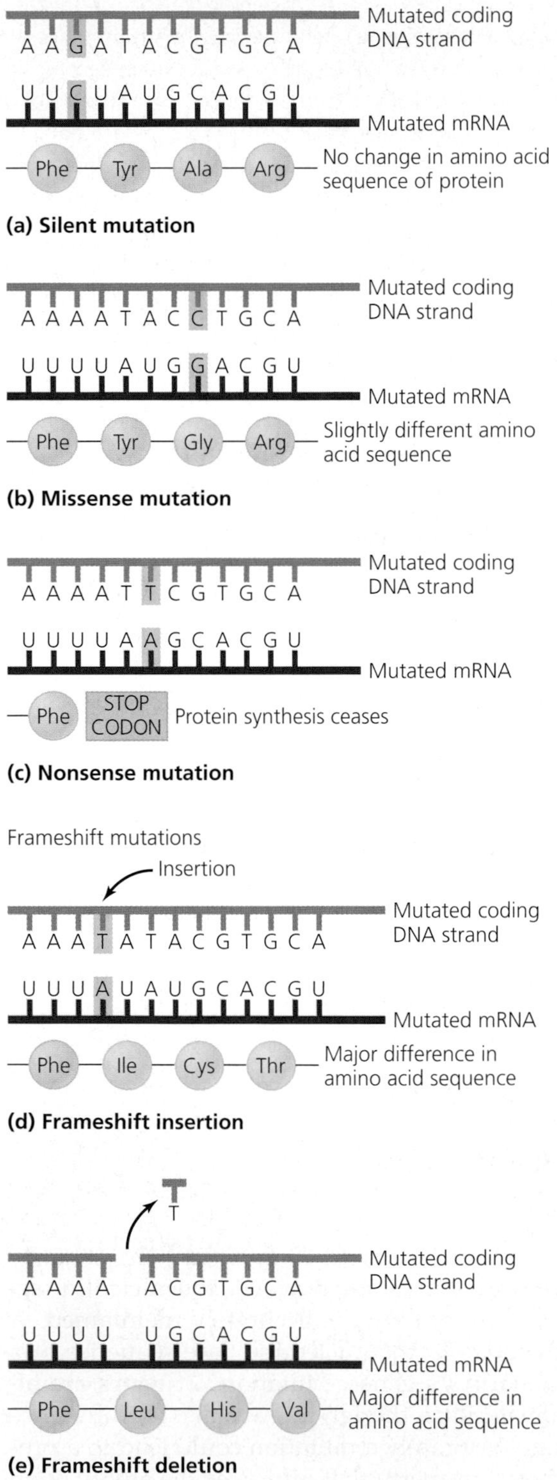

transposition (THE ELK ATE CAT). Such mutations and even larger deletions and insertions are *gross mutations*.

Effects of Mutations

Learning Objective

✓ List three effects of mutations.

Some base-pair substitutions produce **silent mutations** because the substitution does not change the amino acid sequence due to the redundancy of the genetic code **(Figure 7.20a).** For example, when the DNA triplet AAA is changed to AAG, the mRNA codon will be changed from UUU to UUC; however, because both codons specify the amino acid phenylalanine, there is no change in the phenotype—the mutation is silent because it affects the genotype only.

Of greater concern are substitutions that change a codon for one amino acid into a codon for a different amino acid. A change in a nucleotide sequence resulting in a codon that specifies a different amino acid is called a **missense mutation (Figure 7.20b);** what gets transcribed and translated makes sense, but not the right sense. The effect of missense mutations depends on where in the protein the different amino acid occurs. When the different amino is in a critical region of a protein, the protein becomes nonfunctional; however, when the different amino acid is in a less important region, the mutation has no adverse effect.

A third type of mutation occurs when a base-pair substitution changes an amino acid codon into a stop codon. This is called a **nonsense mutation (Figure 7.20c).** Nearly all nonsense mutations result in nonfunctional proteins.

Frameshift mutations (that is, insertions or deletions) typically result in drastic missense and nonsense mutations **(Figure 7.20d, e),** except when the insertion or deletion is very close to the end of a gene.

Table 7.5 summarizes the types and effects of point mutations.

CRITICAL THINKING

Find the codon UGU in Figure 7.9. What DNA nucleotide triplet codes for this codon? Identify a base-pair substitution that would produce a silent mutation at this codon. Identify a base-pair substitution that would result in a missense mutation at this codon. Identify a base-pair substitution that would produce a nonsense mutation at this codon.

Table 7.5 The Types of Point Mutations and Their Effects

Type of Point Mutation	Description	Effects
Substitution	Mismatching of nucleotides or replacement of one base pair by another	*Silent mutation* if change results in redundant codon, as amino acid sequence in polypeptide is not changed. *Missense mutation* if change results in codon for a different amino acid; effect depends on location of different amino acid in polypeptide. *Nonsense mutation* if codon for an amino acid is changed to a stop codon.
Frameshift (insertion)	Addition of one or a few nucleotide pairs creates new sequence of codons	Missense and nonsense mutations
Frameshift (deletion)	Removal of one or a few nucleotide pairs creates new sequence of codons	Missense and nonsense mutations

Mutagens

Learning Objectives

- ✓ Discuss how different types of radiation cause mutations in a genome.
- ✓ Describe three kinds of chemical mutagens and their effects.

Mutations occur naturally during the life of an organism. These *spontaneous mutations* result from errors in replication and repair as well as from *recombination* in which relatively long stretches of DNA move among chromosomes, plasmids, and viruses, introducing frameshift mutations. For example, we have seen that mismatched base pairing during DNA replication results in one error in every billion (10^9) base pairs. Since an average gene has 10^3 base pairs, about one of every 10^6 (million) genes contains an error. Further, though cells have repair mechanisms to reduce the effect of mutations, the repair process itself can introduce additional errors. Mutations can also be induced by many physical or chemical agents called **mutagens** (myū′ tă-jenz), which include radiation and several types of DNA-altering chemicals.

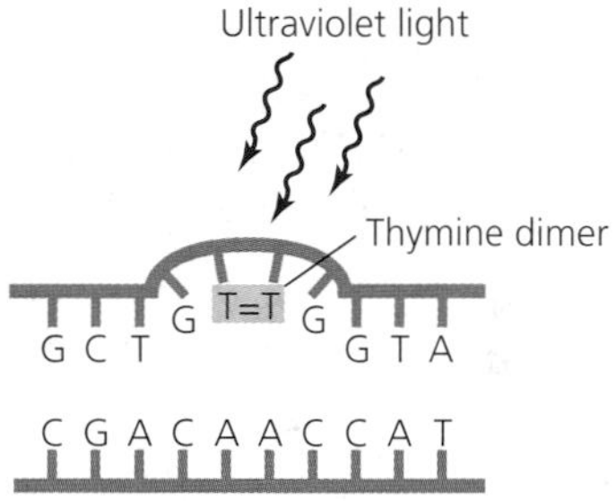

▲ ***Figure 7.21***
A thymine dimer. Ultraviolet light causes adjacent thymine bases to covalently bond to each other, which prevents hydrogen bonding with adenine bases in the complementary strand. The resulting distortion of the sugar-phosphate backbone prevents proper replication and transcription.

Radiation

In the 1920s, Hermann Muller (1890–1967) discovered that *X-rays* increased phenotypic variability in fruit flies by causing mutations. *Gamma rays* also damage DNA. As discussed in Chapter 9, X-rays and gamma rays are *ionizing radiation;* that is, they energize electrons in atoms, causing some of the electrons to escape from their atoms. These free electrons strike other atoms, producing ions that can react with the structure of DNA, producing mutations. More seriously, electrons and ions can break the covalent bonds between the sugars and phosphates of a DNA backbone, causing physical breaks in chromosomes and complete loss of cellular control.

Nonionizing radiation in the form of *ultraviolet (UV) light* is also mutagenic because it causes adjacent thymine bases to covalently bond to one another, forming **thymine dimers (Figure 7.21).** The presence of dimers prevents hydrogen bonding with the adenine nucleotides in the complementary strand, distorts the sugar-phosphate backbone, and prevents proper replication and transcription. Cells have several methods of repairing thymine dimers (discussed shortly).

Chemical Mutagens

Here we consider three of the many basic types of mutagenic chemicals.

Nucleotide Analogs **Nucleotide analogs** are compounds that are structurally similar to normal nucleotides (**Figure 7.22a** on page 224). When nucleotide analogs are available to replicating cells, they may be incorporated into DNA in place of normal nucleotides, where their structural differences either inhibit nucleic acid polymerases or result in mismatched base pairing. When, for example, thymine is replaced by 5′-bromouracil, it pairs with the wrong complement—it pairs with guanine rather than adenine, resulting in a point mutation **(Figure 7.22b).**

Nucleotide analogs make potent antiviral and anticancer drugs because viruses and cancer cells typically

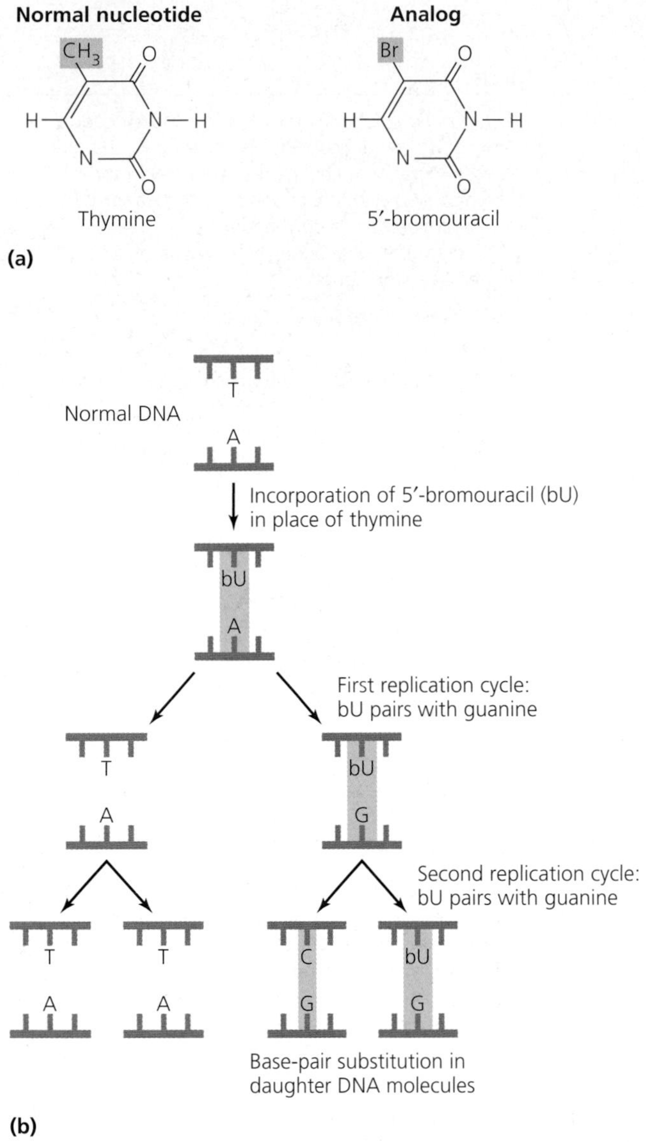

▲ ***Figure 7.22***
The structure and effects of nucleotide analogs. **(a)** The structures of thymine and its nucleotide analog, 5′-bromouracil (bU). **(b)** When 5′-bromouracil is incorporated into DNA, the result is a point mutation. A single replication cycle results in one normal DNA molecule and one mutated molecule in which bU pairs with guanine. After a second replication cycle, a complete base-pair substitution has occurred in one of the four DNA molecules: CG has been substituted for TA.

replicate faster than normal cells. The use of nucleotide analogs as antimicrobial agents is discussed in Chapter 10.

Nucleotide-Altering Chemicals Some chemical mutagens alter the structure of nucleotides. For example, *nitrous acid* (H—O—N=O) removes the amine group of adenine, converting it into a guanine analog. When a cell replicates DNA containing this analog, an AT base pair is changed to a GC base pair in one daughter molecule—a base-pair-substitution mutation.

Another group of nucleotide-altering chemicals are *aflatoxins,* which are produced by *Aspergillus* (as-per-jil′ŭs) molds growing on grains and nuts. Aflatoxins convert guanine nucleotides into thymine nucleotides, so that a GC base pair is converted to a TA base pair, resulting in missense mutations and possibly cancer.

Frameshift Mutagens Still other mutagenic chemical agents insert or delete nucleotide base pairs, resulting in frameshift mutations. Examples of frameshift mutagens are *benzopyrene,* which is found in smoke; *ethidium bromide,* which is used to stain DNA; and *acridine,* one of a class of dyes commonly used as mutagens in genetic research. These chemicals are exactly the right size to slip between adjoining nucleotides in DNA, producing a bulge in the molecule **(Figure 7.23).** When DNA polymerase copies the misshapen strands, one or more base pairs may be inserted or deleted in the daughter strand.

Frequency of Mutation

Learning Objective

✓ Discuss the relative frequency of deleterious and useful mutations.

Mutations are rare events. If they were not, organisms could not live or effectively reproduce themselves. As we have seen, about one of every million (10^6) genes contains an error. Mutagens typically increase the mutation rate by a factor of 10–1000 times; that is, mutagens induce an error in one of every 10^3 to 10^5 genes.

Most mutations are deleterious because they code for nonfunctional proteins or stop transcription entirely. Cells without functional proteins cannot metabolize, therefore many deleterious mutations are removed from the population when the cells die. Rarely, however, a cell acquires a beneficial mutation that allows it to survive, reproduce, and pass the mutation to its descendents. This change in gene frequency is the basis of evolution. For example, a bacterium might randomly acquire a mutation that confers resistance to an antibiotic. In an environment containing the antibiotic, cells without resistance die, but the mutated cell survives and reproduces. As long as the antibiotic is present, cells with such a mutation have an advantage over cells without the mutation.

DNA Repair

Learning Objective

✓ Describe light and dark repair of thymine dimers, base excision repair, mismatch repair, and the SOS response.

We have seen that although a mutation might rarely convey an advantage, most mutations are deleterious. To respond to the dangers mutations pose, cells have numerous methods for repairing damaged DNA, including light and dark repair of thymine dimers, base excision repair, mismatch repair, and an SOS response.

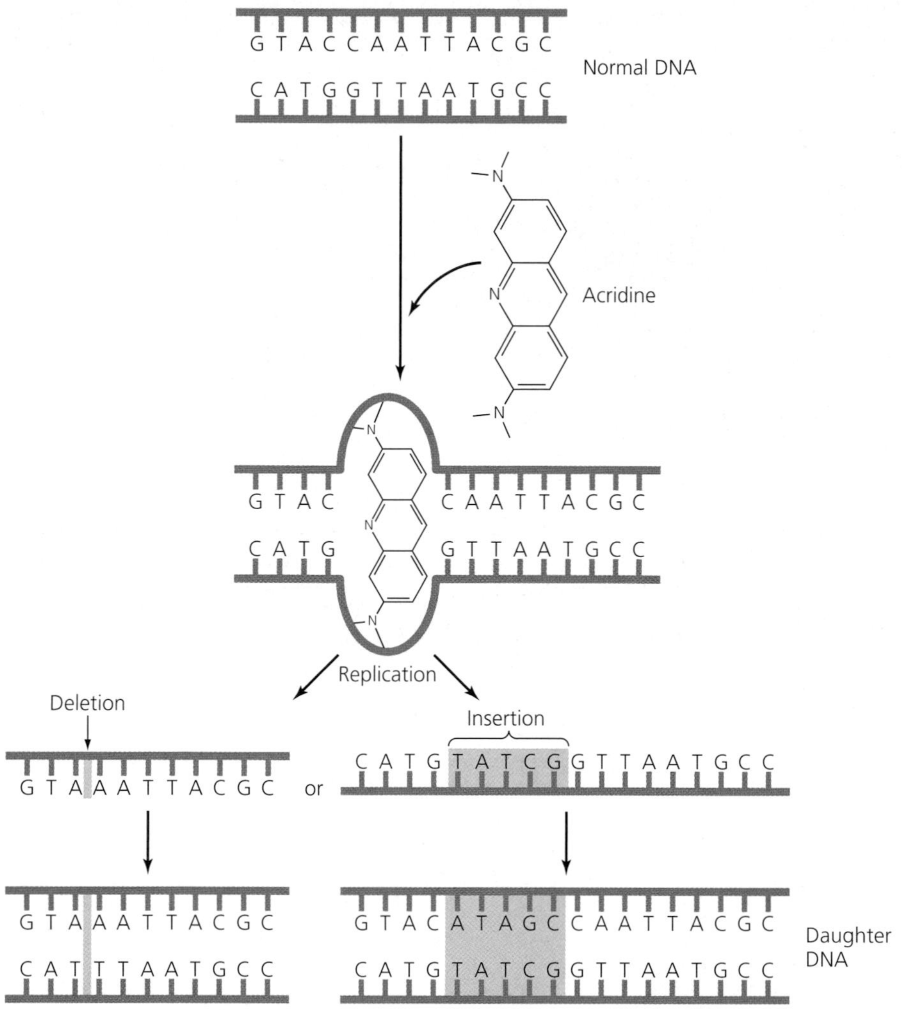

Figure 7.23

The action of a frameshift mutagen. When DNA polymerase passes the bulge caused by the insertion of acridine between the two DNA strands, it incorrectly synthesizes a daughter strand, which contains either a deletion or an insertion mutation.

Repair of Thymine Dimers

Thymine dimers resulting from exposure to UV light constitute the most common type of mutation. Many prokaryotes contain *DNA photolyase,* an enzyme that is activated by visible light to break the bonds between adjoining thymine nucleotides, reversing the mutation and restoring the original DNA sequence (**Figure 7.24a** on page 226). This so-called **light repair** mechanism is advantageous for the prokaryote, but it presents a difficulty to scientists studying UV-induced mutations—they must keep such bacterial strains in the dark, or the mutants will revert to their previous form.

So-called **dark repair** involves a different repair enzyme—one that doesn't require light. Dark repair enzymes cut the damaged section of DNA from the molecule, creating a gap that is repaired by DNA polymerase and DNA ligase **(Figure 7.24b).** Though called dark repair, this mechanism operates either in light or in the dark.

Base-Excision Repair

Sometimes DNA polymerase incorporates an incorrect nucleotide during DNA replication. If the proofreading function of DNA polymerase does not repair the error, cells may use another enzyme system in a process called **base-excision repair.** This enzyme system excises the erroneous base, and then DNA polymerase fills in the gap **(Figure 7.24c).** A typical excision repair is the replacement of uracil, which is not supposed to be in DNA, with thymine.

Mismatch Repair

A similar repair mechanism is called **mismatch repair.** Mismatch repair enzymes scan newly synthesized DNA looking for mismatched bases, which they remove and replace **(Figure 7.24d).** How does the mismatch repair system determine which strand to repair? If it chose randomly, 50% of the time it would choose the wrong strand and introduce mutations.

Mismatch repair enzymes can distinguish between a new DNA strand and an old strand because old strands are methylated. Recognition of an error as far as 1000 base pairs away from an unmethylated portion of DNA triggers the mismatch repair enzymes. Once a new DNA strand is methylated, mismatch repair enzymes cannot correct any errors that remain.

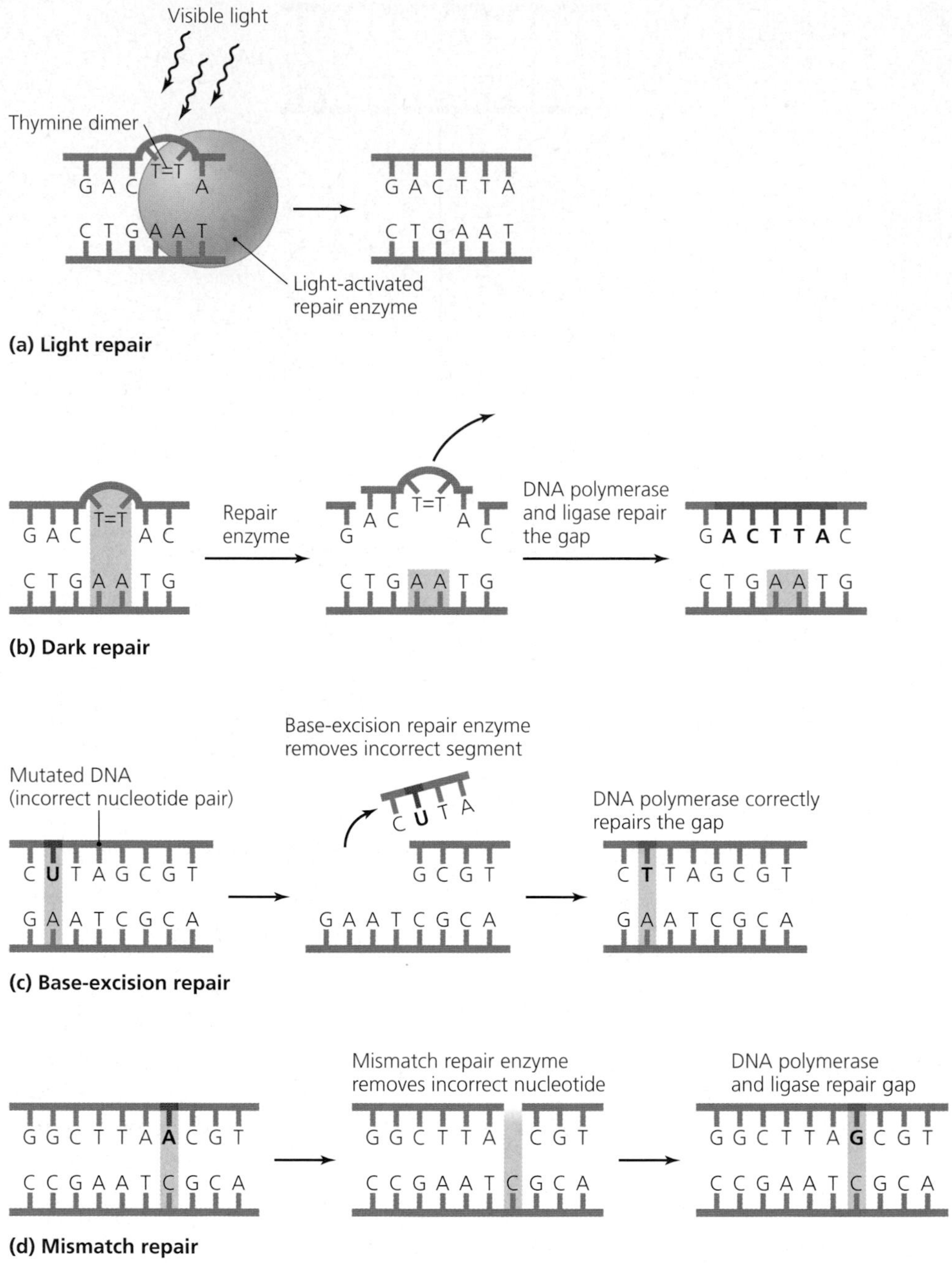

Figure 7.24 ➤

DNA repair mechanisms. **(a)** Light repair of thymine dimers. A light-activated enzyme breaks thymine-to-thymine bonds. **(b)** Dark repair of thymine dimers. After the repair enzyme removes the entire damaged section from one strand of DNA, DNA polymerase and DNA ligase repair the breach. This mechanism is called dark repair because it does not require light; in fact, it operates in either light or darkness. **(c)** Base-excision repair, in which enzymes remove a segment with an incorrect base and DNA polymerase fills the gap. **(d)** Mismatch repair, which involves total excision of an incorrect nucleotide.

SOS Response

Sometimes damage to DNA is so extreme that regular repair mechanisms cannot cope with the damage. In such cases, bacteria resort to what geneticists call an **SOS response** involving a variety of processes, such as the production of a novel DNA polymerase capable of copying less-than-perfect DNA. This polymerase replicates DNA with little regard to the base sequence of the template strand. Of course, this introduces many new and potentially fatal mutations, but presumably SOS repair allows a few offspring of these bacteria to survive.

Identifying Mutants, Mutagens, and Carcinogens

Learning Objectives

- ✓ Contrast the positive and negative selection techniques for isolating mutants.
- ✓ Describe the Ames test, and discuss its use in discovering carcinogens.

If a cell does not successfully repair a mutation, it and its descendents are called **mutants.** In contrast, cells normally found in nature (in the wild) are called **wild type** cells.

Scientists distinguish mutants from wild type cells by observing or testing for altered phenotypes. Since mutations are rare, and nonfatal mutations are even rarer, mutants can easily be "lost in the crowd." Therefore, researchers have developed methods to recognize mutants amidst their wild type neighbors.

Positive Selection

Positive selection involves selecting a mutant by eliminating wild type phenotypes **(Figure 7.25).** Assume, for example, that researchers want to isolate bacterial mutants that are resistant to penicillin from a liquid culture. To do so they spread the liquid medium, which contains mostly penicillin-sensitive cells, but also the few penicillin-resistant mutants, onto a plate containing medium that includes penicillin; only the penicillin-resistant mutants multiply on this medium and produce visible colonies **(Figure 7.25a).**

When a mutagenic agent is added to the liquid culture, it increases the number of mutants **(Figure 7.25b).** While the researchers are isolating mutants, they can also determine the rate of mutation by comparing the number of mutant colonies formed after use of the mutagen with the number formed before treatment. The rate of mutation can be calculated as follows.

$$\frac{\text{number of colonies seen with use of mutagen} - \text{number of colonies seen without use of mutagen}}{\text{number of colonies seen without the use of mutagen}} \times 100\%$$

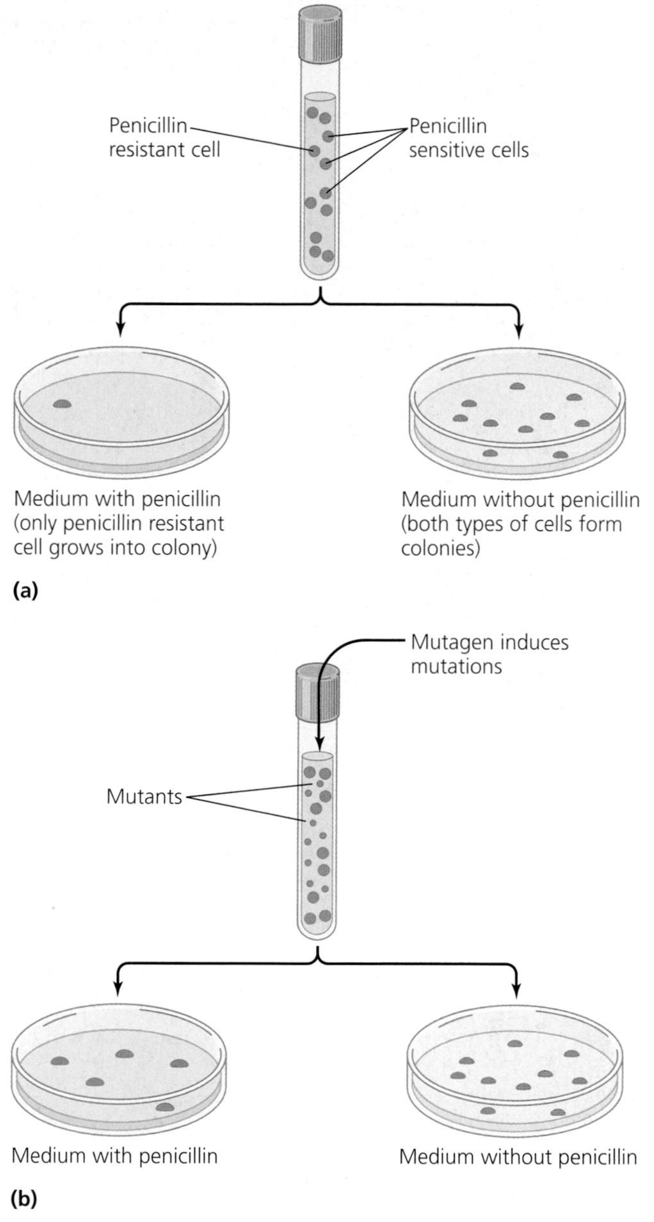

▲ ***Figure 7.25***

Positive selection of mutants. Only mutants that are resistant to penicillin can survive on the plate containing the antibiotic. **(a)** Normally a population includes very few mutants. **(b)** Introduction of a mutagen increases the number of mutants and thus the number of colonies that grow in the presence of penicillin. *What is the rate of mutation induced by the mutagen in (b)?*

Figure 7.25 $\frac{5-1}{1} \times 100\% = 400\%$

Negative (Indirect) Selection

An organism with nutritional requirements that differ from those of its wild-type phenotype is known as an *auxotroph* (awk′sō-trōf).[6] For example, a mutant bacterium that has lost the ability to synthesize tryptophan is auxotrophic for this amino acid—it must acquire tryptophan from the environment. Obviously, if a researcher attempts to grow tryptophan auxotrophs on media lacking tryptophan, the bacteria will be unable to synthesize all its proteins and will die. Therefore, to isolate such auxotrophs we must use a technique called **negative (indirect) selection.** The process by which a researcher uses negative selection to culture a tryptophan auxotroph is as follows (**Figure 7.26** on page 228):

1. The researcher inoculates a sample of a bacterial suspension containing potential mutants onto a plate containing complete media (including tryptophan). The sample is diluted such that the plate receives only about 100 cells.
2. Both auxotrophs and wild-type cells reproduce and form colonies on the plate, but the colonies are indistinguishable.
3. The researcher transfers cells from all the colonies on the plate to a sterile velvet pad by pressing the pad onto the plate.
4. The researcher inoculates two new plates—one containing tryptophan, the other lacking tryptophan—by pressing the pad onto each of them. This technique is called *replica plating*.
5. After the plates have incubated for several hours, the researcher compares the two replica plates. Tryptophan

[6]From Greek *auxein*, meaning to increase, and *trophe*, meaning nutrition.

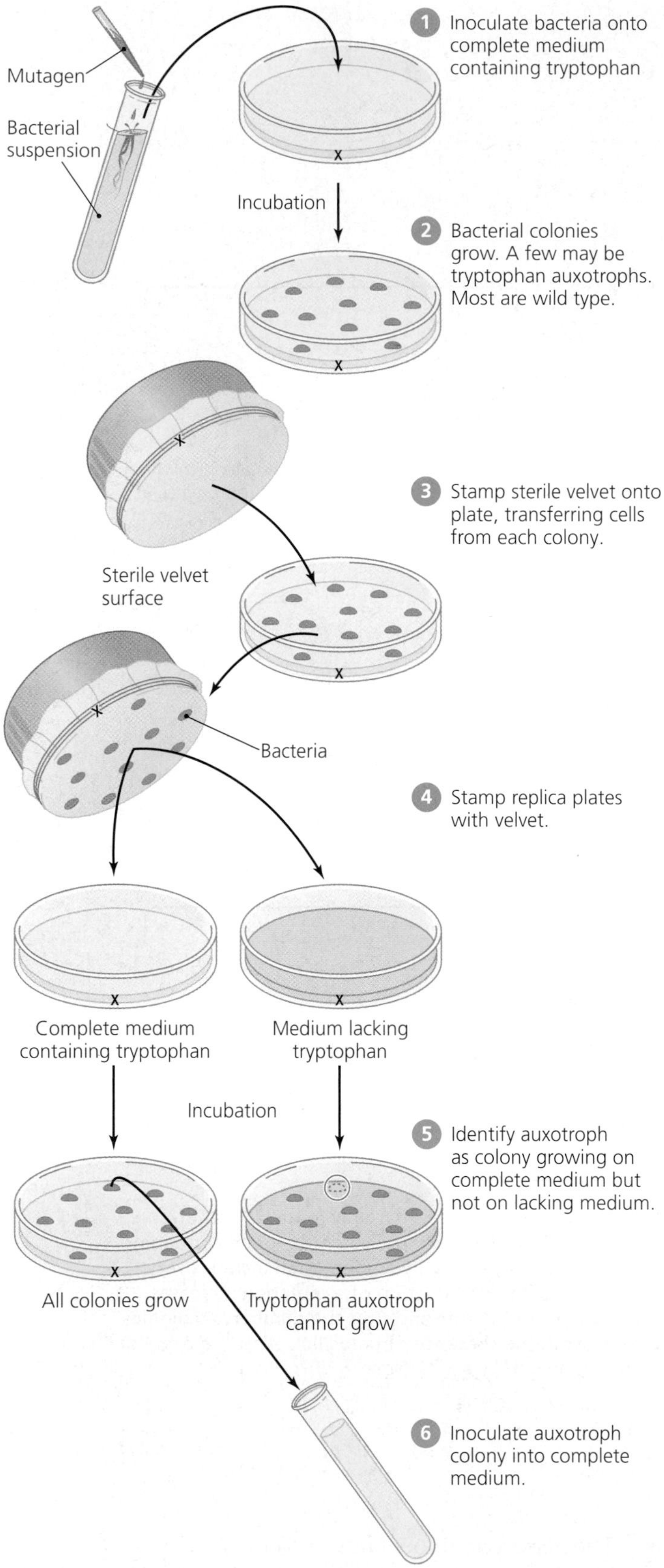
Mutagen
Bacterial suspension
1 Inoculate bacteria onto complete medium containing tryptophan
Incubation
2 Bacterial colonies grow. A few may be tryptophan auxotrophs. Most are wild type.
3 Stamp sterile velvet onto plate, transferring cells from each colony.
Sterile velvet surface
Bacteria
4 Stamp replica plates with velvet.
Complete medium containing tryptophan
Medium lacking tryptophan
Incubation
5 Identify auxotroph as colony growing on complete medium but not on lacking medium.
All colonies grow
Tryptophan auxotroph cannot grow
6 Inoculate auxotroph colony into complete medium.

Figure 7.26

The use of negative (indirect) selection to isolate a tryptophan auxotroph. The plates are marked (in this case, with an X) so that their orientation can be maintained throughout the procedure. Researchers may have to inoculate hundreds of such plates to identify a single mutant.

auxotrophs growing on medium containing tryptophan are revealed by the absence of a corresponding colony on the plate lacking tryptophan.

6 The researcher takes cells of the auxotroph colony from the replica plate and inoculates them into a complete medium. The auxotroph is now isolated.

The Ames Test for Identifying Carcinogens

Numerous chemicals in food, the workplace, and the environment in general have been suspected of being **carcinogenic** (kar′si-nō-jen′ik) mutagens; that is, of causing mutations that result in cancer. Because animal tests to prove that they are indeed carcinogenic are expensive and time-consuming, researchers use a fast and inexpensive method for screening mutagens called an **Ames test,** which is named for its inventor, Bruce Ames (1928–). Interpretation of an Ames test includes the assumption that any given mutagen is a potential carcinogen.

An Ames test uses mutant *Salmonella* bacteria possessing a point mutation that prevents the synthesis of the amino acid histidine; in other words, they are histidine auxotrophs, indicated by the abbreviation *his*$^{-}$. To perform the test, an investigator mixes his^{-} mutants with liver extract and the substance suspected to be a mutagen **(Figure 7.27).** The presence of liver extract simulates the conditions in the body under which liver enzymes can turn innocuous chemicals into mutagens. The researcher then spreads the treated bacteria on a solid medium lacking histidine. If the suspected substance does in fact cause mutations, some of the mutations will reverse the effect of the original mutation, producing revertant cells (designated his^{+}) that have regained the ability to synthesize histidine and thus can survive on a medium lacking histidine. Thus, the presence of colonies during an Ames test reveals that the suspected substance is mutagenic in *Salmonella.*

Given that DNA in all cells is very similar, the ability of a chemical to cause mutations in *Salmonella* indicates that it is likely to cause mutations in humans as well. Note, however, that a positive Ames test does not prove that the chemical is a carcinogen; it only shows that the substance or a derivative of it is mutagenic in *Salmonella.* To prove that the substance can in fact cause cancer, scientists must test it in laboratory animals, a process that usually is warranted only if a chemical is mutagenic in *Salmonella.* Ames testing reduces the cost and time that would be required to assay every chemical for carcinogenicity in animals.

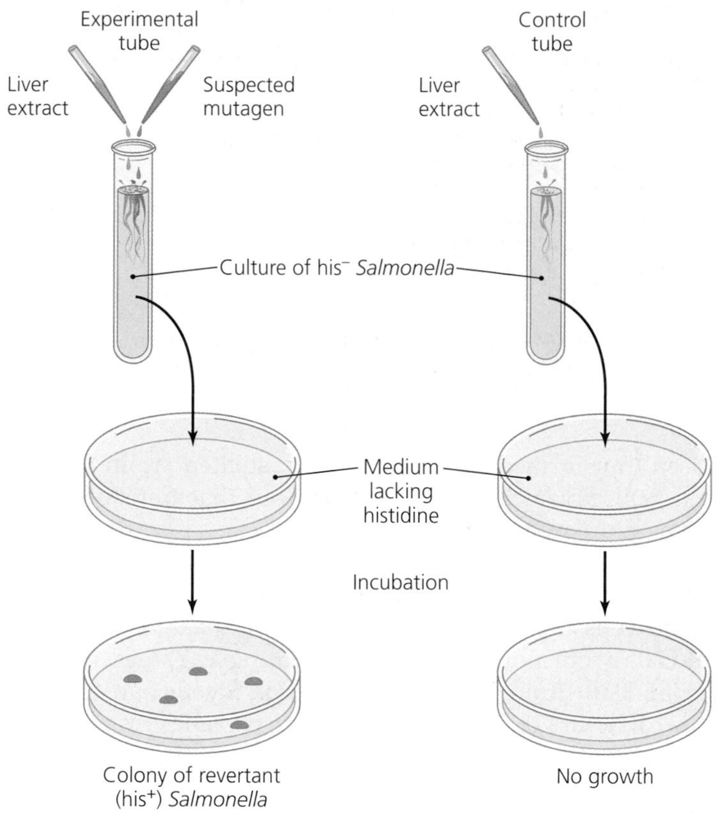

▲ ***Figure 7.27***

The Ames test. A mixture containing his$^-$ *Salmonella* mutants, rat liver extract, and the suspected mutagen is inoculated onto a plate lacking histidine. Colonies will form only if a mutagen reverses the his$^-$ mutation, producing revertant his$^+$ organisms with the ability to synthesize histidine. A control tube that lacks the suspected mutagen demonstrates that reversion did not occur in the absence of the mutagen. *What is the purpose of liver extract in an Ames test?*

Figure 7.27 Liver extract simulates conditions in the body by providing enzymes that may degrade harmless substances into mutagens.

Genetic Recombination and Transfer

Learning Objective

✓ Define genetic recombination.

Genetic recombination refers to the exchange between two DNA molecules of segments that are composed of identical or nearly identical nucleotide sequences called *homologous sequences.* Scientists have discovered a number of molecular mechanisms for genetic recombination, one of which is illustrated simply in **Figure 7.28a** on page 230. In this type of recombination, enzymes nick one strand of DNA at the homologous sequence, and another enzyme inserts the nicked strand into the second DNA molecule. Ligase then reconnects the strands in new combinations, and the molecules resolve themselves into novel molecules. Such DNA molecules that contain new arrangements of nucleotide sequences (and the cells that contain them) are called **recombinants.** Scientists first observed recombinants in eukaryotes during *crossing over*—a process in which portions of homologous chromosomes are recombined during the formation of gametes **(Figure 7.28b).** Chapter 12 discusses crossing over and the formation of gametes in more detail.

Horizontal Gene Transfer Among Prokaryotes

Learning Objectives

✓ Contrast vertical gene transfer with horizontal gene transfer.

✓ Explain the roles of an F factor, F$^+$ cells, and Hfr cells in bacterial conjugation.

✓ Describe the structures and actions of simple and complex transposons.

✓ Compare and contrast crossing over, transformation, transduction, and conjugation.

As we have discussed, both prokaryotes and eukaryotes replicate their genomes and supply copies to their descendants. This is known as *vertical gene transfer*—the passing of genes to the next generation. In addition, many prokaryotes can acquire genes from other microbes of the same generation—a process termed **horizontal gene transfer.** In horizontal gene transfer, a **donor cell** contributes part of its genome to a **recipient cell,** which may be a different species or even different genus from the donor. Typically, the recipient cell inserts part of the donor's DNA into its own chromosome, becoming a recombinant cell. Cellular enzymes then usually degrade the remaining unincorporated DNA. Horizontal gene transfer is a rare event, typically occurring in less than 1% of a population.

Here we consider the three types of horizontal gene transfer: *transformation, transduction,* and *bacterial conjugation.*

Transformation

In **transformation,** a recipient cell takes up DNA from the environment, such DNA as might be released by dead organisms. Frederick Griffith (1879–1941) discovered this process in 1928 while attempting to develop a vaccine for pneumonia caused by *Streptococcus pneumoniae* (strep-tō-kok'ŭs nū-mō' nē-ē), though the organism was called *Diplococcus pneumoniae* (dip'lō-kok'ŭs nū-mō' nē-ē) at that time. Griffith worked with two strains of *Streptococcus.* The first strain has a protective capsule that enables it to escape a body's defensive white blood cells; thus, these encapsulated cells cause deadly pneumonia when injected into mice (**Figure 7.29a** on page 231). (They are called *strain S* because they form *smooth* colonies on an agar surface.) The application of heat kills these encapsulated cells and renders them harmless when injected into mice **(Figure 7.29b).** In contrast, cells of the second strain (called *strain R* because they form *rough* colonies) are mutants that cannot make a capsule. The unencapsulated cells of strain R do not cause

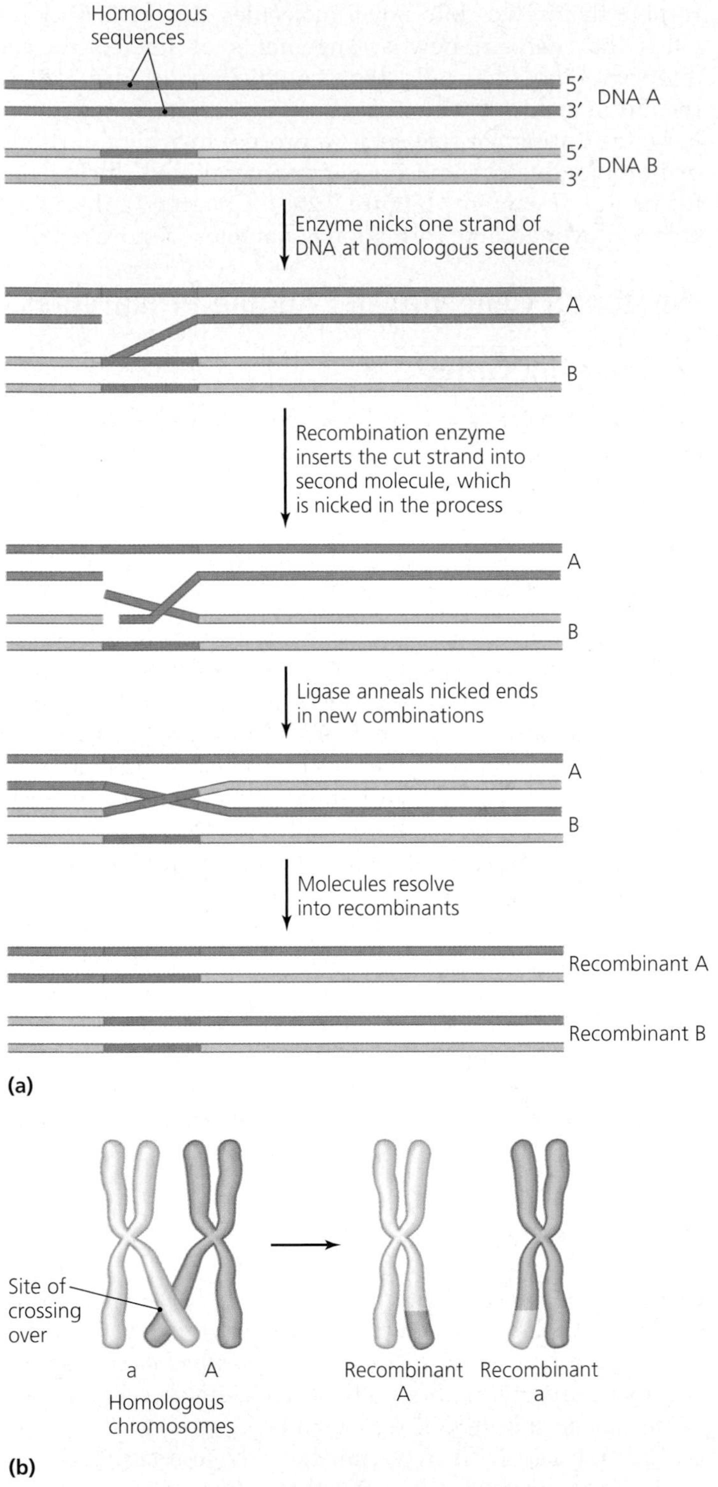

Figure 7.28

Genetic recombination. **(a)** Simplified depiction of one type of recombination between two DNA molecules. After an enzyme nicks one strand (here, strand A), a recombination enzyme rearranges the strands, and ligase seals the gaps to form recombinant molecules. **(b)** Crossing over, a process by which recombination routinely occurs between homologous chromosomes in eukaryotes. *What is the function of ligase during DNA replication?*

Figure 7.28 *Ligase functions to anneal Okazaki fragments during replication of the lagging strand.*

disease **(Figure 7.29c)** because a mouse's defensive white blood cells quickly devour them.

Griffith discovered that when he injected both heat-killed strain S and living strain R into a mouse, the mouse died from pneumonia, even though neither of the injected strains was harmful when administered alone **(Figure 7.29d).** Further, and most significantly, Griffith isolated numerous living, *encapsulated* cells from the dead mouse. He realized that harmless, unencapsulated strain R bacteria had been transformed into deadly, encapsulated strain S bacteria. Subsequent investigations showed that transformation also occurs *in vitro* **(Figure 7.29e).**

The fact that the living encapsulated cells retrieved at the end of this experiment outnumbered the dead encapsulated cells injected at the beginning indicated that strain R cells were not merely appropriating capsules released from dead strain S cells. Instead, strain R cells had acquired the capability of producing their own capsules by assimilating capsule-coding genes released from strain S cells. In 1944, Oswald Avery, Colin MacLeod, and Maclyn McCarty extracted various chemicals from S cells and determined that the transforming agent was DNA. This discovery was one of the conclusive pieces of proof that DNA is the genetic material of cells.

Cells that have the ability to take up DNA from their environment are said to be **competent.** Competence results from alterations in the cell wall and cytoplasmic membrane that allow DNA to enter the cell. Natural competency occurs in only a few types of bacteria, including pathogens in the genera *Streptococcus, Haemophilus* (hē-mof′i-lŭs), *Neisseria, Bacillus* (ba-sil′ŭs), *Staphylococcus* (staf′i-lō-kok′ŭs), and *Pseudomonas* (soo-dō-mō′ nas). Scientists can also generate competency artificially in *Escherichia* and other bacteria by manipulating the temperature and salt content of the medium. Because competent cells take up DNA from any donor genome, competency and transformation are important tools in genetic engineering.

Transduction

A second method of horizontal gene transfer, called **transduction,** involves the transfer of DNA from one cell to another via a replicating virus. Transduction can occur either between prokaryotic cells or between eukaryotic cells; it is limited only by the availability of a virus capable of infecting both donor and recipient cells. Here we will consider transduction in bacteria.

A virus that infects bacteria is called a **bacteriophage** or **phage** (fāj).[7] The process by which a phage participates in transduction is depicted in **Figure 7.30** on page 232. To repli-

---

[7]From Greek *phagein*, meaning to eat.

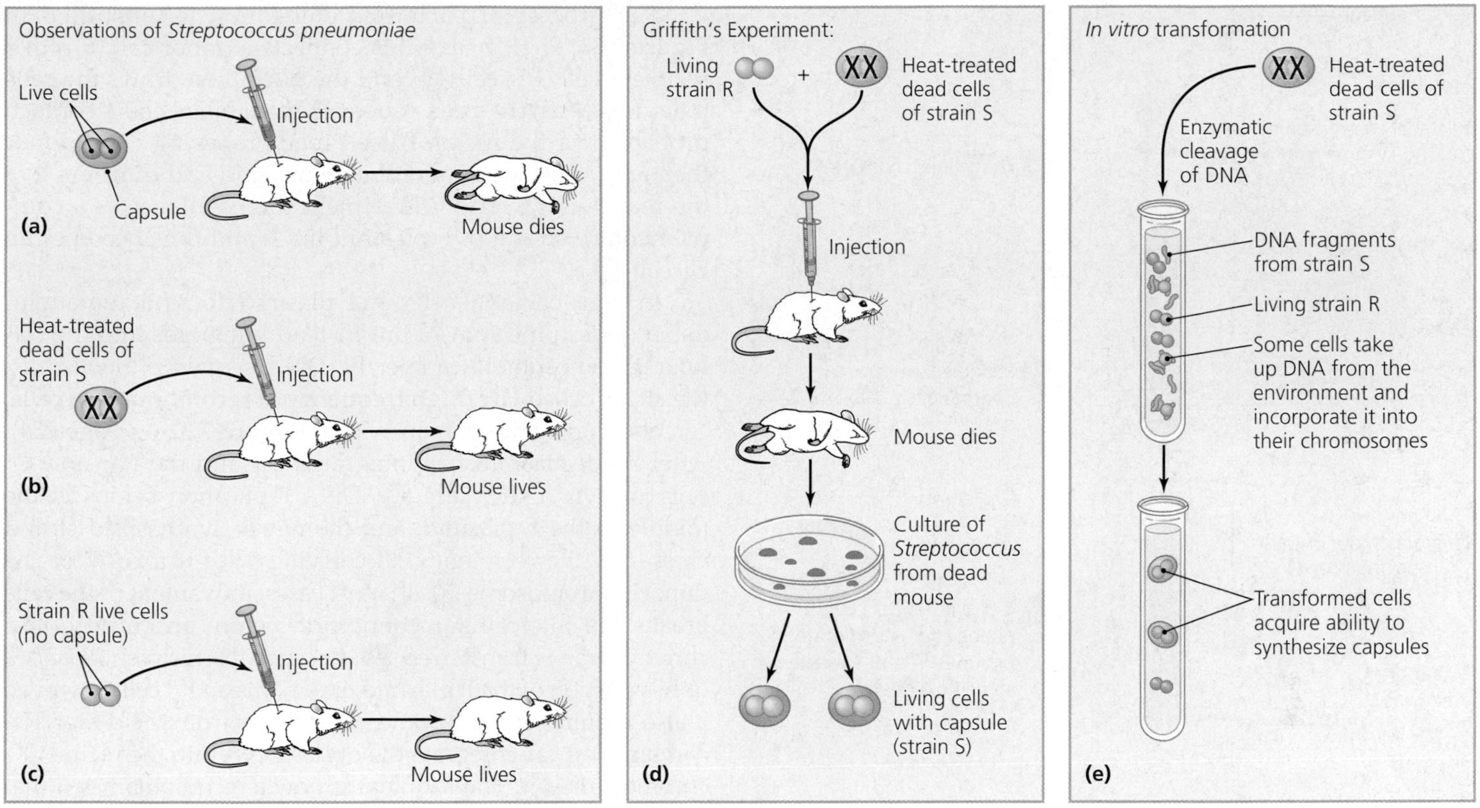

▲ ***Figure 7.29***

Transformation in *Streptococcus pneumoniae*. Griffith's observations revealed that **(a)** encapsulated strain S killed mice, **(b)** heating renders strain S harmless to mice, and **(c)** unencapsulated strain R did not harm mice. In Griffith's experiment **(d),** a mouse injected concurrently with killed strain S and live strain R (each harmless) died and was found to contain numerous living, encapsulated bacteria. **(e)** A demonstration that transformation of R cells to S cells also occurs *in vitro.*

cate, a bacteriophage attaches to a bacterial host cell and injects its genome into the cell ①. Phage enzymes, translated by cellular ribosomes, degrade the cell's DNA ②. The phage genome now controls the cell's functions and directs it to synthesize new phage DNA and phage proteins. Normally, phage proteins assemble around phage DNA to form new phage particles, but some phages mistakenly incorporate remaining fragments of bacterial DNA to form **transducing phages** ③. Eventually the host cell lyses, releasing daughter and transducing phages. Transduction occurs when a transducing phage injects donor DNA into a new host cell (the recipient) ④. The recipient host cell incorporates the donated DNA into its chromosome by recombination ⑤.

In *generalized transduction*, the transducing phage carries a random DNA segment from a donor host cell's chromosome or plasmids to a recipient host cell. The transduction is not limited to a particular DNA sequence. In *specialized transduction,* only certain host sequences are transferred (along with phage DNA). In nature, specialized transduction is important in transferring genes encoding for certain bacterial toxins—including those responsible for diphtheria, scarlet fever, and the bloody, life-threatening diarrhea caused by *E. coli* O157:H7—into cells that would otherwise be harmless. Chapter 8 discusses the use of specialized transduction to intentionally insert genes into cells.

Bacterial Conjugation

A third method of genetic transfer in bacteria is **conjugation.**[8] Unlike the donor cells in transformation and transduction, a donor cell in conjugation remains alive. Further, conjugation requires physical contact between donor and recipient cells. Scientists discovered conjugation between cells of *E. coli,* and it is best understood in this species. Thus, the remainder of our discussion will focus on conjugation in this bacterium.

Conjugation is mediated by **conjugation pili** (pīlī, singular: *pilus*), also called *sex pili,*[9] which are proteinaceous, rodlike structures extending from the surface of a cell. The gene coding for conjugation pili is located on a plasmid

[8]From Latin *conjugatus*, meaning yoked together.

[9]Bacterial conjugation may resemble intercourse, but because no exchange of gametes occurs, it is not true sex.

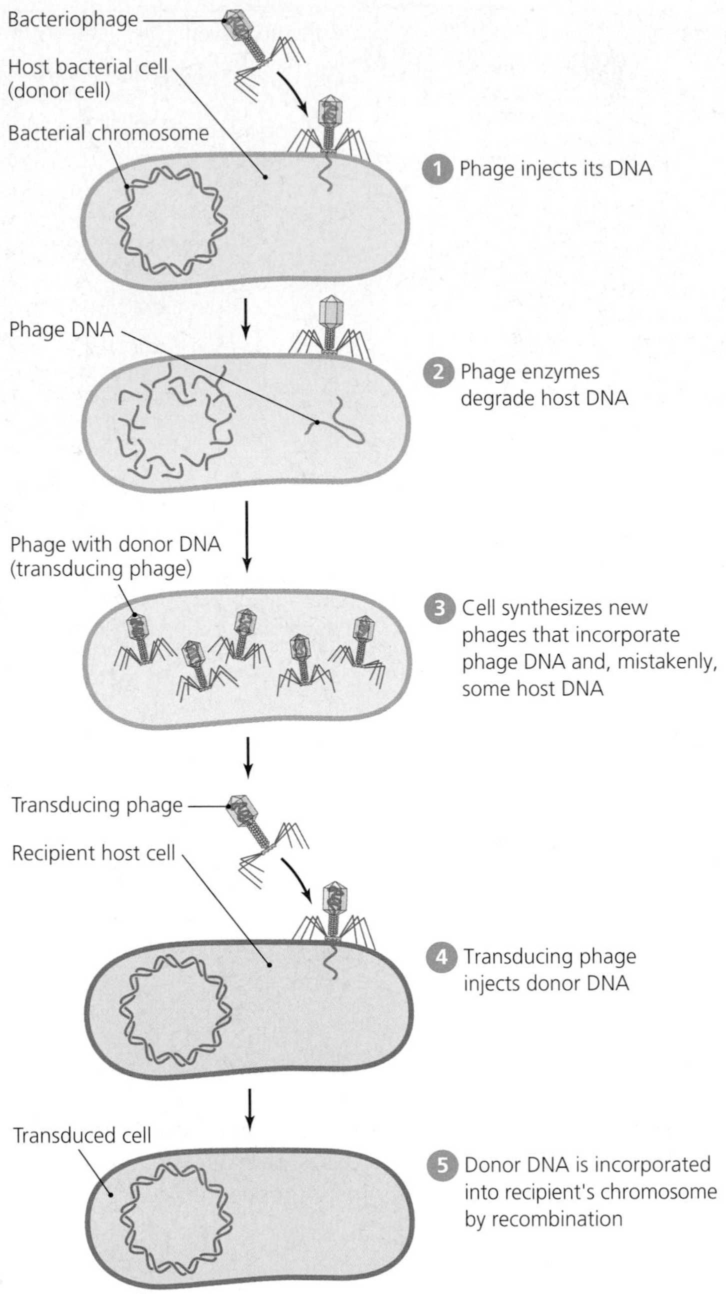
Bacteriophage
Host bacterial cell (donor cell)
Bacterial chromosome
1 Phage injects its DNA
Phage DNA
2 Phage enzymes degrade host DNA
Phage with donor DNA (transducing phage)
3 Cell synthesizes new phages that incorporate phage DNA and, mistakenly, some host DNA
Transducing phage
Recipient host cell
4 Transducing phage injects donor DNA
Transduced cell
5 Donor DNA is incorporated into recipient's chromosome by recombination

▲ ***Figure 7.30***

Transduction. After a virus called a bacteriophage (phage) attaches to a host bacterial cell, it injects its genome into the cell and directs the cell to synthesize new phages. During assembly of new phages, some host DNA may be incorporated, forming transducing phages, which subsequently carry donor DNA to a recipient host cell.

called an **F (fertility) plasmid** or **F factor.** (Recall that a plasmid is a small, circular, extrachromosomal, molecule of DNA; see Figure 7.2b.) Cells that contain an F plasmid are called F^+ cells, and they serve as donors during conjugation. Recipient cells are F^-; that is, they lack an F plasmid and therefore have no conjugation pili.

The process of bacterial conjugation is illustrated in **Figure 7.31.** First, a sex pilus connects a donor cell (F^+) to a recipient cell (F^-) cell (1), and the pilus then draws the cells together. After the cells touch, they stabilize their contact, probably via the fusion of cell membranes (2). The donor then replicates one strand of its F plasmid and transfers it to the recipient (3). The F^- recipient then synthesizes a complementary strand of F plasmid DNA and thus becomes an F^+ cell (4).

In some bacterial cells, an F plasmid does not remain independent in the cytosol but instead integrates into the cellular chromosome at a specific DNA sequence. Such cells, which are called **Hfr (high frequency of recombination) cells,** can also conjugate with an F^- cell (**Figure 7.32** on page 234). After the F plasmid has integrated (1) and the Hfr and F^- cells join via a sex pilus (2), DNA replication begins in the middle of the F plasmid, and the newly synthesized strand crosses to the recipient cell, carrying with it a copy of the donor's chromosome (3). In most cases, movement of the cells breaks the intercellular connection before an entire donor chromosome is transferred (4). Because the recipient receives only a portion of the F plasmid, it remains an F^- cell; however, it also acquires some chromosomal genes from the donor. Recombination can integrate the donor DNA into the recipient's chromosome (5). The recipient is now a recombinant cell that contains its own genes as well as some donor genes.

Because the F plasmid integrates into chromosomes at only a few locations, the order in which genes are transferred is always the same. Scientists have produced maps of the genes on bacterial chromosomes by noting the time, in minutes, required for a particular gene to transfer from donor cells to recipient cells (**Figure 7.33** on page 235).

Conjugation occurs in several species of bacteria, which can be quite promiscuous; that is, conjugation can occur among bacteria of widely varying kinds and even between a bacterium and a yeast cell or between a bacterium and a plant cell. For example, the crown gall bacterium, *Agrobacterium,* transfers some of its genes into the chromosome of its host plant.

The natural transfer of genes by conjugation among diverse organisms heightens some scientists' concerns about the spread of R plasmids among pathogens. These scientists note that antibiotic resistance developed by one pathogen can spread to other pathogens.

Table 7.6 on page 235 summarizes the mechanisms of horizontal gene transfer in bacteria.

Transposons and Transposition

Transposons[10] are segments of DNA, 700–40,000 bp in length, that transpose (move) themselves from one location in a DNA molecule to another location in the same or different molecule. The result of the action of a transposon is

[10]From *transposable* elements.

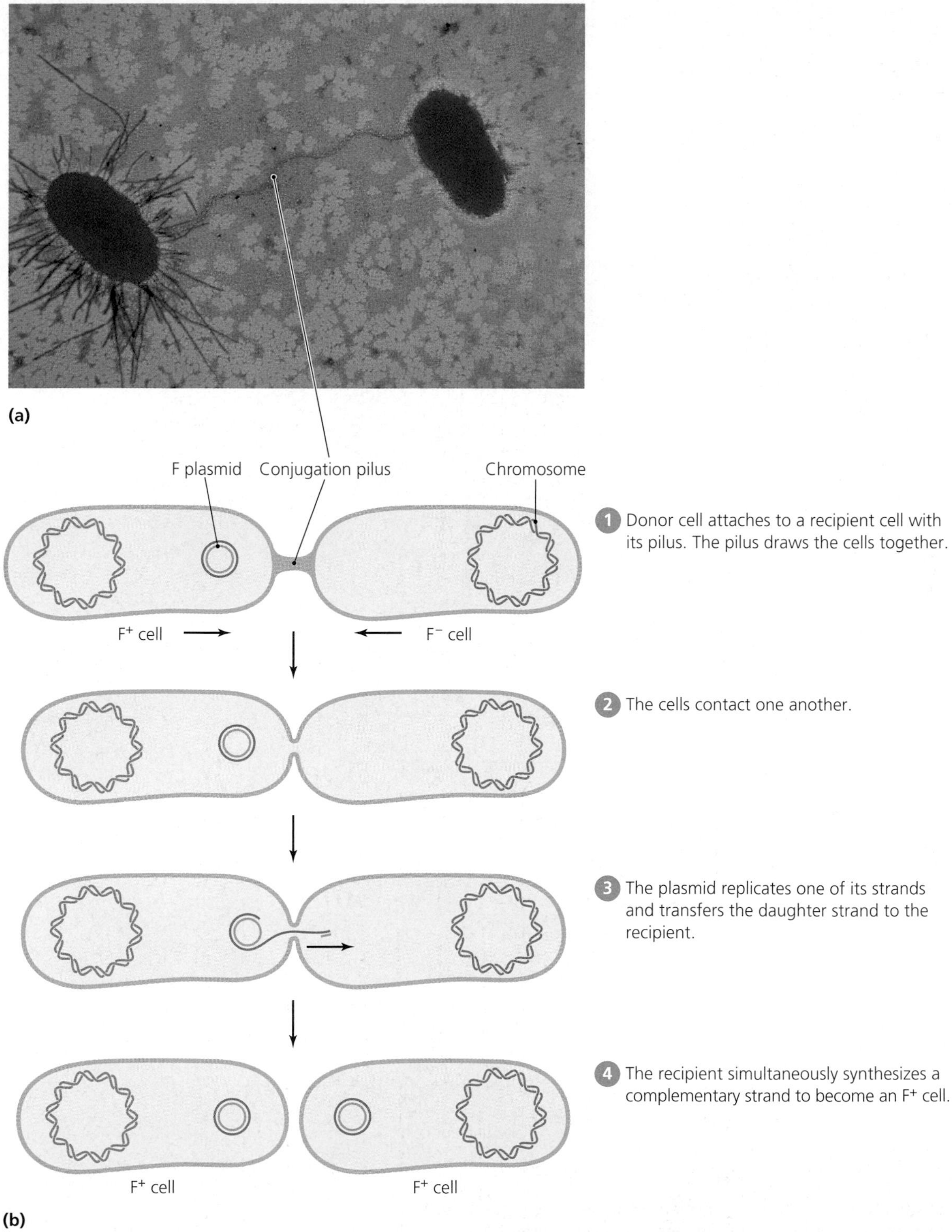
(a)
F plasmid
Conjugation pilus
Chromosome
F⁺ cell
F⁻ cell
1 Donor cell attaches to a recipient cell with its pilus. The pilus draws the cells together.
2 The cells contact one another.
3 The plasmid replicates one of its strands and transfers the daughter strand to the recipient.
4 The recipient simultaneously synthesizes a complementary strand to become an F⁺ cell.
F⁺ cell
F⁺ cell
(b)

▲ ***Figure 7.31***

Bacterial conjugation, a process in which a conjugation pilus connecting two cells mediates the transfer of DNA between the cells. **(a)** Two *E. coli* cells are connected by a conjugation pilus. **(b)** Artist's rendition of the process.

termed *transposition;* in effect it is a kind of frameshift insertion (see Figure 7.20d). This "illegitimate" recombination does not need a region of homology, unlike other recombination events.

The American geneticist Barbara McClintock discovered these so-called "jumping genes" through a painstaking analysis of the colors of the kernels of corn (**Figure 7.34** on page 235). She discovered that the genes for kernel color were turned on and off by the insertion of transposons. Subsequent research has shown that transposons are found in many, if not all,

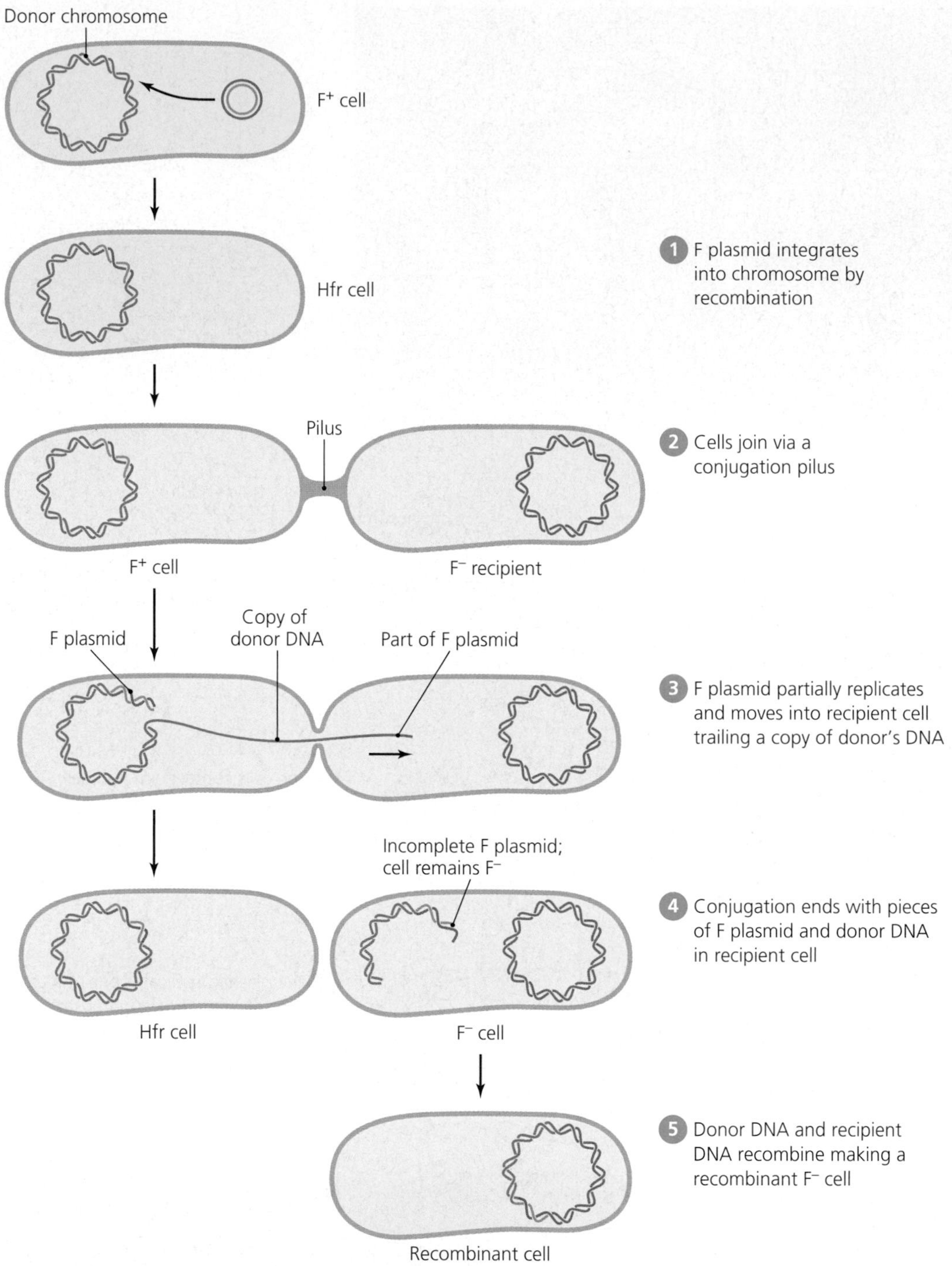
Donor chromosome
F+ cell
Hfr cell
1 F plasmid integrates into chromosome by recombination
Pilus
2 Cells join via a conjugation pilus
F+ cell
F− recipient
F plasmid
Copy of donor DNA
Part of F plasmid
3 F plasmid partially replicates and moves into recipient cell trailing a copy of donor's DNA
Incomplete F plasmid; cell remains F−
4 Conjugation ends with pieces of F plasmid and donor DNA in recipient cell
Hfr cell
F− cell
5 Donor DNA and recipient DNA recombine making a recombinant F− cell
Recombinant cell

▲ ***Figure 7.32***

Conjugation involving an Hfr cell, which is formed when an F^+ cell integrates its F plasmid into its chromosome. Hfr cells donate a partial copy of their DNA and a portion of the F plasmid to a recipient, which is rendered a recombinant cell but remains F^-.

prokaryotes, eukaryotes, and viruses. Fortunately, while transposons are common, transposition and the frameshift mutations it causes are relatively rare occurrences.

Transposons vary in their nucleotide sequences, but all of them contain palindromic sequences at each end. A *palindrome*[11] is a word, phrase, or sentence that has the same sequence of letters when read backward or forward—for example, "Madam, I'm Adam." In genetics, a palindrome is a region of DNA in which the sequence of nucleotides is

[11]From Greek *palin*, meaning again, and *dramein*, meaning to run.

Table 7.6 Natural Mechanisms of Horizontal Genetic Transfer in Bacteria

Mechanism	Requirements	State of Donor	State of Recipient
Transformation	Free DNA in the environment and a competent recipient	Dead	Living
Transduction	Bacteriophage	Killed by bacteriophage	Living
Conjugation	Cell-to-cell contact and F plasmid (either in cytosol or incorporated into chromosome of donor)	Living	Living

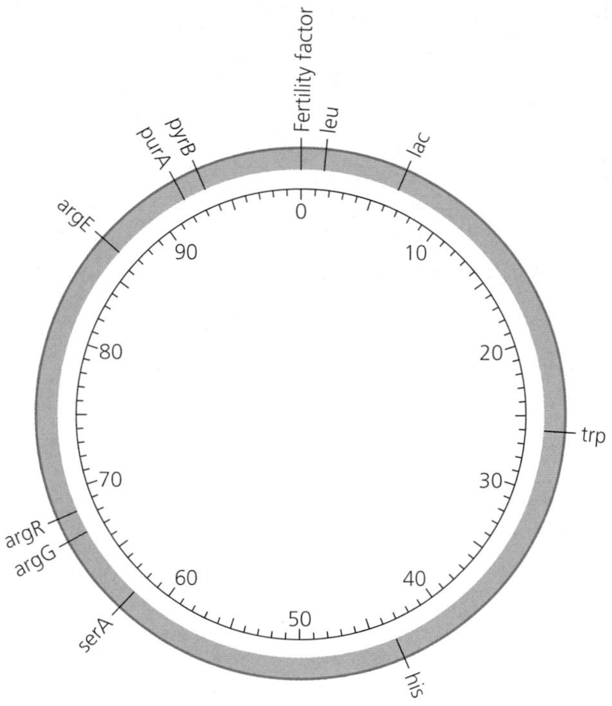

▲ ***Figure 7.33***

A genetic map. This map of the circular genome of *E. coli* is divided into 100 minutes because that is the time required to transfer an entire chromosome from an Hfr donor cell to a recipient cell under standard laboratory conditions. The map shows the locations of a few genes, including the *lac* operon (6 minutes) and the *trp* operon (26 minutes). Insertion of the F factor occurs at 0 minutes. *How long does it take for the genes of the leucine biosynthetic pathway (leu) to pass from the donor cell to the recipient cell? How long for the transfer of the histidine (his) operon?*

Figure 7.33 *It takes 1.5 minutes for the genes of the leucine biosynthetic pathway (leu) to pass from the donor cell to the recipient, and 43.5 minutes for the transfer of the histidine operon.*

▲ ***Figure 7.34***

Variation in color of the kernels of corn, which was instrumental in the discovery of transposons. Barbara McClintock discovered that changes in the inherited colors of kernels were caused by the movement of transposons into and out of genes (in this case, the genes involved in the synthesis of kernel pigments), a process called transposition.

identical to an inverted sequence in the complementary strand. For example, GAATTC is the palindrome of CTTAAG. Such a palindromic sequence is also known as an **inverted repeat (IR).**

The simplest transposons, called **insertion sequences (IS),** consist of no more than two inverted repeats and a gene that encodes the enzyme *transposase* (**Figure 7.35a** on page 236). Transposase recognizes its own inverted repeat in a target site, and then it cuts the DNA at that site and inserts the transposon (or a copy of it) into the DNA molecule **(Figure 7.35b).** Such transposition also produces a duplicate copy of the target site.

Complex transposons also contain one or more genes not connected with transposition, such as genes for antibiotic resistance **(Figure 7.35c).** *R factors,* which are plasmids that convey resistance to antibiotics, often contain transposons. R factors are of great clinical concern because they spread antibiotic resistance to innumerable pathogens.

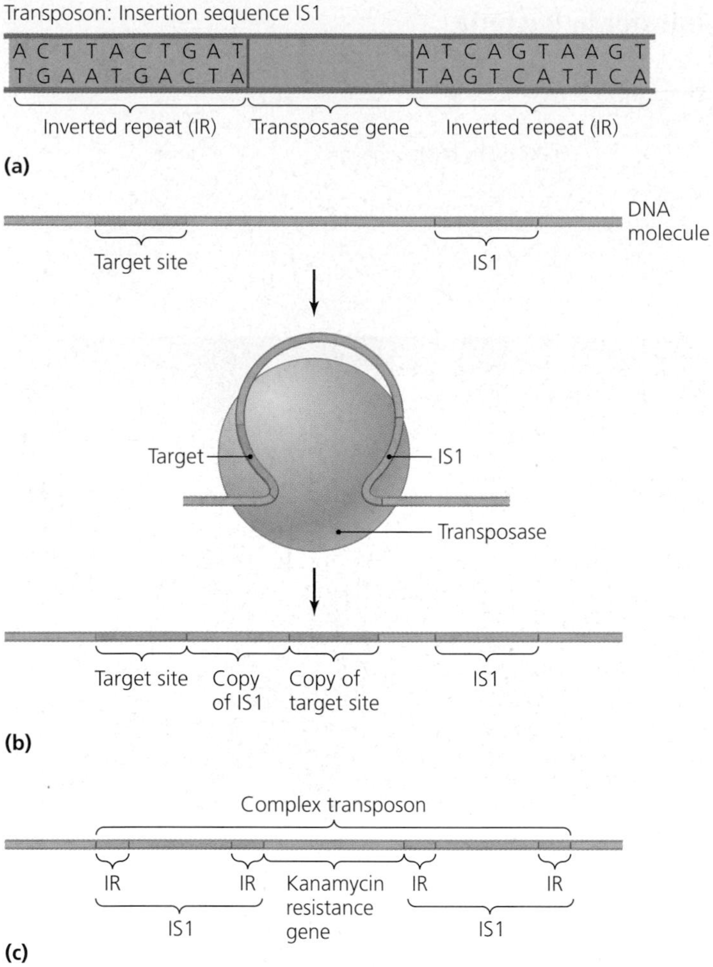

Figure 7.35
Transposons. **(a)** A simple transposon, or insertion sequence. Shown here is Insertion Sequence 1 (IS1), which consists of a gene for the enzyme transposase bounded by identical (though inverted) repeats of nucleotides. **(b)** Transposase recognizes its target site elsewhere in the same DNA molecule, as shown here, or in a different DNA molecule; then it moves the transposon (or, as in the case of IS1, a copy of the transposon) to its new site. The target site is duplicated in the process. **(c)** A complex transposon, which contains genes not related to transposition. Shown here is Tn5, which consists of a gene for kanamycin resistance between two IS1 transposons.

CRITICAL THINKING

Suppose you are a scientist who wants to insert into your dog a gene that encodes a protein that protects dogs from heartworms. A dog's cells are not competent, so they cannot take up the gene from the environment; but you have a plasmid, a competent bacterium, and a related (though incompetent) F^+ bacterium that lives as an intracellular parasite in dogs. Describe a possible scenario by which you could use natural processes to genetically alter your dog to be heartworm resistant.

Chapter 8 discusses how scientists have adapted the natural processes of transformation, transduction, conjugation, and transposition to manipulate the genes of organisms.

CHAPTER SUMMARY

The Structure and Replication of Genomes (pp. 200–208)

1. **Genetics** is the study of inheritance and inheritable traits. Genes are composed of specific sequences of nucleotides that code for polypeptides or RNA molecules. A **genome** is the sum of all the genetic material in a cell or virus. Prokaryotic and eukaryotic cells use DNA as their genetic material; some viruses use DNA, and other viruses use RNA.
2. The two strands of DNA are held together by hydrogen bonds between complementary **base pairs (bp)** of nucleotides. Adenine bonds with thymine, and guanine bonds with cytosine.
3. Prokaryotic genomes consist of one (or rarely, two) **chromosomes,** which are typically circular molecules of DNA associated with protein and RNA molecules, localized in a region of the cytosol called the **nucleoid.** Prokaryotic cells may also contain one or more extrachromosomal DNA molecules called **plasmids,** which contain genes that regulate nonessential life functions, such as bacterial conjugation and resistance to antibiotics.
4. In addition to DNA, eukaryotic chromosomes contain proteins called **histones,** arranged as **nucleosomes** (beads of DNA) that clump to form **chromatin fibers.** Eukaryotic cells also contain extranuclear DNA in mitochondria, chloroplasts, and plasmids.
5. DNA replication is semiconservative; that is, each newly synthesized strand of DNA remains associated with one of the parental strands. After helicase unwinds and unzips the original molecule, synthesis of each of the two daughter strands—called the **leading strand** and the **lagging strand**—occurs from its 5′ end to its 3′ end. Synthesis is mediated by enzymes that prime, join, and proofread the pairing of new nucleotides.
6. After DNA replication, **methylation** occurs. Methylation plays several roles in the control of gene expression, the initiation of DNA replication, recognition of a cell's own DNA, and repair.

Gene Function (pp. 208–221)

1. The **genotype** of an organism is the actual set of genes in its genome, whereas its **phenotype** refers to the physical and functional traits expressed by those genes.
2. RNA has several forms: **messenger RNA (mRNA),** which carries genetic information from DNA to a ribosome; **transfer RNA (tRNA),** which carries amino acids to the ribosome; and **ribosomal RNA (rRNA),** which, together with polypeptides, makes up the structure of ribosomes.

3. The **central dogma** of protein synthesis states that genetic information is transferred from DNA to RNA to polypeptides, which function alone or in conjunction as proteins.
4. The transfer of genetic information begins with **transcription** of the genetic code from DNA to RNA, in which **RNA polymerase** links RNA nucleotides that are complementary to genetic sequences in DNA. Transcription begins at a region of DNA called a **promoter** (recognized by RNA polymerase), and ends with a sequence called a **terminator.** Other proteins may assist in termination, or termination may depend solely on the nucleotide sequence of the transcribed RNA.
5. In **translation,** the sequence of genetic information carried by mRNA is used by ribosomes to construct polypeptides with specific amino acid sequences.
6. The genetic code consists of triplets of mRNA nucleotides, called **codons.** These bind with complementary anticodons on transfer RNAs (tRNA), which are molecules that carry specific amino acids. Ribosomal RNA (rRNA) catalyzes the bonding of one amino acid to another to form a polypeptide. The sequence of nucleotides thus codes for the sequence of amino acids.
7. Eukaryotic mRNA is synthesized as pre-messenger RNA. Before translation can occur, the cell removes noncoding **introns** from pre-mRNA and splices together the **exons,** which are the coding sections.
8. A ribosome contains three tRNA binding sides: an **A site** (associated with incoming amino acids), a **P site** (associated with elongation of the polypeptide), and an **E site** from which tRNA exits the ribosome.
9. An **operon** is a series of genes, a **promoter,** and an **operator** sequence controlled by one regulatory gene. The operon model explains gene regulation in prokaryotes.
10. **Inducible operons** are normally "turned off" and are activated by inducers, whereas **repressible operons** are normally "on" and are deactivated by repressors.

Mutations of Genes (pp. 221–229)

1. A **mutation** is a change in the nucleotide base sequence of a genome. **Point mutations** involve a change in one or a few nucleotide base pairs and include **substitutions** and two types of **frameshift mutations: insertions** and **deletions.** Mutations can be categorized by their effects as **silent, missense,** or **nonsense mutations.**
2. Mutations can be introduced by physical or chemical agents called **mutagens.** Physical mutagens include ionizing radiation such as X-rays and gamma rays, and nonionizing ultraviolet light. Ultraviolet light causes adjacent thymine bases to bond to one another to form **thymine dimers.** Mutagenic chemicals include **nucleotide analogs,** chemicals that are structurally similar to nucleotides and can result in mismatched base pairing, and chemicals that insert or delete nucleotide base pairs, producing frameshift mutations.
3. Cells repair damaged DNA via **light repair** and **dark repair** of thymine dimers, **base-excision repair,** and **mismatch repair.** When damage is so extensive that these mechanisms are overwhelmed, bacterial cells may resort to an **SOS response.**
4. Researchers have developed methods to distinguish **mutants,** which carry mutations, from normal **wild type** cells. These methods include **positive selection, negative (indirect) selection,** and the **Ames test,** which is used to identify potential **carcinogens.**

Genetic Recombination and Transfer (pp. 229–236)

1. Organisms acquire new genes through **genetic recombination**, which is the exchange of segments of DNA. **Crossing over** occurs during gamete formation during sexual reproduction in eukaryotes.
2. Vertical gene transfer is the transmission of genes from parents to offspring. In **horizontal gene transfer,** DNA from a **donor cell** is transmitted to a **recipient cell.** A **recombinant cell** results from genetic recombination between donated and recipient DNA. Transformation, transduction, and bacterial conjugation are types of horizontal gene transfer.
3. In **transformation,** a **competent** recipient prokaryote takes up DNA from its environment. Competency is found naturally and can be created artificially in some cells.
4. In **transduction,** a virus such as a bacteriophage carries DNA from a donor cell to a recipient cell. Donor DNA is accidentally incorporated in such **transducing phages.**
5. In **conjugation,** an F^+ bacterium—that is, one containing an **F (fertility) plasmid (factor)**—forms a **conjugation pilus** that attaches to an F^- recipient bacterium. Plasmid genes are transferred to the recipient, which becomes F^+ as a result.
6. **Hfr (high frequency of recombination) cells** result when an F plasmid integrates into a prokaryotic chromosome. Hfr cells form conjugation pili and transfer cellular genes more frequently than normal F^+ cells.
7. **Transposons** are DNA segments that code for the enzyme transposase bounded by palindromic sequences known as **inverted repeats (IR)** at each end. Transposons move among locations in chromosomes in eukaryotes and prokaryotes. The simplest transposons, known as **insertion sequences (IS)**, consist only of inverted repeats and transposase. **Complex transposons** contain other genes as well.

QUESTIONS FOR REVIEW

(Answers to multiple choice and fill in the blanks are on the web, along with additional review questions. Visit www.microbiologyplace.com.)

Multiple Choice

1. Which of the following is most likely the number of base pairs in a bacterial chromosome?
   a. 4,000,000
   b. 4000
   c. 400
   d. 40
2. Which of the following is a true statement concerning prokaryotic chromosomes?
   a. They typically have two or three origins of replication.
   b. They contain single-stranded DNA.
   c. They are located in the cytosol.
   d. They are associated in linear pairs.

3. A plasmid is
   a. a molecule of RNA found in bacterial cells.
   b. distinguished from a chromosome by being circular.
   c. a structure in bacterial cells formed from plasma membrane.
   d. extrachromosomal DNA.

4. Which of the following forms ionic bonds with eukaryotic DNA and stabilizes it?
   a. chromatin
   b. bacteriocin
   c. histone
   d. nucleosome

5. Nucleotides used in the replication of DNA
   a. are energy storage molecules.
   b. are found in four forms, each with a deoxyribose sugar, a phosphate, and a base.
   c. are present in the cytosol of cells as triphosphate nucleotides.
   d. all of the above.

6. Which of the following molecules functions as a "proofreader" for a newly replicated strand of DNA?
   a. DNA polymerase
   b. primase
   c. helicase
   d. ligase

7. The addition of $—CH_3$ to a cytosine nucleotide after DNA replication is called
   a. methylation.
   b. restriction.
   c. transcription.
   d. translation.

8. In translation, the binding site through which discharged tRNA molecules exit is called the
   a. A site.
   b. X site.
   c. P site.
   d. E site.

9. The Ames test
   a. uses auxotrophs and liver extract to reveal potential mutagens.
   b. is time intensive and costly.
   c. involves the isolation of a mutant by eliminating wild type phenotypes with specific media.
   d. proves that suspected chemicals are carcinogenic.

10. Which of the following methods of DNA repair involves enzymes that recognize and correct nucleotide errors in unmethylated strands of DNA?
    a. light repair of thymine dimers
    b. dark repair of thymine dimers
    c. mismatch repair
    d. SOS response

11. Which of the following is *not* a mechanism of natural genetic transfer and recombination?
    a. transduction
    b. transformation
    c. transcription
    d. conjugation

12. Cells that have the ability to take up DNA from their environment are said to be
    a. Hfr cells.
    b. transposing.
    c. genomic.
    d. competent.

13. Which of the following statements is true?
    a. Conjugation requires sex pili extending from the surface of a cell.
    b. Conjugation involves a C factor.
    c. Conjugation is an artificial genetic engineering technique.
    d. Conjugation primarily involves DNA that has been released into the environment from dead organisms.

14. Which of the following are called "jumping genes"?
    a. Hfr cells
    b. transducing phages
    c. palindromic sequences
    d. transposons

15. Although cells P and Q are totally unrelated, cell Q receives DNA from cell P and incorporates this new DNA into its chromosome. This process is
    a. crossing over of P and Q.
    b. vertical gene transfer.
    c. horizontal gene transfer.
    d. transposition.

16. Which of the following is part of each molecule of mRNA?
    a. palindrome
    b. codon
    c. anticodon
    d. base pair

17. A nucleotide is composed of
    a. a five-carbon sugar.
    b. phosphate.
    c. a nitrogenous base.
    d. all of the above

18. In DNA, adenine forms ________ hydrogen bonds with ________.
    a. three/guanine
    b. two/guanine
    c. two/thymine
    d. three/thymine

19. A sequence of nucleotides formed during replication of the lagging DNA strand is
    a. a palindrome.
    b. an Okazaki fragment.
    c. a template strand.
    d. an operon.

20. Which of the following is *not* part of an operon?
    a. operator
    b. promoter
    c. origin
    d. gene

21. Repressible operons are important in regulating prokaryotic
    a. DNA replication.
    b. RNA transcription.
    c. rRNA processing.
    d. sugar catabolism.

22. Transcription produces ________.
   a. DNA molecules
   b. RNA molecules
   c. polypeptides
   d. palindromes

23. Ligase plays a major role in
   a. lagging strand replication.
   b. mRNA processing in eukaryotes.
   c. polypeptide synthesis by ribosomes.
   d. RNA transcription.

24. Before mutations can affect a population they must be
   a. permanent.
   b. inheritable.
   c. beneficial.
   d. all of the above

25. The *trp* operon is repressible. This means it is usually ________ and is directly controlled by ________.
   a. active/an inducer
   b. active/a repressor
   c. inactive/an inducer
   d. inactive/a repressor

Fill in the Blanks

1. The three steps in RNA transcription are ________, ________, and ________.
2. A triplet of mRNA nucleotides that specifies a particular amino acid is called a ________.
3. Three effects of point mutations are ________, ________, and ________.
4. Insertions and deletions in the genetic code are also called ________ mutations.
5. An operon consists of ________, ________, and ________ and is associated with a regulatory gene.
6. In general, ________ operons are inactive until the substrate of their genes is present.
7. A daughter DNA molecule is composed of one original strand and one new strand because DNA replication is ________.
8. A gene for antibiotic resistance can move among bacterial cells by ________, ________, and ________.
9. ________ are nucleotide sequences containing palindromes and genes for proteins that cut DNA strands.
10. ________ ________ is a recombination event that occurs during gamete formation in eukaryotes.
11. ________ RNA carries amino acids.

Short Answer

1. How does the genotype of a bacterium determine its phenotype? Use the terms *gene, mRNA, ribosome*, and *polypeptide* in your answer.
2. List several ways in which eukaryotic messenger RNA differs from prokaryotic mRNA.
3. Compare and contrast introns and exons.
4. Polypeptide synthesis requires large amounts of energy. How do cells regulate synthesis to conserve energy? Describe one specific example.
5. Describe the operon model of gene regulation.
6. How can knowledge of nucleotide analogs be useful to a cancer researcher?
7. Compare and contrast the structure and components of DNA and RNA in prokaryotes.
8. Besides the fact that it synthesizes RNA, how does RNA polymerase differ in function from DNA polymerase?
9. Describe the formation and function of mRNA, rRNA, and tRNA in prokaryotes and eukaryotes.
10. Describe how DNA is packaged in both prokaryotes and eukaryotes.
11. Explain the central dogma of genetics.
12. Compare and contrast the processes of transformation, transduction, and conjugation.
13. Fill in the following table:

Process	Purpose	Beginning Point	Ending Point
Replication			Origin or end of molecule
Transcription		Promoter	
Translation	Synthesis of polypeptides		

CRITICAL THINKING

1. A scientist uses a molecule of DNA composed of nucleotides containing radioactive sugar molecules as a template for replication and transcription in a nonradioactive environment. What percentage of DNA strands will be radioactive after three DNA replication cycles? What percentage of RNA molecules will be radioactive?

2. If molecules of mRNA have the following nucleotide base sequences, what will be the sequence of amino acids in polypeptides synthesized by eukaryotic ribosomes? The genetic code is displayed in Figure 7.9.
   a. AUGGGGAUACGCUACCCC
   b. CCGUACAUGCUAAUCCCU
   c. CCGAUGUAACCUCGAUCC
   d. AUGCGGUCAGCCCCGUGA

3. The drugs ddC and AZT are used to treat AIDS.

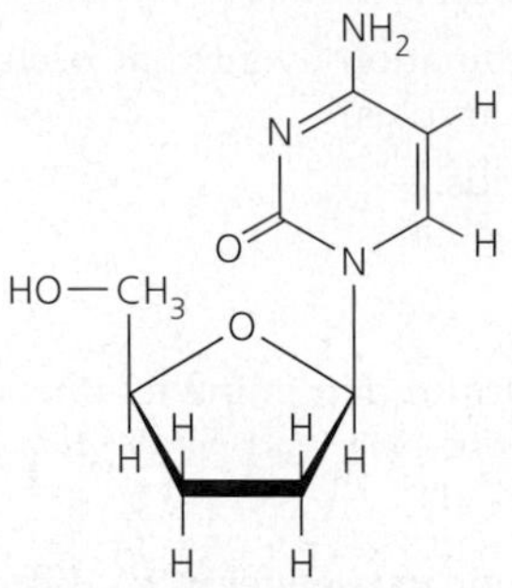

ddc (2′,3′-dideoxycytidine)

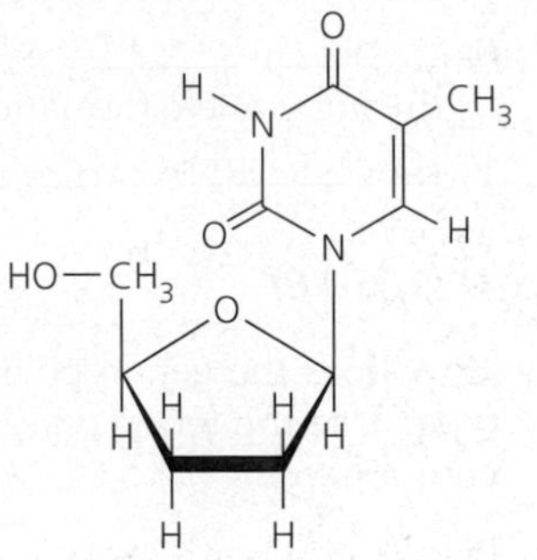

AZT (3′-azido-2′,3′-dideoxythymidine)

Based on their chemical structures, what is their mode of action?

4. Explain why an insertion of three nucleotides is less likely to result in a deleterious effect than an insertion of a single nucleotide.

5. Suppose that *E. coli* sustains a mutation in its gene for the *lac* operon repressor such that the repressor is ineffective. What effect would this have on the bacterium's ability to catabolize lactose? Would the mutant strain have an advantage over wild type cells? Explain your answer.

6. A student claims that nucleotide analogs can be carcinogenic. Another student in the study group insists that nucleotide analogs are sometimes used to treat cancer. Explain why both students are correct.

CHAPTER 8

Recombinant DNA Technology

Recombinant DNA technology affects our lives in many ways. Genetically altered corn, soybeans, or canola are ingredients in over 60% of processed foods in the United States. Scientists now have the ability to take a gene from a bacterium and insert it into a corn plant, enabling the plant to produce a toxin that kills insects but is harmless to humans. Geneticists can insert daffodil and bacterial genes into rice to boost the grain's nutritional content, and they can insert growth-hormone genes into salmon to make the fish grow faster. The possibilities are limitless and intriguing—and frequently controversial, as questions about safety and environmental impact arise. Could crops genetically altered to resist pests spread those resistance genes to weeds? Could genetically modified foods introduce new allergens? What are the effects of these technologies on wildlife populations?

Despite the concerns, recombinant DNA technology holds much promise, not only for the food industry and agriculture, but for medicine as well. This chapter explores the fascinating field of recombinant DNA technology, and the roles of microorganisms within it.

CHAPTER OUTLINE

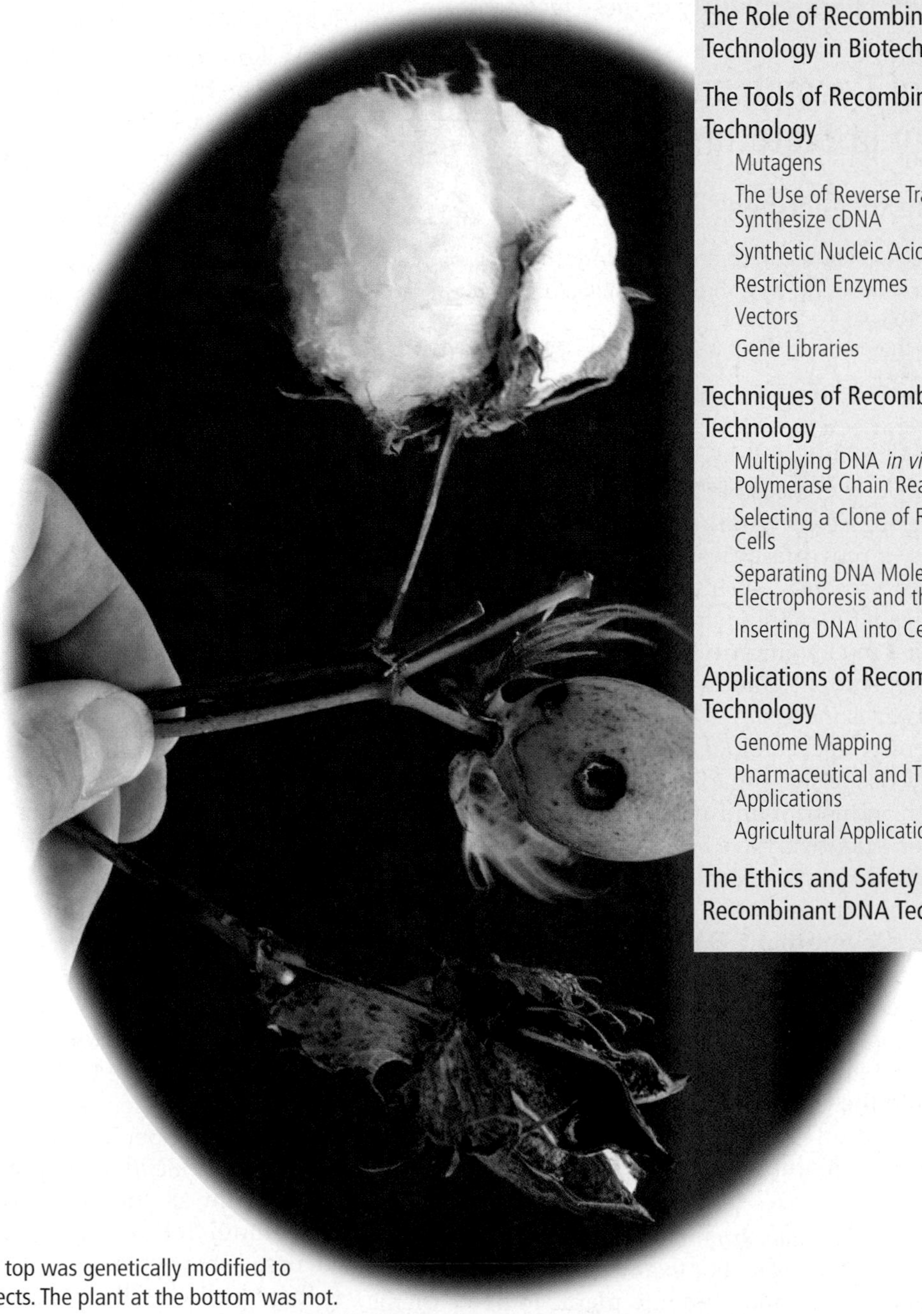

The cotton plant at the top was genetically modified to better resist insects. The plant at the bottom was not.

MicroPrep Pre-Test: *Take the pre-test for this chapter on the web. Visit* ***www.microbiologyplace.com****.*

This chapter examines how genetic researchers have adapted the natural enzymes and processes of DNA recombination, replication, transcription, transformation, transduction, and conjugation to manipulate genes for industrial, medical, and agricultural purposes. Together these techniques are termed *recombinant DNA technology,* commonly called "genetic engineering." We will study the tools and techniques gene researchers use to manipulate DNA and modify genomes for pharmaceutical, therapeutic, and agricultural purposes. We end the chapter with a discussion of the ethics and safety of these techniques.

The Role of Recombinant DNA Technology in Biotechnology

Learning Objectives

- ✓ Define biotechnology and recombinant DNA technology.
- ✓ List several examples of useful products made possible by biotechnology.
- ✓ Identify the three main goals of recombinant DNA technology.

Biotechnology—the use of microorganisms to make practical products—is not a new field. For thousands of years, humans have used microbes to make products such as bread, cheese, soy sauce, and alcohol. During the 20th century, scientists industrialized the natural metabolic reactions of bacteria to make large quantities of acetone, butanol, and antibiotics. More recently, scientists have adapted microorganisms for use in the manufacture of paper, textiles, and vitamins; to assist in cleaning up industrial wastes, oil spills, and radioactive isotopes; and to aid in mining copper, gold, uranium, and other metals.

Until recently, microbiologists were limited to working with naturally occurring organisms and their mutants for achieving such industrial and medical purposes. Since the 1990s, however, scientists have become increasingly adept at intentionally modifying the genomes of organisms, by natural and artificial processes, for a variety of practical purposes. This is **recombinant DNA technology,** and it has expanded the possibilities of biotechnology in ways that seemed like science fiction only a few years ago. Today, scientists isolate specific genes from almost any so-called donor organism, such as a human, a plant, or a bacterium, and insert it into the genome of almost any kind of recipient organism.

Scientists who manipulate genomes have three main goals:

- *To eliminate undesirable phenotypic traits in humans, animals, plants, and microbes.* For example, scientists have inserted genes from microbes into plants to make them resistant to pests or freezing, and since 1999 they have cured children born with a fatal and previously untreatable genetic disorder called severe combined immunodeficiency disease (SCID).
- *To combine beneficial traits of two or more organisms to create valuable new organisms,* such as laboratory animals that mimic human susceptibility to HIV.
- *To create organisms that synthesize products that humans need,* such as vaccines, antibiotics, hormones, and enzymes. For instance, gene therapists have successfully inserted the human gene for insulin into bacteria so that the bacteria synthesize human insulin, which is cheaper and safer than insulin derived from animals.

Recombinant DNA technology is not a single procedure or technique, but rather a collection of tools and techniques scientists use to manipulate the genomes of organisms. In general, they isolate a gene from a cell, manipulate it *in vitro,*[1] and insert it into another organism. **Figure 8.1** illustrates the basic processes involved in recombinant DNA technology.

The Tools of Recombinant DNA Technology

Scientists use a variety of physical agents, naturally occurring enzymes, and synthetic molecules to manipulate genes and genomes. These tools of recombinant DNA technology include *mutagens, reverse transcriptase, synthetic nucleic acids, restriction enzymes,* and *vectors.* Scientists use these molecular tools to create *gene libraries,* which are a time-saving tool for genetic researchers.

Mutagens

Learning Objective

- ✓ Describe how gene researchers use mutagens.

As we saw in Chapter 7, mutagens are physical and chemical agents that produce mutations. Scientists deliberately utilize mutagens to create changes in microbes' genomes so that the microbe's phenotypes are changed. They then select for and culture cells with characteristics considered beneficial for a given biotechnological application. For example, scientists exposed the fungus *Penicillium* to mutagenic agents and then selected strains that produce greater amounts of penicillin. In this manner, they developed a strain of *Penicillium* that secretes over 25 times the amount of penicillin as the strain originally isolated by Alexander Fleming. Today, with recombinant DNA techniques (discussed shortly), researchers can isolate mutated genes rather than dealing with entire organisms.

[1] Latin, meaning within glassware.

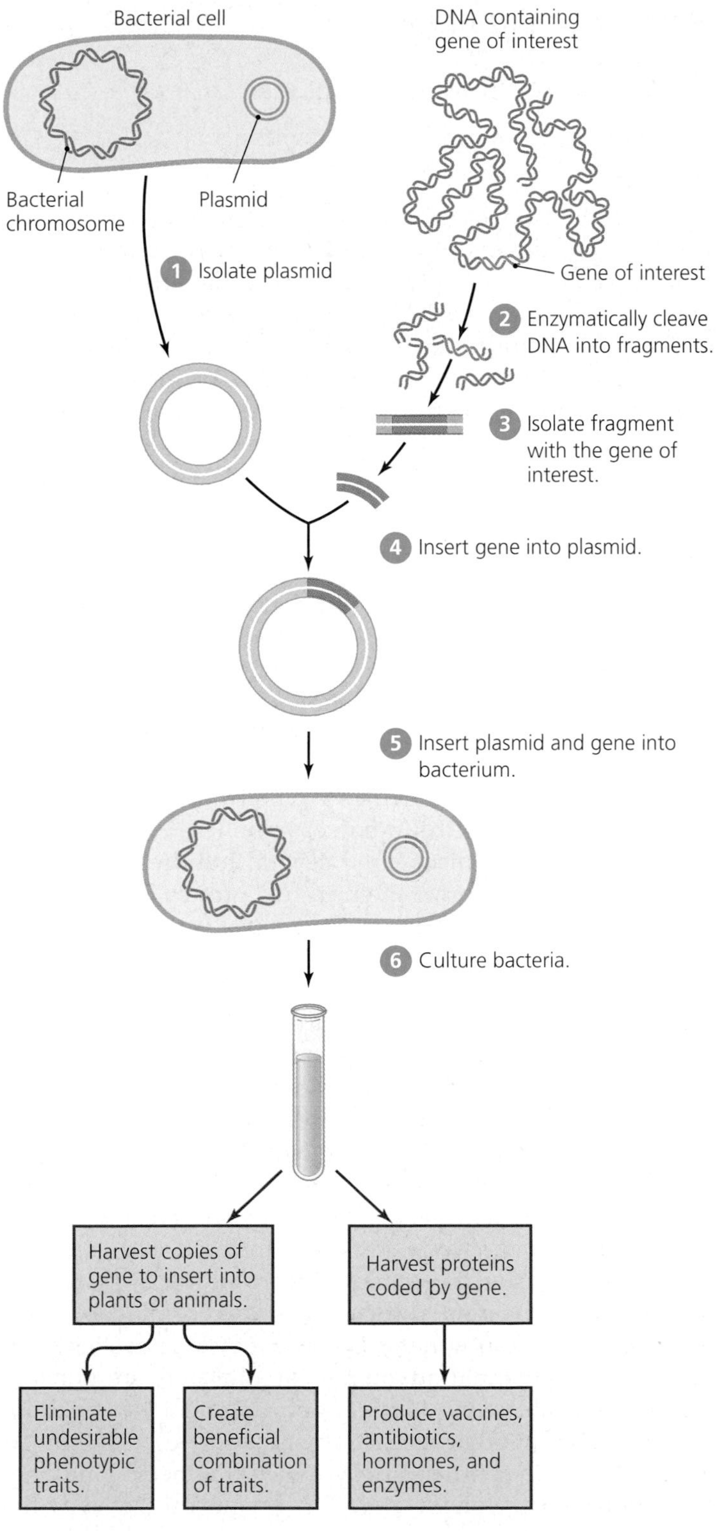

Bacterial cell
DNA containing gene of interest
Bacterial chromosome
Plasmid
1 Isolate plasmid
Gene of interest
2 Enzymatically cleave DNA into fragments.
3 Isolate fragment with the gene of interest.
4 Insert gene into plasmid.
5 Insert plasmid and gene into bacterium.
6 Culture bacteria.
Harvest copies of gene to insert into plants or animals.
Harvest proteins coded by gene.
Eliminate undesirable phenotypic traits.
Create beneficial combination of traits.
Produce vaccines, antibiotics, hormones, and enzymes.

▲ ***Figure 8.1***

Overview of recombinant DNA technology.

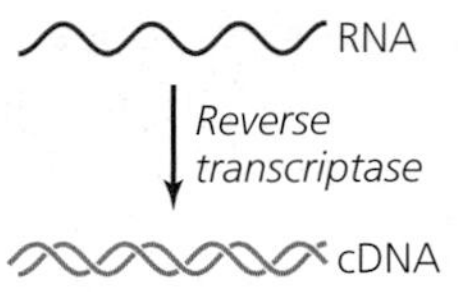

RNA
Reverse transcriptase
cDNA

▲ ***Figure 8.2***

Reverse transcription, in which the viral enzyme reverse transcriptase catalyzes the synthesis of cDNA from an RNA template. The flow of information here—from RNA to DNA—is the opposite of what occurs in cells.

The Use of Reverse Transcriptase to Synthesize cDNA

Learning Objective

✓ Explain the function and use of reverse transcriptase in synthesizing cDNA.

As we discussed in Chapter 7, transcription involves the transmission of genetic information from molecules of DNA to molecules of RNA. The discovery of retroviruses, which have genomes consisting of RNA instead of DNA, led to the discovery of an unusual enzyme—**reverse transcriptase.** Reverse transcriptase in effect creates a flow of genetic information in the opposite direction from what occurs in conventional transcription: It uses an RNA template to transcribe a molecule of DNA, which is called **complementary DNA (cDNA)** because it is complementary to an RNA template **(Figure 8.2).** Because hundreds to millions of copies of mRNA exist for every active gene, it is frequently easier to produce a desired gene by first isolating the mRNA molecules that code for a particular polypeptide and then use reverse transcription to synthesize a cDNA gene from the mRNA template. Further, eukaryotic DNA is not normally expressible by prokaryotic cells, which cannot remove the introns (noncoding sequences) present in eukaryotic pre-mRNA. However, since eukaryotic mRNA has already been processed to remove introns, cDNA produced from it lacks noncoding sequences. Therefore, scientists can successfully insert cDNA into prokaryotic cells, making it possible for the prokaryotes to produce eukaryotic proteins such as human growth factor, insulin, and blood clotting factors.

Synthetic Nucleic Acids

Learning Objectives

✓ Explain how gene researchers synthesize nucleic acids.

✓ Describe three uses of synthetic nucleic acids.

Not only do the enzymes of DNA replication and RNA transcription function *in vivo*;[2] they also function *in vitro*;

[2]Latin, meaning in life (that is, within a cell).

therefore, scientists are able to produce molecules of DNA and RNA in cell-free solutions for genetic research. In fact, scientists have so mechanized the processes of nucleic acid replication and transcription that they can produce molecules of DNA and RNA with any nucleotide sequence; all they must do is enter the desired sequence into a synthesis machine's four-letter keyboard. A computer controls the actual synthesis, using a supply of nucleotides and other required reagents. Nucleic acid synthesis machines synthesize molecules over 100 nucleotides long in a few hours, and scientists can join two or more of these molecules end to end with ligase to create even longer synthetic molecules.

Researchers have used synthetic nucleic acids in many ways, including:

- *Elucidating the genetic code.* Using synthetic molecules of varying nucleotide sequences and observing the amino acids in the resulting polypeptides, scientists elucidated the genetic code. For example, synthetic DNA consisting only of adenine nucleotides yields a polypeptide consisting solely of the amino acid phenylalanine. Therefore, the mRNA codon UUU (transcribed from the DNA triplet AAA) must code for phenylalanine.
- *Creating genes for specific proteins.* Once they know the genetic code and the amino acid sequence of a protein, scientists can create a gene for that protein. In this manner, scientists synthesized a gene for human insulin. Of course, such a synthetic gene may well consist of a nucleotide sequence different from that of its cellular counterpart because of the redundancy in the genetic code (see Table 7.2 on p. 213).
- *Synthesizing DNA and RNA probes to locate specific sequences of nucleotides.* **Probes** are nucleic acid molecules with a specific nucleotide sequence that have been labeled with radioactive or fluorescent chemicals so that their locations can be detected. The use of probes to locate specific sequences of nucleotides is based on the fact that any given nucleotide sequence will preferentially bond to its complementary sequence. Thus a probe constructed of a single strand of DNA with the nucleotide sequence ATGCT will bond to a DNA strand with the sequence TACGA, and the probe's label allows researchers to then detect the complementary site. Probes are essential tools for locating specific DNA sequences such as genes for particular polypeptides.

CRITICAL THINKING

Even though some students correctly synthesize a fluorescent cDNA probe complementary to mRNA for a particular yeast protein, they find that the probe does not attach to any portion of the yeast's genome. Explain why the students' probe does not work.

Restriction Enzymes

Learning Objectives

✓ Explain the source and names of restriction enzymes.
✓ Describe the importance and action of restriction enzymes.

An important development in recombinant DNA technology was the discovery of **restriction enzymes** in bacterial cells. Such enzymes cut DNA molecules and are restricted in their action—they cut DNA only at locations with specific and usually palindromic nucleotide sequences called *restriction sites*. In nature, bacterial cells use restriction enzymes to protect themselves from phages by cutting phage DNA into nonfunctional pieces. (Recall from Chapter 7 that bacterial cells protect their own DNA by methylation of some of their nucleotides, which hides the DNA from the restriction enzymes.)

Researchers name restriction enzymes with three letters (denoting the genus and specific epithet of the source bacterium) and Roman numerals (to indicate the order in which enzymes from the same bacterium were discovered). In some cases, a fourth letter denotes the strain of the bacterium. Thus *Escherichia coli* (esh-ĕ-rik'ē-ă kō'lī) strain R produces the restriction enzymes *Eco*RI and *Eco*RII, and *Hin*dIII is the third restriction enzyme isolated from *Haemophilus influenzae* (hē-mof'i-lŭs in-flu-en'zē) strain Rd.

Scientists have discovered several hundred restriction enzymes, and categorize them in two groups based on the types of cuts they make. The first type, as exemplified by *Eco*RI, makes staggered cuts of the two strands of DNA, producing fragments that terminate in mortise-like *sticky ends,* each of which is composed of up to four nucleotides that form hydrogen bonds with its complementary sticky end **(Figure 8.3a).** Scientists can use these bits of single-stranded DNA to combine pieces of DNA from different organisms into a single recombinant DNA molecule (the enzyme ligase unites the sugar-phosphate backbones of the pieces) **(Figure 8.3b).** Other restriction enzymes, such as *Hin*dII and *Sma*I (from *Serratia marcescens,* sĕ-ră'tē-ă măr-sĕs'enz), cut both strands of DNA at the same point, resulting in *blunt ends* **(Figure 8.3c).** It is more difficult to make recombinant DNA from blunt-ended fragments because they are not sticky, but they have a potential advantage—blunt ends are nonspecific. This enables any blunt-ended fragments, even those produced by different restriction enzymes, to be combined **(Figure 8.3d).** In contrast, sticky-ended fragments bind only to complementary, sticky-ended fragments produced by the same restriction enzyme.

Table 8.1 identifies several restriction enzymes and their target DNA sequences.

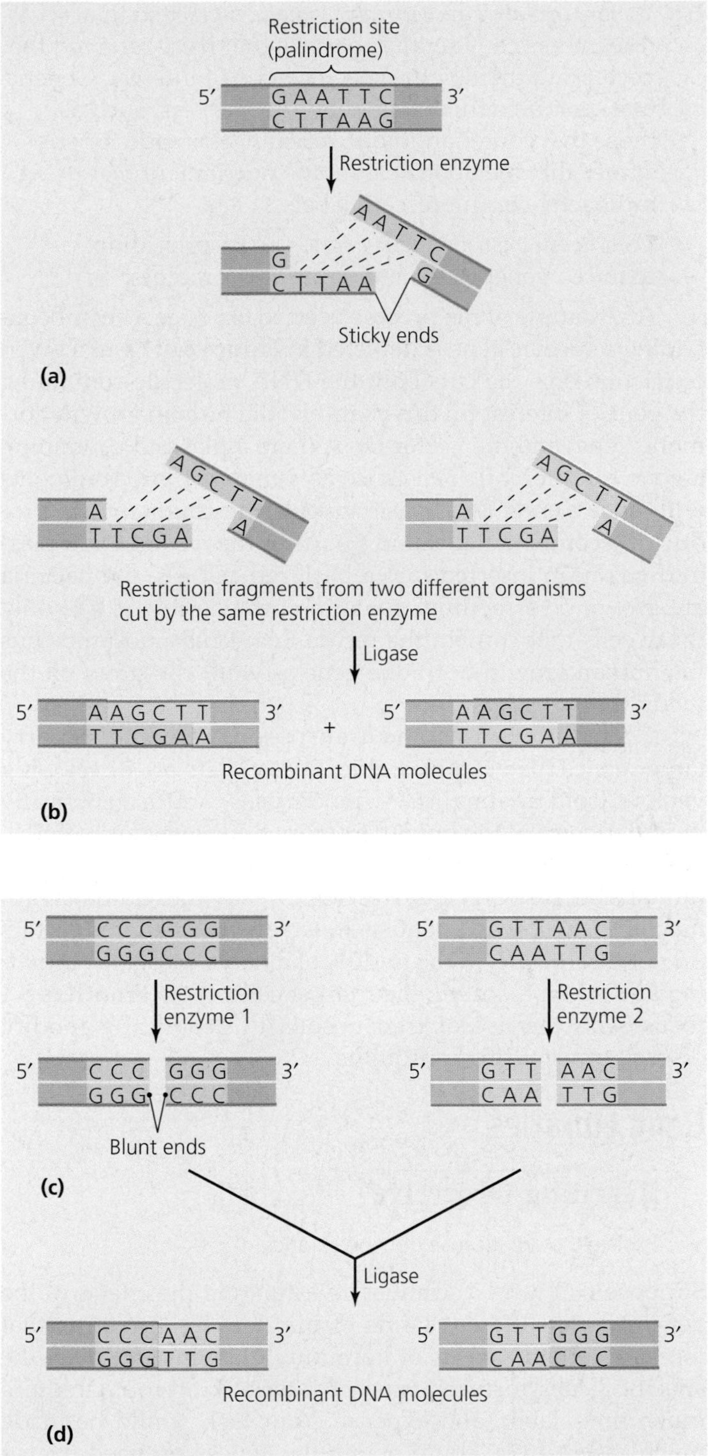

Figure 8.3

Actions of restriction enzymes, which recognize and cut both strands of a DNA molecule at a specific (usually palindromic) restriction site. **(a)** Certain restriction enzymes produce staggered cuts with complementary "sticky ends." **(b)** When two complementary sticky-ended fragments produced by a given restriction enzyme come from different organisms, their bonding (catalyzed by ligase) produces recombinant DNA. **(c)** Other restriction enzymes produce blunt-ended fragments. **(d)** A lack of specificity enables blunt-ended fragments produced by different restriction enzymes to be combined into recombinant DNA. *Which restriction enzymes act at the restriction sites shown?*

Figure 8.3 (a) EcoRI (b) HindIII, and (c) SmaI and HpaI.

Table 8.1 Properties of Some Restriction Enzymes

Enzyme	Bacterial Source	Restriction Site*
*Bam*HI	*Bacillus amyloliquefaciens* H	G↓GATCC CCTAG↑G
*Eco*RI	*Escherichia coli* RY13	G↓AATTC CTTAA↑G
*Eco*RII	*E. coli* R245	CC↓GG GG↑CC
*Hind*II	*Haemophilus influenzae* Rd	GTPy↓PuAC CAPu↑PyTG
*Hind*III	*H. influenzae* Rd	A↓AGCTT TTCGA↑A
*Hin*fI	*H. influenzae* Rf	G↓ANTC CTNA↑G
*Hpa*I	*H. parainfluenzae*	GTT↓AAC CAA↑TTG
*Msp*I	*Moraxella* sp.	CC↓GG GG↑CC
*Sma*I	*Serratia marcescens*	CCC↓GGG GGG↑CCC

*Arrows indicate sites of cleavage; Py = pyrimidine (either T or C); Pu = purine (either A or G); N = any nucleotide (A, T, G, or C).

Vectors

Learning Objective

✓ Define a vector as the term applies to genetic manipulation.

One goal of recombinant DNA technology is to insert a useful gene into a cell so that the cell has a new phenotype—for example, the ability to synthesize a novel protein. To deliver a gene into a cell, researchers use **vectors,** which are nucleic acid molecules such as viral genomes, transposons, and plasmids.

Genetic vectors share several useful properties:

- *Vectors are small enough to manipulate in a laboratory.* Large DNA molecules the size of entire chromosomes are generally too fragile to serve as vectors.
- *Vectors survive inside cells.* Plasmids, which are circular DNA, make good vectors because they are more stable than are linear fragments of DNA, which are typically degraded by cellular enzymes. However, some linear vectors such as transposons and certain viruses insert themselves rapidly into a host's chromosome before they can be degraded.

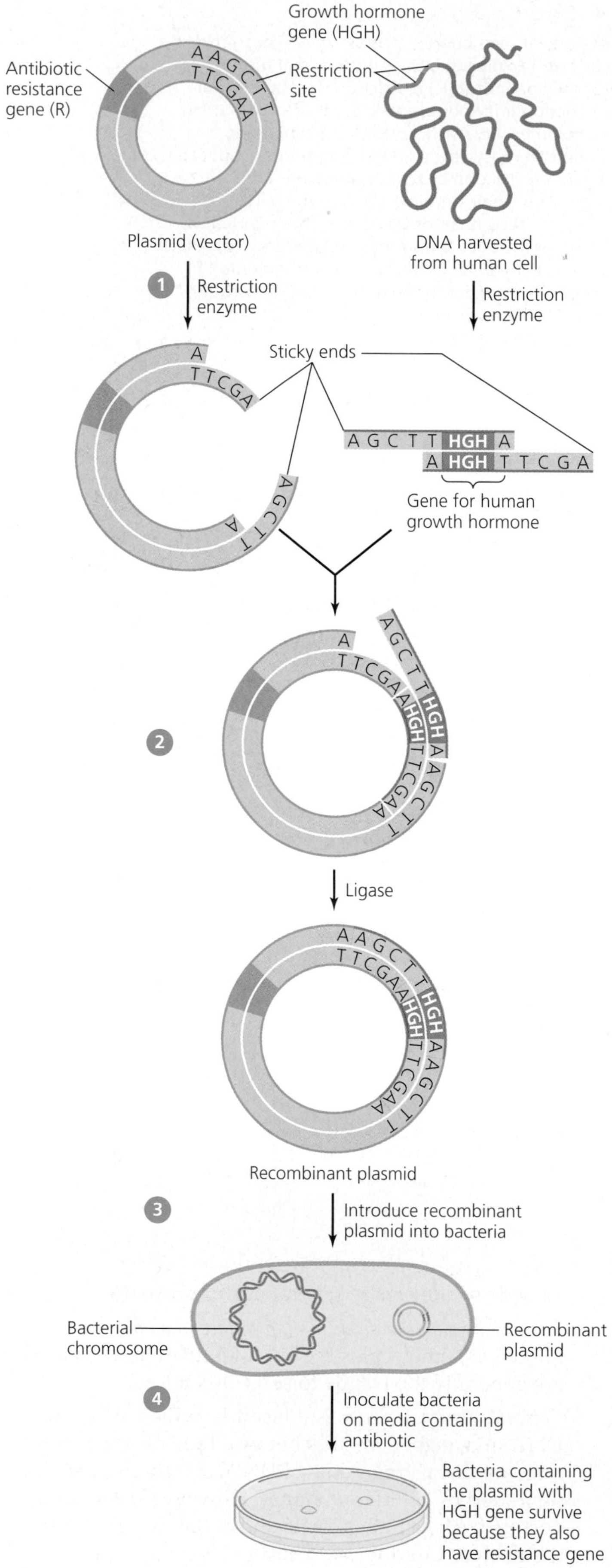

◀ *Figure 8.4*
An example of the process for producing a recombinant vector—in this case, a plasmid. *Which restriction enzyme was used in this example?*

- *Vectors contain a recognizable genetic marker* so that researchers can identify the cells that have received the vector and thereby the specific gene of interest. Genetic markers can either be phenotypic markers, such as those that confer antibiotic resistance or code for enzymes that metabolize a unique nutrient, or *tags* using radioactive or fluorescent labels.
- *Vectors can ensure genetic expression* by providing required genetic elements such as promoters.

An example of the process used to produce a vector containing a specific gene is depicted in **Figure 8.4.** After a given restriction enzyme cuts both the DNA molecule containing the gene of interest (in this example, the human growth hormone gene) and the vector DNA (here a plasmid containing a gene for antibiotic resistance as a marker) into fragments with sticky ends ❶, ligase anneals the fragments to produce a recombinant plasmid ❷. After the recombinant plasmid has been inserted into a bacterial cell ❸, the bacteria are grown on a medium containing the antibiotic ❹; only those cells that contain the recombinant plasmid (and thus the human growth hormone gene as well) can grow on the medium.

Generally, viruses and transposons are able to carry larger genes than can plasmids. Researchers are developing vectors from adenoviruses, poxviruses, and a genetically modified form of the human immunodeficiency virus (HIV). HIV in particular might make an excellent vector for use in humans because HIV inserts itself directly into human chromosomes; however, scientists must ensure that viral vectors do not insert DNA in the middle of a necessary gene, mutating and possibly killing their target cells. **New Frontiers 8.1** focuses on the use of transposons to genetically modify disease-transmitting mosquitos.

Gene Libraries

Learning Objective

✓ Explain the significance of gene libraries.

Suppose you were a scientist investigating the effects of the genes for 24 different kinds of interleukins (proteins that mediate certain aspects of immunity). Having to isolate the specific genes for each type of interleukin would require much time, labor, and expense. Your task would be made much easier if you could obtain the genes you need from a **gene library,** a collection of bacterial or phage clones, each of which contains a portion of the genetic material of interest. In effect, each clone is like one book in a library in that it

Figure 8.4 *HindIII*

New Frontiers 8.1 Building a Better Mosquito

What if scientists could genetically engineer mosquitoes to be incapable of transmitting deadly diseases, such as malaria? Researchers are attempting to do just that. One possible strategy involves using transposons to carry genes into mosquitoes that would boost their production of defensive proteins that destroy pathogens. Mosquitoes naturally produce such proteins, but not in high enough doses to kill off all pathogens. Scientists theorize that if they can genetically engineer and breed mosquitoes that destroy pathogens, these mosquitoes could then be released into the environment to breed with wild mosquitoes, creating subsequent populations that can no longer spread disease.

To date, researchers have successfully used transposons as vectors to insert defense-boosting genes into *Aedes* mosquitoes, which transmit yellow fever. While this is a promising achievement, *Aedes* is only one of over 100 disease-transmitting mosquito species. Recombinant technologies that work on *Aedes* may not necessarily work on, for instance, *Anopheles gambia* mosquitoes, which are the primary carriers of the most deadly form of malaria. In addition, some worry that unleashing transgenic mosquitoes into the wild may have adverse consequences. Given that malaria alone kills over 2 million people each year, however, the quest for a safer mosquito is one that scientists are compelled to continue.

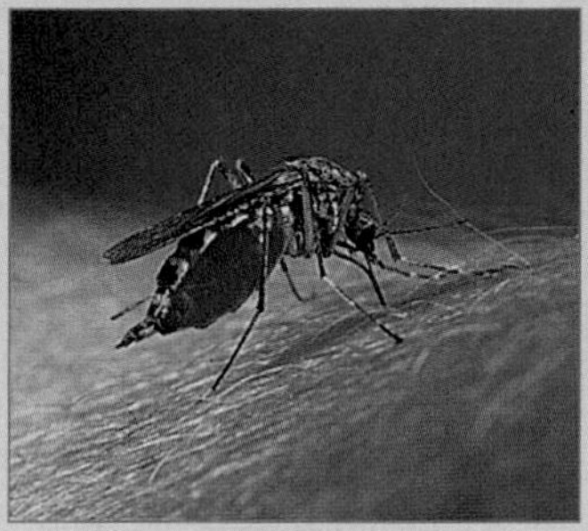

Aedes mosquito

contains one fragment (typically a single gene) of an organism's entire genome. Alternatively, a gene library may contain clones with all the genes of a single chromosome or of the set of cDNA that is complementary to an organism's mRNA.

As depicted in **Figure 8.5** on page 248, genetic researchers can create each of the clones in a gene library by using restriction enzymes to generate fragments of the DNA of interest and then ligase to synthesize recombinant vectors. They insert the vectors into bacterial cells, which are then grown on culture media. Once a scientist isolates a recombinant clone and places it in a gene library, the gene the clone carries becomes available to other investigators, saving them the time and effort required to isolate that gene. Many gene libraries are now commercially available.

Techniques of Recombinant DNA Technology

Scientists use the tools of recombinant DNA technology in a number of basic techniques to multiply, identify, manipulate, isolate, map, and sequence the nucleotides of genes.

Multiplying DNA *in vitro:* The Polymerase Chain Reaction (PCR)

Learning Objective

✓ Describe the purpose and application of the polymerase chain reaction.

The **polymerase chain reaction (PCR)** is a technique by which scientists produce a large number of identical molecules of DNA *in vitro*. Using PCR, researchers start with a single molecule of DNA and generate billions of exact replicas within hours. Such rapid amplification of DNA is critical in a variety of situations. For example, epidemiologists used PCR to amplify the genome of an unknown pathogen that killed 90% of the crows and seven people in New York City in 1999. The large number of identical DNA molecules produced by PCR allowed scientists to determine the nucleotide sequence, which was found to be identical to that of West Nile virus—unknown in the Americas until then. Similarly in 2001, scientists used PCR to amplify DNA from *Bacillus anthracis* (ba-sil'ŭs an-thras'is) spores used in the bioterrorism attacks. The DNA molecules were sequenced and the sequence was compared to those of known *Bacillus* strains in an attempt to identify the source of the spores.

PCR is a repetitive process that alternately separates and replicates the two strands of DNA. Each cycle of PCR consists of the following three steps (**Figure 8.6a** on page 249):

1. *Denaturation*. Exposure to heat (about 94°C) separates the two strands of the target DNA by breaking the hydrogen bonds between base pairs but otherwise leaves the two strands unaltered.
2. *Priming*. After a mixture containing an excess of DNA primers (synthesized such that they are complementary to nucleotide sequences near the ends of the target DNA), DNA polymerase, and an abundance of the four deoxyribonucleotide triphosphates (A, T, G, and C) is added to the target DNA; cooling to about 65°C enables double-stranded DNA to re-form. Because there is an excess of primers, the single strands are more likely to bind to a primer than to one another. The primers provide DNA polymerase with the 3′ hydroxyl group it requires for DNA synthesis.
3. *Extension*. Raising the temperature to about 72°C increases the rate at which DNA polymerase replicates each strand to produce more DNA.

These steps are repeated over and over, so the number of DNA molecules increases exponentially **(Figure 8.6b).** After only 30 cycles—which requires only a few hours to complete—PCR produces over 1 billion identical copies of the original DNA molecule.

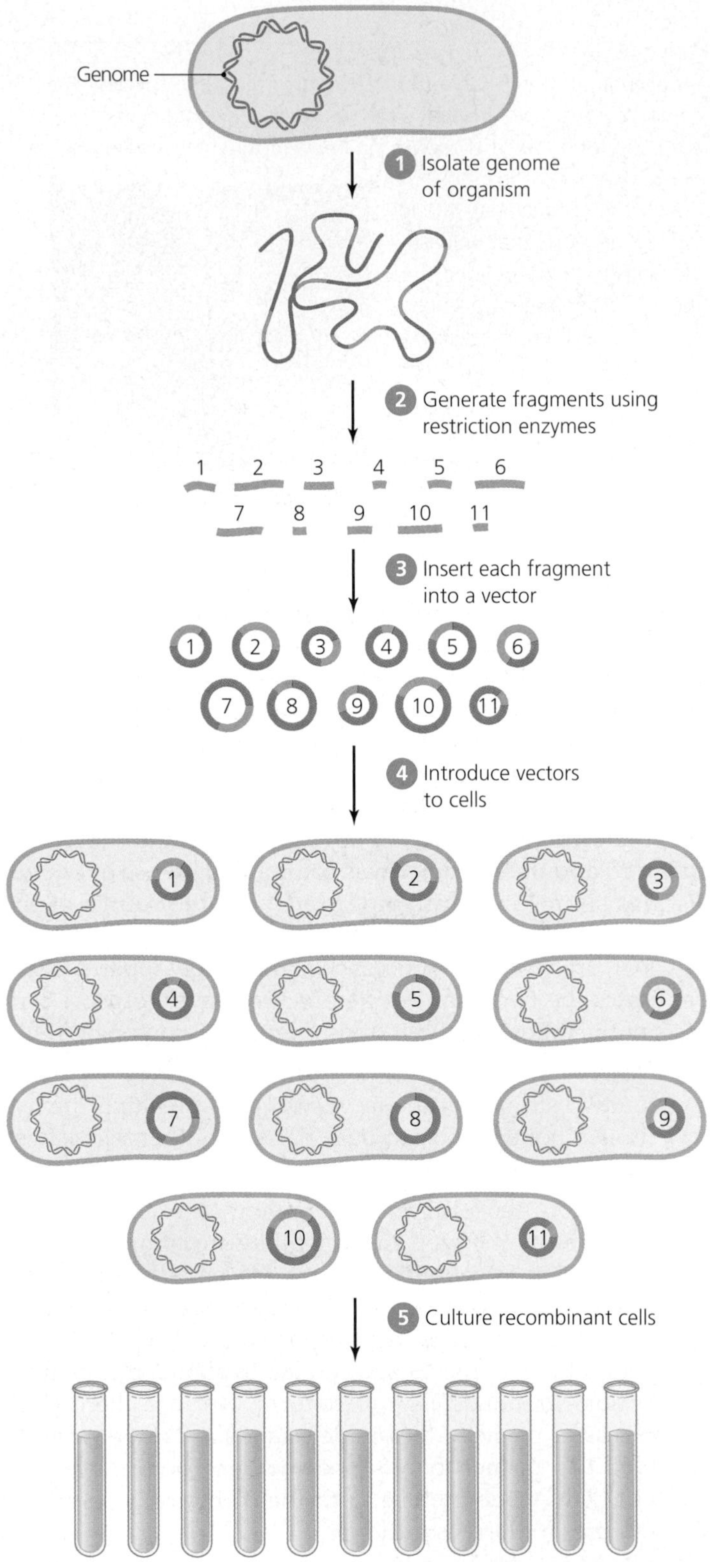

▲ ***Figure 8.5***

Production of a gene library, the population of all cells or phages that together contain all of the genetic material of interest. In this figure, each clone of cells carries a portion of a bacterium's genome.

The process can be automated using a *thermocycler,* a device that automatically performs PCR by continuously cycling all the necessary reagents—DNA polymerase, primers, and triphosphate deoxynucleotides—through the three temperature regimes. A thermocycler uses DNA polymerase derived from hyperthermophilic archaea such as *Thermus aquaticus* (ther'mus a-kwa'ti-cus). This enzyme is not denatured at 94°C, so the machine need not be replenished with DNA polymerase after each cycle.

Selecting a Clone of Recombinant Cells

Learning Objective

✓ Explain how researchers use DNA probes to identify recombinant cells.

Before recombinant DNA technology can have practical application, a scientist must be able to select and isolate recombinant cells that contain particular genes of interest. For example, once researchers have created a gene library, they must find the clone containing the DNA of interest. To do so, scientists use probes—which, you may recall, bind specifically and exclusively to their complementary nucleotide sequences and have either radioactive or fluorescent markers. Researchers then isolate and culture cells that have the radioactive or fluorescent marker, which also aids in identifying the specific location of the genes of interest, as performed in a technique called gel electrophoresis (discussed next).

Separating DNA Molecules: Gel Electrophoresis and the Southern Blot

Learning Objective

✓ Describe the process and use of gel electrophoresis, particularly as it is used in a Southern blot.

Electrophoresis (ē-lek-trō-fōr'ē-sis) is a technique that involves separating molecules based on their electrical charge, size, and shape. In recombinant DNA technology, scientists use **gel electrophoresis** to isolate fragments of DNA molecules that can then be inserted into vectors, multiplied by PCR, or preserved in a gene library.

In gel electrophoresis, DNA molecules, which have an overall negative charge, are drawn through a semisolid gel by an electric current toward the positive electrode within an electrophoresis chamber (**Figure 8.7** on page 250). The gel is typically composed of a purified sugar component of agar, called *agarose,* which in addition to making a more uniform gel than agar, acts as a molecular sieve that retards the movement of DNA fragments down the chamber and separates the fragments by size. Smaller DNA fragments move faster and farther than larger ones. Scientists can determine the size of a fragment by comparing the distance it travels to the distances traveled by standard DNA fragments of known sizes.

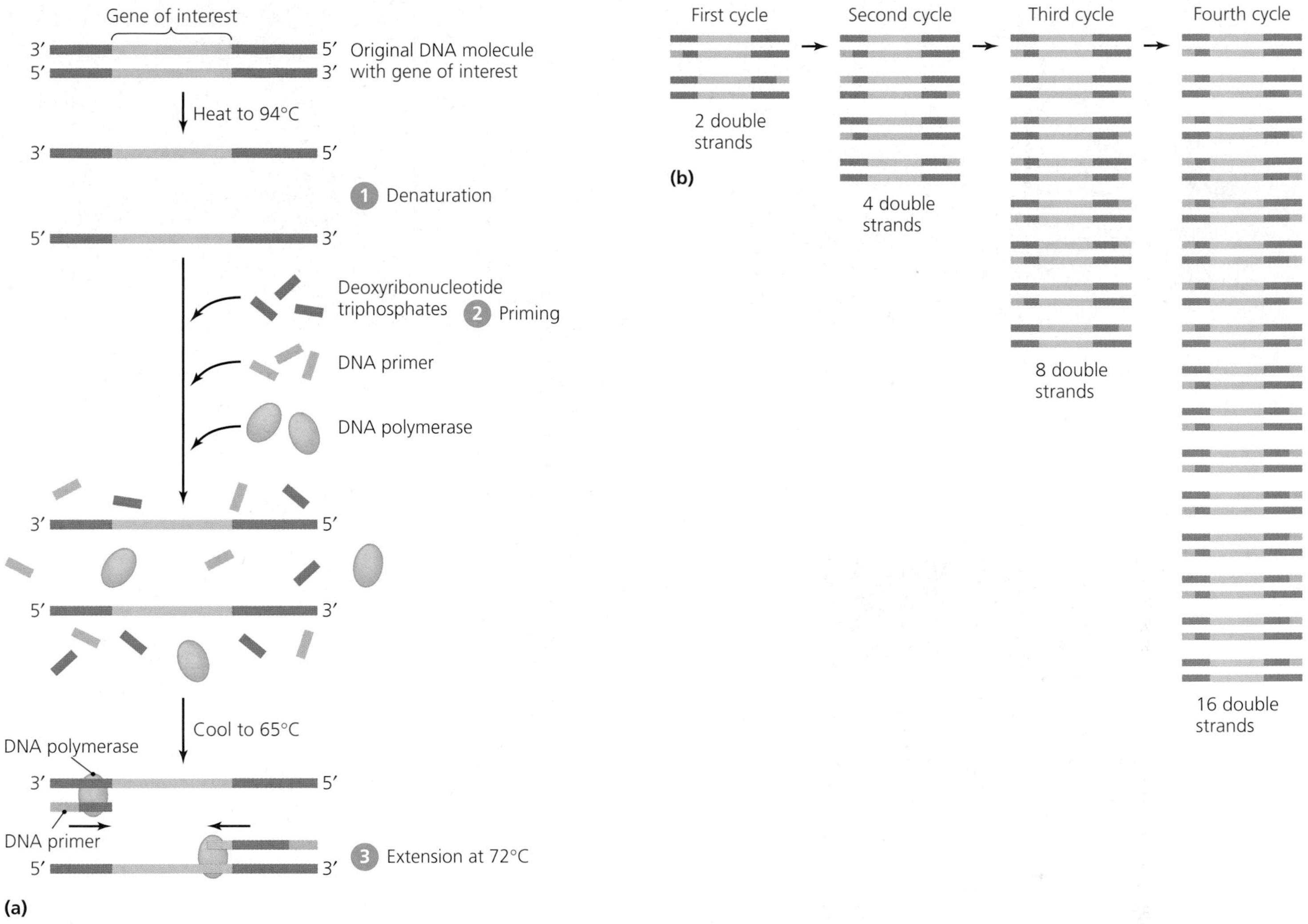

▲ ***Figure 8.6***

The use of the polymerase chain reaction (PCR) to replicate DNA. **(a)** Each cycle of PCR consists of three steps: ① During *denaturation* at 94°C, the hydrogen bonds between DNA strands are broken and the strands separate. ② Cooling to 65°C in the presence of deoxyribonucleotide triphosphates, primers, and DNA polymerase allows *priming* to occur. ③ During *extension* at 72°C, warming speeds the action of DNA polymerase in replicating strands to produce more DNA. **(b)** Each cycle of PCR doubles the amount of DNA; over 1 billion copies of the original DNA molecule are produced by 30 cycles of PCR.

As we have seen, DNA probes allow a researcher to find specific DNA sequences such as genes in a cell. Scientists could also use probes to localize specific sequences in electrophoresis gels, but because gels are flimsy, easily broken, and deform as they dry, it is difficult to probe gels.

In 1975, Ed Southern (1938–) devised a method, called the **Southern blot,** to transfer DNA from agarose gels to nitrocellulose membranes, which are less delicate. The Southern blot technique begins with the procedures of gel electrophoresis just described. Once the DNA fragments have been separated by size, the liquid in the electrophoresis gel is blotted out, the DNA it contains is transferred and bonded to a nitrocellulose membrane, and radioactive probes are used to localize DNA sequences of interest in the membrane (**Figure 8.8** on page 250). A *northern blot* is a similar technique used to detect specific RNA molecules.

Researchers use Southern blots for a variety of purposes, including genetic "fingerprinting" (discussed shortly) and diagnosis of infectious diseases. For example, scientists can detect the presence of genetic sequences unique to hepatitis B virus in a blood sample of an infected patient even before the patient shows symptoms or an immune response.

Scientists also use Southern blotting to demonstrate the incidence and prevalence in an environmental sample of archaea, bacteria, and viruses, particularly those that cannot be cultured. Most microorganisms have never been grown in laboratory; indeed, scientists know them only by unique DNA patterns in electrophoresis gels and Southern blot membranes, called their DNA fingerprints or "signatures." For example, based on the many unique DNA signatures

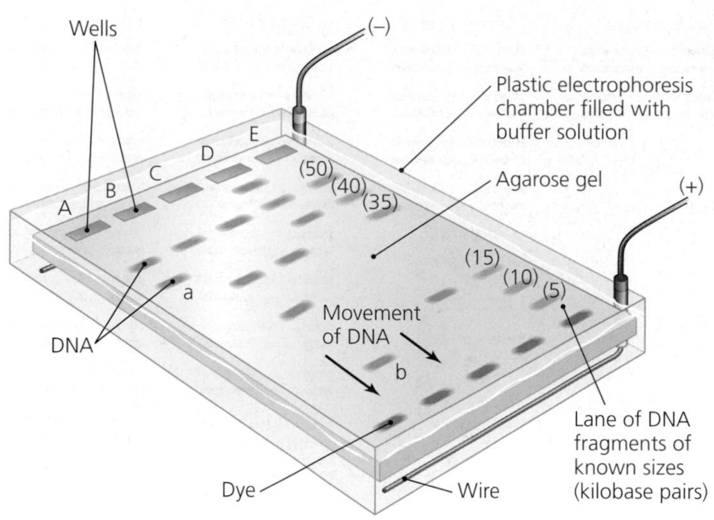

(a)

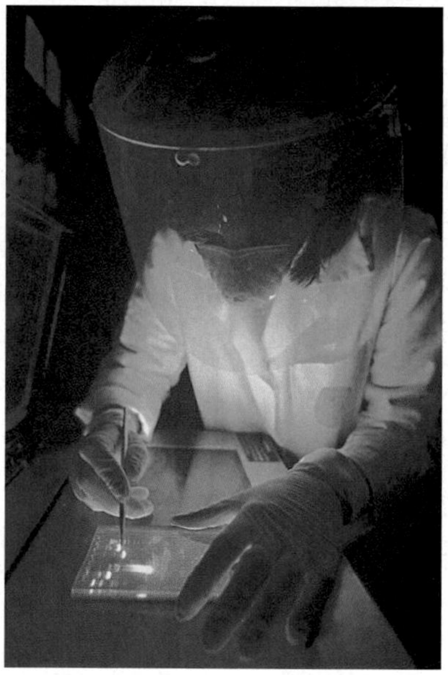

(b)

▲ ***Figure 8.7***

Gel electrophoresis. **(a)** After DNA is cleaved into fragments by restriction enzymes, it is loaded into wells, which are small holes cut into the agarose gel. DNA fragments of known sizes are often loaded into one well (in this case, E) to serve as standards. After the DNA fragments are drawn toward the positive electrode by an electric current, they are stained with a dye. **(b)** Ethidium bromide dye fluoresces under ultraviolet illumination to reveal the locations of DNA within a gel. *Compare the positions of the fragments in lanes A and B to the positions of the fragments of known sizes. What sizes are the fragments labelled a and b?*

Figure 8.7 a–40 kilobase pairs. b–10 kilobase pairs.

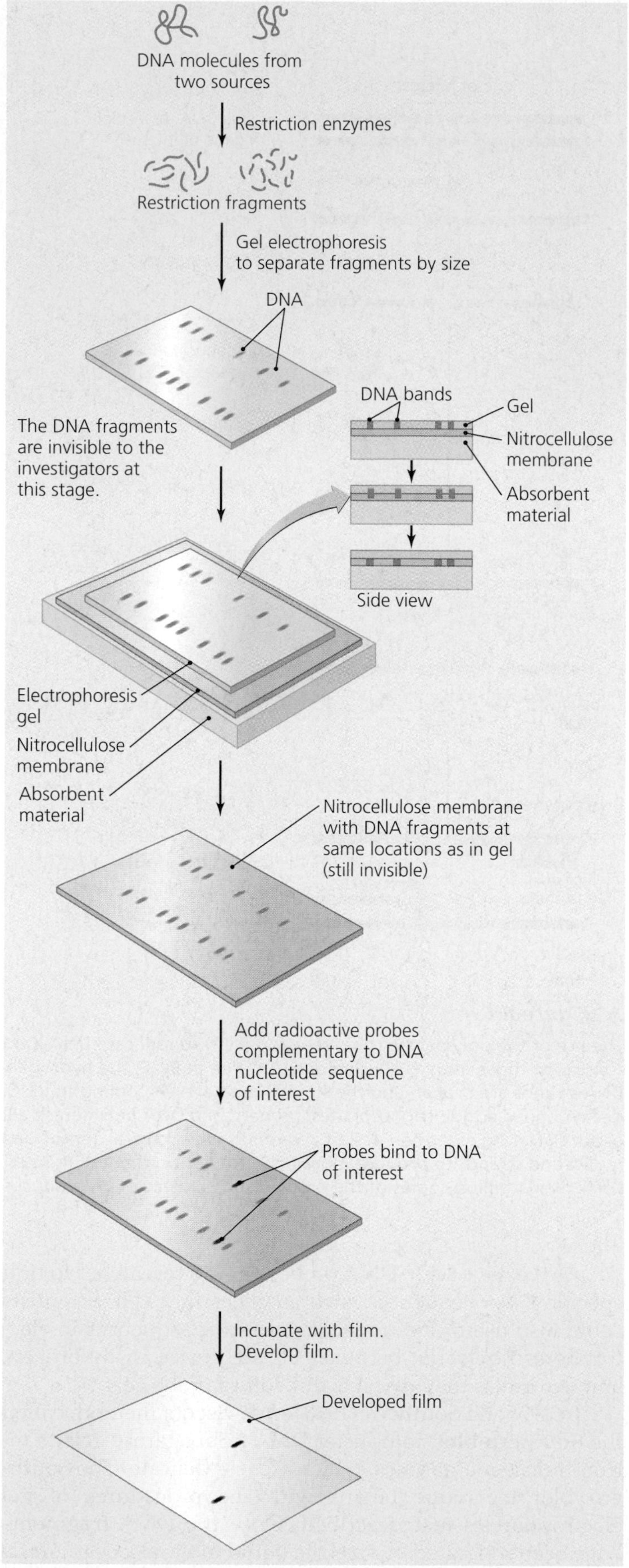

Figure 8.8 ▲

The Southern blot technique, which enables scientists to locate DNA sequences of interest.

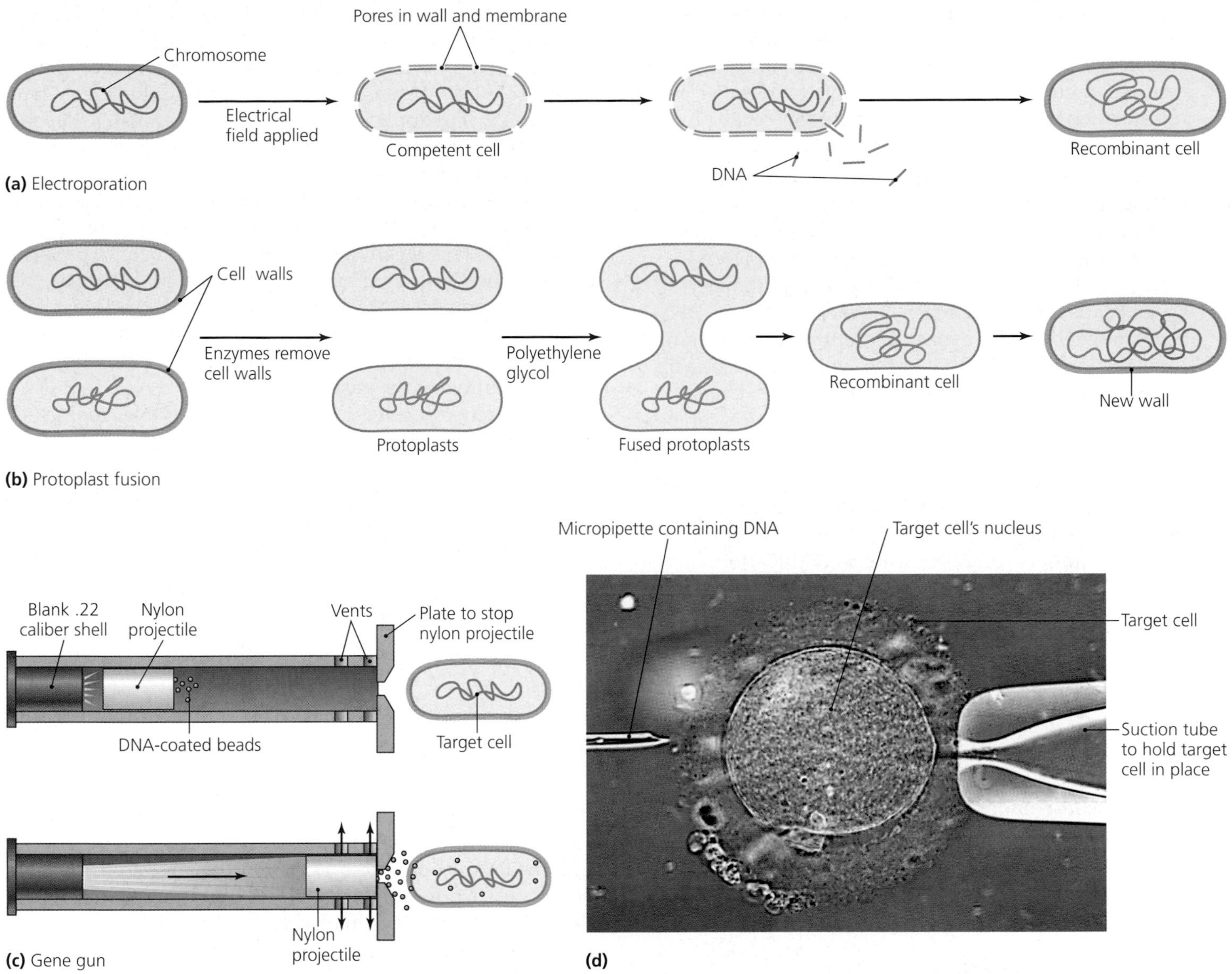

▲ ***Figure 8.9***

Artificial methods of inserting DNA into cells. **(a)** Electroporation, in which an electrical current applied to a cell makes it competent to take up DNA. **(b)** Protoplast fusion, in which enzymes digest cell walls to create protoplasts that fuse at a high rate when treated with polyethylene glycol. **(c)** A gene gun, which fires DNA-coated beads into a cell. **(d)** Microinjection, in which a solution of DNA is introduced into a cell through a micropipette.

isolated from soil that do not match signatures from known microorganisms, scientists calculate that fewer than 1% of soil-dwelling microbes have ever been grown in a laboratory.

Inserting DNA into Cells

Learning Objective

✓ List and explain three artificial techniques for introducing DNA into cells.

A goal of recombinant DNA technology is the insertion of a gene into a cell. In addition to using vectors and the natural methods of transformation, transduction, and conjugation, scientists have developed several artificial methods to introduce DNA into cells, including:

- *Electroporation* **(Figure 8.9a).** Electroporation involves using an electrical current to puncture microscopic holes through a cell's membrane so that DNA can enter the cell from the environment. Electroporation can be used on all types of cells, though the thick-walled cells of fungi and algae must first be converted to *protoplasts,* which are cells whose cell walls have been enzymatically removed. Cells treated by electroporation repair their membranes and cell walls after a time.
- *Protoplast fusion* **(Figure 8.9b).** When protoplasts encounter one another, their cytoplasmic membranes may

fuse to form a single cell that contains the genomes of both "parent" cells. Exposure to polyethylene glycol increases the rate of fusion. The DNA from the two fused cells recombines to form a recombinant molecule. Scientists often use protoplast fusion for the genetic modification of plants.

- *Injection* **(Figure 8.9c** and **d).** Two types of injection are commonly used. Researchers use a *gene gun* powered by a blank .22-caliber cartridge or compressed gas to fire tiny tungsten or gold beads coated with DNA into a target cell. The cell eventually eliminates the inert metal beads. In *microinjection,* a geneticist inserts DNA into a target cell with a glass micropipette having a tip diameter smaller than that of the cell or, in the case of eukaryotes, smaller than that of the nucleus. Unlike electroporation and protoplast fusion, injection can be used on intact tissues such as plant seeds.

In every case, foreign DNA that enters a cell will remain in a cell's progeny only if the DNA is self-replicating, as in the case of plasmid and viral vectors, or if the DNA integrates into a cellular chromosome by recombination.

Applications of Recombinant DNA Technology

The importance of recombinant DNA technology does not lie in the novelty, cleverness, or elegance of its procedures, but in its wide range of applications. In this section we consider how recombinant DNA technology is used to solve various problems and produce products in research, medicine, and agriculture.

Genome Mapping

Learning Objective

✓ Describe genome mapping and genomics, and explain their usefulness.

One application of these tools and techniques is **genome mapping,** which involves locating genes on a nucleic acid molecule. Genome maps provide scientists with useful facts, including information concerning an organism's metabolism and growth characteristics, as well as its potential relatedness to other microbes. For example, in the late 1990s, scientists discovered a virus with a genome map similar to those of certain hepatitis viruses. As a result, they named the new discovery *hepatitis G virus* because it presumably causes hepatitis, though it has not been demonstrated that the virus actually causes hepatitis.

Locating Genes

Until about 1970, scientists identified the specific location of genes on chromosomes by labor-intensive methods. In eukaryotes, the process involved assessing the frequency with which eukaryotic genes remained linked during meiotic crossing over (see Figure 12.1); in bacteria, the process involved timing the transfer of genes during conjugation (see Figure 7.30). Both these techniques are cumbersome, time consuming, and applicable to only some organisms. Recombinant DNA techniques provide simpler and universal methods for genome mapping.

One technique for locating genes, called *restriction fragmentation,* was one of the earliest applications of restriction enzymes. In this technique, which is used for mapping the relative locations of genes in plasmids and viruses, researchers compare DNA fragments resulting from cleavages by several restriction enzymes to determine each fragment's location relative to the others. Consider the following simplified example of the basic procedure.

Researchers cleave identical circular DNA molecules that are 2400 base pairs (2.4 kilobase pairs or kbp) long using three restriction enzymes: *Eco*RI, *Bam*HI, and *Hin*dIII. After they use the enzymes singly and in combinations, they identify the sizes of the fragments by gel electrophoresis. Their results are as follows:

Enzyme	Fragment Size (kilobase pairs)
*Bam*HI	2.4
*Hin*dIII	2.4
*Eco*RI	1.2
*Eco*RI and *Hin*dIII	0.24, 0.96, 1.2
*Eco*RI and *Bam*HI	0.36, 0.84, 1.2

Since either *Hin*dIII or *Bam*HI produces a single fragment, the original DNA must be a circular molecule—cutting a linear molecule always produces more than one fragment. *Eco*RI produces 1.2 kbp fragments, so there must be two restriction sites located on opposite sides of the circle recognized by this enzyme. The *Hin*dIII and *Bam*HI restriction sites must be located in one or the other of the two halves produced by *Eco*RI. **Figure 8.10** illustrates three possible genome maps consistent with these observations. The researchers can determine which of the three is correct by further analysis with other combinations of restriction enzymes. If the researchers know the locations of specific genes on specific fragments, then elucidation of the correct arrangement of the fragments will reveal the relative locations of the genes on the DNA molecule.

Using this method, scientists first completed the entire gene map of a cellular microbe—the bacterium *H. influenzae*—in 1995. Since then, geneticists have elucidated complete gene maps of numerous viral, prokaryotic, and eukaryotic microbes.

CRITICAL THINKING

If the restriction enzymes *Hin*dIII and *Bam*III together produce restriction fragments 1.08 kbp and 1.32 kbp, then which of the three maps shown in Figure 8.10 is correct?

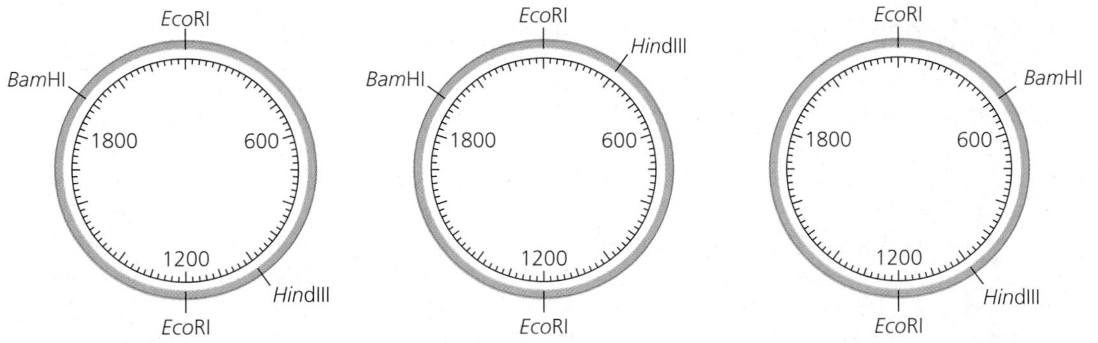

Figure 8.10
Genome mapping using restriction fragmentation. In this technique, a researcher determines the relative locations of restriction sites by an analysis of fragment lengths. Shown are three possible maps corresponding to the actions of the restriction enzymes discussed in the text.

Nucleotide Sequencing

An exciting development in the world of genetics is **genomics,** the sequencing and analysis of the nucleotide bases of genomes. At first, scientists sequenced DNA molecules by selectively cleaving DNA at A, T, G, or C bases, separating the fragments by gel electrophoresis, and mapping the order in which the fragments occur in a complete DNA molecule. Such time-consuming, labor-intensive, and cumbersome sequencing was limited to short DNA molecules such as those of plasmids.

Today, scientists use a faster technique that utilizes cDNA synthesized with nucleotides that have been tagged with four different fluorescent dyes—a different color for each nucleotide base; then an automated DNA sequencer determines the sequence of base colors emitted by the dyes **(Figure 8.11).** Such machines, often running 24 hours a day for months, have sequenced the entire genomes of numerous viruses, bacteria, and eukaryotic organisms. Scientists reached a milestone in 2001 by sequencing the 3 billion nucleotide base pairs that constitute the human genome. **New Frontiers 8.2** on page 254 explains how scientists are even using the nucleotide sequence in the human genome to detect new infectious agents.

Elucidation of the gene sequences of pathogens, particularly those affecting hundreds of millions of people and those with potential bioterrorist uses, is a current priority of researchers. For example, in 2002 scientists finished sequencing the genome of *Plasmodium falciparum* (plaz-mō′dē-ŭm fal-sip′ar-ŭm), one of the protozoa that causes malaria, and hope to use the information to develop novel antimalarial drugs and an effective vaccine. Other researchers are sequencing the genomes of the pathogens that cause anthrax, bubonic plague, and cholera in order to develop more effective therapies and vaccines.

Another use for genomics is to relate DNA sequence data to protein function. For instance, scientists are investigating the genes and proteins of *Deinococcus radiodurans* (dī-nō-kok′ŭs rā-dē-ō-dur′anz), a microorganism that is remarkably resistant to damage of its DNA by radiation.

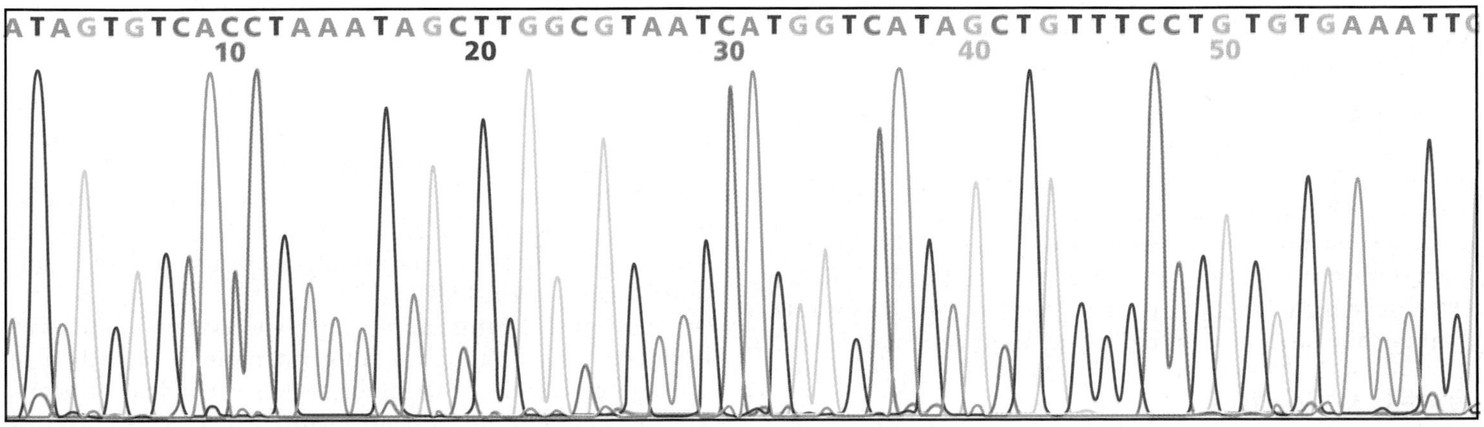

Figure 8.11
Automated DNA sequencing. Each colored line corresponds to a different nucleotide base; each peak indicates the location of a particular base. The lines are continuous because they are drawn by four colored pens onto a moving strip of paper. *What is the nucleotide sequence for the nucleotides in positions 35–45?*

Figure 8.11 TCATAGCTGTT

New Frontiers 8.2 **Using the Human Genome to Detect Disease-Causing Microbes**

Could unidentified microorganisms cause chronic autoimmune diseases such as lupus and multiple sclerosis? When new infectious disease agents emerge—as HIV, Ebola virus, hantavirus, and West Nile virus have in recent decades—how can we identify the microbes? Scientists have recently developed an intriguing new approach to investigating these questions: using the human genome.

The Human Genome Project was begun in 1990 with two ambitious goals: identifying all the genes in human DNA, and determining the sequences of the 3 billion base pairs that make up that DNA. In June 2000, in an announcement that made headlines worldwide, scientists proclaimed they had completed a working draft of the human genome. This was followed in February 2001 by reports containing the revelation that humans likely have only 26,000 to 40,000 genes—only two or three times the number possessed by the fruit fly *Drosophila melanogaster*, which has about 13,000 genes. In comparison, pathogenic *E. coli* has 5,416 genes.

Equipped with DNA sequence data compiled by the Human Genome Project, scientists at the Dana-Farber Cancer Institute have developed a technique called "computational subtraction," in which a portion of DNA from a diseased human tissue sample is sequenced and then compared against the sequence of DNA in the entire human genome. The matches are then eliminated, and what is "left over" is DNA that can then be studied to determine if it is that of a novel microorganism.

Scientists tested this technique by using a tissue sample known to contain a type of cancer-causing human papillomavirus. The method worked—after the matches were eliminated, what was "left over" contained DNA known to be that of the papillomavirus. The hope is that using "computational subtraction" scientists might detect the DNA of either an as-yet-unidentified microbial agent responsible for mysterious chronic illnesses such as lupus, or of newly emerging viruses.

Such studies may lead to methods of reversing genetic damage in cancer patients undergoing radiation therapy. Researchers are also investigating the genetic basis of the enzymes of psychrophiles, which are microorganisms that thrive at temperatures below 20°C. Such enzymes have potential applications in food processing and in the manufacture of drugs.

Table 8.2 summarizes the tools and techniques of recombinant DNA technology.

Pharmaceutical and Therapeutic Applications

Learning Objectives

- ✓ Describe six potential medical applications of recombinant DNA technology.
- ✓ Describe the steps and uses of genetic fingerprinting.
- ✓ Define gene therapy.

Researchers now supplement traditional biotechnology with recombinant DNA technology to produce a variety of pharmaceutical and therapeutic substances, and to perform a host of medically important tasks. Next we explore the use of recombinant DNA technology to synthesize selected proteins, produce vaccines, screen for genetic diseases, match DNA specimens to the organisms from which they came, treat genetic illnesses, and aid in organ transplantation.

Protein Synthesis

Scientists have inserted synthetic genes for insulin, for interferon (a natural antiviral chemical), and for other proteins into bacteria and yeast cells so that the cells synthesize these proteins in vast quantities. In the past, such proteins were isolated from donated blood or from animals—labor-intensive processes that carry the risk of inducing allergies or of transferring pathogens such as hepatitis B and HIV. "Genetically engineered" proteins are safer and less expensive than their naturally occurring counterparts.

Vaccines

Vaccines contain *antigens*—foreign substances such as weakened bacteria, viruses, and toxins that stimulate the body's immune system to respond to and subsequently remember these foreign materials. In effect, a vaccine primes the immune system to respond quickly and effectively when confronted with pathogens and their toxins. However, the use of some vaccines entails a risk—they may cause the disease they are designed to prevent.

Scientists now use recombinant DNA technology to produce safer vaccines. Once they have inserted the gene that codes for a pathogen's antigenic polypeptides into a vector, they can inject the recombinant vector or the proteins it produces into a patient. Thus the patient's immune system is exposed to a so-called subunit of the pathogen—one of the pathogen's antigens—but not to the pathogen itself. Such **subunit vaccines** are especially useful in safely protecting against pathogens that either cannot be cultured or cause incurable fatal diseases. Hepatitis B vaccine is an example of a successful subunit vaccine. Scientists are also pursuing subunit vaccines against HIV.

A promising future approach to vaccination involves introducing genes coding for antigenic proteins of pathogens into common fruits or vegetables such as bananas or beans. The immune systems of people eating such altered produce would be exposed to the pathogen's antigens and theoretically would develop immunologic memory against the pathogen. Such a vaccine would have the advantages of being painless and easy to administer, and vaccination would not require a visit to a health care provider. **New Frontiers 8.3** focuses on such a vaccine that scientists have developed for cattle to protect them against a disease commonly known as "shipping fever."

Table 8.2 Tools and Techniques of Recombinant DNA Technology

Tool or Technique	Description	Potential Application
Mutagen	Chemical or physical agent that creates mutations	Creating novel genotypes and phenotypes
Reverse transcriptase	Enzyme from RNA retrovirus that synthesizes cDNA from an RNA template	Synthesizing a gene using an mRNA template
Synthetic nucleic acid	DNA molecule prepared *in vitro*	Creating DNA probes to localize genes within a genome
Restriction enzyme	Bacterial enzyme that cleaves DNA at specific sites	Creating recombinant DNA by joining fragments
Vector	Transposon, plasmid, or virus that carries DNA into cells	Altering genome of a cell
Gene library	Collection of cells or viruses, each of which carries a portion of a given organism's genome	Providing a ready source of genetic material
Polymerase chain reaction (PCR)	Produces multiple copies of a DNA molecule	Multiplying DNA for various applications
Gel electrophoresis	Uses electrical charge to separate molecules according to their size	Separating DNA fragments for the Southern blot
Electroporation	Uses electrical current to make cells competent	Inserting a novel gene into a cell
Protoplast fusion	Fuses two cells to create recombinants	Inserting a novel gene into a cell
Gene gun	Blasts genes into target cells	Inserting a novel gene into a cell
Microinjection	Uses micropipette to inject genes into cells	Inserting a novel gene into a cell
Nucleic acid probes	RNA or DNA molecules labeled with radioactive or fluorescent tags	Localizing specific genes in a Southern blot
Southern blot	Localizes specific DNA sequences on a stable membrane	Identifying a strain of pathogen
Genome mapping	Uses restriction enzymes to locate relative positions of restriction sites	Locating genes in an organism's genome
DNA sequencing	Determines the sequence of nucleotide bases in DNA	Comparing genomes of organisms

New Frontiers 8.3 Vaccines on the Menu

Wouldn't it be great if instead of receiving painful flu shots we could be vaccinated simply by eating an antigen-laced chocolate bar? We aren't quite there yet, but scientists are making progress in developing genetically modified oral vaccines.

In one case, researchers are currently testing a plant-based oral vaccine to protect cattle against a disease called pneumonic pasteurellosis, commonly known as "shipping fever" because cattle experience this stress-induced disease while being shipped from cow/calf operations to feedlots. During this respiratory illness, caused by the bacterium *Mannheimia* (formerly *Pasteurella*) *haemolytica,* diseased cattle experience decreased appetite, fever, and nasal discharge. An injectable vaccine against the disease is available, but it is costly and labor intensive, and the injection is so stressful for cattle that it may actually contribute to the problem.

Enter Reggie Lo and his colleagues at the University of Guelph in Canada, who have successfully inserted a particular gene from *M. haemolytica* into white clover, so that the plant expresses one of the bacterium's proteins. The idea is that when cattle are fed the genetically altered clover, the bacterial protein acts as an antigen, triggering a protective immune response against *M. haemolytica*. This vaccine is easy to prepare, requires minimal processing, and remains stable in dried clover hay for at least several days. It also takes advantage of the fact that cattle regularly regurgitate their food and chew it repeatedly—allowing immune cells in their digestive tract to come into maximum contact with the protein. To date, the vaccine has worked successfully in laboratory animals; field trials with cattle began in 2002.

White clover

Table 8.3 Some Products of Recombinant DNA Technology Used in Medicine

Product	Modified Cell	Uses of Product
Interferons	*Escherichia coli, Saccharomyces cerevisiae*	To treat cancer, multiple sclerosis, chronic granulomatous disease, hepatitis, and warts
Interleukins	*E. coli*	To enhance immunity
Tumor necrosis factor	*E. coli*	In cancer therapy
Erythropoietin	Mammalian cell culture	To stimulate red blood cell formation; to treat anemia
Tissue plasminogen activating factor	Mammalian cell culture	To dissolve blood clots
Human insulin	*E. coli*	For diabetes therapy
Taxol	*E. coli*	In ovarian cancer therapy
Factor VIII	Mammalian cell culture	In hemophilia therapy
Macrophage colony stimulating factor	*E. coli, S. cerevisiae*	To stimulate bone marrow to produce more white blood cells; to counteract side effects of cancer treatment
Relaxin	*E. coli*	To ease childbirth
Human growth hormone	*E. coli*	To correct childhood deficiency of growth hormone
Hepatitis B vaccine	Carried on a plasmid of *S. cerevisiae*	To stimulate immunity against hepatitis B virus

Another type of vaccination involves producing a recombinant plasmid carrying a gene from a pathogen and injecting the plasmid into a human, whose body then synthesizes polypeptides characteristic of the pathogen. The polypeptides stimulate immunologic memory within the human body, readying it to mount a vigorous immune response and prevent infection should it subsequently be exposed to the real pathogen. Clinical trials of such a vaccine against malaria have shown some promise.

Table 8.3 lists some products of recombinant DNA technology with medical applications.

Genetic Screening

Genetic mutations cause some diseases such as inherited forms of breast cancer and Huntington's disease. Laboratory technicians use the Southern blot procedure to screen patients, prospective parents, and fetuses for such mutant genes. This procedure, called *genetic screening,* can also identify viral DNA sequences in a patient's blood or other tissues. For instance, genetic screening can identify HIV in a patient's cells even before the patient shows any other sign of infection.

DNA Fingerprinting

Medical laboratory technicians and forensic investigators use gel electrophoresis and Southern blotting for so-called **genetic fingerprinting** or **DNA fingerprinting**—identifying individuals or organisms by their unique DNA sequences.

DNA fingerprinting involves procuring a sample of DNA, making multiple copies of it via PCR, cutting the copies with restriction enzymes, and separating the fragments by gel electrophoresis to produce a unique pattern. The process is analogous to standard fingerprinting in that the pattern resulting from a particular DNA sample is unique, and it must be compared to patterns produced from other DNA molecules **(Figure 8.12),** much like a standard fingerprint must be compared to known fingerprints. For example, the patterns from DNA collected at a crime scene either matches or does not match a suspect's or victim's DNA; or the pattern from an environmental sample matches or does not match patterns from known organisms. Genetic fingerprinting is used to determine paternity; to connect blood, semen, or even single skin cells to a particular crime suspect; to identify badly damaged human remains; and to identify pathogens.

Gene Therapy

An exciting use of recombinant DNA technology is **gene therapy,** in which missing or defective genes are replaced with normal copies. Theoretically, a physician could remove a few genetically defective cells—for example, cells that produce a defective protein—from a patient, insert normal genes, and replace the cells into the patient—curing the disease. Alternatively, plasmid or viral vectors could deliver genes directly to target cells within a patient.

Unfortunately, gene therapy has proven difficult in practice because of unexpected results. Specifically, some patients' immune systems react out of control to the presence of vectors, resulting in the death of the patient. Nevertheless, doctors have successfully treated patients for severe combined immunodeficiency disease. Other diseases that may respond well to gene therapy are cystic fibrosis, sickle cell anemia, and some types of hemophilia and diabetes.

Xenotransplants

Xenotransplants[3] are animal cells, tissues, or organs introduced into the human body. For years physicians have performed xenotransplants such as using valves from pig hearts to repair severely damaged human hearts; however, recombinant DNA technology may expand the possibilities. It is theoretically feasible to insert functional human genes into animals to direct them to produce organs and tissues for transplantation into humans. For example, scientists could induce pigs to produce human-like cytoplasmic membrane proteins so that the pigs' organs would not be rejected as foreign tissue by a transplant recipient.

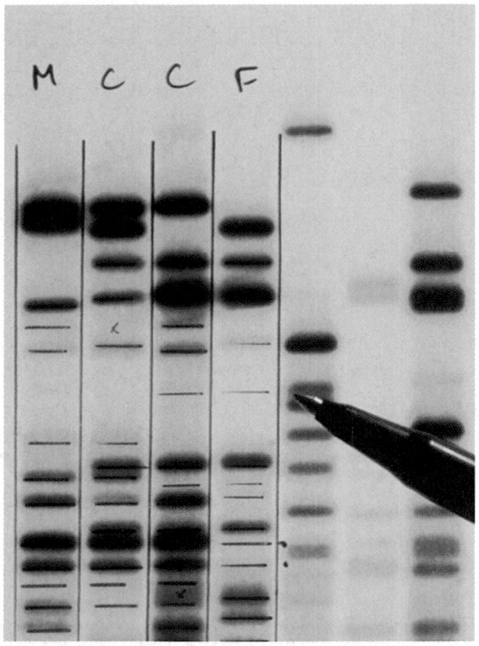

▲ ***Figure 8.12***

DNA fingerprinting. Shown here is a partial X-ray of bands of DNA from four family members: a mother (M), father (F), and two children (C's). Both children share some bands with each parent, proving they are indeed related. A similar process can be used to compare DNA bands from microbial specimens in order to identify a particular specimen.

Agricultural Applications

Learning Objective

✓ Identify five agricultural applications of recombinant DNA technology.

Recombinant DNA technology has been applied to the realm of agriculture to produce **transgenic** organisms—recombinant plants and animals that have been altered for specific purposes by the addition of genes from other organisms. The purposes for which transgenic organisms have been produced are many and varied and include herbicide resistance, tolerance to salty soils, resistance to freezing and pests, and improvements in nutritional value and yield.

Herbicide Resistance

The biodegradable herbicide *glyphosate* (Roundup™) normally kills all plants—weeds and crops alike—by blocking an enzyme that is essential for the synthesis of several amino acids. After scientists discovered and isolated a *Salmonella* bacterial gene that conveys resistance to glyphosate, they produced transgenic crop plants containing the gene. As a result of this application of recombinant DNA technology, farmers can now apply glyphosate to a field of transgenic plants to kill weeds without damaging the crop. An added benefit is that farmers do not need to till the soil to suppress weeds during the growing season, reducing soil erosion. In 2001, over 50% of the soybeans grown in the United States were genetically modified in this manner to be "Roundup™ ready."

Salt Tolerance

Years of irrigation have resulted in excessive salt buildup in farmland throughout the world, rendering the land useless for farming. Though salt-tolerant plants that can grow under these conditions exist, they are not edible.

Scientists have now successfully removed the gene for salt tolerance and inserted it into tomato and canola plants to create food crops that can grow in soil so salty it would poison normal crops. Not only do such transgenic plants survive and produce fruit; they also remove salt from the soil, restoring it and making it suitable to grow unmodified crops as well. Researchers are now attempting to insert the gene for salt tolerance into wheat and corn.

Freeze Resistance

Ice crystals more readily form when bacterial proteins (from natural bacteria present in a field) are available as crystallization nuclei. Scientists have modified strains of the bacterium *Pseudomonas* (soo-dō-mō'nas) with a gene for a polypeptide that prevents ice crystals from forming. Crops sprayed with genetically modified bacteria can tolerate mild freezes, so the farmers no longer lose their crops to unseasonable cold snaps.

Pest Resistance

The bacterium *Bacillus thuringiensis* (ba-sil'ŭs thur-in-jē-en'sis) produces a protein that, when modified by enzymes in the intestinal tracts of insects, becomes *Bt-toxin*. Bt-toxin binds to receptors lining the insect's digestive tract and causes the tissue to dissolve. Unlike many insecticides, Bt-toxin is naturally occurring, harmful only to insects, and biodegradable. Farmers, particularly organic growers, have used Bt-toxin for over 30 years to reduce insect damage to their crops.

[3]From Greek *xenos,* meaning stranger.

Highlight 8.1 Does Genetically Modified Corn Harm Monarch Butterflies?

Controversy often surrounds genetically-modified foods. For example, can pollen from genetically modified corn plants harm monarch caterpillars feeding on nearby milkweed leaves? A 1999 Cornell University study found that under laboratory conditions, the dusting of milkweed leaves with pollen from corn modified with the Bt-toxin gene did indeed stunt or kill monarch caterpillars feeding on the leaves. However, a 2001 study found that in and around actual Bt-toxin cornfields, the amount of pollen present on milkweed leaves is unlikely to reach the levels of pollen used in the Cornell study. Moreover, some researchers argue that butterflies may actually be safer in fields of so-called Bt-corn because they are then not subjected to the chemical pesticides otherwise used. Further research is needed to get the complete story.

Monarch caterpillar on milkweed leaf dusted with Bt-pollen.

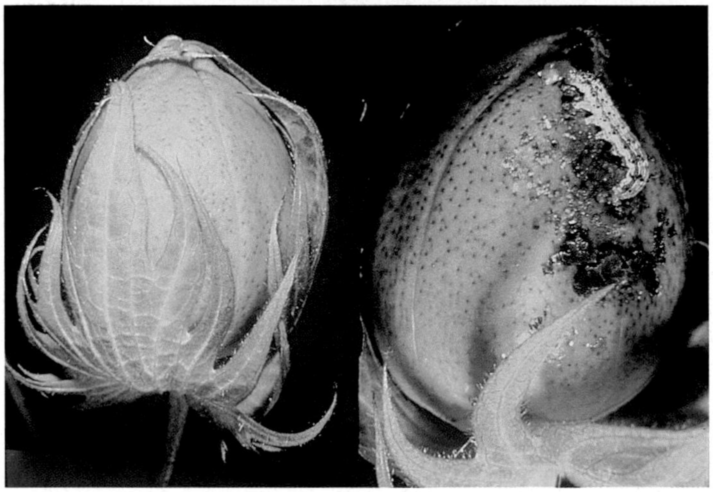

▲ *Figure 8.13*

The cotton boll on the left was genetically modified to resist boll weevils; the boll on the right was not.

Now, the gene for Bt-toxin has been inserted into a variety of crop plants, including potatoes, cotton, and corn, so that they produce Bt-toxin for themselves. Pests feeding on such plants are killed **(Figure 8.13)** while humans and most animals that eat them are unharmed. See **Highlight 8.1** for a discussion of the effects of Bt-toxin on monarch butterflies.

Improvements in Nutritional Value and Yield

Genetic researchers have increased crop and animal yields in several ways. For example, MacGregor tomatoes remain firm after harvest because the gene for the enzyme that breaks down pectin has been suppressed. This allows farmers to let the tomatoes ripen on the vine, and it increases the tomatoes' shelf life.

Another example involves bovine growth hormone (BGH), which when injected into cattle enables them to more rapidly gain weight, have meat with a reduced fat content, and produce 10% more milk. Though BGH can be derived from animal tissue, it is more economical to insert the BGH gene into bacteria so that they produce the hormone, which is then purified and injected into farm and ranch animals.

In yet another application, scientists have improved the nutritional value of rice by adding a gene for β-carotene, which is a precursor to vitamin A. Vitamin A is required for human embryonic development and for vision in adults, and it is an important antioxidant that plays a role in ameliorating cancer and atherosclerosis (hardening of arteries).

Recombinant DNA technology has progressed to the point that scientists are now considering transplanting genes coding for entire metabolic pathways, rather than merely genes encoding simple proteins. For instance, researchers are attempting to transfer into corn and rice plants all the genes for the enzymes and structures some bacteria use to convert atmospheric nitrogen into nitrogenous fertilizer, in effect allowing the recombinant plants to produce their own fertilizer.

Recombinant DNA tools and techniques allow scientists to examine, compare, and manipulate the genomes of microorganisms, plants, animals, and humans for a variety of purposes including gene mapping and forensic, medical, and agricultural applications. However, as with many scientific advances, concerns arise about the ethics and safety of genetic manipulations. The next section deals with these issues.

The Ethics and Safety of Recombinant DNA Technology

Learning Objective

✓ Discuss the pros and cons concerning the safety and ethics of recombinant DNA technology.

The procedures of recombinant DNA technology provide the opportunity to transfer genes among unrelated organisms, even among organisms in different kingdoms, but how safe and ethical are they? "Super-broccoli," "frankenfood," "biological Russian roulette," and "designer humans" are some of the terms opponents use to denigrate gene therapy and transgenic agricultural products. Some oppenents question the ethics of raising genetically altered animals solely for producing tissues and organs for human use. They contend that this exemplifies a supremacist view—the view that humans are of greater intrinsic value than animals. Other critics of transgenic crops and animals correctly state that the long-term effects of transgenic manipulations are unknown, and that unforeseen problems arise from every new technology and procedure. Recombinant DNA technology may burden society with complex, as yet unforeseen regulatory, administrative, financial, legal, social, and environmental problems.

Critics also argue that natural genetic transfer by processes such as transformation and transduction could deliver genes from transgenic plants and animals into other organisms. For example, if a herbicide-resistant plant cross-pollinates with a related weed species, we might be cursed with a weed that is nearly impossible to kill. Opponents further express concern that transgenic organisms could trigger allergies or cause harmless organisms to become pathogenic. Therefore, some opponents of recombinant DNA technology desire a ban on all genetically modified agricultural products.

The U.S. National Academy of Sciences, the U.S. National Research Council, and 81 research projects conducted between 1985 and 2001 by the European Union have not revealed any risks to human health or the environment from genetically modified agricultural products beyond the usual uncertainties inherent in conventional plant breeding. In fact, the European Union concluded in 2001 that "the use of more precise technology and the greater regulatory scrutiny probably make them [genetically modified foods] even safer than conventional plants and foods."

As the debate continues, governments continue to impose standards on laboratories involved in recombinant DNA technology. These are intended to prevent the accidental release of altered organisms or exposure of laboratory workers to potential dangers. Additionally, genetic researchers often design organisms to lack a vital gene so that they cannot survive for long outside of a laboratory.

Unfortunately, biologists can apply the procedures used to create beneficial crops and animals to create biological weapons that are more infective and more resistant to treatment than their natural counterparts are, though international treaties prohibit the development of biological weapons. Nevertheless, *B. anthracis* spores were used in bioterrorist attacks in the United States in 2001, though, thankfully, the strain utilized was not genetically altered to realize its deadliest potential.

Emergent recombinant DNA technologies raise numerous other ethical issues. Should people be routinely screened for diseases that are untreatable or fatal? Who should pay for these procedures: individuals, employers, prospective employers, insurance companies, HMOs, government agencies? What rights do individuals have to genetic privacy? If entities other than individuals pay the costs involved in genetic screening, should those entities have access to *all* the genetic information that results? Should businesses be allowed to have patents on and make profits from any living organisms they have genetically altered? Should governments be allowed to require genetic screening and then force genetic manipulations on individuals to correct so-called "genetic abnormalities" that some claim are the bases of criminality, manic depression, risk-taking behavior, and alcoholism? Should HMOs, physicians, or the government demand genetic screening and then refuse to provide services related to the birth or care of supposedly "defective" children?

We as a society will have to confront these and other ethical considerations as the genomic revolution continues to affect people's lives in many unpredictable ways.

CHAPTER SUMMARY

The Role of Recombinant DNA Technology in Biotechnology (p. 242)

1. **Biotechnology** is the use of microorganisms to make useful products. Historically these have included bread, wine, beer, and cheese.
2. **Recombinant DNA technology** is a new type of biotechnology in which scientists change the genotypes and phenotypes of organisms to benefit humans.

The Tools of Recombinant DNA Technology (pp. 242–247)

1. The tools of recombinant DNA technology include mutagens, reverse transcriptase, synthetic nucleic acids, restriction enzymes, vectors, and gene libraries.
2. **Mutagens** are chemical and physical agents used to create changes in a microbe's genome to effect desired changes in the microbe's phenotype.

3. The enzyme **reverse transcriptase** transcribes DNA from an RNA template; genetic researchers use reverse transcriptase to make complementary DNA **(cDNA).**
4. Scientists used synthetic nucleic acids to elucidate the genetic code, and they now use them to create genes for specific proteins and to synthesize DNA and RNA **probes** labeled with radioactive or fluorescent markers.
5. **Restriction enzymes** cut DNA at specific (usually palindromic) nucleotide sequences and are used to produce recombinant DNA molecules.
6. In recombinant DNA technology, a **vector** is a small DNA molecule (such as a viral genome, transposon, or plasmid) that carries a particular gene and a recognizable genetic marker into a cell.
7. A **gene library** is a collection of bacterial or phage clones, each of which carries a fragment (typically a single gene) of an organism's genome.

Techniques of Recombinant DNA Technology (pp. 247–252)

1. The **polymerase chain reaction (PCR)** allows researchers to replicate molecules of DNA rapidly.
2. **Gel electrophoresis** is a technique for separating molecules (including fragments of nucleic acids) by size, shape, and electrical charge.
3. The **Southern blot** technique allows researchers to stabilize specific DNA sequences from an electrophoresis gel and then localize them using DNA dyes or probes.
4. Geneticists artificially insert DNA into cells by electroporation, protoplast fusion, and injection.

Applications of Recombinant DNA Technology (pp. 252–259)

1. **Genomics** is the sequencing **(genetic mapping),** analysis, and comparison of genomes. Genetic sequencing has been speeded up by an automated machine that distinguishes among flourescent dyes attached to each type of nucleotide base.
2. Scientists synthesize **subunit vaccines** by introducing genes for a pathogen's polypeptides into cells or viruses. When the cells, the viruses, or the polypeptides they produce are injected into a human, the body's immune system is exposed to and reacts against relatively harmless antigens instead of the potentially harmful pathogen.
3. **Genetic screening** can detect certain inherited diseases before a patient shows any sign of the disease.
4. **Genetic fingerprinting (DNA fingerprinting),** which identifies unique sequences of DNA, is used in paternity investigations, crime scene forensics, diagnostic microbiology, and epidemiology.
5. **Gene therapy** cures various diseases by replacing defective genes with normal genes.
6. In **xenotransplants** involving recombinant DNA technology, human genes are inserted into animals to produce cells, tissues, or organs that are then introduced into the human body.
7. **Transgenic** plants and animals have been genetically altered by the inclusion of genes from other organisms.
8. Agricultural uses of recombinant DNA technology include advances in herbicide resistance, salt tolerance, freeze resistance, and pest resistance, as well as improvements in nutritional value and yield.
9. Among the ethical and safety issues surrounding recombinant DNA technology are concerns over the accidental release of altered organisms into the environment, the ethics of altering animals for human use, and the potential for creating genetically modified biological weapons.

QUESTIONS FOR REVIEW

(Answers to multiple choice are on the web, along with additional review questions. Visit www.microbiologyplace.com.)

Multiple Choice

1. Which of the following statements is true concerning recombinant DNA technology?
   a. It will replace biotechnology in the future.
   b. It is a single technique for genetic manipulation.
   c. It is useful in manipulating genotypes but not phenotypes.
   d. It involves modification of an organism's genome.
2. A gene synthesized from an RNA template is
   a. reverse transcriptase.
   b. complementary DNA.
   c. recombinant DNA.
   d. probe DNA.
3. After scientists exposed cultures of *Penicillium* to agents X, Y, and Z, they examined the type and amount of penicillin produced by the altered fungi to find the one that is most effective. Agents X, Y, and Z were probably
   a. recombinant cells.
   b. competent.
   c. mutagens.
   d. phages.
4. Which of the following is *false* concerning vectors in recombinant DNA technology?
   a. Vectors are small enough to manipulate outside a cell.
   b. Vectors contain a recognizable genetic marker.
   c. Vectors survive inside cells.
   d. Vectors must contain genes for self-replication.
5. Which recombinant DNA technique is used to replicate copies of a DNA molecule?
   a. PCR
   b. gel electrophoresis
   c. electroporation
   d. Southern blotting
6. Which of the following techniques is used regularly in the study of genomics?
   a. Clones are selected using a vector with two genetic markers.
   b. Genes are inserted to produce an antigenic protein from a pathogen.
   c. Fluorescent nucleotide bases are sequenced.
   d. Defective organs are replaced with those made in animal hosts.

7. Which application of recombinant DNA technology involves the production of a distinct pattern of DNA fragments on a gel?
   a. genetic fingerprinting
   b. gene therapy
   c. genetic screening
   d. protein synthesis

True/False

Indicate which of the following are true and which are false. Rewrite any false statements to make them true.

___ 1. Restriction enzymes inhibit the movement of DNA.

___ 2. Restriction enzymes act at specific nucleotide sequences within a double-stranded DNA molecule.

___ 3. A thermocycler separates molecules based on their size, shape, and electrical charge.

___ 4. Protoplast fusion is often used in the genetic modification of plants.

Short Answer

1. Describe three artificial methods of introducing DNA into cells.
2. Why is cloning a practical technique for medical researchers?
3. Describe three ways scientists use synthetic nucleic acids.
4. Describe a gene library and its usefulness.

CRITICAL THINKING

1. Examine the restriction sites listed in Table 8.1. Which restriction enzymes produce restriction fragments with sticky ends? Which produce fragments with blunt ends?
2. A cancer-inducing virus, HTLV-1, inserts itself into a human chromosome, where it remains. How can a laboratory technician prove that a patient is infected with HTLV-1?
3. A thermocycler uses DNA polymerase from hyperthermophilic prokaryotes, but it cannot use DNA polymerase derived from *Escherichia coli*. Why not?

CHAPTER 9

Controlling Microbial Growth in the Environment

CHAPTER OUTLINE

In response to the anthrax attacks in the mail during the fall of 2001, the United States Postal Service undertook an extraordinary safety measure: It began irradiating mail at selected postal facilities.

Irradiation systems use high-voltage electron beams to kill bacteria—including endospores of *Bacillus anthracis*, the causative agent of anthrax—that may be present in pieces of mail. The beams of electrons work by disrupting bacterial DNA. The same technology is commonly used to sterilize medical equipment and to irradiate fresh produce; it has proven effective against *Bacillus, Salmonella, Escherichia*, and other bacterial contaminants. Contrary to a common misconception, irradiation does not cause mail or food to become radioactive.

In this chapter we will learn not only about irradiation, but also the wide variety of other methods used to control microorganisms in our environment.

Following the anthrax attacks, postal service employees began wearing protective gloves and mail was irradiated at high-risk facilities.

MicroPrep Pre-Test: *Take the pre-test for this chapter on the web. Visit* ***www.microbiologyplace.com****.*

The control of microbes in health care facilities, in laboratories, and at home is a significant and practical aspect of microbiology. In this chapter we will study the terminology and principles of microbial control, consider the factors affecting the efficacy of microbial control, and examine the control of microorganisms and viruses by various chemical and physical means. One important aspect of microbial control—the use of antimicrobial drugs to assist the body's defenses against pathogens—will be considered in Chapter 10.

Basic Principles of Microbial Control

Scientists, health care professionals, researchers, and government workers should use precise terminology in reference to microbial control in the environment. In the following sections we will discuss the terminology of microbial control, examine the concept of microbial death rates, and discuss the action of antimicrobial agents.

Terminology of Microbial Control

Learning Objectives

- ✓ Contrast sterilization, disinfection, and antisepsis, and describe their practical uses.
- ✓ Contrast the terms *degerming, sanitization,* and *pasteurization*.
- ✓ Compare the effects of "-static" versus "-cidal" control agents on microbial growth.

It is important for microbiologists, health care workers, and others to use correct terminology for describing microbial control. While many of these terms are familiar to the general public, they are often misused.

In its strictest sense, **sterilization** refers to the removal or destruction of *all* microbes, including viruses and bacterial endospores, in or on an object. (The term does not apply to *prions,* which are infectious proteins, because standard sterilizing techniques do not destroy them; in fact, scientists do not know the minimum treatment required to destroy prions.)

In practical terms, sterilization indicates only the eradication of harmful microorganisms and viruses; some innocuous microbes may still be present and viable in an environment that is considered sterile. For instance, *commercial sterilization* of canned food does not kill all hyperthermophilic microbes, but they cannot grow and spoil the food at ambient temperatures, so they are of no practical concern. Likewise, some hyperthermophiles may survive sterilization by laboratory methods (discussed shortly), but they are of no practical concern to technicians because they cannot grow or reproduce at normal laboratory temperatures.

The term **aseptic**[1] (ā-sep'tik) describes an environment or procedure that is free of contamination by *pathogens.* For example, vegetables and fruit juices are available in aseptic packaging, and surgeons and laboratory technicians use aseptic techniques to avoid contaminating a surgical field or laboratory equipment.

Disinfection[2] refers to the use of physical or chemical agents known as **disinfectants,** including ultraviolet light, heat, alcohol, and bleach, to inhibit or destroy microorganisms, especially pathogens. Unlike sterilization, disinfection does not guarantee that all pathogens are eliminated; indeed, disinfectants alone cannot inhibit endospores or some viruses. Further, the term *disinfection* is used only when discussing treatment of inanimate objects. When a chemical is used on skin or other tissue, the process is called **antisepsis**[3] (an-tē-sep'sis), and the chemical is called an **antiseptic.** Antiseptics and disinfectants often have the same components, but disinfectants are more concentrated or can be left on a surface for longer periods of time. Of course, some disinfectants, such as steam or concentrated bleach, are not suitable for use as antiseptics.

Degerming is the removal of microbes from a surface by scrubbing, such as when you wash your hands or a nurse prepares an area of skin for an injection. Though chemicals such as soap or alcohol are commonly used during degerming, the action of thoroughly scrubbing the surface is usually more important than the chemical in removing microbes.

Sanitization[4] is the process of disinfecting places and utensils used by the public to reduce the number of pathogenic microbes to meet accepted public health standards. For example, steam, high-pressure hot water, and scrubbing are used to sanitize restaurant utensils and dishes, and chemicals are used to sanitize public toilets. Thus, the only difference between *disinfecting* dishes at home in a dishwasher and *sanitizing* dishes in a restaurant is the arena—private versus public—in which the activity takes place.

Pasteurization[5] is the use of heat to kill pathogens and reduce the number of spoilage microorganisms in food and beverages. Milk, fruit juices, wine, and beer are commonly pasteurized.

So far, we have seen that there are two major types of microbial control—sterilization, which is the elimination of *all* microbes, and disinfection, which is the destruction of vegetative (nonspore) cells and many viruses. Modifications of disinfection include antisepsis, degerming, sanitization, and pasteurization. Some scientists and clinicians apply these terms only to pathogenic microorganisms.

Additionally, scientists and health care professionals use the suffixes *-stasis/-static*[6] to indicate that a chemical or physical agent inhibits microbial metabolism and growth, but doesn't necessarily kill microbes. Thus, refrigeration is bacteriostatic for most bacterial species; it inhibits their growth, but they can resume metabolism when the optimal temperature is restored. By contrast, words ending in *-cide/-cidal*[7] refer to agents that destroy or permanently

[1]From Greek *a,* meaning not, and *sepsis,* meaning decay.
[2]From Latin *dis,* meaning reversal, and *inficere,* meaning to corrupt.
[3]From Greek *anti,* meaning against, and *sepsis,* meaning putrefaction.
[4]From Latin *sanitas,* meaning healthy.
[5]Named for Louis Pasteur, inventor of the process.
[6]Greek, meaning to stand—that is, to remain relatively unchanged.
[7]From Latin *cidium,* meaning a slaying.

Table 9.1 Terminology of Microbial Control

Term	Definition	Examples	Comments
Antisepsis	Reduction in the number of microorganisms and viruses, particularly potential pathogens, on living tissue	Iodine; alcohol	Antiseptics are frequently disinfectants whose strength has been reduced to make them safe for living tissues.
Aseptic	Refers to an environment or procedure free of pathogenic contaminants	Preparation of surgical field; hand washing; flame sterilization of laboratory equipment	Scientists, laboratory technicians, and health care workers routinely follow standardized aseptic techniques.
-cide **-cidal**	Suffixes indicating destruction of a type of microbe	Bactericide; fungicide; germicide; virucide	Germicides include ethylene oxide, propylene oxide, and aldehydes.
Degerming	Removal of microbes by mechanical means	Hand washing; alcohol swabbing at site of injection	Chemicals play a secondary role to the mechanical removal of microbes.
Disinfection	Destruction of most microorganisms and viruses on nonliving tissue	Phenolics; alcohols; aldehydes; surfactants	The term is used primarily in relation to pathogens.
Pasteurization	Use of heat to destroy pathogens and reduce the number of spoilage microorganisms in foods and beverages	Pasteurized milk and fruit juices	Heat treatment is brief to reduce alteration of taste and nutrients; microbes still remain and eventually cause spoilage.
Sanitization	Removal of pathogens from objects to meet public health standards	Washing tableware in scalding water in restaurants	Standards of sanitization vary among governmental jurisdictions.
-stasis **-static**	Suffixes indicating inhibition, but not complete destruction, of a type of microbe	Bacteriostatic; fungistatic; virustatic	Germistatic agents include some chemicals, refrigeration, and freezing.
Sterilization	Destruction of all microorganisms and viruses in or on an object	Preparation of microbiological culture media and canned food	Typically achieved by steam under pressure, incineration, or ethylene oxide gas

inactivate a particular type of microbe; *virucides* inactivate viruses, *bactericides* kill bacteria, and *fungicides* kill fungal hyphae, spores, and yeasts. *Germicides* are chemical agents that destroy pathogenic microorganisms in general.

Table 9.1 summarizes the terminology used to describe the control of microbial growth.

CRITICAL THINKING

A student inoculates *Escherichia coli* into two test tubes containing the same sterile liquid medium, except the first tube also contains a drop of a chemical with an antimicrobial effect. After 24 hours of incubation, the first tube remains clear while the second tube becomes cloudy with bacteria. Design an experiment to determine if this amount of the antimicrobial chemical is *bacteriostatic* or *bactericidal* against *E. coli.*

Microbial Death Rates

Learning Objective

✓ Define microbial death rate, and describe its significance in microbial control.

Scientists define **microbial death** as the permanent loss of reproductive ability under ideal environmental conditions. One technique for evaluating the efficacy of an antimicrobial agent is to calculate the **microbial death rate,** which is usually found to be constant over time for any particular microorganism under a particular set of conditions **(Figure 9.1).** Suppose for example that a scientist treats a broth containing 1 billion (10^9) microbes with an agent that kills 90% of them in 1 minute. The most susceptible cells die first, leaving 100 million (10^8) hardier cells after the first minute. After another minute of treatment, another 90% die, leaving 10 million (10^7) cells that have even greater resistance to and require longer exposure to the agent before they die. Notice that in this case, each full minute decreases the number of living cells tenfold (that is, by one logarithm). The broth will be sterile when all the cells are dead. When these results are plotted on a semilogarithmic graph—in which the y-axis is logarithmic, and the x-axis is arithmetic—the plot of microbial death rate is a straight line; that is, the microbial death rate is constant.

Action of Antimicrobial Agents

Learning Objective

✓ Describe how antimicrobial agents act against cell walls, cytoplasmic membranes, proteins, and nucleic acids.

There are many types of chemical and physical microbial controls, but their modes of action fall into two basic categories: those that disrupt the integrity of cells by adversely altering their cell walls or cytoplasmic membranes, and

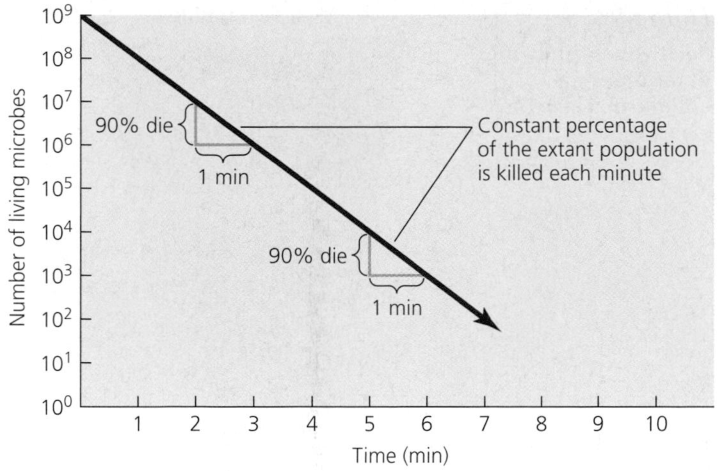

▲ ***Figure 9.1***

A plot of microbial death rate. Microbicidal agents do not simultaneously kill all cells. Rather they kill a constant percentage of cells over time—in this case 90% per minute. On this semilogarithmic graph, a constant death rate is indicated by a straight line. *How many minutes are required for sterilization in this case?*

Figure 9.1 *For this microbe under these conditions, this microbicidal agent requires 9 minutes to achieve sterilization.*

those that interrupt cellular metabolism and reproduction by interfering with the structures of proteins and nucleic acids.

Alteration of Cell Walls and Membranes

As we have seen, a cell wall maintains cellular integrity by counteracting the effects of osmosis when the cell is in a hypotonic solution. If the wall is disrupted by physical or chemical agents, it no longer prevents the cell from bursting as water moves into it by osmosis.

Beneath a cell wall, the cytoplasmic membrane essentially acts as a bag that contains the cytoplasm and controls the passage of chemicals into and out of the cell. Extensive damage to a membrane's proteins or phospholipids by any physical or chemical agent allows the cellular contents to leak out—which, if not immediately repaired, causes death.

In enveloped viruses, the envelope is a membrane composed of proteins and phospholipids that is responsible for the attachment of the virus to its target cell; thus damage to the envelope by physical or chemical agents fatally interrupts viral replication. The lack of an envelope in nonenveloped viruses accounts for their greater tolerance of harsh environmental conditions, including antimicrobial agents.

Damage to Proteins and Nucleic Acids

Proteins in cells act as regulatory compounds in cellular metabolism, function as enzymes in most metabolic reactions, and form structural components in membranes and cytoplasm. As we have seen, a protein's function depends on an exact three-dimensional shape, which is maintained by hydrogen and disulfide bonds between amino acids. When these bonds are broken by extreme heat or certain chemicals, the protein's shape changes (see Figure 5.9). Such *denatured* proteins cease to function, bringing about cellular death.

Chemicals, radiation, and heat can also alter and even destroy nucleic acids. Given that the genes of a cell or virus are composed of nucleic acids, disruption of these molecules can produce fatal mutations. Additionally, that portion of a ribosome that actually catalyses the synthesis of proteins is a *ribozyme*—that is, an enzymatic RNA molecule—so physical or chemical agents that interfere with nucleic acids also stop protein synthesis.

CRITICAL THINKING

Would you expect Gram-negative bacteria or Gram-positive bacteria to be more susceptible to antimicrobial chemicals that act against cell walls? Explain your answer, which you should base solely upon the nature of the cells' walls (see Figure 3.13).

Scientists and health care workers have at their disposal many chemical and physical agents to control microbial growth and activity. In the next section we will consider the factors and conditions that must be considered in choosing a particular control method, as well as some ways for evaluating a method's effectiveness.

The Selection of Microbial Control Methods

Ideally, agents used for the control of microbes should be inexpensive, fast-acting, and stable during storage. Further, a perfect agent would control the growth and reproduction of every type of microbe while being harmless to humans, animals, and objects. Unfortunately, such ideal products and procedures do not exist—every agent and procedure has limitations and disadvantages. In the next two subsections we consider the factors that affect the efficacy of antimicrobial methods and some ways to evaluate disinfectants.

Factors Affecting the Efficacy of Antimicrobial Methods

Learning Objective

- ✓ List factors to consider in selecting a microbial control method.
- ✓ Identify the three most-resistant groups of microbes, and explain why they are resistant to many antimicrobial agents.
- ✓ Discuss environmental conditions that can affect the effectiveness of antimicrobial agents.

In each situation, microbiologists, laboratory personnel, and medical staff must consider at least three factors: the nature of the sites to be treated, the degree of susceptibility of the microbes involved, and the environmental conditions that pertain.

Site to Be Treated

In many cases, the choice of an antimicrobial method depends on the nature of the site to be treated. For example, harsh chemicals and extreme heat cannot be used on humans, animals, and fragile objects such as artificial heart valves and plastic utensils. Moreover, when performing medical procedures, medical personnel must choose a method and level of microbial control based on the site of the procedure, because the site greatly affects the potential for subsequent infection. For example, the use of medical instruments that penetrate the outer defenses of the body such as needles and scalpels carries a greater potential for infection, so they must be sterilized, whereas items that contact only the surface of a mucous membrane or the skin may be disinfected. In the latter case sterilization is required only if the patient is immunocompromised.

Relative Susceptibility of Microorganisms

Though microbial death rate is usually constant for a particular agent acting against a single microbe, death rates do vary—sometimes dramatically—among microorganisms and viruses. Generally these microbes fall along a continuum from most susceptible to most resistant to antimicrobial agents. For example, *enveloped* viruses, such as HIV, are more susceptible to antimicrobial agents and heat than are *nonenveloped viruses*, such as polio virus (see Figures 13.1 and 13.7), because viral envelopes are more easily disrupted than the protein coats of nonenveloped viruses. The relative susceptibility of microbes to antimicrobial agents is illustrated in **Figure 9.2.**

Often, scientists and medical personnel select a method to kill the hardiest microorganisms present, assuming that such a treatment will kill more fragile microbes as well. The most resistant microbes include the following:

- Bacterial endospores. The endospores of *Bacillus* (ba-sil'ŭs) and *Clostridium* (klos-trid'ē-ŭm) are the most resistant forms of life. They can survive environmental extremes of temperature and acidity, and many chemical disinfectants. For example, endospores can survive more than 20 years in 70% alcohol, and scientists have recovered viable endospores that were embalmed with Egyptian mummies thousands of years ago.
- Species of *Mycobacterium* (mī'kō-bak-tēr'ē-ŭm). The cell walls of members of this genus, such as *Mycobacterium tuberculosis*, contain large amounts of waxy lipids. The wax allows these bacteria to survive drying and protects them from most water-based chemicals; therefore, medical personnel must use strong disinfectants or heat to treat whatever comes into contact with tuberculosis patients, including utensils, equipment, and rooms.
- Cysts of protozoa. A protozoan cyst's wall prevents entry of most disinfectants, protects against drying, and shields against radiation and heat.

Figure 9.2 ➤ Relative susceptibilities of microbes to antimicrobial agents.

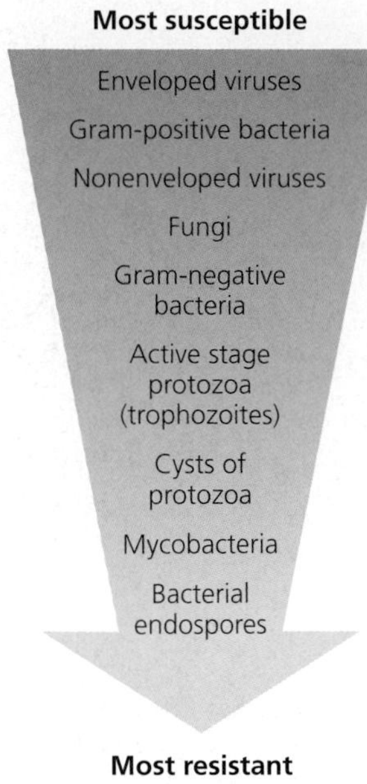

The effectiveness of germicides can be classified as high, intermediate, or low depending on their effectiveness in inactivating or destroying microorganisms on medical and dental instruments that cannot be sterilized with heat. *High-level germicides* kill all pathogens, including bacterial endospores. Health care professionals use them to sterilize invasive instruments such as catheters, implants, and parts of heart-lung machines. *Intermediate-level germicides* kill fungal spores, protozoan cysts, viruses, and pathogenic bacteria, but not bacterial endospores. They are used to disinfect instruments that come in contact with mucous membranes but are noninvasive, such as respiratory equipment and endoscopes. *Low-level germicides* eliminate vegetative bacteria, fungi, protozoa, and some viruses; they are used to disinfect items that only contact the skin of patients, such as furniture and electrodes.

Environmental Conditions

Temperature and pH affect microbial death rates and the efficacy of antimicrobial methods. Warm disinfectants, for example, generally work better than cool ones because chemicals react faster at higher temperatures **(Figure 9.3).** Some chemical disinfectants such as sodium hypochlorite ($NaOCl_2$, household chlorine bleach) are more effective at low pH. The antimicrobial effect of heat is also enhanced by acidic conditions.

Organic materials such as fat, feces, vomit, blood, and the intercellular secretions of biofilms interfere with the penetration of heat, chemicals, and some forms of radiation, and in some cases these materials inactivate chemical disinfectants. For this reason, it is important to clean objects be-

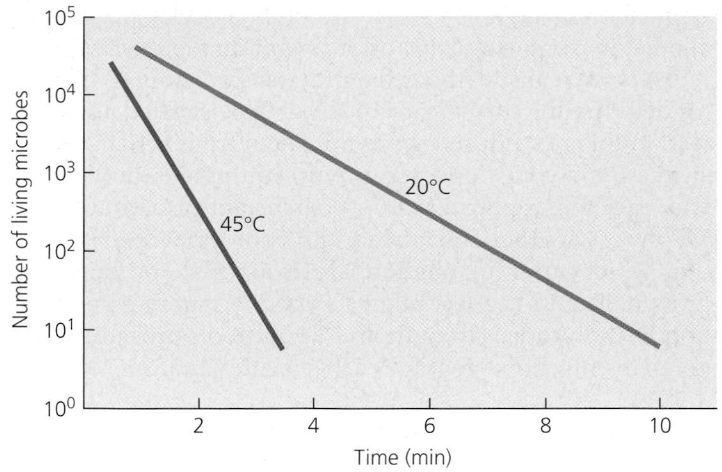

▲ ***Figure 9.3***

Effect of temperature on the efficacy of an antimicrobial chemical. This semilogarithmic graph shows that the microbial death rate is higher at higher temperatures; to kill the same number of microbes, this disinfectant required only 4 minutes at 45°C, but 12 minutes at 20°C.

fore sterilization or disinfection so that antimicrobial agents can thoroughly contact all the object's surfaces.

Methods for Evaluating Disinfectants and Antiseptics

Learning Objective

✓ Compare and contrast four methods used to measure the effectiveness of disinfectants and antiseptics.

With few exceptions, higher concentrations and fresher solutions of a disinfectant are more effective than more dilute, older solutions. We have also seen that longer exposure times ensure the deaths of more microorganisms. However, anyone using disinfectants must consider whether higher concentrations and longer exposures may damage an object or injure a patient.

Scientists have developed several methods to measure the efficacy of antimicrobial agents. These include the phenol coefficient, the use-dilution test, the disk diffusion method, and the in-use test.

Phenol Coefficient

Recall from Chapter 1 that Lister introduced the use of phenol (also known as carbolic acid) as an antiseptic during surgery in the late 1800s. Since then, researchers have evaluated the efficacy of various disinfectants and antiseptics by calculating a ratio that compares a given agent's ability to control microbes to that of phenol under standardized conditions. This ratio is referred to as the **phenol coefficient.** A phenol coefficient greater than 1.0 indicates that an agent is more effective than phenol, and the larger the ratio, the greater the effectiveness. For example, *chloramine*, a mixture of chlorine and ammonia, has a phenol coefficient of 133.0 when used against the bacterium *Staphylococcus aureus* (staf'i-lō-kok'us ô'rē-us), and a phenol coefficient of 100.0 when used against *Salmonella typhi* (sal'mŏ-nel'ă tī'fē). This indicates that chloramine is at least 133 times more effective than phenol against *Staphylococcus* but only 100 times more effective against *Salmonella*. Measurement of an agent's phenol coefficient has been replaced by newer methods because scientists have developed disinfectants and antiseptics much more effective than phenol.

CRITICAL THINKING

What is the phenol coefficient of phenol when used against *Staphylococcus*?

Use-Dilution Test

Another method for measuring the efficacy of disinfectants and antiseptics against specific microbes is the **use-dilution test.** In this test, a researcher dips several metal cylinders into broth cultures of bacteria and briefly dries them at 37°C. The bacteria used in the standard test are *Pseudomonas aeruginosa* (soo-dō-mō'nas ā-ru-ji-nō'sa), *Salmonella cholerasuis* (sal'mŏ-nel'ă kol-er-a-su'is), and *S. aureus*. The researcher then immerses each contaminated cylinder into a different dilution of the disinfectants being evaluated. After 10 minutes, each cylinder is removed, rinsed with water to remove excess chemical, and placed into a fresh tube of sterile medium for 48 hours of incubation. The most effective agent is the one that entirely prevents microbial growth at the highest dilution.

The use-dilution test is the current standard test, though it was developed several decades ago, before the appearance of many of today's pathogens, including hepatitis C virus, HIV, and antibiotic-resistant bacteria and protozoa. Moreover, the disinfectants in use at the time were far less powerful than many used today. Some government agencies have expressed concern that the test is neither accurate, reliable, nor relevant; therefore, the American Official Analytical Chemists are developing a new standard procedure.

Disk-Diffusion Method

The **disk-diffusion method,** which is also known as the *Kirby-Bauer method,* is used in teaching laboratories to demonstrate the effectiveness of disinfectants and antiseptics. An investigator soaks sterile paper disks with the agents to be studied and then places them on the surface of a solid medium that has been heavily inoculated with a test microorganism. After incubation overnight, a *zone of inhibition*—an area where bacteria do not grow—is visible around disks containing effective disinfectants (**Figure 9.4** on page 268). A larger zone of inhibition may indicate greater effectiveness

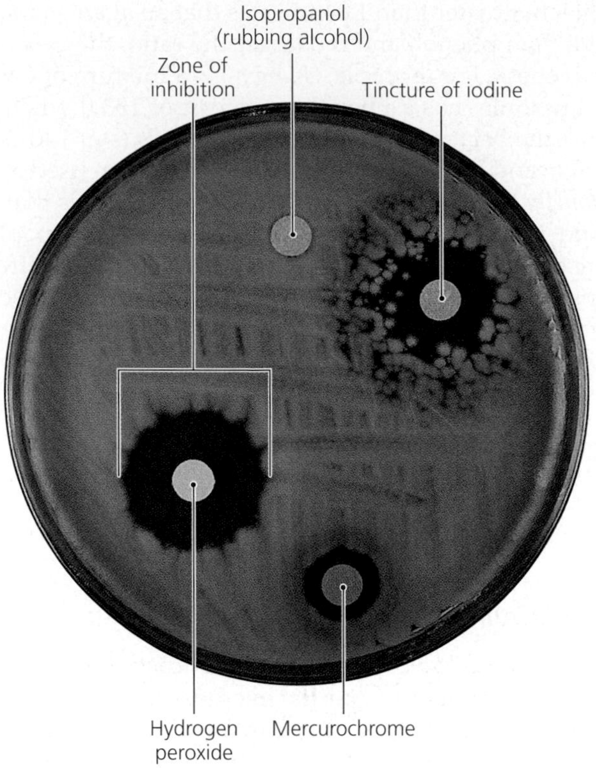

▲ ***Figure 9.4***

Results of the disk diffusion method to determine effectiveness of antimicrobial chemicals. The plate was inoculated with *Bacillus cereus* before disks soaked with antiseptics were placed over the inoculum. Alcohol had no effect, perhaps because it evaporated. *Which antiseptic was most effective against this bacterium under these conditions?*

Figure 9.4 Because it had the largest zone of inhibition, hydrogen peroxide was the most effective antiseptic.

against the test microbe, but it may only indicate that a chemical is more stable or diffuses more rapidly in agar than another. Of course, a disinfectant that is effective against one microorganism may have little or no effect against other microorganisms.

This method has some disadvantages. For example, an antimicrobial agent's molecular size and solubility in water influence its ability to diffuse through agar, affecting the size of the zone of inhibition. Tables relating the sizes of zones of inhibition to the degree of microbial resistance are available for many chemicals.

In-Use Test

Though phenol coefficient, use-dilution, and disk-diffusion tests can be useful for initial screening of disinfectants, they can also be misleading. All three types of evaluation are measures of effectiveness under controlled conditions against one, or at most a few, species of microbes, but disinfectants are generally used in various environments against a diverse population of organisms that are often associated with one another in complex *biofilms* affording mutual protection.

A more realistic (though more time-consuming) method for determining the efficacy of a chemical is called an **in-use test.** In this procedure, swabs are taken from actual objects, such as emergency operating room equipment, both before and after the application of a disinfectant or an antiseptic. The swabs are then inoculated into appropriate growth media, which after incubation are examined for microbial growth. The in-use test allows a more accurate determination of the proper strength and application procedure of a given disinfection agent for each specific situation.

Now that we have studied the terminology and general principles of microbial control, we turn our attention to the actual physical and chemical agents available to scientists, medical personnel, and the general public to control microbial growth.

Physical Methods of Microbial Control

Learning Objective

✓ Describe five types of physical methods of microbial control.

Physical methods of microbial control include exposure of the microbes to extremes of heat and cold, desiccation, filtration, osmotic pressure, and radiation.

Heat-Related Methods

Learning Objectives

✓ Discuss the advantages and disadvantages of using moist heat in an autoclave and dry heat in an oven for sterilization.

✓ Explain the use of *Bacillus stearothermophilus* endospores in sterilization techniques.

✓ Explain the importance of pasteurization, and describe three different pasteurization methods.

Heat is one of the older and more common means of microbial control. As we have seen, high temperatures denature proteins, interfere with the integrity of cytoplasmic membranes and cell walls, and disrupt the function and structure of nucleic acids. Heat can be used for sterilization, in which case all cells and viruses are deactivated, or for practical "sterilization," as in the commercial sterilization of canned goods. In commercial sterilization, endospores of hyperthermophilic prokaryotes remain viable but are harmless because they cannot grow at the normal (room) temperatures in which canned foods are stored.

Though microorganisms vary in their susceptibility to heat, it can be an important agent of microbial control. As a result, scientists have developed concepts and terminology to convey these differences in susceptibility. **Thermal death point** is the lowest temperature that kills all cells in a broth in 10 minutes, while **thermal death time** is the time it takes to completely sterilize a particular volume of liquid at a set temperature.

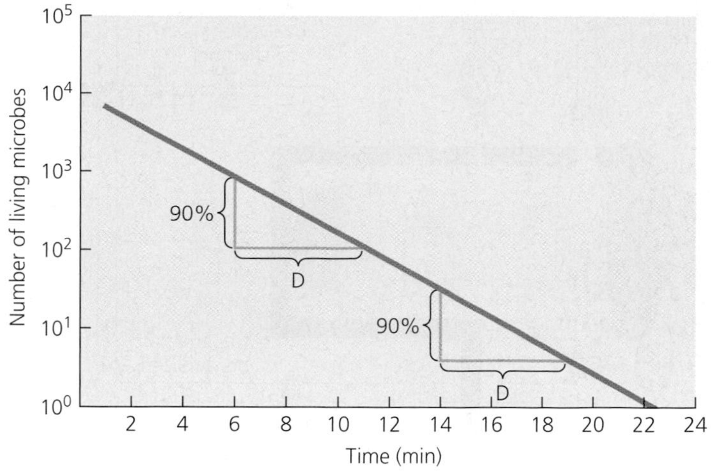

▲ *Figure 9.5*

Decimal reduction time (D) as a measure of microbial death rate. D is defined as the time it takes to kill 90% of a microbial population. Note that D is a constant that is independent of the initial density of the population. *What is the decimal reduction time of this heat treatment against this organism? What is the thermal death time?*

Figure 9.5 *D = 5 minutes; thermal death time = 22.5 minutes.*

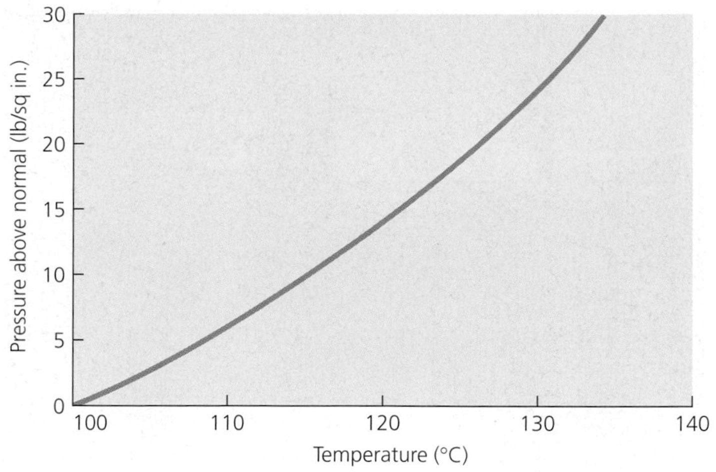

▲ *Figure 9.6*

The relationship between temperature and pressure. Note that higher temperatures—and in consequence, greater antimicrobial action—are associated with higher pressures. *Ultrahigh-temperature pasteurization of milk requires a temperature of 134°C; what pressure must be applied to the milk to achieve this temperature?*

Figure 9.6 *29 psi.*

As we have discussed, cell death occurs logarithmically (see Figure 9.1). When measuring the effectiveness of heat sterilization, researchers calculate the **decimal reduction time (D),** which is the time required to destroy 90% of the microbes in a sample **(Figure 9.5).** This concept is especially useful to food processors because they must heat foods to eliminate the endospores of *Clostridium botulinum* (klos-trid'ē-ŭm bo-tū-lī'num), which could germinate and produce botulism toxin inside sealed cans. The standard in food processing is to apply heat such that a population of 10^{12} *C. botulinum* endospores is reduced to 10^0 (1) endospore (a 12-fold reduction), which leaves only a very small chance that any particular can of food contains even a single endospore. Researchers have calculated that the D value for *C. botulinum* endospores at 121°C is 0.204 minute, so it takes 2.5 minutes (0.204 × 12) to reduce 10^{12} endospores to 1 endospore.

Moist Heat

Moist heat, which is commonly used to disinfect, sanitize, sterilize, and pasteurize, kills cells by denaturing proteins and destroying cytoplasmic membranes. Moist heat is more effective in microbial control than dry heat because water is a better conductor of heat than air. This is easily demonstrated in your kitchen—you can safely stick your hand into an oven at 350°F for a few moments, but putting it into boiling water at 212°F would burn you severely. The first method we consider for controlling microbes using moist heat is boiling.

Boiling Boiling kills the vegetative cells of bacteria and fungi, the trophozoites of protozoa, and most viruses within 10 minutes at sea level. Contrary to popular belief, water at a rapid boil is no hotter than that at a slow boil; boiling water at normal atmospheric pressure cannot exceed boiling temperature (100°C at sea level) because escaping steam carries excess heat away. Therefore, it is impossible to boil something more quickly simply by applying more heat; the added heat is carried away by the escaping steam. Boiling *time* is the critical factor. Further, it is important to realize that water boils at lower temperatures at higher elevations because atmospheric pressure is lower; thus a longer boiling time is required in Denver than in Los Angeles to get the same antimicrobial effect.

Bacterial endospores, protozoan cysts, and some viruses (such as hepatitis viruses) can survive boiling at sea level for many minutes or even hours. In fact, because bacterial endospores can withstand boiling for more than 20 hours, boiling is effective for sanitizing restaurant tableware or disinfecting baby bottles, but it is not recommended when true sterilization is required.

Autoclaving Achieving true sterilization using heat requires higher temperatures than that of boiling water. To achieve the required higher temperatures, pressure is applied to boiling water to prevent the escape of heat in steam. The reason that applying pressure succeeds in achieving sterilization is that the temperature at which water boils (and steam is formed) increases as pressure increases **(Figure 9.6).** In a kitchen, food can be cooked at a higher

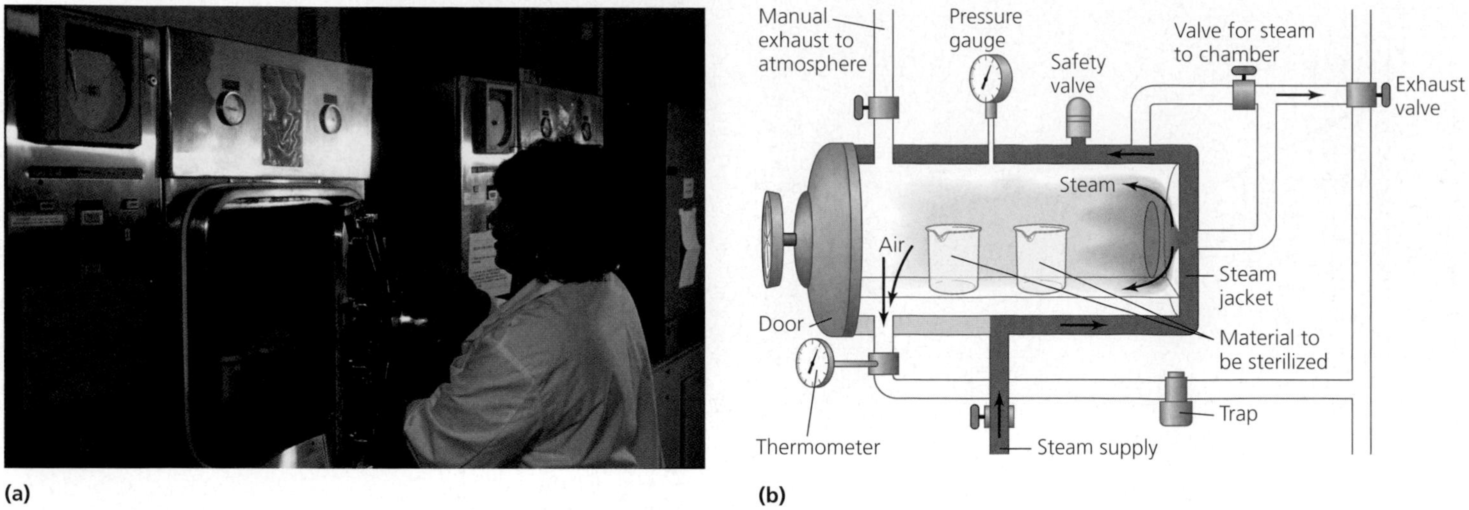

▲ *Figure 9.7*
An autoclave. **(a)** A photo of a laboratory autoclave. **(b)** A schematic of an autoclave, showing how it functions.

temperature—and thus much faster—by using a pressure cooker; in a laboratory, scientists and medical personnel routinely use a piece of equipment called an autoclave to sterilize chemicals and objects that can tolerate moist heat. Of course, alternate techniques (discussed shortly) must be used for items such as some plastics and vitamins that are damaged by heat or water.

An **autoclave** consists of a pressure chamber, pipes to introduce and evacuate steam, valves to remove air and control pressure, and pressure and temperature gauges to monitor the procedure **(Figure 9.7).** As steam enters an autoclave chamber, it forces air out, raises the temperature of the contents, and increases the pressure, until a set temperature and pressure are reached.

Scientists have determined that a temperature of 121°C, which requires the addition of 15 pounds per square inch (psi)[8] of pressure above that of normal air pressure (see Figure 9.6), destroys all microbes in a small volume in about 10 minutes. Typically, an autoclave holds the pressure and temperature for 15 minutes to provide a margin of safety. The presence of large volumes to be sterilized slows the process because it requires more time for heat to penetrate. Thus, it requires more time to sterilize 1 liter of fluid in a flask than the same volume of fluid distributed into smaller tubes. Autoclaving requires extra time to sterilize solid substances such as canned meat because it takes longer for heat to penetrate to their centers.

Sterilization in an autoclave requires that steam be able to contact all liquids and surfaces that might be contaminated with microbes; therefore, solid objects must be wrapped in porous cloth or paper, not sealed in plastic or aluminum foil, which are impermeable to steam. Containers of liquids must be sealed loosely enough to allow steam to escape. Additionally, steam must be able to circulate freely, and all air must be forced out by steam. Since steam is lighter than air, it cannot force air from the bottom of an empty vessel; therefore, empty containers must be tipped so that air can flow out of them.

Scientists use several means to ensure that an autoclave has sterilized its contents. A common one is a chemical that changes color when the proper combination of temperature and time have been reached. Often such a color indicator is impressed in a pattern on tape or paper so that the word *sterile* or a pattern or design appears **(Figure 9.8a).** Another technique uses plastic beads that melt when proper conditions are met.

A biological indicator of sterility uses endospores of the bacterium *Bacillus stearothermophilus* (ba-sil′ŭs ste-rō-ther-ma′fil-ŭs) impregnated into tape. After autoclaving, the tape is aseptically inoculated into sterile broth. If no bacterial growth appears, the original material is considered sterile. In a variation on this technique, the endospores are on a strip in one compartment of a vial that also includes a growth medium containing a pH color indicator. After autoclaving, a barrier between the two compartments is broken, putting the endospores into contact with the medium **(Figure 9.8b).** In this case, the absence of a color change indicates sterility.

CRITICAL THINKING

Where within an autoclave should you place a sterilization indicator? Explain your reasoning.

Pasteurization In Chapter 1 we learned that Louis Pasteur developed a method of heating beer and wine just enough to destroy the microorganisms that cause spoilage

[8]The Standard International (SI) equivalent of 1 psi is 1.01×10^5 pascals.

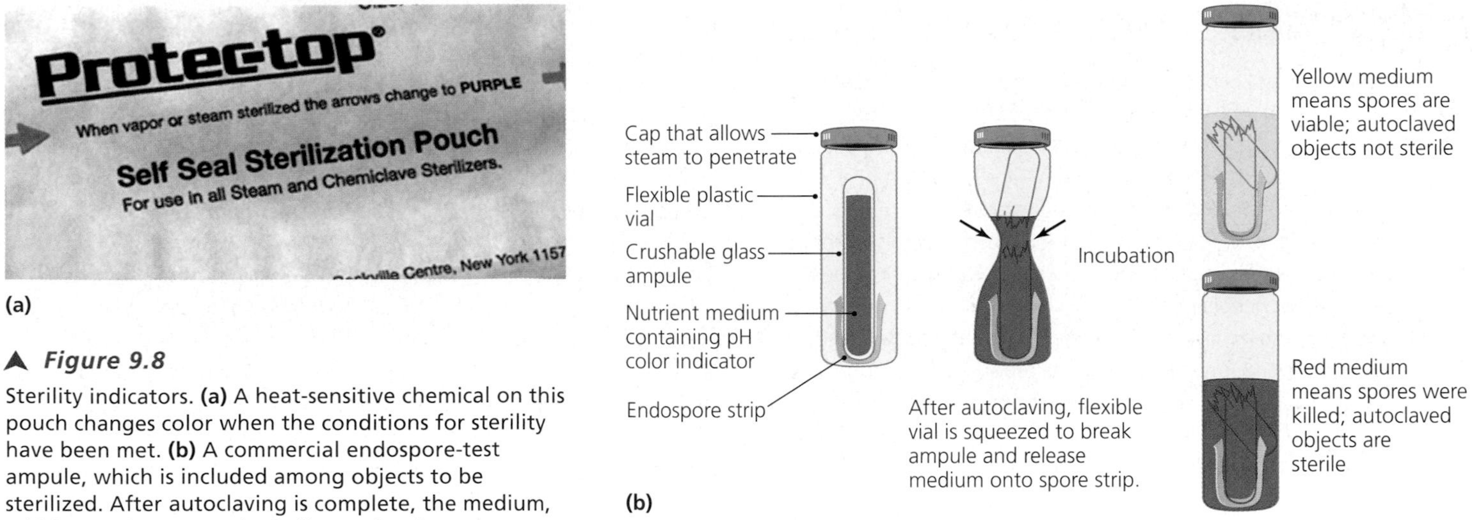

▲ ***Figure 9.8***

Sterility indicators. **(a)** A heat-sensitive chemical on this pouch changes color when the conditions for sterility have been met. **(b)** A commercial endospore-test ampule, which is included among objects to be sterilized. After autoclaving is complete, the medium, which contains a pH color indicator, is released onto the endospore strip by breaking the ampule. If the endospores are still alive, their metabolic wastes lower the pH, changing the color of the medium.

without raising the temperature so much that the taste was ruined. Today, pasteurization is also used to kill pathogens in milk, ice cream, yogurt, and fruit juices. *Brucella melitensis* (brū-sel'lă me-li-ten'sis), *Mycobacterium bovis* (mī'kō-bak-tēr'ē-ŭm bō'vis), and *E. coli*, the causative agents of undulant fever, bovine tuberculosis, and one kind of diarrhea respectively, are controlled in this manner.

Pasteurization is not sterilization. *Thermoduric* and *thermophilic*—heat-tolerant and heat-loving—prokaryotes survive pasteurization, but they do not cause spoilage over the relatively short times during which properly refrigerated and pasteurized foods are stored before consumption. In addition, such prokaryotes are generally not pathogenic.

The combination of time and temperature required for effective pasteurization varies with the product. Because milk is the most familiar pasteurized product, we will consider the pasteurization of milk in some detail. Historically, milk was pasteurized by the *batch method* for 30 minutes at 63°C, but most milk processors today use a high-temperature, short-time method known as *flash pasteurization,* in which milk flows through heated tubes that raise its temperature to 72°C for only 15 seconds. This treatment effectively destroys all pathogens. *Ultrahigh-temperature pasteurization* heats the milk to 134°C for only 1 second, but some consumers claim it adversely affects the taste.

Highlight 9.1 on page 272 describes an outbreak of food poisoning on the West Coast that was caused when a popular brand of unpasteurized apple juice became contaminated with *E. coli.*

Ultrahigh-Temperature Sterilization The dairy industry and other food processors also use *ultrahigh-temperature sterilization,* which involves flash heating milk or other liquids to rid them of all living microbes. The process involves passing the liquid through superheated steam at 140°C for 1 second, and then cooling it rapidly. Treated liquids can be stored indefinitely at room temperature without microbial spoilage, though after months of storage chemical degradation results in flavor changes. Small packages of dairy creamer served in restaurants are often sterilized by the ultrahigh-temperature method. **Table 9.2** summarizes the dairy industry's use of moist heat for controlling microbes in milk.

Table 9.2 Moist Heat Treatments of Milk

Process	Treatment
Historical (batch) pasteurization	63°C for 30 minutes
Flash pasteurization	72°C for 15 seconds
Ultrahigh-temperature pasteurization	134°C for 1 second
Ultrahigh-temperature sterilization	140°C for 1 second

Dry Heat

For substances such as powders and oils that cannot be sterilized by boiling or with steam, or for materials that can be damaged by repeated exposure to steam (such as most metal objects), sterilization can be achieved by the use of dry heat, as occurs in an oven. Treatment with dry heat is also a routine part of standard aseptic laboratory procedure; each time laboratory technicians open or close a test tube or flask, they heat the lip of the container in the flame of a Bunsen burner.

Hot air is an effective sterilizing agent because it both denatures proteins and fosters the oxidation of metabolic and structural chemicals; however, in order to sterilize, dry heat requires higher temperatures for longer times than moist heat because dry heat penetrates more slowly. For instance, whereas an autoclave needs only 15 minutes to sterilize an object at 121°C, an oven at the same temperature requires at least 16 hours to achieve sterility. Scientists

Highlight 9.1 Odwalla's Road to Pasteurization

An apple a day may keep the doctor away, but a batch of bad apples can be deadly. In the fall of 1996, a terrible outbreak of food poisoning occurred in several western states and Canada. Over 60 people fell ill; children were particularly vulnerable, and one 16-month-old girl died. The symptoms included diarrhea, cramping, vomiting, and in the worst cases, a severe kidney disorder called hemolytic uremic syndrome. The culprit: unpasteurized Odwalla brand apple juice contaminated with *Escherichia coli* O157:H7 strain.

At the time, most major juice companies already pasteurized their juice. However, Odwalla had built its reputation on producing the freshest juice possible, and had never pasteurized its products out of concern that the process would make their juice taste bland and strip away nutrients. The company relied on other methods of preventing microbial contamination, such as sorting out rotten fruit and cleaning fruit with sanitizing chemicals. Despite the sorting methods in place, a batch of decayed apples was unintentionally allowed through the production process. It was ultimately linked to the contaminated apple juice. Though no one can be certain, it is believed that the initial *E.coli* contamination occurred when the apples fell from a tree and came into contact with animal feces containing the bacteria. Pasteurization would have killed the *E.coli* present in the resulting juice product.

Odwalla accepted responsibility for the outbreak and conducted an extensive review of its quality assurance policies. The company concluded that pasteurization was the only fail-safe method of preventing *E. coli* contamination in juice, and revised its manufacturing processes accordingly. Not all companies have followed suit, however. Many firms in the apple cider industry remain reluctant to pasteurize, contending that fresh, unpasteurized cider is safe for most adults, and that the risk of *E. coli* infection is far greater in hamburgers and seafood. But for young children, the elderly, those with compromised immune systems, and cautious adults everywhere, the only safe apple drink is the one that's been sufficiently heated (to 72°C for at least 15 seconds.)

typically use higher temperatures—171°C for 1 hour, or 160°C for 2 hours—to sterilize objects in an oven, but objects made of rubber, paper, and many types of plastic oxidize rapidly (combust) under these conditions.

Complete incineration is the ultimate means of sterilization. As part of standard aseptic technique in microbiological laboratories, inoculating loops are sterilized by heating them in the flame of a Bunsen burner or with an electric heating coil until they glow red (about 1500°C). Health care workers incinerate contaminated dressings, bags, and paper cups; field epidemiologists incinerate the carcasses of animals that have diseases such as anthrax or bovine spongiform encephalopathy (mad cow disease).

Refrigeration and Freezing

Learning Objective

✓ Describe the use and importance of refrigeration and freezing in limiting microbial growth.

In many situations, particularly in food preparation and storage, the most convenient method of microbial control is either refrigeration between 0°C and 7°C or freezing (temperatures below 0°C). These processes decrease microbial metabolism, growth, and reproduction because chemical reactions occur more slowly at low temperatures, and because liquid water is not available at subzero temperatures. Note, however, that psychrophilic microbes can multiply in refrigerated food and spoil its taste and suitability for consumption.

Refrigeration halts the growth of most pathogens, which are predominantly mesophiles. Notable exceptions are the bacteria *Listeria* (lis-tēr-ē-ă), which can reproduce to dangerous levels in refrigerated food, and *Yersinia* (yer-sin'ē-ă), which can multiply in refrigerated blood products and be passed on to blood recipients.

Slow freezing, during which ice crystals have time to form and puncture cell membranes, is more effective than quick freezing in inhibiting microbial metabolism, though microorganisms also vary in their susceptibility to freezing. Whereas the cysts of tapeworms and the roundworm larvae that causes trichinosis perish after several days in frozen meat, many vegetative bacterial cells, bacterial endospores, and viruses can survive subfreezing temperatures for years. In fact, scientists store many bacteria and viruses in low-temperature freezers at −30°C to −80°C and are able to reconstitute the microbes into viable populations by warming them in media containing proper nutrients. Therefore, care must be taken in thawing and cooking frozen food, because it can still contain many pathogenic microbes.

Highlight 9.2 describes how freezing is but one method of microbial control used in the preparation of sushi.

Desiccation and Lyophilization

Learning Objective

✓ Compare and contrast desiccation and lyophilization.

Desiccation or drying has been used for thousands of years to preserve such foods as fruits, peas, beans, grain, nuts, and yeast **(Figure 9.9).** Desiccation inhibits microbial growth because metabolism requires liquid water. Drying inhibits the spread of most pathogens, including the bacteria that cause syphilis, gonorrhea, and the more common forms of bacterial pneumonia and diarrhea. However, as with freezing, drying is microbistatic; it does not kill all microbes. For example, most molds can grow on dried raisins and apricots, which have as little as 16% water content.

Figure 9.9

The use of desiccation as a means of preserving foods. In this time-honored practice, drying inhibits microbial growth in the food by removing the water microbes needed for metabolism.

Highlight 9.2 Sushi: A Case Study in Microbial Control

Sushi—it either makes you squirm or salivate! Although technically the term refers to rice, it's generally understood to refer to bite-sized slices of raw fish served with rice. Long a staple of Japanese cuisine, sushi has become popular throughout the United States. But isn't eating raw fish dangerous? What methods of microbial control are applied in the preparation of sushi?

It's true that raw fish can contain harmful microorganisms. Such parasites as the kind of roundworms called anasakids are commonly found in fish and can cause gastrointestinal symptoms. Accordingly, before fish can be served raw, the Food and Drug Administration requires that it be frozen at −20°C (−4°F) for 7 days, or at −37°C (−35°F) for 15 hours. Unfortunately, freezing does *not* kill all bacterial or viral pathogens. Consumers should be aware that they always assume some risk of food poisoning caused by such bacteria as *S. aureus, E. coli,* and species of *Salmonella* and *Vibrio* whenever they eat any kind of raw seafood, including raw oysters, seviche, or carpaccio.

Even though diners tend to fixate on the safety of raw fish, cooked rice left sitting at room temperature is also vulnerable to the growth of pathogens. To counter this potential problem, sushi rice is prepared with vinegar, which acidifies the rice. At pH values below

4.6, rice becomes too acidic to support the growth of pathogens. Sushi bars can also reduce the risk posed by pathogens by keeping restaurant temperatures cool. Additionally, it is believed that wasabi, the fiery horseradish-like green paste commonly eaten with sushi, contains antimicrobial properties, although its mechanism of action is not well understood.

With these antimicrobial precautions in place, the vast majority of diners consume sushi safely meal after meal (although pregnant women and people with compromised immune systems should avoid all raw seafood). When properly prepared, sushi is beautiful, low in calories, and a source of heart-healthy omega 3-fatty acids—and very delicious.

Scientists use **lyophilization** (lī-of'i-li-zā'shŭn), a technique combining freezing and drying, to preserve microbes and other cells for many years. In this process, scientists instantly freeze a culture in liquid nitrogen or frozen carbon dioxide (dry ice); then they subject it to a vacuum that removes frozen water through a process called sublimation, in which the water is transformed directly from a solid to a gas. Lyophilization prevents the formation of large damaging ice crystals. Although not all cells survive, enough are viable to enable the culture to be reconstituted many years later.

CRITICAL THINKING

Why is liquid water necessary for microbial metabolism?

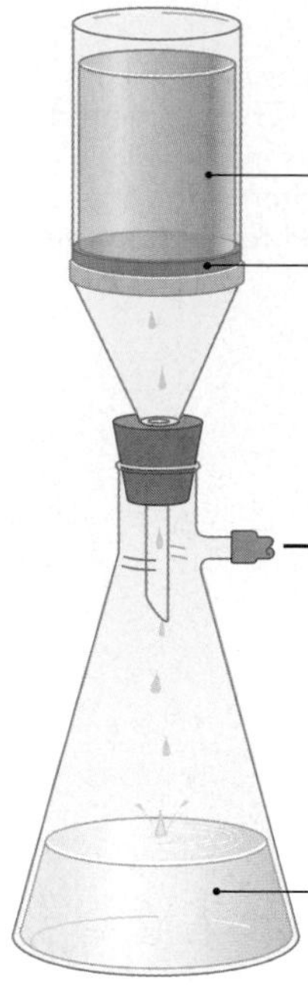

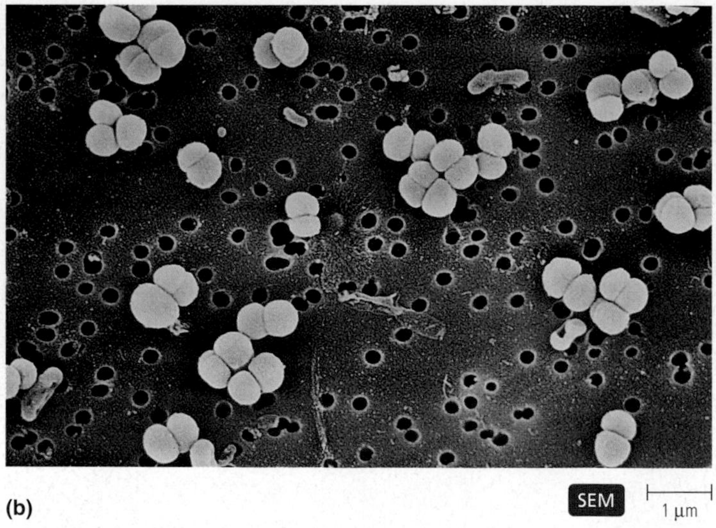

Figure 9.10

Filtration equipment used for microbial control. **(a)** Assembly for sterilization by vacuum filtration. **(b)** Membrane filters composed of various substances and with pores of various sizes can be used to trap diverse microbes.

Filtration

Learning Objective

✓ Describe the use of filters for disinfection and sterilization.

Filtration is the passage of a fluid (either a liquid or a gas) through a sieve designed to trap particles—in this case, cells or viruses—and separate them from the fluid **(Figure 9.10).** Researchers often use a vacuum to assist the movement of fluid through the filter. Filtration traps microbes larger than the pore size, allowing smaller microbes to pass through. In the late 1800s, filters were able to trap cells, but their pores were too large to trap the pathogens of such diseases as rabies and measles. These pathogens were thus named *filterable viruses*, which today has been shortened to *viruses*[9]. Now, filters with pores small enough to trap even viruses are available, so filtration can be used to sterilize such heat-sensitive materials as ophthalmic solutions, antibiotics, vaccines, liquid vitamins, enzymes, and culture media.

Over the years, filters have been constructed from porcelain, glass, cotton, asbestos, and diatomaceous earth, a substance composed of the innumerable glass-like cell walls of single-celled algae called diatoms. Scientists today typically use thin (only 0.1 mm thick), circular **membrane filters** manufactured of nitrocellulose or plastic and containing specific pore sizes ranging from 25 μm to less than 0.01 μm in diameter (see Figure 9.10b). The pores of the latter filters are small enough to trap small viruses and even some large protein molecules. Microbiologists also use filtration to estimate the number of microbes in a fluid by counting the number deposited on the filter after passing a given volume through the filter (see Figure 6.21). **Table 9.3** lists some pore sizes of membrane filters and the microbes they allow through.

Health care and laboratory workers routinely use filtration to prevent airborne contamination by microbes. The use of surgical masks prevents contamination of the environment by microbes exhaled by medical personnel, and cotton plugs in culture vessels prevents contamination of cultures by airborne microbes. Additionally, *high-efficiency particulate air (HEPA) filters* are crucial parts of biological safety cabinets **(Figure 9.11),** and HEPA filters are mounted in the air ducts of some operating rooms, rooms occupied by patients with airborne diseases such as tuberculosis, and rooms of immunocompromised patients such as burn victims and AIDS patients.

[9]Latin, meaning poisons.

CRITICAL THINKING

A virologist needs to remove all bacteria from a solution containing viruses without removing the viruses. What size membrane filter should the scientist use?

Osmotic Pressure

Learning Objective

✓ Discuss the use of hypertonic solutions in microbial control.

Another ancient method of microbial control is the use of high concentrations of salt or sugar in foods to inhibit

Table 9.3 Membrane Filters

Pore Size (μm)	Microbes that Pass Through
0.01	Prions
0.025	Smallest viruses and prions
0.22	Small viruses, pliable bacteria (mycoplasmas, rickettsias, chlamydias, some spirochetes), and prions
0.45	Viruses, small bacteria, and prions
1.2	Bacteria, viruses, and prions
3	Protozoa, smallest algae, bacteria, viruses, and prions
5	Yeasts, protozoa, algae, bacteria, viruses, and prions

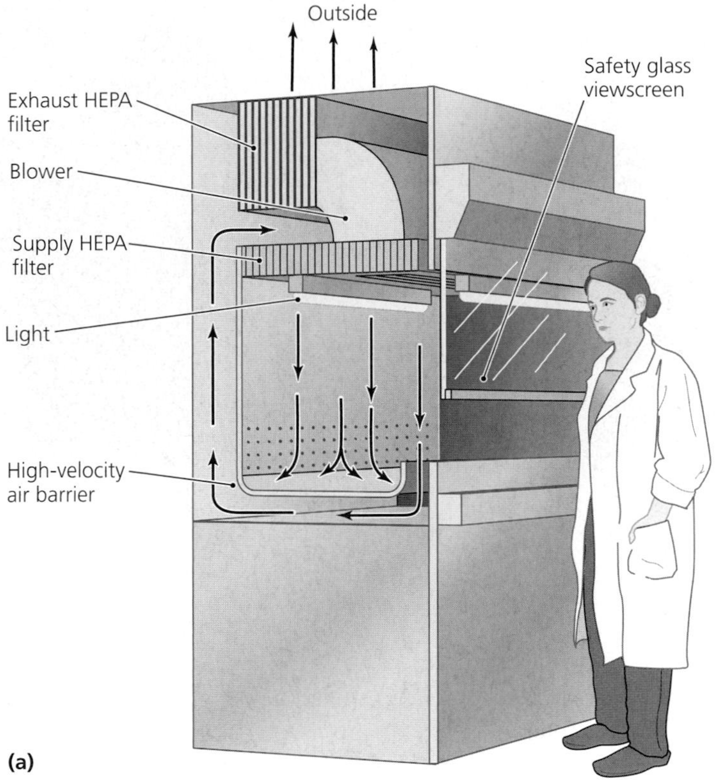

▲ ***Figure 9.11***

The roles of high-efficiency particulate air (HEPA) filters in laminar flow biological safety cabinets. HEPA filters protect workers from exposure to microbes (by maintaining a barrier of filtered air across the opening of the cabinet) and the laboratory environment (by filtering microbes out of exhaust air). HEPA filters are also used in air ducts in operating rooms and in the rooms of highly contagious or immunocompromised patients.

microbial growth by **osmotic pressure.** As we saw in Chapter 3, osmosis is the net movement of water across a semipermeable membrane (such as a cytoplasmic membrane) from an area of higher water concentration to an area of lower water concentration. Cells in a hypertonic solution lose water, whereas cells in a hypotonic solution tend to gain water. When a cell is in an environment with a high salt or sugar concentration, the water in the cell is drawn out by osmosis, and the cell desiccates (see Figure 3.18b). The removal of water inhibits cellular metabolism because enzymes are fully functional only in aqueous environments. Thus, honey, jerky, jams, jellies, salted fish, and some types of pickles are preserved from most microbial attacks.

Fungi have a greater ability than bacteria to tolerate hypertonic environments with little moisture, which explains why jelly in your refrigerator may grow a colony of *Penicillium* (pen-i-sil′ē-ŭm) mold but is not likely to grow the bacterium *Salmonella*.

Radiation

Learning Objective

✓ Differentiate between ionizing radiation and nonionizing radiation as they relate to microbial control.

Another physical method of microbial control is the use of radiation. There are two types of **radiation:** particulate radiation and electromagnetic radiation. Particulate radiation consists of high-speed subatomic particles, such as electrons, that have been freed from their atoms. Electromagnetic radiation can be defined as energy without mass traveling in waves at the speed of light (3×10^5 km/sec). Electromagnetic energy is released from atoms that have undergone internal changes. The *wavelength* of electromagnetic radiation, defined as the distance between two crests of a wave, ranges from very short cosmic and gamma rays, through X rays, ultraviolet light, and visible light, to long infrared rays, and finally to very long radio waves (see Figure 4.1). Though they are particles, electrons also have a wave nature with wavelengths that are even shorter than gamma rays.

The shorter the wavelength of an electromagnetic wave, the more energy it carries and the farther it can penetrate. Therefore, shorter wavelength radiation is more suitable for microbial control than longer wavelength radiation, which carries less energy and is less penetrating. Scientists describe all types of radiation as either *ionizing* or *nonionizing* according to its effects on the chemicals within cells.

Ionizing Radiation

Electron beams, gamma rays, and X rays, all of which have wavelengths shorter than 1 nm, are **ionizing radiation** because when they strike molecules, they have sufficient energy to create ions by ejecting electrons from atoms. Such ions disrupt hydrogen bonding, oxidize double covalent

(a) (b)

Figure 9.12

Irradiated food. **(a)** A demonstration of the increased shelf life of food achieved by ionizing radiation. **(b)** A radura, the symbol used in the United States to label irradiated foods.

bonds, and create highly reactive hydroxide ions (see Chapter 2). These ions in turn denature other molecules, particularly DNA, causing fatal mutations and cell death.

Electron beams are produced by *cathode ray machines* that operate like an analog television tube. Electron beams are highly energetic and therefore very effective in killing microbes in just a few seconds, but they do not penetrate deep into matter. Thus they cannot sterilize thick objects or objects coated with large amounts of organic matter. They are used to sterilize some sliced meats, microbiological plastic ware, and dental and medical supplies such as gloves, syringes, and suturing material.

Gamma rays, which are emitted by some radioactive elements such as radioactive cobalt, penetrate much farther than electron beams but require hours to kill microbes. The FDA has approved the use of gamma irradiation for microbial control in meats, spices, and fresh fruits and vegetables **(Figure 9.12).** Irradiation with gamma rays kills not only microbes, but also the larvae and eggs of insects; it also kills the cells of fruits and vegetables, preventing microbial spoilage and overripening.

Consumers have been reluctant to accept irradiated food. A number of reasons have been cited, including fear that radiation makes food radioactive, and claims that it changes the taste and nutritive value of foods or produces potentially carcinogenic (cancer-causing) chemicals. Supporters of irradiation reply that gamma radiation passes through food and cannot make it radioactive any more than a dental X ray produces radioactive teeth, and they cite numerous studies which concluded that irradiated foods are tasty, nutritious, and safe.

X rays travel the farthest through matter, but they have less energy than gamma rays and require a prohibitive amount of time to make them practical for microbial control.

Nonionizing Radiation

Electromagnetic radiation with a wavelength greater than 1 nm does not have enough energy to force electrons out of orbit. However, such radiation does contain enough energy to excite electrons and cause them to make new covalent bonds, which can affect the three-dimensional structure of proteins and nucleic acids.

Ultraviolet (UV) light, visible light, infrared radiation, and radio waves are nonionizing radiation. Of these, only UV light has sufficient energy to be a practical antimicrobial agent. Visible light and microwaves, which are extremely short wavelength radio waves, have little value in microbial control, though microwaves heat food, which can be inhibitory if the food gets hot enough.

UV light with a wavelength of 260 nm is specifically absorbed by adjacent thymine nucleotide bases in DNA, causing them to form covalent bonds with each other rather than forming hydrogen bonds with adenine bases in the complementary DNA strand (see Figure 7.21). As discussed in Chapter 7, such *thymine dimers* distort the shape of DNA, making it impossible for the cell to accurately transcribe or replicate its genetic material. If thymine dimers remain uncorrected, an affected cell may die.

The effectiveness of UV irradiation is tempered by the fact that UV light does not penetrate well. UV light is therefore suitable primarily for disinfecting air, transparent fluids, and the surfaces of objects such as barber's sheers and

Table 9.4 Physical Methods of Microbial Control

Method	Conditions	Action	Representative Use(s)
Moist heat			
Boiling	10 min at 100°C	Denatures proteins and destroys membranes	Disinfection of baby bottles and sanitization of restaurant cookware and tableware
Autoclaving (pressure cooking)	15 min at 121°C		Autoclave: sterilization of medical and laboratory supplies that can tolerate heat and moisture; pressure cooker: sterilization of canned food
Pasteurization	15 sec at 72°C		Destruction of all pathogens and most spoilage microbes in dairy products, fruit juices, beer, and wine
Ultrahigh-temperature sterilization	1–3 sec at 140°C		Sterilization of dairy products
Dry heat			
Hot air	2 h at 160°C or 1 h at 171°C	Denatures proteins, destroys membranes, oxidizes metabolic compounds	Sterilization of water-sensitive materials such as powders, oils, and metals
Incineration	1 sec at more than 1000°C		Sterilization of inoculating loops, flammable contaminated medical waste, and diseased carcasses
Refrigeration	0–7°C	Inhibits metabolism	Preservation of food
Freezing	<0°C	Inhibits metabolism	Long-term preservation of foods, drugs, and cultures
Desiccation (drying)	Varies with amount of water to be removed	Inhibits metabolism	Preservation of food
Lyophilization (freeze drying)	-196°C for a few minutes while drying	Inhibits metabolism	Long-term storage of bacterial cultures
Filtration	Filter retains microbes	Physically separates microbes from air and liquids	Sterilization of air and heat-sensitive ophthalmic and enzymatic solutions, vaccines, and antibiotics
Osmotic pressure	Exposure to hypertonic solutions	Inhibits metabolism	Preservation of food
Ionizing radiation (electron beams, gamma rays, X rays)	Seconds to hours of exposure (depending on wavelength of radiation)	Destroys DNA	Sterilization of medical and laboratory equipment and preservation of food
Nonionizing radiation (ultraviolet light)	Irradiation with 260 nm wavelength radiation	Formation of thymine dimers inhibits DNA transcription and replication	Disinfection and sterilization of surfaces and of transparent fluids and gases

operating tables. Recently, some cities have begun using UV irradiation in sewage treatment. By passing wastewater past banks of UV lights, they reduce the number of bacteria without using chlorine, which can damage the environment.

Table 9.4 summarizes the physical methods of microbial control discussed in the previous pages.

CRITICAL THINKING

During the bioterrorist attack in which anthrax endospores were sent through the mail in the fall of 2001, one news commentator suggested that people should iron all their incoming mail with a regular household iron as a means of destroying endospores. Would you agree that this is a good way to disinfect mail? Explain your answer. Which disinfectant methods would be both more effective and practical?

Chemical Methods of Microbial Control

Learning Objective

✓ Compare and contrast eight major types of antimicrobial chemicals, and discuss the positive and negative aspects of each.

Although physical agents are sometimes used for disinfection, antisepsis, and preservation, more often chemical agents are used for these purposes. As we have seen, chemical agents act to adversely affect microbes' cell walls, cytoplasmic membranes, proteins, or DNA. And as was also the case for physical agents, the effect of a chemical agent varies with temperature, length of exposure, and the amount of contaminating organic matter in the environment; it also varies with pH, concentration, and freshness of the chemical. Chemical agents tend to destroy or inhibit the growth of

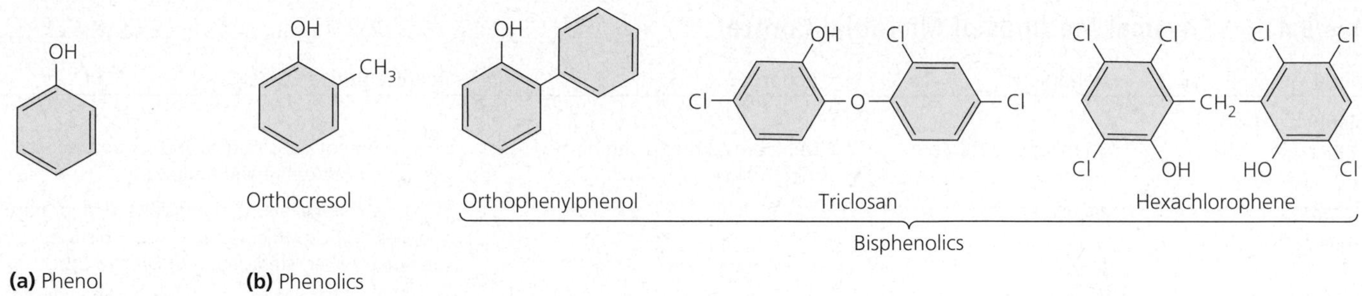

▲ ***Figure 9.13***

Phenol and phenolics. **(a)** Phenol, a naturally occurring molecule that is also called carbolic acid. **(b)** Phenolics, which are compounds synthesized from phenol, have greater antimicrobial efficacy with fewer side effects. Bisphenols are paired, covalently linked phenolics.

enveloped viruses and the vegetative cells of bacteria, fungi, and protozoa more than fungal spores, protozoan cysts, or bacterial endospores. The latter are particularly resistant to chemical agents, as was demonstrated by numerous failed attempts to decontaminate the United States Senate office building of anthrax endospores sent there by bioterrorists in 2001.

In the following sections we discuss eight major categories of antimicrobial chemicals used as antiseptics and disinfectants: *phenols, alcohols, halogens, oxidizing agents, surfactants, heavy metals, aldehydes*, and *gaseous agents*. Some chemical agents combine one or more of these. Additionally, researchers and food processors sometimes use antimicrobics—substances normally used to treat diseases—as disinfectants.

Phenol and Phenolics

Learning Objective

✓ Distinguish between phenol and the types of phenolics, and discuss their action as antimicrobial agents.

In 1867, Dr. Joseph Lister began using phenol **(Figure 9.13a)** to reduce infection during surgery. As stated previously, the efficacy of phenol remains one standard to which the actions of other antimicrobial agents can be compared.

Phenolics are compounds derived from phenol molecules that have been chemically modified by the addition of halogens or organic functional groups **(Figure 9.13b).** For instance, chlorinated phenolics contain one or more atoms of chlorine and have enhanced antimicrobial action and a less annoying odor than phenol. Natural oils such as pine and clove oils are also phenolics and can be used as antiseptics.

Bisphenolics are composed of two covalently linked phenolics. Two examples of bisphenolics are *orthophenylphenol*, which is the active ingredient in the disinfectant Lysol®, and *triclosan*, which is incorporated into numerous consumer products, including garbage bags, diapers, and cutting boards.

Phenol and phenolics are intermediate- to low-level disinfectants that denature proteins and disrupt cell membranes in a wide variety of pathogens. They are effective even in the presence of contaminating organic material such as vomit, pus, saliva, and feces, and they remain active on surfaces for a prolonged time. For these reasons, phenolics are commonly used in health care settings, laboratories, and households.

Negative aspects of phenolics include their disagreeable odor and possible side effects; for example, phenolics irritate the skin of some individuals. *Hexachlorophene* (see Figure 9.13b), which was once a popular household bisphenolic, was found to cause brain damage in infants. Now it is available only by prescription and is used in nurseries only in response to severe staphylococcal contamination.

Alcohols

Learning Objective

✓ Discuss the action of alcohols as antimicrobial agents, and explain why 70% to 90% alcohols are more effective than pure alcohols.

Alcohols are bactericidal, fungicidal, and virucidal against enveloped viruses; however, they are not effective against fungal spores or bacterial endospores. Alcohols are considered intermediate-level disinfectants. Commonly used alcohols include rubbing alcohol (isopropanol) and drinking alcohol (ethanol):

$$CH_3\text{–}CH(OH)\text{–}CH_3 \qquad CH_3\text{–}CH_2OH$$

Isopropanol   Ethanol

Isopropanol is slightly superior to ethanol as a disinfectant and antiseptic. *Tinctures* (tingk'chūrs), which are solutions of other antimicrobial chemicals in alcohol, are often more effective than the same antimicrobials dissolved in water.

Alcohols denature proteins and disrupt cytoplasmic membranes. Surprisingly, pure alcohol is not an effective antimicrobial agent because the denaturation of proteins requires water; therefore, 70% to 90% alcohol is typically used to control microbes. Alcohols evaporate rapidly, which is advantageous in that they leave no residue, but disadvantageous in that they may not contact microbes long enough to be effective (see Figure 9.4). Swabbing the skin with alcohol prior to an injection removes more microbes by physical action (degerming) than by chemical action.

Halogens

Learning Objective

✓ Discuss the types and uses of halogen-containing antimicrobial agents.

Halogens are the four very reactive, nonmetallic chemical elements: iodine, chlorine, bromine, and fluorine. Halogens are intermediate-level antimicrobial chemicals that are effective against vegetative bacterial and fungal cells, fungal spores, some bacterial endospores and protozoan cysts, and many viruses. Halogens are used both alone and combined with other elements in organic and inorganic compounds. Scientists do not know exactly how halogens exert their antimicrobial effect, but it is thought that they damage enzymes via oxidation or by covalently linking to them, denaturing them.

Iodine is a well-known antiseptic. In the past, backpackers and campers disinfected water with iodine tablets, but experience has shown that protozoan cysts can survive iodine treatment unless the iodine concentration is so great that the water is undrinkable. Knowledgeable campers now filter stream and lake water or carry bottled water, even though standard filtration does not remove toxins from water. Activated charcoal filters remove most toxins.

Medically, iodine is used either as a tincture or as an *iodophor*, which is an iodine-containing organic compound that slowly releases iodine. Iodophors have the advantage of being long-lasting, nonstaining, and nonirritating to the skin. Betadine® **(Figure 9.14)** is an example of an iodophor used in medical institutions to prepare skin for surgery and injections and to treat burns.

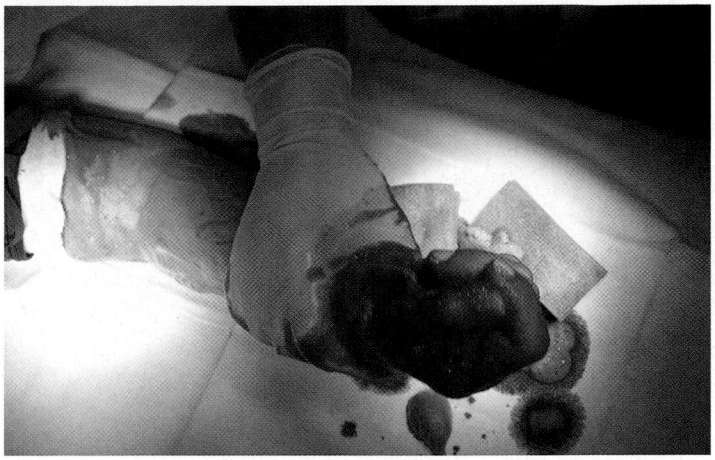

▲ *Figure 9.14*
The use of Betadine®, an iodophor, as a degerming antiseptic in preparation for drawing blood.

Municipalities commonly use *chlorine* in its elemental form (Cl_2) to treat drinking water supplies, swimming pools, and wastewater from sewage treatment plants. Compounds containing chlorine are also effective disinfectants. Examples include *sodium hypochlorite,* which is the active ingredient in chlorine bleach, and *calcium hypochlorite.* The dairy industry and restaurants use these compounds to disinfect utensils and the medical field uses them to disinfect hemodialysis systems. Household bleach diluted by adding two drops to a liter of water can be used in an emergency to make water safer to drink, but it does not kill all protozoan cysts, bacterial endospores, or viruses. *Chlorine dioxide* (ClO_2) is a gas that can be used to disinfect large spaces; for example, it was used in the federal office buildings contaminated with anthrax spores following the bioterrorism attack that occurred in the fall of 2001. Chloramines—chemical combinations of chlorine and ammonia—are used in wound dressings, as skin antiseptics, and in some municipal water supplies. Chloramines are less effective antimicrobial agents than other forms of chlorine, but they release chlorine slowly and are thus longer lasting, and they don't appear to form carcinogenic compounds.

Bromine is an effective disinfectant in hot tubs because it evaporates more slowly than chlorine at high temperatures.

Oxidizing Agents

Learning Objective

✓ Describe the use and action of oxidizing agents in microbial control.

Peroxides, ozone, and *peracetic acid* kill microbes by oxidizing their enzymes, thereby preventing metabolism. Oxidizing agents are high-level disinfectants and antiseptics that are particularly effective against anaerobic microorganisms contaminating deep wounds; they work by releasing hydroxyl radicals, which kill anaerobes. Therefore, health care workers use oxidizing agents to kill anaerobes in deep puncture wounds.

Hydrogen peroxide is a common household chemical that can disinfect and even sterilize the surfaces of inanimate objects such as contact lenses, but is often mistakenly used to treat open wounds. Hydrogen peroxide does not make a good antiseptic for open wounds because *catalase*—an enzyme released from damaged human cells—quickly neutralizes it by breaking it down into water and oxygen gas, which can be seen as escaping bubbles. Though aerobes and facultative anaerobes on inanimate surfaces also contain catalase, the volume of peroxide used as a disinfectant overwhelms the enzyme, making hydrogen peroxide a useful disinfectant. Food processors use hot hydrogen peroxide to sterilize packages such as juice boxes.

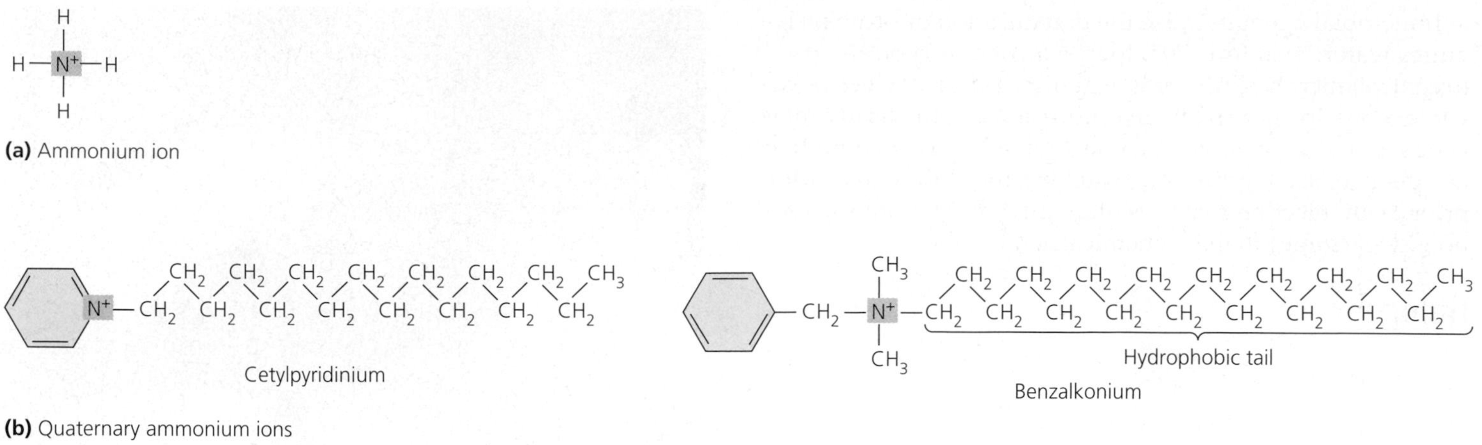

▲ ***Figure 9.15***

Quaternary ammonium compounds (quats). Quats are surfactants in which the hydrogen atoms of an ammonium ion **(a)** are replaced by other functional groups **(b)**.

Ozone (O_3) is a reactive form of oxygen that is generated when molecular oxygen (O_2) is subjected to electrical discharge. Ozone gives air its "fresh smell" after a thunderstorm. Some Canadian and European municipalities treat their drinking water with ozone rather than chlorine, because ozone does not produce carcinogenic by-products like chlorine can. Ozone is also a more effective antimicrobial agent than chlorine, but it is more expensive, and it is difficult to maintain an effective concentration of ozone in water.

Peracetic acid is an extremely effective sporicide that can be used to sterilize surfaces. Because it is not adversely affected by organic contaminants, and it leaves no toxic residue, food processors and medical personnel use peracetic acid to sterilize equipment.

Surfactants

Learning Objective

✓ Define surfactants and describe their antimicrobial action.

Surfactants are "surface active" chemicals. One of the ways surfactants act is to reduce the surface tension of solvents such as water by decreasing the attraction among solvent molecules. One result of this reduction in surface tension is that the solvent becomes more effective at dissolving solute molecules.

Two common surfactants involved in microbial control are soaps and detergents. One end of a soap molecule is hydrophobic because it is composed of fatty acids, and the other end is hydrophilic and negatively charged. When soap is used to wash skin, for instance, the hydrophobic ends of soap molecules are effective at breaking oily deposits (such as secretions from oil glands) into tiny droplets, and the hydrophilic ends attract water molecules; the result is that the tiny droplets of oily material—and any bacteria they harbor—are more easily dissolved in and washed away by water. Thus soaps by themselves are good degerming agents, but poor antimicrobial agents; when household soaps are antiseptic, it is largely because they contain antimicrobial chemicals.

Synthetic **detergents** are positively charged organic surfactants that are more soluble in water than soaps. The most popular detergents for microbial control are **quaternary ammonium compounds** or **quats**, which are composed of an ammonium cation (NH_4^+) in which the hydrogen atoms are replaced by other functional groups or hydrocarbon chains **(Figure 9.15).** Quats are not only antimicrobial; they are colorless, tasteless, and harmless to humans (except at high concentrations), making them ideal for many industrial and medical applications. If your mouthwash foams, it probably contains a quaternary ammonium compound. Examples of quats are Zephiran® (benzalkonium chloride) and Cepacol® (cetylpyridinium chloride).

Quats function by disrupting cellular membranes such that affected cells lose essential internal ions such as potassium ions (K^+). Quats are bactericidal (particularly against Gram-positive bacteria), fungicidal, and virucidal against enveloped-viruses; but they are not effective against nonenveloped viruses, mycobacteria, or endospores. The action of quaternary ammonium compounds is retarded by organic contaminants, and they are deactivated by soaps. Some pathogens, such as *P. aeruginosa*, actually thrive in quats; therefore, quats are classified as low-level disinfectants.

Heavy Metals

Learning Objective

✓ Define heavy metals, give several examples, and describe their use in microbial control.

The ions of high-molecular-weight metals such as arsenic, zinc, mercury, silver, and copper are antimicrobial because they combine with sulfur atoms in molecules of cysteine, an amino acid. Such bonding alters the three-dimensional shapes of proteins, inhibiting or eliminating their function. Heavy metal ions, as they are called, are low-level bacteriostatic and fungistatic agents, and with few exceptions their use has been superceded by more effective antimicrobial agents. **Figure 9.16** illustrates the effectiveness of heavy metals in inhibiting bacterial reproduction on a petri plate.

At one time, many states required that the eyes of newborns be treated with a cream containing 1% *silver nitrate* (AgNO₃) to prevent blindness caused by *Neisseria gonorrhoeae* (nī-sē′rē-ă go-nor-rē′ē), which can enter a baby's eyes while it passes through an infected birth canal. Today, silver nitrate has largely been displaced by antibiotic ointments, because they are less irritating and are also effective against *Chlamydia trachomatis* (kla-mid′ē-ă tra-kō′ma-tis) and other pathogens. Silver still plays an antimicrobial role in some surgical dressings, burn creams, and catheters.

For several decades, drug companies have used *thimerosal*, a mercury-containing compound, to preserve vaccines. In 1999, the U.S. Public Health Service recommended that alternatives be used because mercury is a metabolic poison.

Copper, which interferes with chlorophyll, is used to control algal growth in reservoirs, fish tanks, swimming pools, and water storage tanks. In the absence of organic contaminants, copper is an effective algicide in concentrations as low as 1 ppm (part per million). In addition to copper, zinc and mercury are used to control mildew in paint.

CRITICAL THINKING

What common household antiseptic contains a heavy metal as its active ingredient?

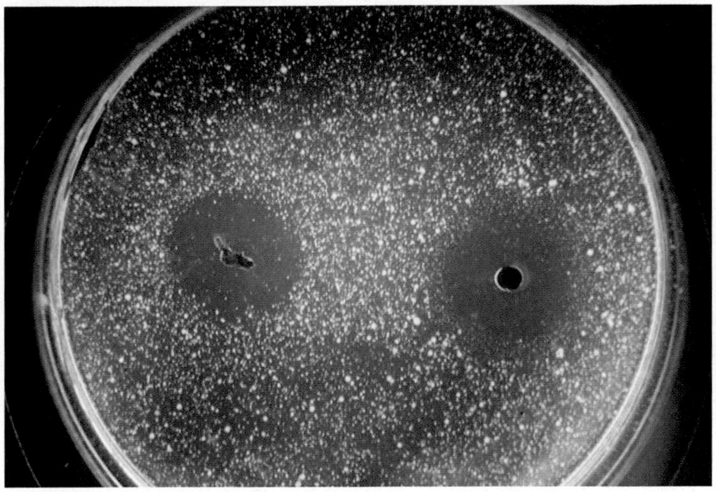

▲ *Figure 9.16*

The effect of heavy metal ions on bacterial growth. The zones of inhibition formed because metal ions inhibit bacterial reproduction through their effects on protein function.

Aldehydes

Learning Objective

✓ Compare and contrast formaldehyde and glutaraldehyde as antimicrobial agents.

Aldehydes are compounds containing terminal —CHO groups (see Table 2.3 on p. 40). *Glutaraldehyde,* which is a liquid, and *formaldehyde,* which is a gas, are highly reactive chemicals with the following structural formulas:

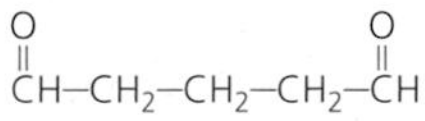

$$\underset{\text{Glutaraldehyde}}{\overset{O}{\overset{\|}{C}}H-CH_2-CH_2-CH_2-\overset{O}{\overset{\|}{C}}H} \qquad \underset{\text{Formaldehyde}}{H\overset{O}{\overset{\|}{C}}H}$$

Aldehydes function in microbial control by cross-linking amino, hydroxyl, sulfhydryl, and carboxyl organic functional groups thereby denaturing proteins and inactivating nucleic acids.

Hospital personnel and scientists use 2% solutions of glutaraldehyde to kill bacteria, viruses, and fungi; a 10-minute treatment effectively disinfects most objects, including medical and dental equipment. When the time of exposure is increased to 10 hours, glutaraldehyde sterilizes.

Although glutaraldehyde is less irritating and more effective than formaldehyde, it is more expensive as well.

Morticians and health care workers use formaldehyde dissolved in water to make a 37% solution called *formalin.* They use formalin for embalming and to disinfect isolation rooms, exhaust cabinets, surgical instruments, and reusable kidney dialysis machines. Formaldehyde must be handled with care because it irritates mucous membranes and is carcinogenic.

Gaseous Agents

Learning Objective

✓ Describe the advantages and disadvantages of gaseous agents of microbial control.

Many items, such as heart-lung machine components, sutures, plastic laboratory ware, mattresses, pillows, artificial heart valves, catheters, electronic equipment, and dried or powdered foods, cannot be sterilized easily with heat or water-soluble chemicals; nor is irradiation always practical for large or bulky items. However, they can be sterilized within a closed chamber containing highly reactive microbicidal and sporicidal gases such as *ethylene oxide, propylene oxide,* and *beta-propiolactone:*

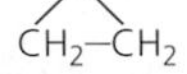

$$\underset{\text{Ethylene oxide}}{\overset{O}{CH_2-CH_2}} \qquad \underset{\text{Propylene oxide}}{\overset{O}{CH_3-CH-CH_2}}$$

$$\underset{\text{Beta-propiolactone}}{\overset{CH_2-CH_2}{\underset{O\;-\;C\diagdown_O}{|\qquad\quad|}}}$$

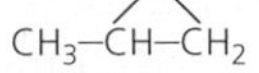

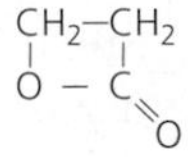

Table 9.5 Chemical Methods of Microbial Control

Method	Action(s)	Level of Activity	Use
Phenol (carbolic acid)	Denatures proteins and disrupts cell membranes	Intermediate to low	Original surgical antiseptic; now replaced by less odorous and injurious phenolics
Phenolics (chemically altered phenol; bisphenols are composed of a pair of linked phenolics)	Denature proteins and disrupt cell membranes	Intermediate to low	Disinfectants and antiseptics
Alcohols	Denature proteins and disrupt cell membranes	Intermediate	Disinfectants, antiseptics, and as a solvent in tinctures
Halogens (iodine, chlorine, bromine,and fluorine)	Presumably denature proteins	Intermediate	Disinfectants, antiseptics, and water purification
Oxidizing agents (peroxides, ozone, and peracetic acid)	Denature proteins by oxidation	High	Disinfectants, antiseptics for deep wounds, water purification, and sterilization of food-processing and medical equipment
Surfactants (soaps and detergents)	Decrease surface tension of water and disrupt cell membranes	Low	Soaps: degerming; detergents: antiseptic
Heavy metals (arsenic, zinc, mercury, silver, and copper, etc.)	Denature proteins	Low	Fungistats in paints; silver nitrate cream: surgical dressings, burn creams, and catheters; copper: algicide in water reservoirs, swimming pools, and aquariums
Aldehydes (glutaraldehyde and formaldehyde)	Denature proteins	High	Disinfectant and embalming fluid
Gaseous agents (ethylene oxide, propylene oxide, and beta-propiolactone)	Denature proteins	High	Sterilization of heat- and water-sensitive objects
Antimicrobials	Act against cell walls, cell membranes, protein synthesis, and DNA transcription and replication	Intermediate to low	Disinfectants and treatment of infectious diseases

These gases rapidly penetrate paper and plastic wraps and diffuse into every crack. Over time (usually 4–18 hours), they denature proteins and DNA by cross-linking organic functional groups, thereby killing everything they contact without harming inanimate objects.

Ethylene oxide is frequently used as a gaseous sterilizing agent in hospitals and dental offices, and NASA uses the gas to sterilize spacecraft designed to land on other worlds, lest they accidentally export earthly microbes. Large hospitals often use ethylene oxide chambers, which are similar in appearance to autoclaves, to sterilize instruments and equipment sensitive to heat.

Despite their advantages, gaseous agents are far from perfect: They can be extremely hazardous to the people using them. Because they are often highly explosive, they must be administered by combining them with 10% to 20% nitrogen gas or carbon dioxide. Moreover, they are extremely poisonous, so workers must extensively flush sterilized objects with air to remove every trace of the gas (which adds to the time required to use them). Finally, gaseous agents, especially beta-propiolactone, are potentially carcinogenic.

Antimicrobials

Learning Objective

✓ Describe the types of antimicrobials and their use in environmental control of microorganisms.

Antimicrobials include antibiotics, semisynthetics, and synthetics. *Antibiotics* are antimicrobial chemicals produced naturally by microorganisms. When scientists chemically modify an antibiotic the agent is called a *semisynthetic*. Scientists have also developed wholly synthetic antimicrobial drugs. The main difference between these antimicrobials

Highlight 9.3 Should You Use Antibacterial Soaps?

Walk through the soap aisle at your local supermarket, and it's likely you will find numerous products advertised as "antimicrobial" or "antibacterial." In fact, it's likely you already have such products in your home and use them every day. But do they really work better than plain, regular soaps? And is it possible that such products may actually do more harm than good?

The answers are less clear than you might think. Although soaps containing antimicrobial drugs are indeed more effective than plain soaps in reducing the presence of microorganisms, there is growing concern that overuse of antimicrobials may actually contribute to the evolution of drug-resistant microorganisms.

Antibacterial soaps kill off weaker bacteria, leaving stronger, more-resistant strains to multiply. Dr. Stuart Levy of Tufts University and some other researchers even contend that overuse of antimicrobial products may be contributing to an increase in hard-to-treat infections caused by the stronger bacteria that antimicrobials allegedly leave behind. Levy's research has focused in particular on triclosan, a common bisphenolic found in a wide range of household products. In tests on *E. coli*, Levy found that triclosan targeted a specific bacterial gene and contends that this specificity raises the potential for the development of resistant bacterial strains. To date, this hypothesis has not been proven.

Levy's critics point to the absence of any scientific evidence that antimicrobial products pose a health threat. They note that triclosan has been used for decades in cleaning products without any sign of encouraging excessive growth of resistant strains. The CDC and the FDA have taken a cautious stance in the controversy, acknowledging the benefits of antimicrobial soaps in certain circumstances while agreeing that further research into the question is needed.

Who knew cleanliness could be so complicated? And in the meantime, what kind of soap should you use? Experts can't seem to agree on a single guideline, but the CDC recommends using mild, regular soap and washing in warm running water for at least 10–15 seconds in most cases, reserving antimicrobial soaps for limited applications: for use in handling food, for the care of newborns, and for the care of high-risk patients by health care workers.

and the chemical agents we have discussed in this chapter is that antimicrobials are typically used for treatment of disease, and not for environmental control of microbes. Nevertheless, some antimicrobials *are* used for antimicrobial control outside the body. For example, the antimicrobials *nisin* and *natamycin* are used respectively to reduce bacterial and fungal growth in cheese. Chapter 10 discusses in more detail the nature and use of antimicrobials to treat infectious diseases.

Table 9.5 summarizes the chemical methods of microbial control discussed in this chapter.

Development of Resistant Microbes

Many scientists are concerned that Americans have become overly preoccupied with antisepsis and disinfection, as evidenced by the proliferation of products containing antiseptic and disinfecting chemicals. For example, one can now buy hand soap, shampoo, toothpaste, hand lotion, foot pads for shoes, deodorants, and bath sponges that contain antiseptics, as well as kitty, litter, cutting boards, scrubbing pads, garbage bags, children's toys, and laundry detergents that contain disinfectants. There is little evidence that the extensive use of such products adds to human or animal health, but it does promote the development of strains of microbes resistant to antimicrobial chemicals because while susceptible cells die, resistant cells remain to proliferate. Scientists have already isolated strains of pathogenic bacteria, including *Mycobacterium tuberculosis, P. aeruginosa, E. coli,* and *S. aureus,* that are less susceptible to common disinfectants and antiseptics.

Highlight 9.3 discusses a controversy regarding the use of antibacterial soap.

CHAPTER SUMMARY

Basic Principles of Microbial Control (pp. 263–265)

1. **Sterilization** is the eradication of microorganisms and viruses; the term is not usually applied to the destruction of prions.
2. **Antisepsis** is the inhibition/killing of microorganisms (particularly pathogens) on skin or tissue by the use of a chemical **antiseptic,** whereas **disinfection** refers to the use of agents to inhibit microbes on inanimate objects.
3. **Degerming** refers to the removal of microbes from a surface by scrubbing.
4. **Sanitization** is the reduction of a prescribed number of pathogens from surfaces and utensils in public settings.
5. **Pasteurization** is a process using heat to kill pathogens and control microbes that cause spoilage of food and beverages.
6. The suffixes -stasis and -static indicate that an antimicrobial agent inhibits microbes, whereas the suffixes -cide and -cidal indicate that the agent kills or permanently inactivates a particular type of microbe.
7. **Microbial** death is the permanent loss of reproductive capacity. **Microbial death rate** measures the efficacy of an antimicrobial agent.

8. Antimicrobial agents destroy microbes either by altering their cell walls and membranes or by interrupting their metabolism and reproduction via interference with proteins and nucleic acids.

The Selection of Microbial Control Methods (pp. 265–268)

1. Factors affecting the efficacy of antimicrobial methods include the site to be treated, the relative susceptibility of microorganisms, and environmental conditions.
2. Four methods for evaluating the effectiveness of a disinfectant or antiseptic are the **phenol coefficient,** which compares the agent's efficacy to that of phenol; the **use-dilution test;** the **disk-diffusion method** (also known as the Kirby-Bauer method); and the **in-use test,** which provides a more accurate determination of efficacy under real-life conditions.

Physical Methods of Microbial Control (pp. 268–277)

1. **Thermal death point** is the lowest temperature that kills all cells in a broth in 10 minutes, whereas **thermal death time** is the time it takes to completely sterilize a particular volume of liquid at a set temperature. **Decimal reduction time (D)** is the time required to destroy 90% of the microbes in a sample.
2. An **autoclave** uses steam heat under pressure to sterilize chemicals and objects that can tolerate moist heat.
3. Pasteurization, a method of heating foods to kill pathogens and control spoilage organisms without altering the quality of the food, can be achieved by several methods: the historical (batch) method, flash pasteurization, and ultrahigh-temperature pasteurization. The methods differ in their combinations of temperature and time of exposure.
4. Under certain circumstances, microbes can be controlled using ultrahigh temperature sterilization, dry heat sterilization, incineration, refrigeration, or freezing.
5. Antimicrobial methods involving drying are **desiccation,** used to preserve food, and **lyophilization** (freeze drying), used for the long-term preservation of cells or microbes.
6. When used as a microbial control method, **filtration** is the passage of air or a liquid through a material that traps and removes microbes. Some **membrane filters** have pores small enough to trap the smallest viruses and some large protein molecules. HEPA (high-efficiency particulate air) filters remove microbes and particles from air.
7. The high **osmotic pressure** exerted by hypertonic solutions of salt or sugar can preserve foods such as jerky and jams by removing the water from microbes that they need to carry out their metabolic functions.
8. **Radiation** includes high-speed subatomic particles and even faster waves of electromagnetic energy released from atoms. **Ionizing radiation** (electromagnetic radiation with wavelengths shorter than 1 nm) creates ions that create effects leading to the denaturation of important molecules and cell death. **Nonionizing radiation,** which has wavelengths longer than 1 nm, is less effective in microbial control; however, UV light causes thymine dimers, which can kill affected cells.

Chemical Methods of Microbial Control (pp. 277–283)

1. **Phenolics,** which are chemically modified phenol molecules, are intermediate- to low-level disinfectants that denature proteins and disrupt cell membranes in a wide variety of pathogens.
2. **Alcohols** are intermediate-level disinfectants that denature proteins and disrupt cell membranes; they are used either as 70% to 90% aqueous solutions or in a tincture, which is a combination of an alcohol and another antimicrobial chemical.
3. **Halogens** (iodine, chlorine, bromine, and fluorine) are used as intermediate-level disinfectants and antiseptics to kill microbes in water or on medical instruments or skin. Although their exact mode of action is unknown, they are believed to denature proteins.
4. **Oxidizing agents** such as hydrogen peroxide, ozone, and peracetic acid are high-level disinfectants and antiseptics that release oxygen radicals, which are toxic to many microbes, especially anaerobes.
5. **Surfactants** include soaps, which act primarily to break up oils during degerming, and detergents such as **quaternary ammonium compounds (quats),** which are low-level disinfectants.
6. **Heavy metal** ions such as arsenic, silver, mercury, copper, and zinc are low-level disinfectants that denature proteins. For most applications they have been superceded by less-toxic alternatives.
7. **Aldehydes** are high-level disinfectants that cross-link organic functional groups in proteins and DNA. A 2% solution of glutaraldehyde and a 37% aqueous solution of formaldehyde (called formalin) are used to disinfect or sterilize medical or dental equipment and in embalming fluid.
8. **Gaseous agents** of microbial control, which include ethylene oxide, propylene oxide, and beta-propiolactone, are high-level disinfecting agents used to sterilize heat-sensitive equipment and large objects. These gases are explosive and potentially carcinogenic.
9. **Antimicrobials,** which include antibiotics, semisynthetics, and synthetics, are compounds that are typically used to treat diseases but can also function as intermediate-level disinfectants.

QUESTIONS FOR REVIEW

(Answers to multiple choice questions are on the web, along with additional review questions. Visit www.microbiologyplace.com.)

Multiple Choice

1. Which of the following statements provides the definition of sterilization in *practical* terms?
   a. Sterilization eliminates all organisms and their spores or endospores.
   b. Sterilization eliminates harmful microorganisms and viruses.
   c. Sterilization eliminates prions.
   d. Sterilization eliminates hyperthermophiles.

2. Which of the following substances or processes kills microorganisms on laboratory surfaces?
   a. antiseptics
   b. disinfectants
   c. degermers
   d. pasteurization

3. Which of the following terms best describes the disinfecting of cafeteria plates?
   a. pasteurization
   b. antisepsis
   c. sterilization
   d. sanitization

4. The microbial death rate is used to measure
   a. the efficiency of a detergent.
   b. the efficiency of an antiseptic.
   c. the efficiency of sanitization techniques.
   d. all of the above

5. Which of the following statements is true concerning the selection of an antimicrobial agent?
   a. An ideal antimicrobial agent is stable during storage.
   b. An ideal antimicrobial agent is fast acting.
   c. Ideal microbial agents do not exist.
   d. all of the above

6. The spores of which organism are used as a biological indicator of sterilization?
   a. *Bacillus stearothermophilus*
   b. *Salmonella typhi*
   c. *Mycobacterium tuberculosis*
   d. *Staphylococcus aureus*

7. A company that manufactures an antimicrobial cleaner for kitchen counters claims that its product is effective when used in a 50% water solution. By what means might scientists have been able to verify this statement?
   a. disk-diffusion test
   b. phenol coefficient
   c. filter paper test
   d. use-dilution test

8. Which of the following items functions most like an autoclave?
   a. a boiling pan
   b. an incinerator
   c. a desiccator
   d. a pressure cooker

9. The preservation of beef jerky from microbial growth relies on which method of microbial control?
   a. filtration
   b. lyophilization
   c. osmotic pressure
   d. radiation

10. Which of the following types of radiation is more widely used as an antimicrobial technique?
    a. electron beams
    b. visible light waves
    c. radio waves
    d. microwaves

11. Which of the following substances would most effectively inhibit anaerobes in a deep wound?
    a. phenol
    b. hexachlorophene
    c. isopropanol
    d. hydrogen peroxide

12. Which of the following adjectives best describes a surgical procedure that is free of microbial contaminants?
    a. disinfected
    b. sanitized
    c. degermed
    d. aseptic

13. Sterile paper disks are soaked in one of three disinfectants: P, Q, and R. The disks are placed on medium inoculated with organism X and incubated. The following chart indicates the results:

Disinfectants	Zone of inhibition
P	13 mm
Q	4.5 mm
R	10 mm

    Assuming the disinfectants diffuse at the same rates and are equally stable, which disinfectant was most effective against organism X?
    a. disinfectant P
    b. disinfectant Q
    c. disinfectant R

14. A sample of *E. coli* has been subjected to heat for a specified time, and 90% of the cells have been destroyed. Which of the following terms best describes this event?
    a. thermal death point
    b. thermal death time
    c. decimal reduction time
    d. none of the above

15. Which of the following substances is least toxic to humans?
    a. carbolic acid
    b. glutaraldehyde
    c. hydrogen peroxide
    d. formalin

16. Which of the following chemicals is active against bacterial endospores?
    a. copper ions
    b. ethylene oxide
    c. ethanol
    d. triclosan

17. Which of the following disinfectants acts against cell membranes?
    a. phenol
    b. peracetic acid
    c. silver nitrate
    d. glutaraldehyde

18. Which of the following disinfectants contains alcohol?
    a. an iodophor
    b. a quat
    c. formalin
    d. a tincture of bromine

Short Answer

1. Describe three types of microbes that are extremely resistant to antimicrobial treatment, and explain why they are resistant.
2. Compare and contrast the four tests that have been developed to measure the efficacy of disinfectants.
3. Why is it necessary to use strong disinfectants in areas exposed to tuberculosis patients?
4. Why do warm disinfectant chemicals generally work better than cool ones?
5. Why are Gram-negative bacteria more susceptible to heat than Gram-positive bacteria?
6. Describe five physical methods of microbial control.
7. What is the difference between thermal death point and thermal death time?
8. Defend the following statement: "Pasteurization is not sterilization."
9. What antimicrobial agents would you recommend for making drinking water safe in your community?
10. Compare and contrast the action of alcohols, halogens, and oxidizing agents in controlling microbial growth.
11. Hyperthermophilic prokaryotes may remain viable in canned goods after commercial sterilization. Why is this situation not dangerous to consumers?
12. Why are alcohols more effective in a 70% solution than in a 100% solution?
13. Contrast the structures and actions of soaps and quats.
14. What are some advantages and disadvantages of using ionizing radiation to sterilize food?
15. How can campers effectively treat stream water to remove pathogenic protozoa, bacteria, and viruses?

CRITICAL THINKING

1. In what ways might it be argued that the widespread commercial use of antiseptics and disinfectants has hurt rather than helped American health?
2. Is desiccation the only antimicrobial effect operating when grapes are dried in the sun to make raisins? Explain.
3. How long would it take to reduce a population of 100 trillion (10^{14}) bacteria to 10 viable cells if the D value of the treatment is 3 minutes?
4. Some potentially pathogenic bacteria and fungi, including strains of *Enterococcus, Staphylococcus, Candida,* and *Aspergillus,* can survive for 1–3 months on a variety of materials found in hospitals, including the synthetic material of privacy curtains, the cotton blends of scrub suits, nurses' clothes, lab coats, and plastics from splash aprons and computer keyboards. What steps could hospital personnel take to reduce the spread of these pathogens to susceptible patients?
5. In the spring of 2001, California health officials identified 23 patients infected with *Salmonella* who ate uncooked alfalfa sprouts. Twenty-one patients developed acute diarrhea, and three had to be hospitalized. Investigators discovered that the sprout producer had subjected the contaminated sprouts to a heat treatment followed by a 15-minute soak in calcium hypochlorite. Based on the discussion in this chapter, what types of heat treatment are available to producers of alfalfa sprouts? Why were the antimicrobial methods not effective in preventing salmonellosis? What other precautions could either the producer or consumers have taken?
6. An over-the-counter medicated foot powder contains camphor, eucalyptus oil, lemon oil, and zinc oxide. Only one of the ingredients is antimicrobial. Which one? How does it act against fungi?

CHAPTER 10

Controlling Microbial Growth in the Body: Antimicrobial Drugs

Meet *Staphylococcus aureus,* a common bacterium that is the number one cause of hospital-acquired infections in the United States. The particular strain of *S. aureus* shown here, however, is remarkable in some very important and alarming ways. Not only is it resistant to methicillin, the antimicrobial that is traditionally used to treat staphylococcal infections; it is also resistant to vancomycin, a drug that was long considered the last line of defense against methicillin-resistant *S. aureus.* Although other drugs may be used to fight it, this "superbug" strain of *S. aureus* leaves patients few options for defense—and indeed it can cause death.

How do antimicrobial drugs work? Why do they work on some microorganisms and not on others? What can be done about the increasing problem of superbugs that resist a wide variety of existing drugs? This chapter focuses on the chemical control of pathogens in the body.

CHAPTER OUTLINE

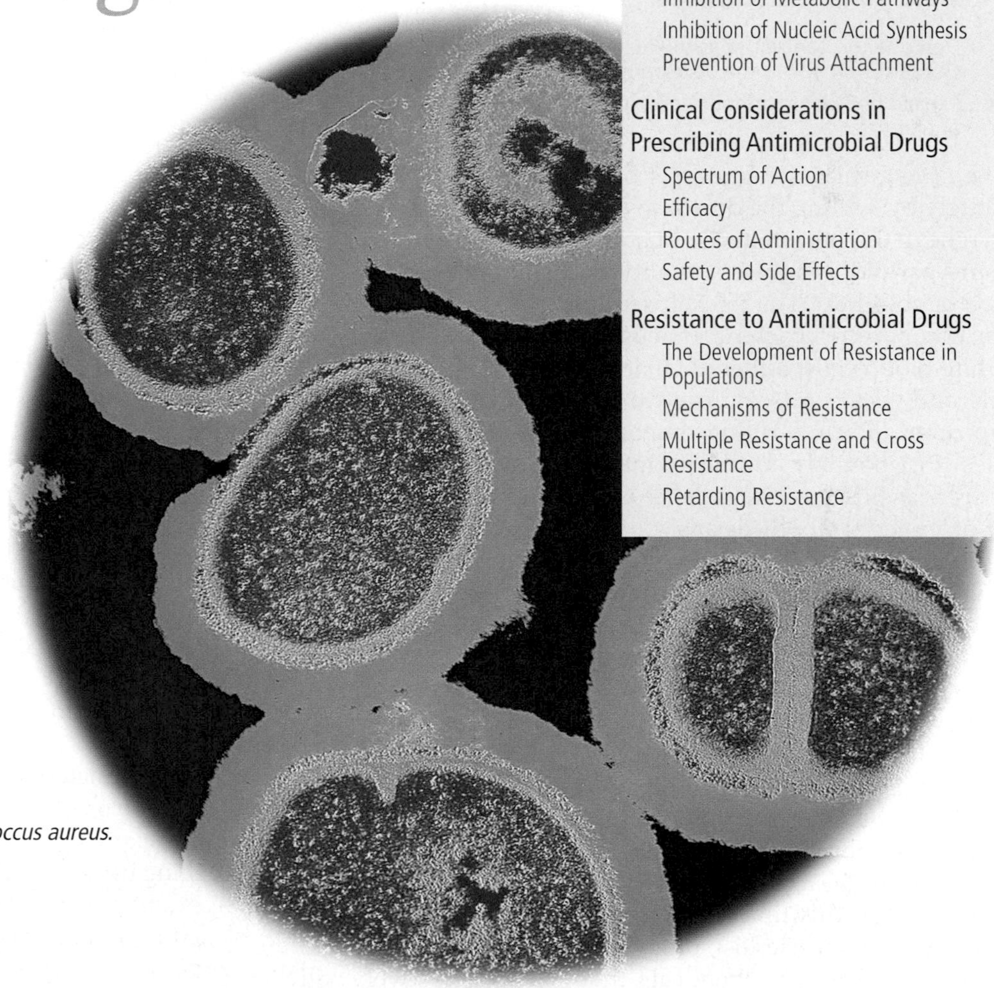

Vancomycin-resistant *Staphylococcus aureus.*

MicroPrep Pre-Test: *Take the pre-test for this chapter on the web.*
Visit ***www.microbiologyplace.com.***

Chemicals that affect physiology in any manner, such as caffeine, alcohol, and tobacco, are called *drugs*. Drugs that act against diseases are called *chemotherapeutic agents*. Examples include insulin, anticancer drugs, and drugs for treating infections—called **antimicrobial agents (antimicrobials),** the subject of this chapter.

In the pages that follow we'll examine the mechanisms by which antimicrobial agents act, the factors that must be considered in the use of antimicrobials, and several issues surrounding resistance to antimicrobial agents among microorganisms. First, however, we begin with a brief history of antimicrobial chemotherapy.

The History of Antimicrobial Agents

Learning Objectives

- ✓ Describe the contributions of Paul Ehrlich, Alexander Fleming, and Gerhard Domagk in the development of antimicrobials.
- ✓ Explain how semisynthetic and synthetic antimicrobials differ from antibiotics.

The little girl lay struggling to breathe as her parents stood mutely by, willing the doctor to do something—anything—to relieve the symptoms that had so quickly consumed their four-year-old daughter's vitality. Sadly, there was little the doctor could do. The thick "pseudomembrane" of diphtheria, composed of bacteria, mucus, blood clotting factors, and white blood cells, adhered tenaciously to her pharynx, tonsils, and vocal cords. He knew that trying to remove it could rip open the underlying mucous membrane, resulting in bleeding, possibly additional infections, and death. In 1902, there was little medical science could offer for the treatment of diphtheria; all physicians could do was wait and hope.

At the beginning of the 20th century, much of medicine involved diagnosing illness, describing its expected course, and telling family members either how long a patient might be sick, or when they might expect her to die. Even though physicians and scientists had recently accepted the germ theory of disease and knew the causes of many diseases, very little could be done to inhibit pathogens, including *Corynebacterium diphtheriae* (kŏ-rī'nē-bak-tēr'ē-ŭm dif-thi'rē-ē), and alter the course of infections. In fact, one-third of children born in the early 1900s died from infectious diseases before the age of five.

It was at this time that Paul Ehrlich (1854–1915), a visionary German scientist, proposed the term *chemotherapy* to describe the use of chemicals that would selectively kill pathogens while having little or no effect on a patient. He wrote of "magic bullets" that would bind to receptors on germs to bring about their death while ignoring host cells, which lacked the receptor molecules.

Ehrlich's search for antimicrobial agents resulted in the discovery of one arsenic compound that killed trypanosome parasites, and another that worked against the bacterial agent of syphilis. A few years later, in 1929, the British bacteriologist Alexander Fleming (1881–1955) reported the antibacterial action of penicillin released from *Penicillium* (pen-i-sil'ē-ŭm) mold **(Figure 10.1).** Fleming coined the term **antibiotics** to describe antimicrobial agents that are produced naturally by an organism. (See **Highlight 10.1** for a discussion of why microorganisms secrete antibiotics.) In common usage today "antibiotic" means "an antibacterial agent," excluding agents with antiviral and antifungal activity.

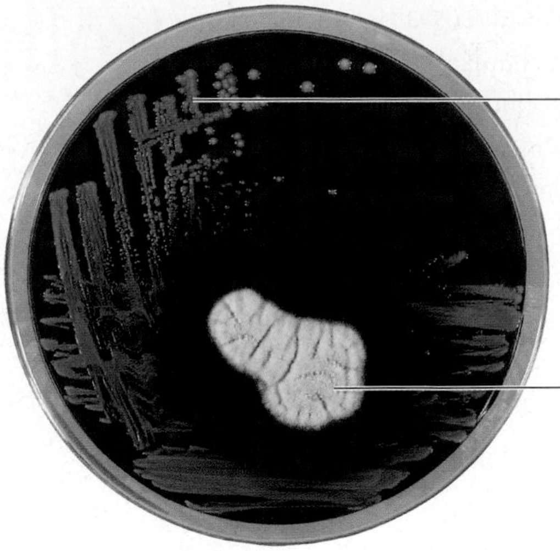

▲ ***Figure 10.1***

Antibiotic effect of the mold *Penicillium chrysogenum*. Alexander Fleming observed that this mold secretes penicillin, which inhibits the growth of bacteria, as is apparent with *Staphylococcus aureus* growing on this blood agar plate.

Though arsenic compounds and penicillin were discovered first, they were not the first antimicrobials in widespread use because Ehrlich's arsenic compounds are toxic to humans, and because penicillin was not available in large enough quantities to be useful until the late 1940s. Instead, *sulfanilamide,* discovered in 1932 by the German chemist Gerhard Domagk (1895–1964), was the first practical antimicrobial agent efficacious in treating a wide array of bacterial infections.

Other microorganisms besides the mold *Penicillium* are sources of useful antimicrobials, most notably the fungus *Cephalosporium* (sef'ă-lō-spor'ē-um) and species of soil-dwelling bacteria in the genera *Bacillus* (ba-sil'ŭs) and *Streptomyces* (strep-tō-mī'sēz). Additionally, by altering the chemical structure of antibiotics, scientists produce **semisynthetics**—drugs that are more effective, longer lasting, or easier to administer than the naturally occurring antibiotics. Antimicrobials that are completely synthesized in a laboratory are called **synthetics.** More than half of all antibiotics and semisynthetics are derived from species of *Streptomyces*.

Table 10.1 provides a partial list of common antibiotics and semisynthetics and their sources.

CRITICAL THINKING

Why aren't antibiotics effective against the common cold?

Highlight 10.1 Why Do Microbes Make Antibiotics?

We all know that antibiotics benefit humans, but what good are they to the microorganisms that secrete them? From the viewpoint of evolutionary theory, the answer may seem obvious: Antibiotics are weapons that confer an advantage to the secreting organisms in their struggle for survival. In reality, however, the answer is not so simple.

Antibiotics are members of an extremely diverse group of metabolic products known as *secondary metabolites,* which typically are complex organic molecules that are not essential for normal cell growth and reproduction and are produced only *after* an organism has established itself in its environment. The production of secondary metabolites results in a metabolic cost for the cell; that is, antibiotic production consumes energy and raw materials that could be used for growth and reproduction. Tetracycline, for example, is the end result of 72 separate enzymatic steps, and erythromycin requires 25 different chemical reactions—none of which contribute to *Streptomyces*'s normal growth or reproduction. Therefore, the question can be modified: Of what use are metabolically "expensive" antibiotics to organisms that are already secure in their environment?

Adding to the conundrum is the fact that antimicrobials against bacteria have never been discovered in natural soil at high enough concentrations to be inhibitory to *any* neighboring cell. For example, it is almost impossible to detect tetracyclines, aminoglycosides, chloramphenicol, or other antibiotics produced by *Streptomyces*, except when the bacteria are grown in a laboratory. Production of antibiotics in inconsequential quantities hardly seems to represent a significant adaptive edge.

Some scientists have suggested that antibiotics are evolutionary vestiges—leftovers of metabolic pathways that were once useful but no longer have a significant role. Others point out that evolutionary theory suggests that there would be tremendous selective pressure against the slightest continued manufacture of complex antibiotics if they truly have little purpose for the microorganism. Further research is required before we may satisfactorily answer the question, if we ever can.

Table 10.1 Sources of Some Common Antibiotics and Semisynthetics

Microorganism	Antimicrobial
Fungi	
Penicillium chrysogenum	Penicillin
Penicillium griseofulvin	Griseofulvin
Cephalosporium spp.	Cephalothin
Bacteria	
Bacillus licheniformis	Bacitracin
Bacillus polymyxa	Polymyxin
Micromonospora purpurea	Gentamicin
Streptomyces griseus	Streptomycin
Streptomyces fradiae	Neomycin
Streptomyces aureofaciens	Tetracycline
Streptomyces orientalis	Vancomycin
Streptomyces venezuelae	Chloramphenicol
Streptomyces erythraeus	Erythromycin
Streptomyces mediterranei	Rifampin
Streptomyces nodosus	Amphotericin B
Streptomyces avermitilis	Ivermectin

Mechanisms of Antimicrobial Action

Learning Objectives

- ✓ Explain the principle of selective toxicity.
- ✓ List six mechanisms by which antimicrobial drugs affect the growth of pathogens.

As Ehrlich foresaw, the key to successful chemotherapy against microbes is **selective toxicity;** that is, an effective antimicrobial agent must be more toxic to a pathogen than to the pathogen's host. Selective toxicity is possible because of differences in structure or metabolism between the pathogen and its host. Typically, the more differences, the easier it is to discover or create an effective antimicrobial agent.

Because there are many differences between the structure and metabolism of pathogenic bacteria and their eukaryotic hosts, antibacterial drugs constitute the greatest number and diversity of antimicrobial agents. Similarly, fewer antifungal, antiprotozoan, and anthelmintic drugs are available because fungi, protozoa, and helminths—like their animal and human hosts—are eukaryotic and thus share many common features. The number of effective antiviral drugs is even more limited, despite major differences in structure, because viruses utilize their host cells' enzymes and ribosomes to metabolize and replicate. Therefore, drugs that are effective against viral replication are likely toxic to the host as well.

Although they can have a variety of effects on pathogens, antimicrobial drugs can be categorized into several general groups according to their mechanisms of action (**Figure 10.2** on page 290):

- Drugs that inhibit cell wall synthesis. These drugs are selectively toxic to certain fungal or bacterial cells, which have cell walls, but not to animals, which lack cell walls.
- Drugs that inhibit protein synthesis (translation) by targeting the differences between prokaryotic and eukaryotic ribosomes.

Figure 10.2 ➤
Mechanisms of action of antibacterial drugs. Also listed are representative drugs for each type of action.

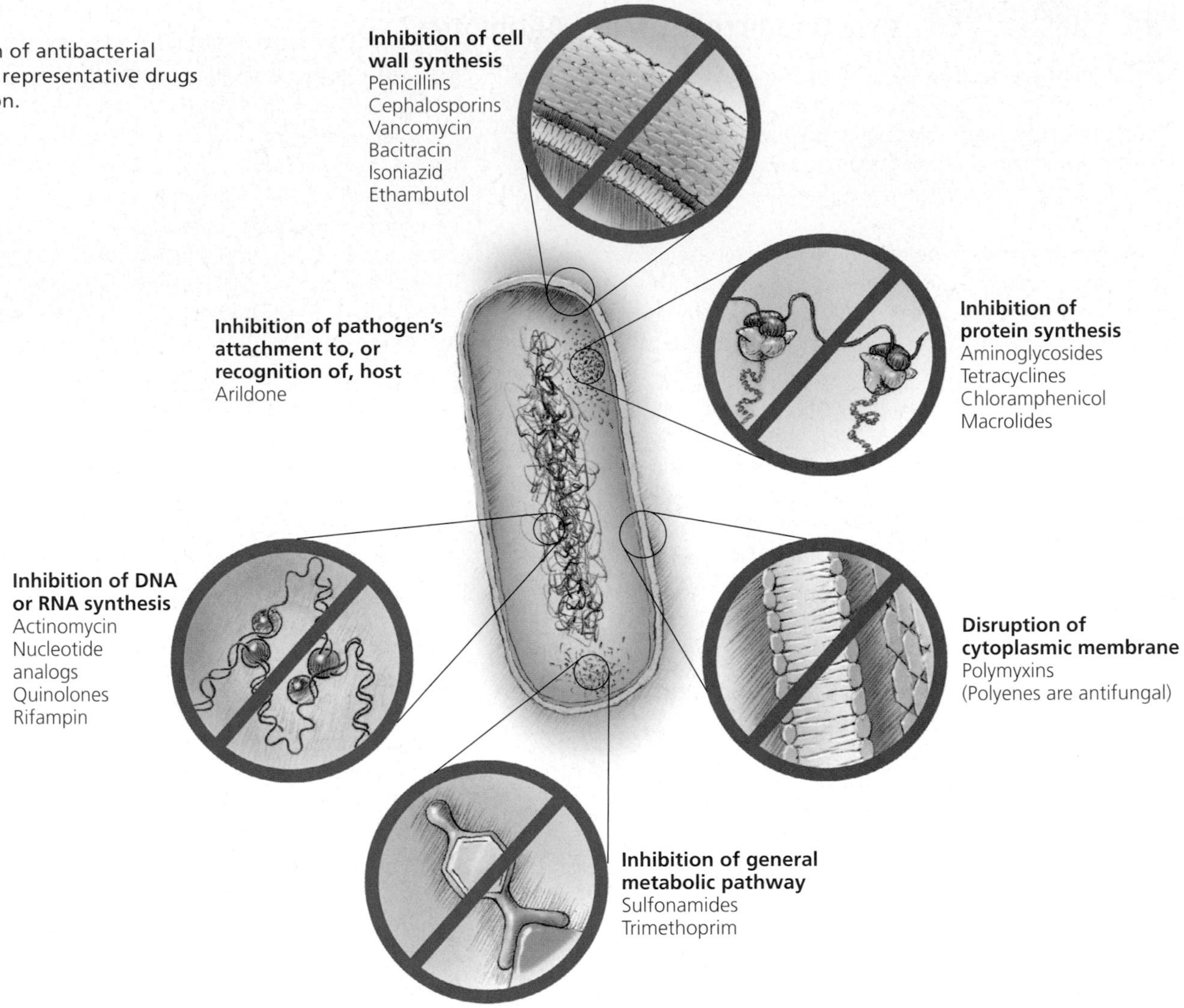

- Drugs that disrupt the cytoplasmic membrane.
- Drugs that inhibit general metabolic pathways.
- Drugs that inhibit nucleic acid synthesis.
- Drugs that block a pathogen's recognition of or attachment to its host.

In the following sections we examine these mechanisms in turn.

Inhibition of Cell Wall Synthesis

Learning Objective

✓ Describe the actions and give examples of drugs that affect the cell walls of bacteria.

A cell wall protects a cell from the effects of osmotic pressure. The major structural component of a bacterial cell wall is its peptidoglycan layer. As we discussed in Chapter 3, peptidoglycan is a single, huge macromolecule composed of polysaccharide chains of alternating N-acetylglucosamine (NAG) and N-acetylmuramic acid (NAM) molecules that are cross-linked by short peptide chains extending between NAM subunits (see Figure 3.12). To enlarge or divide, a cell must synthesize more peptidoglycan by adding new NAG and NAM subunits to existing NAG–NAM chains, and the new NAM subunits must then be bonded to neighboring NAM subunits **(Figure 10.3a and b).**

The most common antibacterial agents act by preventing the cross-linkage of NAM subunits. Most prominent among these drugs are **beta-lactams,** such as penicillins and cephalosporins, which are antimicrobials whose functional portions are called *beta-lactam (β-lactam) rings* **(Figure 10.3c).** Beta-lactams inhibit peptidoglycan formation by irreversibly binding to the enzymes that cross-link NAM subunits **(Figure 10.3d).** In the absence of correctly formed peptidoglycan, growing bacterial cells have weakened cell walls that are less resistant to the effects of osmotic pressure. The underlying cytoplasmic membrane bulges through the weakened portions of cell wall as water moves into the cell, and eventually the cell lyses **(Figure 10.3e).**

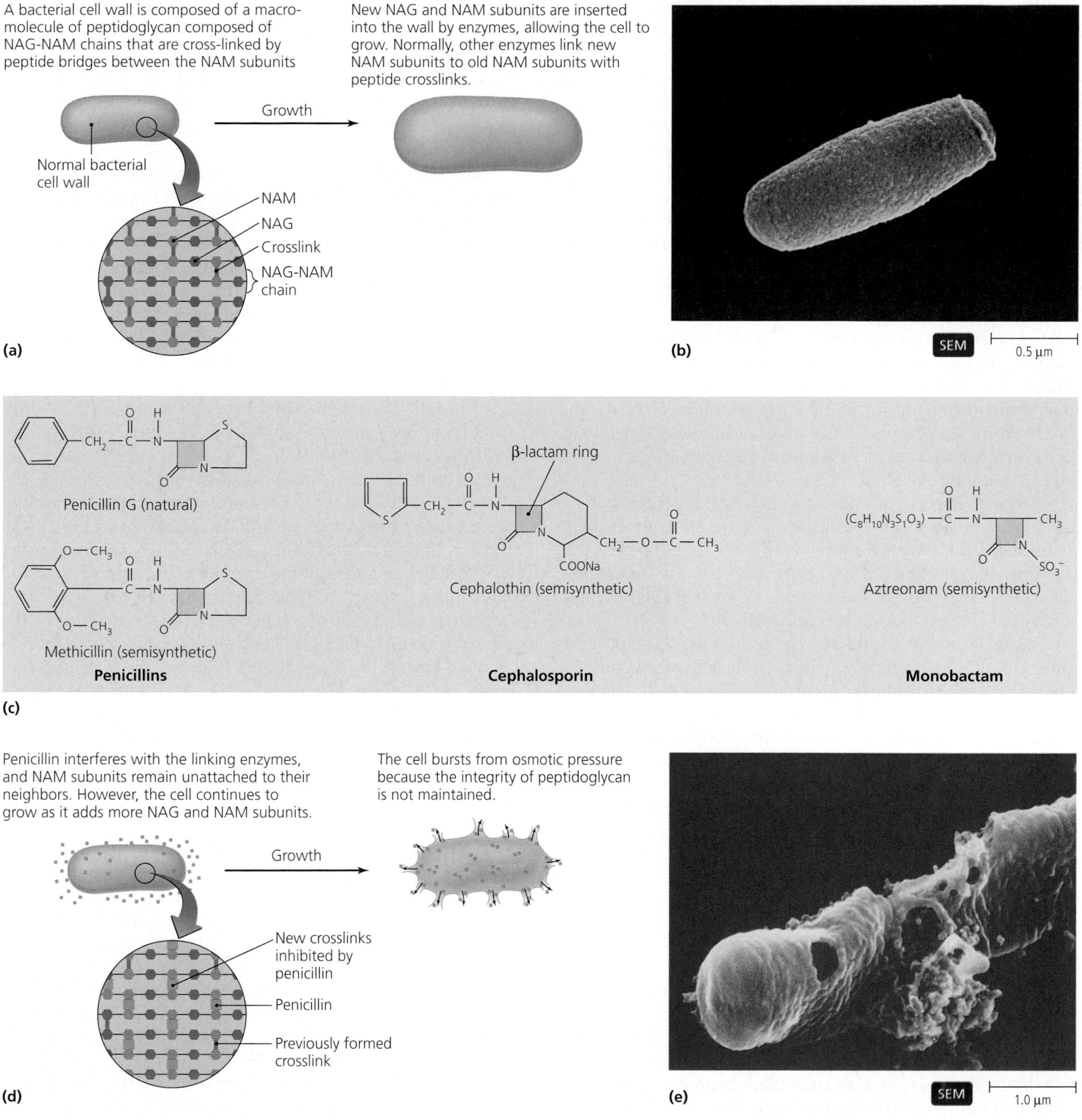
A bacterial cell wall is composed of a macromolecule of peptidoglycan composed of NAG-NAM chains that are cross-linked by peptide bridges between the NAM subunits
New NAG and NAM subunits are inserted into the wall by enzymes, allowing the cell to grow. Normally, other enzymes link new NAM subunits to old NAM subunits with peptide crosslinks.
Growth
Normal bacterial cell wall
NAM
NAG
Crosslink
NAG-NAM chain
(a)
(b)
SEM
0.5 μm
Penicillin G (natural)
Methicillin (semisynthetic)
Penicillins
β-lactam ring
Cephalothin (semisynthetic)
Cephalosporin
Aztreonam (semisynthetic)
Monobactam
(c)
Penicillin interferes with the linking enzymes, and NAM subunits remain unattached to their neighbors. However, the cell continues to grow as it adds more NAG and NAM subunits.
The cell bursts from osmotic pressure because the integrity of peptidoglycan is not maintained.
Growth
New crosslinks inhibited by penicillin
Penicillin
Previously formed crosslink
(d)
(e)
SEM
1.0 μm

▲ ***Figure 10.3***

Bacterial cell wall synthesis, and the inhibitory effects of beta-lactams on it. **(a)** A schematic depiction of a normal peptidoglycan cell wall showing NAG–NAM chains and cross-linked NAM subunits. **(b)** A scanning electron micrograph of a bacterial cell with a normal cell wall. **(c)** Structural formulas of some beta-lactam drugs. Their functional portion is the β-lactam ring. **(d)** A schematic depiction of the effect of penicillin on peptidoglycan in preventing NAM–NAM cross-links. **(e)** A scanning electron micrograph of bacterial lysis due to the effects of penicillin. *After a beta-lactam weakens a peptidoglycan molecule by preventing NAM–NAM cross-linkages, what force actually kills an affected bacterial cell?*

Figure 10.3 *Cells lyse due to osmotic movement of water into the cell.*

Chemists have made alterations to natural beta-lactams, such as penicillin G, to create semisynthetic derivatives such as methicillin and cephalothin (see Figure 10.3c), which are more stable in the acidic environment of the stomach, more readily absorbed in the intestinal tract, less susceptible to deactivation by bacterial enzymes, or more active against more types of bacteria. The simplest beta-lactams are *monobactams,* which are seldom used because they are effective only against aerobic Gram-negative bacteria.

Other antimicrobials such as **vancomycin** (van-kō-mī'sin), which is obtained from *Streptomyces orientalis,* and **cycloserine,** a semisynthetic, disrupt cell wall formation in a different manner. They directly interfere with particular alanine–alanine bridges that link the NAM subunits in many Gram-positive bacteria. Those bacteria that lack alanine–alanine crossbridges are naturally resistant to these drugs. Still another drug that prevents cell wall formation, **bacitracin** (bas-i-trā'sin), blocks the secretion of NAG and NAM from the cytoplasm. Like beta-lactams, vancomycin, cycloserine, and bacitracin result in cell lysis due to the effects of osmotic pressure.

Since all these drugs prevent bacteria from *increasing* the amount of cell wall material but have no effect on existing peptidoglycan, they are effective only on bacterial cells that are growing or reproducing; dormant cells are unaffected. Of course, they do not harm plant or animal cells because such cells lack peptidoglycan cell walls.

Bacteria of the genus *Mycobacterium* (mī'kō-bak-tēr'ē-ŭm), notably the agents of leprosy and tuberculosis, are characterized by unique, complex cell walls that have a layer of arabinogalactan-mycolic acid in addition to the usual peptidoglycan of prokaryotic cells. **Isoniazid** (ī-sō-nī'ă-zid, or **INH**[1]) and **ethambutol** (eth-am'boo-tol) disrupt the formation of this extra layer. Mycobacteria typically only reproduce every 12–24 hours, in part because of the complexity of their cell walls, so antimicrobial agents that act against mycobacteria must be administered for months or even years to be effective. It is often difficult to ensure that patients continue such a long regimen of treatment.

Inhibition of Protein Synthesis

As we discussed in Chapter 2, cells use proteins for structure and regulation, as enzymes in metabolism, and as channels and pumps to move materials across cell membranes. Thus, a consistent supply of proteins is vital for the active life of a cell. Given that all cells use ribosomes to translate proteins using information from messenger RNA templates, it is not immediately obvious that drugs could selectively target differences related to protein synthesis. Recall, however, that prokaryotic ribosomes differ from eukaryotic ribosomes in structure and size: Prokaryotic ribosomes are 70S and composed of 30S and 50S subunits, whereas eukaryotic ribosomes are 80S with 60S and 40S subunits.

Many antimicrobial agents take advantage of the differences between ribosomes to selectively target bacterial protein translation without significantly affecting eukaryotes. Note, however, that because some of these drugs affect eukaryotic mitochondria, which also contain 70S ribosomes like those of prokaryotes, such drugs may be harmful to animals and humans.

Understanding the actions of antimicrobials that inhibit protein synthesis requires an understanding of the process of translation, because various parts of ribosomes are the targets of antimicrobial drugs. Recall from the discussion of translation in Chapter 7 (see p. 212) that both the 30S and 50S subunits of a prokaryotic ribosome play a role in the initiation of protein synthesis, in codon recognition, and in the docking of tRNA–amino acid complexes, and that the 50S subunit contains the enzymatic portion that actually forms peptide bonds.

Among the antimicrobials that target the 30S ribosomal subunit are aminoglycosides and tetracyclines. **Aminoglycosides** (am'i-nō-glī'kō-sīds), such as *streptomycin, amikacin, tobramycin,* and *gentamicin,* change the shape of the 30S subunit, making it impossible for the ribosome to read the codons of mRNA correctly **(Figure 10.4a). Tetracyclines** (tet-ră-sī'klēns) block the tRNA docking site, which then prevents the incorporation of additional amino acids into a growing polypeptide **(Figure 10.4b).**

Other antimicrobials interfere with the function of the 50S subunit. *Chloramphenicol* and similar drugs block the enzymatic site of the 50S subunit **(Figure 10.4c),** which prevents translation. *Clindamycin* and antimicrobial agents called **macrolides** (mak'rō-līds), including *erythromycin,* bind to a different portion of the 50S subunit, preventing movement of the ribosome from one codon to the next **(Figure 10.4d);** as a result, translation is frozen and protein synthesis is halted.

Disruption of Cytoplasmic Membranes

Some antimicrobial drugs disrupt the cytoplasmic membrane of targeted cells, often by becoming incorporated into the membrane and damaging its integrity. This is the mechanism of action of a group of drugs called **polyenes** (pol-ē-ēns'). One polyene called *amphotericin B* **(Figure 10.5a)** is fungicidal because it attaches to *ergosterol,* a lipid constituent of fungal membranes **(Figure 10.5b),** in the process disrupting the membrane and causing lysis of the cell. The cytoplasmic membranes of humans are somewhat susceptible to amphotericin B because they contain cholesterol, which is similar to ergosterol, though cholesterol does not bind amphotericin B as well as does ergosterol.

Most bacterial membranes lack sterols, so these bacteria are naturally resistant to amphotericin B; however, there are other agents that disrupt bacterial membranes. An example of these antibacterial agents is *polymyxin,* produced by *Bacillus polymyxa* (ba-sil'ŭs po-lē-miks'ă). Polymyxin is effective against Gram-negative bacteria, particularly *Pseudomonas* (soo-dō-mō'nas), but because it is toxic to human kidneys it

[1]From *isonicotinic acid hydrazide,* the correct chemical name for isoniazid.

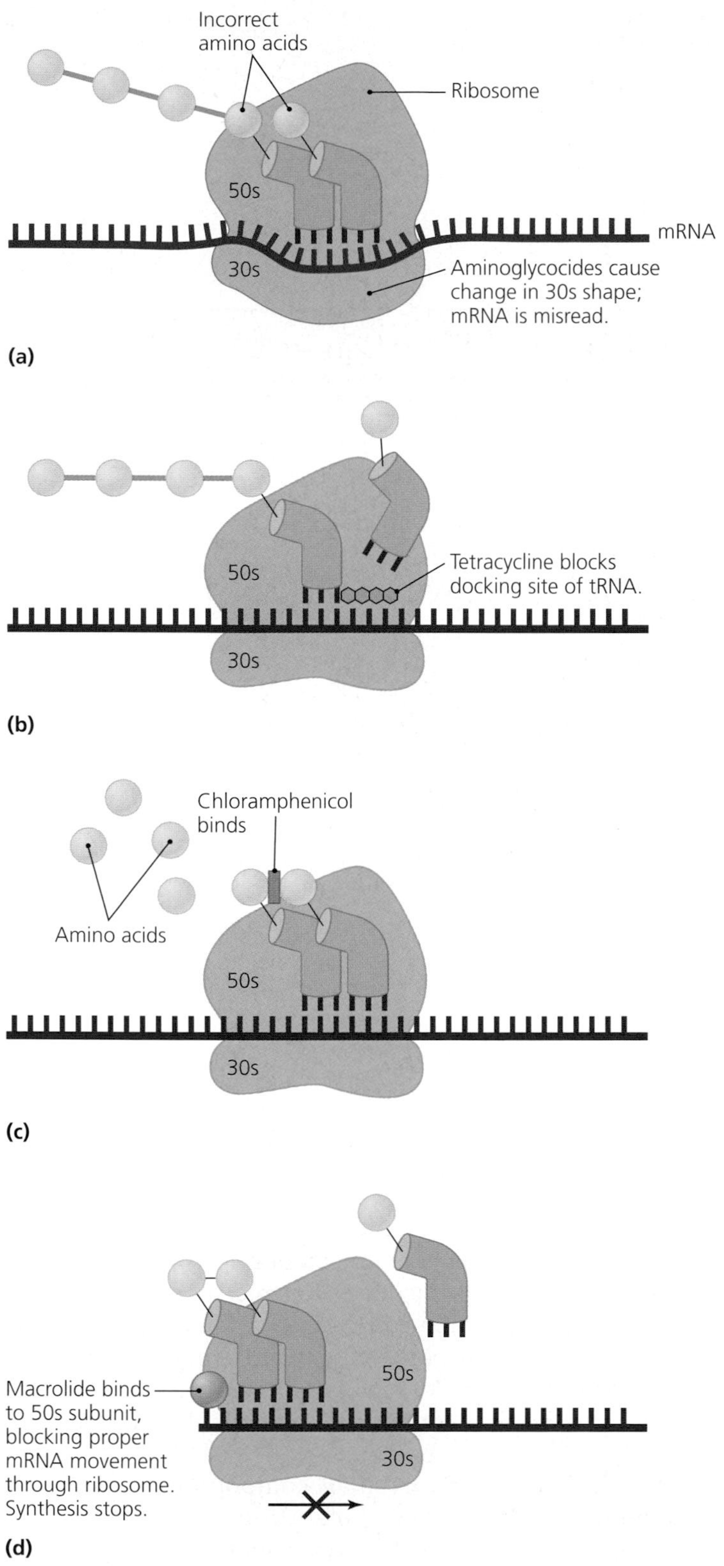

◀ ***Figure 10.4***

The mechanisms by which antimicrobials inhibit protein synthesis by targeting prokaryotic ribosomes. **(a)** Aminoglycosides change the shape of the 30S subunit, causing incorrect pairing of tRNA anticodons with mRNA codons. **(b)** Tetracyclines block the tRNA docking site on the 30S subunit, preventing protein elongation. **(c)** Chloramphenicol blocks enzymatic activity of the 50S subunit, preventing the formation of peptide bonds between amino acids. **(d)** Macrolides bind to the 50S subunit, preventing movement of the ribosome along the mRNA. *Which tRNA anticodon should align with the codon CUG?*

Figure 10.4 *Anticodon GAC is the correct complement for the codon CUG.*

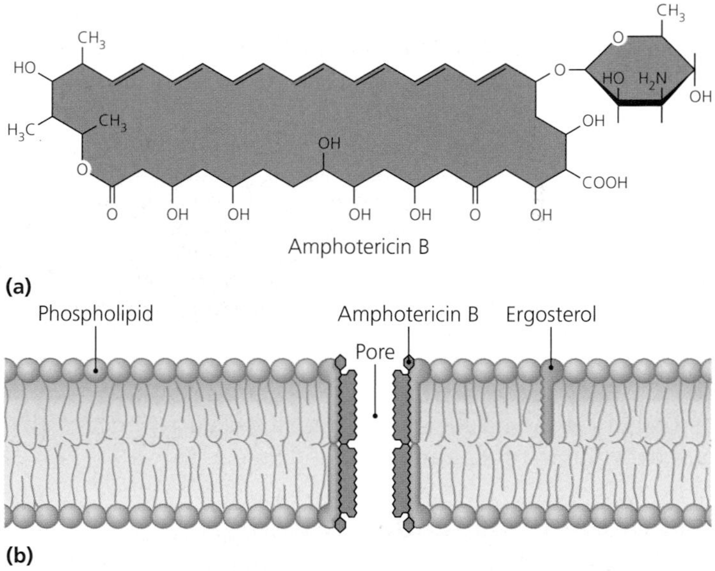

▲ ***Figure 10.5***

Disruption of the cytoplasmic membrane by the antifungal amphotericin B. **(a)** The structure of amphotericin B. **(b)** The proposed action of amphotericin B. The drug binds to molecules of ergosterol, which then congregate, forming a pore.

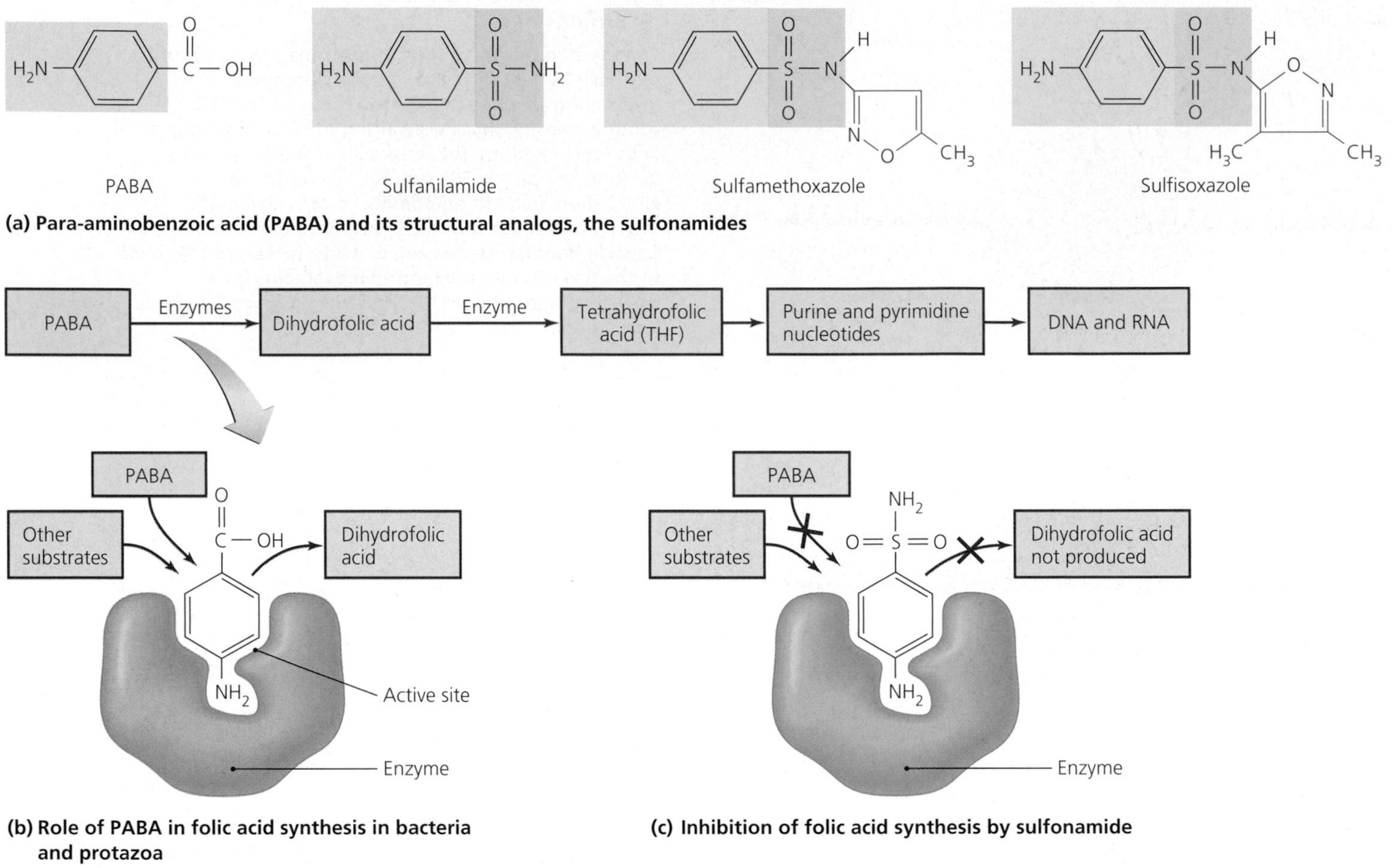

▲ ***Figure 10.6***

The antimetabolic action of sulfonamides in inhibiting nucleic acid synthesis. **(a)** Para-aminobenzoic acid (PABA) and representative members of its structural analogs, the sulfonamides. The analogous portions of the compounds are shaded. **(b)** The metabolic pathway in bacteria and protozoa by which folic acid is synthesized from PABA. **(c)** The inhibition of folic acid synthesis by the presence of a sulfonamide, which deactivates the enzyme by binding irreversibly to the enzyme's active site.

is usually reserved for use against external pathogens that are resistant to other antibacterial drugs.

Inhibition of Metabolic Pathways

As we discussed in Chapter 5, metabolism can be defined simply as the sum of all chemical reactions that take place within an organism. Whereas most living things share certain metabolic reactions—for example, glycolysis—other chemical reactions are unique to certain organisms. Whenever differences exist between the metabolic processes of a pathogen and its host, *antimetabolic agents* can be effective.

A variety of kinds of antimetabolic agents are available, including heavy metals (such as arsenic, mercury, and antimony), which inactivate enzymes; agents that rid the body of parasitic worms by paralyzing them; drugs that block the activation of viruses; and metabolic antagonists such as sulfanilamide, the first commercially available antimicrobial agent.

Sulfanilamide and similar compounds, called **sulfonamides,** act as antimetabolic drugs because they are *structural analogs* of—that is, are chemically very similar to—*para-aminobenzoic acid* (*PABA;* **Figure 10.6a**), a compound that is crucial in the anabolic reactions in the synthesis of nucleotides required for DNA and RNA synthesis. Many organisms, including some pathogens, enzymatically convert PABA into dihydrofolic acid, and dihydrofolic acid into tetrahydrofolic acid (THF), a form of folic acid that is then used as a coenzyme in the synthesis of purine and pyrimidine nucleotides **(Figure 10.6b).** As analogs of PABA, sulfonamides compete with PABA molecules for the active site of the enzyme involved in the production of dihydrofolic acid **(Figure 10.6c),** which leads to a decrease in the production of THF, and thus of DNA and RNA, thereby slowing the production of proteins. Thus the end result of sulfonamide competition with PABA is the cessation of cell metabolism, which leads to cell death.

Another antimetabolic agent, *trimethoprim,* also interferes with nucleic acid synthesis. However, instead of binding to the enzyme that converts PABA to dihydrofolic acid, trimethoprim binds to the enzyme involved in the conversion of dihydrofolic acid to THF, the next step in this metabolic pathway.

Note that humans do not synthesize THF from PABA, as some bacteria and protozoa do; instead, we take simple folic acids found in our diets and convert them into THF. As a result, human metabolism is unaffected by sulfonamides.

Antiviral agents can also target the unique aspects of the metabolism of viruses. After attachment to a host cell, many viruses must penetrate the cell's membrane and be uncoated to release viral genetic instructions and assume control of the cell's metabolic machinery. Some viruses of eukaryotes are uncoated as a result of the acidic environment within phagolysosomes. *Amantadine, rimantadine,* and weak organic bases can neutralize the acid of phagolysosomes and thereby prevent viral uncoating; thus, these are antiviral drugs. Amantadine is used exclusively to prevent infections by influenza type A virus.

CRITICAL THINKING

It would be impractical and expensive for every American to take amantadine during the entire flu season to prevent influenza infections. For what group of people might amantadine prophylaxis be cost effective?

Inhibition of Nucleic Acid Synthesis

As we saw in Chapter 7, the nucleic acids DNA and RNA are built from purine and pyrimidine nucleotides and are critical to the survival of cells. Several drugs function by blocking either the replication of DNA or its transcription into RNA.

Because only slight differences exist between the DNA of prokaryotes and eukaryotes, drugs that affect DNA replication often act against both types of cells. For example, *actinomycin* binds to DNA and effectively blocks DNA synthesis and RNA transcription not only in bacterial pathogens, but in their hosts as well. Generally, drugs of this kind are not used to treat infections, though they are used in research of DNA replication and may be used judiciously to slow replication of cancer cells.

Other compounds that can act as antimicrobials by interfering with the function of nucleic acids are called **nucleotide analogs** because of the compounds' structural similarities to the normal nucleotide building blocks of nucleic acids (**Figure 10.7** on page 296). The structures of certain nucleotide analogs enable them to be incorporated into the DNA or RNA of pathogens, where they distort the shapes of the nucleic acid molecules and prevent further replication, transcription, or translation. Nucleotide analogs are most often used against viruses because viral DNA polymerases are tens to hundreds of times more likely to incorporate nonfunctional nucleotides into nucleic acids than are host polymerase enzymes. Additionally, viral nucleic acid synthesis is usually more rapid than nucleic acid synthesis in cells. These characteristics make viruses more susceptible to nucleotide analogs than their hosts are, though nucleotide analogs are also effective against rapidly dividing cancer cells.

The synthetic drugs called *quinolones* and *fluoroquinolones* are unusual because they are active against prokaryotic DNA specifically. These antibacterial agents inhibit *DNA gyrase,* an enzyme necessary for correct coiling and uncoiling of replicating bacterial DNA; they typically have little effect on eukaryotes or viruses. See **Highlight 10.2** on page 297 for a closer look at ciprofloxacin (sip-rō-floks'ă-sin), a quinolone drug that is prescribed for respiratory and urinary tract infections as well as for suspected cases of anthrax.

Other antimicrobial agents function by binding to and inhibiting the action of RNA polymerases during the synthesis of RNA from a DNA template. Several drugs, including *rifampin* (rif'am-pin), bind more readily to prokaryotic RNA polymerase than to eukaryotic RNA polymerase; as a result, rifampin is more toxic to prokaryotes than to their eukaryotic hosts. Rifampin is used primarily against *Mycobacterium tuberculosis* and other pathogens that metabolize slowly and thus are less susceptible to antimicrobials targeting active metabolic processes and protein synthesis.

Clofazimine binds to the DNA of *Mycobacterium leprae,* the causative agent of leprosy, and prevents normal replication and transcription. It is also used to treat tuberculosis and other mycobacterial infections.

Prevention of Virus Attachment

Many pathogens, particularly viruses, must attach to their host's cells via the chemical interaction between attachment proteins on the pathogen and complementary receptor proteins on a host cell. Attachment of viruses can be blocked by peptide and sugar analogs of either attachment or receptor proteins. When these sites are blocked by analogs, viruses can neither attach to nor enter their host cells. The use of such substances, called *attachment antagonists,* is still in the developmental stage. For example, small peptide chains from HIV attachment proteins may prove effective at preventing infection by HIV. *Arildone,* an antagonist of the receptor of polioviruses and some cold viruses, blocks attachment of these viruses and is being evaluated as a deterrent to infections.

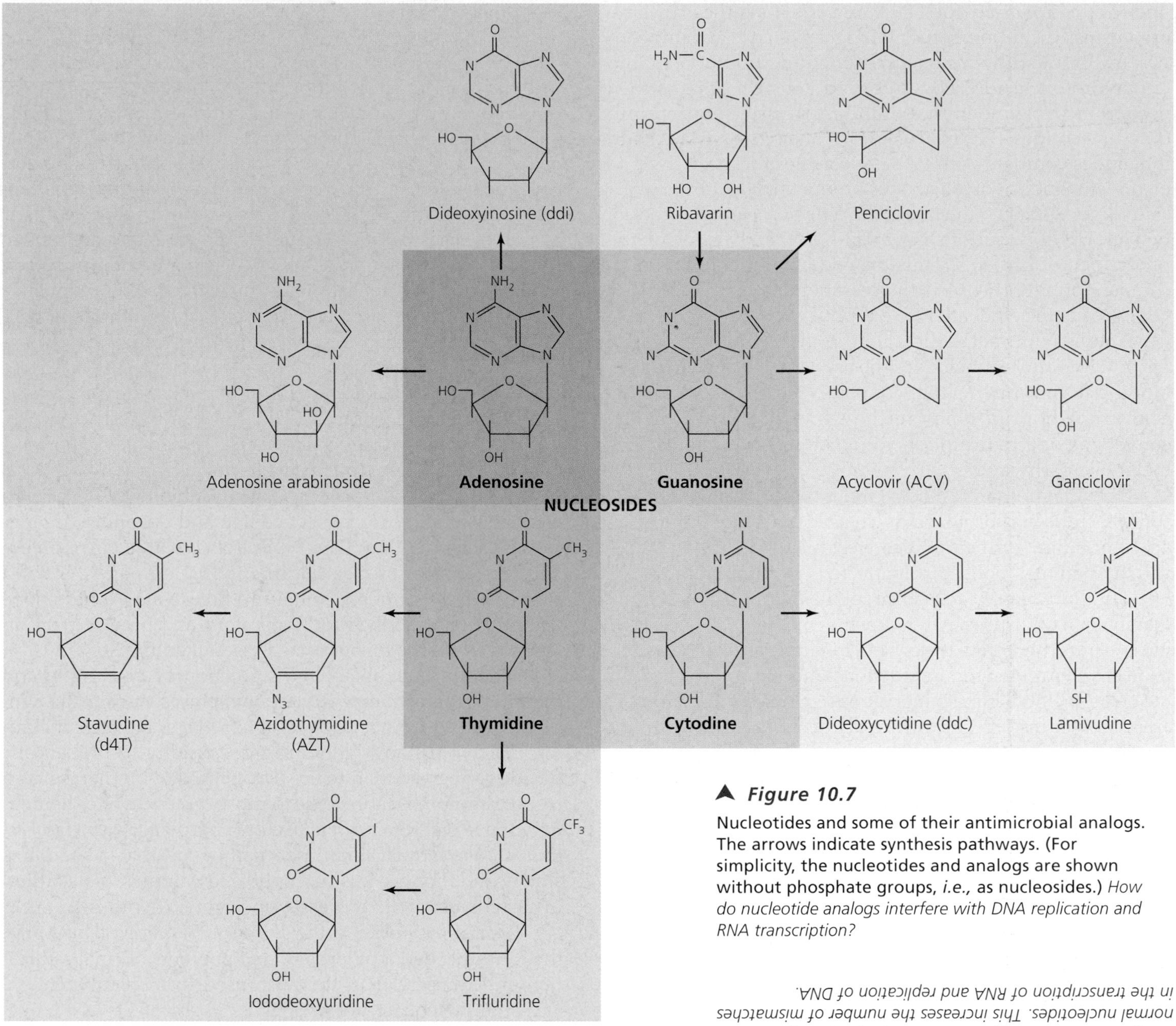

▲ ***Figure 10.7***

Nucleotides and some of their antimicrobial analogs. The arrows indicate synthesis pathways. (For simplicity, the nucleotides and analogs are shown without phosphate groups, *i.e.*, as nucleosides.) *How do nucleotide analogs interfere with DNA replication and RNA transcription?*

Figure 10.7 Because of the distortions they cause in nucleic acids, nucleotide analogs do not form proper base pairs with normal nucleotides. This increases the number of mismatches in the transcription of RNA and replication of DNA.

Clinical Considerations in Prescribing Antimicrobial Drugs

Even though antibiotics are produced commonly by many fungi and bacteria, most of these chemicals are not effective for treating diseases because they are toxic to humans and animals, are too expensive, are produced in minute quantities, or lack potency. The ideal antimicrobial agent to treat an infection or disease would be one that is:

- readily available
- inexpensive
- chemically stable (so that it can be transported easily and stored for long periods of time)
- easily administered
- nontoxic and nonallergenic
- selectively toxic against a wide range of pathogens

No antimicrobial agent has all of these qualities, so doctors and medical laboratory technicians must evaluate antimicrobials with respect to several characteristics: the range of pathogens against which they are effective, called their spectrum of action; their efficacy, including the dosages required to be effective; the routes they can be administered; and their overall safety, and the side effects they produce. We consider

Highlight 10.2 Ciprofloxacin and Anthrax

Ciprofloxacin hydroxide (commercially known as "Cipro") is a synthetic antibiotic used to treat a variety of bacterial infections. In the fall of 2001, following the anthrax attacks through the U.S. mail, Cipro made headlines for its efficacy against *Bacillus anthracis*, the bacterium that causes anthrax. Thousands of postal workers were given Cipro as a preventative measure against inhalational anthrax, and the U.S. government bought 100 million Cipro tablets from a drug manufacturer for an emergency stockpile. Demand for the drug surged as frightened consumers sought Cipro in doctor's offices, pharmacies, and over the Internet, spurring concerns that the drug was being oversold and misused.

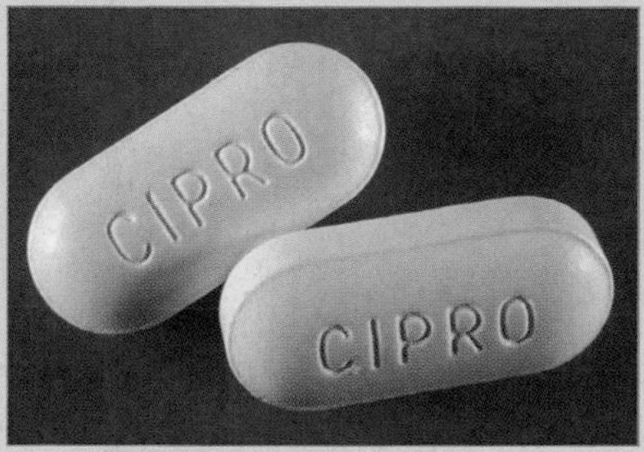

Cipro works by inhibiting the DNA synthesis of bacteria. Specifically, it targets an enzyme called DNA gyrase (topoisomerase II), which is responsible for the coiling and uncoiling of bacterial DNA. Cipro thus interferes with the ability of the bacterial cell to wind and unwind its DNA, which consequently affects the cell's ability to reproduce. It is not the only drug that has proven effective against *B. anthracis*—other antibiotics, including penicillin, can also work—but because it is a relatively new drug, bioterrorists are less likely to have developed strains of *B. anthracis* that can resist it.

Cipro is a powerful drug that should never be taken without the close supervision of a physician. Its side effects can be serious and include vomiting, diarrhea, headaches, dizziness, and central nervous system problems. Indeed, the Centers for Disease Control and Prevention discourages its use by anyone who has not been exposed to anthrax or who has not had the drug prescribed for a medical condition. Random, widespread use of Cipro is not only potentially harmful to individuals; it can also pose public health problems by increasing the chances that a drug-resistant form of *B. anthracis* will develop.

each of these characteristics of microbial agents in the following sections.

Spectrum of Action

Learning Objective

✓ Distinguish between narrow-spectrum drugs and broad-spectrum drugs in terms of their targets and side effects.

The number of different kinds of pathogens a drug acts against is known as its **spectrum of action** (**Figure 10.8** on page 298); drugs that work against only a few kinds of pathogens are **narrow-spectrum drugs,** whereas those that are effective against many different kinds of pathogens are **broad-spectrum drugs.** For instance, because erythromycin acts against Gram-negative bacteria, Gram-positive bacteria, and chlamydias and rickettsias, it is considered a broad-spectrum antibiotic. In contrast, penicillin cannot easily penetrate the outer membranes of Gram-negative bacteria to reach and prevent the formation of their peptidoglycan cell walls, so its efficacy is largely limited to Gram-positive bacteria; thus penicillin has a narrower spectrum of action than erythromycin, which acts against protein synthesis.

The use of broad-spectrum antimicrobials is not always as desirable as it might seem. Besides killing pathogens, broad-spectrum antimicrobials can also open the door to serious secondary infections by transient pathogens or *superinfections* by members of the normal microbiota unaffected by the antimicrobial. This results because the killing of normal microbiota reduces *microbial antagonism,* the competition between normal microbes and pathogens for nutrients and space that reinforces the body's defense by limiting the ability of pathogens to colonize the skin and mucous membranes. Thus a woman using erythromycin to treat strep throat (a bacterial disease) could develop vaginitis resulting from the excessive growth of *Candida albicans* (kan'did-ă al'bi-kanz), a yeast that is unaffected by erythromycin and is freed from microbial antagonism when an antibiotic kills normal bacteria in the vagina.

Efficacy

Learning Objective

✓ Compare and contrast *Kirby-Bauer, Etest,® MIC, and MBC tests.*

To effectively treat infectious diseases, physicians must know which antimicrobial agent is most effective against a particular pathogen. To ascertain the efficacy of antimicrobials, microbiologists conduct a variety of tests, including diffusion susceptibility tests, the minimum inhibitory concentration test, and the minimum bacteriocidal concentration test.

Diffusion Susceptibility Test

Diffusion susceptibility tests, also known as *Kirby-Bauer tests,* are a variation of the *disk-diffusion method* of determining the effectiveness of disinfectants discussed in Chapter 9 (p. 267). Diffusion susceptibility tests, which are simple, inexpensive, and widely used, involve uniformly inoculating a Petri plate with a standardized amount of the pathogen in question. Then small disks of paper containing standard concentrations of the drugs to be tested are firmly arranged on the surface of the plate. The plate is incubated, and the

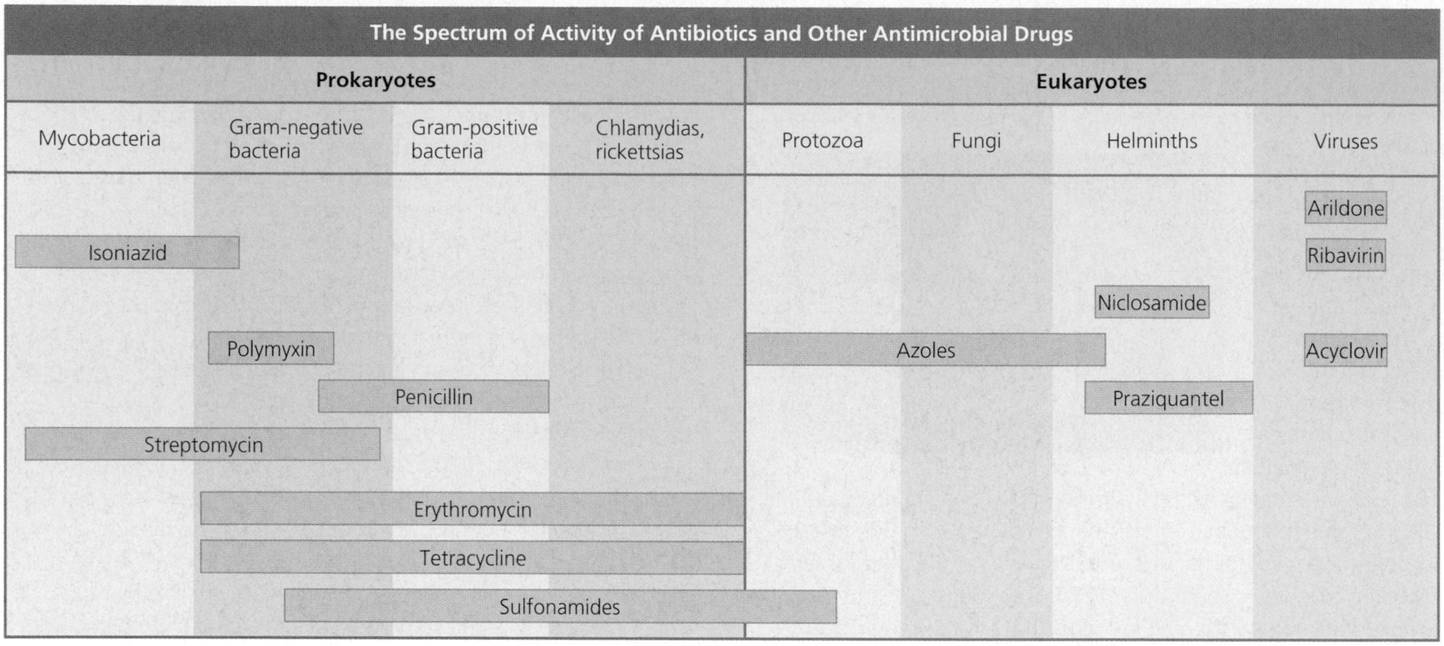

▲ ***Figure 10.8***

Spectrum of action for selected antimicrobial agents. The more kinds of pathogens a drug affects, the broader its spectrum of action.

bacteria grow and reproduce everywhere but the areas where an effective antimicrobial drug diffuses through the agar. After incubation, the plates are examined for the presence of a **zone of inhibition**—that is, a clear area surrounding the disk where the microbe does not grow **(Figure 10.9).** A zone of inhibition is measured as the diameter (to the closest millimeter) of the clear region.

Generally, the larger the zone of inhibition, the more effective that drug is, though the size of the zone also depends on the rate of diffusion of the antimicrobial agent—lower-molecular-weight drugs generally diffuse more quickly than those with higher molecular weights. The size of a zone of inhibition can be compared to a standard table for that particular drug. Diffusion susceptibility tests enable scientists to classify pathogens as *susceptible, intermediate,* or *resistant* to each drug.

CRITICAL THINKING

Sometimes it is not possible to conduct a Kirby-Bauer or other susceptibility test, either because of a lack of time or an inability to access the bacteria (in an inner ear infection, for instance). How could a physician select an appropriate therapeutic agent in such cases?

Minimum Inhibitory Concentration (MIC) Test

Once scientists identify an effective antimicrobial agent, they quantitatively express its potency as a **minimum inhibitory concentration (MIC).** As the name suggests, the MIC is the smallest amount of the drug that will inhibit growth and reproduction of the pathogen. The MIC

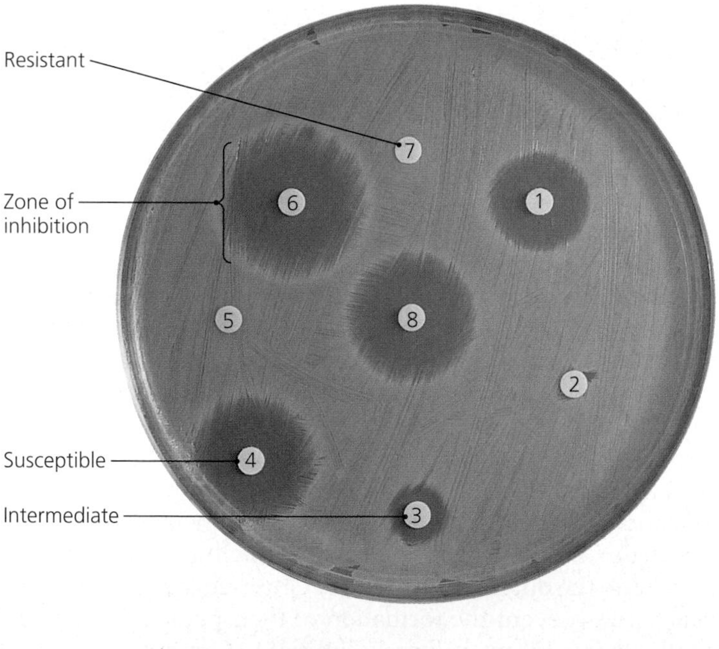

▲ ***Figure 10.9***

Zones of inhibition in a diffusion susceptibility (Kirby-Bauer) test. In general, the larger the zone of inhibition around disks, which are impregnated with an antimicrobial agent, the more effective that antimicrobial is against the organism growing on the plate. The organism is classified as either susceptible, intermediate, or resistant to the antimicrobials tested based on the sizes of the zones of inhibition. *If all eight of these antimicrobial agents diffuse at the same rate and are equally safe and easily administered, which one would be the "drug of choice" for treating this pathogen?*

Figure 10.9 *Drug 6 has the largest zone of inhibition, so all other things being equal, it is the drug of choice.*

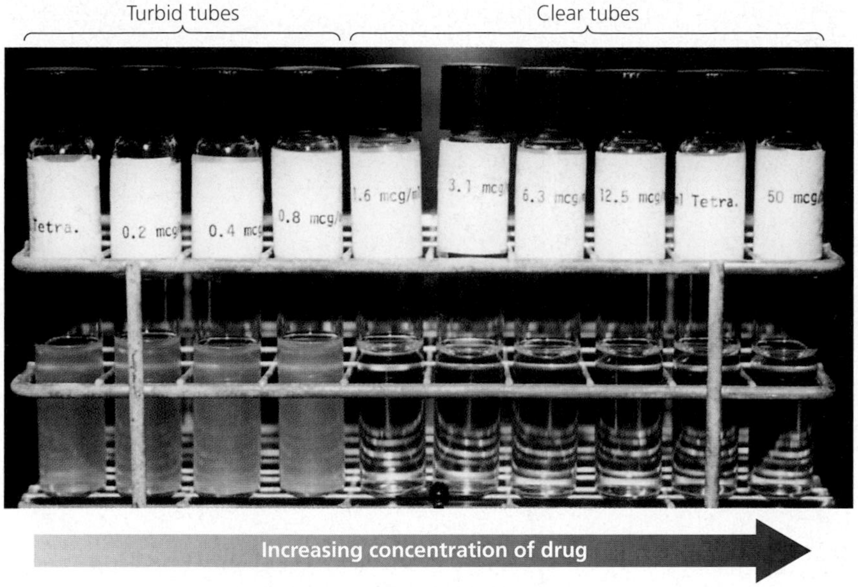

Figure 10.10

Minimum inhibitory concentration (MIC) test in test tubes. *What is the MIC for the drug acting against this bacterium?*

Figure 10.10 The MIC is 1.6 mcg/ml.

is often determined via a **broth dilution test,** in which a standardized amount of bacteria is added to serial dilutions of antimicrobial agents in tubes or wells containing broth. After incubation, turbidity (cloudiness) indicates bacterial growth; lack of turbidity indicates that the bacteria were either inhibited or killed by the antimicrobial agent **(Figure 10.10).**

There are many benefits of dilution tests over diffusion tests. For instance, many dilution tests can be conducted simultaneously in wells, and the entire process can be automated, with turbidity measured by special scanners connected to computers.

Another test that determines minimum inhibitory concentration combines aspects of an MIC test and a diffusion susceptibility test. This test, called an **Etest,**®[2] involves placing a plastic strip containing a gradient of the antimicrobial agent being tested on a plate uniformly inoculated with the organism of interest **(Figure 10.11).** After incubation, an elliptical zone of inhibition indicates antimicrobial activity, and the minimum inhibitory concentration can be noted where the zone of inhibition intersects a scale printed on the strip. The advantage of an Etest® over a standard diffusion susceptibility test is that only a single plate need be inoculated to determine both susceptibility and MIC.

Figure 10.11

An Etest,® which combines aspects of Kirby-Bauer and MIC tests. The plastic strip contains a gradient of the antimicrobial agent of interest. The MIC is estimated to be the concentration printed on the strip where the zone of inhibition intersects the strip. In this example, the MIC is 1.5 μg/ml.

Minimum Bactericidal Concentration (MBC) Test

An extension of the MIC test is a **minimum bactericidal concentration (MBC) test.** In an MBC test, samples taken from clear MIC tubes (or alternatively, from zones of inhibition from a series of diffusion susceptibility tests) are transferred to plates containing a drug-free growth medium (**Figure 10.12** on page 300). The appearance of bacterial growth in these subcultures after appropriate incubation indicates that at least some bacterial cells survived that concentration of the antimicrobial drug and were able to grow and multiply once placed in a drug-free medium. Any drug concentration at which growth occurs in subculture is *bacteriostatic,* not *bactericidal,* for that bacterium. The lowest

[2]The name Etest® has no specific origin.

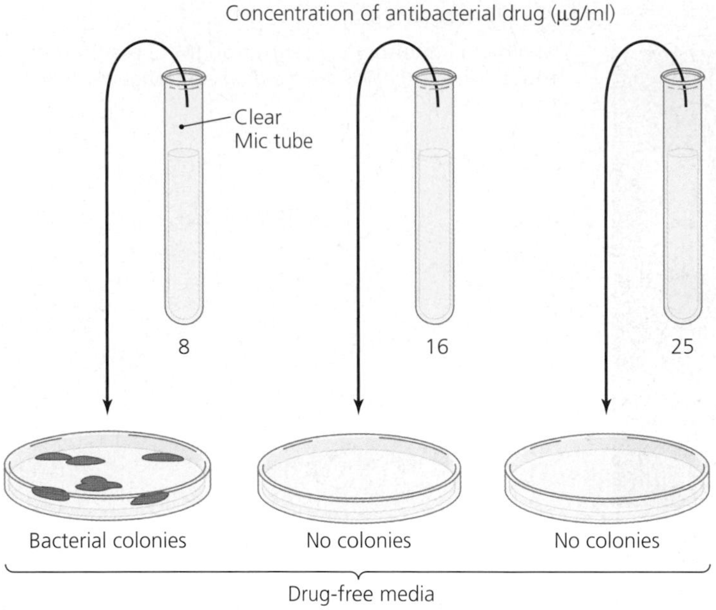

Figure 10.12

A minimum bactericidal concentration (MBC) test. In this test, plates containing a drug-free growth medium are inoculated with samples taken from zones of inhibition or from clear MIC tubes. After incubation, growth of bacterial colonies on a plate indicates that that concentration of antimicrobial drug (in this case, 8 μg/ml) is bacteriostatic. The lowest concentration for which no bacterial growth occurs on the plate is the minimum bactericidal concentration; in this case, the MBC is 16 μg/ml.

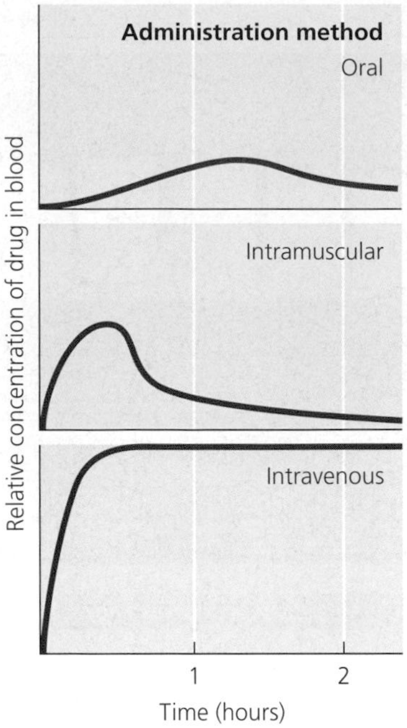

Figure 10.13

The effect of route of administration on blood levels of a chemotherapeutic agent. Although intravenous (IV) and intramuscular (IM) administration achieve higher drug concentrations in the blood, oral administration has the advantage of simplicity.

concentration of drug for which no growth occurs in the subcultures is the minimum bactericidal concentration (MBC).

Routes of Administration

Learning Objective

✓ Discuss the advantages and disadvantages of the different routes of administration of antimicrobial drugs.

An adequate amount of an antimicrobial agent must reach a site of infection if it is to be effective. For external infections such as athlete's foot, drugs can be applied directly. This is known as *topical* or *local* administration. For internal infections, drugs can be administered orally, *intramuscularly (IM)*, or *intravenously (IV)*. Each route has advantages and disadvantages.

Even though the oral route is simplest (it requires no needles and is self-administered), the drug concentrations it achieves are lower than occur via other routes of administration **(Figure 10.13).** Further, because patients need not rely on a health care provider to administer the drug properly, they do not always follow prescribed timetables for taking oral medications.

IM administration via a hypodermic needle allows a drug to diffuse slowly into the many blood vessels within muscle tissue, but the concentration of the drug in the blood is never as high as that achieved by IV administration, which delivers the drug directly into the bloodstream through either a needle or a catheter (a plastic or rubber tube). Even though the amount of the drug in the blood is initially very high for the IV route, the concentration rapidly diminishes as the liver and kidneys remove the drug from the circulation, unless the drug is continuously administered.

In addition to route of administration, physicians must consider how antimicrobial agents will be distributed to infected tissues by the blood. For example, an agent removed rapidly from the blood by the kidneys might be the drug of choice for a bladder infection but would not be chosen to treat an infection of the heart. Finally, given that blood vessels in the brain, spinal cord, and eye are almost impermeable to many antimicrobial agents (because of the so-called blood-brain barrier), infections in these organs are often difficult to treat.

Safety and Side Effects

Learning Objective

✓ Identify three main categories of side effects of antimicrobial therapy.

Another aspect of chemotherapy that physicians must consider is the possibility of side effects. These fall into three main categories—toxicity, allergies, and disruption of normal microbiota—though some side effects are benign.

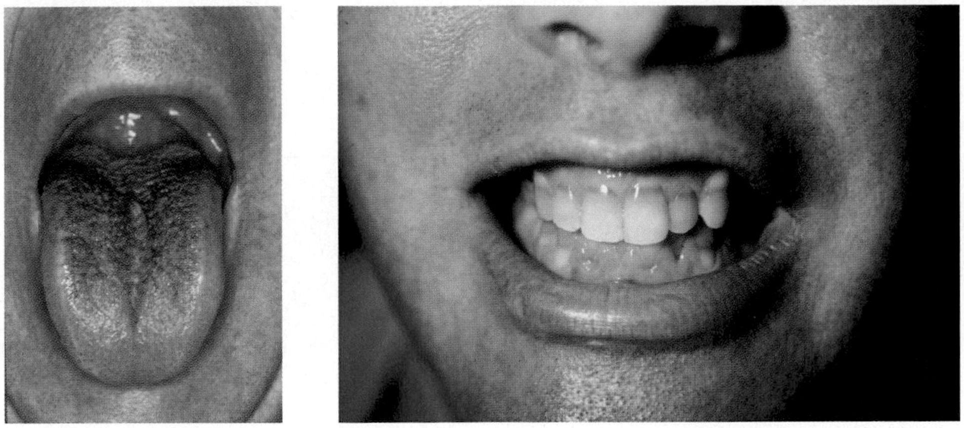

(a) (b)

Figure 10.14

Some side effects resulting from toxicity of antimicrobial agents. **(a)** "Black hairy tongue," caused by the antiprotozoan drug metronidazole (Flagyl®). **(b)** Discoloration and damage of tooth enamel caused by tetracycline. *Who should avoid taking tetracycline?*

Figure 10.14 Pregnant women and children should not use tetracycline.

Toxicity

Though antimicrobial drugs are ideally selectively toxic against microbes and harmless to humans, many in fact have toxic side effects. The exact cause of many adverse reactions is poorly understood, but drugs may be toxic to the kidneys, the liver, or nerves. For example, polymyxin and aminoglycosides can be fatally toxic to kidneys. Not all toxic side effects are so serious. *Metronidazole (Flagyl®),* an antiprotozoan drug, may cause a harmless temporary condition called "black hairy tongue," which results when the breakdown products of hemoglobin accumulate in the papillae of the tongue **(Figure 10.14a).**

Doctors must be especially careful when prescribing drugs for pregnant women, as many drugs that are safe for adults can have adverse affects when absorbed by a fetus. For instance, tetracyclines form complexes with calcium that can become incorporated into bones and developing teeth, causing malformation of the skull and staining of tooth enamel **(Figure 10.14b).**

Allergies

In addition to toxicity, some drugs trigger allergic immune responses in sensitive patients. Although relatively rare, such reactions may be life threatening, especially in an immediate, violent reaction called *anaphylactic shock*. For example, about 0.1% of Americans have an anaphylactic reaction to penicillin, which results in approximately 300 deaths per year. However, not every allergy to an antimicrobial agent is so serious. Recent studies indicate that patients with mild allergies to penicillin frequently lose their sensitivity to it over time. Thus, an initial mild reaction to penicillin need not preclude its use in treating future infections. Chapter 18 discusses allergies in more detail.

Disruption of Normal Microbiota

As we have seen, drugs that disrupt normal microbiota and their microbial antagonism of opportunistic pathogens may result in secondary infections. In instances when a member of the normal microbiota is not affected by a drug, it can overgrow, causing a superinfection. For example, long-term use of broad-spectrum antimicrobials often results in explosions in the growth rate of *Candida albicans* in the vagina (vaginitis) or mouth (thrush), and the multiplication of *Clostridium difficile* (klos-trid′ē-ŭm dif′fi-sil) in the colon that causes a potentially fatal condition called *pseudomembranous colitis*. Such secondary infections are of greatest concern for hospitalized patients, who are often not only debilitated, but also more likely to be exposed to pathogens with resistance to antimicrobial drugs—the topic of the next section.

Resistance to Antimicrobial Drugs

Among the major challenges facing microbiologists today are the problems presented by pathogens that are resistant to antimicrobial agents. In the sections that follow we will examine the development of resistant populations of pathogens, the mechanisms by which pathogens are resistant to antimicrobials, and some ways that resistance can be retarded.

The Development of Resistance in Populations

Learning Objectives

- ✓ Describe how populations of resistant microbes can arise.
- ✓ Describe the relationship between R-plasmids and resistant cells.

Not all pathogens are equally sensitive to a given therapeutic agent; a population may contain a few organisms that are

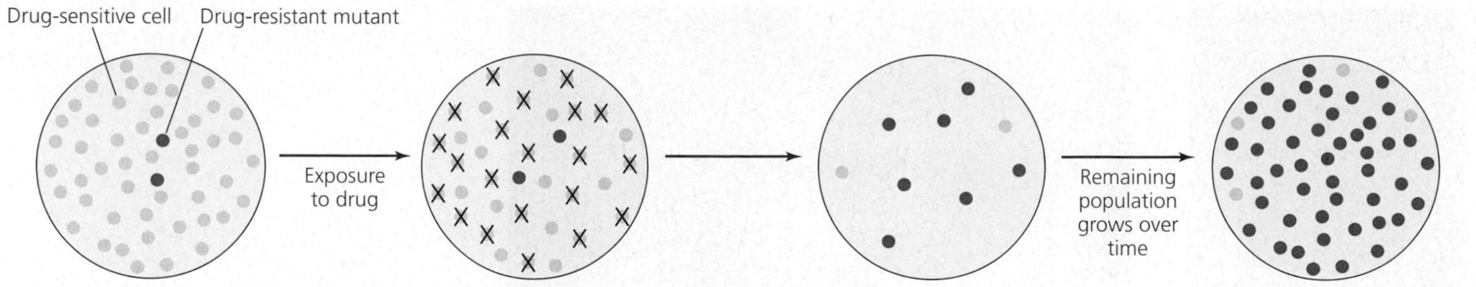

(a) Population of microbial cells **(b)** Sensitive cells inhibited by exposure to drug **(c)** Most cells are now resistant

▲ ***Figure 10.15***

The development of a resistant strain of bacteria. **(a)** A bacterial population contains both drug-sensitive and drug-resistant cells, although sensitive cells constitute the vast majority of the population. **(b)** Exposure to an antimicrobial drug inhibits the sensitive cells; so long as the drug is present, reduced competition from sensitive cells facilitates the multiplication of resistant cells. **(c)** Eventually resistant cells constitute the majority of the population. *Why do resistant strains of bacteria more often develop in hospitals and nursing homes than in college dormitories?*

Figure 10.15 Resistant strains are more likely to develop in hospitals and other health care facilities because the extensive use of antimicrobial agents in those places inhibits the growth of sensitive strains and selects for the growth of resistant strains.

naturally either partially or completely resistant. Among bacteria, individual cells can acquire such resistance in two ways: through new mutations of chromosomal genes, or by acquiring resistance genes on extrachromosomal pieces of DNA called **R-plasmids** (or *R-factors*) via the processes of transformation, transduction, or conjugation (see Chapter 7). We focus here on the development of resistance in populations of bacteria, but resistance to antimicrobials is known to occur among viruses as well—including HIV, as discussed in **New Frontiers 10.1.**

The process by which a resistant strain of bacteria develops is depicted in **Figure 10.15.** In the absence of an antimicrobial drug, resistant cells are usually less efficient than their normal neighbors because they must expend extra energy to maintain resistance genes and proteins. Under these circumstances, resistant cells remain the minority in a population because they reproduce more slowly. However, when an antimicrobial agent is present, the majority of cells (which are sensitive to the antimicrobial) are inhibited or die while the resistant cells continue to grow and multiply, often more rapidly because they then face less competition. The result is that resistant cells soon replace the sensitive cells as the majority in the population. It should be noted that the presence of the chemotherapeutic agent does not *produce* resistance, but instead facilitates the replication of resistant cells that were already present in the population.

CRITICAL THINKING

Enterococcus faecium (en'ter-ō-kok'ŭs fē-sē-um) is frequently resistant to vancomycin. Why might this be of concern in a hospital setting in terms of developing resistant strains of *other* genera of bacteria?

Mechanisms of Resistance

Learning Objective

✓ List five ways by which microorganisms can be resistant to antimicrobial drugs.

Pathogens are resistant to antimicrobial drugs by one of five types of mechanisms:

1. Resistant bacteria may produce an enzyme that destroys or deactivates the drug. This common mode of resistance is exemplified by **β-lactamases** (penicillinases), enzymes that break the β-lactam rings of penicillin and similar molecules, rendering them inactive **(Figure 10.16).** Over 200 different lactamases have been identified. Frequently their genes are located on R-plasmids.
2. Resistant pathogens may slow or prevent the entry of the drug into the cell. This mechanism typically involves changes in the structure or electrical charge of the cytoplasmic membrane proteins that constitute channels or pores. Such proteins in the outer membranes of Gram-negative bacteria are called *porins* (see Figure 3.13b). Altered pore proteins result from mutations in chromosomal genes. Resistance against tetracycline and penicillin are known to occur via this mechanism.
3. Resistant cells may alter the receptor for the drug so that it either cannot attach to or binds less effectively to its target. This form of resistance is often seen against antimetabolites (such as sulfonamides) and against drugs (such as erythromycin) that thwart protein translation.

New Frontiers 10.1 The AIDS "Cocktail" and the Problem of Drug Resistance

One of the most important weapons in the fight against AIDS is a drug regimen known as the "AIDS cocktail." This "cocktail" is a combination of three drugs: a *protease inhibitor* and two kinds of *reverse transcriptase inhibitors*, including zidovudin (commonly known as AZT). Reverse transcriptase inhibitors interfere with enzymes that HIV (the virus that causes AIDS) needs very early in its replication cycle. The protease inhibitor, which revolutionized AIDS treatment upon its introduction in 1995, inhibits an enzyme that HIV needs near the end of its replication cycle. Between 1995 and 2000, this potent "cocktail" of drugs effectively reduced the death rate of HIV patients by more than 50%.

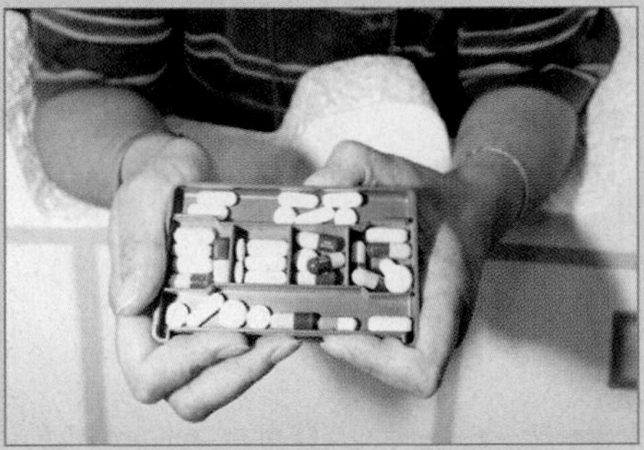

The "cocktail" has its side effects, including anemia, muscle weakness, pneumonia, and bronchitis. It is also an expensive and challenging regimen to maintain; some drug combinations require patients to take up to 35 pills each day. The biggest problem, however, is the emergence of strains of HIV that are resistant to even the powerful drugs that make up the "cocktail." According to a recent study, the number of new patients infected with such drug-resistant strains of HIV is rising—particularly in areas such as San Francisco, where the AIDS "cocktail" has been widely used for many years and the virus has had a greater opportunity to develop resistance.

As a result of the emergence of resistant strains of HIV, researchers are hard at work exploring new anti-HIV agents, including ones that can interfere with other stages of the HIV life cycle. Thus, whereas existing drugs focus on HIV's replication stage, which occurs after the virus has already invaded a cell, new drugs may include "entry inhibitors" that focus on preventing HIV from invading cells in the first place. Early studies of such agents are showing promise, fueling hopes that a new generation of AIDS drugs is about to enter the fray.

References: Stephenson, J. 2002. Researchers Explore New Anti-HIV Agents. *Journal of the American Medical Association* 287:1635–1637.

Trachtenberg, J. D., M. A. Sande. 2002. Emerging Resistance to Nonnucleoside Reverse Transcriptase Inhibitors: A Warning and a Challenge. *Journal of the American Medical Association* 288:239.

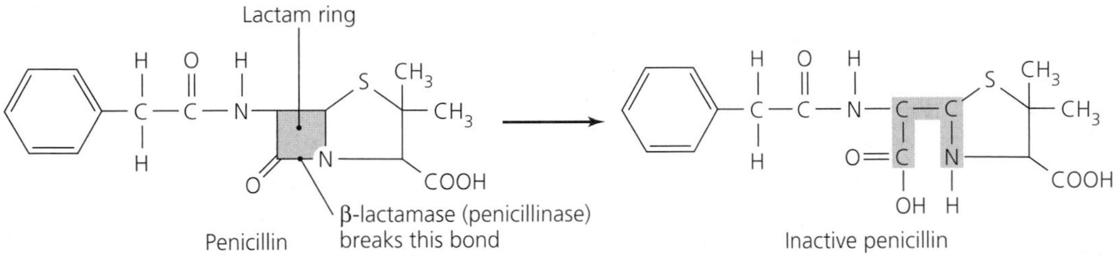

▲ ***Figure 10.16***
How β-lactamase (penicillinase) renders penicillin inactive. The enzyme acts by breaking a bond in the β-lactam ring, the functional portion of the drug.

4. Resistant cells may alter their metabolic chemistry, or they may abandon the sensitive metabolic step altogether. For example, a cell may become resistant to a drug by producing more enzyme molecules for the affected metabolic pathway, effectively reducing the power of the drug. Alternatively, cells become resistant to sulfonamides by abandoning the synthesis of folic acid, absorbing it from the environment instead.
5. Resistant cells may pump the drug out of the cell before the drug can act.

Multiple Resistance and Cross Resistance

Learning Objective

✓ Define cross resistance, and distinguish it from multiple resistance.

A given pathogen can acquire resistance to more than one drug at a time, especially when resistance is conferred by R-plasmids, which are exchanged among bacterial cells. Such multiresistant strains of bacteria frequently develop in hospitals and nursing homes, where the constant use of many kinds of antimicrobial agents eliminates sensitive cells and encourages the development of resistant strains.

Pathogens that are resistant to most antimicrobial agents are sometimes called *superbugs*. Superbug strains of *Staphylococcus, Streptococcus* (strep-tō-kok'ŭs), *Enterococcus* (en'ter-ō-kok'ŭs), *Pseudomonas,* and *Plasmodium* (plaz-mō'dē-ŭm), (the protozoan that causes malaria) pose unique problems for health care professionals, who must treat infected patients without effective antimicrobials while taking extra care to protect themselves and other patients from infection.

Resistance to one antimicrobial agent may confer resistance to similar drugs, a phenomenon called **cross resistance.** Cross resistance typically occurs when drugs are similar in structure. For example, resistance to one aminoglycoside drug, such as streptomycin, may confer resistance to similar aminoglycoside drugs.

Retarding Resistance

Learning Objective

✓ Describe four ways that development of resistance can be retarded.

The development of resistant populations of pathogens can be averted in at least four ways. First, sufficiently high concentrations of the drug can be maintained in a patient's body for a long enough time to kill all sensitive cells and inhibit others long enough for the body's defenses to defeat them. Discontinuing a drug before all of the pathogens have been neutralized promotes the development of resistant strains. For this reason, it is important that patients finish their entire antimicrobial prescription and resist the temptation to "save some for another day."

A second way to avert resistance is to use antimicrobial agents in combination so that pathogens resistant to one drug will be killed by the second, and vice versa. Additionally, one drug sometimes enhances the effect of a second drug in a process called **synergism** (sin'er-jizm) **(Figure 10.17).** In one example of a synergistic drug combination, the inhibition of cell wall formation by penicillin makes it easier for streptomycin molecules to enter bacteria and interfere with protein synthesis. Synergism can also result from combining an antimicrobial drug and a chemical, as occurs when *clavulanic acid* enhances the effect of penicillin by deactivating β-lactamase.

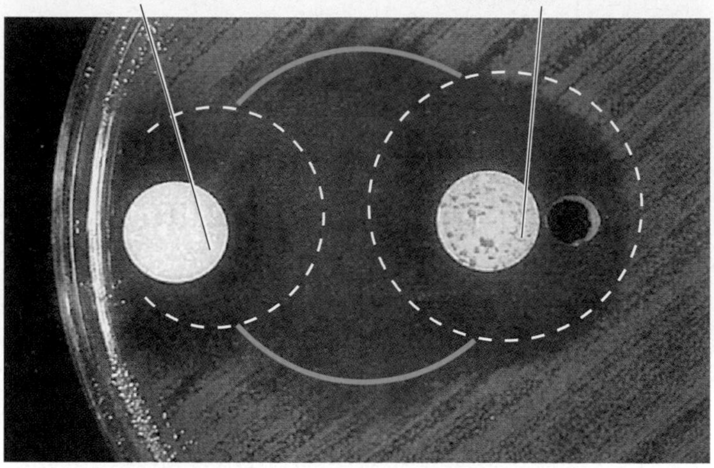

▲ ***Figure 10.17***

An example of synergism between two antimicrobial agents. The portion of the zone of inhibition outlined in green represents the synergistic enhancement of antimicrobial activity beyond the activities of the individual drugs (outlined in white).

Not all drugs act synergistically; some combinations of drugs can be *antagonistic*—interfering with each other. For example, drugs that slow bacterial growth are antagonistic to the action of penicillin, which acts only against growing and dividing cells.

A third way to reduce the development of resistance is to limit the use of antimicrobials to necessary cases. Unfortunately, many antimicrobial agents are used indiscriminately, in both developed countries and in less-developed regions where many are available without a physician's prescription. In the United States, an estimated 50% of prescriptions for antibacterial agents to treat sore throats, and 30% of prescriptions for ear infections, are inappropriate because the diseases are viral in nature. Likewise, because antibacterial drugs have no effect on cold and flu viruses, 100% of antibacterial prescriptions for treating these diseases are superfluous. As discussed previously, the use of antimicrobial agents encourages the reproduction of resistant bacteria by limiting the growth of sensitive cells; therefore, inappropriate use of such drugs increases the likelihood that resistant strains of bacteria will multiply.

Finally, scientists can combat resistant strains by developing new variations of existing drugs, in some cases by adding novel side chains to the original molecule. In this way, scientists develop semisynthetic *second-generation* drugs. If resistance develops to these drugs, *third-generation* drugs may be developed to replace them. Many health care professionals and research scientists are concerned about how long drug developers can stay ahead of the development of resistance by pathogens.

Selected antimicrobial agents, their modes of action, clinical considerations, and other features are summarized in **Tables 10.2** through **10.6.** Particular use of antimicrobial drugs against specific pathogens is covered in the relevant chapters of this book.

Table 10.2 Antibacterial Drugs

Drug	Description and Mode of Action	Clinical Considerations	Resistance
Antibacterial Drugs that Inhibit Cell Wall Synthesis			
Beta-lactams Representative natural penicillins: Penicillin G Penicillin V Representative semisynthetic penicillins: Methicillin Ampicillin Amoxicillin Representative natural cephalosporin: Cephalothin Representative semisynthetic cephalosporins: Cefaclor Cephalexin Monobactam	Large number of natural and semisynthetic derivatives from *Penicillium* (penicillins) and *Cephalosporium* (cephalosporins, *e.g.*, cephalothin); bind to, and deactivate, the enzyme that cross links the NAM subunits of peptidoglycan Monobactams have only a single ring instead of the two rings seen in other beta-lactams	**Spectrum of Action:** Natural drugs have limited action against most G− bacteria because they do not readily cross the outer membrane; semisynthetics have broader spectra of action Monobactams have a limited spectrum of action, affecting only aerobic, G− bacteria **Route of Administration:** Penicillin V, a few cephalosporins (*e.g.*, cephalexin), and monobactams: oral; penicillin G, and many semisynthetics (*e.g.*, methicillin, ampicillin, carbenicillin, cephalothin): IM or IV **Adverse Effects:** Allergic reactions against beta-lactams are seen in some adults; monobactams are least allergenic	Develops in three ways: • G− bacteria change their outer membrane structure to prevent entrance of the drug • modify the enzyme so that the drug no longer binds to it • synthesize beta-lactamases that cleave the functional lactam ring of the drug; genes for lactamases are often carried on R-plasmids
Vancomycin	Produced by *Streptomyces orientalis*; directly interferes with the formation of alanine–alanine crossbridges between NAM subunits	**Spectrum of Action:** Effective against most G+ bacteria, but generally reserved for use against strains resistant to other drugs such as methicillin-resistant *Staphylococcus aureus* (MRSA) **Route of Administration:** IV **Adverse Effects:** Damage to ears and kidneys, allergic reactions	G− bacteria are naturally resistant because the drug is too large to pass through the outer membrane; some G+ bacteria (*e.g.*, *Ersipelothrix*, *Lactobacillus*) are naturally resistant because they do not form alanine–alanine bonds between NAM subunits
Cycloserine	Analog of alanine that interferes with the formation of alanine–alanine crossbridges between NAM subunits	**Spectrum of Action:** Some G+ bacteria, mycobacteria **Route of Administration:** Oral **Adverse Effects:** Toxic to nervous system, producing depression, aggression, confusion, and headache	Some G+ bacteria are naturally resistant.
Bacitracin	Isolated from *Bacillus licheniformis* growing on a patient named Tracy; appears to have three modes of action: • interference with the movement of peptidoglycan precursors through the bacterial cell membrane to the cell wall • inhibition of RNA transcription • damage to the bacterial cytoplasmic membrane The latter two modes of action have not been proven definitely	**Spectrum of Action:** G + bacteria **Route of Administration:** Topical **Adverse Effects:** Toxic to kidneys	Resistance most often involves changes in bacterial cell membranes that prevent bacitracin from entering the cell
Isoniazid (Isonicotinic acid hydrazide, INH)	Blocks the gene for an enzyme that forms mycolic acid; analog of the vitamins nicotinamide and pyridoxine	**Spectrum of Action:** Mycobacteria, including *M. tuberculosis* and *M. leprae* **Route of Administration:** Oral **Adverse Effects:** Occasionally toxic to liver	Resistance is due to random mutations of bacterial chromosomes that result in reduced drug uptake or alteration of target sites

Table 10.2 *(continued)*

Drug	Description and Mode of Action	Clinical Considerations	Resistance
Ethambutol	Prevents the formation of mycolic acid; used in combination with other antimycobacterial drugs	**Spectrum of Action:** Mycobacteria, including *M. tuberculosis* and *M. leprae* **Route of Administration:** Oral **Adverse Effects:** None	Resistance is due to random mutations of bacterial chromosomes that result in reduced drug uptake or alteration of target sites
Antibacterial Drugs that Inhibit Protein Synthesis			
Aminoglycosides Representatives: Kanamycin Streptomycin Gentamicin Amikacin Tobramycin	Compounds in which two or more amino sugars are linked with glycosidic bonds; were originally isolated from species of the bacterial genera *Streptomyces* and *Micromonospora*; inhibit protein synthesis by irreversibly binding to the 30S subunit of prokaryotic ribosomes. This either causes the ribosome to mistranslate mRNA, producing aberrant proteins, or causes premature release of the ribosome from mRNA, which stops synthesis	**Spectrum of Action:** Broad: effective against both G+ and G− bacteria **Route of Administration:** IV; do not traverse blood-brain barrier **Adverse Effects:** May be toxic to kidneys or auditory nerves, causing deafness	Because uptake of these drugs is oxygen-dependent, anaerobic bacteria are naturally resistant; aerobic bacteria alter membrane pores to prevent uptake, or synthesize enzymes that alter or degrade the drug once it enters; rarely, bacteria alter the binding site on the ribosome
Tetracyclines Representatives: Tetracycline Doxycycline Trimocycline	Composed of four hexagonal rings with various side groups; prevent tRNA molecules, which carry amino acids, from binding to ribosomes at the 30S subunit's docking site	**Spectrum of Action:** Most are broad: effective against many G+ and G− bacteria as well as against bacteria that lack cell walls, such as *Mycoplasma* **Route of Administration:** Oral, cross poorly into brain **Adverse Effects:** Nausea, diarrhea, sensitivity to light; forms complexes with calcium, which stains developing teeth and adversely affects the strength and shape of bones	Develops in three ways. Bacteria: • alter chromosomes in gene for pores in outer membrane, preventing drug from entering cell • alter binding site on the ribosome to allow tRNA to bind even in presence of drug • actively pump drug from cell
Chloramphenicol	Rarely used drug that prevents prokaryotic ribosomes from moving along mRNA by binding to their 50S subunits	**Spectrum of Action:** Broad, but rarely used except in treatment of typhoid fever **Route of Administration:** Oral; traverses blood-brain barrier **Adverse Effects:** In 1 of 24,000 patients, causes aplastic anemia, a potentially fatal condition in which blood cells fail to form; can also cause neurologic damage	Develops via gene carried on an R-plasmid that codes for an enzyme that deactivates drug
Macrolides Representative: Erythromycin Clarithromycin Azithromycin	Group of antimicrobials typified by a macrocyclic lactone ring; the most prescribed is erythromycin, which is produced by *Streptomyces erythraeus*; act by binding to the 50S subunit of prokaryotic ribosomes and preventing the elongation of the nascent protein	**Spectrum of Action:** Effective against G+ and a few G− bacteria **Route of Administration:** Oral; do not traverse blood-brain barrier **Adverse Effects:** Nausea, mild gastrointestinal pain, vomiting	Develops via changes in ribosomal RNA that prevent drugs from binding, or via R-plasmid genes coding for the production of macrolide-digesting enzymes; resistance genes are same as those of lincosamides
Clindamycin	Binds to 50S ribosomal subunit and stops protein elongation	**Spectrum of Action:** Effective against G+ and anaerobic G− bacteria **Route of Administration:** Oral or IV; does not traverse blood-brain barrier **Adverse Effects:** Gastrointestinal distress, including nausea, diarrhea, vomiting, and pain	Develops via changes in ribosomal structure that prevent drug from binding; resistance genes are same as those of aminoglycosides

Table 10.2 *(continued)*

Drug	Description and Mode of Action	Clinical Considerations	Resistance
Antibacterial Drugs that Alter Cell Membranes			
Polymyxin	Produced by *Bacillus polymyxa*; destroy cytoplasmic membranes of susceptible cells	Spectrum of Action: Effective against G− bacteria, particularly *Pseudomonas* Route of Administration: Topical Adverse Effects: Toxic to kidneys	Results from changes in cell membrane that prohibit entrance of the drug
Antibacterial Drugs that Are Antimetabolites			
Sulfonamides	Synthetic drug; first produced as a dye; is an analog of PABA that binds irreversibly to enzyme that produces folic acid; synergistic with trimethoprim	Spectrum of Action: Broad: effective against G+ and G− bacteria and some protozoa and fungi; however, resistance is widespread Route of Administration: Oral Adverse Effects: Rare: allergic reactions, anemia, jaundice, mental retardation of fetus if administered in last trimester of pregnancy	*Pseudomonas* is naturally resistant due to permeability barriers; cells that require folic acid as a vitamin are also naturally resistant; chromosomal mutations result in lowered affinity for the drug
Trimethoprim	Blocks second metabolic step in the formation of folic acid from PABA; synergistic with sulfonamides	Spectrum of Action: Broad: effective against G+ and G− bacteria and some protozoa and fungi; however, resistance is widespread Route of Administration: Oral Adverse Effects: Allergic reactions in some patients	*Pseudomonas* is naturally resistant due to permeability barriers; cells that require folic acid as a vitamin are also naturally resistant; chromosomal mutations result in lowered affinity for the drug
Antibacterial Drugs that Inhibit Nucleic Acid Synthesis			
Quinolones Nalidixic acid Norfloxacin Ciprofloxacin	Synthetic agents that inhibit DNA gyrase, which is needed to correctly replicate bacterial DNA; penetrate cytoplasm of cells	Spectrum of Action: Broad: G+ and G− bacteria are affected Route of Administration: Oral Adverse Effects: None of major significance	Results from chromosomal mutations that lower affinity of drug or reduce uptake
Rifampin	Semisynthetic derivative of drug produced by *Streptomyces mediterranei* that binds to bacterial RNA polymerase, preventing transcription of RNA; used with other antimicrobial bacterial drugs	Spectrum of Action: Bacteriostatic against aerobic G+ bacteria; bactericidal against mycobacteria Route of Administration: Oral Adverse Effects: None of major significance	Results from chromosomal mutation that alters binding site on enzyme; G− bacteria are naturally resistant due to poor uptake
Clofazimine	Binds to mycobacterial DNA preventing replication and transcription	Spectrum of Action: Mycobacteria, especially *M. tuberculosis* and *M. leprae* Route of Administration: Oral Adverse Effects: Diarrhea, discoloration of skin and eyes	None

Table 10.3 Antiviral Drugs

Drug	Description and Mode of Action	Clinical Considerations	Resistance
Attachment Antagonists			
Arildone	Block attachment molecule on host cell or pathogen	**Spectrum of Action:** Picornaviruses (*e.g.*, poliovirus, some cold viruses) **Route of Administration:** Not approved for antiviral use by FDA **Adverse Effects:** None	None reported
Antiviral Drugs that Inhibit Viral Uncoating			
Amantadine **Rimantadine**	Neutralize acid environment within phagolysosomes that is necessary for viral uncoating	**Spectrum of Action:** Influenza A virus **Route of Administration:** Oral; rimantadine is approved for adults only **Adverse Effects:** Toxic to central nervous system; results in nervousness, irritability, insomnia, and blurred vision	None reported
Antiviral Drugs that Inhibit Nucleic Acid Synthesis			
Acyclovir (ACV) **Ganciclovir**	Phosphorylation by virally coded kinase enzyme activates the drug; inhibits DNA and RNA synthesis	**Spectrum of Action:** Viruses that code for kinase enzymes: herpes, Epstein-Barr, cytomegalovirus, varicella virus **Route of Administration:** Oral **Adverse Effects:** None	Mutations in genes for kinase enzymes may render them ineffective at drug activation
Ribavirin	Phosphorylation by virally coded kinase enzyme activates the drug; inhibits DNA and RNA synthesis; viral DNA polymerase more likely to incorporate the drugs	**Spectrum of Action:** Respiratory syncytial, hepatitis C, influenza A, measles, some hemorrhagic fever viruses **Route of Administration:** Oral, aerosol, IV **Adverse Effects:** Perhaps harmful to developing fetus	None
Adenosine arabinoside	Phosphorylation by cell-coded kinase enzyme activates the drug: inhibits DNA synthesis; viral DNA polymerase more likely to incorporate the drugs than human DNA polymerase	**Spectrum of Action:** Herpes virus **Route of Administration:** IV **Adverse Effects:** Fatal to host cells that incorporate the drug into cellular DNA	Results from mutation of viral DNA polymerase
Nucleotide analogs Representative (see also Figure 10.7): Azidothymidine (AZT)	Phosphorylation by cell-coded kinase enzyme activates these drugs: inhibits DNA synthesis; viral reverse transcriptase more likely to incorporate these drugs; used in conjunction with protease inhibitor	**Spectrum of Action:** HIV **Route of Administration:** Oral **Adverse Effects:** Nausea, bone marrow toxicity	Results from mutation of viral reverse transcriptase
Antiviral Drugs that Inhibit Viral Proteins			
Protease inhibitors	Computer-assisted modeling of protease enzyme, which is unique to HIV, allowed the creation of drugs that block the active site; used in conjunction with drugs active against nucleic acid synthesis	**Spectrum of Action:** HIV **Route of Administration:** Oral **Adverse Effects:** None	Result from mutation of protease gene

Table 10.4 Antifungal Drugs

Drug	Description and Mode of Action	Clinical Considerations	Resistance
Antifungal Drugs that Inhibit Cell Membrane			
Polyenes Representatives: Amphotericin B Nystatin	Associate with molecules of ergosterol, forming a pore through the fungal membrane, which leads to leakage of essential ions from the cell; amphotericin B is produced by *Streptomyces nodosus*	**Spectrum of Action:** Fungi **Route of Administration:** Amphotericin B: IV; nystatin: topical **Adverse Effects:** Chills, vomiting, fever	None
Azoles Representatives: Miconazole Ketoconazole	Antifungal action due to inhibition of ergosterol synthesis	**Spectrum of Action:** Fungi, G+ bacteria, and parasitic protozoa **Route of Administration:** Topical, IV **Adverse Effects:** Possibly causes cancer in humans	None
Other Antifungal Drugs			
5-fluorocytosine	Fungi, but not mammals, have an enzyme that converts this drug into 5-fluorouracil, an analog of uracil that inhibits RNA function	**Spectrum of Action:** *Candida, Cryptococcus, Aspergillus* **Route of Administration:** Oral **Adverse Effects:** None	Develops from mutations in the genes for enzymes necessary for utilization of uracil
Griseofulvin	Isolated from *Penicillium griseofulvum*; deactivates tubulin, preventing cytokinesis and segregation of chromosomes during mitosis	**Spectrum of Action:** Molds of ringworm (tinea) **Route of Administration:** Topical, oral **Adverse Effects:** None	None

Table 10.5 Antihelminthic Drugs

Drug	Description and Mode of Action	Clinical Considerations	Resistance
Mebendazole **Thiabendazole**	Drugs appear to inhibit some enzymes and inhibit microtubule function, thus preventing cytokinesis and segregation of chromosomes during mitosis	**Spectrum of Action:** Helminths, *Giardia* (protozoan) **Route of Administration:** Oral **Adverse Effects:** None	None
Pyrantel pamoate **Diethylcarbamazine**	Bind to neurotransmitter receptors, causing complete muscular contraction of helminths	**Spectrum of Action:** Nematodes **Route of Administration:** Oral **Adverse Effects:** None	None
Praziquantel	Attracts calcium ions, which are required for muscular contraction; induces complete muscular contraction in helminths	**Spectrum of Action:** Cestodes, trematodes **Route of Administration:** Oral **Adverse Effects:** None	None
Piperazine **Ivermectin**	Produce flaccid paralysis by blocking neurotransmitters	**Spectrum of Action:** Nematodes **Route of Administration:** Oral **Adverse Effects:** Adverse allergic reactions may result from antigens of dead worms	None
Niclosamide	Prevents production of ATP by oxidative phosphorylation	**Spectrum of Action:** Cestodes **Route of Administration:** Oral **Adverse Effects:** None; humans do not absorb the drug	None
Suramin	Inhibits specific enzymes in some nematodes	**Spectrum of Action:** Nematodes **Route of Administration:** Oral **Adverse Effects:** None	None

Table 10.6 Antiprotozoan Drugs

Drug	Description and Mode of Action	Clinical Considerations	Resistance
Heavy metals (Hg, As, Cr, Sb) Representatives: Salvarsan Antimony sodium gluconic acid	Deactivate enzymes by breaking hydrogen bonds necessary for effective tertiary structure; drugs containing arsenic were the first recognized chemotherapeutic agents that were selectively toxic	Spectrum of Action: Metabolically active cells Route of Administration: Topical, oral Adverse Effects: Toxic to active cells, such as those of the brain, kidney, liver, and bone marrow	None
Quinines Representatives: Natural quinine Semisynthetic quinines: Chloroquine Primaquine	Natural and semisynthetic drugs derived from the bark of cinchona tree: inhibits DNA synthesis in malaria parasites	Spectrum of Action: *Plasmodium* Route of Administration: Oral Adverse Effects: Allergic reactions, visual disturbances	Results from presence of quinine pumps that remove the drug from the cells
Sulfonamides	Synthetic drug; first produced as a dye; is an analog of PABA that binds irreversibly to enzyme that produced folic acid; synergistic with trimethoprim	Spectrum of Action: Broad: effective against G+ and G− bacteria and some protozoa and fungi cells; however, resistance is widespread Route of Administration: Oral Adverse Effects: Rare: Allergic reactions, anemia, jaundice, mental retardation of fetus if administered in last trimester of pregnancy	Cells that require folic acid as a vitamin are also naturally resistant; chromosomal mutations result in lowered affinity for the drug
Trimethoprim	Blocks second metabolic step in the formation of folic acid from PABA; synergistic with sulfonamides	Spectrum of Action: Broad: effective against G+ and G− bacteria and some protozoa and fungi; however, resistance is widespread Route of Administration: Oral Adverse Effects: Allergic reactions in some patients	Cells that require folic acid as a vitamin are naturally resistant; chromosomal mutations result in lowered affinity for the drug
Nitroimidazoles Representative: Metronidazole	Mode of action unknown	Spectrum of Action: *Trichomonas, Giardia,* and *Entamoeba* Route of Administration: Oral Adverse Effects: Metronidazole suspected to cause cancer	None

CHAPTER SUMMARY

The History of Antimicrobial Agents (pp. 288–289)

1. Chemotherapeutic agents are chemicals used to treat diseases. Among them are **antimicrobial agents (antimicrobials),** which include **antibiotics** (biologically produced agents), **semisynthetics** (chemically modified antibiotics), and **synthetic** agents.

Mechanisms of Antimicrobial Action (pp. 289–295)

1. Successful chemotherapy against microbes is based on **selective toxicity;** that is, using antimicrobial agents that are more toxic to pathogens than to the patient.
2. Antimicrobial drugs affect pathogens by inhibiting cell wall synthesis, inhibiting the translation of proteins, disrupting cytoplasmic membranes, inhibiting general metabolic pathways, inhibiting the replication of DNA, blocking the attachment of viruses to their hosts, and by blocking a pathogen's recognition of its host.
3. Beta-lactams—penicillins, cephalosporins, and monobactams—have a functional **lactam ring.** They prevent bacteria from cross-linking NAM subunits of peptidoglycan in the bacterial cell wall during growth. **Vancomycin** and **cycloserine** also disrupt cell wall formation in many Gram-positive bacteria.

Bacitracin blocks NAG and NAM secretion from the cytoplasm. **Isoniazid (INH)** and **ethambutol** block mycolic acid synthesis in the walls of mycobacteria.

4. Antimicrobial agents that inhibit protein synthesis include **aminoglycosides** and **tetracyclines,** which inhibit functions of the 30S ribosomal subunit, and chloramphenicol and **macrolides,** which inhibit 50S subunits.
5. Amphotericin B and polymyxin disrupt the cytoplasmic membranes of fungi and Gram-negative bacteria respectively.
6. **Sulfonamides** are **structural analogs** of para-aminobenzoic acid (PABA), a chemical needed by some microorganisms but not by humans. The substitution of sulfonamides in the metabolic pathway leading to nucleic acid synthesis kills those organisms. Trimethoprim is another antimetabolite that blocks this pathway.
7. Drugs that inhibit nucleic acid replication in pathogens include actinomycin, **nucleotide analogs,** fluoroquinolones, quinolones, and rifampin.

Clinical Considerations in Prescribing Antimicrobial Drugs (pp. 296–301)

1. Chemotherapeutic agents have a **spectrum of action** and may be classed as either **narrow-spectrum drugs** or **broad-spectrum drugs** depending on the number of kinds of pathogens they affect.
2. **Diffusion susceptibility tests** such as the Kirby-Bauer test reveal which drug is most effective against a particular pathogen; in general, the larger the **zone of inhibition** around a drug-soaked disk on a Petri plate, the more effective the drug.
3. The **minimum inhibitory concentration (MIC),** usually determined by either a **broth dilution test** or an **Etest,**® is the smallest amount of a drug that will inhibit a pathogen.
4. A **minimum bactericidal concentration (MBC) test** ascertains whether a drug is bacteriostatic and the lowest concentration of a drug that is bactericidal.
5. In choosing antimicrobials, physicians must consider how a drug is best administered, orally, intramuscularly, or intravenously—and possible side effects, including toxicity and allergic responses.

Resistance to Antimicrobial Drugs (pp. 301–304)

1. Some members of a pathogenic population may develop resistance to a drug due to extra DNA pieces called **R-plasmids** or to the mutation of genes. Microorganisms may resist a drug by producing enzymes such as **β-lactamase** that deactivate the drug, by inducing changes in the cell membrane that prevent entry of the drug, by altering the drug's receptor to prevent its binding, by altering the cell's metabolic pathways, or by pumping the drug out of the cell.
2. **Cross resistance** occurs when resistance to one chemotherapeutic agent confers resistance to similar drugs.
3. **Synergism** describes the interplay between drugs that results in efficacy that exceeds the efficacy of either drug alone. Some drug combinations are antagonistic.

QUESTIONS FOR REVIEW

(Answers to multiple choice questions are on the web, along with additional review questions. Visit www.microbiologyplace.com.)

Multiple Choice

1. Diffusion and dilution tests that expose pathogens to antimicrobials are designed to
   a. determine the spectrum of action of a drug.
   b. determine which drug is most effective against a particular pathogen.
   c. determine the amount of a drug to use against a particular pathogen.
   d. both b and c
2. In a Kirby-Bauer susceptibility test, the presence of a zone of inhibition around disks containing antimicrobial agents indicates
   a. that the microbe does not grow in the presence of the agents.
   b. that the microbe grows well in the presence of the agents.
   c. the smallest amount of the agent that will inhibit the growth of the microbe.
   d. the minimum amount of an agent that kills the microbe in question.
3. The key to successful chemotherapy is
   a. selective toxicity.
   b. the diffusion test.
   c. the minimum inhibitory concentration test.
   d. the spectrum of action.
4. Which of the following statements is relevant in explaining why sulfonamides are effective?
   a. Sulfonamides attach to sterol lipids in the pathogen, disrupt the membranes, and lyse the cells.
   b. Sulfonamides prevent the incorporation of amino acids into polypeptide chains.
   c. Humans and microbes use folic acid and PABA differently in their metabolism.
   d. Sulfonamides inhibit DNA replication in both pathogens and human cells.
5. Cross resistance is
   a. the deactivation of an antimicrobial agent by a bacterial enzyme.
   b. alteration of the resistant cells so that an antimicrobial agent cannot attach.
   c. the mutation of genes that affect the cell membrane channels so that antimicrobial agents cannot cross into the cell's interior.
   d. resistance to antimicrobial agent P because of its similarity to antimicrobial agent Q.
6. Superbugs
   a. are resistant to most antimicrobial agents.
   b. can be found in the genus *Plasmodium.*
   c. frequently develop in hospitals.
   d. all of the above

7. Which of the following is most closely associated with a β-lactam ring?
   a. penicillin
   b. vancomycin
   c. bacitracin
   d. isoniazid
8. Drugs that act against protein synthesis include
   a. beta-lactams.
   b. trimethoprim.
   c. polymyxin.
   d. aminoglycosides.
9. Which of the following statements is *false* concerning antiviral drugs?
   a. Macrolide drugs block attachment sites on the host cell wall and prevent viruses from entering.
   b. Drugs that increase the acidity of phagolysosomes prevent viral uncoating and thus prevent viral genetic "takeover."
   c. Nucleotide analogs in antiviral drugs can be used to stop viral replication.
   d. Drugs containing protease inhibitors retard viral growth by blocking the production of essential viral proteins.
10. PABA is
   a. a substrate used in the production of penicillin.
   b. a type of β-lactamase.
   c. molecularly similar to cephalosporins.
   d. a substrate used to synthesize folic acid.

Short Answer

1. What characteristics would an ideal chemotherapeutic agent have? Which drug fits these qualities?
2. Contrast narrow-spectrum and broad-spectrum drugs. Which are more effective?
3. Why is the fact that drug Z destroys the NAM portions of a cell's wall structure an important factor in considering the drug for chemotherapy?
4. Given that both human cells and pathogens synthesize proteins at ribosomal sites, how can antimicrobial agents that target this process be safe to use in humans?
5. Support or refute the following statement: "Antimicrobial agents produce resistant cells."
6. Given that resistant strains of pathogens are a concern to the general health of a population, what can be done to prevent their development?
7. Why are antiviral drugs difficult to develop?
8. A man has been given a broad-spectrum antibiotic for his stomach ulcer. What unintended consequences could arise from this therapy?

CRITICAL THINKING

1. AIDS is often treated with a "cocktail" of several antiviral agents at once. Why is the cocktail more effective than a single agent? What is a physician trying to prevent by prescribing several drugs at once?
2. How does *Penicillium* escape the effects of the penicillin it secretes?
3. How might a colony of *Bacillus licheniformis* escape the effects of its own bacitracin?
4. Fewer than 1% of known antibiotics have any practical value in treatment of disease. Why is this so?
5. In the summer issue of *News of the Lepidopterist's Society* in 2000, a recommendation was made to moth and butterfly collectors to use antibiotics to combat disease in the young of these insects. What are the possible ramifications for human health of such usage of antibiotics?
6. Even though aminoglycosides such as gentamicin can cause deafness, there are still times when they are the best choice for treating some infections. What laboratory test would a clinical scientist use to show that gentamicin is the best choice to treat a particular *Pseudomonas* infection?
7. Your pregnant neighbor has a sore throat and tells you that she is taking some tetracycline she had left over from a previous infection. Give two reasons why her decision is a poor one.
8. Acyclovir has replaced adenosine arabinoside as treatment for herpes infections. Compare the ways these drugs are activated (see Table 10.3). Why is acyclovir a better choice?
9. Why might amphotericin B affect the kidneys more than other human organs?
10. *Clostridium difficile* is an obligate anaerobic opportunistic pathogen. It can cause serious, life-threatening pseudomembranous colitis in patients undergoing long-term antimicrobial therapy. Why might aminoglycosides be used to treat this disease more often than tetracyclines?

CHAPTER 11

Characterizing and Classifying Prokaryotes

In 1980, microbiologist Tony Walsby startled the scientific world by announcing the discovery of something no one had seen before: rectangular-shaped bacteria. These novel prokaryotes were named *Haloarcula*, meaning "salt box," because of their unusual boxy shape and their habitat, the salt-encrusted pools of Egypt's Sinai peninsula. Until this extraordinary find, scientists had only seen spherical, rod-shaped, or spiral prokaryotes. *Haloarcula* was a surprise, one that forced scientists to reconsider some of their assumptions about prokaryotic life forms.

More recently, we have discovered species of archaea that may offer clues to whether or not there is life on other planets. And among the millions of unknown microorganisms that live on Earth, there may very well exist entirely new categories of life. The vast and diverse world of prokaryotes continues to surprise and amaze us.

CHAPTER OUTLINE

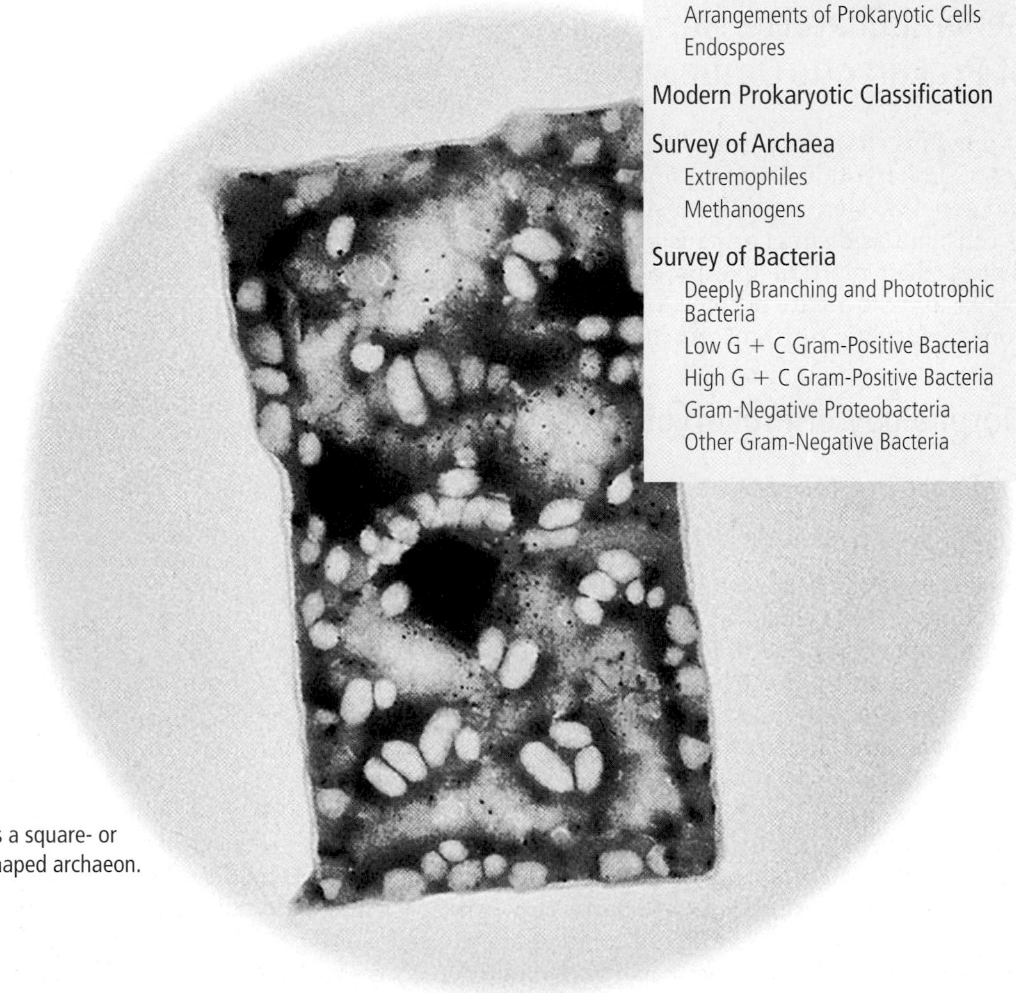

Haloarcula is a square- or rectangular-shaped archaeon.

MicroPrep Pre-Test: *Take the pre-test for this chapter on the web. Visit* ***www.microbiologyplace.com.***

Prokaryotes are by far the most diverse group of organisms. They thrive in various habitats: from Antarctic glaciers to thermal hot springs, from the colons of animals to the cytoplasm of other prokaryotes, from distilled water to supersaturated brine, and from disinfectant solutions to basalt rocks thousands of meters below the Earth's surface. In part because of such great diversity, only a very few prokaryotes have enzymes, toxins, or cellular structures that enable them to colonize humans and cause disease. In this chapter we will begin by examining general prokaryotic characteristics and conclude with a survey of specific prokaryotic taxa. We will briefly mention human pathogens throughout the chapter.

General Characteristics of Prokaryotic Organisms

In previous chapters we considered the general characteristics of prokaryotic cells, including their metabolism, growth, and genetics. Here we will consider prokaryotes not merely as cells, but as distinct organisms, focusing on their cellular shapes, their reproductive processes, their spatial arrangements, and the ability of some to survive unfavorable conditions by forming resistant structures within themselves.

Morphology of Prokaryotic Cells

Learning Objective

✓ Identify six basic shapes of prokaryotic cells.

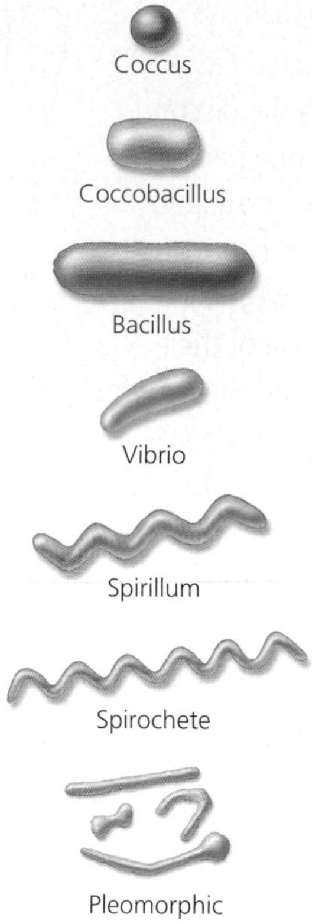

▲ ***Figure 11.1***

Typical prokaryotic morphologies. *What is the difference between a spirillum and a spirochete?*

Figure 11.1 Generally, spirilla are stiff and have fewer than a dozen coils, whereas spirochetes are flexible and have numerous coils.

Prokaryotic cells exist in a variety of shapes, or morphologies **(Figure 11.1).** The three basic shapes are **cocci** (kok′sī, roughly spherical), **bacilli** (bă-sil′ī, rod-shaped), and **spirals.** Cocci are not all perfectly spherical; for example, there are pointed, kidney-shaped, and oval cocci. Similarly, bacilli vary in shape; for example, some bacilli are pointed, spindle-shaped, or threadlike (filamentous). Spiral-shaped prokaryotes are either **spirilla,** which are stiff, or **spirochetes** (spī′rō-kētz), which are flexible. Slightly curved rods are **vibrios,** and the term **coccobacillus** is used to describe cells that are intermediate in shape between cocci and bacilli; that is, when it is difficult to ascertain if a cell is an elongated coccus or a short bacillus. In addition to these basic shapes, there are star-shaped, triangular, and square prokaryotes, as well as prokaryotes that are **pleomorphic**[1] (plē-ō-mōr′ fik); that is, they vary in shape and size (see Figure 11.11).

Reproduction of Prokaryotic Cells

Learning Objectives

✓ List three common types of reproduction in prokaryotes.
✓ Describe snapping division as a type of binary fission.

All prokaryotes reproduce asexually; none reproduce sexually. The most common method of asexual reproduction is **binary fission,** which proceeds as follows **(Figure 11.2):** 1 The cell replicates its DNA; each DNA molecule is attached to the cytoplasmic membrane. 2 The cell grows; as the cytoplasmic membrane elongates, it moves the daughter molecules of DNA apart. 3 The cell forms a cross wall, invaginating the cytoplasmic membrane. 4 The cross wall completely divides daughter cells. 5 The daughter cells may or may not separate. The parental cell disappears with the formation of progeny.

A variation of binary fission called **snapping division** occurs in some Gram-positive species **(Figure 11.3).** In snapping division, only the inner portion of the cell wall is deposited across the dividing cell. The thickening of this new transverse wall puts tension on the outer layer of the old cell wall, which still holds the two cells together. Eventually, as the tension increases, the outer wall breaks at its weakest

[1]From Greek *pleon*, meaning more and *morphe*, meaning form.

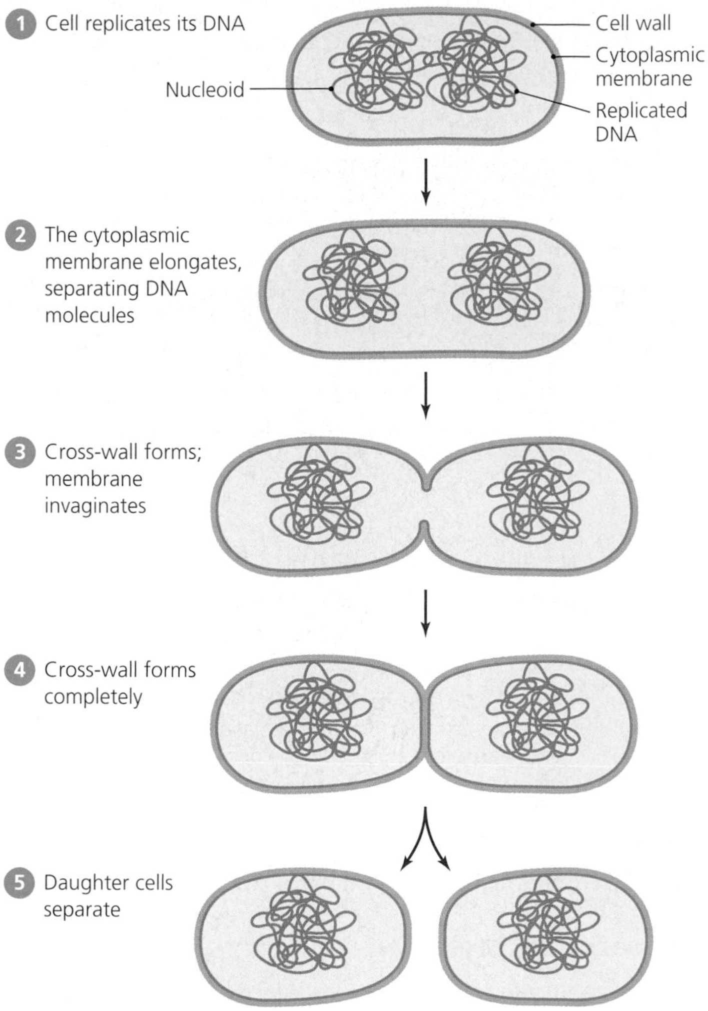

▲ ***Figure 11.2***

Binary fission. The cell replicates its DNA, elongates, forms a cross-wall, and divides into two equal-sized daughter cells.

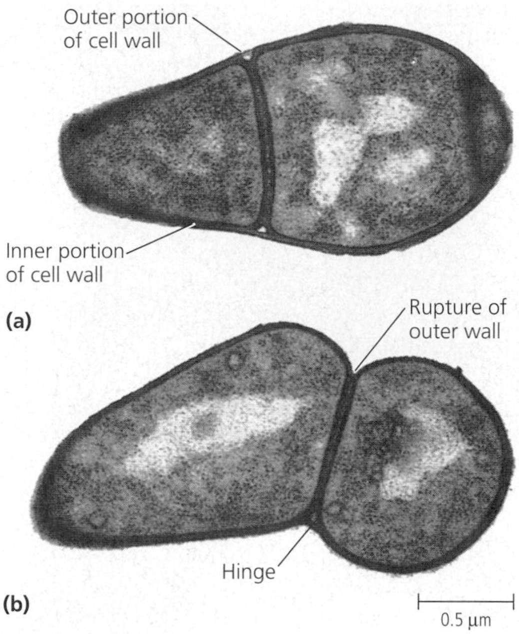

▲ ***Figure 11.3***

Snapping division, a variation of binary fission. **(a)** Only the inner portion of the cell wall forms a cross wall. **(b)** As the daughter cells grow, tension snaps the outer portion of the cell wall, leaving the daughter cells connected by a hinge of old cell wall material.

point with a snapping movement that tears it most of the way around. The daughter cells then remain hanging together, held at an angle by a small remnant of the original outer wall that acts like a hinge.

A few prokaryotes have other methods of reproduction. The parental cell retains its identity during and after these methods. The *actinomycetes* (ak′ti-nō-mī-sētz) produce reproductive cells called **spores** at the ends of their filamentous cells (**Figure 11.4** on page 316). Each spore can develop into a clone of the original organism. Some *cyanobacteria* reproduce by fragmentation into small motile filaments that glide away from the parental strand. Still other prokaryotes reproduce by **budding,** in which an outgrowth of the original cell (a bud) receives a copy of the genetic material and enlarges. Eventually the bud is cut off from the parental cell, typically while it is still quite small (**Figure 11.5** on page 316).

Arrangements of Prokaryotic Cells

Learning Objective

✓ Draw and label five arrangements of prokaryotes.

The arrangements of prokaryotic cells result from two aspects of division during binary fission: the planes in which cells divide, and whether or not daughter cells separate completely or remain attached to each other. Thus cocci that remain attached in pairs are **diplococci** (**Figure 11.6a** on page 317), and long chains of cocci are called **streptococci**[2] **(Figure 11.6b).** Some cocci divide in two planes and remain attached to form **tetrads (Figure 11.6c);** others divide in three planes to form cuboidal packets called **sarcinae**[3] (sar′si-nī) **(Figure 11.6d).** Clusters called **staphylococci**[4] (staf′i-lō-kok-sī), which look like bunches of grapes, form when the planes of cell division are random **(Figure 11.6e).**

Bacilli are less varied in their arrangements than cocci because bacilli divide transversely—that is, perpendicular to the long axis. Daughter bacilli may separate to become single cells or stay attached as either pairs or chains (**Figure 11.7a–c** on page 317). Because the cells of *Corynebacterium diphtheriae* (kŏ-rī′nē-bak-tēr′ē-ŭm dif-thi′rē-ē), the

[2]From Greek *streptos*, meaning twisted, because long chains tend to twist.
[3]Latin for bundles.
[4]From Greek *staphyle*, meaning bunch of grapes.

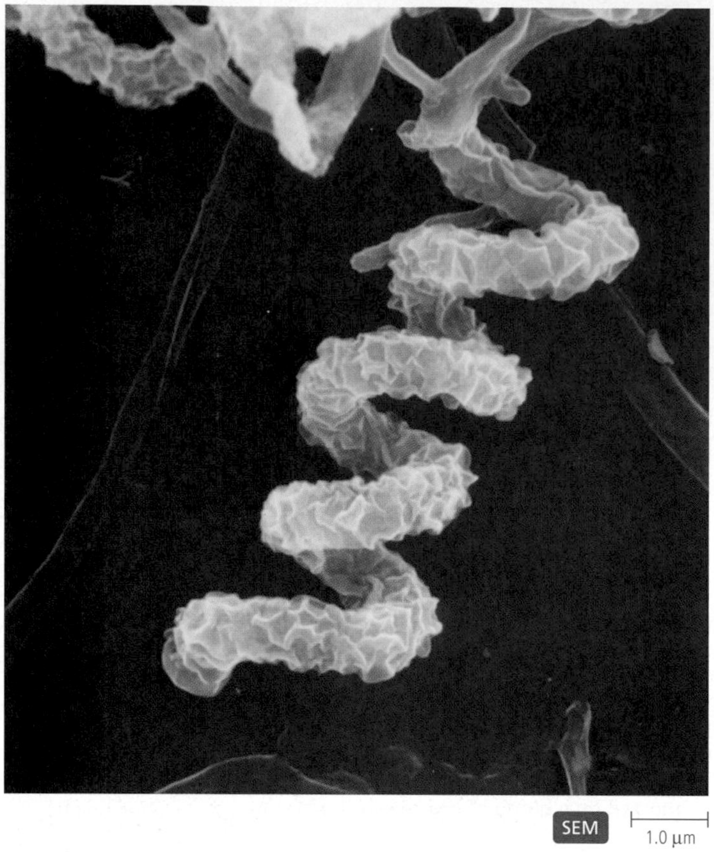

▲ *Figure 11.4*

Prokaryotic reproductive structures. Spores of actinomycetes grow from filamentous vegetative cells.

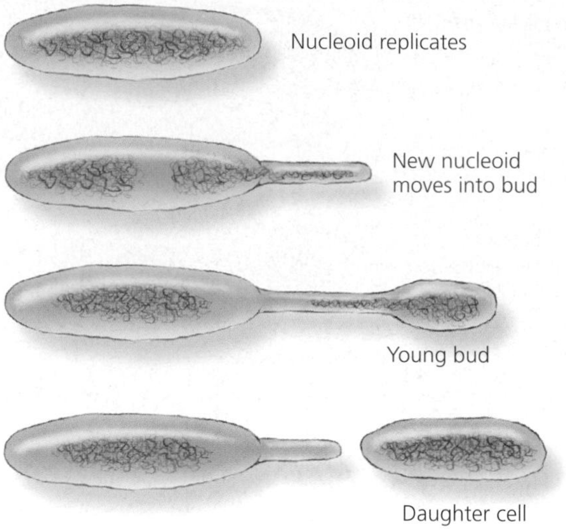

▲ *Figure 11.5*

Budding. *How does budding differ from binary fission?*

Figure 11.5 In binary fission, the parent cell disappears with the formation of two equal-sized offspring; in contrast, a bud is often much smaller than its parent, and the parent remains to produce more buds.

causative agent of diphtheria, divide by snapping division, the daughter cells remain attached to form V-shapes and a side-by-side arrangement called a **palisade**[5] **(Figure 11.7d).**

Note that the same word can be used to refer either to a general shape and/or arrangement or to a specific genus. Thus the characteristic shape of the genus *Bacillus* (ba-sil'ŭs) is a bacillus, a rod-shaped bacterium, and the characteristic arrangement of bacteria in the genus *Sarcina* (sar'si-nă) is cuboidal. In such potentially confusing cases the meaning can be distinguished because genus names are always capitalized and italicized. In other cases the genus name uses the singular form while the arrangement uses the plural form; thus streptococci—spherical cells arranged in a chain—are characteristic of the genus *Streptococcus* (strep'tō-kok'ŭs), and staphylococci—spherical cells arranged in grape-like clusters—are characteristic of the genus *Staphylococcus* (staf' i-lō-kok'ŭs).

[5]From Latin *palus*, meaning stake, referring to a fence made of adjoining stakes.

Endospores

Learning Objective

✓ Describe the formation and function of bacterial endospores.

The Gram-positive bacteria *Bacillus* and *Clostridium* (klos-trid'ē-ŭm) are characterized by the ability to produce unique structures called **endospores,** which are important for several reasons, including their durability and potential pathogenicity (discussed shortly). Though some scientists refer to endospores simply as "spores," endospores should not be confused with the reproductive spores of actinomycetes, algae, and fungi. Each single bacterial cell, called a *vegetative* cell to distinguish it from an endospore, transforms into only one endospore, which then germinates to grow into only one vegetative cell; therefore, endospores are not reproductive structures. Instead, endospores constitute a defensive strategy against hostile or unfavorable conditions.

A vegetative cell normally transforms itself into an endospore only when one or more nutrients (such as carbon or nitrogen) are in limited supply. The process of endospore formation, called *sporulation,* requires 8 to 10 hours and proceeds in seven steps (**Figure 11.8** on page 318). During the process, two membranes, a thick layer of peptidoglycan, and a spore coat form around a copy of the genome and a

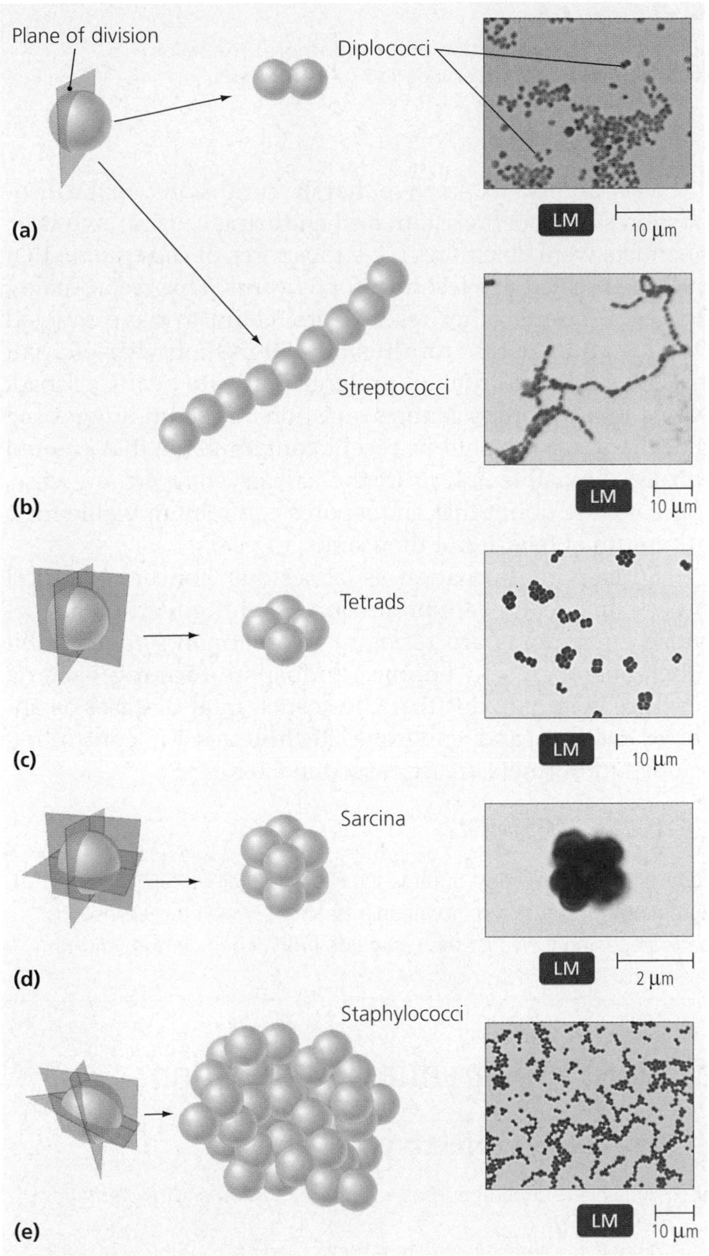

▲ ***Figure 11.6***

Arrangements of cocci. **(a)** The diplococci of *Neisseria gonorrheae.* **(b)** The streptococci of *Streptococcus pyogenes.* **(c)** A tetrad, in this case of *Micrococcus luteus.* **(d)** A sarcina of *Sarcina maxima.* **(e)** The staphylococci of *Staphylococcus aureus.*

small portion of cytoplasm. The cell deposits large quantities of dipicolinic acid, calcium, and DNA-binding proteins within the endospore while removing most of the water. Depending on the species, a cell forms an endospore either *centrally, subterminally* (near one end), or *terminally* (at one end) (**Figure 11.9** on page 319). Sometimes an endospore is so large it swells the original cell.

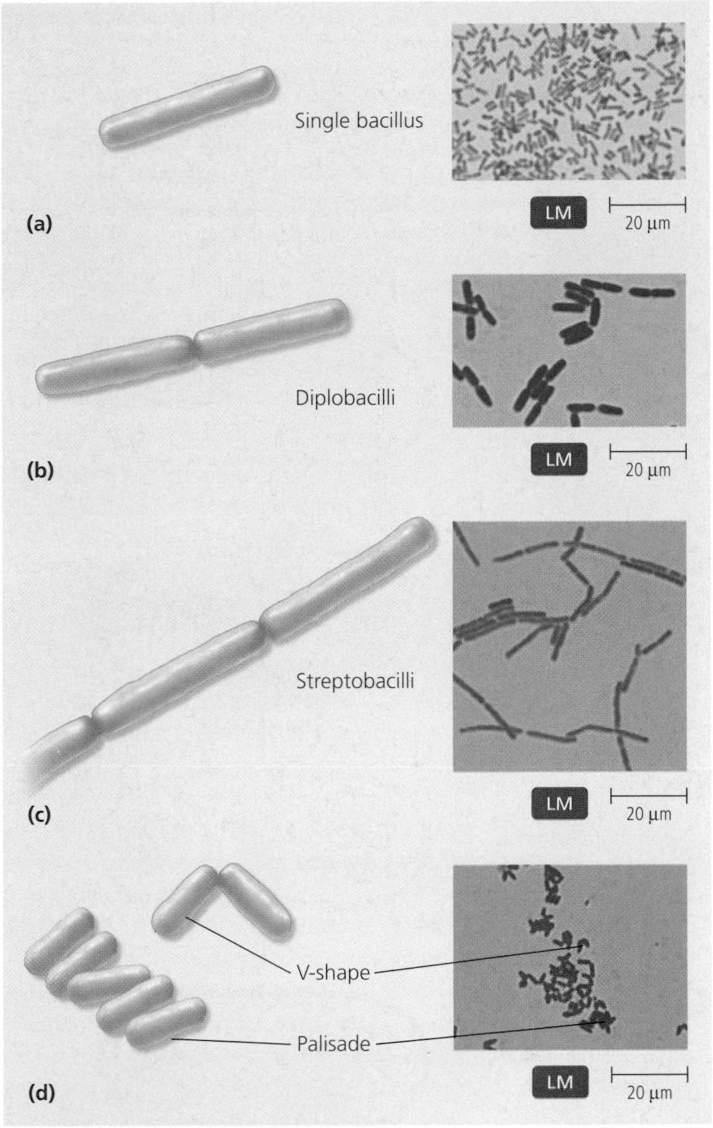

▲ ***Figure 11.7***

Arrangements of bacilli. **(a)** A single bacillus of *Escherichia coli.* **(b)** Diplobacilli in a young culture of *Bacillus cereus.* **(c)** Streptobacilli in an older culture of *Bacillus cereus.* **(d)** V-shapes and a palisade of *Corynebacterium diphtheriae.*

Endospores are extremely resistant to drying, heat, radiation, and lethal chemicals. For example, they remain alive in boiling water for several hours; are unharmed by alcohol, peroxide, bleach, and other toxic chemicals; and can tolerate over 400 rad of radiation, which is more than four times the dose that is lethal to humans. Endospores are stable resting stages that barely metabolize—they are essentially in a state of suspended animation—and they germinate only when conditions improve. Scientists do not know how endospores are able to resist harsh conditions, but it appears that the double membrane, spore coats, dipicolinic acid, calcium, and DNA-binding proteins may serve to stabilize DNA and proteins, protecting them from adverse conditions.

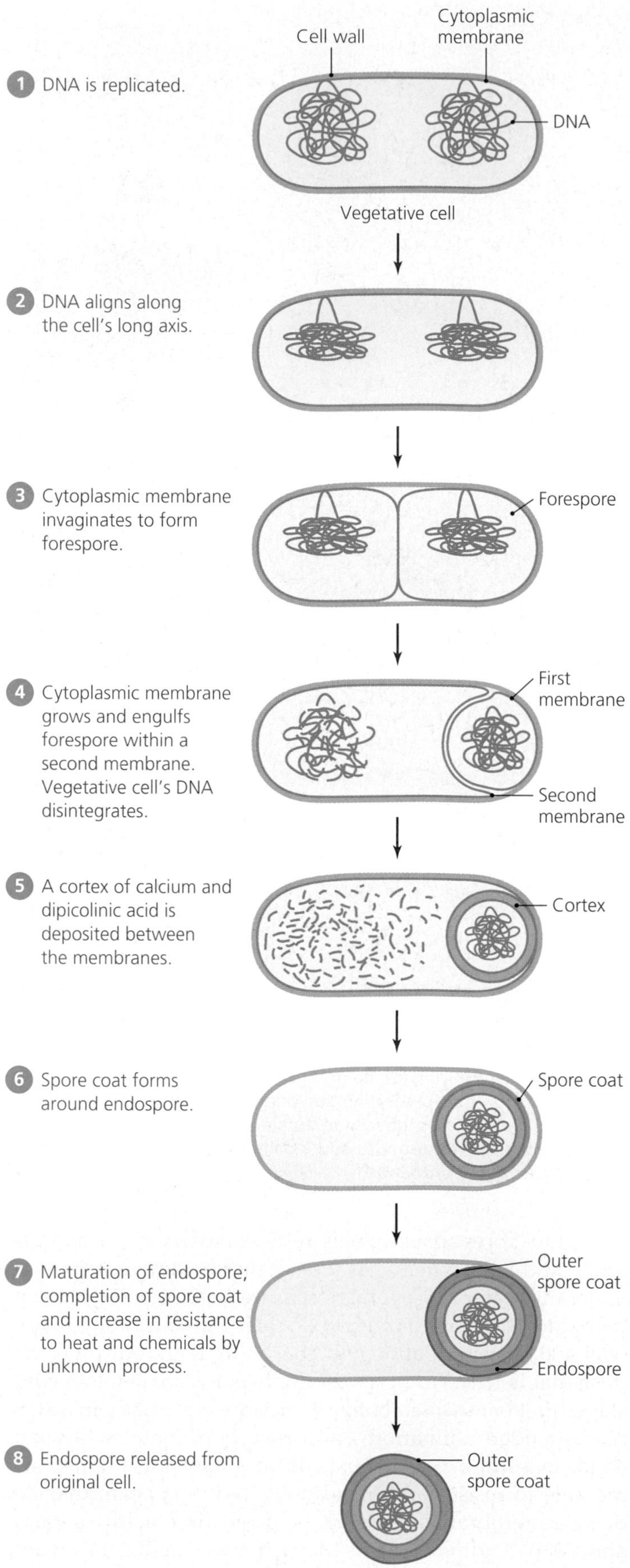

Cell wall
Cytoplasmic membrane
1 DNA is replicated.
DNA
Vegetative cell
2 DNA aligns along the cell's long axis.
3 Cytoplasmic membrane invaginates to form forespore.
Forespore
4 Cytoplasmic membrane grows and engulfs forespore within a second membrane. Vegetative cell's DNA disintegrates.
First membrane
Second membrane
5 A cortex of calcium and dipicolinic acid is deposited between the membranes.
Cortex
6 Spore coat forms around endospore.
Spore coat
7 Maturation of endospore; completion of spore coat and increase in resistance to heat and chemicals by unknown process.
Outer spore coat
Endospore
8 Endospore released from original cell.
Outer spore coat

◀ ***Figure 11.8***
Sporulation, the formation of an endospore. The seven steps depicted occur over a period of 8–10 hours.

The ability to survive harsh conditions makes endospores the most resistant and enduring cells. In one case, scientists were able to revive endospores of *Clostridium* that had been sealed in a test tube for 34 years. This record pales, however, beside other researchers' claim to have revived *Bacillus* endospores from inside 250-million-year-old salt crystals retrieved from an underground site near Carlsbad, New Mexico. Some scientists question this claim, suggesting that the bacteria might be recent contaminants that entered through invisible cracks in the salt crystals. In any case, there is little doubt that endospores can remain viable for a minimum of tens, if not thousands, of years.

Endospore formation is a serious concern to food processors, health care professionals, and governments because endospores are resistant to treatments that inhibit other microbes, and because endospore-forming bacteria produce deadly toxins that cause such fatal diseases as anthrax, tetanus, and gangrene. Techniques for controlling endospore-formers are discussed in Chapter 9.

CRITICAL THINKING

Following the bioterrorist anthrax attacks in the fall of 2001, a news commentator suggested that people steam their mail for 30 seconds before opening it. Would this technique protect people from anthrax infections? Why or why not?

Modern Prokaryotic Classification

Learning Objectives

✓ Explain the general purpose of *Bergey's Manual of Systematic Bacteriology.*

✓ Discuss the veracity and limitations of any taxonomic scheme.

As we saw in Chapter 4, scientists called *taxonomists* group similar organisms into categories called *taxa.* At one time the smallest taxa of prokaryotes (that is, genera and species) were based solely on the characteristics we considered in the previous section, especially morphology and arrangement. More recently, the classification of living things has been based more on genetic relatedness. Accordingly, modern taxonomists have placed all organisms into three *domains*—Archaea, Bacteria, and Eukarya—which are the largest, most inclusive taxa and have essentially replaced kingdoms. Each domain is divided into phyla primarily on the basis of similarities in DNA, RNA, and protein sequences.

We previously noted as well that the vast majority of prokaryotes—perhaps as many as 99%, and probably millions of species—have never been isolated or cultured and are known only from their rRNA "fingerprints"; that is, they

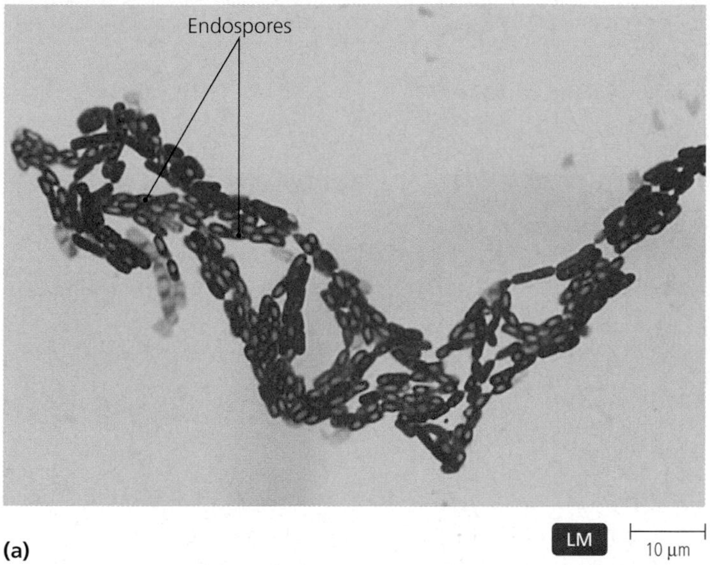

(a)

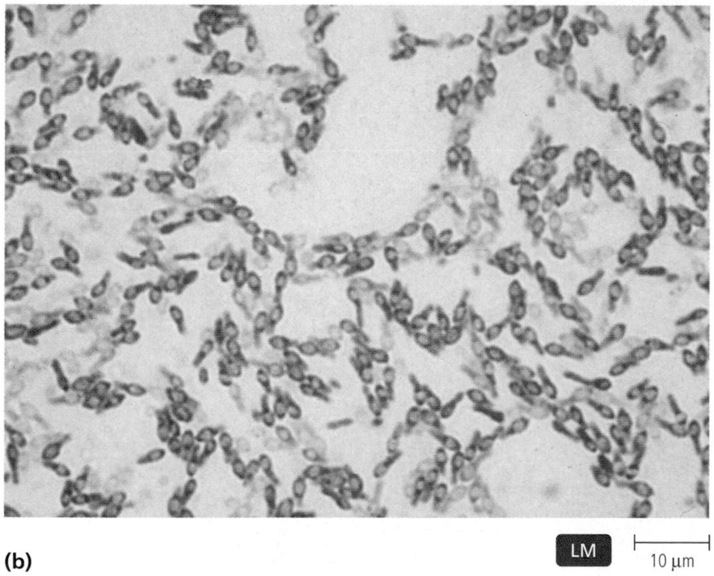

(b)

▲ *Figure 11.9*

Locations of endospores. **(a)** Central endospores of *Bacillus.* **(b)** Terminal endospores of *Clostridium botulinum.* The enlarged endospore has swollen the cell that produced it.

are known only from sequences of rRNA that do not match any known RNA sequences. In light of this information, taxonomists now construct modern classification schemes of prokaryotes based on the relative similarities of rRNA sequences found in various prokaryotic groups (**Figure 11.10** on page 320).

Perhaps the most authoritative reference in modern prokaryotic systematics is *Bergey's Manual of Systematic Bacteriology,* which classifies prokaryotes into 25 phyla—two in Archaea and 23 in Bacteria. When completed, the five volumes of the second edition of *Bergey's Manual* will discuss the great diversity of prokaryotes based in large part (but not exclusively) on their possible evolutionary relationships as reflected in their rRNA sequences. (As of 2003, only Volume I had been published.)

Our examination of prokaryotic diversity in this text is for the most part organized to reflect the taxonomic scheme that will appear in the second edition of *Bergey's Manual,* but it's important to note that as authoritative as it is, *Bergey's Manual* is not an "official" list of prokaryotic taxa. The reason is that taxonomy is partly a matter of opinion and judgment, and not all taxonomists agree. Because there is room in taxonomy for honest differences of opinion, and because legitimately differing views often change as more information is uncovered and examined, *Bergey's Manual* is merely a consensus of experts at a given time. More information about *Bergey's Manual* can be found in Appendix C.

CRITICAL THINKING

A scientist stated that *Bergey's Manual* would never truly be finished. Support her statement.

In the following sections we examine representative prokaryotes in each major phylum. We begin our exploration of prokaryotic diversity with a survey of Archaea.

Survey of Archaea

Learning Objective

✓ Identify the common features of microbes in the domain Archaea.

Scientists originally identified archaea as a distinct type of prokaryotes on the basis of unique rRNA sequences. Archaea also share other common features that distinguish them from bacteria:

- Archaea lack peptidoglycan in their cell walls
- Their cell membrane lipids have branched hydrocarbon chains
- The intial amino acid in their polypeptide chains, coded by the AUG start codon, is methionine (as in eukaryotes and in contrast to the N-formylmethionine used by bacteria).

These and other features of archaea, bacteria, and eukaryotes are compared and contrasted in Chapter 4.

Archaea are currently classified in two phyla—Crenarchaeota and Euryarchaeota—based primarily upon rRNA sequences (see Figure 11.10).

Archaea reproduce by binary fission, budding, or fragmentation. Most archaeal cells are cocci, bacilli, or spiral forms, but some unusual shapes also exist (**Figure 11.11** on page 321). Archaeal cell walls vary among taxa and are

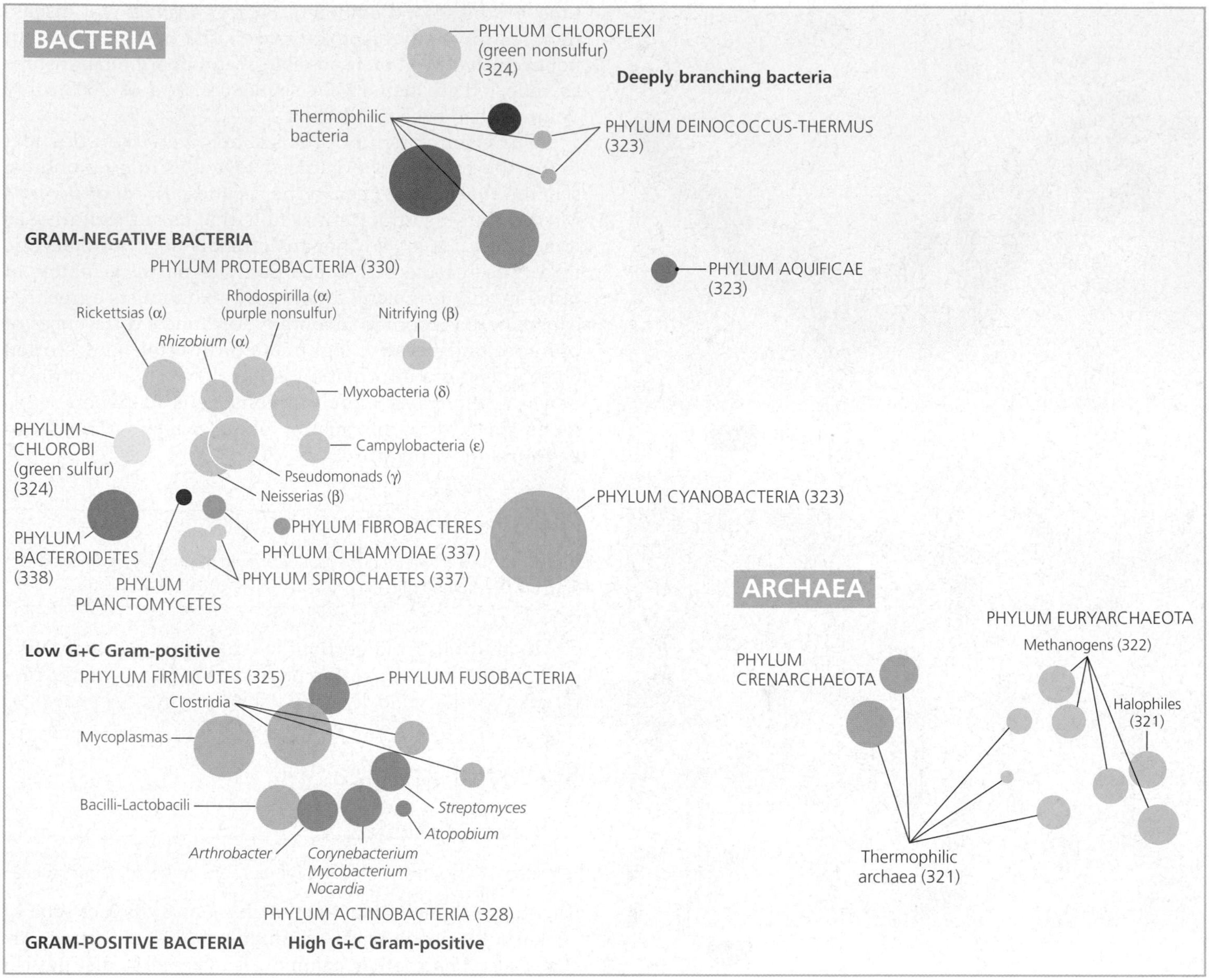

▲ ***Figure 11.10***

A taxonomic scheme for prokaryotes based on their relatedness according to rRNA sequences. The closer together the disks, the more similar are the rRNA sequences of the organisms within the group; the sizes of disks are proportional to the number of sequences known for that group. Note that archaea are distinctly separate from bacteria. The discussion in this chapter is largely based on the scheme depicted in this figure; the numbers in parentheses correspond to the page in this book on which the discussion of that group begins. (Adapted from *Road Map to Bergey's.* 2002, Bergey's Manual Trust.)

composed of a variety of compounds, including proteins, glycoproteins, lipoproteins, and polysaccharides; the feature they have in common is a lack of peptidoglycan. Another interesting feature of archaea is that not one of them is known to cause diseases in humans or animals.

Though most archaea live in moderate environmental conditions, the most noted archaea are *methanogens,* organisms that generate methane gas (discussed shortly), and *extremophiles,* which we discuss next.

Extremophiles

Learning Objective

✓ Compare and contrast the two kinds of extremophiles discussed in this section.

Extremophiles are microbes that require extreme conditions of temperature, pH, and/or salinity to survive. Prominent among the extremophiles are *thermophiles* and *halophiles.*

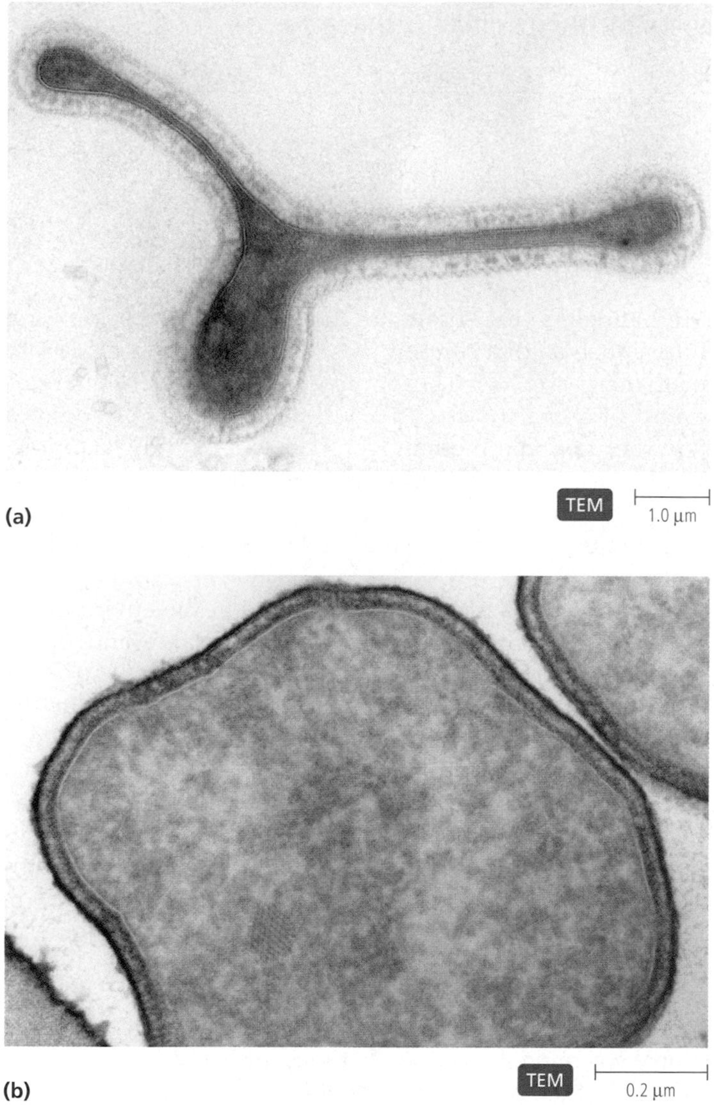

▲ ***Figure 11.11***

Thermophiles. **(a)** *Pyrodictium,* which has disc-shaped cells with filamentous extensions.
(b) *Sulfolobus,* which is pleomorphic.

Thermophiles

As discussed in Chapter 6, **thermophiles**[6] are prokaryotes whose DNA, RNA, cytoplasmic membranes, and proteins require them to live at temperatures over 45°C because these cellular components do not function properly at lower temperatures. Prokaryotes that require temperatures over 80°C are called **hyperthermophiles.** Most thermophilic archaea are in the phylum Crenarchaeota, though some thermophilic species are also found in the phylum Euryarchaeota.

Three representative genera of thermophiles are *Acidanus, Pyrodictium,* and *Sulfolobus* (see Figure 11.11). These microorganisms live in sulfur-rich, acidic hot springs such as those found in Yellowstone National Park **(Figure 11.12)** and similar terrestrial volcanic habitats. The cells of *Pyrodictium,* which lives in deep-sea hydrothermal vents, are irregular disks with elongated protein tubules that attach them to grains of sulfur, which they use as final electron acceptors in respiration. **Table 11.1** on page 322 summarizes the characteristics of these three genera.

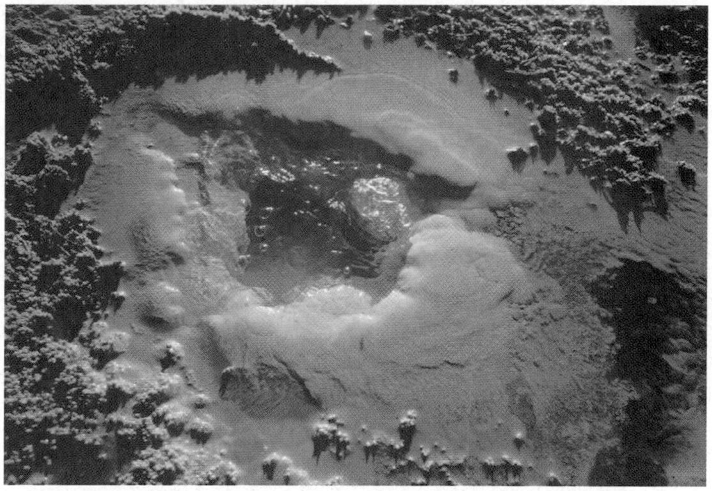

▲ ***Figure 11.12***

A typical habitat of hyperthermophilic archaea: a sulfur-rich hot spring in Yellowstone National Park.

Scientists use thermophiles and their enzymes in recombinant DNA technology applications because thermophiles' cellular structure and enzymes are stable and functional at temperatures that denature most proteins and nucleic acids and kill other cells. As discussed in Chapter 8, DNA polymerase from hyperthermophilic archaea makes possible the automated amplification of DNA in a thermocycler. Heat-stable enzymes are also ideal for many industrial applications, including their use as additives in laundry detergents.

Halophiles

Halophiles[7] are classified in the phylum Euryarchaeota based on rRNA sequences. They inhabit extremely saline habitats such as the Dead Sea, the Great Salt Lake, and solar evaporation ponds used to concentrate salt for use in seasoning and for the production of fertilizer (**Figure 11.13** on page 322). Halophiles can also colonize and spoil such foods as salted fish, sausages, and pork.

The distinctive characteristic of halophiles is their absolute dependence on a concentration of NaCl greater than 9% (that is, a 1.5 molar solution) to maintain the integrity of their cell walls. Most halophiles grow and reproduce within an optimum range of 17–23% NaCl, and many species can survive in a saturated saline solution (35% NaCl). Halophiles contain red to orange pigments that probably

---

[6]From Greek *thermos,* meaning heat, and *philos,* meaning love.
[7]From Greek *halos,* meaning salt.

Table 11.1 Characteristics of Three Representative Genera of Thermophilic Archaea

Genus	Morphology	Optimum Temperature	Optimum pH	Oxygen Requirement
Sulfolobus	Lobed coccus	75°C	2.5	Obligate aerobe
Acidanus	Coccus	88°C	2	Facultative anaerobe
Pyrodictium	Disk-shaped with filaments	105°C	6	Anaerobe

▲ ***Figure 11.13***

One habitat of halophiles: the highly saline water in solar evaporation ponds near San Francisco as seen in an aerial photo. Halophiles often contain red to orange pigments to protect them from intense solar energy.

play a role in protecting them from the intense visible and ultraviolet light.

The most studied halophile is *Halobacterium salinarium* (hā'lō-bak-tēr'ē-ŭm sal-ē-nar'ē-um), which is unusual because even though it uses light energy to drive the synthesis of ATP, it does not synthesize organic compounds from CO_2. Further, *Halobacterium* lacks photosynthetic pigments—chlorophylls and bacteriochlorophylls. Instead, it synthesizes purple proteins, called **bacteriorhodopsins** (bak-tēr'ē-ō-rō-dop'sinz), that absorb light energy and use it to pump protons (hydrogen ions) across the cytoplasmic membrane to establish a proton gradient. (Recall from Chapter 5 that cells use the energy of proton gradients to produce ATP via chemiosmosis.) *Halobacterium* also rotates its flagella with energy from the proton gradient to position itself at the proper water depth for maximum light absorption.

Methanogens

Learning Objective

✓ List at least four significant roles played by methanogens in the environment.

Methanogens are obligate anaerobes in the phylum Euryarchaeota that convert CO_2, H_2, and organic acids into methane gas (CH_4). These microbes constitute the largest group of archaea. Most species are mesophilic, though a few thermophilic methanogens are known. For example, *Methanopyrus*[8] has an optimum growth temperature of 98°C and grows in 110°C seawater around submarine hydrothermal vents. Scientists have also discovered halophilic methanogens.

Methanogens play significant roles in the environment by converting organic wastes in pond, lake, and ocean sediments into methane. Other methanogens such as *Methanobacterium* (meth'a-nō-bak-tēr'ē-ŭm) living in the colons of animals are one of the primary sources of environmental methane. Methanogens dwelling in the intestinal tract of a cow, for example, can produce 400 liters of methane a day. Methanogens have produced about 10 trillion tons of methane—twice the known amount of oil, natural gas, and coal combined—that lies buried in the mud on the ocean floor. Sometimes the production of methane in swamps and bogs is so great that bubbles rise to the surface as "swamp gas." Methane is a so-called *greenhouse gas;* that is, methane in the atmosphere traps heat, which adds to global warming. It is about 25 times more potent as a greenhouse gas than carbon dioxide. If all the methane trapped in ocean sediments were released, it would wreak havoc with the world's climate.

Methanogens also have useful industrial applications. An important step in sewage treatment is the digestion of sludge by methanogens, and some sewage treatment plants burn the methane to heat buildings and generate electricity.

Though we have concentrated our discussion on extremophiles and methanogens, many archaea live in more moderate habitats. For example, archaea make up about a third of the prokaryotic biomass in coastal Antarctic water, providing food for marine animals.

CRITICAL THINKING

A scientist who discovers a prokaryote living in a hot spring at 100°C suspects that it belongs to the archaea. Why does he think it might be archaeal? How could he prove that it is not bacterial?

[8]From Greek *pyrus,* meaning fire.

Survey of Bacteria

As we noted previously, our survey of prokaryotes in this chapter reflects the classification scheme that will be featured in the as-yet incomplete second edition of *Bergey's Manual*. Whereas the classification scheme for bacteria in the first edition of the *Manual* emphasized morphology, Gram reaction, and biochemical characteristics, the second edition will largely base its classification of bacteria on differences in their 16S rRNA sequences. The result is a scheme that includes 23 bacterial phyla grouped in a way that generally reflects our current understanding of the genetic relationships among bacteria. We begin our survey of bacteria by considering the deeply branching and phototrophic bacteria.

Deeply Branching and Phototrophic Bacteria

Learning Objectives

✓ Provide a rationale for the name "deeply branching bacteria."

✓ Explain the function of heterocysts in terms of both photosynthesis and nitrogen fixation.

Deeply Branching Bacteria

The **deeply branching bacteria** are so named because their rRNA sequences and growth characteristics lead scientists to conclude that these organisms are similar to the earliest bacteria; that is, they branched off the "tree of life" at an early stage. For example, the deeply branching bacteria are autotrophic, and early organisms must have been autotrophs because heterotrophs by definition must derive their carbon from autotrophs. Further, many of the deeply branching bacteria live in habitats similar to those some scientists think existed on the early Earth—hot, acidic, anaerobic, and exposed to intense ultraviolet radiation from the sun.

One representative of these microbes—the Gram-negative, microaerophilic *Aquifex*, a bacterium in the Phylum Aquificae—is considered to represent the earliest branch of bacteria. It is chemoautotrophic, hyperthermophilic, and anaerobic, deriving energy and carbon from inorganic sources in very hot habitats containing little oxygen.

Another representative of deeply branching bacteria is *Deinococcus* (dīn-ō-kok'ŭs), in the proposed Phylum "Deinococcus-Thermus" which grows as tetrads of Gram-positive cocci. Interestingly, the cell wall of *Deinococcus* has an outer membrane similar to that of Gram-negative bacteria, but the cells stain purple like typical Gram-positive microbes. *Deinococcus* is extremely resistant to radiation because of the way it packages its DNA and the presence of radiation-absorbing pigments and unique lipids within its membranes. Even when exposed to 5 million rad of radiation, which is enough energy to shatter its chromosome into hundreds of fragments, its enzymes can repair the damage. Not surprisingly, researchers have isolated *Deinococcus* from sites severely contaminated with radioactive wastes.

Phototrophic Bacteria

Phototrophic bacteria acquire the energy needed for anabolism by absorbing light with pigments located in thylakoids called *photosynthetic lamellae*. Of course, being prokaryotes, they lack the membrane-bound thylakoids seen in eukaryotic chloroplasts.

Most phototrophic bacteria are also autotrophic—they produce organic compounds from carbon dioxide. Phototrophs are a diverse group of microbes that are taxonomically confusing. Based on their pigments and their source of electrons for photosynthesis, phototrophic bacteria can be divided into the following five groups (**Table 11.2** on page 325):

- Blue-green bacteria (cyanobacteria)
- Green sulfur bacteria
- Green nonsulfur bacteria
- Purple sulfur bacteria
- Purple nonsulfur bacteria

Although these organisms are classified into four phyla, we consider all of them together because of their common phototrophic metabolism.

Cyanobacteria The organisms known as **cyanobacteria** are Gram-negative phototrophs that vary greatly in shape, size, and method of reproduction. They range in size from 1 μm to 10 μm in diameter and are either coccoid or disc shaped. Coccal forms can be single or arranged in pairs, tetrads, chains, or sheets (**Figure 11.14a and b** on page 324); disc-shaped forms are often tightly appressed end to end to form filaments, which can be either straight, branched, or helical and are frequently contained in a gelatinous glycocalyx called a *sheath* **(Figure 11.14c).** Some filamentous cyanobacteria are motile, moving along surfaces by *gliding*. Cyanobacteria generally reproduce by binary fission, with some species also reproducing by motile fragments or by thick-walled spores called akinetes (ā-kin-ēts'; see Figure 11.14a).

Like plants and algae, cyanobacteria utilize chlorophyll *a* and generate oxygen (are oxygenic) during photosynthesis:

$$12H_2O + 6CO_2 \xrightarrow{\text{light}} C_6H_{12}O_6 + 6H_2O + 6O_2 \qquad (1)$$

For this reason, cyanobacteria were formerly called *blue-green algae;* however, the name *cyanobacteria* properly emphasizes their true prokaryotic nature.

Photosynthesis by cyanobacteria is thought to have transformed the anaerobic atmosphere of the early Earth into our oxygen-containing one, and according to the endosymbiotic theory, chloroplasts developed from cyanobacteria. Indeed, chloroplasts and cyanobacteria have similar rRNA and structures such as 70S ribosomes and photosynthetic membranes.

Nitrogen is an essential element in proteins and nucleic acids. Though nitrogen constitutes about 79% of the atmosphere, relatively few organisms can utilize this gas. A few prokaryotes—species of filamentous cyanobacteria and

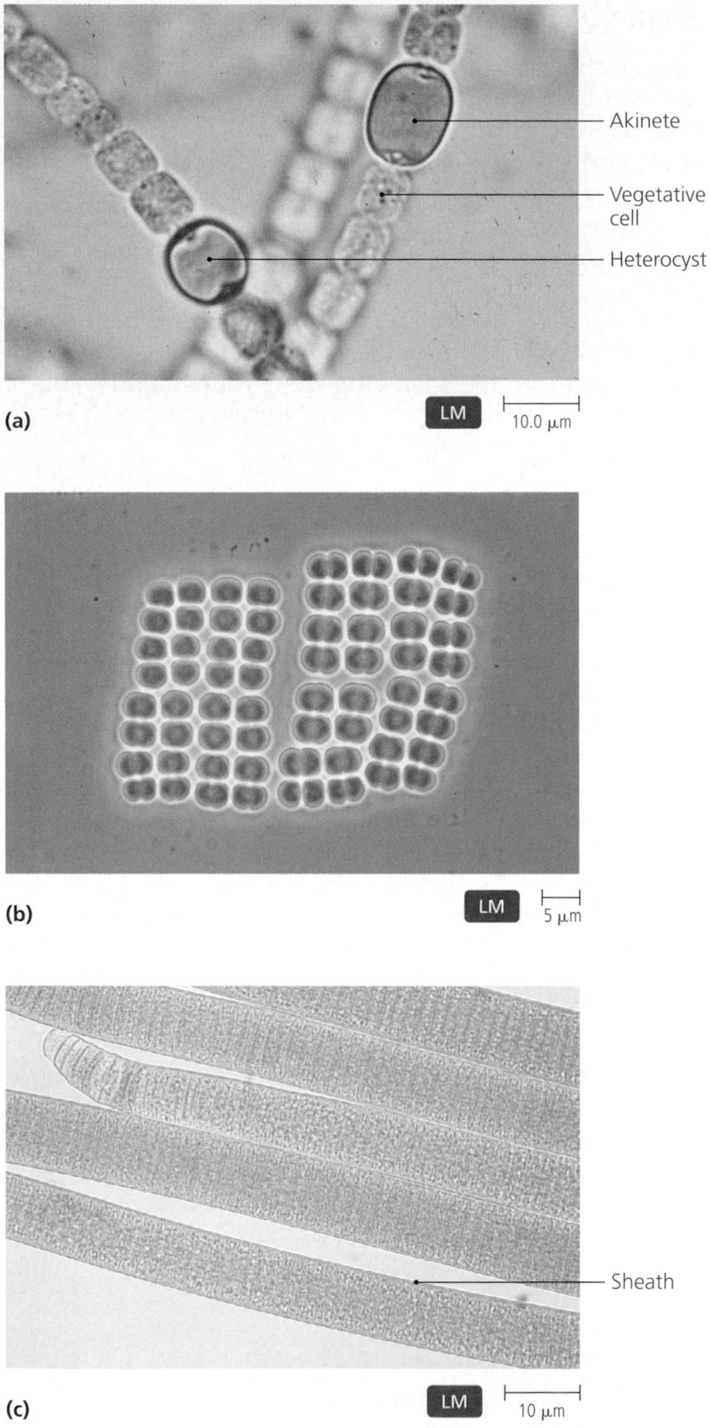

Figure 11.14

Examples of cyanobacteria with different growth habits. **(a)** *Anabaena*, which grows as a filament of cocci with differentiated cells. Heterocysts fix nitrogen; akinetes are reproductive cells. **(b)** *Merismopedia*, which grows as a flat sheet of cocci surrounded by a gelatinous glycocalyx. **(c)** *Oscillatoria*, which forms a filament of tightly appressed disc-shaped cells.

proteobacteria (discussed later in the chapter)—reduce nitrogen gas (N_2) to ammonia (NH_3) via a process called **nitrogen fixation.** Nitrogen fixation is essential for life on Earth because nitrogen-fixers are not only able to enrich their own growth, but also provide nitrogen in a usable form to other organisms.

Because the enzyme responsible for nitrogen fixation is inhibited by oxygen, nitrogen-fixing cyanobacteria are faced with a problem—how to segregate nitrogen fixation, which is inhibited by oxygen, from oxygenic photosynthesis, which produces oxygen. Nitrogen-fixing cyanobacteria solve this problem in one of two ways. Most cyanobacteria isolate the enzymes of nitrogen fixation in specialized, thick-walled, nonphotosynthetic cells called **heterocysts** (see Figure 11.14a). Heterocysts transport reduced nitrogen to neighboring cells in exchange for glucose. Other cyanobacteria photosynthesize during daylight hours and fix nitrogen at night, thereby separating nitrogen fixation from photosynthesis in time rather than in space.

Green and Purple Phototrophic Bacteria Green and purple bacteria differ from plants, algae, and cyanobacteria in two ways: They use *bacteriochlorophylls* for photosynthesis instead of chlorophyll *a,* and they are *anoxygenic;* that is, they do not generate oxygen during photosynthesis. Green and purple phototrophic bacteria commonly inhabit anaerobic, hydrogen sulfide–rich muds at the bottoms of ponds and lakes. These microbes are not necessarily green and purple in color; rather, the terms refer to pigments in some of the better-known members of the groups.

As previously indicated, the green and purple phototrophic bacteria include both sulfur and nonsulfur forms. Whereas nonsulfur bacteria derive electrons for the reduction of CO_2 from organic compounds such as carbohydrates and organic acids, sulfur bacteria derive electrons from the oxidation of hydrogen sulfide to sulfur, as follows:

$$12H_2S + 6CO_2 \xrightarrow{\text{light}} C_6H_{12}O_6 + 6H_2O + 12S \qquad (2)$$

Green sulfur bacteria deposit the resultant sulfur outside their cells, whereas purple sulfur bacteria deposit sulfur within their cells **(Figure 11.15).**

At the beginning of the 20th century, a prominent question in biology concerned the origin of the oxygen released by photosynthetic plants. It was initially thought that oxygen was derived from carbon dioxide, but a comparison of photosynthesis in cyanobacteria (Equation 1) with that in sulfur bacteria (Equation 2) provided evidence that free oxygen is derived from water.

Whereas green sulfur bacteria are placed in Phylum Chlorobi, green nonsulfur bacteria are members of Phylum Chloroflexi (see Figure 11.10). The purple bacteria (both sulfur and nonsulfur) are placed in three classes of Phylum Proteobacteria (see Table 11.2 and Figure 11.10), which is composed of Gram-negative bacteria and is discussed shortly. First, however, we turn our attention to various groups of Gram-positive bacteria, and to a different characteristic of microbes that is used in the classification of Gram-positive bacteria.

Table 11.2 Characteristics of the Major Groups of Phototrophic Bacteria

Phylum	Cyanobacteria	Chlorobi	Chloroflexi	Proteobacteria	Proteobacteria
Class	Cyanobacteria	Chlorobia	Chloroflexi	"γ-proteobacteria"[a]	"α-proteobacteria" and one genus in "β-proteobacteria"
Common name(s)	Blue-green bacteria ("blue-green algae")	Green sulfur bacteria	Green nonsulfur bacteria	Purple sulfur bacteria	Purple nonsulfur bacteria
Major photosynthetic pigments	Chlorophyll *a*	Bacteriochlorophyll *a* plus *c, d,* or *e*	Bacteriochlorophylls *a* and *c*	Bacteriochlorophyll *a* or *b*	Bacteriochlorophyll *a* or *b*
Types of photosynthesis	Oxygenic	Anoxygenic	Anoxygenic	Anoxygenic	Anoxygenic
Electron donor in photosynthesis	H_2O	H_2, H_2S, or S	Organic compounds	H_2, H_2S, or S	Organic compounds
Sulfur deposition	None	Outside of cell	None	Inside of cell	None
Nitrogen fixation	Some species	None	None	None	None
Motility	Nonmotile or gliding	Nonmotile	Gliding	Motile with polar or peritrichous flagella	Nonmotile or motile with polar flagella

[a]The names of classes in quotations are not officially recognized.

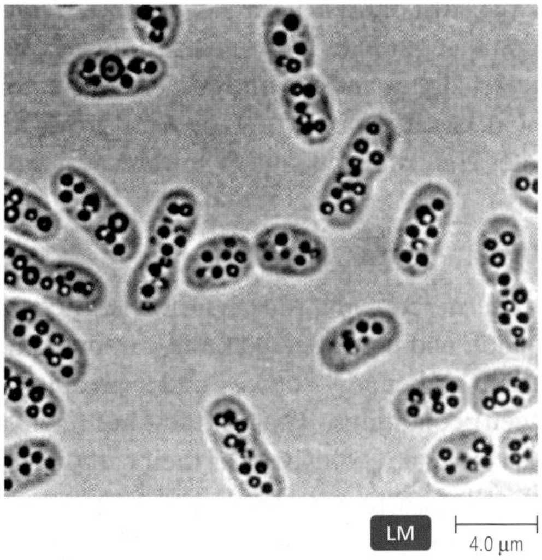

▲ ***Figure 11.15***

Deposits of sulfur within purple sulfur bacteria in the genus *Chromatium.* These bacteria oxidize H_2S to produce the granules of elemental sulfur evident in this photomicrograph. *Where do green sulfur bacteria deposit sulfur grains?*

Figure 11.15 Green sulfur bacteria deposit sulfur grains externally.

Taxonomists have come to use the *G + C ratio*—the percentage of all base pairs in a genome that are guanine–cytosine base pairs—as a useful criterion in classifying microbes. Bacteria with G + C ratios below 50% are considered "low G + C bacteria"; the remainder are considered "high G + C bacteria." Because taxonomists have discovered that Gram-positive bacteria with low G + C ratios have similar sequences in their 16S rRNA, and that those with high G + C ratios also have rRNA sequences in common, they have assigned low G + C bacteria and high G + C bacteria to different phyla. We discuss the low G + C bacteria first.

Low G + C Gram-Positive Bacteria

Learning Objectives

- ✓ Discuss the lack of cell walls in mycoplasmas.
- ✓ Identify significant beneficial or detrimental effects of the genera *Clostridium, Bacillus, Listeria, Lactobacillus, Sreptococcus,* and *Staphylococcus.*

The low G + C Gram-positive bacteria are classified within Phylum **Firmicutes** (fer-mik′ū-tēz), which includes three groups: clostridia, mycoplasmas, and low G + C Gram-positive bacilli and cocci. Next we consider these three groups in turn.

Clostridia

Clostridia are rod-shaped, obligate anaerobes, many of which from endospores. The group is named for the genus

Highlight 11.1 Botulinum Toxin: A Killer and a Healer

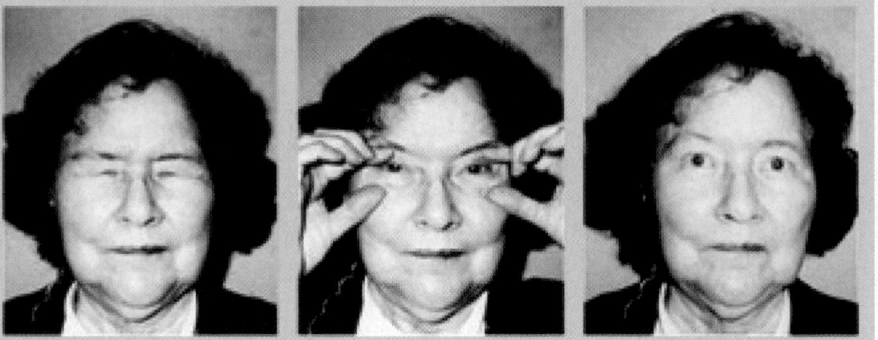

Clostridium botulinum produces botulinum toxins, some of the deadliest toxins known. When absorbed in the body, botulism toxins interfere with the release of acetylcholine, the neurotransmitter that signals muscles to contract. As a result, muscle cells cannot contract, and a slow but progressive paralysis spreads throughout the body. Death occurs when paralysis of respiratory muscles results in respiratory failure. This disease is called botulism.

Amazingly, one toxin known as type A botulism toxin is being increasingly used as part of a cosmetic procedure to erase frown lines, crow's feet, and other facial wrinkles. Purified type A botulinum toxin is marketed as Botox, extremely small doses of which are injected into those facial muscles that cause skin wrinkles, paralyzing or weakening them. Such treatments only last approximately 3–6 months and must be repeated in order to maintain the desired effects.

The same type A botulinum toxin is also injected into the appropriate muscles to treat two eye conditions caused by abnormal muscle contractions: *strabismus,* the inability of the eyes to maintain binocular vision due to an imbalance between eye muscles, and *bleparospasm,* a condition caused by involuntary spasmodic contractions of facial muscles surrounding the eye. The photos in this box show the effects of the toxin on a patient with bleparospasm. The photo at left shows the patient before treatment; she is unable to open her eyes. The middle photo shows the patient, still prior to treatment, forcing her eyes open with her fingers. The photo at right shows the patient after treatment; she is now able to keep her eyes open because the toxin has suppressed the excessive contractions of the treated muscles.

Clostridium[9] (klos-trid'ē-ŭm), which is important both in medicine—in large part because its members produce potent toxins that cause a variety of diseases in humans—and in industry because their endospores enable them to survive harsh conditions, including many types of disinfection and antisepsis. Examples of clostridia include *C. tetani* (which causes tetanus), *C. perfringens* (gangrene), *C. botulinum* (botulism), and *C. difficile* (severe diarrhea). **Highlight 11.1** describes how a deadly toxin produced by *C. botulinum* has been put to use for cosmetic purposes.

The giant bacterium *Epulopiscium* (ep'ū-lō-pis-sē-ŭm) is an unusual relative of *Clostridium.* As discussed in Highlight 3.2 on page 59, *Epulopiscium* is large enough to be seen without the use of a microscope and has numerous short flagella, which make it appear ciliated. For these reasons it was first considered eukaryotic. *Epulopiscium* is also unusual in that it does not reproduce by binary fission but instead forms new cells within the parent cell and releases them through a slit in the parent's cell wall.

Other microbes related to *Clostridium* include sulfate-reducing microbes, which produce H_2S from elemental sulfur during anaerobic respiration, and *Veillonella* (vī-lō-nel'ă), a genus of anaerobic cocci that live as part of the biofilm (plaque) that forms on the teeth of warm-blooded animals. *Veillonella* is unusual because even though it has a typical Gram-positive cell wall structure, it has a negative Gram reaction—it stains pink.

Mycoplasmas

A second class of low G + C bacteria are the **mycoplasmas**[10] (mī'kō-plaz'mas). These facultative or obligate anaerobes lack cell walls, which means they stain pink when Gram stained. Indeed, until their nucleic acid sequences proved their similarity to Gram-positive organisms, mycoplasmas were classified in a phylum of Gram-negative microbes, instead of in Phylum Firmicutes with other low G + C Gram-positive bacteria.

Mycoplasmas are able to survive without cell walls because they colonize osmotically protected habitats such as animal and human bodies, and because they have tough cytoplasmic membranes, many of which contain lipids called *sterols* that give the membranes strength and rigidity. Because they lack cell walls, they are pleomorphic. They were named "mycoplasmas" because their filamentous forms resemble the filaments of fungi. Mycoplasmas have diameters ranging from 0.2 μm to 0.8 μm, making them the smallest free-living cells, and many mycoplasmas have a terminal structure that is used to attach them to eukaryotic cells and that gives the bacteria a pear-like shape. They require organic growth factors, such as cholesterol, fatty acids, vitamins, amino acids, and nucleotides, which they either acquire from their host or which must be added to laboratory media. When growing on solid media, most species form a distinctive "fried egg" appearance because cells in the center

[9]From Greek *kloster,* meaning spindle.
[10]From Greek *mycos,* meaning fungus, and *plassein,* meaning to mold.

of the colony grow into the agar while those around the perimeter only spread across the surface **(Figure 11.16).**

In animals, mycoplasmas colonize mucous membranes of the respiratory and urinary tracts and are associated with pneumonia and urinary tract infections.

Gram-Positive Bacilli and Cocci

A third group of low G + C Gram-positive organisms is composed of bacilli and cocci that are significant in environmental, industrial, and health care settings. Among the genera in this group are *Bacillus, Listeria* (lis-tēr'ē-ă), *Lactobacillus* (lak'tō-ba-sil'ŭs), *Streptococcus, Enterococcus,* and *Staphylococcus.*

Bacillus The genus *Bacillus* includes endospore-forming aerobes and facultative anaerobes that typically move by means of peritrichous flagella. Numerous species of *Bacillus* are common in soil.

A few species of *Bacillus* are directly beneficial to humans. *Bacillus thuringiensis* (thur-in-jē-en'sis) is beneficial to farmers and gardeners. During sporulation, this bacterium produces a crystalline protein that is toxic to caterpillars that ingest it **(Figure 11.17).** Gardeners spray *Bt toxin,* as preparations of the bacterium and toxin are known, on plants to protect them from caterpillars. Scientists have achieved the same effect, without the need of spraying, by introducing the gene for Bt toxin into plants' chromosomes. Other beneficial species of *Bacillus* include *B. polymyxa* and *B. licheniformis,* which synthesize the antibiotics polymyxin and bacitracin, respectively. Chapter 10 discusses the production and effects of antibiotics in more detail.

Bacillus anthracis (an-thrā'sis), which causes anthrax, gained notoriety in 2001 as an agent of bioterrorism. Its endospores are either inhaled or enter the body through breaks in the skin. When they germinate, the vegetative cells produce toxins that kill surrounding tissues. Untreated *cutaneous anthrax* is fatal in 20% of patients; untreated *pulmonary anthrax* is generally 100% fatal.

Endospores of *B. cereus* (se'rē-us) occasionally contaminate rice and cause disease in people who ingest them. These heat-resistant endospores survive cooking, germinate in the intestinal tract, and produce a toxin that induces nausea, vomiting, and abdominal cramping. As discussed in Chapter 9, refrigeration prevents the excessive growth of these and other contaminates.

Listeria Another pathogenic low G + C Gram-positive rod is *Listeria monocytogenes* (mo-nō-sī-to'je-nēz), which can contaminate milk and meat products. This microbe, which does not produce endospores, is notable because it continues to reproduce under refrigeration, and it can survive inside phagocytic white blood cells. *Listeria* rarely causes disease in adults, but it kills fetuses when the organism crosses the placenta from their infected mothers. It also causes meningitis[11] and bacteremia[12] when it infects immunocompromised patients such as the aged and patients with AIDS, cancer, or diabetes.

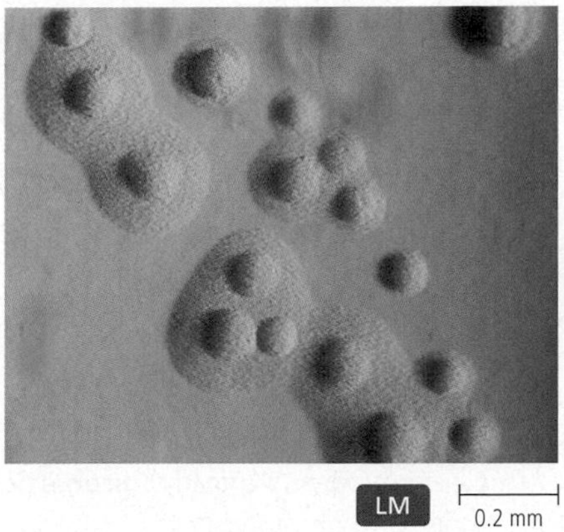

▲ ***Figure 11.16***

The distinctive "fried egg" appearance of *Mycoplasma* colonies growing on an agar surface. This visual feature is unique to this group of bacteria.

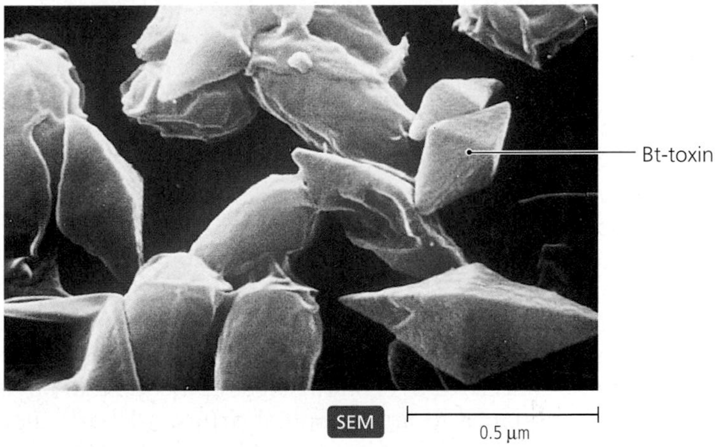

▲ ***Figure 11.17***

Crystals of Bt toxin, produced by the endospore-forming *Bacillus thuringiensis.* The crystalline protein kills caterpillars that ingest it.

Lactobacillus Organisms in the genus *Lactobacillus* are nonspore-forming rods normally found growing in the human mouth, stomach, intestinal tract, and vagina. These organisms rarely cause disease; instead, they protect the body by inhibiting the growth of pathogens—a situation called *microbial antagonism.* Lactobacilli are used in industry in the

[11] Inflammation of the membranes covering the brain and spinal cord; from Greek *meninx,* meaning membrane.

[12] The presence of bacteria, particularly those that produce disease symptoms, in the blood.

New Frontiers 11.1 *Lactobacillus*: A Probiotic to Watch

For over a century, people have eaten yogurt and other dairy products containing live bacterial cultures, believing this would benefit their health. As it turns out, they may have been on to something. Recent studies have shown that the ingestion of some microorganisms—in particular, *Lactobacillus,* the same bacterium that converts milk into cheese and yogurt—can reduce symptoms of diarrhea in children, especially in cases caused by rotaviruses. Scientists are also studying the use of *Lactobacillus* in relieving antibiotic-associated diarrhea, milk allergies, and certain respiratory infections. Indeed, *Lactobacillus* is the focus of many studies in the growing field of **probiotics,** the administration of microorganisms to promote health.

What makes *Lactobacillus* so promising as a medical therapy? The mechanisms are not completely understood, but one of the central ideas is that *Lactobacillus* (which lives naturally and harmlessly in our intestines and other parts of the body) competes with harmful microorganisms for resources, keeping pathogenic microbes in check. It may also interact with the immune system in ways that can help alleviate milk allergies. Further evidence suggests that *Lactobacillus* can induce the expression of certain proteins that protect the intestines against harmful bacteria and viruses. In addition, their natural presence in dairy products makes them easy to administer to patients.

Much more research remains to be done, but the preliminary findings are encouraging. In the exciting field of probiotics, *Lactobacillus* is the star on center stage.

Source: Friedrich, M.J. 2000. A bit of culture for children: Probiotics may improve health and fight disease. *Journal of the American Medical Association* 284(11): 1365–1366.

production of yogurt, buttermilk, pickles, and sauerkraut. **New Frontiers 11.1** discusses the growing role of *Lactobacillus* in promoting human health.

Streptococcus* and *Enterococcus The genera *Streptococcus* and *Enterococcus* are diverse groups of Gram-positive cocci associated in pairs and chains (see Figure 11.6b). They produce numerous human diseases, including pharyngitis (strep throat), scarlet fever, impetigo, fetal meningitis, wound infections, pneumonia, and diseases of the inner ear, skin, blood, and kidneys. In recent years, health care providers have become concerned over strains of streptococci that have developed resistance to antimicrobial drugs. Of particular concern are so called "flesh-eating," drug-resistant streptococci that produce toxins that destroy muscle and fat tissue.

Staphylococcus Among the common inhabitants of humans is *Staphylococcus aureus*[13] (aw'rē-ŭs), which is typically found growing harmlessly on the skin in clusters (see Figure 11.6e). A variety of toxins and enzymes allow some strains of *S. aureus* to invade the body and cause such diseases as bacteremia, pneumonia, wound infections, food poisoning, toxic shock syndrome, and diseases of the joints, bones, heart, and blood.

The characteristics of the low G + C Gram-positive bacteria are summarized in **Table 11.3.** To summarize, these bacteria, which are classified into three classes within Phylum Firmicutes, include the anaerobic endospore-forming rod *Clostridium;* the pleomorphic *Mycoplasma;* the aerobic and facultative aerobic endospore-forming rod *Bacillus,* the nonendospore-forming rods *Listeria* and *Lactobacillus,* and the cocci *Streptococcus* and *Staphylococcus.*

Next we consider Gram-positive bacteria that have high G + C ratios.

[13]From Latin *aurum, meaning* gold, because it produces yellow pigments.

High G + C Gram-Positive Bacteria

Learning Objectives

- ✓ Explain the slow growth of *Mycobacterium.*
- ✓ Identify significant beneficial or detrimental properties of the genera *Corynebacterium, Mycobacterium, Actinomyces, Nocardia,* and *Streptomyces.*

Taxonomists classify Gram-positive bacteria with a G + C ratio greater than 50% in the phylum Actinobacteria, which includes species with rod-shaped cells (many of which are significant human pathogens) and filamentous bacteria, which resemble fungi in their growth habit and in the production of reproductive spores. Here we will discuss briefly some prominent high G + C Gram-positive bacteria.

Corynebacterium

Members of the genus *Corynebacterium* (kŏ-rī'nē-bak-tēr'ē-ŭm) are pleomorphic—though generally rod-shaped—aerobes and facultative anaerobes. They reproduce by snapping division, which often causes the cells to form V-shapes and palisades (see Figure 11.7d). Corynebacteria are also characterized by their stores of phosphate within inclusions called **metachromatic granules,** which stain differently from the rest of the cytoplasm when the cells are stained with the blue dyes methylene blue or toluidine blue. The best known species is *Corynebacterium diphtheriae,* which causes diphtheria.

Mycobacterium

The genus *Mycobacterium* (mī'kō-bak-tēr'ē-um) is composed of aerobic species that are slightly curved to straight rods that sometimes form filaments. Mycobacteria grow very slowly, often requiring a month or more to form a visible

Table 11.3 Characteristics of Selected Gram-Positive Bacteria

Phylum/Class	G + C Ratio	Representative Genera	Special Characteristics	Diseases
Firmicutes				
"Clostridia"[a]	Low (less than 50%)	*Clostridium*	Obligate anaerobic rods; endospore formers	Tetanus Botulism Gangrene Severe diarrhea
		Epulopiscium	Giant rods	
		Veillonella	Part of oral biofilm on human teeth; stain like Gram-negative (pink)	Dental caries
Firmicutes				
Mollicutes	Low (less than 50%)	*Mycoplasma*	Lack cell wall; pleomorphic; smallest free-living cells; stain like Gram-negative (pink)	Pneumonia Urinary tract infections
Firmicutes				
"Bacilli"	Low (less than 50%)	*Bacillus*	Facultative anaerobic rods; endospore formers	Anthrax
		Listeria	Contaminates of dairy products	Listeriosis
		Lactobacillus	Produce yogurt, buttermilk, pickles, sauerkraut	Rare blood infections
		Streptococcus	Cocci in chains	Strep throat, scarlet fever, and others
		Staphylococcus	Cocci in clusters	Bacteremia, food poisoning, and others
Actinobacteria				
Actinobacteria	High (greater than 50%)	*Corynebacterium*	Snapping division; metachromatic granules in cytoplasm	Diphtheria
		Mycobacterium	Waxy cell walls (mycolic acid)	Tuberculosis and meningitis
		Actinomyces	Filaments	Actinomycosis
		Nocardia	Filaments; degrade pollutants	Lesions
		Streptomyces	Produce antibiotics	Rare sinus infections

[a]The names of classes in quotations are not officially recognized.

colony on an agar surface. Their slow growth is partly due to the time and energy required to enrich their cell walls with high concentrations of long carbon-chain waxes called **mycolic acids,** which make the cells resistant to desiccation and to staining with water-based dyes. As discussed in Chapter 4, microbiologists developed the *acid-fast stain* for mycobacteria because they are difficult to stain by standard staining techniques such as the Gram stain. Though some mycobacteria are free-living, the most prominent species are pathogens of animals and humans, including *Mycobacterium tuberculosis* (too-ber'kū-lō-sis) and *Mycobacterium leprae* (lep'rī), which cause tuberculosis and leprosy, respectively. Mycobacteria should not be confused with the low G + C mycoplasmas discussed earlier.

Actinomycetes

Actinomycetes (ak'ti-nō-mī-sētz) are high G + C Gram-positive bacteria that form branching filaments resembling fungi (**Figure 11.18** on page 330). Of course, in contrast to fungi, the filaments of actinomycetes are composed of prokaryotic cells. As we have seen, some actinomycetes also resemble fungi in the production of chains of reproductive spores at the ends of their filaments (see Figure 11.4). These spores should not be confused with endospores, which are

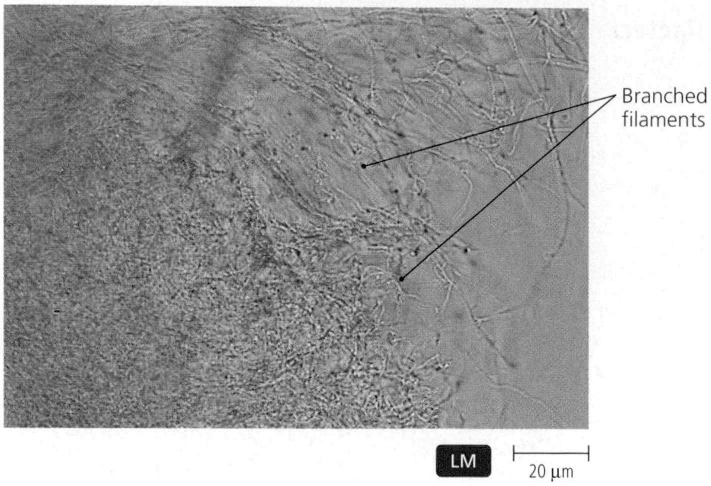

▲ ***Figure 11.18***
The branching filaments of actinomycetes. This photograph shows filaments of *Streptomyces* sp. at the edge of a colony growing on agar. *How do the filaments of actinomycetes compare to the filaments of fungi?*

Figure 11.18 *Filaments of actinomycetes are thinner than those of fungi, and they are composed of prokaryotic cells.*

resting stages and not reproductive cells. Actinomycetes may cause disease, particularly in immunocompromised patients. Among the important actinomycete genera are *Actinomyces* (which gives this group its name), *Nocardia*, and *Streptomyces*.

Actinomyces Species of *Actinomyces* (ak′ti-nō-mī′sēz) are facultative capneic[14] filaments that are normal inhabitants of the mucous membranes lining the oral cavity and throats of humans. *Actinomyces israelii* (is-rā′el-ē-ē) growing in humans destroys tissue to form abscesses and can spread throughout the abdomen, consuming every vital organ.

Nocardia Species of *Nocardia* (nō-kar′dē-ă) are soil- and water-dwelling aerobes that typically form aerial and subterranean filaments that make them resemble fungi. *Nocardia* is notable because it can degrade many pollutants of landfills, lakes, and streams, including waxes, petroleum hydrocarbons, detergents, benzene, polychlorinated biphenyls (PCBs), pesticides, and rubber. Some species of *Nocardia* cause lesions in humans.

Streptomyces Bacteria in the genus *Streptomyces* (strep-tō-mī′sēz) are important in several realms. Ecologically, they recycle nutrients in the soil by degrading a number of carbohydrates, including cellulose, lignin (the woody part of plants), chitin (the skeletal material of insects and crustaceans), latex, aromatic chemicals (organic compounds containing a benzene ring), and keratin (the protein that forms hair, nails, and horns). The metabolic by-products of *Streptomyces* give soil its musty smell. Medically, *Streptomyces* produce most of the important antibiotics, including chloramphenicol, erythromycin, and tetracycline, as discussed in Chapter 10.

Table 11.3 on page 329 includes a summary of the characteristics of the genera of high G + C Gram-positive bacteria (Phylum Actinobacteria) discussed in this chapter.

CRITICAL THINKING

In Chapter 4, we discussed the identification of microbes using dichotomous taxonomic keys (see page 121). Design a key for all the genera of Gram-positive bacteria listed in Table 11.3 (p. 329).

To this point we have discussed archaea, deeply branching bacteria, phototrophic bacteria, and Gram-positive bacteria. Now it's time to turn our attention to the Gram-negative bacteria that are grouped together within the phylum Proteobacteria.

Gram-Negative Proteobacteria

The phylum **Proteobacteria**[15] constitutes the largest and most diverse group of bacteria (see Figure 11.10). Though they have a variety of shapes, reproductive strategies, and nutritional types, they are all Gram-negative and share common 16S rRNA nucleotide sequences. The G + C ratio of Gram-negative species is not critical in delineating taxa of most Gram-negative organisms, so we will not consider this characteristic in our discussion of Gram-negative bacteria.

There are five distinct classes of proteobacteria, designated by the first five letters of the Greek alphabet—α, β, γ, δ, and ε (alpha, beta, gamma, delta, and epsilon). These classes are distinguished by minor differences in their rRNA sequences. Here we will focus our attention on species with novel characteristics as well as species with practical importance.

"Alphaproteobacteria"

Learning Objective

✓ Describe the appearance and function of prosthecae in alphaproteobacteria.

Alphaproteobacteria are typically aerobes capable of growing at very low nutrient levels. Many have unusual methods of metabolism, as we will see shortly. They may be rods, curved rods, spirals, coccobacilli, or pleomorphic in shape. Many species have unusual extensions called *prosthecae* (pros-thē′kē), which are composed of cytoplasm surrounded by the cytoplasmic membrane and cell wall **(Figure 11.19).** They use prosthecae for attachment and to increase surface area for nutrient absorption. Some prosthecate species produce buds at the ends of the extensions.

[14]Meaning they grow best with a relatively high concentration of carbon dioxide.
[15]Named for the Greek god *Proteus*, who could assume many shapes.

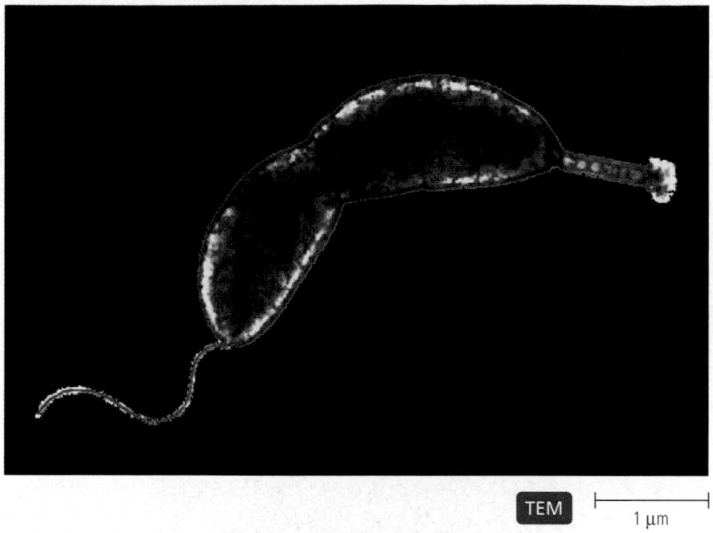

▲ ***Figure 11.19***

A prostheca, an extension of an α-proteobacterial cell that increases surface area for absorbing nutrients and serves as an organ of attachment. This prosthecate bacterium, *Hyphomicrobium facilis,* also produces buds from its prostheca.

▲ ***Figure 11.20***

Nodules on soybean roots stimulated by the growth of *Rhizobium,* a nitrogen-fixing α-proteobacterium.

Nitrogen Fixers Two genera of nitrogen fixers in the class α-proteobacteria—*Azospirillum* (ā-zō-spī′ril-ŭm) and *Rhizobium* (rī-zō′bē-ŭm)—are important in agriculture. They grow in association with the roots of plants, where they make atmospheric nitrogen (N_2) available to the plants as ammonia (NH_3), so-called fixed nitrogen. *Azospirillum* associates with the outer surfaces of roots of tropical grasses, such as sugar cane. In addition to supplying nitrogen to the grass, this bacterium also releases chemicals that stimulate the plant to produce numerous root hairs, which increase a root's surface area and thus its uptake of nutrients. *Rhizobium* grows within the roots of leguminous plants such as peas, beans, and clover, stimulating the formation of nodules on their roots **(Figure 11.20).** *Rhizobium* cells within the nodules make ammonia available to the plant, encouraging growth. Scientists are actively seeking ways to successfully insert the genes of nitrogen fixation into plants such as corn, which require large amounts of nitrogen.

Nitrifying Bacteria As discussed in Chapter 6, chemoautotrophic organisms derive their carbon from carbon dioxide, and their energy from inorganic chemicals. Those that derive electrons from the oxidation of nitrogenous compounds are called **nitrifying bacteria.** These microbes are important in the environment and in agriculture because they convert reduced nitrogen compounds, such as ammonia (NH_3), ammonium (NH_4^+), and nitrite (NO_2^-), into nitrate (NO_3^-)—a process called **nitrification.** Nitrate moves more easily through soil than reduced nitrogenous compounds and is thereby more available to plants. Nitrifying α-proteobacteria oxidize ammonia or ammonium and are in the genus *Nitrobacter.*

CRITICAL THINKING

Contrast the processes of nitrogen fixation and nitrification.

Purple Nonsulfur Phototrophs With one exception (the β-proteobacterium *Rhodocyclus*), purple nonsulfur phototrophs are classified as α-proteobacteria. Purple nonsulfur bacteria grow in the upper layer of mud at the bottoms of lakes and ponds. As we discussed earlier, they harvest light as an energy source by using bacteriochlorophylls, and they do not generate oxygen during photosynthesis. Morphologically, they may be rods, curved rods, or spirals; some species are prosthecate. Refer to Table 11.2 (p. 325) to review the characteristics of all phototrophic bacteria.

Pathogenic Alphaproteobacteria Notable pathogens among the α-proteobacteria include *Rickettsia* (ri-ket′sē-ă), *Brucella* (broo-sel′lă), and *Ehrlichia* (er-lik′ē-ă).

Rickettsia is a genus of small, Gram-negative, aerobic rods that live and reproduce by binary fission inside mammalian cells. They cause a number of human diseases, including typhus and Rocky Mountain spotted fever. Rickettsias cannot use glucose as a nutrient; instead they oxidize amino acids and Krebs cycle intermediates such as glutamic acid and succinic acid. For this reason they are obliged to live within other cells where these nutrients are produced.

Brucella is a coccobacillus that causes *brucellosis,* a disease of mammals characterized by spontaneous abortions and sterility in animals. In contrast, infected humans suffer chills, sweating, fatigue, and fever. *Brucella* is notable because

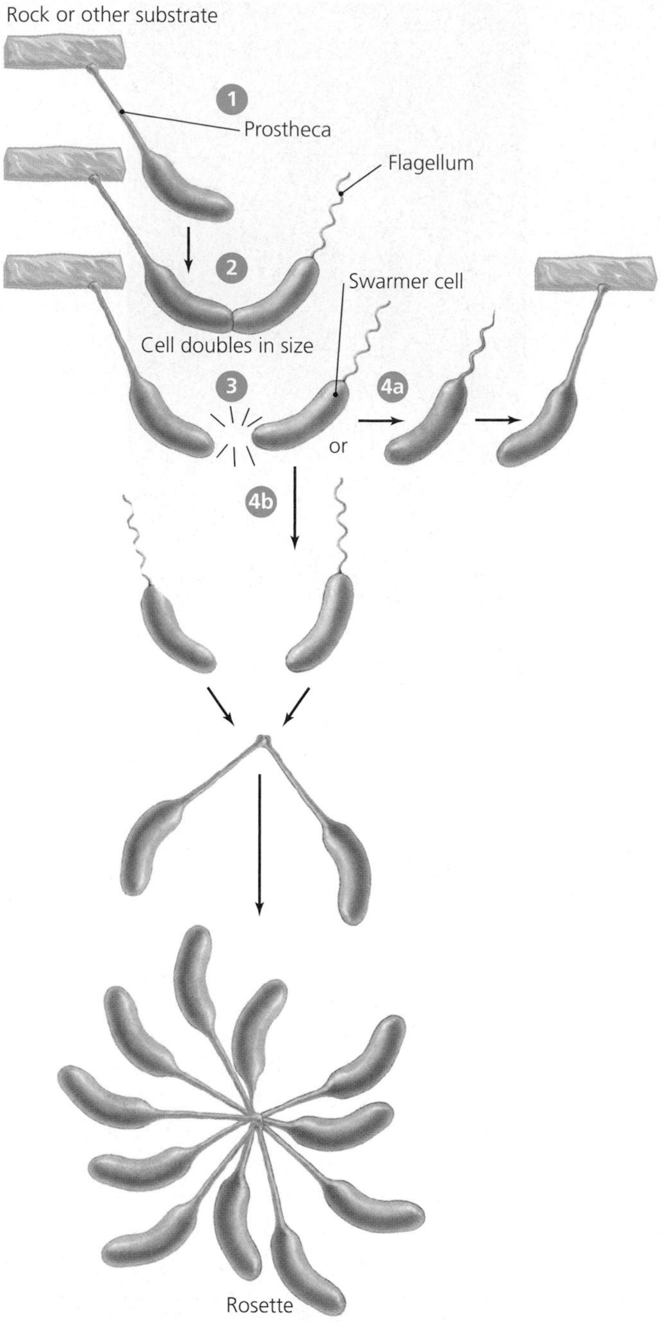

Figure 11.21

Growth and reproduction of *Caulobacter.* ① A cell attached to a substrate by its prostheca. ② Growth and production of an apical flagellum. ③ Division by asymmetrical binary fission to produce a flagellated daughter cell called a swarmer cell. ④a Attachment of a swarmer cell to a substrate by a prostheca that replaces the flagellum. ④b Attachment of a swarmer cell to other cells to form a rosette.

it survives phagocytosis by white blood cells—normally an important step in the body's defense against disease.

Ehrlichia is a tick-borne α-proteobacterium that causes disease in humans by living within white blood cells. Symptoms of *ehrlichiosis* include high fever, headache, muscle pain, and *leukopenia,* an abnormal decrease in the number of white blood cells.

Figure 11.22

A plant gall resulting from infection by the α-proteobacterium *Agrobacterium.* The gall is formed by the proliferation of undifferentiated plant cells caused by the expression of a plant growth hormone gene inserted into a plant chromosome by a plasmid carried by the infecting *Agrobacterium* cells. The gall cells synthesize nutrients for the bacteria.

Other Alphaproteobacteria Other α-proteobacteria are important in industry and the environment. For example, *Acetobacter* (as′ē-tō-bak-ter) and *Gluconobacter* (gloo-kon′ō-bak-ter) are used to synthesize acetic acid (vinegar). *Caulobacter* (kaw-lō-bak-ter) is a common prosthecate rod-shaped microbe that inhabits nutrient-poor seawater and fresh water; it can also be found in laboratory water baths.

Caulobacter has a unique reproductive strategy **(Figure 11.21).** A cell that has attached to a substrate with its prostheca ① grows until it has doubled in size, at which time it produces a flagellum at its apex ②; it then divides by asymmetric binary fission ③. The flagellated daughter cell, which is called a swarmer cell, then swims away. The swarmer cell can either attach to a substrate with a new prostheca that replaces the flagellum ④a or to other swarmer cells to form a rosette of cells ④b. The process of reproduction repeats about every two hours.

Scientists are very interested in the potential usefulness of another α-proteobacterium, *Agrobacterium* (a′grō-bak-tēr-ē-ŭm), which infects plants to form tumors called *galls* **(Figure 11.22).** The bacterium inserts a plasmid, which carries a gene for a plant growth hormone, into a chromosome of the plant. The growth hormone causes the cells of the plant to proliferate into a gall and to produce nutrients for the bacterium. Scientists have discovered that they can insert almost any DNA sequence into the plasmid, making it an ideal vector for genetic manipulation of plants.

Characteristics of selected members of the alphaproteobacteria are listed in **Table 11.4** on page 334.

"Betaproteobacteria"

Learning Objective

✓ Name three pathogenic and three useful betaproteobacterial species.

Betaproteobacteria are another diverse group of Gram-negative bacteria that thrive in habitats with low levels of nutrients. They differ from α-proteobacteria in their rRNA sequences, though metabolically the two groups overlap. One example of the metabolic overlap of the β-proteobacteria with the α-proteobacteria is *Nitrosomonas* (nī-trō-sō-mō'nas), an important nitrifying bacterium of soils; instead of oxidizing ammonia or ammonium to nitrate, however, it oxidizes nitrite to nitrate. In this section we will discuss a few of the interesting β-proteobacteria.

Pathogenic Betaproteobacteria Species of *Neisseria* (nī-sē'rē-ă) are Gram-negative diplococci that inhabit the mucous membranes of mammals and cause such diseases as gonorrhea, meningitis, pelvic inflammatory disease, and inflammation of the cervix, pharynx, and external lining of the eye.

Other pathogenic β-proteobacteria include *Bordetella* (bōr-dĕ-tel'ă), which is the cause of pertussis (whooping cough), and *Burkholderia* (burk-hol-der'ē-ă), which recycles numerous organic compounds in nature. *Burkholderia* commonly colonizes moist environmental surfaces (including laboratory and medical equipment) and the respiratory passages of patients with cystic fibrosis.

Other Betaproteobacteria Members of the genus *Thiobacillus* (thī-ō-ba-sil'ŭs) are colorless sulfur bacteria that are important in recycling sulfur in the environment by oxidizing hydrogen sulfide (H_2S) or elemental sulfur (S^0) to sulfate (SO_4^{2-}). Miners use *Thiobacillus* to leach metals from low-grade ore, though the bacterium does cause extensive pollution when it releases metals and acid from mine wastes.

Sewage treatment supervisors are interested in *Zoogloea* (zō'ō-glē-ă) and *Sphaerotilus* (sfēr-ōl'til-us), two genera that form *flocs*—slimy, tangled masses of bacteria and organic matter in sewage. *Zoogloea* forms compact flocs that settle to the bottom of treatment tanks and assist in the purification process. *Sphaerotilus,* in contrast, forms loose flocs that do not settle and thus impede the proper flow of waste through a treatment plant.

Spirillum (spī-ril'ŭm) is a Gram-negative, spiral-shaped β-proteobacterium whose stained polar flagella are easily visible with a light microscope (see Figure 3.6c). *Spirillum minus* causes rat-bite fever.

The characteristics of the genera of β-proteobacteria discussed in this section are summarized in Table 11.4 on page 334.

"Gammaproteobacteria"

Learning Objective

✓ Describe the gammaproteobacteria.

The **gammaproteobacteria** make up the largest and most diverse class of proteobacteria; almost every shape, arrangement of cells, metabolic type, and reproductive strategy is represented in this group. Ribosomal studies indicate that γ-proteobacteria can be divided into several subgroups:

- Purple sulfur bacteria
- Intracellular pathogens
- Methane oxidizers
- Facultative anaerobes that utilize Emdem-Meyerhof glycolysis and the pentose phosphate pathway
- Pseudomonads, which are aerobes that catabolize carbohydrates by the Entner-Doudoroff and pentose phosphate pathways

Here we will examine some representatives of these groups.

Purple Sulfur Bacteria Whereas the purple *non*sulfur bacteria are distributed among the α- and β-proteobacteria, **purple sulfur bacteria** are γ-proteobacteria. Purple sulfur bacteria are obligate anaerobes that oxidize hydrogen sulfide to sulfur, which they deposit as internal granules. They are found in sulfur-rich zones in lakes, bogs, and oceans. Some species form intimate relationships with marine worms, covering the body of a worm like strands of hair **(Figure 11.23).**

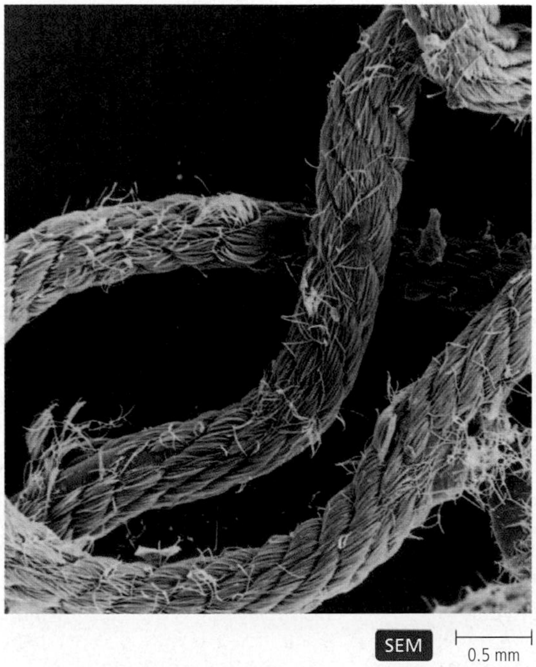

▲ ***Figure 11.23***
Purple sulfur bacteria covering the surface of a nematode (round worm). The presence of so many bacteria produces the appearance of a woven rope.

Table 11.4 Characteristics of Selected Gram-Negative Bacteria

Phylum/Class	Representative Members	Special Characteristics	Diseases
Proteobacteria			
"α-proteobacteria"[a]	*Azospirillum*	Nitrogen fixers	
	Rhizobium	Nitrogen fixers	
	Nitrobacter	Nitrifying bacteria	
	Purple nonsulfur bacteria	Anoxygenic phototrophs	
	Rickettsia	Intracellular pathogens	Typhus and Rocky Mountain spotted fever
	Brucella	Coccobacilli	Brucellosis
	Ehrlichia	Live inside white blood cells	Leukopenia
	Acetobacter, Glucanobacter	Synthesize acetic acid	
	Caulobacter	Prosthecate bacteria	
	Agrobacterium	Cause galls in plants; vectors for gene transfer in plants	
Proteobacteria			
"β-proteobacteria"	*Nitrosomonas*	Nitrifying bacteria	
	Neisseria	Diplococci	Gonorrhea and meningitis
	Bordetella		Pertussis
	Burkholderia		Lung infection of cystic fibrosis patients
	Thiobacillus	Colorless sulfur bacteria	
	Zoogloea	Used in sewage treatment	
	Sphaerotilus	Block sewage treatment pipes	
	Spirillum	Amphitrichous polar flagella	Rat bite fever
Proteobacteria			
"γ-proteobacteria"	Purple sulfur bacteria		
	Legionella	Intracellular pathogens	
	Coxiella	Intracellular pathogens	
	Methane oxidizers		
	Glycolytic facultative anaerobes	Facultative anaerobes that catabolize carbohydrates via glycolysis and the pentose phosphate pathway	See Table 11.5 on page 336
	Pseudomonas	Aerobes that catabolize carbohydrates via Entner-Doudoroff and pentose phosphate pathways	Urinary tract infections, external otitis
	Azotobacter *Azomonas*	Nitrogen fixers not associated with plant roots	
Proteobacteria			
"δ-proteobacteria"	*Desulfovibrio*	Sulfate reducers	
	Bdellovibrio	Pathogens of Gram-negative bacteria	
	Myxobacteria	Reproduce by forming differentiated fruiting bodies	
Proteobacteria			
"ε-proteobacteria"	*Campylobacter*	Curved rods	Gastroenteritis
	Helicobacter	Spirals	Gastric ulcers
Chlamydiae			
"Chlamydiae"	*Chlamydia*	Intracellular pathogens; lack peptidoglycan	Neonatal blindness and lymphogranuloma venereum

[a]The names of classes in quotations are not officially recognized.

Table 11.4 *(continued)*

Phylum/Class	Representative Members	Special Characteristics	Diseases
Spirochaetes			
"Spirochaetes"	*Treponema*	Motile by axial filaments	Syphilis
	Borrelia	Motile by axial filaments	Lyme disease
Bacteroidetes			
"Bacteroides"	*Bacteroides*	Anaerobes that live in animal colons	Abdominal infections
	Cytophaga	Digest complex polysaccharides	

Intracellular Pathogens Organisms in the genera *Legionella* (lē-jŭ-nel'lă) and *Coxiella* (kok-sē-el'ă) are pathogens of humans that avoid digestion by white blood cells, which are normally part of a body's defense; in fact, they thrive inside these defensive cells. *Legionella* derives energy from the metabolism of amino acids, which are more prevalent inside cells than outside. *Coxiella* grows best at low pH, such as is found in the phagolysomes of white blood cells. The bacteria in these genera cause Legionnaire's disease and Q fever, respectively.

Methane Oxidizers Organisms that are **methane oxidizers** are Gram-negative bacteria that utilize methane both as a carbon and as an energy source. Like methanogens, which are a type of archaea, methane oxidizers inhabit anaerobic environments worldwide, growing just above the anaerobic layers that contain methanogens. Although methane is one of the so-called greenhouse gases that retains heat in the atmosphere, methane oxidizers digest most of the methane in their local environment before it can adversely affect the world's climate.

Glycolytic Facultative Anaerobes The largest group of γ-proteobacteria is composed of Gram-negative, facultatively anaerobic rods that catabolize carbohydrates by glycolysis and the pentose phosphate pathway. This group, which is divided into three families (**Table 11.5** on page 336), contains numerous human pathogens. Members of the family Enterobacteriaceae, including *Escherichia coli* (esh-ĕ-rik'ē-ă kō'lī), are frequently used for laboratory studies of metabolism, genetics, and recombinant DNA technology.

Pseudomonads Bacteria called **pseudomonads** are Gram-negative, aerobic, flagellated, straight to slightly curved rods that catabolize carbohydrates by the Entner-Doudoroff and pentose phosphate pathways. These organisms are noted for their ability to break down numerous organic compounds. Many of them are important pathogens of humans and animals and are involved in the spoilage of refrigerated milk, eggs, and meat because they can grow and catabolize proteins and lipids at 4°C. The pseudomonad group is named for its most important genus: *Pseudomonas* (soo-dō-mō's'nas; **Figure 11.24**), which causes diseases such as urinary tract infections and external otitis (swimmers' ear). Other pseudomonads such as *Azotobacter* (ā-zō-tō-bak'ter) and *Azomonas* (ā-zō-mō'nas) are soil dwelling, nonpathogenic nitrogen fixers; however, in contrast to nitrogen-fixing α-proteobacteria, these γ-proteobacteria do not associate with the roots of plants.

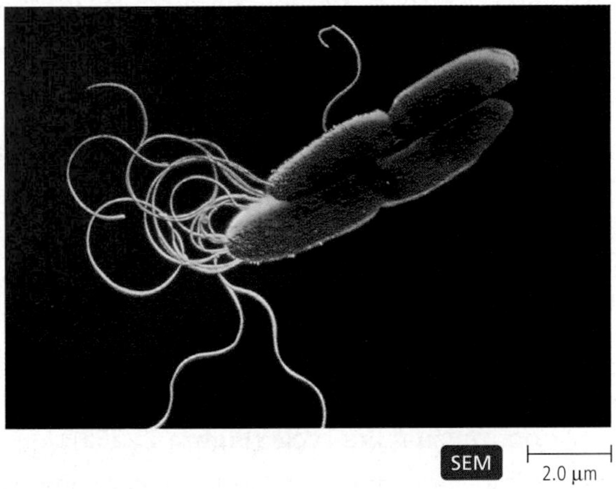

▲ ***Figure 11.24***
Pseudomonas and its characteristic polar flagella.

The characteristics of the members of the γ-proteobacteria discussed in this section are summarized in Table 11.4.

"Deltaproteobacteria"

Learning Objective

✓ List several members of deltaproteobacteria.

The **deltaproteobacteria** are not a large assemblage, but like other proteobacteria they include a wide variety of metabolic types. *Desulfovibrio* (dē'sul-fō-vib-rē-ō) is a sulfate-reducing microbe that is important in the sulfur cycle. It is also an important member of bacterial communities living in the sediments of polluted streams and sewage treatment lagoons, where its presence is often apparent by the odor of

Table 11.5 Representative Glycolytic Facultative Anaerobes of the Class "Gammaproteobacteria"

Family	Special Characteristics	Representative Genera	Typical Human Diseases
Enterobacteriaceae	Straight rods; oxidase negative; peritrichous flagella or nonmotile	*Escherichia*	Gastroenteritis
		Enterobacter	(Rarely pathogenic)
		Serratia	(Rarely pathogenic)
		Salmonella	Enteritis
		Proteus	Urinary tract infection
		Shigella	Shigellosis
		Yersinia	Plague
		Klebsiella	Pneumonia
Vibrionaceae	Vibrios; oxidase positive; polar flagella	*Vibrio*	Cholera
Pasteurellaceae	Cocci or straight rods; oxidase positive; nonmotile	*Haemophilus*	Meningitis in children

hydrogen sulfide that it releases during anaerobic respiration. Hydrogen sulfide reacts with iron to form iron sulfide, so sulfate-reducing bacteria play a primary role in the corrosion of iron pipes in heating systems, sewer lines, and other structures.

Bdellovibrio (del-lō-vib′rē-ō) is not pathogenic to eukaryotes, but instead attacks other Gram-negative bacteria. It has a complex and unusual life cycle **(Figure 11.25a):**

1. A free *Bdellovibrio* swims rapidly through the medium until it collides with a Gram-negative bacterium.
2. It rapidly drills through the cell wall of its prey by secreting hydrolytic enzymes and rotating in excess of 100 revolutions per second **(Figure 11.25b).**
3. Once inside, *Bdellovibrio* lives in the periplasmic space—the space between the cytoplasmic membrane and the cell wall. It kills its host by disrupting the host's cytoplasmic membrane and inhibiting DNA, RNA, and protein synthesis.
4. The invading bacterium uses the nutrients released from its dying prey and grows into a long filament.
5. Eventually, the filament divides into many smaller cells, each of which, when released from the dead cell, produces a flagellum, and swims off to repeat the process. Multiple fission to produce many offspring is a rare form of reproduction.

Myxobacteria are Gram-negative, aerobic, soil-dwelling bacteria with a unique life cycle for prokaryotes in that individuals cooperate to produce differentiated reproductive structures. The life cycle of myxobacteria can be summarized as follows (**Figure 11.26a** on page 338):

1. Vegetative myxobacteria glide on slime trails through their environment, digesting yeasts and other bacteria or scavenging nutrients released from dead cells. When nutrients and cells are plentiful, the myxobacteria divide by binary fission; when nutrients are depleted, however, they aggregate into a mound of cells.
2. Myxobacteria within the mound differentiate to form a macroscopic **fruiting body** ranging in height from 50 μm to 700 μm.
3. Some cells within the fruiting body develop into dormant *myxospores* that are enclosed within walled structures called *sporangia* (singular: *sporangium:* **Figure 11.26b**).
4. The sporangia release the myxospores, which can resist desiccation and nutrient deprivation for a decade or more.
5. When nutrients are again plentiful, the myxospores germinate and become vegetative cells.

Myxobacteria live worldwide in soils that have decaying plant material or animal dung. Though certain species live in the arctic and others in the tropics, most myxobacteria live in temperate regions.

CRITICAL THINKING

What do the names *Desulfovibrio* and *Bdellovibrio* tell you about the morphology of these δ-proteobacteria?

"Epsilonproteobacteria"

Epsilonproteobacteria are Gram-negative rods, vibrios, or spirals. The most important genera are *Campylobacter* (kam′pi-lō-bak′ter), which causes blood poisoning and inflammation of the intestinal tract and *Helicobacter* (hel′ĭ-kō-bak′ter), which causes ulcers.

The characteristics of the δ- and ϵ-proteobacteria are summarized in Table 11.4.

CRITICAL THINKING

Design a dichotomous key for the proteobacteria discussed in this chapter.

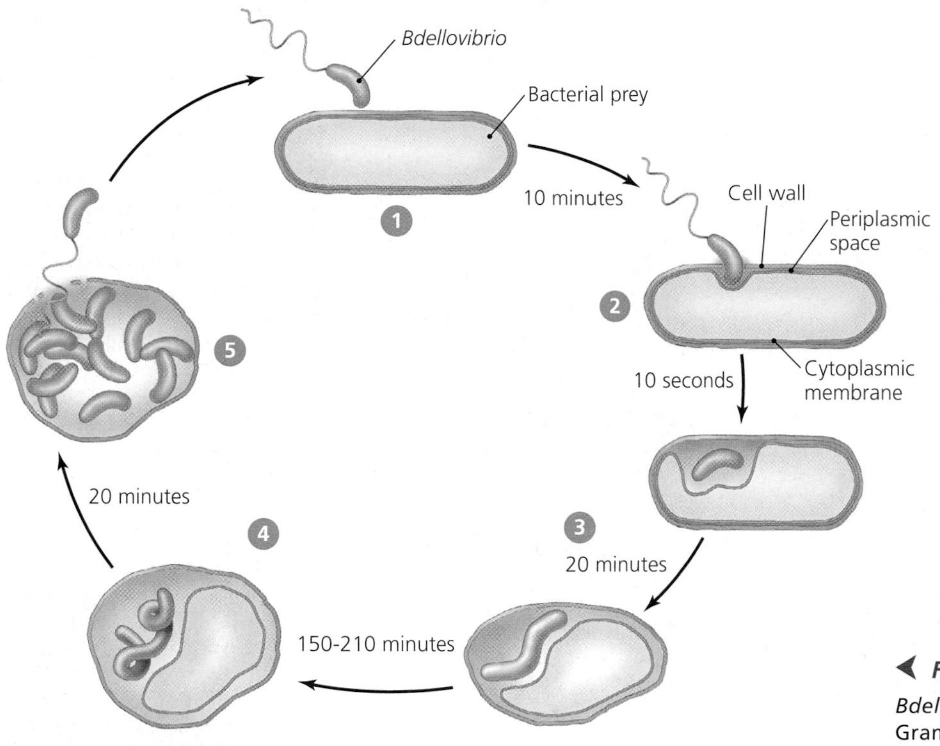

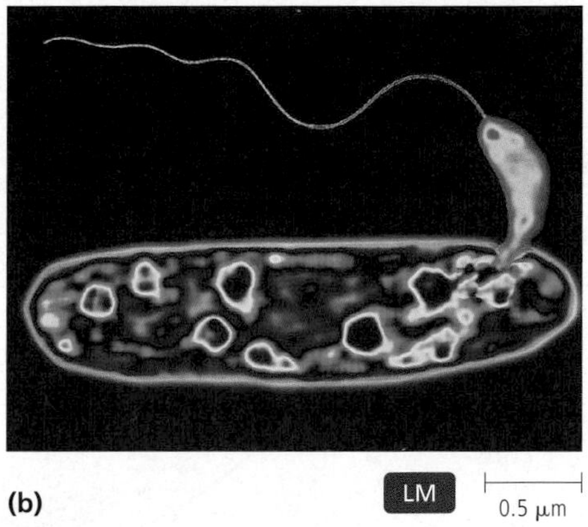

Figure 11.25
Bdellovibrio, a Gram-negative pathogen of other Gram-negative bacteria. **(a)** Life cycle, including the elapsed times between events. **(b)** *Bdellovibrio* invading the periplasmic space of its host.

Other Gram-Negative Bacteria

Learning Objectives

- ✓ Describe the unique features of chlamydias and spirochetes.
- ✓ Describe the ecological importance of bacteroids.

In the final section of this chapter we consider an assortment of Gram-negative bacteria that are classified in the second edition of *Bergey's Manual* into nine phyla that are grouped together for convenience, not because of genetic relatedness. Species in six of the nine phyla are of relatively minor importance, and here we discuss representatives from the three phyla that are either of particular ecological concern or significantly affect human health: the chlamydias (Phylum Chlamydiae), the spirochetes (Phylum Spirochaetes), and the bacteroids (Phylum Bacteroidetes) (see Figure 11.10).

Chlamydias

Microorganisms called **chlamydias** (kla-mid′ē-ăz) are small, Gram-negative cocci that grow and reproduce only within the cells of mammals, birds, and a few invertebrates. The smallest chlamydias—0.2 μm in diameter—are smaller than the largest viruses (poxviruses); however, in contrast to viruses, chlamydias have both DNA and RNA, cytoplasmic membranes, functioning ribosomes, reproduction by binary fission, and metabolic pathways. Like other Gram-negative prokaryotes, chlamydias have two membranes, but in contrast they lack peptidoglycan in their cell wall.

Because chlamydias and rickettsias share an obvious characteristic—both have a requirement for intracellular life—these two types of organisms were grouped together in a single taxon in the first edition of *Bergey's Manual*. Now, however, the rickettsias are typically classified with the γ-proteobacteria, and the chlamydias are in their own phylum. Organisms in the genus *Chlamydia* are the most studied of the chlamydias because they cause neonatal blindness, pneumonia, and a sexually transmitted disease called *lymphogranuloma venereum;* in fact, they are the most common sexually transmitted bacteria in the United States.

Chlamydias have a unique method of reproduction. After invading a host cell, a chlamydial cell forms an **initial body** (also called a *reticulate body*). The initial body grows and undergoes repeated binary fissions until the host cell is filled with reticulate bodies. These then change into **elementary bodies,** which contain electron-dense material and have rigid outer boundaries formed when they cross-link their two membranes with disulfide bonds. Each elementary body, which is relatively resistant to drying, is an infective stage. When the host cell dies, the elementary bodies are released to drift until they contact and attach to other host cells, triggering their own phagocytosis by the new host cells. Once inside a host cell, elementary bodies transform back into initial bodies, and the cycle repeats.

Spirochetes

Spirochetes (spī′rō-kētz) are unique helical bacteria that are motile by means of axial filaments—flagellarlike structures that lie in the periplasmic space (see Figure 3.7). When the

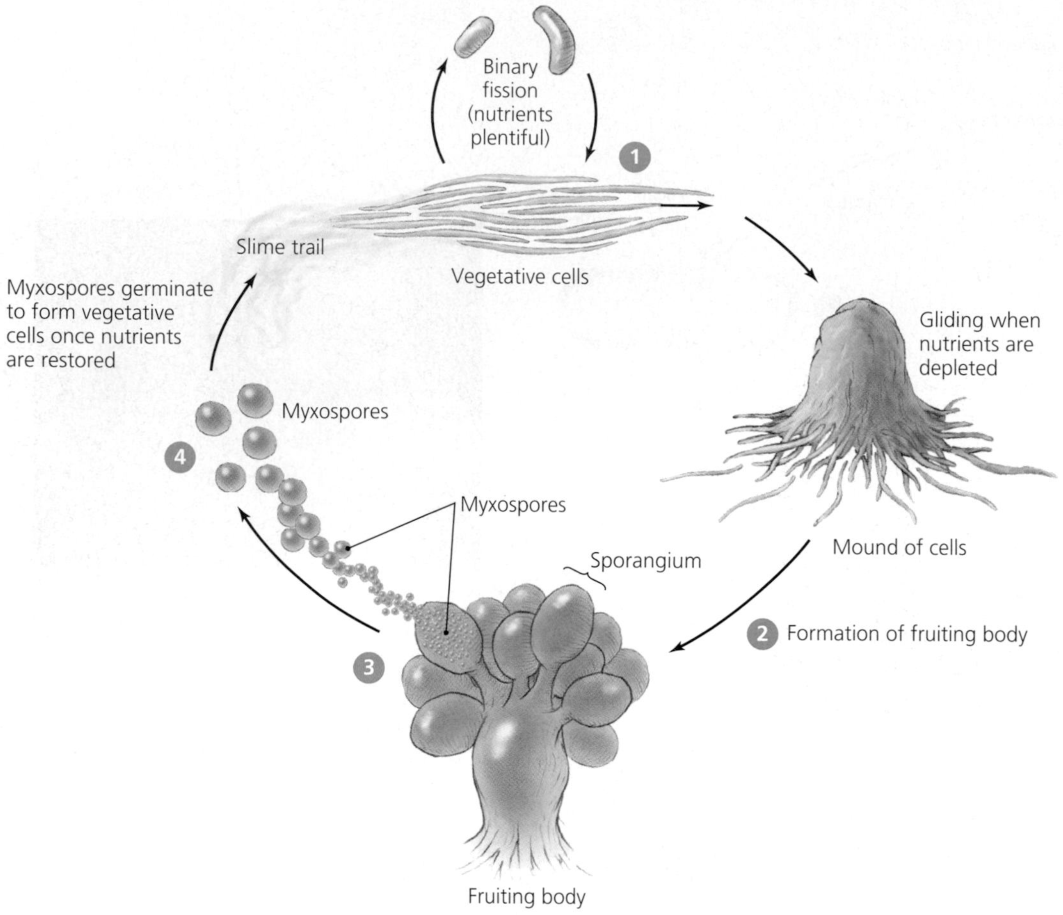

(a)

(b)

Figure 11.26
(a) Myxobacteria life cycle. ① Vegetative myxobacteria divide by binary fission when nutrients are plentiful. They aggregate in mounds when nutrients are depleted. ② The mound forms fruiting bodies. ③ Some cells in a fruiting body develop into dormant myxospores within a sporangium. ④ The sporangia release the myxospores, which remain dormant until nutrients are again plentiful. **(b)** Fruiting body.

axial filaments rotate, the entire cell corkscrews through the medium.

Spirochetes have a variety of types of metabolism and live in diverse habitats. They are frequently isolated from the human mouth, marine environments, moist soil, and the surfaces of protozoa that live in termites' guts. In the latter case, they may coat the protozoan so thickly that they look, and act, like cilia. The spirochetes *Treponema* (trep-ō-nē'mă) and *Borrelia* (bō-rē'lē-ă) cause syphilis and Lyme disease, respectively, in humans.

Bacteroids

Bacteroids are yet another diverse group of Gram-negative microbes that are grouped together on the basis of similarities in their rRNA nucleotide sequences. The group is named for *Bacteroides* (bak-ter-oy'dēz), a genus of obligately anaerobic rods that normally inhabit the digestive tracts of humans and animals. Bacteroids assist in digestion by catabolizing substances such as cellulose and other complex carbohydrates that are indigestible by mammals. About 30% of the bacteria isolated from human feces are *Bacteroides*. Some species of *Bacteroides* cause abdominal, pelvic, blood, and other infections; in fact, they are the most common anaerobic human pathogen.

Bacteroids in the genus *Cytophaga* (sī-tof'ă-gă) are aquatic, gliding, rod-shaped aerobes with pointed ends. These bacteria degrade complex polysaccharides such as agar, pectin, chitin, and even cellulose, so they cause damage to wooden boats and piers; they also play an important role in the degradation of raw sewage. Their ability to glide allows these bacteria to position themselves at sites with optimum nutrients, pH, temperature, and oxygen levels. Organisms in *Cytophaga* differ from other gliding bacteria, such as cyanobacteria and myxobacteria, in that they are nonphotosynthetic and do not form fruiting bodies.

The characteristics of these groups of Gram-negative bacteria are listed in Table 11.4.

CRITICAL THINKING

Design a key for the genera of Gram-negative bacteria listed in Table 11.4.

CHAPTER SUMMARY

General Characteristics of Prokaryotic Organisms (pp. 314–318)

1. Three basic shapes of prokaryotic cells are spherical **cocci,** rod-shaped **bacilli,** and **spirals.** Spirals may be stiff **(spirilla)** or flexible **(spirochetes).**
2. Other variations in shapes include **vibrios** (slighty curved rods), **coccobacilli** (intermediate to cocci and bacilli), and **pleomorphic** (variable shape and size).
3. Cocci may typically be found in groups, including long chains **(streptococci),** pairs **(diplococci),** cuboidal packets **(sarcinae),** and clusters **(staphylococci).**
4. Bacilli are found singly, in pairs, in chains, or in a folded **palisade** arrangement.
5. Prokaryotes reproduce asexually by **binary fission, snapping division** (a type of binary fission), **spore** formation, and **budding.**
6. Environmentally resistant **endospores** are produced within vegetative cells of the Gram-positive genera *Bacillus* and *Clostridium.* Depending on the species in which they are formed, the endospores may be terminal, subterminal, or centrally located.

Modern Prokaryotic Classification (pp. 318–319)

1. Living things are now typically classified into three domains—Archaea, Bacteria, and Eukarya—based largely on genetic relatedness.
2. The most authoritative reference in modern prokaryotic systematics is *Bergey's Manual of Systematic Bacteriology,* which classifies prokaryotes into two phyla in Archaea and 23 phyla in Bacteria. The organization of this text's survey of prokaryotes largely follows the classification scheme proposed in the second edition of *Bergey's Manual.*

Survey of Archaea (pp. 319–322)

1. The domain Archaea includes **extremophiles,** microbes that require extreme conditions of temperature, pH, and/or salinity to survive.
2. **Thermophiles** and **hyperthermophiles** (in the phyla Crenarchaeota and Euryarchaeota) live at temperatures over 45°C and 80°C, respectively, because their DNA, membranes, and proteins do not function properly at lower temperatures.
3. **Halophiles** (Phylum Euryarchaeota) depend on high concentrations of salt to keep their cell walls intact. The halophile *Halobacterium salinarium* synthesizes purple proteins called **bacteriorhodopsins** that harvest light energy to synthesize ATP.
4. **Methanogens** (Phylum Euryarchaeota) are obligate anaerobes that produce methane gas and are useful in sewage treatment.

Survey of Bacteria (pp. 323–338)

1. **Deeply branching bacteria** have rRNA sequences thought to be similar to those of earliest bacteria. They are autotrophic and live in hot, acidic, and anaerobic environments, often with intense exposure to sun.
2. Phototrophic bacteria trap light energy with photosynthetic lamellae. The five groups of phototrophic bacteria are **cyanobacteria,** green sulfur bacteria, green nonsulfur bacteria, purple sulfur bacteria, and purple nonsulfur bacteria.
3. Many cyanobacteria reduce atmospheric N_2 to NH_3 via a process called **nitrogen fixation.** Cyanobacteria must separate (in either time or space) the metabolic pathways of nitrogen fixation from those of oxygenic photosynthesis because nitrogen fixation is inhibited by the oxygen generated during photosynthesis. Many cyanobacteria fix nitrogen in thick-walled cells called **heterocysts.**
4. Green and purple bacteria use bacteriochlorophylls for anoxygenic photosynthesis. Nonsulfur forms derive electrons from organic compounds; sulfur forms derive electrons from H_2S.
5. The phylum **Firmicutes** contains bacteria with a G + C ratio (the proportion of all base pairs that are guanine–cytosine base pairs) of less than 50%. Firmicutes includes clostridia, mycoplasmas, and low G + C cocci and bacilli.
6. Clostridia include the genus *Clostridium* (pathogenic bacteria that cause gangrene, tetanus, botulism, and diarrhea), *Epulopiscium* (which is large enough to be seen without a microscope), and *Veillonella* (often found in dental plaque).
7. **Mycoplasmas** are Gram-negative, pleomorphic, facultative anaerobes and obligate anaerobes that lack cell walls; they are frequently associated with pneumonia and urinary tract infections.
8. Low G + C Gram-positive bacilli and cocci important to human health and industry include *Bacillus* (which contains species that cause anthrax and food poisoning, and includes beneficial Bt-toxin bacteria), *Listeria* (which cause bacteremia and meningitis), *Lactobacillus* (used to produce yogurt and pickles), *Streptococcus* (which causes strep throat and other diseases), and *Staphylococcus* (which causes a number of human diseases).
9. High G + C bacteria (*Corynebacterium, Mycobacterium,* and actinomycetes) are classified in Phylum Actinobacteria.
10. Bacteria in *Corynebacterium* store phosphates in **metachromatic granules;** *C. diphtheriae* causes diphtheria.
11. Members of the genus *Mycobacterium,* including species that cause tuberculosis and leprosy, grow slowly, and have unique, resistant cell walls containing waxy **mycolic acids.**
12. **Actinomycetes** resemble fungi in that they produce spores and form filaments; this group includes *Actinomyces* (normally found in human mouths), *Nocardia* (useful in degradation of pollutants), and *Streptomyces* (produces important antibiotics).
13. Phylum **Proteobacteria** is a very large group of Gram-negative bacteria divided into five classes—the alpha-, beta-, gamma-, delta-, and epsilonproteobacteria.
14. The **alphaproteobacteria** include a variety of aerobes, many of which have unusual attachment extensions of the cell called prosthecae. *Azospirillum* and *Rhizobium* are nitrogen fixers that are important in agriculture.

15. Some members of the alphaproteobacteria are **nitrifying bacteria,** which oxidize NH_3 to NO_3 via a process called **nitrification.** Nitrifying alphaproteobacteria are in the genus *Nitrobacter.*
16. Most purple nonsulfur phototrophs are alphaproteobacteria.
17. Pathogenic α-proteobacteria include *Rickettsia* (some cause typhus and Rocky Mountain spotted fever), *Brucella* (brucellosis), and *Ehrlichia* (ehrlichiosis).
18. There are many beneficial α-proteobacteria, including *Acetobacter* and *Gluconobacter,* both of which are used to synthesize acetic acid. *Caulobacter* is of interest in reproductive studies, and *Agrobacterium* is used in genetic recombination in plants.
19. The **betaproteobacteria** include the nitrifying *Nitrosomonas* and pathogenic species such as *Neisseria* (gonorrhea), *Bordetella* (whooping cough), and *Burkholderia* (which colonizes the lungs of cystic fibrosis patients).
20. Other β-proteobacteria include *Thiobacillus* (ecologically important), *Zoogloea* (useful in sewage treatment), *Sphaerotilus* (hampers sewage treatment), and *Spirillum.*
21. The **gammaproteobacteria** constitute the largest class of proteobacteria; they include purple sulfur bacteria, intracellular pathogens, **methane oxidizers,** facultative anaerobes that utilize glycolysis and the pentose phosphate pathway, and pseudomonads.
22. Both *Legionella* and *Coxiella* are intracellular, pathogenic γ-proteobacteria.
23. Numerous human pathogens are facultatively anaerobic γ-proteobacteria that catabolize carbohydrates by glycolysis.
24. **Pseudomonads,** including pathogenic *Pseudomonas* and nitrogen-fixing *Azotobacter* and *Azomonas,* utilize the Entner-Doudoroff and pentose phosphate pathways for catabolism of glucose.
25. The **deltaproteobacteria** include *Desulfovibrio* (important in the sulfur cycle and in corrosion of pipes), *Bdellovibrio* (pathogenic to bacteria), and **myxobacteria.** The latter form stalked **fruiting bodies** containing resistant, dormant myxospores.
26. The **epsilonproteobacteria** include some important human pathogens, including *Campylobacter* and *Helicobacter.*
27. Chlamydias are Gram-negative cocci typified by the genus *Chlamydia;* they cause neonatal blindness, pneumonia, and a sexually transmitted disease. Within a host cell, chlamydias form **initial bodies,** which change into smaller **elementary bodies** released when the host cell dies.
28. **Spirochetes** are helical bacteria that live in diverse environments. *Treponema* (syphilis) and *Borrelia* (Lyme disease) are important spirochetes.
29. **Bacteroids** include *Bacteroides,* an obligate anaerobic rod that inhabits the digestive tract, and *Cytophaga,* an aerobic rod that degrades wood and raw sewage.

QUESTIONS FOR REVIEW

(Answers to multiple choice and matching questions are on the web, along with additional review questions. Visit www.microbiologyplace.com.)

Modified True-False

For each of the following statements that is true, write "true" in the blank. For each statement that is false, write the word(s) that should be substituted for the italicized word(s) to make the statement correct.

___ 1. All prokaryotes reproduce *sexually.*

___ 2. A *bacillus* is a bacterium with a slightly curved rod shape.

___ 3. If you were to view *staphylococci,* you should expect to see clusters of cells.

___ 4. *Initial* bodies are stable resting stages that do not metabolize but will germinate when conditions improve.

___ 5. Archaea are classified into phyla based primarily on *tRNA sequences.*

___ 6. *Halophiles* inhabit extremely saline habitats such as the Great Salt Lake.

___ 7. Pigments located in thylakoids in phototrophic bacteria trap *light* energy for metabolic processes.

___ 8. Most *cyanobacteria* form heterocysts in which nitrogen fixation occurs.

___ 9. A giant bacterium that is large enough to be seen without a microscope is *Veillonella.*

___10. When environmental nutrients are depleted, myxobacteria aggregate in mounds to form fruiting bodies.

Matching

Match the bacterium on the left with the term with which it is most closely associated.

___ 1. *Bacillus anthracis*	A. wood damage
___ 2. *Veillonella*	B. dental plaque
___ 3. *Clostridium perfringens*	C. gangrene
___ 4. *Clostridium botulinum*	D. botulism poisoning
___ 5. *Bacillus licheniformis*	E. anthrax
___ 6. *Streptococcus*	F. lymphogranuloma venereum
___ 7. *Streptomyces*	G. leprosy
___ 8. *Corynebacterium*	H. tetracycline
___ 9. *Glucanobacter*	I. vinegar
___ 10. *Bordetella*	J. yogurt
___ 11. *Zoogloea*	K. impetigo
___ 12. *Azotobacter*	L. bacitracin
___ 13. *Desulfovibrio*	M. iron pipe corrosion
___ 14. *Chlamydia*	N. pertussis
___ 15. *Cytophaga*	O. nitrogen fixation
	P. floc formation
	Q. diphtheria

Multiple Choice

1. The type of reproduction in prokaryotes that results in a palisade arrangement of cells is called
   a. pleomorphic division.
   b. endospore formation.
   c. snapping division.
   d. binary fission.

2. The thick-walled reproductive spores produced in the middle of cyanobacterial filaments are called
   a. akinetes.
   b. terminal endospores.
   c. metachromatic granules.
   d. myxospores.

3. Which of the following terms best describes stiff, spiral-shaped prokaryotic cells?
   a. cocci
   b. bacilli
   c. spirilla
   d. spirochetes

4. Endospores
   a. can remain alive for decades.
   b. can remain alive in boiling water.
   c. live in a state of suspended animation.
   d. all of the above

5. *Halobacterium salinarium* is distinctive because
   a. it is absolutely dependent on high salt concentrations to maintain its cell wall.
   b. it is found in terrestrial volcanic habitats.
   c. it photosynthesizes without chlorophyll.
   d. it can survive 5 million rad of radiation.

6. Photosynthetic bacteria that also fix nitrogen are
   a. mycoplasmas.
   b. spirilla.
   c. bacteroids.
   d. cyanobacteria.

7. Which genus is the most common anaerobic human pathogen?
   a. *Bacteroides*
   b. *Spirochetes*
   c. *Chlamydias*
   d. *Methanopyrus*

Short Answer

1. Whereas the first edition of *Bergey's Manual* relied on morphological and biochemical characteristics to classify microbes, the new edition focuses on ribosomal RNA sequences. List several other criteria for grouping and classifying bacteria.
2. What are extremophiles? Describe two kinds and give examples.
3. Name and describe three types of bacteria mentioned in this chapter that "glide."
4. Name three groups of low G + C Gram-positive bacteria.
5. Sketch and label six shapes of bacteria.
6. A student was memorizing the arrangements of bacteria and noticed that there are more arrangements for cocci than for bacilli. Why might this be so?
7. How are bacterial endospores different from the spores of actinomycetes, algae, and fungi?
8. How is *Agrobacterium* used in recombinant DNA technology?
9. Name and describe five distinct classes of Phylum Proteobacteria.
10. Explain why organisms formerly known as blue-green algae are now called cyanobacteria.
11. Louis Pasteur said, "The role of the infinitely small in nature is infinitely large." Explain what he meant by using examples of the roles of microorganisms in health, industry, and the environment.

CRITICAL THINKING

1. A microbiology student described "deeply branching bacteria" as having a branched filamentous growth habit akin to *Streptomyces*. Do you agree with this description? Why or why not?
2. Iron oxide (rust) forms when iron is exposed to oxygen, particularly in the presence of water. Nevertheless, iron pipes typically corrode more quickly when they are buried in moist *anaerobic* soil than when they are buried in soil containing oxygen. Explain why this is the case.
3. Why is it that Gram-positive species don't have axial filaments?
4. Even though *Clostridium* is strictly an anaerobic bacterium, it can be isolated easily from the surface of your skin. Explain how this can be.

CHAPTER 12

Characterizing and Classifying Eukaryotes

CHAPTER OUTLINE

The mushrooms in this photo constitute only the tips of what may be the largest living organism on Earth—a fungus known as *Armillaria ostoyae* that lives 3 feet underground in Oregon's Malheur National Forest. Also known as the honey mushroom, it is 3.5 miles long, covers an area equivalent to 1665 football fields, and is estimated to be at least 2400 years old!

Amazingly, this "humongous fungus" began life as a microscopic spore. During its lifespan of over two millennia, it has used root-like structures called rhizomorphic hyphae to draw water and nutrients from tree roots. Despite its gigantic proportions, what is visible to us are ordinary mushrooms.

Fungi are eukaryotes—a vast category that also includes protozoa, algae, parasitic helminths, as well as all plants and animals. In this chapter we will take a closer look at the eukaryotic organisms of microbiological importance.

Mushrooms of *Armillaria ostoyae*

MicroPrep Pre-Test: *Take the pre-test for this chapter on the web. Visit* ***www.microbiologyplace.com****.*

Eukaryotic microbes include a fascinating and almost bewilderingly diverse assemblage. Eukaryotic microbes include unicellular and multicellular protozoa,[1] fungi,[2] algae,[3] water molds, and slime molds. Additionally, microbiologists study parasitic helminths[4] (worms) because they have microscopic stages and they study arthropod vectors because they are intimately involved in the transmission of microbial pathogens. Eukaryotes include both human pathogens and organisms that are vital for human life. For example, one kind of marine algae called diatoms, and a kind of protozoa called dinoflagellates (dī″nō-flaj′ĕ-lātz), provide the basis for the oceans' food chains and produce most of the world's oxygen. Eukaryotic fungi produce penicillin, and tiny bakers' and brewers' yeasts are essential for making bread and alcoholic beverages.

Among the 20 most frequent microbial causes of death worldwide, six are eukaryotic, including the agents of malaria, African sleeping sickness, and amoebic dysentery. *Pneumocystis* pneumonia, toxoplasmosis, and cryptosporidosis—common afflictions of AIDS patients—are all caused by eukaryotic pathogens. A recently discovered dinoflagellate *(Pfiesteria)* appears to secrete volatile toxins that induce hours of vomiting and fits of rage, and even persistent nerve damage, in humans that inhale them.

In previous chapters we discussed characteristics of *cells*—their metabolism, growth, and genetics. In this chapter we discuss four major groups of eukaryotic *organisms* of importance in microbiology—protozoa (single-celled "animals"), fungi, algae, and the water molds and slime molds—and conclude with a brief discussion of the relationship of parasitic helminths and vectors to microbiology.

We begin by discussing general features of eukaryotic reproduction and classification; the following sections survey some representative members of microbiologically important eukaryotic groups, focusing on beneficial, environmentally significant, and unusual species.

General Characteristics of Eukaryotic Organisms

Our discussion of the general characteristics of eukaryotes begins with a survey of the events in eukaryotic reproduction; then we consider some aspects of the complex matter of classifying the great variety of eukaryotic organisms.

Reproduction in Eukaryotes

Learning Objectives

- ✓ State four reasons why eukaryotic reproduction is more complex than prokaryotic reproduction.
- ✓ Describe the phases of mitosis, mentioning chromosomes, chromatids, centromeres, and spindle.
- ✓ Contrast meiosis with mitosis, mentioning homologous chromosomes, tetrads, and crossing over.
- ✓ Distinguish among nuclear division, cytokinesis, and schizogony.

As discussed in Chapter 3, a unique characteristic of living things is the ability to reproduce themselves. Prokaryotic reproduction typically involves replication of DNA and binary fission of the cytoplasm to produce two identical offspring. Reproduction of eukaryotes is more complicated and varied than reproduction in prokaryotes for a number of reasons:

- Most of the DNA in eukaryotes is packaged with histone proteins as *chromosomes* in the form of *chromatin* (krō′ma-tin) *fibers* located within nuclei. The remaining DNA in eukaryotic cells is found in mitochondria and chloroplasts, organelles that reproduce by binary fission in a manner similar to prokaryotic reproduction. In this chapter we will discuss only the nuclear portion of eukaryotic genomes.
- Eukaryotes have a variety of methods of asexual reproduction, including binary fission, budding, fragmentation, spore formation, and *schizogony* (ski-zog′ō-nē) (discussed later).
- Many eukaryotes reproduce sexually—that is, via a process that involves the formation of sexual cells called *gametes,* and the subsequent fusion of two gametes to form a cell called a *zygote.*
- Additionally, algae, fungi, and some protozoa reproduce both sexually and asexually. (Animals generally reproduce only one way or the other.)

Every form of eukaryotic reproduction involves two types of division: nuclear division and cytoplasmic division (also called cytokinesis). After we discuss the various aspects of these two types of cell division, we will consider a special type of eukaryotic reproduction called schizogony.

Nuclear Division

Typically, a eukaryotic nucleus has either one or two complete copies of the chromosomal portion of a cell's genome. A nucleus with a single copy of each chromosome is called a **haploid**[5] or $1n$ nucleus, and one with two sets of chromosomes is a **diploid**[6] or $2n$ nucleus. Generally, each organism has a consistent number of chromosomes. For example, each haploid cell of the brewer's yeast, *Saccharomyces cerevisiae*[7] (sak″ă-rō-mī′sēz se-ri-vis′ē-ē), has 16 chromosomes.

Whereas the cells of most fungi, many algae, and some protozoa are haploid, the cells of most plants and animals and the remaining fungi, algae, and protozoa are diploid. Typically, gametes are haploid, and a zygote (formed from the union of gametes) is diploid.

[1]From Greek *protos,* meaning first, and *zoion,* meaning animal.
[2]Plural of Latin *fungus,* meaning mushroom.
[3]Plural of Latin *alga,* meaning seaweed.
[4]From Greek *helmins,* meaning worm.
[5]From Greek *haploos,* meaning single.
[6]From Greek *diploos,* meaning double.
[7]From Greek *sakcharon,* meaning sugar, and *mykes,* meaning fungus, and Latin *cerevisiae,* meaning beer.

A cell divides its nucleus so as to pass a copy of its chromosomal DNA to each of its descendants, so that each new generation has the necessary genetic instructions to carry on life. There are two types of nuclear division—*mitosis* (mī-tō′sis) and *meiosis* (mī-ō′sis).

Mitosis Cells have two main stages in their life cycle: a stage called *interphase,*[8] during which cells grow and eventually duplicate their DNA, and a stage during which the cell's nucleus divides. In the type of nuclear division called **mitosis,**[9] which begins after the cell has duplicated its DNA such that there are two exact DNA copies (see Figure 7.5), the cell partitions its replicated DNA equally between two nuclei. Thus mitosis maintains the ploidy of the parent nucleus; that is, a haploid nucleus that undergoes mitosis forms two haploid nuclei, and a diploid nucleus that undergoes mitosis produces two diploid nuclei.

Mitosis has four phases: prophase,[10] metaphase,[11] anaphase,[12] and telophase.[13] The events of mitosis proceed as follows **(Figure 12.1a):**

1. **Prophase.** The cell condenses the DNA molecules into visible threads called *chromatids* (krō′mă-tidz). Two identical chromatids, sister DNA molecules, are joined together in a region called a *centromere* to form one chromosome. Also during prophase, a set of microtubules is constructed in the cytosol to form a *spindle*. In most cells, the nuclear envelope disintegrates during prophase so that mitosis occurs freely in the cytosol; however, many fungi and some unicellular microbes (for example, diatoms and dinoflagellates) maintain their nuclear envelopes so that mitosis occurs within a nucleus.
2. **Metaphase.** The chromosomes line up in the middle of the cell and attach to microtubules of the spindle near their centromeres.
3. **Anaphase.** Sister chromatids separate and crawl along the microtubules toward opposite poles of the spindle. Each chromatid is now called a chromosome.
4. **Telophase.** The cell restores its chromosomes to their less compact, nonmitotic state, and nuclear envelopes form around the daughter nuclei. Though a cell may divide during telophase, as shown in Figure 12.1a, mitosis is nuclear division, not cell division.

Though certain specific events distinguish each of the four phases of mitosis, the phases are not discrete steps; that is, mitosis is a continuous process, and there are no clear boundaries between succeeding phases—one phase leads seamlessly to the next. For example, late anaphase and early telophase are indistinguishable.

Students sometimes confuse the terms *chromosome* and *chromatid,* in part because early microscopists used the word *chromosome* for two different things. During prophase and metaphase, a chromosome consists of two chromatids (DNA molecules) joined at a centromere. However, during anaphase and telophase, the chromatids separate, and each chromatid is then called a chromosome. In other words, a "chromosome" is a pair of chromatids during the first two phases, while "chromatid" and "chromosome" are synonymous terms during the latter two phases of mitosis.

Figure 12.1 ➤
The two kinds of nuclear division: mitosis and meiosis. The events occurring during each of the numbered phases are described in the text. **(a)** Mitosis, in which the number of chromosomes (ploidy) in the daughter nuclei is the same as in the parent nucleus. Even though cell division (cytokinesis) may occur simultaneously with mitosis (as shown here), mitosis is not cell division. **(b)** Meiosis, which results in four nuclei, each with half the number of chromosomes as the parent nucleus.

Meiosis In contrast to mitosis, **meiosis**[14] is nuclear division that involves the partitioning of chromatids into four nuclei such that each nucleus receives only half the original amount of DNA. Thus, diploid nuclei use meiosis to produce haploid daughter nuclei. Meiosis is a necessary condition for sexual reproduction (in which nuclei from two different cells fuse to form a single nucleus), because if cells lacked meiosis, with each nuclear fusion to form a zygote the number of chromosomes doubles, and their number would soon become unmanageable.

Meiosis occurs in two stages known as *meiosis I* and *meiosis II* **(Figure 12.1b).** As in mitosis, each stage has four phases, named prophase, metaphase, anaphase, and telophase. The events in meiosis as they occur in a diploid nucleus proceed as follows:

1. Early prophase I (prophase of meiosis I). As with mitosis, DNA replication during interphase has resulted in pairs of identical chromatids, forming chromosomes. But now an additional pairing occurs: *homologous chromosomes*—that is, chromosomes carrying similar or identical genetic sequences—line up side by side. Because these are prophase chromosomes, each of them consists of two identical chromatids; therefore, four DNA molecules are involved in this pairing. An aligned pair of homologous chromosomes is known as a *tetrad*.
2. Late prophase I. Once tetrads have formed, the homologous chromosomes exchange sections of DNA in a random fashion via a process called *crossing over*. This results in recombinations of their DNA. It is because of meiotic crossing over that the offspring produced by sexual reproduction have different genetic makeups from their siblings. Prophase I can last for days or longer.
3. Metaphase I. Tetrads align in the center of the cell and attach to spindle microtubules. Metaphase I differs from

---

[8]Latin, meaning between phases.
[9]From Greek *mitos,* meaning thread, after the threadlike appearance of chromosomes during nuclear division.
[10]From Greek *pro,* meaning before, and *phasis,* meaning appearance.
[11]From Greek *meta,* meaning in the middle.
[12]From Greek *ana,* meaning back.
[13]From Greek *telos,* meaning end.
[14]From Greek *meioun,* meaning to make smaller.

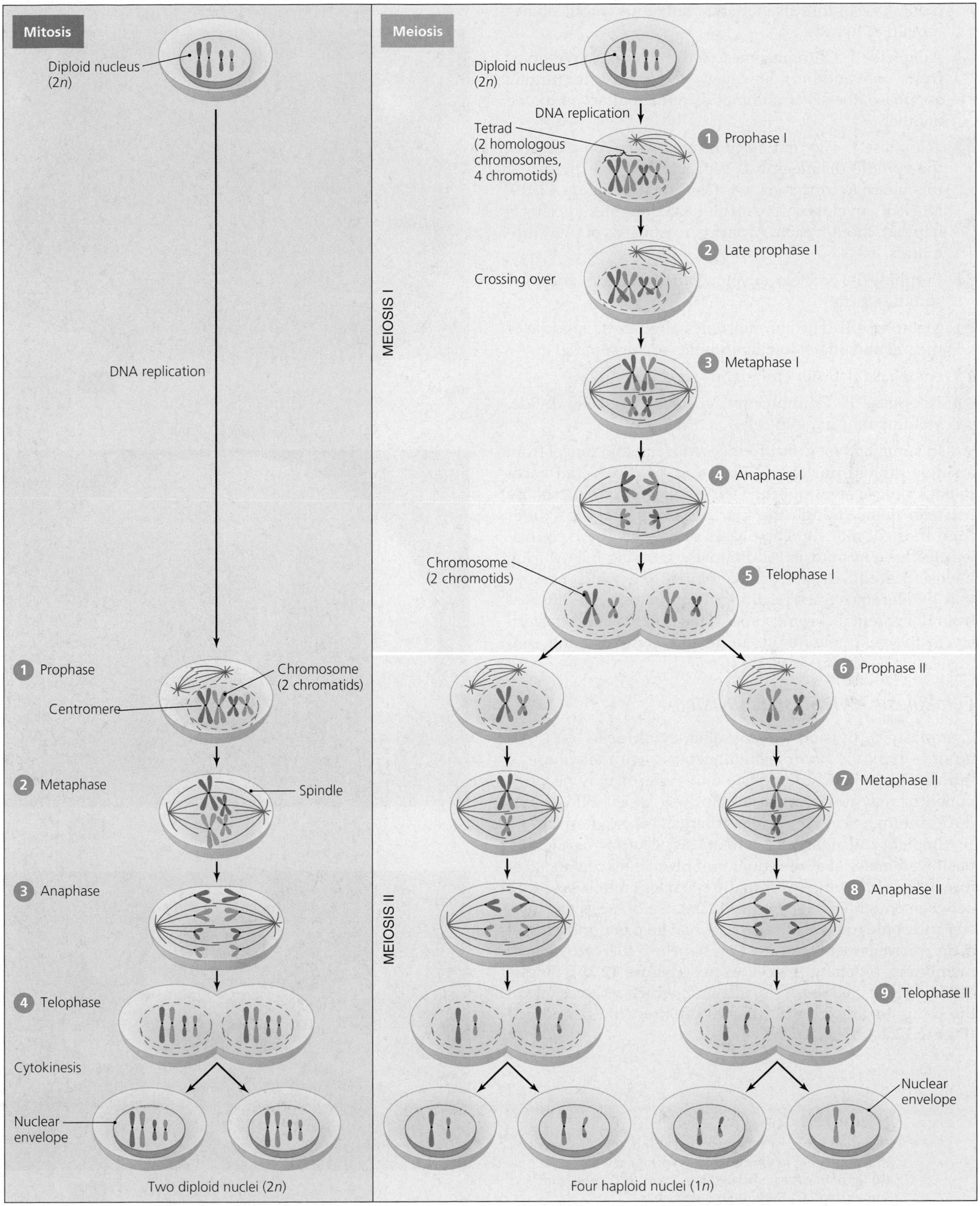
Mitosis
Diploid nucleus (2n)
DNA replication
1 Prophase
Chromosome (2 chromatids)
Centromere
2 Metaphase
Spindle
3 Anaphase
4 Telophase
Cytokinesis
Nuclear envelope
Two diploid nuclei (2n)
Meiosis
Diploid nucleus (2n)
DNA replication
Tetrad (2 homologous chromosomes, 4 chromotids)
1 Prophase I
2 Late prophase I
Crossing over
MEIOSIS I
3 Metaphase I
4 Anaphase I
Chromosome (2 chromotids)
5 Telophase I
6 Prophase II
7 Metaphase II
MEIOSIS II
8 Anaphase II
9 Telophase II
Nuclear envelope
Four haploid nuclei (1n)

metaphase of mitosis in that homologous chromosomes remain as tetrads.

4 Anaphase I. Chromosomes of the tetrads move apart from one another; however, in contrast to mitotic anaphase, the sister chromatids remain attached to one another.

5 Telophase I. The first stage of meiosis is completed as the spindle disintegrates. Typically, the cell divides at this phase to form two cells (as shown in Figure 12.1b). Nuclear envelopes may form. Each daughter nucleus is haploid, though each chromosome consists of two chromatids.

6 Prophase II. Nuclear envelopes disintegrate, and new spindles form.

7 Metaphase II. The chromosomes align in the middle of the cell and attach to microtubules of the spindle.

8 Anaphase II. Sister chromatids separate.

9 Telophase II. Daughter nuclei form. The cells divide, yielding four haploid cells.

In summary, meiosis produces four haploid nuclei from a single diploid nucleus. Meiosis can be considered back-to-back mitoses without the DNA replication of interphase between them, though the four phases of meiosis I differ from those of mitosis. The phases of meiosis II are equivalent to those in mitosis. Additionally, crossing over during meiosis I produces genetic recombinations, which ensures that the chromosomes resulting from meiosis are different from the parental chromosomes. This provides genetic variety in the next generation. **Table 12.1** compares and contrasts mitosis and meiosis.

Cytokinesis (Cytoplasmic Division)

Cytoplasmic division—also called **cytokinesis** (sī″tō-ki-nē′sis)—typically occurs simultaneously with telophase of mitosis, though in some algae and fungi it may be postponed or may not occur at all. In these cases, mitosis produces multinucleate cells called **coenocytes** (sē′nō-sītz).

In plant and algal cells, cytokinesis occurs as vesicles deposit wall material at the equatorial plane between nuclei to form a *cell plate*, which eventually becomes a transverse wall between daughter cells **(Figure 12.2a).** Cytokinesis of protozoa and some fungal cells occurs when an equatorial ring of actin microfilaments contracts just below the cytoplasmic membrane, pinching the cell in two **(Figure 12.2b).** Single-celled fungi called *yeasts*, form a bud, which receives one of the daughter nuclei and pinches off from the parent cell **(Figure 12.2c).**

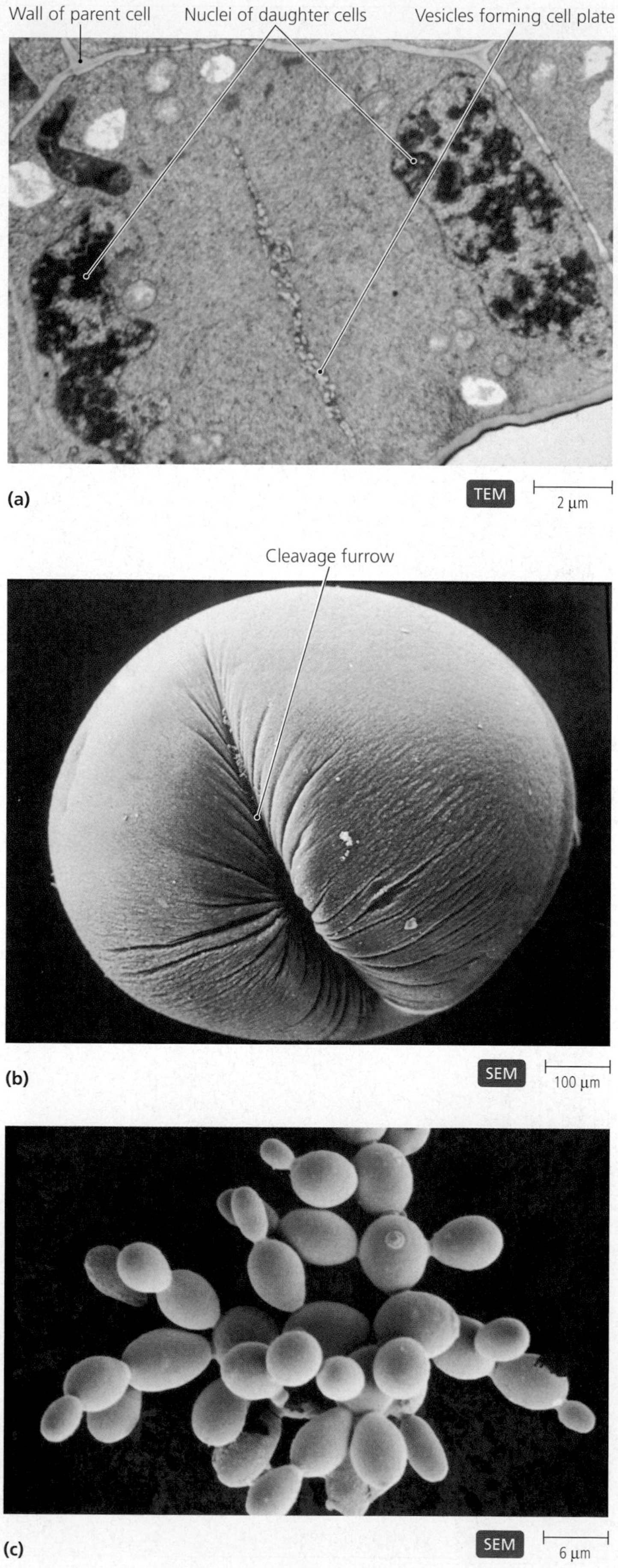

Figure 12.2 ➤

Different types of cytoplasmic division. **(a)** Cytokinesis in a plant cell, in which vesicles form a cell plate. **(b)** Cytokinesis as it occurs in animals, protozoa, and some fungi. **(c)** Budding in yeast cells.

Table 12.1 Characteristics of the Two Types of Nuclear Division

	Mitosis	Meiosis
DNA replication	During interphase, before nuclear division	During interphase, before meiosis I begins
Phases	Prophase, metaphase, anaphase, telophase	Meiosis I—prophase I, metaphase I, anaphase I, telophase I Meiosis II—prophase II, metaphase II, anaphase II, telophase II
Formation of tetrads (alignment of homologous chromosomes)	Does not occur	Early in prophase I
Crossing over	Does not occur	Following formation of tetrads during prophase I
Number of accompanying cytoplasmic divisions that may occur	One	Two
Resulting nuclei	Two nuclei with same ploidy as the original	Four nuclei with half the ploidy of the original

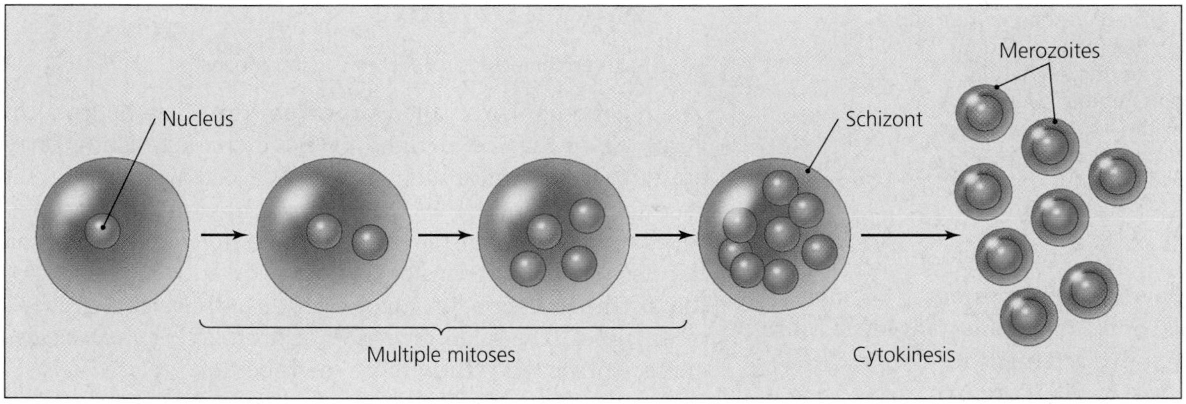

▲ ***Figure 12.3***
Schizogony. Sequential mitoses without intervening cytokineses produce a multinucleate schizont, which later undergoes cytokinesis to produce many daughter cells.

Schizogony

The protozoan *Plasmodium* (plaz-mō′dē-ŭm), which causes malaria, reproduces asexually within red blood cells and liver cells via a special type of reproduction called **schizogony** (ski-zog′ō-nē; **Figure 12.3**). In schizogony, multiple mitoses form a multinucleate **schizont** (skiz′ont); only then does cytokinesis occur, simultaneously releasing numerous uninucleate daughter cells called *merozoites* (mer-ō-zō′ītz). The body of infected hosts responds to the release of huge numbers of merozoites with the cyclic fever and chills characteristic of malaria.

The Classification of Eukaryotic Organisms

Learning Objectives

- ✓ Briefly describe the major groups of eukaryotes as they were first classified in the late 18th century, and as they were classified in the late 20th century.
- ✓ List some of the problems involved in the classification of protists in particular.

Historically, the classification of many of the eukaryotic microbes has been fraught with difficulty and characterized by change. Since the late 18th century, when Linnaeus first classified living things, until near the end of the 20th century, taxonomists grouped organisms together largely according to readily observable structural traits. Linnaeus classified unicellular algae and fungi as plants, and he classified protozoa—as the name suggests—as animals (**Figure 12.4a** on page 348). By the late 20th century, some taxonomists, including Whittaker, had placed fungi in their own kingdom and grouped protozoa and algae together within the kingdom Protista **(Figure 12.4b);** other taxonomists, however, kept the green algae in the kingdom Plantae.

This scheme is troublesome, in part because the kingdom Protista includes both large multicellular algae (such as kelps) and nonphotosynthetic unicellular protozoa. Adding to the confusion is that taxonomists who classify plants and fungi use the term *divisions* to refer to the same taxonomic level that zoologists call *phyla*.

More recently, many taxonomists have abandoned classification schemes that are so strongly grounded in large-scale

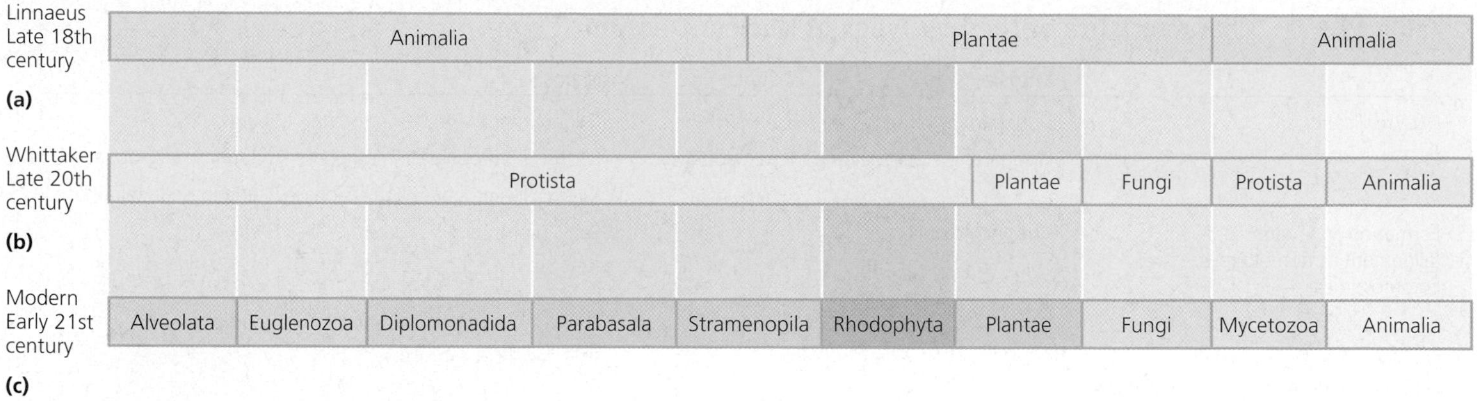

▲ ***Figure 12.4***

The changing classification of eukaryotes over the centuries. **(a)** In the late 18th century, Linnaeus classified all organisms as either plants or animals. **(b)** In the late 20th century, Whittaker placed fungi in their own group and established a new kingdom Protista. **(c)** Today, microbial eukaryotes are classified into numerous kingdoms based largely on their genetic relatedness. Though not all taxonomists would agree about every detail of this scheme, it forms the basis for the discussion of eukaryotic organisms in this chapter.

structural similarities in favor of schemes based on characteristics that are observable only at the molecular level. More specifically, modern taxonomy attempts to clarify the genetic relatedness of microbes, to place them in groups based on similarities in nucleotide sequences and cellular ultrastructure as revealed by electron microscopy. One of the most evident results of such taxonomic studies is that most schemes no longer include a group of "protists"; instead, the latest studies suggest that such eukaryotic microbes belong in several kingdoms.

Though no one classification scheme has garnered universal support, and more thorough understanding based on new information will almost certainly dictate changes, many taxonomists favor a scheme similar to the one shown in **Figure 12.4c.** In this scheme, on which the discussions of eukaryotic microbes in this chapter are largely based, the organisms we commonly refer to as protozoa are classified in the kingdoms Alveolata, Euglenozoa, Diplomonadida, and Parabasala; fungi are in the kingdom Fungi; algae are distributed among the kingdoms Stramenopila, Rhodophyta, and Plantae; water molds join the algae in Kingdom Stramenopila, and slime molds are in Kingdom Mycetozoa. As we study the eukaryotic microbes discussed in this chapter, bear in mind that because the relationships among eukaryotic microbes are not fully understood, not all taxonomists would completely agree with this scheme, and that new information will shape future alterations in the taxonomy of eukaryotic microbes.

We begin our survey of eukaryotic microbes with the diverse group of organisms commonly known as protozoa.

Protozoa

Learning Objective

✓ List three characteristics shared by all protozoa.

The microorganisms called **protozoa** (prō-tō-zō'ă) are a diverse group that are defined by three characteristics: They are eukaryotic, unicellular, and lack a cell wall. Note that "protozoa" is not a currently accepted taxon. With the exception of one subgroup (called apicomplexans), protozoa are also motile by means of cilia, flagella, and/or pseudopodia. By these criteria, protozoa include a diverse assemblage of microbes. The scientific study of protozoa is *protozoology,* and scientists who study these microbes are *protozoologists.*

In the following sections we discuss the distribution, morphology, nutrition, reproduction, and classification of various groups of protozoa.

Distribution of Protozoa

Protozoa require moist environments because their lack of a cell wall subjects them to desiccation. Therefore, most species live worldwide in ponds, streams, lakes, and oceans, where they are critical members of the *plankton*—free-living, drifting organisms that form the basis of aquatic food chains. Some protozoa live in moist soil, beach sand, and decaying organic matter, and a very few are pathogens—that is, disease-causing microbes, of animals and humans.

Morphology of Protozoa

Though protozoa have most of the features of eukaryotic cells discussed in Chapter 3 and illustrated in Figure 3.2, this group of eukaryotic microbes is characterized by great morphologic diversity. Indeed, taxonomists once used the variety in a prominent morphological feature—their locomotory structures—to classify the protozoa as flagellates, ciliates, or protozoa having pseudopodia. These features no longer figure prominently in the taxonomic classification of protozoa because the presence of a given locomotory feature may not indicate evolutionary relatedness.

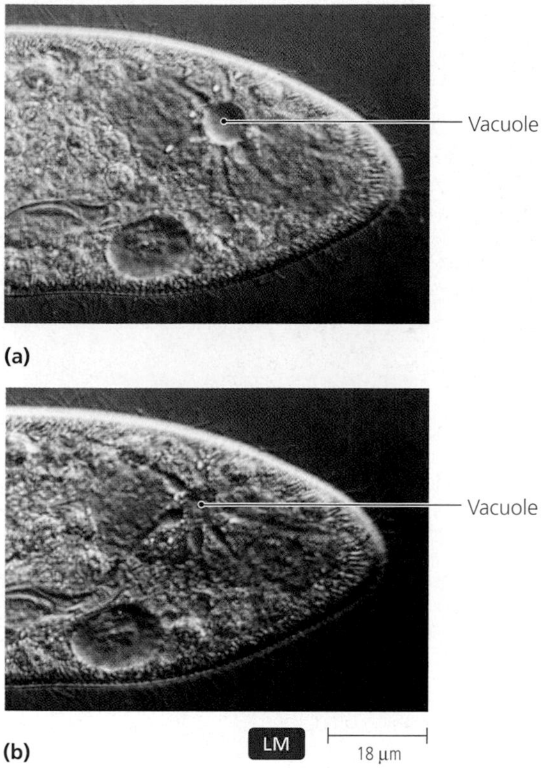

▲ ***Figure 12.5***

A contractile vacuole, a prominent morphological feature of many protozoa such as *Paramecium*. **(a)** A vacuole being filled by water that entered the cell via osmosis. **(b)** A vacuole that is contracting to pump water out of the cell. *Is the environment in this case hypertonic, hypotonic, or isotonic to the cell? Explain.*

Figure 12.5 *The environment is hypotonic to the cell; water moves down its concentration gradient—into the cell—by osmosis.*

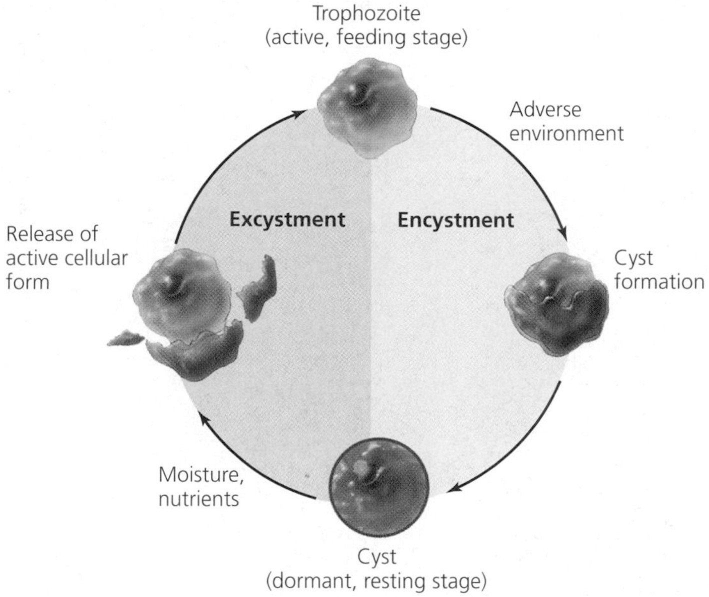

▲ ***Figure 12.6***

The cyclical production of trophozoite and cyst stages in protozoa in response to environmental conditions. All protozoa have a trophozoite, but only some protozoa form cysts. *Why is cyst production a common feature of parasitic intestinal protozoa but not of parasitic blood-borne protozoa?*

Figure 12.6 *Cysts allow intestinal protozoa to survive in the relatively dry conditions outside the body and infect a new host; blood-borne protozoa are typically transmitted by arthropod vectors and thus need not form resistant cysts.*

Some ciliates (*e.g., Paramecium*) have two nuclei: often a larger *macronucleus,* which contains many copies of the genome (often more than 50*n*) and controls metabolism, growth, and sexual reproduction, and a smaller *micronucleus,* which is involved in genetic recombination, sexual reproduction, and regeneration of macronuclei.

Protozoa also show variety in the number and kind of mitochondria they contain. Several groups completely lack mitochondria, while all the others have mitochondria with discoid or tubular cristae rather than the platelike cristae seen in animals, plants, fungi, and many algae. Additionally, some protozoa have *contractile vacuoles* that actively pump water from the cells, protecting them from osmotic lysis **(Figure 12.5).**

All free-living aquatic and pathogenic protozoa exist as a motile feeding stage called a **trophozoite** (trof-ō-zō'īt), and many have a hardy resting stage called a **cyst,** which is characterized by a thick capsule and a low metabolic rate. Cysts of protozoa are not reproductive structures, because one trophozoite forms one cyst, which later becomes one trophozoite **(Figure 12.6).** Such cysts allow intestinal protozoa to pass from one host to another and to survive harsh environmental conditions such as desiccation, nutrient deficiency, extremes of pH and temperature, and lack of oxygen.

Nutrition of Protozoa

Most protozoa are chemoheterotrophic, obtaining nutrients by phagocytizing bacteria, decaying organic matter, other protozoa, or the tissues of a host; a few protozoa absorb nutrients from the surrounding water. Because the protozoa called dinoflagellates and euglenoids (discussed shortly) are photoautotrophic, botanists historically classified them with the algae rather than as protozoa.

Reproduction in Protozoa

Most protozoa reproduce asexually only, by binary fission or schizogony; a few protozoa also have sexual reproduction in which two individuals exchange genetic material. Some sexually reproducing protozoa become **gametocytes** (gametes) that fuse to form a diploid zygote. Ciliates, such as *Paramecium,* reproduce sexually via a complex process called *conjugation,* which involves the coupling of two compatible

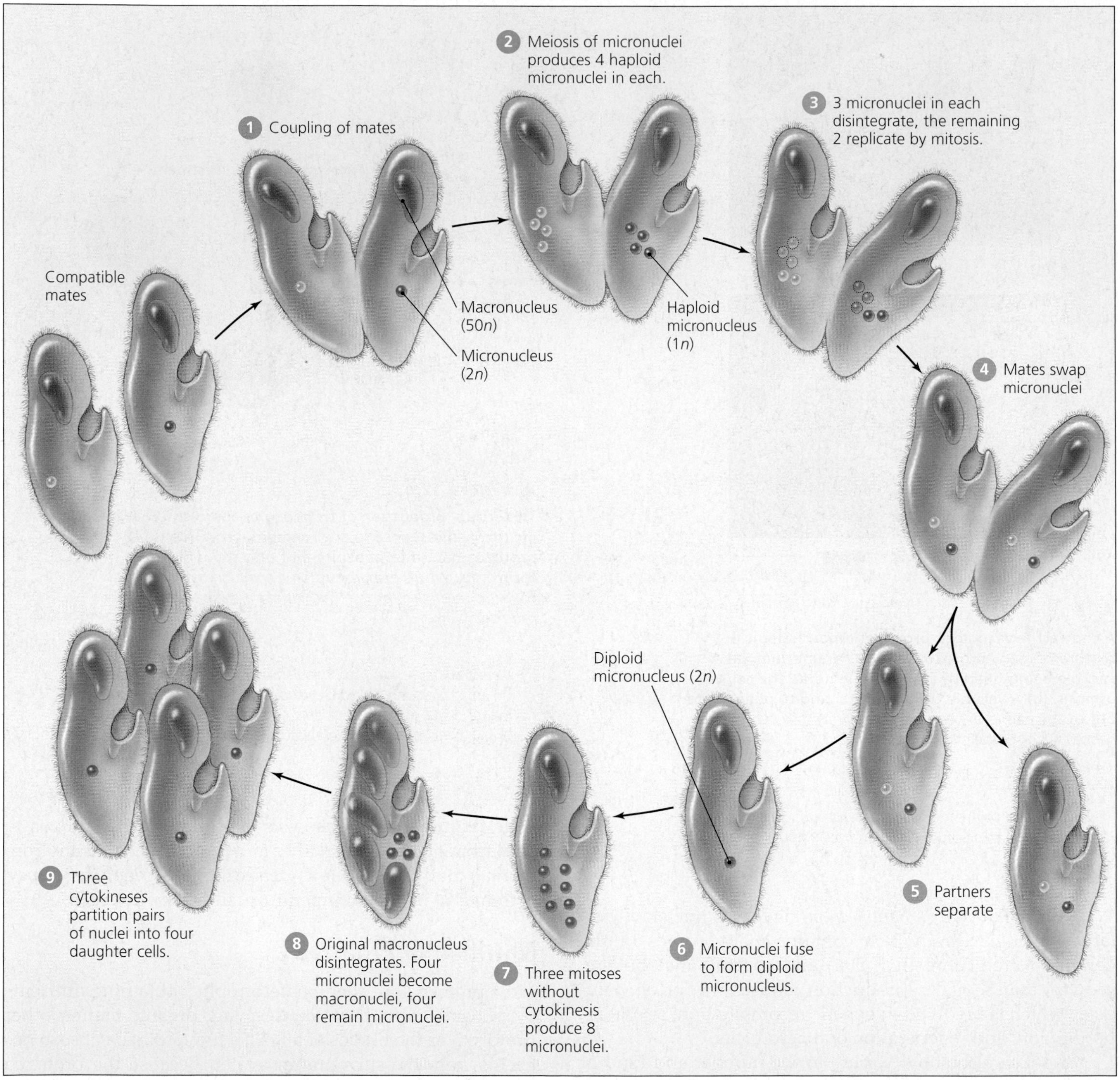

Compatible mates
1 Coupling of mates
Macronucleus (50n)
Micronucleus (2n)
2 Meiosis of micronuclei produces 4 haploid micronuclei in each.
Haploid micronucleus (1n)
3 3 micronuclei in each disintegrate, the remaining 2 replicate by mitosis.
4 Mates swap micronuclei
5 Partners separate
Diploid micronucleus (2n)
6 Micronuclei fuse to form diploid micronucleus.
7 Three mitoses without cytokinesis produce 8 micronuclei.
8 Original macronucleus disintegrates. Four micronuclei become macronuclei, four remain micronuclei.
9 Three cytokineses partition pairs of nuclei into four daughter cells.

▲ ***Figure 12.7***

Sexual reproduction via conjugation in the ciliate *Paramecium*.

mating cells, meiosis of diploid micronuclei, exchange of haploid micronuclei between the coupled cells, uncoupling of the cells, fusion of haploid micronuclei to form a diploid micronucleus, three mitoses of the micronucleus to form eight micronuclei, disintegration of the macronucleus and the subsequent formation of a new macronucleus from four micronuclei, and three cytokineses to produce four daughter cells **(Figure 12.7).**

Classification of Protozoa

Learning Objectives

- ✓ Discuss the reasons for the many different taxonomic schemes for protozoa.
- ✓ Compare and contrast the three types of alveolates.
- ✓ Name one distinguishing characteristic of the apicomplexans.

- ✓ Explain why dinoflagellates and euglenoids were historically classified as both protozoa and algae.
- ✓ Compare and contrast amoeboid movement and euglenoid movement.
- ✓ Identify several features of a typical euglenoid.
- ✓ Describe three types of archaezoa.

As we have seen, over two centuries ago Linnaeus classified protozoa as animals, and in 1959 Whittaker grouped protozoa (and algae as well) into Kingdom Protista in his four-kingdom scheme for eukaryotic classification (see Figure 12.4a and b). Furthermore, some taxonomists divided the protozoa into four groups based on the organisms' mode of locomotion: Sarcodina (which are motile by means of pseudopodia), Mastigophora (flagella), Ciliophora (cilia), and Sporozoa (which are nonmotile). Other taxonomists lumped the first two groups together into a single group called Sarcomastigophora. Grouping of the protozoa according to locomotory features is still in common usage for many practical applications.

Many taxonomists today, however, recognize that these schemes do not reflect genetic relationships, either between protozoa and other organisms or among protozoa. Accordingly, these taxonomists continue to revise and refine the classification of protozoa based on nucleotide sequencing and features made visible by electron microscopy. One such phylogenetic scheme classifies protozoa into the four taxa shown at the left of Figure 12.4c, which different scientists consider kingdoms, subkingdoms, or phyla.

In the following sections we will briefly discuss members of the four groups of protozoa, formed largely according to similarities in nucleotide sequences and ultrastructure, that are shown in Figure 12.4c. We begin with the alveolates.

Alveolates

Alveolates (al-vē′ō-lātz) are protozoa with small membrane-bound cavities called *alveoli*[15] (al-vē′ō-lī) beneath their cell surfaces **(Figure 12.8).** Scientists do not know the purpose of alveoli. Alveolates share at least one other characteristic—tubular mitochondrial cristae. This group is further divided into three subgroups: *ciliates* and *apicomplexans,* which are chemoheterotrophic, and *dinoflagellates,* which are photoautotrophic.

Ciliates As their name indicates, **ciliates** (sil′ē-āts) have cilia by which they either move themselves or move water past their cell surfaces. (The structure and function of cilia were discussed on page 77.) Some ciliates are covered with cilia, whereas others have only a few isolated tufts. All ciliates have two nuclei—one macronucleus and one micronucleus. Some taxonomists consider them the sole members of Phylum Ciliophora.

Notable ciliates include *Vorticella* (vōr-ti-sel′ă), whose apical cilia create a whirlpool-like current to direct food into its "mouth"; *Balantidium* (bal-an-tid′ē-ŭm), which is the only ciliate pathogenic to humans (see Figure 3.23b); and the carnivorous *Didinium* (dī-di′nē-ŭm), which phagocytizes other protozoa such as the well-known pond-water ciliate *Paramecium* **(Figure 12.9).**

Apicomplexans The alveolates called **apicomplexans** (ap-i-kom-plek′sănz) are all pathogens of animals. The name of this group refers to the *complex* of special intracellular organelles, located at the *apices* of the infective stages of

[15]Latin, meaning small hollows.

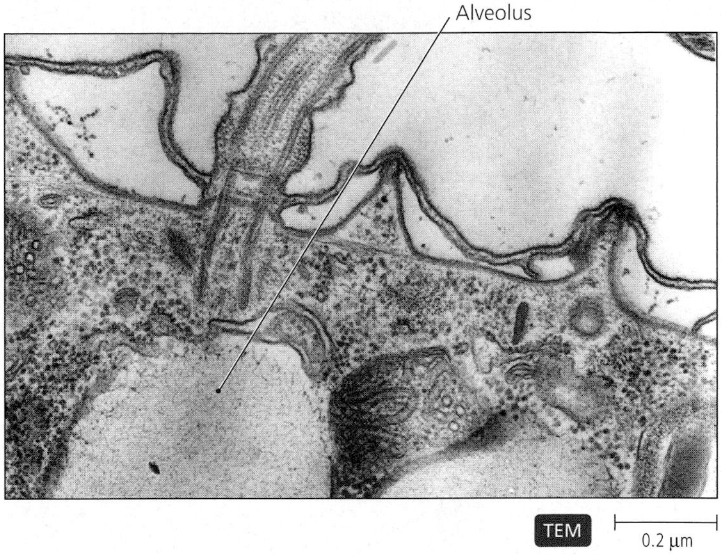

▲ **Figure 12.8**
An alveolus, a membrane-bound structure found in some protozoa. Even though the alveolus' function is not yet known, it is present in eukaryotic microbes with similar rRNA sequences, indicating genetic relatedness and forming the basis of the group of eukaryotes called alveolates.

▲ **Figure 12.9**
A predatory ciliate, *Didinium*, devouring another ciliate, *Paramecium*.

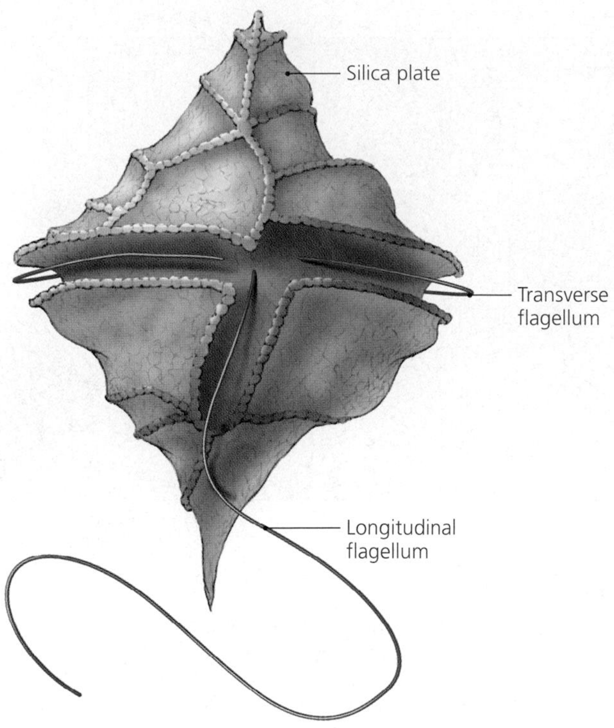

▲ *Figure 12.10*

Peridinium, a motile armored dinoflagellate with two flagella. The transverse flagellum spins the cell; the longitudinal flagellum propels the cell forward.

these microbes, that enables them to penetrate host cells. Examples of apicomplexans are *Plasmodium, Babesia* (bă-bē′zē-ă), and *Toxoplasma* (tok-sō-plaz′mă), which cause malaria, anemia, and toxoplasmosis, respectively.

Dinoflagellates The group of alveolates called **dinoflagellates** are unicellular microbes that have photosynthetic pigments such as carotene and chlorophylls a, c_1, and c_2. Like many plants and algae, their food reserves are starch and oil, and their cell walls are composed of cellulose. Many dinoflagellates are armored with protective plates of *silica* (SiO_2), which is the mineral found in quartz and opal. Even though botanists have historically classified the dinoflagellates as algae (in the division Pyrrhophyta) because of their plantlike features, 18S rRNA sequences and the presence of alveoli indicate that dinoflagellates are more closely related to ciliates and apicomplexans than they are to either plants or algae. Interestingly, unlike other eukaryotic chromosomes, dinoflagellate chromosomes lack histone proteins.

Dinoflagellates make up a large proportion of freshwater and marine plankton. Motile dinoflagellates have two flagella of unequal length **(Figure 12.10).** The transverse flagellum wraps around the equator of the cell in a groove in the cell wall, and its beat causes the cell to spin; the second flagellum extends posteriorly and propels the cell forward.

Many dinoflagellates are bioluminescent—that is, able to produce light via metabolic reactions. When luminescent dinoflagellates are present in large numbers, the ocean water lights up with every crashing wave, passing ship, or jumping fish. Other dinoflagellates produce a red pigment, and their abundance in marine water is called a **red tide.**

Some dinoflagellates, such as *Gymnodinium* (jīm-nō-din′ē-um) and *Gonyaulax* (gon-ē-aw′laks), produce *neurotoxins*—poisons that act against the nervous system. In 1991, over a billion fish died from dinoflagellate poisoning off the coast of North Carolina. Dinoflagellates can also poison humans who eat shellfish that have concentrated the toxins of planktonic dinoflagellates they have ingested.

The neurotoxin of another dinoflagellate, *Pfiesteria*[16] (fes-ter′ē-ă), may be even more potent: It has been claimed the toxin poisons people who merely handle infected fish or breathe air laden with the microbes and that *Pfiesteria* toxin may also cause memory loss, confusion, headache, respiratory difficulties, skin rash, muscle cramps, diarrhea, nausea, and vomiting. The Centers for Disease Control and Prevention (CDC) calls such poisoning *possible estuary-associated syndrome (PEAS).*

Amoebae

The evolutionary history of the group of protozoa we consider next is unclear, and its place within the taxonomic scheme depicted in Figure 12.4c is also unclear. We discuss these protozoa here, even though some members more properly belong in the Euglenozoa—they have disc-shaped mitochondrial cristae and share similar 18S rRNA sequences with *euglenoids* (discussed next).

The organisms called **amoebae**[17] are protozoa that have two characteristics in common: They move and feed by means of pseudopodia (see Figures 3.27 and 3.28), and they lack mitochondria. Beyond these two characteristics, and the fact that they all reproduce via binary fission (an unusual aspect of which is explored in **New Frontiers 12.1**), amoebae exhibit little uniformity. Though many amoebae lack any regular shape, others have loose-fitting protein shells with holes through which they extend pseudopodia to feed and move around; some species of amoebae form cysts. And though a few types are pathogenic in animals and humans, most species live in freshwater, seawater, or soil.

Radiolarians and *foraminifera* are armored, primarily marine amoebae with shells made of silica and calcium carbonate, respectively **(Figure 12.11).** Over 90% of known foraminiferans are fossil species, some of which may form layers of limestone hundreds of meters thick. The great pyramids of Giza outside Cairo, Egypt, are built of foraminiferan limestone. Geologists correlate the ages of sedimentary rocks from different parts of the world by finding identical foraminiferan fossils embedded in them.

---

[16]Named for dinoflagellate biologist Lois Pfiester.

[17]From Greek *ameibein,* meaning to change.

New Frontiers 12.1 Amoeba Midwives

Amoebae divide by binary fission, during which a cell splits into two daughter cells. Occasionally, however, an amoeba fails to divide completely, and the two daughter cells remain attached, as has been observed in recent studies of *Entamoeba invadens,* an amoeba found in the guts of snakes. When stalled in the division process, these amoebae appear to signal for the assistance of neighboring amoebae. Like a midwife helping with the birthing process, an assisting amoeba will advance toward a dividing amoeba and barge into it—completing the separation of the two daughter cells in a fascinating example of microbial cooperation.

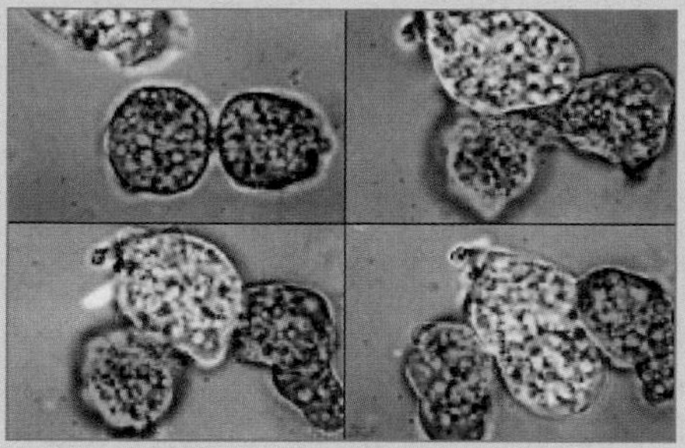

An "assisting" amoeba (yellow) barges into a dividing amoeba (blue).

Source: Biron, D., Libros, P., Sagi, D., Mirelman, D., and Moses, E. 2001. 'Midwives' Assist Dividing Amoebae. *Nature* 410:430.

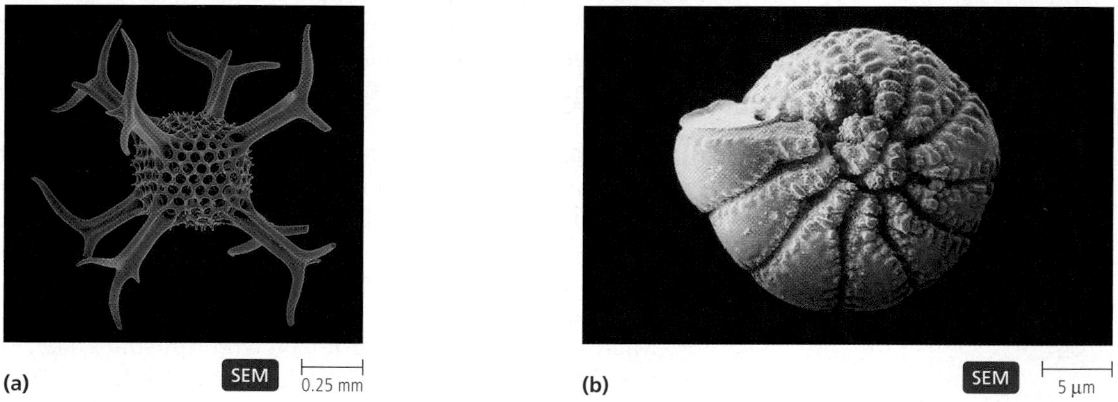

Figure 12.11 Armored marine amoebae. The pseudopodia of these protozoa (not visible) extend through holes in the shells. **(a)** Radiolarians have ornate shells of silica. **(b)** Foraminifera have multichambered shells of calcium carbonate.

The normally free-living amoebae *Naegleria* (nā-glē′rē-ă) and *Acanthamoeba* (ă-kan-thă-mē′bă) cause diseases of the brain in humans and animals that swim in water containing these amoebae. Other amoebae, such as *Entamoeba* (ent-ă-mē′bă), always live inside animals, where they produce potentially fatal amoebic dysentery.

Euglenozoa

Part of the reason that taxonomists established the kingdom Protista in the 1960s was to create a "dumping ground" for *euglenoids,* eukaryotic microbes that share certain characteristics of both plants and animals. More recently, based on similar 18S rRNA sequences and the presence of mitochondria with disc-shaped cristae, some taxonomists have created a new taxon: kingdom Euglenozoa (see Figure 12.4c). The euglenozoa include some of the free-living amoebae (*Naegleria* and *Acanthamoeba*), the euglenoids, and some flagellated protozoa called *kinetoplastids*. We discussed amoebae in the previous section, so we now turn our attention to the euglenoids and kinetoplastids.

Euglenoids The group of euglenozoa called **euglenoids,** which are named for the genus *Euglena* (**Figure 12.12a** on page 354), are photoautotrophic, unicellular microbes with chloroplasts containing light-absorbing pigments—chlorophyll *a*, chlorophyll *b*, and carotene. For this reason, botanists historically classified euglenoids in the division Euglenophyta of the kingdom Plantae. However, one reason for not including euglenoids with plants is that euglenoids store food as a unique polysaccharide called *paramylon* instead of as starch. Euglenoids are similar to animals in that they lack cell walls, have flagella, are chemoheterotrophic phagocytes (in the dark), and move by using their flagella as well as by flowing, contracting, and expanding their cytoplasm. Such a squirming movement, which is similar to amoeboid movement but does not involve pseudopodia, is called *euglenoid movement*.

A euglenoid has a semi-rigid, proteinaceous, helical *pellicle* that underlies its cytoplasmic membrane and helps maintain its shape. Typically each euglenoid also has a red "eyespot," which plays a role in positive phototaxis—movement toward light. Euglenoids reproduce by mitosis followed by longitudinal cytokinesis. They form cysts when exposed to harsh conditions.

Kinetoplastids Euglenozoa called **kinetoplastids (Figure 12.12b)** each have a single large mitochondrion that

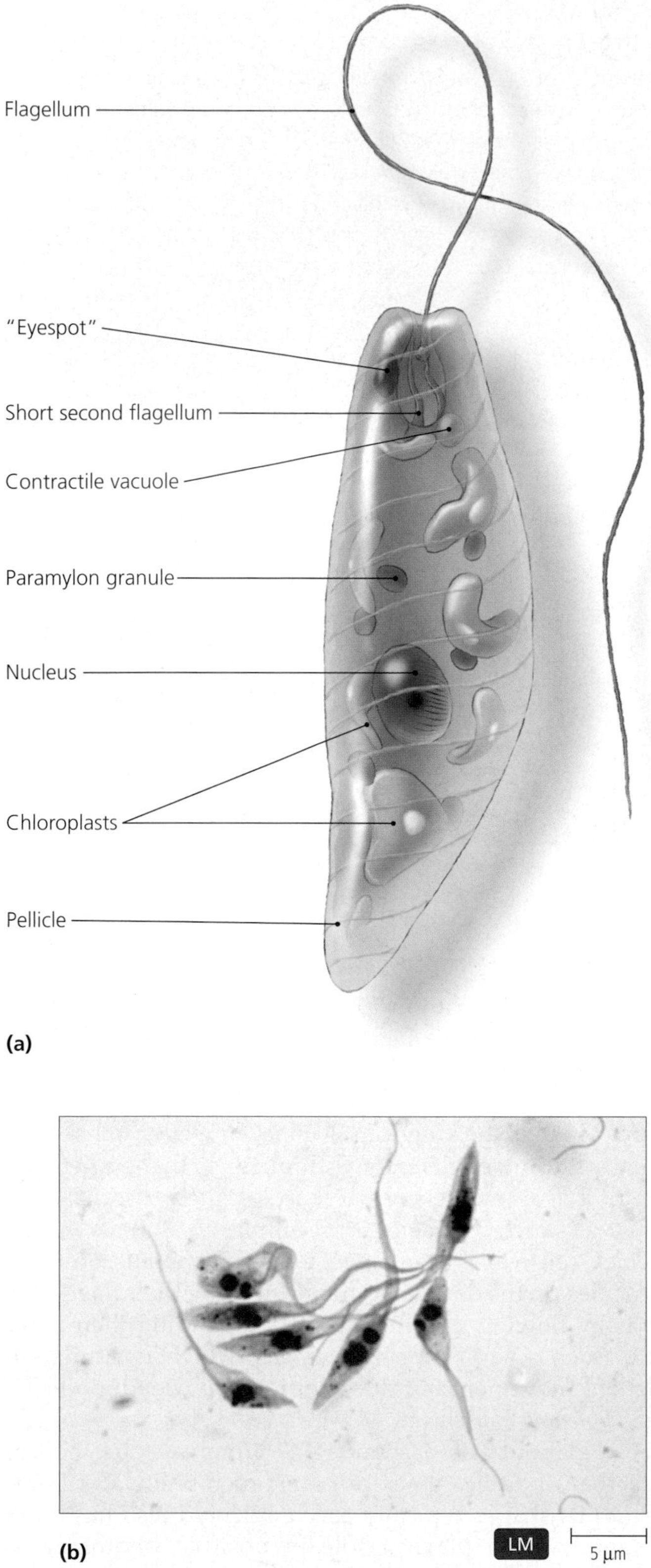

▲ ***Figure 12.12***

Two representatives of the kingdom Euglenozoa. **(a)** The euglenoid *Euglena.* Euglenoids have characteristics that are similar to both plants and animals. **(b)** The kinetoplastid *Trypanosoma cruzi.* *What is the function of the "eyespot" in* Euglena?

Figure 12.12 The "eyespot" functions in positive phototaxis.

contains a unique region of mitochondrial DNA called a *kinetoplast*. Kinetoplastids live inside animals, and some are pathogenic. Among the latter is the genus *Trypanosoma*, certain species of which cause African sleeping sickness and Chagas' disease—potentially fatal diseases of mammals, including humans.

Archaezoa

Because members of the group of eukaryotic microbes called **archaezoa**[18] lack mitochondria, Golgi bodies, and peroxisomes, biologists once thought these organisms were descended from ancient eukaryotes that had not yet phagocytized the prokaryotic ancestors of mitochondria. More recently, however, geneticists have discovered mitochondrial genes in the chromosomes of archaezoa, a finding that suggests that archaezoa might be descended from typical eukaryotes that somehow lost their organelles. In the following sections we discuss the two taxa of archaezoa—Diplomonadida and Parabasala.

Diplomonadida The taxon Diplomonadida includes diplomonads and microsporidia. Diplomonads have two equal-sized nuclei and multiple flagella. A prominent example is *Giardia* (jē-ar′dē-ă), a diarrhea-causing pathogen of animals and humans that is spread to new hosts when they ingest resistant cysts.

Microsporidia are small protozoa with unusual polar filaments. They are obligatory intracellular parasites; that is, organisms that must live within their host's cells. Microsporidia spread from host to host as small resistant cysts called "spores." An example is *Nosema* (nō-sē′mă), which is parasitic on insects such as silkworms and honeybees. The Environmental Protection Agency has approved one species of *Nosema* as a biological control agent for grasshoppers. Seven genera of microsporidia, including *Nosema* and *Microsporidium*, are known to cause diseases in immunocompromised patients.

Parabasala Parabasalids also lack mitochondria, but each has a single nucleus and a *parabasal body*, which is a Golgi-like structure. *Trychonympha* (trik-ō-nimf′ă), a parabasalid with prodigious flagella **(Figure 12.13),** inhabits the guts of termites, where it assists in the digestion of wood. Another well-known parabasalid is *Trichomonas* (trik-ō-mō′nas), which lives in the human vagina. When the normally acidic pH of the vagina is raised, *Trichomonas* proliferates and causes severe inflammation that can lead to sterility. It is spread by sexual intercourse and is usually asymptomatic in males.

In summary, protozoa are a heterogeneous collection of single-celled, mostly chemoheterotrophic organisms that lack a cell wall, and whose relationships with one another and

[18] From Greek *arche,* meaning beginning, and *zoion,* meaning animal.

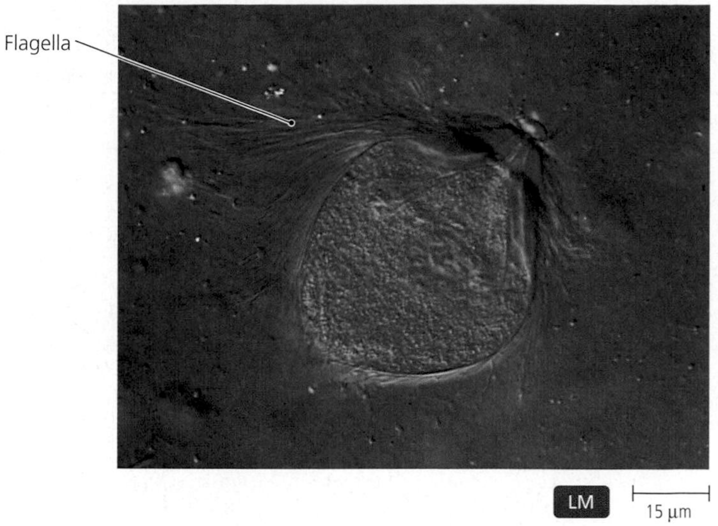

▲ ***Figure 12.13***

Trychonympha, a parabasalid with prodigious flagella.
What evidence suggests that the progenitors of archaezoa possessed mitochondria?

Figure 12.13 Scientists have discovered DNA sequences associated with mitochondrial proteins in the chromosomes of archaezoa.

with other eukaryotic organisms is still unclear. **Table 12.2** on page 356 summarizes the incredible diversity of these microbes.

We next turn our attention to another group of chemoheterotrophs: the fungi, which chiefly differ from the protozoa in that they have cell walls.

CRITICAL THINKING

In Chapter 4 (see p. 121), we discussed identifying microbes using dichotomous taxonomic keys. Design a key for the genera of protozoa discussed in this section.

Fungi

Learning Objective

✓ Cite at least three characteristics that distinguish fungi from other groups of eukaryotes.

Organisms in the kingdom **Fungi** (fŭn′jī; see Figure 12.4c), such as molds, mushrooms, and yeasts, are like most protozoa in that they are chemoheterotrophic; however, unlike protozoa they have cell walls, which typically are composed of a strong, flexible, nitrogenous polysaccharide called **chitin.** (The chitin in fungi is chemically identical to that in the exoskeletons of insects and other arthropods such as grasshoppers, lobsters, and crabs.) Fungi differ from plants in that they lack chlorophyll and do not perform photosynthesis; they differ from animals by having cell walls, although genetic sequencing of fungal and animal genomes has shown that fungi and animals are related. The study of fungi is *mycology*,[19] and scientists who study fungi are *mycologists*.

The Significance of Fungi

Learning Objective

✓ List five ways in which fungi are beneficial.

Fungi are extremely beneficial microorganisms. In nature, they decompose dead organisms (particularly plants) and recycle their nutrients. Additionally, the roots of about 90% of vascular plants form *mycorrhizae*,[20] which are beneficial associations with fungi that assist the plants to absorb water and dissolved minerals.

Humans use fungi for food (mushrooms and truffles), in religious ceremonies (because of their hallucinogenic properties), and in the manufacture of foods and beverages, including bread, alcoholic beverages, citric acid (the basis of the soft drink industry), soy sauce, and some cheeses. Fungi also produce antibiotics such as penicillin and cephalosporin; the immunosuppressive drug *cyclosporine*, which makes organ transplants possible; and *mevinic acids*, which are cholesterol-reducing agents.

Fungi are also important research tools in the study of metabolism, growth, and development, and in genetics and biotechnology. For instance, based on their work with *Neurospora* (noo-ros′pōr-ă) in the 1950s, George Beadle (1903–1999) and Edward Tatum (1909–1975) developed their Nobel Prize–winning theory that one gene codes for one enzyme. Because of similar research, *Saccharomyces* (brewer's yeast) is the best understood eukaryote and the first eukaryote to have its entire genome sequenced.

But not all fungi are beneficial—about 30% of known fungal species produce **mycoses** (mī-kō′sēz), which are fungal diseases of plants, animals, and humans. For example, Dutch elm disease is a fungal disease of elm trees, and athlete's foot is a fungal disease of humans. Because fungi tolerate concentrations of salt, acid, and sugar that inhibit bacteria, fungi are responsible for the spoilage of fruit, pickles, jams, and jellies exposed to air.

In the following sections we will consider the basic characteristics of fungal morphology, nutrition, and reproduction before turning to a brief survey of the major groups of fungi.

Morphology of Fungi

Learning Objective

✓ Distinguish among septate hyphae, aseptate hyphae, and mycelia.

[19]From Greek *mykes*, meaning mushroom, and *logos*, meaning discourse.
[20]From Greek *rhiza*, meaning root.

Table 12.2 Characteristics of Protozoa

Category	Distinguishing Feature(s)	Representative Genera Mentioned in the Text
Alveolates	Alveoli (membrane-bound cavities underlying cytoplasmic membrane); tubular mitochondrial cristae	
Ciliates	Cilia	*Balantidium* *Paramecium* *Didinium*
Apicomplexans	Apical complex of organelles	*Babesia* *Plasmodium* *Toxoplasma*
Dinoflagellates	Photosynthesis; two flagella	*Gymnodinium* *Gonyaulax* *Pfiesteria*
Amoebae	Pseudopodia; lack mitochondria	
Foraminifera		
Radiolarians		
Free-living types		*Naegleria* *Acanthamoeba*
Parasitic types		*Entamoeba*
Euglenozoa	Flagella; disc-shaped mitochondrial cristae	
Euglenoids	Photosynthesis; pellicle	*Euglena*
Kinetoplastids	Single large mitochondrion with DNA localized in kinetoplast	*Trypanosoma*
Archaezoa		
Diplomonadida	Lack mitochondria, Golgi bodies, and peroxisomes	
Diplomonads	Two equal-sized nuclei; multiple flagella	*Giardia*
Microsporidia	Polar filaments	*Nosema*
Parabasala		
Parabasalids	Single nucleus	*Trichonympha* *Trichomonas*

The vegetative (nonreproductive) body of a fungus is called its **thallus**[21] (plural: *thalli*). The morphology of fungal thalli is variable. The thalli of *yeasts* are small, globular, and composed of a single cell **(Figure 12.14a),** whereas the thalli of *molds* are large and composed of long, branched, tubular filaments called **hyphae**[22] **(Figure 12.14b).** Hyphae are either **septate** (divided into cells by crosswalls called *septa*[23]) or **aseptate** (not divided by septa; **Figure 12.14c**). Aseptate hyphae are also coenocytic (multinucleate).

In response to environmental conditions such as temperature or carbon dioxide concentration, some fungi produce both yeastlike thalli and moldlike thalli **(Figure 12.14d);** fungi that produce two types of thalli are said to be **dimorphic** (which means "two-shaped"). Many medically important fungi are dimorphic, including *Histoplasma capsulatum* (his-tō-plaz′mă kap-soo-lā′tŭm), which causes a respiratory disease called histoplasmosis, and *Coccidioides immitis* (kok-sid-ē-oy′dēz im′mi-tis), which causes a flulike disease called *coccidioidomycosis* (kok-sid-ē-oy′dō-mī-kō′sis).

The thallus of a mold is composed of hyphae intertwined to form a tangled mass called a **mycelium** (plural: *mycelia;* **Figure 12.15** on page 358). Mycelia are typically subterranean and thus usually escape our notice, though they can be very large. In fact, as mentioned in the chapter opener, the largest organisms on Earth are fungi in the genus *Armillaria*, the mycelia of which can spread through thousands of acres of forest to a depth of several feet and weigh many hundreds of tons. (In contrast, blue whales, the largest living animals, weigh only about 150 tons.) *Fruiting bodies*, such as puffballs and mushrooms, are the reproductive structures of molds and are only small visible extensions of vast underground mycelia.

Nutrition of Fungi

Fungi acquire nutrients by absorption; that is, they secrete catabolic enzymes outside their thalli to break large organic molecules into smaller molecules, which they then transport into their thalli. Most fungi are **saprobes**[24] (sap′rōbz)—they absorb nutrients from the remnants of dead organisms—but some species trap and kill microscopic soil-dwelling

[21] From Greek *thallos*, meaning young shoot.
[22] From Greek *hyphe*, meaning weaving or web.
[23] Latin, meaning partitions or fences.
[24] From Greek *sapros*, meaning rotten, and *bios*, meaning life.

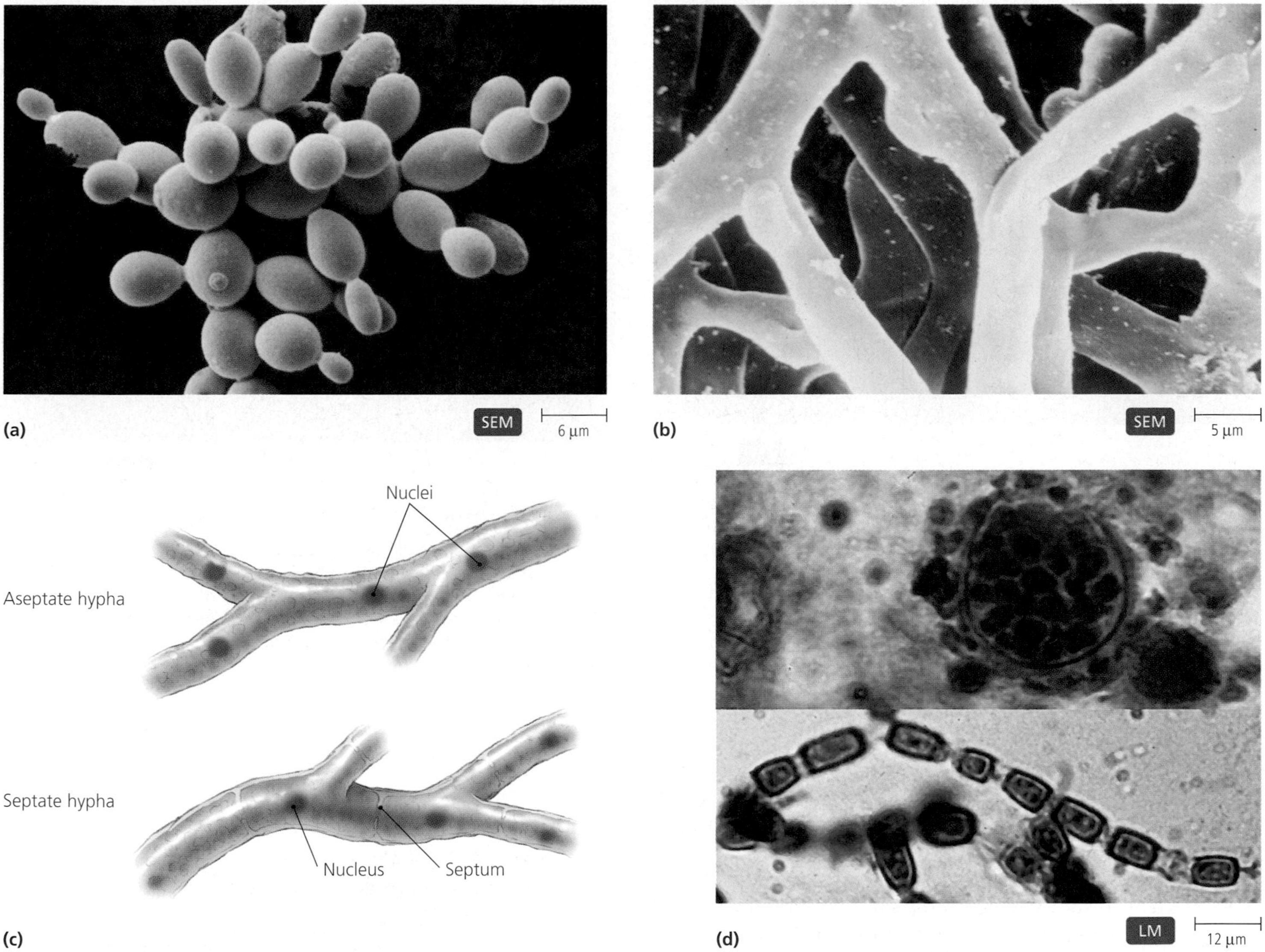

Figure 12.14

Fungal morphology. **(a)** The thalli of *Saccharomyces* (bakers' or brewers' yeast), which are unicellular and spherical to irregularly oval in shape. **(b)** The thallus of a filamentous fungus, which is composed of cylindrical hyphae. **(c)** Examples of septate and aseptate hyphae. **(d)** The thalli of a dimorphic fungus, *Mucor rouxii*, showing both yeastlike and moldlike growth in response to local environmental conditions.

nematodes (worms) (**Figure 12.16** on page 358). Fungi that derive their nutrients from living plants and animals usually have modified hyphae called **haustoria**[25] (haw-stō′rē-ă), which penetrate the tissue of the host to withdraw nutrients. Absorptive nutrition is important in the role that fungi play as decomposers and recyclers of organic waste. Cytoplasmic streaming frequently transports nutrients and organelles, including nuclei, throughout a mycelium. Streaming between cells of septate mycelia occurs through pores in the septa.

Most fungi are aerobic, though many yeasts (for example, *Saccharomyces*) are facultative anaerobes that obtain energy from fermentation, such as occurs in the reactions that produce alcohol. Anaerobic fungi are found in the digestive systems of many herbivores, such as cattle and deer, where they assist in the catabolism of plant material.

Reproduction in Fungi

Learning Objectives

- ✓ Describe asexual reproduction in fungi.
- ✓ List three basic types of asexual spores found in molds.

[25]From Latin *haustor,* meaning someone who draws water from a well.

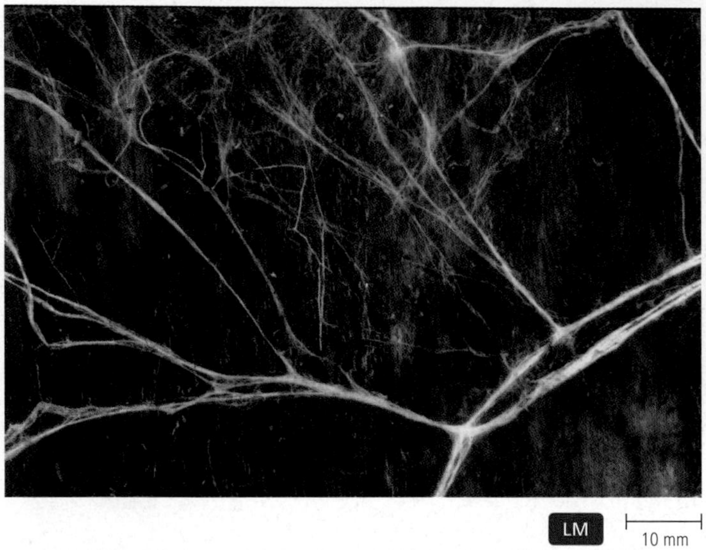

▲ ***Figure 12.15***
A fungal mycelium growing on wood.

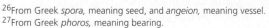

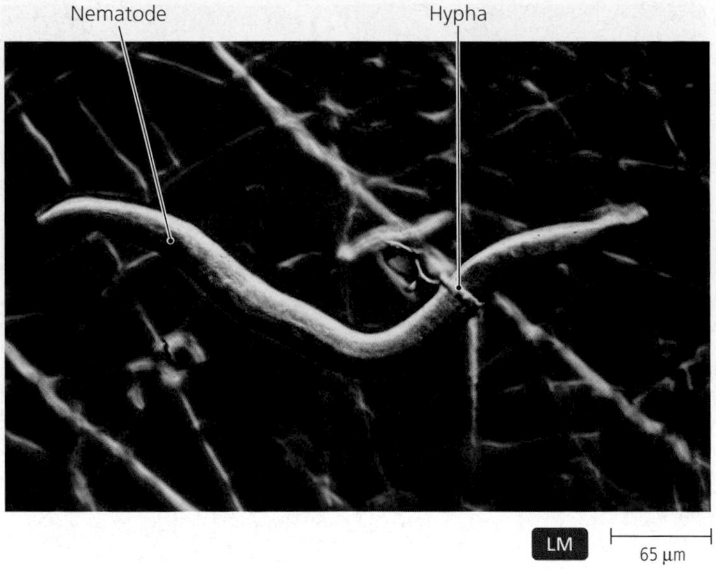

▲ ***Figure 12.16***
Predation of a nematode by the fungus *Arthrobotrys.* The fungus produces special looped hyphae that constrict when the worm contacts the inside of the loop. The fungus secretes enzymes that digest the nematode and then absorbs the resulting nutrients. *What is the more typical mode of nutrition found in fungi?*

Figure 12.16 Most fungi are saprobic.

Whereas all fungi have some means of asexual reproduction involving mitosis followed by cytokinesis, most fungi also reproduce sexually. In the next subsections we briefly examine asexual and sexual reproduction in fungi.

Budding and Asexual Spore Formation

Yeasts typically bud in a manner similar to prokaryotic budding. Following mitosis, one daughter nucleus is sequestered in a small bleb of cytoplasm that is isolated from the parent cell by the formation of a new wall (see Figure 12.14a). In some species, especially *Candida albicans* (kan′did-ă al′bi-kanz), which causes human oral thrush and vaginal yeast infections, a series of buds remain attached to one another and to the parent cell, forming a long filament called a *pseudohypha. Candida* invades human tissues by means of such pseudohyphae, which can penetrate intercellular cracks.

Filamentous fungi reproduce asexually by producing lightweight spores, which enable the fungi to disperse vast distances on the wind. Researchers have isolated fungal spores from wind currents many miles above the surface of the earth. Scientists categorize the asexual spores of molds according to their mode of development:

- *Sporangiospores* form inside a sac called a *sporangium,*[26] which is often borne on a spore-bearing stalk, called a *sporangiophore,*[27] at either the tips or sides of hyphae **(Figure 12.17a).**
- *Chlamydospores* form with a thickened cell wall inside hyphae **(Figure 12.17b).**
- *Conidiospores* (also called *conidia*) are produced at the tips or sides of hyphae, but not within a sac. There are many types of conidia, including *arthroconidia,* which develop from hyphae that fragment into individual spores; *blastoconidia,* which form as buds, and others that develop in chains on stalks called *conidiophores* **(Figure 12.17c).**

Medical lab technologists use the presence and type of asexual spores in clinical samples to identify many fungal pathogens.

Sexual Spore Formation

Scientists designate fungal mating types as "+" and "−" rather than as male and female because their thalli are morphologically indistinguishable. The process of sexual reproduction in fungi has four basic steps (**Figure 12.18** on page 360):

1. Haploid (n) cells from a + thallus and a − thallus fuse to form a *dikaryon,* a cell containing both + and − nuclei. The dikaryotic stage is neither diploid nor haploid, but instead is designated ($n + n$).
2. After a period of time that typically ranges from hours to years but can be centuries, a pair of nuclei within the dikaryon fuse to form one diploid ($2n$) nucleus.
3. Meiosis of diploid nucleus restores the haploid state.
4. The haploid nuclei are partitioned into + and − spores.

Fungi differ in the ways they form dikaryons and in the site at which meiosis occurs.

[26]From Greek *spora,* meaning seed, and *angeion,* meaning vessel.
[27]From Greek *phoros,* meaning bearing.

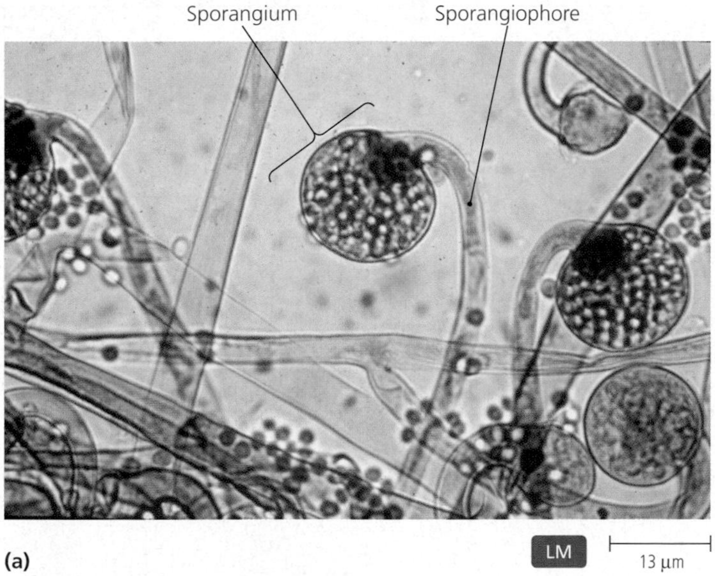

(a)

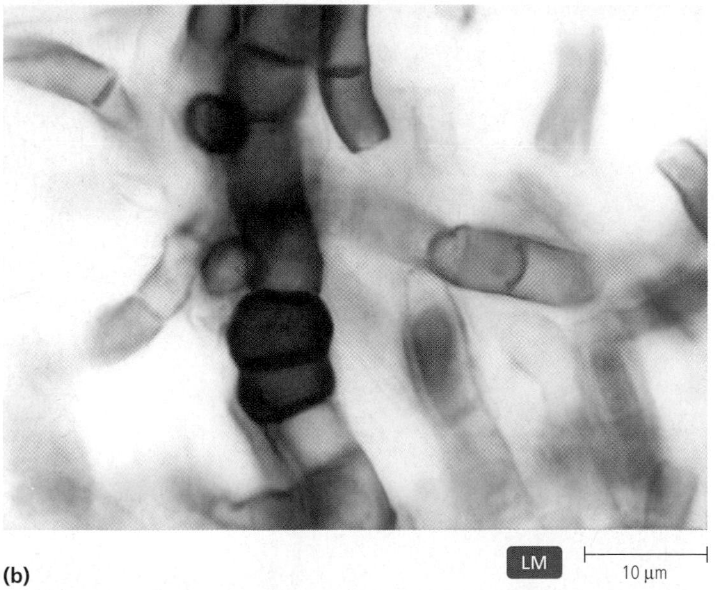

(b)

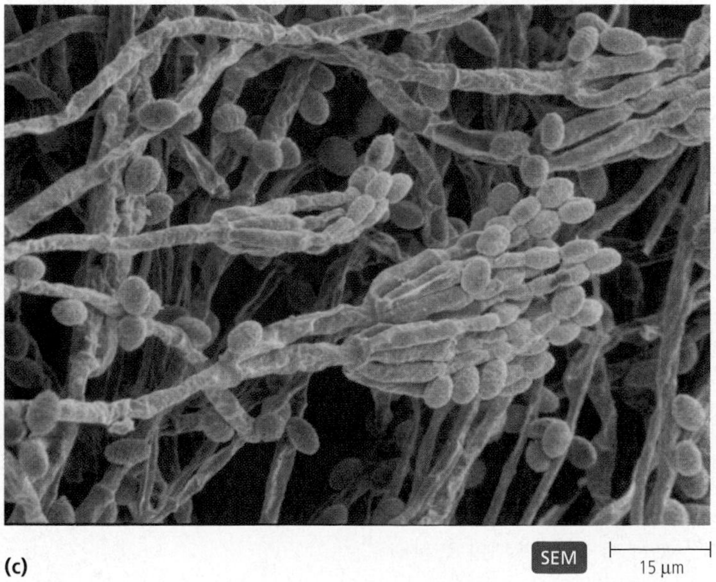

(c)

◀ ***Figure 12.17***

Representative asexual spores of molds. **(a)** Sporangiospores, which develop within a sac called a sporangium that is borne on a sporangiophore. **(b)** Chlamydospores, which are thick-walled spores that form inside hyphae. **(c)** The conidiospores (conidia) of *Penicillium*, which develop on conidiophores at the ends of hyphae. *How do conidia differ from sporangiospores?*

Figure 12.17 Conidia are never enclosed in a sac.

CRITICAL THINKING

Fungi tend to reproduce sexually when nutrients are limited or other conditions are unfavorable, but reproduce asexually when conditions are more ideal. Why is this a successful strategy?

Classification of Fungi

Learning Objectives

- ✓ Compare and contrast the three divisions of fungi with respect to the formation of sexual spores.
- ✓ Describe the deuteromycetes, and explain why this group no longer constitutes a formal taxon.
- ✓ List several beneficial roles or functions of lichens.

In the following sections we will consider the four major subgroups into which taxonomists traditionally divided the kingdom Fungi seen in Figure 12.4c. Three of these subgroups, which are taxons called *divisions* that are equivalent to phyla in kingdom Animalia, are based on the type of sexual spore produced (divisions Zygomycota, Ascomycota, and Basidiomycota); the fourth (the deuteromycetes) was a repository of fungi for which no sexual stage is known. We begin by considering the Zygomycota.

Division Zygomycota

Fungi in the division **Zygomycota** are coenocytic molds called *zygomycetes* (zī'gō-mī-sēts). Of the approximately 1100 species known, most are saprobes; the rest are obligate parasites of insects and other fungi.

Figure 12.19 on page 361 illustrates the life cycle of a typical zygomycete: the black bread mold *Rhizopus* (rī-zō'pŭs) *nigricans*.[28] Zygomycetes reproduce asexually via sporangiospores (steps 1 to 4 in Figure 12.19), but the distinctive feature of zygomycetes is the formation of sexual structures called **zygospores.** Zygospores of *R. nigricans* are black, rough-walled structures that develop from a zygote produced by the fusion of two nuclei within a short-lived dikaryon formed by the fusion of sexually compatible hyphae (see steps 5 to 8 in Figure 12.19). Like other fungal

[28]From Greek *rhiza,* meaning root, and *pous,* meaning foot, and Latin *niger,* meaning black.

Figure 12.18 ➤
The process of sexual reproduction in fungi. The steps in the process are described in the text.

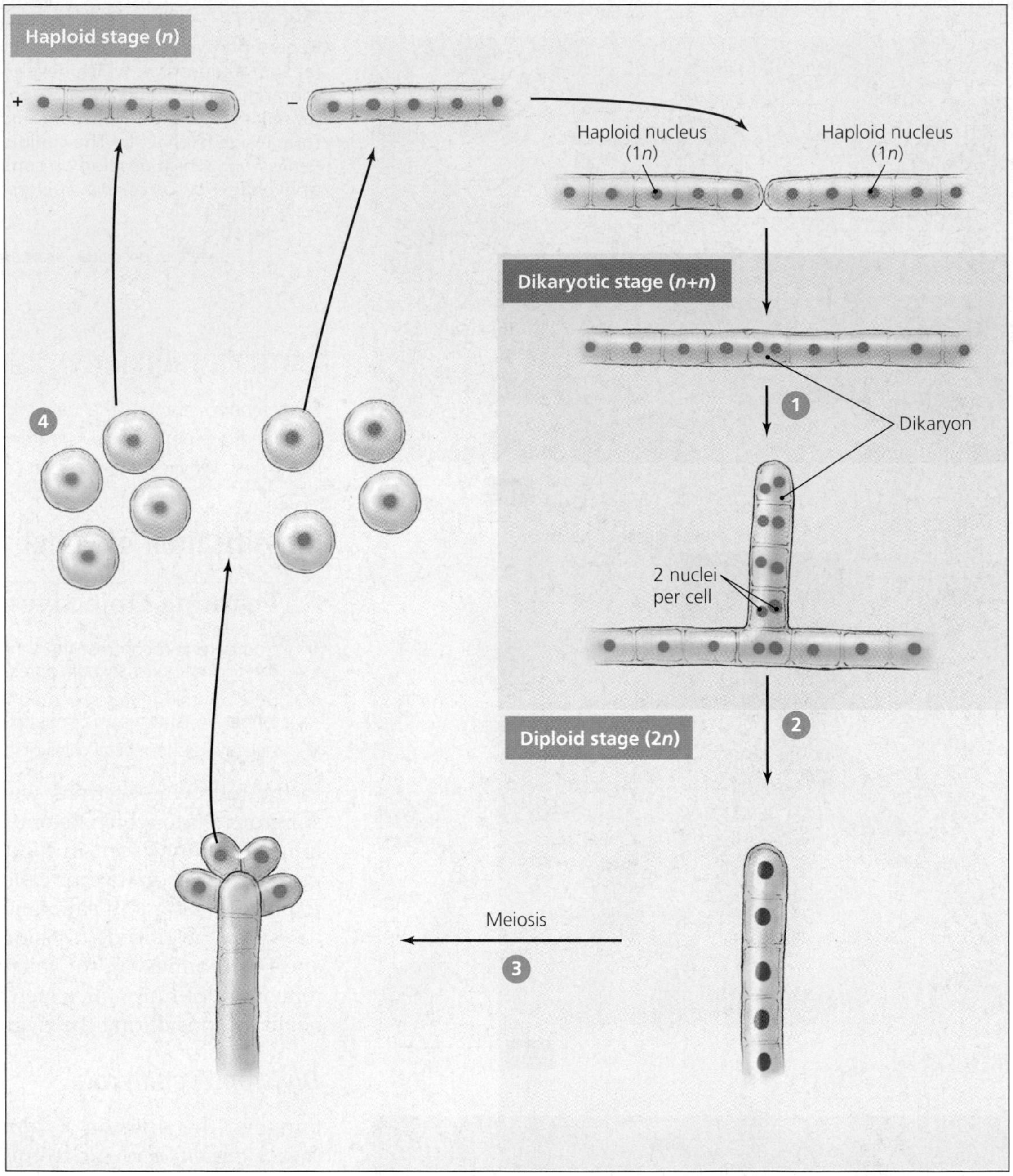

spores, zygospores can withstand desiccation and other harsh environmental conditions.

Each zygospore undergoes meiosis without cytokinesis, but only one of the four meiotic nuclei survives. A zygospore germinates, producing an asexual sporangium that produces numerous haploid spores via mitosis and cell divisions. The sporangium releases its spores, each of which germinates to produce either a + or a − mycelium. This completes the life cycle.

Division Ascomycota

The division **Ascomycota** contains about 32,000 known species of molds and yeasts that are characterized by the formation of haploid **ascospores** within sacs called **asci.**[29] Asci occur in fruiting bodies called *ascocarps*, which have various shapes **(Figure 12.20).** Ascomycetes (as′kō-mī-sēts), as they are called, also reproduce asexually by conidiospores, as illustrated for a representative ascomycete (**Figure 12.21,** steps 1 to 4 on page 362).

Also Figure 12.21 (steps 5 to 11) illustrates the sexual reproduction of an ascomycete, which proceeds as follows:

5 Multinucleate, hyphal tips of opposite mating types fuse to form a dikaryon.

[29]From Greek *askus,* meaning wineskin.

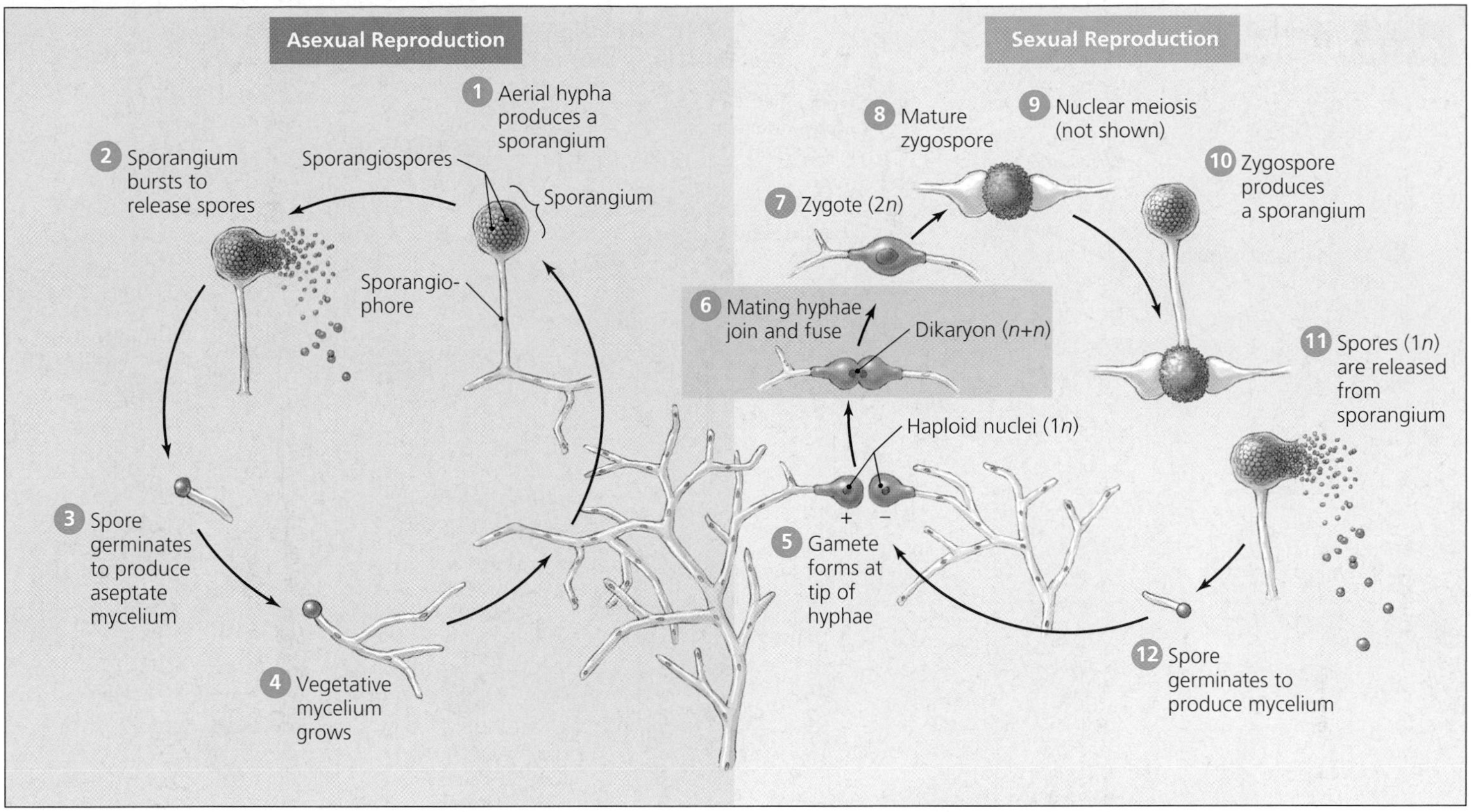
Asexual Reproduction
Sexual Reproduction
1 Aerial hypha produces a sporangium
2 Sporangium bursts to release spores
Sporangiospores
Sporangium
Sporangio-phore
3 Spore germinates to produce aseptate mycelium
4 Vegetative mycelium grows
5 Gamete forms at tip of hyphae
6 Mating hyphae join and fuse
Dikaryon (n+n)
Haploid nuclei (1n)
+
−
7 Zygote (2n)
8 Mature zygospore
9 Nuclear meiosis (not shown)
10 Zygospore produces a sporangium
11 Spores (1n) are released from sporangium
12 Spore germinates to produce mycelium

▲ **Figure 12.19**

Life cycle of the zygomycete *Rhizopus*. During the asexual cycle, the fungus reproduces via sporangiospores that germinate and produce hyphae. In the sexual cycle, the tips of + and − hyphae fuse and form a diploid zygospore that matures, undergoes meiosis, germinates, and produces a sporangium containing haploid sporangiospores, which germinate to form new mycelia.

6. The dikaryon reproduces to form hyphae whose cells are all dikaryontic.
7. In the dikaryon, opposite mating type nuclei fuse to form a diploid nucleus.
8. Each diploid nucleus undergoes meiosis and cytokinesis to form four haploid cells within an ascus.
9. Each haploid cell may undergo mitosis and cytokinesis to form two haploid ascospores. Thus, the process produces eight ascospores, which line up inside the ascus.
10. The asci open to release their ascospores.
11. The ascospores germinate to produce + and − hyphae.

Ascomycetes are familiar and economically important fungi. For example, most of the fungi that spoil food are ascomycetes. This group also includes plant pathogens such as the causative agents of Dutch elm disease and chestnut blight, which have almost eliminated their host trees in many parts of the United States. *Claviceps purpurea* (klav′i-seps poor-poo′rē-ă) growing on grain produces *lysergic acid diethylamide (LSD)*, which causes abortions in cattle and hallucinations in humans.

1 cm

▲ **Figure 12.20**

Ascocarps (fruiting bodies) of the common morel, *Morchella esculenta*, a delectable edible ascomycete. The pits visible in this photograph are lined with asci, sacs that contain numerous ascospores.

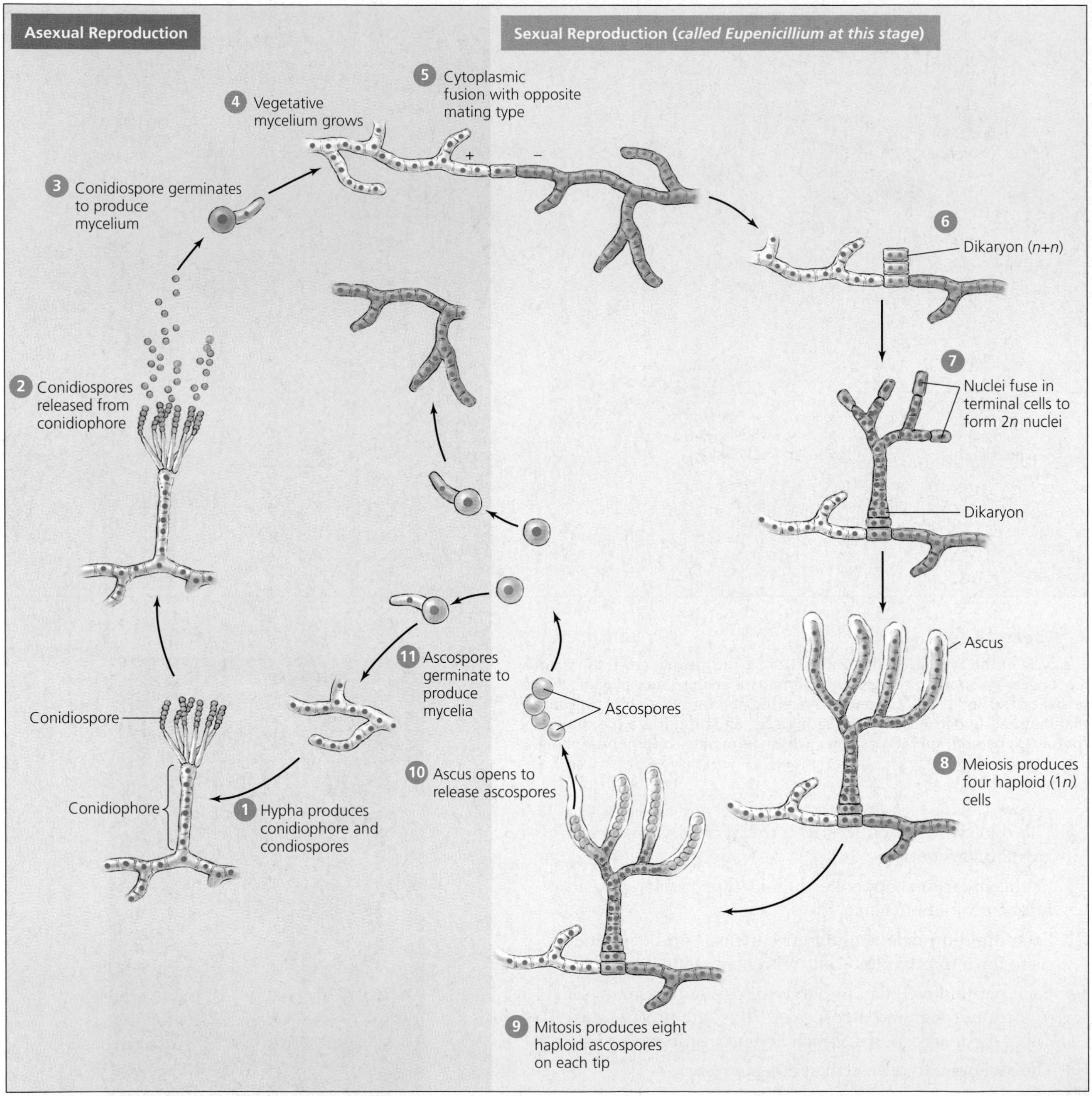

Asexual Reproduction
Sexual Reproduction (*called Eupenicillium at this stage*)
1 Hypha produces conidiophore and condiospores
Conidiospore
Conidiophore
2 Conidiospores released from conidiophore
3 Conidiospore germinates to produce mycelium
4 Vegetative mycelium grows
5 Cytoplasmic fusion with opposite mating type
+
–
6 Dikaryon (*n*+*n*)
7 Nuclei fuse in terminal cells to form 2*n* nuclei
Dikaryon
Ascus
8 Meiosis produces four haploid (1*n*) cells
9 Mitosis produces eight haploid ascospores on each tip
10 Ascus opens to release ascospores
Ascospores
11 Ascospores germinate to produce mycelia

▲ ***Figure 12.21***
Life cycle of an ascomycete.

On the other hand, many ascomycetes are beneficial. For example, *Penicillium* (pen-i-sil′ē-ŭm) mold is the source of penicillin; *Saccharomyces*, which ferments sugar to produce alcohol and carbon dioxide gas, is the basis of the baking and brewing industries; and *truffles* (varieties of *Tuber*) grow as mycorrhizae in association with oak and beech trees to form delectable culinary delights (see **Highlight 12.1**). Another ascomycete, the pink bread mold *Neurospora*, has been an important tool in genetics and biochemistry. Many ascomycetes partner with green algae or cyanobacteria to form *lichens*, which are discussed in more detail shortly.

Highlight 12.1 Truffles: The Gourmet Fungi

Truffles—rare, intensely flavored mushrooms that grow underground—are one of the most luxurious foods on Earth, selling at more than $500 per pound. There are many different varieties of truffles. The most coveted include *Tuber melanosporum,* a black truffle that is also known as the "black diamond," and *Tuber magnatum pico,* a white truffle that can sell for up to $1800 per pound! Because truffles are very difficult to find, truffle hunters often use pigs and dogs trained to sniff them out. (Dogs are preferred, because pigs are more likely to eat the truffles.) The underground habitat of truffles requires them to form symbiotic relationships with trees (for nutrients) and with animals such as squirrels and chipmunks (for spore dispersal; the animals dig up the truffles and thus help spread the spores to other locations).

Incidentally, chocolate truffles only derive their name from their prized fungal counterparts!

Tuber melanosporum

Division Basidiomycota

A walk through fields and woods in most parts of the world may reveal mushrooms, puffballs, stinkhorns, jelly fungi, bird's nest fungi, or bracket fungi, all of which are the visible fruiting bodies of the almost 22,000 known species of fungi in the division **Basidiomycota**. Poisonous mushrooms are sometimes called *toadstools,* but there is no sure way to always distinguish between an edible "mushroom" and a poisonous "toadstool" except by eating them—a truly risky practice!

Mushrooms and other fruiting bodies of *basidiomycetes* (ba-sid'ē-ō-mi-sēts) are called **basidiocarps (Figure 12.22).** The entire structure of a basidiocarp consists of tightly woven hyphae that extend into multiple, often club-shaped projections called *basidia,* the ends of which produce sexual *basidiospores* (typically four on each basidium). **Figure 12.23** on page 364 illustrates the life cycle of *Amanita muscaria* (am-ă-nī'tă mus-ka'rē-ă), a poisonous mushroom.

Besides the edible mushrooms—most notably, the cultivated *Agaricus* and *Cortinellis*—basidiomycetes affect humans in several ways. Most basidiomycetes are important decomposers that digest chemicals such as cellulose and lignin in dead plants and return nutrients to the soil. Many mushrooms produce hallucinatory chemicals or toxins. An example of the latter is the extremely poisonous "Death-cap mushroom," *Amanita* (see Figure 12.23). The basidiomycete yeast *Cryptococcus neoformans* is the leading cause of fungal meningitis. Other basidiomycetes are *rusts* and *smuts,* which cause millions of dollars in crop loss each year.

Deuteromycetes

As noted previously, the divisions Zygomycota, Ascomycota, and Basidiomycota were based on type of sexual spore produced. But because scientists have not observed sexual reproduction in all fungi, taxonomists in the middle of the 20th century created the division **Deuteromycota** to contain the heterogeneous collection of fungi whose sexual stages are unknown—either because they do not produce sexual spores, or because their sexual spores have not been observed. More recently, however, the identification of rRNA sequences has revealed that most deuteromycetes in fact

▲ ***Figure 12.22***

Basidiocarps (fruiting bodies) of the bird's nest fungus, *Crucibulum*. The familiar shapes of mushrooms are also basidiocarps of extensive mycelia.

belong in the division Ascomycota, and thus modern taxonomists have abandoned Deuteromycota as a formal taxon. Nevertheless, many medical laboratory technologists, health care practitioners, and scientists continue to refer to "deuteromycetes" because it is a traditional name.

Most deuteromycetes are either terrestrial saprobes, pathogens of plants, or pathogens of other fungi. Several genera are pathogenic to humans, including the filamentous genus *Trichophyton* (tri-kof'i-tŏn), one of the causes of ringworm, and the dimorphic fungus *Histoplasma capsulatum,* which causes a flu-like illness called *histoplasmosis*. Chapter 22 discusses these and other pathogenic fungi in more detail.

Lichens

A discussion of fungi is incomplete without considering **lichens,** which are partnerships between fungi and photosynthetic microbes—either green algae (discussed shortly) or, more commonly, cyanobacteria. The hyphae of the fungus

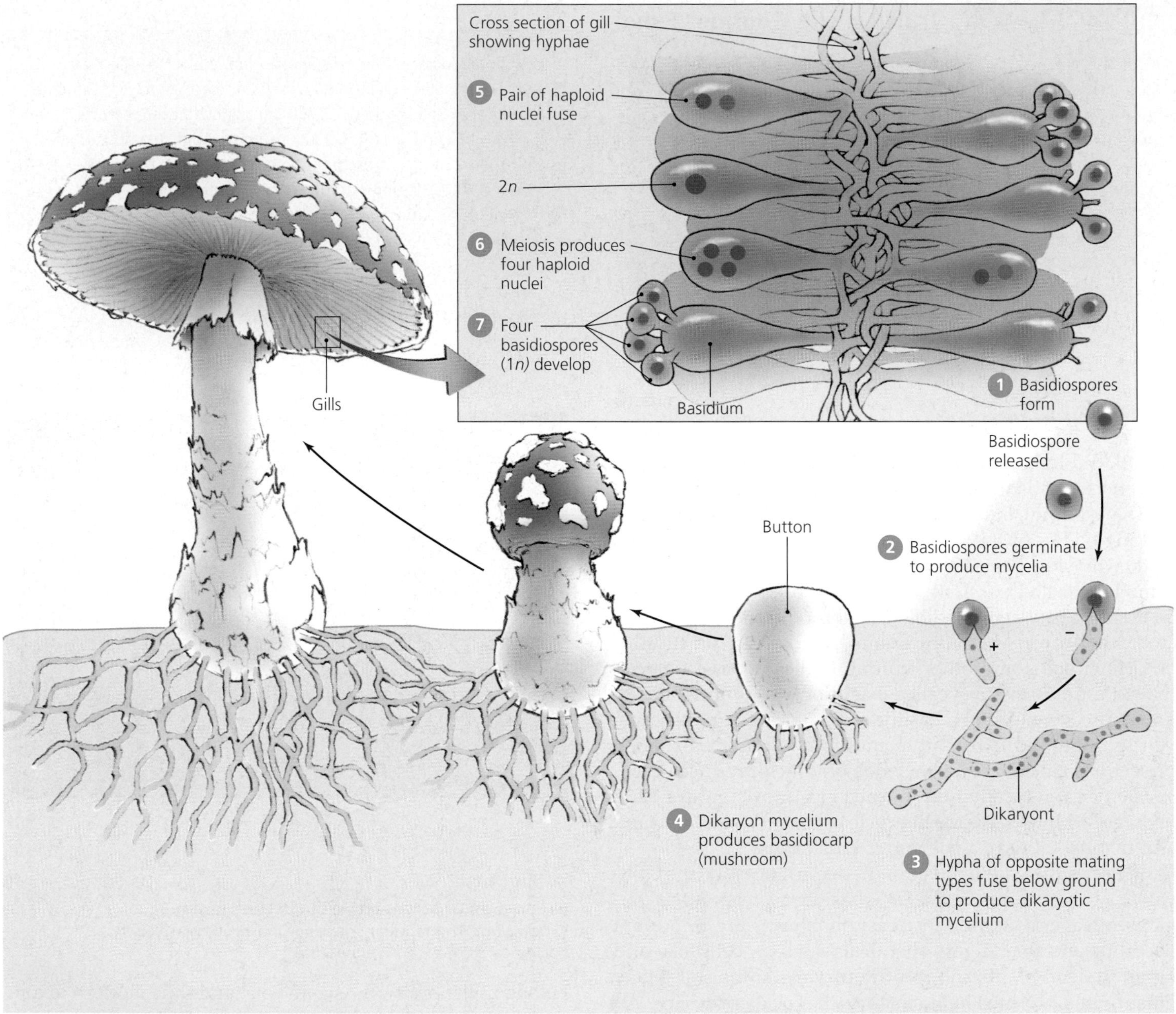

▲ ***Figure 12.23***

Life cycle of *Amanita muscaria*, a poisonous basidiomycete. *How can a novice distinguish between edible and poisonous mushrooms?*

Figure 12.23 *There is no sure way to distinguish between edible and poisonous mushrooms except by assuming the risk of eating them.*

in a lichen, which is usually an ascomycete, surround the photosynthetic cells **(Figure 12.24)** and provide them nutrients, water, and protection from desiccation and harsh light. In return, each alga or cyanobacterium provides the fungus with products of photosynthesis—carbohydrates and oxygen. In some lichens, the phototroph releases 60% of its carbohydrates to the fungus.

The partnership in a lichen is not always mutually beneficial; in some lichens, the fungus produces haustoria that penetrate and kill the photosynthetic member. Such lichens are maintained only because the phototroph's cells reproduce faster than the fungus can devour them.

The fungus of a lichen reproduces by spores, which must germinate and develop into hyphae that capture an appropriate alga or cyanobacterium. Alternatively, bits of lichen called *soredia,* which contain both phototrophs and fungal hyphae, break away and establish a new lichen when they reach a suitable substrate.

▼ Figure 12.24

Makeup of a lichen. The hyphae of a fungus (most commonly an ascomycete) constitute the major portion of the thallus of the lichen; cells of the photosynthetic member of the lichen are concentrated near the lichen's surface. Soredia, which are bits of fungus surrounding phototrophic cells, are the means by which some lichens propagate themselves.

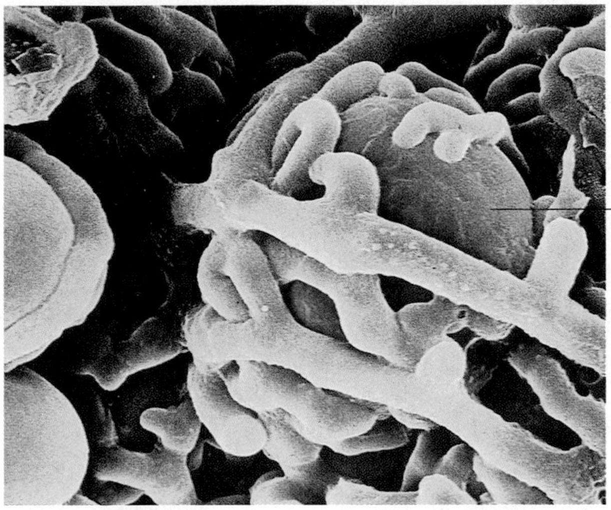

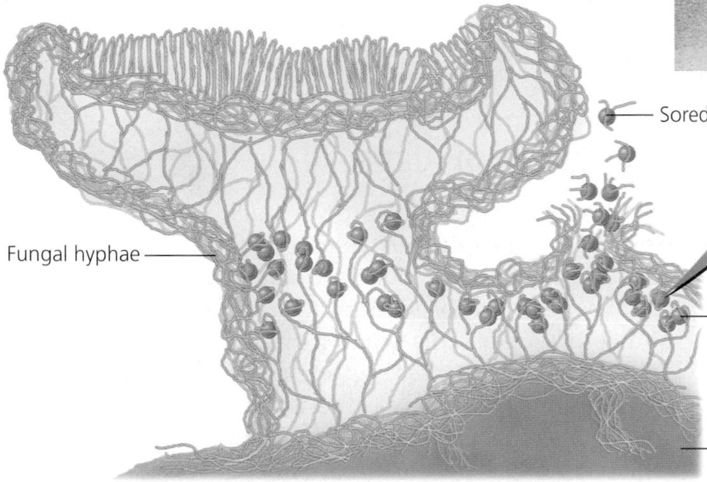

Scientists have identified over 14,000 species of lichens, which are named for their fungal half. Lichens are abundant throughout the world, particularly in pristine unpolluted habitats, growing on soil, rocks, leaves, tree bark, other lichens, and even the backs of tortoises. Indeed, lichens grow in almost every habitat—from high-elevation alpine tundra to submerged rocks on the ocean's shores, from frozen Antarctic soil to hot desert climes. The only unpolluted places where lichens do not consistently grow are in the depths of the oceans and the dark world of caves—after all, lichens require light.

Lichens occur in three basic shapes **(Figure 12.25):**

- *Crustose* lichens grow appressed to their substrates and may extend into the substrate for several millimeters.
- *Foliose* lichens are leaflike, with margins that grow free from the substrate.
- *Fruticose* lichens are either erect or hanging cylinders.

Lichens are important agents in the creation of soil from weathered rocks, and lichens containing nitrogen-fixing cyanobacteria often provide nitrogen to nutrient-poor environments. Many animals eat lichens; for example, reindeer and caribou subsist primarily on lichens throughout the winter. Birds use lichens for nesting materials, and some insects camouflage themselves with bits of living lichen. Humans also use lichens in the production of foods, dyes, clothing, perfumes, and medicines, and as sensitive living assays for monitoring air pollution because lichens will not grow in polluted environments.

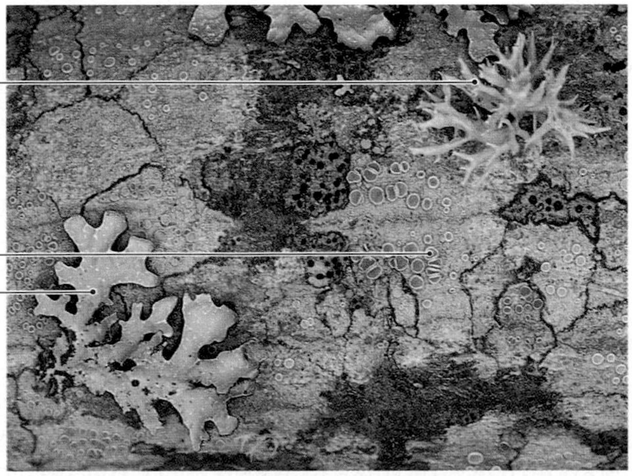

▲ Figure 12.25

Gross morphology of lichens. Crustose forms are flat and tightly joined to the substrate, foliose forms are leaflike with free margins, and fruticose forms are either erect (as shown) or pendulant.

Table 12.3 Characteristics of Fungi

Group	Type of Sexual Spore	Comments	Representative Genera
Division Zygomycota	Zygospores	Coenocytic (aseptate)	*Rhizopus*
Division Ascomycota	Ascospores	Septate; some associated with cyanobacteria or green algae to form lichens	*Claviceps* *Saccharomyces* *Penicillium* *Tuber* *Neurospora*
Division Basidiomycota	Basidiospores	Septate	*Agaricus* *Cortinellis* *Cryptococcus* *Amanita*
Deuteromycetes	None known	Septate	*Histoplasma* *Trichophyton*

In the preceding sections, we have seen that fungi are chemoheterotrophic yeasts and molds that function primarily to decompose and degrade dead organisms. Some fungi are pathogenic, and others associate with algae or cyanobacteria to form lichens. All fungi reproduce asexually either by budding or via asexual spores, and most fungi also produce sexual spores, by which taxonomists classify them. **Table 12.3** summarizes the characteristics of fungi.

CRITICAL THINKING

Design a key for the genera of fungi discussed in this section.

Algae

Learning Objective

✓ Describe the distinguishing characteristic of algae.

The Romans used the word *alga* (al′ga) to refer to any simple aquatic plant, particularly one found in marine habitats. Their usage thus included the organisms we recognize as algae (al′jē) and cyanobacteria, as well as sea grasses and other aquatic plants. Today, the word *algae* properly refers to simple, eukaryotic, phototrophic organisms that, like plants, carry out oxygenic photosynthesis using chlorophyll *a*. Algae differ from plants such as sea grass in having sexual reproductive structures in which every cell becomes a gamete. In plants, by contrast, a portion of the reproductive structure always remains vegetative.

Algae are not a unified group; rather, they differ widely in distribution, morphology, reproduction, and biochemical traits. Moreover, the word *algae* is not synonymous with any taxon; in the taxonomic scheme shown in Figure 12.4c, algae can be found in the kingdoms Stramenopila, Rhodophyta, and Plantae. The study of algae is **phycology**,[30] and the scientists that study them are called *phycologists*.

[30]From Greek *phykos*, meaning seaweed, and *logos*, meaning discourse.

Distribution of Algae

Even though some algae grow in such diverse habitats as in soil and ice, in intimate association with fungi as lichens, and on plants; most algae are aquatic, living in the *photic zone* (penetrated by sunlight) of fresh, brackish, and salt bodies of water. This watery environment both provides some benefits and presents some difficulties for photosynthetic organisms. Whereas the water contains sufficient dissolved chemicals to provide nutrients for an entire large alga, the water also differentially absorbs longer wavelengths of light (including red light), so only shorter (blue) wavelengths penetrate more than a meter below the surface. This is problematic for algae because their primary photosynthetic pigment, chlorophyll *a*, captures red light. Thus to grow in deeper waters, algae must have *accessory photosynthetic pigments* that trap the energy of penetrating, short-wavelength light and pass that energy to chlorophyll *a*. Members of the group of algae known as red algae, for example, contain *phycoerythrin*, a red pigment that absorbs blue light, enabling red algae to inhabit even the deepest parts of the photic zone.

Morphology of Algae

Algae are either unicellular, colonial, or have simple multicellular bodies called thalli, which are commonly composed of branched filaments or sheets. The thalli of large marine algae, commonly called seaweeds, can be relatively complex, with branched *holdfasts* to anchor them to rocks, stemlike *stipes*, and leaflike *blades*. The thalli of many of the larger marine algae are buoyed in the water by gas-filled bulbs called *pneumocysts* (see Figure 12.29). Though the thalli of some marine algae can surpass land plants in length, they lack the well-developed transport systems common to vascular plants.

Reproduction in Algae

Learning Objective

✓ Describe the alternation of generations in algae.

In unicellular algae, asexual reproduction involves mitosis followed by cytokinesis. In unicellular algae that reproduce sexually, each algal cell acts as a gamete and fuses with another such gamete to form a zygote, which then undergoes meiosis to return to the haploid state.

Multicellular algae reproduce asexually by fragmentation, in which each piece of a parent alga develops into a new individual, and by motile or nonmotile asexual spores. As noted previously, in multicellular algae that reproduce sexually, every cell in the reproductive structures of the alga becomes a gamete—a feature that distinguishes algae from all other photosynthetic eukaryotes.

Many multicellular algae reproduce sexually with an **alternation of generations** of haploid and diploid individuals (**Figure 12.26** on page 368). In such life cycles, diploid individuals undergo meiosis to produce male and female haploid spores that develop into haploid male and female thalli, which may look identical to the diploid thallus. Each of these thalli produces gametes that fuse to form a zygote, which grows into a new diploid thallus. Both haploid and diploid thalli may reproduce asexually as well.

Classification of Algae

Learning Objectives

✓ List four divisions of algae and describe the distinguishing characteristics of each.

✓ List several economic benefits derived from algae.

The classification of algae is not yet settled. Historically, taxonomists have used differences in photosynthetic pigments, storage products, and cell wall composition to classify algae into several groups that are named for the colors of their photosynthetic pigments: green algae, red algae, golden algae, yellow-green algae, and brown algae. The following sections present algal divisions—less inclusive groups than the kingdoms shown in Figure 12.4c. We begin with the green algae of the division Chlorophyta.

Division Chlorophyta (Green Algae)

Chlorophyta[31] are green algae that share numerous characteristics with plants—they have chlorophylls *a* and *c*, use sugar and starch as food reserves, and have cell walls composed of cellulose. In addition, the 18S rRNA sequences of green algae and plants are comparable. Because of these similarities, the green algae are often considered to be the progenitors of plants, and in some taxonomic schemes the Chlorophyta are placed in the kingdom Plantae.

Most green algae are unicellular or filamentous (see Figure 1.6a) and live in freshwater ponds, lakes, and pools, where they form green to yellow scum. Some multicellular forms grow in the marine intertidal zone—that is, in the region exposed to air during low tide.

Prototheca (prō-tō-thē′kă) is an unusual green alga in that it lacks pigments, making it colorless. This chemoheterotrophic alga causes a skin rash in sensitive individuals. *Codium* is a member of a group of marine green algae that do not form crosswalls after mitosis; thus, its entire thallus is a single, large, multinucleate cell. Some Polynesians dry and grind *Codium* for use as seasoning pepper. The green alga *Trebouxia* is the most common alga found associated with fungi in lichens.

CRITICAL THINKING

Since *Prototheca* is colorless, how do scientists know that it is really a green alga?

Division Rhodophyta (Red Algae)

The algae of the division **Rhodophyta**[32], which have been placed in the kingdoms Plantae and Protista, are now in their own taxon—Rhodophyta (see Figure 12.4c). They are characterized by the red accessory pigment **phycoerythrin**, the storage molecule glycogen (also known as *floridean*[33] *starch*), cell walls of **agar** or **carrageenan** (kar-ă-gē′nan), and nonmotile male gametes called *spermatia*. As previously noted, phycoerythrin allows red algae to absorb short-wavelength blue light and photosynthesize at depths greater than 100 m. Because the relative proportions of phycoerythrin and chlorophyll *a* vary, red algae range in color from green to black in the intertidal zone to red in deeper water (**Figure 12.27** on page 369). Most red algae are marine, though a few freshwater genera are known.

The gel-like polysaccharides agar and carrageenan, once they have been isolated from red algae such as *Gelidium* (jel-i′dē-ŭm) and *Chondrus* (kon′drŭs), are used as thickening agents for the production of solid microbiological media and numerous consumer products, including ice cream, toothpaste, syrup, salad dressings, and snack foods.

Division Chrysophyta (Golden Algae, Yellow-Green Algae, and Diatoms)

Though the division **Chrysophyta**[34] is a group of algae that are diverse with respect to cell wall composition and pigments, it is unified in using the polysaccharide *chrysolaminarin* as a storage product. (Some also store oils.) Many modern taxonomists group these algae with the brown algae and water molds (discussed shortly) in the kingdom Stramenopila (see Figure 12.4c) based on similarities in nucleotide sequences. Whereas some chrysophytes lack cell walls, others have ornate external coverings such as scales or plates. **Diatoms** (dī′ă-tomz) are unique in having cell

[31]From Greek *chloros,* meaning green, and *phyton,* meaning plant.
[32]From Greek *rhodon,* meaning rose.
[33]Named for a taxon of red algae, *Florideophycidae.*
[34]From Greek *chrysos,* meaning gold.

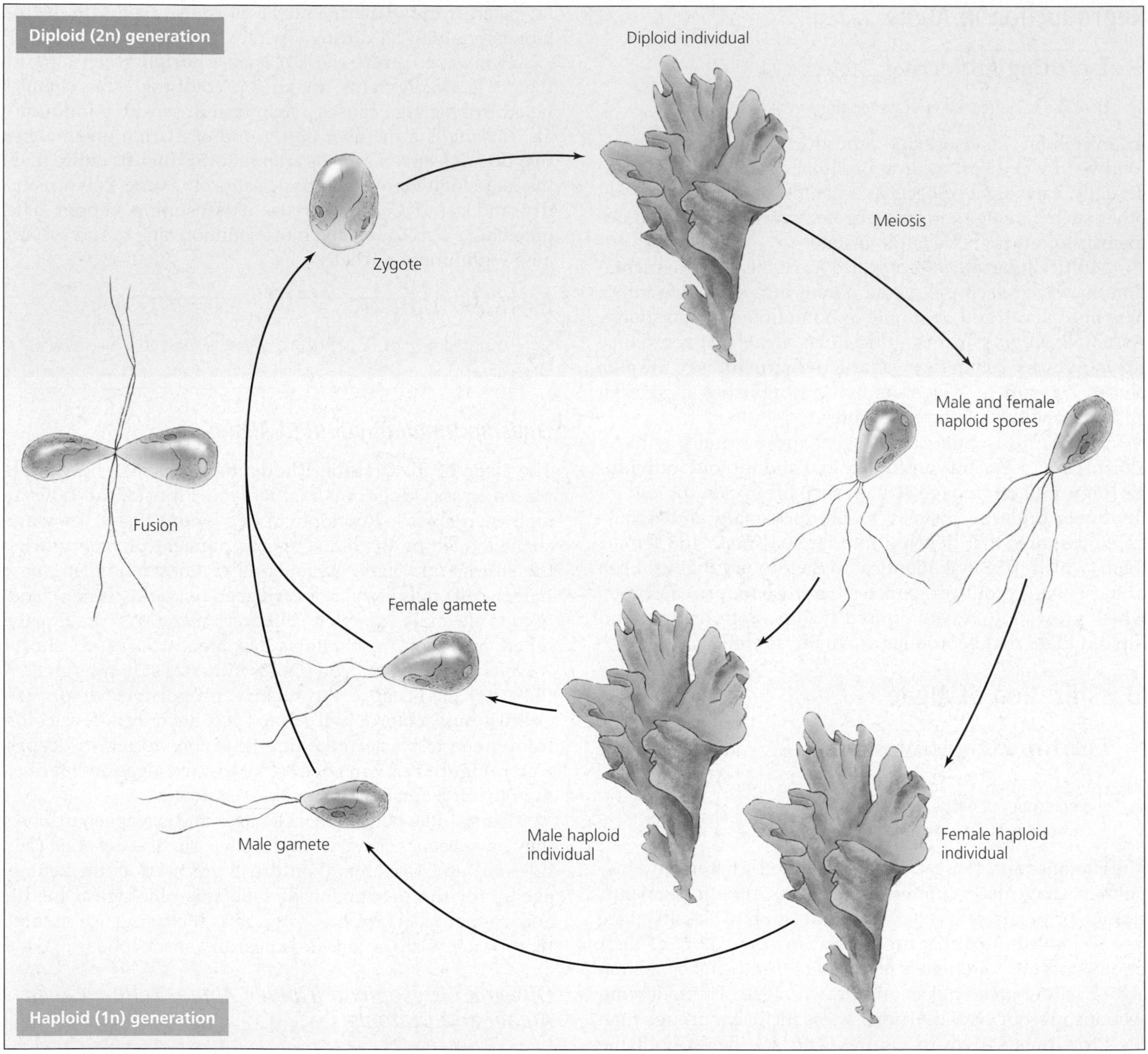

▲ ***Figure 12.26***

Alternation of generations in algae, as occurs in *Ulva*. A diploid individual meiotically produces haploid spores that germinate and grow into male and female thalli. These haploid algae produce gametes that fuse to form a diploid zygote, which grows into a new diploid individual. Each generation can also reproduce asexually by fragmentation and by spores.

walls composed of two halves called *frustules* that fit together like a Petri dish **(Figure 12.28).**

Most chrysophytes are unicellular or colonial. All chrysophytes contain more orange-colored *carotene* pigment than they do chlorophyll, which accounts for the common names of two major classes of chrysophytes—golden algae and yellow-green algae.

Diatoms are a major component of marine *phytoplankton*—free-floating photosynthetic microorganisms that form the basis of food chains in the oceans. Further, because of

▲ *Figure 12.27*
Red alga.

▲ *Figure 12.28*
Diatom. Diatoms have frustules, composed of silica and cellulose, that fit together like a Petri dish.

their enormous number, diatoms are the major source of the world's oxygen. The frustules of diatoms are composed of silica and contain minute holes for the exchange of gases, nutrients, and wastes with the environment. Organic gardeners use *diatomaceous earth,* composed of innumerable frustules of dead diatoms, as a pesticide against harmful insects and worms. Diatomaceous earth is also used in polishing compounds, detergents, paint removers, and as a component of firebrick, soundproofing products, and reflective paints.

Division Phaeophyta (Brown Algae)

The **Phaeophyta**[35] are in the kingdom Stramenopila (see Figure 12.4c) along with chrysophytes and diatoms. They

▲ *Figure 12.29*
The giant kelp *Macrocystis*, a brown alga. A kelp's blades are kept afloat by pneumocysts.

have chlorophylls *a* and *c*, carotene, and brown pigments called *xanthophylls* (zan′thō-fils). Depending on the relative amounts of these pigments, brown algae may appear dark brown, tan, yellow-brown, greenish brown, or green. Most brown algae are marine organisms, and some of the giant kelps, such as *Macrocystis* **(Figure 12.29),** rival the tallest trees in length, though not in girth. Brown algae produce gametes and spores that are motile by means of two flagella—one tinsel-like and one whiplike (**Figure 12.30** on page 370).

Brown algae use the polysaccharide *laminarin* and oils as food reserves and have cell walls composed of cellulose and **alginic acid** (alginate). Alginic acid is used in numerous foods as a thickening agent and emulsifier; brewers use it to maintain a foamy head on beer.

In summary, algae are unicellular and multicellular photoautotrophs characterized by sexual reproductive structures in which every cell becomes a gamete. The colors produced by the combination of their primary and accessory photosynthetic pigments give them their common names and provide the basis of at least one classification scheme. **Table 12.4** on page 371 summarizes the characteristics of the major groups of algae; the dinoflagellates and euglenoids are included in it because botanists often classify these phototrophic protozoa as algae.

CRITICAL THINKING

Design a key for the genera of algae discussed in this section.

Water Molds and Slime Molds

The microbes commonly known as *water molds* and *slime molds* were once classified as fungi because they resemble filamentous fungi in having finely branched filaments; however, neither water molds nor slime molds are true fungi.

[35]From Greek *phaeo,* meaning brown.

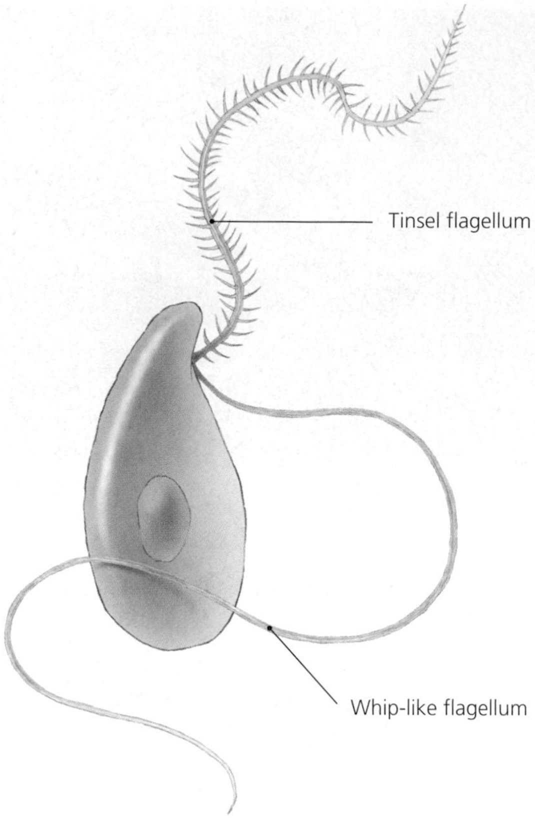

▲ ***Figure 12.30***
The two types of flagella on the sperm of the brown alga *Fucus*.

Water Molds

Learning Objective

✓ List four ways in which water molds differ from true fungi.

Water molds differ from fungi in the following ways:

- They have tubular cristae in their mitochondria.
- They have cell walls of cellulose instead of chitin.
- Their spores have two flagella—one whiplike and one tinsel-like.
- They have true diploid thalli rather than haploid or dikaryon thalli.

As is the case with many eukaryotic microorganisms, taxonomists do not agree on the placement of water molds in classification schemes. Because they often have tinsel-like flagella and certain similarities in rRNA sequence, many scientists think water molds should be grouped together with diatoms, other chrysophytes, and brown algae in the taxon Stramenopila (see Figure 12.4c); other mycologists prefer to classify the water molds in their own kingdom or phylum.

Water molds decompose dead animals and return nutrients to the environment **(Figure 12.31).** Some species are

▲ ***Figure 12.31***
An example of the important role of water molds in recycling organic nutrients in aquatic habitats.

detrimental pathogens of crops such as grapes, tobacco, and potato. In 1845, the water mold *Phytophthora* (fī-tof′tho-ră) was accidentally introduced into Ireland and devastated the potato crop, causing the great famine that killed over 1 million people and forced a greater number to immigrate to the United States and Canada.

Slime Molds

Learning Objectives

✓ List two ways in which slime molds differ from true fungi.
✓ Contrast plasmodial (acellular) and cellular slime molds.

Slime molds differ from true fungi in two main ways:

- They lack cell walls, more closely resembling the amoebae in this regard.
- They are phagocytic rather than absorptive in their nutrition.

Once again, not all taxonomists agree on the classification of slime molds. Some scientists categorize slime molds in a phylum of the kingdom Protista; others classify them in a special division of the kingdom Fungi; still others classify them in a kingdom of their own—kingdom Mycetozoa (see Figure 12.4c).

Species in the two groups of slime molds—plasmodial (acellular) and cellular slime molds—differ based in their morphology, reproduction, and 18S rRNA sequences. Slime molds are important to humans primarily as excellent laboratory systems for the study of developmental and molecular biology.

Plasmodial (Acellular) Slime Molds

Plasmodial slime molds, also known as *acellular* slime molds (*e.g., Physarum*), exist as streaming, coenocytic, colorful filaments of cytoplasm that creep like amoebae through forest litter, feeding by phagocytizing organic debris and bacteria. The thallus, called a *plasmodium* (not to be confused with the agent of malaria—*Plasmodium*), may contain

Table 12.4 Characteristics of Various Algae

Group (Common Name)	Pigments	Storage Product (s)	Cell Wall Component(s)	Habitat	Representative Genera
Chlorophyta (green algae)	Chlorophylls *a* and *b*, carotene, xanthophylls	Sugar, starch	Cellulose	Fresh, brackish, and salt water; terrestrial	*Spirogyra* *Prototheca* *Codium* *Trebouxia*
Rhodophyta (red algae)	Chlorophyll *a*, phycoerythrin, phycocyanin, xanthophylls	Glycogen (floridean starch)	Agar, carrageenan	Mostly salt water	*Chondrus* *Gelidium*
Chrysophyta (golden algae, yellow-green algae, diatoms)	Chlorophylls *a*, c_1 and c_2; carotene, xanthophylls	Chrysolaminarin	Cellulose, silica, calcium carbonate	Fresh, brackish, and salt water; terrestrial; ice	*Navicula*
Phaeophyta (brown algae)	Chlorophylls *a* and *c*, xanthophylls	Laminarin, oils	Cellulose, alginic acid	Brackish and salt water	*Macrocystis*
Pyrrhophyta (dinoflagellates)	Chlorophylls *a*, c_1, and c_2, carotene	Starch, oils	Cellulose	Fresh, brackish, and salt water	*Gymnodinium* *Gonyaulax* *Pfiesteria*
Euglenophyta (euglenoids)	Chlorophylls *a* and *b*, carotene	Paramylon, oils, sugar	Absent	Fresh, brackish, and salt water; terrestrial	*Euglena*

millions of diploid nuclei and cover many square centimeters (**Figure 12.32a** on page 372). Nutrients are distributed throughout the plasmodium by cytoplasmic streaming.

When food or water are in short supply, the plasmodium divides into individual masses of cytoplasm, each of which produces a stalked sporangium. Meiosis occurs within the sporangia to generate haploid spores. These spores germinate to produce *myxamoebae*, which look and act like protozoan amoebae; in the presence of water, however, myxamoebae produce flagella and swim about. (When the water disappears, they become amoeboid again.)

Compatible myxamoebae of opposite mating types fuse to form a diploid zygote. The nucleus of the zygote undergoes numerous mitoses—without cytokineses—to form a new coenocytic plasmodium.

Cellular Slime Molds

Cellular slime molds, such as *Dictyostelium*, exist as individual haploid myxamoeba that phagocytize bacteria, yeasts, dung, and decaying vegetation. **Figure 12.32b** illustrates their life cycle, all of which is haploid—there is no diploid phase. Myxamoebae reproduce by mitosis and cytokinesis when food is abundant; however, in scarcity, some secrete cyclic adenosine monophosphate (cAMP), which acts as a chemotactic attractant for other myxamoebae. The myxamoebae congregate into a sluglike *pseudoplasmodium*, which can migrate for several days. Unlike the true plasmodium of acellular slime molds, the cells of a pseudoplasmodial slug retain their individuality and can be separated mechanically.

Some cells of the pseudoplasmodium form a stalked sporangium; the remaining cells climb the stalk and become spores. In contrast to the spores of plasmodial slime molds, the spores of cellular slime molds do not result from meiosis and are not enclosed in a common wall.

CRITICAL THINKING

Why are cellular slime molds called "cellular"?

In summary, water molds and slime molds are not true molds; they differ from fungi in many aspects of structure, reproduction, and nutrition.

Other Eukaryotes of Microbiological Interest: Parasitic Helminths and Vectors

Learning Objectives

- ✓ Explain why microbiologists study large organisms such as parasitic worms and arthropod vectors.
- ✓ Describe the principal arthropod vectors of human pathogens and give one example of a disease transmitted by each.

Microbiologists are also interested in two other groups of eukaryotes, although they are not microorganisms. The first group are the parasitic **helminths,** commonly called parasitic worms. Microbiologists became interested in parasitic helminths because they observed the microscopic infective and diagnostic stages of the helminths—usually eggs or larvae (immature forms)—in samples of blood, feces, and urine. Thus microbiologists study parasitic helminths in

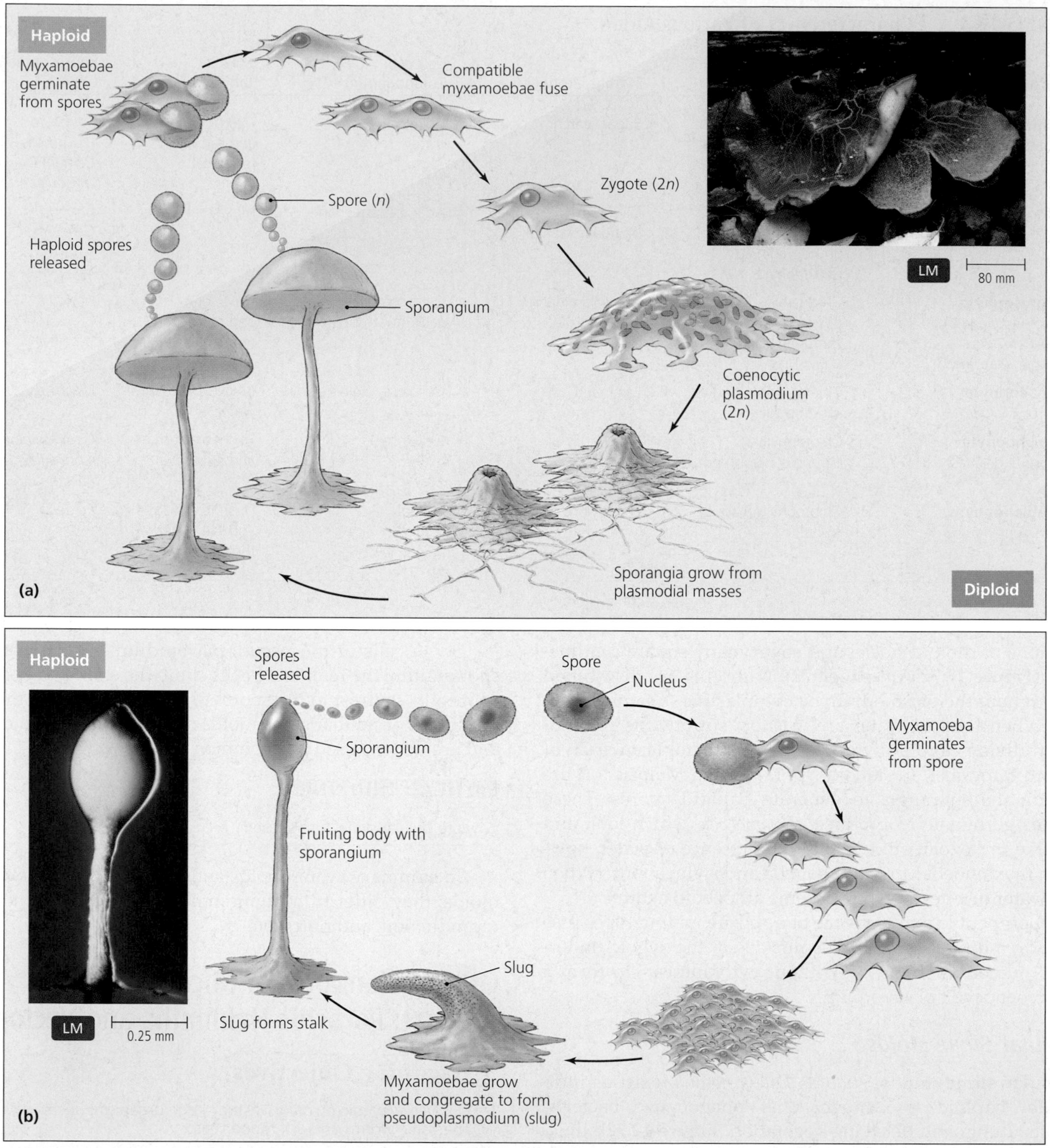

▲ ***Figure 12.32***

Life cycles of slime molds. **(a)** A plasmodial (acellular) slime mold. The coenocytic plasmodium feeds on organic debris and bacteria as it creeps through its environment. Under adverse conditions, it produces sporangia that undergo meiosis to produce haploid spores. Compatible spores fuse to form a zygote that undergoes multiple mitoses, but not cytokinesis, to form a new plasmodium. **(b)** A cellular slime mold. Individual myxamoebae congregate during times of starvation to form a multicellular slug, which produces a stalked sporangium. The sporangium releases spores that germinate when conditions are more favorable. All phases are haploid.

part because they must distinguish the parasites' microscopic forms from other microbes. Chapter 14 examines parasitism and other relationships that exist among microbes and other organisms.

Microbiologists are also interested in **arthropod vectors**—animals with segmented bodies, hard external skeletons, and jointed legs that carry pathogens. Some arthropods are *biological vectors,* meaning that they also serve as hosts for microbial pathogens; others are *mechanical vectors*—they merely carry pathogens. Arthropods are small and produce numerous offspring; therefore, eliminating their role as vectors is an insurmountable task.

There are two types of arthropod vectors—arachnids and insects.

Arachnids

All adult **arachnids** (ă-rak′nidz) have eight legs. Ticks and mites are the arachnid vectors; though spiders are seemingly more common, they do not transmit diseases. Ticks and mites resemble each other morphologically, having disc-shaped bodies, no heads, and a six-leg juvenile stage. Ticks are roughly pea-sized **(Figure 12.33a),** whereas mites (commonly known as chiggers) are usually the size of sand grains.

Hard ticks—those with hard backs—are the most important arachnid vectors. They are distributed worldwide and serve as vectors for bacterial, viral, and protozoan diseases. Some tick-borne diseases are Lyme disease, Rocky Mountain spotted fever, and tularemia.

Parasitic mites also live around the world, wherever humans and animals coexist. A mite species transmits scrub typhus.

Insects

As adults, all **insects** have three pairs of legs and three body regions—head, thorax (chest), and abdomen, though insects are far from uniform in appearance. Some have biting mouthparts, whereas others have sucking mouthparts; some have two wings, some four, and others are wingless. The fact that many insects can fly has epidemiological implications—they have broader home ranges than nonflying insects and may migrate, making control more difficult.

Insects account for the greatest number of vectors. They include fleas, lice, flies, and true bugs. Most insects are found on a host only when they are actively feeding; the only exception being lice, which may spend their entire lives in association with a single individual.

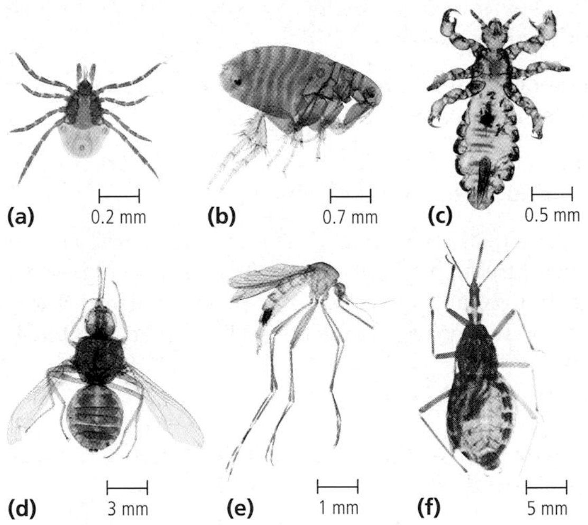

Figure 12.33

Representative arthropod vectors. Arachnid vectors include **(a)** ticks; insect vectors include **(b)** fleas, **(c)** lice, **(d)** true flies, such as this tsetse fly, **(e)** mosquitos, and **(f)** true bugs (e.g. *triatoma* shown here).

Fleas are small, vertically flattened, wingless insects **(Figure 12.33b)** that are found worldwide, though some species have limited ranges. Most are not encountered by humans, though a few species do feed on humans. The most significant microbial disease transmitted by fleas is plague, transmitted by rat fleas.

Lice are horizontally flat, soft-bodied, wingless insects **(Figure 12.33c)** found worldwide. They live in clothing and bedding, and move onto humans to feed. Lice are most common among the poor and those living in severely overcrowded communities. They are involved in epidemic outbreaks of typhus.

Many different species of flies, which have two wings, are found around the world. Bloodsucking flies may transmit disease. For example, tsetse flies **(Figure 12.33d)** vector African sleeping sickness. Mosquitoes, which are the most common arthropod vectors, are a morphologically distinct type of fly. They have elongated bodies, long antennae, long legs, and a long proboscis for feeding on blood **(Figure 12.33e).** Mosquitoes are found worldwide, but particular species are geographically limited. They carry pathogens of malaria, yellow fever, dengue fever, filariasis, viral encephalitis, and other diseases.

Kissing bugs—so called because they preferentially take nocturnal blood meals near the mouth of their hosts—are relatively large, winged bugs with cone-shaped heads and wide abdomens **(Figure 12.33f).** Some transmit Chagas' disease.

CHAPTER SUMMARY

General Characteristics of Eukaryotic Organisms (pp. 343–348)

1. A typical eukaryotic nucleus may be **haploid** (having a single copy of each chromosome) or **diploid** (having two copies). It divides by **mitosis** in four phases—**prophase, metaphase, anaphase,** and **telophase**—resulting in two nuclei with the same ploidy as the original.
2. **Meiosis** is nuclear division that results in four nuclei with half the ploidy of the original.

3. A cell's cytoplasm divides by **cytokinesis** either during or after nuclear division.
4. **Coenocytes** are multinucleate cells resulting from repeated mitoses but postponed or no cytokinesis.
5. Some microbes undergo multiple mitoses by **schizogony** to form a multinucleate **schizont**, which then undergoes cytokinesis.
6. The classification of eukaryotic microbes is problematic and has changed frequently. Historical schemes based on similarities in morphology and chemistry have been replaced with schemes based on nucleotide sequences and ultrastructural features.

Protozoa (pp. 348–355)

1. **Protozoa** are eukaryotic, unicellular organisms that lack cell walls. Most of them are chemoheterotrophs.
2. A motile **trophozoite** is the feeding stage of a typical protozoan; a **cyst,** a resting stage that is resilient to environmental changes, is formed by some protozoa.
3. A few protozoa undergo sexual reproduction by forming **gametocytes** that fuse to form a **zygote.**
4. Protozoa may be classified into four groups: **alveolates, euglenozoa, diplomonads,** and **parabasalids. Amoebae,** protozoa that move and feed with pseudopodia, are found in several kingdoms.
5. Alveolates, with cavities called alveoli beneath their cell surfaces, include **ciliates** (characterized by cilia), **apicomplexans** (all are pathogenic), and **dinoflagellates** (responsible for **red tides**).
6. Unicellular flagellated **euglenoids** are euglenozoans that store food as paramylon, lack cell walls, and have eyespots used in positive phototaxis. Because they exhibit characteristics of both animals and plants, they are a taxonomic problem.
7. **Kinetoplastids** are euglenozoans with a single, large, apical mitochondrion that contains a kinetoplast, which is a region of DNA.
8. Members of the Diplomonadida lack mitochondria, Golgi bodies, and peroxisomes; they include diplomonads (*e.g., Giardia*) and microsporidia (*e.g., Nosema*).
9. Parabasalids (*e.g., Trichomonas*) are characterized by a Golgi-like structure called a parabasal body.

Fungi (pp. 355–366)

1. **Fungi** (studied by mycologists) are chemoheterotrophic eukaryotes with cell walls that are usually composed of **chitin.**
2. Most fungi are beneficial, but some produce **mycoses** (fungal diseases).
3. The nonreproductive body of a filamentous fungus (mold) or yeast (unicellular fungus) is a **thallus.** Mold thalli are composed of tubular filaments called **hyphae.**
4. Hyphae are described as either **septate** or **aseptate** depending on the presence of crosswalls. A **mycelium** is a tangled mass of hyphae.
5. A **dimorphic** fungus has either type of thallus, depending on environmental conditions.
6. Most fungi are **saprobes**—they absorb nutrients by **absorption** from dead organisms; others get nutrients from living organisms using **haustoria** that penetrate host tissues.
7. Fungi reproduce asexually both by budding and via asexual spores, which are categorized according to their mode of development. Most fungi also reproduce sexually via spores.
8. Fungi in the division **Zygomycota** produce rough-walled **zygospores.**
9. Fungi in the division **Ascomycota,** a group of economically important fungi, produce **ascospores** within sacs called **asci.**
10. Fungi of the division **Basidiomycota,** including mushrooms, puffballs, and bracket fungi, produce **basidiospores** at the ends of **basidia.**
11. **Deuteromycota** is an informal grouping of fungi having no known sexual stage.
12. **Lichens** are economically and environmentally important organisms composed of fungi living in partnership with photosynthetic microbes, either green algae or cyanobacteria.

Algae (pp. 366–369)

1. Algae (studied by phycologists) typically reproduce by an **alternation of generations** in which a haploid thallus alternates with a diploid thallus.
2. Large algae have multicellular thalli with stemlike **stipes,** leaflike **blades,** and **holdfasts** that attach them to substrates.
3. Division **Chlorophyta** contains green algae, which are similar to vascular plants.
4. **Rhodophyta,** red algae, contain the pigment **phycoerythrin,** the storage molecule floridean starch, and cell walls of **agar** or **carrageenan,** substances used as thickening agents.
5. **Chrysophyta**—the golden algae, yellow-green algae, and **diatoms**—contain chrysolaminarin as a storage product. The cell walls of diatoms, which are made of silica, are arranged in nesting halves called **frustules.**
6. **Phaeophyta,** brown algae, contain xanthophylls, laminarin, and oils. They have cell walls composed of cellulose and **alginic acid,** which is another thickening agent. A brown algal spore is motile by means of one tinsel flagellum and one whiplike flagellum.

Water Molds and Slime Molds (pp. 369–371)

1. **Water molds** have tubular cristae in their mitochondria, cell walls of cellulose, spores having two different flagella, and diploid thalli. They are placed in the kingdom Stramenopila along with diatoms, other chrysophytes, and brown algae.
2. **Slime molds** lack cell walls and are phagocytic in their nutrition. They are similar to both amoebae and fungi.
3. **Plasmodial** (acellular) **slime molds** are composed of multinucleate cytoplasm. **Cellular slime molds** are composed of myxamoebae that phagocytize bacteria and yeasts.

Other Eukaryotes of Microbiological Interest: Parasitic Helminths and Vectors (pp. 371–373)

1. Parasitic helminths are significant to microbiologists because their infective stages are usually microscopic.
2. Microbiologists also study **arthropod vectors,** which are **arachnids** and **insects** that carry pathogens. Biological vectors also serve as hosts for the pathogens; mechanical vectors merely carry pathogens.
3. Arachnids, which have eight legs as adults, include ticks and mites. Insects, which have six legs as adults, include fleas, lice, flies (including mosquitoes), and bugs. Mosquitoes are the most common arthropod vectors.

QUESTIONS FOR REVIEW

(Answers to multiple choice, fill in the blanks, and matching questions are on the web, along with additional review questions. Visit www.microbiologyplace.com.)

Multiple Choice

1. Haploid genomes
   a. contain one set of chromosomes.
   b. contain two sets of chromosomes.
   c. are found in zygotes.
   d. are found in the cytosol of eukaryotic organisms.
2. Which of the following sequences reflects the correct order of events in mitosis?
   a. telophase, anaphase, metaphase, prophase
   b. prophase, anaphase, metaphase, telophase
   c. telophase, prophase, metaphase, anaphase
   d. prophase, metaphase, anaphase, telophase
3. Which of the following statements accurately describes prophase?
   a. The cell appears to have a line of chromosomes across the mid-region.
   b. The nuclear envelope becomes visible.
   c. The cell constructs microtubules to form a spindle.
   d. Chromatids separate and become known as chromosomes.
4. Multiple nuclear divisions without cytoplasmic divisions result in cells called
   a. mycoses. c. haustoria.
   b. coenocytes. d. pseudohypha.
5. Branched, tubular filaments with crosswalls found in large fungi are
   a. septate hyphae. c. aseptate haustoria.
   b. aseptate hyphae. d. dimorphic mycelia.
6. The type of asexual fungal spore that buds from vegetative hyphae is called a
   a. sporangiospore. c. blastospore.
   b. conidiospore. d. chlamydospore.
7. A phycologist studies which of the following?
   a. Classification of eukaryotes
   b. Alternation of generations in algae
   c. Rusts, smuts, and yeasts
   d. Parasitic worms
8. The stemlike portion of a seaweed is called its
   a. thallus. c. stipe.
   b. holdfast. d. blade.
9. Carrageenan is found in the cell walls of which group of algae?
   a. Red algae c. Dinoflagellates
   b. Green algae d. Yellow-green algae
10. Chrysolaminarin is a storage product found in which group of microbes?
    a. Dinoflagellates c. Golden algae
    b. Euglenoids d. Brown algae
11. Which of the following features characterizes dinoflagellates?
    a. Laminarin and oils as food reserves
    b. Protective plates of silica in their cell walls
    c. Chlorophylls *a* and *c*, carotene, and xanthophylls
    d. Paramylon as a food storage molecule
12. Water molds differ from true fungi in which of the following ways?
    a. They have tubular cristae in their mitochondria.
    b. Their cell walls are made of chitin instead of cellulose.
    c. Their spores are encased in silica shells.
    d. Their thalli are haploid rather than diploid.
13. The motile feeding stage of protozoan is called
    a. an apicomplexa. c. a cyst.
    b. a gametocyte. d. a trophozoite.

Matching

1. ___	Mitosis	A. Cytoplasmic division
2. ___	Meiosis	B. Diploid nuclei producing haploid nuclei
3. ___	Homologous chromosomes	C. Results in genetic variation
4. ___	Crossing over	D. Carry similar genes
5. ___	Cytokinesis	E. Diploid nuclei producing diploid nuclei

1. ___	Chitin	A. Fungal cell wall component
2. ___	Basidiospore	B. Fungus + alga
3. ___	Zygospore	C. Fungal body
4. ___	Thallus	D. Fungal spore formed in a sac
5. ___	Ascospore	E. Diploid fungal zygote with a thick wall
6. ___	Lichen	F. Fungal spore formed in club-shaped hyphae

1. ___	Chlorophyta	A. Yellow-green algae
2. ___	Rhodophyta	B. Green algae
3. ___	Chrysophyta	C. Brown algae
4. ___	Phaeophyta	D. Red algae

Short Answer

1. Compare and contrast the following closely related terms:
   Chromatid and chromosome
   Mitosis and meiosis II
   Hypha and mycelium
   Algal thallus and fungal thallus
   Water molds and slime molds
2. How do fungi transport nutrients?
3. How are lichens useful in environmental protection studies?
4. What are the taxonomic challenges in classifying euglenoids?
5. List several economic benefits of algae.
6. Why are relatively large animals such as parasitic worms studied in microbiology?
7. Why are microbiologists interested in macroscopic ticks, fleas, lice, and mosquitoes?
8. Name two ways that slime molds differ from true fungi.
9. What is the role of rRNA sequencing in the classification of eukaryotic microbes?
10. Describe the nuclear divisions that produce eight ascospores in an ascus.

Fill in the Blanks

1. The study of protozoa is called ________.
2. The study of fungi is called ________.
3. The study of algae is called ________.
4. Fungal diseases are called ________.

CRITICAL THINKING

1. How are cysts of protozoa similar to bacterial endospores? How are they different?
2. The host of a home improvement show suggests periodically emptying a package of yeast into a drain leading to a septic tank. Explain why this would be beneficial.
3. Why doesn't penicillin act against any of the pathogens discussed in this chapter?
4. How can one distinguish between a filamentous fungus and a colorless alga?
5. Why do scientists as a group spend more time and money studying protozoa than they do studying algae?
6. Why are there more antibacterial drugs than antifungal drugs?
7. Which type of metabolic pathways are present in protozoa that lack mitochondria (amoebae and archaezoa)? Which metabolic pathways are absent?
8. Mycologists are certain that none of the filamentous deuteromycetes will be shown to make zygospores. How can they be certain when the sexual stages of deuteromycetes are still unknown?
9. Twenty years ago, *Pneumocystis carinii*, a pathogen of immunocompromised patients that causes pneumonia, was classified as a protozoan because it is a chemoheterotroph that lacks a cell wall. However, taxonomists today classify *Pneumocystis* as a fungus. Why do you think this pathogen has been reclassified from a protozoan to a fungus?

CHAPTER 13

Characterizing and Classifying Viruses, Viroids, and Prions

They may look like aliens landing on another world, but the structures in this photo are actually bacteriophages, viruses that attach to and infect bacteria. Because they are relatively inexpensive and easy to grow, bacteriophages have been the focus of many studies on virus behavior; indeed, much of our knowledge of viral biology has been attained by studying them. Bacteriophages have also inspired renewed interest in the scientific community for their potential as weapons against drug-resistant bacteria.

This chapter is an introduction to the characteristics and behavior of viruses and other pathogenic particles called viroids and prions—how they infect cells, how they multiply, and how they differ from cellular pathogens.

CHAPTER OUTLINE

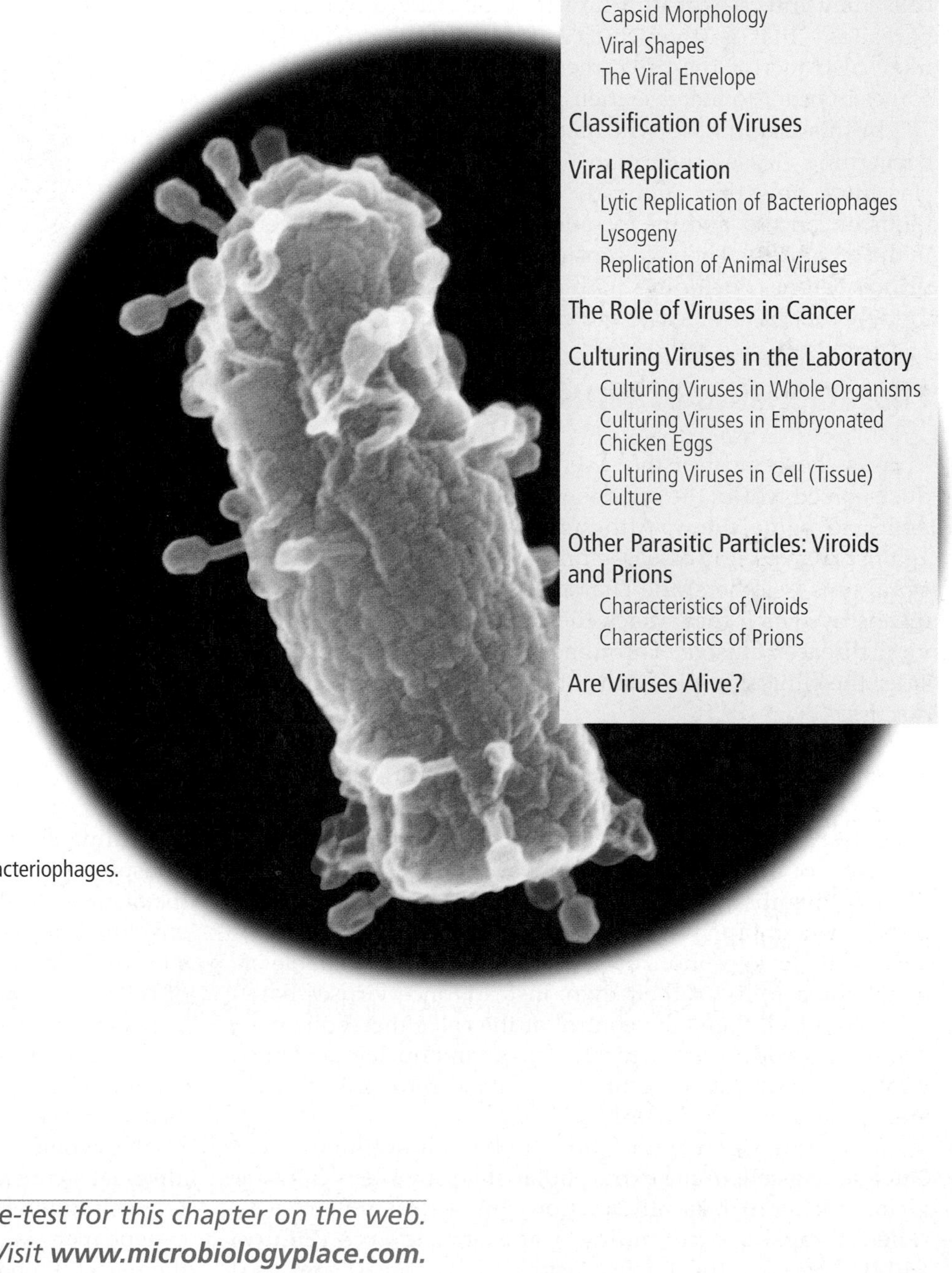

Esherichia coli infected by bacteriophages.

MicroPrep Pre-Test: *Take the pre-test for this chapter on the web. Visit* ***www.microbiologyplace.com.***

Not all pathogens are eukaryotic or prokaryotic cells. Many infections of humans, animals, and plants, and even of bacteria, are caused by **acellular** (noncellular) agents, including viruses and other parasitic particles called viroids and prions. Although these agents are like some eukaryotic and prokaryotic microbes in that they can cause disease when they invade susceptible cells, they are simple compared to a cell, lacking cell membranes and being composed of only a few organic molecules. In addition to lacking a cellular structure, they lack most of the characteristics of life described in Chapter 3: They cannot carry out any metabolic pathway, they can neither grow nor respond to the environment, and they cannot reproduce independently but instead must utilize the chemical and structural components of the cells they infect. Not only are they obligate intracellular parasites—that is, they must invade a living cell—they must also recruit the cell's metabolic chemicals and ribosomes in order to increase their numbers.

In this chapter we will first examine a range of topics concerning viruses: their many characteristics, how they are classified, how they replicate, the role they play in some kinds of cancers, and the techniques used to maintain them in the laboratory. Then we consider the nature of viroids and prions before concluding with further discussion concerning whether or not viruses are alive.

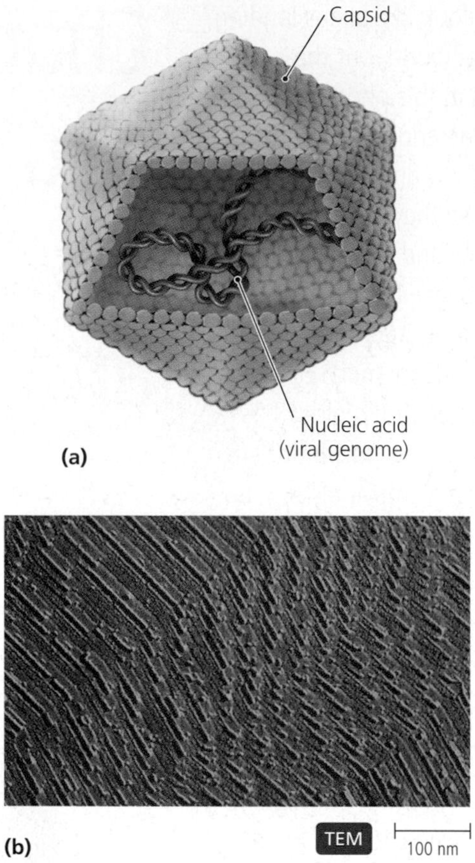

▲ *Figure 13.1*

Virions, complete virus particles, include a nucleic acid, a capsid, and in some cases an envelope. **(a)** A drawing of a nonenveloped polyhedral virus containing DNA. **(b)** A transmission electron micrograph of crystallized virions. Like many chemicals and unlike cells, some viruses can form crystals.

Characteristics of Viruses

Viruses cause most of the diseases that still plague the industrialized world: the common cold, influenza, herpes, and AIDS, to name a few. Although we have immunizations against many viral diseases and are adept at treating the symptoms of others, the characteristics of viruses and the means by which they attack their hosts can make cures for viral diseases elusive. Throughout this section, try to consider the clinical implications of the viral characteristics we discuss.

Let's begin by looking at the characteristics viruses have in common. A **virus** is a miniscule, acellular, infectious agent having one or several pieces of nucleic acid—either DNA or RNA, but never both. The nucleic acid is the genetic material (genome) of the virus. Being acellular, viruses have no cytoplasmic membrane (though, as we will see, some viruses possess a membrane-like *envelope*). Viruses also lack cytosol, and, with one exception, organelles. They are not capable of metabolic activity of their own; instead, once viruses have invaded a cell, they take control of the cell's metabolic machinery to produce more molecules of viral nucleic acid and viral proteins, which are then assembled into new viruses via a process we will examine shortly.

Viruses have an extracellular and an intracellular state. Outside of a cell, in the extracellular state, a virus is called a **virion** (vir'ē-on). Basically, a virion consists of a protein coat, called a **capsid,** surrounding a nucleic acid core **(Figure 13.1a).** Together the nucleic acid and its capsid are also called a *nucleocapsid,* which in many cases can crystallize like crystalline chemicals **(Figure 13.1b).** Some virions have a phospholipid membrane called an envelope surrounding the nucleocapsid. The outermost layer of a virion (capsid or envelope) provides a virus both protection and recognition sites that bind to complementary chemicals on the surfaces of cells. These are involved when a virus penetrates a cell. Envelopes typically fuse with the cell's membrane and the virus moves into the cell. Once inside, the intracellular state is initiated; the capsid is removed, and the virus exists solely as nucleic acid.

Now that we have examined ways in which viruses are alike, let's consider the characteristics that are used to distinguish different viral groups. Viruses differ in the type of genetic material they contain, the kinds of cells they attack, their size, the nature of their capsid coat, their shapes, and the presence or absence of an envelope.

Genetic Material of Viruses

Learning Objective

✓ Discuss viral genomes in terms of dsDNA, ssDNA, ssRNA, dsRNA, and number of segments of nucleic acid.

Viruses show more variety in the nature of their genomes than do cells. Whereas the genome of every cell is double-stranded DNA, the genome of a virus may be either DNA or RNA, but never both. The primary way in which scientists categorize and classify viruses is based on the type of genetic material that makes up the viral genome.

Some viral genomes, such as those of herpes virus and chickenpox virus, are double-stranded DNA (dsDNA), like the genomes of cells. Other viruses use either single-stranded RNA (ssRNA), single-stranded DNA (ssDNA), or double-stranded RNA (dsRNA) as their genomes. These molecules never function as the genome of any cell—in fact, ssDNA and dsRNA are almost nonexistent in cells. Further, a viral genome may be either linear and composed of several molecules of nucleic acid, as in eukaryotic cells, or circular and singular, as in most prokaryotic cells. For example, the genome of the influenza (flu) virus is composed of eight linear segments of single-stranded RNA, whereas the genome of poliovirus is one molecule of single-stranded RNA.

Viral genomes are much smaller than the genomes of cells. Whereas the genome of the smallest chlamydial bacterium has almost 1000 genes, the genome of the bacteriophage MS2 has only three genes. **Figure 13.2** compares the genome of bacteriophage T7, which contains 56 genes, with the genome of the bacterium *Escherichia coli* (esh-ĕ-rik′ē-ă kō′lī), which contains over 4000 genes.

CRITICAL THINKING

Some viral genomes, composed of single-stranded RNA, act as mRNA. What advantage might these viruses have over other kinds of viruses?

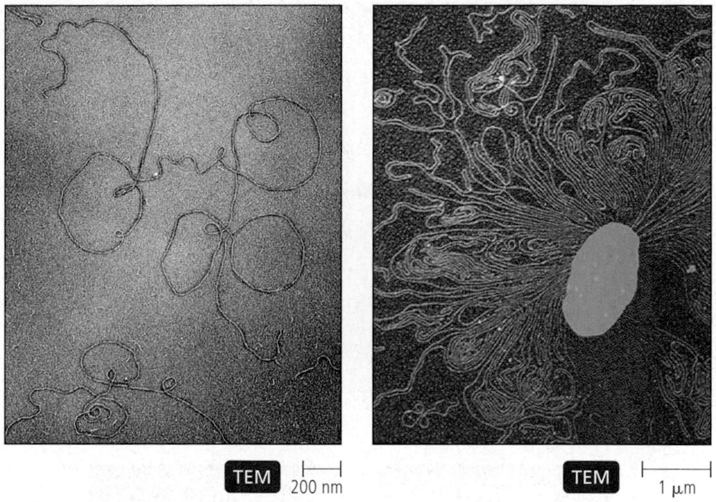

▲ ***Figure 13.2***
The relative sizes of the genomes of *E. coli* (on the right) and bacteriophage T7 on the left. The *E. coli* genome was released by rupture of the bacterium.

Hosts of Viruses

Learning Objectives

✓ Explain how viruses are specific for their host cells.
✓ Compare and contrast viruses of fungi, plants, animals, and bacteria.

Most viruses infect only particular host's cells. This specificity is due to the specific affinity of viral surface proteins or glycoproteins for complementary proteins or glycoproteins on the surface of the host cell. Viruses may be so specific they infect not only a particular host, but a particular kind of cell in that host. For example, HIV (human immunodeficiency virus, the agent that causes AIDS) specifically attacks helper T lymphocytes (a type of white blood cell) in humans and has no effect on, say, human muscle cells or bone cells. By contrast, some viruses are *generalists;* they infect many kinds of cells in many different hosts. An example of a generalist virus is rabies, which can infect most mammals, from humans to bats.

All types of organisms are susceptible to viral attack. There are viruses that infect plant, bacterial, and animal cells (**Figure 13.3** on page 380), and even fungal cells. Most viral research and scientific study has focused on bacterial and animal viruses. As we have seen, a virus that infects bacteria is referred to as a **bacteriophage** (bak-tēr′ē-ō-fāj), or simply a **phage** (fāj). (**Highlight 13.1** on page 381 discusses the potential use of bacteriophages as an alternative to antibiotics.) We will return our attention to bacteriophages and animal viruses later in this chapter.

Viruses of plants are less well known than bacterial and animal viruses, even though viruses were originally identified and isolated from tobacco plants. Plant viruses infect many food crops, including corn, beans, sugarcane, tobacco, and potatoes, resulting in billions of dollars in losses each year. Viruses of plants are introduced into plant cells either through abrasions of the cell wall or by plant parasites such as nematodes and aphids. Plant viruses follow the replication cycle discussed below for animal viruses.

Fungal viruses have been little studied and are not well known. We do know that fungal viruses are different from animal and bacterial viruses in that fungal viruses exist only within cells; that is, they have no extracellular state. Presumably, fungal viruses cannot penetrate the thick fungal cell wall. However, because fusion of cells is typically a part of the fungal life cycle, viral infections can easily be propagated by the fusion of an infected fungal cell with an uninfected one.

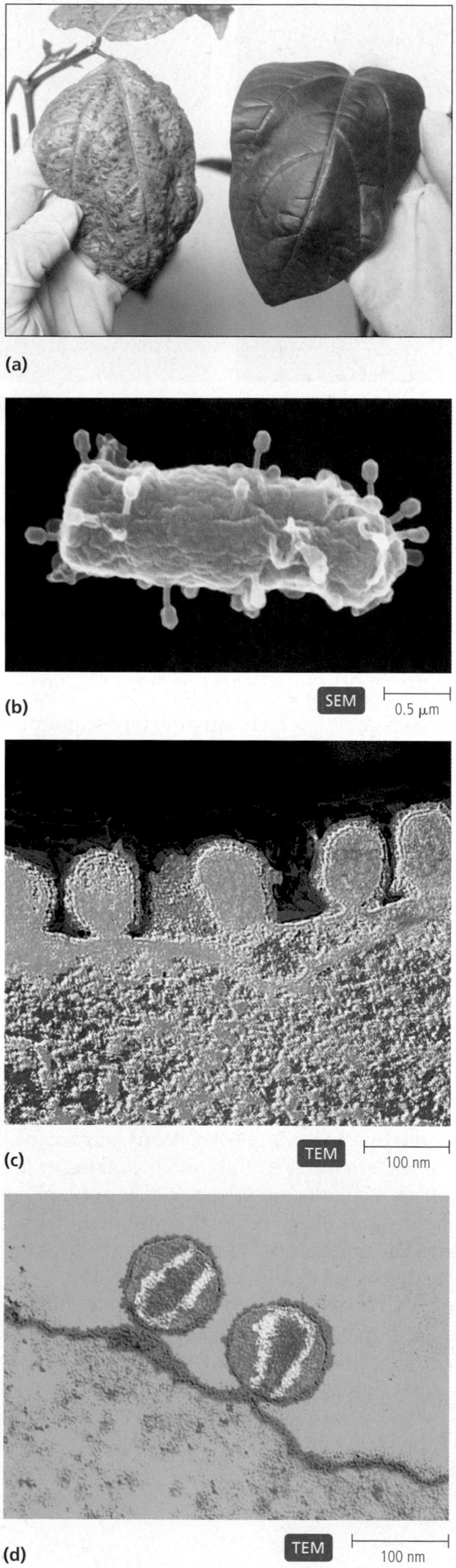

Figure 13.3

Some examples of plant, bacterial, and human hosts of viral infections. **(a)** Left, a tobacco leaf infected with tobacco mosaic virus, the first virus discovered. **(b)** A bacterial cell under attack by bacteriophages. **(c)** Influenza viruses budding from an infected cell. **(d)** A human white blood cell's cytoplasmic membrane, to which two enveloped human immunodeficiency viruses (HIV) are attached.

Sizes of Viruses

In the late 1800s, scientists hypothesized that the cause of many diseases, including polio and smallpox, was an agent smaller than a bacterium. They named these tiny agents "viruses," from the Latin word for "poison." Viruses are so small that most cannot be seen by light microscopy. One hundred million polioviruses could fit on the period at the end of this sentence. The smallest viruses have a diameter of 10 nm, whereas the largest are approximately 300 nm in diameter, which is about the size of the smallest bacterial cell. **Figure 13.4** compares the sizes of selected viruses to *E. coli* and a human red blood cell.

In 1892, Russian microbiologist Dmitri Ivanowsky first demonstrated that viruses are acellular with an experiment designed to elucidate the cause of tobacco mosaic disease. He filtered the sap of infected tobacco plants through a porcelain filter fine enough to trap even the smallest of bacterial cells. The viruses, however, were not trapped, but instead passed though the filter with the liquid, which remained infectious to tobacco plants. This experiment proved the existence of an acellular disease-causing entity smaller than a bacterium. Tobacco mosaic virus (TMV) was isolated and characterized by an American chemist, Wendell Stanley, in 1935. Shortly thereafter, the invention of electron microscopy allowed TMV and other viruses to be seen for the first time.

Capsid Morphology

Learning Objective

✓ Discuss the structure and function of the viral capsid.

As we have seen, viruses have capsids—protein coats that provide both protection for the viral nucleic acid and a means by which many viruses attach to a host's cells. The capsid of a virus is composed of proteinaceous subunits called **capsomeres** (or *capsomers*). Some capsomeres (and therefore some capsids) are composed of only a single type of protein, whereas others are composed of several different protein subunits. Recall that a viral nucleic acid surrounded by its capsid is termed a *nucleocapsid*.

CRITICAL THINKING

In some viruses, the capsomeres act enzymatically as well as structurally. What advantage might this provide the virus?

Highlight 13.1 Prescription Bacteriophages?

As more and more bacteria become resistant to antibiotics, interest in the use of bacteriophages to fight bacterial pathogens has increased. One recent study by Rockefeller University researchers, for example, found that enzymes coded by bacteriophages were able to eliminate the presence of *Streptococcus pneumoniae* (a leading cause of illness in young, elderly, and immunocompromised patients) from mucous membranes in mice. The theory is that using a nasal spray containing these enzymes could help prevent certain infections before they even begin. Other scientists propose using intact phages.

There are several potential advantages to the use of phages over antibiotics. Unlike antibiotics, phages target specific bacteria, so phages can be effective without also destroying "good" bacteria in the normally present microbiota. (The destruction of "good" bacteria is linked to the diarrhea suffered by some users of antibiotics.) Furthermore, phages are inexpensive, easy to produce, effective in small doses (because they replicate), have no known deleterious effects on a patient's cells or metabolism, and are unlikely to contribute to antibiotic resistance—all the more reason to further investigate this intriguing alternative to the use of antibiotics.

Reference: Loeffler, J. M., D. Nelson, V. A. Fischetti, 2001. Rapid killing of *Streptococcus pneumoniae* with a bacteriophage cell wall hydrolase. *Science* 294:2170–2172.

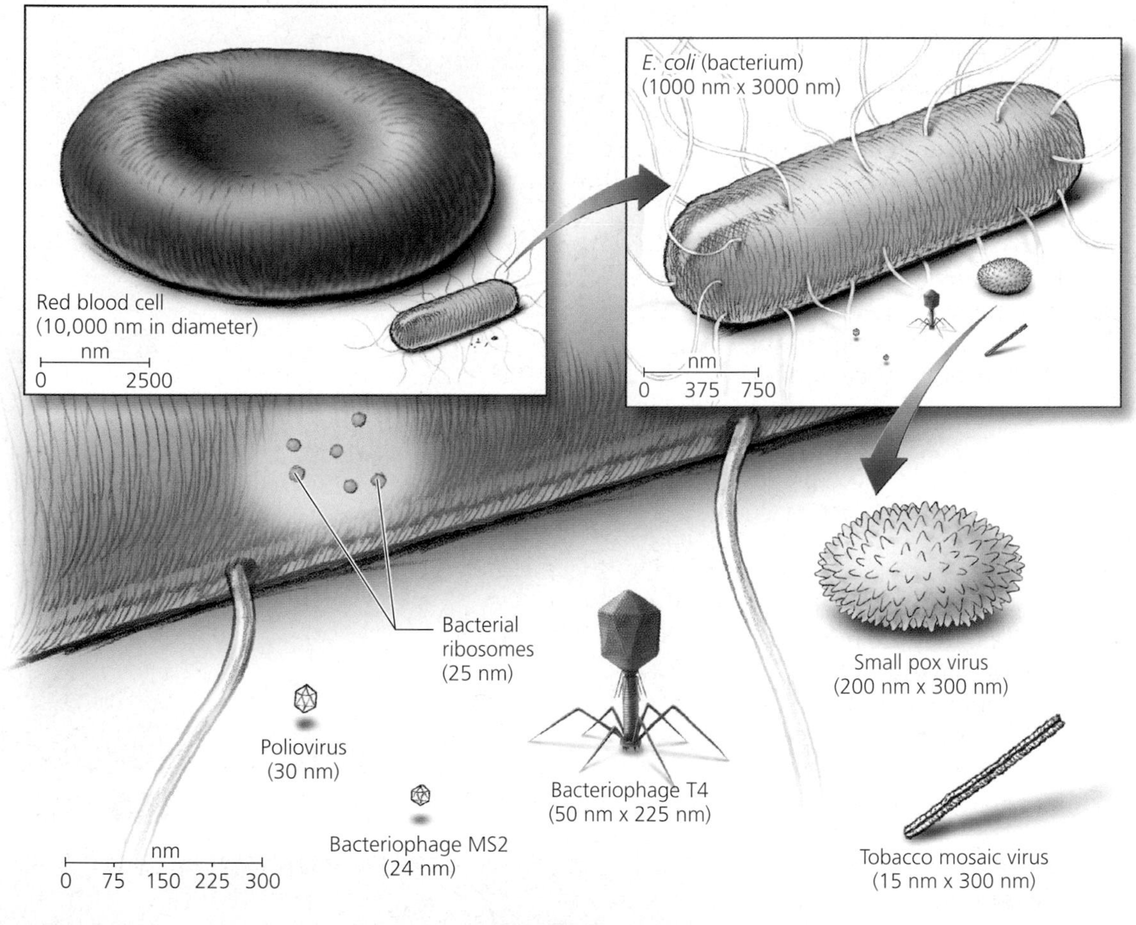

▲ ***Figure 13.4***

The sizes of selected viruses as compared to a bacterium, *E. coli,* and a human red blood cell. *How can viruses be so small and yet still be pathogenic?*

Figure 13.4 *Viruses utilize a host cell's enzymes, organelles, and membranes to replicate.*

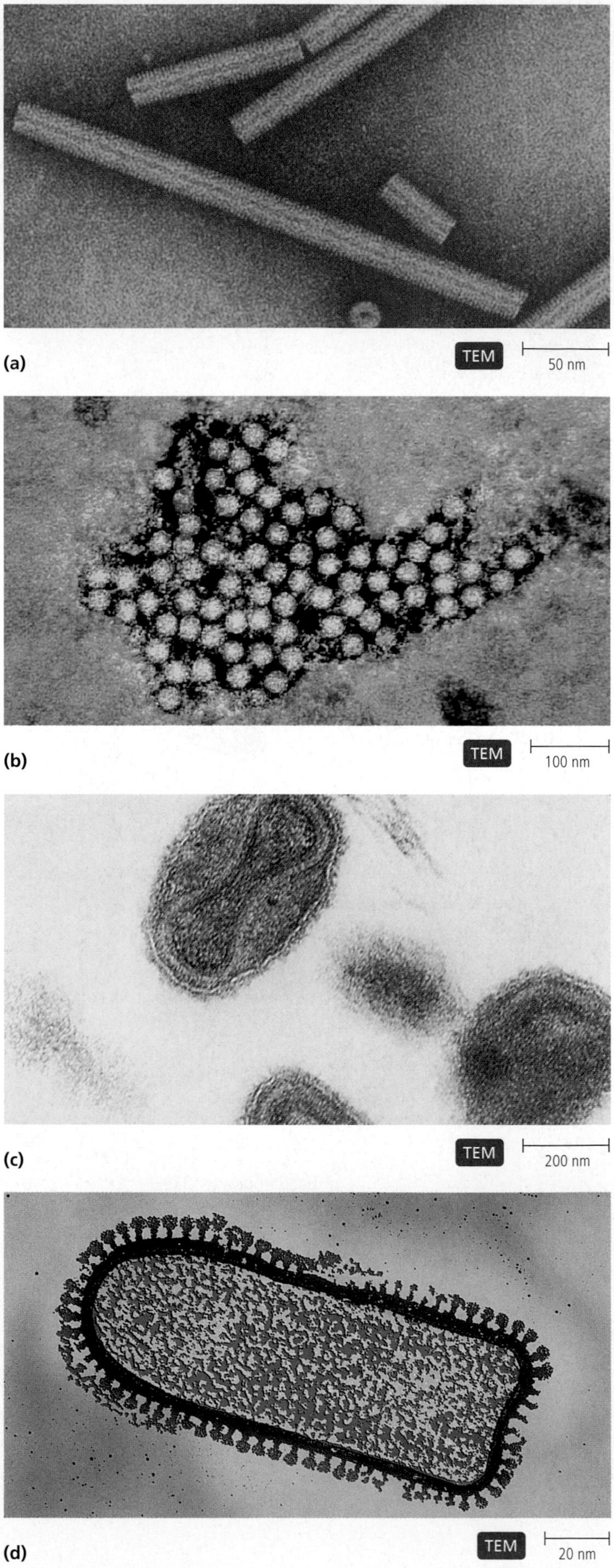

Figure 13.5 ➤
The shapes of virions. **(a)** A helical virus, tobacco mosaic virus. The tubular shape of the capsid results from the tight arrangement of several rows of helical capsomeres. **(b)** Polyhedral virions of a virus that causes the common cold. **(c)** Complex virions of smallpox virus. **(d)** The complex shape of rabies virus, which results from the shapes of the capsid and bullet-shaped envelope surrounding it.

Viral Shapes

The shape of the virion is another characteristic by which viruses are classified. There are three basic types of viral shapes: helical, polyhedral, and complex **(Figure 13.5).** The capsid of a helical virus is composed of capsomeres that bond together in a spiral fashion to form a tube around the nucleic acid. The capsid of a polyhedral virus is roughly spherical, with a shape similar to a geodesic dome. The most common type of polyhedral capsid is an icosahedron, which has 20 sides.

Complex viruses have capsids of many different shapes that do not readily fit into either of the other two categories. An example of a complex virus is smallpox virus, which has several covering layers (including lipid) and no easily identifiable capsid. The complex shapes of many bacteriophages include icosahedral heads, which contain the genome, attached to helical tails with tail fibers. The complex capsids of bacteriophages somewhat resemble NASA's lunar lander **(Figure 13.6).**

The Viral Envelope

Learning Objective

✓ Discuss the origin, structure, and function of the viral envelope.

All viruses lack cell membranes (after all, they are not cells), but some, particularly animal viruses, have a membrane similar in composition to a cytoplasmic membrane surrounding their capsids. Such a membrane is called an **envelope,** and thus a virus with a membrane is an **enveloped virion (Figure 13.7);** a virion without an envelope is called a *nonenveloped* or *naked virion.*

Enveloped viruses acquire their envelope from the host cell during viral replication or release (discussed shortly). Indeed, the envelope of a virus is a portion of the membrane system of a host cell. Like a cytoplasmic membrane, a viral envelope is composed of a phospholipid bilayer and proteins. Some of the proteins are virally-coded glycoproteins, which appear as spikes protruding outward from the envelope's surface (see Figure 13.5d). Host DNA carries the genetic code required for the assembly of the phospholipids and some of the proteins in the envelope, while the viral genome specifies the other integral proteins.

The envelope's proteins and glycoproteins often play a role in the recognition of host cells. A viral envelope does not perform the other physiological roles of a cytoplasmic membrane such as endocytosis or active transport.

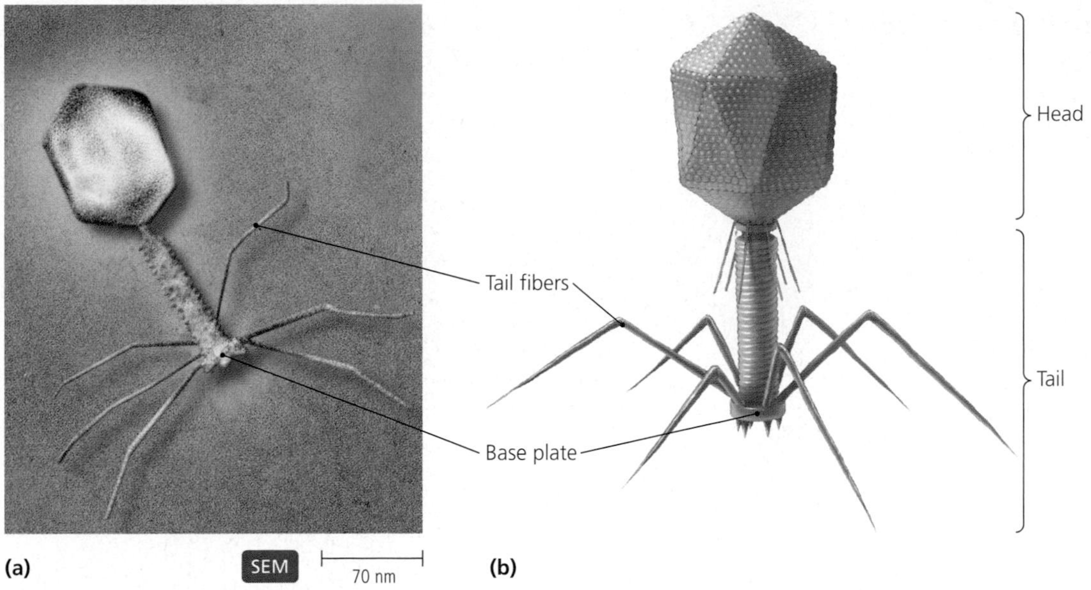

Figure 13.6
The complex shape of bacteriophage T4, which includes an icosahedral head and an ornate tail that enables viral attachment and penetration.

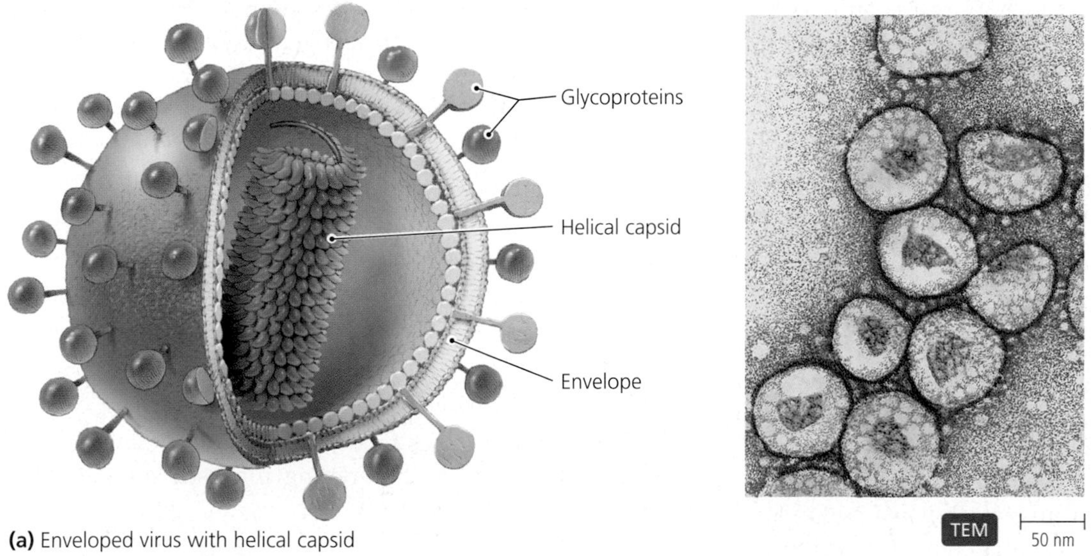

(a) Enveloped virus with helical capsid

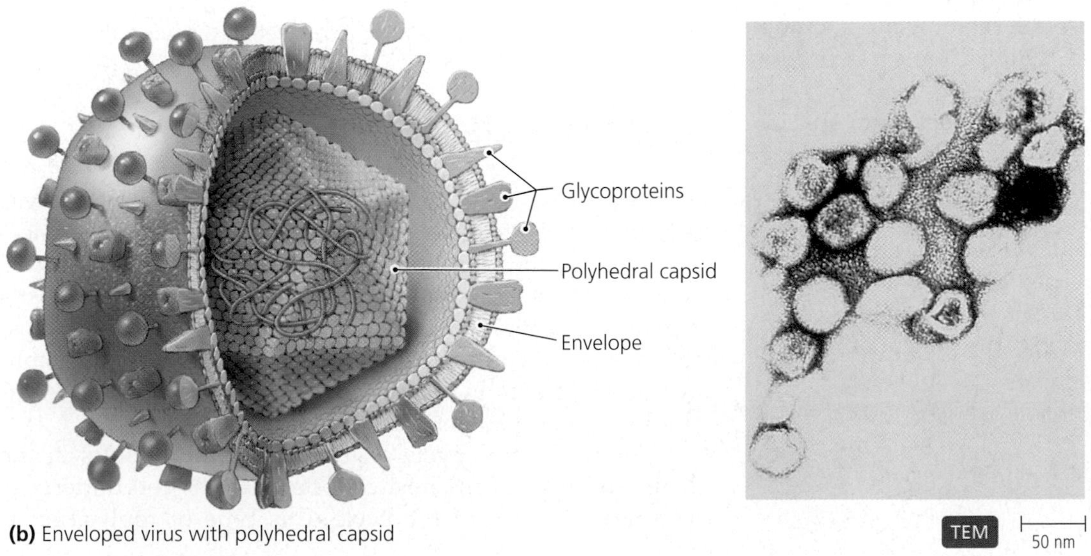

(b) Enveloped virus with polyhedral capsid

Figure 13.7
Enveloped virions. **(a)** Artist's rendition and electron micrograph of a coronavirus, an enveloped virus with a helical capsid. **(b)** Artist's rendition and electron micrograph of a togavirus, an enveloped virus with a polyhedral capsid.

Table 13.1 The Novel Properties of Viruses

Viruses	Cells
Are inert macromolecules outside of a cell, but become active inside a cell	Metabolize on their own
Do not divide or grow	Divide and grow
Acellular	Cellular
Obligate intracellular parasites	Most free-living
Contain either DNA or RNA, never both	Contain both DNA and RNA
Genome can be dsDNA, ssDNA, dsRNA, or ssRNA	Genome is dsDNA
Ultramicroscopic in size, ranging from 10 nm to 300 nm	200 nm to 12 cm in diameter
Have a proteinaceous capsid around genome; some have an envelope around the capsid	Surrounded by a phospholipid membrane and often a cell wall
Replicate in an assembly-line manner using the enzymes and organelles of a host cell	Self-replicating by asexual and/or sexual means

The novel properties of viruses, and how those properties differ from the corresponding characteristics of cells, are summarized in **Table 13.1.** Next we turn our attention to the criteria by which viruses are classified.

Classification of Viruses

Learning Objective

✓ List the characteristics by which viruses are classified.

The International Committee on Taxonomy of Viruses (ICTV) was established in 1966 to provide a single taxonomic scheme for viral classification and identification. Viruses are categorized by their type of nucleic acid, presence of an envelope, shape, and size. For most viruses, families are the highest taxonomic groups established so far by the ICTV; viruses are assigned to certain genera within these families. Kingdoms, divisions, and classes have not been determined for most viruses because the relationships among viruses are not well understood. Only three viral orders have been established.

Family names are typically derived either from special characteristics of viruses within the family or from the name of an important member of the family. For example, family *Picornaviridae* contains very small[1] RNA viruses, and *Hepadnaviridae* contains a DNA virus that causes hepatitis A. Family *Herpesviridae* is named for herpes simplex, a virus that can cause genital herpes. **Table 13.2** lists the major families of human viruses, grouped according to the type of nucleic acid each contains.

[1] Pico means one trillionth.

At this time, Latinized names are not given to viruses at the species level. Instead, species names for viruses are their common English designations. Accordingly, the taxonomy for two important viral pathogens, HIV and rabies virus, is as follows:

	HIV	*Rabies virus*
Order:	—	Mononegavirales
Family:	*Retroviradae*	*Rhabdoviridae*
Genus:	*Lentivirus* (len′ti-vī-rŭs)	*Lyssavirus* (lis′ă-vī-rŭs)
Species:	Human immuno-deficiency virus	Rabies virus

As you know, viruses cause a variety of human diseases, from mild colds to life-threatening illnesses such as AIDS and Ebola hemorrhagic fever.

CRITICAL THINKING

Why has it been difficult to develop a complete taxonomy for viruses?

Viral Replication

As previously noted, viruses cannot reproduce themselves because they have neither the genes for all the enzymes necessary for replication nor functional ribosomes for protein synthesis. Instead, viruses are dependent on their host's organelles and enzymes to produce new virions. Once a host cell falls under control of a viral genome, it is forced to replicate viral genetic material and translate new proteins, including viral capsomeres and viral enzymes.

The replication cycle of a virus usually results in the death and lysis of the host cell. Because the cell undergoes *lysis* near the end of the cycle, this type of replication is

Table 13.2 Families of Human Viruses

Family	Strand Type	Representative Genera (Diseases)
DNA Viruses		
Poxviridae	Double	*Orthopoxvirus* (smallpox)
Herpesviridae	Double	*Simplexvirus,* Herpes type 1 (fever blisters, respiratory infections), Herpes type 2 (genital infections); *Varicellovirus* (chicken pox); *Lymphocryptovirus,* Epstein-Barr virus (infectious mononucleosis, Burkitt's lymphoma); *Cytomegalovirus* (birth defects); *Roseolavirus* (roseola)
Papillomaviridae	Double	*Papillomavirus* (benign tumors, warts, cervical and penile cancers)
Polyomaviridae	Double	*Polyomavirus* (progressive multifocal leukoencephalopathy)
Adenoviridae	Double	*Mastadenovirus* (conjunctivitis, respiratory infections)
Hepadnaviridae	Partial single and partial double	*Hepadnavirus* (hepatitis B)
Parvoviridae	Single	*Erythrovirus* (erythema infectiosum)
RNA Viruses		
Picornaviridae	Single, +[a]	*Enterovirus* (polio); *Hepatovirus* (hepatitis A); *Rhinovirus* (common cold)
Caliciviridae	Single, +	*Norovirus* (gastroenteritis)
Astroviridae	Single, +	*Astrovirus* (gastroenteritis)
Hepeviridae	Single, +	*Hepevirus* (hepatitis E)
Togaviridae	Single, +	*Alphavirus* (encephalitis); *Rubivirus* (rubella)
Flaviviridae	Single, +	*Flavivirus* (yellow fever); Japanese encephalitis virus; *Hepacivirus* (hepatitis C)
Coronaviridae	Single, +	*Coronavirus* (common cold, severe acute respiratory syndrome)
Retroviridae	Single, +, segmented	Human T cell leukemia virus; *Lentivirus* (AIDS)
Orthomyxoviridae	Single, −[b], segmented	*Influenzavirus* (flu)
Paramyxoviridae	Single, −	*Paramyxovirus* (colds, respiratory infections); *Pneumovirus* (pneumonia, common cold); *Morbillivirus* (measles); *Rubulavirus* (mumps)
Rhabdoviridae	Single, −	*Lyssavirus* (rabies)
Bunyaviridae	Single, −, segmented	*Bunyavirus* (California encephalitis virus); *Hantavirus*
Filoviridae	Single, −	*Filovirus* (Ebola); Marburg viruses (hemorrhagic fever)
Arenaviridae	Single, −, segmented	*Lassavirus* (hemorrhagic fever)
Reoviridae	Double, segmented	*Orbivirus* (encephalitis); *Rotavirus* (diarrhea); *Coltivirus* (Colorado tick fever)

[a] +RNA is equivalent to mRNA, *i.e.,* instructs ribosomes in protein translations.
[b] −RNA is complementary to mRNA; it cannot be directly translated.

termed **lytic replication.** In general, a lytic replication cycle consists of the following five stages:

- **Attachment** of the virion to the host cell
- **Entry** of the virion or its genome into the host cell
- **Synthesis** of new nucleic acids and viral proteins by the host cell's enzymes and ribosomes
- **Assembly** of new virions within the host cell
- **Release** of the new virions from the host cell

In the following sections we will examine the events that occur in the replication of bacteriophages and animal viruses. We will begin with lytic replication in bacteriophages, turn to a modification of replication (called lysogenic replication) in bacteriophages, and then consider the replication of animal viruses.

Lytic Replication of Bacteriophages

Learning Objective

✓ Sketch and describe the five stages of the lytic replication cycle as it typically occurs in bacteriophages.

Bacteriophages make excellent tools for the general study of viruses because they are easier and less expensive to culture than animal or human viruses. Studies of bacteriophages revealed the basics of viral biology.

Here we will examine the replication of a much-studied dsDNA bacteriophage of *E. coli* called *type 4* or *T4*. T4 virions are complex, having the polyhedral heads and helical tails seen in many bacteriophages (see Figure 13.6). We begin with the first stage of replication: attachment.

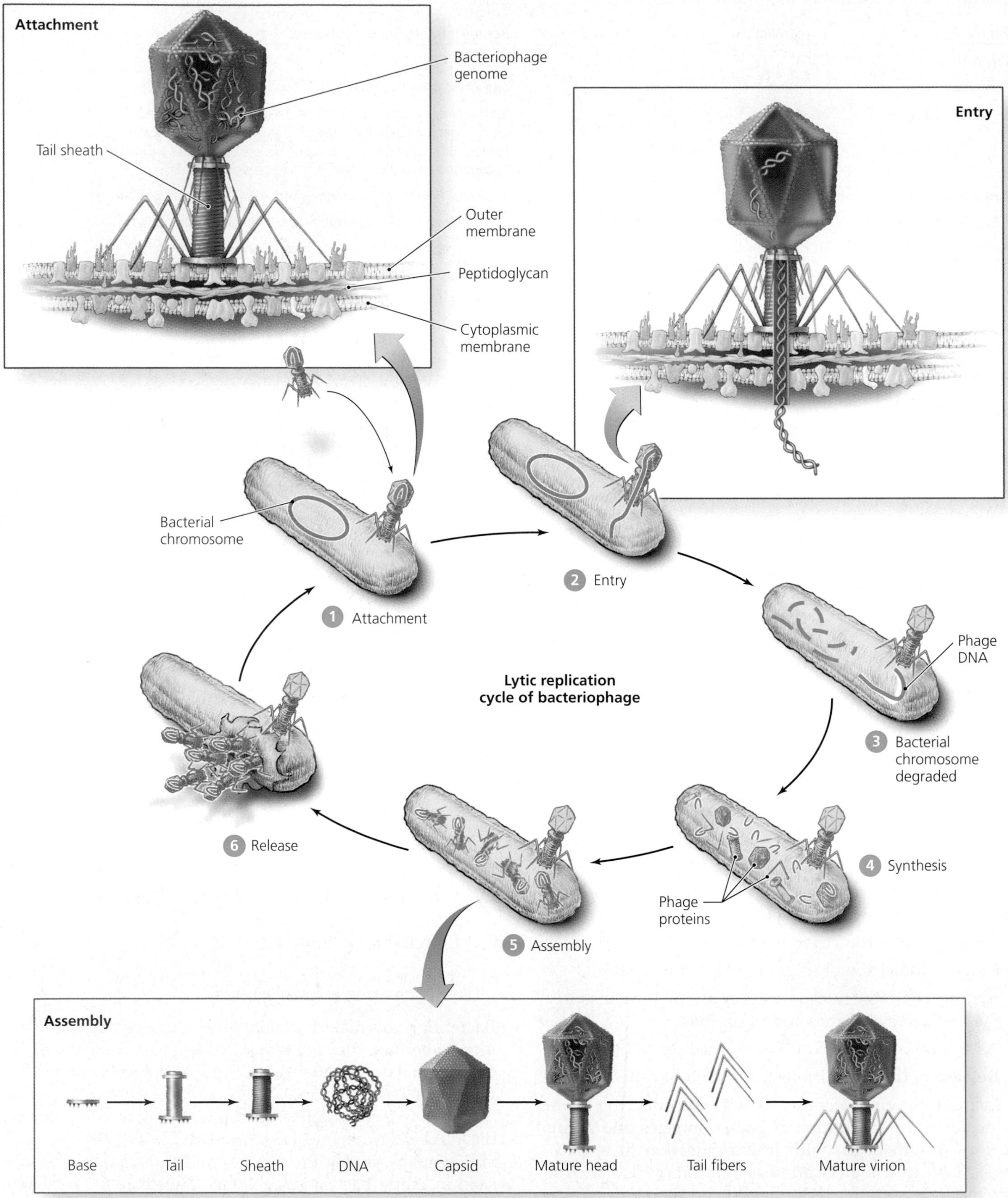
Attachment
Bacteriophage genome
Tail sheath
Outer membrane
Peptidoglycan
Cytoplasmic membrane
Entry
Bacterial chromosome
1 Attachment
2 Entry
Phage DNA
Lytic replication cycle of bacteriophage
3 Bacterial chromosome degraded
4 Synthesis
Phage proteins
5 Assembly
6 Release
Assembly
Base
Tail
Sheath
DNA
Capsid
Mature head
Tail fibers
Mature virion

◀ *Figure 13.8*

The lytic replication cycle in bacteriophages. The phage shown in this illustration is T4, and the bacterium shown is *E. coli.* The circular bacterial chromosome is represented diagrammatically; in reality it would be much longer.

Attachment ①

Because virions are nonmotile, contact with a bacterium occurs by purely random collision, brought about as Brownian motion and currents move virions through the environment. As is the case for many bacteriophages, the structures responsible for the attachment of T4 to its host bacterium are its tail fibers **(Figure 13.8).** Attachment is dependent on the chemical attraction and precise fit between attachment proteins on the phage's tail fibers and complementary receptor proteins on the surface of the host's cell wall. The specificity of the attachment proteins for the receptors ensures that the virus will attach only to *E. coli*. Bacteriophages may attach to receptor proteins on bacterial cell walls, flagella, or pili.

Entry ②

Now that phage T4 has attached to the bacterium's cell wall, it must still overcome the formidable barrier posed by the cell wall and cell membrane if it is to enter the cell. T4 overcomes this obstacle in an elegant way. Upon contact with *E. coli,* T4 releases *lysozyme* (lī'sō-zīm), a protein enzyme carried within the capsid that weakens the peptidoglycan of the cell wall. The phage's tail sheath then contracts, which forces an internal hollow tube within the tail through the cell wall and membrane, much as a hypodermic needle penetrates the skin (see Figure 13.8). The phage genome then moves through the tube and into the bacterium. The empty capsid, having performed its task, is left on the outside of the cell looking like an abandoned spacecraft.

New Frontiers 13.1 describes recent research that reveals the pressure inside the capsid of another phage named phi29 is about 2200 pounds per square inch—comparable to the pressure inside a scuba tank. Such pressure may be used to inject the phage's genome into its host bacterium. In the absence of such high pressure, a phage's genome diffuses into a host cell.

Synthesis ③–④

After entry, viral enzymes (either carried within the capsid or coded by viral genes and made by the bacterium) degrade the bacterial DNA into its constituent nucleotides. As a result, the bacterium stops synthesizing its own molecules and begins synthesizing new viruses under control of the viral genome.

For dsDNA viruses like T4, protein synthesis is straightforward and similar to cellular transcription and translation, except that mRNA is transcribed from viral DNA instead of cellular DNA. Translation by the host cell's ribosomes results in viral proteins, including head capsomeres, components of the tail, DNA polymerase (which replicates viral DNA), and lysozyme (which weakens the bacterial cell wall from within, enabling the virions to leave the cell once they have been assembled).

Assembly ⑤

Scientists do not understand completely how phages are assembled inside a host cell, but it appears that as capsomeres accumulate within the cell, they spontaneously attach to one another to form new capsid heads. Likewise, tails assemble and attach to heads and tail fibers attach to tails to form

New Frontiers 13.1 High-Pressure Viruses

Imagine the pressure inside an unopened bottle of champagne. Now consider that the DNA inside some viruses can be packed at an internal pressure as high as 10 times that of champagne pressures!

To reach this conclusion, researchers at the University of California at Berkeley and the University of Minnesota used an ingenious method of measuring the force required to package DNA into an empty capsid of bacteriophage phi29. They glued a capsid to a plastic bead held stationary with suction through a micropipette. They then positioned a molecule of DNA glued to a second bead near the capsid. A laser light was focused onto the second bead.

When they added ATP to the system, a capsid enzyme stuffed the DNA into the capsid, which pulled the two beads closer together and deflected the laser light. By measuring the amount of deflection, the researchers could measure the force on the bead and thus on the DNA attached to it. At first, the DNA was packaged smoothly at a rate of 100 base pairs per second, but as the capsid filled with DNA, the pressure inside it increased, and the process slowed. By measuring the force and calculating the volume of the capsid, researchers concluded that the pressure inside the head is 2200 pounds per square inch. This tremendous pressure may help some viruses inject their DNA into cells. Among the viruses suspected of packing their DNA in this way are those that cause herpes, chicken pox, shingles, and other human diseases.

Further research in this area may reveal ways to halt the process—ways that may result in the development of drugs designed to interfere with the ability of these viruses to cause disease.

1. Sanders, R. October 18, 2001. Molecular motor powerful enough to pack DNA into viruses at greater than champagne pressures. University of California at Berkeley, press release.

2. Smith, D. E., S. J. Tans, S. B. Smith, S. Grimes, D. L. Anderson, C. Bustamante. 2001. The bacteriophage ϕ29 portal motor can package DNA against a large internal force. *Nature* 413:748–752.

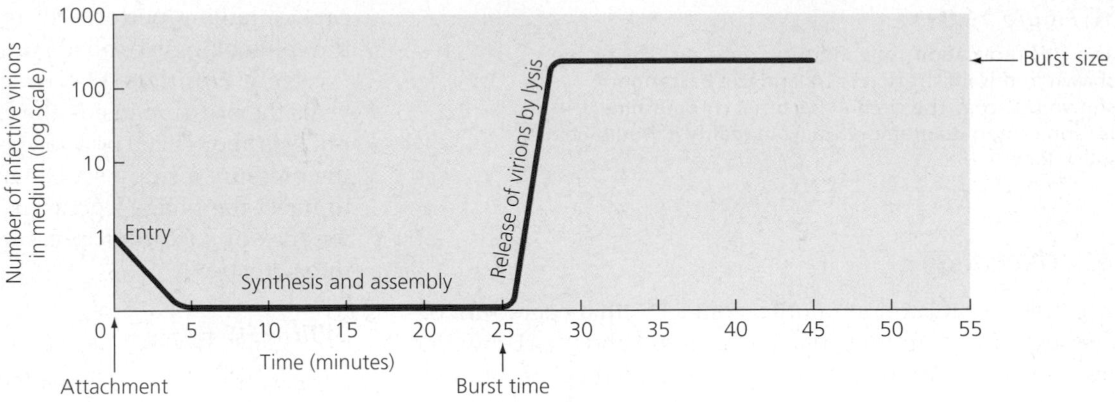

Figure 13.9 Virion abundance over time during a single replication cycle in lytic viruses. New virions are not observed in the culture medium until synthesis, assembly, and release (lysis) are complete, at which time (the burst time) the new virions are released all at once. Burst size is the number of new virions released per lysed host cell.

mature virions (see Figure 13.8). Such capsid assembly is a spontaneous process, requiring little or no enzymatic activity. For many years it was assumed that all capsids formed around a genome in just such a spontaneous manner. However, recent research (including that described in New Frontiers 13.1) has shown that for some viruses, enzymes insert the genome into the capsid after it has assembled. This process resembles stuffing a strand of cooked spaghetti into a matchbox through a single hole.

Sometimes a capsid assembles around leftover pieces of host DNA instead of viral DNA. A virion formed in this manner is still able to attach to a new host by means of its tail fibers, but instead of inserting phage DNA it transfers DNA from the first host into a new host. This process, known as *transduction,* was described in Chapter 8.

Release 6

Newly assembled virions are released from the cell as the lysozyme completes its work on the cell wall and the bacteria disintegrate. Areas of disintegrating bacterial cells in a lawn of bacteria in a Petri plate looks as if the lawn were being eaten, and it was this appearance that prompted early scientists to give the name *bacteriophage,* "bacterial eater," to these viruses.

For phage T4, the process of lytic replication takes about 25 minutes and can produce as many as about 100–200 new virions for each bacterial cell lysed **(Figure 13.9).** For any phage undergoing lytic replication, the period of time required to complete the entire process, from attachment to release, is called the *burst time,* and the number of new virions released from each lysed bacterial cell is called the *burst size.*

CRITICAL THINKING

If a colony of 1.5 billion *E. coli* cells were infected with a single phage T4, and each lytic replication cycle of the phage produced 200 new phages, how many replication cycles would it take for T4 phages to overwhelm the entire bacterial colony? (Assume for the sake of simplicity that every phage completes its replication cycle in a different cell, and that the bacteria themselves do not reproduce.)

Lysogeny

Learning Objective

✓ Compare and contrast the lysogenic replication cycle of viruses with the lytic cycle and with latency.

Not all viruses follow the lytic pattern of phage T4 we just examined. Some bacteriophages have a modified replication cycle in which infected host cells grow and reproduce normally for many generations before they lyse. The additional portion of such a replication cycle is called a **lysogenic replication cycle** or **lysogeny** (lī-soj′ĕ-nē), and the phages are called **lysogenic phages** or **temperate phages.**

We will examine lysogenic replication as it occurs in much-studied temperate phage, *lambda phage,* which is another parasite of *E. coli.* A lambda phage has a linear molecule of dsDNA in a complex capsid consisting of an icosahedral head attached to a tail that lacks tail fibers **(Figure 13.10).**

Figure 13.11 illustrates lysogeny with lambda phage. First, the virion randomly contacts an *E. coli* cell and attaches to it via its tail 1. The viral DNA enters the cell, just as occurs with phage T4, but the host cell's DNA is not destroyed, and the phage's genome does not immediately assume control of the cell. Instead, the virus remains inactive. Such an inactive bacteriophage is called a **prophage** (prō′fāj) 2. A prophage remains inactive by coding for a protein that suppresses prophage genes. A side effect of this repressor protein is that it renders the bacterium resistant to additional infection by other viruses of the same type.

Another difference between the lysogenic cycle and the lytic cycle is that a prophage is inserted into the DNA of the bacterium, becoming a physical part of the bacterial chromosome. For DNA viruses like lambda phage, this is a simple process of fusing two pieces of DNA: One piece of DNA, the virus, is fused to another piece of DNA, the chromosome of the cell. Every time the cell replicates its chromosome, the prophage is also replicated 3. All daughter cells of a lysogenic cell are thus infected with the quiescent virus. A prophage and its descendents may remain a part of the bacterial chromosome for generations.

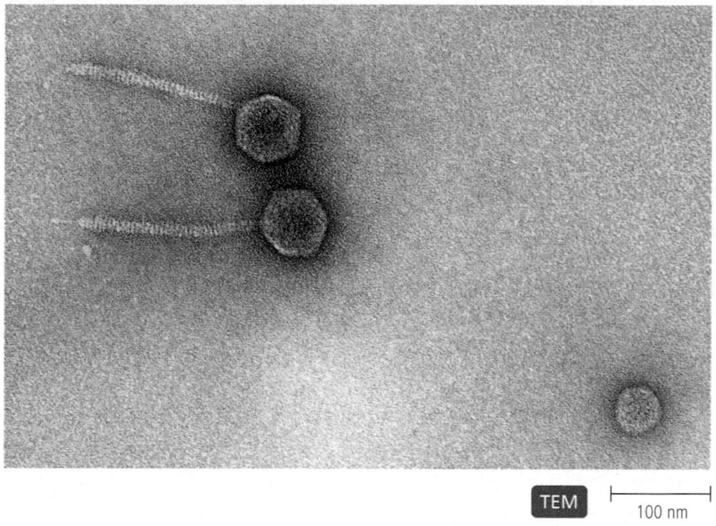

◀ ***Figure 13.10***

Bacteriophage lambda. Note the absence of tail fibers.
Phage T4 attaches by means of molecules on its tail fibers. How does lambda phage, which lacks fibers, attach?

Figure 13.10 *Lambda has attachment molecules at the end of its tail rather than on its tail fibers.*

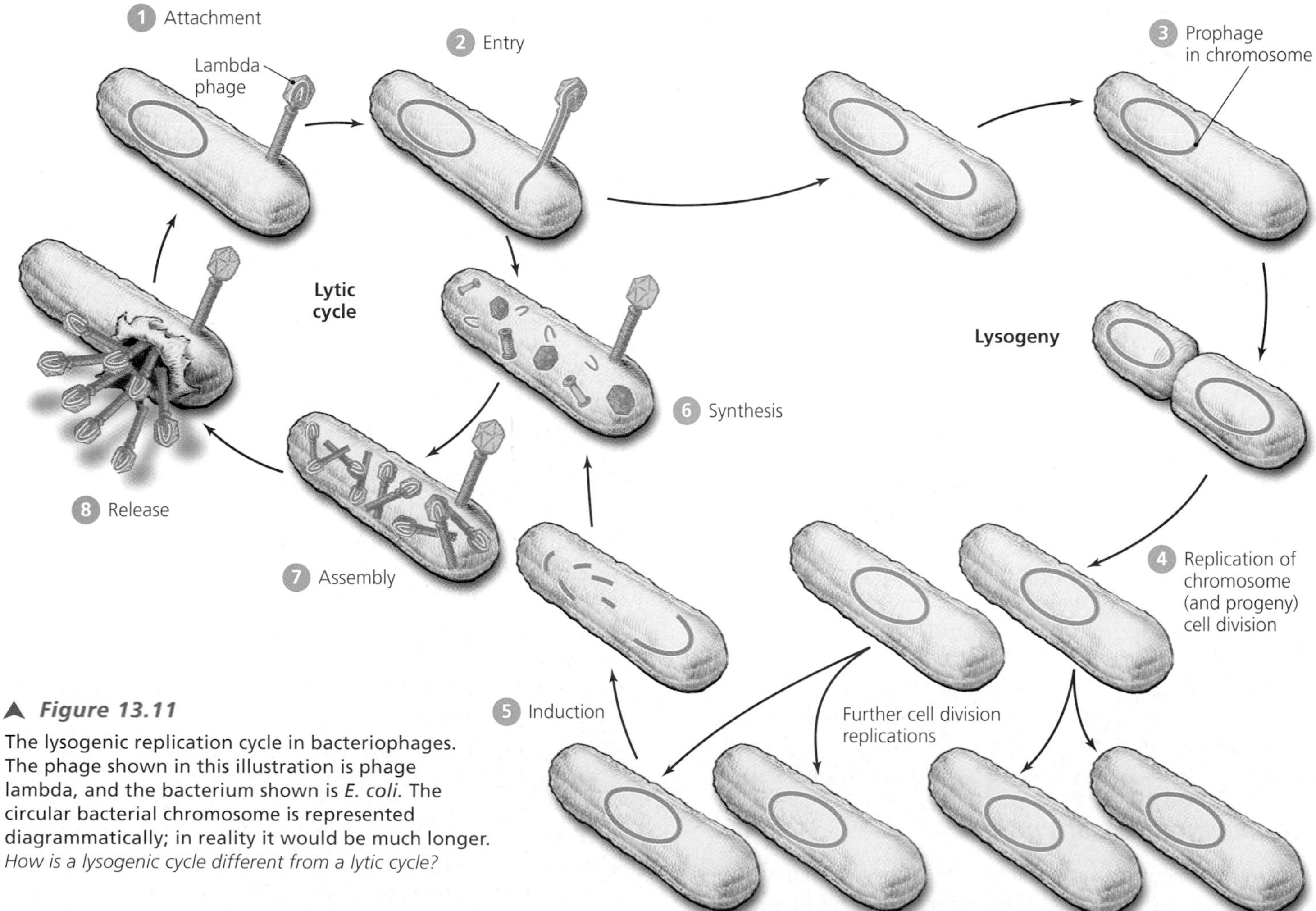

▲ ***Figure 13.11***

The lysogenic replication cycle in bacteriophages. The phage shown in this illustration is phage lambda, and the bacterium shown is *E. coli.* The circular bacterial chromosome is represented diagrammatically; in reality it would be much longer.
How is a lysogenic cycle different from a lytic cycle?

Figure 13.11 *In a lysogenic cycle, the virus is inserted into the bacterial chromosome, and it is replicated and passed on to all daughter cells until it is induced to leave the chromosome; a lytic cycle is a replication cycle that results in cell death.*

At some later time the prophage may be excised from the chromosome by recombination or some other genetic event and reenters the lytic phase. The process whereby a prophage is excised from the host chromosome is called **induction** (5). Inductive agents are typically the same physical and chemical agents that damage DNA molecules, including ultraviolet light, X-rays, and carcinogenic chemicals.

After induction, the lytic steps of synthesis (6), assembly (7), and release (8) resume from the point at which they stopped. The cell becomes filled with virions and breaks open.

Bacteriophages T4 and lambda demonstrate two replication strategies that are typical for many DNA viruses. RNA viruses and enveloped viruses present variations on the lytic and lysogenic cycles we have examined. We will examine some of these variations as they occur with animal viruses.

CRITICAL THINKING

What differences would you expect in the replication cycles of RNA phages from those of DNA phages? (Hints: Think about the processes of transcription, translation, and replication of nucleic acids. Also, note that RNA cannot be inserted into a DNA molecule.)

Replication of Animal Viruses

Learning Objectives

- ✓ Explain the differences between bacteriophage replication and animal viral replication.
- ✓ Compare and contrast the replication and synthesis of DNA, −RNA, and +RNA viruses.
- ✓ Compare and contrast the release of viral particles by lysis and budding.

Animal viruses have the same basic replication pathway as bacteriophages—that is, attachment, entry, synthesis, assembly, and release. However, there are significant differences in the replication of animal viruses that result in part from the presence of envelopes around some of the viruses, and in part from the eukaryotic nature of animal cells as well as their lack of a cell wall.

In this section we will examine the replication processes that are shared by DNA and RNA animal viruses, compare these processes with those of bacteriophages, and discuss how the synthesis of DNA and RNA viruses differ.

Attachment of Animal Viruses

As with bacteriophages, attachment of an animal virus is dependent on the chemical attraction and exact fit between proteins or glycoproteins on the virion and complementary protein or glycoprotein receptors on the animal cell's plasma membrane. Unlike many bacteriophages, animal viruses lack both tails and tail fibers. Instead, animal viruses typically have glycoprotein spikes or other attachment molecules on their capsids or envelopes.

Entry and Uncoating of Animal Viruses

Animal viruses enter a host cell shortly after attachment. Even though entry of animal viruses is not as well understood as for bacteriophages, there appear to be at least three different mechanisms: direct penetration (for naked viruses), and membrane fusion and phagocytosis (for enveloped viruses).

In naked viruses, as with phages, the genome enters the host cell while the capsid remains on the cell's surface **(Figure 13.12a).** Exactly how the genome enters the host cell remains a mystery. Poliovirus is an example of a virion that infects host cells via direct penetration.

In enveloped viruses, by contrast, the entire capsid and its contents (including the genome) enter the host cell. In a few viruses, including the measles and mumps viruses, the viral envelope and the host cell membrane fuse, releasing the capsid into the cell's cytoplasm and leaving the envelope glycoproteins as part of the cell membrane **(Figure 13.12b).** Most enveloped viruses, however, enter host cells after attachment of the virus to receptor molecules on the cell's surface stimulates the cell to phagocytize the entire virus **(Figure 13.12c).** Herpes viruses enter human host cells via phagocytosis.

For those enveloped viruses that penetrate the host cell with their capsids intact, the capsids must be removed to release their genomes before the viruses can continue to replicate. The removal of a viral capsid within a host cell is called **uncoating,** a process that remains poorly understood. It apparently occurs via different means in different viruses; some viruses are uncoated within vesicles by cellular enzymes, whereas others are uncoated by enzymes within the cell's cytosol.

Synthesis of Animal Viruses

Synthesis of animal viruses also differs from synthesis of bacteriophages. Each type of animal virus requires a different strategy for synthesis that depends on the kind of nucleic acid involved—whether it is DNA or RNA, and whether it is double stranded or single stranded. As we discuss the synthesis and assembly of each type of animal virus, consider the following two questions:

- How is mRNA (needed for the translation of viral proteins) synthesized?
- What molecule serves as a template for nucleic acid replication?

dsDNA Viruses Synthesis of new double-stranded DNA (dsDNA) virions is similar to the normal replication of DNA and translation of proteins in cells. After uncoating, the genomes of most dsDNA viruses enter the nucleus of the cell, where cellular enzymes replicate the viral genome in the same manner as host dsDNA is normally replicated—using each strand of viral DNA as a template for its complement. After messenger RNA is transcribed from viral DNA in the nucleus, and capsomere proteins are made in the

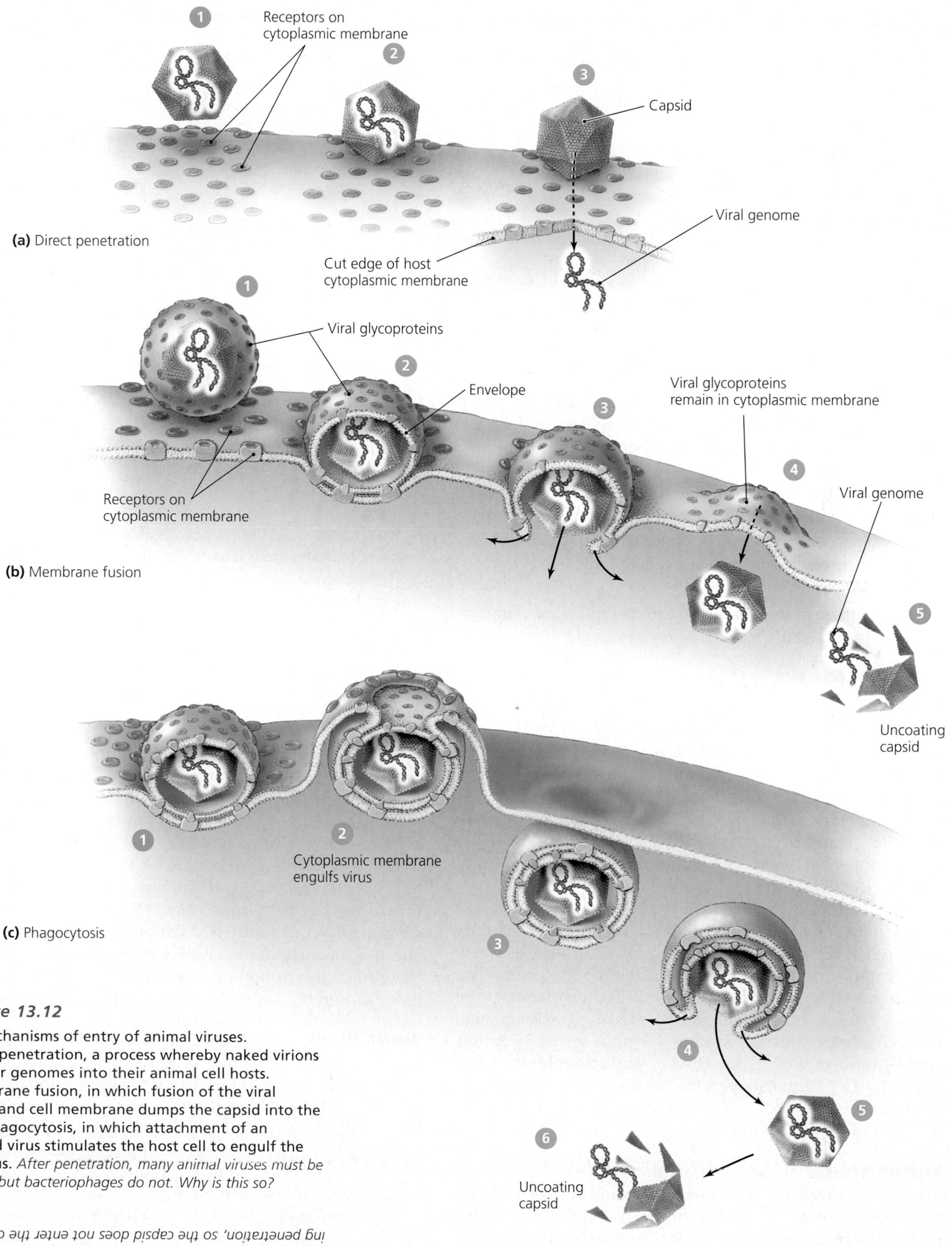

▲ ***Figure 13.12***

Three mechanisms of entry of animal viruses.
(a) Direct penetration, a process whereby naked virions inject their genomes into their animal cell hosts.
(b) Membrane fusion, in which fusion of the viral envelope and cell membrane dumps the capsid into the cell. **(c)** Phagocytosis, in which attachment of an enveloped virus stimulates the host cell to engulf the entire virus. *After penetration, many animal viruses must be uncoated, but bacteriophages do not. Why is this so?*

Figure 13.12 Generally bacteriophages inject their DNA during penetration, so the capsid does not enter the cell.

Table 13.3 Synthesis Strategies of Animal Viruses

Genome	How is mRNA Synthesized?	What Molecule is the Template for Genome Replication?
dsDNA	By RNA polymerase (in nucleus or cytoplasm of cell)	Each strand of DNA serves as template for its complement (except for hepatitis B, which synthesizes RNA to act as the template for new DNA)
ssDNA	By RNA polymerase (in nucleus of cell)	Complementary strand of DNA is synthesized to act as template
+ssRNA	Genome acts as mRNA	−RNA is synthesized to act as template
+ssRNA (*Retroviridae*)	DNA is synthesized from RNA by reverse transcriptase; mRNA is transcribed from DNA by RNA polymerase	DNA
−ssRNA	By RNA-dependent RNA transcriptase	+RNA (mRNA)
dsRNA	Positive strand of genome acts as mRNA	Each strand of genome acts as template for its complement

cytoplasm by host ribosomes, capsomeres enter the cell's nucleus, where new virions spontaneously assemble. This method of replication is seen with herpes and papilloma (wart) viruses.

There are two exceptions to this regimen of dsDNA viruses:

- Every part of a poxvirus is synthesized and assembled in the cytoplasm of the host's cell; the nucleus is not involved.
- The genome of hepatitis B viruses is replicated using an RNA intermediary instead of replicating DNA from DNA.

ssDNA Viruses The only significant human viruses with genomes composed of single-stranded DNA (ssDNA) are parvoviruses (see Table 13.2). When a parvovirus enters the nucleus of a host cell, host enzymes produce a new strand of DNA complementary to the viral genome. This complementary strand binds to the ssDNA of the virus to form a dsDNA molecule. Transcription of mRNA, replication of new ssDNA, and viral assembly then follow the DNA virus pattern just described.

As previously noted, RNA is not used as genetic material in cells, so it follows that the synthesis of RNA viruses must differ significantly from typical cellular processes, and from the replication of DNA viruses as well. There are four types of RNA viruses: positive single-stranded RNA (designated +ssRNA), retroviruses (a kind of +ssRNA virus), negative single-stranded RNA (−ssRNA), and double-stranded (dsRNA). The synthesis process for these RNA viruses is varied and sometimes rather complex. We start with the synthesis of +ssRNA viruses.

+ssRNA Viruses Viral single-stranded RNA that can act directly as mRNA is called **positive strand RNA (+RNA).** An example of a +ssRNA virus is poliovirus. In many +ssRNA viruses, a complementary **negative strand RNA (−RNA)** is transcribed from the +ssRNA genome by viral RNA polymerase; −RNA then serves as the template for the transcription of +ssRNA genomes. Such transcription of RNA from RNA is unique to viruses; no cell can transcribe RNA from RNA.

Retroviruses Unlike other +ssRNA viruses, the +ssRNA viruses called **retroviruses** do not use their genome as mRNA. Instead, retroviruses use a DNA intermediary that is transcribed from +RNA by *reverse transcriptase,* an enzyme carried within the capsid. This DNA intermediary then serves as the template for the synthesis of additional +RNA molecules, which act both as mRNA and as genomes for new virions. Human immunodeficiency virus (HIV) is a prominent retrovirus.

−ssRNA Viruses Other single-stranded RNA virions are −ssRNA viruses, which must overcome a unique problem. In order to synthesize a protein, a ribosome can only utilize mRNA (*i.e.,* +RNA), because −RNA is not recognized by ribosomes. The virus overcomes this problem by carrying within its capsid an enzyme, *RNA-dependent RNA transcriptase,* which is released into the host cell's cytoplasm during uncoating and then transcribes +RNA molecules from the virus's −RNA genome. Translation of proteins can then occur as usual. The transcribed +RNA also serves as a template for transcription of additional copies of −RNA. Diseases caused by −ssRNA viruses include rabies and flu.

dsRNA Viruses Viruses that have double-stranded RNA use yet another method of synthesis. The positive strand of the molecule serves as mRNA for the translation of proteins, one of which is a different RNA polymerase that transcribes more dsRNA. Each strand of RNA acts as a template for transcription of its opposite, which is reminiscent of DNA replication in cells. Double-stranded RNA viruses cause most cases of diarrhea in infants.

The various strategies by which animal viruses are synthesized are summarized in Table 13.3.

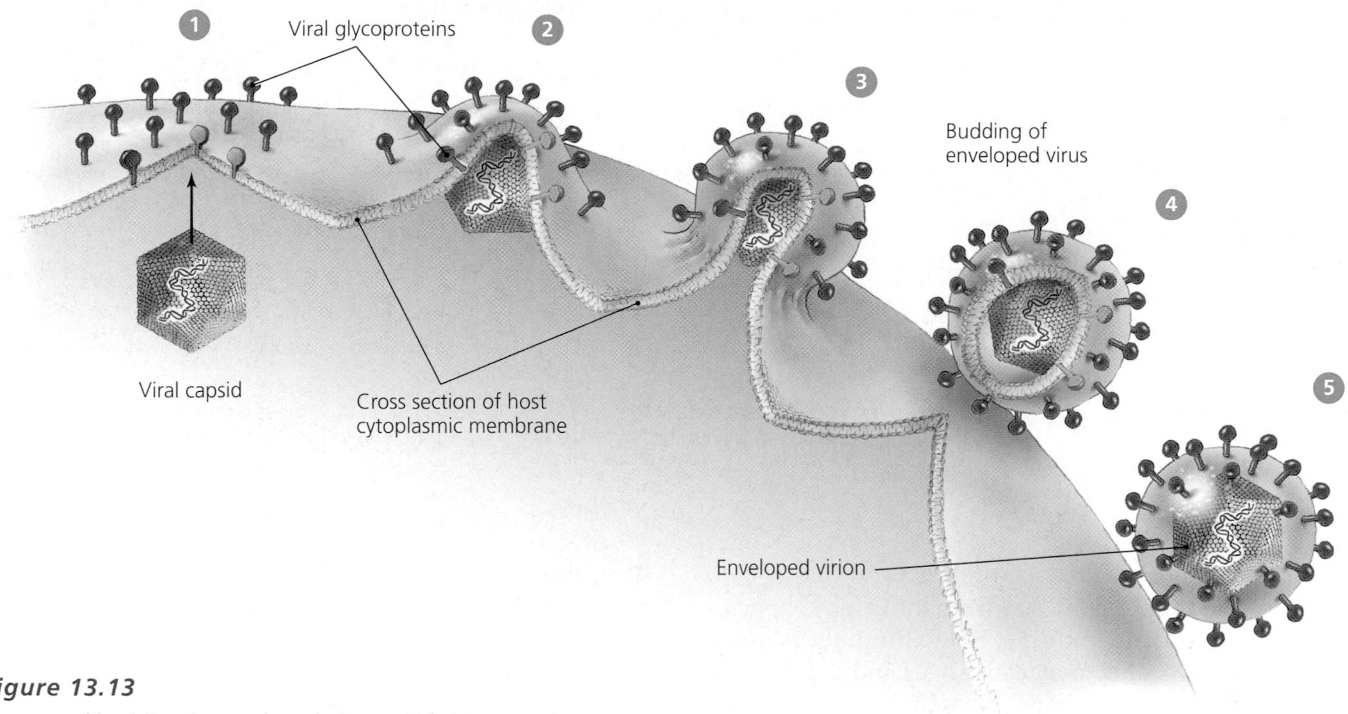

▲ ***Figure 13.13***
The process of budding in enveloped viruses. *What term describes a nonenveloped virus?*

Figure 13.13 Naked.

CRITICAL THINKING

Although many +ssRNA viruses use their genome directly as mRNA, retroviruses do not. Instead, their +RNA is transcribed into DNA by reverse transcriptase. What advantage do retroviruses gain by using reverse transcriptase?

Assembly and Release of Animal Viruses

As with bacteriophages, once the components of animal viruses are synthesized, they are assembled into virions that are then released from the host cell. Most DNA viruses assemble in and are released from the nucleus into the cytosol, whereas most RNA viruses develop solely in the cytoplasm. The number of viruses produced and released depends on both the type of virus and the size and initial health of the host cell.

Replication of animal viruses takes more time than replication of bacteriophages. Herpesviruses, for example, require almost 24 hours to replicate, as compared to 25 minutes for bacteriophage T4.

Enveloped animal viruses are often released via a process called **budding (Figure 13.13).** As virions are assembled, they are extruded through one of the cell's membranes—the nuclear membrane, the endoplasmic reticulum, or the cytoplasmic membrane. Each virion acquires a portion of membrane, which becomes the viral envelope. During synthesis, some viral glycoproteins are inserted into cellular membranes, and these proteins become the glycoprotein spikes on the surface of the viral envelope.

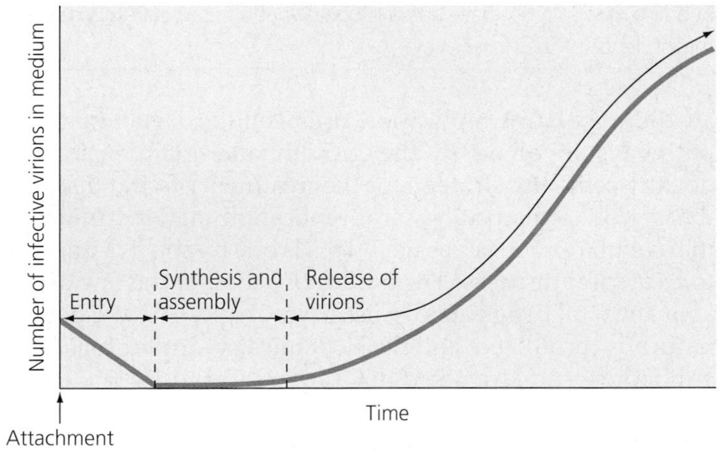

▲ ***Figure 13.14***
A generalized curve of virion abundance for persistent infections by budding enveloped viruses. Because the curve does not represent any actual infection, units for the graph's axes are omitted.

Because the host cell is not quickly lysed, as occurs in bacteriophage replication, budding allows an infected cell to remain alive for some time. Infections with enveloped viruses in which host cells shed viruses slowly and relatively steadily are called *persistent infections;* a curve showing virus abundance over time during a persistent infection lacks the burst of new virions seen in lytic replication cycles **(Figure 13.14).**

Table 13.4 A Comparison of Bacteriophage and Animal Virus Replication

	Bacteriophage	Animal Virus
Attachment	Proteins on tails attach to proteins on cell wall	Spikes, capsids, or envelope proteins attach to proteins or glycoproteins on cell membrane
Penetration	Genome is injected into cell or diffuses into cell	Capsid enters cell by direct penetration, fusion, or phagocytosis
Uncoating	None	Removal of capsid by cell enzymes
Site of synthesis	In cytoplasm	RNA viruses in cytoplasm; most DNA viruses in nucleus
Site of assembly	In cytoplasm	RNA viruses in cytoplasm; most DNA viruses in nucleus
Mechanism of release	Lysis	Naked virions: exocytosis or lysis; enveloped virions: budding
Nature of chronic infection	Lysogeny, always incorporated into host chromosome, may leave host chromosome	Latency, with or without incorporation into host DNA, incorporation is permanent

Naked animal viruses may be released in one of two ways: They may either be extruded from the cell by exocytosis, in a manner similar to budding, but without the acquisition of an envelope, or they may cause lysis and death of the cell, reminiscent of bacteriophage release.

CRITICAL THINKING

If an enveloped virus were somehow released from a cell without budding, it would not have an envelope. What effect would this have on the virulence of the virus? Why?

Because viral replication uses cellular structures and pathways involved in the growth and maintenance of healthy cells, any strategy for the treatment of viral diseases that involves disrupting viral replication may disrupt normal cellular processes as well. This is one reason it is difficult to treat viral diseases. The modes of action of the few available antiviral drugs are discussed in Chapter 10; the body's naturally produced antiviral chemicals—interferons and antibodies—are discussed in Chapters 15 and 16.

Latency of Animal Viruses

Some animal viruses, including chickenpox and herpes viruses, may remain dormant in cells in a process known as **latency;** the viruses involved in latency are called **latent viruses** or **proviruses.** Latency may be prolonged for years with no viral activity, signs, or symptoms. Though latency is similar to lysogeny as seen with bacteriophages, there are differences. Some latent viruses do not become incorporated into the chromosomes of their host cells, whereas lysogenic phages always do.

On the other hand, some animal viruses (*e.g.,* HIV) are more like lysogenic phages in that they do become integrated into a host chromosome as a **provirus.** However, when a provirus is incorporated into its host DNA, the condition is permanent; induction never occurs, as it can in lysogeny. Thus an incorporated provirus becomes a permanent, physical part of the host's chromosome, and all descendants of the infected cell will carry the provirus.

An important feature of HIV infections is that proviruses become a permanent part of the DNA of the host's white blood cells. But given that RNA cannot be incorporated directly into a DNA molecule, how does the ssRNA of HIV become a provirus incorporated into the DNA of its host cell? HIV can become a permanent part of a host's DNA because it, like all retroviruses, carries reverse transcriptase, which transcribes the genetic information of the +RNA molecule to a DNA molecule—which *can* become incorporated into the host cell's genome.

CRITICAL THINKING

A latent virus that is incorporated into a host cell's chromosome is never induced; that is, it never emerges from the host cell's chromosome to become a free virus. Given that it cannot emerge from the host cell's chromosome, can such a latent virus be considered "safe"? Why or why not?

A comparison of the features of the replication of bacteriophages and animal viruses is presented in **Table 13.4.** Next we turn our attention to the part viruses can play in cancer, beginning with a brief consideration of the terminology needed to understand the basic nature of cancer.

The Role of Viruses in Cancer

Learning Objectives

- ✓ Define the terms *neoplasia, tumor, benign, malignant, cancer,* and *metastasis.*
- ✓ Explain in simple terms how a cell may become cancerous, with special reference to the role of viruses.

Under normal conditions, the division of cells in a mature multicellular animal is under strict genetic control; that is, the animal's genes dictate that some types of cells can no longer divide at all, and that those that can divide are prevented from unlimited division. In this genetic control, either genes for cell division are "turned off," or genes that inhibit division are "turned on," or some combination of both these genetic events occur. However, if something upsets the genetic control, cells begin to divide uncontrollably. This phenomenon of uncontrolled cell division in a multicellular animal is called **neoplasia**[2] (nē-ō-plā'zē-ă). Cells undergoing neoplasia are said to be neoplastic, and a mass of neoplastic cells is a **tumor.**

Some tumors are **benign;** that is, they remain in one place and are not generally harmful, although occasionally such noninvasive tumors are painful and rob adjacent normal cells of space and nutrients. However, some tumors are **malignant;** they invade neighboring tissues and even travel throughout the body and invade other organs and tissues to produce new tumors—a process called **metastasis** (mĕ-tas'tă-sis). Malignant tumors are also called **cancers.** Cancers rob normal cells of space and nutrients and cause pain; in some kinds of cancer, malignant cells derange the function of the affected tissues, until eventually the body can no longer withstand the loss of normal function and dies.

Several theories have been proposed to explain the development of cancers, and the role viruses play. These theories revolve around the presence of *protooncogenes* (prō-tō-ong'kō-jēnz), genes that play a role in cell division. So long as the protooncogenes are repressed, no cancer results. However, activity of oncogenes (their name when they are active) or inactivation of oncogene repressors can cause cancer to develop. In most cases, several genetic changes must occur before cancer develops. Put another way, "multiple hits" to the genome must occur for cancer to result **(Figure 13.15).**

A variety of environmental factors contribute to the inhibition of oncogene repressors and the activation of oncogenes. Ultraviolet light, radiation, certain chemicals called *carcinogens* (kar-si'nō-jenz), and viruses have all been implicated in the development of cancer.

Viruses cause cancers in several ways. Some viruses carry copies of oncogenes as part of their genomes; other viruses stimulate oncogenes already present in the host; still other viruses interfere with normal tumor repression when they insert (as a provirus) into a repressor gene.

That viruses cause some animal cancers is well established. In the first decade of the 1900s, virologist F. Peyton Rous (1879–1970) proved that viruses induce cancer in chickens. Though several DNA and RNA viruses are known to cause about 15% of human cancers, the link between viruses and other human cancers has been more difficult to document. Among the virally induced cancers in humans are Burkitt's lymphoma, Hodgkin's disease, Kaposi's sarcoma, and cervical cancer. DNA viruses in the families *Adenoviridae, Herpesviridae, Hepadnaviridae, Papillomaviridae,* and *Polyomaviridae,* and two RNA viruses in the family *Retroviridae,* cause these and other human cancers.

[2]From Greek *neo,* meaning new, and *plassein,* meaning to mold.

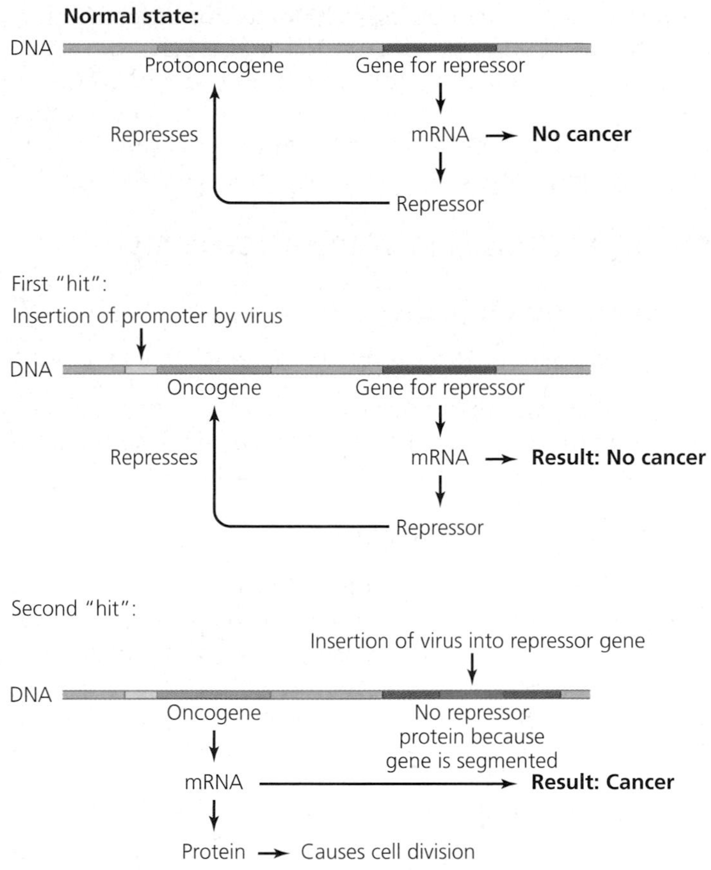

▲ ***Figure 13.15***

The oncogene theory of the induction of cancer in humans. The theory suggests that more than one "hit" to the DNA (that is, any change or mutation), whether caused by a virus (as shown here) or various physical or chemical agents, is required to induce cancer.

CRITICAL THINKING

Why are DNA viruses more likely to cause neoplasias than are RNA viruses?

Culturing Viruses in the Laboratory

Learning Objectives

- ✓ Describe some ethical and practical difficulties to overcome in culturing viruses.
- ✓ Describe three types of media used for culturing viruses.

Viruses must be cultured in order to conduct research and develop vaccines and treatments, but because viruses cannot metabolize or replicate by themselves, they cannot be grown in standard microbiological broths or on agar plates. Instead, they must be cultured inside suitable host cells, a requirement that complicates the detection, identification, and characterization of viruses. Virologists have developed three types of media for culturing viruses: media consisting of whole organisms (bacteria, plants, or animals), embryonated (fertilized) eggs, and cell cultures. We begin by considering the culture of viruses in whole organisms.

Culturing Viruses in Whole Organisms

Learning Objective

✓ Explain the use of a plaque assay in culturing viruses in bacteria.

In the following subsections we consider the use of bacterial cells as a virus culture medium before considering the issues involved in growing viruses in living animals.

Culturing Viruses in Bacteria

Most of our knowledge of viral replication has been derived from research on bacteriophages, which are relatively easy to culture because bacteria are easily grown and maintained. Phages can be grown in bacteria maintained in either liquid cultures or on agar plates. In the latter case, bacteria and phages are mixed with warm (liquid) nutrient agar and poured in a thin layer across the surface of an agar plate. During incubation, bacteria infected by phages lyse and release new phages that infect nearby bacteria, while uninfected bacteria grow and reproduce normally. After incubation, the appearance of the plate includes a uniform bacterial lawn interrupted by clear zones called **plaques,** which are areas where phages have lysed the bacteria **(Figure 13.16).** Such plates enable the estimation of phage numbers via a technique called **plaque assay** (aš'sā), in which each plaque corresponds to a single phage in the original bacterium/virus mixture.

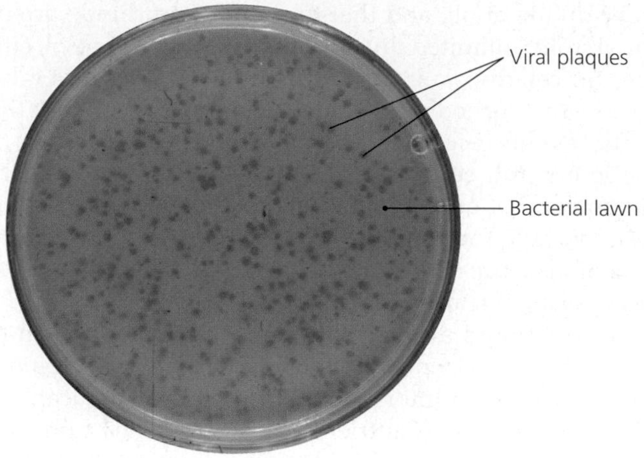

▲ ***Figure 13.16***
Viral plaques in a lawn of bacterial growth on the surface of an agar plate. *What is the cause of viral plaques?*

Figure 13.16 *Each plaque is an area in a bacterial lawn where bacteria have succumbed to phage infections.*

Culturing Viruses in Plants and Animals

Most experiments designed to study disease processes and immune responses, including those involving viruses, require the use of living organisms. Plant and animal viruses can be grown in laboratory plants and animals. Recall that the first discovery and isolation of a virus was the discovery of tobacco mosaic virus in tobacco plants. Rats, mice, guinea pigs, rabbits, pigs, and primates have been used to culture and study animal viruses.

However, maintaining laboratory animals can be difficult and expensive, and this practice raises ethical issues for some. Growing viruses that infect only humans raises additional ethical complications. Therefore, scientists have developed alternative ways of culturing animal and human viruses using fertilized chicken eggs or cell cultures.

Culturing Viruses in Embryonated Chicken Eggs

Chicken eggs are a useful culture medium for viruses because they are inexpensive, among the largest of cells, free of contaminating microbes, and contain a nourishing yolk (which makes them self-sufficient). Most suitable for culturing viruses are chicken eggs that have been fertilized and thus contain a developing embryo, the tissues of which (called membranes, which should not be confused with cellular membranes) provide ideal inoculation sites for growing viruses **(Figure 13.17).** Samples of virus can be injected into embryonated eggs at the sites that are best suited for the virus's maintenance and replication.

Vaccines against some viruses can also be prepared in egg cultures. You may have been asked if you are allergic to eggs before you received such a vaccine, because egg protein may remain as a contaminant in the vaccine.

Culturing Viruses in Cell (Tissue) Culture

Learning Objective

✓ Compare and contrast diploid cell culture and continuous cell culture.

Viruses can also be grown in **cell culture,** which consists of cells isolated from an organism and grown on the surface of a medium or in broth **(Figure 13.18).** Such cultures became practical when antibiotics provided a way to limit the growth of contaminating bacteria. Cell culture can be less expensive than maintaining research animals, plants, or eggs, and it avoids some of the moral problems associated with experiments performed on animals and humans. Cell cultures are sometimes called *tissue cultures,* but the term *cell culture* is more accurate because only a single type of cell is

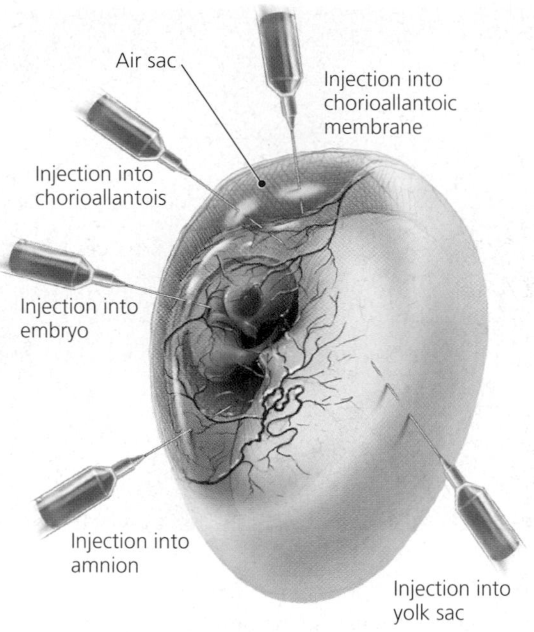

▲ ***Figure 13.17***

Inoculation sites for the culture of viruses in embryonated chicken eggs. *Why are eggs often used to grow animal viruses?*

Figure 13.17 *Eggs are large, sterile, self-sufficient cells that contain a number of different sites that are suitable for viral replication.*

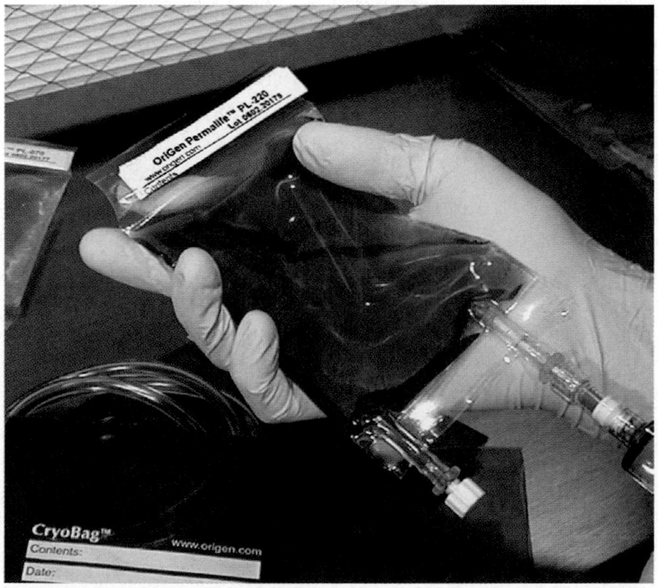

▲ ***Figure 13.18***

An example of cell culture. The bag contains a colored nutrient medium for growing cells in which viruses can be cultured.

used in a culture. (By definition, a tissue is composed of at least two kinds of cells.)

Cell cultures are of two types. The first type, **diploid cell cultures,** are created from embryonic animal, plant, or human cells that have been isolated and provided appropriate growth conditions. The cells in diploid cell culture generally last no more than about 100 generations (cell divisions) before they die.

The second type of culture, **continuous cell cultures,** are more long lasting because they are derived from tumor cells. Recall that a characteristic of neoplastic cells is that they divide relentlessly, providing a never-ending supply of new cells. One of the more famous continuous cell cultures is of HeLa cells, derived from a woman named *He*nrietta *La*cks who died of cervical cancer in 1951. Though she is dead, Mrs. Lacks's cells live on in laboratories throughout the world.

It is interesting that HeLa cells have lost some of their original characteristics. For example, they are no longer diploid because they have lost many chromosomes. HeLa cells provide a semi-standard[3] human tissue culture medium for studies on cell metabolism, aging, and (of course) viral infection.

CRITICAL THINKING

HIV replicates only in certain types of human cells, and one early problem in AIDS research was culturing those cells. How do you think scientists are now able to culture HIV?

Other Parasitic Particles: Viroids and Prions

Viruses are not the only submicroscopic entities capable of causing disorders within cells. In this section we will consider the characteristics of two molecular particles that infect cells: viroids and prions.

Characteristics of Viroids

Learning Objectives

- ✓ Define and describe viroids.
- ✓ Compare and contrast viroids and viruses.

Viroids are extremely small, circular pieces of RNA that are infectious and pathogenic in plants (**Figure 13.19** on page 398). Viroids are similar to an RNA virus except that they lack a capsid surrounding the nucleic acid. Even though they are circular, viroids may appear linear because of hydrogen bonding within the molecule. Several plant diseases, including some of chrysanthemum, potato, cucumber,

[3]HeLa cells are "semi-standard" because different strains have lost different chromosomes, and mutations have occurred over the years. Thus, HeLa cells in one laboratory may be slightly different from HeLa cells in another laboratory.

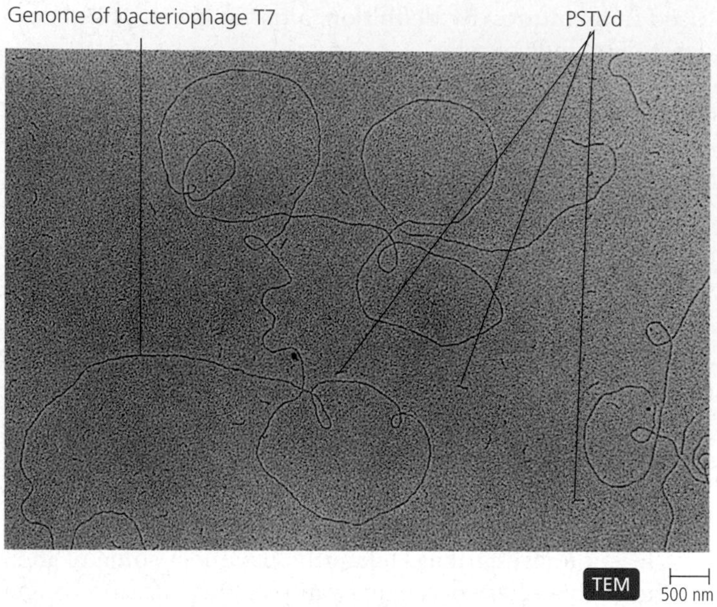

▲ ***Figure 13.19***

The RNA strand of the small potato spindle tuber viroid (PSTV). Also shown for comparison is the longer DNA genome of bacteriophage T7. *How are viroids similar to and different from viruses?*

Figure 13.19 *Viroids are similar to certain viruses in that they are infectious and contain a single strand of RNA; they are different from viruses in that they lack a proteinaceous capsid.*

coconut, and avocado, are caused by viroids, including the stunting shown in **Figure 13.20.**

Viroid-like agents—infectious, pathogenic RNA particles that lack capsids but do not infect plants—affect some fungi. (They are not called viroids because they do not infect plants.) No animal diseases are known to be caused by viroid-like molecules, though the possibility exists that they may be responsible for some diseases in humans.

Characteristics of Prions

Learning Objectives

- ✓ Define and describe prions, including their replication process.
- ✓ Compare and contrast prions and viruses.
- ✓ List four diseases caused by prions.

In 1982, Stanley Prusiner discovered a proteinaceous infectious agent that was different from any other known infectious agent in that it lacked nucleic acid. Prusiner named these new agents of disease **prions** (prē′onz), for *pro*teinaceous *in*fective particles. Before his discovery, the diseases now known to be caused by prions were thought to be caused by what were known as "slow viruses," which were so named because of the long delay between infection and the onset of signs and symptoms. Through convincing ex-

▲ ***Figure 13.20***

One effect of viroids on plants. The potatoes at right are stunted as the result of infection with PSTV viroids.

periments, Prusiner and his colleagues showed that prions are not viruses because they lack any nucleic acid, but their work was not without controversy.

For years some scientists resisted the concept of prions because particles that lack any nucleic acid violate the "universal" rule of protein synthesis—that proteins are translated from a molecule of mRNA, which is a copy of a portion of a DNA molecule. So, argued these detractors, if Prusiner's "infectious proteins" lack nucleic acid, how can they carry the information required to replicate themselves?

Prusiner ascertained that prions are composed of a single protein, called PrP, and that all mammals contain a gene that codes for the primary sequence of amino acids in PrP. Moreover, he determined that the sequence of amino acids in PrP is such that the protein can fold into at least two stable tertiary structures: a normal, functional structure that folds with several α-helices, called *cellular PrP,* and the disease-causing form having β-pleated sheets, called *prion PrP* **(Figure 13.21).**

Prusiner determined that prion PrP is able to convert normal cellular PrP into prion PrP by inducing a conformational change in the shape of cellular PrP. In other words, cellular PrP exposed to prion PrP refolds to form prion PrP. As cellular ribosomes synthesize more cellular PrP, the prions already present cause these new molecules to refold, becoming more prions. Thus, the rule of protein synthesis is maintained, because prions do not code for new prions, but instead convert extant cellular PrP into prions. For his work, Prusiner was awarded the Nobel Prize in 1997.

Why don't prions develop in all mammals, given that all mammals have PrP? Under normal circumstances, it appears that other nearby proteins and polysaccharides force PrP into the correct (cellular) shape. Only excess PrP production or mutations in the PrP gene result in the initial

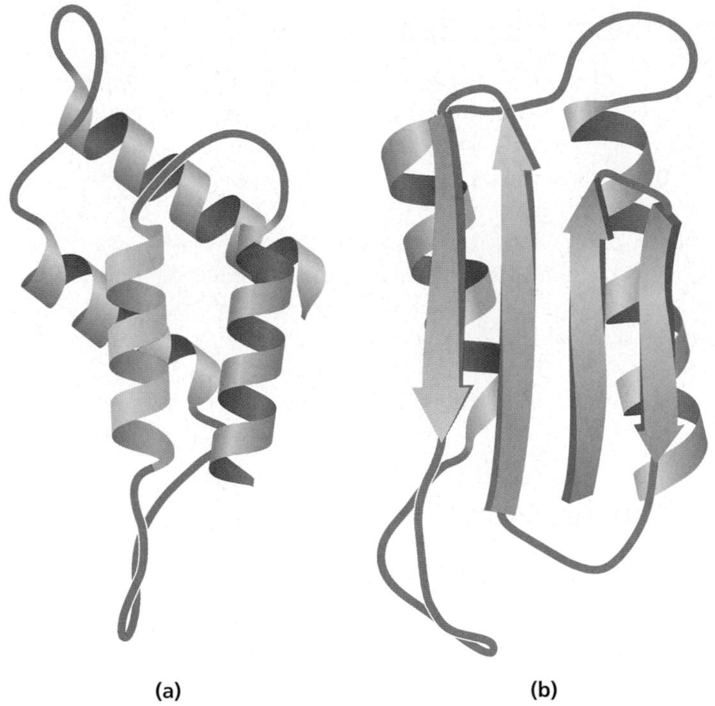

▲ *Figure 13.21*

The two stable, three-dimensional forms of prion protein (PrP). **(a)** Normal functional PrP, which has a preponderance of alpha helices. **(b)** Prion PrP, which has the same amino acid sequence but is folded to produce a preponderance of beta-pleated sheets.

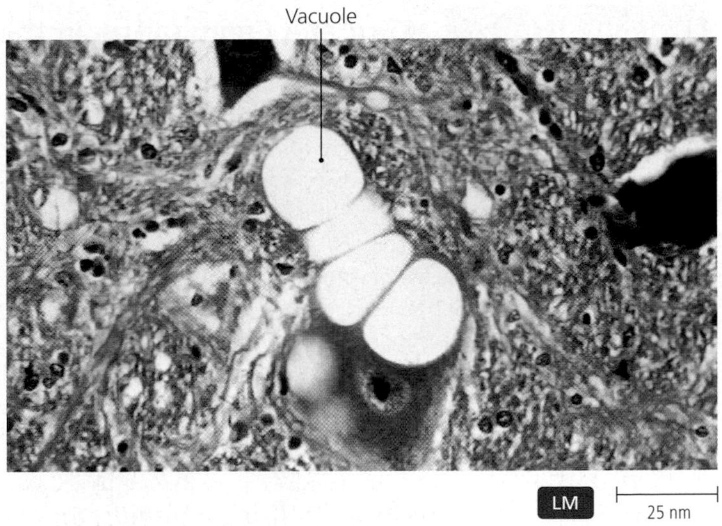

▲ *Figure 13.22*

A brain showing the large vacuoles and spongy appearance typical in prion-induced diseases. Shown here is the brain of a sheep with the prion disease called scrapie.

formation of prion PrP. For instance, it has been determined that human cellular PrP can misfold only if it contains methionine as the 129th amino acid. About 40% of humans have this type of cellular PrP and are thus susceptible to prion disease.

All known prion diseases involve fatal neurological degeneration, the deposition of fibrils in the brain, and the loss of brain matter such that eventually large vacuoles form. The latter process results in a characteristic spongy appearance **(Figure 13.22),** which is why as a group, prion diseases are called *spongiform encephalopathies* (spŭn'ji-fōrm en-sef'ă-lop'ă-thēz). The way in which prion PrP causes these changes in the brain is unknown.

Prions are known to cause several spongiform encephalopathies of animals and of humans, including *bovine spongiform encephalitis* (*BSE*, so-called "mad cow disease"), *scrapie* in sheep, and *Creutzfeldt-Jakob disease (CJD)* in humans. **Highlight 13.2** on page 400 describes *kuru* (koo'roo), a spongiform encephalopathy that has been eliminated. All these diseases are transmitted through either the ingestion of infected tissue, transplants of infected tissue, or contact of mucous membranes or skin abrasions with infected tissue.

Prions are not destroyed by normal cooking or sterilization, though they are destroyed by incineration. There is no treatment for any prion disease, though scientists have determined that the antimalarial drug quinacrine and the antipsychotic drug chlorpromazine forestall spongiform encephalopathy in prion-infected mice. Human trials of these drugs are ongoing.

PrP proteins of different species are slightly different in primary structure, and at one time it was thought unlikely that prions of one species could cross over into another species to cause disease. However, an epidemic of BSE in Great Britain in the late 1980s and in France in the 2000s resulted in the spread of prions not only to other cattle, but also to humans who ingested infected beef. To prevent further infection, the European Union has banned the use of animal-derived protein supplements in animal feed. Unfortunately, this step was too late to prevent the spread of bovine prions to more than 100 people who have developed fatal *variant Creutzfeldt-Jakob disease*.[4]

Prions composed of different proteins may lie behind other muscular and neuronal degenerative diseases, such as Alzheimer's disease, Parkinson's disease, and amyotrophic lateral sclerosis (ALS, or "Lou Gehrig's disease").

CRITICAL THINKING

Why did most scientists initially resist the idea of an infectious, self-replicating protein?

In this chapter we have seen that humans, animals, plants, fungi, and bacteria are susceptible to infection by acellular pathogens: viruses, viroids, and prions. **Table 13.5** on page 400 summarizes the differences and similarities between these pathogenic agents and bacterial pathogens.

[4]"Variant" because it is derived from BSE prions in cattle, as opposed to the regular form of CJD, which is a genetic disease of humans.

Highlight 13.2 Kuru and Cannibalism in the Fore Tribe of Papua New Guinea

Between 1957 and 1968, over 1100 members of the Fore tribe in the highlands of Papua New Guinea died from a frightening disease. This condition, called kuru—the Fore word for trembling or shivering—resulted in tremors, loss of the ability to stand or walk, visual impairment, mental deterioration, and eventually death. Initially the cause of kuru was a mystery to researchers, who were intrigued by the observation that kuru affected women and children at significantly higher rates than men. Subsequent research has since attributed kuru to prions acquired when tribe members handled and ate the brains of deceased relatives during mourning rituals. The high rates in women and children afflicted with the disease are consistent with this finding, as they were the primary participants in cannibalistic mourning rituals, in which men rarely participated.

Due to the influence of missionaries and government decree, cannibalism in Papua New Guinea ceased in the late 1950s, and cases of kuru have since vanished as a result. Elsewhere, however, other spongiform encephalopathies remain—including bovine spongiform encephalitis ("mad-cow disease") and Creutzfeldt-Jakob disease in humans.

Table 13.5 Comparison of Viruses, Viroids, and Prions to Bacterial Cells

	Bacteria	Viruses	Viroids	Prions
Width	200–2000 nm	10–300 nm	2 nm	5 nm
Length	200–550,000 nm	20–800 nm	40–130 nm	5 nm
Nucleic acid?	Both DNA and RNA	Either DNA or RNA, never both	RNA only	None
Protein?	Present	Present	Absent	Present (PrP)
Cellular?	Yes	No	No	No
Cytoplasmic membrane?	Present	Absent (though some viruses do have a membranous envelope)	Absent	Absent
Ribosomes?	Present	Absent (with one exception)	Absent	Absent
Growth?	Present	Absent	Absent	Absent
Self-replicating?	Yes	No	No	Yes; transform PrP protein already present in cell
Responsiveness?	Present	Some bacteriophages respond to a host cell by injecting their genomes	Absent	Absent
Metabolism?	Present	Absent	Absent	Absent

Are Viruses Alive?

Now that we have studied the characteristics and replication processes of viruses, let's return to a question first posed in Chapter 3: Are viruses alive?

To be able to wrestle with the answer, we must first recall the characteristics of life: growth, self-reproduction, responsiveness, and the ability to metabolize, all within structures called cells. According to these criteria, viruses seem to lack the qualities of living things, prompting some scientists to consider them nothing more than complex pathogenic chemicals. For other scientists, however, at least three observations—that viruses use sophisticated methods to invade cells, have the means of taking control of their host cells, and possess genomes containing instructions for replicating themselves—indicate that viruses are the ultimate parasites, because they use cells to make more viruses. According to this viewpoint, viruses are the least complex living entities.

In any case, viruses are right on the threshold of life—outside of cells they do not appear to be alive, but within cells they direct the synthesis and assembly required to make copies of themselves.

What do you think? Are viruses alive?

CHAPTER SUMMARY

1. **Acellular** disease-causing agents, which lack cell structure and cannot metabolize, grow, reproduce, or respond to their environment, include viruses, viroids, and prions.

Characteristics of Viruses (pp. 378–384)

1. A **virus** is a tiny infectious agent with nucleic acid surrounded by proteinaceous **capsomeres** that form a coat called a **capsid.**

A virus exists in an extracellular state and an intracellular state. A **virion** is a complete viral particle, including a nucleic acid and a capsid, outside of a cell.

2. The genomes of viruses includes either DNA or RNA, but never both. Viral genomes may be dsDNA, ssDNA, dsRNA, or ssRNA. They may exist as linear or circular and singular or multiple molecules of nucleic acid, depending on the type of virus.
3. A **bacteriophage** (or **phage**) is a virus that infects a bacterial cell.
4. Virions can have a membranous **envelope** or be naked—that is, without an envelope.

Classification of Viruses (p. 384)

1. Viruses are classified based on type of nucleic acid, presence of an envelope, shape, and size.
2. The International Committee on Taxonomy of Viruses (ICTV) has recognized viral family and genus names. With the exception of three orders, higher taxa are not established.

Viral Replication (pp. 384–394)

1. Viruses depend on random contact with a specific host cell type for replication. Typically a virus in a cell proceeds with a **lytic replication cycle** with five stages: **attachment, entry, synthesis, assembly**, and **release.**
2. Once attachment has been made between virion and host cell, the nucleic acid enters the cell. With phages, only the nucleic acid enters the host cell. With animal viruses, the entire virion often enters the cell, where the capsid is then removed in a process called **uncoating.**
3. Within the host cell, the viral nucleic acid directs synthesis of more viruses using metabolic enzymes and ribosomes of the host cell.
4. Assembly of synthesized virions occurs in the host cell, typically as capsomeres surround replicated or transcribed nucleic acids to form new virions.
5. Virions are released from the host cell either by lysis of the host cell (seen with phages and animal viruses) or by the extrusion of enveloped virions through the host's cell membrane (called **budding**), a process seen only with certain animal viruses. If budding continues over time, the infection is a persistent infection. An envelope is derived from a cell membrane.
6. **Temperate (lysogenic)** bacteriophages enter a bacterial cell and remain inactive, in a process called **lysogeny** or a **lysogenic cycle.** Such inactive phages are called **prophages** and are inserted into the chromosome of the cell and passed to its daughter cells. At some point in the generations that follow, a prophage may be excised from the chromosome in a process known as **induction.** At that point the prophage again becomes a lytic virus.
7. In a similar process called **latency,** an animal virus remains inactive in a cell, possibly for years. A **latent** virus is also known as a **provirus.** A provirus that has become incorporated into a host's chromosome remains there.
8. dsDNA viruses act like cellular DNA in transcription and replication.
9. Some ssRNA viruses have **positive strand RNA (+RNA),** which can be directly translated by ribosomes to synthesize protein. From the +RNA, complementary **negative strand RNA (−RNA)** is transcribed to serve as a template for more +RNA.
10. **Retroviruses,** such as HIV, are +ssRNA viruses that carry an enzyme, *reverse transcriptase,* to transcribe DNA from their RNA. This reverse process (DNA transcribed from RNA) is reflected in the name *retrovirus*.
11. −ssRNA viruses carry an RNA-dependent RNA transcriptase for transcribing mRNA from the −RNA genome so that protein can then be translated. Transcription of RNA from RNA is not found in cells.
12. When **double-stranded RNA (dsRNA)** functions as the genome, one strand of the RNA molecule functions as the genome, and the other strand functions as a template for RNA replication.

The Role of Viruses in Cancer (pp. 394–395)

1. **Neoplasia** is uncontrolled cellular reproduction in a multicellular animal. A mass of neoplastic cells, called a **tumor,** may be relatively harmless **(benign)** or invasive **(malignant).** Malignant tumors are also called **cancer. Metastasis** describes the spreading of malignant tumors. Environmental factors or oncogenic viruses may cause neoplasia.

Culturing Viruses in the Laboratory (pp. 395–397)

1. In the laboratory, viruses must be cultured inside whole organisms, in embryonated chicken eggs, or in cell cultures because they cannot metabolize or replicate alone.
2. When a mixture of bacteria and phages is grown on an agar plate, bacteria infected with phages lyse, producing clear areas called **plaques** on the bacterial lawn. A technique called **plaque assay** enables the estimation of phage numbers.
3. Viruses can be grown in two types of **cell cultures.** Whereas **diploid cell cultures** last about 100 generations, **continuous cell cultures,** derived from cancer cells, last longer.

Other Parasitic Particles: Viroids and Prions (pp. 397–400)

1. **Viroids** are small circular pieces of RNA with no capsid that infect and cause disease in plants. Similar pathogenic RNA molecules are known from fungi.
2. **Prions** are infectious protein particles that lack nucleic acids and replicate by converting similar normal proteins into new prions. Diseases caused by prions are spongiform encephalopathies, which involve fatal neurological degeneration.

Are Viruses Alive? (p. 400)

1. Outside of cells, viruses do not appear to be alive, but within cells, they exhibit lifelike qualities such as the ability to replicate themselves.

QUESTIONS FOR REVIEW

(Answers to multiple choice and matching are on the web, along with additional review questions. Visit www.microbiologyplace.com.)

Multiple Choice

1. Which of the following is *not* an acellular agent?
   a. viroid
   b. virus
   c. rickettsia
   d. prion
2. Which of the following statements is true?
   a. Viruses contain both DNA and RNA in specific proportions.
   b. Viruses are capable of metabolism.
   c. Viruses lack a cell membrane.
   d. Viruses grow in response to their environmental conditions.
3. A virus that is specific for a bacterial host is called a
   a. phage.
   b. prion.
   c. virion.
   d. viroid.
4. A naked virus
   a. has no membranous envelope.
   b. has injected its DNA or RNA into the host cell.
   c. is devoid of capsomeres.
   d. is one that is unattached to a host cell.
5. Which of the following statements is *false*?
   a. Viruses may have circular DNA.
   b. ssDNA and dsRNA are found in bacteria more often than in viruses.
   c. Viral DNA may be linear.
   d. Viruses may have DNA or RNA, but not both.
6. When a cell is infected with an enveloped virus and sheds viruses slowly over time, this infection is
   a. called a lytic infection.
   b. a prophage cycle.
   c. called a persistent infection.
   d. caused by a quiescent virus.
7. Another name for a complete virus is
   a. virion.
   b. viroid.
   c. prion.
   d. capsid.
8. Which of the following viruses can be latent?
   a. HIV
   b. chicken pox
   c. herpes
   d. all of the above
9. Which of the following is *not* a criterion for specific family classification of viruses?
   a. the type of nucleic acid present
   b. envelope structure
   c. capsid type
   d. lipid composition

Matching

Place the letter of the appropriate description in the correct blank in the left column.

___ 1. uncoating	A. dormant virus in a eukaryotic cell
___ 2. prophage	B. a virus that infects a bacterium
___ 3. retrovirus	C. DNA transcribed from RNA
___ 4. bacteriophage	D. protein coat of virus
___ 5. capsid	E. a membrane outside a viral capsid
___ 6. envelope	F. complete viral particle
___ 7. virion	G. inactive virus within bacterial cell
___ 8. provirus	H. removal of capsomeres from a virion
___ 9. benign tumor	I. invasive neoplastic cells
	J. harmless neoplastic cells

Short Answer

1. Compare and contrast a bacterium and a virus by writing either "Present" or "Absent" for each of the following structures.

Structure	Bacterium	Virus
Cell membrane	________	________
Functional ribosomes	________	________
Cytoplasm	________	________
Nucleic acid	________	________
Nuclear membrane	________	________

2. Describe the five phases of a generalized lytic replication cycle.
3. Why is it difficult to treat viral infections?
4. Describe four different ways that viral nucleic acid can enter a host cell.
5. Contrast lysis and budding as means of release of virions from a host cell.
6. What is the difference between a virion and a virus particle?
7. How is a provirus like a prophage? How is it different?
8. Describe lysogeny.

CRITICAL THINKING

1. Many viruses have a single-stranded genome, but it has been observed that larger viruses usually have a double-stranded genome. What reasonable explanation can you offer for this observation?
2. What are the advantages and disadvantages to bacteriophages of the lytic and lysogenic reproductive strategies?
3. How are computer viruses similar to biological viruses? Are computer viruses alive? Why or why not?
4. Compare and contrast lysogeny by a prophage and latency by a provirus.

CHAPTER 14

Infection, Infectious Diseases, and Epidemiology

CHAPTER OUTLINE

Fearing SARS, residents of Hong Kong don face masks.

During the winter of 2002–2003, a mysterious new viral illness began appearing in China's Guangdong province. Characterized by fever, shortness of breath, and atypical pneumonia, the highly infectious illness spread rapidly, particularly among food handlers and hospital workers tending to the already stricken. Initially confined to mainland China, the disease is believed to have begun its worldwide spread in February 2003, when a physician who had contracted the disease in Guangdong traveled to Hong Kong. During his stay in Hong Kong's Metropole Hotel, the physician passed the disease on to at least twelve other hotel guests, including travelers who continued on to Canada, Singapore, and Vietnam, unknowingly advancing the disease's spread to these countries. By March 2003, the disease had a name—severe acute respiratory syndrome (SARS)—and the World Health Organization (WHO) issued a global alert, triggering quarantines, travel cancellations, and other measures intended to curb the spread of this frightening new, sometimes-fatal disease.

How do microorganisms infect and cause disease in people? And how do experts investigate the transmission and spread of infectious diseases in populations? Read on.

MicroPrep Pre-Test: *Take the pre-test for this chapter on the web. Visit* ***www.microbiologyplace.com****.*

In previous chapters we have discussed characteristics of microorganisms and their metabolism. We examined ways microbes can be controlled, viewed, identified, and used for our benefit. Now it is time to examine how microbes directly affect our bodies. In this discussion, continually bear in mind that most microorganisms are neither harmful nor expressly beneficial to humans—they live their lives, and we live ours. Relatively few microbes either benefit us or harm us.

In this chapter we will first examine the general types of relationships that microbes can have with the bodies of their hosts. Then we will discuss how microbes enter and attach to their hosts; the nature of microbial diseases, called *infectious diseases;* and how microbes leave hosts to become available to enter new hosts. Next we will explore the sources of infectious diseases for humans, and the ways that infectious diseases are spread among hosts. Finally, we will consider *epidemiology,* the study of the occurrence and spread of diseases within groups of humans, and the methods by which we can limit the spread of pathogens within society.

▲ ***Figure 14.1***

Wood-eating termites in the genus *Reticulitermes* (shown above) cannot digest the cellulose in wood, but the protozoan *Trichonympha,* which lives in their intestines, can digest cellulose. The two organisms maintain a mutualistic relationship that is crucial for the life of the termite. *What benefits accrue to each of the symbionts in this mutalistic relationship?*

Figure 14.1 The termite gets digested cellulose from the protozoa, which get a constant supply of wood pulp to digest.

Symbiotic Relationships Between Microbes and Their Hosts

Symbiosis (sim-bī-o'sis) means "to live together." Each of us has symbiotic relationships with countless microorganisms, although often we are completely unaware of them. The 10 trillion or so cells in your body provide homes for more than 100 trillion bacteria and fungi and a huge number of viruses. We begin this section by considering the types of relationships between microbes and their hosts.

Types of Symbiosis

Learning Objectives

- ✓ Distinguish among the types of symbiosis, listing them in order from most beneficial to most harmful to the host.
- ✓ Describe the relationships among the terms parasite, host, and pathogen.

Scientists recognize three primary types of symbiosis: mutualism, commensalism, and parasitism.

Mutualism

One common type of symbiosis is **mutualism** (mū'tū-ăl-izm). In a mutualistic relationship between two organisms, both members benefit from their interaction. For example, bacteria in your colon receive a warm, moist, nutrient-rich environment in which to thrive, while you absorb vitamins released from the bacteria when they die. In this example, the relationship is beneficial to microbe and human alike but is not required by either. The bacteria could live elsewhere, and you could get vitamins from your diet.

Some mutualistic relationships provide such important benefits that one or both of the parties cannot live without the other. For instance, termites, which cannot digest the cellulose in wood by themselves, would die without a mutualistic relationship with colonies of wood-digesting protozoa living in their intestines **(Figure 14.1).** When the protozoa break down the wood pulp within the termite intestines, the termites share in the nutrients released from the wood. **Highlight 14.1** on p. 406 describes another relationship that benefits two diverse organisms.

Commensalism

In **commensalism** (kŏ-men'săl-izm), a second type of symbiosis, one member of the relationship benefits without significantly affecting the other. For example, Staphylococcus (staf'i-lō-kok'us) growing on the skin typically causes no measureable harm to a person, though an absolute example of commensalism is difficult to prove because the host may experience some unobserved benefits.

Parasitism

Of concern to health care professionals is a third type of symbiosis called **parasitism**[1] (par'ă-si-tizm). A **parasite** derives

[1] From Greek *parasitos,* meaning one who eats at the table of another.

Highlight 14.1 A Tale of Two Organisms

An unusual relationship exists between the marine unsegmented roundworm (nematode) *Eubostrichus parasitiferus,* and the rod-shaped, aerobic, sulfur-oxidizing bacteria that colonize its surface. The bacteria attach themselves to the nematode and completely cover the outside of the worm, producing a rope-like appearance.

This relationship is mutualistic because both symbionts benefit: The worm's mobility moves the bacteria through the marine sediments that contain their nutrients, while the worm in effect cultivates its own food supply by providing a habitat for the bacteria, its main (perhaps only) source of food.

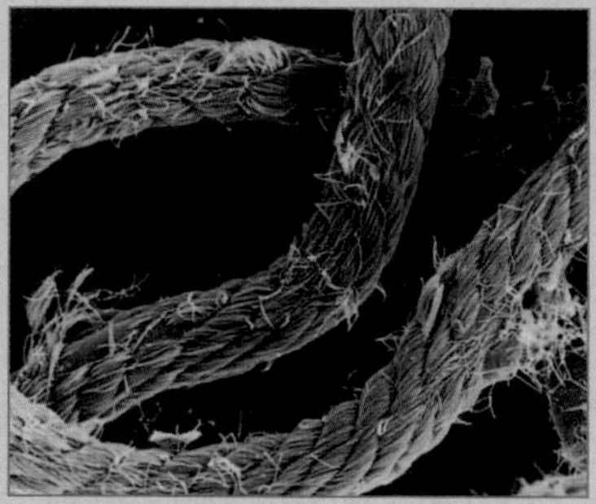

Eubostrichus parasitiferus covered by bacteria

Table 14.1 The Three Types of Symbiotic Relationships

	Organism 1	Organism 2	Example
Mutualism	Benefits	Benefits	Bacteria in human colon
Commensalism	Benefits	Neither benefits nor is harmed	Tapeworm in human intestine
Parasitism	Benefits	Is harmed	Tuberculosis bacteria in human lung

benefit from its **host** while harming it, though some hosts sustain only slight damage. In the most severe cases, a parasite kills its host, in the process destroying its own home. Any parasite that causes disease is called a **pathogen** (path′ō-jen).

A variety of protozoa, fungi, bacteria, and viruses are microscopic parasites of humans. Larger parasites of humans include parasitic worms and biting arthropods,[2] including mites (chiggers), ticks, mosquitos, fleas, and flies. Biting arthropods are discussed as transmitters of disease later in this chapter.

The three types of symbiosis are summarized in **Table 14.1.**

CRITICAL THINKING

Corals are small colonial animals that feed by filtering small microbes from seawater in tropical oceans worldwide. Biologists have discovered that the cells of most (if not all) corals are hosts to microscopic algae called zooxanthellae. Design an experiment to ascertain whether corals and zooxanthellae coexist in a mutual, commensal, or parasitic relationship.

Normal Microbiota in Hosts

Learning Objective

✓ Describe the normal microbiota, including resident and transient members.

Even though many parts of your body are *axenic*[3] environments—that is, sites that are free of any microbes—other parts of your body are home to millions of mutualistic and commensal symbionts. Each square centimeter of your skin, for example, contains about 3 million bacteria. The organisms that colonize the surfaces of the body without normally causing disease constitute the body's **normal microbiota (Figure 14.2),** also sometimes called the *normal flora*[4] or the *indigenous microbiota.* The normal microbiota are of two main types: resident microbiota and transient microbiota.

Resident Microbiota

Resident microbiota remain a part of the normal microbiota of a person throughout life. These organisms are found on the skin and on the mucous membranes of the digestive tract, upper respiratory tract, distal[5] portion of the urethra,[6] and vagina. **Table 14.2** on page 408 illustrates these environments and lists many of the resident bacteria, fungi, and protozoa that live there. Most of the resident microbiota are commensal; that is, they feed on excreted cellular wastes and dead cells without causing harm.

Transient Microbiota

Transient microbiota remain in the body for only a few hours, days, or months before disappearing. They are found in the same locations as the resident members of the normal mi-

[2]From Greek *arthros,* meaning jointed, and *pod,* meaning foot.
[3]From Greek *a,* meaning no, and *xenos,* meaning foreigner.
[4]This usage of the term *flora,* which literally means plants, derives from the fact that bacteria and fungi were considered plants in early taxonomic schemes. *Microbiota* is preferable because it properly applies to all microbes (whether eukaryotic or prokaryotic) and to viruses.
[5]*Distal* refers to the end of some structure, as opposed to *proximal,* which refers to the beginning of a structure.
[6]The tube that empties the bladder.

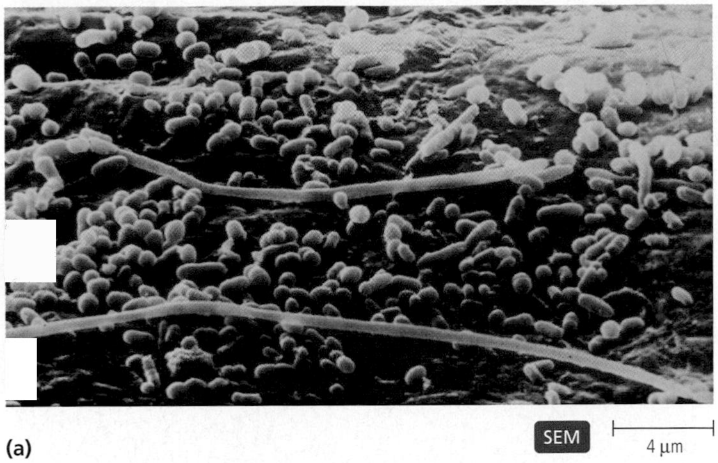

(a)

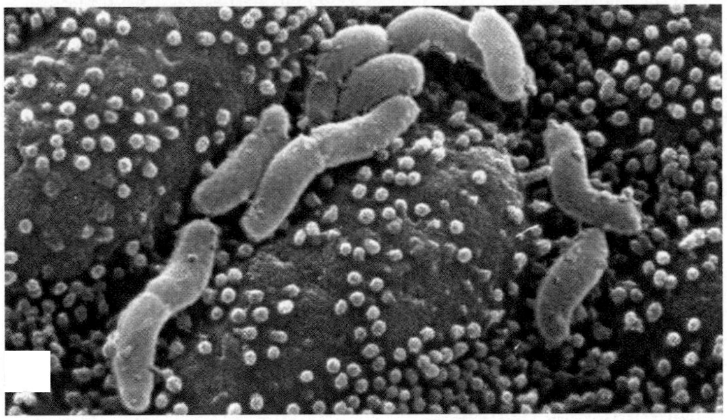

(b)

▲ *Figure 14.2*

Two examples of the normal microbiota of humans. **(a)** Bacteria on the surface of the skin. **(b)** Bacteria in the colon. *What types of symbiotic relationships do the bacteria in parts (a) and (b) have with their host?*

Figure 14.2 *Bacteria on the surface of the skin are commensals, though the host may derive some benefit in that the normal microbiota inhibit the growth of pathogens; the bacteria in the colon are mutualistic with the host.*

crobiota but cannot persist because of competition from other microorganisms, elimination by the body's defense cells, or chemical and physical changes in the body that dislodge them.

Acquisition of Normal Microbiota

You developed in your mother's womb without a normal microbiota because you had surrounded yourself with an amniotic membrane and fluid (bag of waters), which generally kept microorganisms at bay, and because your mother's uterus was essentially an axenic environment.

Your normal microbiota began to develop when your surrounding amniotic membrane ruptured and microorganisms came in contact with you during birth. Microbes entered your mouth and nose as you passed through the birth canal, and your first breath was loaded with microorganisms that quickly established themselves in your upper respiratory tract. Your first meals provided the progenitors of resident microbiota for your colon, while *Staphylococcus epidermidis* (ep-i-der'mī-dis) and other microbes transferred from the skin of both the medical staff and your parents began to colonize your skin. Although you continue to add to your transient microbiota, most of the resident microbiota was initially established during your first months of life.

How Normal Microbiota Become Opportunistic Pathogens

Learning Objective

✓ Describe three conditions that create opportunities for normal microbiota to cause disease.

Under ordinary circumstances, normal microbiota do not cause disease. However, these same microbes may become harmful if an opportunity to do so arises. In this case, normal microbiota become **opportunistic pathogens** or *opportunists.* Conditions that create opportunities for pathogens include the following:

1. *Immune suppression.* Anything that suppresses the body's immune system—including disease, malnutrition, emotional or physical stress, extremes of age (either very young or very old), the use of radiation or chemotherapy to combat cancer, or the use of immunosuppressive drugs in transplant patients—can enable normal microbiota to become opportunistic pathogens. AIDS patients often die from opportunistic infections that are typically controlled by a healthy immune system because HIV infection suppresses immune system function.
2. *Changes in the normal microbiota.* Normal microbiota use nutrients, take up space, and release toxic waste products, all of which make it less likely that arriving pathogens can compete well enough to become established and produce disease. This situation is known as **microbial antagonism** or **microbial competition.** However, changes in the relative abundance of normal microbiota, for whatever reason, may allow a member of the normal microbiota to become an opportunistic pathogen and thrive. For example, when a woman must undergo long-term antimicrobial treatment for a bacterial blood infection, the antimicrobial may also kill normal bacterial microbiota in the vagina. In the absence of competition from bacteria, *Candida albicans* (kan'did-ă al-bi'kanz), a yeast and also a member of the normal vaginal microbiota, grows prolifically, producing an opportunistic vaginal yeast infection. Other conditions

Table 14.2 Resident Microbiota[a]

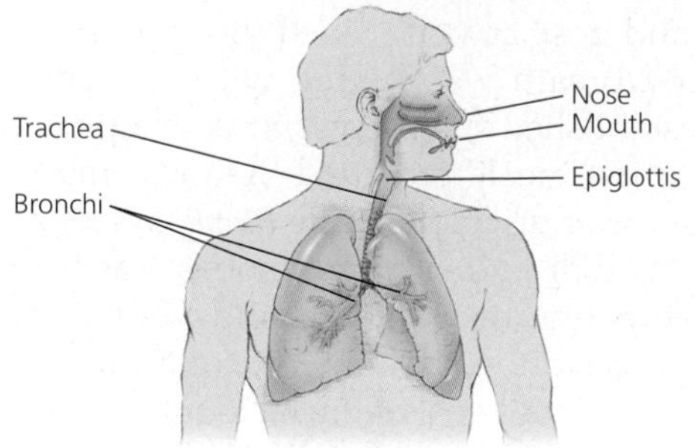

Upper Respiratory Tract

Genera

Staphylococcus, Streptococcus, Moraxella, Haemophilus, Lactobacillus, Veillonella, Fusobacterium, Candida (fungus)

Notes

The nose is cooler than the rest of the respiratory system and has some unique microbiota. The trachea and bronchi have a sparse microbiota compared to the nose and mouth. The alveoli of the lungs, which are too small to see at this magnification, have no natural microbiota.

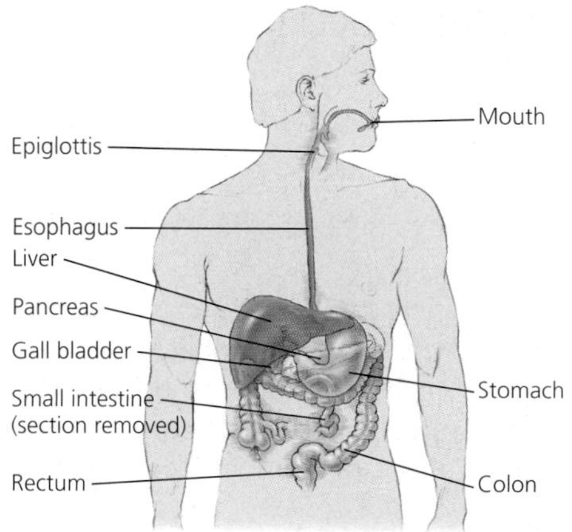

Upper Digestive Tract

Genera

Lactobacillus, Haemophilus, Actinomyces, Bacteroides, Treponema, Neisseria, Corynebacterium, Entamoeba (protozoan), *Trichomonas* (protozoan)

Notes

Microbes colonize surfaces of teeth, gingiva, lining of cheeks, and pharynx, and are found in saliva in large numbers. Dozens of species have never been identified.

Lower Digestive Tract

Genera

Bacteroides, Fusobacterium, Escherichia, Lactobacillus, Clostridium, Bifidobacterium, Enterococcus, Proteus, Shigella, Candida (fungus), *Entamoeba* (protozoan), *Trichomonas* (protozoan)

Notes

Bacteria are mostly strict anaerobes, though some facultative anaerobes are also resident.

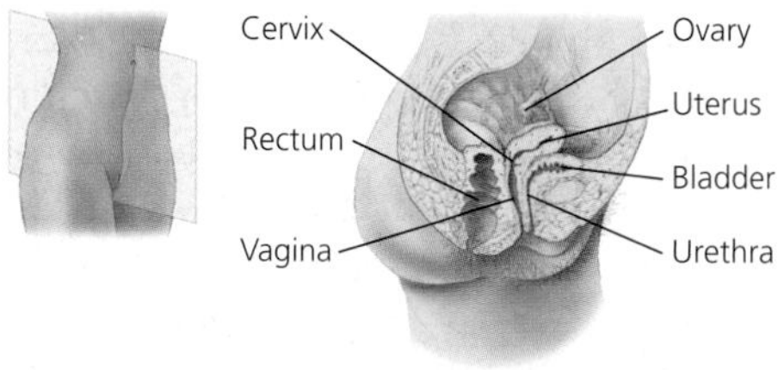

Female Urinary and Reproductive Systems

Genera

Lactobacillus, Streptococcus, Staphylococcus, Bacteroides, Clostridium, Candida (fungus), *Trichomonas* (protozoan)

Notes

Microbiota change as acidity in the vagina changes during menstrual cycle. The flow of urine prevents extensive colonization of the urethra.

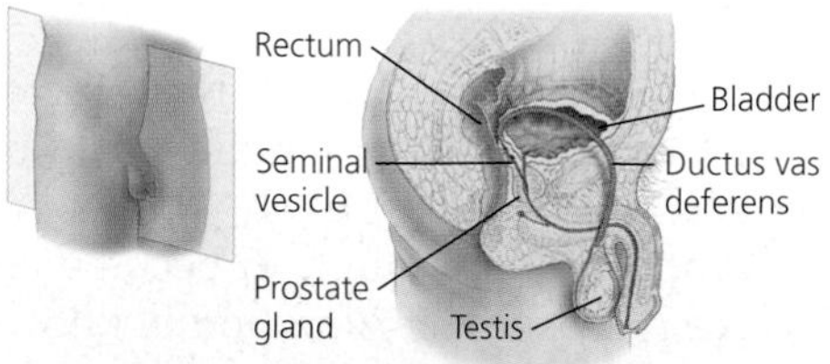

Male Urinary and Reproductive Systems

Genera

Staphylococcus, Streptococcus, Mycobacterium, Bacteroides, Fusobacterium, Peptostreptococcus

Notes

The flow of urine prevents extensive colonization of the urethra.

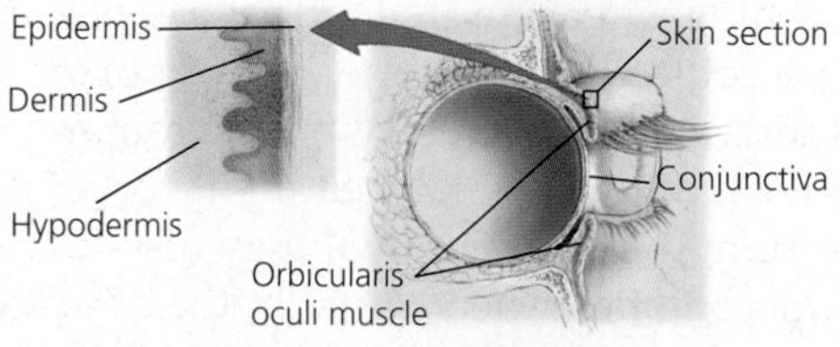

Eyes and Skin

Genera

Skin: *Propionibacterium, Staphylococcus, Corynebacterium, Micrococcus, Malassezia* (fungus), *Candida* (fungus)

Conjunctiva: *Staphylococcus*

Notes

Microbiota live on the outer, dead layers of the skin and in hair follicles and pores of glands. The deeper layers (dermis and hypodermis) are axenic.

Tears wash most microbiota from the eyes, so there are few compared to the skin.

[a]Genera are bacteria unless noted.

that can disrupt the normal microbiota are hormonal changes, stress, changes in diet, and exposure to overwhelming numbers of pathogens.

3. *Introduction of a member of the normal microbiota into an unusual site in the body.* As we have seen, normal microbiota are present only in certain body sites, and each site has only certain species of microbiota. If a member of the normal microbiota in one site is introduced into a site it normally does not inhabit, the organism may become an opportunistic pathogen. In the colon, for example, *Escherichia coli* (esh-ĕ-rik′ē-ă kō′lī) is mutualistic, but should it enter the urethra it becomes an opportunist that can produce a disease.

Now that we have considered the relationships of the normal microbiota to their hosts, we turn our attention to the movement of microbes into new hosts.

The Movement of Microbes into Hosts: Infection

In this section we examine the events that occur when hosts are exposed to microbes, the sites at which microbes can gain entry into hosts, and how entering microbes become established in new hosts.

Exposure to Microbes: Contamination and Infection

Learning Objective

✓ Describe the relationship between contamination and infection.

In the context of the interaction between microbes and their hosts, **contamination** refers to the mere presence of microbes in or on the body. Some microbial contaminants reach the body in food, drink, or the air, whereas others are introduced via wounds, biting arthropods, or sexual intercourse. Several outcomes of contamination by microbes are possible. Some microbial contaminants remain where they first contacted the body (such as the skin or mucous membranes) without causing harm and subsequently become part of the resident microbiota; other microbial contaminants remain on the body for only a short time, as part of the transient microbiota. Still others overcome the body's external defenses, multiply, and become established in the body; such a successful invasion of the body by a pathogen is called an **infection.**

Next we consider in greater detail the various sites through which pathogens gain entry into the body.

Portals of Entry

Learning Objective

✓ Identify and describe the portals through which pathogens invade the body.

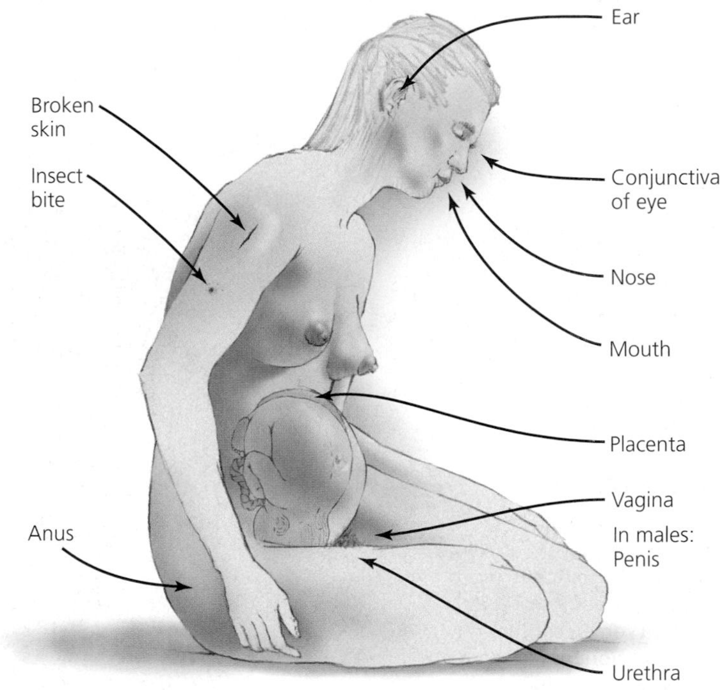

▲ *Figure 14.3*
Portals of entry for invading pathogens. The portals include the skin, placenta, conjunctiva, and mucous membranes of the respiratory, gastrointestinal, urinary, and reproductive tracts.

The sites through which pathogens enter the body can be likened to the great gates or portals of a castle, because those sites constitute the routes by which microbes gain entry. Pathogens thus enter the body at several sites, called **portals of entry (Figure 14.3),** which are of four major types: the skin, the mucous membranes, the placenta, and the so-called *parenteral route* (which is not a portal per se, but a way of circumventing the usual portals). Next we consider each type of portal of entry in turn.

Skin

Because the outer layer of skin is composed of relatively thick layers of tightly packed, dead, dry cells (**Figure 14.4** on p. 410), it forms a formidable barrier to most pathogens so long as it remains intact. Still, some pathogens can enter the body through natural openings in the skin, such as hair follicles and sweat glands; abrasions, cuts, and scrapes open the skin to infection by contaminants. Additionally, the larvae of some parasitic worms are capable of burrowing through the skin to reach underlying tissues, and some fungi can digest the dead outer layers of skin to gain entry to deeper, moister sites.

Mucous Membranes

Other portals of entry for pathogens are the *mucous membranes,* which line all the body cavities that are open to the outside world. They include the linings of the respiratory,

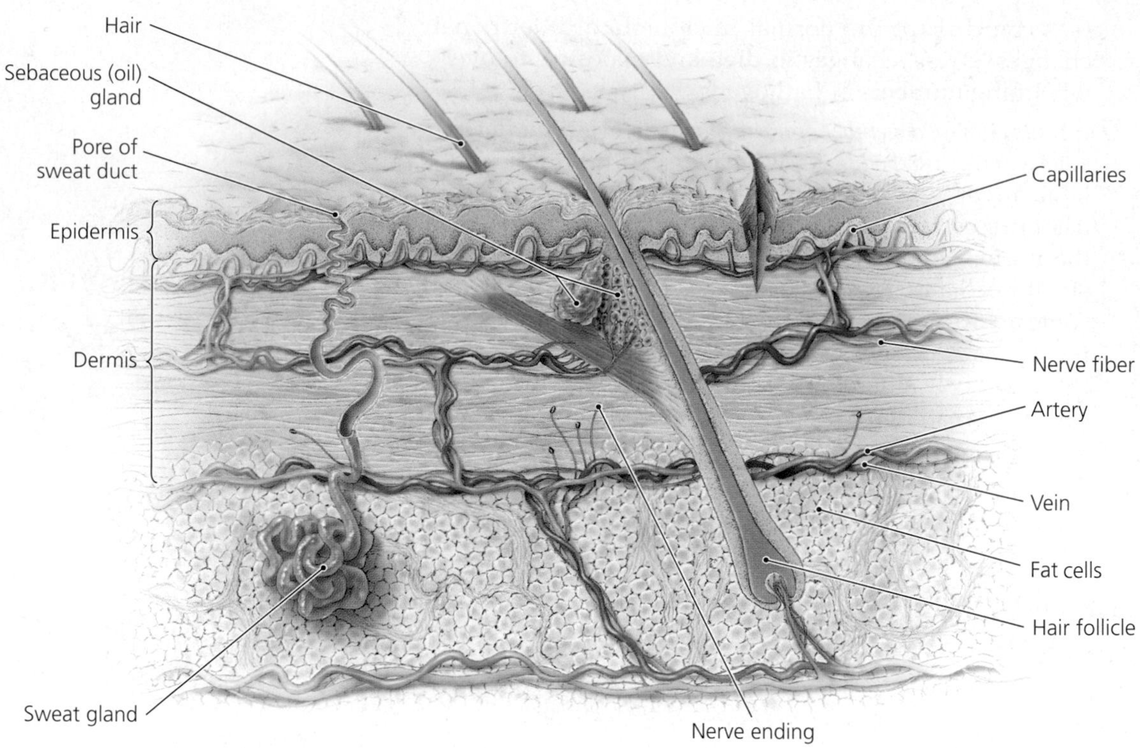

Figure 14.4 A cross-section of skin, showing the layers of cells that constitute a barrier to most microbes so long as the skin remains intact. Some pathogens can enter the body through hair follicles and the ducts of sweat glands.

gastrointestinal, urinary, and reproductive tracts, as well as the *conjunctiva* (kon-jŭnk-tī′vă), the thin membrane covering the surface of the eyeball and the underside of the eyelids. Like the skin, mucous membranes are composed of tightly packed cells; but unlike the skin, mucous membranes are relatively thin, moist, and warm, and their cells are living. Therefore, pathogens find mucous membranes more hospitable and often easier portals of entry.

The respiratory tract is the most frequently used portal of entry. Pathogens enter the mouth and nose in the air, on dust particles, and in drops of moisture. The bacteria that cause whooping cough, diphtheria, pneumonia, strep throat, and meningitis, as well as some fungi, viruses, and protozoa, enter through the respiratory tract.

Surprisingly, many viruses enter the respiratory tract via the eyes. They are introduced onto the conjunctiva by contaminated fingers and are washed into the nasal cavity with tears. Cold and influenza viruses typically enter the body in this manner.

Some parasitic protozoa, helminths, bacteria, and viruses infect the body through the gastrointestinal mucous membranes. These parasites are able to survive the acidic pH of the stomach and the digestive juices of the intestinal tract.

Placenta

The developing embryo forms an organ, called the *placenta,* through which it obtains nutrients from the mother. The placenta is in such intimate contact with the wall of the mother's uterus that nutrients and wastes can diffuse between the blood vessels of the developing child and of the mother, but because the two blood supplies do not actually contact each other, the placenta typically forms an effective barrier to most pathogens. However, in about 2% of pregnancies, pathogens cross the placenta and infect the embryo, or fetus, sometimes causing spontaneous abortion, birth defects, or premature birth. Some pathogens that can cross the placenta are listed in **Table 14.3.**

The Parenteral Route

The **parenteral** (pă-ren′ter-ăl) **route** is not technically a portal of entry, but instead a means by which the portals of entry are sometimes circumvented. To enter the body by the parental route, pathogens must be deposited directly into tissues beneath the skin or mucous membranes, such as occurs in punctures (by a nail, thorn, or hypodermic needle), cuts, bites, stab wounds, deep abrasions, or surgery.

The Role of Adhesion in Infection

Learning Objectives

- ✓ List the types of adhesion factors and the roles they play in infection.
- ✓ Explain how a biofilm may facilitate contamination and infection.

After entering the body, symbionts must adhere to cells if they are to be successful in establishing colonies. The process by which microorganisms attach themselves to cells is called **adhesion** (ad-hē′zhŭn). To accomplish adhesion, pathogens use **adhesion factors**, which are either specialized structures or attachment proteins. Examples of specialized structures that serve as adhesion factors are adhesion

Table 14.3 Some Pathogens that Cross the Placenta

	Pathogen	Condition in the Adult	Effect on Embryo or Fetus
Protozoan	*Toxoplasma gondii*	Toxoplasmosis	Abortion, epilepsy, encephalitis, microcephaly, mental retardation, blindness, anemia, jaundice, rash, pneumonia, diarrhea, hypothermia, deafness
Bacteria	*Treponema pallidum*	Syphilis	Abortion, multiorgan birth defects, syphilis
	Listeria monocytogenes	Listeriosis	Granulomatosis infantiseptica (nodular inflammatory lesions and infant blood poisoning), death
DNA viruses	*Cytomegalovirus*	Usually asymptomatic	Deafness, microcephaly, mental retardation
	Parvovirus B19	Erythema infectiosum	Abortion
RNA viruses	*Lentivirus* (HIV)	AIDS	Immunosuppression (AIDS)
	Rubivirus rubella	German measles	Severe birth defects or death

disks in some protozoa and suckers and hooks in some helminths. In contrast, viruses and many bacteria have surface lipoprotein and glycoprotein molecules called *ligands* that enable them to bind to complementary receptors on host cells **(Figure 14.5).** Ligands are also called *adhesins* on bacteria and *attachment proteins* on viruses. Adhesins are found on fimbriae, flagella, and glycocalyces of many pathogenic bacteria. Receptor molecules on host cells are typically glycoproteins containing mannose and galactose. If ligands or their receptors can be changed or blocked, infection can often be prevented.

The specific interaction of adhesins and receptors with chemicals on host cells often determines the specificity of pathogens for particular hosts. For example, *Neisseria gonorrhoeae* (nī-sē′rē-ă go-nōr-rē′ē) has adhesins on its fimbriae that adhere only to cells lining the urethra and vagina of humans. Thus, this pathogen cannot affect other hosts.

Some bacteria, including *Bordetella* (bōr-dĕ-tel′ă), the cause of whooping cough, have more than one type of adhesin, and other pathogens can change their adhesins over time. This helps the pathogen evade the body's immune system and allows the pathogen to attack more than one kind of cell. Bacterial cells and viruses that have lost the ability to make adhesins or attachment proteins—whether as the result of some genetic change (mutation) or exposure to certain physical or chemical agents (as occurs in the production of some vaccines)—become harmless, or **avirulent** (ā-vir′ū-lent).

Some bacterial pathogens do not attach to host cells directly, but instead interact with each other to form a sticky web of bacteria and polysaccharides called a *biofilm,* which adheres to a surface within a host. A prominent example of a biofilm is dental plaque (**Figure 14.6** on p. 412), which contains the bacteria that cause dental caries (tooth decay).

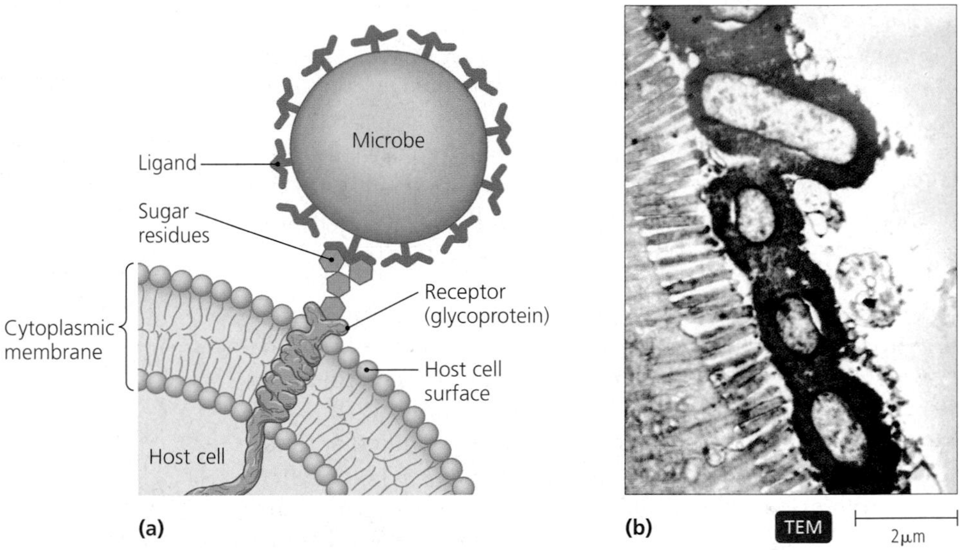

▲ *Figure 14.5*
The adhesion of pathogens to host cells. **(a)** An artist's rendition of the attachment of a microbial ligand to a complementary surface receptor on a host cell. **(b)** A photomicrograph of cells of a pathogenic strain of *E. coli* attached to the mucous membrane of the intestine. Although the thick glycocalyces of the bacteria are visible, the ligands are too small to be seen at this magnification.

CRITICAL THINKING

If a mutation occurred in *Escherichia coli* that deleted the gene for an adhesin, what effect might it have on the ability of *E. coli* to cause urinary tract infections?

The Nature of Infectious Disease

Learning Objective

✓ Compare and contrast the terms *infection, disease, morbidity, pathogenicity,* and *virulence*.

Most infections succumb to the body's defenses (discussed in detail in Chapters 15 and 16), but some infectious agents not only evade the defenses and multiply, but also affect

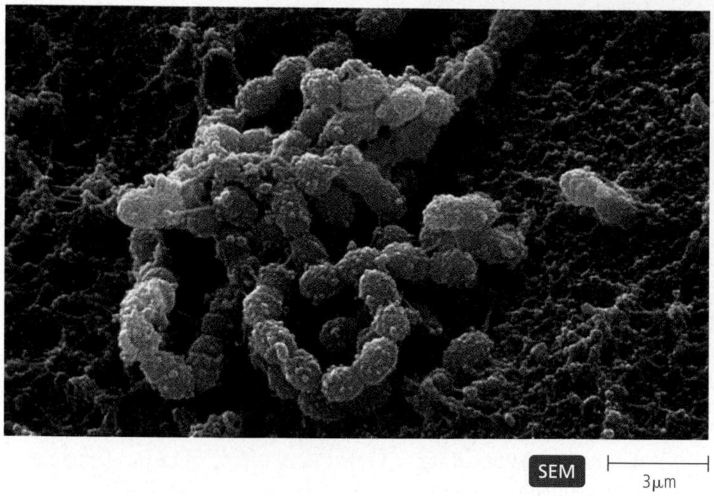

▲ *Figure 14.6*
Dental plaque, a biofilm of bacteria that adhere to one another and to teeth by means of sticky capsules, slime layers, and fimbriae.

body function. By definition, all parasites injure their hosts. When the injury is significant enough to interfere with the normal functioning of the body, the result is termed **disease.** Thus, disease, also known as **morbidity,** is any change from a state of health.

Infection and disease are not the same thing. Infection is the invasion of a pathogen; disease results only if the pathogen multiplies sufficiently to adversely affect the body. Some illustrations will help to clarify this point. While caring for a patient, a nurse may become contaminated with the bacterium *Staphylococcus aureus* (staf′i-lō-kok′ŭs o′rē-ŭs). If the pathogen is able to gain access to his body, perhaps through a break in his skin, he will become infected. Then, if the multiplication of *Staphylococcus* results in the production of a boil, the nurse will have a disease. Another example: A drug addict who shares needles may become contaminated and subsequently infected with HIV without experiencing visible change in bodily function for years. Such a person is infected but not yet diseased.

Manifestations of Disease: Symptoms, Signs, and Syndromes

Learning Objective

✓ Contrast symptoms, signs, and syndromes.

Diseases may become manifest in different ways. **Symptoms** are *subjective* characteristics of a disease that can be felt by the patient alone, whereas **signs** are *objective* manifestations of disease that can be observed or measured by others. Symptoms include pain, headache, dizziness, and fatigue; signs include swelling, rash, redness, and fever. Sometimes pairs of symptoms and signs reflect the same underlying cause: Nausea is a symptom, and vomiting a sign; chills are a symptom, whereas shivering is a sign. Note, however, that even though signs and symptoms may have the same underlying cause, they need not occur together. Thus, for example, a viral infection of the brain may result in both a headache (a symptom) and the presence of viruses in the cerebrospinal fluid (a sign), but viruses in the cerebrospinal fluid are not invariably accompanied by headache. Symptoms and signs are used in conjunction with laboratory tests to make diagnoses. Some typical symptoms and signs are listed in **Table 14.4.**

Table 14.4 Typical Manifestations of Disease

Symptoms (sensed by the patient)	Signs (detected or measured by an observer)
Pain, nausea, headache, chills, sore throat, fatigue or lethargy (sluggishness, tiredness), malaise (discomfort), itching, abdominal cramps	Swelling, rash or redness, vomiting, diarrhea, fever, pus formation, anemia, leukocytosis/leukopenia (increase/decrease in the number of circulating white blood cells), bubo (swollen lymph node), tachycardia/bradycardia (increase/decrease in heart rate)

A **syndrome** is a group of symptoms and signs that collectively characterizes a particular disease or abnormal condition. For example, acquired immunodeficiency syndrome (AIDS) is characterized by malaise, loss of certain white blood cells, diarrhea, weight loss, pneumonia, toxoplasmosis, and tuberculosis.

Some infections go unnoticed because they have no symptoms. Such cases are **asymptomatic** or **subclinical** infections. Note that even though asymptomatic infections by definition lack symptoms, in some cases certain signs may still be detected if the proper tests are performed. Thus, for example, leukocytosis (an excess of white blood cells) may be detected in a blood sample from an individual who feels completely healthy.

Table 14.5 lists some prefixes and suffixes used to describe various aspects of diseases and syndromes.

Causation of Disease: Etiology

Learning Objectives

✓ Define etiology.

✓ List Koch's postulates, explain their function, and describe their limitations.

Even though our focus in this chapter is infectious disease, it's obvious that not all diseases result from infections. Some diseases are *hereditary,* which means they are genetically transmitted from parents to offspring. Other diseases, called *congenital diseases,* are diseases that are present at birth, regardless of the cause (whether hereditary, environmental, or infectious). Still other diseases are classified as *degenerative, nutritional, endocrine, mental, immunological,* or *neoplastic.*[7]

[7] Neoplasms are tumors, which may either remain in one place (benign tumors) or spread (cancers).

Table 14.5 Terminology of Disease

Term	Meaning	Example
carcino-	cancer	carcinogenic, give rise to cancer
col-, colo-	colon	colitis, inflammation of the colon
dermato-	skin	dermatitis, inflammation of the skin
-emia	pertaining to the blood	viremia, viruses in the blood
endo-	inside	endocarditis
-gen, gen-	give rise to	pathogen, give rise to disease
-genesis	development	pathogenesis, development of disease
hepat-	liver	hepatitis, inflammation of the liver
idio-	unknown	idiopathic, pertaining to a disease of unknown cause
-itis	inflammation of a structure	meningitis, inflammation of the meninges (covering of the brain); endocarditis, inflammation of the lining of the heart
-oma	tumor	melanoma, cancer of melanocytes (pigment cells of the skin)
-osis	condition of	toxoplasmosis, being infected with *Toxoplasma*
-patho	abnormal	pathology, study of disease
septi-	literally *rotting,* refers to presence of pathogens	septicemia, pathogens in the blood
terato-	defects	teratogenic, causing birth defects
tox-	poison	toxin, harmful compound

The various categories of diseases are described in **Table 14.6** on p. 415. Note that some diseases can fall into more than one category. For example, liver cancer, which is usually associated with infection by hepatitis B virus, can be classified as both neoplastic and infectious.

The study of the cause of a disease is called **etiology** (ē-tē-ol'ō-jē). Because our focus here and in subsequent chapters is on infectious diseases, we next examine how microbiologists investigate the causation of infectious diseases, beginning with a reexamination of the work of Robert Koch, previously described in Chapter 1.

Using Koch's Postulates

In our modern world, we take for granted the idea that specific microbes cause specific diseases, but this has not always been the case. In the past, disease was thought to result from a variety of causes, including bad air, imbalances in body fluids, or spiritual forces. In the 19th century, Louis Pasteur, Robert Koch, and other microbiologists proposed the **germ theory of disease,** which states that disease is caused by infections of pathogenic microorganisms (at the time, called *germs*).

But which pathogen causes a specific disease? How can we distinguish between the pathogen that causes a disease and all the other biological agents (fungi, bacteria, protozoa, and viruses) that are part of the normal microbiota and are in effect "innocent bystanders"?

Koch developed a series of essential conditions, or *postulates,* that scientists must demonstrate or satisfy to prove that a particular microbe was pathogenic and caused a particular disease. Using his postulates, Koch proved that *Bacillus* (ba-sil'ŭs an-thra'sis) causes anthrax, and that *Mycobacterium tuberculosis* (mī-kō-bak-tēr'ē-ŭm too-ber-kyu-lō'sis) causes tuberculosis.

In order to prove that a given infectious agent causes a given disease, a scientist must satisfy all of **Koch's postulates** (**Figure 14.7** on p. 414):

1. The suspected pathogen must be present in every case of the disease.
2. That pathogen must be isolated and grown in pure culture.
3. The cultured pathogen must cause the disease when it is inoculated into a healthy, susceptible experimental host.
4. The same pathogen must be reisolated from the diseased experimental host.

It is critical that all the postulates be satisfied in order. The mere presence of a pathogen does not prove that it causes a disease. Although Koch's postulates have been used to prove the cause of many infectious diseases in humans, animals, and plants, inadequate attention to the postulates has resulted in incorrect conclusions concerning disease causation. For instance, in the early 1900s *Haemophilus influenzae* (hē-mof'i-lŭs in-flū-en'zē) was found in the respiratory systems of flu victims and identified as the causative agent of influenza. Later, flu victims that lacked *H. pneumoniae* in their lungs were discovered. This discovery violated the first postulate, so *H. influenzae* cannot be the cause of flu. Today we know that an RNA virus causes flu, and that *H. influenzae* was part of the normal microbiota of

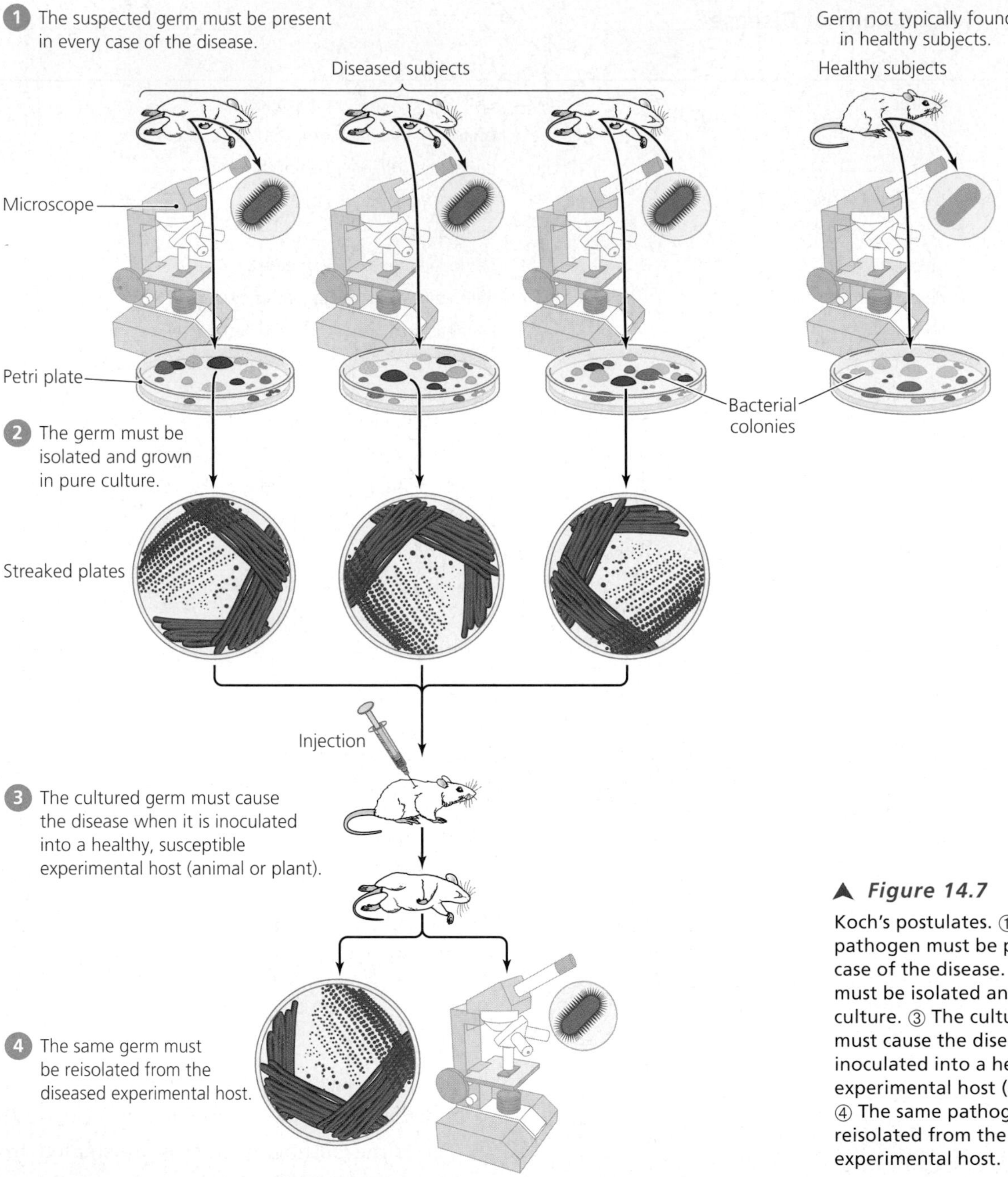
1 The suspected germ must be present in every case of the disease.
Germ not typically found in healthy subjects.
Diseased subjects
Healthy subjects
Microscope
Petri plate
Bacterial colonies
2 The germ must be isolated and grown in pure culture.
Streaked plates
Injection
3 The cultured germ must cause the disease when it is inoculated into a healthy, susceptible experimental host (animal or plant).
4 The same germ must be reisolated from the diseased experimental host.

▲ ***Figure 14.7***

Koch's postulates. ① The suspected pathogen must be present in every case of the disease. ② The pathogen must be isolated and grown in pure culture. ③ The cultured pathogen must cause the disease when it is inoculated into a healthy, susceptible experimental host (animal or plant). ④ The same pathogen must be reisolated from the diseased experimental host.

those early flu patients. (Later studies, correctly using Koch's postulates, showed that *H. influenzae* can cause an often-fatal meningitis in children.)

Exceptions to Koch's Postulates

Clearly, Koch's postulates are a cornerstone of the etiology of infectious diseases, but using them is not always feasible, for the following reasons:

- Some pathogens cannot be cultured in the laboratory. For example, pathogenic strains of *Mycobacterium leprae* (mī-kō-bak-tēr′ē-ŭm lep′rē), which causes leprosy, have never been grown on laboratory media.
- Some diseases are caused by a combination of pathogens, or by a combination of a pathogen and physical, environmental, or genetic cofactors. In such cases, the pathogen alone is avirulent, but when accompanied by another pathogen or the appropriate cofactor, disease results. For example, liver cancer can result when liver cells are infected by both the hepatitis B and hepatitis D viruses, but seldom when only one of the viruses infects the cells.
- Ethical considerations prevent applying Koch's postulates to diseases and pathogens that occur in humans only. In such cases the third postulate, which involves

Table 14.6 Categories of Diseases[a]

	Description	Examples
Hereditary	Caused by errors in the genetic code received from parents	Sickle-cell anemia, diabetes mellitus, Down syndrome
Congenital	Anatomical and physiological (structural and functional) defects present at birth; caused by drugs (legal and illegal), X-ray exposure, or infections	Fetal alcohol syndrome, deafness from rubella infection
Degenerative	Result from aging	Renal failure, age-related far-sightedness
Nutritional	Result from lack of some essential nutrients in diet	Kwashiorkor, rickets
Endocrine (hormonal)	Due to excesses or deficiencies of hormones	Dwarfism
Mental	Emotional or psychosomatic	Skin rash, gastrointestinal distress
Immunological	Hyperactive or hypoactive immunity	Allergies, autoimmune diseases, agammaglobulinemia
Neoplastic (tumor)	Abnormal cell growth	Benign tumors, cancers
Infectious	Caused by an infectious agent	Colds, influenza, herpes infections
Iatrogenic[b]	Caused by medical treatment or procedures; are a subgroup of hospital acquired diseases	Surgical error, yeast vaginitis resulting from antimicrobial therapy
Idiopathic[c]	Unknown cause	Alzheimer's disease, multiple sclerosis

[a]Some diseases may fall in more than one category.
[b]From Greek *iatros*, meaning physician.
[c]From Greek *idiotes*, meaning ignorant person, and *pathos*, meaning disease.

inoculation of a healthy susceptible host, cannot be satisfied within ethical boundaries. For this reason, scientists have never attempted to apply Koch's postulates to proving that HIV causes AIDS; however, observations of fetuses that were naturally exposed to HIV by their infected mothers, as well as accidentally infected health care workers, have in effect satisfied the third postulate.

Additionally, some circumstances may make satisfying Koch's postulates difficult:

- It is not possible to establish a single cause for such infectious diseases as pneumonia, meningitis, and hepatitis, because the names of these diseases refer to conditions that can be caused by more than one pathogen. For these diseases, laboratory technicians must identify the etiologic agent involved in any given case.
- Some pathogens have been ignored. For example, gastric ulcers were long thought to be caused by excessive production of stomach acid in response to stress, but the majority of such ulcers are now known to be caused by a long-overlooked bacterium, *Helicobacter pylori* (hel'i-kō-bak-ter pī'lō-rē).

If Koch's postulates cannot be applied to a disease condition for whatever reason, how can we positively know the causative agent of a disease? *Epidemiological* studies, discussed later in this chapter, can give statistical support to causation theories, but not absolute proof. For example, some researchers propose that the bacterium *Chlamydia pneumoniae* (kla-mid'ē-ă noo-mō'nē-ē) causes most cases of arteriosclerosis on the basis of the bacterium's presence in most patients with the disease. Debate among scientists about such cases fosters a continued drive for knowledge and discovery. For example, we are still searching for the causes of Parkinson's disease, multiple sclerosis, and Alzheimer's disease.

Virulence Factors of Infectious Agents

Learning Objective

✓ Explain how microbial extracellular enzymes, toxins, adhesion factors, and antiphagocytic factors affect virulence.

Microbiologists characterize the disease-related capabilities of microbes by using two related terms. The ability of a microorganism to cause disease is termed **pathogenicity** (path'ō-jĕ-nis'i-tē), and the *degree* of pathogenicity is termed **virulence.** In other words, virulence is the relative ability of a pathogen to infect a host and cause disease. Organisms can be placed along a virulence continuum (**Figure 14.8** on p. 416); highly virulent organisms almost always cause disease, while less virulent organisms (including opportunistic pathogens) cause disease only in weakened hosts or when present in overwhelming numbers.

Pathogens have a variety of traits that enable them to enter a host, adhere to host cells, gain access to nutrients, and escape detection or removal by the immune system. These traits are collectively called **virulence factors.** Virulent pathogens have one or more virulence factors that nonvirulent microbes lack. We discussed adhesion factors previously; now we examine three other virulence factors: extracellular enzymes, toxins, and antiphagocytic factors.

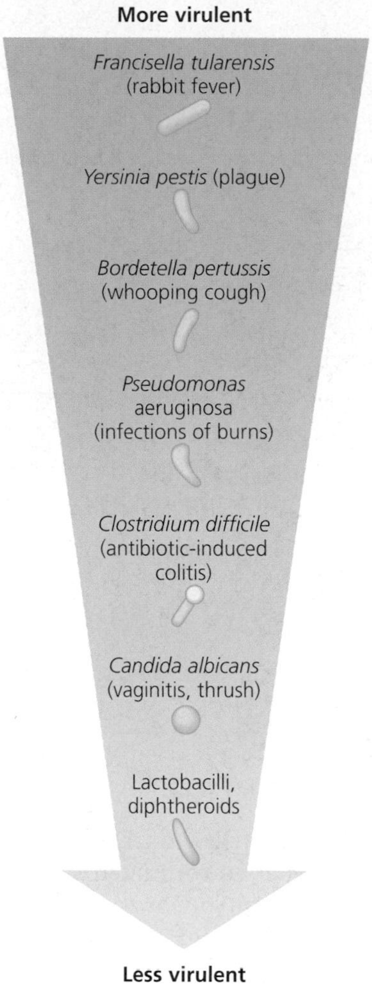

Figure 14.8
Relative virulence of some bacterial pathogens.

Extracellular Enzymes

Many pathogens secrete enzymes that enable them to dissolve structural chemicals in the body and thereby maintain an infection, invade further, and avoid body defenses. **Figure 14.9a** illustrates the action of some extracellular enzymes of bacteria:

- Hyaluronidase and collagenase degrade specific molecules to enable bacteria to invade deeper tissues. Hyaluronidase digests hyaluronic acid, the "glue" that holds animal cells together, and collagenase breaks down collagen, the body's chief structural protein.
- Coagulase coagulates blood proteins, providing a "hiding place" for bacteria within a clot.
- Kinases such as staphylokinase and streptokinase digest blood clots, allowing subsequent invasion of damaged tissues.

Many bacteria with these enzymes are virulent; mutant strains of the same species that have defective genes for these extracellular enzymes are usually avirulent.

Figure 14.9
Virulence factors. **(a)** Extracellular enzymes. Hyaluronidase and collagenase digest structural materials in the body. Coagulase in effect "camouflages" bacteria inside a blood clot, whereas kinases digest clots to release bacteria. **(b)** Toxins. Exotoxins (including cytotoxin, shown here) are released from living pathogens and harm neighboring cells. Endotoxin is released from many dead Gram-negative bacteria and can trigger widespread disruption of normal bodily functions. **(c)** Antiphagocytic factors. Capsules, one antiphagocytic factor, can prevent phagocytosis or stop digestion by a phagocyte. *How can bacteria prevent a phagocytic cell from digesting them once they have been engulfed by pseudopodia?*

Figure 14.9 *Some bacteria secrete a chemical that prevents the fusion of lysosomes with the phagosome containing the bacteria.*

Toxins

Toxins are chemicals produced by pathogens that either harm tissues or trigger host immune responses that cause damage. The distinction between extracellular enzymes and toxins is not always clear, because many enzymes are toxic, and many toxins have enzymatic action. In a condition called **toxemia** (tok-sē′mē-ă), toxins enter the bloodstream and are carried to other parts of the body, including sites that may be far removed from the site of infection. There are two types of toxin: exotoxins and endotoxin.

Exotoxins Many microorganisms secrete **exotoxins** that are central to their pathogenicity in that they destroy host cells or interfere with host metabolism. Exotoxins are of three principal types: (1) *cytotoxins,* which kill host cells in general or affect their function **(Figure 14.9b);** (2) *neurotoxins,* which specifically interfere with nerve cell function; and (3) *enterotoxins,* which affect cells lining the gastrointestinal tract. Examples of pathogenic bacteria that secrete exotoxins are the clostridia bacteria that cause gangrene, botulism, and tetanus; pathogenic strains of *Staphylococcus aureus* that cause food poisoning and other ailments; and diarrhea-causing *E. coli.* Some fungi and marine dinoflagellates (protozoa) also secrete exotoxins. Specific exotoxins are discussed in the chapters that cover specific pathogens.

The body protects itself with **antitoxins,** which are protective molecules called *antibodies* that bind to specific toxins and neutralize them. Health care workers stimulate the production of antitoxins by administering immunizations composed of *toxoids,* which are toxins that have been treated with heat, formaldehyde, chlorine, or other chemicals to make them nontoxic but still capable of stimulating the production of antibodies. Antibodies, toxoids, and immunizations are discussed further in Chapter 17.

Endotoxin Recall from Chapter 3 that Gram-negative bacteria have an outer (cell wall) membrane composed of lipopolysaccharide, phospholipids, and proteins (see Figure 3.15). **Endotoxin,** also called **lipid A,** is the lipid portion of the membrane's lipopolysaccharide.

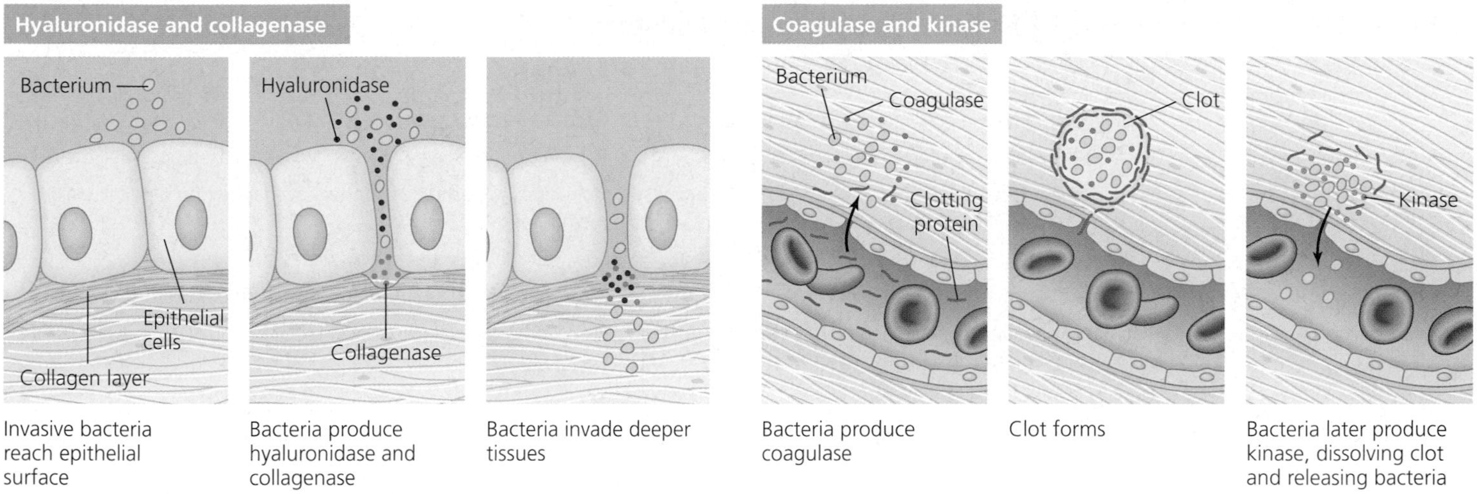
Hyaluronidase and collagenase
Bacterium
Epithelial cells
Collagen layer
Hyaluronidase
Collagenase
Invasive bacteria reach epithelial surface
Bacteria produce hyaluronidase and collagenase
Bacteria invade deeper tissues
Coagulase and kinase
Bacterium
Coagulase
Clotting protein
Clot
Kinase
Bacteria produce coagulase
Clot forms
Bacteria later produce kinase, dissolving clot and releasing bacteria

(a) Extracellular enzymes

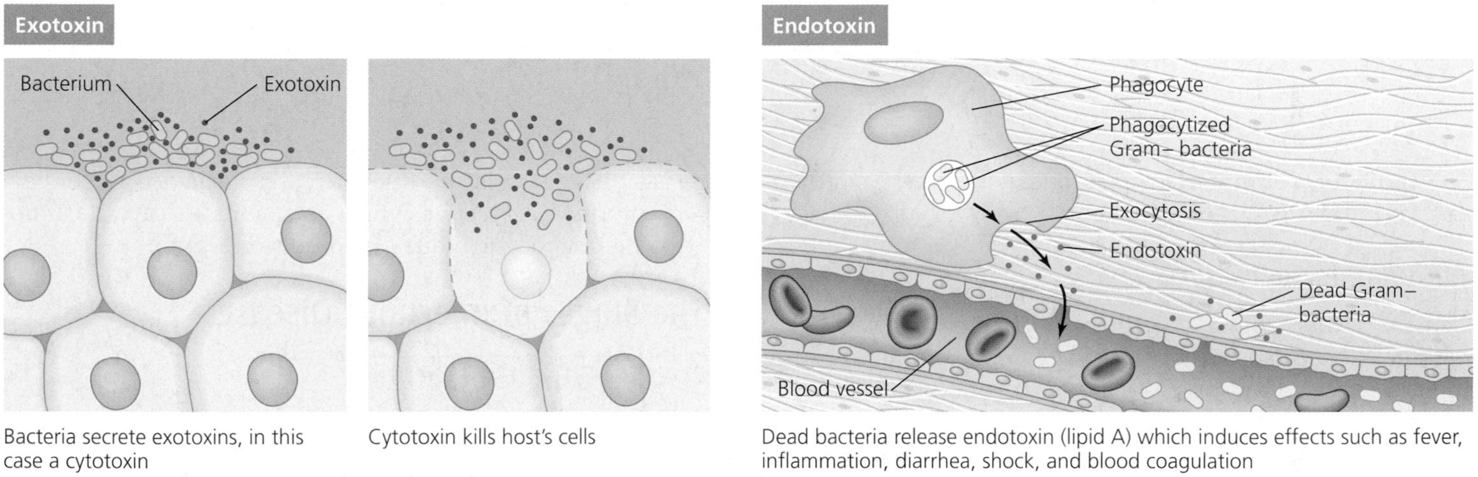
Exotoxin
Bacterium
Exotoxin
Bacteria secrete exotoxins, in this case a cytotoxin
Cytotoxin kills host's cells
Endotoxin
Phagocyte
Phagocytized Gram– bacteria
Exocytosis
Endotoxin
Dead Gram– bacteria
Blood vessel
Dead bacteria release endotoxin (lipid A) which induces effects such as fever, inflammation, diarrhea, shock, and blood coagulation

(b) Toxins

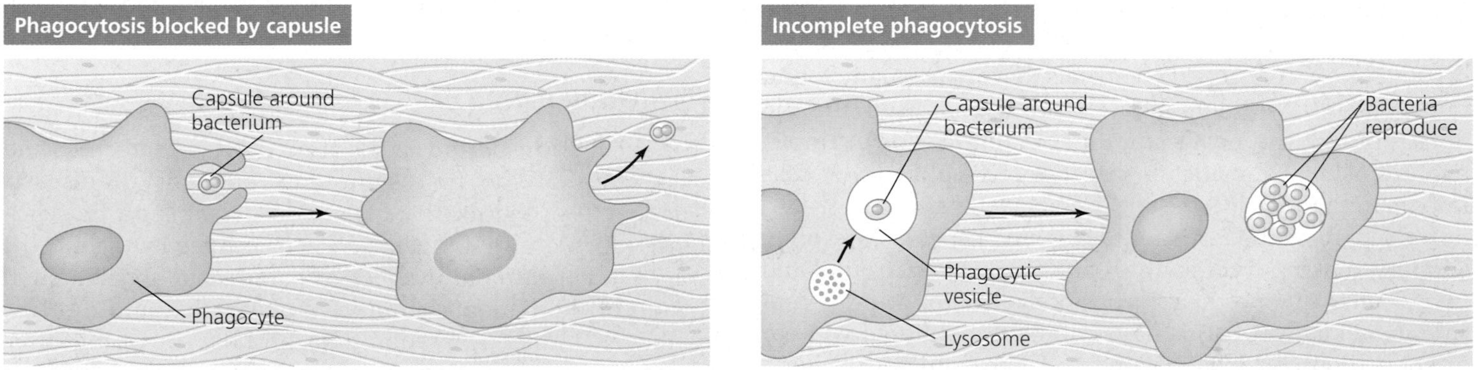
Phagocytosis blocked by capsule
Capsule around bacterium
Phagocyte
Incomplete phagocytosis
Capsule around bacterium
Phagocytic vesicle
Lysosome
Bacteria reproduce

(c) Antiphagocytic factors

Endotoxin is released only when bacterial cells die naturally or are digested by phagocytic cells such as macrophages (see Figure 14.9b). Many types of lipid A stimulate the body to release chemicals that cause fever, inflammation, diarrhea, hemorrhaging, shock, and blood coagulation. Although infections by bacteria that produce exotoxins are generally more serious than infections with Gram-negative bacteria, most Gram-negative pathogens can be potentially life threatening because the release of endotoxin from dead bacteria can produce serious, systemic effects in the host.

Table 14.7 A Comparison of Bacterial Exotoxins and Endotoxin

	Exotoxins	Endotoxin
Source	Mainly Gram-positive and Gram-negative bacteria	Gram-negative bacteria
Relation to bacteria	Metabolic product secreted from living cell	Portion of outer (cell wall) membrane released upon cell death
Chemical nature	Protein or short peptide	Lipid (lipid A) portion of lipopolysaccharide of outer (cell wall) membrane
Toxicity	High	Low, but may be fatal in high doses
Heat stability	Unstable at temperatures above 60°C	Stable for up to 1 hour at autoclave temperatures (121°C)
Effect on host	Variable depending on source; may be cytotoxin, neurotoxin, enterotoxin	Fever, lethargy, malaise, shock, blood coagulation
Fever producing?	No	Yes
Antigenicity[a]	Strong: stimulates antitoxin (antibody) production	Weak
Toxoid formation for immunization?	By treatment with heat or fomaldehyde	Not feasible
Representative diseases	Botulism, tetanus, gas gangrene, diphtheria, cholera, plague, staphylococcal food poisoning	Typhoid fever, tularemia, endotoxic shock, urinary tract infections, meningococcal meningitis

[a]Refers to the ability of a chemical to trigger a specific immune response, particularly the formation of antibodies.

Table 14.7 summarizes differences between exotoxins and endotoxin.

Antiphagocytic Factors

Typically, the longer a pathogen remains in a host, the greater the damage and the more severe the disease. To limit the extent and duration of infections, the body's phagocytic cells, such as the white blood cells called macrophages, engulf and remove invading pathogens. Here we consider some virulence factors related to the evasion of phagocytosis, beginning with bacterial capsules.

Capsules The capsules of many pathogenic bacteria (see Figure 3.4a) are effective virulence factors because many capsules are composed of chemicals that are normally found in the body (including polysaccharides); as a result, they do not stimulate the host's immune response. For example, hyaluronic acid capsules in effect deceive phagocytic cells into treating them and the enclosed bacteria as if they were a normal part of the body. Additionally, capsules are often slippery, making it difficult for phagocytes to surround and phagocytize them—their pseudopodia cannot grip the capsule, much as wet hands have difficulty holding a wet bar of soap **(Figure 14.9c).** Even when encapsulated bacteria are successfully engulfed, they may retard digestion long enough to reproduce inside the phagocyte.

Antiphagocytic Chemicals Some bacteria, including those that cause gonorrhea, produce chemicals that prevent the fusion of lysosomes with phagocytic vesicles, which allows the bacteria to survive inside of phagocytes (see Figure 14.9c). *Streptococcus pyogenes* (strep-tō-kok'ŭs pī-aj'en-ēz) produces a protein on its cell wall and fimbriae, called *M protein,* that resists phagocytosis and thus increases virulence. Other bacteria produce *leukocidins,* which are chemicals capable of destroying phagocytic white blood cells outright.

The Stages of Infectious Diseases

Learning Objective

✓ List and describe the five typical stages of infectious diseases.

Following contamination and infection, a definite sequence of events called the **disease process** occurs. Many infectious diseases have five stages following infection: an incubation period, the prodromal period, illness, decline, and convalescence **(Figure 14.10).**

Incubation Period

The **incubation period** is the time between infection and occurrence of the first symptoms or signs of disease. The length of the incubation period depends on the virulence (degree of pathogenicity) of the infective agent, the infective dose (initial number of pathogens), the state and health of the patient's immune system, the nature of the pathogen and its generation time, and the site of infection. Some diseases have typical incubation periods, whereas for others the incubation period varies considerably. **Table 14.8** lists incubation periods for selected diseases.

Prodromal Period

The **prodromal**[8] (prō-drō'măl) **period** is a short time of generalized, mild symptoms (such as malaise and muscle aches)

[8]From Greek *prodromos,* meaning forerunner.

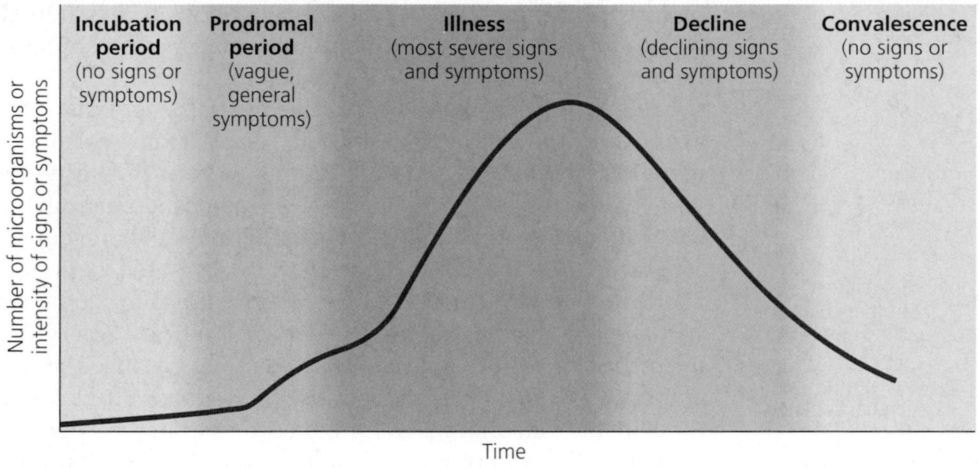

Figure 14.10 The stages of infectious diseases. Not every stage occurs in every disease.

Table 14.8 Incubation Periods of Selected Infectious Diseases

Disease	Incubation Period
Salmonella food-borne infection	<1 day
Influenza	1 day
Cholera	2 days
Genital herpes	5 days
Tetanus	3–21 days
Syphilis	2–4 weeks
Hepatitis B	1–6 months
AIDS	1–>8 years
Leprosy	10–>30 years

that precedes illness. Not all infectious diseases have a prodromal stage.

Illness

Illness is the most severe stage of an infectious disease. Signs and symptoms are most evident during this time. The patient's immune system has not yet fully responded to the pathogens, and their presence is harming the body. This stage is usually when a physician first sees the patient.

Decline

During the period of **decline,** the body gradually returns to normal as the patient's immune response and/or medical treatment vanquish the pathogens. Fever and other signs and symptoms subside. Normally the immune response and its products (such as antibodies in the blood) peak during this stage. If the patient doesn't recover, then the disease is fatal.

Convalescence

During **convalescence** (kon-vă-les′ens), the patient recovers from the illness, and tissues and systems are repaired and returned to normal. The length of a convalescent period depends on the amount of damage, the nature of the pathogen, the site of infection, and the overall health of the patient. Thus, whereas recovery from staphylococcal food poisoning may take a day, recovery from Lyme disease may take years.

A patient is likely to be infectious during every stage of disease. Even though most of us may realize we are infective during the symptomatic periods, many people are unaware that infections can be spread during incubation and convalescence as well. For example, a patient who no longer has any obvious herpes sores is still capable of transmitting herpesviruses. Good aseptic technique can often limit the spread of pathogens from recovering patients.

The Movement of Pathogens Out of Hosts: Portals of Exit

Just as infections occur through portals of entry, so pathogens must leave infected patients through **portals of exit** in order to infect others (**Figure 14.11** on p. 420). Many portals of exit are essentially identical to portals of entry. However, pathogens often exit hosts in materials that the body secretes or excretes. Thus, pathogens may leave hosts in secretions (ear wax, tears, nasal secretions, saliva, sputum, and respiratory droplets), in blood (via arthropod bites, hypodermic needles, or wounds), in vaginal secretions or semen, in milk produced by the mammary glands, and in excreted bodily wastes (feces and urine). As we will see, health care personnel must consider the portals of entry and exit in their efforts to understand and control the spread of diseases within populations.

Sources of Infectious Diseases in Humans

Learning Objective

✓ Describe three types of reservoirs of infection in humans.

Once pathogens of humans have left the body through the various portals of exit, most cannot survive for long in the

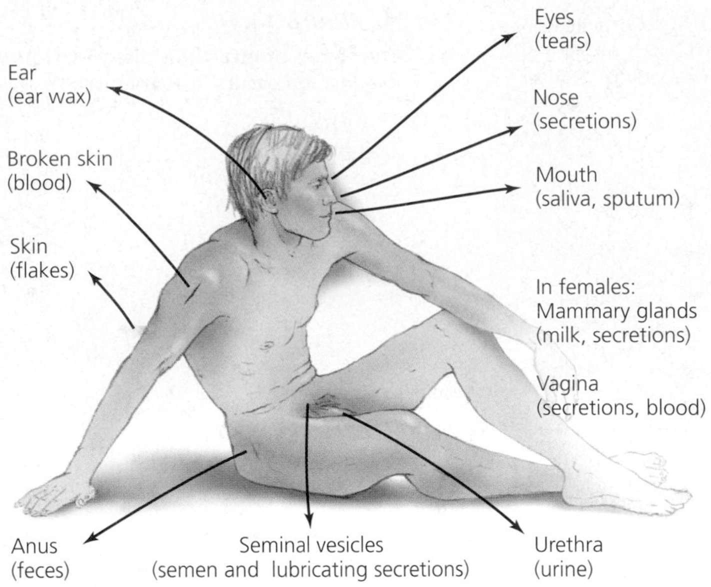

▲ ***Figure 14.11***

Portals of exit. Many portals of exit are the same sites as portals of entry. Pathogens often leave the body via bodily secretions and excretions produced at those sites.

relatively harsh conditions they encounter outside their hosts. Thus if these pathogens are to enter new hosts, they must survive in some site from which they can infect new hosts. These sites where pathogens are maintained as a source of infection are called **reservoirs of infection.** Next we consider three types of reservoirs: animal reservoirs, human carriers, and nonliving reservoirs.

Animal Reservoirs

Many pathogens that normally infect either domesticated or sylvatic[9] (wild) animals can also affect humans. The more similar an animal's physiology is to human physiology, the more likely its pathogens could affect human health. Diseases that are naturally spread from their usual animal hosts to humans are called **zoonoses** (zō-ō-nō′sēz) or **zoonotic diseases**. Over 150 zoonoses have been identified throughout the world. Well-known examples include yellow fever, anthrax, bubonic plague, and rabies.

Humans may acquire zoonoses from animal reservoirs via a number of routes, including various types of direct contact with animals and their wastes, by eating animals, and via bloodsucking arthropods. (Modes of transmission of infectious diseases are discussed in the next section.) Human infections with zoonoses are difficult to eradicate because extensive animal reservoirs are often involved. The larger the animal reservoir (that is, the greater the number of infected animals) and the greater the contact between humans and the animals, the more difficult it is to control the spread of the disease to humans. This is especially true when the animal reservoir consists of both sylvatic and domesticated animals. In the case of rabies, for example, the disease typically spreads from a sylvatic reservoir (often foxes, bats, and skunks) to domestic pets (especially dogs), from which humans may be infected. The wild animals constitute a reservoir for the rabies virus, but because domestic pets can be vaccinated, transmission to humans can be largely prevented.

Table 14.9 provides a brief view of some common zoonoses. Humans are usually dead-end hosts for zoonotic pathogens—that is, they do not act as significant reservoirs for the reinfection of animal hosts—largely because the circumstances under which zoonoses are transmitted favor movement from animals to humans, and not in the opposite direction. For example, animals do not often eat humans these days, and animals less frequently have contact with human wastes than humans have contact with animal wastes. Zoonotic diseases that are transmitted via the bites of blood-sucking arthropods are most likely to be transmitted back to animal hosts.

Human Carriers

Experience tells you that humans with active diseases are important reservoirs of infection for other humans. What may not be so obvious is that some infected people remain both asymptomatic and infective for years. This is true of syphilis and AIDS, for example. Whereas some of these **carriers** incubate the pathogen in their body and eventually develop the disease, others remain a continued source of infection without ever becoming sick. An example of such a carrier is given in **Case Study 14.1.**

Nonliving Reservoirs

Soil, water, and food can be **nonliving reservoirs** of infection. Soil, especially if fecally contaminated, can harbor *Clostridium* (klos-trid′ē-ŭm) bacteria, which cause botulism, tetanus, and other diseases. Water can be contaminated with

Case Study 14.1 A Deadly Carrier

In 1937, a man employed to lay water pipes in Croyden, England, was found to be the source of a severe epidemic of typhoid fever. The man—who was an asymptomatic carrier of *Salmonella typhi*, the bacterium that causes typhoid—habitually urinated at his job site, in the process contaminating the town's water supply with bacteria residing in his bladder. Over 300 cases of typhoid fever developed, and 43 people died, before the man was identified as the carrier.

1. How do you think health officials were able to identify the source of this typhoid epidemic?
2. Given that antibiotics were not generally available in 1937, how could health officials end the epidemic short of removing the man from the job site?

[9]From Latin *sylva*, meaning woodland.

Table 14.9 Some Common Zoonoses

Disease	Causative Agent	Animal Reservoir	Mode of Transmission
Helminthic (worms)			
Tapeworm infestation	*Dipylidium caninum*	Dogs	Ingestion of larvae transmitted in dog saliva
Trichinosis	*Trichinella spiralis*	Pigs, bears	Ingestion of undercooked pork or bear meat containing larvae
Protozoan			
Malaria	*Plasmodium* spp.	Monkeys	Bite of *Anopheles* mosquito
Toxoplasmosis	*Toxoplasma gondii*	Cats and other animals	Ingestion of contaminated meat, inhalation of pathogen, direct contact with infected tissues
Fungal			
Ringworm	*Trichophyton* sp. *Microsporum* sp. *Epidermophyton* sp.	Domestic animals	Direct contact
Bacterial			
Anthrax	*Bacillus anthracis*	Domestic livestock	Direct contact with infected animals, inhalation
Bubonic plague	*Yersinia pestis*	Rodents	Flea bites
Lyme disease	*Borrelia burgdorferi*	Deer	Tick bites
Salmonellosis	*Salmonella* spp.	Birds, rodents, reptiles	Ingestion of fecally contaminated water or food
Typhus	*Rickettsia prowazeki*	Rodents	Louse bites
Viral			
Rabies	*Lyssavirus* sp.	Bats, skunks, foxes, dogs	Bite of infected animal
Hantavirus pulmonary syndrome	*Hantavirus*. sp.	Deer mice	Inhalation of viruses in dried feces and urine
Yellow fever	*Flavivirus* sp.	Monkeys	Bite of *Aedes* mosquito

feces and urine containing parasitic worm eggs, pathogenic protozoa, bacteria, and viruses. Meats and vegetables can also harbor pathogens. Milk can contain many pathogens, which is why it is routinely pasteurized in the United States.

Modes of Infectious Disease Transmission

By definition, an infectious disease agent must be transmitted from either a reservoir or a portal of exit to another host's portal of entry. Transmission can occur by numerous modes that are somewhat arbitrarily categorized into three groups: contact transmission, vehicle transmission, and vector transmission (**Table 14.10** on p. 422).

Contact Transmission

Contact transmission is the spread of pathogens from one host to another by direct contact, indirect contact, or respiratory droplets.

Direct contact transmission, including *person-to-person spread*, typically involves body contact between hosts. Touching, kissing, and sexual intercourse between humans are involved in the transmission of such diseases as warts, herpes, and gonorrhea. Touching, biting, or scratching can transmit zoonoses such as rabies, ringworm, and tularemia from an animal reservoir to a human. The transfer of pathogens from an infected mother to a developing baby across the placenta is another form of direct contact transmission. Direct transmission within a single individual can also occur if the person transfers pathogens from a portal of exit directly to a portal of entry—as occurs, for example, when a person with poor personal hygiene unthinkingly places a finger contaminated with fecal pathogens into the mouth.

Indirect contact transmission occurs when pathogens are spread from one host to another by **fomites** (fōm'i-tēz), which are inanimate objects that are inadvertently used to transfer pathogens to new hosts. Fomites include needles, toothbrushes, paper tissues, toys, money, diapers, drinking glasses, bed sheets, medical equipment, and other objects that can harbor or transmit pathogens. Contaminated needles are a major source of infection of the hepatitis B and AIDS viruses.

Droplet transmission is a third type of contact transmission. Pathogens can be transmitted within *droplet nuclei* (mucus droplets) that exit the body during exhaling, coughing, and sneezing (**Figure 14.12** on p. 422). Pathogens such as cold and flu viruses may be spread in this manner. If pathogens travel more than 1 meter in respiratory droplets, it is considered to be *airborne transmission* (discussed shortly), rather than contact transmission.

Table 14.10 Modes of Disease Transmission

Mode of Transmission	Diseases Spread Include:
Contact Transmission	
Direct Contact: *e.g.* handshaking, kissing, sex, bites	Cutaneous anthrax, genital warts, gonorrhea, herpes, rabies, staphylococcus infections, syphilis
Indirect Contact: *e.g.* drinking glasses, toothbrushes, toys, punctures, droplets from sneezing and coughing (within one meter)	Common cold, enterovirus infections, influenza, measles, Q fever, pneumonia, tetanus, whooping cough
Vehicle Transmission	
Airborne: *e.g.* dust particles	Chicken pox, coccidiomycosis, histoplasmosis, influenza, measles, pulmonary anthrax, tuberculosis
Waterborne: *e.g.* streams, swimming pools	*Campylobacter* infections, cholera, *Giardia* diarrhea
Foodborne: *e.g.* poultry, seafood, meat	Food poisoning (botulism, staphylococcal); hepatitis A, listeriosis, tapeworms, toxoplasmosis, typhoid fever
Vector Transmission	
Mechanical: *e.g.* (on insect bodies) flies, roaches	*E. coli* diarrhea, salmonellosis, trachoma
Biological: *e.g.* lice, mites, mosquitoes, ticks	Chagas' disease, Lyme disease, malaria, plague, Rocky Mountain spotted fever, typhus fever, yellow fever

Vehicle Transmission

Vehicle transmission is the spread of pathogens via air, drinking water, and food, as well as bodily fluids being handled outside the body.

Airborne transmission involves the spread of pathogens to the respiratory mucous membranes of a new host via an **aerosol** (ār′ō-sol), a cloud of small droplets and solid particles suspended in the air. Aerosols may contain pathogens either on dust or inside droplets. (Recall that transmission via droplet nuclei that travel less than 1 meter is considered to be a form of direct contact transmission.) Aerosols can come from sneezing and coughing, or they can be generated by such things as air conditioning systems, sweeping, mopping, changing clothes or bed linens, or even from flaming inoculating loops in microbiological labs. Dust particles can carry *Staphylococcus*, *Streptococcus*, and hantavirus, whereas measles virus and tuberculosis bacilli can be transmitted in dried, airborne droplets. Fungal spores of *Histoplasma* and *Coccidioides* are typically inhaled.

Waterborne transmission is important in the spread of many gastrointestinal diseases, including giardiasis, amoebic dysentery, and cholera. Note that water can act as a reservoir as well as a vehicle of infection. **Fecal-oral infection** is a major source of disease in the world. Some waterborne pathogens, such as *Schistosoma* (skis-tō-sō′mă) worms and enteroviruses, are shed in the feces, enter through the gastrointestinal mucous membrane or skin, and subsequently cause disease elsewhere in the body.

Food-borne transmission involves pathogens in and on foods that are poorly processed, undercooked, or poorly refrigerated. Foods may be contaminated with normal microbiota (*e.g., E. coli* and *S. aureus*), with zoonotic pathogens

▲ *Figure 14.12*

Droplet transmission—in this case, droplets being propelled, primarily from the mouth, during a sneeze. By convention, such transmission is considered contact transmission only if droplets transmit pathogens to a new host within 1 meter of their source. *By what portal of entry does airborne transmission most likely occur?*

Figure 14.12 The most likely portal of entry of airborne pathogens is the respiratory mucous membrane.

such as *Mycobacterium bovis* (mī-kō-bak-tēr′ē-ŭm bō′vis) and *Toxoplasma* (tok-sō-plaz′mă), and with parasitic worms that alternate between human and animal hosts. Contamination of food with feces and pathogens such as hepatitis A virus is another kind of fecal-oral transmission. Because milk is

Case Study 14.2 Unusual Transmission of West Nile Virus

Although West Nile virus is primarily transmitted through mosquito bites, it can also be transmitted through organ transplantation. On August 1, 2002, four organs harvested from a single donor were transplanted into four patients. Within 17 days of transplantation, all four recipients became sick with West Nile virus-associated illnesses. Three of the patients developed meningoencephalitis, an inflammation of the brain and of the membranes surrounding the brain and spinal cord; the fourth patient developed West Nile virus fever. The organ donor's plasma tested positive for West Nile virus, but the source of the donor's infection could not be identified. Complicating matters, the organ donor had received blood products from 63 different blood donors. As this book went to press, an investigation into the source of the donor's infection remained ongoing.

1. In addition to patients who had received organs from the infected donor, what other patient populations might be at risk of West Nile virus infection?
2. As a scientist, what questions would you seek to answer about the infected donor?
3. What precautions would you advise instituting in order to prevent transmission of West Nile virus infection via blood transfusion and organ transplantation?

Source: *Morbidity and Mortality Weekly Report* 51(37): 833–836. 2002.

particularly rich in nutrients that microorganisms use (protein, lipids, vitamins, and sugars), it would be associated with the transmission of many diseases from infected animals and milk handlers if it were not properly pasteurized.

Because blood, urine, saliva, and other bodily fluids can contain pathogens, everyone—but especially health care workers—must take precautions when handling these fluids in order to prevent **bodily fluid transmission.** Special care must be taken to prevent any bodily fluid, which should be considered potentially contaminated with pathogens, from contacting the conjunctiva or any breaks in the skin or mucous membranes. Examples of diseases that can be transmitted via bodily fluids are AIDS, hepatitis, and herpes, which as we have seen can also be transmitted via direct contact. **Case Study 14.2** examines the transmission of West Nile virus via the blood of an infected organ donor.

Vector Transmission

Vectors are animals (typically arthropods) that transmit diseases from one host to another. Vectors can be either biological vectors or mechanical vectors.

Biological vectors not only transmit pathogens; they also serve as hosts for the multiplication of the pathogen during some stage of the pathogen's life cycle. The biological vectors of diseases affecting humans are typically biting arthropods, including mosquitoes, ticks, lice, fleas, blood-sucking flies, blood-sucking bugs, and mites. After the pathogens replicate within a biological vector, often in its gut or salivary gland, the pathogens enter a new host through a bite; either the bite site becomes contaminated with the vector's feces, or the bite directly introduces the pathogens into the new host.

Mechanical vectors are not required as hosts by the pathogens they transmit; they only passively carry pathogens to new hosts on their feet or other body parts. Mechanical vectors such as houseflies and cockroaches may introduce pathogens such as *Salmonella* (sal'mō-nel'ă) and *Shigella* (shē-gel'lă) into drinking water and food or onto the skin.

Table 14.11 on p. 424 lists some arthropod vectors and the diseases they transmit.

Classification of Infectious Diseases

Learning Objectives

✓ Describe the basis for each of the various classification schemes of infectious diseases.
✓ Distinguish among acute, subacute, chronic, and latent diseases.
✓ Distinguish among communicable, contagious, and noncommunicable infectious diseases.

Infectious diseases can be classified in a number of ways. No one way is "*the* correct way"; each has its own advantages. One scheme groups infectious diseases according to the body systems they affect. The appendix lists diseases organized in this way. A difficulty with this method of classification is that many diseases involve more than one organ system. For example, AIDS may begin as either a sexually transmitted infection (reproductive system) or a parenteral infection of the blood (cardiovascular system). It then becomes an infection of the lymphatic system as viruses invade lymphocytes. Finally, the syndrome involves diseases and degeneration of the respiratory, nervous, digestive, and lymphatic systems.

Infectious diseases may also be classified based on the taxonomic groups of their causative agents. Every disease (not just infectious diseases) can also be classified according to its longevity and severity. If a disease develops rapidly but lasts only a short time, it is called an **acute disease.** An example is the common cold. In contrast, **chronic diseases** develop slowly (usually with less severe symptoms) and are continual or recurrent. Infectious mononucleosis, hepatitis C, tuberculosis, and leprosy are chronic diseases. **Subacute diseases** have durations and severities that lie somewhere between acute and chronic.

Table 14.11 Selected Arthropod Vectors

	Disease	Causative Agent
Biological Vector		
Mosquitoes		
Anopheles, Aedes	Malaria	*Plasmodium* spp.
	Yellow Fever	*Flavivirus* sp.
	Elephantiasis	*Wuchereria bancrofti*
	Dengue	*Flavivirus* sp.
	Viral encephalitis	*Alphavirus* spp.
Ticks		
Ixodes, Dermacentor	Lyme disease	*Borrelia burgdorferi*
	Rocky Mountain spotted fever	*Rickettsia rickettsii*
Fleas		
Xenopsylla	Bubonic plague	*Yersinia pestis*
	Endemic typhus	*Rickettsia prowazekii*
Lice		
Pediculus	Epidemic typhus	*Rickettsia typhi*
Blood-sucking flies		
Simulium, Glossina	African sleeping sickness	*Trypanosoma brucei*
	River blindness	*Onchocerca volvulus*
Blood-sucking bug		
Triatoma	Chagas' disease	*Trypanosoma cruzi*
Mites (chiggers)		
Leptotrombidium	Scrub typhus	*Orienta tsutsugamushi*
Mechanical Vectors		
Houseflies		
Musca	Food-borne infections	*Shigella* spp., *Salmonella* spp., *Escherichia coli*
Cockroaches		
Perplaneta, Blattella, Supella	Food-borne infections	*Shigella* spp., *Salmonella* spp., *Escherichia coli*

Subacute sclerosing panencephalitis is a complication of some measles infections. **Latent diseases** are those in which a pathogen remains inactive for a long period of time before becoming active. Herpes is an example of a latent disease.

When an infectious disease comes from another infected host, either directly or indirectly, it is a **communicable** disease. Influenza, herpes, and tuberculosis are examples of communicable diseases. If a communicable disease is easily transmitted between hosts, as is the case for chickenpox or measles, it is also called a **contagious** disease. **Noncommunicable** diseases arise outside of hosts or from normal microbiota. In other words, they are not spread from one host to another, and diseased patients are not a source of contamination for others. Tooth decay, acne, and tetanus are examples of noncommunicable diseases. **Table 14.12** defines these and other terms used to classify infectious diseases.

Yet another way in which all infectious diseases may be classified is by the effects they have on populations, rather than on individuals. Is a certain disease consistently found in a given group of people or geographic area? Under what circumstances is it more prevalent than normal in a given geographic area? How prevalent is "normal"? How is the disease transmitted throughout a population? We next examine issues related to diseases at the population level.

Epidemiology of Infectious Diseases

Learning Objective

✓ Define epidemiology.

Our discussion so far has centered on the negative impact of microorganisms on *individuals*. Now we turn our attention to the effects of pathogens on *populations*. **Epidemiology**[10] (ep-i-dē-mē-ol′ō-jē) is the study of where and when diseases occur, and how they are transmitted within populations. During the 20th century, epidemiologists expanded the scope of their work beyond infectious diseases to also consider injuries and deaths related to automobile and fireworks accidents, cigarette smoking, lead poisoning, and other causes; however, we will limit our discussion primarily to the epidemiology of infectious diseases.

Frequency of Disease

Learning Objectives

✓ Contrast between incidence and prevalence.
✓ Differentiate among the terms *endemic, sporadic, epidemic,* and *pandemic*.

[10]From Greek *epidemios,* meaning among the people, and logos, meaning study of.

Table 14.12 Terms Used to Classify Infectious Diseases

Term	Definition
Acute disease	Disease in which symptoms develop rapidly and that runs its course quickly
Chronic disease	Disease with usually mild symptoms that develop slowly and last a long time
Subacute disease	Disease with time course and symptoms between acute and chronic
Asymptomatic disease	Disease without symptoms
Latent disease	Disease that appears a long time after infection
Communicable disease	Disease transmitted from one host to another
Contagious disease	Communicable disease that is easily spread
Noncommunicable disease	Disease arising from outside of hosts or from opportunistic pathogen
Local infection	Infection confined to a small region of the body
Systemic infection	Widespread infection in many systems of the body; often travels in the blood or lymph
Focal infection	Infection that serves as a source of pathogens for infections at other sites in the body
Primary infection	Initial infection within a given patient
Secondary infection	Infections that follow a primary infection; often by opportunistic pathogens

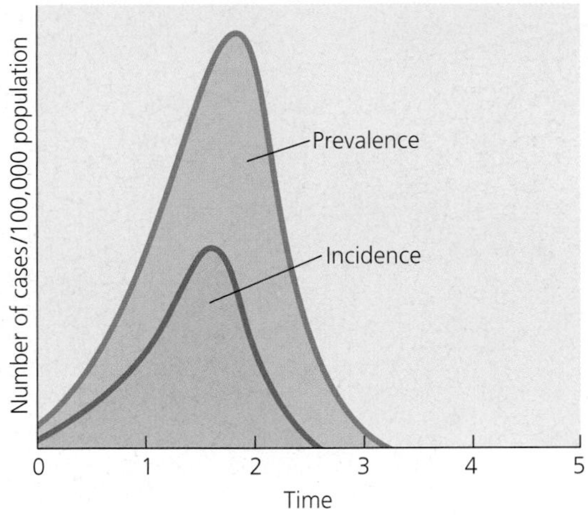

▲ *Figure 14.13*

Curves representing the incidence (orange) and the prevalence (blue) of a given disease over time. Both measures are expressed as a ratio of cases to a selected portion of the population at risk. *Why can the incidence of a disease never exceed the prevalence of that disease?*

Figure 14.13 Because prevalence includes all cases, both old and new, prevalence must always be larger than incidence.

Epidemiologists keep track of the occurrence of diseases by using two measures: incidence and prevalence. **Incidence** is the number of *new* cases of a disease in a given area or population during a given period of time; **prevalence** is the *total number* of cases, both new and already existing, in a given area or population during a given period of time. Both measures are typically expressed as a ratio of the number of cases divided by the number of people at risk:

$$\text{Incidence} = \frac{\text{No. of new cases}}{\text{No. of people at risk}}$$

$$\text{Prevalence} = \frac{\text{No. of old and new cases}}{\text{No. of people at risk}}$$

Thus, for example, the reported number of new cases of tuberculosis in the United States in 2002 was 11,878, and the population at risk was about 289,939,000, so the incidence of tuberculosis was about 4.10 people. However, the prevalence of tuberculosis in 2002 was about 6.87/100,000 people because over 8,000 patients who got the disease prior to 2002 still had the disease. **Figure 14.13** illustrates the relationships between incidence and prevalence.

Epidemiologists report their data in many ways including maps, graphs, charts, and tables (**Figure 14.14** on page 426). Why do they report their data in so many different ways? A variety of formats enables epidemiologists to observe patterns that may give clues about the causes of or ways to prevent diseases. For example, shigellosis (an acute bacterial infection characterized by diarrhea, abdominal pain, and fever) had an overall prevalence of 85 cases per million in the U.S. population in 2000, but when the data are reported by age group, it becomes apparent that the highest prevalence (350 cases per million) occurs in young children (see Figure 14.14c on p. 426).

The occurrence of a disease can also be considered in terms of a combination of frequency and geographic distribution (**Figure 14.15** on p. 427). A disease that normally occurs continually (at moderately regular intervals) at a relatively stable frequency within a given population or geographical area is said to be **endemic**[11] to that population or region. A disease is considered **sporadic** when only a few scattered cases occur within an area or population. Whenever a disease occurs at a greater frequency than is usual for an area or population, the disease is said to be **epidemic** within that area or population.

The commonly held belief that a disease must infect thousands or millions to be considered an epidemic is mistaken. The time period and the number of cases necessary for an outbreak of disease to be classified as an epidemic are not specified; the important fact is that there are more cases than historical statistics indicate are expected. For example,

[11] From Greek *endemos,* meaning native.

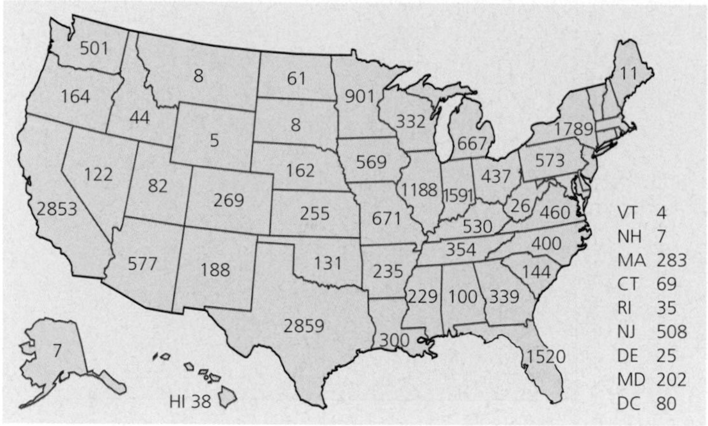

(a)

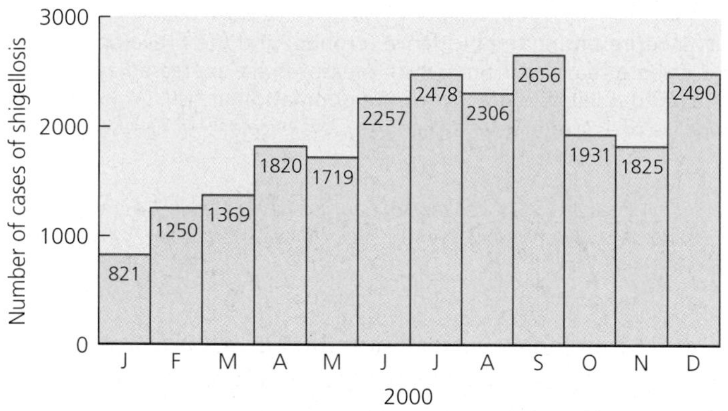

(b)

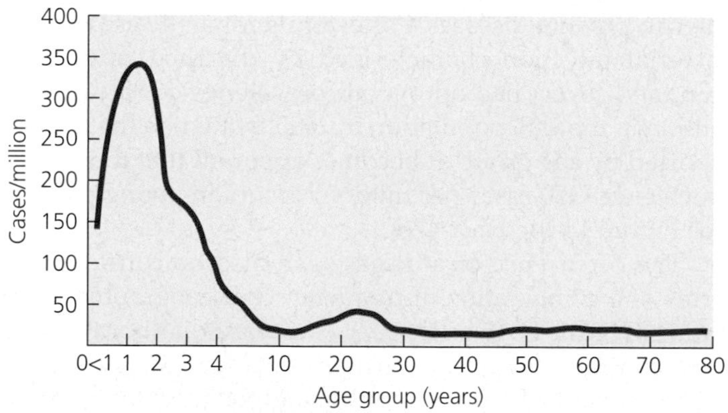

(c)

▲ ***Figure 14.14***

Epidemiologists report data in a number of ways. Here incidence of shigellosis in the U.S. in 2000 is presented **(a)** on a map by state, **(b)** by month, and **(c)** as cases per million by age group. The latter graph reveals that shigellosis in young children is much more prevalent (as high as 350 cases per million) than the prevalance in the U.S. population as a whole, a fact not apparent unless the data are sorted by age.

less than 200 cases of hantavirus pulmonary syndrome occurred in the Four Corners area[12] of the United States in 1993, but because they were the first cases of this disease ever reported for the region, the outbreak was considered an epidemic. At the same time and place, thousands of cases of flu occurred, but there was no flu epidemic because the number of cases observed did not exceed the number expected. **Figure 14.16** illustrates how epidemics are defined according to the number of expected cases, and not according to the absolute number of cases.

If an epidemic occurs simultaneously on more than one continent, it is referred to as a **pandemic** (see Figure 14.15d). AIDS is pandemic worldwide.

Obviously, for disease prevalence to be classified as either endemic, sporadic, epidemic, or pandemic, good records must be kept for each region and population. From such records, incidence and prevalence can be calculated, and then changes in these data can be noted. Health departments at the local and state levels require doctors and hospitals to report certain infectious diseases. Some are also nationally notifiable; that is, their occurrence must be reported to the Centers for Disease Control and Prevention (CDC) in Atlanta, Georgia, which is the headquarters and clearinghouse for national epidemiological research. Nationally notifiable diseases are listed in **Table 14.13** on page 429. Each week the CDC reports the number of cases of all nationally notifiable diseases in the *Morbidity and Mortality Weekly Report (MMWR)* (**Figure 14.17** on p. 428).

Epidemiological Studies

Learning Objective

✓ Explain three approaches epidemiologists use to study diseases in populations.

Epidemiologists conduct research to study the dynamics of diseases in populations by taking three different approaches, called descriptive, analytical, and experimental epidemiology.

Descriptive Epidemiology

Descriptive epidemiology involves the careful tabulation of data concerning a disease. Relevant information includes the location and time of cases of the disease, as well as information about the patients, including age, gender, occupation, health history, and socioeconomic status. Because the time course and chains of transmission of a disease are an important part of descriptive epidemiology, epidemiologists strive to identify the **index case** (the first case) of the disease in a given area or population. Sometimes it is difficult or impossible to identify the index case because the patient has recovered, moved, or died.

[12]Four Corners is the geographic region surrounding the point at which Arizona, Nevada, New Mexico, and Colorado meet.

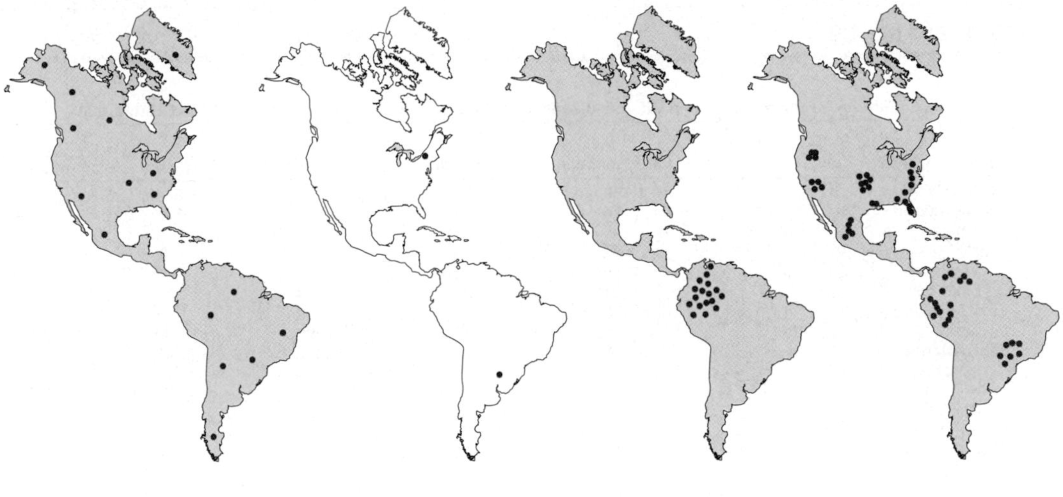

(a) Endemic disease (b) Sporadic (c) Epidemic (d) Pandemic

Figure 14.15 ➤

Illustrations of the different terms for the occurrence of disease. The colored areas indicate the normal range for the disease, and dots represent new cases of the disease. **(a)** An endemic disease, which is normally present in a region. **(b)** A sporadic disease, which occurs irregularly and infrequently. **(c)** An epidemic disease, which is present in greater frequency than is usual. **(d)** A pandemic disease, which is an epidemic disease occurring on more than one continent at a given time. *Regardless of where you live, name a disease that is endemic, one that is sporadic, and one that is epidemic in your state.*

Figure 14.15 *Some possible answers: Flu is endemic in every state; tuberculosis is sporadic in most states; AIDS is epidemic in every state.*

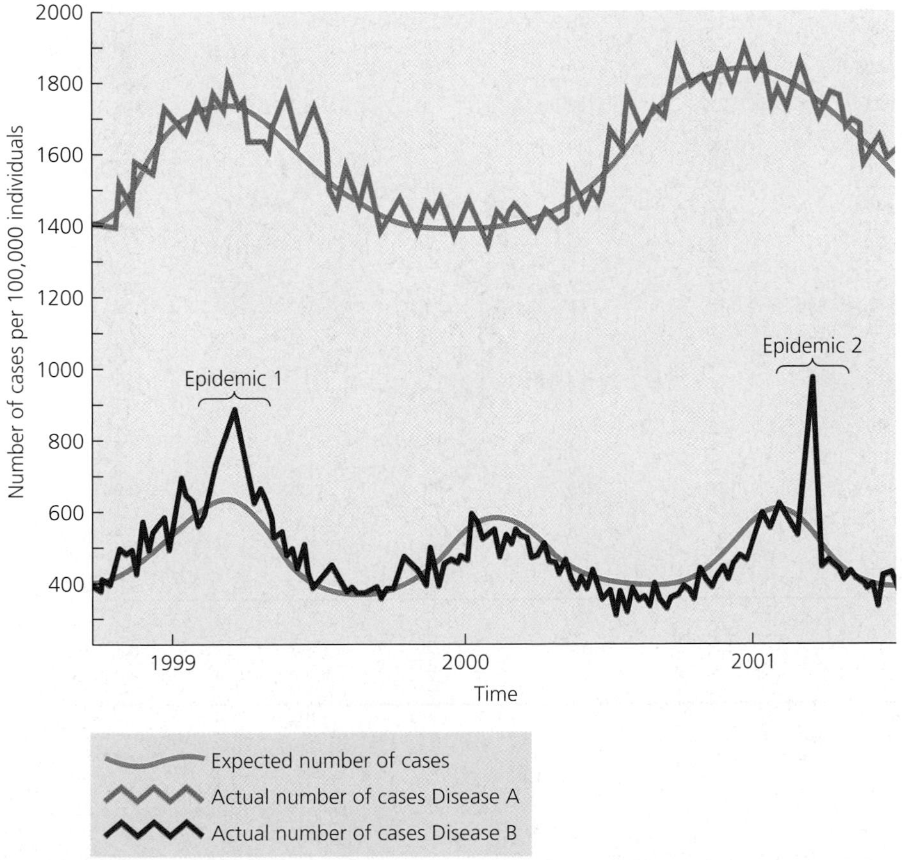

◀ ***Figure 14.16***

Two graphs that demonstrate the independence between the absolute number of cases and a disease's designation as an epidemic. Even though the number of cases of disease A always exceeds the number of cases of disease B, only disease B is ever considered epidemic—at those times when the number of cases observed exceeds the number of cases expected. *How can scientists know the normal prevalence of a given disease?*

Figure 14.16 *Scientists can record every case of the disease so that they will have a "baseline" prevalence, which then becomes the expected prevalence for each disease.*

Vol. 51 / No. 32 MMWR 727

TABLE. (*Continued*) **Reported cases of notifiable diseases, by geographic division and area — United States, 2001**

Reporting area	Legionellosis	Listeriosis	Lyme disease	Malaria	Measles		Meningococcal disease
					Indigenous	Imported*	
United States	**1,168**	**612**	**17,027**	**1,544**	**62**	**54**	**2,326**
New England	**74**	**57**	**5,526**	**107**	**4**	**1**	**113**
Maine	8	2	108	5	—	—	8
N.H.	12	4	129	2	—	—	14
Vt.	5	3	18	1	1	—	7
Mass.	21	30	1,164	53	2	1	57
R.I.	13	3	510	16	—	—	7
Conn.	15	15	3,597	30	1	—	20
Mid. Atlantic	**285**	**119**	**8,907**	**440**	**7**	**13**	**256**
Upstate N.Y.	82	36	4,020	76	—	4	72
N.Y. City	43	26	63	250	3	4	42
N.J.	24	20	2,020	65	—	1	43
Pa.	136	37	2,804	49	4	4	99
E.N. Central	**316**	**88**	**720**	**177**	**—**	**10**	**359**
Ohio	143	17	44	27	—	3	89
Ind.	23	8	26	19	—	4	47
Ill.	24	24	32	71	—	3	88
Mich.	82	25	21	40	—	—	83
Wis.	44	14	597	20	—	—	52
W.N. Central	**55**	**22**	**540**	**77**	**2**	**4**	**173**
Minn.	15	4	461	45	2	2	29
Iowa	8	2	36	9	—	—	31
Mo.	22	10	37	15	—	2	58
N. Dak.	1	—	—	—	—	—	8
S. Dak.	3	—	—	—	—	—	5
Nebr.	5	1	4	2	—	—	27
Kans.	1	5	2	6	—	—	15
S. Atlantic	**223**	**77**	**1,039**	**317**	**3**	**2**	**383**
Del.	12	NN	152	2	—	—	6
Md.	32	16	608	112	2	1	42
D.C.	8	—	17	13	—	—	—
Va.	39	15	156	55	1	—	46
W. Va	NN	6	16	1	—	—	15
N.C.	11	NA	41	19	—	—	63
S.C.	15	5	6	9	—	—	33
Ga.	12	16	—	45	—	1	57
Fla.	94	19	43	61	—	—	121
E.S. Central	**63**	**23**	**72**	**38**	**2**	**—**	**144**
Ky.	14	7	23	14	2	—	27
Tenn.	32	9	31	14	—	—	63
Ala.	13	7	10	6	—	—	35
Miss.	4	—	8	4	—	—	19
W.S. Central	**31**	**34**	**87**	**91**	**—**	**1**	**336**
Ark.	—	1	4	3	—	—	25
La.	7	—	8	6	—	—	78
Okla.	7	2	—	5	—	—	32
Tex.	17	31	75	77	—	1	201
Mountain	**57**	**38**	**15**	**68**	**1**	**1**	**102**
Mont.	—	—	—	3	—	—	4
Idaho	3	1	5	4	—	1	8
Wyo.	3	2	1	1	—	—	5
Colo.	16	9	—	25	—	—	37
N. Mex.	3	7	1	3	—	—	11
Ariz.	21	10	3	19	1	—	21
Utah	7	2	1	4	—	—	8
Nev.	4	7	4	9	—	—	8
Pacific	**64**	**154**	**121**	**229**	**43**	**22**	**460**
Wash.	10	14	9	19	13	2	71
Oreg	NN	12	15	17	3	—	63
Calif.	48	122	95	179	25	15	310
Alaska	1	—	2	1	—	—	3
Hawaii	5	6	—	13	2	5	13
Guam	—	—	—	1	—	—	—
P.R.	2	—	—	6	1	—	9
V.I.	NA	NA	NA	NA	NA	NA	NA
American Samoa	—	—	—	—	—	—	3
C.N.M.I.	—	—	—	—	—	—	—

NA: Not available NN: Not notifiable —: No reported cases

* Imported cases include only those resulting from importation from other countries.

◀ ***Figure 14.17***

A page from the CDC's *Morbidity and Mortality Weekly Report (MMWR)*, which reports epidemiological data state by state for the current week, for the current year to date, and for the previous year to date. The *MMWR* also publishes reports on epidemiological case studies. This page shows the incidence of six diseases in one week in 2001.

An early systematic, epidemiologic study was Dr. John Snow's descriptive study of a cholera outbreak in London in 1854. By carefully mapping the locations of the cholera cases in a particular part of the city, Snow found that the cases were clustered around the Broad Street water pump (**Figure 14.18** on p. 430). This distribution of cases, plus the voluminous watery diarrhea of cholera patients, suggested that the disease was spread via contamination of drinking water by sewage.

Analytical Epidemiology

Analytical epidemiology investigates a disease in detail, including analysis of data acquired in descriptive epidemiological studies, to determine the probable cause, mode of transmission, and possible means of prevention of the disease. Analytical epidemiology may be used in situations where it is not ethical to apply Koch's postulates. Thus, even though Koch's third postulate has never been fulfilled in the case of AIDS (because it is unethical to intentionally inoculate a human with HIV), analytical epidemiological studies indicate that HIV causes AIDS and that it is transmitted primarily sexually.

Often analytical studies are *retrospective;* that is, they attempt to identify causation and mode of transmission after an outbreak has occurred. Epidemiologists compare a group of people who had the disease with a group who did not. The groups are carefully matched by factors such as gender, environment, and/or diet, and then compared to determine which pathogens and factors may play a role in morbidity.

Experimental Epidemiology

Experimental epidemiology involves testing a hypothesis concerning the cause of a disease. The application of Koch's postulates is an example of experimental epidemiology. Experimental epidemiology also involves studies to test a hypothesis resulting from an analytical study such as the efficacy of a preventative measure or certain treatment. For example, analytical epidemiological studies have suggested that the bacterium *Chlamydia pneumoniae* causes arteriosclerosis, resulting in heart attacks. Some scientists have hypothesized that antibacterial drugs could prevent heart attacks by killing the causative agent. To test this hypothesis the scientists administered either the antimicrobial drug azithromycin or a medicinally inactive placebo to 7,700 patients, all of whom had a history of heart disease and were infected with *C. pneumoniae.* The researchers observed no significant difference in the number of heart attacks suffered by patients in the two groups over a two-year period. Thus an experimental epidemiological study disproved a hypothesis suggested by an analytical epidemiological analysis.

Case Study 14.3 on page 431 illustrates the work of epidemiologists.

Table 14.13 Nationally Notifiable Infectious Diseases[a]

Acquired immunodeficiency syndrome (AIDS)
Anthrax
Botulism
Brucellosis
Chancroid
Chlamydia trachomatis, genital infections
Cholera
Coccidioidomycosis
Cryptosporidiosis
Cyclosporiasis
Diphtheria
Ehrlichiosis
Encephalitis/meningitis, arboviral
Enterohemorrhagic *E. coli*
Giardiasis
Gonorrhea
Haemophilus influenzae, invasive disease
Hansen disease (leprosy)
Hantavirus pulmonary syndrome
Hemolytic uremic syndrome, postdiarrheal
Hepatitis A
Hepatitis B
Hepatitis C
HIV infection
Legionellosis
Listeriosis
Lyme disease
Malaria
Measles
Meningococcal disease
Mumps
Pertussis
Plague
Poliomyelitis, paralytic
Psittacosis
Q fever
Rabies, animal
Rabies, human
Rocky Mountain spotted fever
Rubella
Rubella, congenital syndrome
Salmonellosis
Shigellosis
Streptococcal disease, invasive, Group A
Streptococcal toxic shock syndrome
Streptococcus pneumoniae, drug resistant
Streptococcus pneumoniae, invasive in children
Syphilis
Syphilis, congenital
Tetanus
Toxic-shock syndrome
Trichinosis
Tuberculosis
Tularemia
Typhoid fever
Varicella (deaths only)
West Nile encephalitis/meningitis
Yellow fever

[a]Diseases for which hospitals, physicians, and other health care workers are required to report cases to state health departments, who then forward the data to the CDC.

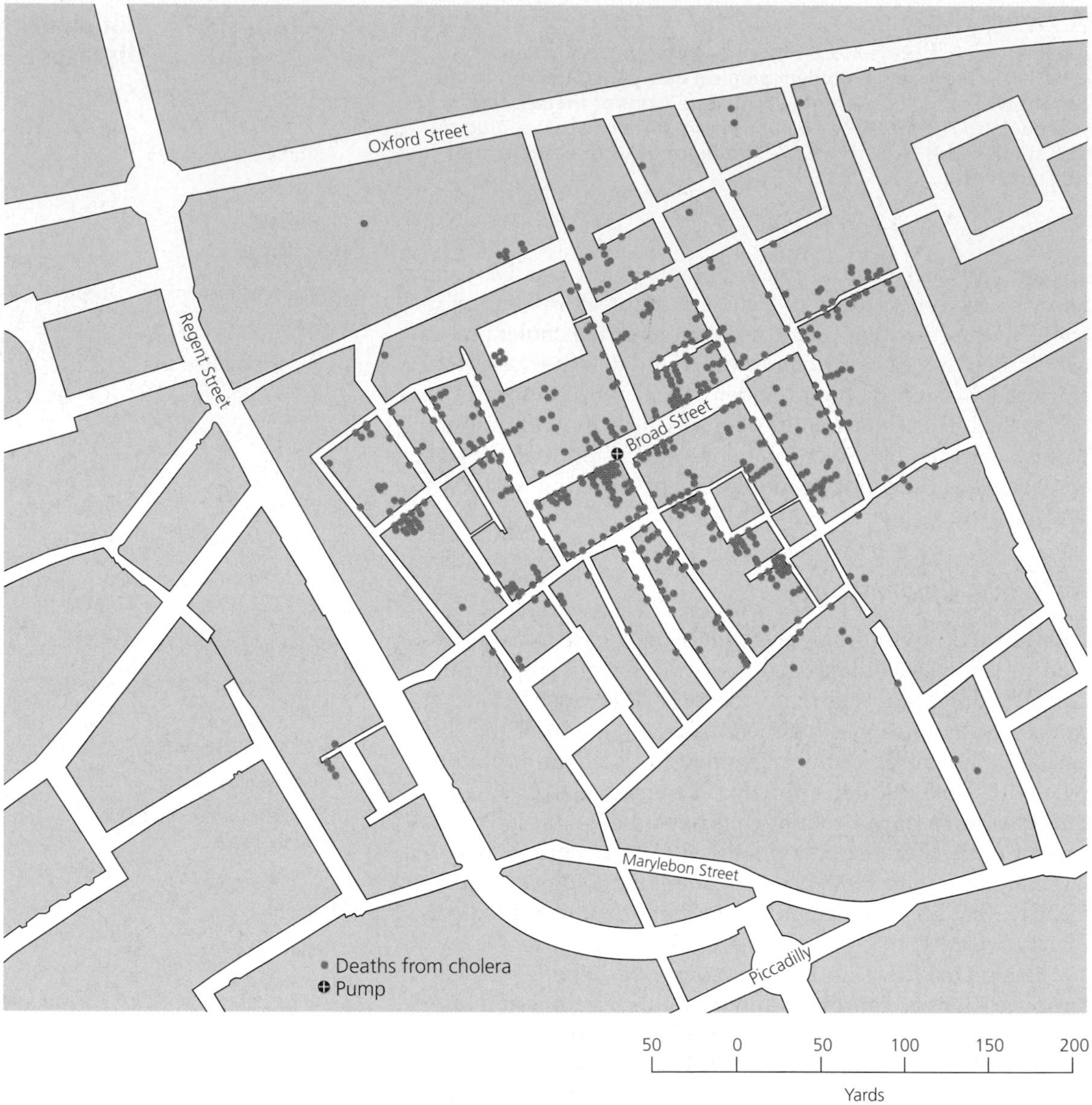

Figure 14.18 ➤ A map showing cholera deaths in a section of London, 1854. From the map he compiled, Dr. John Snow correctly deduced that the Broad Street pump was the source of this cholera epidemic. Snow's work was a landmark in epidemiologic research.

Hospital Epidemiology: Nosocomial Infections

Learning Objectives

- ✓ Explain how nosocomial infections differ from other infections.
- ✓ Describe the factors that influence the development of nosocomal infections.
- ✓ Describe three types of nosocomial infections and how they may be prevented.

Of special concern to epidemiologists and health care workers are nosocomial[13] (nos-ō-kō′mē-ăl) infections and nosocomial diseases. **Nosocomial infections** are those acquired by patients while they are in health care facilities, including hospitals, dental offices, nursing homes, and doctor's waiting rooms. The CDC estimates that about 10% of American patients—over 2 million people—acquire a nosocomial infection each year. **Nosocomial diseases**—those acquired in a health care setting—increase the duration and cost of medical care and result in some 90,000 deaths annually.

Types of Nosocomial Infections

When most people think of nosocomial infections, what likely comes to mind are **exogenous** (eks-oj′ĕ-nŭs) infections, which are caused by pathogens acquired from the health care environment. After all, hospitals are filled with sick people shedding pathogens from every type of portal of exit. However, we have seen that members of the normal microbiota can become opportunistic pathogens as a result of hospitalization or medical treatments such as chemotherapy. Such opportunists cause **endogenous** (en-doj′ĕ-nŭs) nosocomial infectious; that is, they arise from normal microbiota within the patient that become pathogenic because of factors within the health care setting.

[13]From Greek *nosokomeian,* meaning hospital.

Case Study 14.3 *Legionella* in the Produce Aisle

On October 31, 1989, the Lousiana state health department received reports of 33 cases of Legionnaire's disease in the town of Bogalusa (population 16,000). Legionnaire's disease, or legionellosis, is a potentially fatal respiratory disease caused by the growth of a bacterium, *Legionella pneumophila,* in the lungs of patients. The bacterium enters humans via the respiratory portal in aerosols produced by cooling towers, air conditioners, whirlpool baths, showers, humidifiers, and respiratory therapy equipment.

Epidemiologists began trying to ascertain the source of Bogalusa's Legionnaire's disease outbreak by interviewing the victims and their relatives to develop complete histories, and to identify areas of commonality among the victims that were lacking among nonvictims. Victims reported a range of ages, occupations, hobbies, religions, and types and locations of dwellings. No significant differences were identified among the lifestyles, ages, or smoking habits of victims and nonvictims, but one curious fact was discovered: All the victims did their grocery shopping at the same store. However, healthy individuals also shopped at that store.

The air conditioning system of the grocery store proved to be free of *Legionella,* but the vegetable misting machine did not. The strain of *Legionella* isolated from the misting machine was identical to the strain recovered from the lungs of the victims.

1. *Would this outbreak be classified as endemic, epidemic, or pandemic?*
2. *Was this a descriptive, analytical, or experimental epidemiological study?*
3. *Knowing the epidemiology and causative agent of Legionnaire's disease, what questions would you have asked of the victims or of their surviving relatives?*
4. *What, as an epidemiologist, would you have examined at the store?*
5. *How did the victims become contaminated? Why didn't everyone who bought vegetables at the store get legionellosis? What could the owners of the store do to limit or prevent future infections?*

Source: *Morbidity and Mortality Weekly Report 39*(7):108–109. 1990.

Iatrogenic (ī-at-rō-jen′ik) **infections** (literally meaning "doctor induced" infections) are a subset of nosocomial infections that ironically are the direct result of modern medical procedures such as the use of catheters, invasive diagnostic procedures, and surgery.

Superinfections may also result from the use of antimicrobial drugs that, by inhibiting some resident microbiota, allow others to thrive in the absence of competition. For instance, long-term antimicrobial therapy to inhibit a bacterial infection may allow *Clostridium difficile* (klos-trid′ē-ŭm dif′fi-sil), a transient microbe of the colon, to grow excessively and cause a painful condition called pseudomembranous colitis. Such superinfections are not limited to health care settings.

Nosocomial infections most often occur in the urinary, respiratory, cardiovascular, and integumentary (skin) systems, though surgical wounds can become infected and result in nosocomial infections in any part of the body.

Factors Influencing Nosocomial Infections

Nosocomial infections arise from the interaction of several factors in the health care environment, which include the following (**Figure 14.19** on p. 432):

- Exposure to numerous pathogens present in the health care setting, including many that are resistant to antimicrobial agents
- The weakened immune systems of patients who are ill, making them more susceptible to opportunistic pathogens
- Transmission of pathogens among patients, and health care workers—from staff and visitors, to patients, and even from one patient to another via activities of staff members (including invasive procedures and other iatrogenic factors)

Control of Nosocomial Infections

Reduction in the incidence of nosocomial infections involves procedures and precautions designed to reduce the influencing factors, especially the transfer of pathogens within the health care setting. These include disinfection; medical asepsis (including good housekeeping, hand asepsis, bathing, sanitary handling of food, proper hygienic attitudes toward bodily functions, and precautionary measures to avoid the spread of pathogens among patients); surgical asepsis and sterile procedures (including thorough cleansing of the surgical field, use of sterile instruments, and use of sterile gloves, gowns, caps, and masks); isolation of particularly contagious and particularly susceptible patients; and establishment of a nosocomial infection control committee charged with surveillance of nosocomial diseases and review of control measures.

The CDC has instituted the following Universal Precautions to limit occupational exposure to pathogens and reduce the transfer of pathogens to patients:

- Do not sheath used needles, scalpels, or other sharp instruments. Instead, immediately dispose of them in proper "sharps" containers.
- Use protective barriers to prevent exposure to blood, body fluids containing visible blood, and other fluids such as saliva and phlegm.
- Immediately and thoroughly wash hands and other skin surfaces that are contaminated by blood or other fluids.

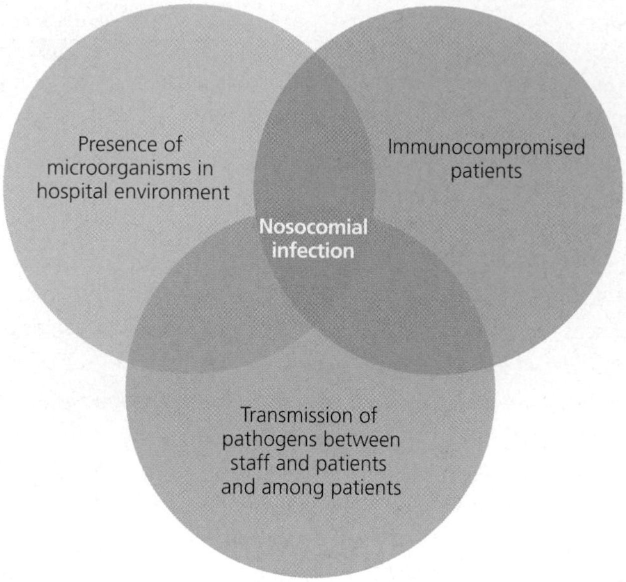

▲ ***Figure 14.19***
The interplay of factors that result in nosocomial infections. Although nosocomial infections can result from any one of the factors shown, most nosocomial infections are the product of the interaction of all three factors.

Numerous studies have shown that the single most effective way to reduce nosocomial infections is effective hand antisepsis by all medical and support staff. In one study, deaths from nosocomial infections were reduced by over 50% when hospital personnel followed strict guidelines about washing their hands frequently.

Epidemiology and Public Health

Learning Objective

✓ List three ways public health agencies work to limit the spread of diseases.

As you have likely realized by now, epidemiologists gather information concerning the spread of disease within populations so that they can take steps to reduce the number of cases and improve the health of individuals within a community. In the following sections we will examine how the various public health agencies share epidemiological data, facilitate the interruption of disease transmission, and educate the public about public health issues. Public health agencies also implement immunization programs; immunization is discussed in Chapter 17.

The Sharing of Data Among Public Health Organizations

Numerous agencies at the local, state, national, and global levels work together with the entire spectrum of health care personnel to promote public health. By submitting reports on cases of disease to public officials, practicing physicians can subsequently learn of current disease trends. Additionally, public health agencies often provide physicians with laboratory and diagnostic assistance.

City and county health departments report data on disease incidence to state agencies. Because state laws govern disease reporting, state agencies play a vital role in epidemiological studies. Most states publish a bulletin similar to the MMWR and assist local health departments and medical practitioners with diagnostic testing for diseases such as rabies and Lyme disease.

Data collected by the states is forwarded to the CDC, which is but one branch of the United States Public Health Service, our national public health agency. In addition to epidemiological studies, the CDC and other branches of the Public Health Service conduct research in disease etiology and prevention, make recommendations concerning immunization schedules, and work with public health organizations of other countries.

The World Health Organization (WHO) coordinates efforts to improve public health throughout the world, particularly in poorer countries, and the WHO has undertaken ambitious projects to eradicate such diseases as smallpox, polio, measles, and mumps. Some other current campaigns involve AIDS education, malaria control, and childhood immunization programs in poor countries.

The Role of Public Health Agencies in Interrupting Disease Transmission

As we have seen, pathogens can be transmitted in air, food, and water, as well as by vectors and via fomites. Public health agencies work to limit disease transmission by enforcing standards of cleanliness in water and food supplies, and by working to reduce the numbers of vectors and reservoirs available for transmitting diseases.

A water supply that is **potable** (fit to drink) is vital to good health. Organisms that cause dysentery, cholera, and typhoid fever are just some of the pathogens that can be spread in water contaminated by sewage. Filtration and chlorination processes are used to reduce the number of pathogens in water supplies. Local, state, and national agencies work to ensure that water supplies remain clean and healthful by monitoring both sewage treatment facilities and the water supply.

Food can harbor infective larvae of parasitic worms, infective stages of protozoa, bacteria, and viruses. National and state health officials ensure the safety of the food supply by enforcing standards in the use of canning, pasteurization, irradiation, and chemical preservatives, and by insisting that food preparers and handlers wash their hands and use sanitized utensils. The Department of Agriculture also provides for the inspection of meats for the presence of pathogens such as *E. coli* and *Trichinella* (trik'i-nel'ă).

Milk is an especially rich food, and the same nutrients that nourish us also facilitate the growth of many microorganisms. In the past, contaminated milk has been responsible for epidemics of tuberculosis, brucellosis, typhoid fever, scarlet fever, and diphtheria. Today, public health agencies require the pasteurization of milk, and as a result disease transmission via milk has been practically eliminated in the United States.

Individuals should assume responsibility for their own health by washing their hands before and during food preparation, using disinfectants on kitchen surfaces, and by using proper refrigeration and freezing procedures. It is also important to thoroughly cook all meats.

Public health officials also work to control vectors, especially mosquitoes and rodents, by eliminating breeding grounds such as stagnant pools of water and garbage dumps. Insecticides have been used with some success to control insects but fail in the long term because of the development of resistance.

Public Health Education

Diseases that are transmitted through the air or sexually are particularly difficult for public health officials to control. In these cases, individuals must take responsibility for their own health, and health departments can only educate the public to make healthful choices.

Colds and flu remain the most common diseases in the United States because of the ubiquity of the viral pathogens and their mode of transmission in aerosols and via fomites. Health departments can encourage afflicted people to remain at home, use disposable tissues to reduce the spread of viruses, and avoid crowds of coughing, sneezing people. As we have discussed, hand washing is also important in preventing the introduction of cold and flu viruses onto the conjunctiva.

Sexually transmitted diseases such as AIDS, syphilis, gonorrhea, and genital warts are completely preventable if the chain of transmission is interrupted by abstinence or mutually faithful monogamy, and their incidence can be reduced by the use of barrier contraceptives (*e.g.* condoms). Still, we as a society are faced with several epidemics of sexually transmitted diseases. Based on the premise that "an ounce of prevention is worth a pound of cure," public health agencies expend considerable effort in public campaigns to educate individuals to make good choices—those that can result in healthier individuals and greater health for the public at large.

CHAPTER SUMMARY

Symbiotic Relationships Between Microbes and Their Hosts (pp. 405–409)

1. Microbes live with their hosts in **symbiotic** relationships, including **mutualism,** in which both members benefit; **parasitism,** in which a **parasite** benefits while the host is harmed; and **commensalism,** in which one member benefits while the other is relatively unaffected. Any parasite that causes disease is called a **pathogen.**
2. Organisms called **normal microbiota** live in and on the body. Some of these microbes are resident, whereas others are transient.
3. **Opportunistic pathogens** cause disease when the immune system is suppressed, when normal **microbial antagonism (competition)** is affected by certain changes in the body, and when normal microbiota are introduced into an area of the body unusual for that microbe.

The Movement of Microbes into Hosts: Infection (pp. 409–411)

1. Microbial **contamination** refers to the mere presence of microbes in or on the body or object. Microbial contaminants include harmless resident and transient members of the microbiota, as well as pathogens, which after a successful invasion cause an **infection.**
2. **Portals of entry** of pathogens into the body include skin, mucous membranes, the placenta, and the **parenteral route,** by which microbes are directly deposited into deeper tissues.
3. Pathogens can attach to cells—a process called **adhesion**—via a variety of structures or attachment proteins called **adhesion factors.** Some bacteria and viruses lose the ability to make adhesion factors called adhesins and thereby become **avirulent.**
4. Some bacteria interact to produce a sticky web of cells and polysaccharides called a biofilm that adheres to a surface.

The Nature of Infectious Disease (pp. 411–419)

1. **Disease,** also known as **morbidity,** is a sufficiently adverse condition to interfere with normal functioning of the body.
2. **Symptoms** are subjectively felt by a patient, whereas an outside observer can observe **signs.** A **syndrome** is a group of symptoms and signs that collectively characterizes a particular disease or abnormal condition.
3. **Asymptomatic** or **subclinical** infections may go unnoticed due to the absence of symptoms, even though clinical tests might reveal signs of disease.
4. **Etiology** is the study of the cause of a disease.
5. Nineteenth-century microbiologists proposed the **germ theory of disease,** and Robert Koch developed a series of essential conditions called **Koch's postulates** to prove the cause of infectious diseases. Certain circumstances can make the use of these postulates difficult or even impossible.
6. **Pathogenicity** is a microorganism's ability to cause disease; **virulence** is a measure of pathogenicity. **Virulence factors** such as adhesion factors, extracellular enzymes, toxins, and antiphagocytic factors affect the relative ability of a pathogen to infect and cause disease.
7. **Toxemia** is the presence in the blood of poisons called **toxins. Exotoxins** are secreted by pathogens into their environment. **Endotoxin,** also known as **lipid A,** is released from the cell wall of dead and dying Gram-negative bacteria and can cause fatal effects.
8. **Antitoxins** are antibodies the host forms against toxins.

9. The stages of infectious diseases are **incubation period, prodromal period, illness, decline,** and **convalescence.**

The Movement of Pathogens Out of Hosts: Portals of Exit (pp. 419–421)

1. **Portals of exit,** such as the nose, mouth, and urethra, allow pathogens to leave the body and are of interest in studying the spread of disease.
2. Living and nonliving continuous sources of infectious disease are called **reservoirs of infection.** Animal reservoirs harbor agents of **zoonotic diseases (zoonoses),** which are diseases of animals that may be spread to humans via direct contact with the animal or its waste products, or via an arthropod **vector.** Humans may be asymptomatic **carriers.**
3. **Nonliving reservoirs** of infection include soil, water, and inanimate objects.

Modes of Infectious Disease Transmission (pp. 421–423)

1. **Direct contact transmission** of infectious diseases involves person-to-person spread by bodily contact. **Indirect contact transmission** occurs when inanimate objects (called **fomites**) inadvertently transmit pathogens.
2. **Droplet transmission** (a third type of contact transmission) occurs when pathogens travel in droplets of mucus less than 1 meter to a new host as a result of speaking, coughing, or sneezing.
3. **Vehicle transmission** involves **airborne, waterborne,** and **food-borne** transmission. **Aerosols** are clouds of water droplets that travel more than 1 meter in airborne transmission. **Fecal-oral infection** can result from sewage-contaminated drinking water or from ingesting fecal contaminents.
4. **Biological vectors** are animals, usually biting arthropods, that serve as both host and vector of pathogens. **Mechanical vectors** are not infected by the pathogens they carry.

Classification of Infectious Diseases (pp. 423–424)

1. There are various ways in which infectious disease may be grouped and studied. When grouped by time course and severity, disease may be described as **acute, subacute, chronic,** or **latent.**
2. When an infectious disease comes either directly or indirectly from another host, it is considered a **communicable** disease. If a communicable disease is easily transmitted from a reservoir or patient, it is called a **contagious** disease. **Noncommunicable** diseases arise either from outside of hosts or from normal microbiota.

Epidemiology of Infectious Diseases (pp. 424–433)

1. **Epidemiology** is the study of where and when diseases occur, and how they are transmitted within populations.
2. Epidemiologists track the **incidence** and **prevalence** of a disease, and classify disease outbreaks as **endemic** (usually present), **sporadic** (occasional), **epidemic** (more cases than usual), or **pandemic** (epidemic on more than one continent). The Centers for Disease Control and Prevention (CDC) report these findings weekly in the *Morbidity and Mortality Weekly Report (MMWR).*
3. **Descriptive epidemiology** is the careful recording of data concerning a disease; it often includes detection of the **index case**—the first case of the disease in a given area or population. **Analytical epidemiology** seeks to determine the probable cause of a disease. **Experimental epidemiology** involves testing a hypothesis resulting from analytical studies.
4. **Nosocomial infections** and **nosocomial diseases** are acquired by patients in health care facilities. They may be **exogenous** (acquired from the health care environment), **endogenous** (derived from normal microbiota that become opportunistic while in the hospital setting), or **iatrogenic** (induced by treatment or medical procedures).
5. Health care workers can protect themselves against occupational exposure to pathogens by following the CDC's Universal Precautions.
6. Public health organizations such as the World Health Organization (WHO) use epidemiological data to promulgate rules and standards for clean, **potable** water and safe food, to prevent disease by controlling vectors and animal reservoirs, and to educate people to make healthful choices concerning the prevention of disease.

QUESTIONS FOR REVIEW

(Answers to multiple choice and fill in the blanks are on the web, along with additional review questions. Visit www.microbiologyplace.com.)

Multiple Choice

1. In which type of symbiosis do both members benefit from their interaction?
   a. mutualism
   b. parasitism
   c. commensalism
   d. pathogenesis
2. An axenic environment is one that
   a. exists in the human mouth.
   b. is free of microbes.
   c. exists in the human colon.
   d. both a and b
3. Which of the following is *false* concerning microbial contaminants?
   a. Contaminants may become opportunistic pathogens.
   b. Most microbial contaminants will eventually cause harm.
   c. Contaminants may be a part of the transient microbiota.
   d. Contaminants may be introduced in a mosquito bite.
4. The most frequent portal of entry for pathogens is
   a. the respiratory tract.
   b. the skin.
   c. the conjunctiva.
   d. a cut or wound.

5. The process by which microorganisms attach themselves to cells is
   a. infection.
   b. contamination.
   c. disease.
   d. adhesion.
6. When pathogenic bacterial cells lose the ability to make adhesins, they
   a. become avirulent.
   b. produce endotoxin.
   c. absorb endotoxin.
   d. increase in virulence.
7. Which of the following are most likely to cause disease?
   a. opportunistic pathogens in a weakened host
   b. pathogens lacking the enzyme kinase
   c. pathogens lacking the enzyme collagenase
   d. highly virulent organisms
8. The nature of bacterial capsules
   a. causes widespread blood clotting.
   b. allows phagocytes to readily engulf these bacteria.
   c. affects the virulence of these bacteria.
   d. has no effect on the virulence of bacteria.
9. Which of the following is the correct sequence of events in infectious diseases?
   a. incubation, prodromal period, illness, decline, convalescence
   b. incubation, decline, prodromal period, illness, convalescence
   c. prodromal period, incubation, illness, decline, convalescence
   d. convalescence, prodromal period, incubation, illness, decline
10. A disease in which a pathogen remains inactive for a long period of time before becoming active is termed a(n)
    a. subacute disease.
    b. acute disease.
    c. chronic disease.
    d. latent disease.
11. Which of the following statements is the best definition of a pandemic disease?
    a. It normally occurs in a given geographic area.
    b. It is a disease that occurs more frequently than usual for a geographical area or group of people.
    c. It occurs infrequently at no predictable time scattered over a large area or population.
    d. It is an epidemic that occurs on more than one continent at the same time.
12. Which of the following types of epidemiologists is most like a detective?
    a. a descriptive epidemiologist
    b. an analytical epidemiologist
    c. an experimental epidemiologist
    d. a reservoir epidemiologist
13. Consider the following case. An animal was infected with a virus. A mosquito bit the animal, was contaminated with the virus, and proceeded to bite and infect a person. Which was the vector?
    a. animal
    b. virus
    c. mosquito
    d. person
14. A patient contracted athlete's foot after long-term use of a medication. His physician explained that the malady was directly related to the medication. Such infections are termed
    a. nosocomial infections.
    b. exogenous infections.
    c. iatrogenic infections.
    d. endogenous infections.
15. Which of the following phrases describes a contagious disease?
    a. a disease arising from fomites
    b. a disease that is easily passed from host to host in aerosols
    c. a disease that arises from opportunistic, normal microbiota
    d. both a and b

Fill in the Blanks

1. A parasite that causes disease is called a ________.
2. Infections that may go unnoticed due to the absence of symptoms are called ________ infections.
3. The study of the cause of a disease is ________.
4. The study of where and when diseases occur and how they are transmitted within populations is ________.
5. Diseases that are naturally spread from their usual animal hosts to humans are called ________.
6. Nonliving reservoirs of disease, such as a toothbrush, drinking glass, or needle, are called ________.
7. ________ infections are those acquired by patients and staff while in health care facilities.

Short Answer

1. List three types of symbiotic relationships, and give an example of each.
2. List three conditions that create opportunities for pathogens to become harmful in a human.
3. List three portals through which pathogens may enter the body.
4. List Koch's four postulates, and describe situations in which they may not be applicable.
5. List in the correct sequence the five stages of infectious diseases.
6. Describe three modes of disease transmission.

CRITICAL THINKING

1. Explain why patient Q, a menopausal woman, may have developed gingivitis from normal microbiota.

2. Patient W died of *E. coli* infection after an intestinal puncture. Explain why this microbe, which normally lives in the colon, could kill this patient.

3. Examine the following graphs of the RED epidemic and the BLUE epidemic. Neither RED disease nor BLUE disease is treatable. A person with either disease is ill for only 1 day and recovers fully. Both epidemics began at the same time. Which epidemic affected more people during the first 3 days? What could explain the short time course for epidemic RED? Why was epidemic BLUE longer lasting?

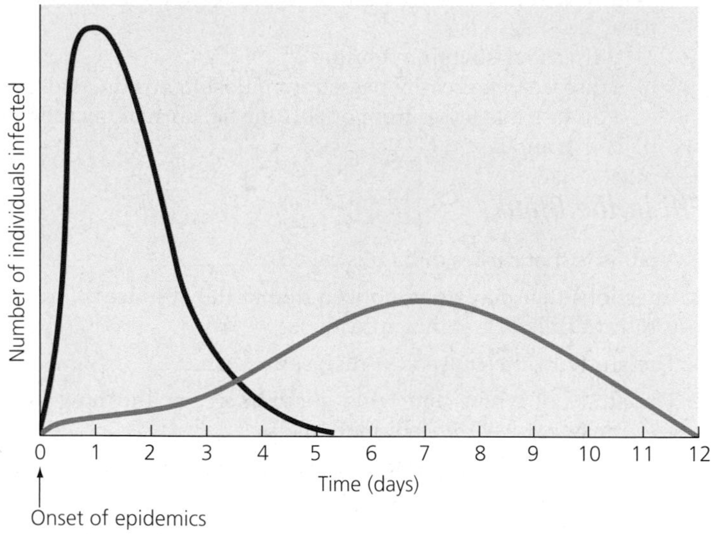

4. A 27-year-old female came to her doctor's office with a widespread rash, fever, malaise, and severe muscle pain. The symptoms had begun with a mild headache 3 days previously. She reported being bitten by a tick 1 week prior to that.

   The doctor correctly diagnosed Rocky Mountain spotted fever (RMSF) and prescribed tetracycline. The rash and other signs and symptoms disappeared in a couple of days, but she continued on her antibiotic therapy for 2 weeks.

   Draw a graph showing the course of disease and the relative numbers of pathogens over time. Label the stages of the disease.

5. Over 30 children younger than 3 years of age developed gastroenteritis after visiting a local water park. These cases represented 44% of the park visitors in this age group on the day in question. No older individuals were affected. The causative agent was determined to be a member of the bacterial genus *Shigella*. The disease resulted from oral transmission to the children.

   Based only on the information given, can you classify this outbreak as an epidemic? Why or why not? If you were an epidemiologist, how would you go about determining which pools in the water park were contaminated? What factors might account for the fact that no older children or adults developed disease? What steps could the park operators take to reduce the chance of future outbreaks of gastroenteritis?

6. Review lichen biology in Chapter 12. Describe the relationship between the photosynthetic member of the lichen and the fungus in terms of the type of symbiosis.

CHAPTER 15

Nonspecific Lines of Defense

Each day the equivalent of a small room full of air enters your respiratory tract through your nose. With that air come dust, smoke, bacteria, viruses, fungi, pollen, soot, fuzz, sand, and more. Acting as a first line of defense, your respiratory mucous membrane uses nose hairs, ciliated epithelium, and mucus to cleanse the inhaled air of pathogens and certain harmful pollutants. Each day you swallow and subsequently digest about a liter of mucus, along with the trapped pathogens and pollutants it contains. Still more mucus clumps around microbes and pollutants to form masses of nasal mucus that may dry out or remain slimy, depending on how rapidly you're breathing and the humidity of the air. Yellowish or greenish mucus contains a large population of trapped bacteria and their waste products.

Mucus is just one example of the body's general, nonspecific defenses against pathogens. The skin, and certain protective cells, chemicals, and processes within the body, are others. In this chapter we will focus on each of these aspects of the body's defenses.

CHAPTER OUTLINE

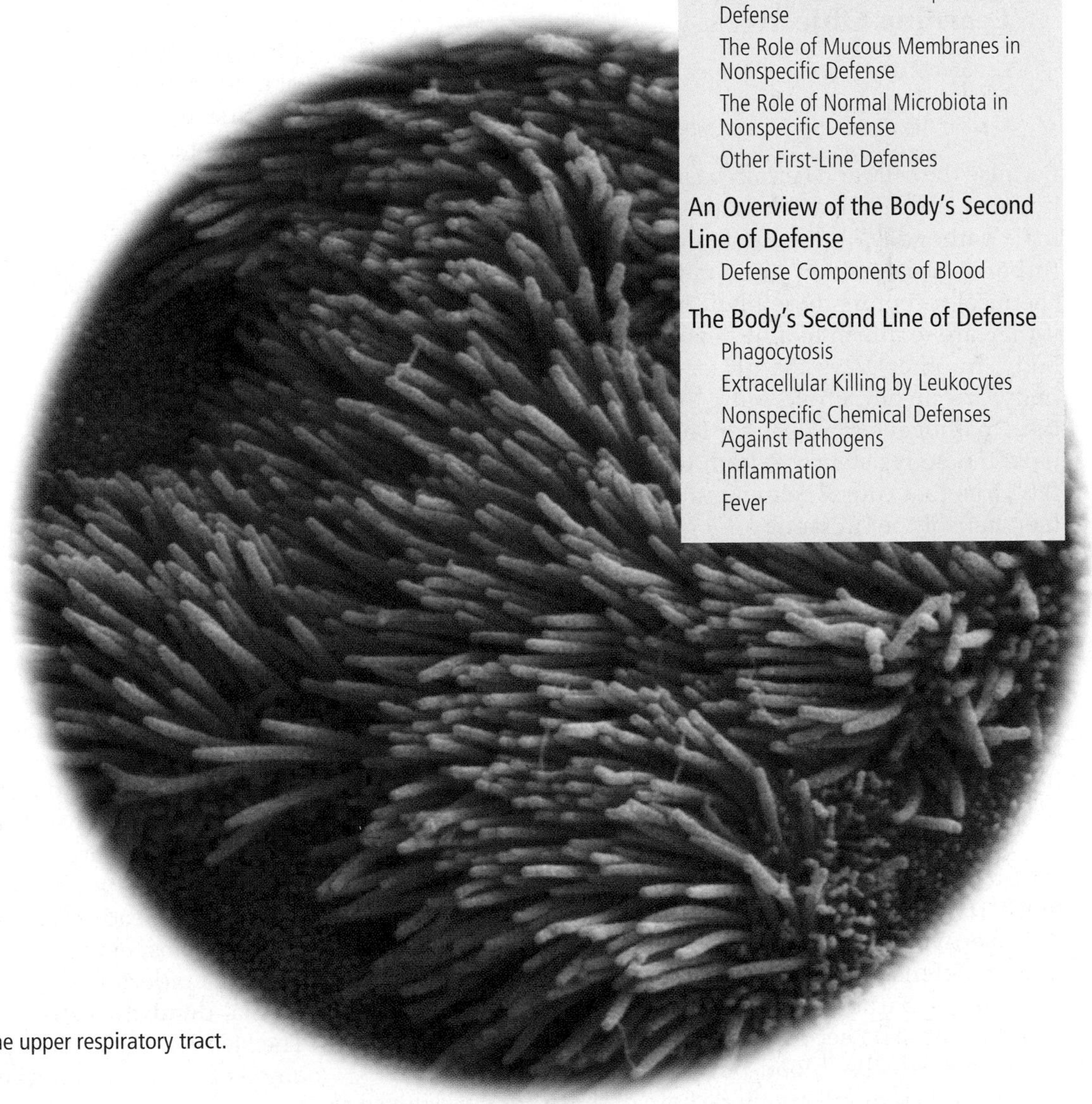

Cilia line the upper respiratory tract.

MicroPrep Pre-Test: *Take the pre-test for this chapter on the web. Visit **www.microbiologyplace.com.***

As we saw in Chapter 14, a pathogen can cause a disease only if it can (1) gain access, either by penetrating the surface of the skin or by entering through some other portal of entry; (2) attach itself to host cells; and (3) evade the body's defense mechanisms long enough to produce harmful changes. In this chapter we will examine the structures, processes, and chemicals that respond in a general way to protect the body from all types of pathogens. Chapter 16 examines the body's specific defense against specific pathogens.

An Overview of the Body's Defenses

Learning Objectives

✓ List and briefly describe the three lines of defense in the human body.

✓ Explain the phrase *nonspecific lines of defense*.

Because the cells and certain basic physiological processes of humans are incompatible with those of most plant and animal pathogens, humans have what is termed **innate (inborn) resistance** to these pathogens. In many cases the chemical receptors these plant or animal pathogens require if they are to attach to a host cell do not exist in the human body; in other cases the pH or temperature of the human body are incompatible with the conditions under which these pathogens can survive. Thus, for example, all humans have innate resistance to both tobacco mosaic virus and to the virus that causes feline immunodeficiency syndrome in members of the cat family.

However, we are confronted every day with a multitude of pathogens that can cause disease in humans. Bacteria, viruses, fungi, protozoa, and parasitic worms come in contact with your body in the air you breathe, the water you drink, the food you eat, and during the contacts you have with other people. Your body must defend itself from these potential pathogens, and in some cases from members of the normal microbiota, which may become opportunistic pathogens.

It is convenient to cluster the structures, cells, and chemicals that act against pathogens into three main lines of defense **(Figure 15.1).** In actuality, however, each line of defense overlaps and reinforces the other two. The first line of defense is chiefly composed of external physical barriers to pathogens, especially the skin and mucous membranes. The second line is internal and is composed of protective cells, blood-borne chemicals, and processes that inactivate or kill invaders. Both of these lines of defense are **nonspecific;** that is, they respond against a wide variety of pathogens, including parasitic worms, protozoa, fungi, bacteria, and viruses. In other words, nonspecific defenses operate in a generalized way independent of the nature of the intruder.

By contrast, the third line of defense is a *specific immune response* that responds against a specific type (species or strain) of pathogen. We will examine immune responses in some detail in Chapter 16, but first we will turn our attention to the two lines of nonspecific defense, the topics of this chapter.

The Body's First Line of Defense

The body's initial line of defense is made up of structures, chemicals, and processes that work together to prevent pathogens from entering the body in the first place. Here we discuss the main components of the first line of defense: the skin and the mucous membranes of the respiratory, digestive, urinary, and reproductive systems. These structures provide a formidable barrier to the entrance of microorganisms. As we discussed in Chapter 14, if these barriers are pierced, broken, or otherwise damaged, they become portals of entry for pathogens. In this section we examine aspects of the first line of defense, including the role of the normal microbiota.

The Role of Skin in Nonspecific Defense

Learning Objective

✓ Identify the physical and chemical aspects of skin that enable it to prevent the entrance of pathogens.

The skin—the organ of the body with the greatest surface area—is composed of two major layers: an outer **epidermis,** and a deeper **dermis,** which contains hair follicles, glands, and nerve endings **(Figure 15.2).** Both the physical structure and the chemical components of skin enable it to act as an effective defense.

The epidermis is composed of multiple layers of tightly packed cells separated by very little extracellular material. It constitutes a physical barrier to most bacteria, fungi, and viruses; very few pathogens can penetrate the layers of epidermal cells unless the skin has been burned, broken, or cut.

The deepest cells of the epidermis continually divide, which pushes their daughter cells toward the surface. As they are pushed toward the surface, the daughter cells flatten and die, and are eventually shed in flakes. Microorganisms that attach to the skin's surface are sloughed off with the flakes of dead cells. **Highlight 15.1** on page 440 describes the fate of lost epidermal cells.

The epidermis also contains phagocytic cells called epidermal **dendritic cells** or *Langerhans* (lahng'ĕr-hahnz) *cells.* The slender, finger-like processes of dendritic cells extend among the surrounding cells, forming an almost continuous network to intercept invaders. These cells both phagocytize pathogens nonspecifically and play a role in specific immunity, as we will see in Chapter 16.

The combination of the barrier function of the epidermis, its continual replacement, and the presence of phagocytic dendritic cells provides significant defense against colonization and infection by pathogens.

The dermis also defends nonspecifically. It contains tough fibers of a protein called collagen. These give the skin

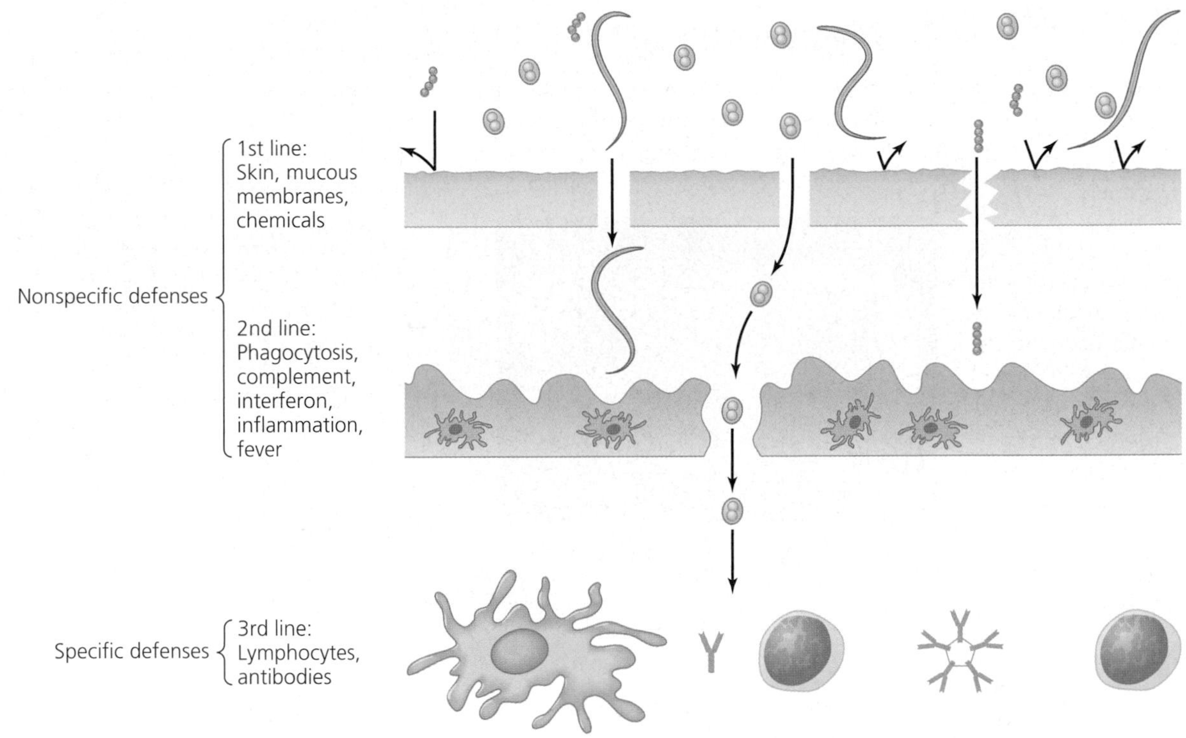

Figure 15.1 ➤
The three lines of defense that protect the body from pathogens.

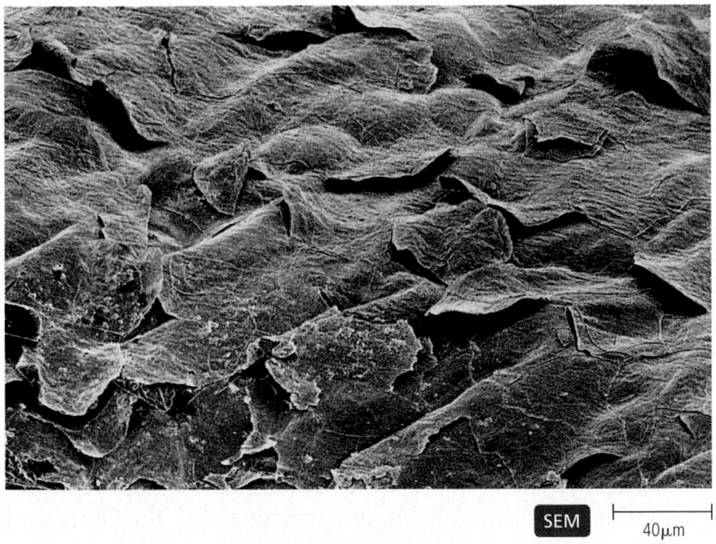

▲ ***Figure 15.2***
A scanning electron micrograph of a section of skin. Note that the epidermal cells are tightly packed, which provides an effective barrier to most microorganisms.

strength and pliability to prevent jabs and scrapes from penetrating the dermis and introducing microorganisms. Blood vessels in the dermis deliver defensive cells and chemicals, which will be discussed shortly.

In addition to its physical structure, the skin has a number of chemical substances that nonspecifically defend against pathogens. Sweat glands secrete perspiration (sweat), which contains salt and lysozyme. Salt draws water osmotically from invading cells, which inhibits their growth and kills them. **Lysozyme** (lī′sō-zīm) is an enzyme that destroys the cell walls of bacteria by cleaving the bonds between the sugar subunits of the walls. Bacteria without cell walls are more susceptible to osmotic shock and digestion by other enzymes within phagocytes (discussed with respect to the second line of defense).

The skin also contains sebaceous (oil) glands that secrete **sebum** (sē′bŭm), an oily substance that not only helps keep the skin pliable and less sensitive to breaking or tearing, but also contains fatty acids that lower the pH of the skin's surface to about pH 4, which is inhibitory to many bacteria.

Although salt, lysozyme, and acidity make the surface of the skin an inhospitable environment for most microorganisms, some bacteria, such as *Staphylococcus epidermidis* (staf′i-lō-kok′ŭs ep-ē-der-mi′dis), find the skin a suitable environment for growth and reproduction (see Figure 14.2a). Such bacteria are particularly abundant in crevices around hairs and in the ducts of glands. Usually they are nonpathogenic, but they may become opportunistic pathogens.

The Role of Mucous Membranes in Nonspecific Defense

Learning Objectives

- ✓ Identify the locations of the body's mucous membranes.
- ✓ Explain how mucous membranes protect the body both physically and chemically.

Mucus-secreting (mucous) membranes, a second part of the first line of defense, line all body cavities that are open to the

Highlight 15.1 What Happens to All That Skin?

Your body sheds tens of thousands of skin flakes every time you walk or move, and you shed at only a slightly lower rate when you stand still. That comes to about 10 billion skin cells per day, or 250 grams (about half a pound) of skin every year! What happens to all that skin?

Much of household dust is skin that you and your housemates have shed as you go about your lives. The skin flakes fall to the rug and upholstery, where they become food for microscopic mites that live sedentary and harmless lives waiting patiently for meals to rain down on them from above. They dwell not only in the rug, but also in your mattress and pillow, and even in the hair follicles of your eyebrows, where they catch skin cells cascading down your forehead.

Dust mite.

By the way, house dust also contains mite feces and mite skeletons, which can trigger allergies. So after reading this chapter, you just might want to pull out the vacuum cleaner.

CRITICAL THINKING

Some strains of *Staphylococcus aureus* produce exfoliative toxin, a chemical that causes portions of the entire outer layer of the skin to be sloughed off in a disease called scalded skin syndrome. Given that cells of the outer layers are going to fall off anyway, why is this disease dangerous?

outside environment. Thus mucous membranes line the lumens[1] of the respiratory, urinary, digestive, and reproductive tracts. Like the skin, mucous membranes act nonspecifically to limit infection both physically and chemically.

Mucous membranes have two distinct layers: an outer covering of superficial (closest to the surface, in this case the lumen) cells called the *epithelium,* and a deeper connective tissue layer that provides mechanical and nutritive support for the epithelium. Epithelial cells of mucous membranes are packed closely together, like those of the epidermis, but they form only a thin tissue. Indeed, in some mucous membranes, the epithelium is only a single cell thick. Unlike surface epidermal cells, surface cells of mucous membranes are alive and play roles in the diffusion of nutrients and oxygen (in the digestive, respiratory, and female reproductive systems) and in the elimination of wastes (in the urinary, respiratory, and female reproductive systems).

The thin epithelium on the surface of a mucous membrane provides a less efficient barrier to the entrance of pathogens than the multiple layers of dead cells found at the skin's surface. So how are microorganisms kept from invading through these thin mucous membranes? In some cases they are not, which is why some mucous membranes, especially those of the respiratory and reproductive systems, can be common portals of entry for pathogens. However, the epithelial cells of mucous membranes are tightly packed to prevent the entry of pathogens. Moreover, epithelial cells are continually shed and then replaced via the cytokinesis of **stem cells,** which are generative cells capable of dividing to form daughter cells of a variety of types; one effect of such shedding is that it carries attached microorganisms away.

In addition, the epithelia of some mucous membranes have still other means of removing pathogens. In the mucous membrane of the trachea, for example, the stem cells produce both **goblet cells,** which secrete a sticky mucus that traps bacteria and other pathogens, and ciliated columnar cells, whose cilia propel the mucus (and the particles and pathogens trapped within it) up from the lungs **(Figure 15.3).** The effect of the action of the cilia is often likened to that of an escalator. Mucus carried into the throat is coughed up and either swallowed or expelled. Because the poisons and tars in tobacco smoke damage cilia, the lungs of smokers are not properly cleared of mucus, so smokers may develop severe coughs as their respiratory tracts attempt to expel excess mucus from the lungs. Smokers also typically succumb to more respiratory pathogens because they are unable to effectively clear pathogens from their lungs.

In addition to these physical actions, some mucous membranes produce chemicals that defend against pathogens. Nasal mucus, for example, contains lysozyme, which chemically destroys bacterial cell walls. **Table 15.1** on page 442 compares the physical and chemical actions of the skin and mucous membranes in the body's first line of defense.

The Role of Normal Microbiota in Nonspecific Defense

As we saw in Chapter 14, the skin and mucous membranes of the body are normally home to a variety of protozoa, fungi, bacteria, and viruses. This normal microbiota plays a role in protecting the body by competing with potential pathogens in a variety of ways, a situation called **microbial antagonism.**

A variety of activities of the normal microbiota make it less likely that a pathogen can compete with them and produce disease. Some microbiota secrete antimicrobial sub-

[1]A lumen is a cavity or channel within any tubular structure or organ.

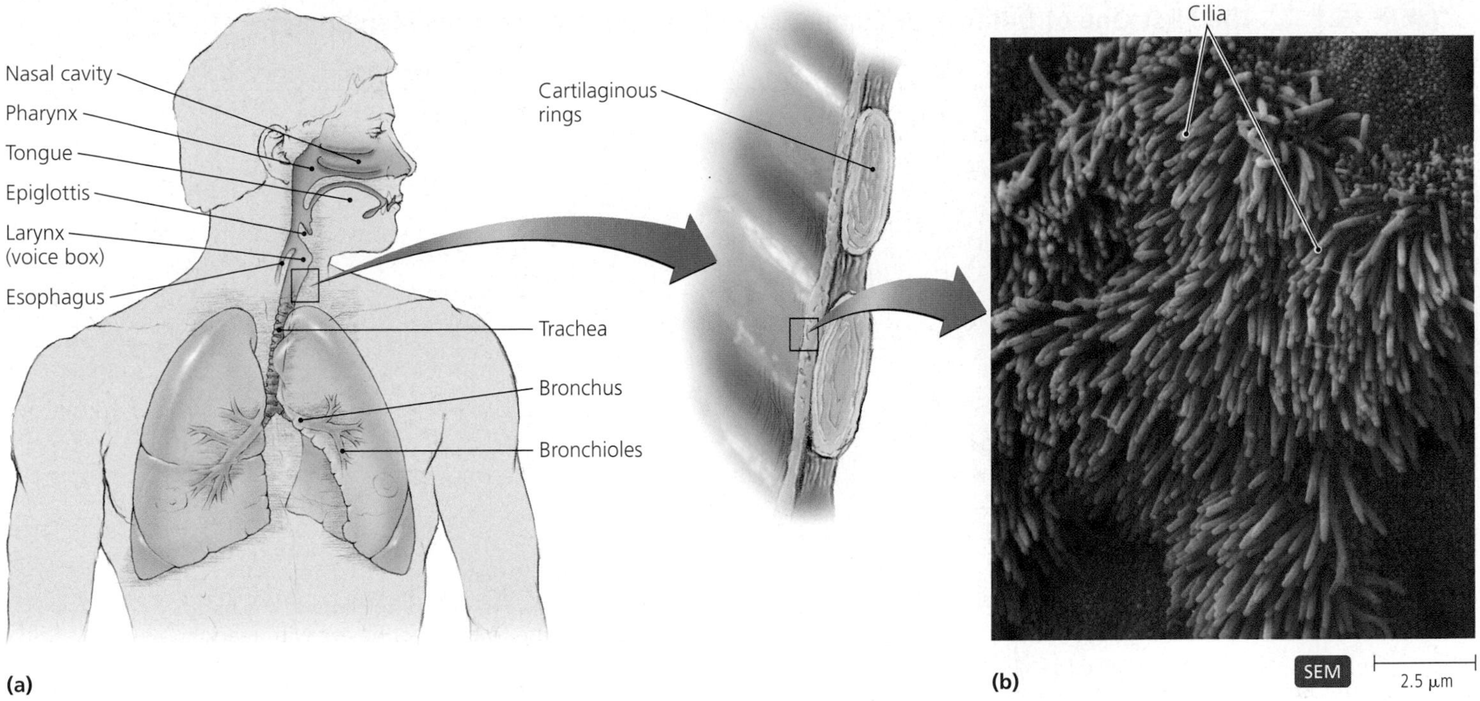

▲ ***Figure 15.3***

The action of the respiratory mucous membrane in expelling pathogens. **(a)** The structure of the respiratory system, which is lined with a mucous membrane. **(b)** The epithelium of the trachea, which contains mucus-secreting goblet cells, and ciliated cells whose cilia propel the mucus (and the microbes trapped within it) up to the larynx for removal. *What is the function of stem cells within the respiratory epithelium?*

Figure 15.3 *Stem cells in the respiratory epithelium undergo cytokinesis to form both ciliated cells and goblet cells to replace those lost during normal shedding.*

stances that limit the growth of potential pathogens. Microbiota also consume nutrients, making them unavailable to pathogens. Additionally, the normal microbiota can change the pH, creating an environment that is favorable for themselves but unfavorable to other microorganisms.

Additionally, the presence of the normal microbiota stimulates the body's second line of defense (discussed shortly). Researchers have observed that animals raised in an *axenic* (ā-zēn′ik) environment—that is, one free of all germs—are slower to defend themselves when exposed to a pathogen.

Finally, the normal microbiota of the intestines improve overall health by providing several vitamins, including biotin and pantothenic acid (vitamin B_5), which are important in glucose metabolism; folic acid, which is essential for the production of the purine and pyrimidine bases of nucleic acids; and vitamin K, which has an important role in blood clotting.

Other First-Line Defenses

Besides the physical barrier of the skin and mucous membranes, other structures hinder microbial invasion. Notable among these is the lacrimal apparatus of the eye, which is a group of structures that produce and drain away tears (**Figure 15.4** on page 442).

Lacrimal glands, located above and to the sides of the eyes, secrete tears into lacrimal gland ducts and onto the surface of the eyes. The tears either evaporate or drain into small lacrimal canals, which carry them into nasolacrimal ducts that empty into the nose. There, the tears join the nasal mucus and flow into the pharynx, where they are swallowed. The blinking action of the eyelids spreads the tears and washes the surfaces of the eyes. Normally, evaporation and flow into the nose balances the flow of tears onto the eye. However, if the eyes are irritated, increased tear production can flood the eyes, and the overflow can carry the irritant away. In addition to their washing action, tears contain lysozyme, which destroys bacteria.

Many other body organs contribute to the first line of defense by secreting chemicals with antimicrobial properties that are secondary to their prime function. For example, whereas stomach acid is primarily present to aid digestion of proteins, it also prevents the growth of many potential pathogens. Likewise, saliva not only contains digestive

Table 15.1 The First Line of Defense: A Comparison of the Skin and Mucous Membranes

	Skin	Mucous Membrane
Number of layers	Many	One to a few
Cells tightly packed?	Yes	Yes
Dead or alive?	Outer layers: dead; inner layers: alive	Alive
Mucus present?	No	With some
Lysozyme present?	Yes	With some
Sebum present?	Yes	No
Cilia present?	No	With some
Constant shedding and replacement of cells?	Yes	Yes

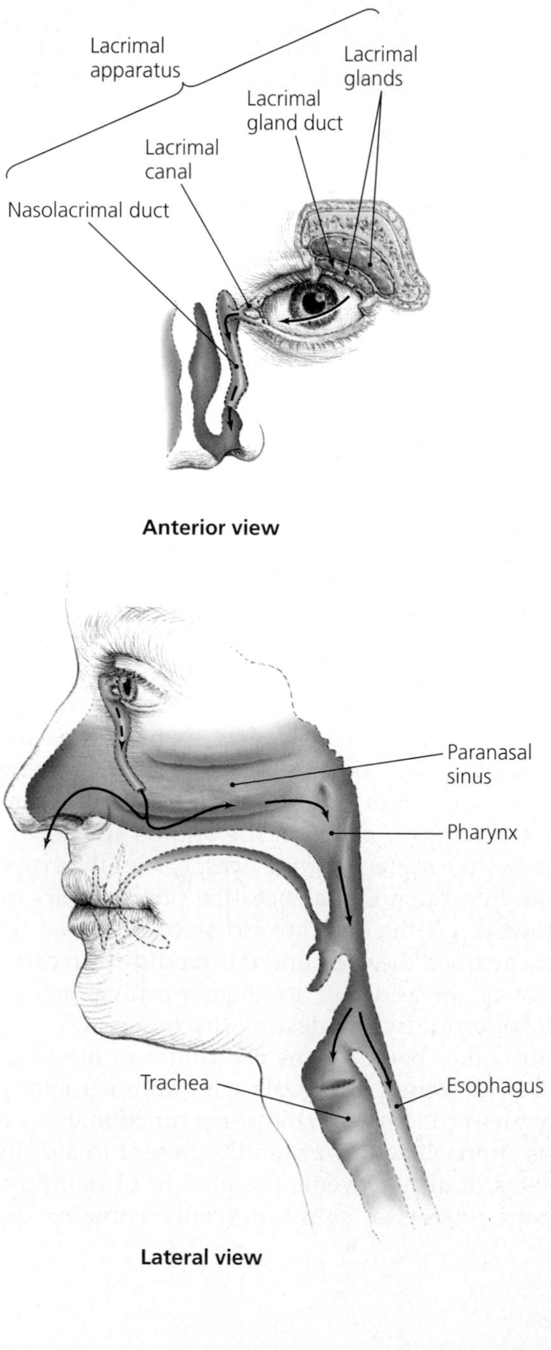

Figure 15.4
The lacrimal apparatus, which functions in the body's first line of defense by bathing the eye with tears. Arrows indicate the route tears take across the eye and into the throat. *What antimicrobial protein is found in tears?*

Figure 15.4 *Tears contain lysozyme, an antimicrobial protein that acts against the peptidoglycan of bacterial cell walls.*

enzymes and provides lubrication for swallowing, it also contains lysozyme and washes microbes from the teeth. The contributions of these and other processes and chemicals to the first line of defense are listed in **Table 15.2.**

An Overview of the Body's Second Line of Defense

Learning Objective

✓ Compare and contrast the body's first and second lines of defense against disease.

When pathogens succeed in penetrating the skin or mucous membranes, the body's second line of defense comes into play. Like the first line of defense, the second line is nonspecific; that is, it operates against a wide variety of pathogens, from parasitic worms to viruses. But unlike the first line of defense, the second line includes no barriers; instead, it is composed of cells (especially phagocytes), antimicrobial chemicals (complement, interferons), and processes (inflammation, fever) (see Figure 15.1). We will consider each of these components of the second line of defense in some detail shortly, but because many of them are either contained in or originate in the blood, we first consider the components of blood.

Table 15.2 Secretions and Activities that Contribute to the First Line of Defense

Secretion/Activity	Function
Digestive System	
Saliva	Washes microbes from teeth, gums, tongue, and palate; contains lysozyme, an antibacterial enzyme
Stomach acid	Digests and/or inhibits microorganisms
Gastroferritin	Sequesters iron during its absorption, making it unavailable for microbial use
Bile	Inhibitory to most microorganisms
Intestinal secretions	Digests and/or inhibits microorganisms
Peristalsis	Moves gastrointestinal contents through GI tract—constantly eliminating potential pathogens
Defecation	Eliminates microorganisms
Vomiting	Eliminates microorganisms
Nervous System	
Tears	Wash eyes; contain lysozyme
Urinary System	
Urine	Contains lysozyme; its acidity inhibits microorganisms; may wash microbes from ureters and urethra during urination
Reproductive System	
Vaginal secretions	Acidity inhibits microorganisms; contains iron-binding proteins that sequester iron, making it unavailable for microbial use
Menstrual flow	Cleanses uterus and vagina
Prostate secretion	Contains iron-binding proteins that sequester iron, making it unavailable for microbial use
Cardiovascular System	
Blood flow	Removes microorganisms from wounds
Coagulation	Prevents entrance of many pathogens
Transferrin	Binds iron for transport, making it unavailable for microbial use

Defense Components of Blood

Learning Objectives

- ✓ Discuss the components of blood and their functions in the body's defense.
- ✓ Explain how macrophages are named.

Blood is a complex liquid tissue composed of cells and portions of cells within a fluid called *plasma*. We begin our discussion of the defense functions of blood by briefly considering plasma.

Plasma

Plasma is mostly water containing electrolytes (ions), dissolved gases, nutrients, and—most relevant to the body's defenses—a variety of proteins. Some plasma proteins are involved in blood clotting, a defense mechanism that both reduces blood loss and the risk of infection. When the clotting factors have been removed from the plasma, as for instance when blood clots, the remaining liquid is called *serum*.

Another group of proteins, called *complement proteins*, is an important part of the second line of defense and is discussed shortly. Still another group of proteins, called *antibodies* or *immunoglobulins*, is a part of specific immunity, the body's third line of defense, which we will explore in Chapter 16.

Defensive Blood Cells: Leukocytes

The cells and cell fragments suspended in the plasma are called **formed elements.** In a process called *hematopoiesis,*[2] blood stem cells located principally in the bone marrow within the hollow cavities of the large bones produce three types of formed elements: **erythrocytes**[3] (ĕ-rith'rō-sītz), **platelets**[4] (plāt'letz), and **leukocytes**[5] (lū'kō-sītz) (**Figure 15.5** on page 444). Erythrocytes, the most numerous of the formed elements, carry oxygen and carbon dioxide in the blood. Platelets, which are pieces of large cells called *megakaryocytes* that have split into small portions of cytoplasm surrounded by cytoplasmic membranes, are involved in blood clotting. Leukocytes, the formed elements that are directly involved in

---

[2]From Greek *haima,* meaning blood, and *poiein,* meaning to make.
[3]From Greek *erythro,* meaning red, and *cytos,* meaning cell.
[4]French for small plates. Platelets are also called thrombocytes, from Greek *thrombos,* meaning lump, and *cytos,* meaning cell, though they are technically not cells, but instead pieces of cells.
[5]From Greek *leuko,* meaning white, and *cytos,* meaning cell.

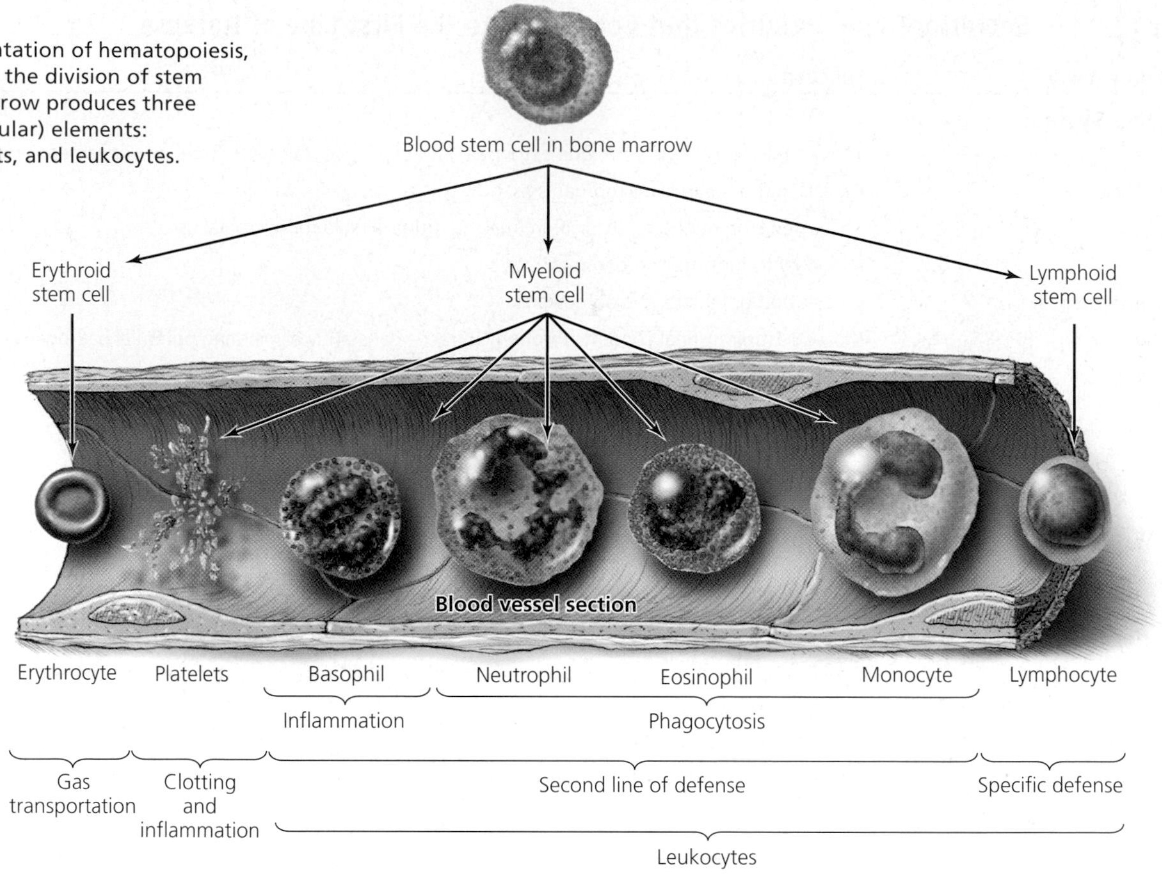

Figure 15.5 A schematic representation of hematopoiesis, the process by which the division of stem cells in the bone marrow produces three types of formed (cellular) elements: erythrocytes, platelets, and leukocytes.

defending the body against invaders, are commonly called white blood cells because they form a whitish layer when the components of blood are separated within a test tube.

Based on their appearance in stained blood smears when viewed under the microscope, leukocytes are divided into two groups: **granulocytes** (gran′ū-lō-sītz) and **agranulocytes** (ă-gran′ū-lō-sītz) **(Figure 15.6).**

Granulocytes have large granules in their cytoplasm that stain different colors depending on the type of granulocyte and the dyes used: **basophils** (bā′sō-fils) stain blue with the basic dye methylene blue; **eosinophils** (ē-ō-sin′ō-fils) stain red to orange with the acidic dye eosin; and **neutrophils** (noo′trō-fils) stain lilac with a mixture of acidic and basic dyes. Both neutrophils (also known as *polymorphonuclear leukocytes*) and eosinophils can phagocytize pathogens, and both can exit the blood to attack invading microbes in the tissues by squeezing between the cells lining capillaries (the smallest blood vessels), a process called **diapedesis**[6] (dī′ă-pĕ-dē′sis) or **emigration (Figure 15.7).** As we will see later in the chapter, eosinophils are also involved in defending the body against parasitic worms and are present in large number during many allergic reactions, though their exact function in allergies is disputed. Basophils can also leave the blood, though they are not phagocytic; instead, they release histamine during inflammation, an aspect of the second line of defense that will be discussed shortly.

The cytoplasm of agranulocytes appears uniform when viewed via light microscopy, though granules do become visible with an electron microscope. Agranulocytes are of two types: **lymphocytes** (lim′fō-sītz), which are the smallest leukocytes and have nuclei that nearly fill the cells, and **monocytes** (mon′ō-sītz), which are large agranulocytes with slightly lobed nuclei. Although most lymphocytes are involved in specific immunity (see Chapter 16), *natural killer (NK) lymphocytes* function in nonspecific defense and thus are discussed later in this chapter. Monocytes leave the blood and mature into **macrophages** (mak′rō-fāj-ĕz), which are phagocytic cells of the second line of defense. Their initial function is to devour foreign objects, including bacteria, fungi, spores, and dust, as well as dead body cells.

Macrophages are named for their location in the body. **Wandering macrophages** leave the blood via diapedesis and perform their scavenger function while travelling throughout the body, including extracellular spaces. Other macrophages are fixed and do not wander. These include the Langerhans cells of the epidermis, **alveolar** (al-vē′ō-lăr) **macrophages**[7] of the lungs, **microglia** (mī-krog′lē-ă) of the central nervous system, and **Küpffer** (kyūp′fĕr) **cells** of the liver. Fixed macrophages generally phagocytize within a specific organ, where they associate with a *reticulum* (network) of fibers that join the organ's cells together **(Figure 15.8).**

[6]From Greek *dia*, meaning through, and *pedan*, meaning to leap.

[7]Alveoli are small pockets at the end of respiratory passages where oxygen and carbon dioxide exchange occurs between the lungs and the blood.

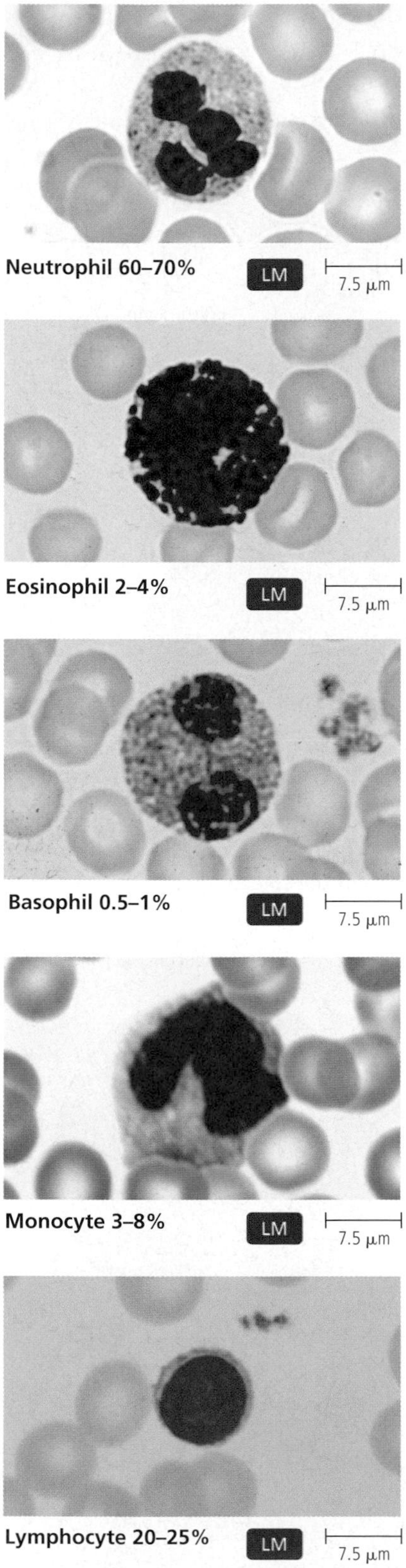

◀ *Figure 15.6*

The appearances of three granulocytes (a neutrophil, an eosinophil, and a basophil) and two agranulocytes (a monocyte and a lymphocyte), as seen in stained blood smears. The numbers are the normal percentages of each cell type among all leukocytes.

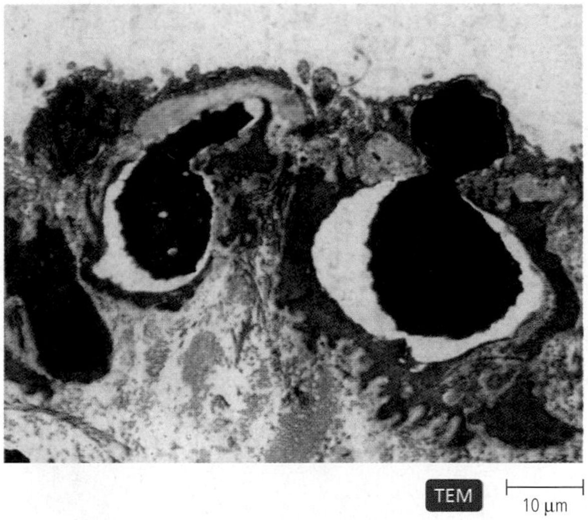

▲ *Figure 15.7*

Diapedesis, the process whereby leukocytes leave intact blood vessels. Shown here is a neutrophil squeezing between the cells lining a capillary on its way to fight an infection in the tissues.

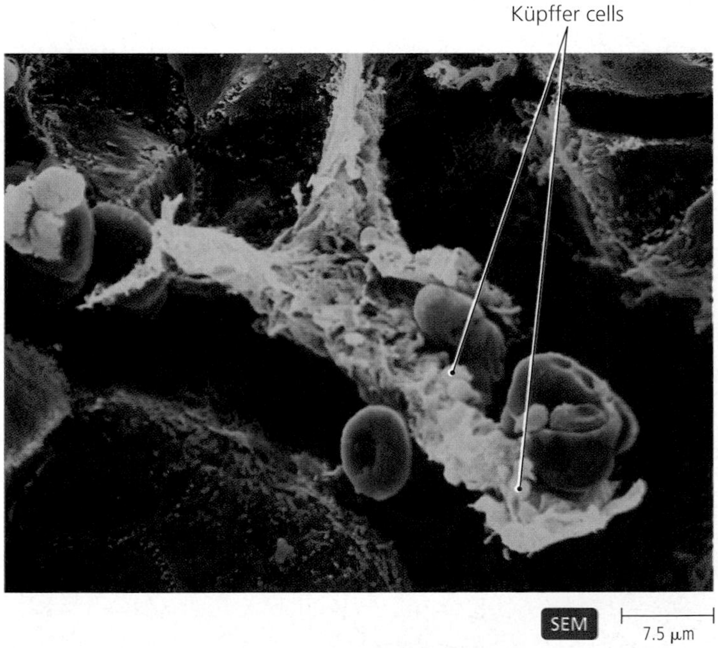

▲ *Figure 15.8*

A Küpffer cell, a fixed macrophage in the liver. Fixed macrophages remain attached to a reticulum of fibers within specific organs, where they survey the surrounding tissue for the presence of pathogens.

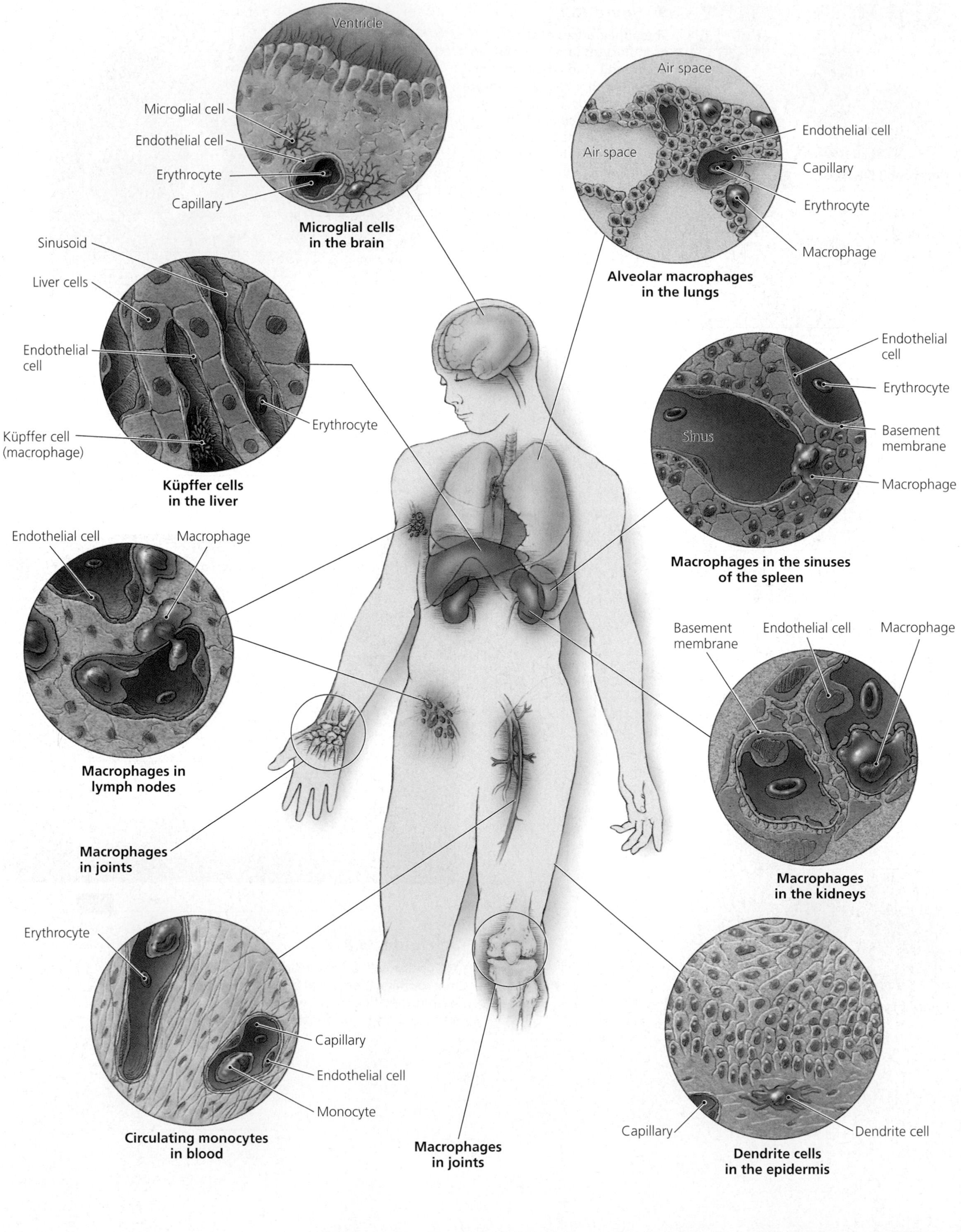
Ventricle
Microglial cell
Endothelial cell
Erythrocyte
Capillary
Microglial cells in the brain
Air space
Air space
Endothelial cell
Capillary
Erythrocyte
Macrophage
Alveolar macrophages in the lungs
Sinusoid
Liver cells
Endothelial cell
Erythrocyte
Küpffer cell (macrophage)
Küpffer cells in the liver
Endothelial cell
Erythrocyte
Sinus
Basement membrane
Macrophage
Macrophages in the sinuses of the spleen
Endothelial cell
Macrophage
Macrophages in lymph nodes
Basement membrane
Endothelial cell
Macrophage
Macrophages in the kidneys
Macrophages in joints
Erythrocyte
Capillary
Endothelial cell
Monocyte
Circulating monocytes in blood
Macrophages in joints
Capillary
Dendrite cell
Dendrite cells in the epidermis

◀ *Figure 15.9*

The mononuclear phagocytic (reticuloendothelial) system. *Why is this system considered a functional system rather than an anatomical (structural) system?*

Figure 15.9 The mononuclear phagocytic system is considered a functional system because it is composed of cells, located in many body systems, that share a common function: phagocytizing invaders.

As a group, macrophages of all types (fixed or wandering), plus monocytes that are attached to *endothelial cells* (which line the heart chambers, blood vessels, and lymphatic[8] vessels), constitute a system called either the **reticuloendothelial** (rē-tik′yū-lō-en-dō-thē′lē-ăl) **system** or the **mononuclear phagocytic system (Figure 15.9).** Though these terms refer to the same thing, they emphasize different aspects of the nature of this system. The first term emphasizes the association of the cells with reticular fibers, whereas the second term emphasizes their microscopic appearance and primary function. The latter name is preferred because the system is primarily a functional system, not an anatomical one.

[8]The lymphatic system is discussed in Chapter 16.

Lab Analysis of Leukocytes Analysis of blood for diagnostic purposes, including white blood cell counts, is one task of medical lab technologists. The proportions of leukocytes, as determined in a **differential white blood cell count,** can serve as a sign of disease. For example, an increase in the percentage of eosinophils can indicate allergies or infection with parasitic worms; bacterial diseases typically exhibit an increase in the number of leukocytes and an increase in the percentage of neutrophils, while viral infections are associated with an increase in the relative number of lymphocytes. The ranges for the normal values for each kind of white blood cell, expressed as a percentage of the total leukocyte population, are shown in Figure 15.6 (on page 445).

CRITICAL THINKING

A medical laboratory technologist argues that granulocytes are a natural group, whereas agranulocytes are an artificial grouping. Based on Figure 15.5, do you agree or disagree with the lab tech? What evidence can you cite to justify your conclusion?

Case Study 15.1 Evaluating a Complete Blood and Differential Leukocyte Count (CBC) Profile

An important diagnostic tool is a count of the number and relative percentages of each type of blood cell in a given volume of blood. Values can increase or decrease as the body responds to infections by bacteria and viruses, as well as to cancer. Consider this portion of a Complete Blood and Differential Leukocyte Count (CBC) profile for a patient named Dorothy. Study the profile and then answer the question concerning her condition on September 7.

1. Which blood cells are at abnormally low levels in Dorothy's blood?
2. Which differential leukocyte counts are outside their normal range?
3. Which of the three lines of defense will be most adversely affected in this patient?

Patient: Dorothy

CBC Profile

CH		9/7/00 09:05	7/3/00 14:36	Normal range of values for women
WBC		6.5	6.9	3.6-10.8 K/μL
RBC	L	3.94	3.62	4.20-5.40 m/Ul
HGB	L	11.7	9.9	12.0-16.0 g/dL
HCT	L	34.8	30.6	37.0-47.0%
MCV		88.4	84.5	81.0-99.0 fL
MCH		29.6	27.3	27.0-31.0 pg
MCHC		33.5	32.3	32.0-36.0 g/dL
RDW	H	15.5	17.0	11.6-14.5%
PLT	H	432	422	130-400 K/μL
MPV	L	6.4	5.7	7.8-11.0 fL
NE%	H	76.3	78.1	37.0-75.0%
LY%	L	10.0	7.1	20.5-51.1%
MO%		10.6	12.7	1.7-15.0%
EO%		2.9	1.6	0.0-10.0%
BA%	L	0.2	0.5	1.0-8.0%
NE#		4.9	5.4	1.4-6.5 K/μL
LY#	L	0.7	0.5	1.2-3.4 K/μL
MO#		0.7	0.9	0.1-0.9 K/μL
EO#		0.2	0.1	0.0-0.7 K/μL
BA#		0.0	0.0	0.0-0.2 K/μL
Plat		Adequate	Adequate	Adequate
Promylo				0-0%

Coulter counter determines the # of each kind of cell (per μL) → WBC, RBC

Number of platelets → PLT

Percentage of leukocytes of each type → NE%, LY%, MO%, EO%, BA%

Actual number of each type of leukocyte → NE#, LY#, MO#, EO#, BA#

Values from two dates

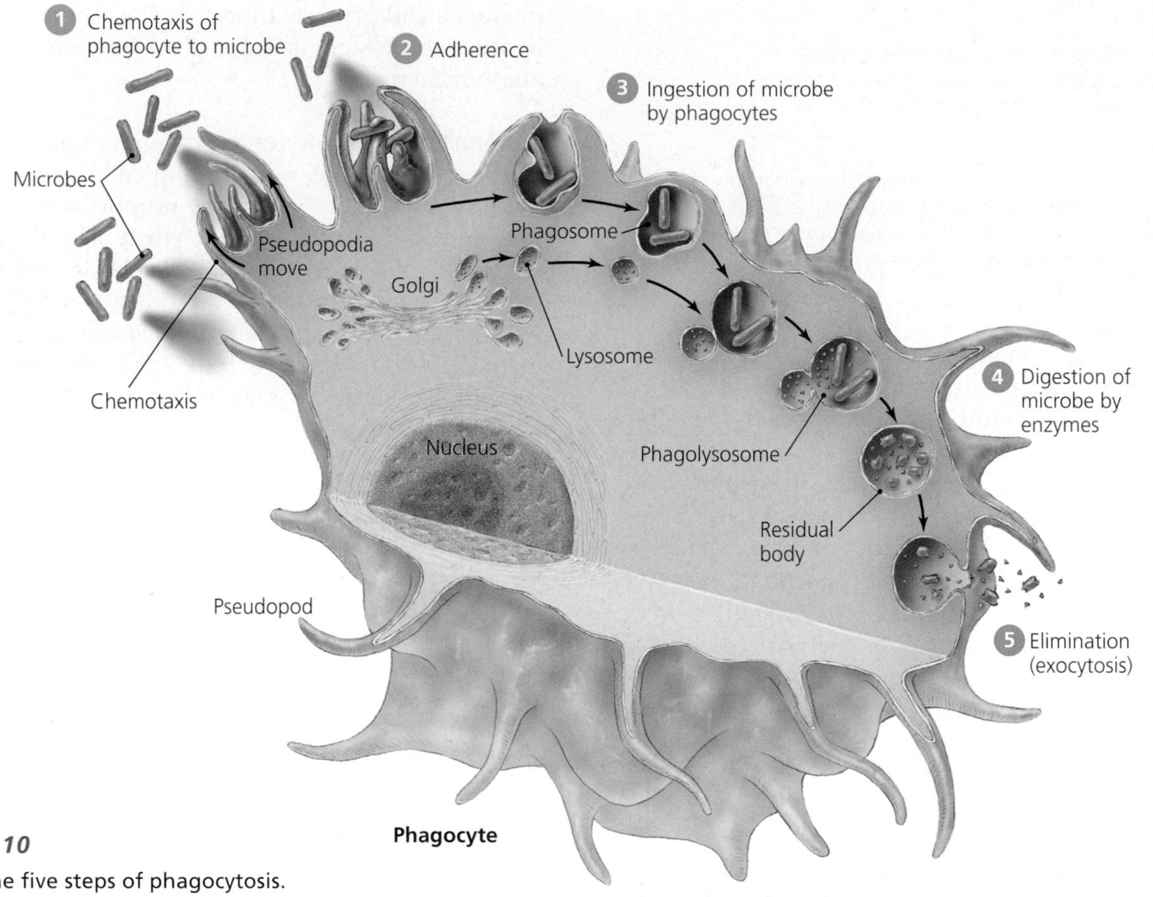

▲ ***Figure 15.10***
The events in the five steps of phagocytosis.

The Body's Second Line of Defense

Now that we have some background concerning the defensive properties of plasma components and leukocytes, we turn our attention to the various components of the body's second line of defense: phagocytosis, extracellular killing by leukocytes, nonspecific chemical defenses, inflammation, and fever.

Phagocytosis

Learning Objective

✓ Name and describe the five stages of phagocytosis.

In Chapter 3 we examined phagocytosis as a way by which a cell can transport substances across the cytoplasmic membrane and into the cytoplasm. Here we consider how those cells of the body that are capable of phagocytosis—collectively known as **phagocytes** (fag'ō-sītz), most of which are leukocytes or their derivatives—play a role against pathogens that get past the body's first line of defense.

Although phagocytosis was discovered over a century ago **(see Highlight 15.2)**, it is a complex process that is still not completely understood. For the purposes of our discussion, we will divide the continuous process of phagocytosis into five steps: (1) chemotaxis, (2) adherence, (3) ingestion, (4) digestion, and (5) elimination **(Figure 15.10).**

Chemotaxis

Recall from Chapter 3 that *chemotaxis* is movement of a cell either toward a chemical stimulus (positive chemotaxis) or away from a chemical stimulus (negative chemotaxis). In the case of phagocytes, positive chemotaxis involves the use of *pseudopodia* (sū-dō-pō'dē-ă) to move toward microorganisms at the site of an infection (step 1 in Figure 15.10). Chemicals that attract phagocytic leukocytes include microbial waste products and secretions, components of damaged tissues and white blood cells, and **chemotactic** (kem-ō-tak'tik) **factors.** Chemotactic factors include peptides derived from complement (discussed later in this chapter) and chemicals called **cytokines** (sī'tō-kīnz) (or *chemokines*), which are released by leukocytes already at a site of infection and are considered in Chapter 16.

Adherence

After arriving at the site of an infection, phagocytes attach to microorganisms through the binding of complementary chemicals such as glycoproteins found on the membranes of cells (step 2 in Figure 15.10). This process is called **adherence.**

Highlight 15.2 The Discovery of Phagocytes

During the latter part of the 19th century, when microbiology was coming of age in the labs of Louis Pasteur and Robert Koch, a Russian microbiologist Elie Metchnikoff (1845–1916) surprised the scientific world with the announcement that animals were protected from pathogens by the action of wandering cells he called *phagocytes.* Metchnikoff had observed these cells moving in the bodies of starfish larvae by means of pseudopodia, and he hypothesized that because they moved like free-living *Amoeba,* they might also eat like *Amoeba,* phagocytizing invasive pathogens.

He tested his hypothesis by stabbing a starfish larva with a rose thorn from his garden. The next morning he observed that thousands of phagocytes had migrated to the site of infection and were actively devouring germs. He subsequently demonstrated the same phenomenon in other animals, and with these discoveries pointed the way toward understanding both what we now call the second line of defense and cellular immunity, which is part of the body's specific defense.

In 1908, Metchnikoff received a Nobel Prize in Medicine or Physiology for his discoveries.

Some bacteria have virulence factors, such as M protein of *Streptococcus pyogenes* (strep-tō-kok'ŭs pī-aj'en-ēz), or slippery capsules that hinder adherence of phagocytes and thereby increase the virulence of the bacteria. Such bacteria are more readily phagocytized if they are pushed up against a surface such as connective tissue, the wall of a blood vessel, or a blood clot.

All pathogens are more readily phagocytized if they are first covered with either the antimicrobial proteins of complement (discussed later), or the specific antimicrobial proteins called antibodies (discussed in Chapter 16). This coating process is called **opsonization,**[9] and the proteins are called **opsonins.**

Ingestion

After phagocytes adhere to pathogens, they extend *pseudopodia* to surround the microbe (step 3 in Figure 15.10, and **Figure 15.11**). The encompassed microbe is internalized as the pseudopodia fuse to form a sac called a **phagosome.**

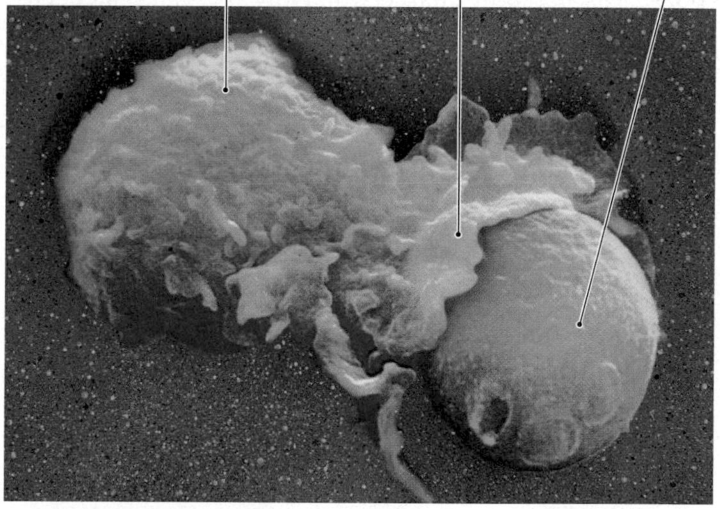

Figure 15.11
A phagocyte (in this case, a macrophage) in the process of ingesting a yeast cell by surrounding it with pseudopodia to form a phagosome.

Digestion

Digestion begins when lysosomes within the phagocyte fuse with the newly formed phagosomes to form **phagolysosomes** (fag-ŏ-lī'sō-sōmz), or digestive vesicles (step 4 in Figure 15.10). Lysosomes contain digestive enzymes and other antimicrobial substances in an environment with a pH of about 4.0 due to the presence of lactic acid. Among the 30 or so different digestive enzymes within lysosomes are lipases, proteases, nucleases, and a variety of others. Not surprisingly, they are most active at acidic pH.

Lysosomes may also contain toxic oxygen derivatives, including singlet oxygen ($^1O_2$), hydrogen peroxide (H_2O_2), nitric acid (NO), superoxide radicals (O_2^-), hydroxyl radical ($OH^·$), and hypochlorite ions (OCl^-). (Hypochlorite is the ingredient in household bleach that accounts for both its bleaching action and its antimicrobial action.) All of these substances damage proteins and lipids in cell membranes, allowing digestive enzymes access to the contents of microbial cells.

Digestion of most pathogens is complete within 30 minutes, though some bacteria contain virulence factors (such as M protein or waxy cell walls) that resist digestion. After digestion, a phagolysosome is known as a *residual body.*

Elimination

Digestion is not always complete, and phagocytes eliminate undigested remnants of microorganisms via *exocytosis,* a process that is essentially the reverse of ingestion (step 5 in Figure 15.10). Some microbial components may remain attached to the cytoplasmic membranes of some phagocytes, a

[9]From Greek *opsonein,* meaning to supply food, and *izein,* meaning to cause; thus, loosely, "to prepare for dinner".

Highlight 15.3 The Fight for Iron

Most cells require iron for metabolism: It is a component of cytochromes of the electron transport chain and functions as an enzyme cofactor. Because iron is relatively insoluble, in humans it is transported through the blood and to cells that need it by a transport protein called *transferrin*. When transferrin-iron complexes reach cells with receptors for transferrin, the binding of the protein to the receptor stimulates the cell to take up the iron via endocytosis. Excess iron is stored within liver cells bound to another protein called *ferritin*.

Though the main function of iron-binding compounds is transporting and storing iron, they play a secondary, defensive role—sequestering iron so that it is unavailable to microorganisms.

Some bacteria respond to a shortage of iron by secreting their own iron-binding proteins called *siderophores*. Because siderophores have a greater affinity for iron than does transferrin, bacteria that produce them in effect steal iron from the body. In response, the body produces *lactoferrin*, which retakes iron from the bacteria due to its even greater affinity. Thus the body and the pathogens engage in a kind of chemical "tug-of-war" for the possession of iron.

Neisseria meningitidis, a pathogen that causes meningitis, bypasses this contest altogether by producing receptors for transferrin and plucking iron from the bloodstream as it flows by.

phenomenon that plays a role in the specific immune response (discussed in Chapter 16). How is it that phagocytes destroy invading pathogens, and leave the body's own healthy cells unharmed? There are at least two mechanisms responsible for this:

- Some phagocytes have cytoplasmic membrane receptors for various bacterial surface components lacking on the body's cells, such as cell wall components or flagellar proteins.
- Opsonins such as complement and antibody provide a signal to the phagocyte.

Extracellular Killing by Leukocytes

Learning Objective

✓ Describe the role of eosinophils and NK cells in extracellular killing of microorganisms and parasitic helminths.

Phagocytosis involves killing a pathogen once it has been ingested—that is, once it is inside the phagocyte. In contrast, eosinophils and natural killer cells can accomplish *extracellular killing*.

As discussed earlier, eosinophils can phagocytize; however, this is not their usual mode of attack. Instead, they primarily attack parasitic helminths (worms) by attaching to the worm's surface, often binding to antibodies that themselves have bound to chemicals that are specific to the worm. Once bound, eosinophils secrete extracellular protein toxins onto the surface of the parasite. These weaken the helminth and may even kill it. **Eosinophilia** (ē-ō-sin′ō-fil-e-ă), an abnormally high number of eosinophils in the blood, is often indicative of helminth infection.

Natural killer lymphocytes (or **NK cells**) are another type of nonspecific, defensive leukocyte that works by secreting toxins onto the surfaces of virally infected cells and neoplasms (tumors). NK cells identify and spare normal body cells because the latter express membrane proteins similar to those on the NK cells. The ability to distinguish one's own healthy cells from diseased cells and pathogens is discussed more fully in Chapter 16.

Nonspecific Chemical Defenses Against Pathogens

Learning Objectives

✓ Describe the complement system, including its classical and alternate pathways.

✓ Explain the roles of interferons and defensins in nonspecific defense.

Chemical defenses augment phagocytosis in the second line of defense. The chemicals involved assist phagocytic cells either by directly attacking pathogens or by enhancing other features of nonspecific resistance. Defensive chemicals include lysozyme (which we have already discussed), complement, interferons, and defensins. Additionally, **Highlight 15.3** describes the antimicrobial activity of *transferrin,* a chemical whose primary function is transporting iron.

Complement

Complement—or more properly, the **complement system**—is a set of serum proteins designated numerically according to the order of their discovery. These proteins initially act as opsonins and chemotactic factors, and trigger inflammation and fever. The end result of complement activation is lysis of foreign cells.

Complement is activated in several ways. Here we consider two of them **(Figure 15.12):**

- In the *classical pathway,* the binding of antibodies to foreign antigens activates complement.
- In the *alternate pathway* (also called the *properdin pathway,* after one of the proteins involved), pathogens or pathogenic products (such as bacterial endotoxins and glycoproteins) activate complement.

As Figure 15.12 shows, the two pathways merge, and then complement proteins react with one another in an amplifying sequence of chemical reactions, in which the product of each reaction becomes an enzyme that catalyzes the next reaction many times over. Such reactions are called *cascades* because they progress in a way that can be likened to a

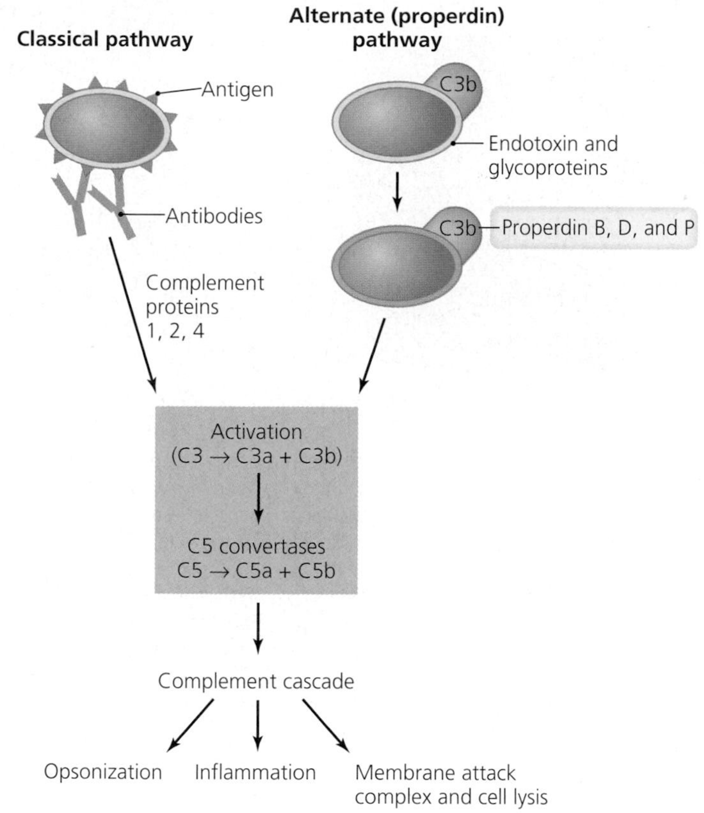

▲ ***Figure 15.12***

The two pathways by which complement is activated. In the classical pathway, the binding of antibodies to antigens activates complement, whereas in the alternate pathway the binding of properdin to endotoxin or glycoproteins in the cell walls of bacteria or fungi activates complement. *How did complement get that name?*

Figure 15.12 Complement proteins add to—or complement—the action of antibodies.

rock avalanche in which one rock dislodges several other rocks, each of which dislodges many others until a whole cascade of rocks is tumbling down the mountain. The products of each step in the complement cascade initiate other reactions, often with wide-ranging effects in the body.

The Classical Pathway Complement got its name from events in the originally discovered "classical" pathway, in which the various complement proteins act nonspecifically to "complement," or act in conjunction with, the action of antibodies. As you study the depiction of the classical complement cascade in **Figure 15.13** on page 452, keep the following concepts in mind:

- Complement enzymes cleave other complement molecules to form *fragments.*
- Most fragments have specific and important roles in achieving the functions of the complement system. Some combine with other complement proteins to form new enzymes; some act to increase vascular permeability, which increases diapedesis; others enhance inflammation; still others are involved as chemotactic factors and in opsonization.
- The end products of the entire cascade are **membrane attack complexes (MAC),** which form circular holes in a pathogen's membrane. The production of numerous MACs (**Figure 15.14** on page 453) leads to lysis in a wide variety of bacterial and eukaryotic pathogens. Gram-negative bacteria are particularly sensitive to the production of MACs via the complement cascade because their outer membranes are exposed. In contrast, a Gram-positive bacterium, which has a thick layer of peptidoglycan overlying its cytoplasmic membrane, is typically resistant to the MAC-induced lytic properties of complement, though they are susceptible to the other effects of the complement cascade.

In addition to its enzymatic roles, fragment C3b acts as an opsonin, and fragments C3a and C5b function as chemotactic factors, attracting phagocytes to the site of infection. C3a and C5b are also inflammatory agents that trigger increased vascular permeability and dilation. The inflammatory roles of these fragments are discussed in more detail shortly.

The Alternate (Properdin) Pathway As previously mentioned, antibodies bound to antigens are necessary for the classical activation of complement, whereas activation of the alternate properdin pathway occurs independent of antibodies. Instead, it is initiated by an interaction between three proteins (called *properdin factors B, D,* and *P*) and the endotoxins and lipopolysaccharides from bacteria and fungi. These molecules from microorganisms stabilize molecules of C3b, which are always found in small quantities in the blood. Stabilized C3b combines with properdin factors B, P, and D to form an enzyme that cleaves many copies of C3 to release many more molecules of C3b. C3b molecules also combine with properdin B to form a second enzyme that cleaves molecules of C5, and the complement cascade then continues as in the classical pathway.

Even though the alternate pathway is less efficient than the classical pathway, it is useful in the early stages of infections by fungi and Gram-negative bacteria—before the specific immune response has created the antibodies needed to activate the classical pathway.

CRITICAL THINKING

A patient has a genetic disorder that makes it impossible for her to synthesize complement protein 8 (C8). Is her complement system nonfunctional? What major effects of complement could still be produced?

Inactivation of Complement We have seen that the complement system is nonspecific, and that MACs can form on any cell's exposed membrane. How do the body's own cells withstand the action of complement? Membrane-bound

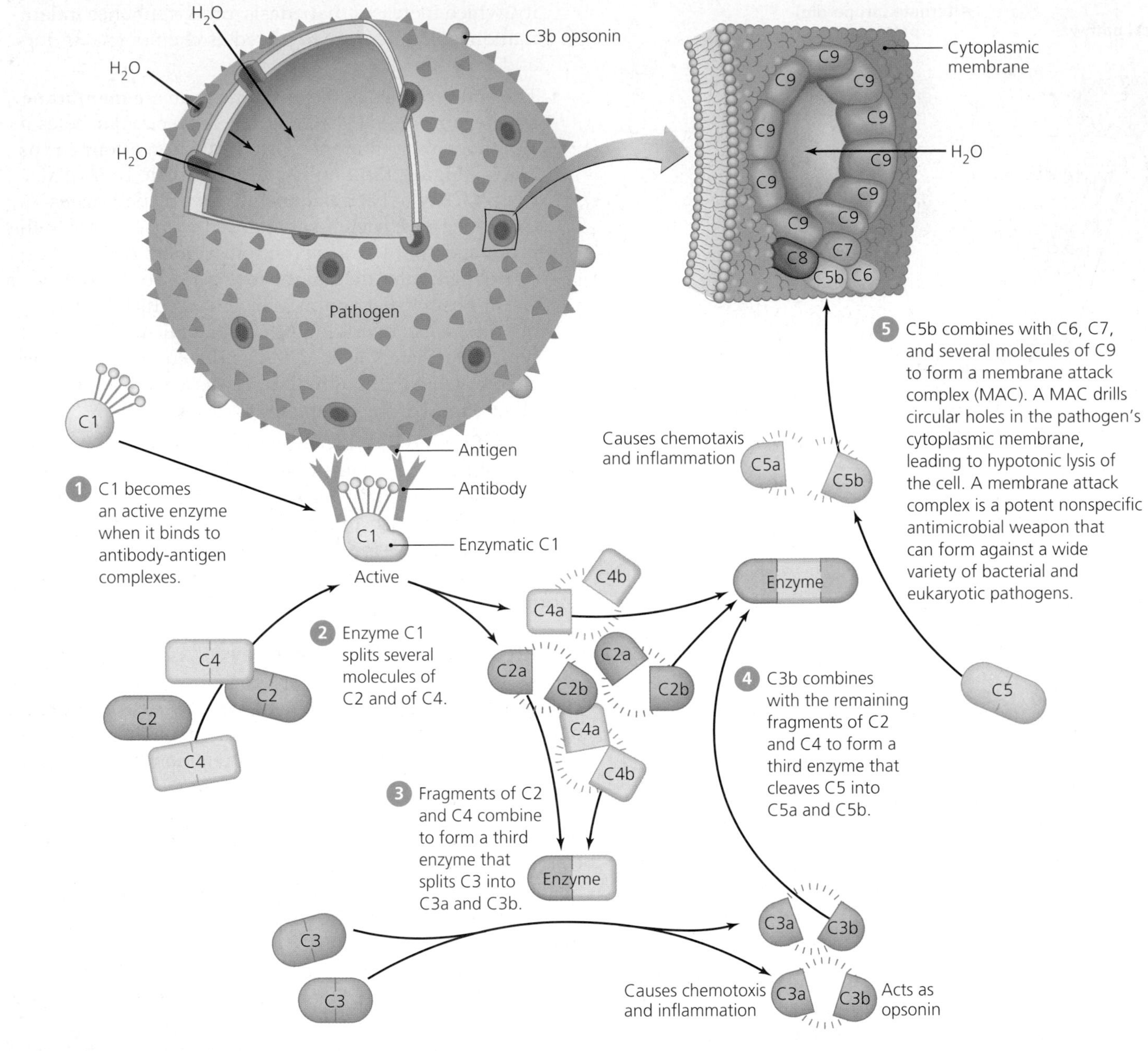

▲ ***Figure 15.13***

The classical complement cascade. *What proteins would be involved if this were the alternate pathway?*

Figure 15.13 Whereas the classical pathway of complement activation involves proteins C1, C2, and C4, the alternate pathway involves properdin factors B, D, and P.

proteins on many cells bind with and break down activated complement proteins, which interrupts the complement cascade before damage can occur. Additionally, because most human cells replace their cell membrane at a very high rate, old membrane and any membrane attack complexes within it are shed or taken in via endocytosis and digested.

Interferons

So far in this chapter we have focused primarily on how the body defends itself against bacteria and eukaryotes. Now we consider how chemicals in the second line of defense act against viral pathogens.

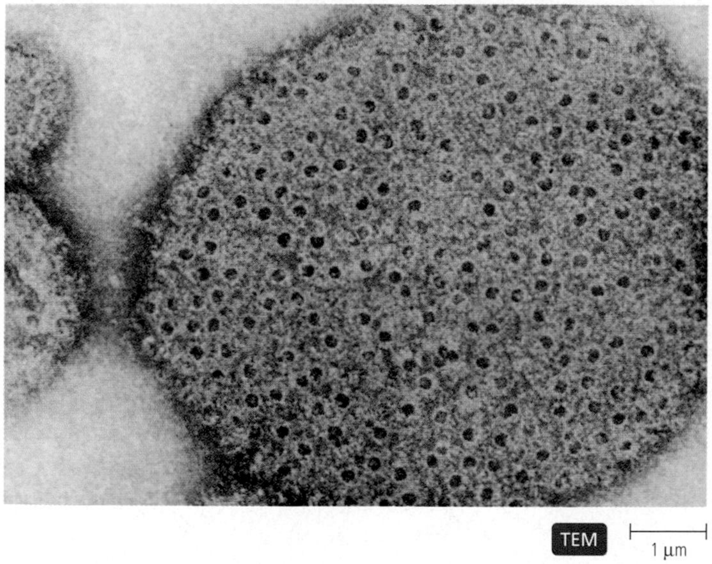

▲ ***Figure 15.14***
Transmission electron micrograph of a cell damaged by numerous punctures produced by membrane attack complexes.

As mentioned in Chapter 3, viruses use a host's metabolic machinery to produce new viruses. For this reason, it is often difficult to interfere with virus replication without also producing deleterious effects on the host. **Interferons** (in-ter-fēr'onz) are protein molecules released by host cells to nonspecifically inhibit the spread of viral infections. Its lack of specificity means that interferon produced against one viral invader protects somewhat against infection by other types of viruses as well.

Interferons are particularly effective against viruses with RNA genomes and are often crucial in reducing their effects; however, interferons also cause the malaise, muscle aches, chills, and fever typically associated with viral infections.

Each of the three basic classes of interferon—alpha, beta, and gamma—is produced by different cell types, and interferons within any given class share certain physical and chemical features though they are specific to the species that produces them. In general, alpha and beta interferons are present early in viral infections, whereas gamma interferon appears somewhat later in the course of infection. Because their actions are identical, we examine alpha- and beta-interferons together before discussing gamma-interferon.

Alpha- and Beta-Interferons Within hours after infection, virally infected monocytes, macrophages, and some lymphocytes secrete small amounts of **alpha-interferons (α-IFN);** similarly, fibroblasts, which are undifferentiated cells in such connective tissues as cartilage, tendon, and bone, secrete small amounts of **beta-interferon (β-IFN)** when infected by viruses. The structures of alpha- and beta-interferons are similar, and their actions are identical.

Interferons do not protect the cells that secrete them—these cells are already infected with viruses. Instead, interferons trigger protective steps in neighboring uninfected cells. Alpha- and beta-interferons bind to interferon receptors on the cytoplasmic membranes of neighboring cells. Such binding triggers the production of **antiviral proteins (AVPs),** which remain inactive within these cells until AVPs bind to double-stranded RNA, a molecule that is common among viruses but generally absent in eukaryotic cells (**Figure 15.15** on page 454).

At least two types of AVPs are produced: *oligoadenylate synthetase,* the action of which results in the degradation of mRNA, and *protein kinase,* which inhibits protein synthesis by ribosomes. Between them, these AVP enzymes essentially destroy the protein production system of a cell, preventing viruses from being replicated. Of course, cellular metabolism is also negatively affected. The antiviral state lasts 3–4 days, which may be long enough for a cell to rid itself of viruses and still a short enough period for the cell to survive without protein production.

Gamma-Interferon The third type of interferon, **gamma-interferon (γ-IFN),** is produced by activated T lymphocytes and NK lymphocytes. Because T lymphocytes are usually produced as part of the specific immune response (see Chapter 16) days after an infection has occurred, gamma-interferon appears later than either alpha- or beta-interferon. Its action in stimulating the activity of macrophages gives γ-IFN its other name: *macrophage activation factor.*

Although gamma-interferon plays a significant role in protecting the body against viral infections, another of its actions—stimulating the phagocytic activity of neutrophils—is also significant. This is particularly evident when one considers *chronic granulomatous disease (CGD),* a congenital condition that results from the inability to synthesize γ-IFN. Because their neutrophils cannot kill bacteria, children with CGD are susceptible to opportunistic bacterial infections, reducing their life expectancy to about 10 years.

Table 15.3 on page 455 summarizes various properties of the three types of interferon in humans.

Interferon Therapy As scientists learned more about the effects of interferons, many thought these proteins might be a "magic bullet" against viral infections. Many types of interferons in all three classes have been produced in laboratories using recombinant DNA technology, in the hopes that antiviral therapy can be improved. Recombinant interferons have two significant advantages over interferons collected from blood serum: They are pure and can be manufactured in large quantities.

The initial enthusiasm that greeted both the discovery of interferon's antiviral role and the production of recombinant interferon was not long lived. Interferons have only a slight

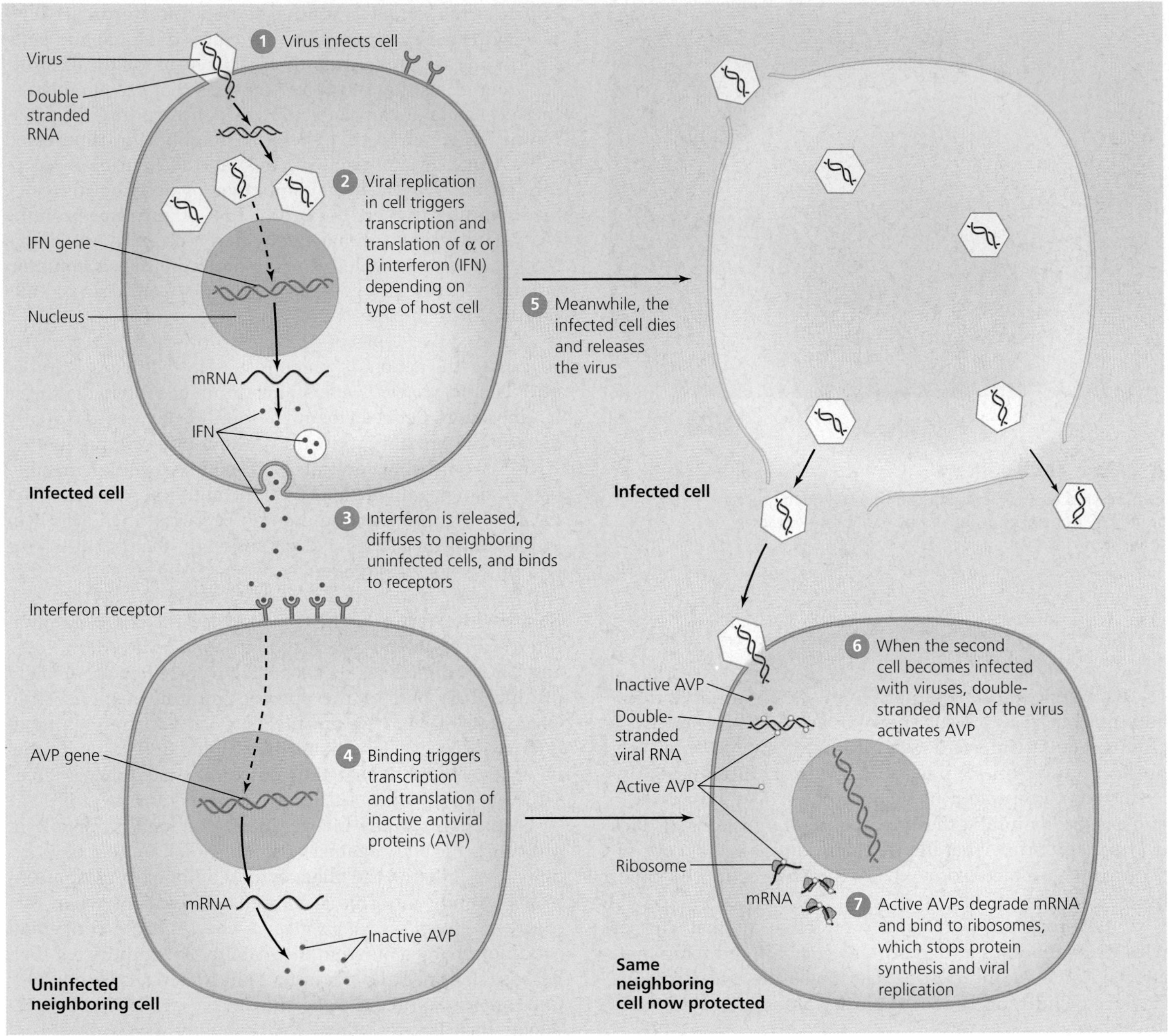

▲ ***Figure 15.15***

The actions of alpha- and beta-interferons.

effect against some major virally induced diseases, such as herpes infections, Kaposi's sarcoma (a cancer of blood vessels primarily affecting AIDS patients), and hepatitis B and C infections. Many viral infections do not respond to interferon therapy at all.

One condition that has responded well to recombinant IFN therapy, however, is chronic granulomatous disease. Injections of γ-IFN enable neutrophils in patients with CGD to kill bacteria normally. Gamma-interferon is not a cure for this condition, but it can be an effective lifelong treatment.

Defensins

Defensins (dē-fen′sinz) are small peptides (about 40 amino acids in length) that function nonspecifically to protect against a broad range of pathogens. They occur in a wide variety of animals, including invertebrates such as insects. Different classes of defensins typically act against a specific group of pathogens (for example, Gram-negative bacteria, Gram-positive bacteria, yeasts, and so on).

Although defensins do not harm the cells of the animals that secrete them, they do act against the cells of pathogens

Table 15.3 The Characteristics of Human Interferons

Property	Alpha-interferon	Beta-interferon	Gamma-interferon
Principal source	Epithelium, leukocytes	Fibroblasts	Activated T lymphocytes and NK lymphocytes
Inducing agent	Viruses	Viruses	Specific immune responses
Action	Stimulates production of antiviral proteins	Stimulates production of antiviral proteins	Stimulates phagocytic activity of macrophages and neutrophils
Molecular weight (kdaltons)	16–23	23	20–25
Glycosylated?	No	Yes	Yes
Stable at pH 2?	Yes	Yes	No
Other names	Leukocyte-IFN	Fibroblast-IFN	Immune-IFN, macrophage activation factor

in a variety of ways. Some defensins punch holes in cytoplasmic membranes (in a fashion similar to complement), some interfere with internal signaling and other metabolic processes, and one interferes with the heat shock protein *DnaK,* which normally preserves the three-dimensional shape of bacterial proteins during heat stress. In humans, defensin production increases in response to the chemicals of inflammation, which we discuss next.

Inflammation

Learning Objective

✓ Discuss the process and benefits of inflammation.

Inflammation is a general, nonspecific response to tissue damage resulting from a variety of causes, including heat, chemicals, ultraviolet light (sunburn), abrasions, cuts, and pathogens. There are two types of inflammation: **Acute inflammation,** which develops quickly and is short lived, is typically beneficial and results in the elimination or resolution of whatever condition precipitated it; whereas **chronic inflammation** develops slowly, lasts a long time, and can cause damage (even death) to tissues, resulting in disease. Both acute and chronic inflammation exhibit similar signs and symptoms, including **rubor** (redness), **calor** (localized heat), **tumor** or *edema* (swelling), and **dolor** (pain). Loss of function may also accompany severe inflammation of either type.

It may not be obvious from this list of signs and symptoms that acute inflammation is beneficial; however, acute inflammation is an important part of the second line of defense because it results in (1) dilation and increased permeability of blood vessels, (2) migration of phagocytes, and (3) tissue repair. Although the chemical details of inflammation are beyond the scope of our study, we now consider these three aspects of acute inflammation.

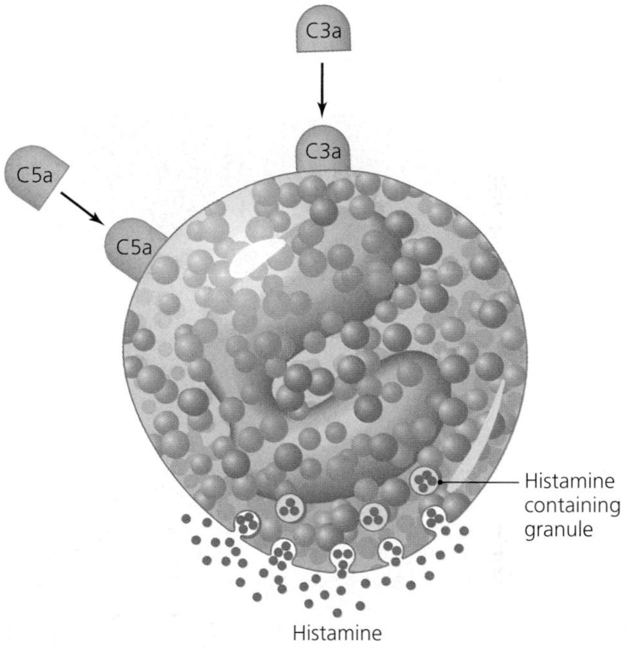

▲ ***Figure 15.16***

The stimulation of inflammation by complement. Complement fragments C3a and C5a bind to platelets, basophils, and mast cells, causing them to release histamine, which in turn stimulates the dilation of small blood vessels.

Dilation and Increased Permeability of Blood Vessels

Part of the body's initial response to an injury or invasion of pathogens is localized dilation (increase in diameter) of blood vessels in the affected region. Damaged cells release various chemicals, including **histamine** (his′tă-mēn), **prostaglandins** (pros-tă-glan′dinz), and **leukotrienes** (loo-kō-trī′ēnz). In addition, basophils, platelets, and specialized cells located in connective tissue—called **mast cells**—release histamine when they are exposed to complement fragments C3a and C5a **(Figure 15.16).** Recall that these complement peptides were cleaved from larger proteins during the complement cascade.

Histamine causes vasodilation of the body's smallest blood vessels (**Figure 15.17** on page 456). Vasodilation results in more blood being delivered to the site of infection, which in turn delivers more phagocytes, oxygen, and nutrients to

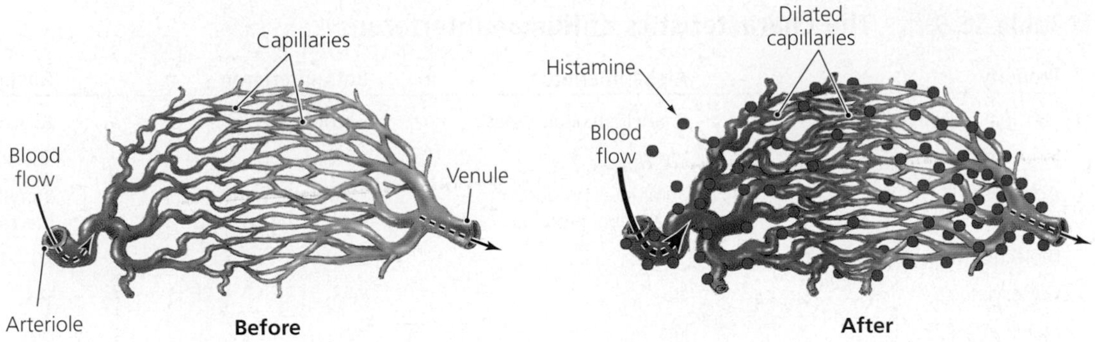

Figure 15.17 ➤
The dilating effect of histamine on small blood vessels. The release of histamine from damaged or infected tissue causes nearby arterioles and capillaries to dilate, which enables more blood to be delivered to the affected site. The increased blood flow causes the reddening and heat associated with inflammation.

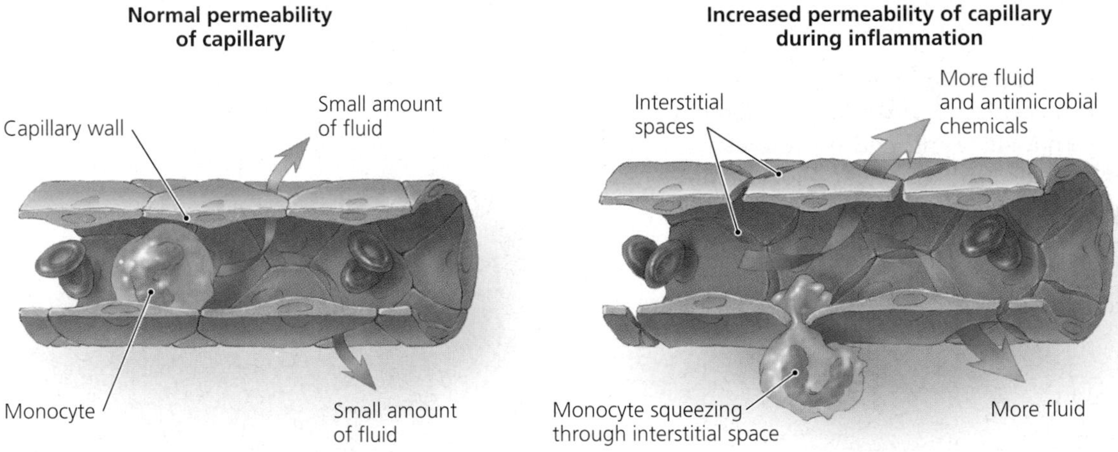

▲ *Figure 15.18*
The production of increased vascular permeability during inflammation. The presence of prostaglandins, leukotrienes, and histamine causes the cells lining capillaries (shown here, but those lining arterioles and venules as well) to pull apart, allowing phagocytes to leave the bloodstream and more easily reach a site of infection. The leakage of fluid and cells causes the edema and pain associated with inflammation.

the site. Prostaglandins, leukotrienes, and (to some extent) histamine also make blood vessels more permeable—that is, they cause cells lining the vessel to contract and pull apart, leaving gaps in the vessel's wall through which macrophages and neutrophils can move into the damaged tissue and fight invaders **(Figure 15.18).** Increased permeability also allows delivery of more blood-borne antimicrobial chemicals to the site.

Dilation of blood vessels in response to histamine results in the redness and localized heat associated with inflammation. At the same time, prostaglandins and leukotrienes cause fluid to leak from the more permeable blood vessels and accumulate in the surrounding tissue, resulting in edema, which is responsible for much of the pain of inflammation as pressure is exerted on nerve endings.

Vasodilation and increased permeability also deliver fibrinogen, the blood's clotting protein. Clots forming at the site of injury or infection wall off the area and help prevent pathogens and their toxins from spreading. One result is the formation of *pus,* a fluid containing dead tissue cells and dead leukocytes in the walled-off area of tissue destruction. Pus may push up toward the surface and erupt, or it may remain isolated in the body, where it is slowly absorbed over a period of days. Such an isolated site of infection is called an **abscess.** Pimples, boils, and pustules are examples of abscesses.

The signs and symptoms of inflammation can be treated with antihistamines, which block histamine receptors on blood vessel walls, or with antiprostaglandins. One of the ways aspirin and ibuprofen reduce pain is by acting as antiprostaglandins.

Migration of Phagocytes

Increased blood flow due to vasodilation delivers monocytes and neutrophils to the site of infection. As they arrive, these leukocytes stick to the walls of the blood vessels in a process called **margination.** They then squeeze between the cells of the vessel's wall (diapedesis) and enter the site of infection, usually within an hour of tissue damage. The phagocytes then destroy the pathogens via phagocytosis.

Figure 15.19 ➤

An overview of the events in inflammation. The process, which is characterized by redness, swelling, heat, and pain, ends with tissue repair. *In general, what types of cells are involved in tissue repair?*

Figure 15.19 *Tissue repair is effected by cells that are capable of cytokinesis and differentiation. If fibroblasts are among them, scar tissue is laid down.*

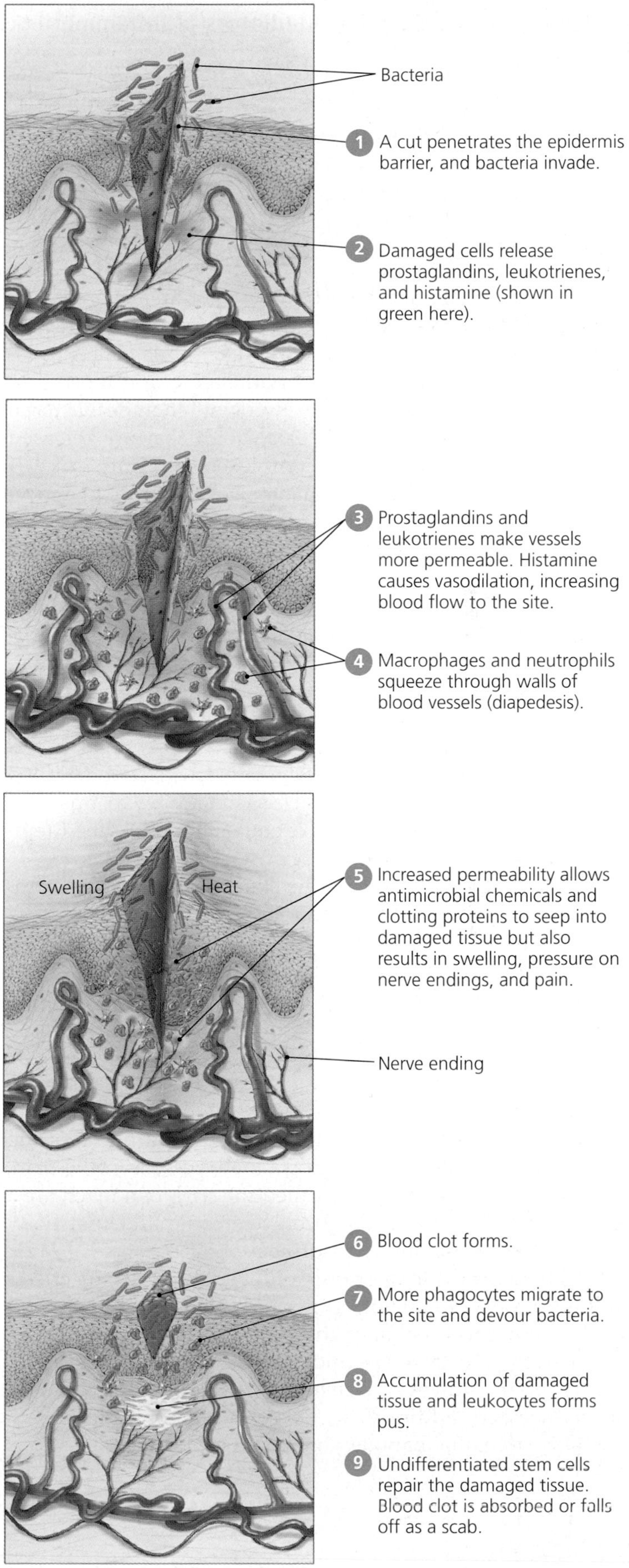

Bacteria
1 A cut penetrates the epidermis barrier, and bacteria invade.
2 Damaged cells release prostaglandins, leukotrienes, and histamine (shown in green here).
3 Prostaglandins and leukotrienes make vessels more permeable. Histamine causes vasodilation, increasing blood flow to the site.
4 Macrophages and neutrophils squeeze through walls of blood vessels (diapedesis).
Swelling
Heat
5 Increased permeability allows antimicrobial chemicals and clotting proteins to seep into damaged tissue but also results in swelling, pressure on nerve endings, and pain.
Nerve ending
6 Blood clot forms.
7 More phagocytes migrate to the site and devour bacteria.
8 Accumulation of damaged tissue and leukocytes forms pus.
9 Undifferentiated stem cells repair the damaged tissue. Blood clot is absorbed or falls off as a scab.

As mentioned previously, phagocytes are attracted to the site of infection by chemotactic factors, including leukotrienes, microbial components and toxins, and C3a and C5a. The first phagocytes to arrive are often neutrophils, which are then followed by monocytes. Once monocytes leave the blood, they change and become wandering macrophages, which are especially active phagocytic cells that devour pathogens, damaged tissue cells, and dead neutrophils. Wandering macrophages are a major component of pus.

Tissue Repair

The final stage of inflammation is tissue repair, which in part involves the delivery of extra nutrients and oxygen to the site. Although not all sites are repairable, areas of the body where cells regularly undergo cytokinesis, such as the skin and mucous membranes, are repaired rapidly.

If the damaged tissue contains undifferentiated stem cells, tissues can be fully restored. For example, a minor skin cut is repaired to such an extent it is no longer visible. However, if cells called *fibroblasts* are involved to a significant extent, scar tissue is formed, inhibiting normal function. Some tissues, such as cardiac muscle and parts of the brain, do not replicate and thus tissue damage cannot be repaired. As a result, these tissues remain damaged following heart attacks and strokes.

An overview of the entire inflammatory process is illustrated in **Figure 15.19. Table 15.4** on page 458 summarizes the chemicals involved in inflammation.

CRITICAL THINKING

While using a microscope to examine a sample of pus from a pimple, Maria observed a large number of macrophages. Is the pus from an early or a late stage of infection? How do you know?

Fever

Learning Objective

✓ Explain the benefits of fever in fighting infection.

Fever is a body temperature above 37°C. Fever augments the beneficial effects of inflammation, but like inflammation it also has unpleasant side effects, including malaise, body aches, and tiredness.

Table 15.4 Chemical Mediators of Inflammation

Vasodilating chemicals	Histamine, serotonin, bradykinin, prostaglandins
Chemotactic factors	Fibrin, collagen, mast cell chemotactic factors, bacterial peptides
Substances with both vasodilating and chemotactic effects	Complement fragments C5a and C3a, interferons, interleukins, leukotrienes, platelet secretions

The hypothalamus, a portion of the brain just above the brain stem, controls the body's internal (core) temperature. Fever results when the presence of chemicals called **pyrogens**[10] (pī′rō-jenz) trigger the hypothalamic "thermostat" to reset at a higher temperature. Pyrogens include bacterial toxins, cytoplasmic contents of bacteria that are released upon lysis, antibody-antigen complexes, and **interleukin 1 (IL-1),** a pyrogen released by phagocytes that have phagocytized bacteria. Although the exact mechanism of fever production is not known, the following discussion and **Figure 15.20** present one possible explanation.

IL-1 produced by phagocytes causes the hypothalamus to secrete prostaglandin, which resets the hypothalamic thermostat by some unknown mechanism. The hypothalamus then communicates the new temperature setting to other parts of the brain, which initiate nerve impulses that produce rapid and repetitive muscle contractions (shivering), an increase in metabolic rate, and constriction of blood vessels of the skin. These processes combine to raise the body's core temperature until it equals the prescribed temperature setting.

Because blood vessels in the skin constrict as fever progresses, one effect of inflammation (vasodilation) is undone. The constricted vessels carry less blood to the skin, causing it to feel cold to the touch, even though the body's core temperature is higher. This symptom is the *chill* associated with fever.

Fever continues as long as IL-1 is present. As an infection comes under control and fewer active phagocytes are involved, the level of IL-1 decreases, the thermostat is reset to 37°C, and the body begins to cool by perspiring, lowering the metabolic rate, and dilating blood vessels in the skin. These processes, collectively called the *crisis* of a fever, are a sign that the infection has been overcome and that body temperature is returning to normal.

The increased temperature of fever enhances the effects of interferons, inhibits the growth of some microorganisms, and is thought to enhance the performance of phagocytes, the activity of cells of specific immunity, and the process of tissue repair. However, if fever is too high, critical proteins are denatured; additionally, nerve impulses are inhibited, resulting in hallucinations, coma, and even death.

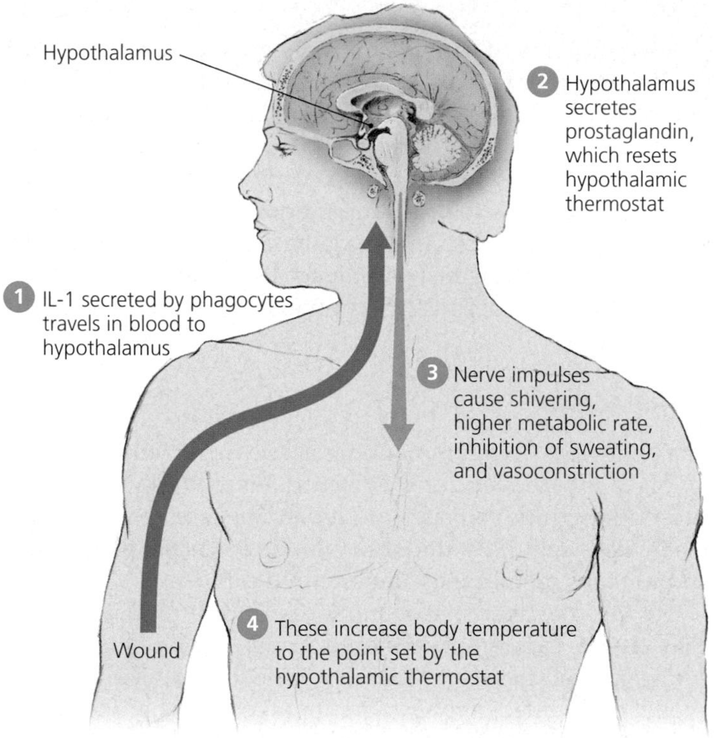
Hypothalamus
2 Hypothalamus secretes prostaglandin, which resets hypothalamic thermostat
1 IL-1 secreted by phagocytes travels in blood to hypothalamus
3 Nerve impulses cause shivering, higher metabolic rate, inhibition of sweating, and vasoconstriction
Wound
4 These increase body temperature to the point set by the hypothalamic thermostat

▲ ***Figure 15.20***
One theoretical explanation for the production of fever in response to infection.

Because of the potential benefits of fever, many doctors recommend that patients refrain from taking fever-reducing drugs unless the fever is prolonged and/or extremely high. Other physicians believe that the benefits of fever are too slight to justify enduring the adverse symptoms.

CRITICAL THINKING

How do drugs such as aspirin and ibuprofen act to reduce fever?

The barriers, cells, chemicals, and processes involved in the body's first two, nonspecific lines of defense are summarized in **Table 15.5.**

[10]From Greek *pyr,* meaning fire, and *genein,* meaning to produce.

Table 15.5 A Summary of Some Nonspecific Components of the First and Second Lines of Defense

First Line	Second Line						
Barriers and Associated Chemicals	Phagocytes	Extracellular Killing	Complement	Interferons	Defensins	Inflammation	Fever
Skin and mucous membranes prevent the entrance of pathogens; chemicals (e.g. sweat, lysozyme, mucus) enhance the protection	Macrophages, neutrophils, and eosinophils ingest and destroy pathogens	Eosinophils and NK lymphocytes kill pathogens without phagocytizing them	Components attract phagocytes, stimulate inflammation, and attack a pathogen's cytoplasmic membrane	Increase resistance of cells to viral infection, slow the spread of disease	Interfere with membranes, internal signaling, metabolism, and heat shock protein	Increases blood flow, capillary permeability, and migration of leukocytes into infected area; walls off infected region; increases local temperature	Mobilizes defenses, accelerates repairs, inhibits pathogens

CHAPTER SUMMARY

An Overview of the Body's Defenses (p. 438)

1. Humans have **innate (inborn) resistance** to certain pathogens, as well as three overlapping lines of defense. The first two lines of defense are **nonspecific** and protect the body against a wide variety of potential pathogens. The third is a specific immune response to particular pathogens.

The Body's First Line of Defense (pp. 438–442)

1. The first line of defense includes the skin, composed of an outer **epidermis** and a deeper **dermis. Dendritic cells** of the epidermis devour pathogens. Sweat glands of the skin produce salty sweat containing an antibacterial protein called **lysozyme. Sebum** is an oily substance of the skin that lowers pH, which deters the growth of many pathogens.
2. The mucous membranes, part of the body's first line of defense, are composed of tightly packed cells often coated with sticky mucus secreted by **goblet cells,** which are produced by **stem cells.**
3. **Microbial antagonism,** the competition between normal microbiota and potential pathogens, also contributes to the body's first line of defense.
4. Tears contain antibacterial lysozyme and also flush invaders from the eyes. Saliva similarly protects the teeth. The low pH of the stomach inhibits most microbes that are swallowed.

An Overview of the Body's Second Line of Defense (pp. 442–447)

1. The second line of defense includes cells (especially **phagocytes**), antimicrobial chemicals (complement, interferons, defensins), and processes (inflammation and fever).
2. Blood is composed of **formed elements** (cells and parts of cells) within a fluid called **plasma.** Serum is that portion of plasma without clotting factors. The formed elements are **erythrocytes** (red blood cells), **leukocytes** (white blood cells), and **platelets.**
3. Based on their appearance in stained blood smears, leukocytes are grouped as either **granulocytes (basophils, eosinophils,** and **neutrophils)** or **agranulocytes (monocytes, lymphocytes).** When monocytes leave the blood they become macrophages.
4. Basophils function to release histamine during inflammation, whereas eosinophils and neutrophils phagocytize pathogens. They exit the capillaries via **diapedesis (emigration).**
5. Macrophages are the principal phagocytic cells of the second line of defense. They are named for their location in the body: **wandering macrophages, alveolar macrophages** (the lungs), **microglia** (the nervous system), and **Küpffer cells** (the liver). The collection of macrophages—along with monocytes attached to the inner layer of blood and lymphatic vessels—are called either the **reticuloendothelial system** or the **mononuclear phagocytic system.**
6. A **differential white blood cell count** is a lab technique that indicates the relative numbers of leukocytes; it can be helpful in diagnosing disease.

The Body's Second Line of Defense (pp. 448–458)

1. **Chemotactic factors,** such as chemicals called **cytokines,** attract phagocytic leukocytes to the site of damage or invasion. The phagocytes attach to pathogens via a process called **adherence.**
2. **Opsonization,** the coating of pathogens by proteins called **opsonins,** makes those pathogens more vulnerable to phagocytes. A phagocyte's pseudopodia then surround the microbe to form a sac called a **phagosome,** which fuses with a lysosome to form a **phagolysosome,** in which the pathogen is digested.
3. Leukocytes can distinguish between the body's normal cells and foreign cells because they have receptor molecules for foreign cells' components or because the foreign cells are opsonized by complement or antibodies.

4. Eosinophils and **natural killer (NK) lymphocytes** attack extracellularly, especially in the case of helminth infections and cancerous cells. Eosinophilia—an abnormally high number of eosinophils in the blood—typically indicates such a helminth infection.
5. The **complement system** is a set of proteins that act as chemotactic attractants, trigger inflammation and fever, and ultimately effect the destruction of foreign cells via the formation of **membrane attack complexes (MAC),** which result in multiple, fatal holes in pathogens' membranes. Complement is activated by a classical pathway involving antibodies and by an alternate pathway triggered by bacterial chemicals.
6. **Interferons** are protein molecules that inhibit the spread of viral infections. **Alpha-interferons** and **beta-interferons,** which are released within hours of infection, trigger **antiviral proteins** to prevent viral reproduction in neighboring cells. **Gamma-interferons,** produced days after initial infection, activate macrophages and neutrophils.
7. **Defensins** are small peptide chains, produced by most animals, that act against a broad range of pathogens.
8. **Acute inflammation** develops quickly and damages pathogens, whereas **chronic inflammation** develops slowly and can cause bodily damage that can lead to disease. Signs and symptoms of inflammation include **rubor** (redness), **calor** (heat), **tumor** (swelling), and **dolor** (pain).
9. Infection causes damaged cells to release chemicals such as **histamine,** which triggers **vasodilation,** and **prostaglandins** and **leukotrienes,** which increase permeability of blood vessels. **Mast cells** also release histamine when exposed to peptides from the complement system. Blood clots may isolate an infected area to form an **abscess** such as a pimple or boil.
10. When leukocytes arrive near the site of infection, they stick to the walls of blood vessels in a process called **margination,** and they then undergo diapedesis to arrive at the site of tissue damage. The increased blood flow of inflammation also brings extra nutrients and oxygen to the infection site to aid in repair.
11. **Fever** results when chemicals called **pyrogens,** including **interleukin 1,** affect the hypothalamus in a way that causes it to reset body temperature at a higher level. The exact process of fever and its control are not fully understood.

QUESTIONS FOR REVIEW

(Answers to multiple choice and matching questions are on the web, along with additional review questions. Visit www.microbiologyplace.com.)

Multiple Choice

1. Fixed phagocytes of the epidermis are called
   a. lysozymes.
   b. Goblet cells.
   c. Küpffer cells.
   d. dendritic cells.
2. Mucus-secreting membranes are found in the
   a. urinary system.
   b. digestive cavity.
   c. respiratory passages.
   d. all of the above
3. The complement system involves
   a. the production of antigens and antibodies.
   b. serum proteins involved in nonspecific defense.
   c. a set of genes that distinguish foreign cells from body cells.
   d. the elimination of undigested remnants of microorganisms.
4. The alternate pathway
   a. is also called the properdin pathway.
   b. involves the cleavage of C5 to form C9.
   c. is more efficient than the classical pathway.
   d. involves recognition of antigens bound to specific antibodies.
5. Complement must be inactivated because if it were not,
   a. viruses could continue to multiply inside host cells using the host's own metabolic machinery.
   b. necessary interferons would not be produced.
   c. protein synthesis would be inhibited, thus halting important cell processes.
   d. it would make holes in the body's own cells.
6. The type of interferon present late in an infection is
   a. alpha-interferon.
   b. beta-interferon.
   c. gamma-interferon.
   d. delta-interferon.
7. Interferons
   a. do not protect the cell that secretes them.
   b. stimulate the activity of neutrophils.
   c. cause muscle aches, chills, and fever.
   d. all of the above
8. In which of the following choices are events in the correct order?
   a. the formation of a MAC, holes in the pathogen's cell membrane, a cascade of reactions, the binding of C1
   b. the binding of C1, a cascade of reactions, the formation of a MAC, holes in the pathogen's cell membrane
   c. a cascade of reactions, the binding of C1, the formation of a MAC, holes in the pathogen's cell membrane
   d. the binding of C1, a cascade of reactions, holes in the pathogen's cell membrane, the formation of a MAC

Matching

In the blank beside each cell, chemical, or process in the left-hand column, write the letter of the line of defense that best applies. Each letter may be used several times.

1. ___ Inflammation	A. First line of defense
2. ___ Monocytes	B. Second line of defense
3. ___ Cleavage of C3	C. Third line of defense
4. ___ Fever	
5. ___ Dendritic cells	
6. ___ Alpha-interferon	
7. ___ Mucous membrane of the digestive tract	
8. ___ Neutrophils	
9. ___ Epidermis	

10. ___ Lysozyme
11. ___ Goblet cells
12. ___ Phagocytes
13. ___ Sebum
14. ___ T lymphocytes
15. ___ Defensins

Write the letter of the description that applies to each of the following.

1. ___ Goblet cell
2. ___ Lysozyme
3. ___ Küpffer cell
4. ___ Dendritic cell
5. ___ Cell from sebaceous gland
6. ___ Bone marrow stem cell
7. ___ Eosinophil
8. ___ Alveolar macrophage
9. ___ Microglia
10. ___ Wandering macrophage

A. Leukocyte that primarily attacks parasitic worms
B. Phagocytic cell in lungs
C. Secrets sebum
D. Devours pathogens in epidermis of skin
E. Breaks bonds in bacterial cell wall
F. Phagocytic cell in central nervous system
G. Phagocytic cell in liver
H. Develops into formed elements of blood
I. Intercellular scavenger
J. Secretes mucus

Modified True/False

Indicate which statements are true. Correct all false statements by writing in the blank a word that, if substituted for the italicized word, would make the statement true.

1. ________ The surface cells of the epidermis of the skin are *alive*.
2. ________ The surface cells of mucous membranes are *alive*.
3. ________ Wandering macrophages experience *diapedesis*.
4. ________ *Monocytes* are immature macrophages.
5. ________ *Lymphocytes* are large agranulocytes.
6. ________ *Phagocytes* exhibit chemotaxis toward a pathogen.
7. ________ Adherence involves the binding between complementary chemicals on a *phagocyte* and on the membrane of a body cell.
8. ________ *Opsonins* result when a phagocyte's pseudopodia surround a microbe and fuse to form a sac.
9. ________ Lysosomes fuse with phagosomes to form *digestive vesicles*.
10. ________ A membrane attack complex drills circular holes in a *macrophage*.
11. ________ Rubor, calor, tumor, and dolor are associated with *fever*.
12. ________ Acute and chronic inflammation exhibit *similar* signs and symptoms.
13. ________ The *hypothalamus* of the brain controls body temperature.
14. ________ Basophils release *leukotrienes* when they are exposed to complement fragments C3a and C5a.
15. ________ Defensins are *phagocytic cells* of the second line of defense.

Short Answer

1. In order for a pathogen to cause disease, what three things must happen?
2. How does a phagocyte "know" it is in contact with a pathogen instead of another body cell?
3. Give three characteristics of the epidermis that make it an intolerable environment for most microorganisms.
4. Other than the different molecules involved, what is the difference between classical and alternate pathways?
5. Describe the classical complement cascade pathway from C1 to the MAC.
6. Why is complement protein C4 "out of order"?

CRITICAL THINKING

1. Neighbor A received a chemical burn on his arm and was instructed by his physician to take an over-the-counter, anti-inflammatory medication for the painful, red, swollen lesions. When Neighbor B suffered pain, redness, and swelling from an infected cut on his foot, he decided to take the same anti-inflammatory drug because his symptoms matched Neighbor A's symptoms. How is Neighbor B's inflammation like that of Neighbor A? How is it different? Is it appropriate for B to medicate his cut with the same medicine A used?
2. What might happen to someone whose body did not produce C3? C5?
3. Duncan, age 65, has had diabetes for 40 years, with resulting damage to the small blood vessels in his feet and toes. His circulation is impaired. How might this condition affect his vulnerability to infection?
4. A patient's chart shows that eosinophils make up 8% of his white blood cells. What does this lead you to suspect? Would your suspicions change if you learned that the patient had spent the previous 3 years as an anthropologist living among an African tribe? What is the normal percentage of eosinophils?
5. There are two kinds of agranulocytes in the blood—monocytes and lymphocytes. Janice noted that monocytes are phagocytic and that lymphocytes are not. She wondered why two agranulocytes would be so different. What facts of hematopoiesis can answer her question?
6. A patient has a genetic disorder that prevents him from synthesizing C8 and C9. What effect would this have on his ability to resist blood-borne Gram-negative and Gram-positive bacteria? What would happen if C3 and C5 fragments were also inactivated?

CHAPTER 16 Specific Defense: The Immune Response

CHAPTER OUTLINE

Elements of Specific Immunity
- Antigens
- The Cells, Tissues, and Organs of the Lymphatic System
- B Lymphocytes (B Cells) and Antibodies
- T Lymphocytes (T Cells)
- Immune System Cytokines

The Body's Preparation for a Specific Immune Response
- Lymphocyte Editing by Clonal Deletion
- The Roles of the Major Histocompatibility Complex
- Antigen Processing

The Humoral Immune Response
- B Cell Activation and Clonal Selection
- Memory B Cells and the Establishment of Immunological Memory

The Cell-Mediated Immune Response
- The Perforin-Granzyme Cytotoxic Pathway
- The CD95 Cytotoxic Pathway
- Memory T Cells
- T Cell Regulation

Types of Acquired Immunity
- Naturally Acquired Active Immunity
- Naturally Acquired Passive Immunity
- Artificially Acquired Active Immunity
- Artificially Acquired Passive Immunity

Imagine that your friend has been bitten on the hand by a dog. The skin is broken, and the wound is deep. He washes the wound with soap and water, but does not use iodine or any of the more powerful antimicrobial agents we discussed in Chapter 9. Within 24 hours, his hand and arm are swollen, red, and painful, and he is feeling seriously ill. His body's inflammatory response is in full force, but it isn't strong enough to overcome the invading microorganisms.

Fortunately, infections like this trigger the body's *specific immunity,* which is far more sophisticated than the nonspecific defense responses discussed in Chapter 15. In this chapter we will explore what happens when cells of the specific immune system—especially lymphocytes—enter the battle against microbial invaders.

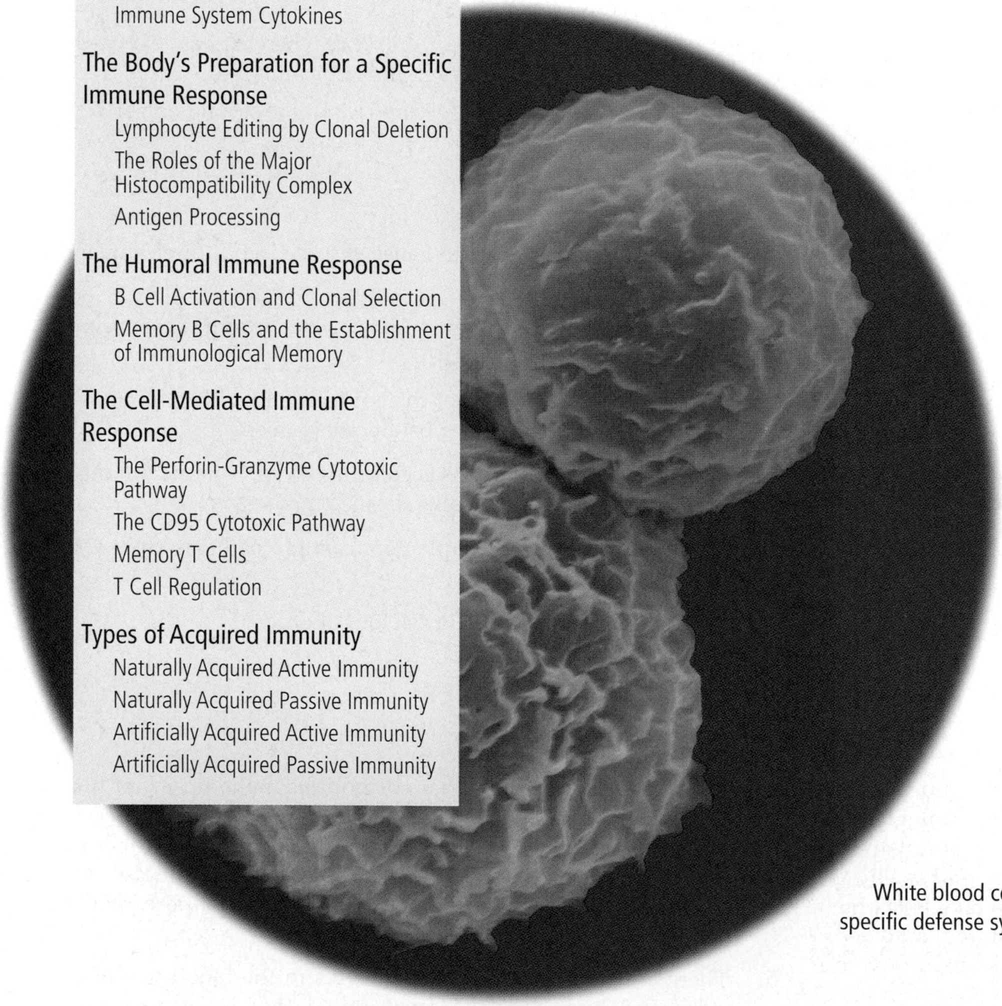

White blood cells called lymphocytes are an important component of the specific defense system.

MicroPrep Pre-Test: *Take the pre-test for this chapter on the web. Visit* ***www.microbiologyplace.com****.*

In Chapter 15 we discussed the first two lines of defense against microbial pathogens. These *nonspecific defenses* include intact skin and mucous membranes, phagocytosis, complement, inflammation, and fever. We saw as well that humans have innate (inborn) resistance against pathogens of plants and certain pathogens of animals because human cells lack surface receptors to which these microbes can attach.

As is evident in the chapter opening vignette, nonspecific defenses do not always offer the ideal approach to defending the body. Although the mechanisms of nonspecific defense are readily available and fast-acting, they do not change greatly in either intensity or effectiveness in response to the great variety of pathogens confronting us—a fever is a fever, whether it is triggered by a mild influenza virus or the deadly Ebola virus. Additionally, we have seen that the inflammatory response can have undesirable, even highly damaging effects; imagine how uncomfortable life would be if the body mounted a full-blown inflammatory response every time even a single bacterium entered the body.

The body augments the mechanisms of nonspecific defense with another line of defense that destroys invaders while "learning" in the process. This third line of defense is a "smart" system that has "memory" about specific pathogens, so that when it encounters them a second time, it responds rapidly and effectively. This response is called **specific immunity**—the body's ability to recognize and defend itself against distinct invaders and their products, whether they are protozoa, fungi, bacteria, viruses, or toxins.

Elements of Specific Immunity

In contrast to innate defenses, specific immunity is *acquired;* that is, even though it is present only in a rudimentary form at birth, it develops as the cells of specific immunity encounter and respond to various pathogens and abnormal body cells. *Immunologists*—scientists who study the cells and chemicals involved in specific immunity—are continually refining and revising our knowledge of specific immune processes as well as revealing entirely new aspects of this third line of defense. In the following sections we will survey aspects of specific immunity. Just as the program at a dramatic presentation introduces the actors and their roles, so the following sections introduce the "cast of characters" involved in specific immunity. First we will examine molecules called *antigens* that trigger specific immune responses, followed by a brief description of the cells, tissues, and organs of specific immunity. Then we will examine in some detail two groups of white blood cells—B lymphocytes and T lymphocytes—that function in specific immunity, and conclude by considering special chemical signals or mediators involved in coordinating a specific immune response.

Antigens

Learning Objectives

- ✓ Identify the characteristics of effective antigens.
- ✓ Distinguish among exogenous antigens, endogenous antigens, and autoantigens.

The body does not direct specific defense against whole bacteria, fungi, protozoa, or viruses, but instead against portions of cells or viruses, and even single molecules that the body recognizes as foreign and worthy of attack. Immunologists call the molecules that trigger a specific immune response **antigens**[1] (an'ti-jenz). Examples of antigens include components of bacterial cell walls, capsules, pili, and flagella, as well as the external and internal proteins of viruses, fungi, and protozoa. Many toxins are also antigenic. Invading microorganisms are not the only source of antigens; for example, food may contain antigens that provoke allergic reactions, and inhaled dust contains mite feces, pollen grains, dander (flakes of skin), and other antigenic particles.

Pathogens (and their antigens) can enter the body via the various portals of entry: through breaks in the skin and mucous membranes of the respiratory, genitourinary, or digestive tracts; through direct injection, as with a snake's bite or a needle administering a vaccine or blood transfusion; or through organ transplants and skin grafts.

Properties of Antigens

Not every molecule is an effective antigen. Among the properties that make certain molecules "more antigenic"—that is, more effective at provoking an immune response—are a molecule's shape, size, and complexity.

The body recognizes antigens by the three-dimensional shapes of regions called **antigenic determinants,** which are also known as *epitopes* (**Figure 16.1a** on page 464). Therefore, stable molecules—those that assume and maintain a definite shape—are more effective antigens than molecules like the protein gelatin, which have changeable shapes.

In general, larger molecules with molecular masses (often called molecular weights) between 5000 and 100,000 daltons are better antigens than smaller ones. The most effective antigens are large foreign macromolecules such as proteins and glycoproteins, but complex carbohydrates and lipids, as well as some bacterial DNA, can be antigenic. Small molecules, especially those with a molecular mass under 5000 daltons, make poor antigens by themselves because they evade detection. Such small molecules, called *haptens,* can become antigenic when bound to larger, carrier molecules (often proteins). For example, proteins in the blood bind to the fungal product, penicillin, which by itself is too small (molecular mass: 302 daltons) to trigger a specific immune response. Bound to a carrier, penicillin can become antigenic and trigger an allergic response in some patients.

[1] From *anti*body *gen*erator; antibodies are specific proteins secreted during a specific immune response.

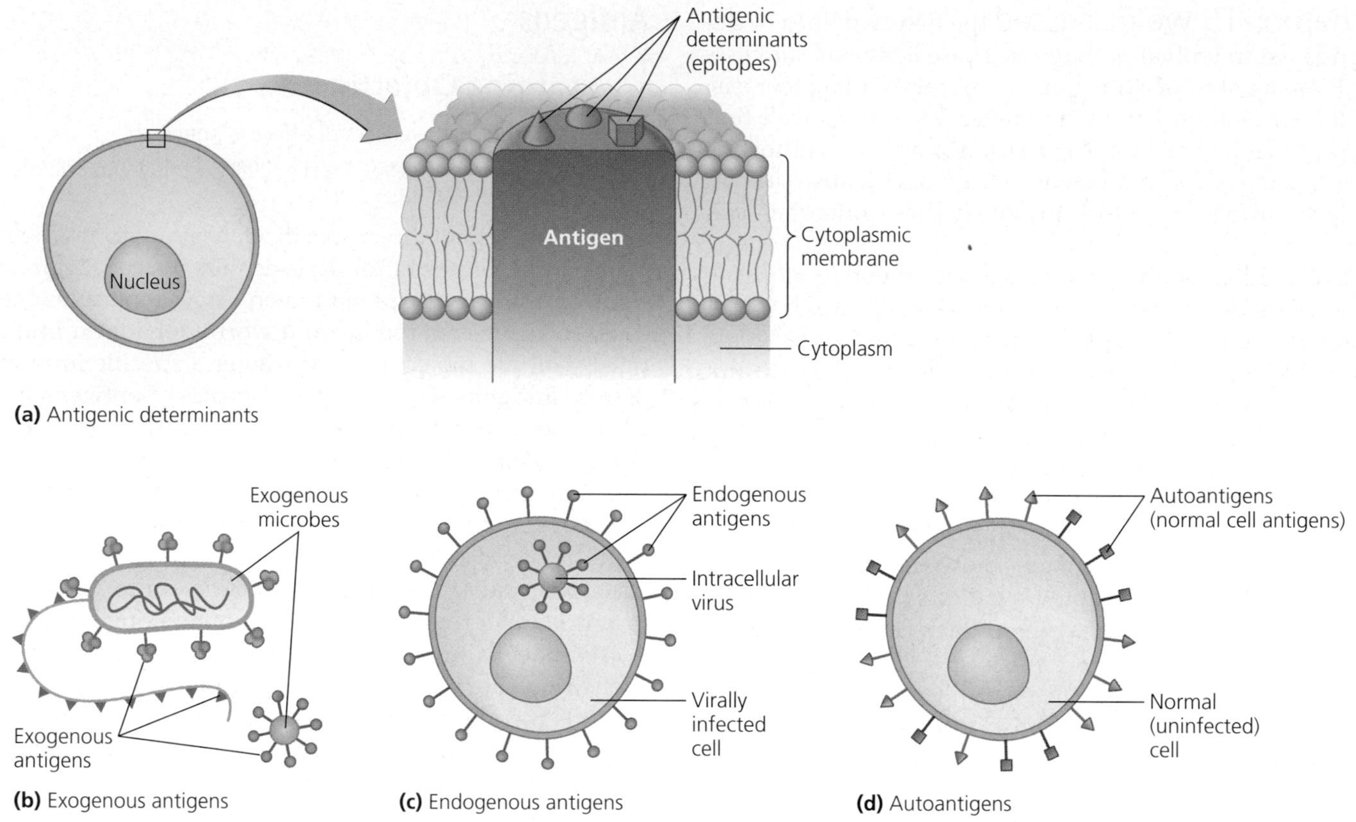

▲ *Figure 16.1*

Antigens, molecules that provoke a specific immune response. **(a)** Antigenic determinants or epitopes, the three-dimensional regions of antigens whose shapes are recognized by cells of the immune system. **(b–d)** Categories of antigens based on their relationship to the body. Exogenous antigens originate from microbes located outside the body's cells; endogenous antigens are produced by intracellular microbes and are typically incorporated into a host cell's cytoplasmic membrane; autoantigens are components of normal, uninfected body cells that are mistakenly recognized as foreign.

Complex molecules make better antigens than simple ones. For example, starch, which is a large polymer of repeating glucose subunits, is not a good antigen despite its very large size because it lacks structural complexity. In contrast, complex molecules, such as glycoproteins and phospholipids, have distinctive shapes and novel combinations of subunits that cells of the immune system recognize as foreign antigenic determinants.

Types of Antigens

Though immunologists categorize antigens in various ways, one especially important way is to group antigens according to their relationship to the body:

- Microbes outside the cells of the body, such as eukaryotic parasites, bacteria, and viruses, are exogenous[2] (eks-oj′en-us) and bear **exogenous antigens (Figure 16.1b).** Exogenous antigens include toxins and other secretions and components of microbial cell walls, membranes, flagella, and pili. Some DNA released from dead bacteria is also antigenic.
- Protozoa, fungi, bacteria, and viruses that reproduce inside a body's cells produce **endogenous**[3] (en-doj′e-nus) **antigens.** The immune system cannot "see" into cells and can respond to endogenous antigens only if the body's cells incorporate such antigens into their cytoplasmic membranes **(Figure 16.1c).**
- **Autoantigens**[4] are antigenic molecules found on an individual's normal (uninfected) cells **(Figure 16.1d).** As we will discuss more fully in a later section, immune cells that recognize autoantigens as foreign are normally eliminated during the development of the immune system. This phenomenon, called *self-tolerance,* prevents the body from mounting an immune response against itself.

In order to understand how the body mounts a specific immune response against antigens, we must first examine

[2]From Greek *exo,* meaning without, and *genein,* meaning to produce.
[3]From Greek *endon,* meaning within.
[4]From Greek *autos,* meaning self.

the lymphatic system, which plays an important role in the production, maturation, and housing of the cells that function in specific immunity.

The Cells, Tissues, and Organs of the Lymphatic System

Learning Objectives

- ✓ Contrast the flow of lymph with the flow of blood.
- ✓ Describe the tissues and organs of the lymphatic system.
- ✓ Describe the importance of red bone marrow, the thymus, lymph nodes, and other lymphoid tissues.

The **lymphatic system** is composed of the *lymphatic vessels,* which conduct the flow of a liquid called *lymph;* and lymphatic cells, tissues, and organs, which are directly involved in specific immunity (**Figure 16.2** on page 466). Taken together, the cells, tissues, and organs of the lymphatic system constitute a surveillance system that screens the tissues of the body—particularly possible points of entry such as the throat and intestinal tract—for foreign antigens. We begin by examining the lymphatic vessels.

The Lymphatic Vessels and the Flow of Lymph

The **lymphatic vessels** form a one-way system that conducts **lymph** (limf)—a colorless, watery liquid similar in composition to blood plasma—from local tissues and returns it to the circulatory system. Indeed, lymph arises from fluid that has leaked out of blood vessels into the surrounding tissues. Lymph is first conducted by remarkably permeable *lymphatic capillaries,* which are located in most parts of the body (exceptions are the bone marrow, brain, and spinal cord) and collect fluid from intercellular spaces. From the lymphatic capillaries, lymph passes into increasingly larger lymphatic vessels until it finally flows via two large *lymphatic ducts* into blood vessels near the heart. Unlike the cardiovascular system, the lymphatic system has no unique pump and is not circular; that is, lymph flows in only one direction—toward the heart. One-way valves ensure that lymph flows only toward the heart as muscular activity squeezes the lymphatic vessels. Located at various points within the system of lymphatic vessels are numerous *lymph nodes* (discussed shortly), which house leukocytes that recognize and attack foreign antigens present in the lymph.

Lymphoid Cells, Organs, and Tissues

The site of production of the cells of specific immunity is the *red bone marrow,* located in the ends of long bones and in the centers of flat bones such as the ribs and hip bones. These sites contain *stem cells,* which are cells that give rise to all types of blood cells, including those that have a central role in specific immunity—**lymphocytes** (lim'fō-sītz). Lymphocytes are the smallest leukocytes and each is characterized by a large, round, central nucleus surrounded by a thin rim of cytoplasm (**Figure 16.3a** on page 467).

Although lymphocytes are indistinguishable microscopically, scientists distinguish two main types, called *B lymphocytes* and *T lymphocytes* (discussed shortly), according to the integral surface glycoproteins present in each lymphocyte's cytoplasmic membrane. These glycoproteins are given internationally accepted designations consisting of a number following the prefix *CD* (for *cluster of differentiation*). These numbers—for example, CD4, CD8, CD26, and CD95—reflect the order in which the glycoproteins were discovered, not the order in which they are produced or function. Scientists have identified about 250 CD molecules, only some of which are present on lymphocytes.

After lymphocytes are produced in the bone marrow, they must undergo a maturation process. The type called B lymphocytes mature within the bone marrow, but T lymphocytes travel to and mature in the *thymus,* located in the chest, just above the heart (see Figure 16.2). The thymus is most active early in life and then decreases in size and activity after puberty, after its function has been largely completed.

Once lymphocytes have arisen and have matured in either the bone marrow or thymus, they migrate to what are often considered secondary lymphoid organs and tissues, including the lymph nodes, the spleen, and other, less-organized accumulations of lymphoid tissue.

Even though lymphocytes can circulate throughout the body in blood and lymph, many remain in lymphoid organs and tissues, where they in effect lie in wait for foreign antigens (see Figure 16.3b). Among these sites are hundreds of **lymph nodes** located throughout the body but concentrated in the cervical (neck), inguinal (groin), axillary (armpit), and abdominal regions (see Figure 16.2). Each lymph node receives lymph from numerous afferent lymphatic vessels and drains lymph into just one or two efferent lymphatic vessels (**Figure 16.4** on page 467). A node consists of a medullary (central) maze of passages that filter the lymph passing through it and houses numerous leukocytes, which survey the lymph for foreign antigens and mount a specific immune response against them. The cortex (outer) portion of a lymph node consists of a tough capsule surrounding *germinal centers,* where some cells of specific immunity replicate.

CRITICAL THINKING

The cross-sectional area of the afferent lymphatic vessels arriving at a lymph node is greater than the cross-sectional area of the efferent lymphatics exiting the lymph node. The result is that lymph moves slowly through a lymph node. What advantage does this provide the body?

The lymphatic system contains other lymphoid tissues and organs, including the spleen, the tonsils, and mucosa-associated lymphatic tissue (MALT; see Figure 16.2). The spleen is similar in structure and function to lymph nodes, except that it filters blood instead of lymph. Thus the spleen removes bacteria, viruses, toxins, and other foreign matter from the blood. It also cleanses the blood of old and damaged

Figure 16.2 ➤
The lymphatic system consists of lymphatic vessels containing lymph, plus various lymphoid structures, including lymph nodes, the thymus, tonsils, and mucosa-associated lymphatic tissue (MALT).

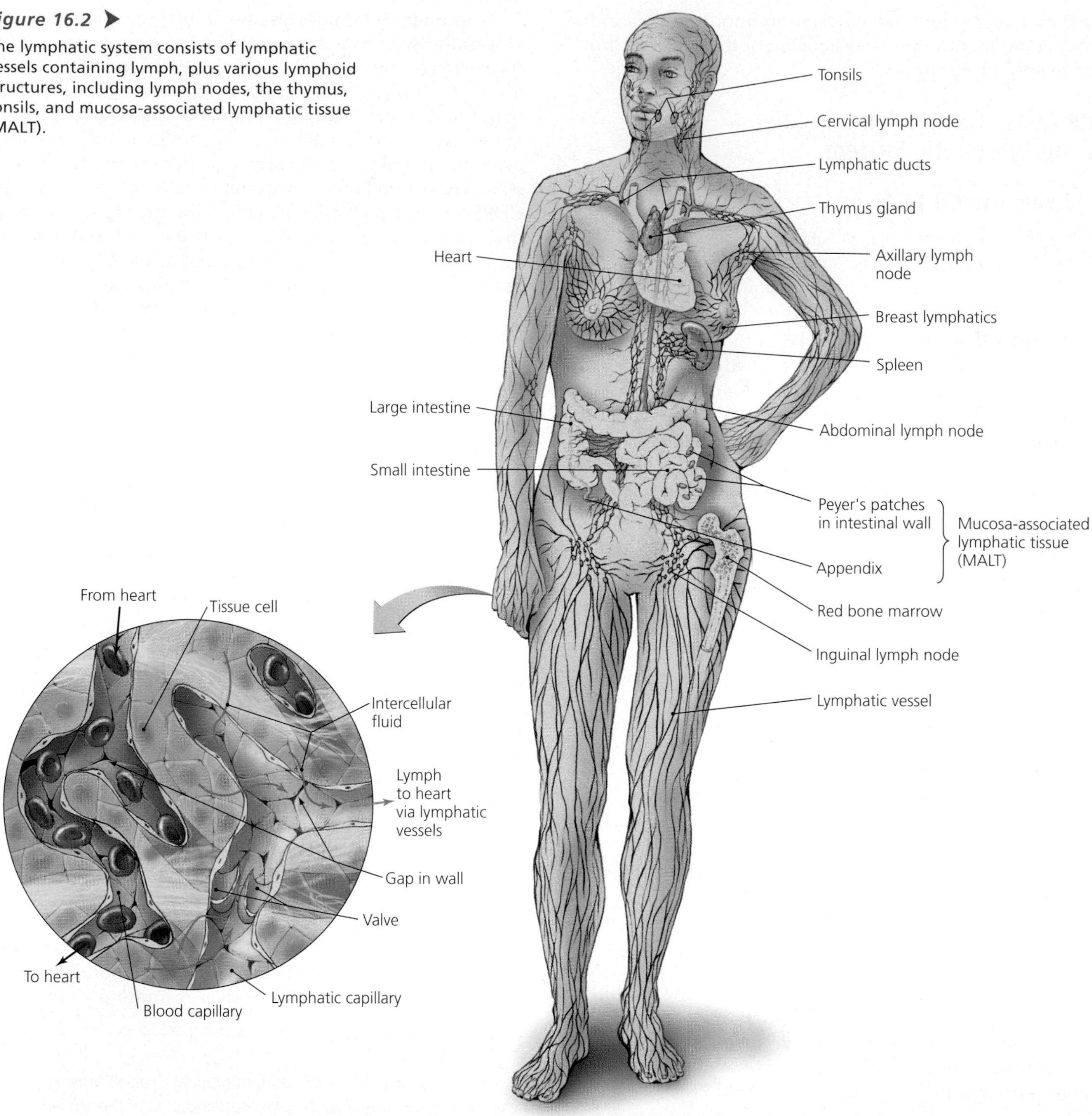

blood cells, stores blood platelets (which are required for the proper clotting of blood), and stores blood components such as iron. The tonsils and MALT lack the tough outer capsules of lymph nodes and spleen, but they function in somewhat the same way by physically trapping foreign particles and microbes. MALT includes the appendix, lymphoid tissue of the respiratory tract, and discrete bits of lymphoid tissue called *Peyer's patches* in the wall of the small intestine. MALT contains leukocytes that defend against foreign antigens entering the blood and the respiratory and digestive tracts.

CRITICAL THINKING

As part of the treatment for some cancers, physicians kill the cancer patients' dividing cells, including the stem cells that produce leukocytes, and then give the patients a bone marrow transplant from a healthy donor. Which cell is the most important cell in such transplanted marrow?

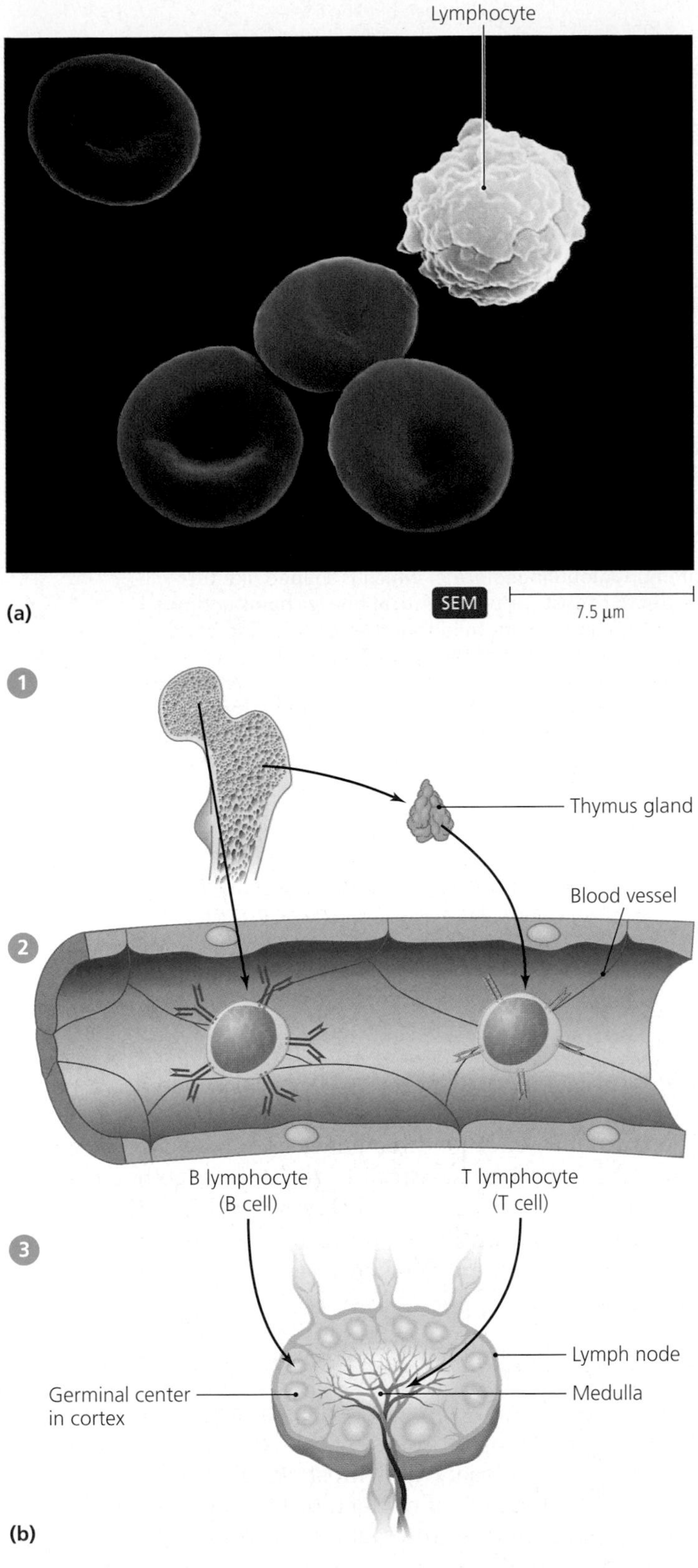

◀ *Figure 16.3*

Lymphocytes, the smallest leukocytes, play a central role in specific immunity. **(a)** A lymphocyte is the smallest leukocyte—about the size of a red blood cell. **(b)** The production, maturation, and deployment of lymphocytes. After lymphocytes are produced in the red bone marrow ①, they mature in either the bone marrow (B cells) or the thymus (T cells) before ② circulating in the blood and lymph or ③ waiting in the lymph nodes (or other lymphoid tissues) for foreign antigens.

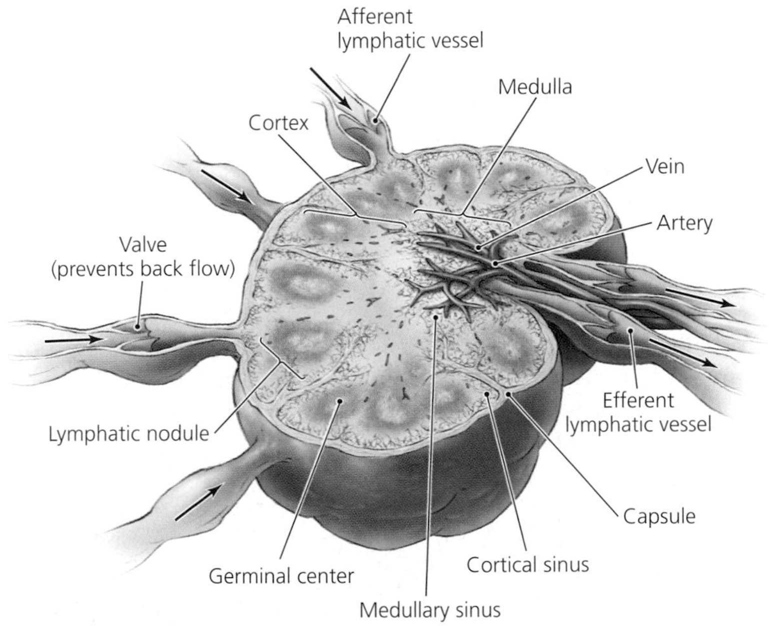

▲ *Figure 16.4*

A lymph node. Lymph enters the node through afferent lymphatic vessels and exits through one or two efferent lymphatic vessels. The outer cortex contains germinal centers, where B lymphocytes proliferate; the inner medulla contains numerous lymphocytes that encounter foreign antigens. A tough capsule surrounds the entire node. *Why do lymph nodes enlarge during an infection?*

Figure 16.4 *During an infection, lymphocytes multiply profusely in lymph nodes. This proliferation, and swelling caused by the secretion of inflammatory chemicals, causes lymph nodes to enlarge.*

We have seen that lymphocytes can either circulate or remain in lymphoid tissues, where they represent a surveillance system that screens the tissues of the body, particularly the portals of entry for foreign antigens. Next we turn our attention to each of the two types of lymphocytes, beginning with B lymphocytes.

B Lymphocytes (B Cells) and Antibodies

Learning Objectives

- ✓ Describe the characteristics of B lymphocytes.
- ✓ Describe the basic structure of an antibody (immunoglobulin) molecule.
- ✓ Contrast the structure and function of the five classes of immunoglobulins.
- ✓ Describe five functions of antibodies.

As we have seen, **B lymphocytes,** which are also called **B cells,** arise in the red bone marrow in adults, and they also mature there. However, the designation "B" does not stand for bone marrow, but instead for the *bursa of Fabricius,* which is a unique lymphoid organ located near the terminus of the intestinal tract in birds. Mammals do not have this bursa, which is where B lymphocytes were originally identified, but the name "B cell" has been retained.

B cells, which are identified by surface proteins (for example, CD21, CD22, and CD35), are found primarily in the spleen, lymph nodes, red bone marrow, and Peyer's patches; only a small percentage of B cells circulate in the blood. The major function of B cells is the secretion of protective antibodies, which we discuss next.

Basic Antibody Structure

B cells that are actively fighting against exogenous antigens are called **plasma cells,** which secrete soluble, proteinaceous antigen binding molecules called **antibodies** or **immunoglobulins** (im'yū-nō-glob'yū-linz) **(Ig)** into the blood or lymph. Historically, bodily fluids such as lymph and blood were called *humors* (the Latin word for moisture), and thus the activity of B cells is said to be a **humoral immune response.**

A basic antibody molecule is formed of four peptide chains—two identical longer chains called *heavy chains,* and two identical shorter *light chains* **(Figure 16.5).** The terms *heavy* and *light* refer to their relative molecular weights. Disulfide bonds—covalent bonds between sulfur atoms in amino acids—link the light chains to the heavy chains in such a way that a basic antibody molecule looks like the letter Y with three regions: a *stem* and two *arms* connected by a flexible *hinge region.*

An antibody stem, also called the $\mathbf{F_C}$ **region** (because it forms a *fragment* that is *crystallizable*), is formed of the lower portions of the two heavy chains. Among all B cells, there are only five types of heavy chains, designated by the Greek letters *gamma, mu, alpha, epsilon,* and *delta.* These five types of heavy chains give their names to five classes of antibodies (immunoglobulins): IgG, IgM, IgA, IgE, and IgD.

The arms of each heavy and each light chain varies in amino acid sequence among B cells, and thus each is called a *variable region.* Though they vary from B cell to B cell, the variable regions of every antibody molecule formed by a given B cell is identical because for antibodies produced by any particular B cell, the two light chain variable regions are

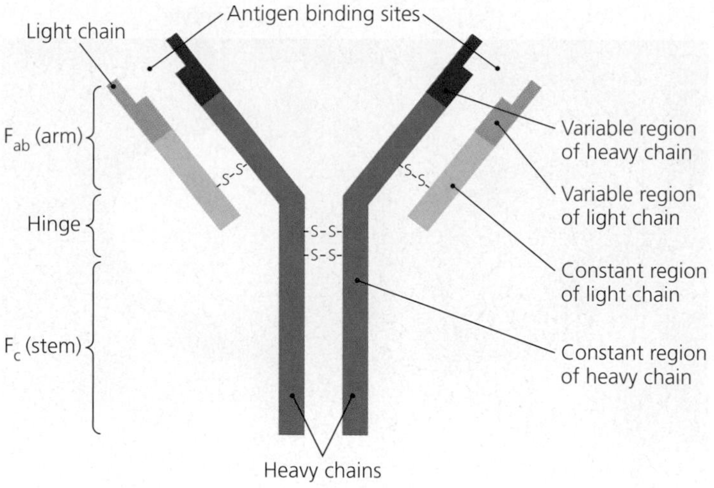

▲ ***Figure 16.5***

Basic immunoglobulin (antibody) structure. Each immunoglobulin molecule, which is shaped like the letter Y, consists of two identical heavy chains and two identical light chains held together by disulfide bonds. Five different kinds of heavy chains form the immunoglobulin's stem (F_C region). The arms terminate in variable regions to form two antigen-binding sites. The hinge region is flexible, allowing the arms to bend in almost any direction. *Why are the two antigen-binding sites on a typical antibody molecule identical?*

Figure 16.5 *Because the amino acid sequences of the two light chains of an antibody molecule are identical, as are the sequences of the two heavy chains, the binding sites—each composed of the ends of one light chain and one heavy chain—must be identical.*

identical and the two heavy chain variable regions are identical. Each B cell randomly chooses genes for variable regions from a set of possible genes and once chosen, makes only those particular variable regions. Together the two variable regions form an **antigen-binding site.** Because the arms of an antibody molecule contain the antigen-binding sites, they are also known as the F_{ab} *regions (fragment, antigen-binding).*

Antibody Function

Antigen-binding sites are complementary to antigenic determinants (epitopes); in fact, the shapes of the two can match so very closely that most water molecules are excluded from the area of contact, producing a strong, noncovalent, hydrophobic interaction. Additionally, hydrogen bonds and other attractions mediate antibody binding to antigenic determinant. Once antibodies are bound to their target antigen, they function in several ways. These include activation of complement, stimulation of inflammation, agglutination, neutralization, and opsonization. Complement activation and inflammation—nonspecific responses of the second line of defense—are discussed in detail in Chapter 15 (see Figures 15.13 and 15.18); here we examine the other antibody functions.

Agglutination Because each basic antibody has two antigen-binding sites, each can attach to two antigenic determinants at once. The result of several immunoglobulin molecules binding with two microbial cells each is **agglutination** (ă-glū-ti-nā'shŭn) or clumping **(Figure 16.6a).** Agglutination hinders the activity of pathogenic organisms and increases the chance that they will be phagocytized.

Neutralization Antibodies can **neutralize** a toxin by binding to a critical portion of the toxin so that it can no longer function against the body. Similarly, antibodies can block attachment molecules on the surfaces of bacteria and viruses such that they cannot adhere to target cells **(Figure 16.6b).**

Opsonization Antibodies act as **opsonins**[5]—molecules that stimulate phagocytosis. Because neutrophils and macrophages have receptors for the F_c regions of IgG molecules, these leukocytes bind to the stems of antibodies. Once the antibodies are so bound, the leukocytes phagocytize them, along with the antigens they carry. Enhanced phagocytosis is called **opsonization** (op'sŏ-nī-zā'shŭn) **(Figure 16.6c).**

CRITICAL THINKING

Two students are studying for an exam on the body's defensive systems. One of them insists that complement is part of the nonspecific second line of defense, but the partner insists that complement is part of a humoral immune response in the third line of defense. How would you explain to them that they are both correct?

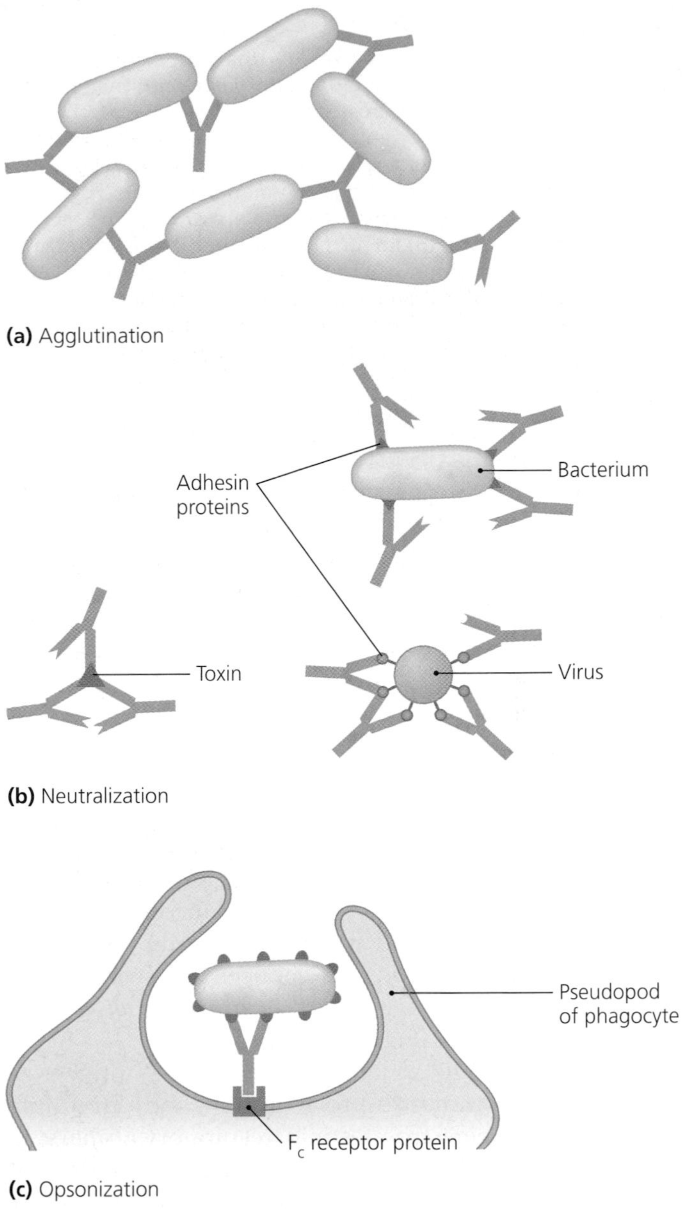

▲ ***Figure 16.6***
Three functions of antibodies. **(a)** Agglutination (clumping). **(b)** Neutralization of toxins and microbes. **(c)** Opsonization. Two other functions of antibodies—activation of complement and participation in inflammation—are discussed in Chapter 15.

Classes of Antibodies (Immunoglobulins)

The threat confronting the immune system can be extremely variable, so it is not surprising that a single type of antibody is not sufficient. Next we consider the structure and functions of the five classes of antibodies that may participate in a humoral immune response; the class involved in any given humoral immune response depends on the type of invading foreign antigens, the portal of entry involved, and the antibody function required.

The most common class of antibody in the blood, accounting for about 85% of serum antibodies, is **immunoglobulin G (IgG).** IgG molecules play a major role in antibody-mediated defense mechanisms, including complement activation, agglutination, opsonization, and neutralization.

Each molecule of IgG is formed from two identical light chains and two identical gamma heavy chains held together with disulfide bonds (**Figure 16.7a** on page 470). Because of its small size, IgG can leave blood vessels more easily than can the other immunoglobulins. This is especially important during inflammation, because it enables IgG to bind to invading pathogens in tissues before they get into the circulatory systems. IgG molecules are also the only type of antibody that can cross the placenta to protect a developing child.

The body produces **immunoglobulin M (IgM),** the second most common antibody class in the blood, during the initial stages of an immune response. An IgM molecule is more than five times larger than a molecule of IgG because it

[5]From Greek *opsonein,* meaning to supply food.

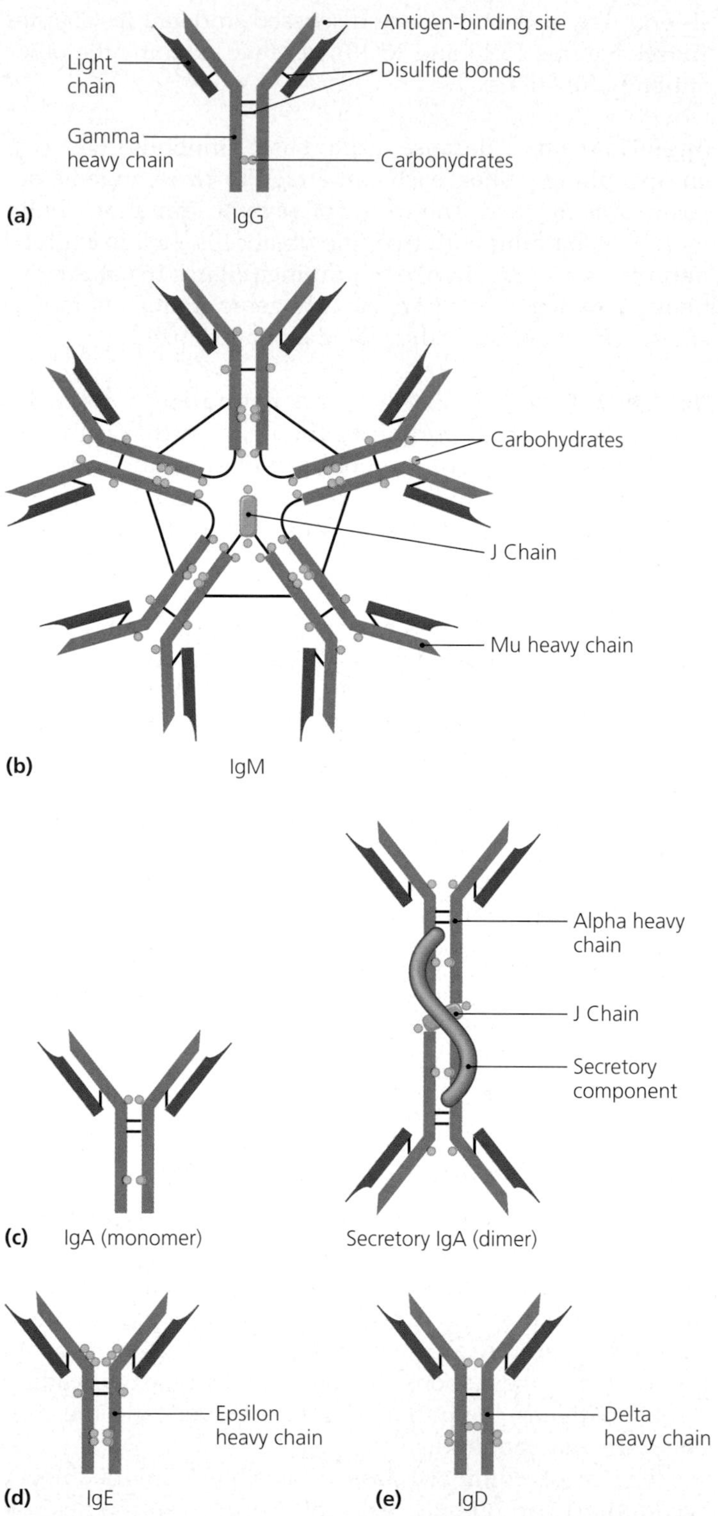

Figure 16.7

Classes of antibodies. **(a)** IgG, the most common immunoglobulin in blood and lymph, and the only class of antibody that crosses the placenta. **(b)** IgM, the largest antibody and the class first secreted during a humoral immune response. **(c)** IgA. Secretory IgA, a dimer composed of two IgA monomers, can cross membrane surfaces due to the presence of the secretory component. **(d)** IgE, which is present in the serum in low concentrations. Its binding to mast cells stimulates the release of histamine, a chemical of inflammation. **(e)** IgD, the rarest class of antibody and an antigen receptor in the membranes of B cells in humans.

is a pentamer, consisting of five Y-shaped subunits linked together in a circular fashion via disulfide bonds and a short polypeptide *joining (J) chain* **(Figure 16.7b).** Each IgM subunit has a conventional immunoglobulin structure, consisting of two light chains and two mu heavy chains. IgM is more efficient than IgG at complement activation, neutralization of virions, and agglutination because of its numerous antigen binding sites.

Immunoglobulin A (IgA), which has alpha heavy chains, is the immunoglobulin most closely associated with various body secretions, including tears and milk. Some of the body's IgA is a monomer with a basic Y shape that circulates in the blood, constituting about 5% of total serum antibody. However, plasma cells in various mucous membranes and in the tear ducts and mammary glands take up IgA monomers from the blood and connect them via a J chain and another short polypeptide (called a *secretory component*) to form **secretory IgA (Figure 16.7c).**

Secretory IgA agglutinates and neutralizes antigens and is of critical importance in protecting the body from infections arising in the gastrointestinal, respiratory, urinary, and reproductive tracts. Secretory component both aids in the transport of secretory IgA across mucous membranes and protects secretory IgA from digestion by intestinal enzymes. Because secretory IgA is secreted into the milk in the mammary glands, it can provide newborns some protection against foreign antigens.

Immunoglobulin E (IgE) is a typical Y-shaped immunoglobulin with two epsilon heavy chains **(Figure 16.7d).** Because it is found in extremely low concentrations in serum (less than 1% of total antibody), it is not important in agglutination, neutralization, or opsonization. Instead, IgE molecules act as signal molecules—they attach to receptors found on certain white blood cells (basophils and mast cells) to trigger the rapid release of histamine, causing inflammation, part of the body's nonspecific defense. As a result, IgE mediates between the specific and nonspecific responses—a process particularly important in allergic diseases. Infections by parasitic worms also stimulate the synthesis of IgE.

Immunoglobulin D (IgD) (Figure 16.7e) is characterized by delta heavy chains. IgD molecules are not secreted, but are membrane-bound antigen receptors on B cells that are often seen during the initial phases of a humoral immune response. IgD is not found in all mammals, and animals that lack IgD show no observable ill effects; therefore, scientists do not know the importance of IgD.

Table 16.1 compares the different classes of immunoglobulins.

Table 16.1 Characteristics of the Five Classes of Immunoglobulins

Name	Function(s)	Structure (Molecular Weight in Daltons)	Location(s)	Percentage in Serum
IgG	Complement activation, agglutination, opsonization, and neutralization; crosses placenta to protect fetus	Monomer (150,000)	Serum, intercellular fluid	85
IgM	Complement activation, agglutination, and neutralization	Pentamer (970,000)	Serum	5–10
IgA	Agglutination and neutralization	Monomer (160,000); dimer (385,000)	External secretions, including milk	5
IgE	Triggers release of histamines from basophils and mast cells	Monomer (188,000)	Serum, mast cell surfaces	<1
IgD	Unknown	Monomer (184,000)	B cell surface	<1

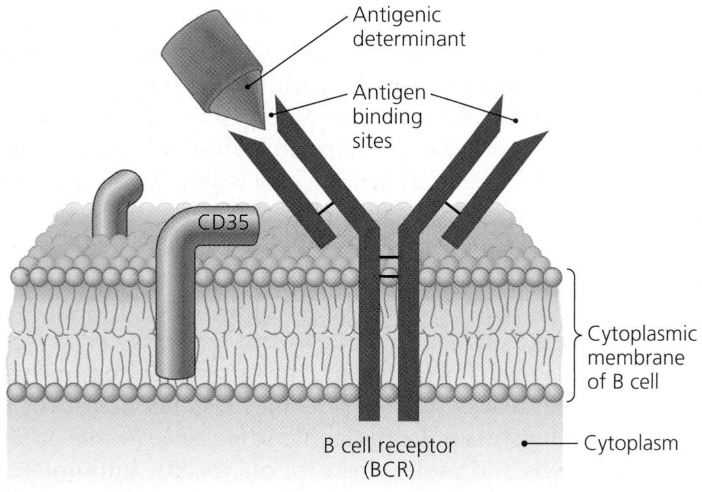

▲ ***Figure 16.8***
Surface structures of B cells. In addition to glycoproteins such as CD35, B cells are characterized by B cell receptors (BCRs), which are symmetrical, Y-shaped antibody molecules with two antigen-binding sites.

B Cell Receptors

The surface of each B lymphocyte is covered with about 250,000 to 500,000 identical copies of **B cell receptor (BCR),** an antibody that remains integral to the cytoplasmic membrane **(Figure 16.8).** Like a secreted antibody molecule, each BCR has two antigen-binding sites identical to the antigen-binding site of the antibodies that particular cell secretes. B lymphocytes do not form BCRs in response to antigens; instead, each cell's randomly generated antibody variable regions determine its specific BCR shape. A BCR is complementary in shape to an antigenic determinant, which the body may or may not encounter. Scientists estimate that each person forms at least 10^{11} different B lymphocytes—each with a distinct set of BCRs! Because there are trillions of lymphocytes, at least one of them fortuitously bears a BCR that is complementary to any given specific antigenic determinant. Thus, for example, each of us has lymphocytes with BCRs complementary to rattlesnake venom proteins, even though it is unlikely that we have been or ever will be bitten by a rattlesnake.

An antigenic molecule (for example, a bacterial protein) typically has numerous antigenic determinants of various shapes and thus will be recognized by many different BCRs; however, it must be emphasized that each B lymphocyte recognizes only one antigenic determinant. Collectively, the cellular receptors on all of an individual's B cells are capable of recognizing millions of different antigenic determinants; thus, the repertoire of BCRs is also the repertoire of B lymphocytes.

In most cases, B cells do not mount a humoral immune response directly. Instead, they respond to most antigens only with the assistance of certain T lymphocytes. Thus, antigens of this type are known as *T-dependent antigens*. We will discuss the details of humoral immunity against T-dependent antigens shortly—after we consider the various T lymphocytes.

T Lymphocytes (T Cells)

Learning Objectives

- ✓ Describe the importance of the thymus to the development of T lymphocytes.
- ✓ Describe the basic characteristics of T lymphocytes.
- ✓ Compare and contrast three types of T cells.
- ✓ Describe five types of cytokines.

An adult's red bone marrow produces lymphocytes that leave the bone marrow to mature in the thymus. In the thymus, each of these cells, called **T lymphocytes** or **T cells,** randomly generates multiple copies of a specific **T cell receptor (TCR)** for its cytoplasmic membrane. A TCR is composed of two different polypeptide chains with a groove between them that acts as an antigen binding site (**Figure 16.9** on page 472). As with BCRs of B cells, each specific TCR recognizes and binds to a complementary antigenic determinant, and there are at least 10^{11} different TCRs—at least one for every possible antigenic shape. T lymphocytes act

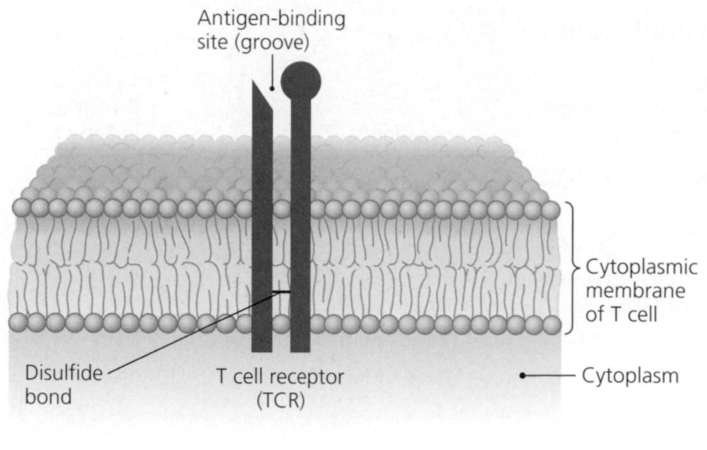

▲ ***Figure 16.9***
A T cell receptor (TCR), an asymmetrical surface molecule composed of two polypeptides containing a single antigen-binding site between them.

directly against endogenous invaders, many of the body's cells that harbor intracellular pathogens, and abnormal body cells such as cancer cells that produce abnormal cell surface proteins. Because T cells act directly against antigens—they do not secrete immunoglobulins—their activity is **cell-mediated immune response.**

After they have matured in the thymus, T cells circulate in the lymph and blood and migrate to the lymph nodes, spleen, and Peyer's patches. They account for about 90% of all lymphocytes in the blood. Based on their surface glycoproteins and their characteristic functions, immunologists recognize three types of T cells: *cytotoxic T cells* and two kinds of *helper T cells*.

Cytotoxic T Cells (T_C Cells)

A **cytotoxic T (T_C) cell** is distinguished by hundreds of copies of its unique TCR as well as the presence of the CD8 cell-surface glycoprotein. The latter gives these lymphocytes their nickname: *T8 cells*. As the name *cytotoxic* T cell implies, these cells directly kill other cells—those infected with viruses and other intracellular pathogens, as well as abnormal cells, such as cancer cells.

Helper T Lymphocytes (T_H Cells)

Immunologists distinguish **helper T (T_H) cells** by the presence of the CD4 glycoprotein; in fact, many medical personnel call these cells *T4 cells*. These cells are called "helpers" because their function is to "help" in regulating the activity of B cells and cytotoxic T cells during an immune response. There are two subpopulations of helper T cells: type 1 helper T cells (T_H1), which assist cytotoxic T cells, and type 2 helper T cells (T_H2), which function in conjunction with B cells. T helper cells secrete various soluble protein messengers, called *cytokines,* that determine which immune response will be activated. (We will consider the types and effects of cytokines shortly.) **Highlight 16.1** describes some of the effects of the destruction of helper T cells in individuals infected with the human immunodeficiency virus (HIV). Immunologists distinguish between T_H1 and T_H2 cells on the basis of their cytokine secretions and by characteristic cell surface proteins: T_H1 cells express CD26 and a cytokine receptor named CCR5,[6] whereas the cytoplasmic membranes of T_H2 cells have CCR3 and CCR4.

Some scientists distinguish a fourth type of T cell, a *suppressor T cell,* which is thought to repress specific immune responses. Other researchers dispute the existence of a specific category of suppressor T cells, contending instead that other regulatory cells and molecules turn off specific immune responses.

Table 16.2 compares and contrasts the features of the various types of lymphocytes.

[6] Immunologists name some cytokine receptors with the prefixes CCR to indicate the presence of two cysteine amino acids in their primary structure.

Highlight 16.1 The Loss of Helper T Cells in AIDS Patients

Neither B cells nor cytotoxic T cells respond effectively to most antigens without the participation of helper T cells, because signals are passed more effectively among leukocytes when the CD4 molecules of helper T cells bind to certain leukocytes.

We also saw in a previous chapter that in order for viruses to enter cells and replicate, they must first attach to a specific protein on their target cells' surface. In the case of human immunodeficiency virus (HIV), the causative agent of AIDS, the surface protein to which it attaches is CD4, and its target cells are helper T cells. Once HIV enters a helper T cell, it, like other viruses, takes over the cell's protein-synthesizing machinery and directs the cell to produce more viruses.

Because they are essential to mounting an effective immune response, the body normally has a surplus of helper T cells—usually about three times as many as it needs. As a result, individuals infected with HIV can typically lose up to about two-thirds of their helper T cells before signs of immune deficiency appear. Cell-mediated immunity is usually affected first; even though B cells require the assistance of helper T cells to function optimally, they can still be stimulated directly by large quantities of antigen. Normal human blood typically contains about three CD4 helper T cells for every two CD8 cytotoxic T cells—a CD4 to CD8 ratio of 3:2. During the course of HIV infection, however, CD4 helper T cells are lost, reducing the CD4 to CD8 ratio such that individuals with full-blown AIDS may have a ratio as low as 0.5:1.

Table 16.2 Characteristics of Selected Lymphocytes

Lymphocyte	Site of Maturation	Representative Cell Surface Glycoproteins	Selected Secretions
B cell	Red bone marrow and germinal center of lymph node	Each has distinct BCR and CD21, CD22, CD35	Antibodies
Helper T cell type 1 (T_H1)	Thymus	Each has distinct TCR and CD4, CD26, CCR5	IL-2, IFN-γ
Helper T cell type 2 (T_H2)	Thymus	Each has distinct TCR and CD4, CCR3, CCR4	IL-4
Cytotoxic T (T_C)	Thymus	Each has distinct TCR and CD8, CD95L	Perforin, granzyme

Immune System Cytokines

Cytokines (sī′tō-kīnz) are soluble regulatory proteins that act as intercellular signals when released by certain body cells including kidney cells, skin cells, and immune cells. Here we are concerned with immune system cytokines, which signal among the various leukocytes. For example, B cells and cytotoxic T cells do not respond fully (if at all) to antigens unless they are first signaled by cytokines.

Many different immune system cytokines are secreted by various leukocytes and affect diverse cells. Many cytokines are redundant; that is, they have almost identical effects. Such complexity has given rise to the concept of a *cytokine network*—a complex web of signals among all the cell types of the immune system. The nomenclature of cytokines is not based on a systematic relationship among them; instead, scientists named cytokines after their cells of origin, their function, and/or the order in which they were discovered. Cytokines of the immune system include the following substances:

- **Interleukins**[7] (in-ter-lū′kinz) **(ILs).** As their name suggests, ILs signal among leukocytes. Immunologists named interleukins sequentially as they were discovered. Currently, scientists have identified about 27 interleukins.
- **Interferons** (in-ter-fēr′onz) **(IFNs).** These antiviral proteins, discussed in Chapter 15, may also act as cytokines. The most important interferon with such a dual function is interferon-gamma (IFN-γ), which is secreted by T_H1 cells.
- **Growth factors.** These proteins stimulate stem cells to divide, ensuring that the body is supplied with sufficient leukocytes of all types.
- **Tumor necrosis**[8] **factors (TNFs).** Macrophages and T cells secrete TNFs to kill tumor cells and to regulate immune responses and inflammation.
- **Chemokines** (kē′mō-kīnz). Chemokines signal leukocytes to rush to a site of inflammation or infection and activate other leukocytes.

To this point, we have examined the "cast of characters" involved in specific immunity—antigens, B cells, antibodies, T cells, and cytokines. Before we consider the detailed processes involved in specific immunity—humoral immune responses and cell-mediated immune responses—we need to consider some initial preparations of the body.

The Body's Preparation for a Specific Immune Response

The body prepares for specific immune responses by killing lymphocytes with receptors complementary to autoantigens, making so-called *major histocompatibility complex* proteins, and processing antigens so that they can be recognized by lymphocytes.

Lymphocyte Editing by Clonal Deletion

Learning Objective

✓ Describe apoptosis, and explain its role in lymphocyte editing by clonal deletion.

Given that lymphocytes randomly generate the shapes of their receptors, every population of maturing lymphocytes includes cells with receptors that will bind normal body components—the autoantigens mentioned earlier. Because it is vitally important that specific immune responses not be directed against autoantigens, the body "edits" lymphocytes via a process by which self-reactive cells are eliminated.

In humans, lymphocyte editing occurs in lymphoid organs: the bone marrow for B lymphocytes (**Figure 16.10** on page 474) and the thymus for T lymphocytes (which is why these cells are so named). During their maturation, lymphocytes are exposed to many autoantigens. Under normal circumstances, when a lymphocyte binds to an autoantigen, that lymphocyte responds by undergoing **apoptosis**[9] (cell suicide). Thus, apoptosis is the critical feature in the development of self-tolerance, because lymphocytes with receptors

[7]From Latin *inter,* meaning between, and Greek *leukos,* meaning white.
[8]From Latin *necare,* meaning to kill.
[9]Greek meaning falling off.

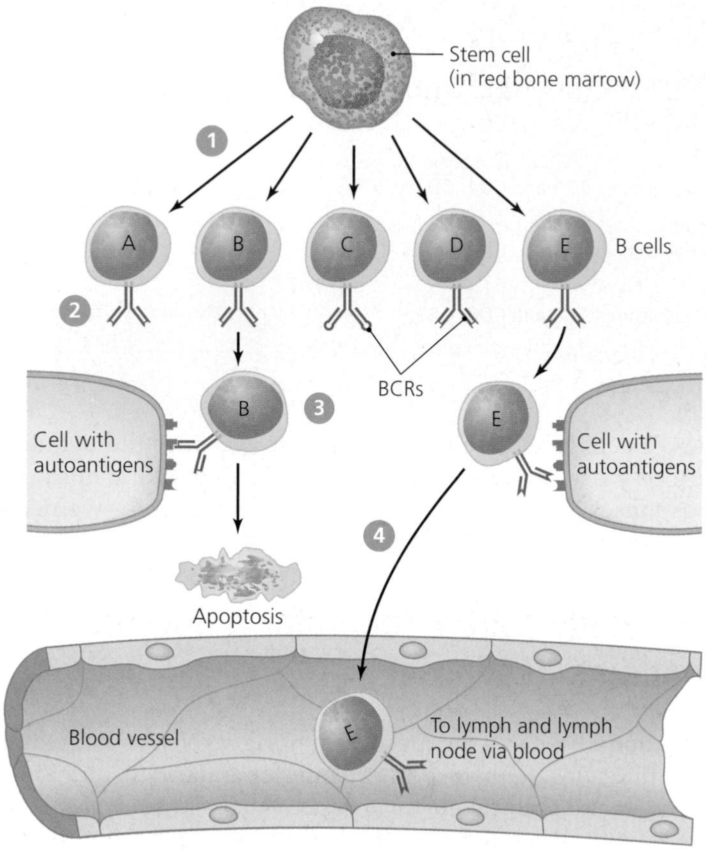

▲ ***Figure 16.10***

Lymphocyte editing (clonal deletion) of B cells. ① Stem cells in the red bone marrow generate a host of lymphocytes. ② Each newly formed B cell randomly generates BCRs with a particular shape. ③ Cells whose BCRs are complementary to some autoantigen bind with that autoantigen, stimulating the cell to undergo apoptosis. Subsequently, an entire set of potential daughter B cells (a clone) that are reactive with the body's own cells are eliminated—thus the term *clonal deletion*. ④ B cells with a BCR that is not complementary to any autoantigen are released from the bone marrow and into the blood. Note that T lymphocytes are also edited via the process of clonal deletion—in their case, in the thymus. *Of the B cells shown, which (besides B) is likely to undergo apoptosis?*

Figure 16.10 The B lymphocyte labeled "C" will also undergo apoptosis.

for autoantigens die before they can reproduce, and their potential daughter cells (clones) are deleted from the repertoire of potential lymphocytes. For this reason, lymphocyte editing is also called **clonal deletion.** Surviving lymphocytes and their descendents respond only to foreign antigens. When self-tolerance is impaired, the result is an *autoimmune disease,* which we examine in Chapter 17.

The Roles of the Major Histocompatibility Complex

Learning Objective

✓ Describe the two classes of major histocompatibility complex (MHC) proteins with regard to their location and function.

When scientists first began grafting organs from one animal into another, they discovered that if the animals were not closely related, the recipients swiftly rejected the grafts. When they analyzed the reason for such rapid rejection, they found that graft recipients were mounting a very strong immune response against a specific group of antigens found on the cells of the graft. When the antigens on the graft's cells were sufficiently dissimilar from the antigens on the host's cells, as seen with unrelated animals, the grafts were rejected. This is how scientists came to understand how the body is able to distinguish self from nonself.

Immunologists named this class of antigens *major histocompatibility*[10] *antigens* to indicate the importance of these molecules in determining the compatibility of tissues in successful grafting. Further research revealed that major histocompatibility antigens are glycoproteins found in the membranes of most cells of vertebrate animals. In humans, these antigens were first identified on leukocytes and were named *human leukocyte antigens (HLA).* Major histocompatibility antigens are coded by a cluster of genes called the **major histocompatibility complex (MHC).** In humans, the MHC is located on each copy of chromosome 6.

Because organ grafting is a modern surgical procedure with no counterpart in nature, scientists reasoned that MHC molecules must have some other "real" function. Indeed, immunologists have determined that MHC proteins in cytoplasmic membranes function to hold and position antigenic determinants for presentation to T cells. Each MHC molecule has an *antigen-binding groove* that lies between two polypeptides; variations in the amino acid sequences of the polypeptides modify the shapes of MHC binding sites and determine which antigenic determinants can be bound and presented.

MHC proteins are of two classes **(Figure 16.11).** Class I MHC molecules are found on the cytoplasmic membranes of all nucleated cells. Thus red blood cells, which do not have nuclei, do not express MHC proteins, while skin and muscles do express MHC class I molecules. In contrast, class II MHC proteins are found only on B lymphocytes and special cells called **antigen presenting cells (APCs).** APCs include monocytes, macrophages, and their functional relatives, such as *dendritic cells,* which are found in lymphoid organs and under the surface of the skin and are characterized by many long, thin cytoplasmic processes called *dendrites.*[11]

The cytoplasmic membrane of an APC has about 100,000 MHC II molecules, which vary among themselves in

[10]From Greek *histos,* meaning tissue, and Latin *compatibilis,* meaning agreeable.
[11]From Greek *dendron,* meaning tree.

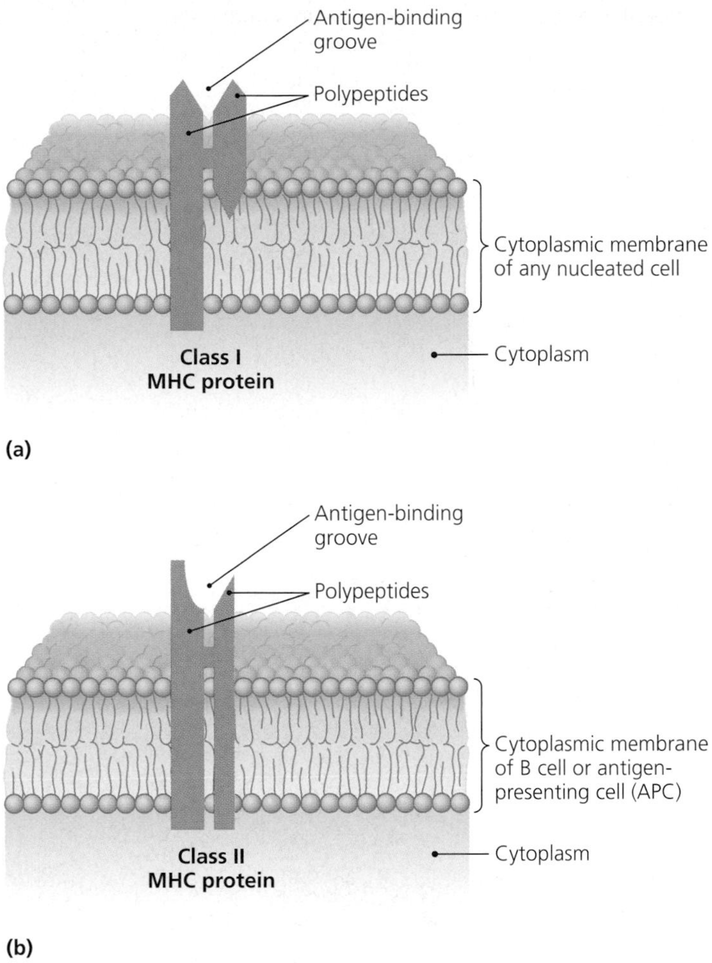

▲ ***Figure 16.11***

The two classes of major histocompatibility complex (MHC) proteins, which are each composed of two polypeptides that form an antigen-binding groove. **(a)** MHC class I glycoproteins, which are found on all cells except red blood cells. **(b)** MHC class II glycoproteins, which are expressed only by B cells and special antigen presenting cells (APCs).

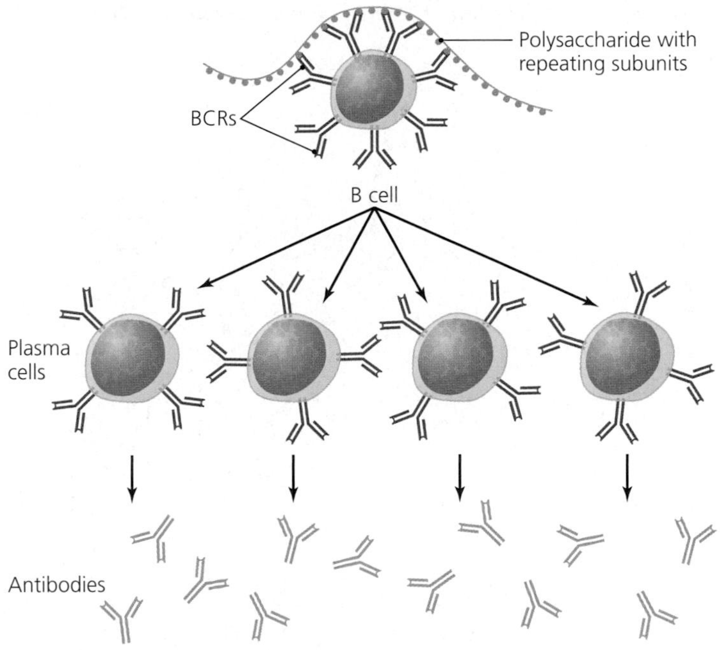

▲ ***Figure 16.12***

The effects of the binding of a T-independent antigen by a B cell. When a molecule with multiple repeating antigenic determinants (such as the polysaccharide shown here) cross-links the BCRs on a B cell, the cell is activated: It proliferates, and its daughter cells become plasma cells that secrete antibodies.

the antigens they can bind. Their diversity is dependent on an individual's genotype. If an antigen fragment cannot be bound to an MHC molecule, it typically does not trigger an immune response. Thus, MHC molecules determine which antigen fragments trigger immune responses.

Antigen Processing

Learning Objectives

- ✓ Contrast T-independent and T-dependent antigens in terms of their structures.
- ✓ Explain the roles of macrophages, dendritic cells, and MHC molecules in antigen processing and antigen presentation.
- ✓ Contrast endogenous antigen processing with exogenous antigen processing.

B cells—but not T cells—can bind directly to large antigens that have numerous, readily accessible, and repeating antigenic determinants such as the repeating polysaccharide subunits of a bacterial capsule; they do not require help. The cross-linking of the repeating subunits to numerous BCRs on a B cell stimulates that cell to proliferate, become a plasma cell, and produce antibodies, all without the help of T cells. Therefore, large antigens with repeating antigenic determinants are called **T-independent antigens (Figure 16.12).**

However, most antigens are smaller and have less accessible antigenic determinants, and B cells targeted against these antigens cannot function without the involvement of helper T cells. These antigens are therefore termed **T-dependent antigens.** Additionally the helper T cells require communication with and assistance from other leukocytes that process the antigen in a way that makes antigenic determinants more accessible. Communication among leukocytes is dependent on MHC glycoproteins and cytokines.

Antigen processing occurs via somewhat different processes according to whether the antigen is exogenous or endogenous. For exogenous antigens such as extracellular pathogens, an APC internalizes the invading pathogen and enzymatically catabolizes the pathogen's antigenic molecules, producing fragments, or antigenic determinants which are contained in a phagolysosome (**Figure 16.13** on

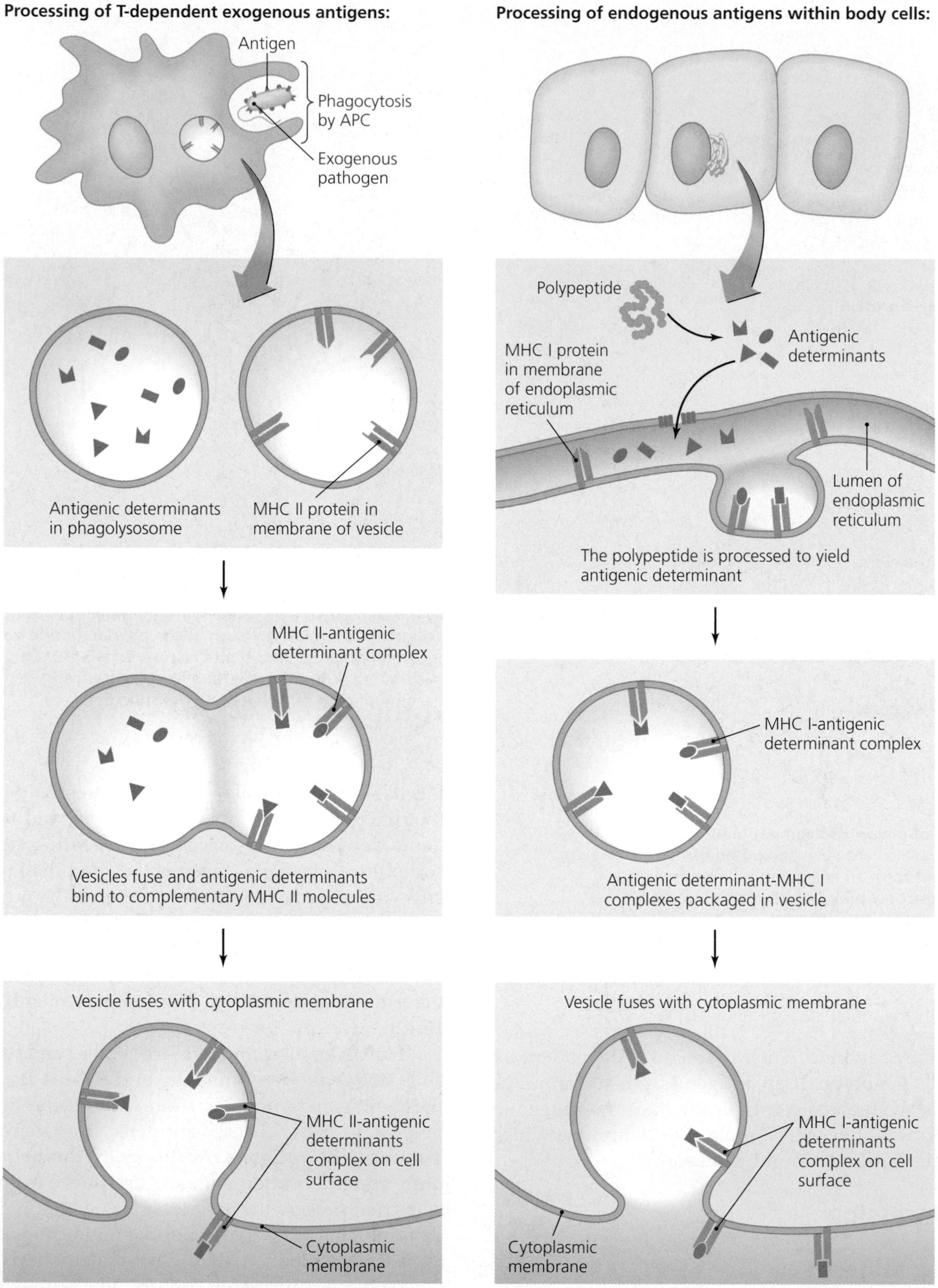

(a) MHC II-antigenic determinant complex displayed on cytoplasmic membrane

(b) MHC I-antigenic determinant complex displayed on cytoplasmic membrane

▲ ***Figure 16.13***

The processing of exogenous and endogenous antigens. **(a)** Exogenous antigens are phagocytized by an APC, which then digests them to release antigenic determinants that are inserted into the antigen-binding grooves of MHC II molecules. The MHC II–antigenic determinant complexes are then displayed on the outside of the APC's cytoplasmic membrane. **(b)** Polypeptides synthesized within a cell are cleaved to release antigenic determinants, which are imported into the endoplasmic reticulum; there they bind to complementary MHC I proteins. The MHC I-antigenic determinant complexes are then packaged in vesicles and exported to the cytoplasmic membrane.

page 476). A vesicle containing MHC class II molecules in its membrane fuses with the phagolysosome and each antigenic determinant then binds to an antigen-binding groove of a complementary MHC class II molecule. The fused vesicle then inserts the complexes of antigenic determinants and MHC class II molecules into the cytoplasmic membrane such that the antigen is presented on the outside.

The endogenous antigens of intracellular bacteria and viruses must also be processed. In fact, some of the new proteins produced within nucleated cells—including proteins produced by intracellular bacteria or viral proteins produced by the cell's metabolic machinery—are catabolized into antigenic determinants and attached to antigen-binding grooves of MHC class I molecules located in endoplasmic reticulum membrane. This membrane is packaged into a vesicle by a Golgi body and then inserted into the cytoplasmic membrane to display the MHC class I-antigenic determinant complex on the cell's surface. All nucleated cells express MHC class I molecules.

Now that we have discussed the organs, cells, secretions, and signaling molecules of the immune system, as well as the preparatory steps of editing, antigen processing, and presentation, we will examine the humoral and cell-mediated immune responses in more detail.

The Humoral Immune Response

As we have discussed, the body mounts humoral immune responses against the antigens of exogenous pathogens. Further, such an immune response against T-dependent antigens requires the assistance of helper T (T_H) cells. The following section examines a humoral immune response in more detail.

B Cell Activation and Clonal Selection

Learning Objectives

✓ Describe the formation and functions of plasma cells and memory B cells.

✓ Describe the steps and effect of clonal selection.

Humoral immunity can be viewed as consisting of the following series of cellular interactions (**Figure 16.14** on page 478), which are mediated by membrane-bound proteins as well as cytokines:

1. **Antigen presentation.** APCs present antigenic determinants to all T_H cells they encounter, but APCs activate only those T_H cells that have binding sites complementary to the presented antigen. Antigen presentation is a random process that depends on chance encounters between APCs and T_H cells. Recall that T_H cells express a unique T cell receptor (TCR) and CD4, and it is CD4 that recognizes and binds to MHC II, which helps stabilize the binding of the antigenic determinant to the TCR.
2. **Differentiation of T_H into T_H2 cells.** In a humoral immune response, APCs signal the T_H cell with IL-1 to become a T_H2 cell.
3. **Clonal selection.** Of all the B cells available, only the B cell with BCRs complementary to the antigenic determinant will be recognized—a process called **clonal selection.**
4. **Activation of B cell.** A complementary T_H2 cell binds to a B cell by recognizing its antigenic determinant–MHC II complex. The T_H cell secretes IL-4, which activates the B cell such that it proliferates rapidly. The offspring of the proliferating B cell differentiate into two types of daughter cells—*memory B cells* (discussed shortly) and *plasma cells.*

The majority of cells produced during B cell proliferation are antibody-secreting **plasma cells,** which are large, activated B cells with an extensive cytoplasm that is rich in rough endoplasmic reticulum. All of the plasma cell offspring of any single activated B cell produce only antibody molecules with active sites identical to the antigen-binding sites of the parent cell's BCR. That is, the antibodies are identical to one another and complementary to the specific antigenic determinant recognized by the parent cell. Each plasma cell produces thousands of molecules of antibody per second. Because of their high activity level, plasma cells are short-lived; they die within a few days of activation, although their antibodies can remain in body fluids for several months, and a clone can persist for years. As discussed previously, antibodies activate complement, trigger inflammation, agglutinate and neutralize antigen, and act as opsonins.

CRITICAL THINKING

What sorts of pathogens would successfully attack a patient with an inability to synthesize B lymphocytes?

Memory B Cells and the Establishment of Immunological Memory

Learning Objective

✓ Contrast primary and memory immune responses.

A small percentage of the cells produced by B cell proliferation do not secrete antibodies but survive as **memory B cells**—that is, long-lived cells with BCRs complementary to the specific antigenic determinant that triggered their production. In contrast to plasma cells, memory cells have BCRs and persist in the lymphoid tissues. They can survive for months or even years, ready to initiate antibody production if the same antigen is encountered again. Let's examine how memory cells provide the basis for immunization to prevent disease, using tetanus immunization as an example.

Because the body produces an enormous variety of B cells (and therefore BCRs), a few T_H cells and B cells bind to and respond to antigens of *tetanus toxoid* (deactivated

Figure 16.14 ➤

A humoral immune response. ① Antigen presentation, in which an APC presents antigen to a complementary T_H cell. ② Differentiation of the T_H cell into a T_H2 cell. ③ Clonal selection, in which the T_H2 cell binds to the B cell clone that recognizes the antigen. ④ Activation of the B cell in response to secretion of IL-4 by the T_H2 cell, which causes the B cell to differentiate into antibody-secreting plasma cells and long-lived memory cells.

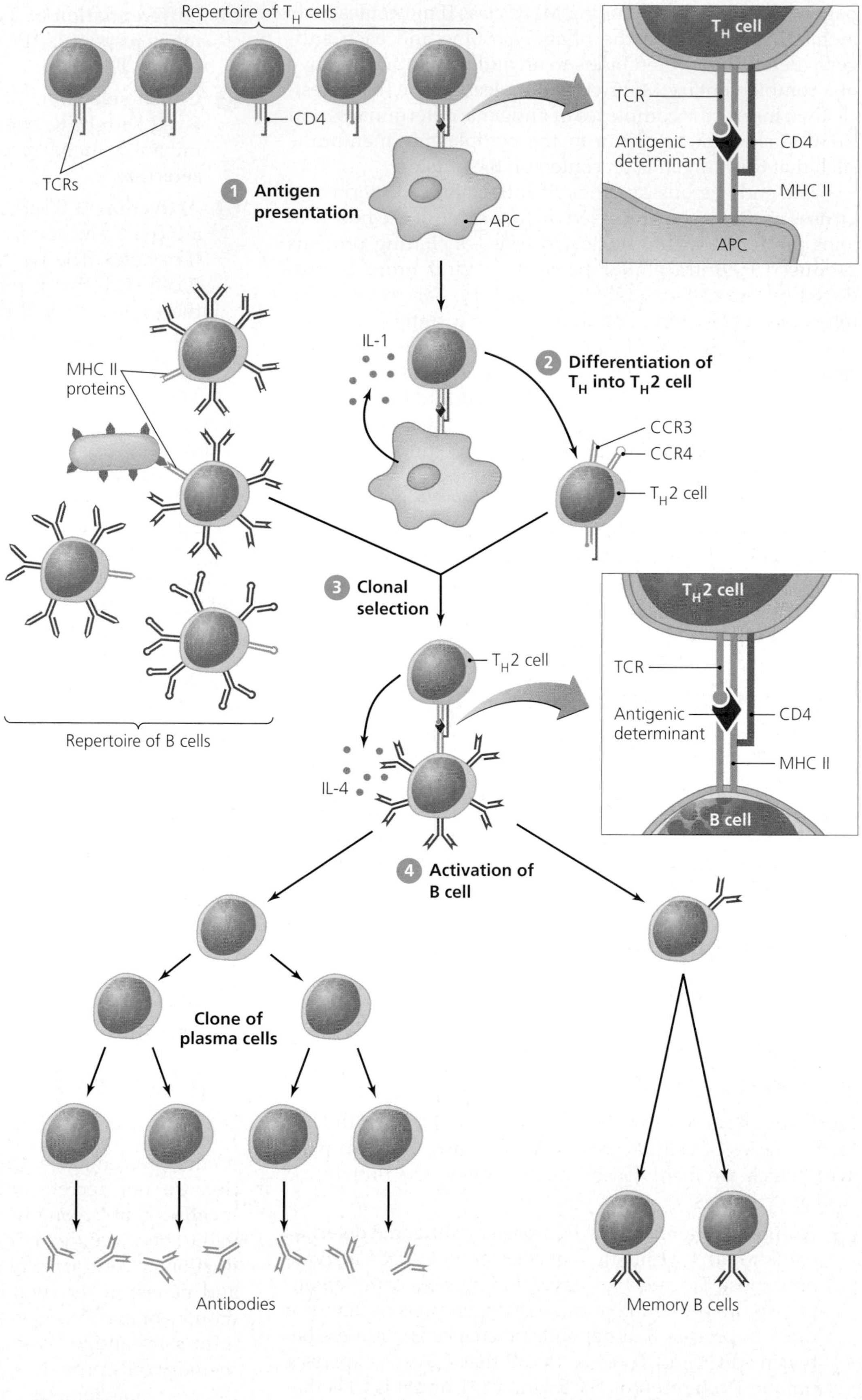

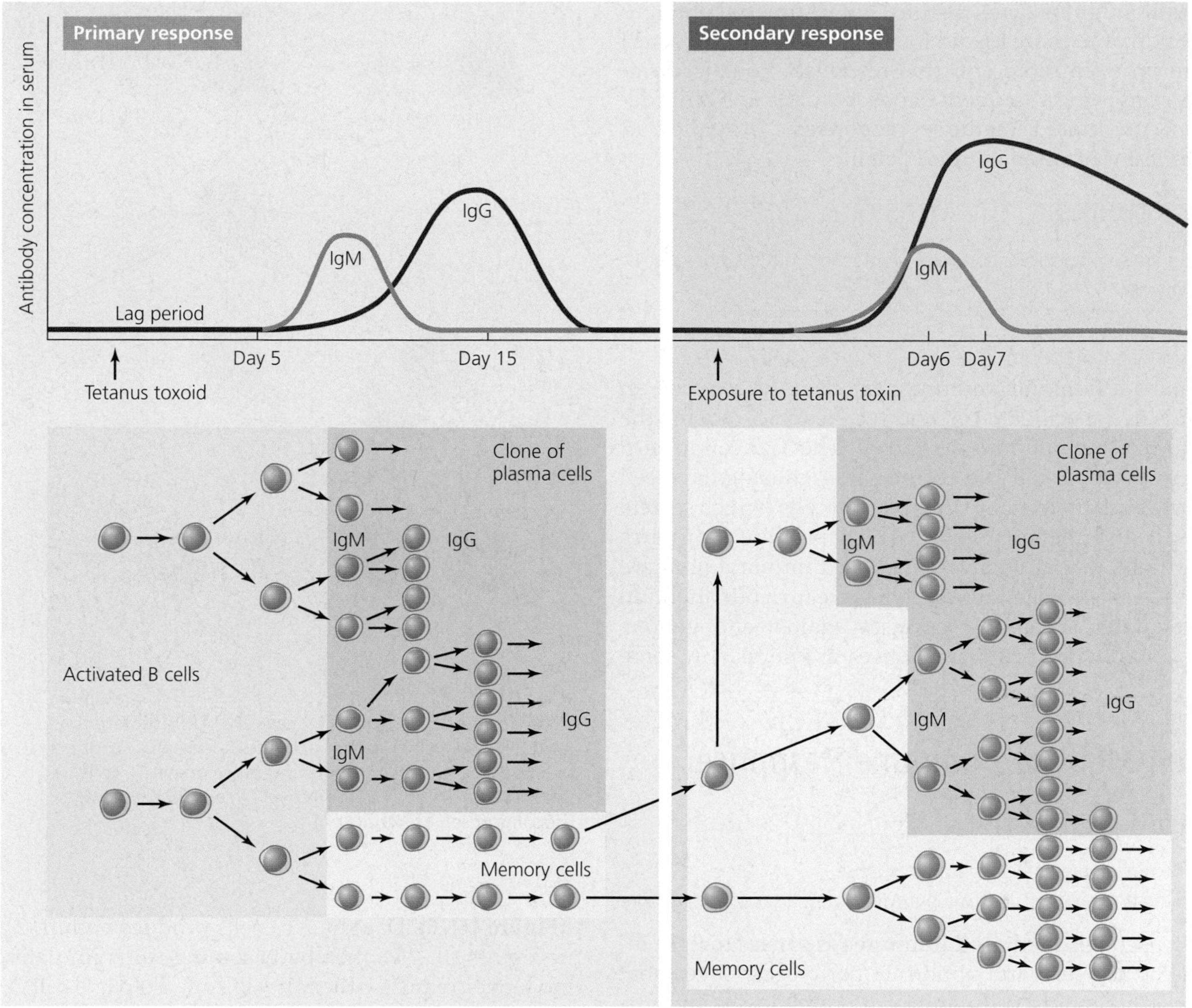

(a) (b)

▲ *Figure 16.15*

The production of primary and secondary humoral immune responses. This example depicts the events following the administration of a tetanus toxoid in immunization. **(a)** Primary response. After the tetanus toxoid is introduced into the body, the body slowly removes the toxoid while producing memory B cells. **(b)** Secondary response. Upon exposure to active tetanus toxin during the course of an infection, memory B cells immediately differentiate into plasma cells and proliferate, producing a response that is faster and results in greater antibody production than occurs in the primary response.

tetanus toxin), which is used for tetanus immunization. In a **primary response (Figure 16.15a),** relatively small amounts of antibodies are produced, and it may take days or weeks before sufficient antibodies are made to completely eliminate the toxoid from the body. Though some antibody molecules may persist for 6 months or more, a primary immune response basically ends when the clone of plasma cells has lived out its normal life span.

Memory B cells, surviving in lymphoid tissue, constitute a reserve of antigen-sensitive cells that become active when there is another exposure to tetanus toxin, perhaps many years later. Thus, tetanus toxin produced during the course of a bacterial infection will encounter a population of memory cells, which proliferate and differentiate rapidly into plasma cells without requiring any interaction with APCs. The newly differentiated plasma cells produce large amounts of antibody within a few days **(Figure 16.15b),** and the tetanus toxin is neutralized before it can cause disease. Because so many more cells are able to recognize and respond to the antigenic determinant, such a **secondary immune response** is much faster and more effective than the initial response.

As you might expect, a third exposure (whether to tetanus toxin or to more toxoid in an immunization booster) results in an even more effective response. Enhanced immune responses to subsequent exposure to the antigenic determinant are called **memory responses.**[12] Chapter 17 discusses immunization in more detail.

CRITICAL THINKING

Plasma cells are vital for protection against infection, but memory B cells are not. Why not?

In summary, a humoral immune response occurs when an APC binds to a specific T_H cell and secretes IL-1, causing the T_H cell to differentiate into a T_H2 cell. The T_H2 cell in turn binds to a specific B cell and secretes IL-4, causing the B cell to multiply and differentiate into plasma cells (which secrete antibodies) and memory cells (which provide long-term protection). Now that we have discussed humoral immune responses against exogenous antigens, we turn our attention to cell-mediated immune responses against endogenous antigens. The two types of responses are similar in some ways.

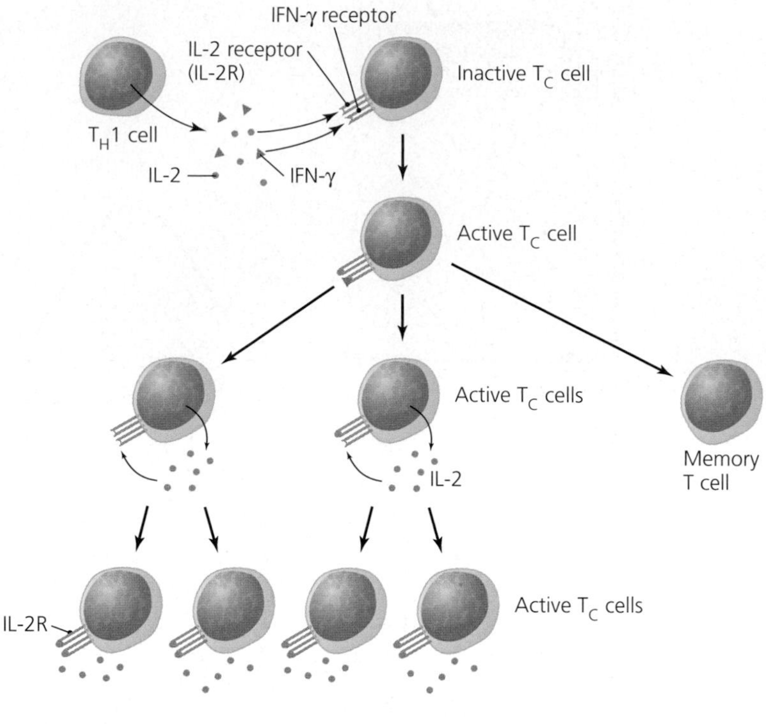

▲ ***Figure 16.16***

Activation of cytotoxic T (T_C) cells. T_H1 cells secrete IL-2 and IFN-γ, which together activate T_C cells. IL-2 triggers active T_C cells to divide, forming more active T_C cells as well as memory T cells. Active T_C cells secrete IL-2, becoming self-stimulatory.

The Cell-Mediated Immune Response

Learning Objectives

✓ Describe a cell-mediated immune response.

✓ Compare and contrast the two pathways of cytotoxic T-cell action.

The body uses **cell-mediated immune responses** to fight intracellular pathogens and abnormal body cells. Given that the most common intracellular invaders are viruses, our discussion of cell-mediated immunity will focus on these pathogens; however, cell-mediated immune responses are also effective against intracellular bacterial pathogens such as *Mycobacterium tuberculosis,* which causes tuberculosis. **New Frontiers 16.1** on page 482 describes an experimental use of T cells to attack one form of cancer.

As previously discussed, when a cell synthesizes certain proteins, it displays fragments of them on its cytoplasmic membrane in the antigen-binding grooves of MHC class I molecules. Thus, when viruses are replicated inside cells, antigenic determinants of viral proteins are displayed on the host cell's surface.

Certain APCs secrete IL-12, which stimulates nearby T_H cells to become T_H1 cells. T_H1 cells secrete IL-2 and IFN-γ, which activate those cytotoxic T (T_C) cells that have TCRs complementary to an antigenic determinant; in fact, IL-2 drives T cell division. Activated T_C cells reproduce to form memory T cells (discussed shortly) and more T_C progeny **(Figure 16.16).** Daughter T_C cells produce both IL-2 and IL-2 receptors (IL-2R), thereby becoming self-stimulating—they no longer require either an APC or a helper T cell.

The initial events in cell-mediated immunity are depicted in **Figure 16.17.** An activated T_C cell binds to an infected cell via its TCR, which is complementary to the antigenic determinant, and via its CD8, which is complementary to the MHC class I protein of the infected cell **(Figure 16.17a).** Next, cytotoxic T cells kill their targets through one of two pathways: The *perforin-granzyme pathway,* which involves the synthesis of special killing proteins, or the *CD95 pathway,* which is mediated through CD95, a cell surface glycoprotein.

The Perforin-Granzyme Cytotoxic Pathway

The cytoplasm of cytotoxic T cells has vesicles containing two key protein molecules called *cytotoxins*—**perforins** (per-fōr-inz) and **granzymes** (gran′zīmz). When a cytotoxic T cell first attaches to its target, these vesicles move toward the region of contact, where they release their cytotoxins. Perforin molecules aggregate into a tubular structure in the infected cell's membrane, forming a channel similar to the MAC channels of complement. Granzymes move through the perforin channel and into the target cell, where they activate apoptotic enzymes **(Figure 16.17b).** Having forced its

[12]Sometimes also called anamnestic responses, from Greek *ana,* meaning again, and *mimneskein,* meaning to call to mind.

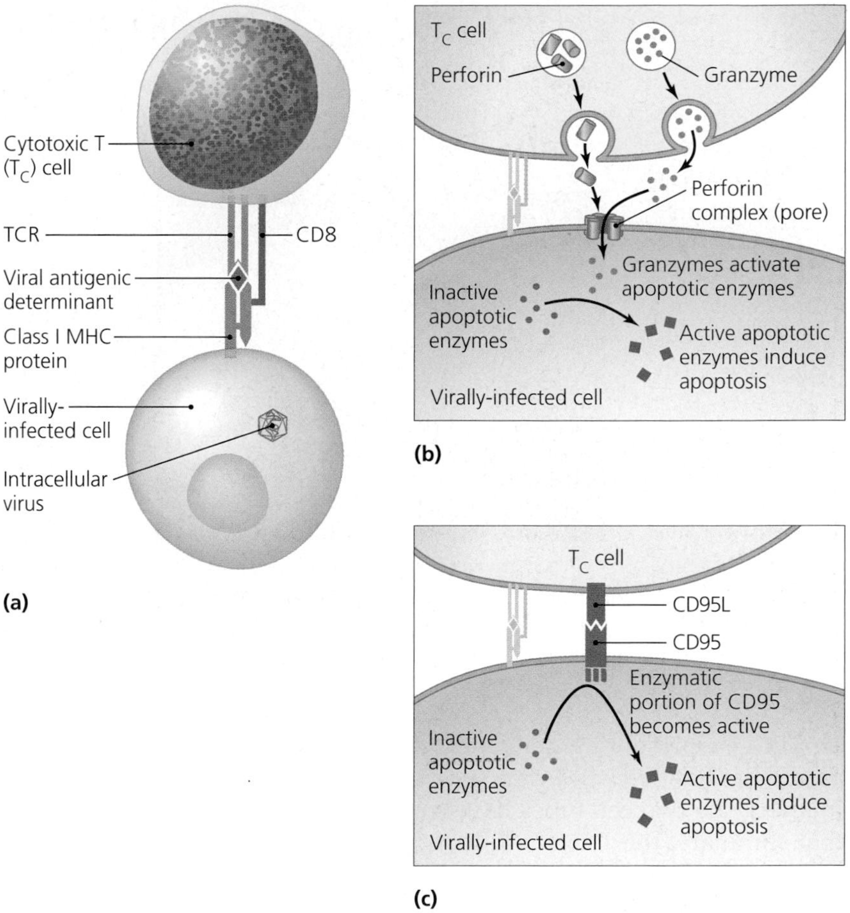

Figure 16.17 ➤
A cell-mediated immune response. **(a)** The binding of a virus-infected cell by a cytotoxic T (T_C) cell. **(b)** The perforin-granzyme cytotoxic pathway. After perforins and granzymes have been released from T_C cell vesicles, granzymes enter the infected cell through the perforin complex pore and activate the enzymes of apoptosis. **(c)** The CD95 cytotoxic pathway. Binding of CD95L on the T_C cell activates the enzymatic portion of the infected cell's CD95 such that apoptosis is induced.

target to commit suicide, the cytotoxic T cell disengages and moves on to another infected cell.

CRITICAL THINKING

Why did scientists give the name "perforin" to this molecule secreted by T_C cells?

The CD95 Cytotoxic Pathway

The **CD95 pathway** of cell-mediated cytotoxicity involves an integral glycoprotein called *CD95* that is present in the cytoplasmic membranes of most body cells. Its receptor, *CD95L,* is present on activated T_C cells. When an activated T_C cell comes into contact with its target, its CD95L binds to CD95 on the target, which then activates enzymes that trigger apoptosis **(Figure 16.17c).**

Memory T Cells

Learning Objective

✓ Describe the establishment of memory T cells.

Just as some B cells become memory cells, so too some activated T cells become **memory T cells.** When stimulated by endogenous antigen and cytokines from T_H1 cells, T cells undergo repeated division and differentiation. Most become cytotoxic T cells and eliminate infected cells, but others simply divide and persist as memory T cells for months or years in lymphoid tissues. If a memory T cell subsequently contacts an antigenic determinant matching its TCR, it responds immediately (without a need for interaction with APCs) and produces cytotoxic T cells.

Because the number of memory T cells is greater than the number of T cells that recognized the antigen during initial exposure, a secondary cell-mediated immune response is much more effective than a primary response. As with humoral immunity, an enhanced cell-mediated immune response upon subsequent exposure to the same antigen is called a memory response.

T Cell Regulation

Learning Objective

✓ Explain the process and significance of the regulation of cell-mediated immunity.

The body carefully regulates cell-mediated immune responses so that T cells do not respond to autoantigens. To prevent this from happening, T cells require additional signals from an antigen presenting cell. If they do not receive these signals, they will not respond. Thus, when a T cell and

New Frontiers 16.1 Attacking Cancer with Lab-Grown T Cells

The body naturally manufactures T cells that attack cancer cells, but it sometimes does not make enough of them to shrink tumors and effectively halt the disease's progress. A recent study from the National Institutes of Health, however, shows that cancer-fighting T cells can be mass-produced in a laboratory and injected into patients, where they survive, multiply, and effectively reduce tumors. The study focused on 13 patients with malignant melanoma, an aggressive form of skin cancer. Researchers took samples of the patients' own cancer-fighting T cells and cloned them in a laboratory until they numbered in the billions. These billions of identical T cells were then injected back into the individuals that first produced them. The results were encouraging. Four of the 13 patients became "virtually cancer free," while tumors were substantially reduced in two other patients.

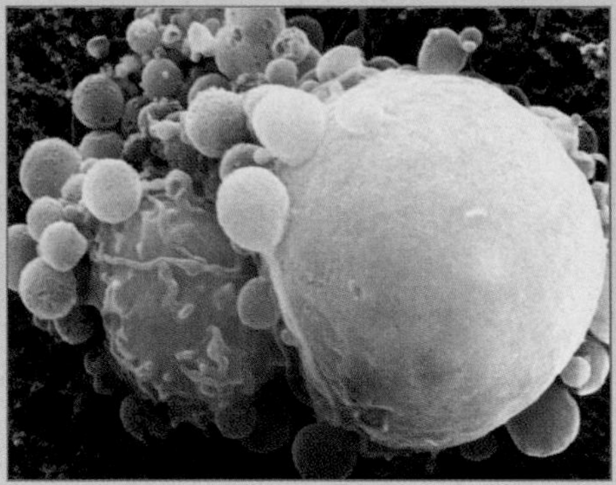

T cells attacking cancer cells.

Such T cell therapy remains highly experimental and is still years away from becoming a generally accepted cancer treatment. Scientists do not yet understand why the therapy works in some patients and not others. In addition, because the T cells attack malignant melanoma tumor antigens that are similar to, or the same as, antigens that are present on certain normal body cells, the therapy can cause autoimmune disease in some patients. To date, the therapy has only been tested against malignant melanoma, but planning to test it against other types of cancer is underway, and some scientists believe that a similar therapy may be effective against viral diseases such as hepatitis.

an antigen presenting cell interact, a specialized contact area, called an *immunological synapse,* forms between them; in effect, the two cell types have a chemical "conversation" that stimulates the T cell to fully respond to the antigen.

If a T cell does not receive the signals required for its activation, it will "shut down" as a precaution against autoimmune responses.

In summary, the body's response to infectious agents seldom relies on one mechanism alone, because this course of action would be far too risky. Therefore, the body typically uses several different mechanisms to combat infections. In the example from the chapter opener, the body's initial response to intruders was inflammation (a nonspecific response), but a specific immune response against the invading microorganisms was also necessary. APCs would phagocytize some of the invaders, process their antigenic determinants, and stimulate both humoral and cell-mediated immune responses, the results of which include the neutralization of toxins and microbial enzymes by antibody; the lysing of bacteria by antibody-activated complement, opsonization by antibodies, and the killing of intracellular bacteria and viruses by cytotoxic T cells. The relative importance of each of these pathways depends on the type of pathogen involved and on the mechanisms by which they cause disease.

CRITICAL THINKING

What sorts of pathogens could successfully attack a patient with an inability to produce T lymphocytes?

Types of Acquired Immunity

Learning Objective

✓ Contrast active versus passive acquired immunity, and naturally acquired versus artificially acquired immunity.

As we have seen, specific immunity is acquired during an individual's life. Immunologists categorize specific immunity as either naturally or artificially acquired. Naturally acquired immunity occurs when the body mounts an immune response against antigens, such as influenza viruses or food antigens, encountered during the course of daily life. Artificial immunity is the body's response to antigens introduced in vaccines, as occurs with immunization against the flu. Immunologists further distinguish acquired immune responses as either *active* or *passive;* that is, the immune system either responds actively to antigens via humoral or cell-mediated responses, or the body passively receives antibodies from another individual. Next we consider each of four types of acquired immunity.

Naturally Acquired Active Immunity

Naturally acquired active immunity occurs when the body responds to exposure to pathogens and environmental antigens by mounting specific immune responses. The body is naturally and actively engaged in its own protection. As we have seen, once an immune response occurs, immunological memory persists—on subsequent exposure to the same antigen, the immune response will be rapid and powerful and often provides the body complete protection.

Table 16.3 A Comparison of the Types of Acquired Immunity

Naturally Acquired		Artificially Acquired	
Active	Passive	Active	Passive
The body responds to antigens that enter naturally, such as during infections	Antibodies are transferred from mother to offspring, either across the placenta (IgG) or in breast milk (secretory IgA)	Health care workers introduce antigens in vaccines; the body responds with humoral or cell-mediated immune responses	Health care workers introduce preformed antibodies (obtained from immune individuals) as antisera and antitoxins into a patient

Naturally Acquired Passive Immunity

Although newborns possess the cells and tissues needed to mount an immune response, they respond slowly to antigens. If required to protect themselves solely via naturally acquired active immunity, they might die of infectious disease before their immune systems were mature enough to respond adequately. However, they are not on their own; in the womb, IgG molecules cross the placenta from the mother's bloodstream to provide protection, and after birth, babies receive secretory IgA in breast milk. Via these two processes, a mother provides her child with antibodies that protect it during its early months. Because the child is not actively producing its own antibodies, this type of protection is known as **naturally acquired passive immunity.**

Artificially Acquired Active Immunity

Physicians induce immunity in their patients by introducing antigens in the form of vaccines. The patients' own immune systems then mount active responses against the foreign antigens, just as if the antigens were part of a naturally acquired pathogen. Such **artificially acquired active immunity** is the basis of immunization, which is discussed in detail in Chapter 17.

Artificially Acquired Passive Immunity

Active immunity usually requires days to weeks to develop fully, and in some cases such a delay can prove detrimental or even fatal. For instance, an active immune response is too slow to protect against rattlesnake venom or infection with hepatitis A virus. Therefore, medical personnel routinely harvest antibodies against toxins and pathogens that are so deadly or so fast-acting that an individual's active immune response is inadequate. They can acquire these antibodies from the blood of immune humans or animals, typically a horse or other large animal. Physicians then inject such *antitoxins* or *antisera* into infected patients to confer **artificially acquired passive immunity.** Chapter 17 discusses this type of treatment in greater detail.

Active immune responses, whether naturally or artificially induced, are advantageous because they result in immunological memory and protection against future infections. However, they are slow-acting. Passive immunity, in which individuals are provided fully formed antibodies, has the advantage of speed but does not confer immunological memory because B and T lymphocytes are not activated. **Table 16.3** summarizes the four types of acquired immunity.

CHAPTER SUMMARY

Elements of Specific Immunity (pp. 463–473)

1. Immunologists study **specific immunity,** which is the ability of a vertebrate to recognize and defend against distinct species or strains of invaders.
2. **Antigens** are substances that trigger specific immune responses. The three-dimensional shape of a region of an antigen that is recognized by the immune system is the **antigenic determinant** (or epitope). Effective antigen molecules are large, complex, stable, degradable, and foreign to their host.
3. **Exogenous** antigens are found on microorganisms that multiply outside the cells of the body; **endogenous** antigens are produced by pathogens multiplying inside the body's cells.
4. Ideally, the body does not attack antigens on the surface of its normal cells, called **autoantigens;** this phenomenon is called self-tolerance.
5. The **lymphatic system** is composed of lymphatic vessels, which conduct the flow of **lymph,** and lymphoid tissues and organs that are directly involved in specific immunity. The latter include lymph nodes, the thymus, the spleen, the tonsils, and mucosa-associated lymphoid tissue (MALT). Cells with a central role in specific immunity, **lymphocytes,** migrate to and persist in various lymphoid organs, where they are available to encounter foreign invaders in the blood and lymph. Lymphocytes originate and mature in the red bone marrow and express characteristic membrane proteins that are designated cluster of differentiation (CD) molecules.
6. **B lymphocytes (B cells),** which mature in the red bone marrow secrete soluble proteins called **antibodies (immunoglobulins, Ig)** that attack exogenous antigens in **humoral immune responses.**

7. Antibodies consist of two **light chains** and two **heavy chains** joined via disulfide bonds to form Y-shaped molecules whose stems, called **F_C regions,** are of five basic classes. Each B cell randomly chooses (once in its life) genes for the upper portion of the "arms" of its antibody molecules; therefore, the upper portions are called variable regions. Together the variable regions of a heavy and light chain form an **antigen-binding site** and the upper portions of antibody molecules are called F_{ab} regions. Each basic antibody molecule has two antigen-binding sites and can potentially bind to two antigenic determinants.
8. Antibodies function in complement activation, inflammation, **agglutination** (clumping), **neutralization** (blocking the action of a toxin or attachment of a pathogen), and **opsonization** (enhanced phagocytosis).
9. The five types of heavy chains in antibody stems distinguish five different classes of antibodies (immunoglobulins). **IgG** is the predominant antibody found in the bloodstream and is largely responsible for defense against invading bacteria. IgG can cross a placenta to protect the fetus. **IgM,** a pentamer with ten antigen-binding sites, is the predominant antibody produced first during a primary humoral response. Two molecules of IgA are attached via J chains and a polypeptide secretory component to produce **secretory IgA,** which is found in milk, tears, and mucous membrane secretions. **IgE** triggers inflammation and allergic reactions. It also functions during helminth infections. Immunoglobulin D **(IgD)** is found in cytoplasmic membranes of some animals.
10. B lymphocytes also express **B-cell receptors (BCRs),** which are antibody-like proteins that are complementary to antigenic determinants but which are inserted into the cytoplasmic membranes of B cells rather than secreted. A BCR has the same shaped binding sites as do the antibodies that can be synthesized by that cell.
11. **T lymphocytes** (T cells) have **T-cell antigen receptors (TCRs),** mature in the thymus, and attack cells that harbor endogenous pathogens during **cell-mediated immune responses.**
12. In cell-mediated immunity, T_H1 cells activate **cytotoxic T cells** (CD8 cells), which then secrete perforins and granzymes that destroy infected or abnormal body cells. Two types of helper T cells—**T_H1** and **T_H2**—are characterized by CD4 and direct cell-mediated and humoral immune responses respectively.
13. **Cytokines** are soluble regulatory proteins that act as intercellular signals to direct activities in immune responses. Cytokines include **interleukin (ILs), interferons (IFNs), growth factors, tumor necrosis factors (TFCs),** and **chemokines.**

The Body's Preparation for a Specific Immune Response (pp. 473–477)

1. Cells with receptors that respond to autoantigens are selectively killed via **apoptosis**—a process known as **clonal deletion.** Only cells that respond to foreign antigens survive to defend the body.
2. The initial step in mounting an immune response is that antigens are captured, ingested, and degraded into antigenic determinants by **antigen presenting cells (APCs)** such as macrophages and dendritic cells. Antigenic determinants are then bound to membrane molecules called **major histocompatibility complex (MHC)** class II molecules, which are inserted to present the antigenic determinant on the outer surface of the APCs' cytoplasmic membrane.

The Humoral Immune Response (pp. 477–480)

1. The combination of antigenic determinant and MHC II proteins activates the helper T cell that has a complementary TCR to become a T_H2 cell. The T_H2 cell secretes the cytokine interleukin-4 (IL-4), which activates the matching B cell to divide and differentiate.
2. **Plasma cells,** which arise from B cells, produce antibodies. Plasma cells live for only a short time but secrete very large amounts of antibodies. B cells that migrate to lymphoid tissues to await a subsequent encounter with the same antigen are called **memory B cells.**
3. The **primary response** to an antigen is slow to develop and of limited effectiveness. When that antigen is encountered a second time, the activation of memory cells ensures that the immune response is rapid and strong. This is a **memory response.**

The Cell-Mediated Immune Response (pp. 480–482)

1. A cell-mediated immune response is directed primarily against infected or abnormal body cells, including virus-infected cells, bacteria-infected cells, some fungal- or protozoan-infected cells, some cancer cells, and foreign cells that enter the body as a result of organ transplantation.
2. Cytotoxic T cells recognize abnormal molecules expressed on the surface of an infected, cancerous, or foreign cell. Cytotoxic T cells function only in the presence of cytokines from T_H1 cells.
3. Cytotoxic T cells destroy their target cells via two pathways: the **perforin-granzyme pathway,** which kills the affected cells by secreting **perforins** and **granzymes,** or the **CD95 pathway,** in which CD95L binds to CD95 on the target cell, which triggers target cell apoptosis. Cytotoxic T cells may also form **memory T cells.**

Types of Acquired Immunity (pp. 482–483)

1. When the body mounts a specific immune response against an infectious agent, the result is called **naturally acquired active immunity.** The passing of maternal IgG to the fetus, or of secretory IgA in milk to a baby, are examples of **naturally acquired passive immunity. Artificially acquired active immunity** is achieved by deliberately injecting someone with antigens in vaccines to provoke an active response, as in the process of immunization. **Artificially acquired passive immunity** involves the administration of preformed antibodies in antitoxins or antisera to a patient.

QUESTIONS FOR REVIEW

(Answers to multiple choice and matching questions are on the web, along with additional review questions. Visit www.microbiologyplace.com.)

Multiple Choice

1. Antibodies function to
   a. directly destroy foreign organ grafts.
   b. mark invading organisms for destruction.
   c. kill intracellular viruses.
   d. promote cytokine synthesis.
   e. stimulate T cell growth.

2. MHC class II molecules bind to ________ and trigger ________.
   a. endogenous antigens, cytotoxic T cells
   b. exogenous antigens, cytotoxic T cells
   c. antibodies, B cells
   d. endogenous antigens, helper T cells
   e. exogenous antigens, helper T cells

3. Rejection of a foreign skin graft is an example of
   a. destruction of virus-infected cells.
   b. tolerance.
   c. antibody-mediated immunity.
   d. a secondary immune response.
   e. a cell-mediated immune response.

4. An autoantigen is
   a. an antigen from bacteria.
   b. a normal body component.
   c. an artificial antigen.
   d. any carbohydrate antigen.
   e. a nucleic acid.

5. Among the key molecules that mediate cell-mediated cytotoxicity are
   a. perforins.
   b. immunoglobulins.
   c. complement.
   d. cytokines.
   e. interferons.

6. Which of the following lymphocytes predominates in blood?
   a. cytotoxic T cells
   b. helper T cells
   c. B cells
   d. memory cells
   e. all are about equally prevalent

7. The major class of immunoglobulin found on the surfaces of the walls of the intestines and airways is
   a. IgG.
   b. IgM.
   c. IgA.
   d. IgE.
   e. IgD.

8. Which cells express MHC class I molecules?
   a. red cells
   b. antigen presenting cells only
   c. neutrophils only
   d. all nucleated cells
   e. dendritic cells only

9. In which of the following sites in the body can B cells be found?
   a. lymph nodes
   b. spleen
   c. red bone marrow
   d. intestinal wall
   e. all of the above

Modified True/False

Mark each statement as either true or false. Rewrite false statements to make them true by changing the italicized words.

1. MHC class II molecules are found on *T cells*.
2. *Apoptosis* is the term used to describe cellular suicide.
3. Lymphocytes with CD8 glycoprotein are *helper* T cells.
4. *Cytotoxic T cells* secrete immunoglobulin.
5. Secretion of antibodies by activated B cells is a form of *cell-mediated* immunity.

Matching

1. Match each cell in the left column with its associated protein from the right column.

___ Plasma cell	A. MHC II molecule
___ Cytotoxic T cell	B. Interleukin-4
___ T_H2 cell	C. Perforin and granzyme
___ Dendritic cell	D. Immunoglobulin

2. Match each type of immunity in the left column with its associated example from the right column.

___ Artificially acquired passive immunity	A. Production of IgE in response to pollen
___ Naturally acquired active immunity	B. Acquisition of maternal antibodies in breast milk
___ Naturally acquired passive immunity	C. Administration of tetanus toxoid
___ Artificially acquired active immunity	D. Administration of antitoxin

Essay Questions

1. When is antigen processing an essential prerequisite for an immune response?
2. Why does the body have both humoral and cell-mediated immune responses?

CRITICAL THINKING

1. Why is it advantageous that the lymphatic system lacks a pump?
2 Contrast nonspecific defenses with specific immunity.
3. What is the benefit to the body of requiring the immune system to process antigen?
4. Scientists can develop genetically deficient strains of mice. Describe the immunological impairments that would result in mice deficient in each of the following: class I MHC, class II MHC, TCR, BCR, IL-2 receptor, or IFN-γ.
5. Human immunodeficiency virus (HIV) preferentially destroys CD4 cells. Specifically, what effect does this have on humoral and cell-mediated immunity?
6. What would happen to a person who failed to make MHC molecules?
7. Why does the body make five different classes of immunoglobulins?
8. Some materials such as metal bone pins and plastic heart valves can be implanted into the body without fear of rejection by the patient's immune system. Why is this? What are the ideal properties of any material that is to be implanted?

Immunization and Immune Testing

CHAPTER 17

A 13-month-old girl is brought into the hospital, coughing and crying. Her mother tells the pediatrician that she had noticed a mucous discharge pooling in the corners of her baby's eyes the previous evening. The pediatrician checks the baby's medical record and notes that she has not yet received her first measles/mumps/rubella (MMR) vaccination. He then swabs the baby's throat, draws a blood sample, and tells the mother that he is going to have the lab perform a "special immune test" to confirm his diagnosis: measles. Because he has seen several children with measles in the past few weeks and the baby's symptoms support the diagnosis, the pediatrician administers an intramuscular injection of antimeasles immunoglobulin. He explains that this injection will protect the child against the most severe form of the illness, which can be fatal. He also schedules her daughter for a routine measles/mumps/rubella vaccination. Mother and daughter return home to rest and recuperate.

How do immunoglobulins and vaccines work? How do immunologic tests aid in the diagnosis of specific diseases? This chapter is an introduction to the important topics of immunization and immune testing.

CHAPTER OUTLINE

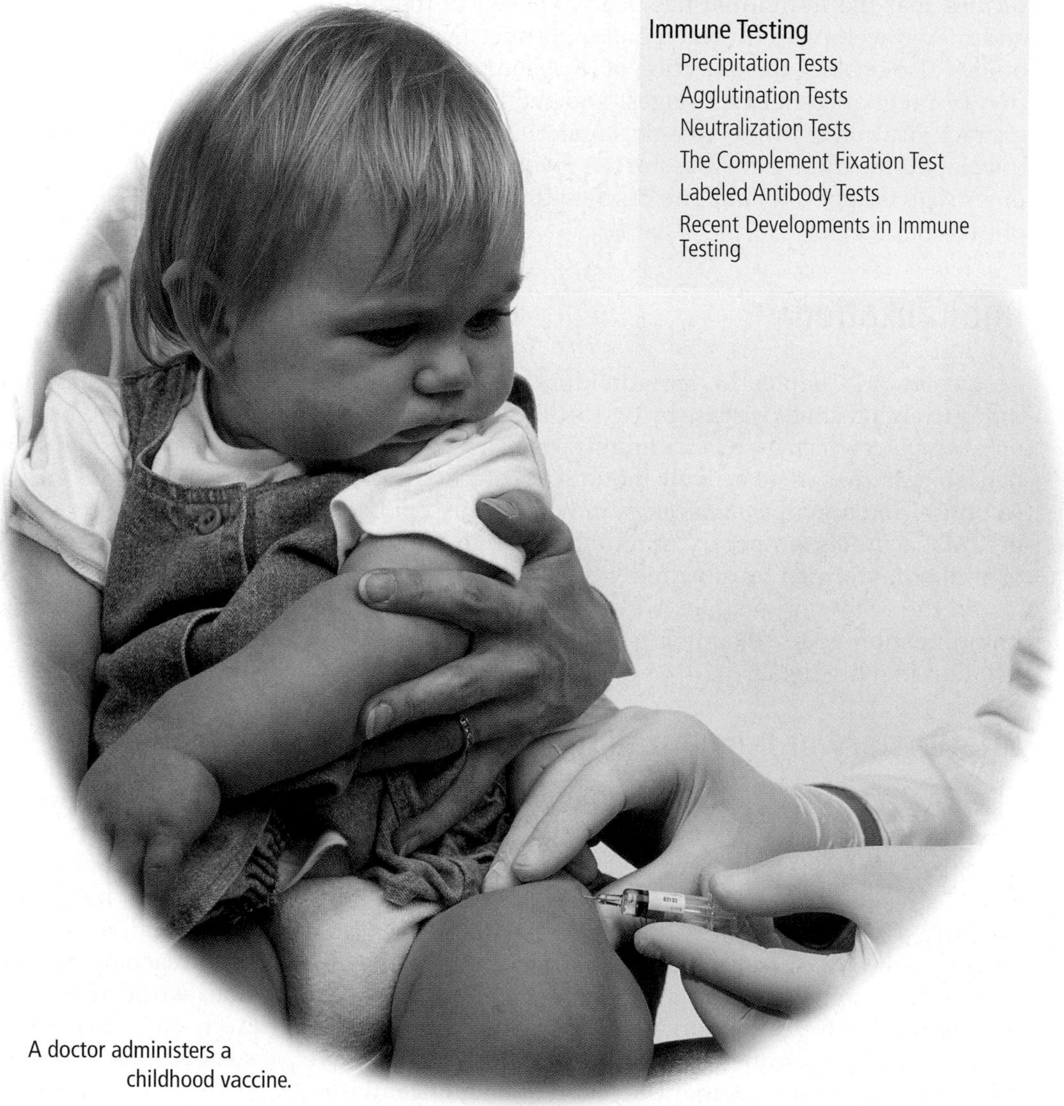

A doctor administers a childhood vaccine.

MicroPrep Pre-Test: *Take the pre-test for this chapter on the web. Visit* ***www.microbiologyplace.com.***

In this chapter we will discuss three applications of immunology: active immunization, or vaccination; passive immunization using immunoglobulins; and immune testing. Vaccination has proven the most efficient and cost-effective method of controlling infectious diseases. Without the use of effective vaccines, millions more people worldwide would suffer each year from potentially fatal infectious diseases, including measles, diphtheria, and pertussis (whooping cough). The administration of immunoglobulins has further reduced morbidity and mortality from certain infectious diseases, such as hepatitis A and yellow fever, in unvaccinated individuals. Medical personnel also make practical use of the immune response as a diagnostic procedure. For example, the detection of antibodies to HIV in a person's blood indicates that the individual has been exposed to that virus and may develop AIDS. The remarkable specificity of antibodies also enables the detection of drugs in urine, recognition of pregnancy at early stages, and the identification or characterization of other biological material. The many tests developed for these purposes are the focus of the discipline of *serology* (sĕ-rol'ō-jē) and are discussed in the second half of this chapter.

Immunization

As we saw in Chapter 16, an individual may be made immune to an infectious disease by two artificial methods: *active immunization,* which involves administering a vaccine to a patient so that the patient actively mounts a protective immune response, and *passive immunization,* in which a susceptible individual acquires temporary immunity through the transfer of antibodies formed by other individuals or animals.

In the following sections we will review the history of immunization before examining active and passive immunization in more detail.

Brief History of Immunization

Learning Objective

✓ Discuss the history of vaccination from the 12th century through the present.

As early as the 12th century, the Chinese noticed that children who recovered from smallpox never contracted the disease a second time. They therefore adopted a policy of deliberately infecting young children with particles of ground smallpox scabs from children who had survived mild cases. By doing so, they succeeded in significantly reducing the population's overall morbidity and mortality from the disease. News of this procedure, called *variolation* (var'ē-ō-lā'shŭn), spread westward through central Asia, and the technique was widely adopted.

Lady Mary Montagu, the wife of the English ambassador to the Ottoman Empire, learned of the procedure, had it performed on her own children, and told others about it upon her return to England in 1721. As a result, variolation came into use in England and in the American colonies; however, although effective and usually successful, it still caused death from smallpox in 1–2% of recipients and in people exposed to recipients, so in time variolation was outlawed.

Thus, when the English physician Edward Jenner demonstrated in 1796 that protection against smallpox could be conferred by inoculation with crusts from a person infected with cowpox—a related but very mild disease—the new technique was adopted with enthusiasm. Because cowpox was also called *vaccinia*[1] (vak-sin'ē-ă), Jenner called the new technique **vaccination** (vak'si-nā'shŭn), and the protective inoculum a **vaccine** (vak-sēn')—terms that are still commonly used to describe all immunizations. For many years thereafter, vaccination against smallpox was widely practiced, even though no one understood how it worked or whether similar techniques could be used for other diseases.

In 1879, Louis Pasteur conducted experiments on the bacterium *Pasteurella multocida* (pas-ter-el'ă mul-tŏ'si-da) and demonstrated that he could make an effective vaccine against this organism (which causes a disease in birds called fowl cholera). Once the basic principle of vaccine manufacture was understood, vaccines against anthrax and rabies rapidly followed. Moreover, once it was discovered that these vaccines provide protection through the actions of antibodies, the technique of transferring protective antibodies to susceptible individuals—that is, passive immunization—was developed soon thereafter.

By the late 1900s, immunologists and health care providers had formulated vaccines that significantly reduced the number of cases of polio, measles, mumps, and rubella in the United States **(Figure 17.1).** Now we also have successful vaccines against many other diseases, including chickenpox, hepatitis A, hepatitis B, and many strains of microbes that cause diarrhea, influenza, and pneumonia. Health care providers, governments, and international organizations working together have rid the world of naturally occurring smallpox and are well on the way to the worldwide eradication of polio, measles, mumps, and rubella. **Highlight 17.1** on page 490 discusses why a vaccine for the common cold does not yet exist.

Even though immunologists have produced vaccines that protect people against many deadly diseases, a variety of political, social, and economic problems prevent vaccines from reaching all those who need them. In developing nations worldwide, over 3 million children still die each year from vaccine-preventable infectious diseases because of socioeconomic and political obstacles. Additionally, some pathogens such as the protozoa of malaria and the virus of AIDS, still frustrate all attempts to develop effective vaccines against them. Furthermore, the existence of vaccine-associated risks—both medical risks (the low but persistent incidence of vaccine-caused diseases) and financial risks (the high costs of developing and producing

[1] From Latin *vacca*, meaning cow.

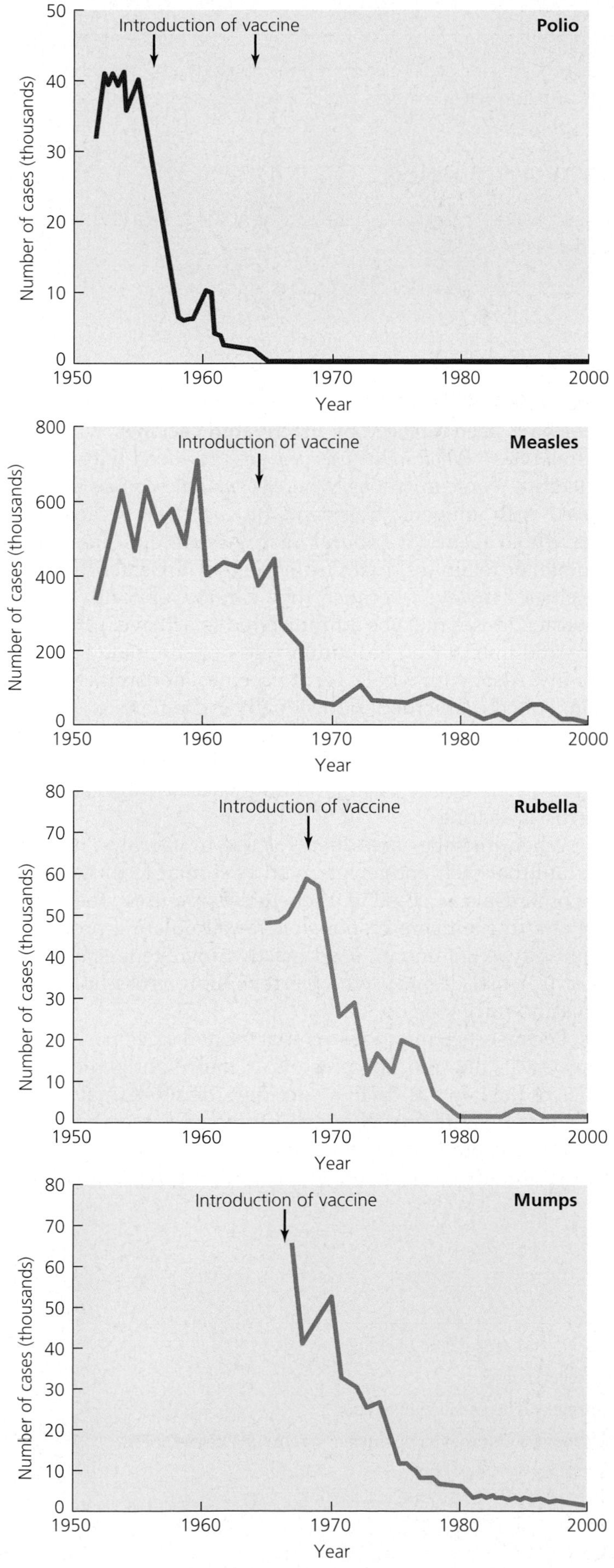

vaccines)—has in recent years discouraged investment in new vaccines. Thus although the history of immunization is marked by stunning advancements in public health, the future of immunization poses immense challenges.

Next we take a closer look at active immunization, commonly known as vaccination.

Active Immunization

Learning Objectives

- ✓ Describe the advantages and disadvantages of the three types of vaccines.
- ✓ Describe three methods by which molecular biology techniques can be used to develop improved vaccines.
- ✓ Delineate the risks and benefits of routine vaccination in healthy populations.

In the following subsections we examine the types of vaccines, the roles of technology in producing modern vaccines, and issues concerning vaccine safety.

Vaccine Types

Scientists are constantly striving to develop vaccines of maximal efficacy and safety, but not all types of vaccines are equally safe or effective. The three general types of vaccines, each of which has its own combination of strengths and weaknesses, are attenuated (live) vaccines, killed (or inactivated) vaccines, and toxoid vaccines.

Attenuated (Live) Vaccines Virulent viruses and organisms are normally not used in vaccines because they cause disease. Instead, immunologists reduce virulence so that, although still living, the pathogens no longer cause disease. The process of reducing virulence is called **attenuation** (ă-ten-ū-ā′shŭn). The most common method for attenuating viruses involves raising them for numerous generations in tissue culture cells, until the viruses lose the ability to produce disease. Thus for rabies viruses, which preferentially attack nerve cells, avirulent viruses for use in vaccines are produced by prolonged tissue culture in rabbit spinal cords, until the viruses' virulence in nerve cells is lost. Bacteria may be made avirulent by culturing them under unusual conditions or by using genetic manipulation.

◀ ***Figure 17.1***

The effect of immunization in reducing the incidence of four infectious diseases in the United States. The two arrows in the graph of polio cases represent the introduction of two different types of vaccine. Only polio has been entirely eradicated in the United States.

Highlight 17.1 Why Isn't There a Cold Vaccine?

We have vaccines for the flu, so why don't we have a vaccine for the common cold? The reason is that whereas strains of only one influenza virus cause flu, over 200 different strains of adenoviruses, coronaviruses, and rhinoviruses are known to cause the common cold, and each of these viruses has its own distinct antigens and antigenic strains making it extremely difficult to create a single vaccine to prevent them all. To further complicate matters, viruses can mutate, resulting in changes in their antigens; with over 200 different cold viruses in existence, such mutations create immense logistical challenges in vaccine development. Fortunately, the common cold typically lasts only a few days and is adequately treated with rest and self-care.

CRITICAL THINKING

Immunologically speaking, is it likely a patient will catch a cold caused by the same virus twice?

Attenuated vaccines—those containing attenuated microbes—are also called *modified live vaccines*. Because they contain living but avirulent organisms or viruses, these vaccines cause very mild infections but no serious disease. Attenuated viruses in such a vaccine infect host cells and replicate; the infected cells then process endogenous viral antigens. As a result, modified live viral vaccines trigger a cell-mediated immune response dominated by type 1 helper T cells (T_H1) and cytotoxic T cells (T_C). Because modified live vaccines contain active microbes, a large number of antigen molecules are available to stimulate an immune response. Further, vaccinated individuals can infect those around them, providing so-called *herd immunity*—that is, immunity beyond the individual receiving the vaccine.

While usually very effective, attenuated vaccines can be hazardous because the modified microbes may retain enough residual virulence to cause disease in immunosuppressed people. Additionally, pregnant women should not receive live vaccines because of the danger that the attenuated pathogen will cross the placenta and harm the fetus. Occasionally, modified viruses actually revert to wild type or mutate to a form that causes persistent infection or disease. For example, in 2000 a polio epidemic in the Dominican Republic and Haiti resulted from the reversion of an attenuated live virus in oral polio vaccine to the virulent poliovirus. For this reason, we no longer use live polio vaccine to immunize children in the United States.

Inactivated (Killed) Vaccines For some diseases live vaccines have been replaced by **inactivated vaccines,** which are of two types: *Whole agent vaccines* are produced with deactivated but whole microbes, whereas *subunit vaccines* are produced with antigenic fragments of microbes. Because neither whole agent nor subunit vaccines can replicate, revert, mutate, or retain residual virulence, they are safer than live vaccines. However, because they cannot replicate, several "booster" doses must be administered to achieve full immunity, and immunized individuals do not stimulate herd immunity. Also, with whole agent vaccines, nonantigenic portions of the microbe occasionally stimulate a painful inflammatory response in some individuals. As a result, pertussis vaccine, formerly available as a whole agent vaccine, is now being replaced with a subunit vaccine called acellular pertussis vaccine.

When microbes are killed for use in vaccines, it is important that their antigens remain as similar to those of living organisms as possible. If chemicals are used, they must not alter the antigens responsible for stimulating protective immunity. A commonly used inactivating agent is *formaldehyde* (fōr-mal′dĕ-hīd), which irreversibly cross-links proteins and nucleic acids.

Because the microbes of inactivated vaccines cannot reproduce, they do not present as many antigenic molecules to the body as do live vaccines; therefore, inactivated vaccines are antigenically weak. To be effective, they must be administered in high doses or in multiple doses, or incorporated with materials called **adjuvants** (ad′joo-

Table 17.1 Some Common Adjuvants

Adjuvant	Effects
Aluminum phosphate (alum)	Slows processing and degradation of antigen
Saponin (soap-like plant product)*	Stimulates T cell responses
Mineral oil*	Slows processing and degradation of antigen
Freund's complete adjuvant (mineral oil containing killed mycobacteria*)	Slows processing and degradation of antigen; stimulates T cell responses

*Considered too toxic for humans; used in vaccinating animals only.

văntz), substances that increase the effective antigenicity of the vaccine. Unfortunately, the use of adjuvants to increase antigenicity may stimulate local inflammation, and high individual doses and multiple dosing increase the risk of producing allergies. **Table 17.1** lists some common adjuvants and their effects in enhancing the efficacy of vaccines.

Because all types of killed vaccines are recognized by the immune system as exogenous antigens, they stimulate a T_H2 response, promoting antibody-mediated (rather than cell-mediated) immunity.

Toxoid Vaccines For some bacterial diseases, notably tetanus and diphtheria, it is more efficacious to induce an immune response against bacterial toxins than against cellular antigens. **Toxoid** (tok'soyd) **vaccines** are chemically or thermally modified toxins that are used in vaccines to stimulate active immunity. As with killed vaccines, toxoids stimulate antibody-mediated immunity. Because toxoids have few antigenic determinants, effective immunization requires multiple childhood doses as well as reinoculations every 10 years for life.

The 2003 immunization schedule recommended by the CDC is presented in **Figure 17.2.** The types of vaccines available for immunizing against each of the diseases in the vaccination schedule, as well as some other available vaccines, are listed in **Table 17.2** on page 492.

Modern Vaccine Technology

Although live, inactivated, and toxoid vaccines have been highly successful in controlling infectious diseases, researchers are always seeking ways to make vaccines more effective, cheaper, and safer. Scientists can use a variety of

CDC Recommended Immunization Schedule – United States, 2003

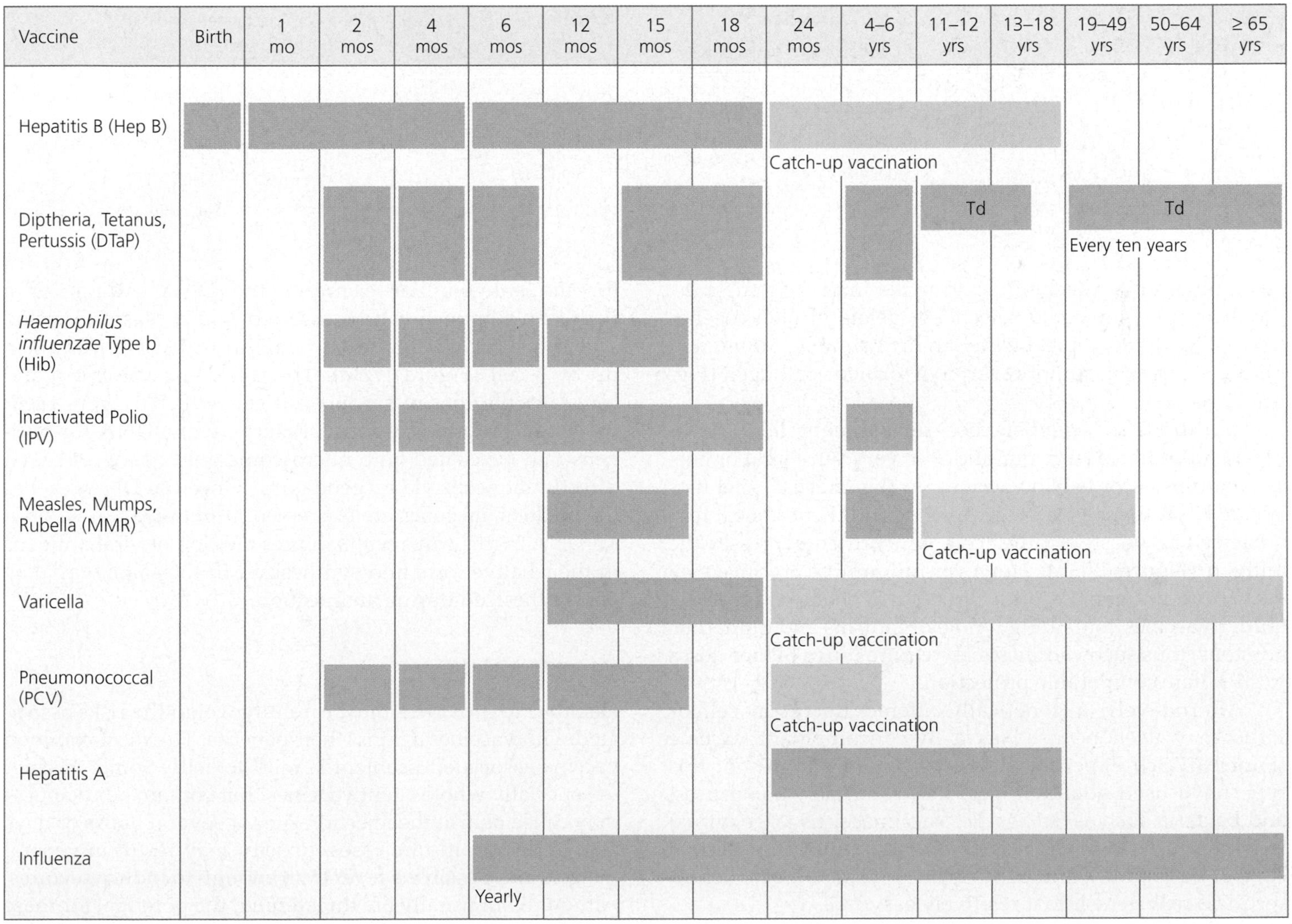
Vaccine
Birth
1 mo
2 mos
4 mos
6 mos
12 mos
15 mos
18 mos
24 mos
4–6 yrs
11–12 yrs
13–18 yrs
19–49 yrs
50–64 yrs
≥ 65 yrs
Hepatitis B (Hep B)
Catch-up vaccination
Diptheria, Tetanus, Pertussis (DTaP)
Td
Td
Every ten years
Haemophilus influenzae Type b (Hib)
Inactivated Polio (IPV)
Measles, Mumps, Rubella (MMR)
Catch-up vaccination
Varicella
Catch-up vaccination
Pneumonococcal (PCV)
Catch-up vaccination
Hepatitis A
Influenza
Yearly

▲ ***Figure 17.2***
The CDC's recommended immunization schedule, 2002.

Table 17.2 Principal Vaccines to Prevent Human Diseases

Vaccine	Disease Agent	Disease	Vaccine Type
Recommended by CDC			
Hepatitis B	Hepatitis B virus	Hepatitis	Inactivated subunit
Diphtheria/	Diphtheria toxin	Diphtheria	Toxoid
tetanus/	Tetanus toxin	Tetanus	Toxoid
acellular pertussis (DTaP)	*Bordetella pertussis*	Whooping cough	Inactivated subunit (inactivated whole also available)
Haemophilus influenzae Type B	*Haemophilus influenzae*	Meningitis, pneumonia	Inactivated subunit
Polio	Poliovirus	Poliomyelitis	Inactivated (attenuated also available)
Measles/	Measles virus	Measles	Attenuated
mumps/	Mumps virus	Mumps	Attenuated
rubella	Rubella virus	Rubella (German measles)	Attenuated
Varicella	Chickenpox virus	Chickenpox	Attenuated
Pneumococcal	*Streptococcus pneumoniae*	Pneumonia	Inactivated subunit
Hepatitis A	Hepatitis A virus	Hepatitis	Inactivated whole
Influenza	Influenza viruses	Flu	Inactivated subunit
Not recommended for general population			
Anthrax	*Bacillus anthracis*	Anthrax	Inactivated whole
BCG (bacillus of Calmette and Guerin)	*Mycobacterium tuberculosis, M. leprae*	Tuberculosis, leprosy	Attenuated
Yellow fever	Yellow fever virus	Yellow fever	Attenuated
Rabies	Rabies virus	Rabies	Inactivated whole
Vaccinia (cowpox)	Smallpox virus	Smallpox	Attenuated

recombinant DNA techniques to make improved vaccines. For example, they could selectively delete virulence genes from a pathogen, producing an irreversibly attenuated microbe, one that cannot revert to a virulent pathogen **(Figure 17.3a).**

Additionally, scientists use recombinant DNA techniques to produce large quantities of very pure viral or bacterial antigens for use in vaccines. In this process, scientists isolate DNA that codes for an antigen and then insert it into a bacterium, yeast, or other cell, which then expresses the antigen **(Figure 17.3b).** However, immunity produced by such pure antigens is often inferior to that produced by cruder vaccines containing whole organisms and numerous antigens; thus such vaccines may require much higher doses to stimulate comparable protection.

Alternatively, a genetically altered microbial cell or virus may itself be used as a live recombinant vaccine **(Figure 17.3c).** Experimental recombinant vaccines of this type have used adenoviruses, herpesviruses, poxviruses, and bacteria such as *Salmonella*. Vaccinia virus (a poxvirus) is widely used because it is easy to administer by dermal scratching or orally, and because its large genome makes inserting a new gene into it relatively easy.

Another innovative method of immunization involves injection not of antigens, but instead of the DNA that codes for the antigen. For example, the DNA coding for a pathogen's antigen can be inserted into a plasmid vector, which is then injected into an individual, where it is taken up by a cell **(Figure 17.3d).** The cell then transcribes and translates the gene to produce antigen, which triggers a cell-mediated immune response. Alternatively, DNA for antigens may be coated onto microscopic gold beads and fired directly into cells via a "gene gun." Once the DNA reaches the nucleus, its genes are expressed. Immunization with viral DNA in this way results in presentation of viral antigens in their native form and synthesized in the same way they are synthesized during an infection.

Vaccine Safety

Health care providers must carefully weigh the risks associated with vaccines against their benefits. The most common vaccine-associated problem is mild toxicity. Some vaccines —especially whole agent vaccines that contain adjuvants— may cause pain at the injection site for several hours or days after injection. In rare cases, toxicity may result in general malaise and possibly a fever high enough to induce seizures. Although not usually life threatening, the potential for these symptoms may be sufficient to discourage people from being vaccinated or having their infants vaccinated.

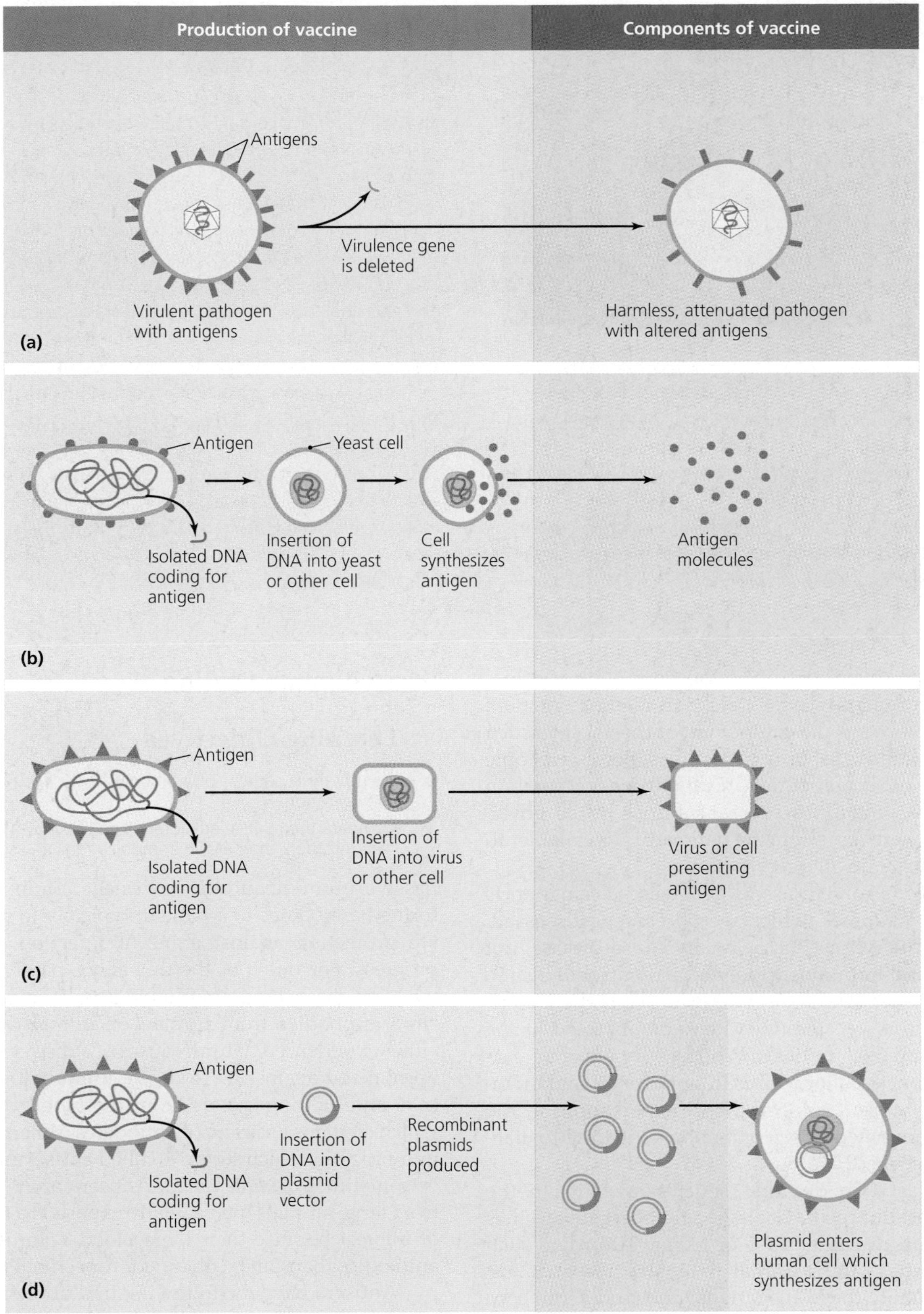

▲ ***Figure 17.3***

Some potential uses of recombinant DNA technology for making improved vaccines. **(a)** Deletion of virulence gene(s) to create an irreversibly attenuated pathogen for use in a vaccine. **(b)** Insertion of an isolated gene that codes for a selected antigen into a cell (in this case, a yeast cell), which then produces large quantities of the antigen for use in a vaccine. **(c)** Insertion of a gene that codes for a selected antigen into a cell or virus (in this case, vaccinia virus), which produces the antigen. The entire recombinant is used in a vaccine. **(d)** Injection of DNA containing a selected gene (in this case, as part of a plasmid) into an individual. Once some of this DNA is incorporated into a cell's genome, that cell synthesizes and processes the antigen, which stimulates a cell-mediated immune response.

Highlight 17.2 Smallpox: To Vaccinate or Not to Vaccinate?

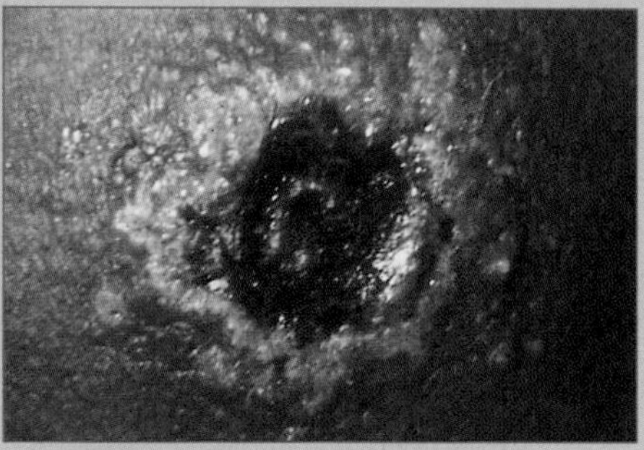

Vaccinia necrosum

Medical personnel in the United States regularly administered the smallpox vaccine to the general public until 1971, at which time the risk of contracting smallpox was deemed too low to justify required vaccinations. Indeed, in 1980 the World Health Assembly declared smallpox successfully eradicated from the natural world. However, recent concerns about the potential use of smallpox virus as an agent of bioterrorism has sparked debate about whether or not citizens should once again be vaccinated against it.

Although safe and effective for most healthy adults, for others the smallpox vaccine, which is attenuated cowpox virus, can result in serious side effects—even death. Individuals with compromised immune systems (such as AIDS patients or cancer patients undergoing chemotherapy) are considered to be at particularly high risk for developing adverse reactions. Pregnant women, infants, and individuals with a history of the skin condition eczema are also considered poor candidates for the vaccine. Though rare, certain otherwise healthy individuals may also develop adverse reactions to the vaccine. The more serious side effects include vaccinia necrosum (characterized by progressive necrosis in the area of vaccination; see the box photo) and encephalitis (inflammation of the brain). Approximately 1 in every 1 million individuals receiving the vaccine for the first time develops a fatal reaction to it. Some experts also fear that vaccinated individuals may pose a risk to immunocompromised people whom they contact.

Is the risk of a bioterrorist smallpox attack great enough to warrant the exposure to the known risks of administering smallpox vaccine to the general population? If you were a public health official, what would you decide?

A much more severe problem associated with vaccination is the risk of anaphylactic shock, an allergic reaction, which may develop to some component of the vaccine, such as egg proteins, adjuvants, or preservatives. Because people are rarely aware of such allergies ahead of time, vaccination recipients should remain for several minutes in the physician's office, where epinephrine is readily available to counter any signs of an allergic reaction.

A third major problem associated with vaccination is that of residual virulence, which we previously discussed. Attenuated vaccine viruses occasionally cause disease, not only in fetuses and immunosuppressed patients, but also in healthy children and adults. A good example is the attenuated oral poliovirus vaccine (OPV) commonly used in the United States until the late 1990s. While a very effective vaccine, it causes clinical poliomyelitis in 1 of every 2 million recipients or their close contacts. Medical personnel in the United States eliminated this problem by switching to inactivated polio vaccine (IPV).

Over the past two decades, several lawsuits in the United States and Europe have alleged that certain vaccines against childhood diseases cause or trigger disorders such as autism, diabetes, and asthma. Extensive research has failed to substantiate these allegations, and to date no conclusive evidence for such a link has been found. Indeed, vaccine manufacturing methods have improved tremendously in recent years, ensuring that modern vaccines are much safer than those in use even a decade ago. The U.S. Food and Drug Administration has established a Vaccine Adverse Event Reporting System for monitoring vaccine safety. **Highlight 17.2** discusses the issues surrounding the administration of smallpox vaccinations to the general public.

Passive Immunization

Learning Objectives

- ✓ Identify three sources of antibodies for use in passive immunization.
- ✓ Compare the relative advantages and disadvantages of active immunization and passive immunization.

Passive immunization involves the administration of preformed antibodies to a patient, typically to provide immediate protection against a recent infection or a disease in progress. For much of the time since passive immunization was first discovered, immunologists have harvested the desired antibodies from human or animal donors that have either experienced natural causes of a disease or were actively immunized against it. To do so, immunologists remove the cells and clotting factors from the blood to acquire *serum*, which contains a variety of antibodies. When used for passive immunization, such serum is called **antiserum** (an-tē-sē'rŭm) or sometimes *immune serum*. Antisera are typically collected from large animals intentionally exposed to the disease agent of interest because their large blood volume contains more antibodies than can be obtained from smaller animals.

Antisera have the following limitations:

- Because they contain antibodies against many different antigens, they are not limited solely to the antigen of interest.
- Repeated injections of antisera collected from a different species can trigger serious allergic responses in recipients. The result is that the recipient mounts an immune response against the animal antibodies found in the antisera.

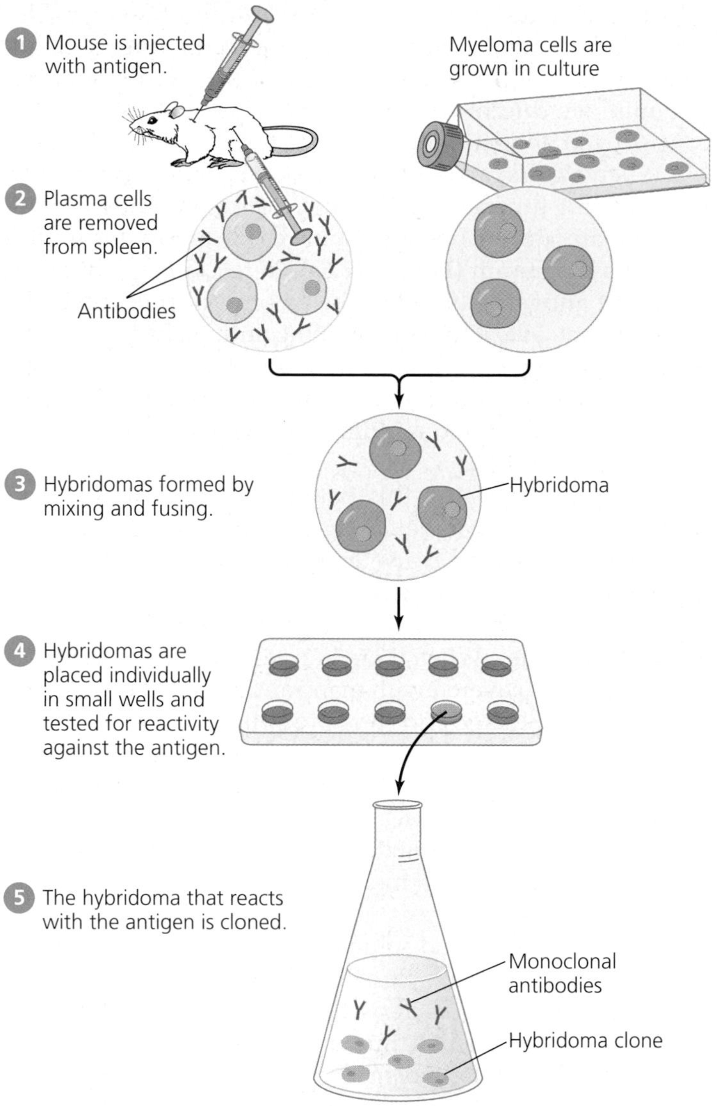

1 Mouse is injected with antigen.
Myeloma cells are grown in culture
2 Plasma cells are removed from spleen.
Antibodies
3 Hybridomas formed by mixing and fusing.
Hybridoma
4 Hybridomas are placed individually in small wells and tested for reactivity against the antigen.
5 The hybridoma that reacts with the antigen is cloned.
Monoclonal antibodies
Hybridoma clone

Figure 17.4

The production of hybridomas. After a laboratory animal is injected with the antigen of interest ①, plasma cells, which of course secrete antibodies, are removed from the animal's spleen and isolated ②. When these plasma cells are fused with cultured cancer cells called myelomas, the results are hybridomas—long-lived tumor cells that produce antibodies while grown in culture③. Once the hybridomas are cultured individually and the hybridoma that produces antibodies against the antigen of interest is identified ④, it is cloned to produce a large number of hybridomas, all of which secrete identical antibodies called monoclonal antibodies ⑤.

- Antisera may be contaminated with viral pathogens.
- The antibodies of antisera are degraded relatively quickly by the recipient. The half-life of such antibodies is about 3 weeks; that is, half the molecules are degraded every 3 weeks.

Scientists have overcome most of the limitations of antisera by developing **hybridomas** (hī-brid-ō'măz), which are tumor cells created by fusing antibody-secreting plasma cells with cancerous plasma cells called *myelomas* (mī-ĕ-lō'măz) **(Figure 17.4).** Each hybridoma divides continuously (because of the cancerous plasma cell component) to produce clones of itself, and each clone secretes large amounts of a single antibody molecule. These identical antibodies are called **monoclonal** (mon-ō-klō'năl) **antibodies** because all of them are secreted by clones originating from a single plasma cell. Once scientists have identified the hybridoma that secretes the antibody complementary to the antigen of interest, they maintain it in tissue culture to produce the antibodies needed for passive immunization.

Physicians use passive immunization when protection against a recent infection or an ongoing disease is needed quickly. Passive immunization provides rapid protection because it does not require the body to mount a response; instead, preformed antibodies are immediately available to bind to antigen, enabling agglutination, neutralization, and opsonization to proceed without delay. For example, if a previously unvaccinated patient has a wound contaminated with *Clostridium tetani* (klos-trid'ē-ŭm te'tan-ē), passive immunization with preformed antibodies against the bacterium and its toxin can provide protection quickly enough to save the patient's life. Antisera directed against toxins are also called *antitoxins* (an-tē-tok'sinz); *antivenin* used to treat snakebites is another antitoxin. In some cases, infections with certain viruses—hepatitis A and B, measles, rabies, rubella, chickenpox, and shingles—are treated with antisera directed against the causative viruses.

Active and passive immunization are used in different circumstances because they provide protection with different characteristics **(Figure 17.5).** As just noted, passive

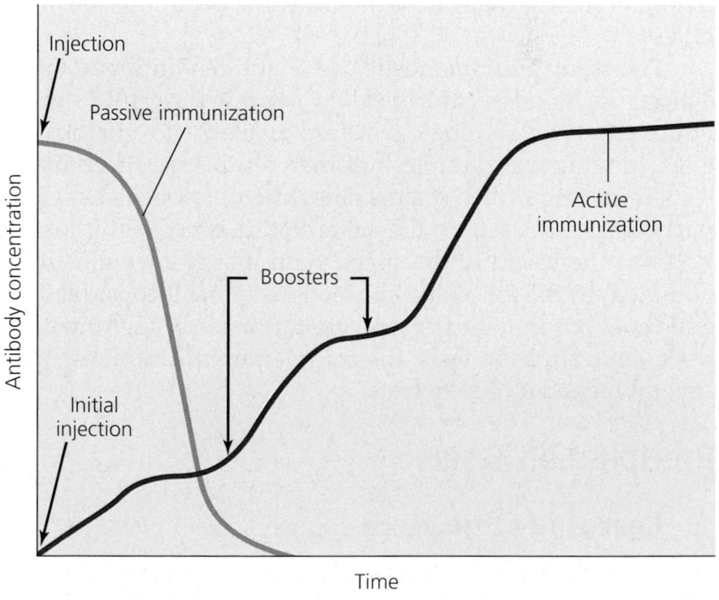

Injection
Passive immunization
Active immunization
Boosters
Antibody concentration
Initial injection
Time

Figure 17.5

The characteristics of immunity produced by active and passive immunization. Whereas passive immunity provides strong and immediate protection, it dissipates relatively quickly. Active immunity takes some time and may require additional booster inoculations to reach protective levels, but it is long lasting and capable of restimulation.

immunization with preformed antibodies is used whenever immediate protection is required. However, because the preformed antibodies are removed rapidly from the blood, the protection they provide wanes, and thus eventually the recipient becomes susceptible again. Whenever long-term protection is the goal, active immunization provides a slower buildup of protection that is capable of restimulation. Thus, when initiated before any exposure to *C. tetani* has occurred, active immunization using a tetanus toxoid develops long-lasting protection that is readily available upon exposure to the toxin.

Immune Testing

Learning Objectives

- ✓ In general terms, compare and contrast precipitation, agglutination, neutralization, complement fixation, and labeled antibody tests.
- ✓ Define serology.

Laboratory scientists use immunological processes in two general diagnostic ways. They may either use known antibodies to detect or identify antigens associated with an infectious agent (such as a particular strain of virus in tissue culture or in body tissues), or they may use selected antigens to detect specific antibodies in a patient's blood, thereby determining whether an individual has been exposed to a specific pathogen. The latter process can be used either to monitor the spread of an infection through a population or to establish a diagnosis in an individual. For example, the presence of antibodies against HIV in an adult's serum is strong evidence that this individual has been infected with HIV.

The study and diagnostic use of antigen-antibody interactions in blood serum is called **serology** (sĕ-rol'ō-jē). A wide variety of serologic tests are available to microbiologists, ranging from simple, automated processes to complex tests requiring skilled technicians. Microbiologists choose a particular test based on the suspected diagnosis, the cost to perform the test, and the speed with which a result can be obtained. In the following subsections we will consider several types of serologic tests: precipitation tests, agglutination tests, neutralization tests, the complement fixation test, and several tagged antibody tests.

Precipitation Tests

Learning Objectives

- ✓ Describe the general principles of precipitation testing.
- ✓ Compare and contrast immunodiffusion and immunoelectrophoresis.
- ✓ Discuss the production of anti-antibodies for immune testing.

One of the simplest of serologic tests relies on the fact that when antigens and antibody are mixed in proper proportions, they form huge lattice-like macromolecular complexes called precipitates. When, for example, a solution of a soluble antigen is mixed with an antiserum containing antibodies against the antigen, the mixture becomes cloudy within a few minutes due to formation of an insoluble precipitate consisting of antigen-antibody complexes.

When a given amount of antibody is added to each of a series of test tubes containing increasing amounts of antigen, the amount of precipitate increases gradually until it reaches a maximum **(Figure 17.6a).** In test tubes containing still more antigen molecules, the amount of precipitate declines; in fact, in test tubes containing antigen in great excess over antibody, no precipitate at all develops. Thus, a graph of the amount of precipitate versus the amount of antigen has a maximum in the middle and lower values on either side **(Figure 17.6b).**

The reasons behind this pattern of precipitation reactions are simple. Complex antigens are generally multivalent—each possesses many antigenic determinants—and antibodies have pairs of active sites and therefore can simultaneously cross-link the same antigenic determinant on two antigen molecules. When there is excess antibody, each antigen molecule is covered with many antibody molecules, preventing cross-linkage and thus precipitation **(Figure 17.6c).** When the reactants are in optimal proportions, the ratio of antigen to antibody is such that cross-linking and lattice formation are extensive. As this lattice grows, it becomes insoluble and eventually precipitates. In mixtures in which antigen is in excess, each antigen molecule is bound to only two antibody molecules. There is no cross-linkage; because these complexes are small and soluble, no precipitation occurs.

In the body, phagocytic cells do not easily remove small immune-complexes (such as those formed in antigen excess). As a result, small immune complexes may be deposited in the joints and in the tiny blood vessels of the kidneys, where they trigger allergic reactions, as discussed in Chapter 18.

Because precipitation requires that antigen and antibody are mixed in optimal proportions, it's not possible to perform a precipitation test by combining just any two solutions containing these reagents. To ensure that the optimal concentrations of antibody and antigen come together, scientists use two techniques involving movement of the molecules through an agar gel: immunodiffusion and immunoelectrophoresis.

Immunodiffusion

In the technique called *immunodiffusion,* double *immunodiffusion,* or an *Ouchterlony test,* a researcher cuts cylindrical holes called *wells* about 5 mm in diameter in an agar plate. One well is filled with a solution of antigen, and the other with a solution of antibodies against the antigen. The two solutions diffuse in all directions out of the wells and into the surrounding agar, and where they meet in optimal proportions, a line of precipitation appears **(Figure 17.7a).** If the solutions contain many different antigens and antibod-

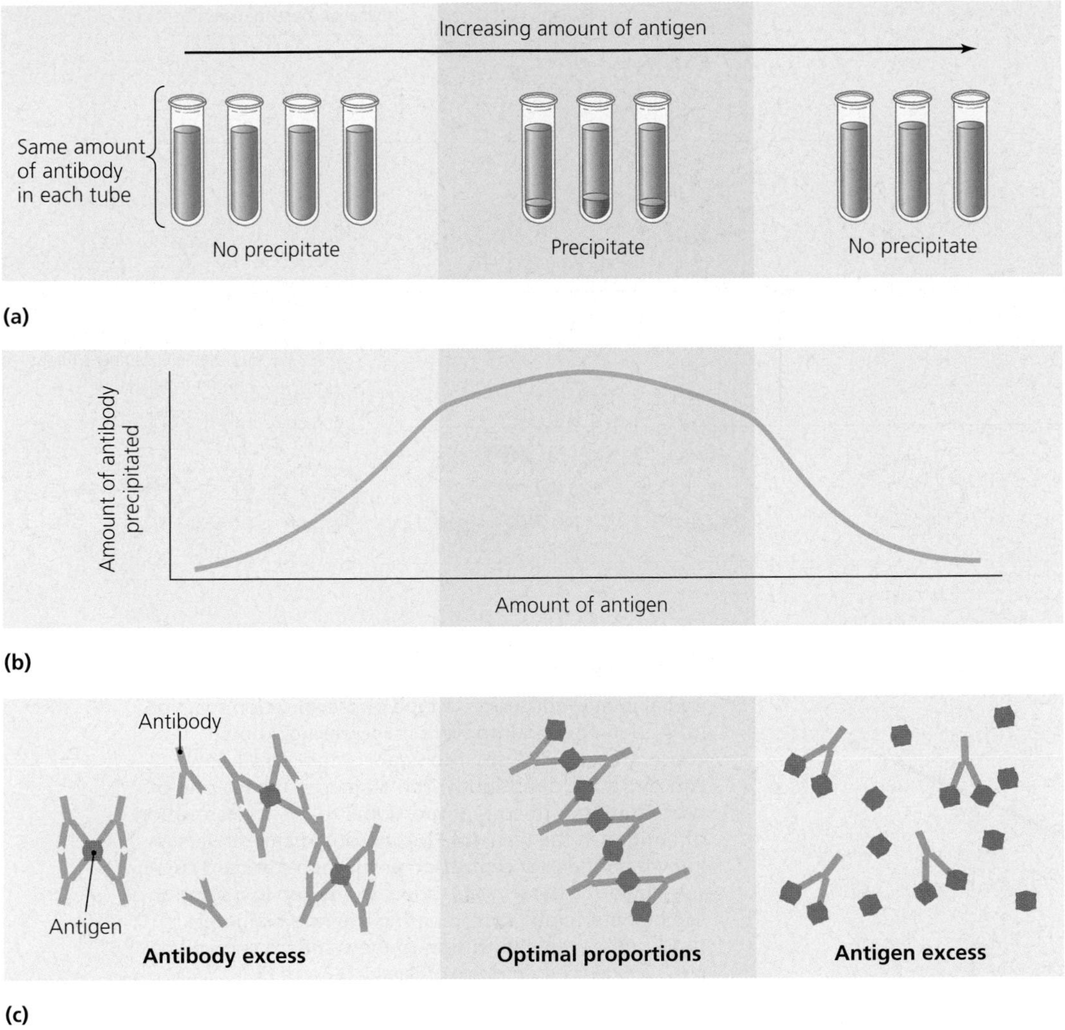

◀ *Figure 17.6*

Characteristics of precipitation reactions. **(a)** When increasing amounts of antigen are placed in tubes containing a constant amount of antibody, precipitate forms only in tubes having moderate amounts of antigen. **(b)** A graph of the amount of precipitate formed versus the amount of antigen present. **(c)** In the presence of excess antigen or antibody, immune complexes are small and soluble; only when antigen and antibody are in optimal proportions do large complexes form and precipitate.

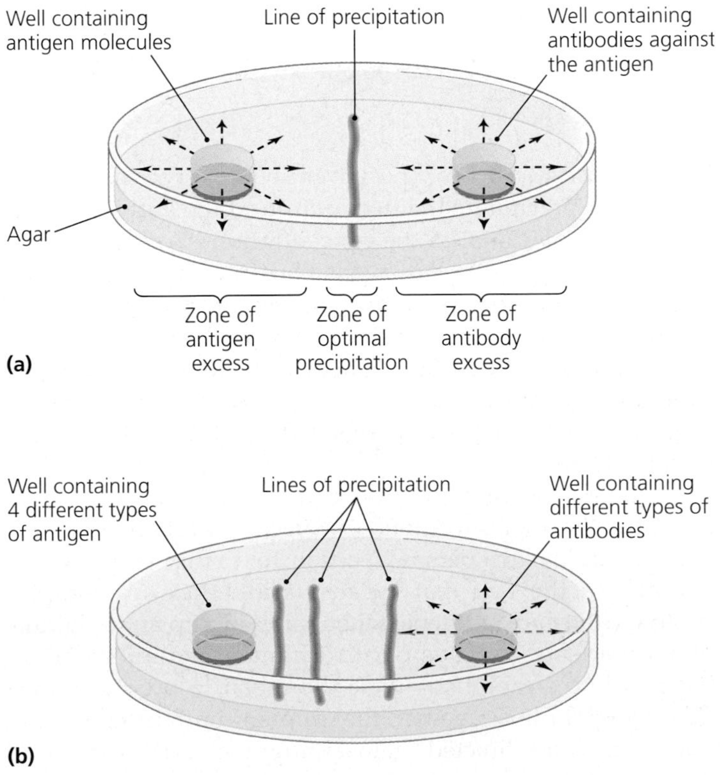

▶ *Figure 17.7*

Immunodiffusion, a type of precipitation reaction. **(a)** Antigen and antibody placed in wells diffuse out and through the agar; where they meet in optimal proportions, a line of precipitation forms. **(b)** When multiple antigens and antibodies are placed in the wells, multiple lines of precipitation mark the sites where different antigen-antibody combinations occurred in optimal proportions. *Why did only three lines of precipitation occur when the antigen well contained four antigens?*

Figure 17.7. None of the antibodies used in the test was complementary to the fourth antigen.

ies, each complementary pair of reactants reaches optimal proportions at different positions, and numerous lines of precipitation are produced—one for each interacting antigen-antibody pair **(Figure 17.7b).**

Another useful variation on precipitation is called **radial immunodiffusion** (rā′dē-ăl im′ū-nō-di-fū′zhŭn) (**Figure 17.8a** on page 498). In this technique, an antigen in solution is allowed to diffuse from a well into agar that contains a

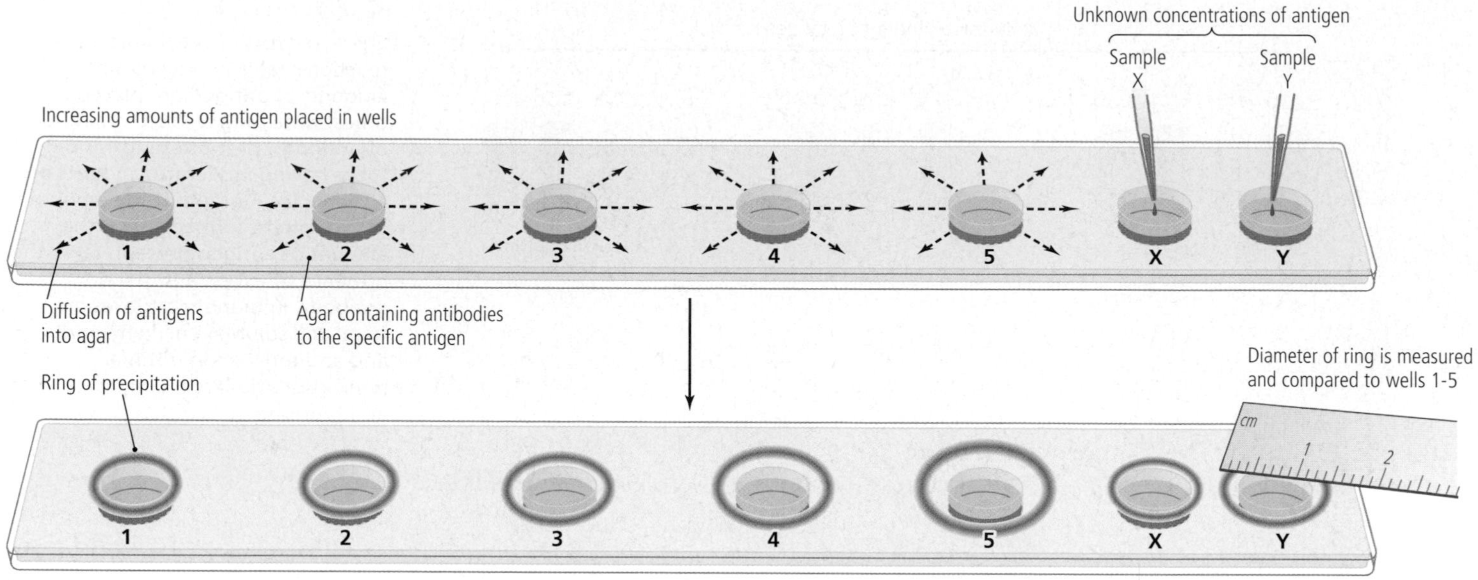

(a)

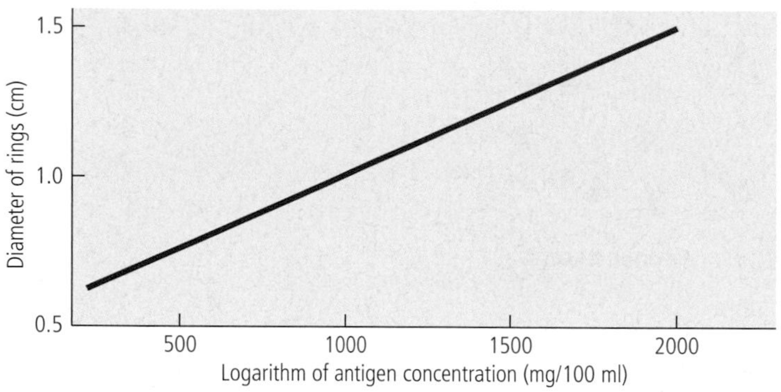

(b)

▲ ***Figure 17.8***

Radial immunodiffusion, a type of precipitation reaction. **(a)** Antigen placed into wells in increasing, known concentrations diffuses into agar containing a known concentration of antibody. The diameter of each ring of precipitation formed is proportional to the concentration of antigen in the well. **(b)** Plotting ring diameter versus known antigen concentration produces a standard curve; subsequently, the concentration of antigen in a solution can be determined by comparing the ring diameter produced by radial diffusion to the standard curve. *What is the antigen concentration of Sample Y?*

Figure 17.8 *Based on the graph in (b), the concentration in Sample Y is about 1900 mg/100 ml.*

known concentration of specific antibodies. The diameter of the ring of precipitate that forms around the antigen well is directly proportional to the concentration of antigen in the well. By using a series of wells containing increasing but known concentrations of antigen, researchers construct a standard curve that relates ring diameter to antigen concentration **(Figure 17.8b)**. Subsequently, the amount of antigen in a solution can be accurately assayed by comparing the ring diameter produced in other radial diffusion tests to the standard line.

Scientists commonly use radial immunodiffusion to measure the concentrations of specific antibodies or immunoglobulins in a person's serum. To do this, they take advantage of the fact that because antibodies are complex proteins, they are antigenic when injected into an individual of another species. Thus purified human immunoglobulins injected into a rabbit stimulate the rabbit to produce rabbit immunoglobulins against the human immunoglobulins. Such antibodies directed against immunoglobulins are called anti-antibodies. So, in a radial immunodiffusion test to measure the concentration of a specific human immunoglobulin, the "antigen" in the test is the human immunoglobulin, and the antibody in the test is rabbit anti-antibody.

It is possible to make anti-antibodies that specifically target a particular class of antibody; for example, anti-IgG anti-antibodies target gamma heavy chains, and anti-IgD anti-antibodies target delta heavy chains. By incorporating anti-IgG, anti-IgA, anti-IgM, anti-IgE, and anti-IgD anti-antibodies into different agar gels and filling the wells of each gel with a sample of an individual's serum, scientists can determine the concentrations of the five antibody classes in a serum.

Immunoelectrophoresis

Although conventional gel-diffusion techniques produce a separate precipitation line for each antigen-antibody combination present, it is sometimes difficult to distinguish all the lines if they form within a small area of gel. In **immunoelectrophoresis** (im′ū-nō-ē-lek′trō-fō-rē′sis), which improves the resolution of an immunodiffusion test, the first step is to separate antigens from one another by *electrophoresis*, which as we saw in Chapter 8 separates molecules on the basis of electrical charge and size by running an electrical current through the gel **(Figure 17.9a).** Positively charged antigens move toward the negative pole, negatively charged antigens move toward the positive pole, and antigens without a charge and very large molecules do not move at all.

After electrophoresis, a trough is cut in the agar parallel to the line of separated antigens. Antiserum placed in the trough diffuses laterally, and where diffusing antibodies encounter their complementary antigens in the correct proportions, curved lines of precipitation form **(Figure 17.9b).** One line of precipitation forms for each pair of complementary reactants.

Immunoelectrophoresis can resolve more than 30 distinct antigens at once. Diagnosticians use the technique either to demonstrate the absence of a normal antigen or to detect the presence of excessive amounts of an antigen—for example, the excess of so-called *prostate specific antigen* in the serum of a patient with prostate cancer.

Agglutination Tests

Learning Objectives

- ✓ Contrast agglutination and precipitation tests.
- ✓ Describe how agglutination is used in immunological testing, including titration.

Not all antigens are soluble proteins that can be precipitated by antibody. Because of their multiple antigen-binding sites, antibodies can also cross-link *particulate* antigens, such as whole bacteria, causing **agglutination** (ă-gloo-ti-nā′shŭn) (clumping). The difference between agglutination and precipitation is that agglutination involves the clumping of insoluble particles, whereas precipitation involves the aggregation of soluble molecules. Agglutination reactions are easy to see and interpret with the unaided eye. IgM antibodies, which have 10 active sites, are more efficient than IgG antibodies (with only two active sites) at causing agglutination. When the particles agglutinated are red blood cells, the reaction is called *hemagglutination* (hē-mă-gloo′ti-nā′shŭn). One use of hemagglutination is to determine blood type in humans. Blood is considered type A if the red blood cells possess surface antigens called A antigen, type B if they possess B antigens, type AB if they possess both antigens, and type O if they have neither antigen. In a hemagglutination reaction to determine blood type **(Figure 17.10),**

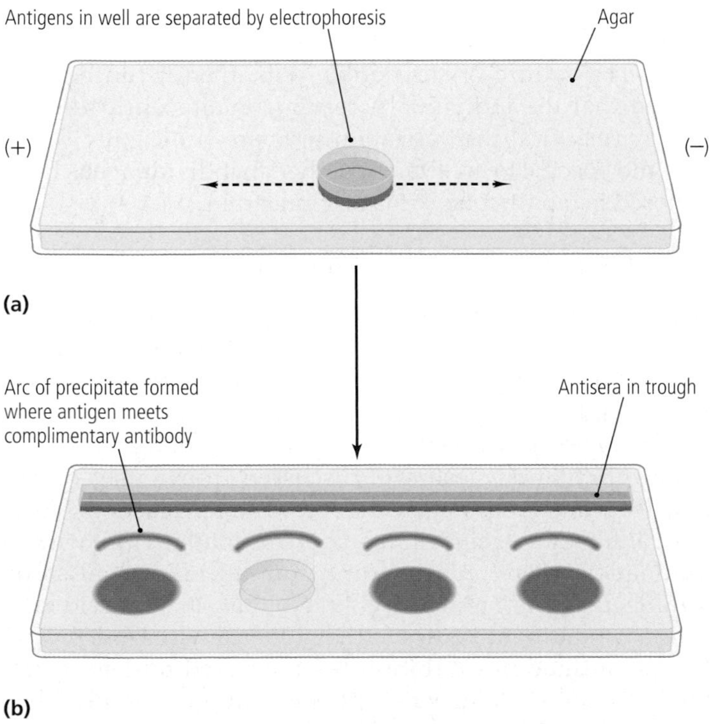

Figure 17.9
Immunoelectrophoresis, a type of precipitation reaction used to identify antigens in a mixture. **(a)** Antigens in a mixture are placed in a well and separated by electrophoresis. **(b)** After a trough running parallel to the direction of the electrical current is cut and filled with antiserum, arcs of precipitate form where complementary antigen and antibody meet in optimal concentrations.

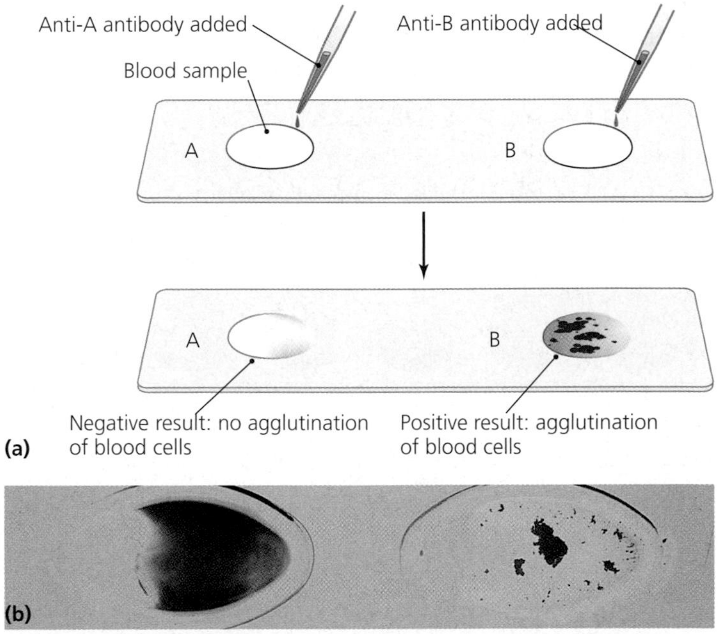

Figure 17.10
The use of hemagglutination to determine blood types in humans. **(a)** Antibodies with active sites that bind to either of two red blood cell surface antigens (antigen A or antigen B) are added to portions of a given blood sample. Where the antibodies react with the surface antigens, the blood cells can be seen to agglutinate or clump together. **(b)** Photo of actual test. *What is the blood type of the person whose blood was used in this hemagglutination reaction?*

Figure 17.10 The individual who donated the blood sample has type B blood.

two portions of a given blood sample are placed on a slide, and anti-A antibodies are added to one portion, and anti-B antibodies to the other; the antibodies agglutinate (clump) those blood cells that possess complementary antigens.

CRITICAL THINKING

Draw a picture showing, at *both the molecular and cellular levels,* IgM agglutinating red blood cells.

Another use of agglutination is in a type of test that determines the concentration of antibodies in a clinical sample. Although the simple *detection* of antibodies is sufficient for many purposes, it is usually more desirable to measure the *amount* of antibodies in serum. By doing so, clinicians can determine whether a patient's antibody levels are rising, as occurs in response to the presence of active infectious disease, or falling, as occurs during the successful conclusion of a fight against an infection. One way of measuring antibody levels in blood sera is by **titration** (tī-trā'shŭn). In titration, the serum being tested is serially diluted, and each dilution is then tested for agglutinating activity **(Figure 17.11).** Eventually, the antibodies in the serum become so dilute that they can no longer cause agglutination. The highest dilution of serum giving a positive reaction is called the **titer** (tī'ter), which provides an estimate of the amount of antibody in that serum. Thus, a serum that must be greatly diluted before agglutination ceases (for example, has a titer of 1000) contains more antibodies than a serum that no longer agglutinates after minimal dilution (has a titer of 10).

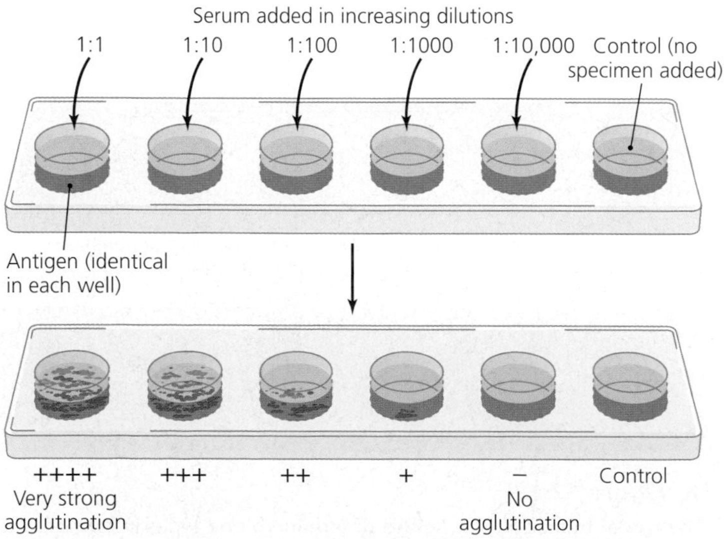

▲ *Figure 17.11*
Titration, the use of agglutination to quantify the amount of antibody in a serum sample. Serial dilutions of serum are added to wells containing a constant amount of antigen. At lower serum dilutions (higher concentrations of antibody), agglutination occurs; at higher serum dilutions, antibody concentration is too low to produce agglutination. The serum's titer is the highest dilution at which agglutination can be detected—in this case, 1:1000.

Neutralization Tests

Learning Objectives

- ✓ Explain the purpose of neutralization tests.
- ✓ Contrast a viral hemagglutination inhibition test with a hemagglutination test.

Neutralization tests work because antibodies can *neutralize* the biological activity of many pathogens and their toxins. For example, combining antibodies against tetanus toxin (which is lethal to mice) with a sample of toxin renders the sample harmless to mice because the antibodies have reacted with and neutralized the toxin. Next we briefly consider two neutralization tests that, although not simple to perform, effectively reveal the biological activity of antibodies.

Viral Neutralization

One neutralization test is **viral neutralization,** which is based on the fact that many viruses introduced into appropriate cell cultures will invade and kill the cells, a phenomenon called a *cytopathic effect*. However, if the viruses are first mixed with specific antibodies against them, their ability to kill culture cells is neutralized. In a viral neutralization test, the lack of cytopathic effects when a mixture containing serum and a known virus is introduced into a cell culture indicates the presence of antibodies against that virus in the serum. For example, if a mixture containing an individual's serum and a sample of hantavirus produces no cytopathic effect in a culture of susceptible cells, then it can be concluded that the individual's serum contains antibodies to hantavirus. Viral neutralization tests are sufficiently sensitive and specific to ascertain whether an individual has been exposed to a particular virus or viral strain.

Viral Hemagglutination Inhibition Test

Because not all viruses are cytopathic, a neutralization test cannot be used to identify all viruses. However, many viruses (including influenza viruses) have surface proteins that naturally clump red blood cells. This natural process, called *viral hemagglutination*, must not be confused with the hemagglutination test we discussed previously—viral hemagglutination is not an antibody-antigen reaction. However, antibodies against influenza virus inhibit viral hemagglutination; therefore, if serum from an individual stops viral hemagglutination, we know that the individual's serum contains antibodies to the influenza virus. Such **viral hemagglutination inhibition tests** are used to detect antibodies against influenza, measles, mumps, and other viruses that naturally agglutinate red blood cells.

The Complement Fixation Test

Learning Objectives

- ✓ Briefly explain the phenomenon that is the basis for a complement fixation test.
- ✓ Describe the use of the complement fixation test.

As we discussed in Chapter 15, activation of the classical complement system by antibody leads to the generation of membrane attack complexes (MACs) that disrupt cytoplasmic membranes (see Figure 15.17). This phenomenon is the basis for the **complement fixation** (kom′plĕ-ment fik-sā′shŭn) **test,** which is used to detect the presence of specific antibodies in an individual's serum. A complement fixation test can detect the presence of small amounts of antibody—amounts too small to detect by agglutination. A complement fixation test is performed in three stages:

1. A serum sample from an individual is heated to 56°C to destroy any naturally present complement.
2. Antigen from the disease agent of interest and a standard amount of complement are added to the serum.
3. Sheep red blood cells (RBCs) and antibodies against sheep RBCs (anti-RBC antibodies) are added to the mixture, which is then examined for the lysis of the RBCs.

Figure 17.12 illustrates the steps in a complement fixation test and depicts the events that occur in both positive and negative complement fixation tests. In a positive test—that is, one in which the individual's serum contains antibodies against the disease agent of interest (Figure 17.12a), the antibodies bind the antigen, and the resulting antigen-antibody complexes bind complement, which is thus no longer available to bind to the anti-RBC antibodies. As a result, the sheep RBCs remain intact, and the solution remains cloudy. In a negative test—that is, one in which the individual's serum lacks antibodies against the disease agent of interest (Figure 17.12b)—the lack of antigen-antibody complexes enables complement to bind to the anti-RBC antibodies (which have attached to the sheep RBCs), triggering lysis of the RBCs by MACs and turning the solution a transparent red.

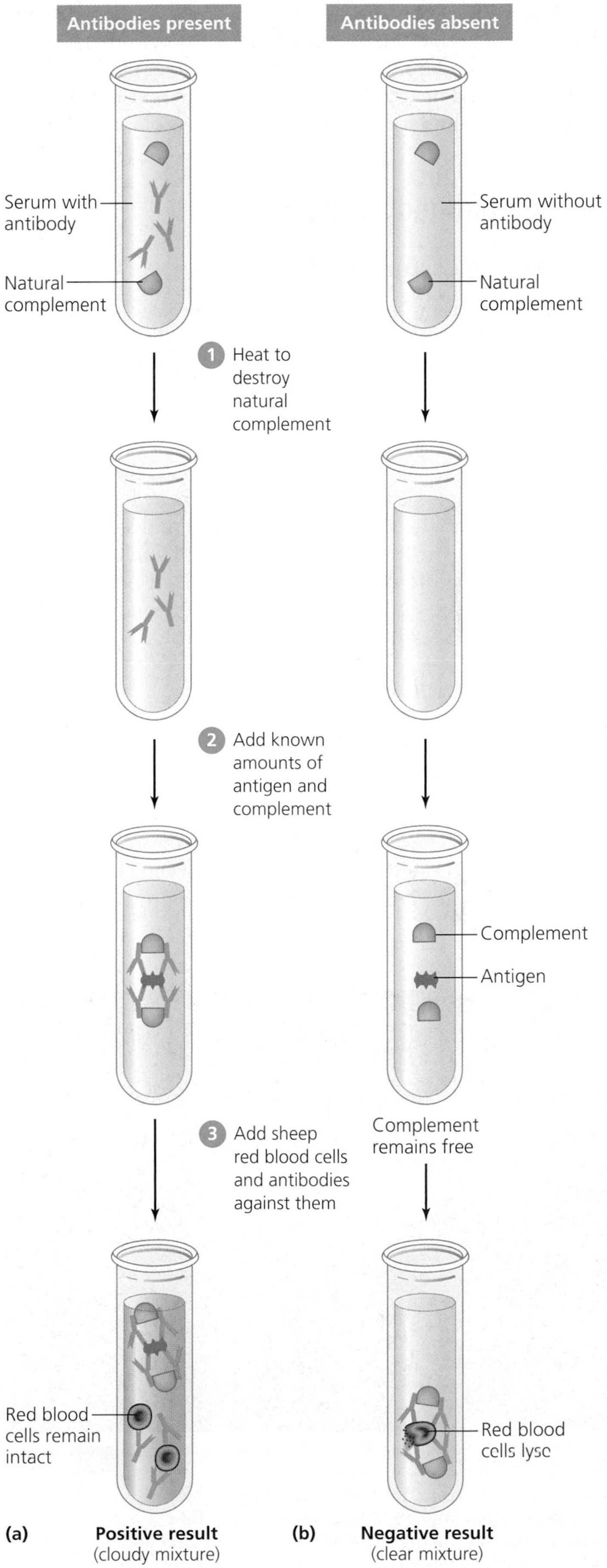

▶ *Figure 17.12*

The complement fixation test. To conduct the test, ① an individual's serum is heated to destroy naturally present complement; ② known amounts of the antigen of interest and complement are added; and ③ sheep red blood cells (RBCs) and antibodies against them (anti-RBC antibodies) are added. **(a)** A positive test. In the presence of antibodies against the antigen, antigen-antibody complexes bind the complement, which then cannot bind with anti-RBC antibodies to lyse the RBCs. **(b)** A negative test. The absence of antibodies (and thus of antigen-antibody complexes) leaves complement free to bind to anti-RBC antibodies attached to the RBCs, which are then lysed. *Why must the serum be heated?*

Figure 17.12 Heat destroys excess complement so that only the standard amount is available.

A complement fixation test called the Wasserman test was once the standard diagnostic test for syphilis. Though the Wasserman test has been replaced by labeled antibody tests (discussed next), complement fixation is still used in diagnosis of some infections of fungi, viruses, and rickettsial bacteria.

Labeled Antibody Tests

Learning Objectives

- ✓ List three tests that use labeled antibodies to detect either antigen or antibodies.
- ✓ Compare and contrast the direct and indirect fluorescent antibody tests, and identify at least three uses for these tests.
- ✓ Compare and contrast the methods, purposes, and advantages of ELISA and the western blot tests.

A different form of serologic testing involves *labeled* (or *tagged*) *antibody tests,* so-named because these tests use antibody molecules that are linked to some molecular "label" that enables them to be detected easily. Labeled antibody tests can be used to detect either antigens or antibodies. In the following subsections we will consider fluorescent antibody tests, ELISA, and the western blot test.

Fluorescent Antibody Tests

Fluorescent dyes are used as labels in several important serologic tests. The most important of these dyes is *fluorescein* (flōr-es′ē-in), which can be chemically linked to an antibody without affecting the antibody's ability to bind antigen. When exposed to fluorescent light (as in a fluorescent microscope), fluorescein glows bright green. Fluorescein-labeled antibodies are used in direct and indirect fluorescent antibody tests.

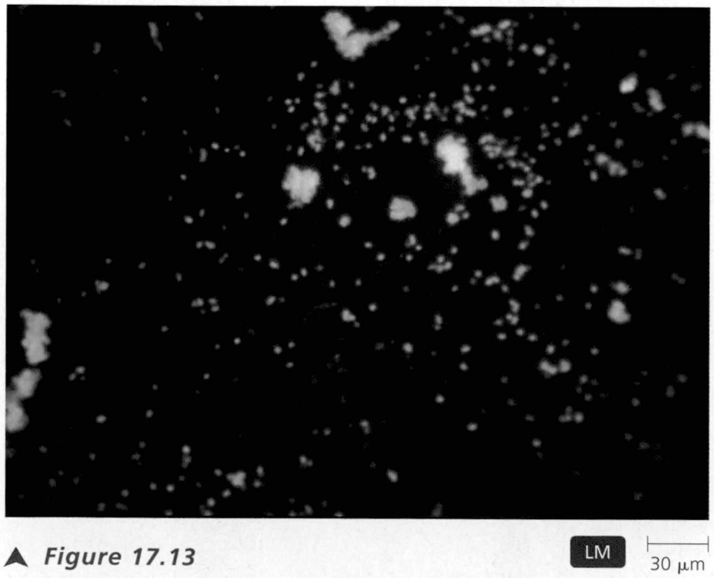

▲ ***Figure 17.13***

The direct fluorescent antibody test. The fluorescence from fluorescein-labeled antibodies against rabies viruses indicates the presence of the viruses in a section of brain tissue.

Direct fluorescent antibody tests identify the presence of antigen in a tissue. The test is straightforward: A scientist floods a tissue sample suspected of containing the antigen with labeled antibody, waits a short time to allow the antibody to bind to the antigen, washes the preparation to remove any unbound antibody, and examines it with a fluorescent microscope. If the suspected antigen is present, labeled antibody will adhere to it, and the scientist will see fluorescence. This is not a quantitative test—the amount of fluorescence observed is not directly related to the amount of antigen present.

Scientists use direct fluorescent antibody tests to identify small numbers of bacteria in patient tissues. This technique is used to detect *Mycobacterium tuberculosis* (mī-kō-bak-tēr′ē-ŭm tū-ber-kyū-lō′sis) in sputum and viruses infecting tissues. In one use, medical laboratory technologists employ a direct fluorescent antibody test to detect the presence of rabies virus in the brain of an infected animal **(Figure 17.13).**

Like direct fluorescent antibody tests, **indirect fluorescent antibody tests** are used to detect antigens in cells or patient tissues, but they can also be used to detect the presence of specific antibodies in an individual's serum via a two-step process **(Figure 17.14a):**

1. After the antigen of interest is fixed to a microscope slide, the individual's serum is added for long enough to allow serum antibodies to bind to the antigen. The serum is then washed off, leaving any specific antibodies bound to the antigen (but not yet visible).
2. Anti-antibodies (in this example, anti-IgG) labeled with fluorescein is added to the slide and binds to the antibodies already bound to the antigen. After washing to remove unbound anti-antibodies, the slide is examined with a fluorescent microscope.

The presence of fluorescence indicates the presence of the labeled anti-antibodies, which are bound to serum antibodies bound to the fixed antigen; thus fluoresence indicates that the individual has serum antibodies against the antigen of interest.

Indirect fluorescent antibody testing is used to detect antibodies against many viruses and some bacterial pathogens, including *Neisseria gonorrhoeae* (nī-sē′rē ă gon-ō-rē′ē), the causative agent of gonorrhea **(Figure 17.14b).**

ELISAs

In another type of labeled antibody test, called an **enzyme-linked immunosorbent assay** (im′ū-nō-sōr′bent as′sā) (ELISA), the label is not a dye, but instead an enzyme that reacts with its substrate to produce a colored product that indicates a positive test. The most common form of ELISA is used to detect the presence (and quantify the abundance) of antibodies in serum. This test, which often takes place in commercially produced polystyrene-coated plates, has five steps (**Figure 17.15** on page 504):

1. Each of the wells in the plate is coated with antigen molecules in solution.

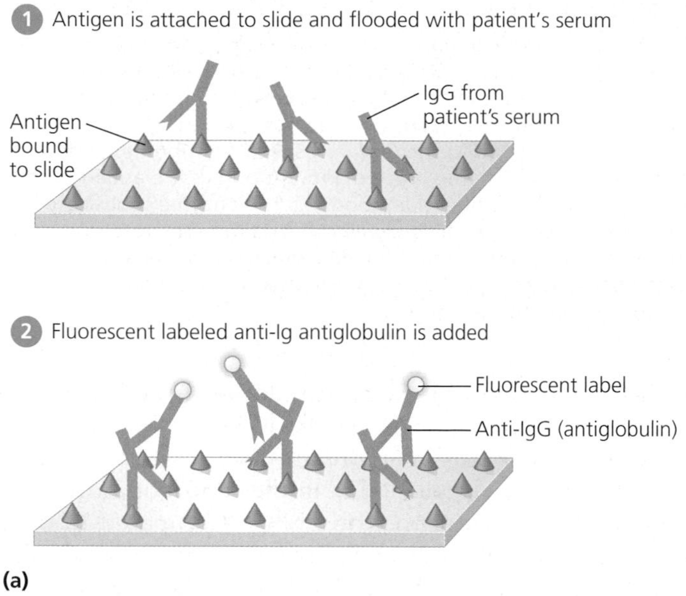

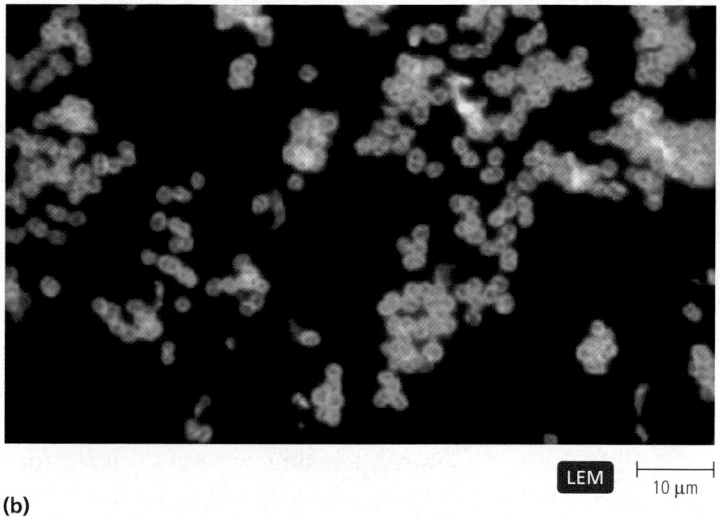

Figure 17.14

The indirect fluorescent antibody test. **(a)** The test procedure. ① Antigen is attached to the slide, which is then flooded with an individual's serum to allow specific antibodies in the serum to bind to the antigen. ② After fluorescein-labeled anti-antibodies are added and then washed off, the slide is examined with a fluorescent microscope. **(b)** A positive indirect antibody test, in which fluorescence indicates the presence of antibodies against *Neisseria gonorrhoeae* (the agent of gonorrhea) in the individual's serum.

2 Excess antigen molecules are washed off, and another protein (such as gelatin) is added to the well to completely coat any of the polystyrene surface not coated with antigen.

3 A sample of each of the sera being tested is added to a separate well. Whenever a serum sample contains antibodies against the antigen, they bind to the antigen affixed to the plate.

4 Anti-antibodies labeled with an enzyme are added to each well.

5 The enzyme's substrate is added to each well. The enzyme and substrate are chosen because their reaction produces products that cause a visible color change.

A positive reaction in a well, indicated by the development of color, can occur only if the labeled anti-antibody has bound to antibodies attached to the antigen of interest. The intensity of the color, which can be estimated visually or measured accurately using a spectrophotometer, is proportional to the amount of antibody present in the serum.

ELISA has become the test of choice for many diagnostic procedures, such as determination of HIV infection, because of its many advantages:

- Like other labeled antibody tests, ELISA can detect either antibody or antigen.
- Unlike some diffusion and fluorescent tests, ELISA can quantify amounts of antigen or antibody.
- ELISAs are easy to perform, are relatively inexpensive, and can simultaneously test many samples quickly; moreover, they lend themselves to efficient automation and can be read easily, either by direct observation or by machine.
- Plates coated with antigen and gelatin can be stored for testing, whenever they are needed.

A modification of the ELISA technique, called an *antibody sandwich ELISA,* is commonly used to detect antigen (**Figure 17.16** on page 504). In testing for the presence of HIV in blood serum, for example, the polystyrene plates are first coated with antibody (instead of antigen). Then the sera from individuals being tested for HIV are added to the wells, and any antigen in the sera will bind to the antibody attached to the well. Finally, each well is flooded with enzyme-labeled antibodies specific to the antigen. The name "antibody sandwich ELISA" refers to the fact that the antigen being tested for is "sandwiched" between two antibody molecules. Such tests can also be used to quantify the amount of antigen in a given sample.

CRITICAL THINKING

A diagnostician used an ELISA to show that a newborn had antibodies against HIV in her blood. However, 6 months later the same test was negative. How can this be?

Western Blot Test

A technique for detecting antibodies against multiple antigens in a complex mixture is a **western blot test.** The name "western blot" is a play on words that refers to the similarity

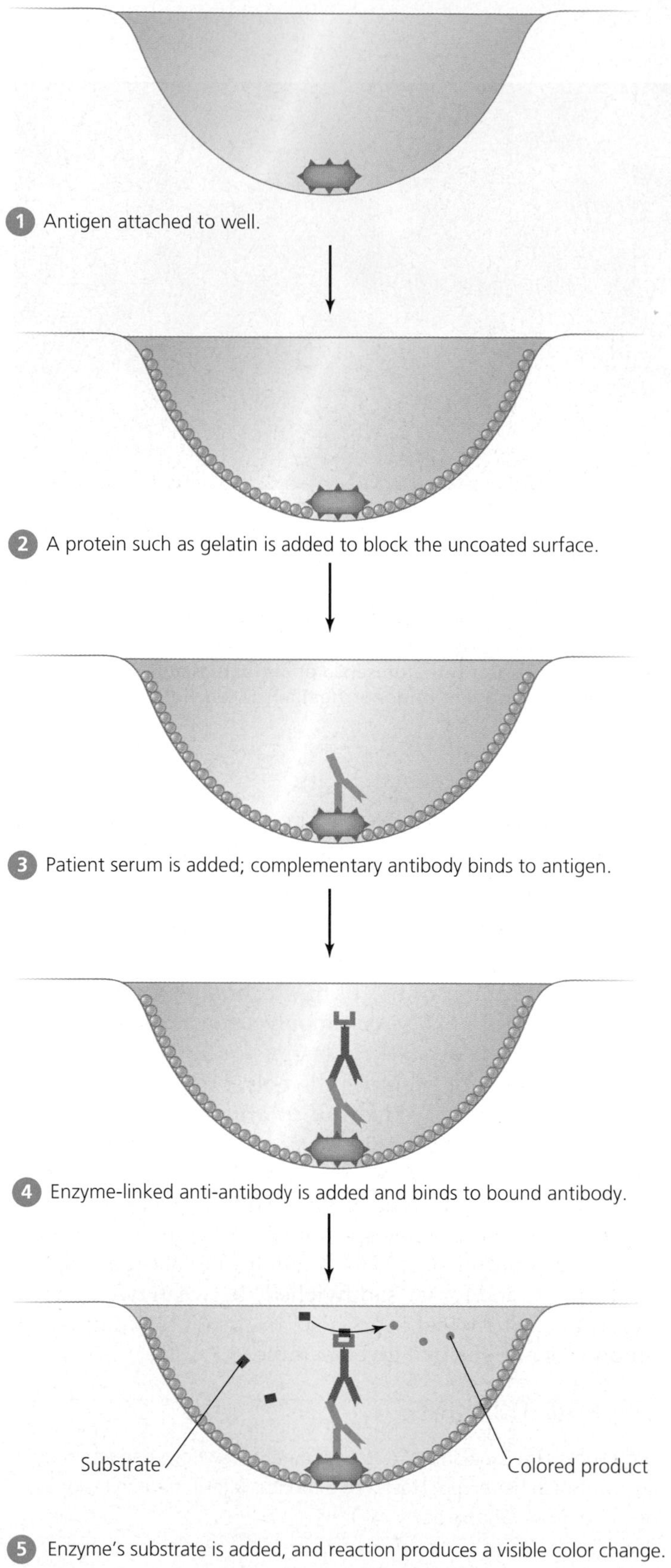

1 Antigen attached to well.
2 A protein such as gelatin is added to block the uncoated surface.
3 Patient serum is added; complementary antibody binds to antigen.
4 Enzyme-linked anti-antibody is added and binds to bound antibody.
Substrate
Colored product
5 Enzyme's substrate is added, and reaction produces a visible color change.

◀ *Figure 17.15*

The enzyme-linked immunosorbent assay (ELISA). Shown is one well in a polystyrene plate. ① Antigen added to the well attaches to it irreversibly. ② After excess antigen is removed by washing, gelatin is added to cover any portion of the well not covered by antigen. ③ Test serum is added to the well; any specific antibodies in it bind to the antigen. ④ Enzyme-labeled anti-antibodies are added to the well and bind to any bound antibody. ⑤ The enzyme's substrate is added to the well, and the enzyme converts the substrate into a colored product; the amount of color, which can be measured via spectrophotometry, is directly proportional to the amount of antibody bound to the antigen.

of this technique to the Southern blot test (see Figure 8.8), named for the man who developed it.

Western blot tests are currently used to verify the presence of antibodies against HIV in the serum of individuals who are antibody-positive by ELISA. Compared to other tests, western blot tests can detect more types of antibodies and are less subject to misinterpretation. A western blot test has three steps **(Figure 17.17a):**

1. Electrophoresis. Antigens in a solution (in this case, HIV proteins) are placed into wells and separated by gel electrophoresis. Each of the proteins in the solution is resolved into a single band, producing a pattern of protein bands.
2. Blotting. Each of the protein band patterns is transferred to an overlying nitrocellulose membrane by absorbing the solution into absorbent paper. The nitrocellulose membrane is then cut into strips.
3. ELISA. Each nitrocellulose strip is incubated with a test solution—in this case, a urine sample and a serum sample from each of three individuals are being tested for antibodies against HIV. After the strips are washed, an enzyme-labeled anti-antibody solution is added for a time; then the strips are washed again and exposed to the enzyme's substrate.

Color develops wherever antibodies against the HIV proteins in the test solutions have bound to the HIV proteins. In this case, the individual tested in strips 1 and 2 is negative for antibodies against HIV, whereas the other two individuals are positive for antibodies against HIV.

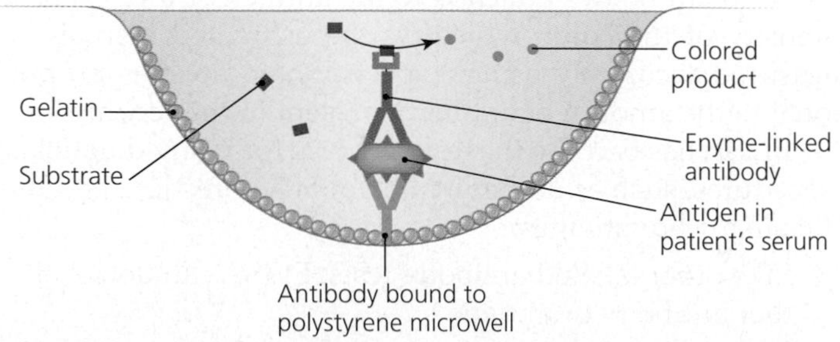

▲ *Figure 17.16*

An antibody sandwich ELISA. Because this variation of ELISA is used to test for the presence of antigen, antibody is attached to the well in the initial step.

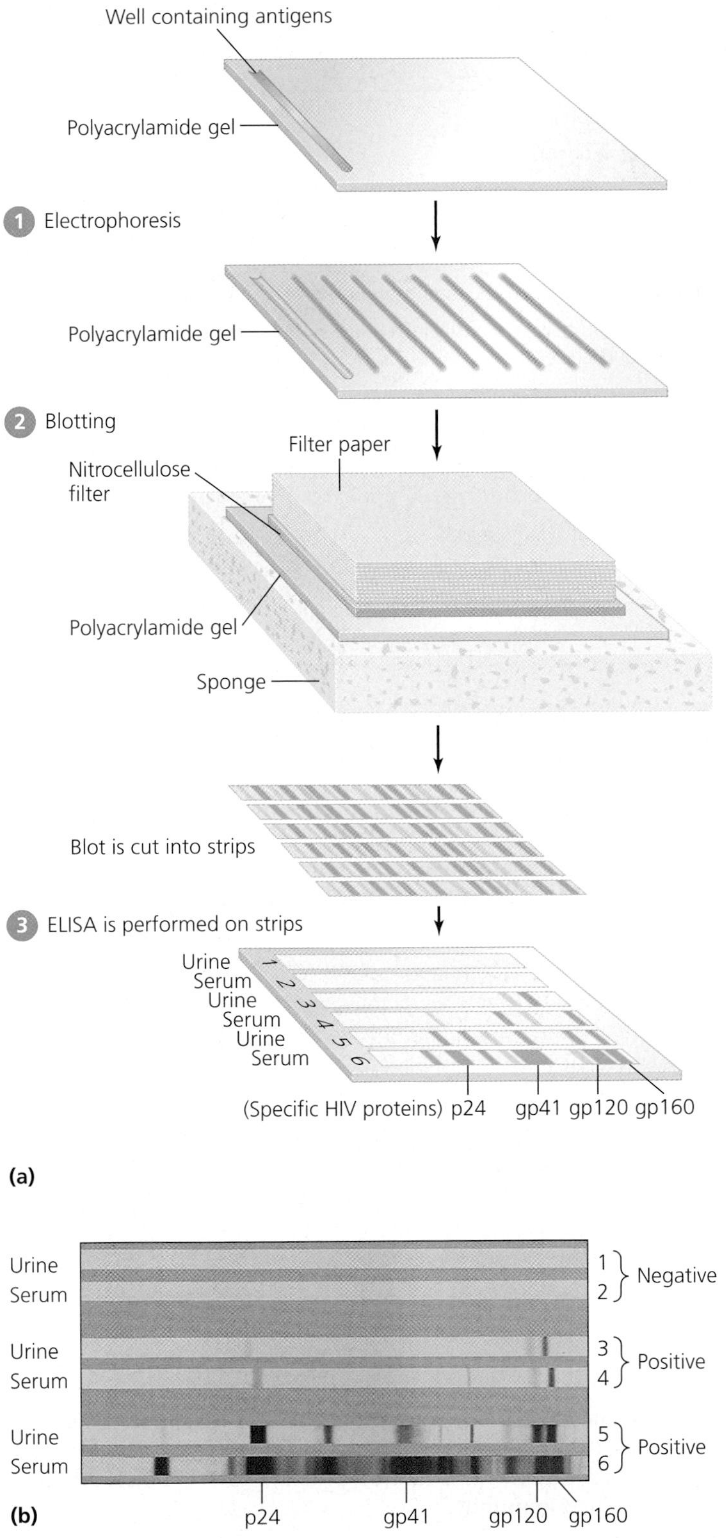

Recent Developments in Immune Testing

Learning Objectives

- ✓ Discuss the benefits of using immunofiltration assays rather than an ELISA.
- ✓ Contrast immunofiltration and immunochromatographic assays.

Recent years have seen the development of simple immunoassays that give clinicians useful results within minutes. The most commonly used of these are immunofiltration and immunochromatography assays. These tests, which generally are not quantitative but simply give a positive or negative result, are very useful in arriving at a preliminary diagnosis.

Immunofiltration (im'ū-nō-fil-trā'shŭn) **assays** are rapid ELISAs based on the use of antibodies bound to a membrane filter rather than to polystyrene plates. Because of the large surface area of a membrane filter, reactions proceed faster and assay times are significantly reduced as compared to a traditional ELISA.

Immunochromatographic (im'ū-nō-krō'mat-ō-graf'ik) **assays** are still faster and easier to read ELISAs. In these systems, an antigen solution (such as infected blood) flowing through a porous strip encounters antibody labeled with either pink colloidal gold or blue colloidal selenium. Where antigen and antibody bind, colored immune complexes form in the fluid, which then flows through a region where the complexes encounter antibody against them, resulting in a clearly visible pink or blue line. These assays are used for pregnancy testing, which test for *human chorionic growth hormone*—a hormone produced only by an embryo or fetus, and for rapid identification of infectious agents such as HIV.

Table 17.3 on page 506 lists the types of immune tests currently used to diagnose some selected bacterial and viral diseases. **New Frontiers 17.1** on page 506 describes a novel approach to detecting pathogens that might hold promise as a new development in immune testing.

Figure 17.17

A western blot, a technique for demonstrating the presence of antibodies against multiple antigens in a complex mixture. **(a)** Steps in a western blot. ① Antigens (in this case, HIV proteins) are separated by gel electrophoresis. ② Separated proteins are transferred to a nitrocellulose membrane. ③ Test solutions (in this case, urine and serum that may or may not contain antibodies against HIV), enzyme-labeled anti-antibody, and the enzyme's substrate are added; color changes are detected wherever antibody in the test solutions has bound to HIV proteins. **(b)** The results of this western blot. The individual tested in the top 2 strips is negative for antibodies to HIV; the other two individuals have antibodies against HIV.

Table 17.3 Immunological Tests and Some of Their Uses

Test	Use
Immunodiffusion (precipitation)	Diagnosis of syphilis, pneumococcal pneumonia
Immunoelectrophoresis (precipitation)	Assay production of particular classes of antibodies
Agglutination	Blood typing; pregnancy testing; diagnosis of salmonellosis, brucellosis, gonorrhea, rickettsial infection, mycoplasma infection, yeast infection, typhoid fever, meningitis caused by *Haemophilus*
Viral neutralization	Diagnosis of infections by specific strains of viruses
Viral hemagglutination inhibition	Diagnosis of viral infections including influenza, measles, mumps, rubella, mononucleosis
Complement fixation	Diagnosis of measles, influenza A, syphilis, rubella, rickettsial infections, scarlet fever, rheumatic fever, infections of respiratory syncytial virus and *Coxiella*
Direct fluorescent antibody	Diagnosis of rabies, infections of group A *Streptococcus*
Indirect fluorescent antibody	Diagnosis of syphilis, mononucleosis
ELISA	Pregnancy testing; presence of drugs in urine; diagnosis of hepatitis A, hepatitis B, rubella; initial diagnosis of HIV infection
Western blot	Verification of infection with HIV, diagnosis of Lyme disease

New Frontiers 17.1 Listening for Pathogens

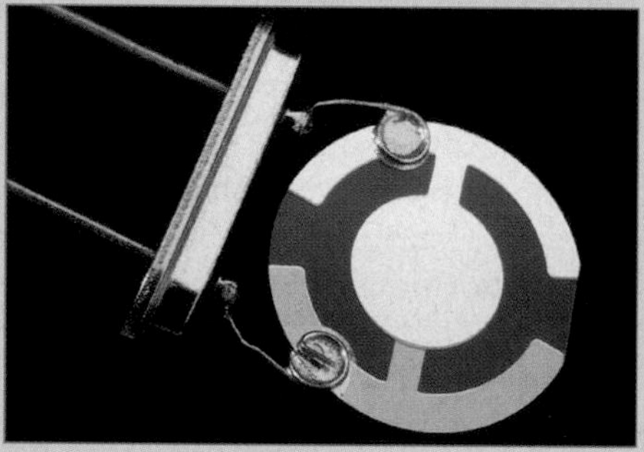

Quartz resonator

Scientists at the University of Cambridge have developed a fascinating new method of detecting the presence of pathogens: by listening for them. They conducted an experiment in which they coated one side of a thin slice of quartz crystal with antibodies against herpes virus, and then introduced herpes viruses, which promptly bound to the antibodies. They then passed an electrical current through the crystal, gradually increasing the voltage until the quartz began to vibrate. They found that the force of the vibration eventually caused the virus and the antibody to break apart, creating a faint popping sound that the crystal registered and converted into electrical signals detectable by researchers.

Why is this important? Because whereas traditional immunological assays detect the presence of antibodies, many of them do not tell us whether a given pathogen is actually present in a patient. The method used by the Cambridge scientists, called *rupture event scanning*, could possibly be adapted to create an instrument that can confirm the presence of a pathogen by its sound. Although herpes virus was used in the Cambridge study, rupture event scanning could work on a wide variety of pathogens, including the causative agents of anthrax, smallpox, foot-and-mouth disease, and Ebola, and might provide quicker, less expensive results than existing serological tests.

CHAPTER SUMMARY

Immunization (pp. 488–496)

1. The first vaccine was developed in the late 1700s by Edward Jenner against smallpox. Vaccines against rabies, anthrax, measles, mumps, rubella, polio and other diseases were developed thereafter.
2. Individuals can be made immune to many infections by either active immunization or passive immunization.
3. Active immunization involves giving antigen in the form of either **attenuated vaccines, inactivated (killed) vaccines,** or **toxoid vaccines.** Pathogens in attenuated vaccines are weakened so that they no longer cause disease, though they are still alive or active and residual virulence may remain a problem. Inactivated vaccines are either whole agent or subunit vaccines and often contain **adjuvants,** which are chemicals added to increase their ability to stimulate active immunity. Toxoid vaccines use modified toxins to stimulate antibody-mediated immunity.
4. Passive immunization involves administration of an **antiserum** containing preformed antibodies.
5. The fusion of myelomas (cancerous plasma cells) with plasma cells results in **hybridomas,** the source of monoclonal antibodies, which can be used in passive immunization.

Immune Testing (pp. 496–506)

1. **Serology** is the study and use of immunologic tests to diagnose and treat disease or identify antibodies or antigens.

2. The simplest of the serologic tests is a precipitation test, in which antigen and antibody meet in optimal proportions to form an insoluble precipitate. Often this test is performed in clear gels, where it is called immunodiffusion.
3. **Radial immunodiffusion** is a quantitative variant of an immunodiffusion test; another variant used to analyze complex antigen mixtures is **immunoelectrophoresis.**
4. **Agglutination tests** involve the clumping of antigenic particles by antibodies. The amount of these antibodies, called the **titer,** is measured by diluting out the serum in a process called **titration.**
5. Antibodies to viruses or toxins can be measured using a **neutralization test,** such as a viral neutralization test. Infection by viruses that naturally agglutinate red blood cells can be demonstrated using a **viral hemagglutination inhibition test.**
6. The **complement fixation test** is a complex assay used to determine the presence of specific antibodies by the ability of the antibodies to nullify the action of complement on sheep red blood cells.
7. Fluorescently–labeled antibodies—those chemically linked to a fluorescent dye such as fluorescein—can be used in a variety of **direct** and **indirect fluorescent antibody** tests. The presence of labeled antibodies is visible through a fluorescent microscope.
8. **Enzyme-linked immunosorbent assays (ELISAs)** are a family of simple tests that can be readily automated and read by machine. These tests are among the most common serologic tests used. A variation of the ELISA is the **western blot test,** which is used to detect antibodies against multiple antigens in a mixture. Together, ELISA and the western blot are used to detect and confirm infection with HIV.
9. **Immunofiltration assays** and **immunochromatographic assays** are very rapid modifications of ELISA tests that can give rapid diagnostic results.

QUESTIONS FOR REVIEW

(Answers to multiple choice, True/False, and matching questions are on the web, along with additional review questions. Visit www.microbiologyplace.com.)

Multiple Choice

1. In order to obtain immediate immunity against tetanus, a patient should receive
   a. an attenuated vaccine of *Clostridium tetani.*
   b. a modified live vaccine of *C. tetani.*
   c. tetanus toxoid.
   d. immunoglobulin against tetanus toxin (antitoxin).
   e. a subunit vaccine against *C. tetani.*
2. Which of the following vaccine types is commonly given with an adjuvant?
   a. an attenuated vaccine
   b. a modified live vaccine
   c. a chemically killed vaccine
   d. an immunoglobulin
   e. an agglutinating antigen
3. Which of the following viruses is widely used in living recombinant vaccines?
   a. coronavirus
   b. vaccinia virus
   c. influenza virus
   d. retrovirus
   e. myxovirus
4. When antigen and antibodies combine, maximal precipitation occurs when
   a. antigen is in excess.
   b. antibody is in excess.
   c. antigen and antibody are at equivalent concentrations.
   d. antigen is added to the antibody.
   e. antibody is added to the antigen.
5. An anti-antibody is used when
   a. an antigen is not precipitating.
   b. an antibody is not agglutinating.
   c. an antibody does not activate complement.
   d. an antigen is insoluble.
   e. the antigen is an antibody.
6. The many different proteins in serum can be analyzed by
   a. an anti-antibody test.
   b. a complement fixation test.
   c. a precipitation test.
   d. an agglutination test.
   e. immunoelectrophoresis.
7. A complement fixation test requires which of the following?
   a. heat-inactivated serum
   b. normal guinea pig serum
   c. peritoneal fluid
   d. heated plasma
   e. antibodies against complement
8. An ELISA uses which of the following reagents?
   a. an enzyme-labeled anti-antibody
   b. a radioactive anti-antibody
   c. a source of complement
   d. an enzyme-labeled antigen
   e. an enzyme-labeled antibody
9. A direct fluorescent antibody test can be used to detect the presence of
   a. hemagglutination
   b. specific antigens
   c. antibodies
   d. complement
   e. precipitated antigen-antibody complexes
10. Which of the following is the best test to detect rabies virus in the brain of a dog?
    a. agglutination
    b. hemagglutination inhibition
    c. virus neutralization
    d. western blot
    e. direct immunofluorescence

True/False

___ 1. Passive immunization provides more prolonged immunity than active immunization.

___ 2. It is standard to attenuate killed virus vaccines.

___ 3. One single serologic test is inadequate for an accurate diagnosis of HIV infection.

___ 4. ELISA is very easily automated.

___ 5. ELISA has basically replaced the western blot test.

Matching

Match the characteristic in the first column with the therapy it most closely describes in the second column. Some choices may be used more than once.

___ 1. Induces rapid onset of immunity	A. Attenuated viral vaccine
___ 2. Induces mainly an antibody response	B. Adjuvant
___ 3. Induces good cell-mediated immunity	C. Subunit vaccine
___ 4. Increases antigenicity	D. Immunoglobulin
___ 5. Uses antigen fragments	E. Residual virulence
___ 6. Uses attenuated microbes	

CRITICAL THINKING

1. Is it ethical to approve the use of a vaccine that causes significant illness in 1% of patients, if it protects immunized survivors against a serious disease?

2. Which is worse: to use a diagnostic test for HIV that may falsely indicate that a patient is not infected (false negative), or to use one that sometimes falsely indicates that a patient is infected (false positive)? Defend your choice.

3. Discuss the importance of costs and technical skill in selecting a practical serologic test. Under what circumstances does automation become important?

4. What body fluids, in addition to blood serum, might be usable for immune testing?

5. Why might a serologic test give a false positive result?

6. Some researchers want to distinguish between B cells and T cells in a mixture of lymphocytes. How could they do this without killing the cells?

CHAPTER 18

Hypersensitivities, Autoimmune Diseases, and Immune Deficiencies

It is summer, and you are hiking along a densely wooded trail wearing shorts and a T-shirt. You spot a beautiful wildflower just a few feet away and wander briefly off the trail to get a better look, wading through knee-high foliage. The next morning, an itchy red rash has developed on both of your legs. Within days, the rash has progressed to painful, oozing blisters. You realize, unhappily, that you have contacted poison ivy and will be treating the blisters with calamine lotion until they heal—typically within 14 to 20 days.

A poison ivy rash is caused by an allergic response to urushiol, a chemical that coats poison ivy leaves. The body's immune system essentially overreacts to the urushiol: It identifies the chemical as a foreign substance and initiates an immune response that results in allergic contact dermatitis (inflammation of the skin). In the absence of an immune response, poison ivy would be harmless. About 85% of the population experiences an allergic reaction when exposed to urushiol.

In this chapter we examine a variety of immune system disorders, including allergies and other immune hypersensitivities, autoimmune diseases, and immune deficiencies.

CHAPTER OUTLINE

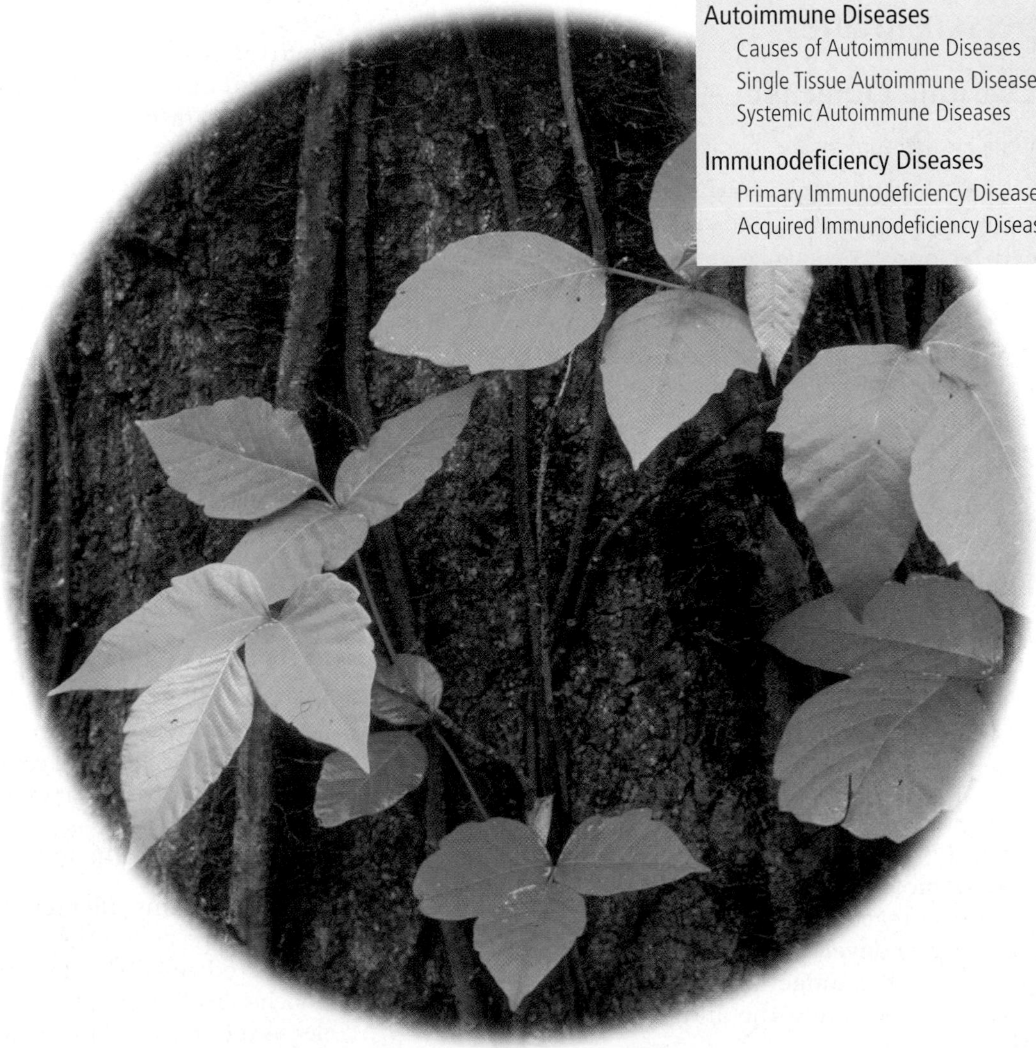

The characteristic foliage of poison ivy with three leaflets.

MicroPrep Pre-Test: *Take the pre-test for this chapter on the web. Visit* ***www.microbiologyplace.com.***

As we have seen, the immune system is an absolutely essential component of the body's defenses. However, if the immune system functions abnormally—either by overreacting or underreacting—the malfunction may cause significant disease, even death. If the immune system functions excessively, the body develops any of a variety of *immune hypersensitivities*. An important example of a hypersensitivity is an allergy, which occurs when the immune system reacts to generally harmless substances such as pollen. Another type of hypersensitivity, graft rejection, occurs when the immune system attacks transplanted tissues. Further, *autoimmune diseases* develop when the immune system attacks the body's own tissues. Finally, diseases that result from a failure of the immune system are called *immunodeficiency diseases*. In this chapter we discuss each of these three categories of abnormalities of the immune system.

Hypersensitivities

Learning Objective

✓ Compare and contrast the four types of hypersensitivity.

Hypersensitivity (hī′per-sen-si-tiv′i-tē) may be defined as any immune response against a foreign antigen that is exaggerated beyond the norm. For example, most people can wear wool, smell perfume, or dust furniture without experiencing itching, wheezing, runny nose, or watery eyes. When these symptoms do occur, the person is said to be experiencing a hypersensitivity response. In the following sections we examine each of the four main types of hypersensitivity response, designated as type I through type IV.

Type I (Immediate) Hypersensitivity

Learning Objectives

✓ Describe the two-part mechanism by which type I hypersensitivity occurs.
✓ Explain the roles of three inflammatory chemicals (mediators) released from mast cell granules.
✓ Describe three disease conditions resulting from type I hypersensitivity mechanisms.

Type I hypersensitivities are either localized or systemic (whole body) reactions that result from the release of inflammatory molecules (such as histamine) in response to an antigen. These reactions are termed *immediate* hypersensitivity because they develop within seconds or minutes following exposure to the antigen. They are also commonly called **allergies**[1] (al′er-jēz), and the antigens that stimulate them are called **allergens.**

In the next two subsections we will examine the two-part mechanism of a type I hypersensitivity reaction: sensitization upon an initial exposure to an allergen, and the degranulation of sensitized cells.

[1]From Greek *allos,* meaning other, and *ergon,* meaning work.

► ***Figure 18.1***
The mechanisms of type I hypersensitivity reaction. **(a)** Sensitization. After normal processing of an antigen (allergen) by an antigen-presenting cell ①, $T_{H}2$ cells are stimulated to secrete IL-4 ②. This cytokine stimulates a B cell ③, which becomes a plasma cell that secretes IgE ④, which then binds to and sensitizes mast cells, basophils, and eosinophils ⑤. **(b)** Degranulation. When the same allergen is subsequently encountered, its binding to the IgE molecules on the surface of sensitized cells triggers rapid degranulation and the release of inflammatory chemicals from the cells.

Sensitization upon Initial Exposure to an Allergen

All of us are exposed to antigens in the environment, against which we typically mount immune responses that result in the production of IgG. But when cytokines (especially IL-4) from type 2 helper T ($T_{H}2$) cells stimulate B cells in *allergic* individuals, the B cells become plasma cells that produce not IgG, but instead IgE **(Figure 18.1a).** Typically, IgE is directed against parasitic worms, which rarely infect most Americans, so IgE is not often found in the blood serum—except in allergic individuals.

The precise reason that only some people produce IgE (and thereby suffer from allergies) is a matter of intense research. Some researchers think that environmental factors—including exposure during infancy and childhood to high levels of indoor antigens such as cigarette smoke, dust mites, and molds—sensitize people, making them hypersensitive (allergic). Other researchers believe that childhood exposure to such antigens is protective, that young children who experience multiple bacterial and viral infections as well as exposure to pet dander (flakes of animal skin) are less likely to develop allergies than those who have been relatively sheltered from infectious diseases and unexposed to pets. **New Frontiers 18.1** on page 513 examines the second hypothesis.

In any case, following initial exposure to allergens, the plasma cells of allergic individuals secrete IgE, which binds very strongly via its stem to three types of defense cells—*mast cells, basophils,* and *eosinophils*—sensitizing these cells to respond to subsequent exposures to the allergen.

Degranulation of Sensitized Cells

When the same allergen subsequently enters the body, it binds to the active sites of IgE molecules on the surfaces of sensitized cells. This binding triggers a cascade of internal biochemical reactions that causes the sensitized cells to release the inflammatory chemicals from their granules into the surrounding intercellular space—an event called *degranulation* (dē-gran-ū-lā′shŭn) **(Figure 18.1b).** It is these inflammatory mediators that generate the characteristic symptoms of type I hypersensitivity reactions: respiratory distress, rhinitis (inflammation of the nasal mucous membranes commonly called "runny nose"), watery eyes, inflammation, and reddening of the skin.

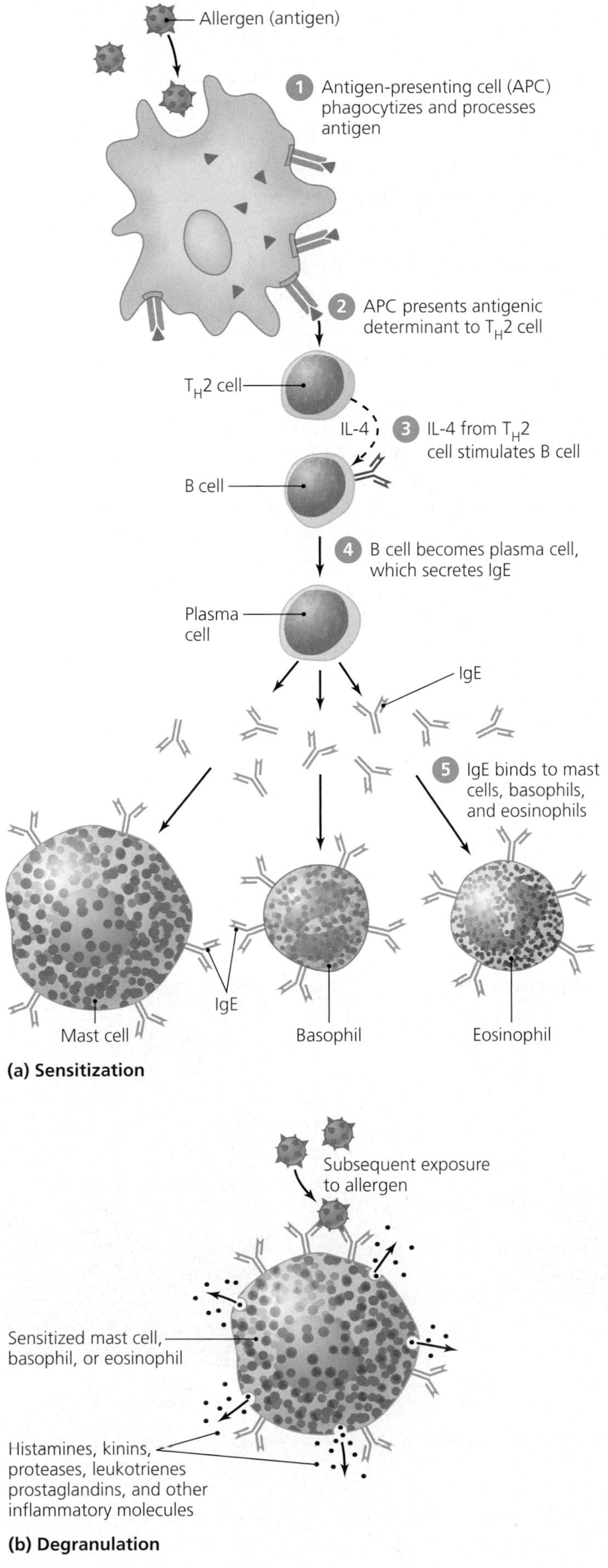

Allergen (antigen)
1 Antigen-presenting cell (APC) phagocytizes and processes antigen
2 APC presents antigenic determinant to T_H2 cell
T_H2 cell
IL-4
3 IL-4 from T_H2 cell stimulates B cell
B cell
4 B cell becomes plasma cell, which secretes IgE
Plasma cell
IgE
5 IgE binds to mast cells, basophils, and eosinophils
IgE
Mast cell
Basophil
Eosinophil
(a) Sensitization
Subsequent exposure to allergen
Sensitized mast cell, basophil, or eosinophil
Histamines, kinins, proteases, leukotrienes prostaglandins, and other inflammatory molecules
(b) Degranulation

The Roles of Degranulating Cells in an Allergic Reaction

Unlike most defensive cells, **mast cells** are not white blood cells; indeed, they are not blood cells at all, but rather are derived from other stem cells and are distributed throughout the body in connective tissue other than blood. These large, round cells are most often found in sites close to body surfaces, including the skin and the walls of the intestines and airways. Their characteristic feature is a cytoplasm packed with very large granules that are loaded with a mixture of very potent inflammatory chemicals.

The most significant of the chemicals released from mast cell granules is **histamine** (his′-tă-mēn), a small molecule related to the amino acid histidine. Histamine stimulates strong contractions in the smooth muscles of the bronchi, gastrointestinal tract, uterus, and bladder, and it also makes blood capillaries dilate (expand) and become leaky. As a result, tissues in which mast cells degranulate become red and swollen. Histamine also stimulates nerve endings, causing itching and pain. Finally, histamine is a potent stimulator of bronchial mucus secretion, tear formation, and salivation.

Other mediators of allergies released by degranulating mast cells include **kinins** (kī′ninz), which are powerful inflammatory chemicals, and **proteases** (prō′tē-ās-ez), enzymes that destroy nearby cells and activate the complement system, which in turn results in the release of still more inflammatory chemicals. Proteases account for more than half the proteins in a mast cell granule. In addition, the binding of allergens to IgE on mast cells activates other enzymes that trigger the production of **leukotrienes** (lū-kō-trī′ēnz) and **prostaglandins** (pros-tă-glan′dinz), lipid molecules that are very potent inflammatory agents. **Table 18.1** on page 512 summarizes the inflammatory molecules released from mast cells.

Basophils, the least numerous type of leukocyte, contain cytoplasmic granules (see Figure 15.6) that stain intensely with basophilic dyes and are filled with inflammatory chemicals similar to those found in mast cells. Sensitized basophils bind IgE and degranulate in the same way as mast cells when they encounter allergens.

The blood and tissues of allergic individuals also accumulate many **eosinophils,** leukocytes that contain numerous cytoplasmic granules (see Figure 15.6) that stain intensely with the red dye eosin and function primarily to destroy parasitic worms. The process during type I hypersensitivity reactions that results in the accumulation of eosinophils in the blood—a condition termed *eosinophilia*—begins with mast cell degranulation, which releases peptides that stimulate the release of eosinophils from the bone marrow. Once in the bloodstream, eosinophils are attracted to the site of mast cell degranulation, where they themselves degranulate. Eosinophil granules contain unique inflammatory mediators and produce large amounts of leukotrienes, which increase vascular permeability and stimulate smooth muscle contraction, thereby contributing greatly to the severity of a hypersensitivity response.

Table 18.1 Inflammatory Molecules Released from Mast Cells

Molecules	Role in Hypersensitivity Reactions
Released during degranulation	
Histamine	Causes smooth muscle contraction, increased vascular permeability, and irritation
Kinins	Cause smooth muscle contraction, inflammation, and irritation
Proteases	Damage tissues and activate complement
Synthesized in response to inflammation	
Leukotrienes	Cause slow, prolonged smooth muscle contraction
Prostaglandins	Some contract smooth muscle; others relax it

Clinical Signs of Localized Allergic Reactions

Type I hypersensitivity reactions are usually mild and localized. The site of the reaction depends on the portal of entry of the antigens. For example, inhaled allergens may provoke a response in the upper respiratory tract commonly known as **hay fever,** a local allergic reaction marked by a profuse watery nasal discharge, sneezing, itchy throat and eyes, and excessive tear production. Mold spores; pollens from grasses, flowering plants, and some trees; and feces and dead bodies of house dust mites, are among the more common allergens **(Figure 18.2).**

If inhaled allergen particles are sufficiently small, they may reach the lungs. The resulting type I hypersensitivity there can cause an episode of severe difficulty in breathing known as **asthma,** characterized by wheezing, coughing, excessive production of a thick, sticky mucus, and constriction of the smooth muscles of the bronchi. Although a localized reaction, asthma can be life threatening: Without medical intervention, the increased mucus and bronchial constriction can quickly cause suffocation.

Some foods also contain powerful allergens, which may provoke diarrhea and other gastrointestinal signs and symptoms. Common food allergens include high-protein foods such as eggs, soy, dairy products, shellfish, and peanuts.

Other allergens—including latex, wool, certain metals, and the venom or saliva of wasps, bees, fire ants, deer flies, fleas, and other stinging or biting insects—may cause a local dermatitis. As a result of the release of histamine and other mediators into nearby skin tissues, plus the ensuing leakage of serum from local blood vessels, the individual suffers round, raised, red areas called *hives* or **urticaria**[2] (er′ti-kar′i-ă; **Figure 18.3**). These lesions are very itchy because histamine irritates local nerve endings.

[2]From Latin *urtica,* meaning nettle.

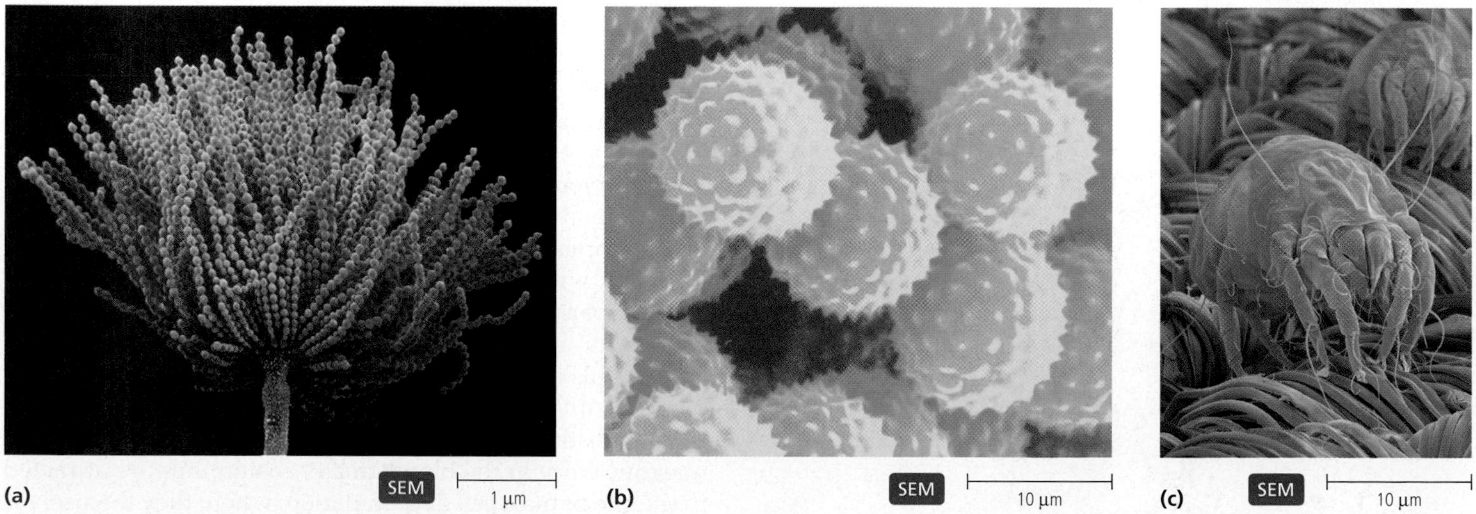

▲ *Figure 18.2*

Some common allergens. **(a)** Spores of the fungus, *Aspergillus flavus.* **(b)** Pollen of *Ambrosia trifida* (ragweed), the most common cause of hay fever in the United States. **(c)** A house dust mite. Mites' fecal pellets and dead bodies may become airborne and are a common cause of hay fever.

New Frontiers 18.1 Can Pets Help Decrease Children's Allergy Risks?

A recent study headed by Dennis Ownby of the Medical College of Georgia suggests that children who grow up in households with two or more cats or dogs may be less likely to develop common allergies than children raised without pets. Researchers suspect that exposure to bacteria harbored by cats and dogs can somehow suppress allergic reactions—not only to pets, but also to other common allergens such as dust mites and grass. Ownby and his team studied 474 children from birth until 6–7 years of age, at which point the children were tested for IgE against common allergens. They found that children exposed to two or more dogs or cats during their first year of life were, on average, 66–77% less likely to have antibodies to common allergens than were children exposed to only one or no pets during the same period. Exactly how the pets and/or their bacteria may suppress the allergic response remains unclear.

Source: Ownby, Dennis R. *et al.* 2002. Exposure to dogs and cats in the first year of life and risk of allergic sensitization at 6 to 7 years of age. *Journal of the American Medical Association* 288(8):963–972.

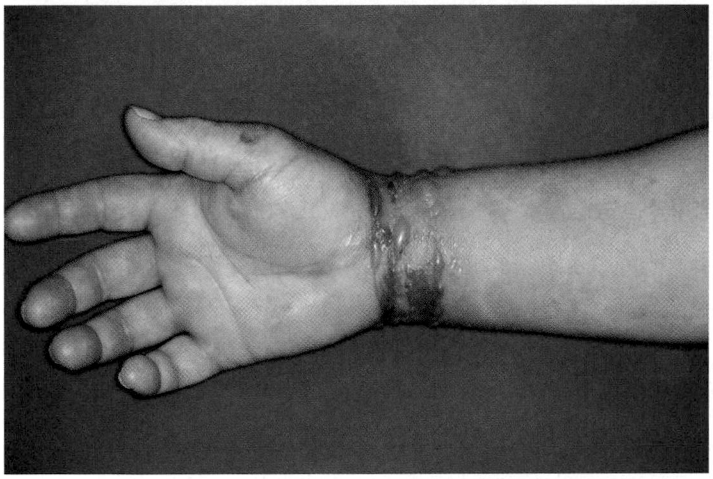

▲ ***Figure 18.3***

Urticaria, the presence of round, red, fluid-filled patches on the skin prompted by the release of histamine in response to an allergen. *What causes fluid to accumulate in the patches seen in urticaria?*

Figure 18.3 The fluid accumulates because histamine from degranulated mast cells causes blood capillaries to become more permeable and leak fluid.

Clinical Signs of Systemic Allergic Reactions

Following a sensitized individual's contact with an allergen, many mast cells may degranulate simultaneously, releasing copious amounts of histamine and other inflammatory mediators into the bloodstream. The release of chemicals may exceed the body's ability to adjust, resulting in **acute anaphylaxis**[3] (an'ă-fī-lak'sis) or **anaphylactic shock.**

[3]From Greek *ana,* meaning away from, and *phylaxis,* meaning protection.

A common cause of acute anaphylaxis is a bee sting in an allergic and sensitized individual. The first sting produces sensitization and the formation of IgE antibodies, and the second sting—which may occur years later—produces an anaphylactic reaction. Other allergens commonly implicated in acute anaphylaxis are certain foods, vaccines, antibiotics such as penicillin, iodine dyes, local anesthetics, certain narcotics such as morphine, and blood products.

The clinical signs of acute anaphylaxis are those of rapid suffocation. Bronchial smooth muscle, which is highly sensitive to histamine, contracts violently. In addition, increased leakage of fluid from blood vessels causes swelling of the larynx and other tissues. The patient also experiences contraction of the smooth muscle of the intestines and bladder. Without prompt administration of *epinephrine* (ep'i-nef'rin) (discussed shortly), an individual in anaphylactic shock may suffocate, collapse, and die within minutes.

Diagnosis of Type I Hypersensitivity

Type I hypersensitivity is commonly diagnosed by injecting a very small quantity of a dilute solution of the allergens being tested into the skin. In most cases the individual is screened for more than a dozen potential allergens simultaneously, using the forearms as injection sites. When the individual tested is sensitive to an allergen, local histamine release will cause redness and swelling at the injection site within a few minutes **(Figure 18.4).**

Prevention of Type I Hypersensitivity

Because many different allergens provoke type I hypersensitivity, prevention begins with *identification* and *avoidance* of the allergens responsible. Air conditioning and avoidance of rural areas during pollen season can reduce upper respiratory allergies provoked by some pollens. Encasing bedding in mite-proof covers, frequent vacuuming, and avoidance of home furnishings that trap dust (such as wall-to-wall carpeting and heavy drapes) can reduce the severity of other

New Frontiers 18.2 When Kissing Triggers Allergic Reactions

Individuals with particularly severe food allergies not only need to watch what they eat—they also need to watch who kisses them. In a recent study from the University of California at Davis, researchers found that out of 316 patients with severe allergies to peanuts, tree nuts, and/or seeds, 20 (about 6%) reported developing allergic reactions after being kissed. In nearly all of the cases, the kisser had eaten nuts to which the allergic individual was sensitive. In some of the cases, the allergic individuals were so sensitive that they developed reactions even though the kissers had brushed their teeth or had consumed the nuts several hours earlier. Food allergists believe that the duration and intensity of a kiss may also be a factor—that is, longer kisses involving exchange of saliva may be more likely to elicit allergic reactions than a quick buss on the cheek. The moral of the story: If you have a severe food allergy, you might want to tell your romantic partner about it!

Source: Hallett, R., L. A. D. Haapanen, and S. S. Teuber. 2002. Food allergies and kissing. *New England Journal of Medicine* 346:1833.

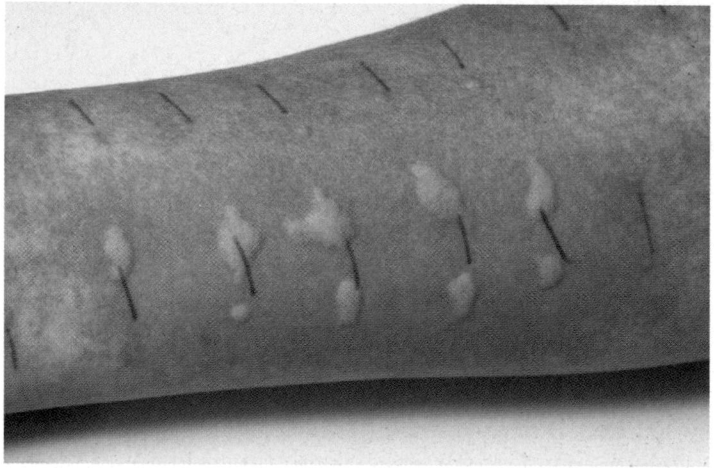

▲ ***Figure 18.4***
Skin tests for diagnosing type I hypersensitivity. The presence of redness and swelling at an injection site indicates sensitivity to the allergen injected at that site.

household allergies. A dehumidifier can reduce mold in a home. The best way for individuals to avoid allergic reactions to pets is to find out if they are allergic before they get a pet. If an allergy develops afterward, they may have to find a new home for the animal.

Food allergens can be identified and then avoided by using a medically supervised elimination diet in which foods are removed one at a time from the diet to see when signs of allergy cease. Peanuts and shellfish are among the foods more commonly implicated in anaphylactic shock; allergic individuals must scrupulously avoid consuming even minute amounts (see **New Frontiers 18.2**). For example, individuals sensitized against shellfish can suffer anaphylactic shock from consuming meat sliced on an unwashed cutting board previously used to prepare shellfish. Some vaccines contain minute amounts of egg protein and should not be given to individuals allergic to eggs.

Health care workers must assess a patient's medical history carefully before administering penicillin, iodine dyes, and other potential allergens. Therapies such as antidotes and respiratory stimulants must be immediately available if a patient's sensitivity status is unknown.

In addition to avoidance of allergens, type I hypersensitivity reactions can be prevented by *immunotherapy* (im'ū-nō-thār'ă-pē), commonly called "allergy shots" or "desensitization," which involves the administration of a series of injections of dilute allergen, usually once a week for many months. It is unclear just how allergy shots work, but they may alter the helper T cell balance from T_H2 cells to T_H1 cells, reducing the production of antibodies, or they may stimulate the production of IgG, which binds antigen before the antigen can react with IgE on mast cells, basophils, and eosinophils. Immunotherapy reduces the severity of allergy symptoms by about 50% in about two-thirds of patients with upper respiratory allergies; however, the series of injections must be repeated every 2–3 years. Immunotherapy is not effective in treating asthma.

Treatment of Type I Hypersensitivity

One way to treat type I hypersensitivity is to administer drugs that specifically counteract the inflammatory mediators released by degranulating cells. Thus **antihistamines** (an-tē-his'tă-mēnz) are administered to either prevent histamine release or its effects. However, because histamine is but one of many mediators released in type I hypersensitivity, antihistamines do not completely eliminate all clinical signs. Indeed, because antihistamines are essentially useless in patients with asthma, asthmatics are typically prescribed an inhalant containing *corticosteroid* (kōr'ti-kō-stēr'oyd) and a *bronchodilator* (brong-kō-dī-lā'ter), which counteract the effects of histamine and the other mediators.

The hormone epinephrine neutralizes many of the lethal mechanisms of anaphylaxis by relaxing smooth muscle and reducing vascular permeability. Epinephrine is thus the drug of choice for the emergency treatment of both severe asthma and anaphylactic shock. Patients who suffer from severe type I hypersensitivities may carry a prescription epinephrine kit so that they can inject themselves before their allergic reactions become life threatening.

Type II (Cytotoxic) Hypersensitivity

Learning Objectives

- ✓ Discuss the mechanisms underlying transfusion reactions.
- ✓ Construct a table comparing the key features of the four blood types in the ABO blood group system.
- ✓ Describe the mechanisms and treatment of hemolytic disease of the newborn.

The second major form of hypersensitivity results when cells are destroyed by an immune response—typically by the combined activities of complement and antibodies (see Figure 15.16). So-called *cytotoxic* (sī-tō-tok′sik) *hypersensitivity* is part of many autoimmune diseases, which we will discuss later in this chapter, but the most significant examples of type II hypersensitivity are the destruction of donor red blood cells following an incompatible blood transfusion, and the destruction of fetal red blood cells in hemolytic disease of the newborn. We begin our discussion by focusing on these two manifestations of type II hypersensitivity.

The ABO System and Transfusion Reactions

Red blood cells, like nucleated cells, have many different glycoprotein and glycolipid molecules on their surface. The surface molecules of red blood cells, called **blood group antigens,** have various functions, including transportation of glucose and ions across the cytoplasmic membrane. Although most of these surface molecules are cell-membrane components, some are soluble molecules found free in serum, saliva, and other body fluids that have simply stuck to red blood cell surfaces.

Blood group antigens vary in complexity. The ABO group, for example, consists of just two antigens arbitrarily given the names A antigen and B antigen. Each person's red blood cells have either A antigen, B antigen, both A antigen and B antigen, or neither antigen. Individuals with neither antigen are said to have blood type O. In contrast, the MN blood group is much more complex, with 29 known antigens.

As you probably know, blood can be readily transfused from one person to another. However, if blood is transfused to an individual with a different blood type, then the donor's blood group antigens stimulate the production of antibodies in the recipient—that bind to and eventually destroy the transfused cells. The result can be a potentially life-threatening *transfusion reaction.*

Transfusion reactions, the most problematic of which involve the ABO group, develop as follows:

- If the recipient has preexisting antibodies to foreign blood group antigens, then the donated blood cells will be destroyed immediately—either the antibody-bound cells will be phagocytized by macrophages and neutrophils, or the antibodies will agglutinate the cells, and complement will rupture them, a process called *hemolysis* **(Figure 18.5).** Hemolysis releases hemoglobin into the bloodstream, which may cause severe kidney damage. At the same time, the membranes of the ruptured blood cells trigger blood clotting within blood vessels, blocking them and causing circulatory failure, fever, difficulty in breathing, coughing, nausea, vomiting, and diarrhea. The liver, which is responsible for breaking down dead red blood cells, becomes overwhelmed. If the patient survives, recovery follows the elimination of all the foreign red blood cells.
- If the recipient has no preexisting antibodies to the foreign blood group antigens, then the transfused cells circulate and function normally, but only for a while—that is, until the recipient's immune system mounts a primary response against the foreign antigens that produces enough antibody to destroy the foreign cells. This happens over a long enough time that the severe symptoms and signs mentioned above do not occur.

To prevent transfusion reactions, laboratory personnel must cross-match blood types between donors and recipients. Thus, before a recipient receives a blood transfusion, the red blood cells and serum of the donor are mixed with the serum and red blood cells of the recipient. If any signs of agglutination are seen, then that blood is not used, and an alternative donor is sought. **Table 18.2** on page 517 presents the characteristics of the ABO blood group, and the compatible donor/recipient matches that can be made.

In most people, production of antibodies against foreign ABO antigens is not stimulated by exposure to foreign blood cells, but instead by exposure to antigenically similar molecules found on a wide range of plants and bacteria. Individuals encounter and make antibodies against these antigens on a daily basis, but only upon receipt of a mismatched blood transfusion are they problematic.

CRITICAL THINKING

During the Gulf War in 2003, an Army corporal with type AB blood received a life-saving blood transfusion from his sergeant who had type O blood. Later the sergeant was involved in a traumatic accident and needed blood desperately. The corporal wanted to help but was told his blood was incompatible. Explain the immunological reasons the corporal could receive blood from, but could not give blood to, the sergeant.

The Rh System and Hemolytic Disease of the Newborn

Many decades ago, researchers discovered the existence of an antigen common to the red blood cells of humans and rhesus monkeys. They called this antigen, which is a cytoplasmic membrane protein that transports anions and glucose across the membrane, *rhesus* (rē′sŭs) *antigen* or **Rh antigen.** Laboratory analysis of human blood samples eventually showed that the Rh antigen is present on the red blood cells of about 85% of humans; that is, about 85% of the population is *Rh positive (Rh+),* and about 15% of the population is *Rh negative (Rh−).*

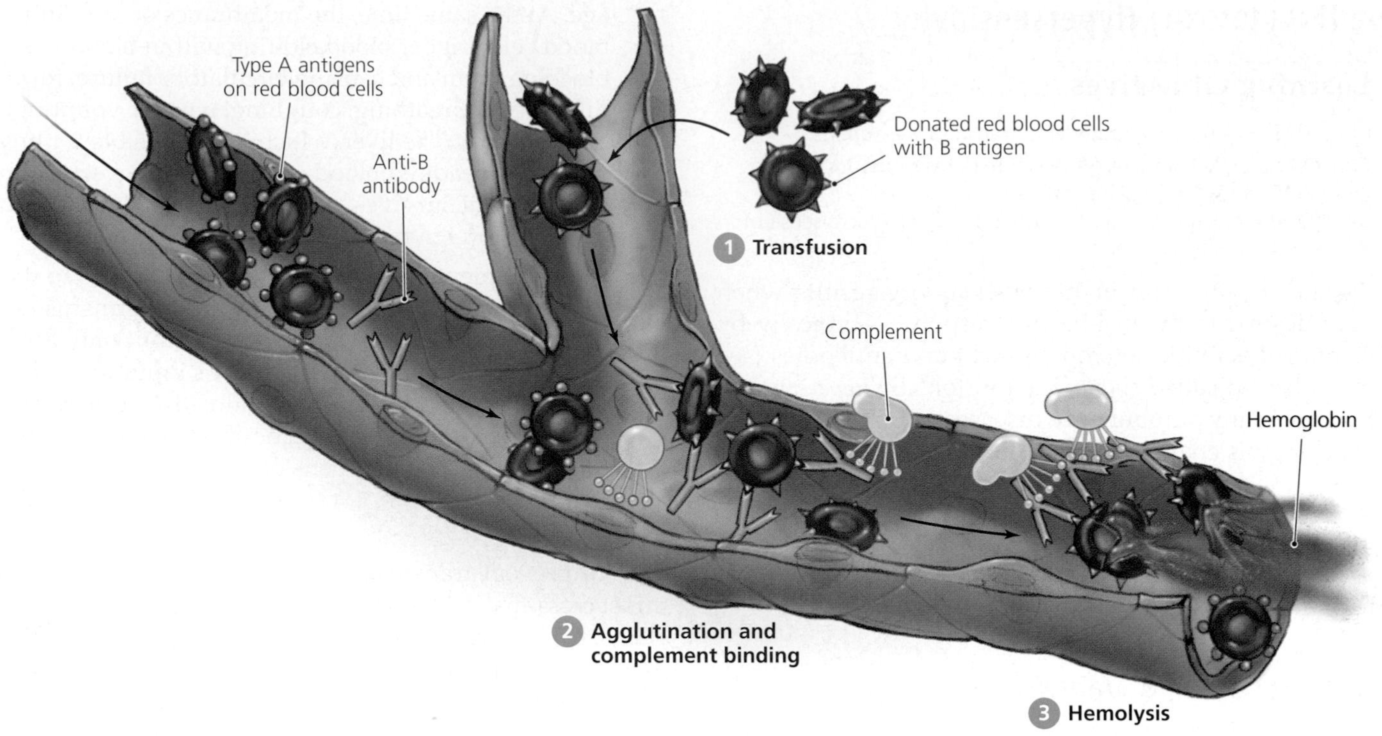

Figure 18.5

Events leading to hemolysis, one of the negative consequences of a transfusion reaction. When antibodies against a foreign ABO antigen combine with the antigen on transfused red blood cells ①, the cells are agglutinated, and the complexes bind complement ②. The resulting hemolysis releases large amounts of hemoglobin into the bloodstream ③, producing several additional negative consequences throughout the body.

In contrast to the situation with ABO antigens, preexisting antibodies against Rh antigen do not occur. Although a transfusion reaction can occur in an Rh negative patient who receives more than one transfusion of Rh positive blood, such a reaction is usually minor because Rh antigen molecules are less abundant than A or B antigens. Instead, the primary problem posed by the Rh system is the risk of **hemolytic** (hē-mō-lit′ik) **disease of the newborn (Figure 18.6).**

This hypersensitivity reaction develops when an Rh− mother is pregnant with her second or subsequent Rh+ fetus (who inherited an Rh gene from its father). Normally, the placenta keeps fetal red blood cells separate from the mother's blood, so fetal cells do not enter the mother's bloodstream. However, in 20–50% of pregnancies—especially during the later weeks of pregnancy, during a clinical or spontaneous abortion, or during childbirth—fetal red blood cells escape into the mother's blood. The mother's immune system recognizes these Rh+ fetal cells as foreign and initiates a humoral immune response by developing antibodies against the Rh antigen. Initially, only IgM antibodies are produced, and because IgM is a very large molecule that cannot cross the placenta, no problems arise during the first pregnancy.

However, IgG antibodies can cross the placenta, so if during subsequent pregnancies the fetus is Rh+, anti-Rh IgG molecules produced by the mother cross the placenta and destroy the fetus's Rh+ red blood cells. Such destruction may be limited, or it may be severe enough—especially in a third or subsequent pregnancy—to cause grave problems. One classic feature of hemolytic disease of the newborn is severe jaundice from excessive *bilirubin* (bil-i-roo′bin), a yellowish blood pigment released during the degradation of hemoglobin from lysed red blood cells. The liver is normally responsible for removing bilirubin, but because of the immaturity of the fetal liver and overload from the hemolytic process, the bilirubin may instead be deposited in the brain, causing severe neurological damage or death during the latter weeks of pregnancy or shortly after the baby's birth.

In the past, when prevention of hemolytic disease of the newborn was impossible, this terrible disease occurred in about 1 of every 300 births. Today, however, physicians can routinely and drastically reduce the prevalence of the disease by administering anti-Rh serum, called *Rhogam,* to Rh-negative pregnant women at 28 weeks gestation and also within 72 hours following abortion, miscarriage, or childbirth to destroy any fetal red cells that may have entered the mother's body. As a result, sensitization of the mother does not occur, and subsequent pregnancies are safer.

Table 18.2 ABO Blood Group Characteristics and Donor/Recipient Matches

ABO Blood Group	ABO Antigen(s) Present	Antibodies Present	Can Donate To	Can Receive From
A	A	Anti-B	A or AB	A or O
B	B	Anti-A	B or AB	B or O
AB	A and B	None	AB	A, B, AB, or O (universal recipient)
O	None	Both anti-A and anti-B	A, B, AB, or O (universal donor)	O

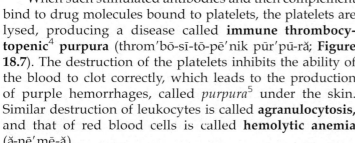

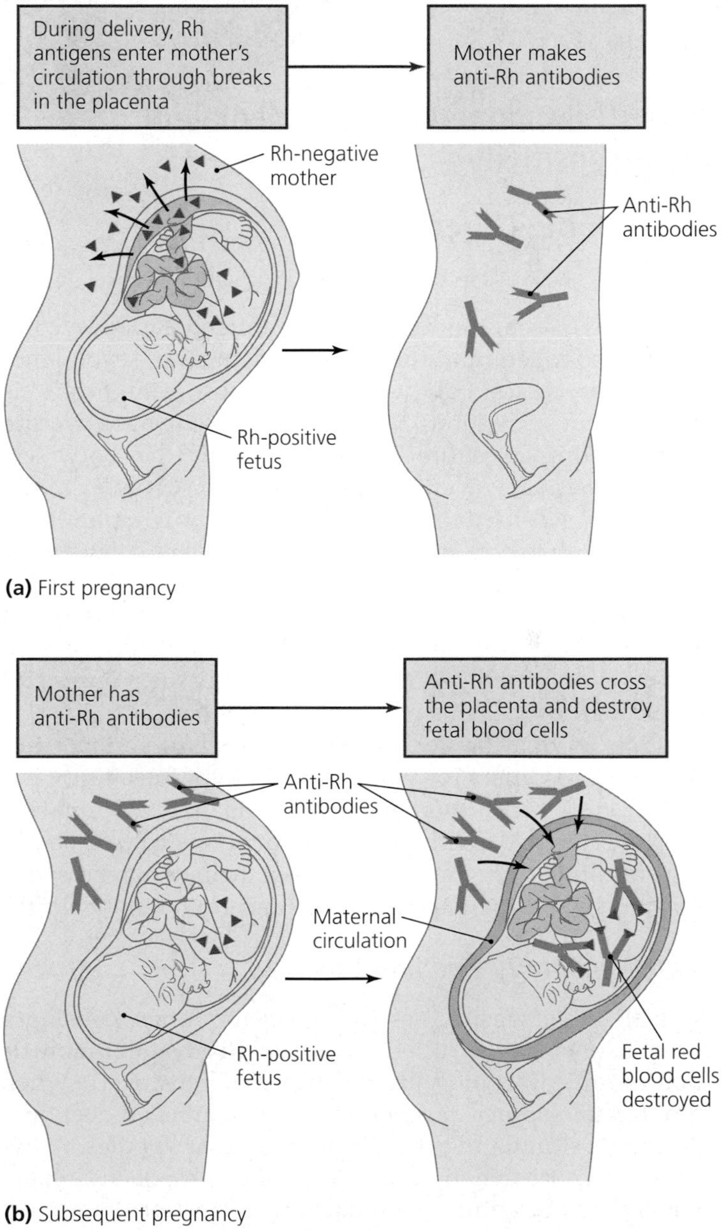

Figure 18.6

Events in the development of hemolytic disease of the newborn. **(a)** During an initial pregnancy, Rh+ red blood cells from the fetus enter the Rh− mother's circulation, often during childbirth. As a result, the mother produces anti-Rh antibodies. **(b)** During a subsequent pregnancy with another Rh+ child, anti-Rh IgG molecules cross the placenta and trigger destruction of the fetus's red blood cells. *How is it possible that the child of an Rh− mother is Rh+?*

Figure 18.6 The father of the child is Rh positive.

Drug-Induced Cytotoxic Reactions

Another kind of type II hypersensitivity involves cytotoxic reactions to drugs. Although the molecules of such drugs as quinine, penicillin, or sulfanilamide are too small to trigger an immune response by themselves, they can be haptens; that is, when they spontaneously bind to blood cells or platelets, they become antigenic and stimulate production of antibodies (a humoral immune response).

When such stimulated antibodies and then complement bind to drug molecules bound to platelets, the platelets are lysed, producing a disease called **immune thrombocytopenic**[4] **purpura** (throm′bō-sī-tō-pē′nik pūr′pū-ră; **Figure 18.7**). The destruction of the platelets inhibits the ability of the blood to clot correctly, which leads to the production of purple hemorrhages, called *purpura*[5] under the skin. Similar destruction of leukocytes is called **agranulocytosis,** and that of red blood cells is called **hemolytic anemia** (ă-nē′mē-ă).

[4] *Thrombocyte* is another name for platelet.

[5] From Latin *porphyra,* meaning purple.

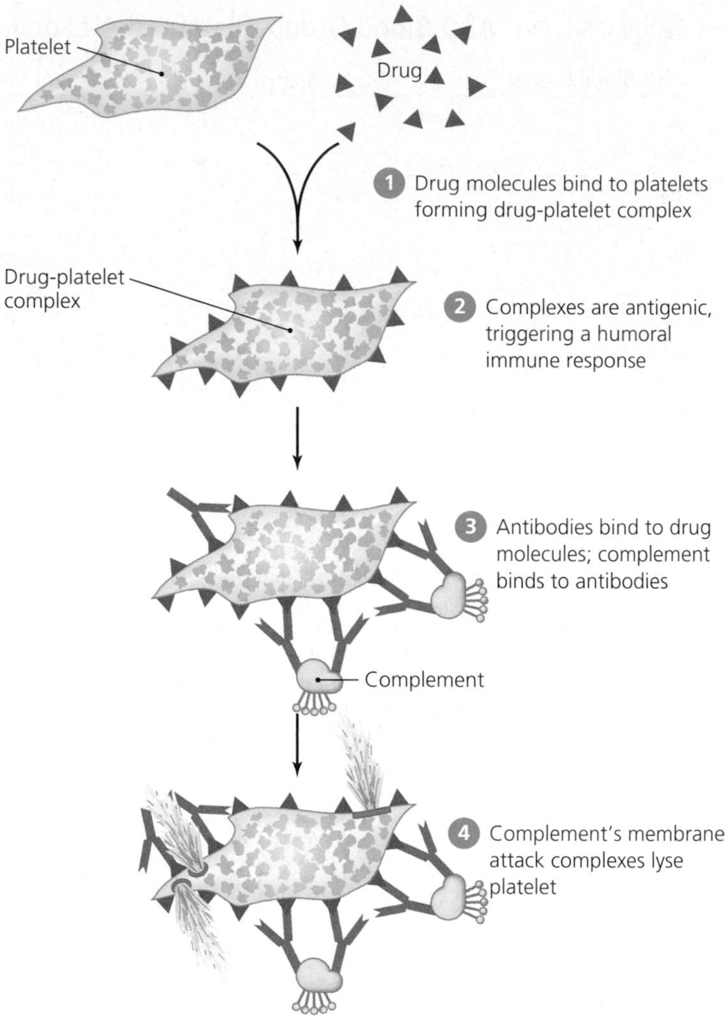
Platelet
Drug
1 Drug molecules bind to platelets forming drug-platelet complex
Drug-platelet complex
2 Complexes are antigenic, triggering a humoral immune response
3 Antibodies bind to drug molecules; complement binds to antibodies
Complement
4 Complement's membrane attack complexes lyse platelet

▲ ***Figure 18.7***

Events in the development of immune thrombocytopenic pupura, a disease resulting from a drug-induced type II (cytotoxic) hypersensitivity reaction. The destruction of platelets through the action of bound antibodies and complement produces the inhibition of blood clotting characteristic of this disease. *What similar disease results from type II (cytotoxic) destruction of red blood cells?*

Figure 18.7 Hemolytic anemia is the type II (cytotoxic) destruction of red blood cells.

Type III (Immune-Complex Mediated) Hypersensitivity

Learning Objective

✓ Outline the basic mechanism of type III hypersensitivity.

As we have seen, the formation of antigen-antibody complexes, also called **immune-complexes,** initiates several molecular processes, of which the most significant is complement activation. Normally, complement-activating immune-complexes are eliminated from the body via phagocytosis. However, in type III (immune-complex mediated) hypersensitivity reactions, the immune complexes are small enough to escape phagocytosis, and so circulate freely in the bloodstream until they become trapped in organs, joints, and tissues (such as the walls of blood vessels). In these sites they trigger mast cells to degranulate, releasing inflammatory chemicals that damage the tissue. **Figure 18.8** illustrates the mechanism of type III reactions.

Later in this chapter we will discuss two systemic autoimmune disorders for which type III hypersensitivity is a factor: systemic lupus erythematosus and rheumatoid arthritis. Here we briefly discuss two more-localized conditions resulting from immune-complex mediated hypersensitivity: hypersensitivity pneumonitis and glomerulonephritis.

Hypersensitivity Pneumonitis

Type III hypersensitivities can affect the lungs, causing a form of pneumonia called **hypersensitivity pneumonitis** (noo-mō-nī'tus). Individuals become sensitized when minute mold spores or other antigens are inhaled deep into the lungs, stimulating the production of antibodies. A hypersensitivity reaction occurs when the subsequent inhalation of the same antigen stimulates the formation of immune complexes that then activate complement.

One form of hypersensitivity pneumonitis, called *farmer's lung,* occurs in farmers chronically exposed to spores from moldy hay. Many other syndromes in humans develop via a similar mechanism and are usually named after the source of the inhaled allergen. Thus, *pigeon breeder's lung* arises following exposure to the dust from pigeon feces, *mushroom grower's lung* is a response to the soil or fungal spores encountered when cultivating mushrooms, and *librarian's lung* results from inhalation of dust from old books.

Glomerulonephritis

Glomerulonephritis (glō-mār'ū-lō-nef-rī'tis) occurs when immune complexes circulating in the bloodstream are deposited in the walls of *glomeruli,* which are networks of minute blood vessels in the kidneys. Immune complexes damage the glomerular cells, leading to enhanced local production of cytokines that stimulate nearby cells to produce more of the intercellular proteins that underlie the cells, impeding blood filtration. Sometimes, the immune complexes are deposited in the center of the glomeruli, where they stimulate local cells to proliferate and compress nearby blood vessels, again interfering with kidney function. The net result is kidney failure; the glomeruli lose their ability to filter wastes from the blood, ultimately resulting in death.

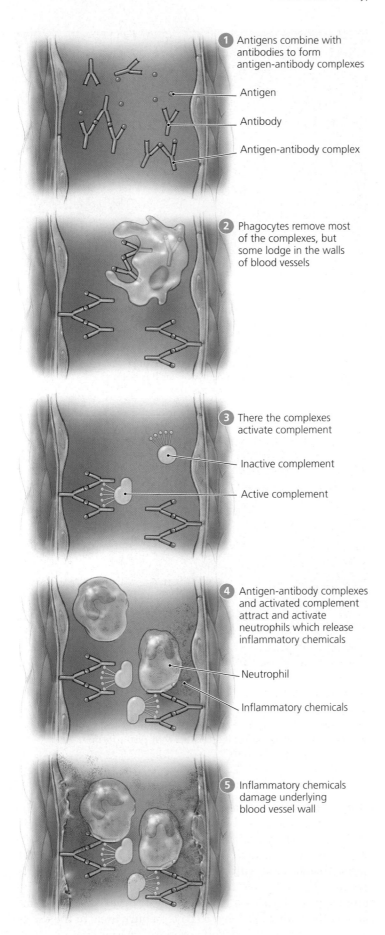

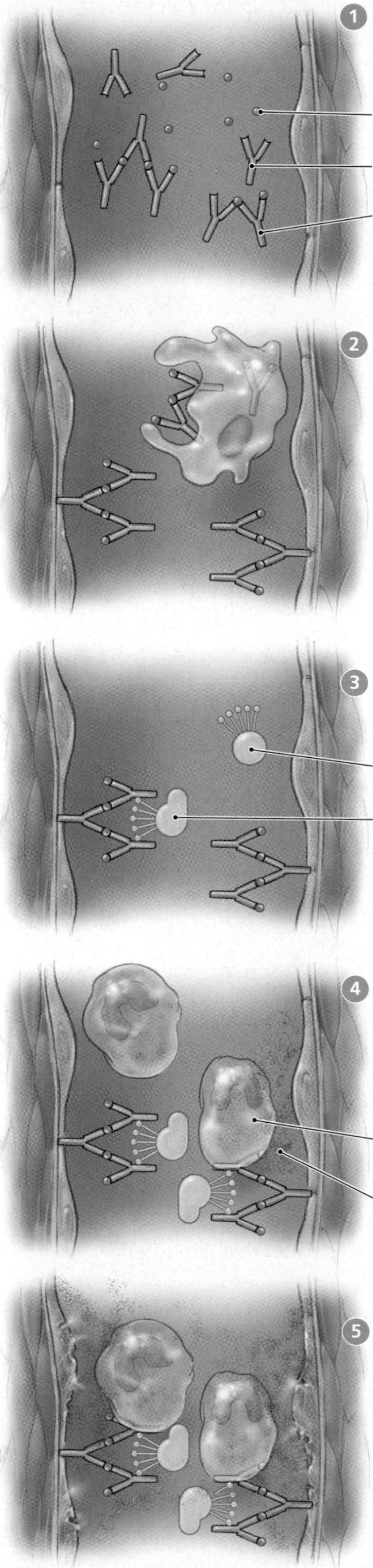
1 Antigens combine with antibodies to form antigen-antibody complexes
Antigen
Antibody
Antigen-antibody complex
2 Phagocytes remove most of the complexes, but some lodge in the walls of blood vessels
3 There the complexes activate complement
Inactive complement
Active complement
4 Antigen-antibody complexes and activated complement attract and activate neutrophils which release inflammatory chemicals
Neutrophil
Inflammatory chemicals
5 Inflammatory chemicals damage underlying blood vessel wall

◀ ***Figure 18.8***

The mechanism of type III (immune-complex mediated) hypersensitivity. After immune complexes form ①, the complexes that are not phagocytized lodge in certain tissues (in this case, the wall of a blood vessel) ②. Trapped complexes then activate complement ③, which attracts and activates neutrophils ④. The release of inflammatory chemicals from the activated neutrophils leads to destruction of tissue ⑤.

Type IV (Delayed or Cell-Mediated) Hypersensitivity

Learning Objectives

- ✓ Outline the mechanism of type IV hypersensitivity.
- ✓ Describe the significance of the tuberculin test.
- ✓ Identify four types of grafts.
- ✓ Explain the concept of privileged sites, and explain why a mother does not reject a fetus even though the uterus is not a privileged site.
- ✓ Compare four types of drugs commonly used to prevent graft rejection.

When certain antigens contact the skin of sensitized individuals, they provoke inflammation that begins to develop at the site only after 12–24 hours. Such **delayed hypersensitivity reactions** result not from the interactions of antigen and antibodies, but rather from interactions among antigen, antigen-presenting cells, and T cells; thus, a type IV reaction is also called *cell-mediated hypersensitivity.* The delay in this cell-mediated response reflects the time it takes for macrophages and T cells to migrate to and proliferate at the site of the antigen. We begin our discussion of type IV reactions by considering two common examples: the tuberculin response and allergic contact dermatitis. Then we discuss two type IV hypersensitivity reactions involving the interactions between the body and tissues grafted to (transplanted into) it—graft rejection, and graft versus host disease—before considering donor-recipient matching and tissue typing.

The Tuberculin Response

The **tuberculin** (too-ber′kū-lin) **response** is an important example of a delayed hypersensitivity reaction in which the skin of an individual exposed to tuberculosis or tuberculosis vaccine reacts to a subcutaneous injection of *tuberculin*—a protein solution obtained from *Mycobacterium tuberculosis* (mī-kō-bak-tēr′ē-ŭm tū-ber-kyū-lō′ sis). Health care providers use tuberculin tests to diagnose contact with antigens of *M. tuberculosis.*

When tuberculin is injected into the skin of a healthy, never infected or vaccinated individual, no response occurs. In contrast, when tuberculin is injected into someone currently or previously infected with *M. tuberculosis* or an individual previously immunized with tuberculosis vaccine, a red hard swelling (10 mm or greater in diameter) indicating a positive tuberculin test develops at the site **(Figure 18.9).** Such inflammation reaches its greatest intensity within 24–72 hours and may persist for several weeks before fading. Microscopic examination of the lesion reveals that it is infiltrated with lymphocytes and macrophages.

A tuberculin response is mediated by memory T cells. When an individual is first infected by *M. tuberculosis,* the resulting cell-mediated immune response generates memory T cells that persist in the body. When a sensitized individual is subsequently injected with tuberculin, phagocytic cells migrate to the site and attract memory T cells, which secrete a mixture of cytokines that attract still more T cells and macrophages, giving rise to a slowly developing inflammation. The macrophages ingest and destroy the injected tuberculin, allowing the tissues eventually to return to normal.

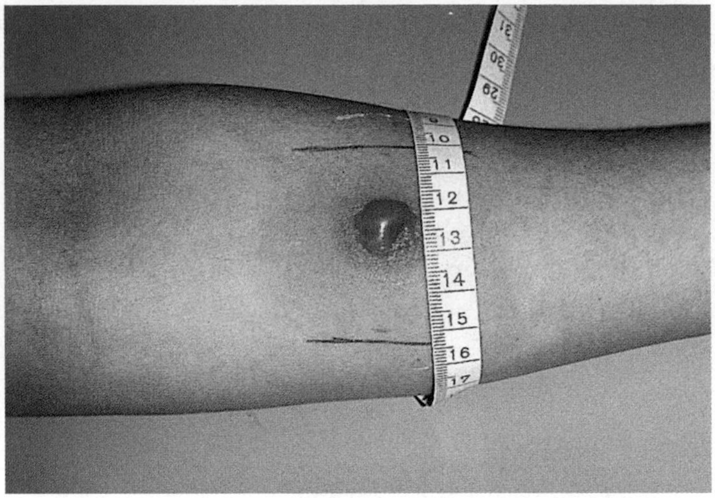

▲ ***Figure 18.9***

A positive tuberculin skin test. The hard, red swelling 10mm or greater in diameter that is characteristic of the tuberculin response indicates that the individual has been vaccinated against *Mycobacterium tuberculosis* or is now or has been previously infected with the bacterium.

Allergic Contact Dermatitis

Urushiol, the oil of poison ivy *(Toxicodendron radicans)* and related plants, is a highly reactive hapten that binds to almost any protein it contacts—including proteins in the skin of anyone who rubs against the plant. The body regards these chemically modified skin proteins as foreign, which triggers a cell-mediated immune response resulting in an intensely irritating skin rash called **allergic contact dermatitis** (der-mă-tī′tis). In severe cases, T_C cells destroy so many skin cells that acellular, fluid-filled blisters develop **(Figure 18.10).**

Other haptens that combine chemically with skin proteins can also induce allergic contact dermatitis. Examples include formaldehyde; some cosmetics, dyes, and drugs; metal ions; and chemicals used in the production of latex for hospital gloves and tubing.

Because T cells mediate allergic contact dermatitis, epinephrine and other drugs used for the treatment of immediate hypersensitivity reactions are ineffective. T cell activities and inflammation can, however, be suppressed by corticosteroid treatment. Good strategies for dealing with exposure to poison ivy include washing the area thoroughly and immediately with a strong soap, and washing all exposed clothes as soon as possible.

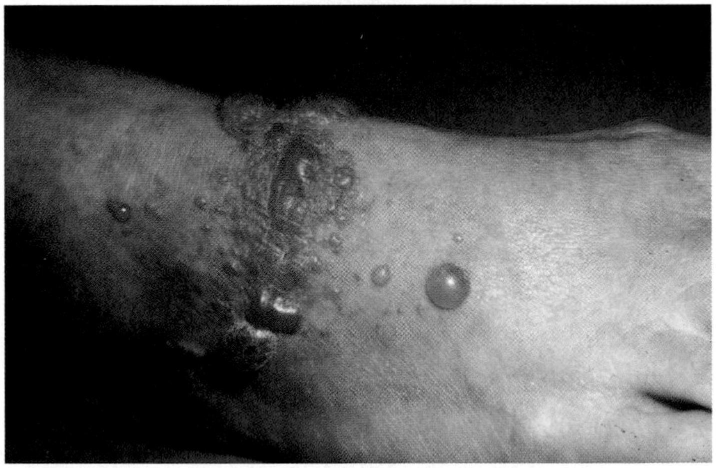

▲ ***Figure 18.10***

Allergic contact dermatitis, a type IV hypersensitivity response—in this case, to exposure to poison ivy. Note the large, acellular, fluid-filled blisters that result from the destruction of skin cells.

Graft Rejection

A special case of type IV hypersensitivity is **graft rejection,** in which the body attacks tissues or organs that have been transplanted. Even though advances in surgical technique enable surgeons to move grafts freely from site to site, grafts perceived as foreign by the recipient undergo rejection, a highly destructive phenomenon that can severely limit the success of organ and tissue transplantation. As discussed in Chapter 16, graft rejection is a normal immune response against foreign major histocompatibility complex (MHC) proteins on the surface of graft cells. The likelihood of graft rejection depends on the degree to which the graft is foreign to the recipient, which in turn is related to the type of graft.

Types of Grafts Scientists name grafts according to the degree of relatedness between the donor and the recipient **(Figure 18.11).** A graft is called an **autograft** (aw′tō-graft) when tissues are moved to a different location within the same individual. Autografts do not trigger immune responses because they do not express foreign antigens. Examples of autografts include the grafting of skin from one area of the body to another to cover a burn area, or the use of a leg vein to bypass blocked coronary arteries.

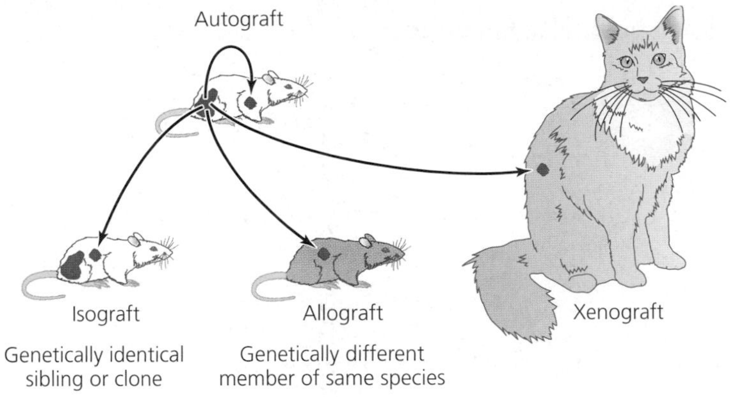

▲ *Figure 18.11*
Types of grafts, which are based on the degree of relatedness between donor and recipient. Autografts are grafts moved from one location to another within a single individual. In isografts, the donor and recipient are either genetically identical siblings or clones. In allografts, the donor and recipient are genetically distinct individuals of the same species. In xenografts, the donor and recipient are of different species.

Isografts (ī′sō-graftz) are grafts transplanted between two genetically identical individuals; that is, between identical siblings or clones. Because these individuals have identical MHC proteins, the immune system of the recipient cannot differentiate between the grafted cells and its own normal body cells. As a result, isografts are not rejected.

Allografts (al′ō-graftz) are grafts transplanted between genetically distinct members of the same species. Most grafts performed in humans are allografts. Because the MHC proteins of the allograft are different from those of the recipient, allografts typically induce a strong type IV hypersensitivity, resulting in graft rejection. Rejection must be suppressed with immunosuppressive drugs if the graft is to survive.

Xenografts (zen′ō-graftz) are grafts transplanted between individuals of different species. Thus, the transplant of a baboon's heart into a human is a xenograft. Because xenografts are usually very different from the tissues of the recipient, both biochemically and immunologically, they usually provoke a rapid, intense rejection that is very difficult to suppress. Therefore, xenografts from mature animals are not commonly used therapeutically.

Privileged Sites Certain sites in the body, including certain parts of the eye such as the cornea, the thymus, the testes, and the brain, are said to be **privileged sites** because grafts involving them are not likely to be rejected. Different sites are privileged for different reasons. The brain lacks lymphatic vessels, and its blood vessel walls are impermeable to lymphocytes such as T cells. The cornea also lacks extensive blood vessels, while the eyes and the testes contain naturally high levels of immunosuppressive molecules. Other sites either lack dendritic cells or express low levels of MHC molecules, so antigen processing does not occur.

Why Fetuses Are Not Rejected The uterus is not a privileged site—allographs placed in the uterine wall experimentally are readily rejected. Given that a fetus has different MHC genes and proteins than those of its mother, why is a fetus in the uterus not rejected like an allograft?

Immunological rejection of a fetus is prevented by the combined activities of many different immunosuppressive mechanisms:

- Early embryos do not express MHC class I or II molecules on the placental layer that is in contact with maternal tissues.
- Cytokines that enhance MHC expression have no effect on placental cells.
- T cells are prevented from functioning in the placenta to reject the fetus. The reason is that certain fetal placental cells produce *indoleamine 2,3-dioxygenase,* an enzyme, which destroys tryptophan, an amino acid that T cells cannot synthesize for themselves. The lack of available tryptophan in their environment prevents T cell division and promotes their destruction via apoptosis.

Donor-Recipient Matching and Tissue Typing

Although it is usually not difficult to ensure that donor and recipient have identical blood groups, MHC compatibility is much harder to achieve. The reason is the very high degree of MHC variability, which ensures that unrelated individuals differ widely in their MHCs. In general, the more closely donor and recipient are related, the smaller will be their MHC difference. Given that in most cases an identical sibling is not available, it is usually preferable that grafts are donated by a parent or sibling possessing MHC antigens similar to those of the recipient.

In practice, of course, closely related donors are not always available. And even though a paired organ (such as a kidney) or a portion of an organ (such as liver tissue) may be available from a living donor, unpaired organs (such as the heart) almost always come from an unrelated cadaver. In such cases, attempts are made to match donor and recipient as closely as possible by means of *tissue typing.* Physicians examine the white cells of potential graft recipients to determine what MHC proteins they have. When a donor organ becomes available, it too is typed. Then the individual whose MHC proteins most closely match those of the donor is chosen to receive the graft. Though a perfect match is rarely achieved, the closer the match, the less intense will be the rejection process, and the greater the chance of successful grafting. A match of 50% or less of the MHC proteins is usually acceptable for most organs, but near absolute matches are required for successful bone marrow transplants.

Graft-versus-Host Disease

Bone marrow allografts are often used as a component of the treatment for cancers of leukocytes (leukemias and lymphomas). In this procedure, physicians use total body

Table 18.3 The Characteristics of the Four Types of Hypersensitivity Reactions

Descriptive	Name	Cause	Time Course	Characteristic Cells Involved
Type 1	Immediate hypersensitivity	IgE on sensitized cells' membranes binds antigen, causing degranulation	Seconds to minutes	Mast cells, basophils, and eosinophils
Type II	Cytotoxic hypersensitivity	Antibodies and complement lyse target cells	Minutes to hours	Red blood cells
Type III	Immune-complex mediated hypersensitivity	Nonphagocytized immune complexes trigger mast cell degranulation	Several hours	Neutrophils
Type IV	Delayed hypersensitivity	T_C cells attack the body's cells	Several days	Activated T cells

irradiation combined with cytotoxic drugs to kill tumor cells in the patient's bone marrow. In the process, the patient's existing leukocytes and all stem cells are also destroyed, which completely eliminates the body's ability to mount any kind of immune response. Physicians then inject the patient with donated stem cells, which migrate to the bones and multiply to produce a new set of leukocytes. Ideally, within a few months, a fully functioning bone marrow is restored.

Unfortunately, when they are transplanted, the donated bone marrow cells may regard the patient's cells as foreign, mounting an immune response against them and giving rise to a condition called **graft-versus-host disease.** If the donor and recipient mainly differ in MHC class I molecules, the grafted T cells attack all of the recipient's tissues, producing especially destructive lesions in the skin and intestine. If the donor and recipient differ mainly in MHC class II molecules, then the grafted T cells attack the antigen-presenting cells of the host, leading to immunosuppression, leaving the recipient vulnerable to infections. The same immunosuppressive drugs used to prevent graft rejection (discussed shortly) can limit graft-versus-host disease.

The major characteristics of the four types of hypersensitivity are summarized in **Table 18.3.** Next we turn our attention to the actions of various types of immunosuppressive drugs used in situations in which the immune system is overactive.

CRITICAL THINKING

A patient arrives at the doctor's office with a rash covering her legs. How could you determine if the rash were a type I or a type IV hypersensitivity?

The Actions of Immunosuppressive Drugs

The development of highly potent immunosuppressive drugs has played a large role in the dramatic success of modern transplantation procedures. Note that these drugs can also be effective in combating certain autoimmune diseases (discussed shortly). In the following discussion we consider four classes of immunosuppressive drugs: corticosteroids, cytotoxic drugs, immunophilins, and lymphocyte-depleting therapies.

Corticosteroids, commonly called *steroids,* have been used as immunosuppressive agents for many years. Corticosteroids such as prednisone (pred'ni-sōn) and methylprednisone suppress the response of T cells to antigen and inhibit such mechanisms as T cell cytotoxicity and cytokine production. These drugs have a much smaller effect on B cell function. Synthetic glucocorticoids also suppress inflammation by inhibiting increases in vascular permeability and vasodilatation, and preventing the production of leukotrienes and prostaglandins.

Cytotoxic (sī-tō-tok'sik) **drugs** inhibit mitosis and cytokinesis (cell division). Given that lymphocyte proliferation is a key feature of specific immunity, blocking cellular reproduction is a potent, although very nonspecific, method of immunosuppression. Among the cytotoxic drugs currently in common use, the following three drugs are noteworthy:

- *Cyclophosphamide* (sī-klō-fos'fă-mīd) cross-links daughter DNA molecules in mitotic cells, preventing their separation and blocking mitosis. It impairs both B cell and T cell responses.
- *Azathioprine* (ā-ză-thī'ō-prēn) is a purine analog; it competes with purines during the synthesis of nucleic acids, thus blocking DNA replication and suppressing both primary and secondary antibody responses.
- *Methotrexate* (meth-ō-trek'sāt) is a folic acid antagonist that inhibits the production of thymidine and purine nucleotides, thereby suppressing DNA replication, RNA transcription, and (indirectly) antibody formation.

Three newly developed cytotoxic drugs used for immunosuppression include *mycophenolate mofetil,* which inhibits purine synthesis, and *brequinar sodium* and *leflunomide,* which inhibit pyrimidine synthesis.

Drugs called **immunophilins** (im'u-nō-fil'inz) constitute a family of potent inhibitors of T cell function. The most important of these drugs is *cyclosporine* (sī-klō-spōr'ēn), a polypeptide derived from fungi. It prevents T cell production of interleukins and interferons, thereby blocking T_H1 responses. Because cyclosporine acts only on activated T cells and has no effect on resting T cells, it is far less toxic than the nonspecific drugs just described. When given to prevent allograft rejection, only activated T cells attacking the graft are suppressed. Because steroids have a similar effect, the combination of corticosteroids and cyclosporine is especially potent and can enhance survival of allografts.

Scientists have developed techniques involving relatively specific *lymphocyte-depleting therapies* in an attempt to reduce the many adverse side effects associated with the use of less-specific immunosuppressive drugs. One technique involves administering an antiserum called *antilymphocyte globulin,* which is specific for lymphocytes. Another, even more specific antilymphocyte technique uses monoclonal antibodies against CD3, which is found only on T cells. An even more specific monoclonal antibody is directed against the interleukin 2 receptor (IL-2R), which is expressed only on activated lymphocytes. Such immunosuppressive therapies are effective in reversing graft rejection, and because they target a narrow range of cells, they produce fewer undesirable side effects than less-specific drugs.

Table 18.4 on page 524 lists some drugs, and their actions, for each of the four classes of the immunosuppressive drugs.

Autoimmune Diseases

Just as today's military must control its arsenal of sophisticated weapons to avert losses of its own soldiers from "friendly fire," the immune system must be carefully regulated so that it does not damage the body's own tissues. However, the immune system does occasionally produce antibodies and cytotoxic T cells that target normal body cells—a phenomenon called *autoimmunity.* Although such responses are not always damaging, they can give rise to **autoimmune diseases,** some of which are life threatening.

Causes of Autoimmune Diseases

Learning Objective

✓ Briefly discuss seven hypotheses concerning the causes of autoimmunity and autoimmune diseases.

Most autoimmune diseases appear to develop spontaneously and at random, with no obvious predisposing cause. Nevertheless, scientists have noted some common features of autoimmune disease. For example, they occur more often in older individuals, and they are also much more common in women than in men, although the reasons for this gender difference are unclear.

Hypotheses to explain the etiology of autoimmunity abound. They include the following:

- Estrogen may stimulate the destruction of tissues by cytotoxic T cells.
- During pregnancy, some maternal cells may cross the placenta and colonize the fetus; because these cells are more likely to survive in a daughter than in a son (because they respond to female hormones), they might trigger an autoimmune disease in a daughter later in her life. Fetal cells may also cross the placenta and trigger autoimmunity in the mother.
- Environmental factors may contribute to the development of autoimmune disorders. Some autoimmune diseases—type I diabetes mellitus and rheumatoid arthritis, for example—develop in a few patients following their recovery from viral infections, though other individuals who develop autoimmune diseases have no history of such viral infections.
- Genetic factors may play a role in autoimmune diseases. MHC genes that in some way promote autoimmunity are found in individuals with autoimmune diseases, whereas MHC genes that somehow protect against autoimmunity may dominate in other individuals. MHC genes may also determine susceptibility to autoimmune disease by regulating the elimination of self-reactive T cells in the thymus.
- Some autoimmune diseases may develop when T cells encounter self-antigens that are normally "hidden" in sites where T cells rarely go. For example, because sperm develop within the testes during puberty, long after the body has selected its T cell population, men may have T cells that recognize their own sperm as foreign. This is normally of little consequence because sperm are sequestered from the blood, but should the testes be injured, T cells may enter the site of damage and mount an autoimmune response against the sperm, resulting in male infertility.
- Infections with a variety of microorganisms may trigger autoimmunity as a result of **molecular mimcry,** which occurs when an infectious agent has an antigenic determinant that is very similar or identical to a self-antigen. In responding to the invader, the body produces *autoantibodies* (antibodies against self-antigens) that damage body tissues. For example, children infected with some strains of group A *Streptococcus pyogenes* (strep-tō-kok'ŭs pī-aj'en-ēz) may produce antibodies to heart muscle and so develop heart disease. Other strains of streptococci trigger the production of antibodies that cross-react with glomerular basement membranes and so cause kidney disease. It is possible that some virally triggered autoimmune diseases may also result from molecular mimicry.
- Other autoimmune responses may result from failure of the normal control mechanisms of the immune system. Thus even though harmful self-reactive T lymphocytes are normally destroyed in the thymus via apoptosis triggered through the cell surface receptor CD95 (see Chapter 16), defects in CD95 can cause autoimmunity by permitting abnormal T cells to survive and cause disease.

Table 18.4 The Four Classes of Immunosuppressive Drugs

Class	Examples	Action
Corticosteroids	Prednisone, methylprednisone	Anti-inflammatory; kills T cells
Cytotoxic drugs	Cyclophosphamide, azathioprine, methotrexate, mycophenolate mofetil, brequinar sodium, leflunomide	Blocks cell division nonspecifically
Immunophilins	Cyclosporine	Blocks T cell responses
Lymphocyte-depleting therapies	Antilymphocyte globulin, monoclonal antibodies	Kills T cells nonspecifically, kills activated T cells

Regardless of the specific mechanism that causes an autoimmune disease, immunologists categorize them into two major categories: single tissue autoimmune diseases, which affect a single organ or tissue, and systemic autoimmune diseases.

Single Tissue Autoimmune Diseases

Learning Objective

✓ Describe a serious autoimmune disease associated with each of the following: blood cells, endocrine glands, and nervous tissue.

Among the many single tissue autoimmune diseases recognized, common examples affect blood cells, endocrine glands, or nervous tissue.

Autoimmunity Affecting Blood Cells

Some patients make autoantibodies to their leukocytes. As a result, they have difficulty combating infections. Other patients make autoantibodies to blood platelets, and as a result their blood will not clot.

As we have seen, individuals with hemolytic anemia produce antibodies against their own red blood cells. These autoantibodies hasten the destruction of the red blood cells, and the patient becomes severely anemic. Different hemolytic anemia patients make antibodies of different classes. Some patients make IgM autoantibodies, which bind to red blood cells and activate the classical complement pathway; the red blood cells are lysed, and degration products from hemoglobin released into the bloodstream are excreted in the urine. Other hemolytic anemia patients make IgG autoantibodies, which are less effective in activating complement and instead serve as opsonins that promote phagocytosis. In this case, red blood cells are removed by macrophages in the liver, spleen, and bone marrow. Even though these patients have no hemoglobin in their urine, they are still severely anemic.

The precise causes of all cases of autoimmune hemolytic anemia are unknown, but some cases follow infections with viruses or treatment with certain drugs, both of which alter the surface of red blood cells such that they are recognized as foreign and trigger an immune response.

Autoimmunity Affecting Endocrine Organs

Other common targets of autoimmune attack are the endocrine (hormone-producing) organs. Patients can develop autoantibodies or produce T cells against cells of the islets of Langerhans within the pancreas, against cells of the thyroid gland, or against cells of other endocrine glands. In most cases, the ensuing autoimmune reaction results in damage to or destruction of the gland, and in hormone deficiencies as endocrine cells are killed.

Immunological attack on the islets of Langerhans results in a loss of the ability to produce the hormone *insulin,* which leads to the development of **type I diabetes mellitus** (dī-ă-bē′tēz mĕ-lī′tēs) (also known as juvenile-onset diabetes). As with other autoimmune diseases, the exact cause of type I diabetes is unknown, but many patients endured a severe viral infection some months before the onset of diabetes. Additionally, some patients are known to have a genetic predisposition to developing type I diabetes that is associated with the possession of certain class I MHC molecules. Some physicians have been successful in delaying the onset of type I diabetes by treating at-risk patients with immunosuppressive drugs before damage to the islets of Langerhans becomes apparent.

An autoimmune response can lead to stimulation rather than to inhibition or destruction of glandular tissue. An example of this is **Graves' disease,** which involves the thyroid gland—a major endocrine gland located in the neck region. Affected patients make autoantibodies that bind to and stimulate receptors on the cytoplasmic membranes of cells in the anterior pituitary gland. The stimulated pituitary cells produce *thyroid-stimulating hormone,* which elicits excessive production of *thyroid hormone* and growth of the thyroid gland. Such patients develop a *goiter* (enlarged thyroid gland), protruding eyes, rapid heartbeat, fatigue, and weight loss despite increased appetite. Physicians treat Graves' disease by surgically removing excess thyroid tissue or by administering immunosuppressive drugs.

Autoimmunity Affecting Nervous Tissue

Of the group of autoimmune diseases affecting nervous tissue, the most important is **multiple sclerosis** (sklĕ-rō′sis) **(MS).** In this disease, cytotoxic T cells attack and destroy the *myelin sheaths* that normally insulate brain and spinal cord neurons and increase the speed of nerve impulses along the length of the neurons. Consequently, MS patients experience deficits in vision, speech, and neuromuscular function that may be quite mild and intermittent, or may ultimately lead to death. Like other autoimmune diseases, MS may be triggered by a viral infection in individuals with certain genetic backgrounds.

Systemic Autoimmune Diseases

Learning Objective

✓ Discuss two important systemic autoimmune diseases.

Two classic examples of autoimmune diseases that affect multiple organs or the entire body are systemic lupus erythematosus and rheumatoid arthritis.

Systemic Lupus Erythematosus

Systemic lupus erythematosus (loo′pŭs er-ĭ-thē′mă-tō-sus) **(SLE),** often shortened to *lupus,* is a generalized immunologic disorder that results from a loss of control of both humoral and cell-mediated immune responses. Affected individuals make autoantibodies against numerous antigens found in normal organs and tissues, giving rise to many different pathological lesions and clinical manifestations. One consistent feature of SLE is the development of autoantibodies against nucleic acids, especially DNA. These autoantibodies combine with free DNA released from dead cells to form immune complexes that are deposited in glomeruli, causing glomerulonephritis and kidney failure. Thus, SLE results from a type III (immune-complex mediated) hypersensitivity reaction. Immune complexes may also be deposited in joints, where they provoke arthritis.

The disease's curious name—systemic lupus erythematosis—stems from two features of the disease: *Systemic* simply reflects the fact that it affects different organs throughout the body; *lupus* is Latin for wolf, and *erythematosus* refers to a redness of the skin, so these latter two words describe the characteristic red, butterfly-shaped rash that develops on the face of many patients, giving them what is sometimes described as a wolf-like appearance (**Figure 18.12**). This rash is caused by deposition of nucleic acid–antibody complexes in the skin, and is worse in skin areas exposed to sunlight.

Although autoantibodies to nucleic acids are characteristic of SLE, many other autoantibodies are also produced. Autoantibodies to red blood cells cause hemolytic anemia; autoantibodies to platelets give rise to bleeding disorders; antilymphocyte antibodies alter immune reactivity; and antibodies against muscle cells provoke muscle inflammation and, in some cases, damage to the heart. Because of its variety of symptoms, SLE can be misdiagnosed.

The cause or causes of lupus are unknown. In experimental animal models, a lupus-like disease can be induced by some drugs, viral infections, or genetic mutations. It is likely that SLE in humans also has many causes. Some patients are known to develop lupus in response to a complement deficiency; the absence of certain complement components prevents the removal of immune complexes from the circulation, leading to their deposition in the kidneys and joints.

Physicians treat lupus with immunosuppressive drugs that reduce autoantibody formation, and with corticosteroids that reduce the inflammation associated with the deposition of immune-complexes.

Rheumatoid Arthritis

Rheumatoid arthritis (roo′mă-toyd ar-thrī′tis) is another crippling, systemic, autoimmune disease resulting from a type III hypersensitivity reaction. This disease commences when B cells in the joints produce autoantibodies against collagen, a protein in cartilage, which commonly covers joint surfaces. The resulting immune-complexes and complement bind to mast cells, which release the inflammatory chemicals in their granules. The resulting inflammation causes the tissues to swell, thicken, and proliferate into the joint. As the altered tissue extends into the joint, the inflammation further erodes and destroys joint cartilage and the neighboring bony structure, until the joints begin to break down and fuse; as a result, affected joints become distorted and lose their range of motion **(Figure 18.13).** The course of rheumatoid arthritis is often intermittent; however, with each recurrence, the lesions and damage get progressively more severe.

As with other autoimmune diseases, the cause or causes of rheumatoid arthritis is not well understood. The fact that there are no animal models (because the disease appears to affect humans only) significantly hinders research on its cause. Many cases demonstrate that rheumatoid arthritis

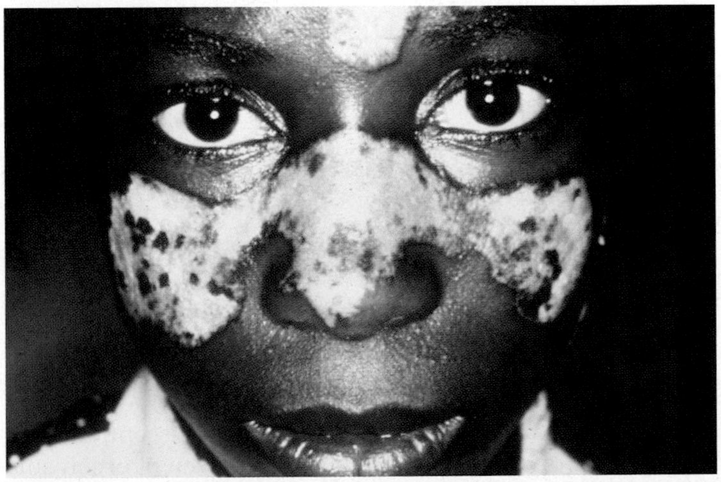

▲ ***Figure 18.12***
The characteristic facial rash of systemic lupus erythematosus. Its shape corresponds to areas most exposed to sunlight, which worsens the condition.

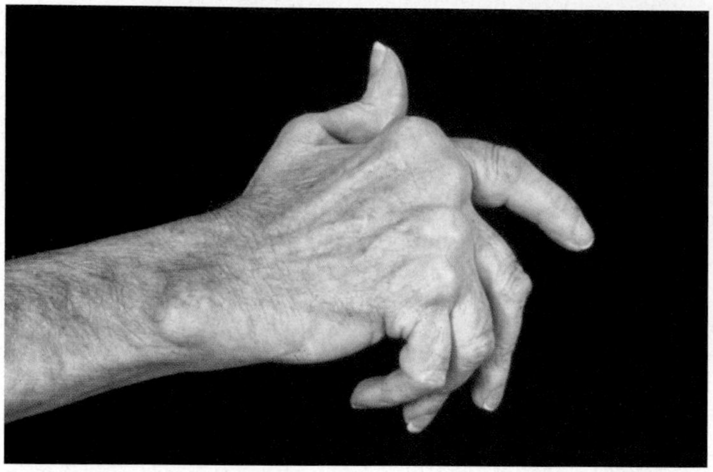

▲ ***Figure 18.13***
The crippling distortion of joints characteristic of rheumatoid arthritis.

commonly follows an infectious disease in a genetically susceptible individual. Possession of certain MHC genes appears to promote susceptibility.

Physicians treat rheumatoid arthritis by administering anti-inflammatory drugs such as corticosteroids to prevent additional joint damage and methotrexate to inhibit the humoral immune response.

CRITICAL THINKING

A 43-year-old woman has been diagnosed with rheumatoid arthritis. Unable to find relief from her symptoms, she seeks treatment from a doctor in another country who injects a daily regimen of antibodies and complement into her afflicted joints. Do you expect the treatment to improve her condition? Explain your reasoning.

Immunodeficiency Diseases

Learning Objective

✓ Differentiate between primary and acquired immunodeficiencies, and cite two diseases associated with each form of immunodeficiency.

You may have noticed that during periods of increased emotional or physical stress, you are more likely to succumb to a cold or a bout of the flu. You are not imagining this phenomenon; stress has long been known to decrease the efficiency of the immune system. Similarly, chronic defects in the immune system typically first become apparent when affected individuals show increased susceptibility to infections of opportunistic pathogens (or even of organisms not normally considered pathogens). Such opportunistic infections, which were discussed in Chapter 14, are the hallmarks of *immunodeficiency* (im'ū-nō-dē-fish'en-sē) diseases, which are conditions resulting from defective immune mechanisms.

Researchers have characterized a large number of immunodeficiency diseases in humans, which are of two general types:

- **Primary immunodeficiency diseases,** which are detectable near birth and develop in infants and young children, result from some genetic or developmental defect.
- **Acquired (secondary) immunodeficiency diseases** develop in later life (in older children, adults, and the elderly) as a direct consequence of some other recognized cause, such as malnutrition, severe stress, or infectious disease.

Next we consider each general type of immunodeficiency disease in turn.

Primary Immunodeficiency Diseases

Many different inherited defects have been identified in all of the body's lines of defense, affecting first and second lines of defense as well as humoral and cell-mediated immune responses.

One of the more important inherited defects in the second line of defense is **chronic granulomatous disease,** in which children have recurrent infections characterized by the development of large masses of inflammatory cells in lymph nodes, lungs, bones, and skin. The reason is that these children have neutrophils that are unable to kill ingested bacteria due to an inherited biochemical defect in one of the polypeptides of *NADPH oxidase,* an enzyme that is crucial in the increased metabolic activity that occurs during phagocytosis.

Most primary immunodeficiencies are associated with defects in the components of the third line of defense—specific immunity. For example, some children fail to develop any lymphoid stem cells whatsoever, and as a result they produce neither B cells nor T cells and cannot mount any sort of immune response. The resulting defects in the immune system cause **severe combined immunodeficiency disease (SCID),** which is discussed in **Highlight 18.1.**

Other children suffer from T cell deficiencies alone. For example, **DiGeorge anomaly** results from a failure of the thymus to develop. In consequence, there are no T cells. The importance of T cells in protecting against viruses is emphasized by the observation that individuals with DiGeorge anomaly generally die of viral infections while remaining resistant to most bacteria. Physicians cure DiGeorge anomaly with a thymic stem cell transplant.

B cell deficiencies also occur in children. The most severe of the B cell deficiencies, called **Bruton-type agammaglobulinemia**[6] (ā-gam'ă-glob'ū-li-nē'mē-ă), is an inherited disease in which affected babies, usually boys, cannot make immunoglobulins. These children experience recurrent bacterial infections but are usually resistant to viral, fungal, and protozoan infections.

[6]Agammaglobulinemia means an absence of gamma globulin (IgG).

Highlight 18.1 SCID: "Bubble Boy" Disease

David Vetter

As a result of the widely publicized case of David Vetter—a Houston, Texas resident commonly known as the "bubble boy" because he lived from birth until age 12 in a sterile, plastic-enclosed environment—*severe combined immunodeficiency disease (SCID)* has been known as "bubble boy" disease since the 1970s. Like all SCID patients, David had no immune system—his body produced neither B cells nor T cells—so the slightest infection could be lethal. As a result, contact with him could occur only through a pair of antiseptic rubber gloves built into one of the walls of his plastic enclosure. At age 12, David underwent an experimental bone marrow transplant in the hope that marrow donated by his older sister would enable him to build up an immune system. Unfortunately, the donated cells turned out to be infected with Epstein-Barr virus (a common virus that causes mononucleosis), ultimately killing David.

Today, there is much more hope for patients with SCID. Virus-free bone marrow transplants are now possible. More recently, some children have been treated via gene therapy, in which the gene for an enzyme (adenine deaminase) missing in SCID patients is inserted, via retroviral vectors, into clones of the patient's bone marrow cells. These genetically altered cells are then returned to the child, where they grow and synthesize sufficient enzyme to normalize the immune response. While successful in several children studied in a French trial, the therapy was halted in September 2002 after one child developed a leukemia-like illness following treatment. Investigations into whether or not the illness was caused by the gene therapy are underway; however, scientists remain hopeful that gene therapy can achieve a cure for SCID—without side effects—in the near future.

Inherited deficiencies of individual immunoglobulin classes are more common than a deficiency of all classes. Among these, IgA deficiency is the most common. Because affected children cannot produce secretory IgA, they experience recurrent infections in the respiratory and gastrointestinal tracts.

CRITICAL THINKING

Two boys have autoimmune diseases: One has Bruton-type agammaglobulinemia, and the other has DiGeorge anomaly. On a camping trip, each boy is stung by a bee and each falls into poison ivy. What hypersensitivity reactions might each boy experience as a result of his camping mishaps?

Table 18.5 summarizes the major primary immunodeficiency diseases.

Acquired Immunodeficiency Diseases

Learning Objectives

- ✓ Describe five acquired conditions that suppress immunity.
- ✓ Define AIDS and differentiate between a disease and a syndrome.
- ✓ Describe HIV, including its structure, possible origin, replication cycle, and transmission.
- ✓ Describe the relationship of the helper T cell population to the course of AIDS.
- ✓ Describe diagnosis, treatment, and prevention of AIDS and list four behaviors that increase the risk of infection with HIV.

Unlike inherited primary immunodeficiency diseases, acquired immunodeficiency diseases affect older individuals who were born with a healthy immune system.

Acquired immunodeficiencies result from a number of causes. In all humans the immune system (but especially T cell production) deteriorates with increasing age; as a result, older individuals normally have less effective immunity, especially cell-mediated immunity, than younger individuals, leading to an increased incidence of both viral diseases and certain types of cancer. Severe emotional or physical stress can also lead to immunodeficiencies by prompting the secretion of increased quantities of corticosteroids, which are toxic to T cells and thus suppress cell-mediated immunity. This is why, for example, cold sores may "break out" on the faces of students during final exams: The causative herpes simplex viruses, which are controlled by a fully functioning cell-mediated immune response, escape immune control in stressed individuals. Malnutrition and certain environmental toxins can also cause acquired immunodeficiency diseases by inhibiting the normal production of B cells and T cells.

Of course, the most significant example of an acquired immunodeficiency is **acquired immunodeficiency syndrome (AIDS).** From the time of its discovery in 1981 among homosexual males in the United States to its emergence as a worldwide pandemic, no affliction has affected modern life as much as AIDS.

Signs and Symptoms of AIDS

AIDS is not a single disease but a **syndrome;** that is, a group of signs and symptoms associated with a common pathology. Currently, epidemiologists define AID syndrome as the presence of several opportunistic or rare infections associated with a positive test showing the presence of human immunodeficiency virus (HIV). The infections include diseases of the skin, such as shingles and herpes **(Figure 18.14a);**

Table 18.5 Primary Immunodeficiency Diseases

Disease	Defect	Manifestation
Chronic granulomatous disease	Ineffective neutrophils	Uncontrolled infections
Severe combined immunodeficiency disease	A lack of T cells and B cells	No resistance to any type of infection, leading to rapid death
Bruton-type agammaglobulinemia	A lack of B cells and thus a lack of immunoglobulins	Death from overwhelming bacterial infections
DiGeorge anomaly	A lack of T cells and thus no cell-mediated immunity	Death from overwhelming viral infections

diseases of the nervous system, including meningitis, toxoplasmosis, and *Cytomegalovirus* (sī-tō-meg'ă-lō-vī'rŭs) disease; diseases of the respiratory system, such as tuberculosis, *Pneumocystis* (nū-mō-sis'tis) pneumonia, histoplasmosis, and coccidioidomycosis; and diseases of the digestive system, including chronic diarrhea, thrush, and oral hairy leukoplakia. A rare cancer of blood vessels called Kaposi's sarcoma is also commonly seen in AIDS patients **(Figure 18.14b).** AIDS often results in dementia during the final stages.

Table 18.6 summarizes these and other diseases associated with AIDS.

AIDS Pathogen and Its Virulence Factors

Human immunodeficiency virus (HIV) causes AIDS. HIV is an enveloped, positive single-stranded RNA (+ssRNA) virus of the type called *retroviruses,* because it uses **reverse transcriptase** to make a DNA copy of its genome. Luc Montagnier (1932–) and his colleagues at the Pasteur Institute discovered HIV in 1983 and originally named it *lymphadenopathy-associated virus;* today, taxonomists classify HIV in the genus *Lentivirus,* family *Retroviridae.* Note that the acronym HIV indicates that it is a virus; therefore, it is redundant and incorrect to call this pathogen "HIV virus." A more appropriate name is "AIDS virus," or simply "HIV."

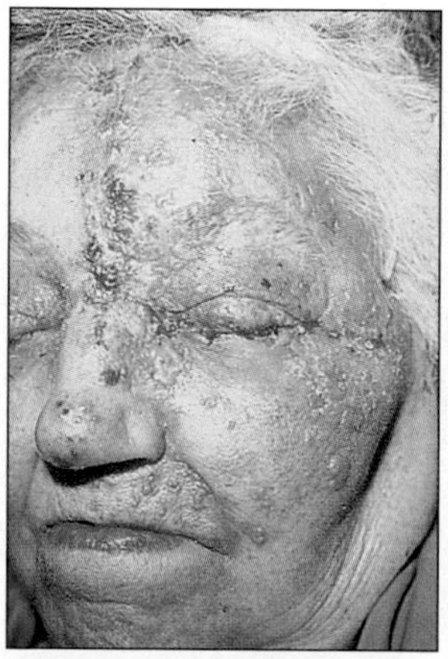

(a)

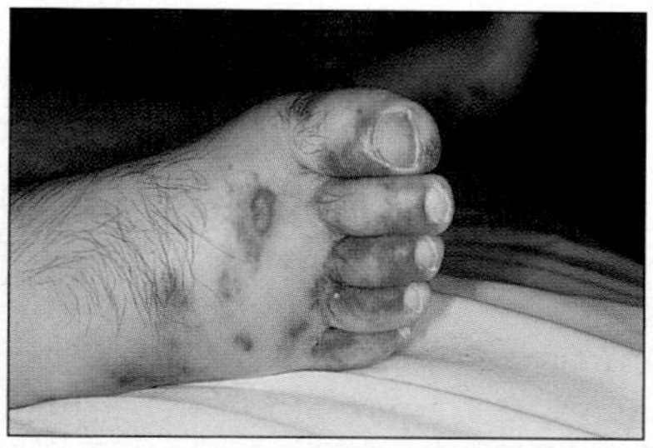

(b)

▲ *Figure 18.14*
Diseases associated with AIDS. **(a)** Disseminated herpes. **(b)** Kaposi's sarcoma, a cancer of blood vessels.

There are two major types of HIV. HIV-1 is more prevalent in the United States and Europe, while HIV-2 is more prevalent in West Africa. HIV-2, which shares about 50% of the nucleic acid sequence of HIV-1, reproduces more slowly than HIV-1. Researchers have studied HIV-1 more, so the following sections focus on this strain.

Structure of HIV HIV is a typical retrovirus in size and shape **(Figure 18.15).** Two antigenic glycoproteins characterize its envelope. The larger glycoprotein, named **gp120**[7], is the primary attachment molecule of HIV. Its antigenicity changes during the course of prolonged infection, making an effective antibody response against it very difficult. The smaller glycoprotein, **gp41,** promotes fusion of the viral envelope to a target cell. The effects of these structural characteristics—antigenic variability and the ability to fuse with host cells—impede immune clearance of HIV from a patient.

Origin of HIV Evidence suggests that HIV arose from mutation of a similar virus—simian immunodeficiency virus (SIV)—found in African chimpanzees; the nucleotide sequence of SIV is similar to that of HIV. Based on mutation rates and the rate of antigenic change in HIV, researchers estimate that HIV emerged in the human population in Africa about 1930. Scientists have identified antibodies against HIV in human blood stored since 1959, though they did not document the first cases of AIDS until 1981.

The relationship among the two types of HIV and SIV is not clear. HIV-1 and HIV-2 may be derived from different strains of SIV, or one type of HIV may be derived from the other. We may never know the exact temporal relationships among the immunodeficiency viruses.

Replication of HIV HIV replicates like a typical animal retrovirus in five steps:

(1) **Attachment.** HIV attaches via gp120 to a molecule on a host cell known as **CD4**[8], which acts as a receptor for

[7]gp120 is short for "glycoprotein with a molecular weight of 120,000 daltons."
[8]Scientists commonly name membrane proteins with letter-number combinations; in this case, the name stands for the fourth discovered cluster of differentiation glycoprotein.

Table 18.6 Opportunistic Infections Associated with AIDS

Disease	Causative agent	Organ primarily affected
Coccidioidomycosis	*Coccidioides* (fungus)	Lung
Cytomegalovirus disease	*Cytomegalovirus*	Brain, liver
Diarrhea (severe and prolonged)	Various bacteria, *Cryptosporidium* (protozoan)	Intestines
Herpes	*Herpesvirus*	Skin
Histoplasmosis	*Histoplasma* (fungus)	Lung
Kaposi's sarcoma	Human herpes virus 8	Blood vessels
Meningitis	*Cryptococcus* (yeast), *Listeria* (bacterium)	Brain and meninges
Oral hairy leukoplakia	*Lymphocryptovirus* (Epstein-Barr virus)	Tongue
Pneumonia	*Pneumocystis* (fungus)	Lung
Shingles	*Varicellovirus*	Skin
Thrush	*Candida* (yeast)	Mouth and tongue, vagina
Toxoplasmosis	*Toxoplasma* (protozoan)	Brain
Tuberculosis	*Mycobacterium*	Lung

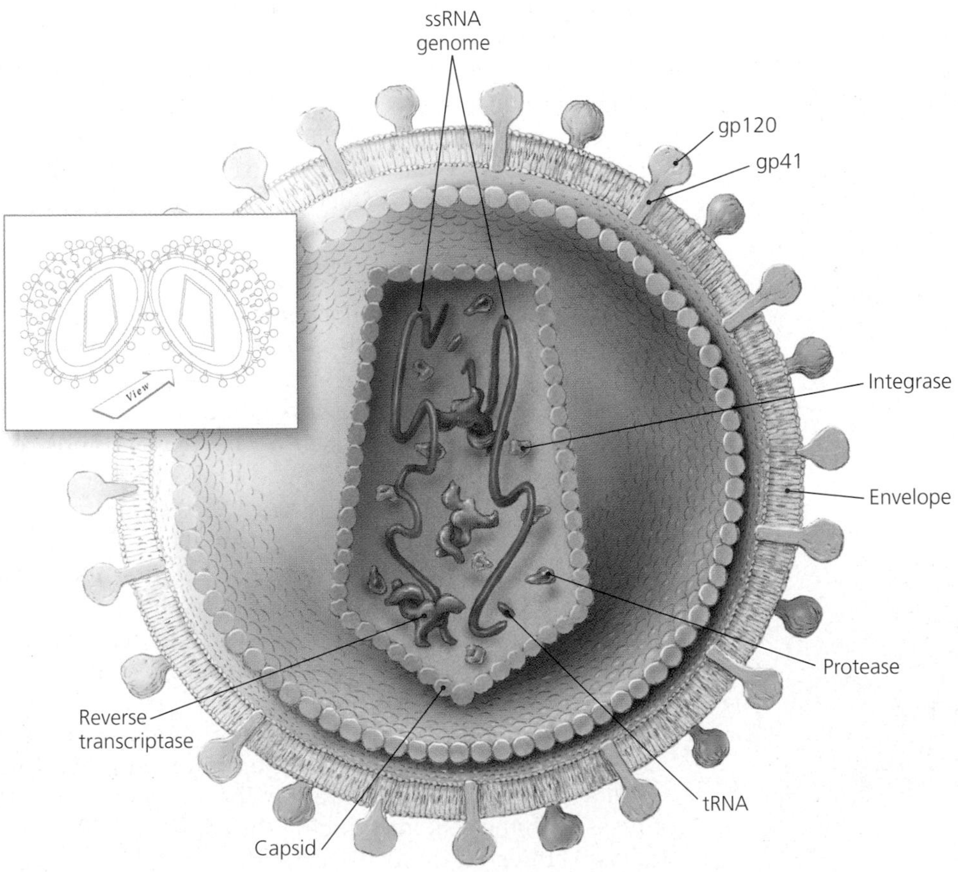

Figure 18.15 Artist's conception of HIV. The virion is about 90 nm in diameter.

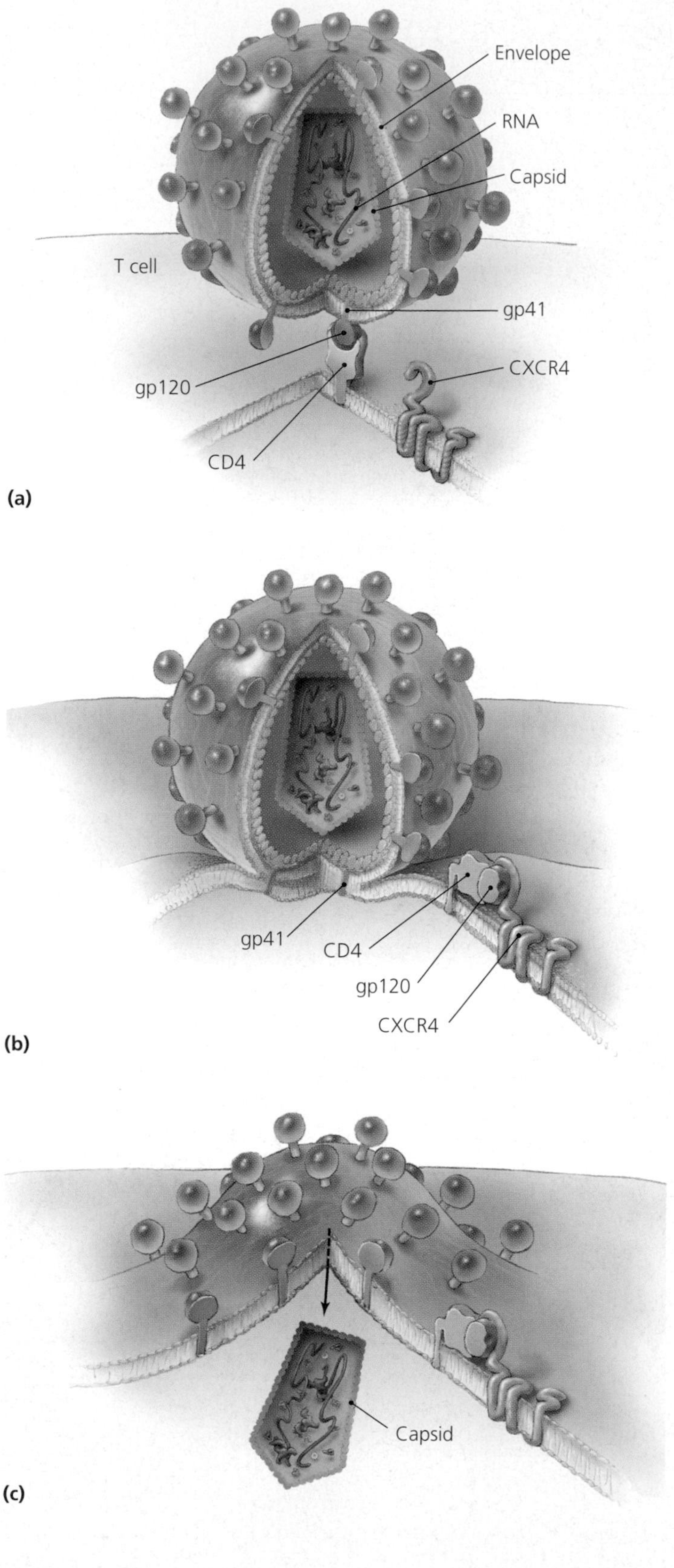

Figure 18.16
The process by which HIV attaches to and enters a host cell. **(a)** Binding of gp120 to a CD4 receptor on the cell membrane. **(b)** Removal of the CD4-gp120 complex, allowing gp41 to attach to the cytoplasmic membrane. **(c)** Fusion of the lipid bilayers, introducing the capsid into the cytoplasm. *Which cells have CXCR4 receptors (as shown here)? Which cells have CCR5 receptors?*

Figure 18.16 *Helper T cells have CXCR4 receptors, whereas cells of macrophage lineage and smooth muscle cells have CCR5 receptors.*

HIV **(Figure 18.16a).** HIV primarily attaches to four kinds of cells: helper T cells; cells of the macrophage lineage, including monocytes, macrophages, and microglia; smooth muscle cells, such as those in arterial walls; and dendritic cells. HIV also rarely infects nerve cells, liver cells, and some epithelial cells.

Additionally, B lymphocytes adhere to HIV that has been covered with complement proteins. Though HIV does not infect these cells, B cells deliver HIV to T cells maturing in lymphoid tissues. Similarly, infected macrophages secrete chemotactic chemicals that attract T cells, which become infected when the macrophages pass the virus to them.

(2) **Entry.** The gp120-CD4 complex binds to another receptor on the cell, called a **chemokine receptor,** which removes gp120 from the virion. Different chemokine receptors exist on various target cells—*CXCR4* on most T cells and *CCR5* on macrophages and smooth muscle cells **(Figure 18.16b).** Finally, once a virion binds to a cell, its envelope contacts the cytoplasmic membrane via gp41 and fuses with it **(Figure 18.16c),** introducing the HIV capsid into the cytoplasm. The envelope remains behind as part of the cell's cytoplasmic membrane. Glycoprotein 41, remaining on the cytoplasmic membrane, induces the cell to fuse to as many as 500 neighboring cells to form a **syncytium** (sin-sish′ē-ŭm), a giant multinucleate cell.

HIV enters dendritic cells in a single step when gp120 attaches to **DC-SIGN,** a receptor on a dendritic cell's cytoplasmic membrane. The envelope of HIV fuses with the cell, introducing the capsid into the cytoplasm. Infected dendritic cells deliver HIV to T cells during antigen presentation.

(3) **Synthesis of DNA.** As with all retroviruses, reverse transcriptase transcribes a double-stranded DNA (dsDNA) copy of its +ssRNA genome by catalyzing three nearly simultaneous reactions as shown in **Figure 18.17:** The enzyme synthesizes a −ssDNA copy of the +ssRNA, forming a ssDNA-ssRNA hybrid. Transfer RNA carried

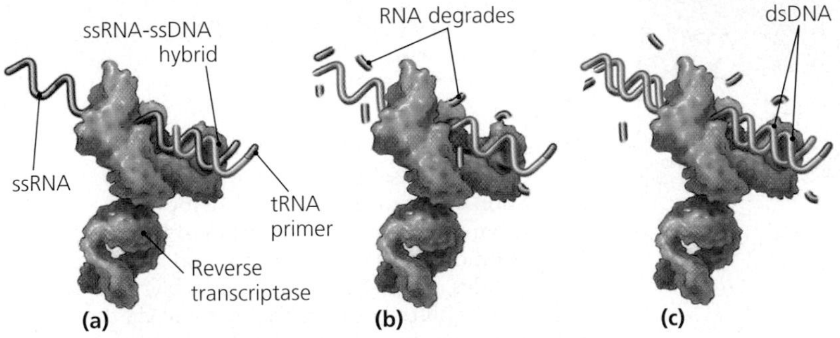

▲ ***Figure 18.17***

Action of reverse transcriptase, depicted here as three distinct steps. **(a)** The enzyme transcribes a complementary −ssDNA molecule from a DNA-RNA hybrid. It uses tRNA brought from its previous cell to use as a primer for transcription. **(b)** The RNA portion of the hybrid is degraded leaving −ssDNA. **(c)** The enzyme synthesizes a complementary +ssDNA strand, forming dsDNA.

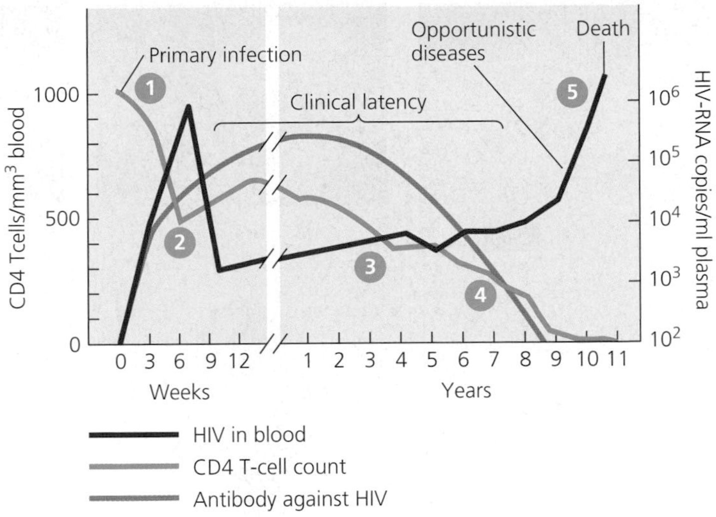

▲ ***Figure 18.18***

The course of AIDS follows the course of helper T cell destruction. The circled numbers correspond to the steps described in the text.

within the capsid from a previous host cell primes this reaction. The enzyme then degrades ssRNA to leave −ssDNA and finally completes the process by synthesizing a complementary strand of DNA, forming dsDNA.

Reverse transcriptase is very error prone, making about five errors per genome, and this generates multiple antigenic variations of HIV. Millions or billions of variants develop in a single patient over the course of the syndrome.

(4) **Integration.** HIV is a latent virus; that is, the dsDNA molecule—known as a *provirus*—enters the nucleus and inserts into a human chromosome by the virion-carried enzyme *integrase*. This occurs within 72 hours of infection. Once integrated, the provirus remains a part of the cellular DNA for life.

(5) **Synthesis of RNA and polypeptides.** An infected cell transcribes and replicates integrated HIV to produce genomic RNA and messenger RNA, which is translated by ribosomes to make viral polypeptides including capsomeres, gp120, gp41, reverse transcriptase, and integrase.

(6) **Release.** Two molecules of genomic RNA along with viral enzymes and two molecules of tRNA (the latter are coded by the host's genome) bud from the cytoplasmic membrane. An infected cell produces 3,000 to 4,000 virions at any one time.

(7) **Assembly.** Virions bud from the cell in a nonvirulent form because their capsids are not assembled and reverse transcriptase is inactive. **Protease** (prō′tē-ās), another viral enzyme packaged in the virion, cleaves a large polypeptide to release reverse transcriptase and capsomeres. This action of protease, which occurs after a virion buds from the cell, allows final assembly of a virulent HIV.

Figure 18.19 on page 532 illustrates the replication cycle of HIV, and **Table 18.7** on page 533 summarizes some of the characteristics that make HIV particularly difficult to combat.

Pathogenesis of AIDS

Only human cells replicate HIV effectively and, as its name indicates, the virus destroys a human's immune system. **Figure 18.18** illustrates the observation that the destruction of helper T cells directly relates to the course of AIDS. Initially, there is a burst of virion production and release from infected cells ①. Fever, fatigue, weight loss, diarrhea, and body aches accompany this primary infection.

The immune system responds by producing antibodies, and the number of free virions plummets (red line) ②. The body and HIV are waging an invisible war with few signs or symptoms. During this period, the body destroys almost a billion virions each day, but the viruses and cytotoxic T cells kill about 100 million CD4 helper T cells. No specific symptoms accompany this stage, and the patient is often unaware of the infection.

Integrated viruses continue to replicate and virions are released into the blood to such an extent that the body cannot adequately replenish the supply of helper T cells. Over the course of 5–10 years, the number of T helper cells declines to a level that severely impairs the immune response (green line) ③. The rate of antibody formation falls precipitously (purple line) as helper T function is lost ④.

HIV production climbs, and the patient dies ⑤. Many of the diseases associated with the loss of immune function in AIDS (see Table 18.6 on page 529) are nonlethal infections in other patients, but AIDS patients cannot effectively resist them. Diseases such as Kaposi's sarcoma, disseminated herpes, toxoplasmosis, and *Pneumocystis* pneumonia occur rarely except in AIDS patients.

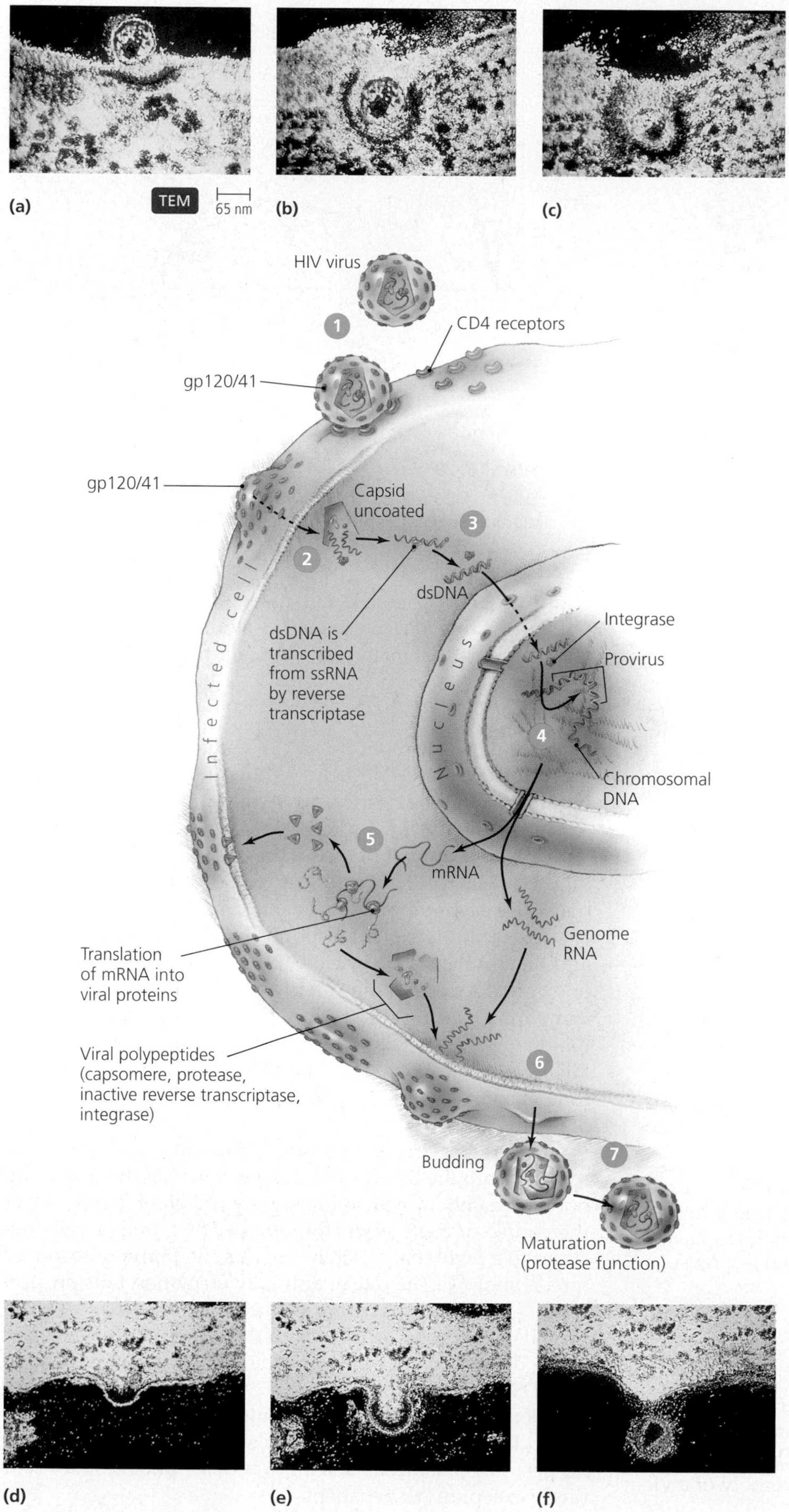

Figure 18.19

The replication cycle of HIV. The artist's rendition depicts seven steps involved in the replication of the virus within a T helper cell: ① attachment, ② entry, ③ synthesis of DNA, ④ integration, ⑤ synthesis of RNA and polypeptides, ⑥ release, and ⑦ assembly, which is completed outside the host cell via the action of protease. Photos **(a)**, **(b)**, and **(c)** show attachment and penetration; photos **(d)**, **(e)**, and **(f)** show the budding and release of a virion. *How does the replication of HIV differ from the replication of bacteriophage T4?*

Figure 18.18 *Bacteriophage T4 is a DNA virus; therefore, it does not have reverse transcription. Further, T4 does not integrate into a bacterium's chromosome; it assembles completely before being released from the cell, it has no envelope, nor does it carry enzymes.*

Table 18.7 Characteristics of HIV That Challenge the Immune System

Characteristic	Effect(s)
Retrovirus with a genome that consists of two copies of +ssRNA	Reassortment of viral genes possible; virus is highly mutable—reverse transcription produces much genetic variation; genome can integrate into host's chromosome
Targets helper T cells especially, but also macrophages, dendritic cells, and muscle cells, and possibly liver, nerve, and epithelial cells	Infects key cells of host's immune system
Antigenic variability	Numerous antigenic variations due to mutations help virus evade host's immune response
Induces formation of syncytia	Increases routes of infection; intracellular site helps virus evade immune detection

Epidemiology of AIDS

Epidemiologists first identified AIDS in young male homosexuals in the United States, but now AIDS is worldwide. The World Health Organization (WHO) estimates as of December 2003 that there are about 40 million HIV-infected individuals worldwide with approximately 14,000 new infections occurring each day. About a third of those infected have already developed AIDS; presumably, most of the rest will eventually succumb to the syndrome. The spread of AIDS in sub-Saharan Africa is particularly horrific (**Figure 18.20** on page 534).

All body secretions of AIDS patients contain HIV; however, the virion typically exists in sufficient concentration to cause infection only in blood, semen, vaginal secretions, and breast milk. Virions may be free or inside infected leukocytes. Infected blood contains 1,000–100,000 virions per milliliter, while semen has 10–50 virions per milliliter. Other secretions have lower concentrations and are less infective than blood or semen.

Infected fluid must be injected into the body or encounter a tear or lesion in the skin or mucous membranes, though such a lesion may be microscopic and not bleed. Because HIV is only about 90 nm in diameter, a break in the protective membranes of the body large enough to allow entry of HIV may be too small to see or feel. Sufficient numbers of virions must be transmitted to target cells to establish an infection, though scientists do not know the exact number of virions needed. The number probably varies with the strain of HIV and the overall health of the patient's immune system.

HIV is transmitted primarily via sexual contact (including vaginal, anal, and oral sex, either homosexual or heterosexual) and intravenous drug use (**Figure 18.21** on page 534). Blood transfusions, organ transplants, hemophiliac therapy, tattooing, and accidental medical needle-sticks also transmit HIV, though rarely. For example, the risk of infection from a needle-stick injury involving an HIV-infected patient is less than 1%. Mothers also transmit HIV to their babies across the placenta and in breast milk—HIV infects approximately one-third of babies born to HIV-positive women.

Behaviors that increase the risk of infection include the following:

- Sexual promiscuity; that is, sex with more than one partner
- Anal intercourse, especially receptive anal intercourse
- Intravenous drug use
- Sexual intercourse with anyone in the previous three categories

The mode of infection is unknown for about 9% of AIDS patients who are often unable or unwilling to answer questions about their sexual and drug use history. The CDC has documented a few cases of casual spread of HIV, including infections from sharing razors and toothbrushes and from mouth-to-mouth kissing. In all cases of casual spread, researchers suspect that small amounts of the donor's blood may have entered abrasions in the recipient's mouth.

Diagnosis, Treatment, and Prevention

Physicians diagnose AIDS based upon unexplained weight loss, fatigue, fever, and fewer than 200 CD4 lymphocytes per microliter (μl)[9] of blood, combined with other signs and symptoms, which vary according to the diseases involved, and the demonstration of antibodies against HIV. Recall that by definition AIDS is the presence of one or more rare diseases and anti-HIV antibodies.

HIV itself is difficult to locate in patient's secretions and blood because it becomes a provirus inserted into chromosomes, and because for years it is kept in check by the immune system. Therefore, serological diagnosis involves detecting antibodies against HIV using ELISA, agglutination, or western blot testing. Most individuals develop antibodies within 6 months of infection, though some remain without detectable antibodies for up to 3 years. A positive test for antibodies does not mean that HIV is currently present or that the patient has or will develop AIDS; it merely indicates that the patient has been exposed to HIV.

A small percentage of infected individuals, called *long-term nonprogressors,* do not develop AIDS even years or decades after infection. It appears that either these individuals are infected with defective virions, they have mutated

[9]The normal value for CD4 cells is 500–700 cells/μl.

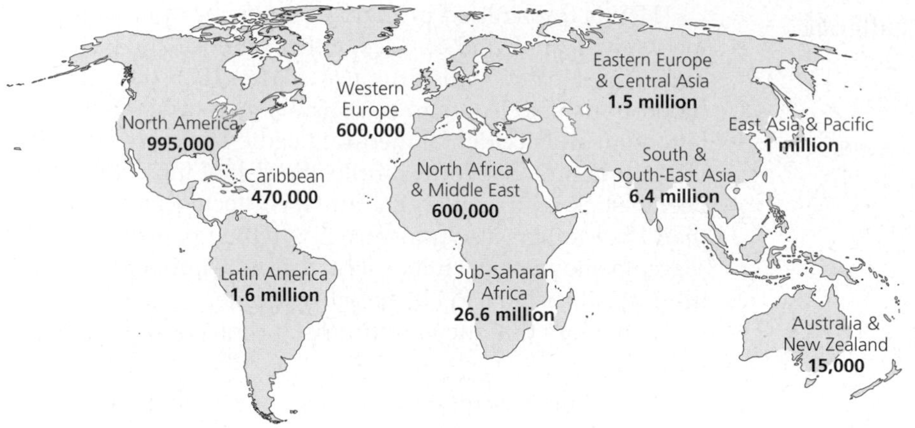

◀ *Figure 18.20*

The global distribution of HIV/AIDS. Figures indicate the estimated number of children and adults living with HIV/AIDS as of the end of 2003. Note the alarmingly high number of people infected in sub-Saharan Africa.

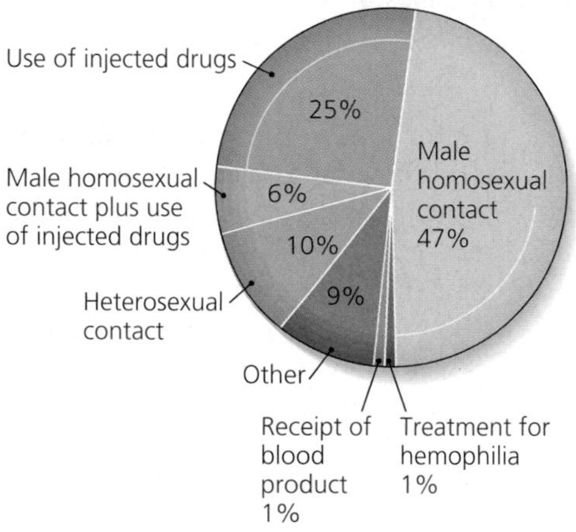

▲ *Figure 18.21*

Modes of HIV transmission in Americans over 12 years of age during 2002. Percentages represent the estimated proportions of HIV infections that result from each type of transmission. (Estimates are rounded to the nearest 1% and thus do not add up to 100%.)

coreceptors (CXCR4 and CCR5), which do not bind effectively to HIV, or they have unusually well-developed immune systems.

Discovering a treatment for AIDS is an area of intense research and development. Currently physicians prescribe **highly active antiretroviral therapy (HAART)**—a "cocktail" of three to four antiviral drugs, including nucleotide analogs, protease inhibitors, and reverse transcriptase inhibitors (see Chapter 10)—to reduce viral replication. HAART is expensive and generally must be taken on a strict schedule. Studies indicate that HAART stops the replication of HIV, because strains of the virus are unlikely to develop resistance to all of the drugs simultaneously. As long as treatment continues, a patient can live a relatively normal life; however, treatment is not a cure because the infection remains. Scientists estimate that it would take 60 years on HAART for all HIV-infected cells to die. In addition to HAART, physicians manage and treat individual diseases associated with AIDS on a case-by-case basis.

Researchers are working to develop methods that will prevent attachment of HIV to target cells, block the entry of HIV into cells, and halt HIV synthesis and release.

Progress in developing a vaccine against HIV has been disappointing. Among the problems that must be overcome in developing an effective vaccine are the following:

- A vaccine must generate both secretory IgA (to prevent sexual transmission and infection), and cytotoxic T lymphocytes (to eliminate infected cells).
- IgG induction, while a necessary part of a vaccine, can actually be detrimental to a patient. Because IgG-viral complexes bind to B cells and also remain infective inside phagocytic cells, a vaccine must stimulate cellular immunity more than humoral immunity so as to kill infected cells.
- HIV is highly mutable, generating antigenic variants that enable it to evade the immune response. Indeed, a patient with untreated AIDS is likely to possess viruses with every possible single-base error in their RNA genomes.
- HIV can spread through syncytia (formed by gp41), thus moving from cell to cell while evading immune surveillance.
- HIV infects and inactivates macrophages, dendritic cells, and helper T cells—cells that combat infections.
- Testing a vaccine presents ethical and medical problems because HIV is a pathogen of humans only. For example, researchers stopped a study in 1994 when HIV infected five volunteers despite their having received an experimental vaccine.

Scientists have developed a vaccine that protects monkeys from SIV disease, but have not been able to translate this success into an effective vaccine for humans. In the meantime, individuals can slow the AIDS epidemic by effecting changes in behavior:

- Abstinence and mutually faithful monogamy between uninfected individuals are the only truly safe forms of sex. Studies have shown that whereas condoms reduce a heterosexual individual's risk of acquiring HIV by about 69%, the benefit of condom usage to a population can be undone by an overall increase in sexual activity that may result from a false sense of security that condoms provide.
- Use of new, clean needles and syringes for all injections, as well as caution in dealing with sharp, potentially contaminated objects, can reduce HIV infection rates. If clean supplies are not available, undiluted bleach or other antiviral agents can deactivate HIV.
- AZT given to pregnant women has reduced transplacental transfer of HIV, though infected breast milk can transfer the virus to infants. HIV-infected mothers should not breast-feed their infants.
- Screening blood, blood products, and organ transplants for HIV and anti-HIV antibodies has virtually eliminated the risk of HIV infection from these sources.
- Proper use of gloves, protective eyewear, and masks can prevent contact with infected blood.

Case Study 18.1 A Case of AIDS

A 25-year-old male was admitted to the hospital with thrush, diarrhea, unexplained weight loss, and difficulty in breathing. Cultures of pulmonary fluid revealed the presence of *Pneumocystis*. The man admitted to being a heroin addict and to sharing needles in a "shooting gallery."

1. What laboratory tests could confirm a diagnosis of AIDS?
2. How did the man most likely acquire an HIV infection?
3. What changes to the man's immune system allowed the opportunistic infections of *Pneumocystis* to arise?
4. What precautions should health care providers take with the patient's blood?

CHAPTER SUMMARY

1. Immunological responses may give rise to inflammatory reactions called **hypersensitivities.** An immunological attack on normal tissues gives rise to autoimmune disorders. A failure of the immune system to function normally may give rise to **immunodeficiency diseases.**

Hypersensitivities (pp. 510–523)

1. Type I hypersensitivity gives rise to **allergies.** Antigens that trigger this response are called **allergens.**
2. Allergies result when allergens bind to IgE molecules that were previously bound to **mast cells, basophils,** and **eosinophils.** This causes the sensitized cells to degranulate and release **histamine, kinins, proteases, leukotrienes,** and **prostaglandins.**
3. Depending on the amount of and site at which these molecules are released, the result can produce various clinical syndromes, including **hay fever, asthma, urticaria** (hives), or various other allergies. When the inflammatory mediators exceed the body's coping mechanisms, **acute anaphylaxis (anaphylactic shock)** may occur. The specific treatment for anaphylaxis is epinephrine.
4. Type I hypersensitivity can be diagnosed by skin testing and can be partially prevented by avoidance of allergens and immunotherapy. Some type I hypersensitivities are treated with **antihistamines.**
5. Type II cytotoxic hypersensitivities, such as incompatible blood transfusions and hemolytic disease of the newborn, result when cells are destroyed by an immune response.
6. Red blood cells have **blood group antigens** on their surface. If incompatible blood is transfused into a recipient, a severe transfusion reaction can result.
7. The most important of the blood group antigens is the **ABO system,** which is largely responsible for transfusion reactions.
8. Approximately 85% of the human population carries **Rh antigens,** which are also found in rhesus monkeys. If an Rh negative pregnant woman is carrying an Rh positive fetus, the fetus may be at risk of **hemolytic disease of the newborn,** in which antibodies made by the mother against the Rh antigens may cross the placenta and destroy the fetus's red blood cells. Rhogam administered to pregnant Rh− women may prevent this disease.
9. Drugs bound to blood platelets may subsequently bind antibodies and complement, causing **immune thrombocytopenic purpura,** in which the platelets lyse.
10. If excessive amounts of **immune-complexes** are deposited in tissues, they can cause significant tissue damage. Immune-complexes deposited in the lung cause a **hypersensitivity pneumonitis,** of which the most common example is farmer's lung. If large amounts of immune-complexes form in the bloodstream, they may be filtered out by the glomeruli of the kidney, causing **glomerulonephritis,** which can result in kidney failure.
11. Type IV hypersensitivity, also known as **delayed hypersensitivity,** is a T cell-mediated inflammatory reaction that is so called because it takes 24–72 hours to reach maximal intensity.

12. A good example of a delayed hypersensitivity reaction is the **tuberculin response,** generated when tuberculin, a protein extract of *Mycobacterium tuberculosis,* is injected into the skin of an individual who has been infected with or vaccinated against *M. tuberculosis.*
13. Another example of a type IV hypersensitivity reaction is **allergic contact dermatitis,** which is T cell-mediated damage to chemically modified skin cells. The best known example is a reaction to poison ivy.
14. An organ or tissue **graft** can be made between different sites within a single individual (an **autograft**), between genetically identical individuals (an **isograft**), between genetically dissimilar individuals (an **allograft**), or between individuals of different species (a **xenograft**). Most surgical organ grafting involves allografts, which, if not treated with immunosuppressive drugs, lead to graft rejection.
15. Some tissues, such as portions of the eye and the testes, are regarded as **privileged sites** because allografts transplanted into these areas are not rejected. Despite being similar to an allograft, a fetus is not rejected by a pregnant mother because a variety of immunosuppressive mechanisms prevent rejection.
16. Commonly used immunosuppressive drugs include **corticosteroids, cytotoxic drugs, immunophilins,** and lymphocyte-depleting therapies, which involve treatment with antibodies against cells of specific immunity.

Autoimmune Diseases (pp. 523–526)

1. **Autoimmune diseases** may result when an individual begins to make autoantibodies or cytotoxic T cells against normal body components.
2. There are many hypotheses concerning the cause of autoimmune disease. One involves **molecular mimicry,** in which microorganisms with antigenic determinants similar to self-antigens trigger autoimmune tissue damage. Others implicate estrogen, pregnancy, and environmental and genetic factors.
3. One group of autoimmune diseases involves only a single organ or cell type. Examples of such diseases include autoimmune hemolytic anemia, **type I diabetes mellitus, Graves' disease,** and **multiple sclerosis.**
4. A second group of autoimmune diseases involves multiple organs or body systems. Examples include **systemic lupus erythematosus (SLE)** and **rheumatoid arthritis,** in which antibodies made against many different autoantigens are part of immune-complexes deposited in tissues. In rheumatoid arthritis, the immune-complexes result in the growth of inflammatory tissue within joints.

Immunodeficiency Diseases (pp. 526–535)

1. **Primary immunodeficiency diseases** result from mutations or developmental anomalies and occur in young children. **Acquired** or **secondary immunodeficiency diseases** result from other known causes such as viral infections, stress, aging, and malnutrition.
2. Examples of primary immunodeficiency diseases include **chronic granulomatous disease,** in which a child's neutrophils are incapable of killing ingested bacteria. Inability to produce both T cells and B cells is called **severe combined immunodeficiency disease.** In **DiGeorge anomaly,** the thymus fails to develop. In **Bruton-type agammaglobulinemia,** B cells fail to function.
3. **Acquired immunodeficiency syndrome (AIDS)** is the most common acquired immunodeficiency disease.
4. A **syndrome** is a complex of signs, symptoms, and diseases with a common cause. AIDS is defined by a presence of a number of opportunistic infections in the presence of antibodies against HIV.
5. **Human immunodeficiency viruses** (HIV-1 and HIV-2) cause AIDS. These retroviruses, which apparently evolved from a chimpanzee virus (SIV), are characterized by glycoproteins gp120 and gp41.
6. The glycoproteins enable attachment via host cytoplasmic membrane proteins **CD4** and **chemokine receptors** on helper T cells or **DC-SIGN** on dendritic cells. The glycoproteins also provide antigenic variability and ability to form a **syncytium** (fused cells).
7. HIV replicates inside macrophages, helper T cells, smooth muscle cells, and dendritic cells. **Reverse transcriptase** synthesizes a dsDNA genome, using the virus' +ssRNA genome as a template. The DNA integrates into a chromosome of the host cell, becoming a latent provirus.
8. The infected cell transcribes and translates the integrated provirus, synthesizing viral genomes and polypeptides, which bud from the cell, forming immature HIV.
9. **Protease** activates HIV by releasing reverse transcriptase and capsomeres within a released immature virion.
10. HIV progressively kills leukocytes, particularly CD4 helper T cells. The body fights this onslaught, but eventually succumbs, allowing opportunistic pathogens to proliferate.
11. Sexual activity and sharing needles are the main ways individuals spread HIV.
12. Physicians diagnose AIDS based on elucidation of characteristic opportunistic infections in the presence of antibodies against HIV.
13. **Highly active antiretroviral therapy (HAART)** or an AIDS "cocktail" is the standard treatment, but it does not eliminate the infection in a timely manner.
14. No effective vaccine against AIDS exists. Prevention of infection primarily depends upon faithful mutual monogamy, proper condom usage, use of clean sterile needles, and screening blood and donated organs for HIV. HIV-positive mothers should not breast-feed their children.

QUESTIONS FOR REVIEW

(Answers to multiple choice, True/False, and matching questions are on the web, along with additional review questions. Visit www.microbiologyplace.com.)

Multiple Choice

1. The immunoglobulin class that mediates type I hypersensitivity is
   a. IgG.
   b. IgM.
   c. IgD.
   d. IgE.
   e. IgA.
2. The major inflammatory mediator released by degranulating mast cells in type I hypersensitivity is
   a. immunoglobulin.
   b. complement.
   c. histamine.
   d. interleukin.
   e. prostaglandin.
3. The major blood group antigens responsible for hemolytic disease of the newborn are
   a. MHC proteins.
   b. MN antigens.
   c. ABO antigens.
   d. rhesus antigen.
   e. type II proteins.
4. Farmer's lung is a hypersensitivity pneumonitis resulting from
   a. a type I hypersensitivity reaction to grass pollen.
   b. a type II hypersensitivity to red cells in the lung.
   c. a type III hypersensitivity to mold spores.
   d. a type IV hypersensitivity to bacterial antigens.
   e. none of the above.
5. A positive tuberculin skin test indicates that a patient not immunized against tuberculosis
   a. is free of tuberculosis.
   b. is shedding *Mycobacterium.*
   c. has been exposed to tuberculosis antigens.
   d. is susceptible to tuberculosis.
   e. is resistant to tuberculosis.
6. Which of the following is an autoimmune disease?
   a. a heart attack
   b. acute anaphylaxis
   c. farmer's lung
   d. graft-versus-host disease
   e. systemic lupus erythematosus
7. When a surgeon conducts a cardiac bypass operation by transplanting a piece of vein from a patient's leg to the same patient's heart, this is
   a. a xenograft.
   b. an autograft.
   c. an allograft.
   d. an isograft.
   e. a coreograft.
8. A deficiency of both B cells and T cells is most likely
   a. a secondary immunodeficiency.
   b. a complex immunodeficiency.
   c. an acquired immunodeficiency.
   d. a primary immunodeficiency.
   e. an induced immunodeficiency.
9. Infection with HIV causes
   a. primary immunodeficiency disease.
   b. acquired hypersensitivity syndrome.
   c. acquired immunodeficiency syndrome.
   d. anaphylactic immunodeficiency diseases.
   e. combined immunodeficiency diseases.
10. An enzyme that activates immature HIV is
    a. reverse transcriptase.
    b. integrase.
    c. gp120.
    d. protease.
    e. CXCR4.

True/False

Indicate whether each statement is true or false. If the statement is false, change the italicized word or phrase to make the statement true.

___ 1. *Cyclosporine* is released by degranulating mast cells.

___ 2. *Type III* hypersensitivity reactions may lead to the development of glomerulonephritis.

___ 3. ABO blood group antigens are found on *nucleated* cells.

___ 4. The tuberculin reaction is a *type I* hypersensitivity.

___ 5. Graft-versus-host disease can follow a bone marrow *isograft.*

Matching

___ 1. Acute anaphylaxis

___ 2. Allergic contact dermatitis

___ 3. Systemic lupus erythematosus

___ 4. Allograft rejection

___ 5. AIDS

___ 6. Graft-versus-host disease

___ 7. Milk allergy

___ 8. Farmer's lung

A. Type I hypersensitivity

B. Type II hypersensitivity

C. Type III hypersensitivity

D. Type IV hypersensitivity

E. Not a hypersensitivity

SHORT ANSWER

1. Why is AIDS more accurately termed a "syndrome" rather than a disease?
2. Why is a child born to an Rh+ mother not susceptible to hemolytic disease of the newborn?
3. Contrast HIV attachment, penetration, and uncoating in a helper T lymphocyte, a macrophage, and a dendritic cell.
4. Why is a patient who produces a large amount of IgE more likely to experience anaphylactic shock than a patient who produces an equal amount of IgG instead?
5. Describe the functions of reverse transcriptase, integrase, and protease in the replication cycle of HIV.
6. Contrast autografts, isografts, allografts, and xenografts.
7. Compare and contrast the functions of four classes of immunosuppressive drugs.

CRITICAL THINKING

1. What possible advantages might an individual gain from making class E antibodies (IgE)?
2. Why can't physicians use skin tests similar to the tuberculin reaction to diagnose other bacterial diseases?
3. In both Graves' disease and juvenile onset diabetes mellitus, autoantibodies are directed against cytoplasmic membrane receptors. Speculate on the clinical consequences of an autoimmune response to estrogen receptors.
4. In general, people with B cell defects acquire numerous bacterial infections, whereas those with T cell defects get viral diseases. Explain why this is so.
5. What types of illnesses cause death in patients with combined immunodeficiencies or AIDS?
6. Because of the severe shortage of organ donors for transplants, many scientists are examining the possibility of using organs from nonhuman species such as pigs. What special clinical problems might be encountered when using these xenografts?
7. Why do drug resistant strains of HIV frequently arise during the course of an infection?

APPENDIX A: Metabolic Pathways

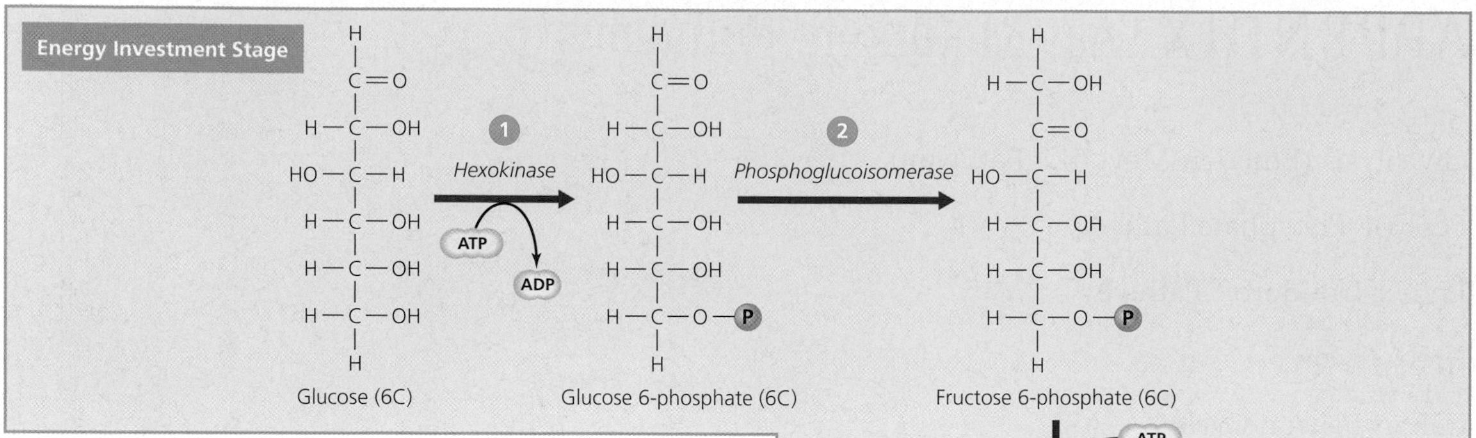

GLYCOLYSIS (Embden-Meyerhof Pathway):

Energy-Investment Stage

Step 1. *Hexokinase* transfers phosphate from ATP to carbon 6 of glucose, forming glucose 6-phosphate—a charged molecule. Since the cytoplasmic membrane is impermeable to ions, phosphorylated glucose cannot diffuse out of a cell.

Step 2. *Phosphoglucoisomerase* rearranges the atoms of glucose to form an isomer—fructose 6-phosphate.

Step 3. Phosphofructokinase invests more energy by adding another phosphate group from ATP to form fructose 1,6-bisphosphate.

Lysis Stage

Step 4. *Aldolase* splits six-carbon fructose 1,6-bisphosphate into three-carbon glyceraldehyde 3-phosphate (G3P) and three-carbon dihydroxyacetone phosphate (DHAP). It is this cleavage that gives glycolysis its name.

Step 5. *Triose phosphate isomerase* rearranges the atoms of DHAP to form another molecule of glyceraldehyde 3-phosphate. From this point, every step of glycolysis occurs twice—once for each of the three-carbon molecules.

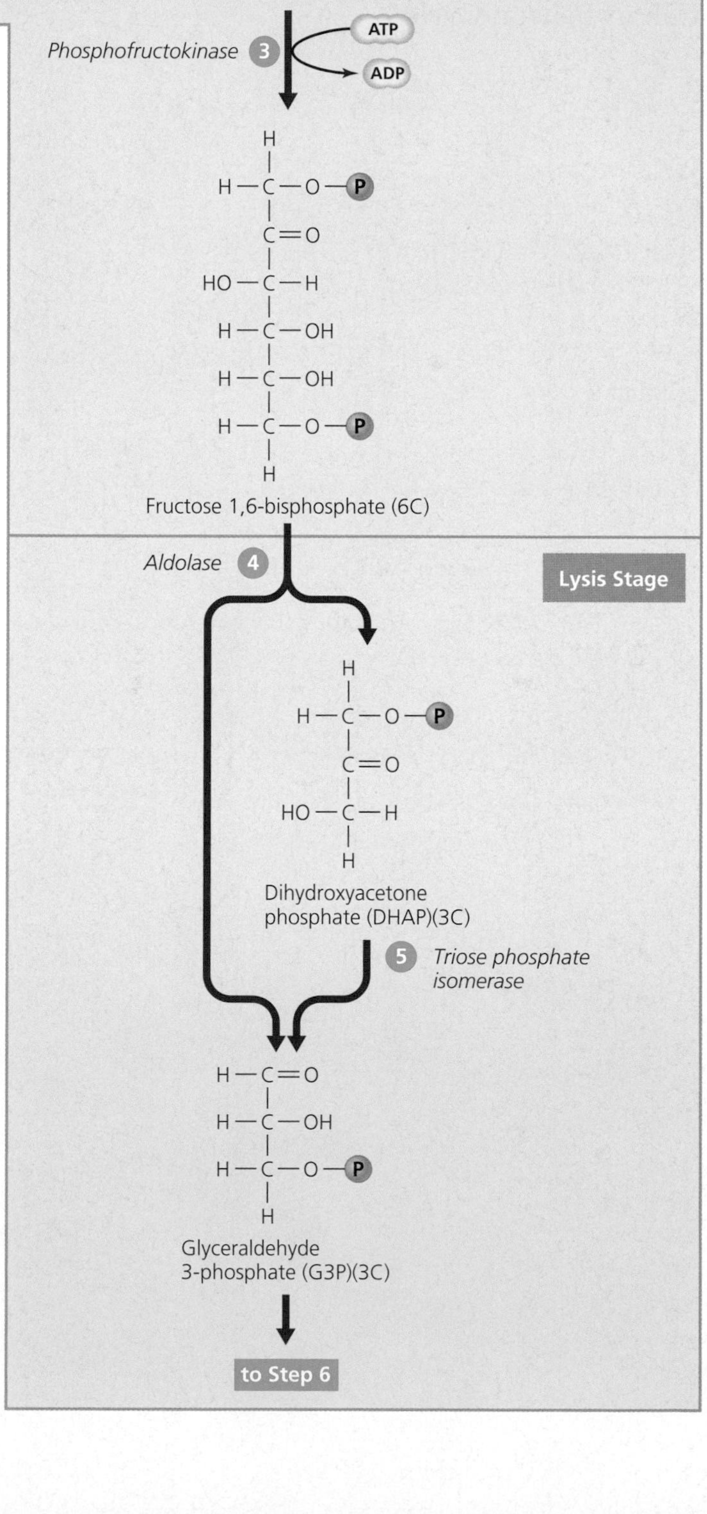

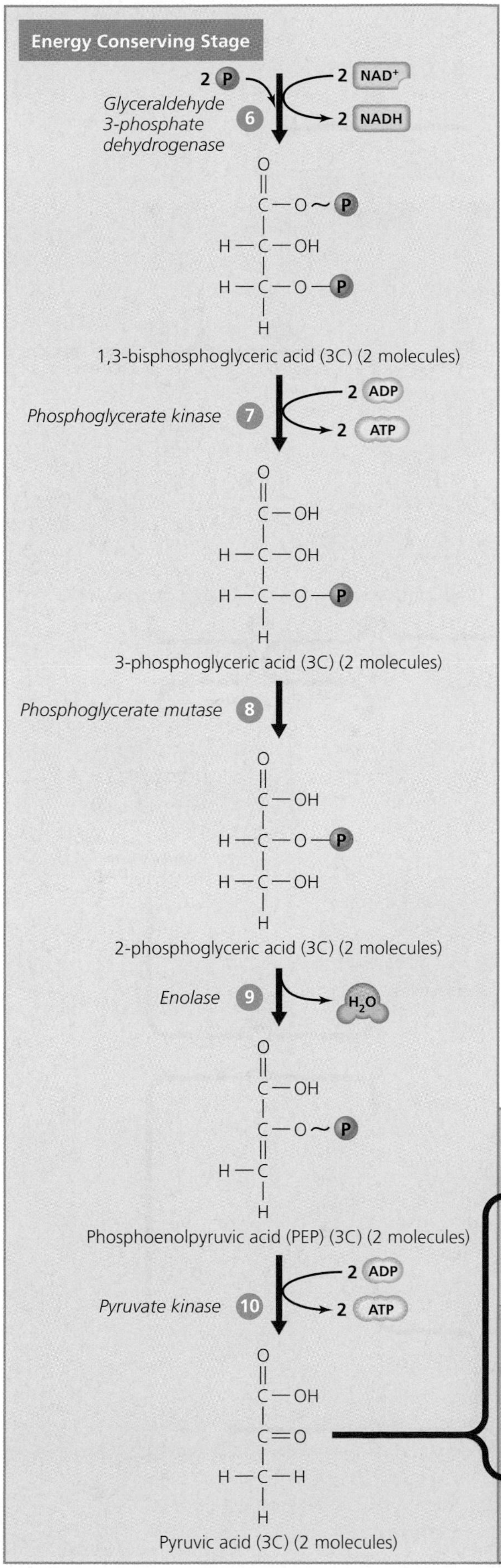

Energy-Conserving Stage

Step 6. *Glyceraldehyde 3-phosphate dehydrogenase* catalyzes a reaction with two parts: (a) It oxidizes G3P, transferring electrons (and hydrogen) to NAD^+ to form NADH and (b) it adds inorganic phosphate from the cytosol to G3P with a high-energy bond to form 1,3-bisphosphoglyceric acid. This two-part reaction is among the more important in glycolysis because it generates a molecule of NADH for each molecule of G3P and because it creates the first high-energy intermediate.

Step 7. *Phosphoglycerate kinase* conserves the energy in the high-energy bonds in two molecules of ATP, yielding 3-phosphoglyceric acid molecules.

Step 8. *Phosphoglycerate mutase* rearranges the atoms to form 2-phosphoglyceric acid.

Step 9. *Enolase* removes a molecule of water from each substrate, forming a double bond and a high-energy bond with phosphate.

Step 10. *Pyruvate kinase* ends glycolysis by transferring energy to ATP, forming pyruvic acid. In the final analysis, two molecules of ATP are invested to yield four molecules of ATP—a net gain of two molecules of ATP—and two molecules of NADH. Pyruvic acid then undergoes respiration (when conditions are aerobic) or fermentation (when conditions are anaerobic).

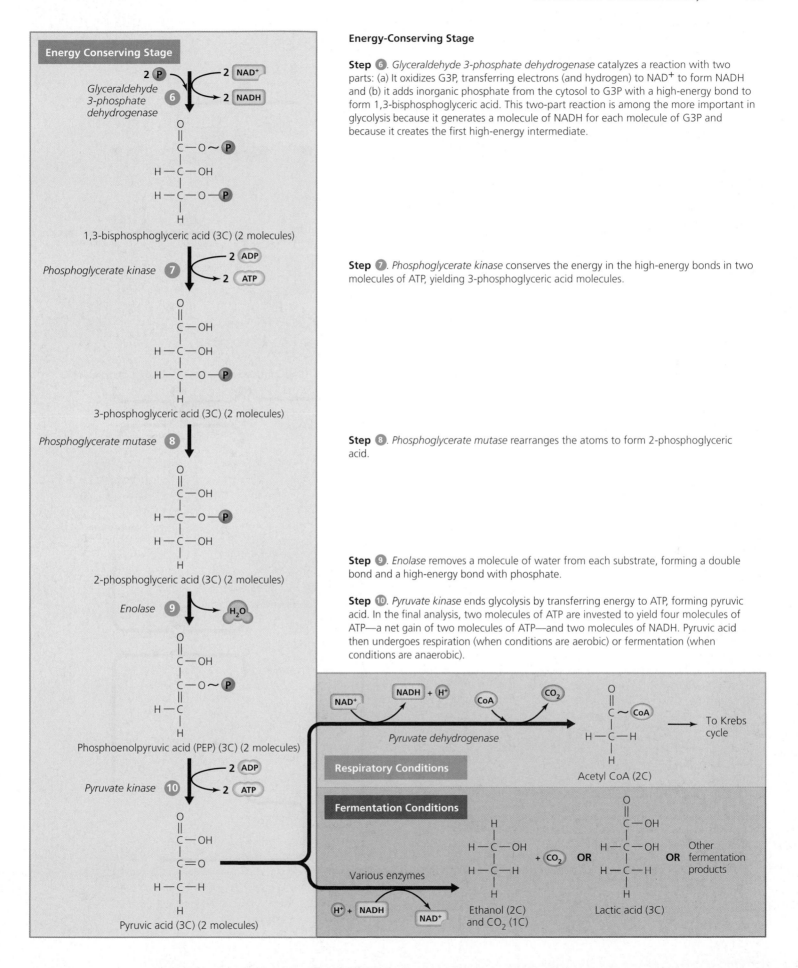

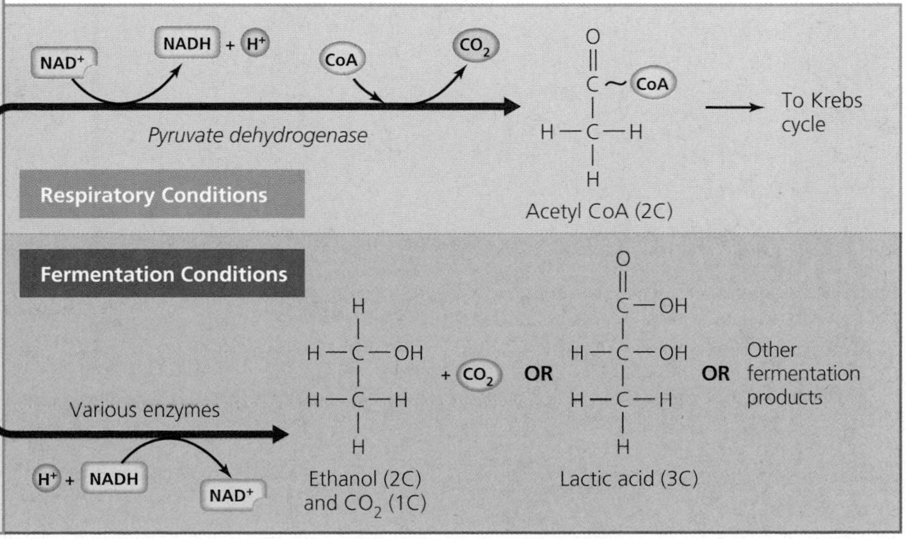

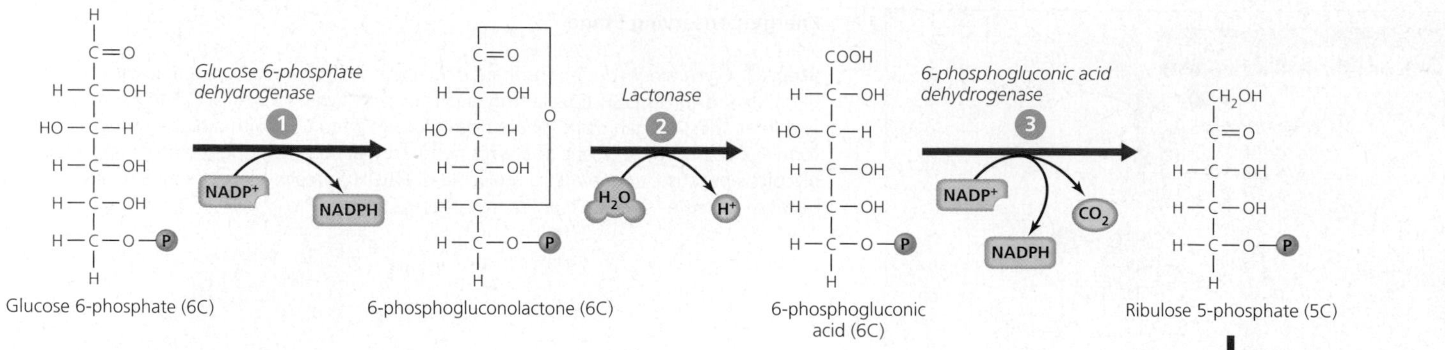

PENTOSE-PHOSPHATE PATHWAY

Step 1. *Glucose 6-phosphate dehydrogenase* oxidizes glucose 6-phosphate to 6-phosphogluconolactone by transferring hydrogen to NADP+, forming NADPH.

Step 2. *Lactonase* removes a water molecule, forming 6-phosphogluconic acid.

Step 3. *6-phosphogluconic acid dehydrogenase* oxidizes this intermediate (transferring hydrogen to another molecule of NADPH) and removes a molecule of carbon dioxide to form ribulose 5-phosphate. This five-carbon phosphorylated sugar is used in the synthesis of nucleotides, certain amino acids, and glucose (via photosynthesis).

Step 4. *Phosphopentose epimerase* converts some molecules of ribulose 5-phosphate to xylulose 5-phosphate.

Step 5. Simultaneously, *phosphopentose isomerase* converts other molecules of ribulose 5-phosphate to ribose 5-phosphate.

Step 6. *Transketolase* catalyzes a reaction in which a two-carbon fragment from xylulose 5-phosphate is transferred to ribose 5-phosphate, forming seven-carbon sedoheptulose 7-phosphate and three-carbon glyceraldehyde 3-phosphate (G3P).

Step 7. *Transaldolase* then transfers a three-carbon fragment from sedoheptulose 7-phosphate to G3P, forming four-carbon erythrose 4-phosphate and six-carbon fructose 6-phosphate.

Step 8. *Transketolase* transfers another two-carbon fragment from another molecule of xylulose 6-phosphate to erythrose 4-phosphate, forming another molecule of fructose 6-phosphate and another molecule of G3P. G3P enters glycolysis at step 6; fructose 6-phosphate can enter glycolysis at step 1 or may be converted into glucose 6-phosphate, which reenters the pentose-phosphate pathway.

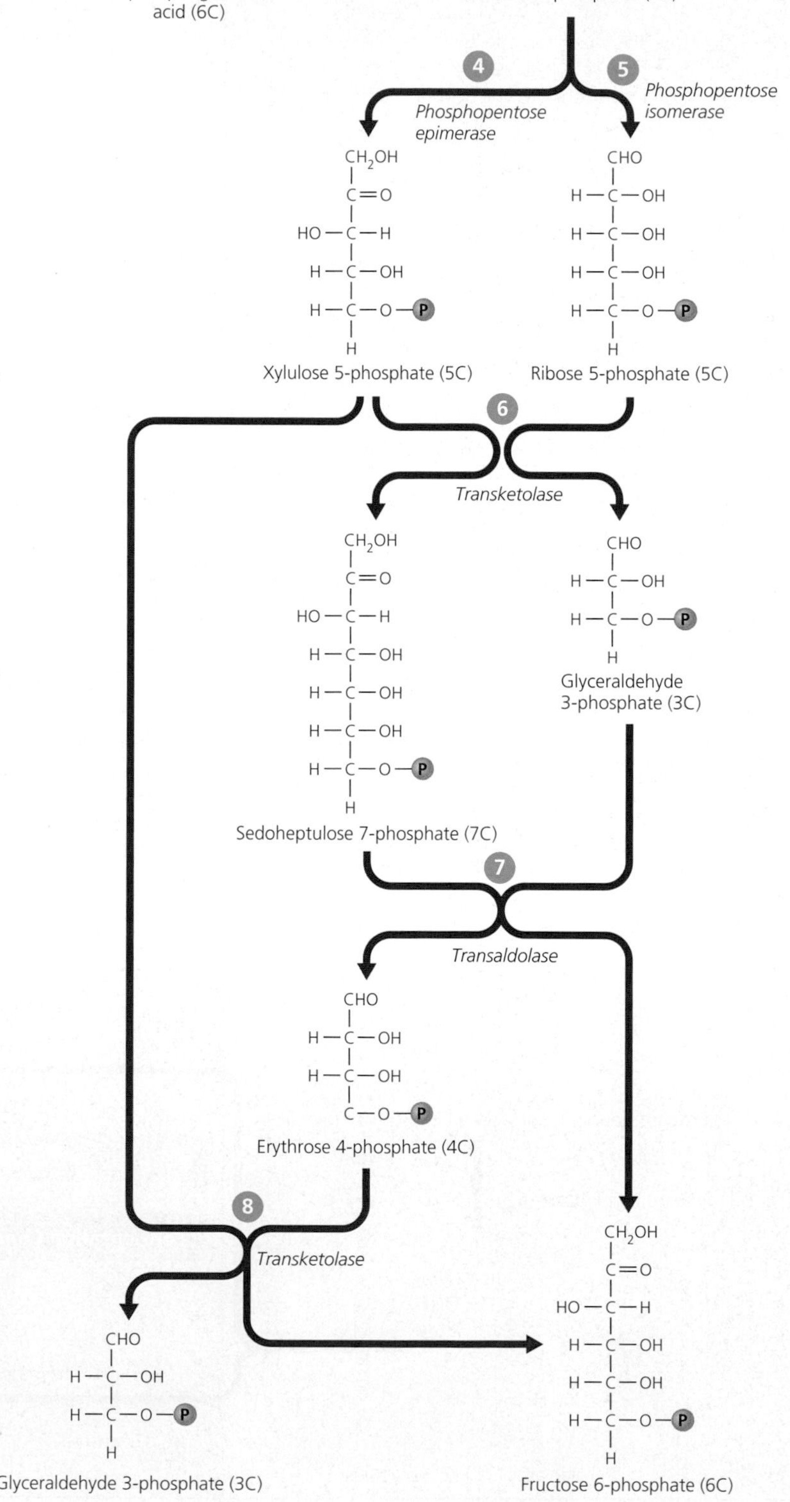

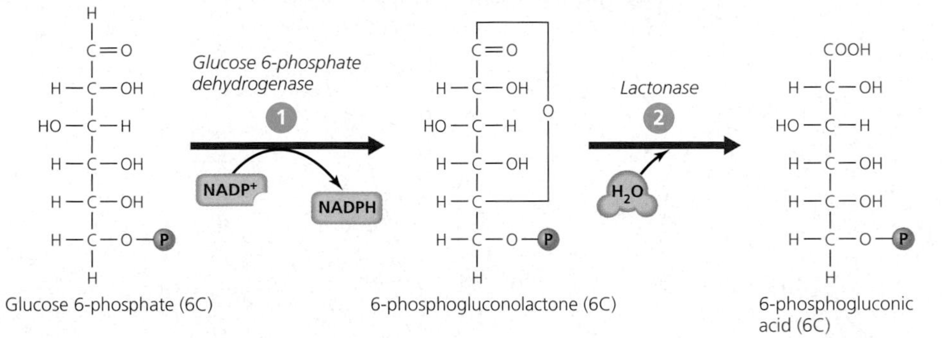

ENTNER-DOUDOROFF PATHWAY

The first two steps of the Entner-Doudoroff pathway are the same as the first two steps of the pentose-phosphate pathway:

Step 1. *Glucose 6-phosphate dehydrogenase* oxidizes glucose 6-phosphate to 6-phosphogluconolactone by transferring hydrogen to $NADP^+$, forming NADPH.

Step 2. As in step two of the pentose phosphate pathway, *lactonase* removes a water molecule, forming 6-phosphogluconic acid.

Step 3. *6-phosphogluconic acid dehydrase* removes a molecule of water to form a six-carbon molecule. (The enzyme in this step should not be confused with the similarly named 6-phosphogluconic acid dehydrogenase of the pentose phosphate pathway.)

Step 4. *Aldolase* splits 2-keto-3-deoxy-6-phosphogluconic acid into two three-carbon compounds: pyruvic acid and glyceraldehyde 3-phosphate (G3P).

Step 5. G3P is converted to pyruvic acid by steps 6 through 10 of Embden-Meyerhof glycolysis.

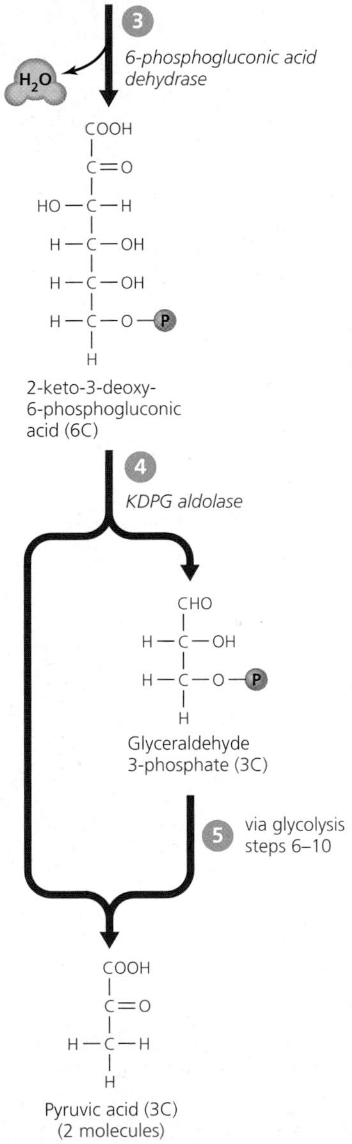

KREBS CYCLE

In a complex reaction with several parts, pyruvate dehydrogenase decarboxylates and oxidizes pyruvic acid (from the Embden-Meyerhof or Entner-Doudoroff pathways) and adds the remaining two-carbon acetic acid to coenzyme-A, forming acetyl-CoA. It is this molecule that enters the Krebs cycle. Recall that for every molecule of glucose that enters glycolysis, two molecules of pyruvic acid are produced, necessitating two sets of Krebs cycle reactions.

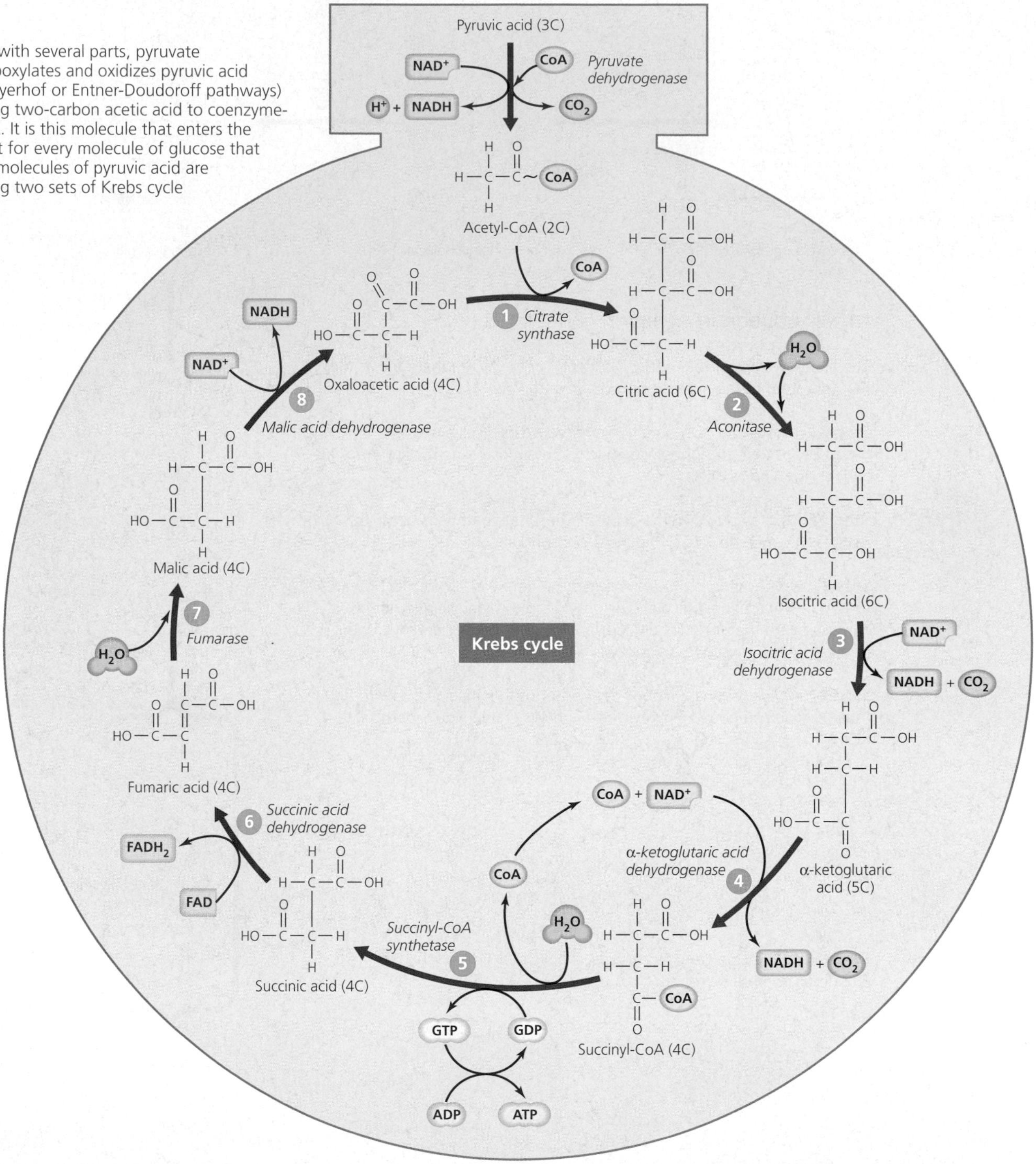

Step 1. *Citrate synthase* adds the two carbons of acetic acid from acetyl-CoA to oxaloacetic acid, forming citric acid. This step gives the Krebs cycle its alternate name—the citric acid cycle.

Step 2. *Aconitase* forms an isomer, isocitric acid, by removing a molecule of water and then adding another back.

Step 3. *Isocitric acid dehydrogenase* oxidizes and decarboxylates six-carbon isocitric acid to five-carbon α-ketoglutaric acid; a molecule of NADH is formed in the process.

Step 4. *α-ketoglutaric acid dehydrogenase* oxidizes and decarboxylates α-ketoglutaric acid to form a four-carbon fragment that is attached to coenzyme-A, forming succinyl-CoA.

Step 5. *Succinyl-CoA synthetase* forms four-carbon succinic acid and simultaneously phosphorylates GDP to form GTP. The latter molecule subsequently phosphorylates ADP to ATP.

Step 6. *Succinic acid dehydrogenase* oxidizes succinic acid to four-carbon fumaric acid, forming a molecule of $FADH_2$.

Step 7. *Fumarase* rearranges the four carbons to form malic acid.

Step 8. *Malic acid dehydrogenase* oxidizes malic acid to reform oxaloacetic acid, completing the cycle and forming another molecule of NADH.

NADH and $FADH_2$ carry electrons to the electron transport chain.

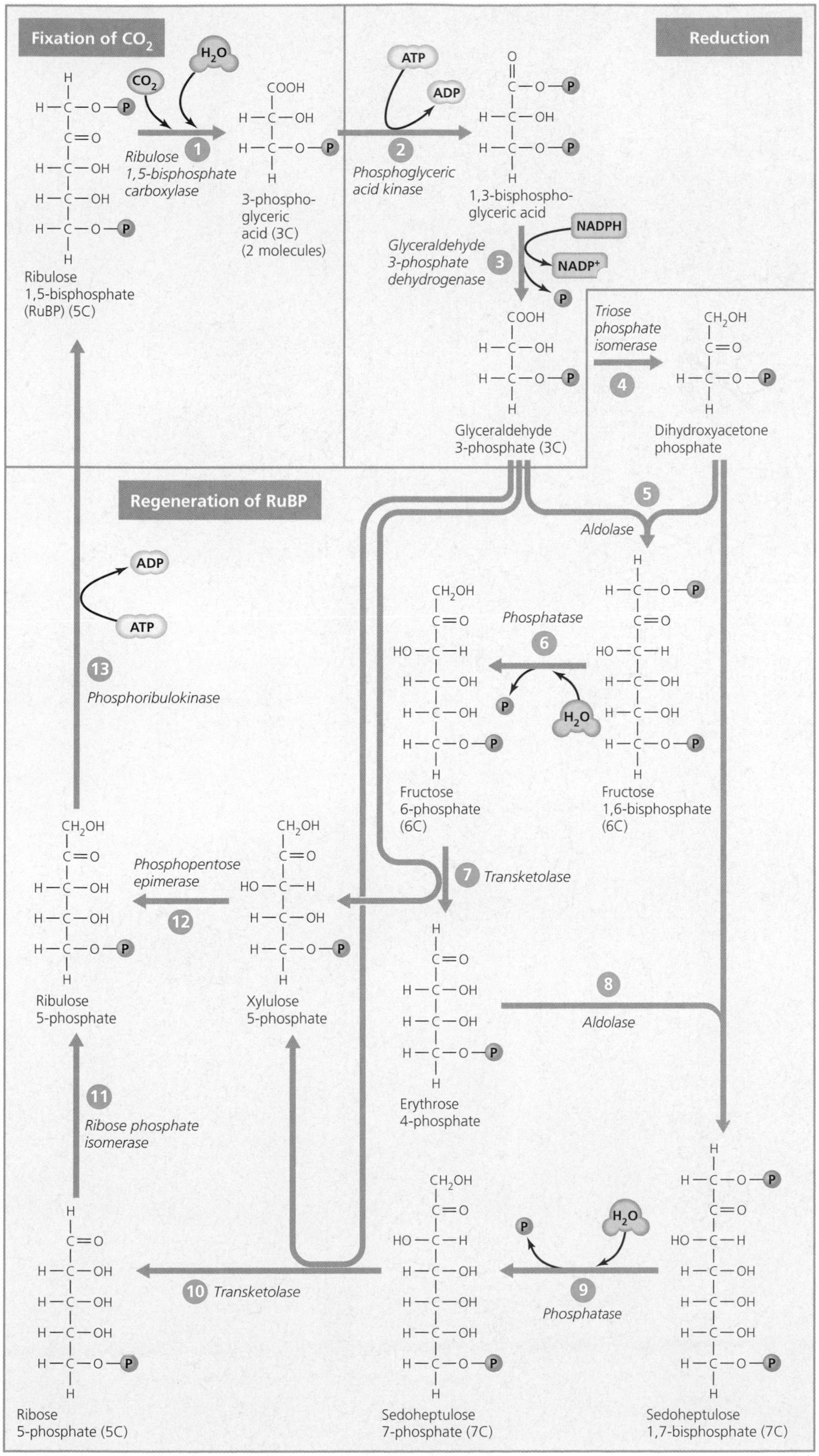

CALVIN-BENSON CYCLE

Fixation of CO_2

Step 1. *Ribulose 1,5-bisphosphate carboxylase* adds CO_2 and H_2O to five-carbon ribulose 1,5-bisphosphate (RuBP), forming a six-carbon intermediate (not shown) that is immediately split into two, three-carbon molecules of 3-phosphoglyceric acid.

Reduction

Step 2. *Phosphoglyceric acid kinase* phosphorylates phosphoglyceric acid at the expense of ATP to form 1,3-bisphosphoglyceric acid.

Step 3. *Glyceraldehyde 3-phosphate dehydrogenase* reduces (using NADPH) and removes one phosphate from 1,3-bisphosphoglyceric acid, forming glyceraldehyde 3-phosphate (G3P).

Regeneration of RuBP

Step 4. *Triose phosphate isomerase* changes some molecules of G3P to dihydroxyacetone phosphate (DHAP).

Step 5. *Aldolase* combines a molecule of G3P and a molecule of DHAP to form fructose 1,6-bisphosphate.

Step 6. *Phosphatase* adds H_2O and removes a phosphate group, forming fructose 6-phosphate.

Step 7. *Transketolase* combines fructose 6-phosphate with a molecule of G3P (from step 3) to form five-carbon xylulose 5-phosphate and four-carbon erythrose 4-phosphate.

Step 8. *Aldolase* rearranges the carbons of erythrose 4-phosphate and of a molecule of DHAP (from step 4), forming sedoheptulose 1,7-bisphosphate.

Step 9. *Phosphatase* adds water and removes a phosphate group to form sedoheptulose 7-phosphate.

Step 10. *Transketolase* combines this molecule with a molecule of G3P (from step 3) to form five-carbon ribose 5-phosphate and five carbon xylulose 5-phosphate.

Step 11. *Ribose 5-phosphate isomerase* rearranges the atoms of ribose 5-phosphate to form its isomer—ribulose 5-phosphate.

Step 12. *Phosphatase epimerase* similarly rearranges the atoms of xylulose 5-phosphate, forming another molecule of ribulose 5-phosphate.

Step 13. *Phosphoribulokinase* phosphorylates ribulose 5-phosphate to regenerate ribulose bisphosphate (RuBP).

For every three molecules of CO_2 that enter the Calvin-Benson cycle (at step 1), one molecule of G3P leaves the cycle to be used for the synthesis of glucose via the reversal of the early reactions of glycolysis.

APPENDIX B: Some Mathematical Considerations in Microbiology

Scientific Notation

Scientific notation (also called exponential notation) is a mathematical device developed to express numbers, particularly very small and very large numbers, in a manner that is convenient and readable. For example, 1.234×10^{-8} is much less cumbersome than 0.00000001234. Similarly, 5.67×10^{17} is more easily read than 567,000,000,000,000,000.

A number expressed in scientific notation is composed of a *coefficient,* which is always a number with only one whole number digit to the left of the decimal place, times ten to some *exponential power*. In the examples above, the coefficients are 1.234 and 5.67 respectively. The exponents are -8 and 17 respectively.

To write a number in scientific notation, move the decimal point either right or left so that there is only one nonzero number to the left of it. For example,

454 becomes 4.54

In this case we moved the decimal point (which is understood to be at the end of a whole number) two places to the left. Since we moved two places left, the exponent will be positive 2, and the scientific notation is

$$4.54 \times 10^2$$

Similarly, 4,500,264.75 becomes 4.50026475×10^6 since we moved the decimal six places to the left.

As you would expect, when we have to move the decimal point to the right, as in 0.00562 (which becomes 5.62), then the exponent will be a negative number, in this case -3; therefore, $0.00562 = 5.62 \times 10^{-3}$.

Logarithms

A **logarithm** is the number of times a number, called the *base number,* must be multiplied by itself to get a certain number. For example, the number 10 must be multiplied by itself three times to get 1000 ($10 \times 10 \times 10$); therefore, the logarithm of 1000 in base 10 is 3. Similarly, the logarithm of 100,000 is 5. Though any number can be used as the base number, in microbiology, base 10 is commonly used. Microbiologists use base 10 logarithms to express pH values and the size of bacterial populations in culture.

When a number is expressed in scientific notation, the logarithm is the exponent when the coefficient is exactly 1. When the coefficient is a number other than 1, the logarithm must be calculated using the logarithm function on a calculator.

Generation Time

When bacteria and other microbes reproduce by binary fission, the number in the population doubles with each division cycle (generation). Such growth is called *logarithmic* (or *exponential*) *growth.*

The number in a population can be calculated as 2 (because each cell produces two offspring) multiplied by itself as many times as there are generations. In other words, the number of cells in a population arising from a single individual is expressed mathematically as

$$2^{\text{number of generations}}$$

A cell dividing for five generations would produce a population of 32 ($2^5 = 2 \times 2 \times 2 \times 2 \times 2 = 32$). When the population begins with more than one individual, then the final population equals

$$\text{Original number of cells} \times 2^{\text{number of generations}}$$

For example, seven cells reproducing for five generations would produce a population of 224 cells ($7 \times 2^5 = 7 \times 2 \times 2 \times 2 \times 2 \times 2 = 224$).

In most cases, microbiologists are not concerned with the number of generations required to produce a given population, but they do want to know the **generation time;** that is, the time required for a bacterial cell to grow and divide. This is also the time required for a population of cells to double in number. To calculate generation time, scientists must first calculate how many generations have been produced, knowing the beginning and ending population sizes. When population sizes are converted to logarithms, the number of generations is calculated as

$$\text{Number of generations} = \frac{\text{logarithm (log) number of cells at the end of reproduction} - \text{log number of cells initially}}{\log 2}$$

Log 2 is used in the formula because each cell produces two offspring each time it divides. $\log 2 = 0.301$.

The number of generations is used to calculate generation time:

$$\text{Generation time (min/generation)} = \frac{60 \text{ minutes} \times \text{number of hours}}{\text{number of generations}}$$

For example, if 100 bacteria multiply to produce a population of 3.28×10^6 cells in seven hours, then the generation time for this bacterium is calculated as follows:

$$\text{Number of generations} = \frac{\text{logarithm (log) number of cells at the end of reproduction} - \text{log number of cells initially}}{\log 2}$$

$$= \frac{\log (3.28 \times 10^6) - \log (100)}{\log 2}$$

$$= 15 \text{ generations}$$

$$\text{Generation time (min/generation)} = \frac{60 \text{ minutes} \times \text{hours}}{\text{number of generations}}$$

$$= \frac{60 \text{ min} \times 7 \text{ h}}{15}$$

$$= 28 \text{ minutes/generation}$$

APPENDIX C: Classification of Bacteria According to *Bergey's Manual**

Domain: Archaea
Phylum I: Crenarchaeota
Class I: Thermoprotei
Order I: Thermoproteales
Family I: Thermoproteaceae
Caldivirga
Pyrobaculum
Thermocladium
Thermoproteus
Family II: Thermofilaceae
Thermofilum
Order II: Desulfurococcales
Family I: Desulfurococcaceae
Acidolobus
Aeropyrum
Desulfurococcus
Igniococcus
Staphylothermus
Stetteria
Sulfophobococcus
Thermodiscus
Thermosphaera
Family II: Pyrodictiaceae
Hyperthermus
Pyrodictium
Pyrolobus
Order III: Sulfolobales
Family I: Sulfolobaceae
Acidianus
Metallosphaera
Stygiolobus
Sulfolobus
Sulfurisphaera
Sulfurococcus
Phylum II: Euryarchaeota
Class I: Methanobacteria
Order I: Methanobacteriales
Family I: Methanobacteriaceae
Methanobacterium
Methanobrevibacter
Methanosphaera
Methanothermobacter
Family II: Methanothermaceae
Methanothermus
Class II: Methanococci
Order I: Methanococcales
Family I: Methanococcaceae
Methanococcus
Methanothermococcus
Family II: Methanocaldococcaceae
Methanocaldococcus
Methanotorris
Order II: Methanomicrobiales
Family I: Methanomicrobiaceae
Methanoculleus
Methanogenium
Methanolacinia
Methanomicrobium
Methanoplanus
Methanofollis
Family II: Methanocorpusculaceae
Methanocorpusculum
Family III: Methanospirillaceae
Methanospirillum
Genera incertae sedis
Methanocalculus
Order III: Methanosarcinales
Family I: Methanosarcinaceae
Methanococcoides
Methanohalobium
Methanohalophilus
Methanolobus
Methanomicrococcus
Methanosarcina
Methanosalsum
Family II: Methanosaetaceae
Methanosaeta
Class III: Halobacteria
Order I: Halobacteriales
Family I: Halobacteriaceae
Haloarcula
Halobacterium
Halobaculum
Halococcus
Haloferax
Halogeometricum
Halorhobdus
Halorubrum
Haloterrigena
Natrialba
Natrinema
Natronobacterium
Natronococcus
Natronomonas
Natronorubrum
Class IV: Thermoplasmata
Order I: Thermoplasmatales
Family I: Thermoplasmataceae
Thermoplasma
Family II: Picrophilaceae
Picrophilus
Family III: Ferroplasmatacea
Ferroplasma
Class V: Thermococci
Order II: Thermococcales
Family I: Thermococcaceae
Palaeococcus
Pyrococcus
Thermococcus
Class VI: Archaeoglobi
Order I: Archaeoglobales
Family I: Archaeoglobaceae
Archaeoglobus
Ferroglobus
Class VII: Methanopyri
Order I: Methanopyrales
Family I: Methanopyraceae
Methanopyrus
Domain: Bacteria
Phylum I: Aquificae
Class I: Aquificae
Order I: Aquificales
Family I: Aquificaceae
Aquifex
Calderobacterium
Hydrogenobacter
Thermocrinis
Genera incertae sedis
Desulfurobacterium
Phylum II: Thermotogae
Class I: Thermotogae
Order I: Thermotogales
Family I: Thermotogaceae
Fervidobacterium
Geotoga
Petrotoga
Thermosipho
Thermotoga
Phylum III: Thermodesulfobacteria
Order I: Thermodesulfobacteriales
Family I: Thermodesulfobacteriaceae
Thermodesulfobacterium

**Bergey's Manual of Systematic Bacteriology,* 2nd ed., 5 vols. (2002), is the reference for classification. *Bergey's Manual of Determinative Bacteriology,* 9th ed. (1994), should be used for identification of culturable bacteria and archaea. For updates to *Bergey's Manual,* visit: http://www.springer-ny.com/bergeysoutline/main.htm

Phylum IV: Deinococcus-Thermus
Class I: Deinococci
Order I: Deinococcales
Family I: Deinococcaceae
Deinococcus
Order II: Thermales
Family I: Thermaceae
Meiothermus
Thermus
Phylum V: Chrysiogenetes
Class I: Chrysiogenetes
Order I: Chrysiogenales
Family I: Chrysiogenaceae
Chrysiogenes
Phylum VI: Chloroflexi
Class I: Chloroflexi
Order I: Chloroflexales
Family I: Chloroflexaceae
Chloroflexus
Chloronema
Heliothrix
Family II: Oscillochloridaceae
Oscillochloris
Order II: Herpetosiphonales
Family I: Herpetosiphonaceae
Herpetosiphon
Phylum VII: Thermomicrobia
Class I: Thermomicrobia
Order I: Thermomicrobiales
Family I: Thermomicrobiaceae
Thermomicrobium
Phylum VIII: Nitrospira
Class I: Nitrospira
Order I: Nitrospirales
Family I: Nitrospiraceae
Leptospirillum
Magnetobacterium
Nitrospira
Thermodesulfovibrio
Phylum IX: Deferribacteres
Class I: Deferribacteres
Order I: Deferribacterales
Family I: Deferrivacter
Deferribacter
Denitrovibrio
Flexistipes
Geovibrio
Genera incertae sedis
Synergistes
Phylum X: Cyanobacteria
Class I: Cyanobacteria
SubSection I
Chamaesiphon
Chroococcus
Cyanobacterium
Cyanobium
Cyanothece
Dactylococcopsis
Gloeobacter
Gloeocapsa
Gloeothece
Microcystis
Prochlorococcus
Prochloron
Synechococcus
Synechocystis
SubSection II
Chroococcidiopsis
Cyanocystis
Dermocarpella
Myxosarcina
Pleurocapsa
Stanieria
Xenococcus
SubSection III
Arthrospira
Borzia
Crinalium
Geitlerinema
Halospirulina
Leptolyngbia
Limnothrix
Lyngbya
Microcoleus
Oscillatoria
Planktothrix
Prochlorothrix
Pseudoanabaena
Spirulina
Starria
Symploca
Trichodesmium
Tychonema
Subsection IV
Anabaena
Anabaenopsis
Aphanizomenon
Calothrix
Cyanospira
Cylindrospermum
Nodularia
Nostoc
Rivularia
Scytonema
Tolypothrix
SubSection V
Chlorogloeopsis
Fischerella
Geitleria
Iyengariella
Nostochopsis
Stigonema
Phylum XI: Chlorobi
Class I: Chlorobia
Order I: Chlorobiales
Family I: Chlorobiaceae
Ancalochloris
Chlorobium
Chloroherpeton
Pelodictyon
Prosthecochloris
Phylum XII: Proteobacteria
Class I: Alpha-Proteobacteria
Order I: Rhodospirillales
Family I: Rhodosprillaceae
Azospirillum
Magnetospirillum
Phaeospirillum
Rhodocista
Rhodospira
Rhodospirillum
Rhodothalassium
Rhodovibrio
Roseospira
Shermanella
Family II: Acetobacteraceae
Acetobacter
Acidiphilium
Acidosphaera
Acidocella
Acidomonas
Asaia
Craurococcus
Gluconacetobacter
Gluconobacter
Paracraurococcus
Rhodopila
Roseococcus
Stella
Zavarzinia
Order II: Rickettsiales
Family I: Rickettsiaceae
Orientia
Rickettsia
Wolbachia
Family II: Ehrlichiaceae
Aegyptianella
Anaplasma
Cowdria
Ehrlichia
Neorickettsia
Xenohaliotis
Family III: Holosporaceae
Caedibacter
Holospora
Lyticum
Odyssella
Polynucleobacter
Pseudocaedibacter
Symbiotes
Tectibacter

Order III: Rhodobacterales
Family I: Rhodobacteraceae
Ahrensia
Amaricoccus
Antarctobacter
Gemmobacter
Hirschia
Hyphomonas
Maricaulis
Methylarcula
Octadecabacter
Paracoccus
Rhodobacter
Rhodovulum
Roseobacter
Roseibium
Roseinatronobacter
Roseovarius
Roseovivax
Rubrimonas
Ruegeria
Sagittula
Staleya
Stappia
Sulfitobacter
Order IV: Sphingomonadales
Family I: Sphingomonodaceae
Blastomonas
Erythrobacter
Erythromicrobium
Erythromonas
Porphyrobacter
Rhizomonas
Sandaracinobacter
Sphingomonas
Zymomonas
Order V: Caulobacterales
Family I: Caulobacteraceae
Asticcacaulis
Brevundimonas
Caulobacter
Phenylobacterium
Order VI: Rhizobiales
Family I: Rhizobiaceae
Agrobacterium
Carbophilus
Chelatobacter
Ensifer
Rhizobium
Sinorhizobium
Family II: Bartonellaceae
Bartonella
Family III: Brucellaceae
Brucella
Mycoplana
Ochrobactrum
Family IV: Phyllobacteriaceae
Allorhizobium
Aminobacter
Aquamicrobium
Defluvibacter
Mesorhizobium
Phyllobacteriaceae
Pseudaminobacter
Family V: Methylocystaceae
Methylocystis
Methylopila
Methylosinus
Family VI: Beijerinckiaceae
Beijerinckia
Chelatococcus
Derxia
Family VII: Bradyrhizobiaceae
Afipia
Agromonas
Blastobacter
Bosea
Bradyrhizobium
Nitrobacter
Oligotropha
Rhodopseudomonas
Family VIII:
Hyphomicrobiaceae
Ancalomicrobium
Ancylobacter
Angulomicrobium
Aquabacter
Azorhizobium
Blastochloris
Devosia
Dichotomicrobium
Filomicrobium
Gemmiger
Hyphomicrobium
Labrys
Methylorhabdus
Pedomicrobium
Prosthecomicrobium
Rhodomicrobium
Rhodoplanes
Seliberia
Starkya
Xanthobacter
Family IX:
Methylobacteriaceae
Methylobacterium
Protomonas
Roseomonas
Family X: Rhodobiaceae
Rhodobium
Class II: Beta-Proteobacteria
Order I: Burkholderiales
Family I: Burkholderiaceae
Burkholderia
Cupriavidus
Lautropia
Pandoraea
Thermothrix
Family II: Ralstoniaceaee
Ralstonia
Family III: Oxalobacteraceae
Duganella
Herbaspirillum
Janthinobacterium
Massilia
Oxalobacter
Telluria
Family IV: Alcaligenaceae
Achromobacter
Alcaligenes
Bordetella
Pelistega
Sutterella
Taylorella
Family V: Comamonadaceae
Acidovorax
Aquabacterium
Brachymonas
Comamonas
Delftia
Hydrogenophaga
Ideonella
Leptothrix
Polaromonas
Rhodoferax
Rubrivivax
Sphaerotilus
Tepidimonas
Thiomonas
Variovorax
Order II: Hydrogenophilales
Family I: Hydrogenophilus
Hydrogenophaga
Thiobacillus
Order III: Methylophilales
Family I: Methylophilaceae
Methylobacillus
Methylophilus
Methylovorus
Order IV: Neisseriales
Family I: Neisseriaceae
Alysiella
Aquaspirillum
Catenococcus
Chromobacterium
Eikenella
Formivibrio
Iodobacter
Kingella
Microvirgula
Neisseria

Prolinoborus
Simonsiella
Vitreoscilla
Vogesella
Order V: Nitrosomonadales
Family I: Nitrosomonadaceae
Nitrosomonas
Nitrosospira
Family II: Spirillaceae
Spirillum
Family III: Gallionellacea
Gallionella
Order VI: Rhodocyclales
Family I: Rhodocyclaceae
Azoarcus
Azonectus
Azospira
Azovibrio
Propionibacter
Propionivibrio
Rhodocyclus
Thauera
Zoogloea
Class III: Gammaproteobacteria
Order I: Chromatiales
Family I: Chromatiaceae
Allochromatium
Amoebobacter
Chromatium
Halochromatium
Halothiobacillus
Isochromatium
Lamprobacter
Lamprocystis
Marichromatium
Nitrosococcus
Pfennigia
Rhabdochromatium
Thermochromatium
Thioalkalicoccus
Thiocapsa
Thiococcus
Thiocystis
Thiodictyon
Thiohalocapsa
Thiolamprovum
Thiopedia
Thiorhodococcus
Thiorhodovibrio
Thiospirillum
Family II: Ectothiorhodospiraceae
Arhodomonas
Ectothiorhodospira
Halorhodospira
Nitrococcus
Thiorhodospira
Order II: Acidothiobacillales
Family I: Acidothiobacillaceae
Acidothiobacillus
Order III: Xanthomonadales
Family I: Xanthomonadaceae
Frateuria
Lureimonas
Lyssobacter
Nevskia
Pseudoxanthomonas
Rhodanobacter
Stenotrophomonas
Xanthomonas
Xylella
Order IV: Cardiobacteriales
Family I: Cardiobacteriaceae
Cardiobacterium
Dichelobacter
Suttonella
Order V: Thiotrichales
Family I: Thiotrichaceae
Achromatium
Beggiatoa
Leucothrix
Macromonas
Thiobacterium
Thiomargarita
Thioploca
Thiospira
Thiothrix
Family II: Piscirickettsiaceae
Cycloclasticus
Hydrogenovibrio
Methylophaga
Piscirickettsia
Thiomicrospira
Family III: Francisellaceae
Francisella
Order VI: Legionellales
Family I: Legionellaceae
Legionella
Family II: Coxiellaceae
Coxiella
Rickettsiella
Order VII: Methylococcales
Family I: Methylococcaceae
Methylobacter
Methylocaldum
Methylococcus
Methylomicrobium
Methylomonas
Methylosphaera
Order VIII: Oceanospirillales
Family I: Oceanospirillaceae
Balneatrix
Fundibacter
Marinomonas
Marinospirillum
Neptunomonas
Oceanospirillum
Family II: Halomonadaceae
Alcanivorax
Carnimonas
Chromohalobacter
Deleya
Halomonas
Zymobacter
Order IX: Pseudomonadales
Family I: Pseudomonadaceae
Azomonas
Azotobacter
Cellvibrio
Chryseomonas
Flavimonas
Lampropedia
Mesophilobacter
Morococcus
Oligella
Pseudomonas
Rhizobacter
Rugamonas
Serpens
Thermoleophilum
Xylophilus
Family II: Moraxellaceae
Acinetobacter
Moraxella
Psychrobacter
Order X: Alteromonadales
Family I: Alteromonadaceae
Allishewanella
Alteromonas
Colwellia
Ferrimonas
Idiomarina
Marinobacter
Marinobacterium
Microbulbifer
Moritella
Pseudoalteromonas
Shewanella
Order XI: Vibrionales
Family I: Vibrionaceae
Allomonas
Enhydrobacter
Listonella
Photobacterium
Salvinivibrio
Vibrio
Order XII: Aeromonadales
Family I: Aeromonadaceae
Aeromonas
Oceanomonas
Tolumonas

Family II: Succinivibrionaceae
Anaerobiospirillumm
Ruminobacter
Succinomonas
Succinivibrio
Order XIII: Enterobacteriales
Family I: Enterobacteriaceae
Alterococcus
Arsenophonus
Brenneria
Buchnera
Budvicia
Buttiauxella
Calymmatobacterium
Cedecea
Citrobacter
Edwardsiella
Enterobacter
Erwinia
Escherichia
Ewingella
Hafnia
Klebsiella
Kluyvera
Leclercia
Leminorella
Moellerella
Morganella
Obesumbacterium
Pantoea
Pectobacterium
Photorhabdus
Plesiomonas
Pragia
Proteus
Providencia
Rahnella
Saccharobacter
Salmonella
Serratia
Shigella
Sodalis
Tatumella
Trabulsiella
Wigglesworthia
Xenorhabdus
Yersinia
Yokenella
Order XIV: Pasteurellales
Family I: Pasteurellaceae
Actinobacillus
Haemophilus
Lonepinella
Pasteurella
Mannheimia
Phocoenobacter
Class VI: Deltaproteobacteria
Order I: Desulfurales
Family I: Desulfurellaceae
Desulfurella
Hippea
Order II: Desulfovibrionales
Family I: Desulfovibrionaceae
Bilophila
Desulfovibrio
Lawsonia
Family II: Desulfomicrobiaceae
Desulfomicrobium
Family III: Desulfonalobiaceae
Desulfohalobium
Desulfomonaas
Desulfonatronovibrio
Order III: Desulfobacterales
Family I: Desulfobacteraceae
Desulfobacter
Desulfobacterium
Desulfobacula
Desulfococcus
Desulfofaba
Desulfofrigus
Desulfonema
Desulfosarcina
Desulfospira
Desulfocella
Desulfotalea
Desulfotignum
Family II: Desulfobulbaceae
Desulfobulbus
Desulfocapsa
Desulfofustis
Desulforhopalus
Family III: Nitrospinaceae
Nitrospina
Desulfobacca
Desulfomonile
Order IV: Desulfuromonadales
Family I: Desulfomonadaceae
Desulfuromonas
Desulfuromusa
Family II: Geobacteraceae
Geobacter
Family III: Pelobacteriaceae
Pelobacter
Melonomonas
Trichlorobacter
Order V: Syntrophobacterales
Family I: Syntrophobacteraceae
Desulfacinum
Syntrophobacter
Desulforhabdus
Desulfovirga
Thermodesulforhabdus
Family II: Syntrophaceae
Smithella
Syntrophus
Order VI: Bdellovibrionales
Family I: Bdellovibrionaceae
Bacteriovorax
Bdellovibrio
Micavibrio
Vampirovibrio
Order VII: Myxococcales
Family I: Myxococcaceae
Angiococcus
Myxococcus
Family II: Archangiaceae
Archangium
Family III: Cystobacteraceae
Cystobacter
Melittangium
Stigmatella
Family IV: Polyangiaceae
Chondromyces
Nannocystis
Polyangium
Class V: Epsilonproteobacteria
Order I: Campylobacterales
Family I: Campylobacteraceae
Arcobacter
Campylobacter
Sulfurospirillum
Thiovulum
Family II: Helicobacteraceae
Helicobacter
Wolinella
Phylum XIII: Firmicutes
Class I: Clostridia
Order I: Clostridiales
Family I: Clostridiaceae
Acetivibrio
Acidaminobacter
Anaerobacter
Caloramator
Clostridium
Coprobacillus
Natronineola
Oxobacter
Sarcina
Sporobacter
Thermobrachium
Thermohalobacter
Tindallia
Family II: Lachnospiraceae
Acetitomaculum
Anaerofilum
Butyrivibrio
Catenibacterium
Catonella
Coprococcus
Johnsonella
Lachnospira

Pseudobutyrivibrio
Roseburia
Ruminococcus
Sporobacterium
Family III: Peptostreptococcace
Filifactor
Fusibacter
Helcococcus
Micromonas
Peptostreptococcus
Tissierella
Family IV: Eubacteriaceae
Acetobacterium
Anaerovorax
Eubacterium
Mogibacterium
Pseudoramibacter
Family V: Peptococcaceae
Anaeroaarcus
Anaerosinus
Anaerovibrio
Carboxydothermus
Centipeda
Dehalobacter
Dendrosporobacter
Desulfitobacterium
Desulfonispora
Desulfosporosinus
Desulfotomaculum
Mitsuokella
Peptococcus
Propionispira
Succinispira
Syntrophobotulus
Thermoterrabacterium
Family VI: Heliobacteriaceae
Heliobacterium
Heliobacillus
Heliophilum
Heliorestis
Family VII: Acidaminocaccaceae
Acetonema
Anaeromonas
Acidaminococcus
Dialister
Megasphaera
Papillibacter
Pectinatus
Phascolarctobacterium
Quinella
Schwartzia
Selenomonas
Sporomusa
Succiniclasticum
Veillonella
Zymophilus
Family VIII: Syntrophomonadaceae
Acetogenium
Aminobacterium
Aminomonas
Anaerobaculum
Anaerobranca
Caldicellulosiruptor
Dethiosulfovibrio
Pelospora
Syntrophomonas
Syntrophospora
Syntrophothermus
Thermoaerobacter
Thermoaerovibrio
Thermohydrogenium
Thermosyntropha
Order II: Thermoanaerobactriales
Family I: Thermoanaerobacteriaceae
Ammonifex
Carboxydobrachium
Coprothermobacter
Moorella
Sporotomaculum
Thermacetogenium
Thermoanaerobacter
Thermoanaerobacterium
Thermoanaerobium
Order III: Haloanaerobiales
Family I: Haloanaerobiaceae
Haloanaerobium
Halocella
Halothermothrix
Natroniella
Family II: Halobacteroidaceae
Acetohalobium
Haloanaerobacter
Halobacteroides
Orenia
Sporohalobacter
Class II: Mollicutes
Order I: Mycoplasmatales
Family I: Mycoplasmataceae
Mycoplasma
Eperythrozoon
Haemobartonella
Ureaplasma
Order II: Entoplasmatales
Family I: Entoplasmatales
Entomoplasma
Mesoplasma
Family II: Spiroplasmataceae
Spiroplasma
Order III: Acholeplasmataceae
Family I: Anaeroplasmataceae
Acholeplasma
Order IV: Anaeroplasmatales
Anaeroplasma
Asteroleplasma
Order V: Incertae sedis
Family I: Erysipelothrichaeceae
Bulleidia
Erysipelothrix
Holdemania
Solobacterium
Class III: Bacilli
Order I: Bacillales
Family I: Bacillaceae
Amphibacillus
Anoxybacillus
Bacillus
Exiguobacterium
Gracilibacillus
Halobacillus
Saccharococus
Salibacillus
Virgibacillus
Family II: Planococcaceae
Filibacter
Kurthia
Planococcus
Sporosarcina
Family III: Caryophanaceae
Caryophanon
Family IV: Listeriaceae
Brochothrix
Listeria
Family V: Staphylococcaceae
Gemella
Macrococcus
Salinicoccus
Staphylococcus
Family VI: Sporolactobacillaceae
Marinococcus
Sporolactobacillus
Family VII: Paenibacillaceae
Ammoniphilus
Aneurinibacillus
Brevibacillus
Oxalophagus
Paenibacillus
Thermicanus
Thermobacillus
Family VIII: Alicyclobacillaceae
Alicyclobacillus
Pasteuria
Sulfobacillus
Family IX: Thermoactinomycetaceae
Thermoactinomyces
Order II: Lactobacillales

Family I: Lactobacillaceae
Lactobacillus
Paralactobacillus
Pediococcus
Family II: Aerococcaceae
Abiotrophia
Aerococcus
Dolosicoccus
Eremococcus
Facklamia
Globicatella
Tetragenococcus
Ignavigranum
Family III: Carnobacteriaceae
Agitococcus
Alloiococcus
Carnobacterium
Desemzia
Dolosigranulum
Granulicatella
Lactosphaera
Trichococcus
Family IV: Enterococcaceae
Atopobacter
Enterococcus
Melissococcus
Tetragenococcus
Vagococcus
Family V: Leuconostocaceae
Leuconostoc
Oenococcus
Weissela
Family VI: Streptococcaceae
Lactococcus
Streptococcus
Family VII: Incertae sedis
Acetoanaerobium
Oscillospira
Syntrophococcus

Phylum XIV: The Actinobacteria
Class I: Actinobacteria
Order I: Acidimicrobiales
Family I: Acidimicrobiaceae
Acidimicrobium
Order II: Rubrobacterales
Family I: Rubrobacteraceae
Rubrobacter
Order III: Coriobacteridae
Family I: Coriobacteriaceae
Atophobium
Collinsella
Coriobacterium
Cryptobacterium
Denitrobacterium
Eggerthella
Slakia
Order IV: Sphaerobacterales
Family I: Sphaerobacteraceae
Sphaerobacter
Order V: Actinomycetales
Suborder: Actinomycineae
Family I: Actinomycetaceae
Actinobaculum
Actinomyces
Arcanobacterium
Mobiluncus
Suborder: Micrococcineae
Family I: Micrococcaceae
Arthrobacter
Kocuria
Micrococcus
Nesterenkonia
Renibacterium
Rothia
Stomatococcus
Family II: Bogoricellaceae
Bogoriella
Family III: Rarobacteriaceae
Rarobacter
Family IV: Sanguibacteraceae
Sanguibacter
Family V: Brevibacteriaceae
Brevibacterium
Family VI: Cellulomonadaceae
Cellulomonas
Oerskovia
Family VII: Dermabacteraceae
Brachybacterium
Dermabacter
Family VIII: Dermatophilaceae
Dermatophilus
Family IX: Dermacoccaceae
Dermacoccus
Demetria
Kytococcus
Family X: Intrasporangiaceae
Intrasporangium
Janibacter
Ornithinicoccus
Ornithinimicrobium
Terrabacter
Terracoccus
Tetrasphaera
Family XI: Jonesiaceae
Jonesia
Family XII: Microbacteriaceae
Agrococcus
Agromyces
Aureobacterium
Clavibacter
Cryobacterium
Curtobacterium
Frigoribacterium
Leifsonia
Leucobacter
Microbacterium
Rathayibacter
Subtercola
Family XIII: Beutenbergiaceae
Beutenbergia
Family XIV: Promicromonosporaceae
Promicromonospora
Suborder: Corynebacterineae
Family I: Corynebacteriaceae
Corynebacterium
Family II: Dietziaceae
Dietzia
Family III: Gordoniaceae
Gordonia
Skermania
Family IV: Mycobacteriaceae
Mycobacterium
Family V: Nocardiaceae
Nocardia
Rhodococus
Family VI: Tsukamurellaceae
Tsukamurella
Family VII: Williamsiaceae
Williamsia
Suborder: Micromonosporaceae
Family I: Micromonosporaceae
Actinoplanes
Catellatospora
Catenuloplanes
Couchioplanes
Dactylosporangium
Micromonospora
Pilimelia
Spirilliplanes
Verrucosispora
Suborder: Propionibacterineae
Family I: Propionibacteriaceae
Luteococcus
Microlunatus
Propionibacterium
Propioniferax
Tessaracoccus
Family II: Nocardioidaceae
Aeromicrobium
Friedmanniella
Hongia
Kribbella
Micropruina
Marmoricola
Nocardiodes
Suborder: Pseudonocardineae
Family I: Pseudonocardiaceae
Actinoalloteichus
Actinopolyspora

Amycolatopsis
Kibdelosporangium
Kutzneria
Prauserella
Pseudonocardia
Saccharomonospora
Saccharopolyspora
Streptoalloteichus
Thermobispora
Thermocrispum
Family II: Actinosynnemataceae
Actinokineospora
Actinosynnema
Lentzea
Saccharothrix
Suborder: Streptomycineae
Family I: Streptomycetaceae
Streptomyces
Streptoverticillium
Suborder: Streptosporangineae
Family I: Streptosporangiaceae
Acrocarpospora
Herbidospora
Microbispora
Microtetraspora
Nonomuraea
Planobispora
Planomonospora
Planopolyspora
Planotetraspora
Streptosporangium
Family II: Nocardiopsaceae
Nocardiopsis
Thermobifida
Family III: Thermomonosporaceae
Actinomadura
Spirillospora
Thermomonospora
Suborder: Frankineae
Family I: Frankiaceae
Frankia
Family II: Geodermatophilaceae
Blastococcus
Geodermatophilus
Modestobacter
Family III: Microsphaeraceae
Microsphaera
Family IV: Sporichthyaceae
Sporichthya
Family V: Acidothermaceae
Acidothermus
Family VI: Kineosporaceae
Cryptosporangium
Kineococcus
Kineosporia
Suborder XI: Glycomycineae
Family I: Glycomycetaceae
Glycomyces
Order VI: Bifidobacteriales
Family I: Bifidobacteriaceae
Bifidobacterium
Falcivibrio
Gardnerella
Family II: Unknown Affiliation
Actinobispora
Actinocorallia
Excellospora
Pelczaria
Turicella

Phylum XV: Planctomycetes
Order I: Planctomycetales
Family I: Planctomycetaceae
Gemmata
Isosphaera
Pirellula
Planctomyces

Phylum XVI: Chlamydiae
Order I: Chlamydiales
Family I: Chlamydiaceae
Chlamydia
Chlamydophila
Family II: Parachlydiaceae
Parachlamydia
Family III: Simkaniaceae
Simkania
Family IV: Waddliaceae
Waddlia

Phylum XVII: Spirochaetes
Class I: Spirochaetes
Order I: Spirochaetales
Family I: Spirochaetaceae
Borrelia
Brevinema
Clevelandina
Cristispira
Diplocalyx
Hollandina
Pillotina
Spirochaeta
Treponema
Family II: Serpulinaceae
Brachyspira
Serpulina
Family III: Leptospiraceae
Leptonema
Leptospira

Phylum XVIII: Fibrobacteres
Class I: Fibrobacteres
Family I: Fibrobacteraceae
Fibrobacter

Phylum XIX: Actinobacter
Family I: Acidobacteriaceae
Acidobacterium
Geothrix
Holophaga

Phylum XX: Bacteroidetes
Class I: Bacteroidetes
Order I: Bacteroidales
Family I: Bacteroidaceae
Acetofilamentum
Acetomicrobium
Acetothermus
Anaerorhabdus
Bacteroides
Megamonas
Family II: Rikenellaceae
Marinilabilia
Rikenella
Family III: Porphyromonadaceae
Dysgonomonas
Porphyromonas
Family IV: Prevotellaceae
Prevotella
Class II: Flavobacteria
Order I: Flavobacteriales
Family I: Flavobacteriaceae
Bergeyella
Capnocytophaga
Cellulophaga
Chryseobacterium
Coenonia
Empedobacter
Flavobacterium
Gelidibacter
Ornithobacterium
Polaribacter
Psychroflexus
Psychroserpens
Riemerella
Saligentibacter
Weeksella
Family II: Myroideaceae
Myroides
Psychromonas
Family III: Blattabacteriaceae
Blattabacterium
Class III: Sphingobacteria
Order I: Sphingobacteriales
Family I: Sphingobacteriaceae
Pedobacter
Sphingobacterium
Family II: Saprospiraceae
Haliscomenobacter
Lewinella
Saprospira
Family III: Flexibacteraceae
Cyclobacterium

Cytophaga
Dyadobacter
Flectobacillus
Flexibacter
Hymenobacter
Meniscus
Microscilla
Runella
Spirosoma
Sporocytophaga

Family IV: Flammeovirgaceae
Flammeovirga
Flexithrix
Persicobacter
Thermonema

Family V: Crenotrichaceae
Chitinophaga
Crenothrix
Rhodothermus
Toxothrix

Phylum XXI: Fusobacteria
Class I: Fusobacteria
Order I: Fusobacteriales
Family I: Fusobacteriaceae
Fusobacterium
Ilyobacter
Leptotrichia
Propionigenium
Sebaldella
Streptobacillus

Genera incertae sedis
Cetobacterium

Phylum XXII: Verrucomicrobia
Class I: Verrucomicrobiae
Order I: Verrucomicrobiales
Family I: Verrucomicrobiaceae
Prosthecobacter
Verrucomicrobium
Family II: Xiphinematobacteriaceae
Xiphinematobacter

Phylum XXIII: Dictyoglomas
Class I: Dictyoglomi
Order I: Dictyoglomales
Family I: Dictyoglomaceae
Dictyoglomus

APPENDIX D: Major Microbial Agents of Disease by Body System Affected

Table D.1 Microbial Disease Agents Affecting the Skin, Mucous Membranes, and Eyes

Bacterial Disease Agent	Target of Infection	Disease(s) Caused	Mode of Transmission/Acquisition	Text Reference:
Staphylococcus aureus	Moist skin folds	Scalded skin syndrome; impetigo; folliculitis; furuncles (boils); carbuncles; toxic shock syndrome	Direct contact between individuals; via fomites; historically, use of highly absorbent tampons	Ch. 19 (pp. 535–537)
Streptococcus pyogenes	Skin and deeper tissues	Pyoderma; erysipelas; necrotizing fasciitis	Respiratory droplets; close contact between individuals	Ch. 19 (pp. 538–540)
Bacillus anthracis	Skin	Cutaneous anthrax	Inhalation of spores; inoculation of spores into a body via break in skin; ingestion of spores	Ch. 19 (pp. 545–547)
Clostridium perfringens	Skin, muscle and connective tissue	Gas gangrene	Physical trauma resulting in introduction of spores into the body	Ch. 19 (pp. 547–548)
Mycobacterium leprae	Peripheral nerve endings; skin cells	Leprosy (Hansen's disease)	Person-to-person contact; inhalation of respiratory droplets; via breaks in skin	Ch. 19 (pp. 559–560)
Propionibacterium acnes	Sebaceous (oil) glands of skin	Acne	Normal microbiota on skin	Ch. 19 (pp. 560–561)
Actinomyces spp.	Mucous membranes	Actinomycosis	Via breaks in mucous membranes resulting from trauma or infection by other pathogens	Ch. 19 (p. 562)
Pasteurella multocida	Skin and local lymph nodes	Pasteurellosis	Animal bites or scratches; inhalation of aerosols from animals	Ch. 20 (p. 582)
Bartonella henselae	Skin and local lymph nodes	"Cat-scratch disease"	Cat scratches and bites	Ch. 20 (p. 584)
Neisseria gonorrhoeae	Cornea	Ophthalmia neonatorum	Gonococcal infection during birth	Ch. 20 (p. 569)
Chlamydia trachomatis	Conjunctiva	Trachoma	Infection during birth; bacteria from genitalia introduced to eyes via fomites or fingers	Ch. 21 (p. 604)

Table D.1 *(continued)*

Fungal Disease Agent	Target of Infection	Disease(s) Caused	Mode of Transmission/Acquisition	Text Reference:
Blastomyces dermatitidis	Lungs, skin	Cutaneous blastomycosis	Inhalation of dust carrying fungal spores, resulting in dissemination from lungs to skin	Ch. 22 (p. 628)
Trichophyton spp. Microsporum spp. Epidermophyton floccosum	Cutaneous tissues (skin, nails; *Trichophyton* also infects hair)	Dermatophytoses (e.g. Tinea pedis, or "athlete's foot"; Tinea cruris, or "jock-itch")	Spores; fungal elements shed in the environment by infected individuals	Ch. 22 (pp. 638–640)
Malassezia furfur	Skin	Pityriasis; folliculitis; seborrheic dermatitis; dandruff	Normal microbiota on skin	Ch. 22 (p. 639)
Fonsecaea pedrosoi; F. compacta; Phialophora verrucosa; Cladophialophora carrionii	Subepidermal tissues in skin	Chromoblastomycoses	Traumatic introduction of fungal elements to subepidermal tissues	Ch. 22 (p. 640)
Madurella, Pseudallescheria, Exophalia, Acremonium	Skin, fascia (lining of muscles), bones of the hands and feet	Mycetomas	Prick wounds or scrapes caused by twigs, thorns, or leaves contaminated with fungi	Ch. 22 (p. 642)
Sporothrix schenckii	Subcutaneous tissues, usually in arms or legs	Sporotrichosis ("rose-gardener's disease")	Thorn pricks, wood splinters	Ch. 22 (p. 642)
Aspergillus spp.	Eyelids, conjunctiva, eye socket	Aspergillosis	Contamination from spores	Ch. 22 (p. 632)

Parasitic Disease Agent	Target of Infection	Disease(s) Caused	Mode of Transmission/Acquisition	Text Reference:
Acanthamoeba	Eye	Keratitis	Through the conjunctiva via abrasions from contact lenses or trauma; inhalation of contaminated water	Ch. 23 (p. 652)
Leishmania	Skin; mucous membranes of the mouth, nose, or soft palate	Cutaneous leishmaniasis; mucocutaneous leishmaniasis	Bite from infected sand fly	Ch. 23 (pp. 654–655)

Table D.1 *(continued)*

Viral Disease Agent	Target of Infection	Disease(s) Caused	Mode of Transmission/Acquisition	Text Reference:
Smallpox virus (Variola)	Skin	Smallpox	Inhalation of droplets or dried crusts	Ch. 24 (pp. 680–682)
Molluscipoxvirus	Skin	Molluscum contagiosum	Person-to-person contact	Ch. 24 (p. 682)
Herpes simplex virus 1	Lips, skin	Fever blisters (cold sores); ocular herpes; whitlow	Via mucous membranes; cut or break in skin	Ch. 24 (pp. 684–687)
Herpes simplex virus 2	Genitals	Genital herpes; whitlow	Sexual transmission	Ch. 24 (pp. 684–687)
Varicella-zoster virus	Respiratory tract, eyes, blood, skin	Chickenpox; shingles	Person-to-person contact	Ch. 24 (pp. 687–689)
Human herpesvirus 6	Skin, lymph nodes	Roseola	Contact with droplets	Ch. 24 (pp. 692–693)
Human papillomavirus	Epithelium of skin or mucous membranes	Warts, verruca	Via direct contact and fomites; sexual transmission	Ch. 24 (pp. 693–695)
Parvovirus B19	Skin (cheeks, arms, thighs, buttock, trunk)	Erythema infectiosum (Fifth disease)	Contact with droplets	Ch. 24 (pp. 699–700)
Coxsackie A virus	Mouth, pharynx, skin	Herpangina; hand-foot-and-mouth-disease; acute hemorrhagic conjunctivitis; viral meningitis	Via fecal-oral route	Ch. 25 (pp. 707–708)
Rubella virus	Skin, respiratory tract, lymph nodes, blood	Rubella	Via respiratory route	Ch. 25 (p. 713)
Measles virus	Respiratory tract, lymph, blood, conjunctiva, urinary tract, blood vessels, lymphatics, central nervous system	Measles, rubeola	Via respiratory droplets	Ch. 25 (pp. 725–727)

Table D.2 Microbial Disease Agents Affecting the Nervous System

Bacterial Disease Agent	Target of Infection	Disease(s) Caused	Mode of Transmission/Acquisition	Text Reference:
Clostridium botulinum	Neuromuscular junction	Botulism	Consumption of toxin from canned foods; ingestion of endospores	Ch. 19 (pp.549–550)
Clostridium tetani	Inhibitory neurons	Tetanus	Introduction of endospores to break in skin, resulting in formation of toxin	Ch. 19 (pp. 550–551)
Mycobacterium leprae	Peripheral nerve endings	Leprosy (Hansen's disease)	Person-to-person contact; inhalation of respiratory droplets	Ch. 19 (pp.559–560)
Neisseria meningitidis	Brain and spinal cord membranes	Meningococcal meningitis; meningococcal septicemia	Person-to-person contact; inhalation of respiratory droplets	Ch. 20 (pp. 569–570)
Haemophilus influenzae	Brain, meninges	Meningitis; infantile arthritis; epiglottitis; subcutaneous tissue inflammation	Person-to-person contact	Ch. 20 (pp. 582)

Fungal Disease Agent	Target of Infection	Disease(s) Caused	Mode of Transmission/Acquisition	Text Reference:
Cryptococcus neoformans	Central nervous system	Cryptococcal meningitis	Inhalation of spores and/or dried yeast forms from contaminated pigeon or chicken droppings	Ch. 22 (p. 635)

Parasitic Disease Agent	Target of Infection	Disease(s) Caused	Mode of Transmission/Acquisition	Text Reference:
Acanthamoeba spp.	Brain	Amoebic encephalitis	Introduction via cuts or scrapes in the skin; inhalation of contaminated water	Ch. 23 (p. 652)
Echinococcus granulosus	Brain, liver, other organs	Hydatid disease	Consumption of food or water contaminated with eggs shed in dogs' feces.	Ch. 23 (p. 666)
Naegleria spp.	Nasal mucosa; brain	Amoebic meningoencephalitis	Inhalation of contaminated water	Ch. 23 (p. 652)
Trypanosoma brucei	Lymphatic, circulatory and central nervous systems	African sleeping sickness	Bite of infected tsetse fly	Ch. 23 (p. 653)
Trypanosoma cruzi	Blood, lymph, spinal fluid, other body cells	Chagas' disease	Feces from infected *Triatoma* bug introduced via bite wound	Ch. 23 (p. 655)

Table D.2 *(continued)*

Viral Disease Agent	Target of Infection	Disease(s) Caused	Mode of Transmission/Acquisition	Text Reference:
JC polyomaviruses	Kidney cells, blood, brain, central nervous system	Progressive multifocal leukoencephalopathy	Via respiratory route	Ch. 24 (p. 695)
Poliovirus	Pharyngeal and intestinal cells, meninges, central nervous system, spinal cord, brain	Poliomyelitis	Contaminated water, food; via fecal-oral route	Ch. 25 (pp. 706–707)
Togavirus	Blood; occasionally brain, liver, skin, and blood vessels	Eastern equine encephalitis (EEE); Western equine encephalitis (WEE); Venezuelan equine encephalitis (VEE)	Arthropod vectors	Ch. 25 (pp. 709–710)
Flavivirus	Blood; occasionally brain, liver, skin, and blood vessels	St. Louis, Japanese, West Nile, and Russian spring-summer encephalitis; Dengue fever; yellow fever	Arthropod vectors	Ch. 25 (pp. 710–713)
Rabies virus	Skeletal muscle cells; neurons; spinal cord; brain; salivary glands	Rabies	Via saliva of infected animals; introduction of virus through break in skin	Ch. 25 (pp. 728–730)
Bunyavirus	Blood, brain, kidney, liver, blood vessels	California encephalitis; Rift Valley fever	Arthropod vectors	Ch. 25 (pp. 735–736)
Human immunodeficiency virus (HIV)	T-helper cells, macrophage cells; smooth muscle cells; dendritic cells	AIDS	Sexual transmission; intravenous drug use; blood transfusion; contact with infected bodily fluid	Ch. 25 (pp. 718–724)
Prions	Brain	Spongiform encephalopathies (Creutzfeldt-Jakob disease, Kuru)	Consumption of infected tissue	Ch. 13 (p. 399)

Table D.3 Microbial Disease Agents Affecting the Cardiovascular and Lymphatic Systems

Bacterial Disease Agent	Target of Infection	Disease(s) Caused	Mode of Transmission/Acquisition	Text Reference:
Staphylococcus aureus	Lining of heart valves; vagina	Endocarditis; staphylococcal toxic shock syndrome	Introduction of pathogen via skin wound or contaminated medical device; historically, use of highly absorbent tampons	Ch. 19 (pp. 535–536)
Streptococcus pyogenes (Group A streptococcus)	Blood, skin, heart valves, muscle	Septicemia; pericarditis; rheumatic fever; scarlet fever; streptococcal toxic shock syndrome	Person-to-person contact; infection of skin or mucous membrane leads to invasion of deeper tissues and organs	Ch. 19 (pp. 538-541)
Bacillus anthracis	Skin, blood, lungs	Anthrax	Inoculation of spores into body via break in skin; ingestion of spores; inhalation of spores	Ch. 19 (p. 545)
Yersinia pestis	Blood	Plague	Bite of infected rat flea	Ch. 20 (pp. 580–581)
Bartonella henselae	Skin, lymph nodes	"Cat-scratch" disease	Infected cat scratches and bites; bites from infected fleas	Ch. 20 (p. 584)
Brucella spp.	Mucous membranes	Brucellosis (undulant fever)	Consumption of contaminated dairy products; contact with animal blood, urine, and placentas via breaks in mucous membranes	Ch. 20 (p. 585)
Francisella tularensis	Skin, lymph nodes	Tularemia	Bite of infected tick or contact with infected animals such as rabbits and muskrats (bacteria can pass through apparently unbroken skin)	Ch. 20 (pp. 588–589)
Rickettsia ricketsii	Endothelial cells lining blood vessels	Rocky Mountain spotted fever	Bite of infected tick	Ch. 21 (p. 599)
Rickettsia prowazekii	Blood vessels	Epidemic typhus	Bite of infected human body louse	Ch. 21 (p. 600)
Ehrlichia chafeensis; Ehrlichia equi	Blood, leukocytes	Ehrlichiosis	Bite of infected tick, including the dog tick and deer tick	Ch. 21 (pp. 600–601)
Borrelia burgdorferi	Blood	Relapsing fever; Lyme disease	Bite of infected *Ixodes* tick	Ch. 21 (pp. 610–611)

Parasitic Disease Agent	Target of Infection	Disease(s) Caused	Mode of Transmission/Acquisition	Text Reference:
Trypanoma cruzi	Blood	Chagas' disease (results in inflammation and damage of heart muscle)	Bite of infected *Triatoma* bugs	Ch. 23 (p. 655)
Leishmania spp.	Liver, spleen, bone marrow, lymph nodes	Leishmaniasis	Bite of infected female sand flies	Ch. 23 (p. 655)
Plasmodium spp.	Blood, liver, kidney	Malaria	Bite of infected female *Anopheles* mosquito	Ch. 23 (pp. 658–659)
Toxoplasma gondii	Lymph nodes	Toxoplasmosis	Ingestion of contaminated undercooked meat; ingestion or inhalation of contaminated soil; contact with infected cats	Ch. 23 (p. 661)
Schistosoma spp.	Blood; liver, lungs, brain	Schistosomiasis; swimmer's itch	Skin contact with contaminated water	Ch. 23 (p. 667)
Wuchereria bancrofti	Lymphatic system and subcutaneous tissues	Filariasis (elephantiasis)	Infected mosquitoes	Ch. 23 (p. 671)

Table D.3 *(continued)*

Viral Disease Agent	Target of Infection	Disease(s) Caused	Mode of Transmission/Acquisition	Text Reference:
Epstein Barr virus	Epithelial cells of the pharynx and parotid salivary glands; blood; B lymphocytes	Burkitt's lymphoma; infectious mononucleosis	Via saliva	Ch. 24 (pp. 689–691)
Cytomegalovirus	Lymph glands, liver	Birth defects	Via bodily secretions such as saliva, mucus, milk, urine, feces, semen, cervical secretions; sexual transmission; birth; blood transfusions; organ transplants; infected needles	Ch. 24 (pp. 691–692)
Coxsackie B virus	Heart muscle, skin	Myocarditis; pericardial infections; pleurodynia ("devil's grip"); viral meningitis	Via fecal-oral route	Ch. 25 (pp. 707–708)
HTLV-1, HTLV-2	Lymphocytes	Adult acute T-cell lymphocytic leukemia (HTLV-1); Hairy-cell leukemia (HTLV-2)	Sexual transmission; blood transfusion, infected needles	Ch. 25 (p. 717)
Marburg virus, Ebola virus	Multiple body cells, especially macrophages and liver cells	Viral hemorrhagic fevers	Person-to-person via contaminated body fluids, primarily blood	Ch. 25 (pp. 730–731)
Hantavirus	Blood, lungs	Hantavirus pulmonary syndrome	Inhalation of virions in dried deer-mouse urine or feces	Ch. 25 (p. 735)

Table D.4 Microbial Disease Agents Affecting the Respiratory System

Bacterial Disease Agent	Target of Infection	Disease(s) Caused	Mode of Transmission/Acquisition	Text Reference:
Streptococcus pyogenes	Pharynx; tongue	Strep throat; scarlet fever	Inhalation of respiratory droplets	Ch. 19 (p. 539)
Streptococcus pneumoniae	Pharynx; lungs; alveoli	Pneumococcal pneumonia	Inhalation of pneumococci from the pharynx into damaged lungs	Ch. 19 (p. 543)
Bacillus anthracis	Lungs	Inhalation anthrax	Inhalation of spores; inoculation of spores into a body via break in skin; ingestion of spores	Ch. 19 (pp. 545–547)
Corynebacterium diphtheriae	Respiratory membranes	Diphtheria	Inhalation of respiratory droplets	Ch. 19 (pp. 554–555)
Mycobacterium tuberculosis	Lungs	Tuberculosis	Inhalation of respiratory droplets	Ch. 19 (pp. 556–557)
Haemophilus influenzae	Lungs, pharynx	Bronchitis; pneumonia	Inhalation of respiratory droplets	Ch. 20 (p. 583)
Bordetella pertussis	Ciliated epithelial cells of the trachea	Whooping cough	Inhalation of respiratory droplets	Ch. 20 (p. 585)
Legionella pneumophila	Lungs	Legionellosis	Inhalation of aerosols such as those produced by showers, misters, hot tubs, vaporizers, whirlpools, air conditioning systems	Ch. 20 (p. 590)
Coxiella burnetti	Lungs, liver	Q fever	Inhalation of infected bodies that become airborne when released from dried tick and animal feces /urine; consumption of contaminated milk	Ch. 20 (p. 590)
Mycoplasma pneumoniae	Lungs; mucous membranes of the respiratory and urinary tracts	Mycoplasmal pneumonia	Nasal secretions among people in close contact	Ch. 21 (pp. 596–597)
Chlamydia pneumonia	Lungs	Bronchitis, pneumonia, sinusitis	Inhalation of respiratory droplets	Ch. 21 (p. 605)
Chlamydia psittaci	Lungs	Pneumonia caused by psittacosis (ornithosis, or "parrot fever")	Inhalation of respiratory droplets	Ch. 21 (p. 606)

Fungal Disease Agent	Target of Infection	Disease(s) Caused	Mode of Transmission/Acquisition	Text Reference:
Blastomyces dermatitidis	Lungs	Pulmonary blastomycosis	Inhalation of spores from dust	Ch. 22 (p. 628)
Coccidioides immitis	Alveolar spaces of lungs	Coccidioidomycosis	Inhalation of spores	Ch. 22 (p. 629)
Histoplasma capsulatum	Alveolar macrophages in lungs	Histoplasmosis	Inhalation of spores near bird droppings	Ch. 22 (p. 630)
Pneumocystis carinii	Lungs	*Pneumocystis* pneumonia	Inhalation of spores	Ch. 22 (p. 636)

Viral Disease Agent	Target of Infection	Disease(s) Caused	Mode of Transmission/Acquisition	Text Reference:
Adenovirus	Respiratory tract	Common cold	Inhalation of respiratory droplets	Ch. 24 (p. 696)
Rhinovirus	Upper respiratory tract	Common cold	Inhalation of respiratory droplets; via fomites; direct person-to-person contact	Ch. 25 (pp. 704–705)
Coronavirus	Respiratory tract	Severe acute respiratory syndrome (SARS); common cold	Inhalation of respiratory droplets	Ch. 25 (p. 715)
Mumps virus	Upper respiratory cells	Mumps	Respiratory secretions	Ch. 25 (pp. 727–728)
Respiratory syncytial virus	Respiratory tract	Respiratory syncytial disease	Via fomites, hands; especially in child day-care centers	Ch. 25 (p. 728)
Influenzavirus	Lungs, respiratory tract	Influenza	Inhalation of airborne viruses; via fomites	Ch. 25 (pp. 731–734)

Table D.5 Microbial Disease Agents Affecting the Digestive System

Bacterial Disease Agent	Target of Infection	Disease(s) Caused	Mode of Transmission/Acquisition	Text Reference:
Staphylococcus aureus	Lower digestive system	Staphylococcal food poisoning	Consumption of contaminated food, such as potato salad, processed meats, ice cream	Ch. 19 (p. 535)
Viridans streptococci	Mouth	Dental caries	Normal microbiota	Ch. 19 (p. 542)
Clostridium perfringens	Lower digestive system	Gastroenteritis	Consumption of contaminated food	Ch. 19 (p. 547)
Escherichia coli	Lower digestive system	Gastroenteritis	Consumption of contaminated food	Ch. 20 (p. 575)
Salmonella enterica Dublin	Lower digestive system	Salmonellosis	Consumption of contaminated food, such as chicken eggs, or inadequately pasteurized milk	Ch. 20 (p. 577)
Salmonella typhi	Lower digestive system	Typhoid fever	Ingestion of contaminated food or water	Ch. 20 (p. 578)
Shigella	Lower digestive system	Shigellosis	Poor personal hygiene; ingestion of bacteria on hands; consumption of contaminated foods	Ch. 20 (p. 579)
Yersinia enterocolitica	Lower digestive system	Gastroenteritis	Consumption of contaminated food or water	Ch. 20 (p. 580)
Vibrio cholerae	Lower digestive system	Cholera	Ingestion of contaminated food or water	Ch. 21 (p. 615)
Vibrio parahemolyticus	Lower digestive system	Gastroenteritis	Ingestion of contaminated shellfish	Ch. 21 (p. 616)
Vibrio vulnificus	Lower digestive system; blood	Gastroenteritis; septicemia	Consumption of contaminated shellfish	Ch. 21 (p. 616)
Campylobacter jejuni	Lower digestive system	Gastroenteritis	Consumption of contaminated poultry	Ch. 21 (p. 616)
Helicobacter pylori	Lower digestive system	Peptic ulcers	Not known for certain; possibly fecal-oral path of infection; cats may transmit disease to humans	Ch. 21 (p. 618)

Fungal Disease Agent	Target of Infection	Disease(s) Caused	Mode of Transmission/Acquisition	Text Reference:
Aspergillus flavus	Liver	Aflatoxin poisoning	Consumption of toxin in food crops; ingestion of contaminated milk	Ch. 22 (p. 643)

Table D.5 *(continued)*

Parasitic Disease Agent	Target of Infection	Disease(s) Caused	Mode of Transmission/Acquisition	Text Reference:
Entamoeba histolytica	Lumen of the intestine, colon, intestinal mucosa	Amoebic dysentery (amoebiasis)	Drinking of water contaminated with feces containing cysts; fecal contamination of hands or food; oral-anal intercourse	Ch. 23 (p. 651)
Giardia intestinalis	Lumen of the intestine; intestinal mucosa; colon	Giardiasis	Ingestion of cysts in contaminated water; contaminated raw foods; contact with feces	Ch. 23 (p. 657)
Cryptosporidium parvum	Intestine	Cryptosporidiosis	Drinking contaminated water; fecal-oral transmission; poor hygiene	Ch. 23 (p. 662)
Cyclospora cayetanensis	Intestine	Cyclospora diarrheal infection	Ingestion of oocytes on contaminated foods and water	Ch. 23 (p. 663)
Taenia saginata	Intestine	Tapeworm (beef)	Ingestion of cysticerci in raw or undercooked beef	Ch. 23 (p. 665)
Taenia solium	Intestine	Tapeworm (pork)	Ingestion of cysticerci in raw or undercooked pork	Ch. 23 (p. 665)
Echinococcus granulosus	Intestine, circulatory system, liver	Hydatid disease	Consumption of food or water contaminated with eggs shed in dogs' feces	Ch. 23 (p. 666)
Necator americanus; Ancylostoma duodenale	Small intestine; intestinal mucosa	Hookworm	Burrow through skin	Ch. 23 (p. 670)
Ascaris lumbricoides	Small intestine; circulatory and lymphatic systems; lungs	Ascariasis	Consumption of contaminated food or water	Ch. 23 (p. 670)
Enterobius vermicularis	Anus	Pinworms	Ingestion of eggs from fingers or food	Ch. 23 (p. 671)

Viral Disease Agent	Target of Infection	Disease(s) Caused	Mode of Transmission/Acquisition	Text Reference:
Hepatitis B	Liver	Serum hepatitis; hepatic cancer	Sexual transmission; contact with infected bodily fluid; contaminated syringes	Ch. 24 (pp. 696–699)
Hepatitis A	Liver cells	Infectious hepatitis	Fecal-oral route	Ch. 25 (p. 709)
Hepatitis C	Liver cells	Chronic hepatitis	Sexual transmission; contaminated syringes	Ch. 25 (p. 709)
Hepatitis D	Liver cells	Delta agent hepatitis	Sexual transmission; contaminated syringes	Ch. 25 (p. 709)
Hepatitis E	Liver cells	Enteric hepatitis	Fecal-oral route	Ch. 25 (p. 709)
Rotavirus	Respiratory and digestive tracts	Infantile gastroenteritis; diarrhea	Fecal-oral route	Ch. 25 (p. 737)
Cytomegalovirus	Lymph glands, liver	Birth defects	Via bodily secretions such as saliva, mucus, milk, urine, feces, semen, cervical secretions; sexual transmission; birth; blood transfusions; organ transplants	Ch. 24 (pp. 691–692)

Table D.6 Microbial Disease Agents Affecting the Urinary and Reproductive Systems

Bacterial Disease Agent	Target of Infection	Disease(s) Caused	Mode of Transmission/Acquisition	Text Reference:
Staphylococcus aureus	Vagina, blood	Toxic shock syndrome	Historically, use of highly absorbent tampons	Ch. 19 (pp. 535–537)
Neisseria gonorrhoeae	Epithelial cells of mucous membranes lining the genital, urinary and digestive tracts	Gonorrhea; pelvic inflammatory disease (PID)	Sexual transmission	Ch. 20 (pp. 568–569)
Escherichia coli	Urethra, bladder	Cystitis (urinary bladder infection); kidney infection	Normal microbiota	Ch. 20 (p. 575)
Haemophilus ducreyi	Genitals	Chancroid	Sexual transmission	Ch. 20 (p. 582)
Mycoplasma genitalium; Ureaplasma urethritis	Urinary and genital tracts	Nongonococcal urethritis (NGU)	Sexual transmission	Ch. 20 (p. 598)
Chlamydia trachomatis	Mucous membranes of urethra, uterus, uterine tubes, anus, rectum; lymph nodes	Lymphogranuloma venereum	Sexual transmission	Ch. 20 (p. 603)
Treponema pallidum	Genitals	Syphilis	Sexual transmission	Ch. 20 (p. 607)
Leptospira interrogans	Skin, mucous membranes, blood, central nervous system, kidneys	Leptospirosis (kidney infection)	Direct contact with urine of infected animals; indirectly via contact with spirochetes in contaminated environments	Ch. 20 (p. 613)
Fungal Disease Agent	**Target of Infection**	**Disease(s) Caused**	**Mode of Transmission/Acquisition**	**Text Reference:**
Candida albicans	Vagina, urinary tract	Candidiasis	Normal microbiota; sexual contact; childbirth	Ch. 22 (p. 633)
Parasitic Disease Agent	**Target of Infection**	**Disease(s) Caused**	**Mode of Transmission/Acquisition**	**Text Reference:**
Trichomonas vaginalis	Vagina	Vaginitis	Sexual transmission	Ch. 23 (p. 658)
Viral Disease Agent	**Target of Infection**	**Disease(s) Caused**	**Mode of Transmission/Acquisition**	**Text Reference:**
Herpes simplex virus 2	Genitals	Genital herpes; whitlow	Sexual transmission	Ch. 24 (pp. 684–687)
Human papillomavirus	Epithelium of skin or mucous membranes	Warts, verruca	Via direct contact and fomites; sexual transmission	Ch. 24 (pp. 693–694)
Hepatitis B	Liver	Serum hepatitis; hepatic cancer	Sexual transmission; contact with infected bodily fluid; contaminated syringes	Ch. 24 (pp. 696–699)
Human immunodeficiency virus (HIV)	T-helper cells, macrophage cells; smooth muscle cells; dendritic cells	AIDS	Sexual transmission; intravenous drug use; blood transfusion; contact with infected bodily fluid	Ch. 25 (pp. 718–724)

Glossary

A site In a ribosome, a binding site which accommodates tRNA delivering an amino acid.
Abscess An isolated site of infection such as a pimple, boil, or pustule.
Abyssal zone In marine habitats, the zone of water beneath the benthic zone, virtually devoid of life except around hydrothermal vents.
Acellular Noncellular.
Acetyl-CoA Combination of two carbon atoms and co-enzyme A.
Acid Compound that dissociates into one or more hydrogen ions and one or more anions.
Acid-fast stain In microscopy, a differential stain used to penetrate waxy cell walls.
Acidic dye In microscopy, an anionic chromophore used to stain alkaline structures. Works most effectively in acidic environments.
Acidophile Microorganism requiring acidic pH.
Acne Skin disorder characterized by presence of whiteheads, blackheads, and in severe cases, cysts; caused by infection with *Propionibacterium acnes*.
Acquired (secondary) immunodeficiency diseases Any of a group of immunodeficiency diseases that develop in older children, adults, and the elderly as a direct consequence of some other recognized cause, such as infectious disease.
Acquired immunodeficiency syndrome (AIDS) Cluster of characteristic signs and symptoms that develop several years after infection with the human immunodeficiency virus (HIV), which destroys helper T cells.
Actinomycetes High G+C Gram-positive bacteria that form branching filaments and produce spores, thus resembling fungi.
Activation energy The amount of energy needed to trigger a chemical reaction.
Active site Functional site of an enzyme, the shape of which is complementary to the shape of the substrate.
Active transport The movement of a substance against its electrochemical gradient via carrier proteins and requiring cell energy from ATP.
Acute anaphylaxis Condition in which the release of inflammatory mediators overwhelms the body's coping mechanisms.
Acute disease Any disease that develops rapidly but lasts only a short time, whether it resolves in convalescence or death.
Acute inflammation Type of inflammation that develops quickly, is short-lived, and is usually beneficial.
Adenine Ring-shaped nitrogenous base found in nucleotides of DNA and RNA.
Adenosine triphosphate (ATP) The primary short-term, recyclable energy molecule fueling cellular reactions.
Adherence Process by which phagocytes attach to microorganisms through the binding of complementary chemicals on the cell membranes.
Adhesion The attachment of microorganisms to host cells.
Adhesion factors A variety of structures or chemical receptors by which microorganisms attach to host cells.
Adjuvant Chemical added to a vaccine to increase its ability to stimulate active immunity.
Aerobic respiration Type of cellular respiration requiring oxygen atoms as final electron acceptors.
Aerosol A cloud of small droplets and solid particles suspended in the air.
Aerotolerant anaerobe Microorganism which prefers anaerobic conditions but can tolerate exposure to low levels of oxygen.
African sleeping sickness Potentially fatal disease caused by a bite from a tsetse fly carrying *Trypanosoma brucei* and characterized by chancre formation at the site of the bite, followed by parasitemia and central nervous system invasion.
Agar Gel-like polysaccharide isolated from red algae and used as thickening agent.
Agglutination Clumping caused when antibodies bind to two antigens, hindering the activity of pathogenic microorganisms and increasing the chance that they will be phagocytized.
Agglutination test In serology, a procedure in which antiserum is mixed with a sample that potentially contains its target antigen.
Agranulocyte Type of leukocyte having a uniform cytoplasm lacking large granules.
Agroterrorism The use of microbes to terrorize humans by destroying the livestock and crops that constitute their food supply.
Airborne transmission Spread of pathogens to the respiratory mucous membranes of a new host via an *aerosol*.
Alcohol Intermediate-level disinfectant that denatures proteins and disrupts cell membranes.
Aldehyde Compound containing terminal –CHO groups used as a high-level disinfectant because it cross-links organic functional groups in proteins and DNA.
Algae Eukaryotic unicellular or multicellular photosynthetic organisms with simple reproductive structures.
Alkalinophile Microorganism requiring alkaline pH environments.
Allergen An antigen that stimulates an allergic response.
Allergic contact dermatitis Type of delayed hypersensitivity reaction in which chemically modified skin proteins trigger a cell-mediated immune response.
Allergy An immediate hypersensitivity response against an antigen.
Allograft Type of graft in which tissues are transplanted from a donor to a genetically dissimilar recipient.
Alpha-interferons Interferons secreted by virally infected monocytes, macrophages, and some lymphocytes within hours after infection.
Alphaproteobacteria Class of aerobic bacteria in the phylum Proteobacteria capable of growing at very low nutrient levels.
Alternation of generations Method of sexual reproduction in which diploid thalli alternate with haploid thalli.
Alveolar macrophage Fixed macrophage of the lungs.
Alveolates Protozoa with small membrane-bound cavities called *alveoli* beneath their cell surfaces.
Ames test Method for screening mutagens that is commonly used to identify potential carcinogens.
Amination Reaction involving the addition of an amine group to a metabolite to make an amino acid.
Amino acid A monomer of protein.
Aminoglycoside Antimicrobial agent that inhibits protein synthesis by changing the shape of the 30S ribosomal subunit.
Ammonification Process by which microorganisms disassemble proteins in soil wastes into their amino acids, which are then converted to ammonia.
Amoeba Protozoan that moves and feeds by pseudopodia, lacks mitochondria, and reproduces via binary fission.
Amoebiasis A mild to severe dysentery that, if invasive, can cause the formation of lesions in the liver, lungs, brain, and other organs; caused by infection with *Entamoeba histolytica*.
Amoebic encephalitis Often fatal inflammation of the brain characterized by headache, altered mental state, and neurological deficit; caused by infection with *Acanthamoeba*.
Amoebic meningoencephalitis Often fatal inflammation of the brain characterized by headache, vomiting, fever, and destruction of neurological tissue; caused by infection with *Naegleria*.
Amphibolic reaction A reversible metabolic reaction; that is, a reaction which can be catabolic or anabolic.
Amphitrichous Term used to describe a cell having flagella at both ends.
Amphotericin B Medication that is widely used to treat mycoses.
Anabolism All of the synthesis reactions in an organism taken together.
Anaerobic respiration Type of cellular respiration not requiring oxygen atoms as final electron acceptors.
Analytical epidemiology Detailed investigation of a disease, including analysis of data to determine the probable cause, mode of transmission, and possible means of prevention.

Anaphase Third stage of mitosis, during which sister chromatids separate and move to opposite poles of the spindle to form chromosomes. Also used for the comparable stage of meiosis.

Anaphylactic shock Condition in which the release of inflammatory mediators overwhelms the body's coping mechanisms, causing suffocation, edema, smooth muscle contraction, and often death.

Anion A negatively charged ion.

Anthrax Gastrointestinal, cutaneous, or pulmonary disease that is usually fatal without aggressive treatment; caused by ingestion, inoculation, or inhalation of spores of *Bacillus anthracis.*

Antibiotic Antimicrobial agent that is produced naturally by an organism.

Antibody (*immunoglobulin*) Proteinaceous antigen-binding molecule secreted by plasma cells.

Antigen Molecule that triggers a specific immune response.

Antigen presenting cell (APC) Monocytes, macrophages, and their close relatives found in lymphoid organs and under the surface of the skin and characterized by many long, thin cytoplasmic processes called *dendrites.*

Antigen-binding site Site formed by the variable regions of a heavy and light chain of an antibody.

Antigenic determinant (*epitope*) The three-dimensional shape of a region of an antigen that is recognized by the immune system.

Antigenic drift Phenomenon that occurs every 2–3 years when a single strain of influenza mutates within a local population.

Antigenic shift Major antigenic change that occurs on average every 10 years and results from the reassortment of genomes from different influenzavirus strains within host cells.

Antihistamines Drugs which specifically neutralize histamine.

Antimicrobial agent Chemotherapeutic agent used to treat microbial infection.

Antimicrobic Any compound used to treat infectious disease; may also function as intermediate-level disinfectant.

Antisepsis The inhibition or killing of microorganisms on skin or tissue by the use of a chemical antiseptic.

Antiseptic Chemical used to inhibit or kill microorganisms on skin or tissue.

Antiserum In serology, blood fluid containing antibodies which binds to the antigens that triggered their production.

Antitoxin Antibodies formed by the host that bind to and protect against toxins.

Antiviral proteins Proteins triggered by alpha- and beta-interferons that prevent viral replication.

Apicomplexans In protozoan taxonomy, group of pathogenic alveolate protozoa characterized by the complex of special intracellular organelles located at the apices of the infective stages of these microbes.

Apoenzyme The protein portion of protein enzymes that is inactive unless bound to one or more cofactors.

Apoptosis Cell suicide.

Applied microbiology Branch of microbiology studying the commercial use of microorganisms.

Arachnid Group of arthropods distinguished by the presence of eight legs, such as spiders, ticks, and mites.

Arborviruses Group of enveloped, positive ssRNA viruses which are transmitted by arthropods.

Archaea (*Archaeon,* sing.) In Woese's taxonomy, domain which includes all prokaryotic cells having archaeal rRNA sequences.

Arenaviruses Group of segmented, negative ssRNA viruses which cause zoonotic diseases.

Arthropod Animal with a segmented body, hard exoskeleton, and jointed legs, including arachnids and insects.

Artificial wetlands Use of successive ponds, marshes, and meadowlands to remove wastes from sewage as the water moves through the wetlands.

Artificially acquired active immunity Type of immunity which occurs when the body receives antigens by injection, as with vaccinations, and mounts a specific immune response.

Artificially acquired passive immunity Type of immunity which occurs when the body receives via injection preformed antibodies in antitoxins or antisera, which can destroy fast-acting and potentially fatal antigens such as rattlesnake venom.

Ascariasis Symptomatic but not typically fatal disease caused by infection with the nematode *Ascaris lumbricoides.*

Ascomycota Division of fungi characterized by the formation of haploid ascospores within sacs called asci.

Ascospore Haploid germinating structure of ascomycetes.

Ascus Sac in which haploid ascospores are formed and from which they are released in fungi of the division Ascomycota.

Aseptate Lacking crosswalls.

Aseptic Characteristic of an environment or procedure that is free of contamination by pathogens.

Aspergillosis Term for several localized and invasive diseases caused by infection with *Aspergillus* species.

Assembly In virology, fourth stage of the lytic replication cycle, in which new virions are assembled in the host cell.

Asthma Hypersensitivity reaction affecting the lungs and characterized by bronchial constriction and excessive mucous production.

Astroviruses Group of small, round enteric viruses which cause diarrhea, typically in children.

Asymptomatic Characteristic of disease that may go unnoticed because of absence of symptoms, even though clinical tests may reveal signs of disease.

Attachment In virology, first stage of the lytic replication cycle, in which the virion attaches to the host cell.

Atom The smallest chemical unit of matter.

Atomic force microscope Type of probe microscope that uses a pointed probe to traverse the surface of a specimen. A laser beam detects vertical movements of the probe which a computer translates to reveal the atomic topography of the specimen.

Atomic mass (*atomic weight*) The sum of the masses of the protons, neutrons, and electrons in an atom.

Atomic number The number of protons in the nucleus of an atom.

ATP synthase Enzyme that phosphorylates ATP in oxidative phosphorylation and photophosphorylation.

Attenuation The process of reducing vaccine virulence.

Attenuated vaccine Inoculum in which pathogens are weakened so that, theoretically, they no longer cause disease; residual virulence can be a problem.

Autoantigens Antigens on the surface of normal body cells.

Autoclave Device that uses steam heat under pressure to sterilize chemicals and objects that can tolerate moist heat.

Autograft Type of graft in which tissues are moved to a different location within the same patient.

Autoimmune disease Any of a group of diseases which result when an individual begins to make autoantibodies or cytotoxic T cells against normal body components.

Autoimmune hemolytic anemia Disease resulting when an individual produces antibodies against his or her own red blood cells.

Avirulent Harmless.

Axial filament In cell morphology, structure composed of rotating endoflagella that allows a spirochete to "corkscrew" through its medium.

B cell receptor (BCR) Antibody integral to the cytoplasmic membrane and expressed by B lymphocytes.

B lymphocyte (*B cell*) Lymphocyte which arises in the red bone marrow in adults and is found primarily in the spleen, lymph nodes, red bone marrow, and Peyer's patches of the intestines and which secretes antibodies.

Bacillus Rod-shaped prokaryotic cell.

Bacitracin Antimicrobial which blocks NAG and NAM secretion from the cytoplasm, prompting cell lysis.

Bacteremia The presence of bacteria in the blood; often caused by infection with *Staphylococcus aureus* or *Streptococcus pneumoniae.*

Bacteria Prokaryotic microorganisms typically having cell walls composed of peptidoglycan. In Woese's taxonomy, domain which includes all prokaryotic cells having bacterial rRNA sequences.

Bacteriophage (*phage*) Virus that infects and usually destroys bacterial cells.

Bacteriorhodopsin Purple protein synthesized by *Halobacterium* that absorbs light energy to synthesize ATP.

Bacteroid Diverse group of Gram-negative

microbes that have similar rRNA nucleotide sequences, e.g., *Bacteroides*.

Balantidiasis A mild gastrointestinal illness caused by *Balantidium coli*.

Barophile Microorganism requiring the extreme hydrostatic pressure found at great depth below the surface of water.

Bartonellosis Often fatal disease characterized by fever, severe anemia, headache, muscle and joint pain, and chronic skin infections; caused by infection with *Bartonella bacilliformis*.

Base Molecule that binds with hydrogen ions when dissolved in water.

Base-excision repair Mechanism by which enzymes excise a section of a DNA strand containing an error, and then DNA polymerase fills in the gap.

Base pair A complementary arrangement of nucleotides in a strand of DNA or RNA. For example, in both DNA and RNA, guanine and cytosine always pair.

Basic dye In microscopy, a cationic chromophore used to stain acidic structures. Works most effectively in alkaline environments.

Basidiocarp Fruiting body of basidiomycetes; includes mushrooms, puffballs, stinkhorns, jelly fungi, bird's nest fungi, and bracket fungi.

Basidiomycota Division of fungi characterized by production of basidiospores and basidiocarps, fruiting bodies such as mushrooms, puffballs, and bracket fungi.

Basophil Type of granulocyte which stains blue with the basic dye methylene blue.

Bejel An oral disease caused by infection with *Treponema pallidum endemicum*.

Benign tumor Mass of neoplastic cells that remains in one place and is not generally harmful.

Benthic zone Bottom zone of freshwater or marine water, devoid of light and nutrients.

Beta-interferons Interferons secreted by virally infected fibroblasts within hours after infection.

Beta-lactam Antimicrobial whose functional portion is composed of beta-lactam rings, which inhibit peptidoglycan formation by irreversibly binding to the enzymes that cross-link NAM subunits.

Beta-lactamase Bacterial enzyme which breaks the beta-lactam rings of penicillin and similar molecules, rendering them inactive.

Beta-oxidation A catabolic process in which enzymes split pairs of hydrogenated carbon atoms from a fatty acid and join them to coenzyme A to form acetyl-CoA.

Betaproteobacteria Diverse group of Gram-negative bacteria in the phylum Proteobacteria capable of growing at very low nutrient levels.

Binary fission The most common method of asexual reproduction of prokaryotes, in which the parental cell disappears with the formation of progeny.

Binomial nomenclature The classification method used in the Linnaean system of taxonomy, which assigns each species both a genus name and a specific epithet.

Biochemical oxygen demand (BOD) A measure of the amount of oxygen aerobic bacteria require to metabolize organic and inorganic nitrogenous wastes in water.

Biochemistry Branch of chemistry which studies the chemical reactions of living things.

Bioconversion The transformation of organic materials into fuels via the use of microorganisms.

Biodiversity The number of species living in a given ecosystem.

Biofilm A slimy, organized system of bacteria on an environmental surface.

Biogeochemical cycling The movement of elements and nutrients from unusable forms to useable forms due to the activities of microorganisms.

Biological vector Biting arthropod or other animal that transmits pathogens and serves as host for the multiplication of the pathogen during some stage of the pathogen's life cycle.

Biomass The quantity of all organisms in a given ecosystem.

Bio-mining The science involved in searching natural populations of soil microbes for useful organisms.

Bioremediation The use of microorganisms to metabolize toxins in the environment to reclaim soils and waterways.

Bioreporter Type of biosensor composed of microbes with innate signaling capabilities.

Biosensor Device that combines bacteria or microbial products such as enzymes with electronic measuring devices to detect other bacteria, bacterial products, or chemical compounds in the environment.

Biosphere The region of Earth inhabited by living organisms.

Biotechnology Branch of microbiology in which microbes are manipulated to manufacture useful products.

Bioterrorism The use of microbes or their toxins to terrorize human populations.

Black piedra Benign scalp disease caused by infection with *Piedraea hortae*.

Blastomycosis Pulmonary disease found in the southeastern United States caused by infection with *Blastomyces dermatidis*.

Blood group antigens The surface molecules of red blood cells.

Bodily fluid transmission Spread of pathogenic microorganisms via blood, urine, saliva, or other bodily fluids.

Botulism Potentially fatal intoxication with botulism toxin; three types include foodborne botulism, infant botulism, and wound botulism.

Broad-spectrum drug Antimicrobial that works against many different kinds of pathogens.

Broth A liquid, nutrient-rich medium used for cultivating microorganisms.

Broth-dilution test Test for determining the minimum inhibitory concentration in which a standardized amount of bacteria is added to serial dilutions of antimicrobial agents in tubes or wells containing broth.

Bruton-type agammaglobulinemia An inherited disease in which affected babies cannot make immunoglobulins and experience recurrent bacterial infections.

Bubonic plague Severe systemic disease, fatal if untreated in 75% of patients, and characterized by fever, tissue necrosis, and the presence of buboes; caused by infection with *Yersinia pestis*.

Budding In prokaryotes, reproductive process in which an outgrowth of the parent cell receives a copy of the genetic material, enlarges, and detaches. In virology, extrusion of enveloped virions through the host's cell membrane.

Buffer A substance, such as a protein, that prevents drastic changes in internal pH.

Bulbar poliomyelitis Infection of the brain stem and medulla resulting in paralysis of muscles in the limbs or respiratory system; caused by infection with poliovirus.

Bunyaviruses Group of zoonotic pathogens that have a segmented genome of three –ssRNA molecules and are transmitted to humans via arthropods.

Burkitt's lymphoma Infectious cancer of the jaw caused by infection with Epstein-Barr virus.

Caliciviruses Group of small, round enteric viruses which cause diarrhea, nausea, and vomiting.

Calor Heat.

Calvin-Benson cycle Stage of photosynthesis in which atmospheric carbon dioxide is fixed and reduced to produce glucose.

Cancer Disease characterized by the presence of one or more malignant tumors.

Candidiasis Term for several opportunistic diseases caused by infection with *Candida* species.

Capnophile Microorganism which grows best with high levels of carbon dioxide in addition to low levels of oxygen.

Capsid A protein coat surrounding the nucleic acid core of a virion.

Capsomere A proteinaceous subunit of a capsid.

Capsule Glycocalyx composed or repeating units of organic chemicals firmly attached to the cell surface.

Capsule stain (see *negative stain*)

Carbohydrate Organic macromolecule consisting of atoms of carbon, hydrogen, and oxygen.

Carbon cycle Biogeochemical cycle in which carbon is cycled in the form of organic molecules.

Carbon fixation The attachment of atmospheric carbon dioxide to ribulose 1,5-bisphosphate (RuBP).

Carbuncle The coalescence of several furuncles extending deep into underlying tissues; caused by infection with *Staphylococcus aureus*.

Carcinogen Chemical capable of causing genetic mutations that result in cancer.

Caries (*cavities*) Tooth decay; caused by viridans streptococci.

Carotenoid Plant pigment which acts as an antioxidant.

Carrageenan Gel-like polysaccharide isolated from red algae and used as thickening agent.
Carrier In human pathology, continuous asymptomatic human source of infection.
Catabolism All of the decomposition reactions in an organism taken together.
Catarrhal phase In pertussis, initial phase lasting 1–2 weeks and characterized by signs and symptoms resembling those of a common cold.
Cation A positively charged ion.
Cat-scratch disease Common and occasionally serious infection in children; characterized by fever and malaise plus localized swelling; caused by infection with *Bartonella henselae*.
CD95 pathway In cell-mediated cytotoxicity, pathway involving CD95 protein which triggers apoptosis of infected cells.
Cell culture Cells isolated from an organism and grown on the surface of a medium or in broth. Viruses can be grown in a cell culture.
Cell wall In prokaryotic and some eukaryotic cells, structural boundary composed of polysaccharide chains that provides shape and support against osmotic pressure.
Cell-mediated immune response Immune response used by T cells to fight intracellular pathogens and abnormal body cells.
Cellular respiration Metabolic process that involves the complete oxidation of substrate molecules and production of ATP via a series of redox reactions.
Cellular slime mold Individual haploid myxamoeba that phagocytizes bacteria, yeasts, dung, and decaying vegetation.
Central dogma In genetics, fundamental description of protein synthesis which states that genetic information is transferred from DNA to RNA to polypeptides, which function alone or in conjunction as proteins.
Centrioles Nonmembranous organelles in animal cells which appear to function in the formation of flagella and cilia and in cell division.
Centrosome Region of a cell containing centrioles.
Cesspool Home equivalent of primary wastewater treatment in which wastes enter a series of porous concrete rings buried underground and are digested by microbes.
Cestodes (*tapeworms*) Group of helminths that are long, flat, and segmented and lack digestive systems.
Chagas' disease Potentially fatal disease caused by a bite from a kissing bug carrying *Trypanosoma cruzi* and characterized by the formation of swellings called *chagomas* at the site of the bite, followed by fever, swollen lymph nodes, myocarditis, organ enlargement, and eventually congestive heart failure.
Chancre Painless red lesion that appears at the site of infection with *Treponema pallidum*, the agent of syphilis.
Chemical bond An interaction between atoms in which electrons are either shared or transferred in such a way as to fill their valence shells.
Chemical fixation In microscopy, a technique which uses methyl alcohol or formalin to attach a smear to a slide.
Chemical reaction The making or breaking of a chemical bond.
Chemiosmosis Use of ion gradients to generate ATP.
Chemoautotroph Microorganism which uses carbon dioxide as a carbon source and catabolizes organic molecules for energy.
Chemoheterotroph Microorganism which uses organic compounds for both energy and carbon.
Chemokine An immune system cytokine which signals leukocytes to rush to the site of inflammation or infection and activate other leukocytes.
Chemotactic factors Peptides derived from complement and chemicals called cytokines which are released by leukocytes at a site of infection and attract phagocytic cells.
Chemotaxis Cell movement that occurs in response to chemical stimulus.
Chemotherapeutic agent Chemical used to treat disease.
Chemotherapy A branch of medical microbiology in which chemicals are studied for their potential to destroy pathogenic microorganisms.
Chitin Strong, flexible nitrogenous polysaccharide found in fungal cell walls and in the exoskeletons of insects and other arthropods.
Chlamydia Any of the Gram-negative pathogenic cocci having both DNA and RNA and growing and reproducing within the cells of mammals, birds, and a few invertebrates.
Chlorophyll Pigment molecule which captures light energy for use in photosynthesis.
Chlorophyta Green-pigmented division of algae that have chlorophylls *a* and *c*, store sugar and starch as food reserves, have cell walls composed of cellulose, and have rRNA sequences similar to plants. Considered the progenitors of plants.
Chloroplast Light-harvesting organelle found in photosynthetic eukaryotes.
Cholera Disease contracted through the ingestion of food and water contaminated with *Vibrio cholerae* and characterized by vomiting and watery diarrhea.
Cholera toxin Exotoxin produced by *Vibrio cholerae* which causes the movement of water out of the intestinal epithelium.
Chromatin Thread-like mass of DNA and associated histone proteins that becomes visible during mitosis as chromosomes.
Chromatin fiber An association of nucleosomes and proteins found within the chromosomes of eukaryotic cells.
Chromoblastomycosis Cutaneous and subcutaneous disease characterized by lesions that can disseminate internally; caused by traumatic introduction of dematiaceous fungi into the skin.
Chromosome In prokaryotes, a typically circular molecule of DNA associated with protein and RNA molecules and localized in a region of the cytosol called the nuceloid. In eukaryotes, chromosomes are thread-like and are most visible during mitosis and meiosis.
Chronic disease Any disease that develops slowly, usually with less severe symptoms, and is continual or recurrent.
Chronic granulomatous disease Primary immunodeficiency disease in which children have recurrent infections characterized by the development of large masses of inflammatory cells in lymph nodes, lungs, bones, and skin.
Chronic inflammation Type of inflammation that develops slowly, lasts a long time, and can cause damage (even death) to tissues, resulting in disease.
Chrysophyta Division of algae including the golden algae, yellow-green algae, and diatoms.
Cilia Short, hairlike, rhythmically motile projections covering some eukaryotic cells.
Ciliate In protozoan taxonomy, group of alveolate protozoa characterized by the presence of cilia in their trophozoite stages.
Class Taxonomical grouping of similar orders of organisms.
Clinical specimen Sample of human material, such as feces or blood, that is examined or tested for the presence of microorganisms.
Clonal deletion Process by which cells with receptors that respond to autoantigens are selectively killed via apoptosis.
Coccidioidomycosis Pulmonary disease found in the southwestern United States caused by infection with *Coccidioides immitis*.
Coccobacillus A prokaryotic cell intermediate in shape between a sphere and a rod, e.g., an elongated coccus.
Coccus Spherical prokaryotic cell.
Codon Triplet of mRNA nucleotides that codes for specific amino acids. For example, AAA is a codon for lysine.
Coenocyte Multinucleate cell resulting from repeated mitosis but postponed or absent cytokinesis.
Coenzyme Organic cofactor.
Cofactor Inorganic ions or organic molecules that are essential for enzyme action.
Cold enrichment Incubation of a specimen in a refrigerator to enhance the growth of cold-tolerant species.
Coliforms Enteric Gram-negative bacteria that ferment lactose to gas and are found in the intestinal tracts of animals and humans.
Colony Visible population of microorganisms living in one place.
Colony-forming unit A single cell or group of related cells used to obtain a pure culture.
Colorado tick fever Mild disease characterized by fever and chills and caused by infection with a coltivirus.
Coltiviruses Group of reoviruses that cause Colorado tick fever, a mild disease involving fever and chills.
Commensalism Symbiotic relationship in which one member benefits without significantly affecting the other.
Communicable disease Any infectious disease that comes either directly or indirectly from another host.

Competence Ability of a prokaryotic cell to take up DNA from the environment.

Competitive inhibitor Inhibitory substance that blocks enzyme activity by blocking active sites.

Complement fixation test A complex assay used to determine the presence of specific antibodies by the ability of the antibodies to nullify the action of complement on sheep red blood cells.

Complement system Set of proteins that act as chemotactic attractants, trigger inflammation and fever, and ultimately effect the destruction of foreign cells.

Complementary DNA (cDNA) DNA synthesized from an mRNA template using reverse transcriptase.

Complex medium Culturing medium that contains nutrients released by the partial digestion of yeast, beef, soy, or other proteins; thus, the exact chemical composition is unknown.

Complex transposon Transposon containing inverted repeats, transposase, and other genes not connected with transposition.

Compound A molecule containing atoms of more than one element.

Compound microscope Microscope using a series of lenses for magnification.

Concentration gradient The difference in concentration of a chemical on the two sides of a membrane. Also called a *chemical gradient*.

Condenser lens In a compound microscope, a lens which directs light through the specimen as well as one or more mirrors which deflect its path.

Condyloma acuminata Large, cauliflower-like genital warts caused by infection with a papillomavirus.

Confocal microscope Type of light microscope which uses ultraviolet lasers to illuminate fluorescent chemicals in a single plane of the specimen.

Congenital syphilis Disease characterized by mental retardation, organ malformation, and in some cases, death of the fetus of a woman infected with *Treponema pallidum*, the agent of syphilis.

Conjugation Method of horizontal gene transfer in which a bacterium containing a fertility plasmid forms a conjugation pilus that attaches and transfers plasmid genes to a recipient.

Conjugation pilus Proteinaceous, rodlike structure extending from the surface of a cell; mediates conjugation.

Contagious disease A communicable disease that is easily transmitted from a reservoir or patient.

Continuous cell culture Type of cell culture created from tumor cells.

Contrast The difference in intensity between two objects, or between an object and its background.

Convalescence In the infectious disease process, final stage during which the patient recovers from the illness and tissues and systems are repaired and return to normal.

Convalescent phase In pertussis, final phase lasting 3–4 weeks during which the ciliated lining of the trachea grows back and frequency of coughing spells diminishes, but secondary bacterial infections may ensue.

Cord factor A cell-wall component of the cell walls of *Mycobacterium tuberculosis* that produces strands of daughter cells, inhibits migration of neutrophils, and is toxic to body cells.

Coronaviruses Group of enveloped ssRNA viruses which cause colds as well as gastroenteritis in children.

Corticosteroids Immunosuppressive agents, including prednisone and methylprednisone, which suppress the response of T cells.

Counterstain In a Gram stain, red stain which provides contrasting color to the primary stain, causing Gram-negative cells to appear pink.

Covalent bond The sharing of a pair of electrons by two atoms.

Coxsackieviruses Group of enteroviruses which cause mild fever, muscle aches, and malaise in humans.

Cristae Folds within the inner membrane of mitochondria that increase its surface area.

Cross resistance Phenomenon in which resistance to one antimicrobial drug confers resistance to similar drugs.

Crossing over Process in which portions of homologous chromosomes are recombined during the formation of gametes.

Croup Inflammation and swelling of the larynx, trachea, and bronchi and a "seal bark" cough caused by infection with a parainfluenza virus.

Cryptococcoses Two diseases often resulting in meningitis and caused by infection with *Cryptococcus neoformans*.

Cryptosporidiosis A gastrointestinal disease caused by infection with *Cryptosporidium parvum*; in humans, characterized by diarrhea and fluid and weight loss; may be fatal in HIV-positive patients.

Culture Act of cultivating microorganisms or the microorganisms that are cultivated.

Cyanobacteria Gram-negative phototrophic bacteria that vary greatly in shape, size, and method of reproduction.

Cyclic photophosphorylation Return of electrons to the original reaction center after passing down an electron transport chain.

Cycloserine Semisynthetic antibiotic used to treat infections with Gram-positive bacteria.

Cyst In protozoan morphology, the hardy resting stage characterized by a thick capsule and a low metabolic rate.

Cytokines Chemicals released by leukocytes at a site of infection.

Cytokinesis Division of a cell's cytoplasm.

Cytoplasm General term used to describe the semiliquid, gelatinous material inside a cell.

Cytoplasmic membrane Membrane found in prokaryotic and eukaryotic cells, and composed of a fluid mosaic of phospholipids and proteins.

Cytosine Ring-shaped nitrogenous base found in nucleotides of DNA and RNA.

Cytoskeleton Internal network of fibers contributing to the basic shape of eukaryotic and rod-shaped prokaryotic cells.

Cytosol The liquid portion of the cytoplasm.

Cytotoxic drugs Group of drugs which inhibit mitosis and cytokinesis.

Cytotoxic T cell (*CD8 cell*) In cell-mediated immune response, type of cell characterized by CD8 cell-surface glycoprotein which secretes perforins and granzymes that destroy infected or abnormal body cells.

Dark-field microscope Microscope used for studying pale or small specimens; deflects light rays so that they miss the objective lens.

Dark repair Mechanism by which enzymes cut damaged DNA sections from a molecule, creating a gap that is repaired by DNA polymerase and DNA ligase.

Deamination Process in which amine groups are split off from amino acids.

Death phase Phase in a growth curve in which the organisms are dying more quickly than they are being replaced by new organisms.

Decimal reduction time The time required to destroy 90% of the microbes in a sample.

Decline In the infectious disease process, period in which the body gradually returns to normal as the patient's immune response and any medical treatments vanquish the pathogens.

Decolorizing agent In a Gram stain, a solution which breaks down the wall of Gram-negative cells, allowing the primary stain to be washed away.

Decomposition reaction A chemical reaction in which the bonds of larger molecules are broken to form smaller atoms, ions, and molecules.

Deep freezing Long-term storage of cultures at temperatures ranging from –50°C to –95°C.

Deeply branching bacteria Prokaryotic autotrophs with rRNA sequences and growth characteristics thought to be similar to those of earliest bacteria.

Defensins Small peptide chains that act against a broad range of pathogens.

Defined medium (synthetic medium) Culturing medium of which the exact chemical composition is known.

Definitive host In the life cycle of parasites, host in which mature and sometimes sexual forms of the parasite are present and usually reproducing.

Degerming The removal of microbes from a surface by scrubbing.

Dehydration synthesis Type of synthesis reaction in which two smaller molecules are joined together by a covalent bond, and a water molecule is formed.

Delayed hypersensitivity reaction (*type IV hypersensitivity*) T cell-mediated inflammatory reaction that takes 24–72 hours to reach maximal intensity.

Deletion Type of mutation in which a nucleotide base pair is deleted.

Deltaproteobacteria Group of proteobacteria that includes *Desulfovibrio*, *Bdellovibrio*, and myxobacteria.

Denaturation Process by which the bonds within an enzyme or other protein break, disrupting the protein's three-dimensional structure.
Dendritic cells (*Langerhans cells*) Cells of the epidermis that devour pathogens.
Dengue fever Self-limiting but extremely painful disease caused by a flavivirus transmitted by *Aedes* mosquitoes.
Dengue hemorrhagic fever Potentially fatal disease involving a hyperimmune response to reinfection with the dengue virus and causing ruptured blood vessels, internal bleeding, and shock.
Denitrification The conversion of nitrate into nitrogen gas by anaerobic respiration.
Deoxyribonucleic acid (**DNA**) Nucleic acid consisting of nucleotides made up of phosphoric acid, a deoxyribose pentose sugar, and an arrangement of the bases adenine, guanine, cytosine, and thymine.
Dermatophytoses Any of a variety of superficial skin, nail, and hair infections caused by dermatophytes.
Dermis The layer of the skin deep to the epidermis and containing hair follicles, glands, and nerve endings.
Descriptive epidemiology The careful recording of data concerning a disease.
Dessication Inhibition of microbial growth by drying.
Deuteromycetes Informal grouping of fungi having no known sexual stage.
Diapedesis (*emigration*) Process whereby leukocytes leave intact blood vessels by squeezing between lining cells.
Diatom Type of algae in the division Chrysophyta which has cell walls made of silica arranged in nesting halves called frustules.
Dichotomous key In taxonomy, method of identifying organisms in which information is arranged in paired statements, only one of which applies to any particular organism.
Differential interference contrast microscope Type of phase microscope which uses prisms to split light beams, giving images a three-dimensional appearance.
Differential medium Culturing medium formulated such that either the presence of visible changes in the medium or differences in the appearances of colonies helps microbiologists differentiate among different kinds of bacteria growing on the medium.
Differential stain In microscopy, a stain using more than one dye so that different structures can be distinguished. The Gram stain is the most commonly used.
Differential white blood cell count Lab technique that indicates the relative numbers of leukocytes.
Diffusion The net movement of a chemical down its concentration gradient.
Diffusion-susceptibility test (Kirby-Bauer test) Simple, inexpensive test widely used to reveal which drug is most effective against a particular pathogen. Procedure involves inoculating a Petri plate uniformly with a standardized amount of the pathogen in question and arranging on the plate disks soaked in the drugs to be tested.
DiGeorge anomaly Failure of the thymus to develop, and thus, absence of T cells.
Dimorphic Having two forms, e.g., dimorphic fungi have both yeast-like and mold-like thalli.
Dinoflagellate In protozoan taxonomy, group of unicellular alveolate protozoa characterized by photosynthetic pigments such as carotene and chlorophylls.
Diphtheria Mild to potentially fatal respiratory disease caused by diphtheria toxin following infection with *Corynebacterium diphtheriae.*
Diplococcus A pair of cocci.
Diploid A nucleus with two copies of each chromosome.
Diploid cell culture Type of cell culture created from embryonic animal, plant, or human cells that have been isolated and provided appropriate growth conditions.
Direct contact transmission Spread of pathogens from one host to another involving body contact between the hosts.
Direct fluorescent antibody test Immune test allowing direct observation of the presence of antigen in a tissue sample flooded with labeled antibody.
Disaccharide Carbohydrate consisting of two monosaccharide molecules joined together.
Disease Any adverse internal condition severe enough to interfere with normal body functioning.
Disease process Definite sequence of events following contamination and infection.
Disinfectant Physical or chemical agent used to inhibit or destroy microorganisms on inanimate objects.
Disinfection The use of physical or chemical agents to inhibit or destroy microorganisms on inanimate objects.
Disk-diffusion method Method of evaluating the effectiveness of a disinfectant or antiseptic in which a zone of inhibition is visible around disks soaked in effective agents. Also called *Kirby-Bauer method.*
Disseminated intravascular coagulation (DIC) The formation of blood clots within blood vessels throughout the body; triggered by lipid A.
Dolor Pain.
Domain Any of three basic types of cell groupings distinguished by Carl Woese, containing or replacing the Linnaean taxon of *kingdom.*
Donor cell In horizontal gene transfer, a cell which contributes part of its genome to a recipient.
Droplet transmission Spread of pathogens from one host to another via droplet nuclei that exit the body during exhaling, coughing, and sneezing.

E site In translation, site at which tRNA exits from the ribosome.
Eastern equine encephalitis (EEE) Potentially fatal infection of the brain caused by a togavirus.
Ebola virus Type of hemorrhagic fever caused by a filamentous virus and fatal in 90% of cases.
Echoviruses (*enteric cytopathic human orphan viruses*) Group of enteroviruses which cause viral meningitis and colds.
Ecosystem All of the organisms living in a particular habitat, and the relationships between the two.
Electrical gradient Voltage across a membrane created by the electrical charges of the chemicals on either side.
Electrochemical gradient The chemical and electrical gradients across a cell membrane.
Electrolyte Any hydrated cation or anion; can conduct electricity through a solution.
Electron A negatively charged subatomic particle.
Electron transport chain Series of redox reactions that pass electrons from one membrane-bound carrier to another, and then to a final electron acceptor.
Electronegativity The attraction of an atom for electrons.
Elek test Immunodiffusion assay used to detect the presence of diphtheria toxin in a fluid sample.
Element Matter that is composed of a single type of atom.
Elementary bodies Infectious stage in the life cycle of chlamydias.
Elephantiasis Enlargement and hardening of tissues, especially in the lower extremities, where lymph has accumulated following infection with *Wuchereria bancrofti.*
Empyema In patients with staphylococcal pneumonia, the presence of pus in the alveoli of the lungs.
Encystment In the life cycle of protozoa, stage in which cysts form in host tissues.
Endemic In epidemiology, a disease that occurs at a relatively stable frequency within a given area or population.
Endemic relapsing fever Disease characterized by recurrent episodes of fever and septicemia separated by symptom-free intervals; caused by infection with any of several types of *Borrelia* carried by ticks in the genus *Ornithodoros.*
Endocarditis Potentially fatal inflammation of the endocardium; typically caused by infection with *Staphylococcus aureus* or *Streptococcus pneumoniae.*
Endocytosis Active transport process, used by some eukaryotic cells, in which pseudopodia surround a substance and move it into the cell.
Endogenous antigen Antigen produced by microorganisms that multiply inside the cells of the body.
Endogenous infection An infection arising within the patient from opportunistic pathogens.
Endoflagellum A special flagellum of spirochetes that spirals tightly around a cell, rather than protruding from it.
Endoplasmic reticulum Netlike arrangement of hollow tubules continuous with the outer membrane of the nuclear envelope and functioning as a transport system.
Endospore Environmentally resistant structure produced by the transformation of a

vegetative cell of the Gram-positive genus *Bacillus* or *Clostridium*.
Endosymbiotic theory Theory proposing that eukaryotes formed from the phagocytosis of small aerobic prokaryotes by larger anaerobic prokaryotes.
Endothermic reaction Any chemical reaction that requires energy.
Endotoxin (*lipid A*) Potentially fatal toxin released from the cell wall of dead and dying Gram-negative bacteria.
Enrichment culture Technique used to enhance the growth of less abundant microorganisms by using a selective medium.
Enterobacteriaceae (enteric bacteria) Family of oxidase negative Gram-negative bacteria which can be pathogenic.
Enteroviruses Group of picornaviruses that are transmitted via the fecal-oral route but cause disease in any of a variety of target organs.
Entner-Doudoroff pathway Series of reactions that catabolize glucose to pyruvic acid using different enzymes from those used in either glycolysis or the pentose phosphate pathway.
Entry In virology, second stage of the lytic replication cycle, in which the virion or its genome enters the host cell.
Envelope In virology, membrane surrounding the viral capsid.
Environmental microbiology Branch of microbiology studying the role of microorganisms in soils, water, and other habitats.
Environmental specimen Sample of material taken from such sources as ponds, soil, and air and tested for the presence of microorganisms.
Enzyme An organic catalyst.
Enzyme-linked immunosorbent assays (ELISAs) A family of simple immune tests that use an enzyme as a label and that can be readily automated and read by machine.
Eosinophil Type of granulocyte which stains red to orange with the acidic dye eosin.
Epidemic In epidemiology, a disease that occurs at a greater than normal frequency for a given area or population.
Epidemic relapsing fever Disease characterized by recurrent episodes of fever and septicemia separated by symptom-free intervals; caused by infection with *Borrelia recurrentis*.
Epidemic typhus Disease characterized by high fever, mental and physical depression, and rash following exposure to lice carrying *Rickettsia prowazekii*.
Epidemiology Study of the occurrence, distribution, and spread of disease in humans.
Epidermis The outermost layer of the skin.
Epsilonproteobacteria Group of Gram-negative rods, vibrios, or spirals in the phylum Proteobacteria.
Erysipelas Pyoderma occurring on the faces of children accompanied by pain and inflammation and caused by infection with group A streptococci.
Erythema infectiosum (*fifth disease*) Harmless red rash occurring in children and caused by infection with B19 virus.
Erythrocyte Red blood cell.
Erythrocytic cycle In the lifecycle of *Plasmodium*, stage during which merozoites infect and cause lysis of erythrocytes.
Etest Test for determining minimum inhibitory concentration in which a plastic strip containing a gradient of the antimicrobial agent being tested is placed on a plate inoculated with the pathogen of interest.
Ethambutol Antimicrobial which blocks mycolic acid synthesis in the walls of mycobacteria.
Etiology The study of the causation of disease.
Euglenoids Euglenozoan protozoans that store food as paramylon, lack cell walls, and have eyespots used in positive phototaxis.
Eukarya In Woese's taxonomy, domain which includes all eukaryotic cells.
Eukaryote Any organism made up of cells containing a nucleus composed of genetic material surrounded by a distinct membrane. Classification includes animals, plants, algae, fungi, and protozoa.
Eutrophication The overgrowth of microorganisms in aquatic systems.
Exchange reaction Type of chemical reaction in which atoms are moved from one molecule to another by means of the breaking and forming of covalent bonds.
Excystment In the life cycle of protozoa, stage following ingestion by the host, in which cysts become trophozoites.
Exocytosis Active transport process, used by some eukaryotic cells, in which vesicles fuse with the cytoplasmic membrane and export their substances from the cell.
Exoerythrocytic cycle In the lifecycle of *Plasmodium*, stage during which infected mosquito infects sporozoites into the blood.
Exogenous antigen Antigen produced by microorganisms that multiply outside the cells of the body.
Exogenous infection An infection caused by pathogens acquired from the health care environment.
Exon Coding sequence of mRNA. Exons are connected to produce a functional mRNA molecule.
Exothermic reaction Any chemical reaction that releases energy.
Exotoxin Toxin secreted by a pathogenic microorganism into its environment.
Experimental epidemiology The testing of hypotheses resulting from analytical epidemiology concerning the cause of a disease.
Extracellular state In virology, state in which a virus exists outside of a cell and is called a virion.
Extremophile Microbe in the domain Archaea that requires extreme conditions of temperature, pH, and/or salinity to survive.

F (fertility) plasmid (F factor) Molecule of DNA coding for conjugation pili. Cells that contain an F plasmid are called F^+ cells and serve as donors during conjugation.
Facilitated diffusion Movement of substances across a cell membrane via protein channels.
Facultative anaerobe Microorganism which can live with or without oxygen.
Family Taxonomical grouping of similar genera of organisms.
Fats Compounds composed of three fatty acid molecules linked to a molecule of glycerol.
F_C region The stem region of an antibody.
Fecal-oral infection Spread of pathogenic microorganisms in feces to the mouth, such as results from drinking sewage-contaminated water.
Feedback inhibition (negative feedback) Method of controlling the action of enzymes in which the end-product of a series of reactions inhibits an enzyme in an earlier part of the pathway.
Fermentation The partial oxidation of sugar to release energy using an organic molecule rather than an electron transport chain as the final electron acceptor.
Fever Body temperature above 37°C.
Fever blisters (*cold sores*) Painful, itchy lesions on the lips; characteristic of infection with herpes simplex virus type 1.
Filariasis Infection of the lymphatic system caused by a filarial nematode.
Filtration The passage of air or liquid through a material that traps and removes microbes.
Fimbriae Sticky, nonmotile, proteinaceous extensions of some bacterial cells that function to adhere cells to one another and to environmental surfaces.
Firmicutes Phylum of bacteria that includes clostridia, mycoplasmas, and low G+C Gram-positive bacilli and cocci.
Flagellates Group of protozoa that possess at least one long flagellum, generally used for movement.
Flagellum A long, whiplike structure, composed of a basal body, hook, and filament, protruding from a cell.
Flavine adenine dinucleotide (**FAD**) Important vitamin-derived electron carrier molecule.
Fluid mosaic model Model describing the arrangement and motion of the proteins within the cytoplasmic membrane.
Fluorescent microscope Type of light microscope which uses an ultraviolet light source to fluoresce objects.
Folliculitis Infection of a hair follicle by *Staphylococcus aureus*.
Fomite Object inadvertently used to transfer pathogens to new hosts, e.g., needle, coin, etc.
Food microbiology The use of microorganisms in food production and the prevention of food-borne illnesses.
Food-borne transmission Spread of pathogenic microorganisms in or on foods that are poorly processed, undercooked, or improperly refrigerated.
Formed elements Cells and cell fragments suspended in blood plasma.
Frameshift mutation Type of mutation in which nucleotide triplets subsequent to an insertion or deletion are displaced, creating new sequences of codons that result in vastly altered polypeptide sequences.

Fungi Eukaryotic organisms lacking cell walls which obtain food from other organisms.
Functional group An arrangement of atoms common to all members of a class of organic molecules, such as the amine group found in all amino acids.
Furuncle A large, painful, nodular extension of folliculitis into surrounding tissue; caused by infection with *Staphylococcus aureus*.

Gametocyte In sexual reproduction of protozoa, cell which fuses with another gametocyte to form a diploid zygote.
Gamma-interferons Interferon produced by T lymphocytes and NK lymphocytes which activates macrophages and neutrophils days after an infection.
Gammaproteobacteria Largest and most diverse class of proteobacteria, including purple sulfur bacteria, methane oxidizers, pseudomonads, and others.
Gaseous agent High-level disinfecting gas used to sterilize heat-sensitive equipment and large objects.
Gel electrophoresis Technique used in recombinant DNA technology to separate molecules by size, shape, and electrical charge.
Gene A specific sequence of nucleotides that code for polypeptides or RNA molecules.
Gene library Collection of bacterial or phage clones, each of which carries a fragment of an organism's genome.
Gene therapy The use of recombinant DNA technology to insert a missing gene or repair a defective gene in human cells.
Generation time Time required for a bacterial cell to grow and divide.
Genetic engineering The manipulation of genes via recombinant DNA technology for practical applications.
Genetic fingerprinting (DNA fingerprinting) Technique which identifies unique sequences of DNA to: determine paternity; connect blood, semen, or skin cells to suspects in criminal investigations; or identify pathogens.
Genetic mapping Application of recombinant DNA technology in which genes are located on a nucleic acid molecule.
Genetic recombination The exchange of segments, typically genes, between two DNA molecules that are composed of homologous nucleotide sequences.
Genetic screening Procedure by which laboratory tests are used to screen patient and fetal DNA for mutant genes.
Genetics The study of inheritance and inheritable traits as expressed in an organism's genetic material.
Genome The sum of all the genetic material in a cell or virus.
Genomics The sequencing, analysis, and comparison of genomes.
Genotype Actual set of genes in an organism's genome.
Genus Taxonomical grouping of similar species of organisms.
Germ theory of disease Hypothesis formulated by Pasteur in 1857 that microorganisms are responsible for disease.
Giardiasis A mild to severe gastrointestinal illness caused by ingestion of cysts of *Giardia intestinalis*.
Glomerulonephritis Deposition of immune complexes in the walls of the glomeruli, networks of minute blood vessels in the kidneys; may result in kidney failure; typically caused by infection with Group A streptococcus.
Glycocalyx Sticky external sheath of prokaryotic and eukaryotic cells.
Glycolysis (Embden-Meyerhof pathway) First step in the catabolism of glucose via respiration and fermentation.
Goblet cells Mucous-secreting cells in the epithelium of the respiratory and gastrointestinal tracts.
Golgi body In eukaryotic cells, a series of flattened, hollow sacs surrounded by phospholipid bilayers and functioning to package large molecules for export in secretory vesicles.
Gonorrhea A common sexually transmitted disease caused by infection with *Neisseria gonorrhoeae*.
gp41 Antigenic HIV glycoprotein that promotes fusion of the viral envelope to a target cell.
gp120 Antigenic glycoprotein that is the primary attachment molecule of HIV.
Grafts Tissues or organs that have been transplanted, whether between sites on the same patient or between a donor and a recipient.
Graft-versus-host disease Disease resulting when donated bone marrow cells mount an immune response against the recipient's cells.
Gram-negative cell Bacterial cell having a wall composed of a thin layer of peptidoglycan, an external membrane of lipopolysaccharide, and a periplasmic space between and appearing pink after the Gram staining procedure.
Gram-positive cell Bacterial cell having a wall composed of a thick layer of peptidoglycan containing teichoic acids and retaining the crystal violet dye used in the Gram staining procedure.
Gram stain Technique for staining microbial samples by applying a series of dyes which leave some microbes purple and others pink. Developed by Christian Gram in 1884.
Granulocyte Type of leukocyte having large granules in the cytoplasm.
Granzyme Protein molecule in the cytoplasm of cytotoxic T cells which causes an infected cell to undergo apoptosis.
Graves' disease Production of autoantibodies that stimulate pituitary cells to secrete thyroid-stimulating hormone, which elicits excessive production of thyroid hormone and growth of the thyroid gland.
Group A streptococcus (*S. pyogenes*) A coccus that produces protein M and a hyaluronic acid capsule, both of which contribute to the pathogenicity of the species.
Group B streptococcus (*S. agalactiae*) A Gram-positive coccus that normally resides in the lower GI, genital, and urinary tracts but can cause disease in newborns.
Group translocation Active process, occurring in some prokaryotes, by which a substance being actively transported across a cell membrane is chemically changed during transport.
Growth An increase in size.
Growth curve Graph that plots the number of bacteria growing in a population over time.
Growth factor Organic chemical such as a vitamin required in very small amounts for microbial metabolism. In immunology, an immune system cytokine that stimulates stem cells to divide, ensuring that the body is supplied with sufficient leukocytes of all types.
Guanine Ring-shaped nitrogenous base found in nucleotides of DNA and RNA.
Gumma Syphilitic lesion that occurs in bones, nervous tissue, or on skin in patients with tertiary syphilis.

Habitat The physical localities in which organisms are found.
Halogen One of the four very reactive, nonmetallic chemical elements: iodine, chlorine, bromine, and fluorine. Used in disinfectants and antiseptics.
Halophile Microorganism requiring a saline environment (greater than 9% NaCl).
Hantavirus pulmonary syndrome Rapid, severe, and often fatal pneumonia caused by infection with a hantavirus.
Hantaviruses Group of bunyaviruses that are transmitted to humans via inhalation of virions in dried deer-mouse urine and cause hantavirus pulmonary syndrome.
Haploid A nucleus with a single copy of each chromosome.
Haustoria Modified hyphae which penetrate the tissue of the host to withdraw nutrients.
Hay fever Allergic reaction localized to the upper respiratory tract and characterized by nasal discharge, sneezing, itchy throat and eyes, and excessive tear production.
Heat fixation In microscopy, a technique which uses the heat from a flame to attach a smear to a slide.
Heavy metal ions The ions of high-molecular-weight metals such as arsenic that are used as antimicrobial agents because they denature proteins. They have largely been replaced because they are also toxic to human cells.
Helminths Multicellular eukaryotic worms, some of which are parasitic.
Helper T cell (T_H cell) In cell-mediated immune response, type of cell characterized by CD4 cell-surface glycoprotein which regulates the activity of B cells and cytotoxic T cells during an immune response.
Hemagglutinin (HA) Component of glycoprotein spikes in the lipid envelope of influenza viruses that help them attach to pulmonary epithelial cells.
Hemolytic disease of the newborn Disease that results when antibodies made by the mother of an Rh negative woman cross the placenta and destroy the red blood cells of an Rh positive fetus.
Hepatitis B Inflammatory condition of the

liver caused by infection with hepatitis B virus and characterized by jaundice.
Hepatitis C Chronic disease that can cause liver damage or hepatic cancer; caused by infection with hepatitis C virus.
Hepatitis D Disease that can cause liver damage or hepatic cancer; caused by infection with hepatitis D virus simultaneously with hepatitis B virus.
Hepatitis E (*enteric hepatitis*) Infection of the liver which is fatal in 20% of infected pregnant women.
Herpes zoster (*shingles*) Extremely painful skin rash caused by reactivation of latent varicella-zoster virus.
Heterocyst Thick-walled nonphotosynthetic cell of cyanobacteria which transports reduced nitrogen to neighboring cells in exchange for glucose.
Hfr (high frequency of recombination) cell Cell containing an F plasmid that is integrated into the prokaryotic chromosome. Hfr cells form conjugation pili and transfer cellular genes more frequently than normal F^+ cells.
Histamine Inflammatory chemical released from damaged cells that causes vasodilation of capillaries.
Histone Globular protein found in eukaryotic chromosomes.
Histoplasmosis Pulmonary, cutaneous, ocular, or systemic disease found in the Ohio River Valley and caused by infection with *Histoplasma capsulatum*.
Holoenzyme The combination of an apoenzyme and its cofactors.
Horizontal gene transfer Process in which a donor cell contributes part of its genome to a recipient cell which may be a different species or genus from the donor.
Human granulocytic ehrlichiosis Disease caused by infection with *Ehrlichia equi* transmitted by ticks, and characterized by leukopenia, rash, malaise, and death in about 10% of untreated cases.
Human immunodeficiency viruses Group of retroviruses which destroy the immune system.
Human monocytic ehrlichiosis Disease caused by infection with *Ehrlichia chaffeensis* transmitted by ticks, and characterized by leukopenia, rash, malaise, and death in about 5% of untreated cases.
Human T-lymphocyte viruses Group of oncogenic retroviruses associated with cancer of lymphocytes.
Humoral immune response The antigen-antibody immune response used by B cells to fight exogenous antigens.
Hybridomas Tumor cells created by fusing antibody-secreting plasma cells with cancerous plasma cells called *myelomas*.
Hydatid Fluid-filled.
Hydatid disease Potentially fatal disease caused by infection with the canine tapeworm *Echinococcus granulosus* and characterized by the presence of fluid-filled cysts in the liver and other tissues.
Hydrogen bond The electrical attraction between a partially charged hydrogen atom and a full or partial negative charge on a different region of the same molecule or another molecule. Hydrogen bonds confer unique properties to water molecules.
Hydrolysis A decomposition reaction in which a covalent bond is broken, and the ionic components of water are added to the products.
Hydrophobia Literally a fear of water; symptom caused by painful swallowing characteristic of rabies infection.
Hydrophobic Insoluble in water.
Hydrophilic Attracted to water.
Hydrothermal vent Vent in marine abyssal zone that spews super-heated, nutrient-rich water.
Hydroxyl radical Most reactive of the toxic forms of oxygen.
Hypersensitivity Any immune response against a foreign antigen that is exaggerated beyond the norm.
Hypersensitivity pneumonitis Type of pneumonia resulting from repeated inhalation of mold spores and other antigens deep into the lungs, stimulating the formation of immune complexes.
Hyperthermophile Microorganism requiring temperatures above 80°C.
Hypertonic Characteristic of a solution having a higher concentration of solutes than another.
Hyphae Long, branched, tubular filaments in the thalli of molds.
Hypotonic Characteristic of a solution having a lower concentration of solutes than another.

Iatrogenic infections A subset of nosocomial infections that are the direct result of a medical procedure or treatment, such as surgery.
Illness In the infectious disease process, the most severe stage, in which signs and symptoms are most evident.
Immune complexes Antigen-antibody complexes.
Immune thrombocytopenic purpura Disease resulting when drugs bound to platelets bind antibodies and complement, causing the platelets to lyse.
Immunochromatographic assay Rapid modification of ELISA test in which an antigen solution flows through a porous strip, encountering labeled antibody; used for pregnancy testing and for rapid identification of infectious agents.
Immunoelectrophoresis Type of immune test used to analyze complex antigen mixtures in which electrophoresis is used to separate antigens before precipitation testing is performed.
Immunofiltration assay Rapid modification of ELISA test using membrane filters rather than polystyrene plates.
Immunoglobulin (Ig) (see *antibody*)
Immunoglobulin A (IgA) The antibody class most commonly associated with various body secretions, including tears and milk. IgA pairs with a secretory component to form *secretory IgA*.
Immunoglobulin D (IgD) A membrane-bound antibody molecule found in some animals as a B cell receptor.
Immunoglobulin E (IgE) Signal antibody molecule which triggers the inflammatory response, particularly in allergic reactions and infections by parasitic worms.
Immunoglobulin G (IgG) The predominant antibody class found in the bloodstream and the primary defender against invading bacteria.
Immunoglobulin M (IgM) The second most common antibody class and the predominant antibody produced first during a primary humoral immune response.
Immunology Study of the body's specific defenses against pathogens.
Immunophilins Immunosuppressive drugs, such as cyclosporine, which inhibit T cell function.
Impetigo Presence of red, pus-filled vesicles on the face and limbs of children; caused by infection with *Staphylococcus aureus*.
Inactivated polio vaccine (IPV) Inoculum developed by Jonas Salk in 1955 for vaccination against poliovirus.
Inactivated vaccine Inoculum containing either whole agents or subunits and often adjuvants.
Incidence In epidemiology, the number of new cases of a disease in a given area or population during a given period of time.
Inclusion Deposited substance such as a lipid, gas vesicle, or magnetite, stored within the cytosol of a prokaryote.
Inclusion bodies In the life cycle of chlamydias, phagosomes full of reticulate bodies.
Incubation period Stage in infectious disease process between infection and occurrence of the first symptoms or signs of disease.
Index case In epidemiology, the first of the disease in a given area or population.
Indirect contact transmission Spread of pathogens from one host to another via inanimate objects called *fomites*.
Indirect fluorescent antibody test Immune test allowing observation through a fluorescent microscope of the presence of antigen in a tissue sample flooded with labeled antibody.
Inducible operon Type of operon that is not normally transcribed and must be activated by inducers.
Induction In virology, excision of a prophage from the host chromosome, at which point the prophage reenters the lytic phase.
Industrial microbiology Branch of microbiology in which microbes are manipulated to manufacture useful products.
Infection Successful invasion of the body by a pathogenic microorganism.
Infection control Branch of microbiology studying the prevention and control of infectious disease.
Infectious mononucleosis (*mono*) Disease characterized by sore throat, fever, fatigue, and enlargement of the spleen and liver; caused by infection with Epstein-Barr virus.
Influenza (*flu*) Infectious disease caused by two species of orthomyxoviruses and characterized by fever, malaise, headache, and myalgia; certain strains can be fatal.
Initial body (*reticulate body*) Reproductive

structure of chlamydias that undergoes repeated binary fissions until the host cell is filled.

Innate (*inborn*) **resistance** Resistance to a pathogen conferred by cellular or physiological incompatibility with the pathogen.

Inoculum Sample of microorganisms.

Inorganic chemical Molecule lacking carbon.

Insertion Type of mutation in which a base pair is inserted into a genetic sequence.

Insertion sequence A simple transposon consisting of no more than two inverted repeats and a gene that encodes the enzyme transposase.

Integrase Enzyme carried by the virions of HIV which allows integration into a human chromosome.

Interferons Protein molecules that inhibit the spread of viral infections.

Interleukin An immune system cytokine that signals among leukocytes.

Interleukin-1 A pyrogen released by phagocytes that have phagocytized bacteria.

Intermediate host In the life cycle of parasites, host in which immature forms of the parasite are present and undergoing various stages of maturation.

Intracellular state In virology, state in which a virus exists solely as nucleic acid within a cell.

Intron Noncoding sequence of mRNA which is removed to make functional mRNA.

In-use test Method of evaluating the effectiveness of a disinfectant or antiseptic which tests efficacy under specific, real-life conditions.

Inverted repeat Palindromic sequence found at each end of a transposon.

Ion An atom or group of atoms that has either a full negative charge or a full positive charge.

Ionic bond A type of bond formed from the attraction of opposite electrical charges. Electrons are not shared

Ionizing radiation Form of radiation that uses wavelengths shorter than 1 nm to strike molecules, creating ions by ejecting electrons from atoms.

Isograft Type of graft in which tissues are moved between genetically identical individuals (identical twins).

Isoniazid (INH) Antimicrobial which blocks mycolic acid synthesis in the walls of mycobacteria.

Isotonic Characteristic of a solution having the same concentration of solutes and water as another.

Isotopes Atoms of a given element that differ only in the number of neutrons they contain.

Ixodes Genus of ticks which transmit Lyme disease.

Kinetoplastid Euglenozoan protozoan with a single large mitochondrion that contains an apical region of mitochondrial DNA and is called a *kinetoplast*.

Kingdom Taxonomical grouping of similar phyla of organisms.

Kinins Powerful inflammatory chemicals released by mast cells.

Koch's postulates A series of steps, elucidated by Robert Koch, that must be taken to prove the cause of any infectious disease.

Koplik's spots Mouth lesions characteristic of measles.

Krebs cycle Series of eight enzymatically catalyzed reactions that transfer stored energy from acetyl-CoA to coenzymes NAD^+ and FAD.

Küppfer cells Fixed macrophages of the liver.

Lag phase Phase in a growth curve in which the organisms are adjusting to their environment.

Lagging strand Daughter strand of DNA synthesized in short segments that are later joined. Synthesis of the lagging strand always moves away from the replication fork, and lags behind synthesis of the leading strand.

Lassa hemorrhagic fever Infectious disease characterized by hemorrhage and caused by infection with an arenavirus.

Latency In virology, process by which an animal virus is not incorporated into the chromosomes of the cell and remains inactive in the cell, possibly for years.

Latent disease Any disease in which a pathogen remains inactive for a long period of time before becoming active.

Latent virus (see *provirus*) An animal virus that remains inactive in a host cell.

Leading strand Daughter strand of DNA synthesized continuously toward the replication fork as a single long chain of nucleotides.

Legionnaires' disease Severe pneumonia caused by infection with *Legionella pneumophila.*

Leishmaniasis Any of three clinical syndromes caused by a bite from a sand fly carrying *Leishmania* and ranging from painless skin ulcers to disfiguring lesions to visceral leishmaniasis, which is systemic and fatal in 95% of untreated cases.

Leprosy Disease caused by infection with *Mycobacterium leprae* that produces either a nonprogressive tuberculoid form or a progressive lepromatous form which destroys tissues, including facial features, digits, and other structures.

Leptospirosis Zoonotic disease contracted by humans upon exposure to infected animals; characterized by pain, headache, and liver and kidney disease; caused by infection with *Leptospira interrogans.*

Leukocyte White blood cell.

Leukotrienes Inflammatory chemical released from damaged cells that increase vascular permeability.

Lichen Organism composed of a fungus living in partnership with photosynthetic microbes, either green algae or cyanobacteria.

Light-dependent reaction Reaction of photosynthesis requiring light.

Light-independent reaction Reaction synthesizing glucose from carbon dioxide and water and not requiring light.

Light repair Mechanism by which prokaryotic DNA photolyase enzyme breaks the bonds between adjoining thymine nucleotides, restoring the original DNA sequence.

Limnetic zone Sunlit, upper layer of freshwater or marine water away from the shore.

Littoral zone Shoreline zone of freshwater or marine water.

Lipid Any of a diverse group of organic macromolecules not composed of monomers and insoluble in water.

Lipid A The lipid component of lipopolysaccharide which is released from dead Gram-negative bacterial cells and can trigger shock and other symptoms in human hosts.

Lipopolysaccharide (**LPS**) Molecule composed of lipid A and polysaccharide found in the external membrane of Gram-negative cell walls.

Lithotroph Microorganism which acquires electrons from inorganic sources.

Log phase Phase in a growth curve in which the population is most actively growing.

Logarithmic (exponential) growth Increase in size of a microbial population in which the number of cells doubles in a fixed interval of time.

Lophotrichous Term used to describe a cell having a tuft of flagella at one end.

Lyme disease Disease carried by ticks infected with *Borrelia burgdorferi* and characterized by a "bull's eye" rash, neurologic and cardiac dysfunction, and severe arthritis.

Lymphatic system Body system composed of lymphatic vessels and lymphoid tissues and organs.

Lymphocyte Type of small agranulocyte which originates and matures in the red bone marrow and has nuclei that nearly fill the cell.

Lymphocytic choriomeningitis (LCM) Type of hemorrhagic fever resulting in flu-like symptoms and caused by infection with an arenavirus.

Lymphogranuloma venereum Sexually transmitted disease caused by infection with *Chlamydia trachomatis* and leading in some cases to proctitis or, in women, pelvic inflammatory disease.

Lyophilization Removal of water from a frozen culture or other substance by means of vacuum pressure. Used for the long-term preservation of cells and foods.

Lysogenic replication cycle (*lysogeny*) Process of viral replication in which a bacteriophage enters a bacterial cell, inserts into the DNA of the host, but remains inactive. The phage is then replicated every time the host cell replicates its chromosome.

Lysosome Vesicle in animal cells that contains digestive enzymes.

Lysozyme Antibacterial protein secreted in sweat.

Lytic replication cycle Process of viral replication consisting of five stages ending with lysis of and release of new virions from the host cell.

Macrolide Antimicrobial agent that inhibits protein synthesis by inhibiting the ribosomal 50S subunits.

Macrophage Mature form of monocyte, and phagocyte of bacteria, fungi, spores, and dust, as well as dead cell bodies.

Macule Any flat, reddened skin lesion; characteristic of early infection with a poxvirus.
Magnification The apparent increase in size of an object viewed via microscopy.
Major histocompatibility complex (MHC) A cluster of genes, located on each copy of chromosome 6 in humans, that codes for membrane-bound glycoproteins called major histocompatibility antigens.
Malaria A mild to potentially fatal disease caused by a bite from an Anopheles mosquito carrying any of four species of *Plasmodium;* characterized by fever, chills, hemorrhage, and potential destruction of brain tissue.
Malignant tumor Mass of neoplastic cells that can invade neighboring tissues and may metastasize to cause tumors in distant organs or tissues.
Marburg virus Type of hemorrhagic fever caused by a filamentous virus and fatal in 90% of cases.
Margination Process by which leukocytes stick to the walls of blood vessels at the site of infection.
Mast cells Specialized cells located in connective tissue that release histamine when they are exposed to complement.
Matter Anything that takes up space and has mass.
Measles (*rubeola*) Contagious disease characterized by fever, sore throat, headache, dry cough, conjunctivitis, and lesions called Koplik's spots; caused by infection with *Morbillivirus*.
Mechanical vector Housefly, cockroach, or other animal that passively carries pathogens to new hosts on its feet or other body parts and is not infected by the pathogens it carries.
Medium A collection of nutrients used for cultivating microorganisms.
Meiosis Nuclear division of diploid eukaryotic cells resulting in four haploid nuclei.
Membrane attack complexes (MAC) The end products of the complement cascade, which form circular holes in a pathogen's membrane.
Membrane filters Thin circles of nitrocellulose or plastic containing specific pore sizes, some small enough to trap viruses.
Membrane filtration Direct method of estimating population size in which a large sample is poured through a filter small enough to trap cells.
Memory B cell B lymphocyte that migrates to lymphoid tissues to await a subsequent encounter with the same antigen.
Memory T cell Type of T cell that persists in lymphoid tissues for months or years awaiting subsequent contact with an antigenic determinant matching its TCR, at which point it produces cytotoxic T cells.
Memory response The rapid and enhanced immune response to a subsequent encounter with a familiar antigen.
Mesophile Microorganism requiring temperatures ranging from 20°C to about 40°C.
Messenger RNA (mRNA) Form of ribonucleic acid which carries genetic information from DNA to a ribosome.
Metabolism The sum of all chemical reactions, both anabolic and catabolic, within an organism.
Metachromatic granules Inclusions of *Corynebacteria* that store phosphate and stain differently from the rest of the cytoplasm.
Metaphase Second stage of mitosis, during which chromosomes line up and attach to microtubules of the spindle. Also used for the comparable stage of meiosis
Metastasis The spreading of malignant cancer cells to nonadjacent organs and tissues, where they produce new tumors.
Methane oxidizer Any Gram-negative bacterium that utilizes methane both as a carbon and as an energy source.
Methanogen Obligate anaerobe that produces methane gas.
Methicillin-resistant *Staphylococcus aureus* (MRSA) Strain of *Staphylococcus aureus* that is resistant to many common antimicrobial drugs, and has emerged as a major nosocomial problem.
Methylation Process in which a cell adds a methyl group to one or two bases that are part of specific nucleotide sequences.
Microaerophile Microorganism which requires low levels of oxygen.
Microbe An organism too small to be seen without a microscope.
Microbial antagonism (*microbial competition*) Normal condition in which established microbiota use up available nutrients and space, reducing the ability of arriving pathogens to colonize.
Microbial contamination The presence of microorganisms in or on the body.
Microbial death Permanent loss of reproductive capacity of a microorganism.
Microbial death rate A measurement of the efficacy of an antimicrobial agent.
Microbial ecology The study of the interactions of microorganisms among themselves and their environment.
Microglia Fixed macrophages of the nervous system.
Micrograph A photograph of a microscopic image.
Microorganism An organism too small to be seen without a microscope.
Microscopy The use of light or electrons to magnify objects.
Minimum bactericidal concentration (MBC) test An extension of the MIC test in which samples taken from clear MIC tubes are transferred to plates containing a drug-free growth medium and monitored for bacterial replication.
Minimum inhibitory concentration (MIC) The smallest amount of a drug that will inhibit a pathogen.
Mismatch repair Mechanism by which enzymes scan newly synthesized, nonmethylated DNA for mismatched bases, removes them, and replaces them.
Missense mutation A substitution in a nucleotide sequence resulting in a codon that specifies a different amino acid: what is transcribed makes sense, but not the right sense.
Mitochondria Spherical to elongated structures found in most eukaryotic cells that produce most of the ATP in the cell.
Mitosis Nuclear division of a eukaryotic cell resulting in two nuclei with the same ploidy as the original.
Mold A typically multicellular fungus that grows as long filaments called *hyphae* and reproduces by means of spores.
Molecular biology Branch of biology combining aspects of biochemistry, cell biology, and genetics to explain cell function at the molecular level, particularly via the use of genome sequencing.
Molecular mimicry Process in which microorganisms with antigenic determinants similar to self-antigens trigger autoimmune tissue damage.
Molecule Two or more atoms held together by chemical bonds.
Molluscum contagiosum Infectious disease caused by *Molluscipoxvirus* which results in tumor-like growths on the skin.
Monoclonal antibodies Identical antibodies secreted by a cell line originating from a single plasma cell.
Monocyte Type of agranulocyte which has slightly lobed nuclei.
Monomer A subunit of a macromolecule such as a protein.
Mononuclear phagocytic system (see *reticuloendothelial system*)
Monosaccharide (*simple sugar*) A monomer of carbohydrate, such as a molecule of glucose.
Monotrichous Term used to describe a cell having a single flagellum.
Morbidity Any change from a state of health.
Mordant In microscopy, a substance that binds to a dye and makes it less soluble.
Most probable number (MPN) method Statistical estimation of the size of a microbial population based upon the dilution of a sample required to eliminate microbial growth.
Multiple sclerosis Autoimmune disease in which cytotoxic T cells attack and destroy the myelin sheath that insulates neurons.
Mumps Disease caused by infection with the mumps virus and characterized by fever, parotitis, pain in swallowing, and in some cases, meningitis or deafness.
Murine typhus (*endemic typhus*) Disease transmitted by fleas and characterized by high fever, headache, chills, muscle pain, and nausea; caused by infection with *Rickettsia typhi*.
Mutagen Physical or chemical agent that introduces a mutation that changes the microbe's phenotype.
Mutant A cell with an unrepaired genetic mutation, or any of its descendents.
Mutation In genetics, change in the nucleotide base sequence of a genome.
Mutualism Symbiotic relationship in which both members benefit from their interaction.
Mycelium Tangled mass of hyphae.
Mycetismus Mushroom poisoning.

Mycetoma Destructive, tumor-like infection of the skin, fascia, and/or bones of the hands or feet caused by mycelial fungi of several genera in the division Ascomycota.
Mycolic acid Long carbon-chain waxy lipid found in the walls of cells in the genus *Mycobacterium* that makes them resistant to dessication and staining with water-based dyes.
Mycology The field of medicine concerned with the diagnosis, management, and prevention of fungal diseases.
Mycosis Fungal disease.
Mycotoxicosis Poisoning caused by eating food contaminated with fungal toxins.
Mycotoxins Secondary metabolites produced by fungi and toxic to humans.
Mycoplasmas Class of low G+C bacteria which lack cytochromes, enzymes of the Krebs cycle, and cell walls and are pleomorphic.
Myxobacteria Gram-negative, aerobic, soil-dwelling bacteria with a unique life cycle including a stage of differentiation into fruiting bodies containing resistant myxospores.

Narrow-spectrum drug Antimicrobial that works against only a few kinds of pathogens.
Natural killer (NK) lymphocyte Type of nonspecific, defensive leukocyte that secretes toxins onto the surfaces of virally infected cells and neoplasms.
Naturally acquired active immunity Type of immunity which occurs when the body responds to exposure to antigens by mounting specific immune responses.
Naturally acquired passive immunity Type of immunity which occurs when a fetus, newborn, or child receives antibodies across the placenta or within breastmilk.
Necrotizing fasciitis Potentially fatal condition marked by toxemia, organ failure, and destruction of muscle and fat tissue following infection with group A streptococci
Negative feedback (see *feedback inhibition*)
Negative selection (*indirect selection*) Process by which auxotrophic mutants are isolated and cultured.
Negative stain (*capsule stain*) In microscopy, a staining technique used primarily to reveal bacterial capsules and involving application of an acidic dye which leaves the specimen colorless and the background stained.
Negative strand RNA (-RNA) Viral single-stranded RNA transcribed from the +ssRNA genome by viral RNA polymerase.
Negri bodies Aggregates of virions in the brains of rabies patients.
Neisseria Genus of pathogenic Gram-negative cocci which has fimbriae, a polysaccharide capsule, and lipooligosaccharide in the cell walls.
Nematodes Group of round, unsegmented helminths with pointed ends that have a complete digestive tract.
Neoplasia Uncontrolled cell division in a multicellular animal.
Neuraminidase (NA) Component of glycoprotein spikes in the lipid envelope of influenza viruses that provides access to cell surfaces by hydrolyzing mucus in the lungs.
Neutralization Antibody function in which the action of a toxin or attachment of a pathogen is blocked.
Neutralization test Immune test that measures the ability of antibodies to neutralize the biological activity of pathogens and toxins.
Neutron An uncharged subatomic particle.
Neutrophil Type of granulocyte which stains lilac with a mixture of acidic and basic dyes.
Neutrophile Microorganism requiring neutral pH.
Nicotinamide adenine dinucleotide (NAD^+) Important vitamin-derived electron carrier molecule.
Nicotinamide adenine dinucleotide phosphate ($NADP^+$) Important vitamin-derived electron carrier molecule.
Nitrification The process by which bacteria convert reduced nitrogen compounds such as ammonia into nitrate, which is more available to plants.
Nitrifying bacteria Chemoautotrophic bacteria that derive electrons from the oxidation of nitrogenous compounds.
Nitrogen cycle Biogeochemical cycle involving nitrogen fixation, ammonification, nitrification, and denitrification.
Nitrogen fixation The conversion of atmospheric nitrogen to ammonia.
Noncommunicable disease An infectious disease that arises from outside of hosts or from normal microbiota.
Noncompetitive inhibitor Inhibitory substance that blocks enzyme activity by binding to an allosteric site located elsewhere on the enzyme.
Noncyclic photophosphorylation The production of ATP by noncyclic electron flow.
Nonionizing radiation Electromagnetic radiation with a wavelength greater than 1 nm.
Nonliving reservoir of infection Soil, water, food, and inanimate object that is a continuous source of infection.
Nonsense mutation A substitution in a nucleotide sequence which causes an amino acid codon to be replaced by a stop codon.
Nonspecific defense Type of defense that protects the body against a wide variety of pathogens and that operates in a generalized way independent of the nature of the intruder.
Normal microbiota Microorganisms that colonize the surfaces of the human body without normally causing disease. They may be resident or transient.
Noroviruses Group of caliciviruses that cause diarrhea.
Nosocomial disease A disease acquired in a health care facility.
Nosocomial infection An infection acquired in a health care facility.
Nuclear envelope Double membrane composed of phospholipid bylayers surrounding a cell nucleus.
Nuclear pores Spaces in the nuclear envelope that function to control the transport of substances through it.
Nucleoid Region of the prokaryotic cytosol containing the cell's chromosome(s).
Nucleolus Specialized region in a cell nucleus where RNA is synthesized.
Nucleoplasm The semiliquid matrix of a cell nucleus.
Nucleosome Bead of DNA bound to histone in a eukaryotic chromosome.
Nucleotide Monomer of a nucleic acid.
Nucleotide analog Compound structurally similar to a normal nucleotide that can be incorporated into DNA, resulting in mismatched base pairing.
Nucleus Spherical to ovoid membranous organelle containing a eukaryotic cell's genetic material.
Numerical aperture The ability of a lens to gather light.
Nutrient Any chemical such as carbon, hydrogen, etc., required for growth of microbial populations.

Objective lens In microscopy, the lens immediately above the object being magnified.
Obligate aerobe Microorganism which requires oxygen as the final electron acceptor of the electron transport chain.
Obligate anaerobe Microorganism which cannot tolerate oxygen and uses a final electron acceptor other than oxygen.
Obligate halophile Microorganism requiring high osmotic pressure.
Ocular herpes (*ophthalmic herpes*) Disorder characterized by conjunctivitis, a gritty feeling in the eye, and pain, and characteristic of latent herpes simplex virus type 1.
Ocular lens In microscopy, the lens closest to the eyes. May be single (*monocular*) or paired (*binocular*).
Operator Regulatory element in an operon where repressor protein binds to stop transcription.
Operon A series of genes, a promoter, and an operator sequence controlled by one regulatory gene. The operon model explains gene regulation in prokaryotes.
Opportunistic pathogens Microorganisms that cause disease when the immune system is suppressed, when microbial antagonism is reduced, or when introduced into an abnormal area of the body.
Opsonin Antimicrobial protein which readies a pathogen for phagocytosis.
Opsonization The coating of pathogens by proteins called *opsonins*, making them more vulnerable to phagocytes.
Optimum growth temperature Temperature at which a microorganism's metabolic activities produce the highest growth rate.
Oral polio vaccine (OPV) Inoculum developed by Albert Sabin in 1961 for vaccination against poliovirus.
Order Taxonomical grouping of similar families of organisms.
Organelle Internal, membrane-bound structure found in eukaryotic cells that acts as a tiny organ to carry out one or more cell functions.

Organic compounds Molecules that contain both carbon and hydrogen atoms.
Organotroph Microorganism which acquires electrons from organic sources.
Ornithosis (*parrot fever*) A respiratory disease of birds that can be transmitted to humans and is caused by infection with *Chlamydia psittaci*.
Osmosis The diffusion of water molecules across a selectively permeable membrane.
Osmotic pressure The pressure exerted across a selectively permeable membrane by the solutes in a solution on one side of the membrane. The osmotic pressure exerted by high-salt or high-sugar solutions can be used to inhibit microbial growth in certain foods.
Osteomyelitis Inflammation of the bone marrow and surrounding bone; caused by infection with *Staphylococcus*.
Otitis media Inflammation of the middle ear; typically caused by *Streptococcus pneumoniae*.
Oxidation-reduction reaction (**redox reaction**) Any metabolic reaction involving the transfer of electrons from an electron donor to an electron acceptor; reactions in which electrons are accepted are called *reduction* reactions, whereas reactions in which electrons are donated are *oxidation* reactions.
Oxidation lagoons Successive wastewaster treatment areas (lagoons) used by farmers and ranchers to treat animal wastes and from which water is released into natural water systems.
Oxidative phosphorylation The use of energy from redox reactions to attach inorganic phosphate to ADP.
Oxidizing agent Antimicrobial agent which releases oxygen radicals.

P site In a ribosome, a binding site which holds a tRNA and the growing polypeptide.
Palisade arrangement In cell morphology, a folded arrangement of bacilli.
Pandemic In epidemiology, the occurrence of an epidemic on more than one continent simultaneously.
Papilloma (*wart*) Benign growth of the epithelium of the skin or mucous membranes.
Papule Any raised, reddened skin lesion that progresses from a macule; characteristic of infection with a poxvirus.
Parabasalid Group of single-celled, animal-like microorganisms that contain a Golgi-like parabasal body.
Paracoccidioidomycosis Pulmonary disease found from southern Mexico to South America caused by infection with *Paracoccidioides brasiliensis*.
Parainfluenza viruses Group of enveloped, negative ssRNA viruses which cause respiratory disease, particularly in children.
Parasite A microbe that derives benefit from its host while harming it or even killing it.
Parasitism Symbiotic relationship in which one organism derives benefit while harming, or even killing, its host.
Parasitology The study of protozoan and helminthic parasites of animal and humans.
Parenteral route A means by which pathogenic microorganisms can be deposited directly into deep tissues of the body, as in puncture wounds and surgery.
Paroxysmal phase In pertussis, second phase lasting 2–4 weeks and characterized by exhaustive coughing spells.
Parvoviruses Group of extremely small, pathogenic ssDNA viruses.
Pasturellaceae Family of gammaproteobacteria, two genera of which—Pasteurella and Haemophilus—are pathogenic.
Pasteurization The use of heat to kill pathogens and reduce the number of spoilage microorganisms in food and beverages.
Pathogen A microorganism capable of causing disease.
Pathogenicity A microorganism's ability to cause disease.
Pelvic inflammatory disease Infection of the uterus and uterine tubes; may be caused by infection with any of several bacteria.
Pentose phosphate pathway Enzymatic formation of phosphorylated pentose sugars from glucose 6-phosphate.
Peptic ulcer Erosion of the mucous membrane of the stomach or duodenum and caused by infection with *Helicobacter pylori*.
Peptide bond A covalent bond between amino acids in proteins.
Peptidoglycan Large, interconnected polysaccharide composed of chains of two alternating sugars and crossbridges of four amino acids. Main component of bacterial cell walls.
Perforin Protein molecule in the cytoplasm of cytotoxic T cells which forms channels (perforations) in an infected cell's membrane.
Periplasmic space In Gram-negative cells, the space between the cell membrane and the outer membrane containing peptidoglycan and periplasm.
Peritrichous Term used to describe a cell having flagella covering the cell surface.
Peroxide Toxic form of oxygen which is detoxified by catalase or peroxidase.
Peroxisome Vesicle found in all eukaryotic cells that degrades poisonous metabolic wastes.
Pertussis (*whooping cough*) Pediatric disease characterized by development of copious mucus, loss of tracheal cilia, and deep "whooping" cough; caused by infection with *Bordetella pertussis*.
Petri plate Dish filled with solid medium used in culturing microorganisms.
pH scale A logarithmic scale used for measuring the concentration of hydrogen ions in a solution.
Phaeohyphomycosis Cutaneous and subcutaneous disease characterized by lesions that can disseminate internally; caused by traumatic introduction of dematiaceous fungi into the skin.
Phaeophyta Brown-pigmented division of algae having cell walls composed of cellulose and alginic acid, a thickening agent.
Phage (see *bacteriophage*.)
Phage typing Method of classifying microorganisms in which unknown bacteria are identified by observing plaques.
Phagocytes Cells such as leukocytes which are capable of phagocytosis.
Phagocytosis Type of endocytosis in which solids are moved into the cell.
Phagolysosome Digestive vesicle formed by the fusing of a lysosome with a phagosome.
Phagosome A sac formed by a phagocyte's pseudopodia.
Pharyngitis (*strep throat*) Inflammation of the throat caused by infection with group A streptococci.
Phase microscope Type of microscope used to examine living microorganisms or fragile specimens.
Phase-contrast microscope Type of phase microscope that produces sharply defined images in which fine structures can be seen in living cells.
Phenol coefficient Method of evaluating the effectiveness of a disinfectant or antiseptic which compares the agent's efficacy to that of phenol.
Phenolic Compound derived from phenol molecules that have been chemically modified to denature proteins and disrupt cell membranes in a wide variety of pathogens.
Phenotype The physical features and functional traits of an organism expressed by genes in the genotype.
Phospholipid Phosphate-containing lipid made up of molecules with two fatty acid chains.
Phospholipid bilayer Bipolar structure of a cytoplasmic membrane.
Photoautotroph Microorganism which requires light energy and uses carbon dioxide as a carbon source.
Photoheterotroph Microorganism which requires light energy and gains nutrients via catabolism of organic compounds.
Photophosphorylation The use of energy from light to attach inorganic phosphate to ADP.
Photosynthesis Process in which light energy is captured by chlorophylls and transferred to ATP and metabolites.
Photosystem Network of light-absorbing chlorophyll molecules and other pigments held within a protein matrix on thylakoids.
Phototaxis Cell movement that occurs in response to light stimulus.
Phylum Taxonomical grouping of similar classes of organisms.
Picornaviruses Family of viruses that contain positive single-stranded RNA and naked polyhedral capsids; many are human pathogens.
Pilus A long, hollow, nonmotile tube of protein that connects two bacterial cells and mediates the transfer of DNA. Also called *conjugation pilus*.
Pinocytosis Type of endocytosis in which liquids are moved into the cell.
Pinta A disfiguring skin disease caused by infection with *Treponema carateum*.
Pityriasis Condition characterized by depigmented or hyperpigmented patches of scaly skin resulting from infection with *Malassezia furfur*.
Plaque In phage typing, the clear region within the bacterial lawn where growth is inhibited.

Plaque assay Technique for estimating phage numbers in which each plaque corresponds to a single phage in the original bacterium/virus mixture.

Plasma The liquid portion of blood.

Plasma cells B cells that are actively fighting against exogenous antigens and secreting antibodies.

Plasmid A small, circular molecule of DNA that replicates independently of the chromosome. Each carries genes for its own replication and often for one or more nonessential functions such as resistance to antibiotics.

Plasmodial slime mold (acellular slime mold) Streaming, coenocytic, colorful filaments of cytoplasm that phagocytize organic debris and bacteria.

Plasmolysis Shriveling of a cell's cytoplasm caused by osmotic pressure.

Platelet Cell fragments involved in blood clotting.

Pleomorphic In cell morphology, term used to describe a variably-shaped prokaryotic cell.

***Pneumocystis* pneumonia (PCP)** Debilitating fungal pneumonia that is the leading cause of death in AIDS patients and is caused by opportunistic infection with *Pneumocystis jiroveci.*

Pneumonia Inflammation of the lungs; typically caused by infection with a strain of *Staphylococcus* or *Streptococcus pneumoniae*.

Pneumonic plague Fever and severe respiratory distress caused by infection of the lungs with *Yersinia pestis*; fatal if untreated in nearly 100% of cases.

Point mutation A genetic mutation affecting only one or a few base pairs in a genome. Point mutations include substitutions, insertions, and deletions.

Polar In cell morphology, pertaining to either end of a cell, e.g., polar flagella.

Polar covalent bond Type of bond in which there is unequal sharing of electrons between atoms with opposite electrical charges.

Poliomyelitis (*polio*) Infection of varied degrees of severity from asymptomatic to crippling and caused by infection with poliovirus.

Polluted Containing microorganisms or chemicals in excess of acceptable values.

Polyenes Group of antimicrobial drugs such as amphotericin B which disrupt the cytoplasmic membrane of targeted cells by becoming incorporated into the membrane and damaging its integrity.

Polymer Repeating chains of covalently linked monomers found in macromolecules.

Polymerase chain reaction (PCR) Technique of recombinant DNA technology that allows researchers to produce a large number of DNA molecules rapidly *in vitro*.

Polyomaviruses Group of opportunistic viruses that infect the kidneys of most people but cause tumors in immunocompromised patients.

Polysaccharide Carbohydrate polymer composed of several to thousands of covalently linked monosaccharides.

Polyunsaturated fat Triglyceride with several double bonds between adjacent carbon atoms in its fatty acids.

Portal of entry Entrance site of pathogenic microorganisms, including the skin, mucous membranes, and placenta.

Portal of exit Exit site of pathogenic microorganisms, including the nose, mouth, and urethra.

Positive selection Process by which mutants are selected by eliminating wild type phenotypes.

Positive strand RNA (+RNA) Viral single-stranded RNA that can act directly as mRNA.

Potable Fit to drink.

Pour plate Method of culturing microorganisms in which colony-forming units are separated from one another using a series of dilutions.

Pox (*pocks*; see *pustule*)

Precursor metabolite Any of twelve molecules typically generated by a catabolic pathway and essential to the synthesis of organic macromolecules in many bacterial cells.

Prevalence In epidemiology, the total number of cases of a disease in a given area or population during a given period of time.

Primary atypical pneumonia So-called *walking pneumonia* characterized by mild respiratory symptoms that last for several weeks; caused by infection with *Mycoplasma pneumoniae.*

Primary immunodeficiency diseases Any of a group of diseases detectable near birth and resulting from a genetic or developmental defect.

Primary response The slow and limited immune response to a first encounter with an unfamiliar antigen.

Primary stain In a Gram stain, the initial stain, which colors all cells purple.

Prion Proteinaceous infectious particle that lacks nucleic acids and replicates by converting similar normal proteins into new prions.

Privileged sites The thymus, the cornea of the eye, and other sites at which grafts are not likely to be rejected.

Probe Nucleic acid molecule with a specific nucleotide sequence that has been labeled with a radioactive or fluorescent chemical so that its location can be detected.

Prodromal period In the infectious disease process, short time of generalized, mild symptoms that precedes illness.

Products The atoms, ions, or molecules that remain after a chemical reaction is complete.

Profundal zone Zone of freshwater or marine water beneath the limnetic zone and above the benthic zone.

Proglottids Body segments of a tapeworm that grow continuously as long as the worm remains attached to its host.

Progressive multifocal leukoencephalopathy (PML) Irreversible and typically fatal neurological disease caused by infection with JC virus in immunocompromised patients.

Prokaryote Any unicellular microorganism that lacks a nucleus. Classification includes bacteria and archaea.

Promoter Region of DNA where transcription begins.

Prophage An inactive bacteriophage.

Prophase First stage of mitosis, during which DNA condenses into chromatids and the spindle apparatus forms. Also used for the comparable stage of meiosis.

Prostaglandins Inflammatory chemicals released from damaged cells that increase vascular permeability.

Protease Enzyme secreted by microorganisms that digests proteins into amino acids outside a microbe's cell wall; in inflammatory reactions, chemicals released by mast cells that activate the complement system; in virology, an internal viral enzyme that makes HIV virulent.

Protein A complex macromolecule consisting of carbon, hydrogen, oxygen, nitrogen, and sulfur and important to many cell functions.

Proteobacteria Phylum of prokaryotes that includes five classes (designated alpha, beta, gamma, delta, and epsilon) of Gram-negative bacteria sharing common 16S rRNA nucleotide sequences.

Proton A positively charged subatomic particle.

Proton gradient Electrochemical gradient of hydrogen ions across a membrane.

Protozoa Single-celled eukaryotes that lack a cell wall and are similar to animals in their nutritional needs and structure.

Provirus Viral DNA that has become incorporated into a host chromosome.

Pseudomonad Any Gram-negative, aerobic, rod-shaped bacterium in the class Gammaproteobacteria that catabolizes carbohydrates by the Entner-Doudoroff and pentose phosphate pathways.

Pseudopodia Movable extensions of the cytoplasm and membrane of some eukaryotic cells.

Psychrophile Microorganism requiring very cold temperatures (below 20°C).

Pure culture (axenic culture) Culture containing cells of only one species of microorganism.

Purple sulfur bacteria Group of gammaproteobacteria which are obligate anaerobes and oxidize hydrogen sulfide to sulfur.

Pustule (*pox*) Any raised, pus-filled skin lesion; characteristic of infection with a poxvirus.

Pyoderma Any confined, pus-producing lesion on the exposed skin of the face, arms, or legs; often caused by infection with group A streptococci.

Pyrogen Chemical that triggers the hypothalamic "thermostat" to reset at a higher temperature, inducing fever.

Q fever Disease characterized by high fever, severe headache, chills, and mild pneumonia, and caused by infection with *Coxiella burnetii*.

Quaternary ammonium compound (**quat**) Detergent antimicrobial which is harmless to humans.

Quorum sensing Process by which bacteria respond to changes in microbial density by utilizing signal and receptor molecules.

Rabies Neuromuscular disease characterized by hydrophobia, seizures, hallucinations, and paralysis, and fatal if untreated, caused by infection with the rabies virus.

Radial immunodiffusion Type of immunodiffusion test in which an antigen solution is allowed to diffuse into agar containing specific concentrations of antibodies, causing a ring of precipitate to form.

Radiation The release of high-speed subatomic particles or waves of electromagnetic energy from atoms.

Reactants The atoms, ions, or molecules that exist at the beginning of a chemical reaction.

Reaction center chlorophyll In a photosystem, a chlorophyll molecule in which electrons excited by light energy are passed to an acceptor molecule which is the initial carrier of an electron transport chain.

Recipient cell In horizontal gene transfer, a cell that receives part of the genome of a donor cell.

Recombinant Cell resulting from genetic recombination between donated and recipient DNA.

Recombinant DNA technology Type of biotechnology in which scientists change the genotypes and phenotypes of organisms to benefit humans in various medical and agricultural applications.

Red tide Abundance of red-pigmented dinoflagellate protozoa in marine water.

Reducing medium Special culturing medium containing compounds that combine with free oxygen and remove it from the medium.

Refrigeration Short-term storage of cultures at 5°C.

Release In virology, final stage of the lytic replication cycle, in which the new virions are released from the host cell, which lyses.

Reoviruses Group of naked, segmented, dsRNA viruses that cause respiratory and gastrointestinal disease.

Repressible operon Type of operon that is continually transcribed until deactivated by repressors.

Reproduction An increase in number.

Reservoir of infection Living or nonliving continuous source of infectious disease.

Resolution The ability to distinguish between objects that are close together.

Responsiveness An ability to respond to environmental stimuli.

Restriction enzyme Enzyme which cuts DNA at specific nucleotide sequences and is used to produce recombinant DNA molecules.

Reticulate bodies (*initial bodies*) Noninfectious stage in the life cycle of chlamydias.

Reticuloendothelial system (*mononuclear phagocytic system*) Collectively, macrophages of all types and monocytes attached to endothelial cells.

Retrovirus Any +ssRNA virus that uses the enzyme reverse trasncriptase carried within its capsid to transcribe DNA from its RNA.

Reverse transcriptase Complex enzyme that allows retroviruses to make dsDNA from RNA templates.

Revolving nosepiece Portion of a compound microscope on which several objective lenses are mounted.

Rh antigens Cytoplasmic membrane proteins common to the red blood cells of 85% of humans as well as rhesus monkeys.

Rheumatic fever A complication of untreated group A streptococcal pharyngitis in which inflammation leads to damage of the heart valves and muscle.

Rheumatoid arthritis A crippling, systemic autoimmune disease, resulting from a type III hypersensitivity reaction, in which B cells in the joints produce autoantibodies against collagen.

Rhinoviruses Group of picornaviruses that cause upper respiratory tract infection and the "common cold."

Rhodophyta Red algae containing the pigment phycoerythrin, the storage molecule floridean starch, and cell walls of agar or carrageenan.

Ribonucleic acid (RNA) Nucleic acid consisting of nucleotides made up of phosphoric acid, a ribose pentose sugar, and an arrangement of the bases adenine, guanine, cytosine, and uracil.

Ribosomal RNA (rRNA) Form of ribonucleic acid which, together with polypeptides, makes up the structure of ribosomes.

Ribosome Nonmembranous organelle found in prokaryotes and eukaryotes that is composed of protein and ribosomal RNA and functions to make proteins.

Ribozyme RNA molecule functioning as an enzyme.

Rickettsias Group of extremely small, Gram-negative, obligate intracellular parasites that appear almost wall-less.

RNA polymerase Enzyme that synthesizes RNA by linking RNA nucleotides that are complementary to genetic sequences in DNA.

Rocky Mountain spotted fever (RMSF) Serious illness caused by infection with *Rickettsia rickettsii* transmitted by ticks and characterized by rash, malaise, petechiae, encephalitis, and death in 5% of cases.

Roseola Endemic illness of children characterized by an abrupt fever, sore throat, enlarged lymph nodes, and faint pink rash; caused by infection with human herpesvirus 6.

Rotaviruses Group of reoviruses that cause a potentially fatal infantile gastroenteritis.

Rough endoplasmic reticulum (RER) Type of endoplasmic reticulum that has ribosomes adhering to its outer surface which produce proteins for transport throughout the cell.

R-plasmid Extrachromosomal piece of DNA containing genes for resistance to antimicrobial drugs.

Rubella (*German measles*) Disease caused by infection with *Rubivirus* resulting in characteristic rash lasting about three days; mild in children but potentially teratogenic to fetuses of infected women.

Rubor Redness

Salmonellosis A serious diarrheal disease usually resulting from consumption of food contaminated with the enteric bacterium *Salmonella*.

Salt A crystalline compound formed by ionic bonding of elements.

Sanitization The process of disinfecting surfaces and utensils used by the public.

Saprobe Fungus that absorbs nutrients from dead organisms.

Sarcina A cuboidal packet of cocci.

Satellite virus A virus, such as hepatitis D virus, that requires glycoproteins coded by another virus to complete its replication cycle.

Saturated fat A triglyceride in which all but the terminal carbon atoms are covalently linked to two hydrogen atoms.

Scalded skin syndrome Reddening and blistering of the skin caused by infection with *Staphylococcus aureus*.

Scanning electron microscope (SEM) Type of electron microscope that uses magnetic fields within a vacuum tube to scan a beam of electrons across a specimen's metal-coated surface.

Scanning tunneling microscope Type of probe microscope in which a tungsten probe passes slightly above the surface of the specimen, revealing surface details at the atomic level.

Scarlet fever Diffuse rash and sloughing of skin caused by infection with group A streptococci.

Schaeffer-Fulton endospore stain In microscopy, staining technique that uses heat to drive a malachite green primary stain into an endospore.

Schistosomiasis A potentially fatal disease caused by infection with a blood fluke in the genus *Schistosoma*; may cause tissue damage in the liver, lungs, brain, or other organs.

Schizogony Special type of asexual reproduction in which the protozoan *Plasmodium* undergoes multiple mitoses to form a multinucleate schizont.

Schizont Multinucleate body formed by the protozoan *Plasmodium* which undergoes cytokinesis to release merozoites.

Scientific method Process by which scientists attempt to prove or disprove hypotheses through observations of the outcomes of carefully controlled experiments.

Scolex Small attachment organ that possesses suckers and/or hooks used to attach a tapeworm to host tissues.

Scrub typhus Disease caused by infection with *Orienta tsutsugamushi* transmitted by mites and characterized by fever, headache, muscle pain, and in a few cases, heart failure and death.

Sebum Oily substance secreted by the sebaceous glands of the skin that lowers pH.

Secretory IgA The combination of IgA and a secretory component, found in tears, mucous membrane secretions, and breastmilk, where it agglutinates and neutralizes antigens.

Secretory vesicle In eukaryotic cells, vesicles containing secretions packaged by the Golgi body which fuse to the cytoplasmic membrane and then release their contents outside the cell via exocytosis.

Selective medium Culturing medium containing substances that either favor the growth

of particular microorganisms or inhibit the growth of unwanted ones.
Selective toxicity Principle by which an effective antimicrobial agent must be more toxic to a pathogen than to the pathogen's host.
Selectively permeable In cell physiology, characteristic of a membrane that allows some substances to cross while preventing the crossing of others.
Semisynthetic antibiotic Antibiotic that has been chemically altered.
Septate Characterized by the presence of crosswalls.
Septic tank The home equivalent of primary wastewater treatment, consisting of a sealed concrete holding tank in which solids settle to the bottom and the effluent flows into a leach field that acts as a filter.
Serology The study and use of immunologic tests to diagnose and treat disease or identify antibodies or antigens.
Severe combined immunodeficiency disease (SCID) Primary immunodeficiency disease in children which affects both T cells and B cells and causes recurrent infections.
Shiga toxin Exotoxin secreted by *Shigella dysenteriae* which stops proteins synthesis in host cells.
Shigellosis A severe form of dysentery caused by any of four species of *Shigella*.
Shingles (see *herpes zoster*)
Signs In human pathology, objective manifestations of a disease that can be observed or measured by others.
Silent mutation Mutation produced by base pair substitution that does not change the amino acid sequence due to the redundancy of the genetic code.
Simple microscope Microscope containing a single magnifying lens.
Simple stain In microscopy, a stain composed of a single dye such as crystal violet.
Singlet oxygen Toxic form of oxygen which is neutralized by pigments called carotenoids.
Sinusitis Inflammation of the nasal sinuses; typically caused by *Streptococcus pneumoniae.*
Slant tube (slant) Test tube containing agar media that solidified while the tube was resting at an angle.
Slime layer Loose, water-soluble glycocalyx.
Slime mold Eukaryotic microbe resembling a filamentous fungus but lacking a cell wall and phagocytizing rather than absorbing nutrients.
Sludge After primary treatment of wastewater, the heavy material remaining at the bottom of settling tanks.
Smallpox Infectious disease eradicated in nature by 1980 and characterized by high fever, malaise, delirium, pustules, and death in about 20% of untreated cases.
Smear In microscopy, the thin film of organisms on the slide.
Smooth endoplasmic reticulum (SER) Type of endoplasmic reticulum that lacks ribosomes and plays a role in lipid synthesis and transport.
Snapping division A variation of binary fission in Gram-positive prokaryotes in which the parent cell's outer wall tears apart with a snapping movement to create the daughter cells.
SOS response Mechanism by which prokaryotic cells with extensive DNA damage use a variety of processes to induce DNA polymerase to copy the damaged DNA.
Southern blot Technique used in recombinant DNA technology that allows researchers to stabilize specific DNA sequences from an electrophoresis gel and then localize them using DNA dyes or probes.
Species Taxonomical category of organisms that can successfully interbreed.
Specific epithet In taxonomy, latter portion of the descriptive name of a species of organisms, describing that species' specific characteristics.
Specific immunity The ability of a vertebrate to recognize and defend against distinct species or strains of invaders.
Spectrum of action The number of different kinds of pathogens a drug acts against.
Spiral In cell morphology, a spiral-shaped prokaryotic cell.
Spirillus A stiff spiral-shaped prokaryotic cell.
Spirochetes Group of helical, Gram-negative bacteria with axial filaments that cause the organism to corkscrew, enabling it to burrow into a host's tissues.
Spoilage Any unwanted change to a food.
Spontaneous generation The theory that living organisms can arise from nonliving matter.
Sporadic In epidemiology, characteristic of a disease that occurs in only a few scattered cases within a given area or population during a given period of time.
Spore Reproductive cell of actinomycetes.
Sporogonic cycle In the lifecycle of *Plasmodium*, stage during which sporozoites are produced in the mosquito's digestive tract,and migrate into the mosquito's salivary glands.
Sporotrichosis Subcutaneous infection usually limited to the arms and legs in which lesions form around the site of infection with *Sporothrix schenckii*.
Staining Coloring microscopy specimens with stains called *dyes*.
Staphylococcus A cluster of cocci.
Starter culture Group of known microorganisms that carry out specific and reproducible fermentation reactions.
Stationary phase Phase in a growth curve in which new organisms are being produced at the same rate at which older organisms are dying.
Stem cells Generative cells capable of dividing to form daughter cells of a variety of types.
Sterile Free of microbial contamination.
Sterilization The eradication of all microorganisms, including bacterial endospores and viruses, although not prions, in or on an object.
Steroid Lipids consisting of four fused carbon rings attached to various side chains and functional groups.
Streak plate Method of culturing microorganisms in which a sterile inoculating loop is used to spread an inoculum across the surface of a solid medium in Petri dishes.
Streptococcal toxic shock syndrome Rare but potentially fatal syndrome characterized by fever, vomiting, diarrhea, pain, and organ failure, and caused by systemic infection with group A streptococci.
Streptococcus A long chain of cocci.
Structural analog Antimicrobial drug that competes with a structurally similar molecule, such as para-aminobenzoic acid (PABA), that is critical to the pathogen's metabolism.
Sty Infection of a hair follicle at the base of an eyelid by *Staphylococcus aureus*.
Subacute disease Any disease that has a duration and severity that lies somewhere between acute and chronic.
Subacute sclerosing panencephalitis (SSPE) Slow, progressive disease of the central nervous system that results in memory loss, muscle spasms, and death several years after infection with a defective measles virus.
Subclinical (see *asymptomatic*).
Substitution Type of mutation in which a nucleotide base pair is replaced.
Substrate The molecule upon which an enzyme acts.
Substrate-level phosphorylation The transfer of phosphate to ADP from another phosphorylated organic compound.
Subunit vaccine Type of immunization developed using recombinant DNA technology which exposes the recipient's immune system to a pathogen's antigens but not the pathogen itself.
Sulfonamide Antimetabolic drug that is a structural analog of para-aminobenzoic acid (PABA).
Superoxide radical Toxic form of oxygen which is detoxified by superoxide dismutase.
Surfactant Chemical which acts to reduce the surface tension of solvents such as water by decreasing the attraction among solvent molecules.
Symbiosis A close association between two or more organisms that is often but not always mutually beneficial.
Symptoms Subjective characteristics of a disease that can be felt by the patient alone.
Syncytium Giant, multinucleated cell formed by fusion of virally infected cell to neighboring cells.
Syndrome A group of symptoms and signs that collectively characterizes a particular disease or abnormal condition.
Synergism Interplay between drugs that results in efficacy that exceeds the efficacy of either drug alone.
Synthesis In virology, third stage of the lytic replication cycle, in which the host cell's metabolic machinery is used to synthesize new nucleic acids and viral proteins.
Synthesis reaction A chemical reaction involving the formation of larger, more complex molecules.
Synthetic drug Antimicrobial that has been completely synthesized in a laboratory.
Syphilis Sexually transmitted disease caused by infection with *Treponema pallidum* and characterized by a painless red lesion called a *chancre*.

Systemic lupus erythematosus (SLE) (*lupus*) A systemic autoimmune disease in which the individual produces autoantibodies against numerous antigens, including nucleic acids.

T cell receptor (TCR) Antigen receptor generated in the cytoplasmic membrane of T lymphocytes.
T lymphocyte (*T cell*) Lymphocyte that matures in the thymus and acts directly against endogenous antigens in cell-mediated immune responses.
Taxa Nonoverlapping groups of organisms sorted on the basis of mutual similarities.
Taxis Cell movement that occurs as a positive or negative response to light or chemicals.
Taxonomic system A system for naming and grouping similar organisms together.
Taxonomy The science of classifying and naming organisms.
Telophase Final stage of mitosis, during which nuclear envelopes form around the daughter nuclei. Also used for the comparable stage of meiosis.
Temperate phage (*lysogenic phage*; see *prophage*) A bacteriophage that enters a host cell but remains inactive.
Teratogenic Characterized by an ability to cause birth defects.
Terminator Region of DNA where transcription ends.
Tetanus Potentially fatal infection with the endospores of *Clostridium tetani*, which produce tetanus toxin, a potent neurotoxin.
Tetracycline Antimicrobial agent that inhibits protein synthesis by blocking the tRNA docking site.
Thallus Nonreproductive body of a filamentous mold or yeast.
Thermal death time The time it takes to completely sterilize a particular volume of liquid at a set temperature.
Thermal death point The lowest temperature that kills all cells in a broth in ten minutes.
Thermophile Microorganism requiring temperatures above 45°C.
Thylakoid In photosynthetic cells, portion of cellular membrane containing light-absorbing photosystems.
Thymine Ring-shaped nitrogenous base found in nucleotides of DNA.
Thymine dimer Mutation caused by nonionizing radiation in the form of ultraviolet light in which adjacent thymine bases covalently bond to one another.
Titer The highest dilution of blood serum giving a positive reaction during titration.
Titration Serial dilution of blood serum to test for agglutination activity.
Total magnification A multiple of the magnification achieved by the objective and ocular lenses of a compound microscope.
Toxemia Presence in the blood of poisons called *toxins*.
Toxic shock syndrome Potentially fatal syndrome characterized by fever, vomiting, red rash, low blood pressure, and loss of sheets of skin, and caused by systemic infection with strains of *Staphylococcus* that produce TSS toxin.
Toxin Chemicals produced by pathogens that either harm tissues or trigger host immune responses that cause damage.
Toxoid vaccine Inoculum using modified toxins to stimulate antibody-mediated immunity.
Toxoplasmosis A disease affecting cats and other animals and caused by infection with *Toxoplasma gondii*; in humans, characterized by mild, febrile symptoms; may be fatal in AIDS patients; transplacental transmission may result in miscarriage, stillbirth, or severe birth defects.
Trace element Element required in very small amounts for microbial metabolism.
Trachoma Serious eye disease caused by *Chlamydia trachomatis*.
Transamination Reaction involving transfer of an amine group from one amino acid to another.
Transcription Process in which the genetic code from DNA is copied as RNA nucleotide sequences.
Transducing phage Virus that transfers bacterial DNA from one bacterium to another.
Transduction Method of horizontal gene transfer in which DNA is transferred from one cell to another via a replicating virus.
Transfer RNA (tRNA) Form of ribonucleic acid which carries amino acids to the ribosome.
Transformation Method of horizontal gene transfer in which a recipient cell takes up DNA from the environment.
Transgenic organism Plant or animal that has been genetically altered by the inclusion of genes from other organisms.
Translation Process in which the sequence of genetic information carried by mRNA is used by ribosomes to construct polypeptides with specific amino acid sequences.
Transmission electron microscope (TEM) Type of electron microscope which generates a beam of electrons that passes through the specimen and produces an image on a fluorescent screen.
Transport medium A special type of medium used to move clinical specimens from one location to another while preserving the relative abundance of organisms and preventing contamination of the specimen or environment.
Transposon Segment of DNA found in many prokaryotes, eukaryotes, and viruses that codes for the enzyme transposase and can move from one location in a DNA molecule to another location in the same or different molecule.
Trematodes (*flukes*) Group of helminths that are flat, leaf-shaped, have incomplete digestive systems, and have oral and ventral suckers.
Trench fever Fever, headache, and pain in the long bones caused by infection with *Bartonella quintana*.
Trophozoite The motile feeding stage of all protozoa.
Tuberculin response Type of delayed hypersensitivity reaction in which the skin of an individual exposed to tuberculosis or tuberculosis vaccine reacts to a subcutaneous injection of tuberculin.
Tuberculin skin test Test for a delayed hypersensitivity reaction to a subcutaneous injection of tuberculin.
Tuberculosis (TB) A respiratory disease caused by infection with *Mycobacterium tuberculosis*; its disseminated form can result in wasting away of the body and death.
Tularemia Zoonotic disease causing fever, chills, malaise, and fatigue, and caused by infection with *Francisella tularensis*.
Tumor In the pathology of cancer, a mass of neoplastic cells. In inflammation, a symptom of swelling (*edema*).
Tumor necrosis factor An immune system cytokine secreted by macrophages and T cells to kill tumor cells and to regulate immune responses and inflammation.
Type I diabetes mellitus Immunologic attack on the islets of Langerhans cells resulting in the inability to produce the hormone insulin.
Typhoid fever Fever, headache, and malaise produced by infection with *Salmonella typhi*; severe infections may cause peritonitis.

Uncoating In animal viruses, the removal of a viral capsid within a host cell.
Unsaturated fat A triglyceride with at least one double bond between adjacent carbon atoms, and thus at least one carbon atom bound to only a single hydrogen atoms.
Uracil Ring-shaped nitrogenous base found in nucleotides of RNA.
Urticaria Hives.
Use-dilution test Method of evaluating the effectiveness of a disinfectant or antiseptic against specific microbes in which the most effective agent is the one that entirely prevents microbial growth at the highest dilution.

Vaccination Active immunization.
Vaccine The inoculum used in active immunization.
Vacuole General term for membranous sac that stores or carries a substance in a cell.
Vaginitis A sexually transmitted disease caused by infection with *Trichomonas vaginalis* and characterized by discharge, vaginal and cervical lesions, and pain.
Valence The combining capacity of an atom.
Vancomycin Antibiotic obtained from *Streptomyces orientalis* and used to treat infections with Gram-positive bacteria.
Varicella (*chickenpox*) Highly infectious disease characterized by fever, malaise, and skin lesions, and caused by infection with varicella-zoster virus.
Variola Common name for the smallpox virus.
Variola major Variant of smallpox virus which causes severe disease with a mortality rate of 20% or higher.
Variola minor Variant of smallpox virus which causes less severe disease and mortality rate of less than 1%.
Vector Nucleic acid molecule such as a viral genome, transposon, or plasmid, that is used to deliver a gene into a cell. In epidemiology, an animal (typically an arthropod) that transmits disease from one host to another.

Vehicle transmission Spread of pathogens via air, drinking water, and food, as well as bodily fluids being handled outside the body.

Venezuelan equine encephalitis (VEE) Potentially fatal infection of the brain caused by a togavirus.

Vesicle General term for membranous sac that stores or carries a substance in a cell; in human pathology, any raised skin lesion filled with clear fluid.

Viable plate count Estimation of the size of a microbial population based upon the number of colonies formed when diluted samples are plated onto agar media.

Vibrio A slightly curved rod-shaped prokaryotic cell.

Viral hemagglutination inhibition test Immune test commonly used to detect antibodies against influenza, measles, and other viruses that naturally agglutinate red blood cells.

Viremia The presence of virions in the blood.

Viridans streptococci Group of alpha-hemolytic streptococci which produce a green pigment when grown on blood media and normally inhabit the mouth, throat, GI, genital, and urinary tract.

Virion A virus outside of a cell, consisting of a proteinaceous capsid surrounding a nucleic acid core.

Viroid Extremely small, circular piece of RNA that is infectious and pathogenic in plants.

Virulence A measure of pathogenicity.

Virulence factors Enzymes, toxins, and other factors that affect the relative ability of a pathogen to infect and cause disease.

Virus Tiny infectious acellular agent with nucleic acid surrounded by proteinaceous capsomeres that form a coat called a capsid.

Wandering macrophage Type of macrophage that leaves the blood via diapedesis to travel to distant sites of infection.

Wastewater (*sewage*) Any water that leaves homes or businesses after being used for washing or flushed from toilets.

Water mold Eukaryotic microbe resembling a filamentous fungus but having tubular cristae in their mitochondria, cell walls of cellulose, two flagella, and true diploid thalli.

Water-borne transmission Spread of pathogenic microorganisms via water.

Wavelength The distance between two corresponding points of a wave.

Wax Alcohol-containing lipid made up of molecules with one fatty-acid chain.

Western blot test Variation of an ELISA test which can detect the presence of antibodies against multiple antigens; used to verify the presence of antibodies against HIV in the serum of individuals who have tested positive by ELISA.

Western equine encephalitis (WEE) Potentially fatal infection of the brain caused by a togavirus.

White piedra Benign scalp disease caused by infection with *Trichosporon beigelii*.

Whitlow Inflamed blister that may result from infection with herpes simplex viruses via a cut or break in the skin.

Wild type cell A cell normally found in nature (in the wild); a non-mutant.

Xenograft Type of graft in which tissues are transplanted between individuals of different species.

Xenotransplant Technique involving recombinant DNA technology in which human genes are inserted into animals to produce cells, tissues, or organs that are then introduced into the human body.

Yaws A skin disease caused by infection with *Treponema pallidum pertenue*.

Yeast A unicellular, typically oval or round fungus that usually reproduces asexually by budding.

Yellow fever Often fatal hemorrhagic disease contracted through a mosquito bite carrying a flavivirus.

Zone of inhibition In a diffusion susceptibility test, a clear area surrounding the drug-soaked disk where the microbe does not grow.

Zoonoses (*zoonotic diseases*) Diseases that are naturally spread from usual animal host to humans.

Zygomycoses Opportunistic fungal infections involving the brain and caused by various genera of fungi classified in the division Zygomycota.

Zygomycota Division of fungi including coenocytic molds called zygomycetes. Most are saprobes.

Zygospore Thick, black, rough-walled sexual structure of zygomycetes that can withstand desiccation and other harsh environmental conditions.

Zygote In sexual reproduction of protozoa, diploid cell formed by the union of gametocytes.

ILLUSTRATION CREDITS

Chapter 1: Figures 1.10, 1.12, 1.14, 1.15 J.B. Woolsey Associates, LLC; Figure 1.13 Precision Graphics.

Chapter 2: 2.1–2.12, 2.15–2.27 Precision Graphics; 2.13, 2.14 J.B. Woolsey Associates, LLC.

Chapter 3: 3.5, 3.7, 3.12, 3.13, 3.14, 3.23, 3.30, 3.31, 3.32, 3.33, 3.36, 3.37 Kenneth Probst; 3.8, 3.11, 3.19, 3.20, 3.29 Precision Graphics; 3.15, 3.16, 3.18, 3.35 Darwen Hennings/Precision Graphics; 3.24 Darwen Hennings.

Chapter 4: 4.1, 4.3, 4.26 Precision Graphics; 4.2, 4.4, 4.5, 4.6, 4.7, 4.8, 4.14, 4.17 J.B. Woolsey Associates, LLC; 4.25, 4.31 Darwen Hennings.

Chapter 5: 5.1–5.20, 5.22–5.25, 5.29–5.34, Highlight Box, Precision Graphics; 5.21, 5.26, 5.27, 5.28 Kenneth Probst.

Chapter 6: 6.1, 6.4a, 6.5, 6.7a, 6.17a, 6.18, 6.19, 6.20 Precision Graphics; 6.3, 6.8a, 6.9a, 6.16, 6.21, 6.22a, 6.23, 6.24, 6.25 J.B. Woolsey Associates, LLC; 6.15 Cassio Lynm.

Chapter 7: 7.3– 7.24, 7.28, 7.30–7.33, 7.35, Highlight Box, Closer Look Box Precision Graphics; 7.25–7.27, 7.29 J.B. Woolsey Associates, LLC.

Chapter 8: 8.1–8.6, 8.8–8.11 Precision Graphics.

Chapter 9: 9.1–9.3, 9.5, 9.6, 9.13, 9.15 Precision Graphics; 9.8, 9.10a, 9.11, J.B. Woolsey Associates, LLC.

Chapter 10: 10.2 Ken Probst; 10.3– 10.8, 10.13, 10.16 Precision Graphics; 10.12, 10.15 J.B. Woolsey Associates, LLC.

Chapter 11: 11.1, 11.5, 11.21, 11.26 Kenneth Probst; 11.2, 11.6–11.8, 11.10, 11.25 Precision Graphics.

Chapter 12: 12.1, 12.3, 12.4 Precision Graphics; 12.6, 12.7, 12.10, 12.12a, 12.14c, 12.18, 12.19, 12.21, 12.23, 12.24, 12.26, 12.30, 12.32 Kenneth Probst.

Chapter 13: 13.1, 13.4, 13.6, 13.7, 13.8, 13.11, 13.12, 13.13, 13.17 Kenneth Probst; 13.9, 13.14, 13.15, 13.21 Precision Graphics.

Chapter 14: Table 14.2, 14.3, 14.4, 14.11 Kenneth Probst; 14.5, 14.8, 14.9, 14.10, 14.13–14.19 Precision Graphics; 14.7 J.B. Woolsey Associates, LLC.

Chapter 15: 15.1, 15.12, 15.13, 15.15, 15.16 Precision Graphics; 15.3, 15.4, 15.5, 15.9, 15.10, 15.18, 15.19, 15.20 Kenneth Probst.

Chapter 16: 16.1, 16.3b, 16.5–16.17 Precision Graphics; 16.2, 16.4 Kenneth Probst.

Chapter 17: 17.1, 17.2, 17.3, 17.5, 17.6, 17.12, 17.14, 17.16, 17.17 Precision Graphics; 17.4, 17.7–17.11 J.B. Woolsey Associates, LLC.

Chapter 18: 18.1, 18.7, 18.8 Precision Graphics; 18.6, 18.9 Cassio Lynm; 18.12 J.B. Woolsey Associates, LLC.

PHOTO CREDITS

Chapter Opener 1 M.I. Walker / Photo Researchers, Inc.; Highlight 1 Center for Disease Control; 1.1 Courtesy of Pfizer, Inc.; 1.2 Alan Shinn; 1.3 Rich Robison, Brigham Young University; 1.4 a Manfred Kage/Peter Arnold; 1.4 b John D. Cunningham/Visuals Unlimited; 1.5 a *Microbiology; A Photographic Atlas for the Laboratory;* Benjamin Cummings 2001; 1.5 b M. Abbey/ Visuals Unlimited; 1.5 c, d Center for Disease Control; 1.6 a M.I. Walker/Photo Researchers, Inc.; 1.6 b Lester V. Berman/ Corbis; 1.7, 1.8 *Microbiology; A Photographic Atlas for the Laboratory;* Benjamin Cummings 2001; 1.9 Lee D. Simon/Photo Researchers, Inc.; 1.11 Louis Pasteur in his library: Oil on canvas, 1889, by Albert Edelfelt / The Granger Collection; 1.16 Culver Pictures, Inc.; 1.17, 1.18 *Microbiology; A Photographic Atlas for the Laboratory;* Benjamin Cummings 2001; 1.19 The Granger Collection, French Colored Engraving, 1883; 1.21 C. James Webb / Phototake.

Chapter Opener 2 NASA/Mark Marten/Photo Researchers, Inc.; Highlight 1 Topham / The Image Works; 2.12 Herman Eisenbeiss Photo Researchers, Inc.

Chapter Opener 3 D. Balkwill, D. Maratea; Highlight 1 NIBSC / Science Photo Library / Photo Researchers, Inc.; Highlight 2 Esther R. Angert / Phototake; New Frontiers 1 David Scharf/ Peter Arnold, Inc.; 3.1 a Dr. Gopal Murti / Photo Researchers, Inc.; 3.1 b Dennis Kunkel/ Phototake; 3.1 c Cabisco/Visuals Unlimited; 3.1 d VU/ Science VU/Visuals Unlimited; 3.4 a Dr. Immo Rantala / Science Photo Library / Photo Researchers, Inc.; 3.4 b Omikron / Photo Researchers, Inc.; 3.6 a Hans Glederbion / Stone; 3.6 b Dr. K.S.Kim / Peter Arnold, Inc.; 3.6 c Ed Reschke/ Peter Arnold, Inc.; 3.6 d A.B. Dowsett / Science Photo Library / Photo Researchers, Inc.; 3.7 a Chris Bjornberg/ Photo Researchers. Inc.; 3.9. 3.10 Dennis Kunkel/ Phototake; 3.21 P. S. Poole; 3.22 Ruth Carallido-Lopez; 3.23 a, b *Microbiology; A Photographic Atlas for the Laboratory;* Benjamin Cummings 2001; 3.23 c Omnikron / Science Library / Photo Researchers, Inc.; 3.25 Robert W. Bauman; 3.26 Don. W. Fawcett/ Visuals Unlimited; 3.27 a–d Mike Abbey/ Visuals Unlimited; 3.28 *Microbiology; A Photographic Atlas for the Laboratory;* Benjamin Cummings 2001; 3.29 b Albert Tousson/ Phototake; 3.30 a David M. Phillipis / Visuals Unlimited; 3.31 D.W. Fawcett / Photo Researchers, Inc.; 3.32 R. Bolender, D. Fawcett / Visuals Unlimited; 3.33 Biophoto Associates / Photo Researchers, Inc.; 3.34 George B. Chapman & Peter K. Chan/Visuals Unlimited; 3.36 D.W. Fawcett / Photo Researchers, Inc.; 3.37 G. Chapman / Visuals Unlimited.

Chapter Opener 4 Robert Bauman; U1 Phillipe Fauchet, Phd /ASM News Volume 68 No.2, 2002; 4.4 Rich Robison, Brigham Young University; 4.4 Charles D. Winter / Photo Researchers, Inc.; 4.6 a–d Rich Robison, Brigham Young University; 4.9 a, b Rich Robison, Brigham Young University; 4.10 Fred Marsik / Visuals Unlimited; 4.11 David Becker / SPL / Photo Researchers, Inc.; 4.12 WM Omerod/Visuals Unlimited; 4.13 Microworks Color / Phototake; 4.15 a Stanley Flegler/Visuals Unlimited; 4.15 b David M. Phillips/Visuals Unlimited; 4.15 c Oliver Meckes/Photo Researchers, Inc.; 4.15 d Stanley Flegler/Visuals Unlimited; 4.16 a, b © 1996 Digital Instruments; 4.17 *Microbiology; A Photographic Atlas for the Laboratory;* Benjamin Cummings 2001; 4.18 a–c Rich Robison, Brigham Young University; 4.19 a–e Rich Robison, Brigham Young University; 4.20 a, b *Microbiology; A Photographic Atlas for the Laboratory;* Benjamin Cummings 2001; 4.21. 4.22 Rich Robison, Brigham Young University; 4.23, 4.24 a–c, 4.27 a, b, 4.28 a, b *Microbiology; A Photographic Atlas for the Laboratory;* Benjamin Cummings 2001; 4.29 a Raymond B. Otero / Visuals Unlimited; 4.30 Courtesy of the Microbial Diseases Laboratory, Berkeley, CA.

Chapter Opener 5 Randall Hyman; Highlight 1 Photo Researchers, Inc.; New Frontiers 1 Dr. Morley Read / Science Photo Library, Photo Researchers, Inc.; 5.6 Ken Eward Biografx.com; 5.26 eos Wanner Photo Researchers, Inc.

Chapter Opener 6 Wayne Armstrong; N1 NASA; U1 Douglas Faulkner Photo Researchers, Inc.; 6.2, 6.4 b *Microbiology; A Photographic Atlas for the Laboratory;* Benjamin Cummings 2001; 6.6 a Gerald & Buff Corsi /Visuals Unlimited; 6.6 b Dr. Ron Hoham; 6.7 *Microbiology; A Photographic Atlas for the Laboratory;* Benjamin Cummings 2001; 6.8 b Raymond B. Otero/Visuals Unlimited; 6.9 b *Microbiology; A Photographic Atlas for the Laboratory;* Benjamin Cummings 2001; 6.10, 6.11 *Microbiology; A Photographic Atlas for the Laboratory;* Benjamin Cummings 2001; 6.12 Rich Robison, Brigham Young University; 6.13, 6.14 a–c *Microbiology; A Photographic Atlas for the Laboratory;* Benjamin Cummings 2001; 6.15b Coy Laboratories, Inc.; 6.17 b Alfred Pasieka / Peter Arnold, Inc.; 6.22 b Pall/Visuals Unlimited; 6.22 c Jack Bosrtak/Visuals Unlimited; 6.25 a Fundamental Photographs; 6.25 b Courtesy of TOPAC.

New Frontiers 1 CDC / Janice Carr; 7.2 a Visuals Unlimited; 7.2 b Huntington Potter and David Dressler; 7.3 a Courtesy Barbara Hamkalo; 7.3 b Courtesy of J. R. Paulsen and U. K. Laemmli, *Cell*, Copyright Cell Press.; 7.3 c G.F. Bahr / Armed Forces Institute of Pathology; 7.3 d G.F. Bahr / Armed Forces Institute of Pathology; 7.16 Visuals Unlimited; 7.31 a Dennis Kunkel / Phototake; 7.34 John A. Anderson Earth Scenes.

Chapter Opener 8 David Nance / ARS / NSDA; Highlight 1 Kent Loeffler; New Frontiers 1 Michael Gadomski/Animal Animals; New Frontiers 3 Michael Gadomski/Animal Animals; 8.7 b M. Schroder, Argus Fotoachiv / Peter Arnold; 8.9 d Sovereign / Phototake; 8.12 D. Parker / Science Photo Library; 8.13 SCIENCE VU/MONSANTO/VU.

Chapter Opener 9 Dianne Bodareff / POOL /AP Wide World Photo; U1 Lois Ellen Frank / Corbis; 9.4 *Microbiology; A Photographic Atlas for the Laboratory;* Benjamin Cummings 2001; 9.7 a Custom Medical; 9.8 Ricahrd Megna/Fundamental Photographs; 9.9 Jonathon Blair/CORBIS; 9.10 b Pall / Visuals Unlimited; 9.12 a Richard Megra / Fundamental Photographs; 9.12 b Courtesy Iowa State University Food Safety Project; 9.14 Richard Megra / Fundamental Photographs; 9.16 Talaro / Visuals Unlimited.

Chapter Opener 10 Dr. Kari Lounatmaa / Science Photo Library; F1 AP Wideworld Photo; Highlight 1 Leonard Lessin/Photo Researchers, Inc.; 10.1 Rich Robison, Brigham Young University; 10.3 Lennart Nilsson / *The Body Victorious,* Delacourt Publishing; 10.3 Lennart Nilsson; 10.9 *Microbiology; A Photographic Atlas for the Laboratory;* Benjamin Cummings 2001; 10.10 Dr. Craig Warner / Albert Einstein College of Medicine; 10.11 *Microbiology; A Photographic Atlas for the Laboratory;* Benjamin Cummings 2001; 10.14 a Barts Medical Library / Phototake; 10.14 b Ken Greer / Visuals Unlimited; 10.17 E. Vercauteren, Department of Microbiology, University Hospital, Antwerp, Belgium.

Chapter Opener 11 Dr. H.W. Jannasch, Woods Hole Oceanographic Institute; Highlight 1 Joseph Janovic, MD. Director Parkinson's Disease Center and Movement Disorders Clinic, Baylor College of Medicine; 11.3 a, b T. A. Krulwich; 11.4 Yoko Takahashi, Kitasato Institute for Life Sciences, Kitasato University, Tokyo; 11.6 a–e *Microbiology; A Photographic Atlas for the Laboratory;* Benjamin Cummings 2001; 11.7 a–d *Microbiology; A Photographic Atlas for the Laboratory;* Benjamin Cummings 2001; 11.9 a, b Rich Robison, Brigham Young University; 11.11 a Dr. Reinhard Rachel; 11.11 b Dr. Reinhard Rachel and K.O. Stetter; 11.12 Stephen Fuller/Peter Arnold, Inc; 11.13 Science VU/NASAGSFC/Visuals Unlimited; 11.14 a Phillip Sze/ Visuals Unlimited; 11.14 b M. Abbey / Visuals Unlimited; 11.14 c Eichelberger / Visuals Unlimited; 11.15 c Paul Johnson / Biological Photo Service; 11.16 Michael Gabridge / Visuals Unlimited; 11.17 J.R. Adams / Visuals Unlimited; 11.18 *Microbiology; A Photographic Atlas for the Laboratory;* Benjamin Cummings 2001; 11.19 Dr. Yves Brun, University of Indiana; 11.20 Wally Eberhart/Visuals Unlimited; 11.22 Leonard Lessin/Peter Arnold, Inc; 11.23 J. Ott; 11.24 Dr KS Kim/ Peter Arnold, Inc.; 11.25 b Alfred Pasieka / Peter Arnold, Inc.; 11.26 Eisenbach / Phototake.

Chapter Opener 12 Courtesy Dr. Greory Filip/Oregon State University; Highlight 1 AFP / Corbis; New Frontiers 1 Weizmann Institute of Science / Nature Magazine; 12.2 a RMF / Visuals Unlimited; 12.2 b David M. Phillips / Visuals Unlimited; 12.2 c Simko / Visuals Unlimited; 12.5 a M. Abbey / Visuals Unlimited; 12.5 b Cabisco / Visuals Unlimited; 12.8 Guy Brugerelle, Universitad Clearmont / Ferrand; 12.9 c Meckes / Ottowas / Eye on Science / Photo Researchers, Inc.; 12.11 a Manfred Kage / Peter Arnold, Inc.; 12.11 b R. Kessell - G. Shih / Visuals Unlimited; 12.12 b Luis De La Maza / Phototake; 12.13 M. Abbey / Visuals Unlimited; 12.14 a Simko / Visuals Unlimited; 12.14 b Gerald van Dyke / Visuals Unlimited; 12.14 d Tom Volk / University of Wisconsin; 12.15 Ken Wagner / Visuals Unlimited; 12.16 Cabisco / Visuals Unlimited; 12.17 a K. Talaro / Visuals Unlimited; 12.17 b Courtesy of Buckman Laboratories

International; 12.17 c David Scharf / Peter Arnold, Inc.; 12.20 Matt Meadows / Peter Arnold, Inc.; 12.22 Ed Reschke / Peter Arnold, Inc.; 12.24 V. Ahmakojian / Visuals Unlimited; 12.25 Fred Rhoades / Mycena Consulting; 12.27 Wim Van Egmond; 12.28 Jamie Woodward; 12.29 Bob Evans / Peter Arnold, Inc.; 12.31 Fred Rhoades / Mycena Consulting; 12.32 Matt Meadows / Peter Arnold, Inc.; 12.32 Cabisco / Visuals Unlimited.

Chapter Opener 13 SPL / Photo Researchers, Inc.; 13.1 b Russell Heere / Visuals Unlimited; 13.2 left Huntington Potter and David Dressler; 13.2 right Courtesy of Theodor O. Diener, Agricultural Research Service, U.S. Department of Agriculture.; 13.3 a David Parker / Science Photo Library; 13.3 b SPL / Photo Researchers, Inc.; 13.3 c Mike Miller / Science Photo Library; 13.3 d Dr. Klaus Boller / Science Photo Library; 13.5 a Harold Fisher / Visuals Unlimited; 13.5 b Science Photo Library / Photo Researchers, Inc.; 13.5 c CDC; 13.5 d Dhris Bjornberg/ Photo Researchers, Inc.; 13.6 a Dr. Harold Fisher / Visuals Unlimited; 13.7 a2 G. Murti / Photo Researchers, Inc.; 13.10 M. Wurtz , Biozentrum / University of Basil / Science Photo Library; 13.16 E. Chow / Visuals Unlimited; 13.18 Courtesy of Origen Biomedical; 13.19 Courtesy of Theodor O. Diener, Agricultural Research Service, U.S. Department of Agriculture; 13.20 H. D. Thurston, Cornell University; 13.22 James King-Holmes / SPL / Photo Researchers, Inc.

Chapter Opener 14 Reuters New Media, Inc. / Corbis; Highlight 1 J. Ott; 14.1 Photo Researchers, Inc.; 14.2 a D.M. Phillips / Visuals Unlimited; 14.2 b Visuals Unlimited; 14.5 W. Costerton; 14.6 David Scharf / Peter Arnold, Inc.; 14.12 Matt Meadows / Peter Arnold, Inc.

Chapter Opener 15 RMF / Visuals Unlimited; Highlight 1 Science Photo Library; 15.2 David Scharf / Peter Arnold, Inc.; 15.3 b RMF / Visuals Unlimited; 15.6 a Carolina Biological / Phototake; 15.6 b Ed Resche / Peter Arnold, Inc.; 15.6 c Ed Resche / Peter Arnold, Inc.; 15.6 d–e J.O. Ballard, M.D.; 15.8 Prof. P. M. Motta, T Fujita, M. Moto / Science Photo Library; 15.11 Biology Media / Photo Researchers, Inc.; 15.14 Dr. Sucharit Bhakdi.

Chapter Opener 16 Dennis Kunkel / Phototake; New Frontiers 1 Science Photo Library; 16.3 Manfred Kage / Peter Arnold, Inc.

Chapter Opener 17 Saturn Stills / Science Photo Library; Highlight 2 Center for Disease Control; New Frontiers 1 Akubio Ltd.; 17.10 b *Microbiology; A Photographic Atlas for the Laboratory;* Benjamin Cummings 2001; 17.13 Visuals Unlimited; 17.14 Maxine Jalbert / Center for Disease Control; 17.17 *Microbiology; A Photographic Atlas for the Laboratory;* Benjamin Cummings 2001.

Chapter Opener 18 E.R. Degginger; Highlight 1 Science VU / Visuals Unlimited; New Frontiers 1 Elizabeth Bauman; 18.2 a NYU Franklin Research Fund / Phototake; 18.2 b CNRI / Science Photo Library; 18.2 c NYU Franklin Research Fund / Phototake; 18.3 a M.F. Brown / Visuals Unlimited; 18.3 b Carolina Biological Supply Company / Phototake; 18.3 c Andrew Syred / Photo Researchers, Inc.; 18.4 ISM / Phototake; 18.5 SUI / Visuals Unlimited; 18.10 Princess Margaret Hospital / Department of Pediatrics; 18.11, 18.13 Ken Greer / Visuals Unlimited; 18.14 SUI / Visuals Unlimited; 19.C1 Dr. Kari Lounatmaa / Science Photo Library.

INDEX

NOTE: A *t* following a page number indicates tabular material, an *f* following a page number indicates a figure, and a *b* following a page number indicates a boxed feature.